2011 7th International Conference - Workshop Compatibility and Power Electronics

(CPE 2011)

Tallinn, Estonia
1 – 3 June 2011

IEEE Catalog Number: CFP11851-PRT
ISBN: 978-1-4244-8806-3

Copyright © 2011 by the Institute of Electrical and Electronic Engineers, Inc
All Rights Reserved

Copyright and Reprint Permissions: Abstracting is permitted with credit to the source. Libraries are permitted to photocopy beyond the limit of U.S. copyright law for private use of patrons those articles in this volume that carry a code at the bottom of the first page, provided the per-copy fee indicated in the code is paid through Copyright Clearance Center, 222 Rosewood Drive, Danvers, MA 01923.

For other copying, reprint or republication permission, write to IEEE Copyrights Manager, IEEE Service Center, 445 Hoes Lane, Piscataway, NJ 08854. All rights reserved.

***This publication is a representation of what appears in the IEEE Digital Libraries. Some format issues inherent in the e-media version may also appear in this print version.**

IEEE Catalog Number: CFP11851-PRT
ISBN 13:   978-1-4244-8806-3

Additional Copies of This Publication Are Available From:

Curran Associates, Inc
57 Morehouse Lane
Red Hook, NY 12571 USA
Phone: (845) 758-0400
Fax: (845) 758-2633
E-mail: curran@proceedings.com
Web: www.proceedings.com

2011 7th International Conference-Workshop Compatibility and Power Electronics (CPE)

TECHNICAL CO-SPONSORSHIP BY IEEE INDUSTRIAL ELECTRONICS SOCIETY

**7TH INTERNATIONAL CONFERENCE-WORKSHOP
JUNE 1-3, 2011 TALLINN ESTONIA**

TABLE OF CONTENT

Power Quality, Alternative Energy and Distributed Systems

HOS-based Virtual Instrument for Power Quality Assessment 1
Juan José González de la Rosa, José María Sierra Fernández, Daniel Ayora Sedeño, Agustín Agüera Pérez, José Carlos Palomares Salas, Antonio Moreno-Muñoz

Support Vector Machine for Power Quality Disturbances Classification using Higher-Order Statistical Features 6
José Carlos Palomares, Agustín Agüera, Juan José González de la Rosa

Reflections on current components according to CPC concept in a single phase circuit with nonsinusoidal voltage and current waveforms and LTI load 11
Marek Hartman

Harmonic Effect in Street Lighting 16
Aurora Gil-de-Castro, Antonio Moreno-Muñoz, Victor Pallares-Lopez, Agustin Agüera-Perez

Deterministic Ethernet synchronism with PTP-base system for synchrophasor in SmartGrid 22
Victor Pallares-Lopez, Antonio Moreno-Muñoz, Juan Jose Gonzalez De La Rosa, Miguel Jesus Gonzalez-Redondo, Rafael Real-Calvo, Aurora Gil de Castro, Francisco Domingo Perez

Neuro-fuzzy Control System for Active Filter with Load Adaptation 28
Oleksandr Husev, Sergey Ivanets, Dmitri Vinnikov

Pollution of High Power Charging Electric Vehicles in Urban Distribution Grids 34
Constantinos Sourkounis, Alexander Broy, Bingchang Ni

Useful lifetime and rational replacement of power transformers 40
Gerards Gavrilovs, Oleg Borscevskis

Tracks of Power Quality Transients in High Order Statistics Spaces 44
Agustín Agüera Pérez, Juan José González de la Rosa, José Carlos Palomares Salas, Aurora Gil de Castro, Antonio Moreno-Muñoz

The through simulation of devices on the basis of the structural linguistic method 50
Andrey Nikiforov

Electronic Synchronous Machine for Dynamic Power Conditioning in Wind Parks 56
Constantinos Sourkounis, Jan Wenske

Electromagnetic Compatibility Test System 62
Francisco Domingo-Perez, Jose-Maria Flores-Arias, Antonio Moreno-Muñoz, Juan J. De la Rosa, Victor Pallares, Aurora Gil-de-Castro, Isabel M. Moreno-Garcia

Integration of Active Power Filters in a Harmonic Load Flow Algorithm for Optimizing Location and Strategy 68
Eva González-Romera, Enrique Romero-Cadaval, Sergio Ruiz-Arranz, María Isabel Milanés-Montero

Generator Overloading owing to Neutral Wire Current 74
Vaclav Kus, Bohumil Skala

Power Analysis of a Multimodular Wind Power System Including PMG, Active Rectifier and VSI 78
Sergey Kharitonov, Sergey Brovanov, Gennady Zinoviev

Operation of Photovoltaic Power Systems with Energy Storage 86
Bernard Jerzy Szymanski, Kamil Kompa, Antoni Dmowski, Lukasz Roslaniec, Jerzy Szymanski

Generalized Proportional-Integral Control for Voltage-Sag Compensation in Dynamic Voltage Restorers 92
Alfonso Parreño Torres, Pedro Roncero-Sánchez, Xavier del Toro García, Vicente Feliu Batlle

Concept of Modular Uninterruptible Power Supply System with Alternative energy storages and sources 98
Andrew Stepanov, Ilya Galkin

NordPoolSpot price pattern analysis for households energy management 103
Aivar Auväärt, Argo Rosin, Nadezhda Belonogova, Denis Lebedev

Power Electronics Interfaces for Low Voltage Distribution Generation – EMC Issues 107
Grzegorz Benysek, Adam Kempski, Robert Smoleński, Marcin Jarnut

Voltage Sag Independent Operation of Induction Motor based on Z-Source Inverter 113
Uroš Flisar, Danijel Vončina, Peter Zajec

Voltage Drops Mitigations Using Flywheel Energy Storage System in Production Lines 119
Ahmad Al-Diab, Constantinos Sourkounis

Modular-Based PFC for Low Power Three-Phase Wind Generator 125
Freddy Flores-Bahamonde, Hugo Valderrama-Blavi, Josep M. Bosque, José A. Barrado, Luís Martínez-Salamero

Frequency Dependent Grid-Impedance Determination with Pulse-Width-Modulation-Signals 131
Michael Jordan, Hauke Langkowski, Trung Do Thanh, Detlef Schulz

Trans-Z-source Inverter with Built-in DC Current Blocking Capacitors 137
Marek Adamowicz, Jaroslaw Guzinski, Natalia Strzelecka, Dmitri Vinnikov

Performance Comparison of SiC Schottky Diodes and Silicon Ultra Fast Recovery Diodes 144
Marek Adamowicz, Sebastian Giziewski, Jedrzej Pietryka, Zbigniew Krzeminski

Power Electronics Controllers for Power Systems

Three-Phase Multilevel Inverter Based on LeBlanc Transformer 150
Pires Vitor, Guerreiro Manuel, Joao Martins, Fernando Silva

Alleviating Bottlenecks in Power Interconnectors by Means of FACTS 155
Rolf Grünbaum, Talivaldis Podins

Angle Estimation using High Frequency Injection for Sliding Mode Controlled Permanent Magnet Synchronous Machines Drives 161
Victor Repecho, Antoni Arias, Domingo Biel, Francesc Guinjoan

A New Criterion for Selecting the Inductors of an Active Power Line Conditioner 167
Pedro González-Castrillo, Enrique Romero-Cadaval, M. Isabel Milanés-Montero, Fermín Barrero-González, Miguel Ángel González-Martínez

Design and Implementation of a Digital Control and Monitoring System for an AC/DC UPS 173
Andriy Palamar, Mykola Karpinskyy, Valery Vodovozov

Three- Phase Regenerative Electronic Load to Test Shunt Power Conditioners 178
Carlos Roncero-Clemente, María Isabel Milanés-Montero, Victor Manuel Miñambres-Marcos, Enrique Romero-Cadaval

High Power, Medium Voltage, Modular Active Power Filtering System 184
Daniel Wojciechowski, Ryszard Strzelecki

Novel and Simple Method for Power Electronics Compensator Design and Optimization 190
Carlos Martinez, Virgilio Valdivia, Javier Lourido, Isabel Quesada, Antonio Lazaro, Andres Barrado

Influence of a High Precision Current Sensor for Improving the Efficiency of PV Power Systems 196
Abdoulkarim Bouabana, Ahmad Al-Diab, Constantinos Sourkounis

Modeling and Analysis of Three-Phase Hybrid Transformer Using Buck-Boost MRC 202
Jacek Kaniewski

Electro-mechanical Energy Conversion

Dynamical Operational Behaviour of the Power Drain of Wind Energy Converters with PMSM Considering Different Current Control Methods.. 208
Constantinos Sourkounis, Alexander Broy

Influence of control parameters of SRM for output characteristics... 214
Krzysztof Tomczuk, Marcin Parchomiuk

Design Guidelines using Selective Harmonic Elimination Advanced Method for DC-AC PWM with the Walsh Transform...................... 220
Jesus Vicente, Rafael Pindado, Inmaculada Martinez

Energy Saving Possibilities in the Industrial Robot IRB 1600 Control... 226
Anton Rassylkin, Hardi Hõimoja, Raivo Teemets

Analysis and Optimization of Sinusoidal Voltage Source Inverter Losses for Variable Output Power Applications.................................. 230
Enrique Romero-Cadaval, Víctor Miñambres-Marcos, María-Isabel Milanés-Montero

Determination of defects of electric motors without their disconnection.. 236
Nikolay Grebchenko, Maria Smirnova, Aleksey Sidorenko, Natalia Ragulina, Ilya Bielchev

Computer Aided Investigation System of Switched Reluctance Motors 242
Vasyl Tkachuk, Marius Klytta

Estimation of Parameters of Induction Motor with Broken Rotor Bars... 246
Danilo Makuc, Klemen Drobnič, Vanja Ambrožič, Damjan Miljavec, Rastko Fišer, Mitja Nemec

PWM-VSI Overmodulation Technique with Carrier Harmonics Reduction for Auxiliary Railway Power Supply 252
Carlos Martinez, Ramon Vazquez, Isabel Quesada, Carlos Lucena, Antonio Lazaro, Andres Barrado

Feasible Solutions Space for the Harmonic Cancellation Technique 258
Isabel Quesada, Carlos Martinez, Carmen Raga, Antonio Lazaro, Andres Barrado, Ramon Vazquez, Ignacio Gonzalez

Fundamental PMSM Model for Estimation of Cogging Torque Harmonic Components 264
Lovrenc Gašparin, Rastko Fišer

Comparison study of full-bridge and reduced switch count three-phase voltage source inverters 270
Roman Grinberg, Francisco Canales, Mikko Paakkinen

Advances in CSI-fed Induction Motor Drives.. 276
Marek Adamowicz, Marcin Morawiec

EMI and ESI Problems

Radiated Magnetic Field of a Low-Frequency Ferrite Rod Antenna.. 283
Alexander Stadler

Common Mode Model of Matrix Converter... 289
Jordi Espina, Antoni Arias, Josep Balcells, Carlos Ortega, Liliana De Lillo, Lee Empringham

Industrial DC/DC Converters in Terms of EMC 295
Zdenek Kubik, Jiri Skala

Application of Databases for Compatibility of Electrical Installation Design 299
Madis Lehtla, Margus Müür, Ivar Kõivastik, Kristjan Müül

Special Power Electronic Systems and Applications

Impact of Component Losses on the Voltage Boost Properties and Efficiency of the qZS-Converter Family 303
Dmitri Vinnikov, Indrek Roasto

Multicell-type Transistor Converter with Combined Control for Micro Resistance Welding 309
Yuriy E. Paerand, Oleksandr F. Bondarenko, Julia V. Bondarenko

High Efficiency Adaptive Boost Converter for LED Drivers 315
Yang Lu, Dariusz Czarkowski, Wieslaw E. Bury

Staff Training for Servicing Special Power Electronic Applications.. 319
Zoja Raud, Valery Vodovozov

DC-DC converter with isolation transformer for traction vehicles application – simulation studies 325
Marcin Parchomiuk, Krzysztof Tomczuk

Power Electric Aiding Controller for Automated Bus Stopping 330
Jorge Godoy, Vicente Milanés, Joshué Pérez, Jorge Villagrá, Carlos González

Analysis of Current Doubler Rectifier Based High Frequency Isolation Stage for Intelligent Transformer........................... 336
Viktor Beldjajev, Indrek Roasto, Dmitri Vinnikov

High Power Li-Ion Battery Charger for Electric Vehicle 342
Alon Kuperman, Udi Levy, Joseph Goren, Aryeh Zafranski, Alex Savernin

Neutral point clamped quasi-impedance-source inverter 348
Silver Ott, Indrek Roasto, Dmitri Vinnikov

New Converter for Interfacing PMSG based Small-Scale Wind Turbine with Residential Power Network 354
Lauris Bisenieks, Dmitri Vinnikov, Ilya Galkin

Energy-Efficient High-Voltage Switch Based on Parallel Connection of IGBT and IGCT........................ 360
Andrei Blinov, Dmitri Vinnikov, Volodymyr Ivakhno

Some Aspects of Blended Learning for Tallinn University of Technology and Tallinn Center of Industrial Education 365
Eduard Brindfeldt, Aleksandr Grinko, Margus Müür

Three-phase Bidirectional Battery Charger for Smart Electric Vehicles 371
Javier Gallardo-Lozano, María Isabel Milanés-Montero, Miguel Angel Guerrero-Martínez, Enrique Romero-Cadaval

Closed Current Loop Regulator for Electric Kart DC Motor Torque Stabilization 377
Kristaps Vitols, Nadav Reinberg, Ilya Galkin, Alvis Sokolovs

Comparative Study of Steady-State Performance of LED Dimmers at Different Modulation Techniques 382
Ilya Galkin, Irena Milashevski, Oleg Teteryonok

Space Vector Modulation with Reduced Switching Losses for Motor Drive Inverters........................ 388
Mikhail Egorov, Valery Vodovozov

Quasi-Z-Source Inverter Based Bi-Directional DC/DC Converter: Analysis of Experimental Results 394
Janis Zakis, Dmitri Vinnikov, Indrek Roasto, Leonids Ribickis

Comparison of two Power Electronic Schemes for 3kW Li-Ion Battery Charger........................ 400
Alexander Suzdalenko, Ilya Galkin

Power Losses Calculation Methodology to Evaluate Inverter Efficiency in Electrical Vehicles........................ 404
Josep Pou, Daniel Osorno, Jordi Zaragoza, Salvador Ceballos, Carles Jaen

Perspectives of Improvement of AC Power Transmission Based on Achievements of Modern Power Electronics........................ 410
Yevgeny I. Sokol, Yury P. Goncharov, Vladimir V. Ivakhno, Vladimir V. Zamaruev, Sergey Y. Krivosheev, Alexander P. Lastovka, Alexander Y. Ivanov, Marina A. Chernetchenko

Parallel resonant inverter with auxiliary AC/DC converter used for induction heating 415
Jan Mucko

Sensorless position identification of permanent magnet motor by zero voltage vector........................ 420
Janusz Wisniewski, Wlodzimierz Koczara

Optimized Cabin Power Supply with a +/- 270 V DC Grid on a Modern Aircraft........................ 425
Johannes Brombach, Arno Lücken, Torben Schröter, Detlef Schulz

Supply System Based on the 18-pulse Diode Rectifier with Coupled Reactors and Integrated with the Series Active Power Filter 429
Piotr Mysiak, Ryszard Strzelecki, Henryk Tunia, Krzysztof Zymmer

High-Voltage LED-Based Efficient Lighting using a Hysteretic Controlled Boost Converter 439
Antonio Leon-Masich, Hugo Valderrama-Blavi, Josep Maria Bosque-Moncusí, Àngel Cid-Pastor, Luís Martínez-Salamero

Motor Cable Effect on the Converter Fed AC Motor Common Mode Current 445
Jaroslaw Luszcz

Enabling Technologies for Matrix Converters in Aerospace Applications........................ 451
Lee Empringham, Liliana de Lillo, Sudarat Khwan-On, Chris Brunson, Pat Wheeler, Jon Clare

Power Electronic Grid-Interface for Renewable Ocean Wave Energy........................ 457
Marian Kazmierkowski, Marek Jasiński

FPGA-Based Implementation by Direct Torque Control of a Permanent Magnet Synchronous Machine........................ 464
Badre Bossoufi, Mohammed Karim, Silviu Ionita, Ahmed Lagrioui

Magnetic Sensor Coil Shape Geometry and Bandwidth Assessment 470
Lauri Kütt, Muhammad Shafiq

2011 7th International Conference-Workshop Compatibility and Power Electronics (CPE)

Honorary Chairmen

Prof. Andres Keevallik
Rector of Tallinn University of Technology (Estonia)

Prof. Peep Sürje
Past Rector of Tallinn University of Technology (Estonia)

General Co-chairs

Dmitri Vinnikov
Tallinn University of Technology (Estonia)

Ryszard Strzelecki
Gdynia Maritime University (Poland)

Prof. Juhan Laugis
Tallinn University of Technology (Estonia)

Technical Program Co-chairs

Dmitri Vinnikov
Tallinn University of Technology (Estonia)

Enrique Romero Cadaval
University of Extremadura (Spain)

Vanja Ambrozic
University of Ljubljana (Slovenia)

Marek Hartman
Gdynia Maritime University (Poland)

Antoni Arias Pujol
Technical University of Catalonia (Spain)

Organization Committee

Taavi Möller
Tallinn University of Technology (Estonia)

Indrek Roasto
Tallinn University of Technology (Estonia)

Eva González Romera
University of Extremadura (Spain)

Daniel Wojciechowski
Gdynia Maritime University (Poland)

Marek Adamowicz
Gdynia Maritime University (Poland)

Technical Program Committee

- **Marek Adamowicz**
  Gdynia Maritime University (Poland)
- **Vanja Ambrožič**
  University of Ljubljana (Slovenia)
- **Antoni Arias Pujol**
  Technical University of Catalonia (Spain)
- **Lucian Asiminoaei**
  Danfoss Drives A/S (Denmark)
- **Andrés Barrado Bautista**
  Universidad Carlos III de Madrid (Spain)
- **Fermín Barrero González**
  Universidad de Extremadura (Spain)
- **Grzegorz Benysek**
  University of Zielona Góra (Poland)
- **Frede Blaabjerg**
  Aalborg University (Denmark)
- **Jon C. Clare**
  The University of Nottingham (United Kingdom)
- **Eugenio Domínguez Amarillo**
  University of Sevilla (Spain)
- **Leopoldo G. Franquelo**
  University of Sevilla (Spain)
- **Ilja Galkin**
  Riga Technical University (Latvia)
- **Juan Jose Gonzalez de la Rosa**
  University of Cadiz (Spain)
- **Eva González Romera**
  Universidad de Extremadura (Spain)
- **Weifeng Guo**
  Harbin Institute of Technology (China)
- **Marek Hartman**
  Gdynia Maritime University (Poland)

- **Carles Jaen**
  Technical University of Catalonia (Spain)
- **Tapani Jokinen**
  Helsinki University of Technology (Finland)
- **Marian-P. Kaźmierkowski**
  Warsaw University of Technology (Poland)
- **Georg Kerber**
  IEEE (Germany)
- **Sergey Kharitonov**
  Novosibirsk State Technical University (Russia)
- **Włodzimierz Koczara**
  Warsaw University of Technology (Poland)
- **Johann Kolar**
  Swiss Ferderal Institute of Technology Zurich (Switzerland)
- **Juhan Laugis**
  Tallinn University of Technology (Estonia)
- **Marco Liserre**
  Polytechnic of Bari (Italy)
- **Jaroslaw Luszcz**
  Gdansk University of Technology (Poland)
- **Mariusz Malinowski**
  Warsaw University of Technology (Poland)
- **Luis Martínez Salamero**
  Universidad Rovira i Virgili (Spain)
- **João Francisco Martins**
  Universidade Nova de Lisboa (Portugal)
- **María Isabel Milanés Montero**
  Universidad de Extremadura (Spain)
- **Antonio Moreno Muñoz**
  University of Córdoba (Spain
- **David Nedeljković**
  University of Ljubljana (Slovenia)

- **Yves Nguegan**
  University of Kassel (Germany)
- **Josep Pou**
  Technical University of Catalonia (Spain)
- **Leonids Ribickis**
  Riga Technical University (Latvia)
- **Enrique Romero Cadaval**
  Universidad de Extremadura (Spain)
- **Juergen Schlabbach**
  University of Applied Sciences (Germany)
- **Robert Smolenski**
  University of Zielona Gora (Poland)
- **Andreas Steimel**
  Ruhr-University Bochum (Germany)
- **Ryszard Strzelecki**
  Gdynia Maritime University (Poland)
- **Paolo Tenti**
  University of Padova (Italy)
- **Xavier del Toro**
  University of Castilla-La Mancha (Spain)
- **Jaime Valencia**
  Universidad de Antioquia (Colombia)
- **Sergio Vázquez**
  University of Sevilla (Spain)
- **Dmitri Vinnikov**
  Tallinn University of Technology (Estonia)
- **Valery Zhuikov**
  National Technical University of Ukraine "KPI" (Ukraine)
- **Gennady Zinoviev**
  Novosibirsk State Technical University (Russia)

Editorial

On behalf of the Organization Committee it is our pleasure to welcome you to the 7th International Conference–Workshop CPE 2011 Compatibility and Power Electronics.

The history of the CPE conference dates back to 1999. From its beginning the objective of this Conference has been to offer a forum of discussion to power electronics specialists. At present, CPE has become a reference for European researchers but it has also developed into a reference for the whole world.

After the very successful CPE 2009 organized two years ago in Badajoz (Extremadura, Spain) this year's conference is being held in the Baltic region of Northern Europe for the first time. The host city of CPE 2011 is Tallinn. As a Hanseatic Town with its history dating back to 1154 Tallinn has been a meeting point for East and West for centuries. Its closeness to all major European capitals and its good infrastructure have made it an ideal venue for any international event. In 2011 Tallinn is the European Capital of Culture, and naturally this highlights the underlying idea for the whole program of CPE 2011.

CPE 2011 is technically co-sponsored by IEEE IES and is organized by Tallinn University of Technology and Gdynia Maritime University, with collaboration of the University of Extremadura and Power Electronic Systems Laboratory of ETH Zurich.

This year, one of the main objectives is to enhance the proved quality of the papers presented at the conference. All papers have been reviewed by at least three referees. 110 papers were sent for consideration and finally 78 of these papers will be presented during the conference. All papers have been scheduled in oral sessions to allow a dynamic exchange among the researchers and a better understanding of the new concepts and ideas presented at the conference.

We hope this conference will provide a good platform for forging international alliances and networking of thoughts.

Finally, we hope that your stay in Tallinn and the social events we have planned will be most delightful.

We would like to express our gratitude to committee members, reviewers and volunteers for their unconditional collaboration and efforts to make CPE 2011 an outstanding and memorable event.

Dmitri Vinnikov
Ryszard Strzelecki
General Co-Chairmen of CPE 2011

Reviewers
List obtained from the IEEE-IES Conference Management System

Antonio J. Marques Cardoso
University of Coimbra, Portugal

Antoni Arias Pujol
Technical University of Catalonia,
Spain

Brayima Dakyo
University of Le Havre, France

Jakub Dawidziuk
Bialystok University of Technology,
Poland

Duarte Valério
IDMEC/IST, TULisbon, Portugal

Eva Gonzalez Romera
University of Extremadura, Spain

Grzegorz Benysek
University of Zielona Gora, Poland

Hardi Hõimoja
Tallinn University of Technology,
Estonia

Jan Iwaszkiewicz
Electrotechnical Institute, Poland

João Francisco A. Martins
Fct/Unl, Portugal

Josef Lutz
Chemnitz University of Technology,
Germany

Kzaimierz Buczek
Rzeszow University of Technology,
Poland

Manuel Lamich
Universitat Politècnica de Catanunya,
Spain

Luis Martinez Salamero
Rovira i Virgili University, Spain

Mariusz Malinowski
Warsaw University of Technology,
Poland

Marek Hartman
Gdynia Maritime University, Poland

Marian-P. Kazmierkowski
Warsaw University of Technology,
Poland

Piotr Chrzan
Gdansk University of Technology,
Poland

Cristian Ilie Pitic
Continental Automotive, Romania

Rastko Fišer
University of Ljubljana, Slovenia

Volker Staudt
University of Bochum, Germany

Toomas Rang
Tallinn University of Technology,
Estonia

Danjel Voncina
University of Ljubljana, Slovenia

Antonio Moreno Muñoz
Universidad de Cordoba, Spain

Josep Balcells
Universitat Politècnica de Catalunya,
Spain

Johannes Brombach
Helmut-Schmidt-University, Germany

Dariusz Czarkowski
NYU-Poly, USA

Valery Vodovozov
Tallinn University of Technology,
Estonia

Frede Blaabjerg
Aalborg University, Denmark

Gennady Zinoviev
NSTU, Russia, Russian Federation

Ahmet Hava
Middle East Technical University,
Turkey

Janis Zakis
Tallinn University of Technology,
Estonia

José Ignacio León
University of Seville, Spain

Josep Maria Bosque Moncusí
Universitat Rovira i Virgili, Spain

Mahesh Krishnamurthy
Illinois Institute of Technology, USA

Lucian Asiminoaei
Danfoss Power Electronics, Denmark

Liping Zheng
Calnetix,Inc, USA

Maria Carmela Di Piazza
CNR - ISSIA, UOS Palermo, Italy

María Isabel Milanés Montero
University of Extremadura, Spain

Jan Mucko
University of Technology and Life
Sciences in Bydgoszcz, Poland

Beata Palczynska
Gdynia Maritime University, Poland

Josep Pou
Technical University of Catalonia,
Spain

Ryszard Strzelecki
Gdynia Maritime University, Poland

Tapani Jokinen
Aalto University, Finland

Valery Vodovozov
Tallinn University of Technology,
Estonia

Zbigniew Fedyczak
University of Zielona Gora, Poland

Andrea Tortella
University of Padova, Italy

Andres Barrado Bautista
Carlos III University, Spain, Spain

Carla Tassoni
Dip. Ing. dell'Informazione, Italy

Detlef Schulz
Helmut-Schmidt-University Hamburg,
Germany

Elmo Pettai
Tallinn University of Technology,
Estonia

Fernando Tadeo
Univ. Valladolid, Spain

Ilya Galkin
Riga Technical University, Latvia

Indrek Roasto
Tallinn University of Technology,
Estonia

Javier Gallardo-Lozano
University of Extremadura, Spain

Jaroslaw Luszcz
Gdansk University of Technology,
Poland

Juergen Schlabbach
University of Applied Sciences
Bielefeld, Germany

Wlodzimierz Koczara
Warsaw University of Technology,
Poland

Alexander P. Lastovka
National technical university
"Kharkov polytechnic institute",
Kharkov, Ukraine

Mihai Cernat
Transilvania University of Brasov,
Romania

Marius Klytta
University of Applied Sciences
Giessen-Friedberg, Germany

Mitja Nemec
University of Ljubljana, Slovenia

Olga Muravleva
Tomsk Polytechnic University,
Russian Federation

Peter Zajec
University of Ljubljana, Slovenia

Robert Smolenski
University of Zielona Gora, Poland

Sandrine Moreau
Poitiers university, France

Tadeusz Piotrowski
Gdynia Maritime University, Poland

Valery Zhuikov
National Technical University,
Ukraine

Zbigniew Krzeminski
Gdansk University of Technology,
Poland

Antonio Lázaro Blanco
Carlos III University of Madrid, Spain

Oleksandr Bondarenko
Donbas State Technical University,
Ukraine

Valery Chrisanov
West Pomeranian Univ. of
Technology in Szczecin, Poland

David Gonzalez Diez
Universitat Politècnica de Catalunya,
Spain

Ali Emadi
Illinois Institute of Technology, USA

J. Fernando A. Silva
TULisbon, IST, Portugal

Hussein A. Kazem
Sohar University, Oman

Jaan Järvik
Tallinn University of Technology,
Estonia

Jose Carpio Ibanez
Uned, Spain

Janusz Nieznanski
Gdansk University of Technology,
Poland

Jaime Valencia
Universidad de Antioquia, Colombia

Johann Kolar
ETH Zurich, Switzerland

Leonids Ribickis
Riga Technical University, Latvia

Marek Adamowicz
Gdynia Maritime University, Poland

Marko Petkovsek
University of Ljubljana, Slovenia

Marek Jasinski
Warsaw University of Technology,
Poland

Piotr Mysiak
Gdynia Maritime University, Poland

Pedro González
University of Extremadura, Spain

Rajesh Gupta
M. N. National Institute of
Technology, India

Sergej Kalaschnikow
Danfoss Drives, USA

Tanel Jalakas
Tallinn University of Technology,
Estonia

Vanja Ambrozic
University of Ljubljana, Slovenia

Francisco Arcega
University of Zaragoza, Spain

Janos Botzheim
Szechenyi Istvan University, Hungary

David Nedeljkovic
University of Ljubljana, Faculty of
Electrical Engineering, Slovenia

Dmitri Vinnikov
Tallinn University of Technology,
Estonia

Enrique Romero Cadaval
University of Extremadura, Spain

ich W. Fuchs
Schristian-Albrechts-University of
Kiel, Germany

H. Y. Kanaan
Saint-Joseph University, Lebanon

Carles Jaen
Technical University of Catalonia,
Spain

Jesús Doval Gandoy
University of Vigo, Spain

Jon C. Clare
Nottingham University, United
Kingdom

Nobuyuki Kasa
Okayama University of Science,
Japan

Krzysztof Zawirski
Poznan University of Technology,
Poland

Marco Liserre
DEE, Poliba, Italy

Maks Berlec
University of Ljubljana, Slovenia

Miguel A. Guerrero Martínez
University of Extremadura, Spain

Juan Carlos Montaño Asquerino
Csic, Spain

Oskars Krievs
Riga Technical University, Latvia

Rafael Pindado
Universitat Politècnica de Catalunya,
Spain

Ivars Rankis
Riga Technical University, Latvia

Andrzej Sikorski
Bialystok Technical University,
Poland

Tarasiuk Tomasz
Gdynia Maritime University, Poland

Victor Miñambres
Universidad de Extremadura, Spain

HOS-based Virtual Instrument for Power Quality Assessment

Juan José González de la Rosa[1], J. M. Sierra Fernández[1], D. Ayora-Sedeño[1], Agustín Agüera-Pérez[1],
J. C. Palomares-Salas[1] and A. Moreno-Muñoz[2]

[1]Research Group PAIDI-TIC-168. Univ. of Cádiz. Electronics. EPSA. Av. Ramón Puyol, S/N. E-11202-Algeciras-Cádiz, (*Spain*)
[2]Research group PAIDI-TIC-168. Univ. of Córdoba. Electronics. Campus de Rabanales. E-14071-Córdoba, (*Spain*)
Email: juanjose.delarosa@uca.es; agustin.aguera@uca.es; josecarlos.palomares@uca.es; amoreno@uco.es

Abstract—This paper describes a virtual instrument for PQ assessment. Conceived to detect transients, sags and swells, the computational nucleus is based in higher-order statistics, which enhance the statistical characterization of the raw data, with the consequent improvement of the instrument's performance. This measurement application is thought to be used by the plant operator previously trained in PQ analysis. The normal operation limit is previously established on empirical expert knowledge. The instrument also allows online visualization of the statistical parameters (variance, skewness and kurtosis). The repeatability is 85% roughly. Results convey the idea of an easy transferable technology, due to its low cost and optimized computational nucleus.

I. INTRODUCTION

Non-linear loads supporting high voltages are very common elements in power systems. Parasitic harmonics generated in this type of loads cause anomalies in the ideal power sine wave, that affect to sensitive equipments, introducing failures in its normal function and consequently provoking losses and reducing the life of many systems; e.g. induction motors [1-4]. Traditionally the analysis of deformation of power waveform was performed via FFT, wavelet transforms and other frequency decomposition system (e.g. calculating the THD).

Switching between lines during normal operation of power systems and switching loads introduce temporal voltage fluctuations (sags and swells), which affect to all interconnected systems, e.g. reductions of motor's torque, resetting electronic systems, turning off them by under voltage or destroy them by over voltage [5]. The traditional measurement procedure implemented in the monitoring equipment is based in RMS value and peak detection of the power waveform [6].

The estimators traditionally used to detect the former anomalies are based in second-order algorithms [7], which are capable of completely characterize Gaussian signals. Nevertheless, the complexity of the electrical anomalies suggests the idea of using statistics of order 3 and more. Without any perturbation, power waveform exhibits a Gaussian behavior; non-Gaussian features provide complete information about the type of fail, generally associated with an impulsive behavior [8].

During the last ten years higher-order statistics (HOS) are being introduced in the measurement PQ algorithms as a

complement with no evidence of practical implementation. We remark the advances in [8], where two types of transients are distinguished via HOS, and [9] associated to PQ and plague detection. Virtual instruments (VI) allow low-cost development and implementation;

This paper describes a VI implementing HOS for PQ targeting. The computational guts are based in second, third and fourth order estimators in the time domain, which sweep the signal under test by means of a sliding window, over which the statistics are calculated. The maximum computational power in HOS calculation is reached by adjusting the window conveniently, as described further. The paper is organized in the following way. Section II summarizes HOS which are implemented in the VI, described in Section III, which gathers the instrument design users' interface and results. Finally, conclusions are drawn in Section IV.

II. HIGHER-ORDER STATISTICS

Higher-order cumulants are used to infer new properties about the data of non-Gaussian processes [9]. In multiple-signal processing it is very common to define the combinational relationship among the cumulants of r stochastic signals, $\{x_i\}_{i \in [1,r]}$, and their moments of order $p, p \leq r$, given by using the *Leonov-Shiryaev* expression, given in Eq. (1) [10]:

$$Cum(x_1,...,x_r) = \sum (-1)^{p-1} \cdot (p-1)! \cdot E\left\{ \prod_{i \in s_1} x_i \right\}$$
$$E\left\{ \prod_{j \in s_2} x_j \right\} \cdots E\left\{ \prod_{k \in s_p} x_k \right\}, \qquad (1)$$

where the addition operator is extended over all the partitions, like one of the form $(s_1, s_2,...,s_p)$, with $p = 1,2,...,r$; and $(1 \leq i \leq p \leq r)$; being s_1 a set belonging to a partition of order p, of the set of integers $1,...,r$.

Let $\{x(t)\}$ be an rth-order stationary random real-valued process. The rth-order cumulant is defined in Eq. (2) as the

978-1-4244-8806-3/11 $26.00 © 2011 IEEE

joint rth-order cumulant of the random variables $x(t), x(t+\tau_1),...,x(t+\tau_{r-1})$,

The expressions in Eq. (4) are measurements of the variance, skewness and kurtosis of the distribution in terms of cumulants al zero lags (the central cumulants). Normalized

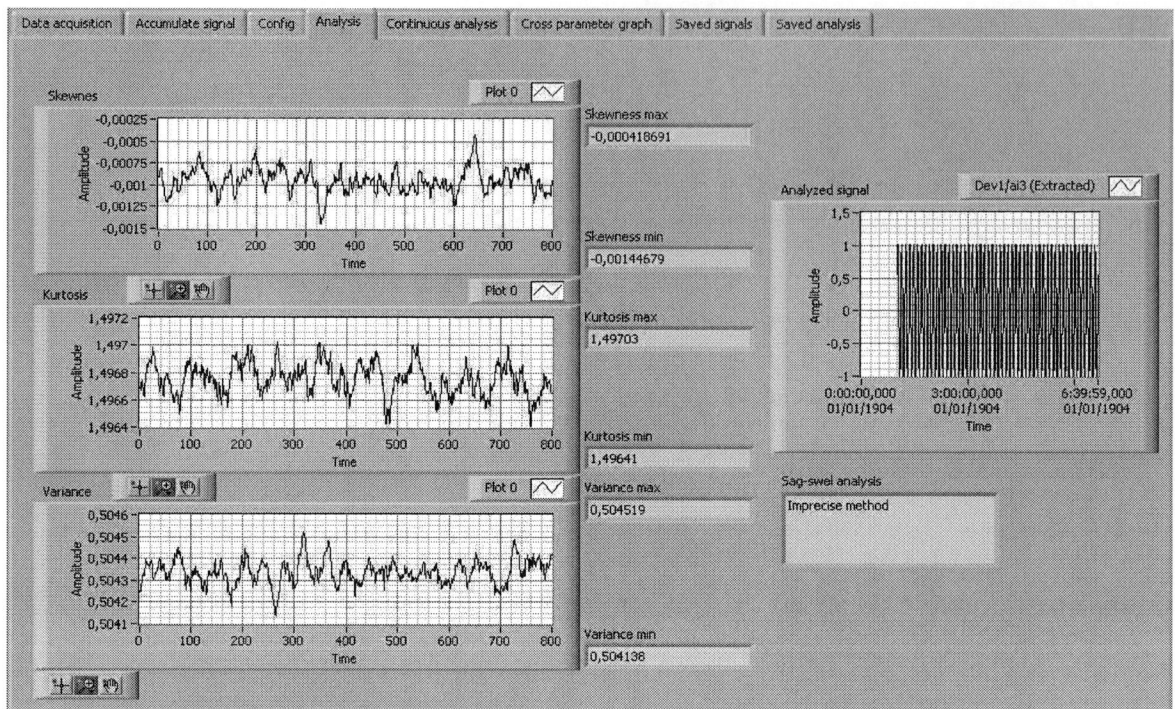

Fig. 1. "Analysis" tab in the virtual instrument.

$$C_{r,x}(\tau_1, \tau_2, ..., \tau_{r-1})$$
$$= Cum[x(t), x(t+\tau_1), ..., x(t+\tau_{r-1})] \qquad (2)$$

The second-, third- and fourth-order cumulants of zero-mean $x(t)$ can be expressed via Eq. (3) [11]:

$$C_{2,x}(\tau) = E\{x(t) \cdot x(t+\tau)\} \qquad (3a)$$

$$C_{3,x}(\tau_1, \tau_2) = E\{x(t) \cdot x(t+\tau_1) \cdot x(t+\tau_2)\} \qquad (3b)$$

$$C_{4,x}(\tau_1, \tau_2, \tau_3)$$
$$= E\{x(t) \cdot x(t+\tau_1) \cdot x(t+\tau_2) \cdot x(t+\tau_3)\}$$
$$- C_{2,x}(\tau_1)C_{2,x}(\tau_2 - \tau_3)$$
$$- C_{2,x}(\tau_2)C_{2,x}(\tau_3 - \tau_1) \qquad (3c)$$
$$- C_{2,x}(\tau_3)C_{2,x}(\tau_1 - \tau_2)$$

By putting $\tau_1 = \tau_2 = \tau_3 = 0$ in Eq. (3), we obtain Eq. (4):

$$\gamma_{2,x} = E\{x^2(t)\} = C_{2,x}(0) \qquad (4a)$$

$$\gamma_{3,x} = E\{x^3(t)\} = C_{3,x}(0,0) \qquad (4b)$$

$$\gamma_{4,x} = E\{x^4(t)\} - 3(\gamma_{2,x})^2 = C_{4,x}(0,0,0) \qquad (4c)$$

kurtosis and skewness are defined as $\gamma_{4,x}/(\gamma_{2,x})^2$ and $\gamma_{3,x}/(\gamma_{2,x})^{3/2}$, respectively. We will use and refer to normalized quantities because they are shift and scale invariant. If $x(t)$ is symmetrically distributed, its skewness is necessarily zero (but not *vice-versa*, almost impossible situations); if $x(t)$ is Gaussian distributed, its kurtosis is necessarily zero (but not *vice versa*). In the section of results, analysis is performed using sliding cumulants, i.e. a moving window in the time domain over which to compute each cumulant (3rd and 4th-order cumulants for zero time-lag).

III. INSTRUMENT DESIGN AND RESULTS

The VI analyzer has been designed in LabVIEW (ver. 8.5). It drives a PCMCIA data acquisition board (DAQ-NI-6036E). Fig. 1 shows instrument's organization, using tabs. "Analysis" is the main tab, shows the analyzed signal (last fragment of real time signal) normalized in "Analyzed signal" graph. Result of HOS analysis is plotted in the "Variance", "Skewness" and "Kurtosis" graphs. These measurement time-series constitute an example of normal operation. The horizontal axis shows the number of the window tested (iteration number within the calculation process). These peaks are evidences of the imperfections associated to the power signal, along with the noise that the acquisition board introduces. The ideal operation steady state and the observed deviation range correspond to the triplet: Variance = 0.5000 ± 0.0045, Skewness = 0.0000 ± 0.0014 and Kurtosis = 1.500 ± 0.003.

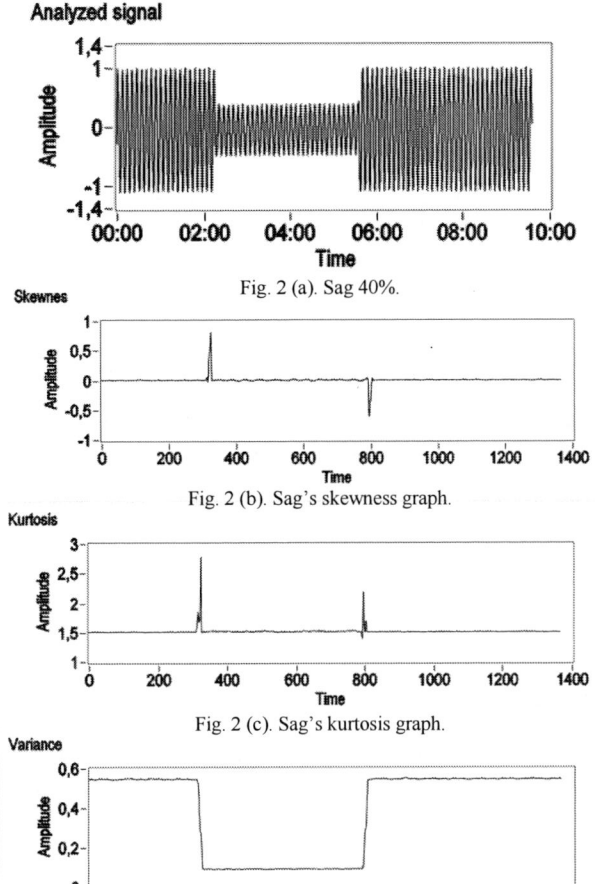

Fig. 2 (a). Sag 40%.

Fig. 2 (b). Sag's skewness graph.

Fig. 2 (c). Sag's kurtosis graph.

Fig. 2 (d). Variance graph: a "valley" targets the sag.

The maxima and minima values of the magnitudes are indicated next to each HOS graph. These values are compared with normal the operation limits, based on empirical experience. If these limits are trespassed, an anomaly is targeted in the signal and the VI save it for a further detailed analysis. The last indicator of this tab, "Sag-swell analysis", shows the variance calculation results. In the case shown in Fig. 1, it is displayed "Imprecise method", which means that data show a deviation less than 12 % (87 % Sag or 112 % Swell).

The rest of the VI tabs are described hereinafter. "Data acquisition" shows the configuration of the DAQ board. "Accumulate signal" shows the intermediate buffer between the acquisition and the analysis, with the objective of keeping data if occurs that the analysis is very time-consuming. The tab "Config" controls the main analysis configuration. The user must indicate a perfect signal in order to lock the normal amplitude, and to modify the window's length, the "slide ratio" (window's shift data), and normal operation limits. In this situation, window's wide is 1 cycle, sliding is 25 data, and the normal operation limits for the variance range from

0.45 to 0.55, -0.02 to 0.02 for skewness and for kurtosis 1.47 to 1.53.

The tab for "Continuous analysis" gathers an historical of HOS analysis graphs. The tab for "Cross parameter graph" allows drawing a graph with some points obtained of previous simulations for different types of anomalies. The VI marks a rectangle over these defects with the associated normal operation limits, and a cross with the actual analysis values. Bi-dimensional graphs depict extremes associated to each statistical parameter, e.g. variance maximum vs. minimum. By watching the graph, the user can conclude if the signal is normal or defective.

Finally, the tabs called "Saved signals" and "Saved analysis" let the user load signals, previously saved for exceeding normal operation limits or within a historical analysis, respectively.

Due to its versatility, the VI is adaptable a wide branch of physical inputs, e.g. it has been tested too with Agilent DSO6012A Digital oscilloscope. It is even possible to join it to a bigger system with power signal data available. Besides, the system's response could be output, for example different levels of alarm, depending on the type and number of defects.

This VI has been tested with simulated data by "Agilent 6811B AC Power Source/Analyzer" instrument. These signals are different situations of Sag, Swell and impulsive transient defect types, modifying different parameters.

Fig. 2 (a) shows a normalized test signal (50 Hz – 230 V_{rms}) with a reduced zone of 50 Hz – 92 V RMS, a sag of 40 %. The normalized defect is 0.4 valued. Fig. 2 (b)-(c)-(d) show the HOS analysis graphs. The skewness sub-figure shows two impulses, around the zero steady state. These peaks are caused by the sudden change of the signal's amplitude. The sign and level of the peaks change depending on different conditions; for example the phase of the sinusoid when the sag starts or finishes changes the sign and the level. Maximum value is 0.808 and minimum -0.623, both exceeding the limits of 0.02 and -0.02, respectively.

Fig. 2 (c) shows two peaks at the same position of Fig. 2 (b), always associated to the amplitude fluctuations. They also behave differently, always over the value 1.5, and always appear in a sag-swell defect. The highest observed peak is 2.768 and the minimum is 1.377 at the starting of the second peak. Again, both exceed normal operation limits, 1.53 and 1.47 respectively.

Fig. 2 (d) is the main graph in sag-swell analysis, the variance graph. Variance is related to the amplitude's square, giving us information about amplitude evolution. The appearance is a zone of less variance during the sag, because the amplitude is a fraction of the normal operation amplitude. Maximum of this graph is 0.547, inside max normal operation limit (0.55), but minimum is 0.085, out of the min limit (0.45). For this concrete sag, all extreme values except one are out of the limits, so the instrument classifies it defective. Result of "Sag-swell analysis" is sag 40 %, which is exactly the defect present in this test, confirming the equipment calibration.

If the defect was swell (the symmetrical case of sag), the situation would be similar. The same peaks in skewness and kurtosis graphs at the defect's start and end are found. The main difference is observed in the variance graph. During the

978-1-4244-8806-3/11 $26.00 © 2011 IEEE

defect, the level is higher than the normal operation value, instead of the reduction associated to sag.

With all this, we can determine a sag defect when there is a zone with less variance (more in swell case), starting and ending with a peak in skewness and kurtosis.

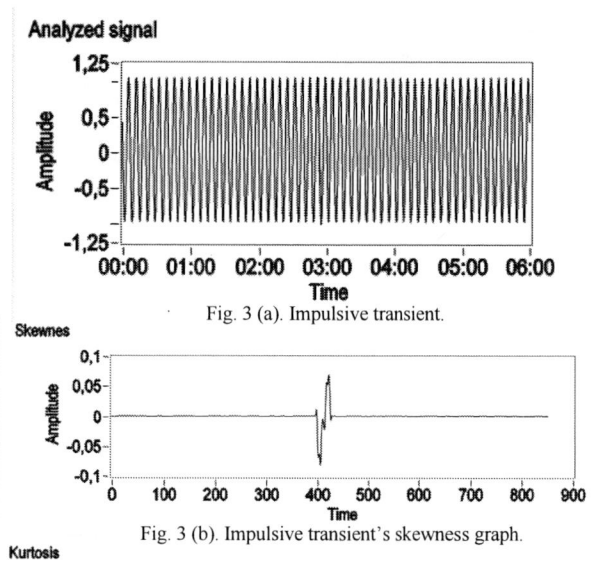

Fig. 3 (a). Impulsive transient.

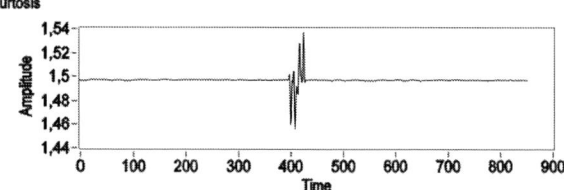

Fig. 3 (b). Impulsive transient's skewness graph.

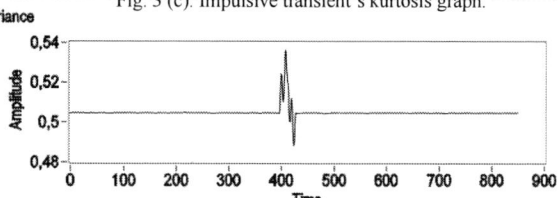

Fig. 3 (c). Impulsive transient's kurtosis graph.

Fig. 3 (d). Impulsive transient's variance graph.

At a first glance the upper sub-figure in Fig. 3, looks like a normal 50 Hz waveform, but it has a buried impulsive transient, starting in phase 170°, with maximum amplitude 30 % of normal amplitude. In this case the defect is not obvious, like the previous one.

Both peaks in Fig. 3 (b) give information about transient. Maximum value is 0.069, and minimum - 0.076, out of limits of 0.02 and -0.02.

Fig. 3 (c) is similar to Fig. 3 (b). It has peaks of similar form and exactly in the same point, created by the same transient. Peaks in the skewness are similar over and less than zero, but kurtosis has a normal value of 1.5, so peaks have approximately the same extension around that value, but always positive. Maximum value is 1.536 and minimum 1.450; they are out of the limits, of 0.53 and 0.47 although maximum is very near to the limit.

Fig. 3(d) is the variance evolution. In this case, there are peaks higher than the normal value, and a little peak lower.

But they are in the same place that previous peaks, exactly in the location of the defect. Maximum value is 0.537 and minimum 0.487, they are in the range of normal operation: 0.55 to 0.45.

"Sag-swell analysis" returns "Imprecise method", because conditions of Sag-swell are no present in this analysis.

In impulsive defects Skewness and Kurtosis are out of normal operation limits, but variance not. One value out of limits is enough to determine the presence of a defect, and in this case we have four.

With all this, we can determine an impulsive transient defect when there are peaks at the same place of all tree HOS graphs, and any of them are out of normal operations limits.

Test signals which have been detected are 93 sag/swell defects and 60 impulsive transient defects. This set of 153 test waveforms represents the 85% of the distorted signals which have been tested. "Sag-swell analysis" only fails in less than 12 % deviation of normal amplitude, from sag 87% to swell 112 %. In sag/swell out of that range, work perfectly. Normal operation limits has been fixed for detect all simulated defects, and ignore normal operation activity of two different signals sources and our electric power source (waveform in our plug).

IV. CONCLUSIONS

This paper has shown the performance of a virtual instrument conceived to detect electrical anomalies within a range pre-defined by the user. Its computational modulus is based in HOS, giving consequently extra information beyond the Gaussian characterization.

This VI has been tested with sags, swells and impulsive transients, detecting successfully the 85% of defective signals. User can adjust all range limits, adapting instrument's sensibility.

Sags are targeted by the presence of a "valley" in the variance graph, as well as a peak in the skewness and kurtosis graph in the starting and finishing the "valley". Swells are detected symmetrically. Impulsive transients are detected by targeting impulses in the same position of the three graphs (skewness, variance and kurtosis).

The VI can operate online and requires an eventual presence of the panelist, with the end of checking the historical of detection hints.

ACKNOWLEDGEMENTS

The authors would like to thank the *Spanish Ministry of Science and Innovation* for funding the research projects TEC2009-08988 and TEC2010-19242-C03-03 (SIDER-HOSAPQ). Our unforgettable thanks to the trust we have from the *Andalusian Government* for funding the Research Group PAIDI-TIC-168 in *Computational Instrumentation and Industrial Electronics-ICEI*.

REFERENCES

[1] H. Mõlder1, T. Vinnal, V. Beldjajev, "Harmonic Losses in Induction Motors Caused by Voltage Waveform Distortions", IEEE Electric Power Quality and Supply Reliability Conference (PQ), 2010, 16-18 June, Kuressaare, Estonia, pp. 143 –150.

[2] Elmoudi, A. ,Lehtonen, M., Nordman, H., "Effect of harmonics on transformers loss of life", IEEE Electrical Insulation, 2006. Conference Record of the 2006, 11-14 June, Toronto, Ont., pp. 408 - 411.

[3] Inan, A., Attar, F. "The life expectancy analysis for an electric motor due to harmonic's" IEEE Electrotechnical Conference, 1998. MELECON 98., 9th Mediterranean, 18-20 May, Tel-Aviv, vol.2 pp. 997 - 999.

[4] P. Caramia. G. Carpinelli. P. Verde, G. Mazzanti. A. Cavallini. G.C. Montanari, "An approach to Life Estimation of Electrical Plant Components in the Presence of Harmonic Distortion" IEEE Ninth International Conference on harmonics and Quality of Power, 2000. Proceedings, 1-4 Oct, Orlando, FL, vol.3 pp. 887 – 892.

[5] Juan A. Zorrilla de San Martín, "Detección automática de perturbaciones en la calidad de la energía eléctrica y clasificación basada en inteligencia artificial", CURSO PEDECIBA Redes Neuronales y Memorias Distribuidas 2007 – Fac. Ciencias.

[6] T. Radil, P.M. Ramos, F.M. Janeiro, A.C. Serra, "PQ Monitoring System for Real-Time Detection and Classification of Disturbances in a Single-Phase Power System", IEEE Transactions on Instrumentation and Measurement, Aug. 2008, Braunschweig, Germany, pp. 1725 – 1733.

[7] Wang Zhan-guo, Shi Huang-shuang, Wang Song, "A New Harmonic Analyzer based on Wavelet and Virtual Instrument", IEEE International Conference on Intelligent Human-Machine Systems and Cybernetics, 2009. IHMSC '09, 26-27 Aug., Hangzhou, Zhejiang, pp. 253 – 255.

[8] Juan José González de la Rosa, Antonio Moreno Muñoz, A. Gallego, R. Piotrkowski, E. Castro, "Higher-order characterization of power quality transients and their classification using competitive layers", Measurement (Ed. Elsevier) 42 (Issue3) (2009) pp. 478–484.

[9] Juan José González de la Rosa, Antonio Moreno Muñoz, "Higher-order cumulants and spectral kurtosis for early detection of subterranean termites," Mechanical Systems and Signal Processing, Volume 22, Issue 2, February 2008, pp. 279-294.

[10] C. L. Nikias, A. P. Petropulu, Higher-Order Spectra Analysis. A Non-Linear Signal Processing Framework, Englewood Cliffs, NJ, Prentice-Hall, 1993.

[11] C. L. Nikias, J. M. Mendel, Signal processing with higher-order spectra, IEEE Signal Processing Magazine (1993) 10–37.

Support Vector Machine for Power Quality Disturbances Classification using Higher-Order Statistical Features

J.C. Palomares-Salas[a,b,1], A. Agüera-Pérez[a,b], J.J.G. de la Rosa[a,b]

Abstract—**Support Vector Machine (*SVM*), which is based on Statistical Learning theory, is a universal machine learning method. This paper proposes the application of *SVM* in classifying to several power quality disturbances. For this purpose, a process based in *HOS* has been realized to extract features that help in classification. In this stage the geometrical pattern established via higher-order statistical measurements is obtained, and this pattern is function of the amplitudes and frequencies of the power quality disturbances associated to the 50-Hz power-line. Once the features are managed will be segmented to form training and test sets and them will be applied in the statistical method used to perform automatic classification of *PQ* disturbances. The result is shown according to correlation and mistake rates.**

I. INTRODUCTION

The increasing pollution of power signals and its impact on the power quality supplied by power plants to customers are pushing forward the development of signal processing tools to monitor and control of Power Quality (*PQ*) anomalies. For this reason, the interest of the research community in *PQ* has dramatically increased over the past decade. Among the most important reasons behind such interest are [1]: (1) the release of electric power markets which generates a competitiveness based on the electric quality of service, (2) the increased complexity of power networks, and, (3) the major sensitivity of modern loads to power quality disturbances. Consequently, monitoring of *PQ* disturbances is essential to offer solutions to industrial and electrical areas.

With the increasing amount of measurement data from *PQ* monitors, it is desirable that analysis, characterization, classification, and compression can be performed automatically [2], [3], [4]. Further, it is desirable to find out the cause of each disturbance, for example, whether a voltage dip is caused by a fault or by some other system event such as motor starting or transformer energizing. For this purpose, companies are putting a lot of efforts in innovation and new technologies because today's equipment, and automated manufacturing devices, are highly sensitive to the power line signal's imperfections (*PQ* events), making the production cost excessive. Malfunctioning not only has to be detected, but also predicted, undoubtedly diagnosed and localized, to identify the cause and prevent the system from a similar shock. This would be reflected *a posteriori* in an enhancement of the industrial production [5], [6].

[a] Research Group PAIDI-TIC-168: Computational Instrumentation and Industrial Electronics (ICEI)
[b] University of Cádiz. Area of Electronics. EPSA. Av. Ramón Puyol S/N. E-11202-Algeciras-Cádiz-Spain
[1] E-mail: josecarlos.palomares@uca.es

Hence, it is evident that a computerized system analysis is essential for the realization of effective and efficient power quality diagnosis systems. In the state of the art there are several classification techniques. Nevertheless, many open question remain, particularly concerning the most effective methods for classifying *PQ* problems.

This paper represents another step in this direction. To start with, many signals are generated by simulating different types of events causing poor power quality. These signals are used as inputs to feature extraction module which is used for selecting a number of representative coefficients of the generated signals. Such representative coefficients are based on higher-order characterization of *PQ* disturbances according to the maxima and minima of the 2nd, 3rd, 4th, 5th and 6th-order cumulants at zero lags (directly related to the variance, skewness, kurtosis, standard moment 5th and 6th-order). Once the features vectors have been obtained, we will apply double crossvalidation method to obtain training and validation sets which will be used in variety of classify techniques. Finally, the correlation and mistake rates of different classification algorithms have been obtained for comparison of them.

The paper is structured as follows. The following Section 2 explains the higher-order statistics. A brief summary of the *SVM* classification technique is shown in Section 3. Finally, the results are presented in Section 4 and conclusions are drawn in Section 5.

II. HIGHER-ORDER STATISTICS

HOS measures are extensions of second-order measures (such as the autocorrelation function and power spectrum) to higher orders. The second-order measures work fine if signals has a Gaussian (Normal) probability density function but many real-life signals are non-Gaussian. Moreover some contributions have demonstrated that *HOS*-based techniques are more appropriate to deal with non-Gaussian processes and nonlinear systems than $2^{n}d$-order-based ones. Remarkable results regarding detection, classification and system identification with cumulants-based techniques have been reported.

Higher-order cumulants are used to infer new properties about the data non-Gaussian process. In multiple-signal processing it is very common to define the combinational relationship among the cumulants of r stochastic signals, $\{x_i\}_{i\in[1,r]}$, and their moments of order $p, p \leq r$, given by using the *Leonov-Shiryaev* formula [7], [8]

$$Cum(x_1, ..., x_r) = \sum (-1)^{p-1} \cdot (p-1)! \cdot E\{\prod_{i \in s_1} x_i\}$$
$$\cdot E\{\prod_{i \in s_2} x_j\} \cdots E\{\prod_{i \in s_p} x_k\}, \tag{1}$$

where the addition operator is extended over all the partitions, like one of the form $(s_1, s_2, \ldots, s_p)$, $p = 1, 2, \cdots, r$; and $(1 \leq i \leq p \leq r)$; being s_i a set belonging to a partition of order p, of the set of integers $1, \ldots, r$.

Let $\{x(t)\}$ be an rth-order stationary random real-valued process. The rth-order cumulant is defined as the joint rth-order cumulant of the random variables $x(t)$, $x(t+\tau_1), \ldots$, $x(t+\tau_{r-1})$,

$$C_{r,x}(\tau_1, \tau_2, \ldots, \tau_{r-1}) = Cum[x(t), x(t+\tau_1), \ldots, x(t+\tau_{r-1})] \tag{2}$$

The second-, third-, fourth-, fifth- and sixth-order cumulants of zero-mean $x(t)$ can be expressed via [9]:

$$C_{2,x}(\tau) = E\{x(t) \cdot x(t + \tau)\} \tag{3a}$$

$$C_{3,x}(\tau_1, \tau_2) = E\{x(t) \cdot x(t + \tau_1) \cdot x(t + \tau_2)\} \tag{3b}$$

$$
\begin{aligned}
C_{4,x}&(\tau_1, \tau_2, \tau_3) \\
&= E\{x(t) \cdot x(t + \tau_1) \cdot x(t + \tau_2) \cdot x(t + \tau_3)\} \\
&\quad - C_{2,x}(\tau_1)C_{2,x}(\tau_2 - \tau_3) \\
&\quad - C_{2,x}(\tau_2)C_{2,x}(\tau_3 - \tau_1) \\
&\quad - C_{2,x}(\tau_3)C_{2,x}(\tau_1 - \tau_2)
\end{aligned}
\tag{3c}
$$

$$
\begin{aligned}
C_{5,x}&(\tau_1, \tau_2, \tau_3, \tau_4) \\
&= E\{x(t) \cdot x(t + \tau_1) \cdot x(t + \tau_2) \cdot x(t + \tau_3) \\
&\quad \cdot x(t + \tau_4)\} \\
&\quad - C_{2,x}(\tau_1)C_{2,x}(\tau_2 - \tau_3 - \tau_4) \\
&\quad - C_{2,x}(\tau_2)C_{2,x}(\tau_3 - \tau_4 - \tau_1) \\
&\quad - C_{2,x}(\tau_3)C_{2,x}(\tau_4 - \tau_1 - \tau_2) \\
&\quad - C_{2,x}(\tau_4)C_{2,x}(\tau_1 - \tau_2 - \tau_3)
\end{aligned}
\tag{3d}
$$

$$
\begin{aligned}
C_{6,x}&(\tau_1, \tau_2, \tau_3, \tau_4, \tau_5) \\
&= E\{x(t) \cdot x(t + \tau_1) \cdot x(t + \tau_2) \cdot x(t + \tau_3) \\
&\quad \cdot x(t + \tau_4) \cdot x(t + \tau_5)\} \\
&\quad - C_{2,x}(\tau_1)C_{2,x}(\tau_2 - \tau_3 - \tau_4 - \tau_5) \\
&\quad - C_{2,x}(\tau_2)C_{2,x}(\tau_3 - \tau_4 - \tau_5 - \tau_1) \\
&\quad - C_{2,x}(\tau_3)C_{2,x}(\tau_4 - \tau_5 - \tau_1 - \tau_2) \\
&\quad - C_{2,x}(\tau_4)C_{2,x}(\tau_5 - \tau_1 - \tau_2 - \tau_3) \\
&\quad - C_{2,x}(\tau_5)C_{2,x}(\tau_1 - \tau_2 - \tau_3 - \tau_4)
\end{aligned}
\tag{3e}
$$

By putting $\tau_1 = \tau_2 = \tau_3 = \tau_4 = \tau_5 = 0$ in Eq. (3), we obtain

$$\gamma_{2,x} = E\{x^2(t)\} = C_{2,x}(0) \tag{4a}$$

$$\gamma_{3,x} = E\{x^3(t)\} = C_{3,x}(0, 0) \tag{4b}$$

$$\gamma_{4,x} = E\{x^4(t)\} - 3(\gamma_{2,x})^2 = C_{4,x}(0, 0, 0) \tag{4c}$$

$$\gamma_{5,x} = E\{x^5(t)\} - 10\gamma_{3,x}\gamma_{2,x} = C_{5,x}(0, 0, 0, 0) \tag{4d}$$

$$
\begin{aligned}
\gamma_{6,x} &= E\{x^6(t)\} - 15\gamma_{4,x}\gamma_{2,x} - 10(\gamma_{3,x})^2 - 15(\gamma_{2,x})^3 \\
&= C_{6,x}(0, 0, 0, 0)
\end{aligned}
\tag{4e}
$$

The expressions in Eq. (4) are measurements of the variance, skewness, kurtosis, fifth-order standard moment, and sixth-order standard moment of the distribution in terms of cumulants at zero lags (the central cumulants).

Normalized sixth-order standard moment, fifth-order standard moment, kurtosis and skewness are defined as $\gamma_{6,x}/(\gamma_{2,x})^3$, $\gamma_{5,x}/(\gamma_{2,x})^{5/2}$, $\gamma_{4,x}/(\gamma_{2,x})^2$ and $\gamma_{3,x}/(\gamma_{2,x})^{3/2}$, respectively. We will use and refer to normalized quantities because they are shift and scale invariant. If $x(t)$ is symmetrically distributed, its skewness is necessarily zero (but not *vice versa*, almost impossible situations); if $x(t)$ is Gaussian distributed, its kurtosis is necessarily zero (but not *vice versa*). In the section IV, results are obtained by using sliding cumulants, i.d. a moving window in the time domain over which to compute the each cumulant (3rd, 4th, 5th and 6th-order cumulants for zero time-lag).

III. SUPPORT VECTOR MACHINE

Support vector machines are a set of related supervised learning methods used for classification and regression. Two main issues of interest in *SVM* classifiers are the generalization performance and the complexity of classifier which they make this method more attractive than many other classification methods. First is due the performance in the test set and not in training set. For a *SVM* classifier, there is a guarantee of the upper error bound on the test set based on statistical learning theory. Complexity of classifiers is a practical implementation issue. For a *SVM*, the complexity of the classifier is associated with the so-called *VC* dimension.

Given a set of training examples, each marked as belonging to one of two categories as it is shown in Fig. 1, a *SVM* training algorithm builds a model that predicts whether a new example falls into one category or the other. More formally, support vector machine introduced by Vapnik [10] uses the concept of support vector (*SV*) methods to construct a optimal separating hyperplane or set of hyperplanes in a high or infinite dimensional space. A good separation is achieved by the hyperplane that has the largest distance to the nearest training data-points of any class (so-called functional margin), since in general the larger the margin the lower the generalization error of the classifier. An *SVM* classifier minimizes the generalization error on the test set under the structural risk minimization (SRM) principle. The process methodology can be found in [11].

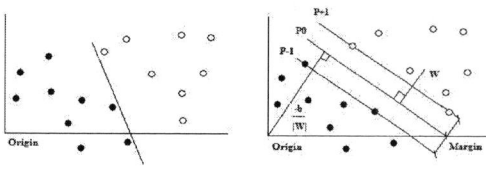

Fig. 1.   Separating hyperplane.

IV. DATA ARRANGEMENT AND EXPERIMENTAL RESULTS

In order to obtain representative signals with the most common *PQ* disturbances, we have used *Matlab R2008b* software to generate them. The generated disturbances have been: sags, swells and oscillatory transients (*OT*). A battery of these signals is generated with the following characteristics: the employed decaying laws are exponential, linear and parabolic, and each disturbance vanishes within less than half a cycle (2*ms*), corresponding to the power line. A sample frequency of 0.1*ms* is chosen and each time-series under test for computation performance contains eight cycles of the power-line. For the *OT* a frequency and amplitude swept have been done. The frequency range is 100 to 4500 to step of 100 Hz and the amplitude range is 0.05 to 0.9 to step of 0.01 V. On other hand, an amplitude swept of −0.9 to 0.9 to step of 0.01 V for sags and swells have been performed. Once the synthetic database has been obtained, we apply the proposed technique. Such technique consists two phases: feature extraction phase and classification phase.

A. Feature extraction phase

As mentioned above, the use of higher-order statistical (*HOS*) based on cumulants seems to be a very promising approach for disturbance detection in voltage signals because they are more appropriate for dealing with Gaussian signals.

The signal analysis is performed using an initial condition which helps to detect disturbances in the sinusoidal power line. This condition states that the sliding window used to extract *HOS* features must enclose an exact number of cycles of the 50-*Hz* power line. In this work the analysis will be based on windows of 0.02 seconds which cover 1 cycle. Thus, displacing the sliding window along a healthy signal, the set of values analyzed by the *HOS* processing is identical, returning an specific constant value. Any disturbance or anomaly in the healthy signal will produce variations from this constant value, thereby revealing its existence.

The first step to detect disturbances in the power line is the definition of a healthy signal. In the Fig. 2 can be observed that a pure sinusoidal signal with a normalized amplitude (healthy signal) is associated to the following constant values for the *HOS* estimators: 0.5 for the variance, 0 for the skewness, −1.5 for the kurtosis, 0 for the fifth-order standard moment and 2.4813 for the sixth-order standard moment. The simple characterization of healthy signals, according to the proposed methodology is this phase, helps to detect any coupled disturbance as a deviation in the *HOS* values.

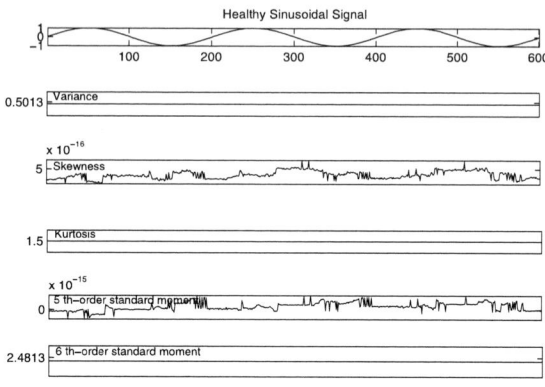

Fig. 2.   Characterization of *HOS* values for healthy sinusoidal signal.

Then, for each register the second-, third-, fourth-, fifth- and sixth-order cumulants are computed according to the following procedure. For a given statistical order the feature extraction algorithm computes a cumulant over the sliding window, and then it jumps to the following starting point (overlapping 98%). Besides, each nth-order cumulant, $Cum_{n,x}[i]$, associated to the ith computation window has been normalized by $Cum_{2,x}[i]^2$, so that it give a real statistical characterization. When the computation sweep along a signal is finished, the maximum and the minimum are calculated for all the sliding windows results. Consequently, each signal is characterized by its second, third, fourth, fifth and sixth-order extremes.

This procedure can be showing in the Fig. 3 where it display the results of the proposed *HOS* analysis performed on a synthetic, which comprises the three studied disturbance kinds. The perturbations introduced in the healthy signal modify the *HOS* values of them, such it can be observed.

B. Classification phase

Once the signal characteristics have been obtained, which will help us in the classification process, a database was synthetically generated. This database consists of 30,600 samples including the three studied disturbance kinds and it is composed as follows: 10,460 *OT* signals, 10,080 sag signals and 10,060 swell signals. Finally we have a matrix with 30,600 samples with three classes (corresponding to perturbation types) and 10 dimensions (corresponding to *HOS* values).

978-1-4244-8806-3/11 $26.00 © 2011 IEEE

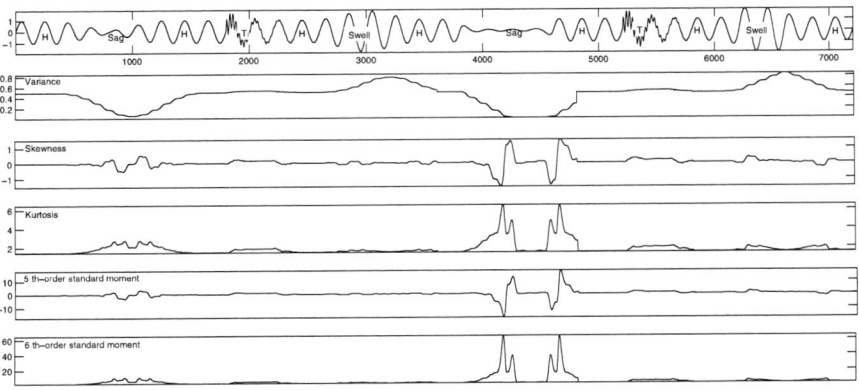

Fig. 3. On site measurement simulation which shows the evolution of the constant values associated to six statistics (variance, skewness, kurtosis, fifth-order standard moment and sixth-order standard moment).

TABLE I
THE CORRELATION AND MISTAKE RATES OF *SVM* METHOD.

	Mean Correct Rate	Mean Error Rate
SVM	0.9994	$6.3667e^{-4}$

The algorithm used to select the training and test sets is common to all these methodologies. This algorithm is the double crossvalidation which selects two sets of available data, one for training and another for test. Then we apply the *SVM* technique as mentioned above in the section III.

- The adjustable parameters of *SVM* are: kernel function that maps the training data into kernel space, the sequential minimal optimization method to find the separating hyperplane and box constraint of $1,000$ for the soft margin. The available kernel functions are: '*linear*', '*quadratic*', '*gaussian radial basis*', '*polynomial*' and '*multilayer perceptron*'. In the Gaussian radial basis function kernel a scale factor (*sigma*) of 0.7 is specified. For the polynomial kernel a $5th$-degree polynomial is chosen and the scale and bias parameters in the multilayer perceptron kernel are selected to $[0.7, -1]$. As in the previous used method, a composed system of three *SVM* networks is used and the correlation rate and mistake rate in each of them were calculated. Again the used training and test sets are selected from the matrix of 10 dimensions. We selected several kernel functions to perform various simulations and the average value of these three obtained values in each experiment have been calculated.

Table I shows the correlation and mistake rates of selected method.

V. CONCLUSION

In this work a detection automatic system and classification of three kinds of *PQ* disturbances has been developed. The primary question considered to carry out the design of a classifier is the choice of appropriate features. Proper selection of features will make a more precise learning process in the automatic systems.

The features extraction phase proposed in this work is built on *HOS*. The obtained statistics in this phase are as follows: variance, skewness, kurtosis, 5 th-order standard moment and 6 th-order standard moment. The signal processing methodology is based on a tuned 50-*Hz* sliding window over which compute each statistical estimator; thus processing the 50-*Hz* events produces constant values to detect the occurrence and the start and end point of events in voltage signals. This tuned window allows signal analysis without preprocessing, i.e. preserving all the information and without introducing nonlinear behaviors.

The *SVM* classifier has shown very satisfactory results using *HOS* features, obtaining an average correlation rate of 99.94%. This is because this one is a robust statistical classification method. Then, we can say that higher-order statistics are appropiated to characterize power quality signals.

ACKNOWLEDGMENT

The authors would like to thank the *Spanish Ministry of Science and Innovation* for funding the research projects $TEC2009-08988$ and $TEC2010-19242-C03-03$ (*SIDER-HOSAPQ*). Our unforgettable thanks to the trust we have from the *Andalusian Government* for funding the Research Group PAIDI-TIC-168 in *Computational Instrumentation and Industrial Electronics-ICEI*.

REFERENCES

[1] M. V. Ribeiro, C. A. G. Marques, C. A. Duque, A. S. Cerqueira, and J. L. R. Pereira, "Detection of disturbances in voltage signals for power quality analysis using *HOS*," *Journal on Advances in Signal Processing*, vol. 2007, no. 1, pp. 1–13, February 2007.
[2] A. K. Khan, "Monitoring power for the future," *Power Engineering Journal*, vol. 15, no. 2, pp. 81–85, April 2001.
[3] M. McGranaghan, "Trends in power quality monitoring," *IEEE Power Engineering Review*, vol. 21, no. 10, pp. 3–9, October 2001.
[4] M. H. J. Bollen, *Understanding Power Quality Problems: Voltage Sags and Interruptions*. New York, NY, USA: Wiley-IEEE Press, 1999.

[5] A. Moreno and *et al*, *Mitigation Technologies in a Distributed Environment*, 1st ed., ser. Power Systems. Springer-Verlag, 2007.

[6] "IEEE Recommended practice for monitoring electric power quality," The Institute of Electrical and Electronics Engineers, Inc., Tech. Rep. IEEE Std. 1159-1995, 1995.

[7] C. L. Nikias and J. M. Mendel, "Signal processing with higher-order spectra," *IEEE Signal Processing Magazine*, vol. 10, no. 3, pp. 10–37, 1993.

[8] J. M. Mendel, "Tutorial on higher-order statistics (spectra) in signal processing and system theory: theorical results and some applications," in *Proceedings of the IEEE*, vol. 79, no. 3, March 1991, pp. 278–305.

[9] A. K. Nandi, *Blind Estimation using Higher-Order Statistics*. Boston: Kluwer Academic Publichers, 1999, vol. 1, no. 1.

[10] V. Vapnik, *The Nature of Statistical Learning Theory*, A. Press, Ed. Springer-Verlag, New York, 1995.

[11] R. G. Brereton and G. R. Lloyd, "Support vector machines for classification and regression," *Analyst*, vol. 135, pp. 230–267, 2010.

Reflections on current components according to CPC concept in a single phase circuit with nonsinusoidal voltage and current waveforms and LTI load

Marek T. Hartman
Faculty of Marine Electrical Engineering
Gdynia Maritime University
Gdynia, Poland
mhartman@am.gdynia.pl

Abstract—**Properties of the current decomposition according to Czarnecki's concept have been analyzed. The paper showed that components of scattered current are either in the phase or in opposition to their corresponding active current harmonics. Each harmonic of the scattered current with the supply voltage generate the active power, however the sum of these powers is equal to zero in period T. It has been proposed to use the sum of active and scattered currents in the compensation procedure of the reactive current according to Czarnecki's concept as well as in scattered current elimination, to get the optimal energy transfer from the source supply to the load.**

Keywords- power theory, Current Physical Components, reactive power compensation, current formatting

I. INTRODUCTION

23 years after Czarnecki had asked his question "What is wrong with Budeanu's concept..." [1], a scientific confirmation of erroneousness of the proposed by Budeanu concept of reactive power and deformation power was obtained [5] and the concept ceased being recommended for the use in engineering practice [10]. Thus, within the area of the power theory in a single-phase circuit, we have at present the following propositions:

- Depenbrock's concept distinguishing a basic harmonic and some others in non-sinusoidal waveforms; on this distinction the definition of power was based. This concept is contained in the latest Standard IEEE [10],
- the concept proposed by Czarnecki, called Current's Physical Components (CPC), which constitutes the original development of the concept proposed by Fryze [3,4] as well as Shepherd and Zakikhani [8],
- the concept proposed by Tenti and Mattavelli, called Conservative Power Theory (CPT), which is the original development of the concept of Fryze and Iliovici [7].

The aim of the Author is to show the properties of Czarnecki's concept, especially to emphasize its usefulness during compensation of reactive current and elimination of scattered current in single-phase circuits with nonsinusoidal voltage and current waveforms and reactive linear load, time invariant (LTI).

II. CPC CONCEPT IN SINGLE – PHASE CIRCUITS WITH LTI LOADS

The basic concept of CPC was presented on the basis of the publication [2], which was the final effect of many earlier publications of Czarnecki.
„ Let us assume that linear time-invariant load, shown in Fig .1 is supplied with voltage

$$u(t) = \sqrt{2}\,\text{Re}\{ \sum_{n \in N} U_n e^{j(n\omega_1 t + \varphi_n)} \} \qquad (1)$$

where: $n = 1.....N$ (DC component is omitted).

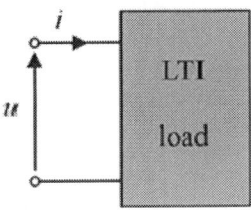

Fig.1 The circuit with nonsinusoidal voltage $u(t)$ and current $i(t)$ with LTI load

The load can be characterized by admittance for harmonic frequencies

$$\underline{Y}_n = G_n + jB_n = Ye^{j\phi_n} \qquad (2)$$

thus, the current can be expressed as

$$i(t) = \sqrt{2}\,\text{Re}\{(G_n + jB_n)U_n e^{j(n\omega_1 t + \varphi_n)} \qquad (3)$$

On the basis of Fryze's concept [3,4], the load current contains the active component , known as the active

978-1-4244-8806-3/11 $26.00 © 2011 IEEE

current $i_a(t)$. It is the current component proportional to the supply voltage $u(t)$, of the minimum value needed for energy permanent transfer to the load with the average rate equal to the active power P. It can be expressed as

$$i_a(t) \triangleq G_e u(t) = \sqrt{2}\, \text{Re}\{ \sum_{n \in N} G_e U_n e^{j(n\omega_1 t + \varphi_n)} \} \qquad (4)$$

Since the active power of resistive load is $P = G\|u\|^2$, such a load is equivalent with respect to the active power P at voltage $u(t)$ to the original load if its conductance is

$$G = G_e \triangleq \frac{P}{\|u\|^2} \qquad (5)$$

It is referred to as *equivalent conductance*. The remaining component of the load current $i(t)$ (3), after subtracting the active current $i_a(t)$ is equal to

$$i(t) - i_a(t) = \sqrt{2}\, \text{Re}\{ \sum_{n \in N} (G_n + jB_n - G_e) U_n e^{j(n\omega_1 t + \varphi_n)} \} \qquad (6)$$

The equation (6) can be decomposed into

$$\sqrt{2}\, \text{Re}\{ \sum_{n \in N} (G_n - G_e) U_n e^{j(n\omega_1 t + \varphi_n)} \triangleq i_s(t) \qquad (7)$$

and

$$\text{Re}\{ \sum_{n \in N} jB_n U_n e^{j(n\omega_1 t + \varphi_n)} \} \triangleq i_r(t) \qquad (8)$$

This decomposition reveals the two entirely different phenomena, which contribute to these useless components of the supply current of LTI load at nonsinusoidal supply voltage:

1. The presence of the current harmonic component shifted by 90^0 with respect to the voltage harmonics. Their sum is the *reactive current* $i_r(t)$. In fact, it is identical with Shepherd and Zakikhani's [8] quadrature current.

2. The difference of the load conductance G_n for harmonic frequency from the load equivalent conductance G_e. The sum of these components creates current $i_s(t)$. Since it occurs when conductance G_n is scattered around the equivalent conductance G_e, this current was called *scattered current* [2]. The presence of the scattered current in the supply current of LTI loads is a new phenomenon revealed by this decomposition.

Thus, the supply current of LTI loads at nonsinusoidal voltage can be expressed as

$$i(t) = i_a(t) + i_s(t) + i_r(t) \qquad (9)$$

The current of (9) contains three components associated with three distinctively different phenomena in the load:

1. Permanent energy conversion: - the active current $i_a(t)$

2. Change of the load conductance G_n with harmonic order: - the scattered current $i_s(t)$

3. Phase-shift between the voltage and current harmonics: - the reactive current $i_r(t)$.

Therefore, these currents were called Current's Physical Components (CPC)"

Czarnecki stressed that "The objective 'physical' does not mean however, that these currents do exist physically. They do not exist as physical entities, but like harmonics,

only as mathematical entities, associated with some physical phenomena in the load.

Czarnecki wrote that the scalar product of periodic components is equal to zero, so the particular components of the load current are mutually orthogonal:

$$(i_a, i_s) \triangleq \frac{1}{T} \int_0^T i_a(t) i_s(t) dt = 0 \qquad (10)$$

$$(i_a, i_r) \triangleq \frac{1}{T} \int_0^T i_a(t) i_r(t) dt = 0 \qquad (11)$$

$$(i_s, i_r) \triangleq \frac{1}{T} \int_0^T i_s(t) i_r(t) dt = 0 \qquad (12)$$

so consequently

$$\|i\|^2 = \|i_a\|^2 + \|i_s\|^2 + \|i_r\|^2 \qquad (13)$$

Multiplying eq. (13) by the square of the voltage rms value, the power equation of LTI loads is obtained

$$S^2 = P^2 + D_s^2 + Q_{S/Z}^2 \qquad (14)$$

where

$D_s \triangleq \|u\|\|i_s\|$ is the scattered power and $Q_{S/Z} \triangleq \|u\|\|i_r\|$ is Shepherd and Zakikhani's reactive power.

III. SOME REMARKS TO CZARNECKI'S CONCEPT

A. Reflection on the scattered current $i_s(t)$

The definition of the equivalent conductance G_e is given by equation (5). On the other hand, the active power P can be calculated as

$$P = \sum_{n \in N} G_n U_n^2 \qquad (15)$$

Based on equations (5) and (15) we can write

$$G_e \triangleq \frac{P}{\|u\|^2} = \frac{\sum_{n \in N} G_n U_n^2}{\sum_{n \in N} U_n^2} = \frac{G_1 U_1^2 + G_2 U_2^2 + .. + G_n U_n^2}{U_1^2 + U_2^2 + .. + U_n^2} \qquad (16)$$

where $G_e \geq 0$ and $G_n \geq 0$ for $n = 1....k.....N$ and

$$(G_1, .., G_k \geq G_e \geq (G_{k+1}, .., G_N) \qquad (17)$$

or

$$(G_1, .., G_k \leq G_e \leq (G_{k+1}, ... G_N) \qquad (18)$$

depends on the load character.

According to relations (17) and (18) "...*conductance* G_n *is scattered around the equivalent conductance* G_e ". Based on equations (4) and (7) one can easily notice that harmonics of the active current and the scattered current have the same phase-shift angles or are shifted by π e.g.

$$i_a(t) = \sqrt{2} G_e \{ U_1 \sin(\omega_1 t + \varphi_1) + U_2 \sin(2\omega_1 t + \varphi_2) + \\ + U_3 \sin(3\omega_1 t + \varphi_3) + ... + U_n \sin(n\omega_1 t + \varphi_n) \} \qquad (19)$$

$$i_s(t) = \sqrt{2}\{(G_1 - G_e)U_1 \sin(\omega_1 t + \varphi_1) +$$
$$+(G_2 - G_e)U_2 \sin(2\omega_1 t + \varphi_2) +$$
$$+(G_3 - G_e)U_3 \sin(3\omega_1 t + \varphi_3) + ... \qquad (20)$$
$$+(G_n - G_e)U_n \sin(n\omega_1 t + \varphi_n)\}$$

Knowing (17) and (18), one can see that a particular harmonic of the scattered current is in phase or in opposition to the same harmonic of the active current and to the supply voltage depending on the load character

$$i_{sn}(t) + i_{an}(t) = \pm\sqrt{2}G_n U_n \sin(n\omega t + \varphi_n) \qquad (21)$$

If $i_{sn}(t) + i_{an}(t) = -\sqrt{2}G_n U_n \sin(n\omega t + \varphi_n)$, the sign minus (-) means that a particular harmonic of the scattered current is in opposition to the same harmonic of the active current (Fig. 2)

a)

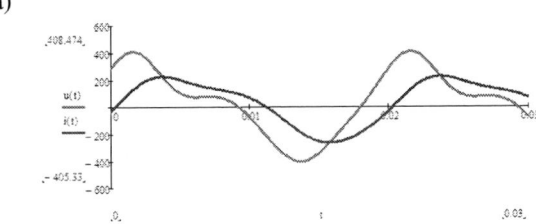

b)

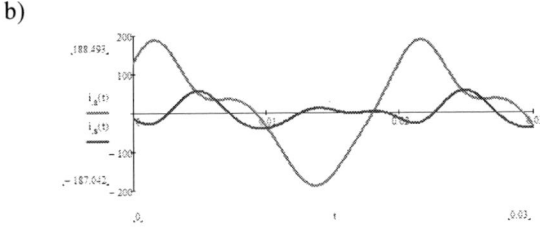

c)

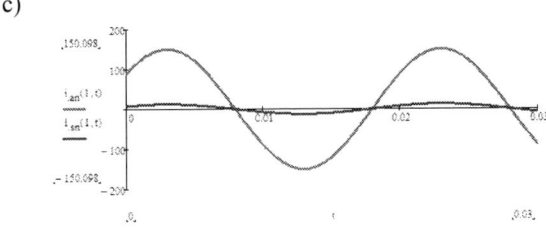

d)

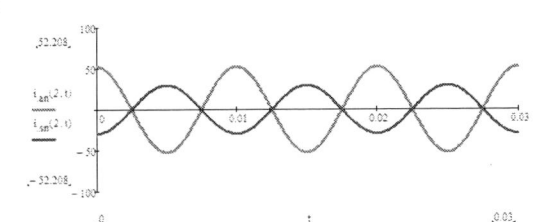

Fig. 2 Example of waveforms: (a) supply voltage and current,(b) the active current and the scattered current,(c) harmonics for n=1 of the active current and the scattered current, (d)) harmonics for n=2 of the active current and the scattered current, in a circuit with LTI load

In the circuits both the current components exist simultaneously and their sum is equal to

$$i_a(t) + i_s(t) = \sqrt{2}\{G_1 U_1 \sin(\omega_1 t + \varphi_1) +$$
$$+G_2 U_2 \sin(2\omega_1 t + \varphi_2) +$$
$$+G_3 U_3 \sin(3\omega_1 t + \varphi_3) + ... \qquad (22)$$
$$+G_n U_n \sin(n\omega_1 t + \varphi_n)\}$$

The mathematical current decomposition proposed by Czarnecki has the specific feature: minus (-) in the scattered current description implicates that each harmonic of the sum of currents ($i_a(t) + i_s(t)$ is in the phase or in opposition to its corresponding supply voltage harmonic. This mathematical combination corresponds to the current flowing through fictitious resistance load of variable resistance for each harmonic. What is more, the sum of these currents transfers the energy of the power equal to the active power P to this load

$$\frac{1}{T}\int_0^T \{i_a(t) + i_s(t)\}u(t)dt = G_1 U_1^2 + G_2 U_2^2 + ... + G_n U_n^2 = P \qquad (23)$$

One can also notice that the scattered current cannot be "generated" by any passive LTI load. As Czarnecki stressed very strongly: this component is a result of mathematical calculation and no physical interpretation can be concluded. However, despite no physical interpretation of the scattered current, this component can be very useful in engineering practice. During the elimination procedure one can remove the scattered current from the line current to generate, by the so called active filter, the current equal to this scattered current. The scattered current components do not influence the active power P but they influence the supply current shape.

B. Reflection on the active current $i_a(t)$ and the scattered current $i_s(t)$.

The proof of orthogonality of the active current and the scattered current given by Czarnecki [2] was based on the identity of equations (5) and (15) defining the active power P. Based on (7), (19) and (20) it can be stated that each harmonic of the same order of the active current (19) is in the phase or is shifted by π in relation to the scattered current (20) e.g.

$$i_{a1}(t) = \sqrt{2}G_e U_1 \sin(\omega_1 t + \varphi_1) \qquad (24)$$
$$i_{s1}(t) = \sqrt{2}(G_1 - G_e)U_1 \sin(\omega_1 t + \varphi_1) \qquad (25)$$

Then, the equation (10) is fulfilled, but the vectors $I_{a1}e^{j\varphi_1}$ and $I_{s1}e^{j\varphi_1}$ representing the instantaneous currents are either in phase or shifted by π depending on the load character (17)(18), as shown in figure 3.

a)

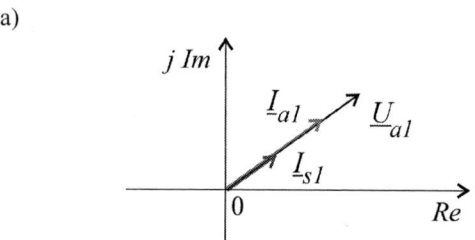

b)

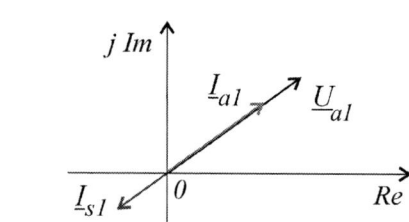

c)

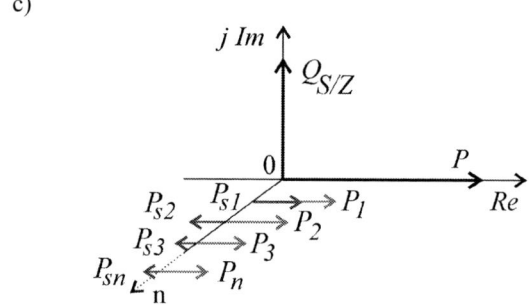

Fig. 3 Arrangement on the complex plane a) vectors $\underline{I}_{a1}$ and $\underline{I}_{s1}$ when $G_1 > G_e$,b) vectors $\underline{I}_{a1}$ and $\underline{I}_{s1}$ when $G_1 < G_e$, c) powers when $G_1 > G_e$ and $G_2, G_3, .G_n < G_e$

C . Reflection on the quadratic power equation and power orthogonality.

On the basis of deduction from part *B* and not assuming that particular components are orthogonal, the equation describing current decomposition $i(t)$ (9) can be written in the following form

$$i(t) = [i_a(t) + i_s(t)] + i_r(t)$$

Consequently, the effective value (rms) of the component $[i_a(t) + i_s(t)]$ is equal

$$\left\| i_a + i_s \right\|^2 = \left\| i_{as} \right\|^2 = \left\| i_a \right\|^2 + 2\left\| \sqrt{i_a i_s} \right\|^2 + \left\| i_s \right\|^2 \quad (26)$$

because

$$\frac{1}{T}\int_0^T 2\, i_a\, i_s dt = \frac{1}{T}\int_0^T \left(\sqrt{2\, i_a\, i_s}\right)^2 dt = \left\| \sqrt{2 i_a\, i_s} \right\|^2 = 2\left\| \sqrt{i_a\, i_s} \right\|^2$$

Since the decomposition components $[i_a(t) + i_s(t)]$ and $i_r(t)$ are orthogonal, the equation (13) has the form

$$\left\| i \right\|^2 = \left\| i_{as} \right\|^2 + \left\| i_r \right\|^2 \quad (27)$$

Multiplying eq.(27) by the square of the voltage rms value, the quadratic power equation of LTI loads is obtained in the form

$$S^2 = P_{as}^2 + Q_{S/Z}^2 \quad (28)$$

Despite formal similarity of the notation of the equation (28) to the concept proposed by Fryze, there exists a basic difference in both these concepts, so:
According to Fryze:

$$P = \left\| i_a \right\| \left\| u \right\| \quad (29)$$

and

$$Q_F = \left\| i_{rF} \right\| \left\| u \right\| = \left\| i_s + i_r \right\| \left\| u \right\| \quad (30)$$

According to (9) and (26)

$$P_{as} = \left\| i_{as} \right\| \left\| u \right\| \quad (31)$$

and

$$Q_{S/Z} = \left\| i_r \right\| \left\| u \right\| \quad (32)$$

On the basis of (26), we have

$$P_{as}^2 = P^2 + D_{as}^2 + D_s^2 \quad (33)$$

where:

$$D_{as}^2 = 2\left\| \sqrt{i_a i_s} \right\|^2 \left\| u \right\|^2 \text{ and } D_s^2 = \left\| i_s \right\|^2 \left\| u \right\|^2$$

C . Reflection on Shepherd and Zakikhani's reactive power $Q_{S/Z}$

Shepherd and Zakikhani's reactive power definition is given by (14) as

$$Q_{S/Z} \triangleq \left\| u \right\| \left\| i_r \right\| \quad (34)$$

where

$$\left\| i_r \right\| = \sqrt{\sum_{n \in N} B_n^2 U_n^2} = \sqrt{\sum_{n \in N} \left(\frac{Q_n}{U_n}\right)^2} \quad (35)$$

The equation (34) corresponding to the reactive (electromagnetic) energy accumulated in the load. The same procedure (but different mathematical equation) corresponding to the reactive energy (power) accumulated in the load can be found as modified Iliovici's proposal[6] or in Tenti's and Mattavelli's [9] concept.

D . Reflection on the transfer energy in the circuits

Multiplying the equation (9) by the sides by the value of supply voltage $u(t)$ and taking into account the direction of "the flow" of the scattered current $i_s(t)$, the equation of the instantaneous power is obtained in the following form

$$p(t) = p_a(t) + p_r(t) + p_s(t) \quad (36)$$

where:
$p_a(t) = i_a(t)u(t)$ is the instantaneous active power,
$p_r(t) = i_r(t)u(t)$ is the instantaneous reactive power,
$p_s(t) = i_s(t)u(t)$ is the instantaneous scattered power.

In the Author opinion, in relation to the load, the powers $p_a(t)$ and $p_r(t)$ are working powers and are characterized by the energy transfer from the supply source to the load . In many real applications the reactive power $p_r(t)$ is the working power because without it some loads would not operate, for instance: electrical machines. The power $p_s(t)$ can be treated as the excessive power, the load of which could not be converted into the working power and which can be successfully returned to the supply source . The equation (35) can be given some engineering

978-1-4244-8806-3/11 $26.00 © 2011 IEEE

interpretation: occurrence of the instantaneous scattered power $p_s(t)$ signifies non-optimal energy transfer from the supply source to the load. The load is not able to convert this energy , [represented by $p_s(t)$] into the working energy represented by the power $p_u(t)$

$$p_u(t) = p_a(t) + p_r(t) \qquad (37)$$

The scattered power $p_s(t)$ can be eliminated and at the same time can facilitate the transfer of the energy from the source to the load without changing the active power P because

$$P = \frac{1}{T}\int_0^T p(t)dt = \frac{1}{T}\int_0^T [p_u(t) + p_s(t)]dt = \frac{1}{T}\int_0^T p_u(t)dt \qquad (38)$$

The Author proposes to name the process of eliminating the scattered power (scattered current) *formatting the power (current).*

IV. CONCLUSIONS

The concept of Czarnecki described in the paper was and is a certain alternative to the other concepts [3,4,7,9,10]. The novelty of the presented analysis is a description of the properties of the sum of the active and scattered currents (22), which allows us to modify the concept of *CPC*. It is suggested that instead of the active component of the supply current one can use the sum of two components: active and scattered at the unchanged value of the active power dissipated in the load. These two components of the supply current remain after compensation of the reactive components of the supply current (the reactive power $Q_{S/Z}$). This constitutes a clue suggesting that we should first perform the compensation of the reactive component , for instance with the help of a reactive element, and then eliminate the scattered current component (the Author calls this process *formatting the current* and not the compensation of the scattered component of the supply current) with the use of the current generated by the so called active filter. The paper shows that the harmonics of the scattered current are either in the phase or in opposition to the harmonics of the active current corresponding to them.

The proposed by Czarnecki concept of power calculation in single-phase circuits with nonsinusoidal current and voltage waveforms and with the load LTI is based mainly on mathematical transformations without justification of all physical phenomena occurring in the circuit.

In the Author's opinion the concept of *CPC or its modified proposition can constitute a very useful algorithm of controlling compensators and active filters* in circuits with nonsinusoidal voltage and current waveforms and LTI load.

ACKNOWLEDGEMENT

The author thanks Tadeusz St. Piotrowski, Ph. D, from Department of Marine Electrical Engineering, Maritime University, Gdynia, Poland for his assistance and fruitful discussion on Czarnecki's concept.

REFERENCES

[1] Czarnecki L.S: What is wrong with Budeanu's Concept of Reactive and Distortion Power and Why it Should be Abandoned, *IEEE Transactions on Instrumentation and Measurement*, vol.IM-36, No.3, Sept.1987

[2] Czarnecki L.S.: Currents' Physical Components (CPC): Fundamental of Power Theory, *Przeglad Elektrotechniczny (Electrical Review)* R84, No.6/2008, p.28-37

[3] Fryze S.: Moc rzeczywista, urojona i pozorna w obwodach elektrycznych o przebiegach odkształconych prądu i napięcia. *Przegląd Elektrotechniczny*, issue.7, pp.192-203; issue.8, pp.225-234, 1931 and issue.22, pp.673-676, 1932 (in Polish)

[4] Fryze S.: Wybrane zagadnienia teoretyczne podstaw elektrotechniki. PWN, Warszawa-Wrocław 1966 (in Polish)

[5] Hartman M.T., Orthogonality of functions describing electric power quantities in Budeanu's concept. *Przegląd Elektrotechniczny (Electrical Review)*, ISSN 0033-2097, R. 87 NR 1/2011, p.14-18.

[6] Hartman M.T.:The integral method to calculate the power states in electrical circuits, *Przeglad Elektrotechniczny (Electrical Review)*, ISSN 0033-2097, R. 86 NR 3/2010, p.194-199

[7] Ilovici M.:Definition of measure de la puissance de l'energie reactives, *France Electr.* 1925, 49-52, 931-954

[8] Sheperd W., Zakikhani P. Suggested definition of reactive power for nonsinusoidal system. *Proc. IEE*, 119, (1972), no.9, 1361-1362

[9] Tenti P., Mattavelli P., Morales Parades H.: Conservative power theory, sequence components and accountability in smart grids. *Przegląd Elektrotechniczny (Electrical Rewiev)*, R.86 NR 6/2010, p.30-37

[10] IEEE Std 1459-2010, "IEEE Trial-Use Standard Definitions for the Measurement of Electric Power Quantities Under Sinusoidal, Nonsinusoidal, Balanced, Or Unbalanced Conditions

Harmonic Effect in Street Lighting

Aurora Gil de Castro*, Antonio Moreno Muñoz*, Víctor Pallarés López*, Agustín Agüera Pérez†
* Research Group PAIDI-TIC-168
University of Córdoba, Area of Electronics, Campus de Rabanales, Leonardo da Vinci building, E-14071 Córdoba, Spain
E-mail: agil@uco.es
† Research Group PAIDI-TIC-168
University of Cádiz, Area of Electronics, Av. Ramón Puyol S/N. E-11202-Algeciras-Cádiz, Spain

Abstract- **In this paper, several high power lighting networks based on high pressures sodium (HPS) lamps with rated lamp powers of 70, 150 and 250 W have been set up. They have also been tested with different levels of power using dimming for a 220 V power supply. These have been chosen because of they are used in public lighting. Each luminaire consists of sodium lamp with dimmable electronic ballast, which can dim the lamp output smoothly and uniformly. This paper focuses on the harmonic characterization of modern HPS lamps connected to dimmable electronic ballasts. Harmonics under full and reduced power for several HPS lamps have been measured. It presents an investigation on the attenuation effect and a proposal of an index called the equivalent lamp index to characterize this effect on HPS lamps. Also it is concluded that using a combination of HPS lamps can result in a reduction of the current distortion.**

I.   INTRODUCTION

The Public Lighting Systems in our cities are a basic and vital service for city councils and other public administration. On the one hand, citizens demand high quality service in accordance with our high development society. On the other hand, a lighting installation is an important energy consumption source that is affected by factors such as regulation and maintenance. The International Energy Agency (IEA) estimated that the potential energy savings by influencing the use and the technology of lighting were at approximately 133–212 TWh/year globally. The corresponding reduction of CO_2 emissions is about 86 to 137 MT/year for IEA countries. Lighting in the IEA countries requires the output of several hundreds of power stations, but consumption can be reduced by 15-20% and more (in the service sector alone) with existing technology. A recent study carried out for the European Commission has shown that between 30% and 50% of electricity used for lighting could be saved investing in energy-efficient lighting systems. In most cases, such investments are not only profitable but they also maintain or improve lighting quality [1]. The main recommendation is that streetlights and other outdoor lighting should be made more efficient as part of a comprehensive strategy to reduce CO_2 emissions including cleaner options for electricity generation, vehicle emissions, more energy efficient buildings, and smart electric meters combined with smart appliances which shift electricity use from peak to off-peak periods [2].

It is highly recommended the suitable election of the lamp. The majority of light sources used in public lighting are HID (High Intensity Discharge) lamps. They can be classified according to their color: orange light (low pressure sodium (LPS) lamps); yellow or orange light (high pressure sodium (HPS) lamps); and white light (high pressure mercury (HPM) lamps, metal halide (MH) lamps and ceramic metal halide (CMH) lamps) [1]. The minimum acceptable requirements for lighting controls are that they provide sufficient light for the users of a space, avoid waste and reduce lighting levels without compromising users' satisfaction and productivity. Therefore, by having an understanding of the lamps, ballasts, luminaires and control options available today as well as the techniques used to develop efficient lighting, lighting can be produced so that it is energy efficient, cost effective and yields a better quality of light [3].

The following measures are recommended for decreasing electrical energy consumption afferent to the public lighting [4]: first, reduction of the luminance level (dimming) on the duration of the hours with reduced traffic, through the decrease of the feeding voltage of the lamps (this adopted measurement allows a decrease in the electrical energy consumption); second, street classification compliant with international standards and the establishment of the light technical parameters based on this classification, just as utilization of performance lamps, and lighting devices; and third, proposal to adopt a special price for the electric energy destined for public lighting, due to the consumption on duration of the night. Another important measure is to encourage the installation of smart dimmable electronic ballasts which must be among other things: able to auto-detect lamp and electrical failures; able to measure and send data such as lamp status, lamp level, power consumption, voltage, current, and power factor; and able to receive switch and dimming commands from a streetlight segment controller.

Thus, the most important savings could be achieved with the installation of a centralized SCADA, which can be used to monitor failed lamps and report their location in addition to the age and condition of every lamp. Consequently, maintenance expenses (materials, routing, labor, etc.) could be minimized by considering the remaining life of nearby lamps that might be replaced during the same service call. Finally, data collected by the SCADA that tracks the hours of illumination for each lamp can be used to claim warranty replacement, establish unbiased product and supplier selection criteria, and validate energy bills for the system. Communicating with lighting controls requires software protocols; however it is difficult to develop a whole system approach to lighting controls compatible to all the components. Usually, communication protocols require a separate set of communication wiring, which adds cost to lighting control systems. Fig. 1 illustrates a schematic of the experimental setup which is based on the LONWorks Power-Line communications (PLC) protocol that supports a large number of media which makes integration with complex traffic management and geographic information systems easier. In this experiment, the protocol connects the segment controller with dimmable electronic ballasts equipped with a LONWorks node.

Nowadays, the main concern is that lamps produce harmonics on lighting networks depending on control gear quality and lamp age. The attenuation effect of many devices has been documented in several studies. Some of them have characterized this effect by using EMTP simulation or experimental tests. Recent researches have shown the harmonic attenuation effect of Compact Fluorescent Lamps (CFLs) [5], [6], [7] and [8] among others. Nevertheless, only a few numbers of authors have decided to study this effect over HPS lamps [9]. This paper presents a good understanding of the HPS lamps harmonic and the attenuation of harmonic currents, based on experimental tests with these kinds of lamps.

The reminder of the article is organised as follows: Section II outlines the background where it is explained the basis of the article. Section III contains a description of the experiments that have been done. Results are shown and explained in section IV. In section V it is analysed the harmonic attenuation in the previous experiments. Section VI concludes.

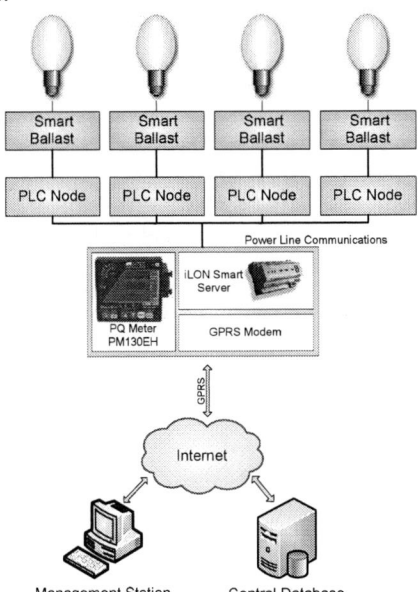

Fig. 1. Monitored Streetlight Networks. Schematic of the experimental setup.

II. BACKGROUND

A number of compatibility problems have occurred in the field of lighting as a result of installing electronic ballasts without understanding how to avoid such problems. Examples of the problems that have occurred include [10]:
- Early failure of ballasts and lamps
- Early failure of occupancy sensors
- Malfunctions of energy management systems
- Malfunctions of centralized clock systems
- Malfunctions of infra-red-based consumer electronic devices
- Malfunctions of personal electronic devices such as a hearing aid.

Lighting also affects the Power Quality (PQ) of the electrical distribution system. PQ is concerned with deviations of the voltage or current from the ideal single-frequency sine wave of constant amplitude and frequency. A consistent set of definitions can be found in [11]. Poor PQ is a concern because it wastes energy, reduces electrical capacity, and can harm equipment and the electrical distribution system itself. Power quality deterioration is due to transient disturbances (voltage sags, voltage swells, impulses, etc.) and steady state disturbances (harmonic distortion, unbalance, flicker). This paper is focused on the second group, and, specifically on harmonic distortion [12].

The main objective must be to provide guidelines for minimizing any PQ impacts resulting from application of energy-saving technologies with regards to lighting. The primary focus is electronic ballast-driven HPS lamps for lighting. However, energy savings is often used as one of the selling features for these devices and customers need to have a clear understanding of the energy-saving potential of these types of technologies.

Concretely, harmonic analysis is a primary matter of PQ assessment. With the widespread use of power electronics equipment and nonlinear loads in industrial, residential and commercial office buildings, the modelling of harmonic sources has become an essential part of harmonic analysis [13]. This paper focuses on analysing harmonics on HPS lamps.

Apart from that, harmonic attenuation refers to the interaction of the load voltage and current distortion [14]. Various research works have shown that a nonlinear load supplied with distorted voltage will inject less harmonic currents than those generated when the load is supplied by undistorted voltage.

Individually, single-phase power electronic-based loads pose no problem to power systems. In total, however, they have the potential to raise harmonic voltages and currents to unacceptably high levels. The two guidelines for modeling the net harmonics currents produced by these loads contemplate [15]: 1. Attenuation due to system impedance and the corresponding voltage distortion that tend to reduce

978-1-4244-8806-3/11 $26.00 © 2011 IEEE

the net harmonic currents produced by these loads; and 2. harmonic current cancellation due to phase angle diversity.

In this paper it is going to evaluate the first one. Upgrading to lighting equipment with clean PQ (high power factor and low harmonic distortion) can improve the power quality of the electrical system. Furthermore, upgrading with higher efficiency and higher power factor lighting equipment can also free up valuable electrical capacity. This benefit alone may justify the cost of a lighting upgrade.

III. OUTDOOR TEST

In order to evaluate PQ, it has done an experiment with lamps and equipment commonly used in up to date street lighting. As seen in Fig. 1, in this experiment the behaviour of four sodium lamps has been logged. The test time period was the same in all cases and environmental and electrical conditions were the actual situations expected. From one up to four lamps were connected to the outlet, each one with different power to achieve in this way a complete range of power from 80 to 700 W. These smart ballasts can be electronically controlled at different dimming level with a voltage control signal from 1V to 10 V.

Electronic circuitry is more energy efficient than conventional ballasts. Usually the required power by each electronic ballast is around 1 W. For high frequency control, the lamps cannot be fully dimmed to extinction, and residual light output and power consumption will appear. However, such system operation may be less noticeable and less annoying to occupants. It was reported that most dimming ballasts could dim lamps to less than 20% of maximum light output.

The monitoring device selected was a portable, stand-alone, 3-phase power quality analyzer [16]. Some of the key monitors requirements included the ability to transfer the surveyed data to an in-house computer program, appropriate numerical storage, and inexpensive and easy to use. The logged data were sent to a computer through Profibus.

As it can be seen in Fig. 1 and Table I, the experiment involved up to four HPS lamps (Philips Lighting) connected to their correspondent electronic ballast [17], [18], [19] and outdoor lamp controller [20].

TABLE I
LAMPS USED IN THE EXPERIMENTS

	LAMP	BALLAST	CONTROLLER
A	Master SON 150 W	SELC 2000 HID 150 W	Candelon Node C100
B	Master SON 150 W	SELC 2000 HID 150 W	Candelon Node C100
C	Master SON 250 W	RomLight Electronic Ballast 250 W	
D	Master SON 70 W	SELC 2000 HID 70 W	Candelon Node C100

Four different tests were done:

Test 1)   Only the case A indicated in Table I.

Test 2)   The cases A and B

Test 3)   The cases A, B and C

Test 4)   The cases A, B, C and D.

Each of the tests was done setting the electronic ballast to three different levels: 50%, 75% and 100%, which means that there are three different lamp power outputs.

IV. RESULTS

Firstly it has been represented for each test, the evolution of the current harmonics depending on the dimming. It can be seen in Fig. 2 that increasing the load will result in decreasing the harmonic current distortion due to the attenuation effect. Additionally, the difference between at half and full load is higher for low-order harmonic currents (3rd, 5th and 7th) than for higher harmonic orders (9th, 11th and 13th). It is noticeable that the tests accomplish the limit value according to IEC 61000-3-2 [21] Class C for all the harmonic orders (PF=0.99).

As in [7], Fig. 2 illustrates that the higher harmonic orders the lower their equivalent power; but this trend changes from 11th harmonic onwards. Nevertheless this doesn't happen neither in [7] nor in [6], because the fluorescent lamps decrease linearly from 3rd harmonic onwards. While the typical 3rd harmonic value in CFLs is near 70% of the fundamental, in this case, not surprisingly, is near 7% of the fundamental in test 1 at full load.

Furthermore, test 1 results in a current total harmonic distortion (THDI) of 7% at full load, that it is close from results obtained in [9], which obtained a THDI from 5.1 to 6.5% with HPS lamps connected to electronic control gear. In contrast, [6] obtained a THDI from 92.3 to 143.69% in the case of CFLs. Accordingly with [22], our loads would be of type 2: medium THDI (60-100%), dominated by harmonic currents of 3, 5, 7 and 9th order, but results show a lower THDI. These results are not surprising if it bears in mind that the ballast is high power factor electronic ballast, and the manufacturer certifies a total harmonic distortion less that 7%.

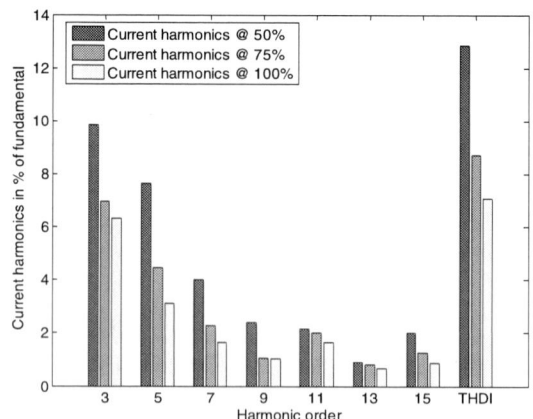

Fig. 2. Evolution in current harmonics and THDI in test 1 at different levels of power (dimming).

However, the study has confirmed other studies as [6] that indicate that similar spectra for all the lamps could be found, but it would rule out the effects of the interaction between the distorted voltage and current if it would suggest a single model to represent a range of lamp types. This is the reason why in this paper it is studied the evolution of harmonic values for the wide range of power. For this, it is mixed the same harmonic value from one to four lamps with their corresponding dimming; and it is shown in Fig. 3 that clearly follows a second grade curve with a confidence bound of 95%.

The same curve is analyzed for all the harmonic orders, and all of them follow the same behavior, so it is going to separate and balance each harmonic order according to their power. In this way it will obtain one curve to each test and each harmonic order with the same range of power, shown in Figs. 4-6.

Fig. 4 as well as showing a drop of current harmonic value the more lamps are connected, the harmonic current value is higher at half load than at full load. This is more representative the lower the number of lamps, because in test 4 the difference between at half and full load is 26%, and in

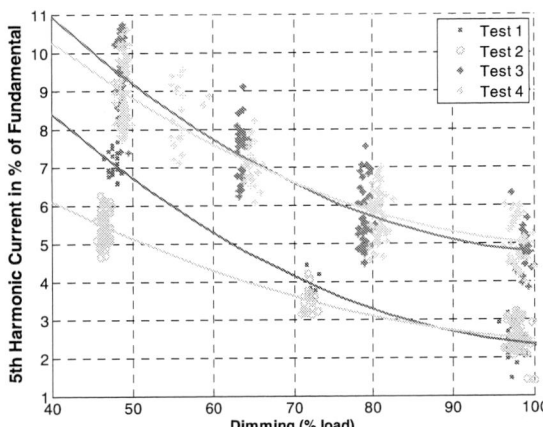

Fig. 5. Evolution of 5th harmonic current with load for tests 1, 2, 3 and 4.

test 1 the difference is 71.7%. Another important result is the higher the power the nearer current harmonic value.

The same is found in [7], where for fluorescent lamps it was observed that increasing the load will result in decreasing the harmonic current distortion.

With respect to 5th harmonic, tests 1 and 2 are below 3 and 4 tests. It should be noted that the 5th harmonic behavior is not the same that harmonics 3rd and 7th behavior, which have the same as it is observed by comparing Fig. 4 and Fig. 6.

In conclusion, it can see that it is produced a decrease on the harmonic value as higher the number of lamps connected. Because it has the value of the harmonic depending on the power, it is going to study the attenuation effect on these lamps.

V. ATTENUATION EFFECT CHARACTERIZATION

Nonlinear loads produce harmonic distortion according to their individual harmonic current spectrum. Traditionally, large single-point harmonics-producing loads have been treated as fixed harmonic current injectors. The same method has been used to predict the harmonic levels in distribution systems caused by large numbers of distributed single-phase

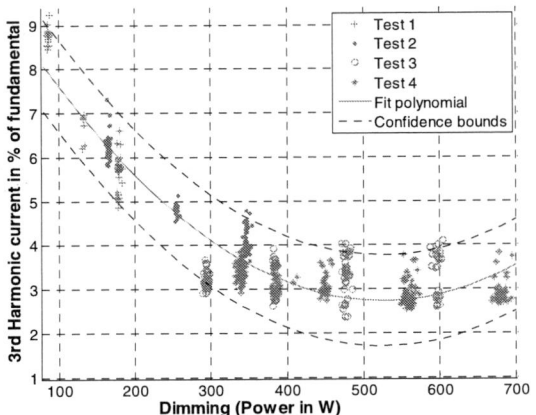

Fig. 3. Reduction in 3rd harmonic current with power for tests 1, 2, 3 and 4.

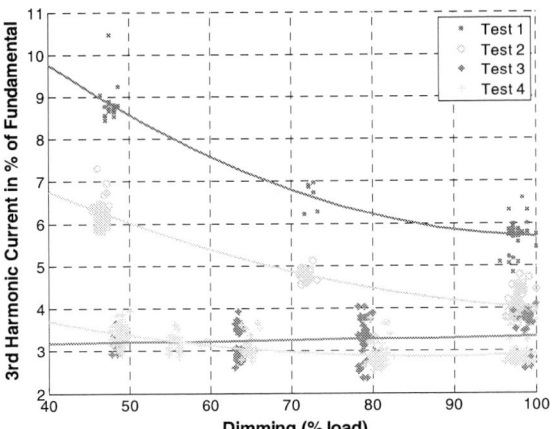

Fig. 4. Evolution of 3rd harmonic current with load for tests 1, 2, 3 and 4.

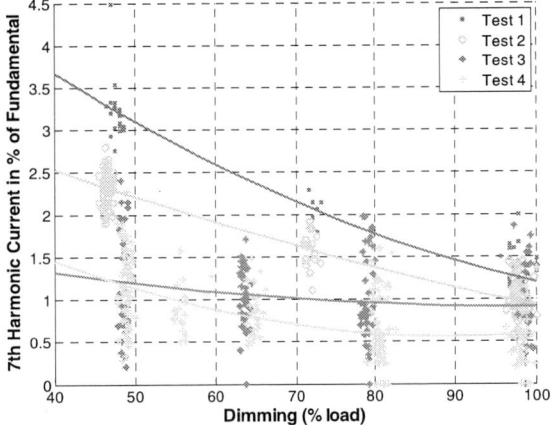

Fig. 6. Evolution of 7th harmonic current with load for tests 1, 2, 3 and 4.

loads, where the typical harmonic current spectrum of one load is scaled in proportion to total load power [15]. However, a large number of a variety of linear and non linear loads connected at the low/medium voltage bus of a distribution transformer, commonly known as the point of common coupling (PCC), really form an aggregate load [22]. Net harmonic current produced by aggregate harmonic loads (AHL) is usually significantly smaller than the algebraic sum of the harmonic currents produced by the individual nonlinear load, mainly due to phase cancellation [22]. Therefore, THDI of the aggregate load (i.e, THDI at the PCC) is influenced by both the participation (fraction) of linear loads into the total demand of the aggregate load as well as composite harmonic current spectra of the AHL. Field measurements have indicated that THDI at the PCC of low-voltage buses typically do not exceed 20% in comparison to THDI of an individual nonlinear load, which ranges between 20%–120%. In this case, a significant reduction in THDI at the PCC can be attributed to the large fraction of linear loads in the power demand of aggregate load and harmonic current cancellation due to phase-angle diversity. This phenomenon is known as attenuation, which refers to the interaction of the voltage and current distortion and it can be as significant as 50% or greater. The assumption of attenuation will usually be valid within customer-owned facilities and should be taken into account when predicting net harmonic levels [15].

Another important characteristic of harmonic currents produced by aggregate harmonic loads is that they are random with a changing average over time. Here it is in the presence of the cancellation related with the same kind of lamps (HPS), although with different power, but dimming with the same levels. In order to represent the attenuation effect of multiple identical loads, it is used in [14] the traditional index given as follows:

$$AF_h = I_h^N / (N \times I_h^1) \qquad (1)$$

where AF_h is the attenuation factor, I_h^N is the resultant current for harmonic h for N units operating in parallel and I_h^1 is the current for harmonic h when N=1.

This formula indicates that, in general, the attenuation due to a shared system impedance is more pronounced for higher-order harmonics, and tends to increase with the number of lamps connected. Although the attenuation factor increases in 13th and 15th harmonics, it is not important since current magnitudes will be negligible in those orders.

By the contrary, [5] and [23] stated that the previous equation doesn't fit in the case of CFL; consequently, they propose the below index, which is obtained by employing 12 CFLs connected to different wire lengths.

$$Neq_h = I_h^N / I_{h\_0}^1 \qquad (2)$$

where Neq_h is the equivalent lamp index, I_h^N is the measured total hth harmonic current for parallel N CFLs and $I_{h\_0}^1$ is the

hth current harmonic produced by one CFL under undistorted voltage supply conditions.

After using both formulas and in the case of sodium lamps, neither the first nor the second is consistent with our data. Because of that, it decided to propose another equation that, based on field experiment could justify the results obtained. It is inspired on the IEC 61000-3-2 [21], where, for devices belonging to D class, the maximum permissible harmonic current is rated with power, and harmonic value depends on the power of the system. It has mixed up the effects produced by dimming lamps in a wide range with the fact that there are four combinations of lamps with different power. Thus, the equivalent lamp index proposed here is the following:

$$Neq_h = I_h^{P_i} / (P_i * I_{h\_0}) \qquad (3)$$

where Neq is the number of equivalent lamps, $I_h^{P_i}$ is the hth total current harmonic measured for P_i power of four lamps and I_{h_0} is the hth current harmonic produced by less power. This index is essentially the ratio of the hth harmonic current produced by P power to the hth harmonic current produced with less power considered weighted with the power in this moment.

This results in a family of Neq_h curves, each representing the attenuation effect at a particular harmonic number. The three sample curves shown in Fig. 7 reveal the consistency of our approximation.

In addition, this figure indicates that a reduction of the harmonic current occurs when the supply voltage becomes more distorted. Harmonic 3rd has higher value than the rest; the higher the power the lower the harmonic value for all the orders; and at the highest power of the load, all of the harmonic values tend to the same one.

VI. CONCLUSIONS

This paper reports the dimming characteristics of large lighting networks based on HPS lamps. It is found that the six types of HPS lamps under test can work smoothly without flickering under dimming conditions. For 70, 150 and 250 W

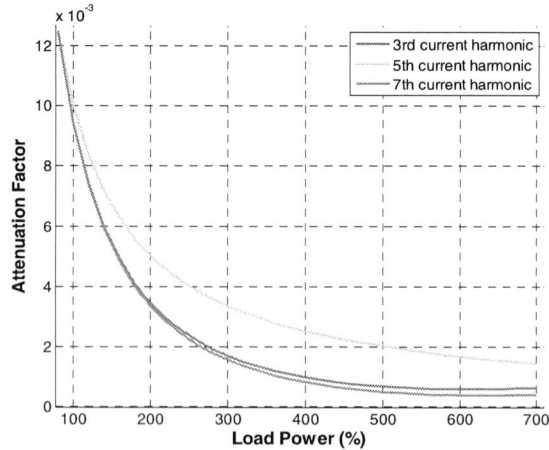

Fig. 7. Attenuation effect of the HPS lamps for 3rd, 5th, and 7th harmonics.

HPS lamps that are commonly used in road lighting systems and are of similar dimensions and structures, their trends for the light output variations with input voltage are similar. There are two types of results: analysis of harmonic according to the level of dimming for all the tests, and analysis of harmonic attenuation by analyzing each harmonic for all the tests. In this way it has the evolution of each harmonic order according to their powers and it also has it according to each test. Both types of result reveal the same phenomenon but present it from different perspectives.

In this paper it has presented a harmonic analysis on four tests that implied four HPS lamps, electronic ballasts and controllers to dim the light. It has compared this analysis with others made by different authors over CFLs. The results obtained are partially similar, i.e., the decrease in the harmonic value as increase the harmonic order, and also the decrease in the value at half load. But the pattern isn't the same because it changes with higher order harmonic.

This study also found that THDI is lower with HPS lamps than with CFLs, and THDI. The harmonic current distortion in a single type load is highly dependent on the loading level because is higher at half load than at full load.

Combining different types of loads can effectively reduce the harmonic current distortion due to attenuation. In order to assess this issue and due to others equations stated by different authors don't fit with our results, it has presented a proposal on an equation that represents the harmonic attenuation effect over these HPS lamps.

Although the technical features of a lighting installation are the first determinant of power efficiency, the ultimate determinant is effective operation management. The operation of lighting installations presents unique features (variation of use periods, lights degradation, environmental aggression, etc.) that, besides geographical dispersion, make the right management difficult.

ACKNOWLEDGMENT

This research was supported by the Company Telvent Energy, Spain, through the project Malaga SmartCity under contract number 12009028. SmartCity's budget is partly financed by the European regional development fund (ERDF) with backing from the Junta de Andalucía and the Ministry of Science and Innovation's Centre for the Development of Industrial Technology. The authors would like to thank the Spanish Ministry of Science and Innovation for funding the research project TEC2009-08988.

REFERENCES

[1] Electricity for More Efficiency: Electric Technologies and their Energy Savings Potential. Union of the Electricity Industry-EURELECTRIC. July 2004

[2] Energy Efficient Streetlights -- Potentials for Reducing Greater Washington's Carbon Footprint --. Prepared by: Robert T. Grow, ACCE Ford Fellow (Greater Washington Board of Trade). March 18, 2008

[3] Ali, N.A.M., Fadzil, S.F.S., Mallya, B.L. Improved illumination levels and energy savings by uplamping technology for office buildings. 2009 *International Association of Computer Science and Information*

Technology - Spring Conference, IACSIT-SC 2009, art. no. 5169423, pp. 598-603.

[4] Ceclan, A., Micu, D.D., Simion, E., and Donca, R. Public lighting systems - an energy saving technique and product.

[5] Nassif, A.B., Xu, W. Characterizing the harmonic attenuation effect of compact fluorescent lamps. 2009. IEEE Transactions on power delivery, 24 (3), pp. 1748-1749.

[6] Nassif, A.B., Acharya, J. An investigation on the harmonic attenuation effect of modern compact fluorescent lamps, 2008, ICHQP 2008: 13th International Conference on Harmonics and Quality of Power, art. no. 4668759

[7] El-Saadany, E.F., Salama, M.M.A. Effect of interactions between voltage and current harmonics on the net harmonic current produced by single phase non-linear loads. 1997 Electric Power Systems Research 40 (3), pp. 155-160

[8] Suárez, J.A., Di Mauro, G.F., Anaut, D.O., Agüero, C.A. Partial cancellation of harmonic currents caused by nonlinear residential loads. 2005 Informacion Tecnologica 16 (3), pp. 63-70.

[9] Manzano, E.R., Carlorosi, M., Tapia Garzón, M. Performance and measurement of power quality due to harmonics from street lighting networks. *2009 International Conference on Renewable Energies and Power Quality (ICREPQ'09)*. ISBN: 978 84 612 8010 8

[10] Power quality guidelines for energy-efficient device application. Consultant report. Prepared By: Electric Power Research Institute. January 2003.

[11] Moreno-Muñoz, A. Power quality. Mitigation Technologies in a distributed environment. 1st edition. London. Springer. 2007.

[12] Salmerón, P., Herrera, R.S., Valles, A.P., Prieto, J. New distortion and unbalance indices based on power quality analyzer measurements. 2009 *IEEE Transactions on Power Delivery* 24 (2), pp. 501-507.

[13] Acarkan, B., Erkan, K. Harmonics modeling and harmonic activity analysis of equipments with switch mode power supply using MATLAB and simulink. 2007 *Proceedings of IEEE International Electric Machines and Drives Conference, IEMDC 2007* 1, art. no. 4270692, pp. 508-513.

[14] Mansoor, A., Grady, W.M., Chowdhury, A.H., Samotyj, M.J. An investigation of harmonic attenuation and diversity among distributed single-phase power electronic loads. 1995. IEEE Transaction Power Delivery, 10 (1), pp. 467-473.

[15] Grady, W.M., Mansoor, A., Fuchs, E.F., Verde, P., Doyle, M. Estimating the net harmonic currents produced by selected distributed single-phase loads: Computers, televisions, and incandescent light dimmers. 2002 Proceedings of the IEEE Power Engineering Society Transmission and Distribution Conference 2, pp. 1090-1094.

[16] PM130EH Plus, SATEC is the Company. Available at: http://www.satec-global.com/eng/products.aspx?product=44

[17] SELC 2000 HID Smart Ballast 150 W. Available at: http://www.selc.ie/media/pdf/SELC2000-1.pdf

[18] ROMlight International Inc. HID Electronic Ballast. Available at: http://www.romlightintl.com/ballasts.cfm

[19] SELC 2000 HID Smart Ballast 70 W. Available at: http://www.selc.ie/media/pdf/SELC2000-2.pdf

[20] SELC CANDELON NODE C100 – Outdoor Lamp Controller. Available at: http://www.selc.ie/media/pdf/C100-10.pdf

[21] IEC 61000-3-2 Standards on Electromagnetic Compatibility (EMC), Part 3, Section 2: Limits for Harmonic Current Emissions, 2005

[22] Au, M.T., Milanović, J.V. Development of stochastic aggregate harmonic load model based on field measurements. 2007 *IEEE Transactions on Power Delivery* 22 (1), pp. 323-330

[23] Yong, J., Chen, L., Nassif, A.B., Xu, W. A frequency-domain harmonic model for compact fluorescent lamps. 2010 *IEEE Transactions on Power Delivery* 25 (2), art. no. 5332254, pp. 1182-1189

Deterministic Ethernet synchronism with PTP-base system for synchrophasor in Smart Grid

Victor Pallares-Lopez[1], Antonio Moreno-Muñoz[1], Miguel Gonzalez-Redondo[1],
Rafael Real-Calvo[1], Isabel Moreno Garcia[1], Aurora Gil de Castro[1], Francisco Domingo Perez[1]
[1] Electronics and Electronic Technology Area, Department of AC, Electronics and TE. University of Cordoba
Cordoba. SPAIN. vpallares@uco.es, amoreno@uco.es

Juan Jose Gonzalez de la Rosa[2],
[2] Electronics Area, Department of ISA, TE and Electronics.
University of Cadiz. Algeciras
Cadiz. SPAIN. juanjose.delarosa@uca.es

Abstract – We propose a new synchronized technique for SmartGrid. A PTP-Based IEEE 1588-2008 Global System has been defined. Specifically for PMUs with synchronism needs up to the microsecond range, for Energy Measurements System (EMS) and for Intelligent Electronics Devices (IEDs). This new Electronic Devices comprise two main technologies; LM3S8962 microcontroller as a slave IEEE 1588-2008, a Single-Board-RIO integrating the acquisition and synchronization. For tasks such as PTP master a NI PCI_1588 card and a Symmetricom's XLI IEEE1588 GrandMaster system for test. For the essays we have defined one experimental system for high precision Synchronism.

I. INTRODUCTION

Synchrophasor measurements are key information needed by system operators to assess the status of the power grid. Using data from Phasor Measurement Units (PMUs), received by phasor data concentrators (PDCs), grid operators will be able to have better visibility of power grid operations and respond to grid disturbances earlier to prevent major blackouts [8].

Two Standards are related to communications of phasor measurement unit (PMU) data and information. IEEE C37.118 was published in 2005 for PMUs. IEC 61850 has been substantially developed for substations but is seen as a key standard for all field equipment operating under both real-time and non-real time applications. The use of IEC 61850 for wide-area communication is already discussed in IEC 61850-90-1 in the context of communication between substations; it is only a small step to use it as well for transmission of PMU data. The models for PMU data need to be defined in IEC 61850. This work seeks integration study with experimentally test [10].

At present, synchronized measurements based on an accurate time reference, e.g. GPS (Global Positioning System), provide the missing link now allowing more efficient use of phasor data [7]. This phasor meters are very geographically dispersed through wide areas and still capture electrical waveforms on a synchronized way with a precision up to the microsecond range. The synchronization requirements are very close to the ones imposed to systems working with a unique clock.

This paper proposes a new synchronized technique for SmartGrid. A PTP-Based IEEE 1588-2008 Global System has been defined and an Experimental PTP-Based System has been developed to provide synchronized on substation for phasor measurements. It implements the Precision Time Protocol (PTP) to perform time stamping for these IEDs.

A NI PCI_1588 card as an experimental feasible Master managed by the virtual instruments implemented in the LabVIEW environment and experimental equipped IED with a PTPd Version 2 slave LM3S8962 microcontroller

For example, synchrophasor Standard [2] imposes critical synchronism requirements. To keep TVE Level-0 (highest) below 1% threshold, highest phasor angle error allowed is 0.57º, on a 50Hz nominal frequency for electrical network (all data from now on, referenced to 50Hz nominal frequency networks). A time error of 10μs corresponds to a phase error of 0.18º. Furthermore, our technical proposal integrates a variety of features in order to reduce to a minimum synchronism errors in the signal sampling and conversion process [16].

On the other hand the IEC61850 is better described as a communication system than a protocol. It includes parts for modeling of the components and the system, description of data types and classes, abstract service definitions, specific mapping for system implementation, and conformance testing.

IEC 61850 v1 only included communication within the substation, so connection to the control center was informational only and did not include 61850 services.

With the completion of section 90-1 in 2009, methods for direct communication with 61850 outside of a single substation became a part of the standard and are fully described and supported.

Integrating IEEE C37.118 with IEC 61850 will help to remove overlaps between the standards, which may impede development of interoperable equipment and systems.

IEEE C37.118 is intended to support applications, for example, protection. IEC 61850 is suitable for system-wide applications that require higher publishing rates [10].

A standards-based approach for time synchronization that addresses the requirements from all applications will support interoperability and facilitate implementation of new Smart Grid applications.

II. HARMONIZATION OF IEEE C37.118 WITH IEC 61850 AND PRECISION TIME SYNCHRONIZATION.

There are significant differences in scope and content of the two standards. IEEE C37.118 includes communication as well as measurement requirements and is also intended to support applications such as protection. IEC 61850 is suitable for system-wide applications that require higher publishing rates. The approach including possible models for PMU data needs to be defined in IEC 61850 [8].

Common time synchronization will be a key for many Smart Grid applications. The IEEE 1588 standard will be a key element to achieve that synchronization and PC37.238 standard is developing for application of PTP to Electric Power [12].

It is possible to use a similar approach for the transmission of PMU and PDC data but the capability needs to be formally defined in IEC 61850. PC37.239 standard defines a Common Format for Event Data Exchange (COMFEDE) for Power Systems [13].

III. PC37.239 FOR COMMON FORMAT FOR EVENT DATA EXCHANGE.

This standard defines a common format for the data files needed for the exchange of various types of power network events in order to facilitate event data integration and analysis from multiple data sources and from different vendor devices. The flexibility provided by digital devices in recording network fault event data in the electric utility industry have generated the need for a standard format for the exchange of data. These data are being used with various devices to enhance and automate the analysis, testing, evaluation, and simulation of power systems and related protection schemes during fault and disturbance conditions. Since each source of data may use a different proprietary format, a common data format is necessary to facilitate the exchange of such data between applications. This will facilitate the use of proprietary data in diverse applications and allow users of one proprietary system to use digital data from other systems [13].

IV. PC37.238 FOR APPLICATION OF PTP TO ELECTRIC POWER

This standard specifies a common profile for use of IEEE 1588-2008 Precision Time Protocol (PTP) in power system protection, control, automation and data communication applications utilizing Ethernet communications architecture [12].

The profile specifies a well-defined subset of IEEE 1588-2008 mechanisms and settings aimed at enabling device interoperability, robust response to network failures, and deterministic control of delivered time quality. It specifies the preferred physical layer (Ethernet), higher level protocol used for PTP message exchange and the PTP protocol configuration parameters. Special attention is given to ensuring consistent and reliable time distribution within substations, between substations, and across wide geographic areas.

V. PTP-BASED GLOBAL SYSTEM FOR SYNCHRONIZED EVENT SMART GRID

Proper time synchronization across Interconnections is a very important function for many reasons. Common time synchronization will be a key for many Smart Grid applications. The IEEE Std 1588™ standard will be a key element to achieve that synchronization. This standard is available to achieve highly accurate synchronization over a communication network. Many applications related to Smart Grid require time synchronization.

At present, phasor measurement units (PMU) can be considered as SMT devices commonly used in power system applications [23]. A significant advantage of using SMT is that all measurement signals are attached with a high-accuracy time stamp, which will facilitate the transition from a conventional measurement system, based on SCADA, towards a more intelligent measurement system using synchronized measurements from geographically distant locations. This feature is essential to develop the SmartGrid concept.

Thus, for synchronizing SmartGrids the use of a PTP-based global system can provide a secure communication channel with a delay that does not compromise the correct operation of the global system. This would imply the advantage of reusing the infrastructure of existing telecommunications networks to transmit synchronism information between PMUs. The Fig.1 shows an example of generic application of PMUs in Smart Grids.

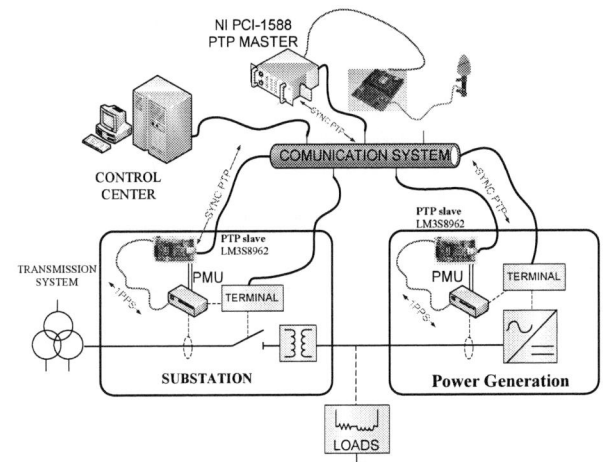

Fig.1 PTP-BASED GLOBAL SYSTEM.

Multiple IEDs sharing data or control commands results in new distribution protection, control and automation functions.

978-1-4244-8806-3/11 $26.00 © 2011 IEEE

This has the potential to supersede and eliminate much of the dedicated control wiring in a substation, plus costly special purpose communication channels between the stations and power network.

This standardization enables the integration of the equipment and systems for controlling the electric power process into complete system solution, which is necessary to support utilities processes. Ensure the interoperability of equipment and systems by providing compatibility between interfaces, protocols and data models. With IEC61850's standardization of data acquisition and description methods, integration efforts are reduced [18].

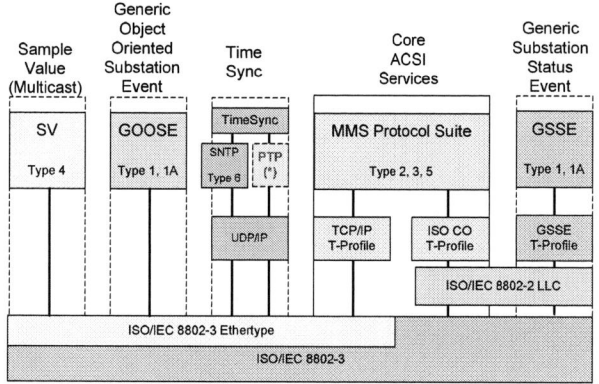

Fig.2- Protocol Mapping Profile.

The data concentration function also requires supporting a wide range of communications protocols. And they should support the newer standard protocols for both IEDs and SCADA masters. Standard protocols such as DNP3™, IEC 60870-5 and IEC61850 (including GOOSE) may be needed now or in the future. When applicable, both serial and LAN formats should be specified. User-friendly features such as configuration templates for all protocols can reduce the configuration time considerably [9].

In addition Network Time Protocol (NTP), Simple Network Time Protocol (SNTP), and the Precision Time Protocol (PTP) may be required to allow time synchronization over the network.

We also study the possibility of adding functionality to transmitting GOOSE messages on an Ethernet network and the integration of the PTP protocol for synchronizing tasks Fig.2 as proposed in the paper [18]. This scheme represents the IEC61850 PROTOCOL MAPPING PROFILE.

The OMICRON IEDScout software was used to detect and subscribe GOOSE messages on the network. Several GOOSE messages that were transmitted on the network were detected by the EDScout software.

VI. EXPERIMENTAL PTP-BASED SYSTEM FOR SYNCHROPHASORS MEASUREMENT

The experimental system is a complete system ready to measure the outside: The system is composed of a PCI1588-Master (GPS), two LM3S8962-Slave and two PMUs. For calibration of each of the experimental Slaves: this system is based in the XLi's Time Interval/Event Time (TIET) feature can be used to measure PTP synchronization across timing networks.

It has been designed to provide highest stability level to the A/D conversion process. System has been over-dimensioned to provide enough room for future analysis, grid management and protection features. Main system components have been selected to comply "Level-0" IEEE 37.118 requirements:

- **Two NI Single-Board RIO 9631**; Simultaneous 6-channel, 16-bit ADC, operating up to 51.2k Samples/s and FPGA synchronism.
- **Two Slave-PTP V2**; A LM3S8962 microcontroller with PTPd soft. Sub-microsecond synchronism error (when directly connected to PTP master).
- **Master PTP V2**; An Industrial-PC with A PCI-1588 as Master PTP V2 and a receptor GPS.

A.1.- Issues with implementation of PCI1588- Master.

Fig.3 shows how the phasor meters units are synchronized with a Master-PTP V2 who acts as the central system. This system is an evolution of "*An IEEE1588-BASED system for synchronized PMUs and protective relaying functions*" [24]. This is equipped with a standard RJ-45 connection for Ethernet communication (up to 100 Mbps), with PFI terminals and with a 10 MHz TCXO. TCXO is characterized by a initial accuracy of ±1.5 PPM, a temperature stability (0° to 55° C) of ±2 PPM and a aging per year of ±1 PPM.

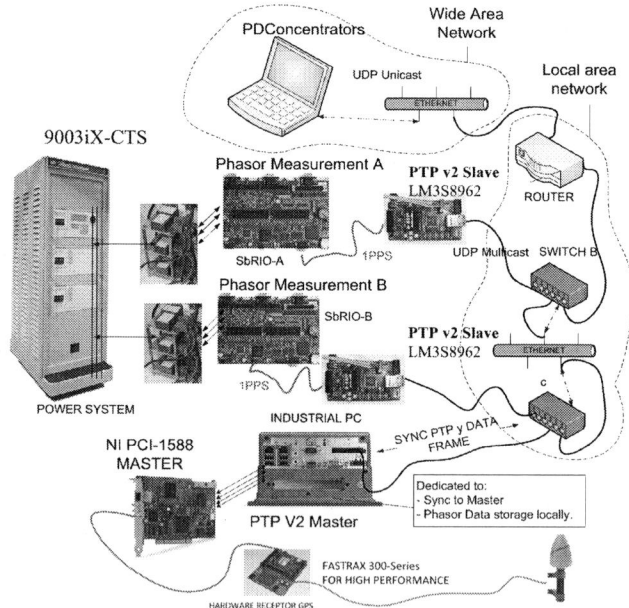

Fig.3- Phasor meters synchronization method.

This Master oversees and manages the synchronism of a set of PTP slaves connected to a standard Ethernet network. PTP Master sends multicast synchronism packets "sync and Delay_req" to every slave unit in order to synchronize their local clocks with the master unit one. PTP master local clock gets stabilized in turn from the GPS signal provided by a

FasTrax IT321 for high performance and Ultra small, low-power and highly sensitive GPS receiver. PTP transceivers provide PPS signal for synchronism of remote units, which use it as a basis for phasor estimation and data transfer [17][22].

A.2.- Issues with design and implementation of ARM- Slave.

The slave is developed by Stellaris LM3S8962 Evaluation Board Layout. The LM3S8962 microcontroller based on the ARM® Cortex™-M3 [15] controller core operating at 50 MHz. The LM3S8962 also features hardware-assisted support for synchronized industrial networks utilizing the IEEE 1588 Precision Time Protocol (PTP) [3]. High precision time stamps can be achieved with the support of specialized hardware interfaces in the physical layer of the network.

The software integrates the PTPd Version 2. The Precision Time Protocol daemon (PTPd) is a complete implementation of the IEEE 1588-2008 specification for a standard non-boundary clock.

A.3.- Synchronism Method.

This LM3S8962 microcontroller provides a PPS signal with the same stability features and TimeStamp with SPI Channel. This signal enables us to re-synchronize sampling Fig.4 and "Data Frame" sending processes by asking PTP master exact PPS timestamp. "Data Frame" messages enclose phasor, time and frequency information.

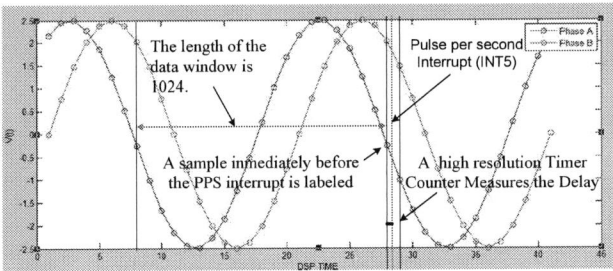

Fig.4 - Synchronism method

The NI-sbRIO-9631 CLK module provides a real-time clock. This clock can be used to measure the passage of time [8], as well as to add timestamp messages to event logs. Both the high resolution times are stored as 64-bit values. The 64-bit counter (FPGA I/O functions) is implemented using a Single-Cycle Timed Loop (SCTL) which guarantees that all operation will execute in one clock pulse at 266MHz internal time-base.

Also we can emphasize that with this system we appraise the deviation in nanoseconds between the pulse per second and the most nearby sample with a timer of high resolution. This displacement adds to the received TimeStamp.

A.4.- Data sampling and processing issues

"Level-0" complaint requirements force us to use high data sampling rates. First estimations lead us to consider a data sampling range among 256 and 1024 samples per cycle (12.8KSamples/s to 51.2KSamples/s).

Data sampling rate must be kept as low as possible, since it considerably increases, the already high computational load required for correlation DFT and FFT phasor estimation methods [11]. When nominal frequency remains constant, phasor estimations for N samples per cycle can be obtained [5] on a continuous basis from a correlation DFT. Between sampling and sampling are calculated each of the terms of the DFT to guarantee the processing in real-time.

An electrical disturbance has effect on several analysis windows, first with the quality factor increasing to reach a maximum and then again decreasing until an acceptable value is reached [5][6].

Voltage and current phasor estimations, timestamp, measured frequency and frequency deviation with regard to analysis window data are informed on a regular basis (25 or 50 data frames per second sending rates). The information is transmitted by a channel UDP Unicast. A central team receives the frames of two experimental IEDs.

VII. GRANDMASTER FOR EXPERIMENTAL TEST

The standard "XLI IEEE1588 GrandMaster Clock [1], as can be seen on Fig.5, provides a complete implementation of a Precise Time Protocol (PTP) "ordinary clock" over a dedicated IEEE 1588 card. The IEEE 1588 card can be configured to operate as a PTP grandmaster or as a PTP slave.

As a PTP grandmaster, the IEEE 1588 card typically synchronizes PTP slaves on the network to International Atomic Time (TAI). The XLi IEEE 1588 Clock derives TAI from the Global Positioning System (GPS). In addition, Symmetricom designed the XLi IEEE 1588 Clock so the user can distribute Coordinated Universal Time (UTC) or user-entered time over PTP.

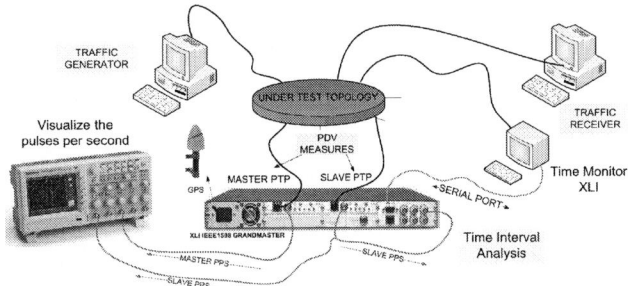

Fig.5- Network Measurement Test Set-up

The XLi's Time Interval/Event Time (TIET) feature can be used to measure PTP synchronization across timing networks. The XLi IEEE 1588 Clock is characterized by the following nominal specification: Frequency Output Accuracy; $<2 \times 10^{-12}$, Frequency/Timing: Allan Deviation, Stability; 1×10^{-9} @ 1sec, 2×10^{-10} @ 1000sec y 1×10^{-12} @ 1 day. The XLi comes with the standard TCVCXO oscillator described below. The stability of the following oscillators is dependent on the reference source (GPS). GPS characterized by Tracking Up to 12 satellites with TRAIM, Position Accuracy Typically < 10m when tracking four (4) satellites,

TRAIM Mask 1µS, 1PPS Accuracy UTC-USNO ±30ns RMS 100ns Peakby a PPS accuracy within 15ns to GPS/UTC.

Synchronization performance depends on several factors, including, but not limited to Slave oscillator quality and PLL control [4], networking equipment, network traffic levels and network topology. System designer generally cannot easily modify slave oscillator and control. However, PTP settings and network design are under the control of the system designer.

Through careful network design, synchronization performance of measurement systems can be maintained. Network characterization is an important step for determining the fitness for high performance synchronization. Two parameters that aids the characterization process are "Packet Delay Variation" (PDV) and Slave PPS Time Error. PDV measures variations in the master to slave packet delay at the physical layer of the network. Measuring slave PPS time error from the hardware-generated PPS signals provides direct observation of master-slave end-to-end synchronization. Errors can be viewed using a frequency counter, oscilloscope or a grandmaster equipped with an integrated time interval measurement input "XLI IEEE 1588".

When system behavior degrades in such a way that TVE threshold is surpassed, phasor estimations cannot be correlated with the ones coming from points with better stability conditions. To prevent or reduce stability issues, alternative network topologies can be evaluated:

Share synchronism and data paths but using only PTP switchs. Existing studies [4][21] demonstrate that with PTP switchs and a flat network topology both communication paths can be unified.

The use of ordinary switches or router should be avoided in critical timing application where sub-microseconds or better accuracy is needed. In these cases Transparent Clocks (TCs) and Boundary Clocks (BCs) should be utilized. This work [20] investigated about advantages from the use of these devices using the IEEE 1588-2008.

VIII. AC/DC POWER SOURCES FOR EXPERIMENTAL TESTS.

An AC/DC power sources 9003iX-CTS with a high performance power analyzer. Applies a waves form precision for two phases simultaneously. The slave unit 1 is implemented in a LM3S8962 microcontroller. The slave unit 2 is equipped with a 3.20 GHz processor with 2GB (2 x 1 GB) 667 MHz DDR2 RAM memory and with NI PCI-1588 PTP. The same slave transmits to the central device the frames as the norm Synchrophasor Standard [2]. The method must emulate the traffic in a substation.

IX. CONCLUSIONS

With experimental PTP-based V1 system [24], substation events could be synchronized within 12 microseconds. TVE requirements [2], limit time error to be lower than 31µs. Nevertheless with a slave PCI-1588 for test recovers its

stability around 150ns. The tests are very similar to [16][19] with a PXI system.

We also study the possibility of adding functionality to transmitting GOOSE messages on an Ethernet network with IEEE 1588-2008 synchronization.

With experimental PTP-based V2 system proposed in this paper we hope to achieve accuracy within some 100 nanoseconds. Some important enhancements [20], among others, are: enablers for increased accuracy higher timestamp resolution, shorter sync intervals, correction field, rapid reconfiguration after network topology changes, fault tolerance, unicast operation and new mappings (for example, PTP directly on Ethernet MAC layer, without IP/UDP).

ACKNOWLEDGEMENTS

This research was supported partially by the Company Telvent Energy, Spain, through the project Malaga SmartCity under contract number 12009028. SmartCity's budget is partly financed by the European regional development fund (ERDF) with backing from the Junta de Andalucía and the Ministry of Science and Innovation's Centre for the Development of Industrial Technology. The authors would like to thank the Spanish Ministry of Industry, Tourism and Trade for funding the Project TSI-020100-2010-484 which partially supports this work. Our unforgettable thanks to the Spanish Ministry of Science and Innovation for funding the research project TEC2010-19242-C03-02.

REFERENCES

[1] XLi IEEE 1588 Grandmaster, http://www.symmttm.com/products_gps_XLI_IEEE_1588.asp

[2] IEEE 37.118-2005. IEEE Standard for Synchrophasors for Power System.

[3] IEEE Std 1588-2008, IEEE Standard for a Precision Clock Synchronization Protocol for Networked Measurement and Control Systems. (2008) The Institute for Electrical and Electronics Engineers, Inc., New York.

[4] Eidson, J.C., Measurement, Control and Communication Using IEEE 1588. (2006). Springer-Verlag London Limited, Lond.

[5] Synchronized Phasor Measurements and Their Applications. A.G. Phadke • J.S. Thorp. EDIT. Springer, 2008. ISBN: 978-0-387-76535-8.

[6] DSP Applications Using C and the TMS320C6x DSK. Rulph Chassaing. EDIT. JOHN WILEY & SONS, INC., 2002. ISBN 0-471-22112-0.

[7] Andrea Carta, Nicola Locci, Carlo Muscas, Member, IEEE, and Sara Sulis, Member, IEEE, "A Flexible GPS-Based System for Synchronized Phasor Measurement in Electric Distribution Networks", IEEE transactions on instrumentation and measurement, VOL. 57, NO. 11, NOVEMBER 2008

[8] NIST Special Publication 1108; "NIST Framework and Roadmap for Smart Grid Interoperability Standards, Release 1.0". Janury 2010.

[9] Erich W. Gunther, "Cybectec Substation Gateways", SmartGridnew.com, Publication Year: Nov 5, 2008.

[10] PAP13 61850 C37118 Harmonize and Synchronization; "Time Synchronization, IEC 61850 Objects/IEEE C37.118 Harmonization (6.1.2, 6.2.2)". 2009

[11] Lee Barford, Member, IEEE, and Jeff Burch. "Fourier Analysis From Networked Measurements Using Time Synchronization". IEEE Transactions on Instrumentation and Measurement. vol. 56, no. 5, october 2007

[12] PC37.238™/D5.6 2 Draft Standard Profile for Use of IEEE 3 Std. 1588 Precision Time Protocol in 4 Power System Applications. February 2011.

[13] IEEE PC37.239™/D05 2 Draft Standard for Common Format for 3 Event Data Exchange (COMFEDE) for 4 Power Systems. May 2010.

[14] The Precision Time Protocol daemon (PTPd). http://ptpd.sourceforge.net/.

[15] Yang, G.H, Wen, B.Y., "A Device for Power Quality Monitoring Based on ARM and DSP", Industrial Electronics and Applications, 2006 1ST IEEE Conference, Publication Year: 2006 , Page(s): 1 – 5.

[16] Marco Lixia, Carlo Muscas, Sara Sulis: "Application of IEEE 1588 to the Measurement of Synchrophasors in Electric Power Systems", ISPCS 2009 International IEEE Symposium on Precision Clock Synchronization for Measurement, Control and Communication Brescia, Italy, October 12-16, 2009

[17] P. Ferrari, A. Flammini, D. Marioli, S. Rinaldi, A. Taroni: "Synchronization of the Probes of a Distributed Instrument for RealTime Ethernet Networks", IEEE ISPCS 2007, 1-3 Oct. 2007, pp. 33-40.

[18] Yang Liu, R. Zivanovic and S. Al-Sarawi, C.Marinescu and R. Cochran, "A Synchronized Event Logger for Substation Topology Processing", Power Engineering Conference, 2009. AUPEC 2009. Australasian Universities Publication Year: 2009 , Page(s): 1 - 6.

[19] Andrea Carta, Nicola Locci, and Carlo Muscas, Member, IEEE: "A PMU for the Measurement of Synchronized Harmonic Phasors in Three-Phase Distribution Networks". IEEE Transactions on instrumentation and measurement, vol. 58, no. 10, october 2009

[20] Jiho Han, Student Member, IEEE, and Deog-Kyoon Jeong, Senior Member, IEEE:" A Practical Implementation of IEEE 1588-2008 Transparent Clock for Distributed Measurement and Control Systems". IEEE transactions on instrumentation and measurement, vol. 59, no. 2, february 2010.

[21] J.-C. Tournier, Xiao Yin, "Improving Reliability of IEEE1588 in Electric Substation Automation", ISPCS 2008 – International IEEE Symposium on Precision Clock Synchronization for Measurement, Control and Communication, Ann Arbor, Michigan, Sept. 22–26, 2008.

[22] C. Na, D. Obradovic, R. L. Scheiterer, G. Steindl, and F. J. Goetz: "Synchronization performance of the precision time protocol in Industrial Automation Networks", IEEE Transactions on Instrumentation and Measurement, Vol. 58, No. 6, June 2009.

[23] Martin, K. E., Hamai, D., Adamiak, M. G., Anderson, S., Begovic, M., Benmouyal, G., et al. (2008). Exploring the IEEE standard C37.118-2005 synchrophasors for power systems. IEEE Transactions on Power Delivery, 23(4), 1805-1811.

[24] Pallarés-López, V., Moreno-Muñoz, A., De La Rosa, J.J.G., Real-Calvo, R. (2010). "An IEEE1588-BASED system for synchronized PMUs and protective relaying functions". SPEEDAM 2010 - International Symposium on Power Electronics, Electrical Drives, Automation and Motion , art. no. 5542201, pp. 593-598.

Neuro-Fuzzy Control System for Active Filter with Load Adaptation

Oleksandr Husev[1,2], Sergey Ivanets[1], Dmitri Vinnikov[2]
Chernihiv State Technological University (Ukraine)[1], Tallinn University of Technology (Estonia)[2]
gsfki@ukr.net

Abstract-**The paper proposes a neuro-fuzzy regulator based control system for parallel active filters, which allows for adaptation of the control mode to changes in the amplitude and spectral composition of the load current. The efficiency of this control system in terms of reactive power compensation quality in the network is demonstrated in a simulation. The main advantage of the proposed control system is in the possibility of its implementation without changes in the compensator circuitry and the switch modulation frequency.**

I.   INTRODUCTION

Increasing requirements for the quality of mains current supplied by the power network lead to the need for devices compensating the reactive power in both the fundamental frequency and higher harmonics. Such devices can be classified into passive filters, active filters (AF), and their combinations, hybrid filters [1], [2]. In systems with the spectral composition of the load current constantly changing over a wide range, a passive filter-based solution becomes inefficient. The principal advantage of active filters is in their ability to compensate a wide spectrum of higher harmonics. The design of all active filters includes a storage element, a converter and a charger. Fig. 1 represents a so-called active power line conditioner (APLC) [1]-[3].

Active filters exist in a large variety of topologies [4]-[10], a specific topology being selected depending on the type of load. In the absence of zero-phase sequence currents to be compensated (a symmetric load or no neutral wire), the above version becomes optimal.

Active filters allow for compensation of a wide spectrum of the load current by means of control, although they are not free of shortcomings - namely, a limited maximum amplitude of reactive power fluctuations and the complexity of implementation [11]-[13].

The constraint on the amplitude of reactive power fluctuations is primarily associated with passive components of the active filter, which determine its nominal load ratings. On the other hand, an APLC may be adapted to load changes by the regulation of the voltage levels on the compensator capacitors [4]. Obviously, the quality of the active filter in this case will depend on the quality of control.

In this case, nonlinearity of the control law substantially limits the applicability of classical regulators. The problem of implementing a more efficient control may be solved by means of artificial intelligence systems: genetic algorithms, ant algorithms, fuzzy systems, or neural networks [14]-[17].

Fuzzy systems present an advantage of being able to implement nonlinear control algorithms. In this case, in the process of tuning the fuzzy regulator, the adjuster operates with the concepts very close to the way of human thinking, which simplifies tuning and understanding the operation of the regulator. Sufficient experimental data and results of simulation exist to facilitate tuning of this regulator. There are many versions of construction and application of such systems [18], [19], where their advantages are in detail described. From other side, disadvantages are obvious: an enormous quantity of experimental data, which encumbers the process of tuning, or the results of the simulations, which, strictly speaking, need the experimental check.

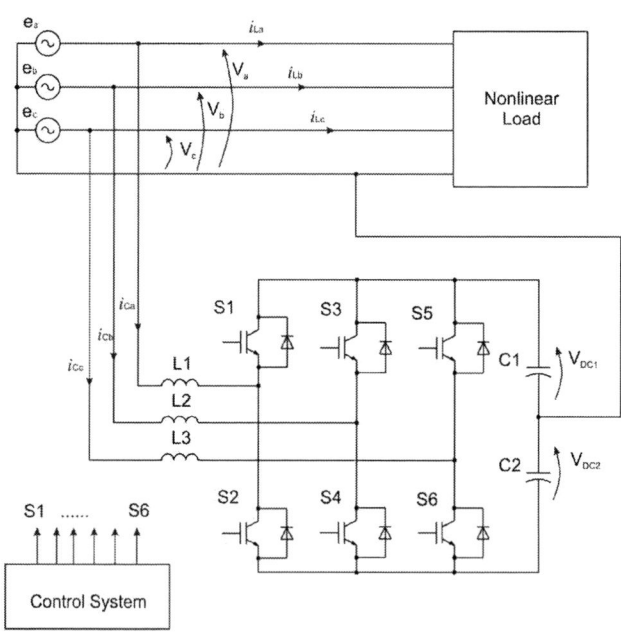

Fig. 1. The structure of APLC.

Learning capabilities of neural networks are also well known and understood [20], and constitute their fundamental advantage. Neuro-fuzzy regulators are combinations of fuzzy logic with neural networks, combining their advantages. In this paper, we propose a control system for an active filter based on a neuro-fuzzy regulator providing adaptation to the changing reactive power in the load.

978-1-4244-8806-3/11 $26.00 © 2011 IEEE

II. CONTROL SYSTEM STRUCTURE

Fig. 2 presents a block diagram of a control system for a parallel three-phase APLC. The control system consists of three basic blocks: a device determining compensator reference currents, a power switch controller, and a capacitor voltage regulator. The information about the load voltage and current in all three phases comes from current and voltage sensors.

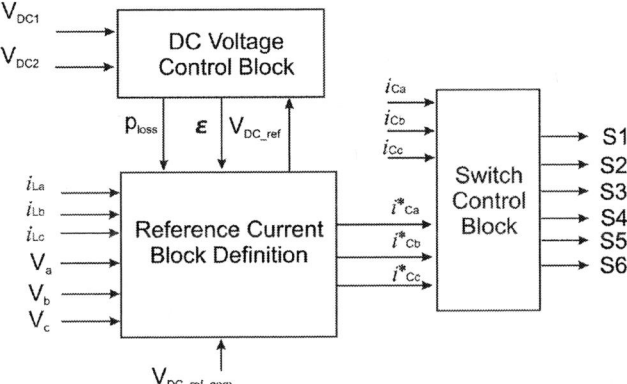

Fig. 2. Block scheme of control system of APLC.

The device determining the reference currents from the instantaneous values of load voltage and current of load calculates the reference currents to be formed by the APLC for complete compensation for reactive power. The control algorithm is based on a modified P-Q- theory of instantaneous power [21], [22].

Fig. 3 presents a block diagram of the regulator controlling the voltage on the capacitors C_1 and C_2, which is used for compensating the ohmic losses in the compensator.

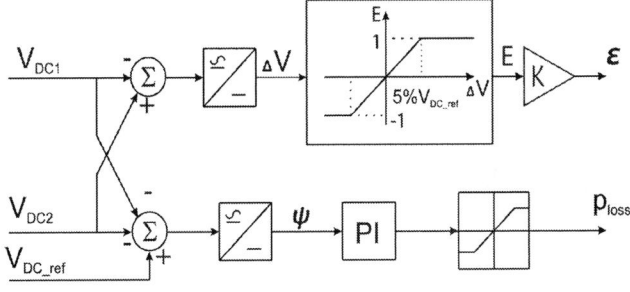

Fig. 3. Block scheme of voltage regulator on the capacities.

The input values of the regulator are the instantaneous voltages on the capacitors (V_{DC1}, V_{DC2}) and the reference voltage V_{DC_ref} equal to the sum of the target values of V_{DC1} and V_{DC2}. The top channel forms the signal ε to equalize the voltages on the capacitors by charging or discharging them. Constant components of the voltages V_{DC1} and V_{DC2} are only used to prevent the influence of voltage ripple on the capacitors on the operation of the unit. The top channel of the regulator covers a ±5% range around the reference voltage due to the use a nonlinear voltage limiter element ("clipper").

The bottom channel controls the voltage levels on the capacitors C_1, C_2 and indirectly affects the losses in the compensator itself. Pi- regulator is used in this channel to maintain the output voltage at the assigned level V_{DC_ref}.

The device determining the compensator reference currents (Fig. 4) is designed according to the instantaneous reactive power theory for a three-phase network with a neutral wire (P-Q theory).

The value of currents of load (i_{La}, i_{Lb}, i_{Lc}), the phase values of voltages (Va, Vb, Vc,), the power of ohmic losses (P_{Loss}) in the compensator, the constant component of current (ε) and the nominal value of voltage on capacitors $V_{DC\_ref\_nom}$ are given to the entrance of block. The currents of load and the phase values of voltages will be given to the block, which determines P-Q components of instantaneous power. Then constant component from the P-constituting instantaneous power is intercepted by the low-pass filter and subtractor. The power of losses in the compensator is added to remained variable P-component of power. With the aid of the block of inverse transformation the currents of compensator (i_{Ca}, i_{Cb}, i_{Cc}) are formed, into which, the direct current of the recharging of capacitors (ε) is added with the aid of the adders. As a result, the currents of compensator (i_{Ca}^*, i_{Cb}^*, i_{Cc}^*) include the reactive current of load, active component for compensating the losses and constant component for regulating voltage across capacitors C_1, C_2.

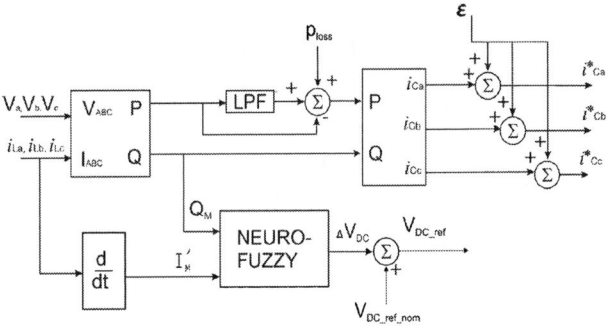

LPF - Low-pass Filter

Fig. 4. Block scheme of the block of the formation of the reference currents APLC.

The adaptation of control block achieves of neuro-fuzzy regulator NEURO-FUZZY. It forms the additional voltage of ΔV_{DC} on the capacitors, which is added to the nominal reference voltage $V_{DC\_ref\_nom}$. NEURO-FUZZY regulator has two inputs: the amplitude of reactive power Q_m and maximum speed of changing in the current of compensator I_{cm}' during the cycle. Operating principle consists in the formation of additional voltage when the input values of the regulator are different from the nominal rating of work of APLC. Basic task with tuning of neuro-fuzzy regulator consists in creating control surface of regulator, which will establish such additional value of voltage, with which the quality of compensation will be the best. This control surface is called optimum.

978-1-4244-8806-3/11 $26.00 © 2011 IEEE

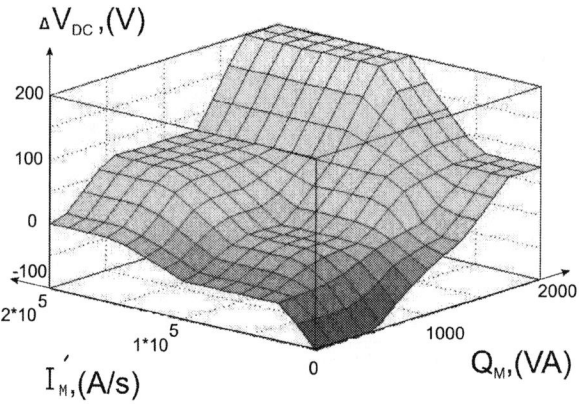

Fig. 6. Control surface of neuro-fuzzy regulator.

The commutation block of power switches is built according to the vector control method. According to the difference between the reference and actual currents of compensator control signal of transistors is formed.

III. NEURO-FUZZY CONTROLLER

The complexity of solution of this problem is determined by the absence of the mathematical apparatus, which would make it possible to obtain the dependence of the quality of compensation on the amplitude of reactive power, maximum value of the current and voltage growth on the capacities. Therefore, the problems of constructing such systems were solved by the simulation and experiment, with using of the obtained results for constructing the neuro-fuzzy regulator.

According to the results of the simulation of work AF in different regimes it is possible to build the table of rules and to establish the parameters of membership functions of NEURO-FUZZY variables in order to obtain the control surface, close to the optimum. The table of rules of neuro-fuzzy regulator (Table I) is comprised for the neuro-fuzzy regulator with two entrances and one output.

TABLE I
NEURO-FUZZY RULES

Q_m \ I'_{cm}	Z	L	M	B	SB
Z	NB	NB	N	Z	P
L	N	N	Z	Z	P
M	N	Z	Z	Z	P
B	N	Z	Z	P	PB
SB	Z	P	P	P	PB

To each variable corresponds five membership functions (Fig. 5). Input variables Q_m and I'_{cm} (Fig. 5(a) and 5(b) respectively) can take the values: Z (zero) – minimum values; L (little) – small values; M (middle) – average values; B (big) – great values; SB (super of big) – maximum values. Output variable ΔV_{DC} (Fig. 4(c)) can have five values: NB (negative of big) – big negative values; N (negative) – negative values; Z (zero) – zero values; P (positive) – positive values; PB (positive of big) – big positive values.

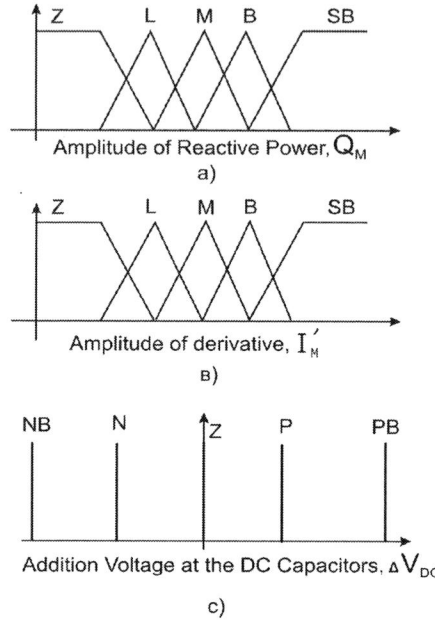

Fig. 5. Membership functions of the neuro-fuzzy regulator.

Neuro-fuzzy regulator is realized on the Takagi-Sugeno algorithm. The adjusting of this regulator concludes in selection of neuron network weight coefficients. The adjusting is possible in several ways. In this case the algorithm of adjusting with the teacher is used.

As the training sample were used the obtained simulation results of work AF in the non-nominal regimes, and also dependences of the total harmonic distortion (THD) of the voltage on the capacity of converter. In Fig. 6 is shown the control surface of the obtained regulator.

The idea of the algorithm of neuro-fuzzy regulator, which is reflected in the table of rules, lies in the fact that reduce the reference voltage (to form the negative value of additional voltage) on the capacitor of compensator with the decrease of the pulsation of reactive power and rate of current changing. It is necessary to increase reference voltage (additional positive voltage), when the value of input variables increases.

IV. SIMULATION RESULTS

The passive components of compensator and the minimum frequency of the commutation of power switches with the vector method of control are calculated according to the method [23], [24]:

978-1-4244-8806-3/11 $26.00 © 2011 IEEE

$$C = \frac{4 \cdot \gamma \cdot Q_m}{K \cdot V_{DC} \cdot \omega \cdot V_m}, \tag{1}$$

$$L = \frac{V_{DC} - V_m}{I'_{cm}}, \tag{2}$$

$$f_{\min} = \frac{V_{DC} - V_m}{L \cdot \Delta i_{\max}}, \tag{3}$$

where:

Q_m – amplitude of the pulsation of reactive power in the three-phase network;

γ – the relative duration of the pulsation of reactive power;

ω – network frequency;

V_m – amplitude value of phase voltage;

V_{DC} – value of voltage on capacitor;

K – ripple factor of voltage on the capacity;

I'_{cm} – maximum value of the rate of current changing;

$f_{\min}$ – minimum frequency of the commutation of power switches;

$\Delta i_{\max}$ – the maximum permissible significance of a deviation of the current of compensator from the given one.

Parameters of load are shown in the Table II.

Knowing the parameters of network and after assigning capacitors voltage DC and ripple factor, from expressions (1) and (2) we determine the value of the passive components of compensator (Table II).

After assigning the tolerant value of a current deviation of the compensator, from expression (3) it is possible to determine the minimum frequency of the power switches commutation. With an increase in the frequency of commutation the coefficient of total harmonic distortions (THD) will be reduced. The selected frequency of the switches commutation in the system is $100\ kHz$. MATLAB is medium of simulation.

TABLE II
THE TABLE OF AF AND POWER NETWORK PARAMETERS

Network parameters	Load parameters	AF determined parameters	AF calculated parameters
$\omega = 2\pi \cdot 50\,Hz$, $V_m = 96\,V.$	$Q_m \approx 1.8\,kVA$, $\gamma = \dfrac{1}{12}$, $I'_{cm} = 4.8 \cdot 10^4 \dfrac{A}{s}$, $P \approx 1\,kVA.$	$V_{DC} = 200\,V$, $K = 0.1$, $\Delta i_{\max} = 0.5\,A.$	$C \approx 1000\,\mu F$, $L \approx 2.2\,mH$, $f_{\min} = 94\,kHz$

Three-phase rectifier is used as the nonlinear load. In the process of simulation the work of control system with different changes in the load was investigated. For example, in Fig. 7 and 8 are shown the waveforms of AF work with the jump of load in the network with absence of neuro-fuzzy regulator in control system.

Amplitude increasing of current in the load occurs at the moment of time $t = 0.3\ s$ (Fig. 7).

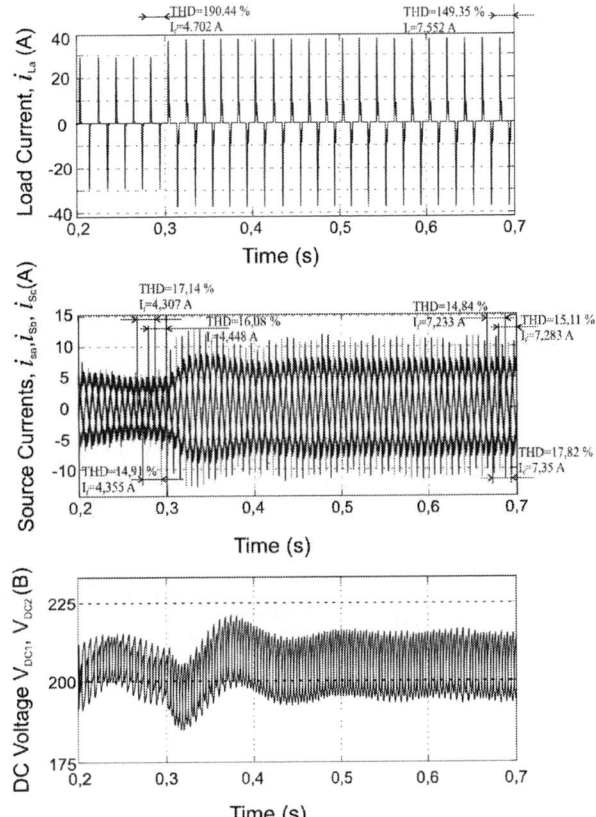

Fig. 7. The waveforms of AF work with the load increasing in the network without the neuro-fuzzy regulator in the control system.

In this case the amplitude value of the fundamental harmonic of current in the load increased from $I_f = 4.7\ A$ to $I_f = 7.6\ A$, and THD of current in the load was reduced from 190.4% to 149.4%. This occurs because the amplitude of high-frequency component, against the background increase in the fundamental harmonic, changes insignificantly. THD of means current also is reduced from 16.1% to 15.1%. In Fig. 8 are shown the waveforms of means current within one cycle before and after the load changing. It is evident that the presence of low-frequency components is noticeable after an increase of the current in the load in the means current.

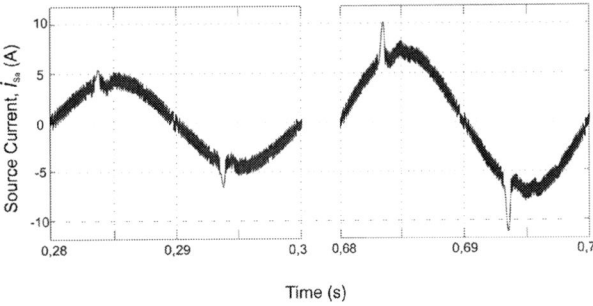

Fig. 8. The waveforms of mean current during one period before and after increasing of the load without the neuro-fuzzy regulator in the control.

In Fig. 9 is shown transient process with load current increasing with the presence of neuro-fuzzy regulator.

In this case neuro-fuzzy regulator forms additional positive incremental voltage on the capacity, as a result the quality of compensation is improved, and the THD is reduced to 11,1% . The waveforms of the means current before and after load increasing are shown in Fig. 10.

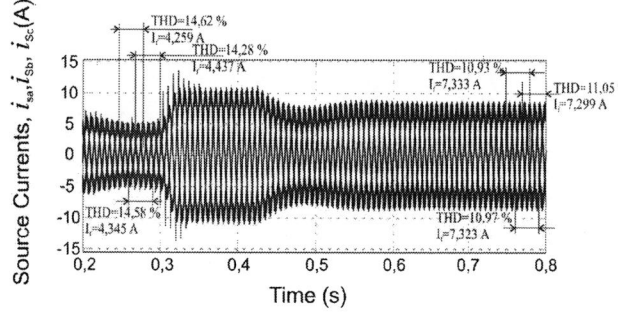

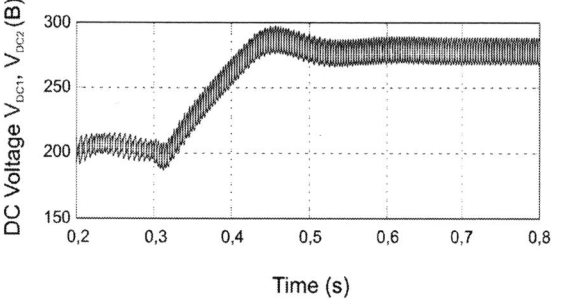

Fig. 9. The waveforms of work AF with the jump of load in the network with the neuro-fuzzy regulator in the control system.

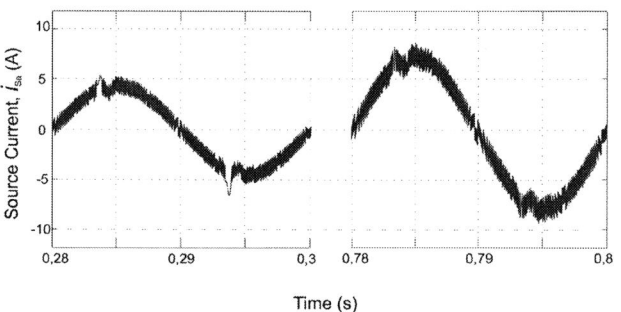

Fig. 10. The waveforms of mean current during one period before and after increasing of the load with the neuro-fuzzy regulator in the control system.

The other simulation results are shown in Table III. Analyzing the obtained results, it is possible to assert that the presence of neuro-fuzzy regulator leads to an improvement in the quality of compensation, with AF constant elements. In this case, the greater range of a change in the current of load, the greater it is necessary to change voltage on the capacity. In particular, an increase of the current of load 1,6 times leads to the need for increasing voltage across capacitor of 1,35 times, which follows from figures 7- 10.

TABLE III
SIMULATION RESULTS

Current step	THD, %			
	Without neuro-fuzzy regulator		With neuro-fuzzy regulator	
	Before step	After step	Before step	After step
4,7 A to 7,6 A	16,1%	15,1%	16,1%	11,1%
4,7 A to 3,1 A	16,1%	19,4%	16,1%	17%

V. CONCLUSIONS

In the work is proposed the new control system of three-phase AF on the basis of neuro-fuzzy regulator, which accomplishes an adaptation of the control mode to changes in the current of load both in the amplitude and by the spectral composition. As a result simulation shows the effectiveness of this control system, in the part of the improvement in the quality of the compensation for reactive power in the network. The basic merit of proposed control system consists in the fact that its adaptation is accomplished without a change in the element base of compensator and frequency of switch modulation on the basis of formation by neuro-fuzzy regulator of the nonlinear law of control AF. It should also be noted that the use of neuro-fuzzy regulator allows constructing of more complex systems with studying in real time.

REFERENCES

[1] H. Akagi, "Trends in active power line conditioners," *IEEE Trans. Power Electron.*, vol. 9, no. 3, pp. 271-350, pp, May 1994.

[2] H. Akagi, "Modern Active Filters and traditional passive filters," *Bulletin of the Polish Academy of sciences, Technical sciences*, vol. 54, no. 3, pp. 255-269, 2006.

[3] H. Akagi, H. Fuijita, "A new power line conditioner for harmonic compensation in power systems," *IEEE Trans. Power Delivery.*, vol. 10, no. 3, pp. 1570-1575, july 1995.

[4] R.L.A. Ribeiro, F. Profumo, "A non standart control strategy for active power filters for unbbalanced conditions of the power mains," *in Conf. Rec. of the 37th Industry Applications Conference (IAS'02),* Pittsburgh (PA, USA), October 2002, vol. 2, pp.896–903.

[5] E. H. Watanabe, M. Aredes, H. Akagi, "The p-q theory for active filter control: some problems and solutions," *Revista Controle & Automacao*, 2004, vol.15, no. 1, pp.78–84.

[6] E.R Cadaval, F.B. Gonzalez, M.I.M. Montero, "Active power line conditioner based on two parallel converters topology," *Compatibility in Power Electronics*, 2005, IEEE, June 1, pp. 134 - 140.

[7] S. Bhattacharya, Po-Tai Cheng, D. M. Divan, "Hybrid solutions for improving passive filter performance in high power application," *IEEE Trans. Industry Applications*, vol. 33, Issue 3, May/June 1997 pp. 732 – 747.

[8] A. Chaghi, A. Guettafi, A. Benoudjit, "Four-Legged Active Power Filter Compensation For a Utility Distribution System," *Journal of Electrical Engineering*, vol. 55, no. 1-2, 2004, pp. 31-35.

[9] J. G. Pinto, R. Pregitzer, L.F.C. Monteiro, J.L. Afonso, "3-phase 4-wire Shunt Active Power Filter with Renewable Energy Interface," *Proc. of the International Conference on Renewable Energies and Power Quality (ICREPQ'07)*, Sevilha (Spain), March 2007.

[10] J. W. Dixon, L. Moran, C. Elgueta, "Some improvements in 81-Level Inverters for Traction Applications," *Proc. of 21th Electric Vehicle Sysmposium (EVS 21)*, Montecarlo (Monaco), April 2005.

[11] H. Akagi, E.H. Watanabe and M. Aredes, "Instantaneous Power Theory and Applications to Power Conditioning," *Wiley-IEEE Press*, April 2007.

[12] D. Nedeljkovic, M. Nemec, K. Drobnic, V. Ambrozic, "Direct current control of active power filter without filter current measurement," *Power Electronics, Electrical Drives, Automation and Motion, SPEEDAM 2008*, International Symposium on, 11-13 June 2008, pp: 72 - 76

[13] H. Kim, F. Blaabjerg, B. Bak-Jensen, "Spectral Analysis of Instantaneous Powers in Singl-Phase and Three-Phase Systems With Use of p-q-r Theory*," IEEE Trans. Power Systems*, vol. 17, no 5, September 2002, pp. 711 – 720.

[14] J. W. Dixon, J. M. Contardo, L. A. Moran, "A Fuzzy controlled, active front-end rectifier with current harmonic filtering characteristics and minimum sensing variables," *IEEE Trans. Power Electronics*, 1999, vol.14, no 4, pp.724–729.

[15] Y.-M. Chen, "Passive filter design using genetic algorithms," *IEEE Trans. Industrial Electronics*, vol. 50, no 1, Feb 2003, pp. 202 - 207.

[16] Y.-M. Chen, R.M. O'Connell, "Active power line conditioner with a neural network control," *IEEE Trans. Industry Applications*, vol. 33, no 4, Jul/Aug 1997, pp.1131 – 1136.

[17] S. H. Ali, N. Reza, "Self tuning Fuzzy PI controller for active filter optimized by Ant colony method," *Power Electronic & Drive Systems & Technologies Conference (PEDSTC)*, 2010 1st, 17-18 Feb., pp 351 – 356.

[18] Y. O. Denisov, S.A. Ivanec, "Impulse stabilization systems of DC voltage with fuzzy and adaptive controller," *Electricity*, Russia, 2007, no. 7, pp.35-39.

[19] C. Sharmeela, M.R. Mohan, G.Uma, J. Baskaran, "Fuzzy logic controller based three-phase shunt active filter for line harmonics reduction," *Journal of Computer Science*, 2007, no 3(2), pp.76–80.

[20] V.P. Voytenko, M.A. Homenko, "Sanitizations of industry neuro controller with improved dynamic," *Technical Electrodynamics. Subject issue*, Ukraine, 2007, part 3., pp. 50-55.

[21] M. Gaiceanu, "Active power compensator of the current harmonics based of the instantaneous power theory," *Department of Electrical engineering*, University of Galati, Domneasca Street 47, 6200-Galati Romania.

[22] F. Couto, J. S. Martins, J. L. Afonso, "Simulation results of a shunt active power filter with controller based on P-Q Theory," *Proc. of International Conference on Renewable Energies and Power Quality (ICREPQ'03)*, Vigo (Spain), April 2003.

[23] S.A. Ivanec, O.O. Husev, A.I. Chub, "Method of choice elements of active line conditioner," *Journal of Chernigov State Technological University*, Ukraine, 2009, no. 40, pp. 223-232.

[24] J. Zakis, D. Vinnikov, J. Laugis, I. Rankis, "Feasibility study of flexible systems for reactive power compensation," *Technical Electrodynamics. Subject Issue*, Ukraine, 2010, part 2, pp. 16-21.

Pollution of High Power Charging Electric Vehicles in Urban Distribution Grids

Constantinos Sourkounis, Bingchang Ni, Alexander Broy
Ruhr-University Bochum, Power Systems Technologie and Power Mechatronics
sourkounis@eele.rub.de, ni@eele.rub.de, broy@eele.rub.de

Abstract- **The currently ongoing discussion about large scale increase of electro mobility supposes a high acceptance of electrical vehicles by the public. To achieve a high acceptance, beyond the ecological aspects, the availability and the short-term flexibility of use are of essential importance. Quick charging leads directly to numerous technical questions regarding the integration of high power charging stations in rural areas. Dynamic reactive power compensation and a short term on demand active power supply on site can be an appropriate approach to realise high power charging with adequate technical efforts. The investigation results of these approaches for intelligent integration of high power charging stations in distribution grids will be discussed. Furthermore, the mains pollution by harmonics will be examined and discussed in the scope of a detailed investigation of the battery charger.**

I. INTRODUCTION

The currently ongoing discussion about large scale increase of electro mobility supposes a high acceptance of electrical vehicles by the public. To achieve a high acceptance, beyond the ecological aspects, the availability and the short-term flexibility of use are of essential importance. Amongst others the acceptance can be increased by very short charging durations for the batteries, e.g. in less than thirty minutes. This is described by the term high power charging.

Unfortunately this approach leads directly to numerous technical questions regarding the integration of high power charging stations in the electrical power distribution grid in conurbations or in rural areas. In most cases the therefore required power capacity of the grids is not available or can not be accessed all the time.

In conurbations it is generally possible to enable high power charging during periods of low load from private households or commercial consumers, controlled by a grid side power management which takes into account the urgency of charging for individual vehicles. Normally, the grid capacity for high power charging in rural areas is not sufficient at any time, despite periods with low grid capacity utilization. In most cases the required power for a high power charging station would demand the whole or even more than the possible grid capacity. In this case, additional technical devices can be installed for grid stabilization. Besides an expensive grid expansion, a dynamic reactive power compensation and a short term on demand active power supply on site can be an appropriate approach to realize high

power charging with adequate technical efforts. The investigation results of these approaches for intelligent integration of high power charging stations in distribution grids will be discussed in this paper.

Furthermore, the mains pollution by harmonics depending on the type of charging devices and their switching frequency will be examined and discussed in the scope of a detailed investigation of the battery charger.

II. DESCRIPTION OF THE OBJECTIVES

Based on the specific energy demand of today's electric vehicles e_F, which is between 120 and 200 Wh/km, and an average driving range s between 120 and 200 km, the required energy amount, calculated according to the formula

$$E_A = \frac{1}{\eta_{LG} \cdot \eta_A} \cdot e_F \cdot s \qquad (1)$$

results in a required capacity of the accumulator between 17.6 kWh and 49 kWh. Thereby, the converter to supply the drive motor is supposed to work with an efficiency of $\eta_{WR} = 0.96$ and the discharge process of the accumulator is supposed to work with an efficiency of $\eta_{AE} = 0.85$. The electric energy which has to be provided by the mains to charge the accumulator has to additionally include the losses of the charging process. These losses are considered by the efficiency of the charger η_{LG} as well as the accumulator η_{AL} during the charging process. The electric energy that hast to be provided by the mains amounts according to

$$E_N = \frac{1}{\eta_{LG} \cdot \eta_{AL} \cdot \eta_{WR} \cdot \eta_{AE}} \cdot e_F \cdot s \qquad (2)$$

between 21.6 kWh and 60 kWh. For the charging of the accumulator within 30 minutes, the required power capacity of the grid connection and the charger for the accumulator is between 45 kW and 120 kW, according to

$$P_N = P_{LG} = \frac{E_N}{t_L} \qquad (3)$$

The required power at charging stations for high power charging results in noteworthy load on the connected urban distribution grid. Hereby, depending on the short circuit power of the urban distribution grid and the coupling point of the charging station respectively, impermissible voltage variations can be caused. Further more, by the elements of the charger for the accumulator switching with very high frequency; harmonics are emitted into the urban distribution grid. Depending on their amplitude and frequency, these harmonics can detract other electrical systems which are connected to the same urban distribution grid. Therefore the mentioned grid perturbation, amongst others, will also be investigated in the scope of this paper.

III. ANALYTICAL DESCRIPTION OF GRID PERTURBATION

As stated before, the grid perturbation of high power charging station for electric vehicles depend on many different characteristics; the short circuit power of the urban distribution grid in relation to the power of the charging station and the topology, the switching frequency and the control method of the accumulator charger. Based on the most commonly used concepts for charging stations, the grid perturbations will be investigated and discussed with analytical methods.

1.   Charger with Mains Commutated Converters

The charging devices with mains commutated converters cause harmonics with the frequencies

$$f_\nu = f_N \cdot (6k \pm 1) \quad \text{with } k = 1, 2, 3, \ldots \tag{4}$$

This results in harmonics with five, seven, eleven, and so on, times the grid frequency f_N.

The amplitude depends on the short circuit power of the grid connection or the grid impedance $Z_{i,N}$ and the size of the additionally installed commutation inductivities respectively. It can be determined with following equation:

$$I_{\nu L} \cdot \sqrt{2} = C_\nu = \sqrt{A_\nu^2 + B_\nu^2}$$

$$\text{with} \quad A_\nu = \frac{1}{\pi} \int_0^{2\pi} i_L(\omega t) \cdot \cos \nu \omega t \cdot d\omega t \tag{5}$$

$$B_\nu = \frac{1}{\pi} \int_0^{2\pi} i_L(\omega t) \cdot \sin \nu \omega t \cdot d\omega t$$

Because the wave shape of the branch current does not depend on the control angle α, the relation between the amplitude of the harmonics and the basic frequency stays constant for any duty cycle. Hereafter, at stationary operation of a mains commutated converter following harmonics can be measured

$$\hat{I}_{\nu;L} = \sqrt{2} \cdot I_{\nu,L} = \frac{\sqrt{2}}{\nu} \cdot I_{1,L} = \frac{\sqrt{2}}{\nu} \cdot \frac{\sqrt{6}}{\pi} \cdot I_d \tag{6}$$

Under consideration of the compatibility level for urban distribution grids described in [2], the unwanted emission of harmonics will be judged. Table 1 shows a list of the compatibility levels for different voltage harmonics.

TABLE 1
COMPATIBILITY LEVEL $U_{\nu,VT}$ FOR SINGLE VOLTAGE HARMONICS IN THE LOW VOLTAGE GRID ACCORDING TO [2].

1Ordinal number ν[-]		compatibility level $u_{\nu,VT}$ [%]
odd, not containing multiples of 3, ordinal numbers	5	26
	7	5
	11	3,5
	13	3
	17	2
	19	1,5
	23	1,5
	25	1,5
	>25	0,2+0,5*25/ν
odd, multiple of 3 ordinal numbers	3	5
	9	1,5
	15	0,3
	21	0,2
even ordinal numbers	2	2
	4	1
	6	0,5
	8	0,5
	10	0,5
	>10	0,2

By calculating the relation between the voltage harmonic u_ν and the relevant compatibility level $u_{\nu,VT}$, a harmonics interference factor B_ν results. It describes the relation between the nominal apparent power $S_{LG,nenn}$ and the short circuit power S_{kV} for charging devices being operated at a particular coupling point of the grid.

$$B_\nu = \frac{u_\nu}{u_{\nu,VT}} = \frac{k_{P\nu} \nu I_{\nu,L}}{u_{\nu,VT}} \cdot \frac{S_{LG,nenn}}{S_{kV}} \tag{7}$$

$$\text{with} \quad \angle Z_{iN} \geq 80°$$

With $k_{P\nu}$ being the harmonics weighting factor (see Table 2). The highest value B_{max}

$$B_{max} = \max\{B_\nu\} \quad \text{for } \nu = 1 \ldots 25 \tag{8}$$

will be considered for the evaluation. According to [2], the necessary $k_{P\nu}$-values to calculate B_ν can be taken from table 2.

TABLE 2
RECOMMENDED VALUES FOR THE HARMONICS WEIGHTING FACTOR k_{Pv} [2].

$\nu3$	3	5	7	11	13	17	19
	0,3	0,1	0,1	0,1	0,1	-	-
	0,4	0,3	0,2	0,1	0,1	0,1	-
	0,6	0,5	0,3	0,2	0,2	0,1	0,1
	0,7	0,7	0,5	0,4	0,4	0,3	0,2
	0,9	0,8	0,7	0,6	0,6	0,5	0,5
	1	1	1	1	1	1	1

The necessary short circuit power of the grid connection point S_{kV} can be determined according to

$$S_{kV} \geq \frac{k_{Pv}\nu I_{vL}}{u_{vVT}} \cdot \frac{S_{LG,nenn}}{B_{max}} \qquad (9)$$

In a similar way, the highest allowable power rating of the charging device at the grid coupling point can be determined.

$$S_{LG,nenn} \leq \frac{u_{vVT}B_{max}}{k_{Pv}\nu I_{vL}} \cdot S_{kV} \qquad (10)$$

The harmonic interference factor is maximally allowed to amount up to $B_{max} = 0.02$, according to [2].

2. Charger with Self Commutated Converters

Charging devices with self-commutated rectifier cause harmonic voltage components of different spectra, contrary to charging devices with mains-commutated rectifier; this is directly dependent on the control method. Three different methods can be used for the control of a mains-operated rectifier taking into account the demand for lower mains pollutions:

- sine-delta modulation
- space vector modulation, and
- current hysteresis procedure.

In case of sine-delta and space vector modulations, a continuous switching frequency f_p can be selected taking into account the semi-oscillation symmetry as multiple of the basic frequency which corresponds to a mains frequency of 50 Hz, so that the inducement of intermediate harmonics is basically avoided. This procedure reduces the demands on the mains-side filter and the height of the short-circuit power at the mains connection point, because the maximal admissible values of the intermediate harmonics are considerably lower than the admissible values of the harmonic components.

Nowadays available electronic switching elements allow for very high switching frequencies as compared with the first harmonic frequency ($f_p/f_1 > 21$), so that the low-frequency harmonic components are suppressed as far as possible. With these procedures, high-frequency harmonic components with ordinal numbers $\nu = f_p/f_1$ occur in packages only. Harmonic

components of order ν can then be determined according to

$$U_{vWR} = \left(1 + 2\sum_{n=1}^{N}(-1)^n \cos \nu\alpha_n\right)\frac{\sqrt{2}}{\nu\pi}U_d$$

$$(11)$$

with $N = \frac{\nu-1}{2}$ and $\nu = \frac{f_P}{f_1}$

[3]. α_n describes the switching angles within a quarter of the cycle duration.

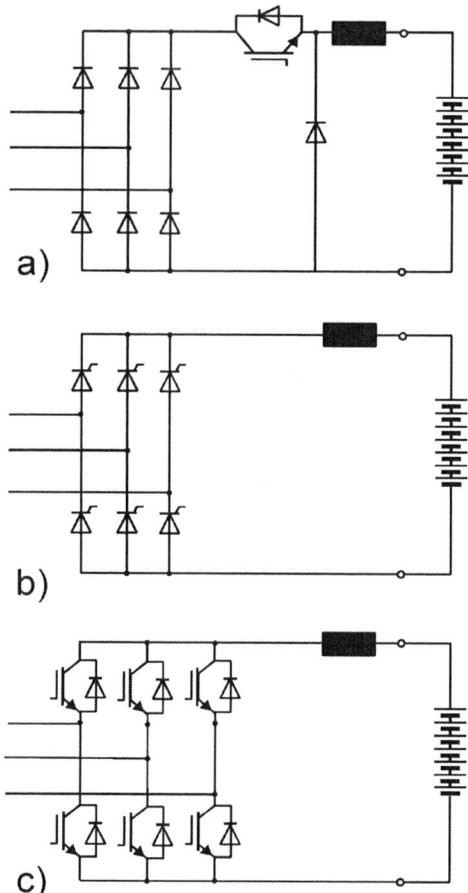

Fig. 1: Circuit topology of charging devices [1].

Different to the hitherto described control methods, a switching frequency is not preset in case of the current hysteresis method. A tolerance band around the desired course of the mains current is defined in this procedure. As long as the actual value of the current remains inside the tolerance band, no switching action happens. When the upper limits are exceeded, the "high-side" electronic switch is switched-off in the phase, and the "low-side" switch is engaged (see Fig. 1c), vice versa when the lower limits of the tolerance band are undercut. Since the desired sine shape of the current features different gradients within one cycle duration, the switching frequency varies accordingly. Furthermore, the switching frequency is directly connected

978-1-4244-8806-3/11 $26.00 © 2011 IEEE

with the effective inductivity in the respective phase on the mains side. The tolerance band determines the maximal harmonic component content of the mains current directly, so that an alteration of the short-circuit power of the circuit or of the internal mains impedance or an alteration of the filter impedance would entail a change of the switching frequency accordingly. This makes it necessary to lay out the inductivities of the mains filter in such that not too high switching frequencies occur. The switching frequency is limited by the resulting switching losses of electronic switches. The average switching frequency can be derived by the following equation:

$$\overline{U}_d - U_B = i\left(R_K + R_{iN}\right) + \left(L_K + L_{iN}\right)\frac{di}{dt}$$

$$\text{with } R_K, R_{iN} \ll \Rightarrow Z_{iN} \approx \omega L_{iN}$$

$$Z_F \approx \omega L_K \qquad (12)$$

$$\Rightarrow \overline{f}_P = \frac{1}{2\Delta t} = \frac{\overline{U}_d - U_B}{2\left(L_K + L_{iN}\right)}$$

From this equation, the dependency of the switching frequency on the internal mains impedance or accordingly the inductivity at the mains connection point and the inductivity of the mains filter, which can be assumed to be an inductance as the first approximation, can be clearly seen.

$$\overline{f}_P \leq \frac{\overline{U}_d - U_B}{2\,L_K} \qquad (13)$$

A site-specific layout is necessary because of the direct interrelation between the harmonic component propagation and the internal impedance of the mains connection. This leads to the necessity to stationary install the mains filter at particular mains connection point or to act on the worst case assumption for the layout of the necessary inductance of the mains filter. The worst case for the layout of the necessary inductance of the mains filter would be to neglect the internal mains impedance. Therefore, the equivalent diagram in Fig. 2 can be taken as a basis for the layout. Based on analytical considerations, a number of different scenarios for high power charging of electric vehicles have been examined by means of simulation.

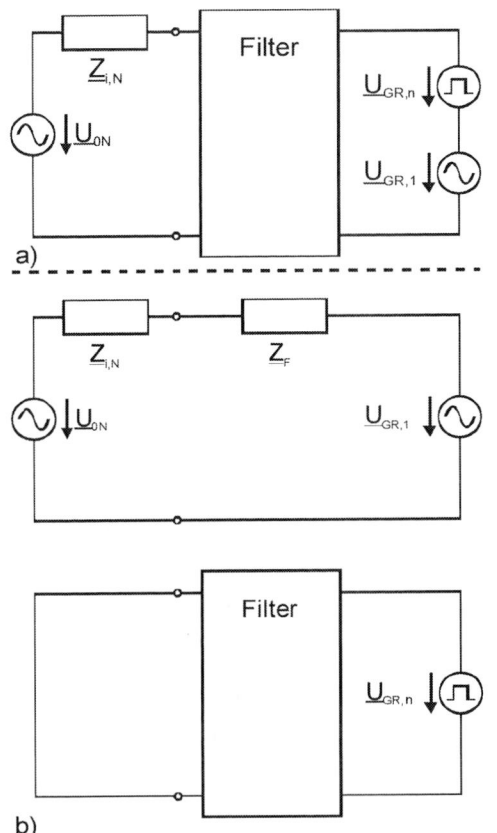

Fig. 2: Equivalent diagram for the calculation of the mains-side filter, a) complete diagram, b) basic oscillation and harmonics diagram.

IV. SIMULATION EXAMINATIONS

A model residential area supplied with electrical energy by a transformer station has been used as a basis for the simulation examinations (see Fig. 3). The residential area consists of 100 accommodation units and features a maximal load of 500 kVA at an average cosine value of 0.92. The high power charging procedure is carried out at times with low net loading. A synchronism factor of 0.2 has been assumed for the examinations. This assumption is based on measured values over one year and featuring synchronism factors of between 0.1 and 0.7.

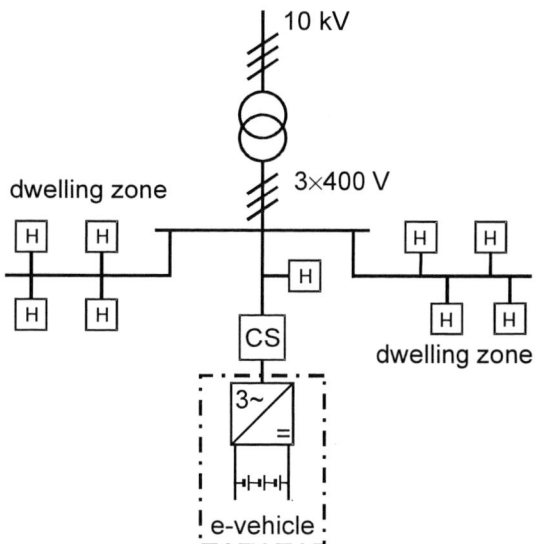

Fig. 3: Grid topology for simulation examinations.

The dynamical behavior of the low-voltage supply grid and the influence of the high-power charging device on the grid were subject of the examinations. Two 50 kW high-power devices have been examined on the basis of a self-commutated rectifier. The applied charging devices differ from each other with respect to the type of control. A sine-delta modulation method and a current hysteresis band method have been used. The sine-delta modulation procedure is a classical control method and features advantages with regard to the spectrum of emitted voltage harmonic components.

At high switching frequencies ($f_p/f_1 > 21$), the first harmonic components occur at a switching frequency with sidebands of lower amplitudes, so that specific higher-order filters can be installed to cope with the expected harmonic component frequencies. A disadvantage is that a direct influence on the current ripple is not possible during operation.

This can, however, be realized with the current hysteresis band method. By presetting of the current hysteresis band, the current ripple at the connection point is determined directly. When this charging unit is operated at a net connection point with different short-circuit powers resp. internal impedance, the switching frequency varies, but not the height of the generated harmonic components.

A first-order filter has been used for these examinations. The reason is that the filters have to be universally applicable. The filters which are part of the charging stations must guarantee compatibility with all kinds of charging devices. The use of higher-order filters in connection with the current hysteresis band method would lead to a malfunction of the self-commutated rectifier or the charging unit respectively. For the integration of the grid filter into the charging device and thus the electric vehicle it is generally possible to adapt the filter order to the type of control. At the same time, oscillatory filters of higher order may lead to voltage variations under certain circumstances when grid connections with high short-circuit powers (stiff grid) are concerned.

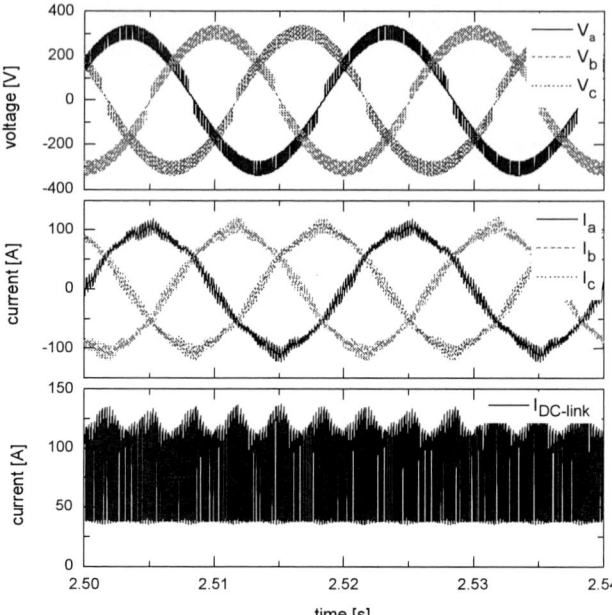

Fig. 4: Voltage and current at the PCC and the dc-link current for sine-delta modulation with 5MVA grid short circuit power.

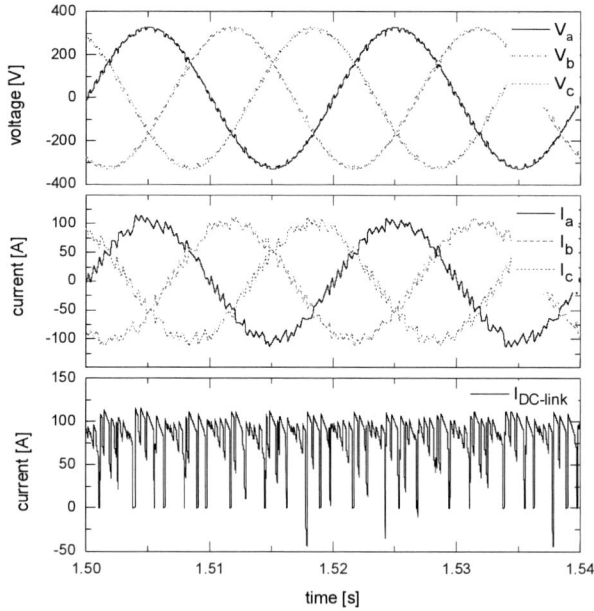

Fig. 5: Voltage and current at the PCC and dc-link current for current hysteresis procedure with 5MVA grid short circuit power.

In fig. 4 and 5 the time curves of the voltage and current at the point of common coupling and the dc-link current are shown.

In the cause of these examinations, the short-circuit power of the grid connection has been varied for a constant grid filter which has been realized on the basis of inductances. Herewith, an operation of the respective charging devices at

978-1-4244-8806-3/11 $26.00 © 2011 IEEE

different charging stations resp. grid connections has been simulated. The simulation results have been evaluated on the basis of the harmonic component spectrum and the voltage drop caused by the charging unit.

The simulation results clearly show correlation between the grid short circuit power and the induced harmonics by high power chargers. A higher short circuit power, and therefore lower internal grid impedance, leads to lower amplitudes of the harmonics. At the same time it can be concluded that a first order filter element, which would be designed for a specific control method and for installation at any grid connection, would not be sufficient to dampen all harmonics. This can be observed, especially for the sine delta modulation control method (see fig. 6). In order to keep the limits according to [2] of 0.2% (here 0.46V) a short circuit power of 100 MVA is needed. This results for a short circuit voltage u_K of 4% in a nominal power of the grid $S_{A,Nenn}$ of 4 MVA.

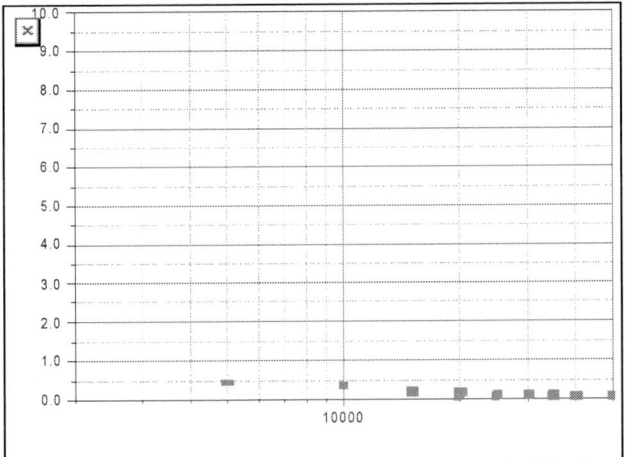

Fig. 6: Harmonics induced by high power charging device with sine delta control

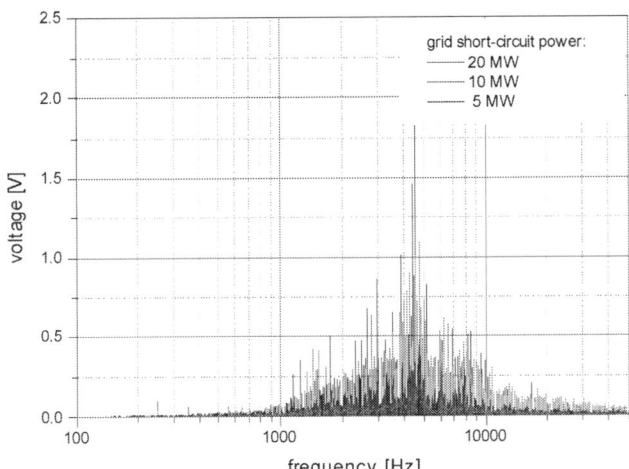

Fig. 7: Harmonics induced by high power charging device with hysteresis control

In contrast, for the current hysteresis control method a grid short circuit power of 13 MVA is sufficient to keep the voltage harmonics within the permitted limits (see fig. 7). This corresponds for a $u_K = 4\%$ to a nominal grid power of 520 kVA. With it, high power charging with 50 kW would be possible in the investigated areas without putting the grid quality at risk.

In principle an increase of the filter inductances will reduce the needed short circuit power and nominal power respectively. But, this leads to a high volume and weight of the filter, whereby noteworthy disadvantages arise for integration into the vehicle. Furthermore the higher investment const are an additional disadvantage.

V. CONCLUSION

The investigations show that depending on the control methods of the charging devices, differently sized grid filters are needed to avoid the spreading of prohibited voltage harmonics. A filter topology is needed which is applicable independent from the grid short circuit power and the internal inductance at the connection point, as well as from the applied control method of the battery charger. At the same time it should be of high order to efficiently dampen the voltage harmonics. For this it is essential to establish generally accepted standards.

VI. REFERENCES

Books:

VII. M. Meyer, "Leistungselektronik", Springer Verlag Berlin 1990
VIII. VDEW, "Grundsätze für die Beurteilung von Netzrückwirkungen", VWEW, Frankfurt, 1992

Periodicals:

IX. T. M. Undeland, N. Mohan, "Overmodulation and Loss Considerations in High-Frequency Modulated Transitorized Induction Motor Drives." IEEE Transactions on Power Electronics Vol. 3 (1988) No. 4
X. W. Kempton and K. Toru; "Electric-drive Vehicles for Peak Power in Japan", Energy Policy 2000, 28(1): 9-18
XI. W. Kempton and J. Tomic; "Vehicle to Grid Implementation: from stabilizing the grid to supporting large-scale renewable energy". Journal of Power Sources Volume 144, Issue 1, 1 June 2005, Pages 280-294.
XII. W. Kempton and A. Dhanju, "Electric Vehicles with V2G: Storage for Large-Scale Wind Power", Windtech International 2 (2), March 2006, pp 18-21.

Papers from Conference Proceedings (Published):

XIII. C. Sourkounis, B. Heusler-Sourkounis, J. Koppe; "Software to Design Autonomous Power Supply Systems Based on Renewable Energies with a Mathematic Optimisation Algorithm Tool" EWEC 2006- European Wind Energy Conference, Athens Greece
XIV. A. Soni, C. S. Ozveren; "Improved Control of Isolated Power System by the Use of Feeding Technique", Proceedings of the 41st International Universities Power Engineering Conference, 2006. UPEC '06. Volume 3, 6-8 Sept. 2006 Page(s):974 – 977

Useful Lifetime and Rational Replacement of Power Transformers

Gerards Gavrilovs, Olegs Borscevskis
Riga Technical University
gerards.gavrilovs@latvenergo.lv, olegs.borscevskis@latvenergo.lv

Abstract--This paper presents main ideas of power transformers maintenance during lifetime and its rational replacement which is based on resolve of technical condition. Replacement of power transformers always affined with costs and requires financial supply. The recalculation of energy systems nominal value, i.e. power of transformers, is following too. There are many affected nuances.

I. INTRODUCTION

Power transformer is the most high-priced electrical equipment in electrical network (transmission and distribution). Its average existing age could be varying between 20 and 40 years. Concepts of power transformer life extension, increasing loading and reduced maintenance costs are often discussed. These ideas at first appearance are contradictory, but all strive for the same results – to reduce operating costs and improve reliability in the delivery of electricity to consumer.

Power transformers are the single largest capital item in substations, comprising almost 55 percent of the total belonged equipment investment. The utility expenditures associated with this investment in acquisition, installation, operation and maintenance typically do not reflect the relevance of this investment. The cost of premature and unexpected failure of one of these assets can be several times more than the initial cost of the power transformer. There is not only the modernization or replacement cost, but also costs associated with failure prevention [1]. Is important to take into account that the technical condition of power transformer base in the time of slow rates of reestablishing and replacement of out-of-resource power transformer can become a source of threat for power safety of the region and of country overall. The useful lifetime extension of power transformers is prior moment in the maintenance. It can be done basis on diagnostics results, modernization effectiveness and using of new technical condition prognosis methods. Certainly, the decision of extension useful lifetime is accepted for each power transformer separately. The conclusion about the extension of power transformer resource should be accepted on the basis of careful inspection of technical condition, estimation of insulation aging, detection and elimination of failures.

The coordination of diagnostic results with power planning survey, give possibility to change transformer on correct one.

II. USEFUL LIFETIME OF POWER TRANSFORMERS

The most appropriate life assessment should concentrate on factors, which directly influence the life of the transformer – insulation life. A relative comparison of the transformers is an alternative to remaining life determination, which involves many uncertain assumptions.

Power transformer useful life involves several mechanisms of degradations. The life of a transformer may be introduced as the change of its condition with time under impact of thermal, electric, dielectric, chemical, electromagnetic and electrodynamics stresses, as well as under the impact of various contamination and aging processes. Technical life of a transformer maybe thought as of several components:

- *"Dielectric Life"*- life span up to critical reduction of dielectric margin of insulation.
- *"Thermal Life"*-time up to critical decomposition of winding conductor's insulation, e.g. degree of polymerization ($DP \leq 250$).
- *"Mechanical life"*-up to critical mechanical weakness and deformation of windings under cumulative stresses of through faults, in-rush currents, vibrations.
- *"Life of accessories"*, especially bushings and load tap changer (LTC), which sometimes can be shorter than complex life of active part of a transformer.

Service experiences give motives to conclude that many failures occur just due to aging phenomena. However statistics available has not exhibited yet correlation between number of failures and transformer years. Average age of failed transformer is still between 20-22 years. Contribution of generation "more than 25" is becoming more weighty, however there has been experienced also a number of failures of new equipment. There is still little information available about the units that have failed primarily due to thermal degradation of insulation material.

In fact *"Dielectric life"* is shorter than *"Thermal life"* due to critical effect of oil aging products resulting in reduction of dielectric withstands strength of oil and degradation of surface strength. The quality of insulation oil has impact on power transformer reliability. It affect on power transformer withstand in maintenance. For new power transformer, its' useful lifetime and quality of construction are expected like is shown in Fig. 1. The withstand stress of a power transformer decreases with time as the insulation system ages. The aging of the insulation system reduces both the mechanical and

978-1-4244-8806-3/11 $26.00 © 2011 IEEE

dielectric withstand of the power transformer. Unfortunately the applied stress tends to increase with added load or reduced protection over time [2].

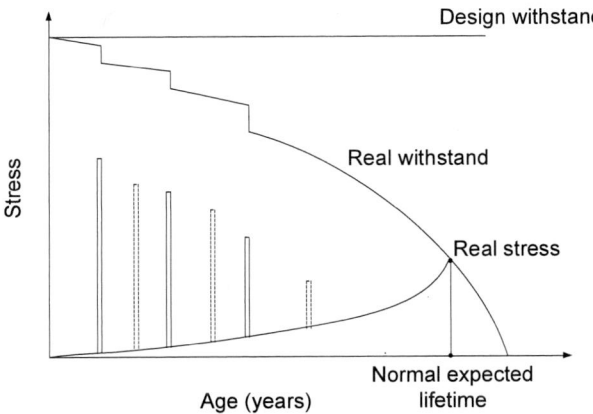

Fig. 1. Power transformer normal expected useful lifetime

Fig. 1 shows the normal useful lifetime of a power transformer which has been able to withstand the periodic accidental faults on the system. Load has grown gradually and withstand decreased due to normal aging. Finally, at the end of lifetime, the power transformer hasn't withstand operating or fault stress, which in many cases is greatly higher than the original designed for stress (real stress zone).

III. TECHNICAL CONDITION DETERMINATION OF POWER TRANSFORMERS AND CASE STUDY

First of all, each specialist must be with a good experience in diagnostic, fault prognosis and schedule of energy system should be able for analyzing and technical data storage. There must be clear difference between real existent situation and theoretic power transformers operating possibilities. The end of lifetime is presented as 95% of accumulated life loss and 5% as remaining design life left [3].

Power transformer age and end of life has a poor correlation. There is example of one power transformers producing year group, which were under diagnostic during maintenance, see Table I.

TABLE I
DIAGNOSTIC RESULTS OF INSULATION CONDITION

Transf. Produc. year	1983		1981		1982	
Meas.No	Rabs	tg	Rabs	tg	Rabs	tg
1	1.66	0.33	1.68	0.46	2	0.32
2	2.17	0.45	2	0.53	2.98	0.29
3	2	0.43	2.5	0.53	2.9	0.42
	1991		1989		1990	
1	1.26	0.8	1.12	0.13	1.23	0.11
2	1.26	0.9	1.14	0.17	1.31	0.11
3	1.27	0.6	1.12	0.23	1.4	0.25

TABLE I (CONTINUATION)
DIAGNOSTIC RESULTS OF INSULATION CONDITION

Transf. Produc. year	1983		1981		1982	
Meas. No.	Rabs	tg	Rabs	tg	Rabs	tg
	1992		1993		1998	
1	1.25	0.3	1.43	0.22	1.15	0.12
2	1.36	0.2	1.7	0.3	1.17	0.14
3	1.34	0.24	1.64	0.25	1.08	0.21
	1996		1994		2000	
1	1.24	0.19	1.29	0.2	1.48	0.13
2	1.15	0.19	1.76	0.27	1.76	0.16
3	1.22	0.21	1.83	0.25	1.51	0.24
	1999		1997		2008	
1	1.13	0.2	1.31	0.3	1.76	0.24
2	1.23	0.2	2.07	0.4	1.59	0.25
3	1.19	0.2	2.06	0.4	1.86	0.32
	2003		2004			
1	1.21	0.22	1.51	0.34		
2	1.23	0.23	2	0.4		
3	1.06	0.23	1.88	0.41		
			2009			
1			1.59	0.49		
2			1.88	0.58		
3			1.81	0.52		

Dielectric losses (loss tangent-tg) and insulation measurements are shown in Table I. It is necessary to detect correlation rate between both technical condition regulated values. For our investigation we took measuring results of power transformer produced in 1981. There are measuring results for main scheme i.e. high voltage winding, Table II.

TABLE II
CONCRETE POWER TRANSFORMER DIAGNOSTIC RESULTS

Measuring year	Rabs	tg
1981	1.68	0.46
1989	1.12	0.13
1993	1.43	0.22
1994	1.29	0.2
1997	1.31	0.3
2004	1.51	0.34
2009	1.59	0.49

978-1-4244-8806-3/11 $26.00 © 2011 IEEE

Fig. 2. is constructed for visual imaging of insulation absorption rate (*Rabs*) and dielectric looses during maintenance.

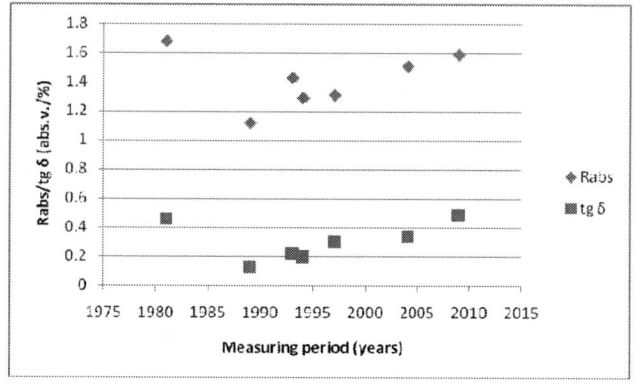

Fig. 2. Insulation absorption rate and dielectric losses

The second measuring point from the beginning has done after power transformer tackle and switching on to energy system.

For checking on practice, in engineering, of correlation theory between two factors we built the Fig. 3.

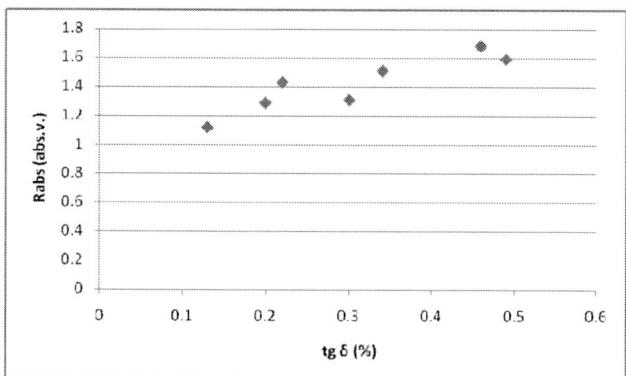

Fig. 3. Correlation between *Rabs* and *tg*

Increased dielectric loses follows with decreased absorption rate of solid insulation and it is the maintenance theoretical rule. It means that between them negative linear correlation has place to be. In our case we have linear positive correlation both those two factors to maintenance time period, and between itself. The last one explains by improvement and modernization works in maintenance of power transformer. *Correlation rate is r= 0.904053*, which indicates that between marks exist intent positive linear relevance. The relevance and density of those technical condition's factors for each power transformer can be different.

The replacement of each power transformer is calculated individually, not chaotically. It is very easy to replace the existing transformer for the new one (the same rated power), but during replacement we must take into account the customers` possible marketable power in the service area of the concrete transformer substations. Example of replaced high power transformers of some energy transmission company is shown in Fig. 4.

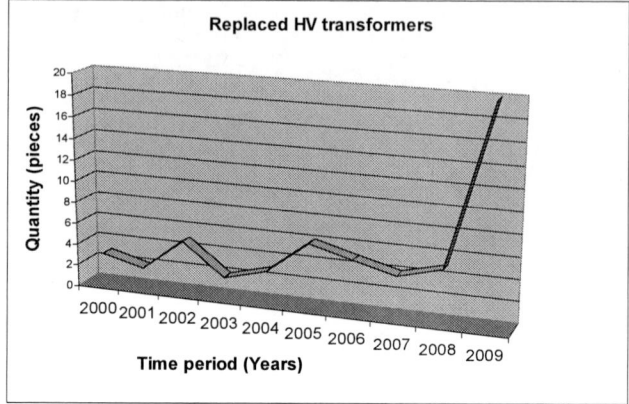

Fig. 4. Replaced high voltage transformers quantity

IV. THEORETICAL PROGNOSIS TRANSFORMER SUBSTATIONS` CAPITAL INVESTMENTS

For rational construction of power system it is necessary to determine the optimal power of substations for each voltage level. This is a complex and laborious task. In paper the choice of transformer substation (*TS*) optimum power for the 110kV network on basis of variants` comparison is made [4].

Total investments of variants K_Σ can be found by expression:

$$K_\Sigma = K_{TS\Sigma} + K_{TP\Sigma} + K_{AL\Sigma} + K_{VL\Sigma} \qquad (1)$$

where $K_{TS\Sigma}$ are capital investments for 110/10 kV *TS* construction;

$K_{AL\Sigma}$ are capital investments of 110 kV cable line for 110/10kV *TS* connection;

$K_{TP\Sigma}$ are capital investments for 10/0.4 kV *TS* construction;

$K_{VL\Sigma}$ are 10 kV cable line capital investments for 10/0.4kV transformer point (*TP*) connection.

Certain functions of the components can be calculated as follows:

$$K_{TS\Sigma} = n_{TS} \cdot K_{TS} \qquad (2)$$

where n_{TS} is the number of 110/10kV TS;

$$
\begin{aligned}
K_{AL\Sigma} &= \lambda_A \cdot A \cdot n_{TS} \cdot K_{ipAL} = \\
&= 1.7 \cdot 1.1 \sqrt{\Pi_{TS}} \cdot n_{TS} \cdot K_{ipAL} = \\
&= 1.9 \sqrt{\Pi_{TS}} \cdot n_{TS} \cdot K_{ipAL}
\end{aligned} \qquad (3)
$$

where λ_A is 110kV network configuration factor (loop-through scheme adopted $\lambda_A = 1.7$);

978-1-4244-8806-3/11 $26.00 © 2011 IEEE

A is the theoretical distance between the 110/10 kV substation adjacent;

K_{ipAL} is 1 km of 110kV cable line construction costs;

$$K_{TP\Sigma} = n_{TP} \cdot K_{TP} \cdot n_{TS} \qquad (4)$$

where n_{TP} is 10/0.4kV TP quantity in one 110/10kV TS service area;

K_{TP} is 10/0.4kV TP cost.

In addition:

$$n_{TP} = \frac{\Pi_{TS}}{\Pi_{TP}} = \frac{S_{r,TS}}{k_{o,TP} \cdot S_{r,TP}} \qquad (5)$$

where Π_{TP} is 10/0.4kV TP service area;

$S_{r,TP}$ is 10/0.4kV TP rated power;

$k_{o,TP}$ is factor of simultaneity at transformers' load maximum, depending of 10/0.4kV TP quantity;

$$K_{VL\Sigma} = n_{TP} \cdot K_{ipVL} \cdot L_{VL} =$$
$$= n_{TP}(K_{ipVL} \cdot \lambda_V \cdot 1.1\sqrt{\Pi_{TP}} \cdot n_{fid}) =$$
$$= n_{TP} \cdot K_{ipVL} \cdot 3.3\sqrt{\Pi_{TP}} \cdot \frac{n \cdot \beta \cdot S_{TS}}{S_{fid}} \qquad (6)$$

where K_{ipVL} is 1 km of 10kV cable line construction costs.;

L_{VL} is length of new 10kV cable lines to connect TP;

λ_V is 10kV network configuration factor (loop-through scheme adopted $\lambda_A = 2.5$);

Π_{TP} is TP service area;

n_{fid} is 10 kV feeder average number from TP;

S_{fid} is 10 kV feeder medium permeability ($S_{fid} = 2.5 MVA$).

The optimum solution of function (1) is equal with the minimum of total investments of variants` comparison. The results of calculations are shown in Fig. 5.

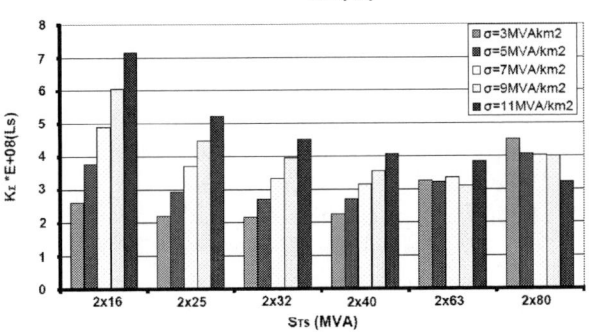

Fig. 5. Total capital investments depending on the transformer substations' power $K_\Sigma = f (S_{TS})$

On the basis of the results (Fig. 5) followed that cities with a load density $\sigma = 3\text{-}5$ MVA/km^2 is economically to use the TS with power in 2x32 MVA, with a load density $\sigma = 5\text{-}9$ MVA/km^2 - 2x40 MVA substation, with load density $\sigma = 9$ MVA/km^2 - 2x63 MVA substation, $\sigma > 11$ MVA/km^2 - 2x80 MVA substation [4].

CONCLUSIONS

First of all, power transformers' owner must to know the situation in power substations and in power system in total: in which circumstances power transformers working, what should expect from overloading. The second, not less important thing for owner is, economical situation in power transformers world marketing, as a result - changes on prices in producing industry of power transformers. All this, before mentioned in paper information, has influence on choosing of accurate maintenance strategy and technical conclusion making. But it is not enough to make decision for remaining lifetime prognosis and extension of useful lifetime. For that reason, technical staff must estimate each power transformer separately, in view of performed modernization works. Technical information and history should be stored. Determination methods of power transformers' technical conditions are normally applied.

The calculation program Microsoft EXCEL and the graphic computer program AutoCAD are offered to automate the calculation process.

The developed method makes it possible to get a rational decision on the development of energy supply circuits of the city in the early stages of the design under conditions of incomplete and uncertain information.

The diagnostics specialists and engineering personal is needed to realize main idea: to extend power transformers useful lifetime with minimum/rational resource involvement and recheck substations rated power needs, in case of necessity.

ACKNOWLEDGMENT

This work has been supported by the European Social Fund within the project „Support for the implementation of doctoral studies at Riga Technical University"

REFERENCES

[1] G. Gavrilovs and S. Vītoliņa. "Technical Condition and remaining lifetime assessment strategies of power transformers," in *Proc. 2010 of the 5th International Conference on Electrical and Control Technologies ECT-2010*, pp. 248-252.

[2] M. A. Franchek and D. J. Woodcock, "Life-cycle considerations of loading transformers above nameplate rating," Weidmann Technical Services Inc., p. 34, 1998.

[3] IEEE Std. C57.91, 1995, "IEEE Guide for Loading Mineral-Oil-Immersed Transformers".

[4] S. Guseva, O. Borscevskis, N. Skobeleva, and L. Kozireva, "Load Determination and Selection of Transformer Substations' Optimal Power for Tasks of Urban Networks' Development," *Power and Electrical Engineering*, vol. 27, no. 4, RTU, Riga, pp. 31-36, 2010.

Tracks of Power Quality Transients in High Order Statistics Spaces

Agustín Agüera Pérez*, Juan José González de la Rosa*, José Carlos Palomares Salas*, Aurora Gil de Castro[†],
Antonio Moreno-Muñoz[†]

*Research Group PAIDI-TIC-168

University of Cádiz. Area of Electronics, Av. Ramón Puyol S/N. E-11202-Algeciras-Cádiz-Spain

Email: agustin.aguera@uca.es

[†]Research Group PAIDI-TIC-168

University of Córdoba, Area of Electronics, Campus de Rabanales, Leonardo da Vinci building, E-14071 Córdoba, Spain

Abstract—This paper deals with the automatic classification of power quality transients according to their amplitudes and frequencies, and following the geometrical pattern previously established via higher-order statistical measurements. The clustering is achieved thanks to the third and fourth-order features associated to these electrical anomalies, which in turn are coupled to the 50-Hz power-line. The main contribution of the paper is the novel finding that the maxima and the minima of these higher-order cumulants distribute according families of curves of constant frequency or constant amplitude. A couple of extremes (min-max) of the higher-order statistical estimator belongs to a couple of curves (constant-frequency and constant-amplitude). The random grouping along each curve reveals the *a priori* hidden geometry, in turn linked to the subjacent electrical anomaly. The 2D (amplitude-frequency) regular surface grid in the input space experiments a transformation to both output spaces, which is developed via the higher-order non-linear mapping.

I. INTRODUCTION

Companies are putting a lot of efforts in innovation and new technologies to monitor and control Power Quality (PQ) anomalies because today's equipment, and automated manufacturing devices, are highly sensitive to the power line signal's imperfections (PQ events), making the production cost excessive. Malfunctioning not only has to be thereby detected, but also predicted and undoubtedly diagnosed, to identify the cause and prevent the system from a similar shock. This would be reflected *a posteriori* in an enhancement of the industrial production [1], [2].

Recent works are bringing a higher-order statistics (HOS) based strategy, dealing with PQ analysis [3], [4], and other fields of Science and Technology [5], [6], [7]. They are based in the following premise. Without perturbation, the 50-Hz of the voltage waveform exhibits a constant statistical behavior (stationarity), generally Gaussian. Deviations can be detected and characterized via HOS; non-Gaussian processes need at least 3rd and 4th-order statistical characterization in order to be completely characterized, because 2nd-order moments and cumulants are not capable of differentiate non-Gaussian events.

Concretely, the problem of differentiating between a transient of long duration named oscillatory (within a signal period) and a short duration transient, or impulsive transient (25 per cent of a cycle), has been outcome under controlled conditions in [8], and the idea of differentiating between healthy signals and signals with transients was pointed out and accomplished in [9]. This problem was previously described in [10] and matches the HOS category, in the following sense. The short transient could also bring the 50-Hz voltage to zero instantly and, generally affects the sinusoid dramatically. By the contrary, the long-duration transient could be considered as a modulating signal (the 50-Hz signal is the carrier), and is associated to load charges [10]. Consequently, given a signal, and once the analysis has been done, we can suggest the PQ type phenomena, and perhaps the origin of the fault in the energy distribution system. Thus, from the basis of these works, this paper conveys the idea of the inverse problem in PQ analysis. That is, given a set features, to guess the rest of the electrical anomaly's characteristics and, occasionally, to target the faulty element in the energy plant.

The present work shows that there is a relationship between the frequency and the amplitude of a concrete transient and its associated higher-order features. This is demonstrated via a mathematical transformation which maps the 2D (amplitude-frequency) input space into two output 2D spaces which relate higher-order features to the frequency and the amplitude. Concretely, the main contribution of the paper is the novel finding that the maxima and the minima of the higher-order cumulants distribute along these families of curves of constant frequency or constant amplitude. A couple of extremes (min-max) of the higher-order statistical estimator belongs to a couple of curves (constant-frequency and constant-amplitude), therefore the image position is given by the cross point of the curves. The random grouping along each curve reveals the *a priori* hidden geometry, in turn linked to the subjacent electrical anomaly.

The paper is structured as follows. The following Section II explains the fundamentals and the importance for PQ monitoring. Higher-Order Statistics are outlined then in Section III. Finally, results are presented in Section IV and conclusions are drawn in Section V.

II. POWER QUALITY CHARACTERIZATION

Categorization of electrical transients based on waveform shapes and their underlying causes (or events) has been studied

978-1-4244-8806-3/11 $26.00 © 2011 IEEE

in [10], and a few previous studies [3], [4] using HOS for feature extraction of electrical signals have shown the possibility of distinguish transients based on details beyond the second-order. In a real-life 50-Hz power line signal, it is very common to find these transients. In Fig. 1 we show an example of anomalous signal, including transients which are not classified between short-duration and long-duration. We show the computation of three higher-order time-domain statistics in order to introduce them qualitatively. The second-order estimator operates as an increase-of-power detector, showing the bumps associated to the increase of power, which in turn are associated to the anomalies of the power-line sine wave, but the third and fourth-order sliding cumulants have to be interpreted further. The most intuitive procedure is to calculate their maxima and minima.

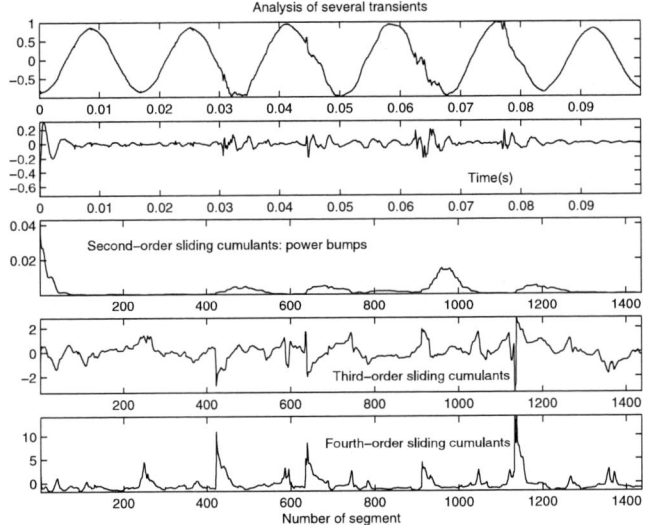

Fig. 1. Several transients in the power line 50-Hz sine wave, and the computation of time-domain statistics. The signal has been previously normalized and high-pass filtered in order to remain with the transients.

Once the foundations of PQ have been settled down, in the following Section we present higher-order statistics in the time-domain in order to present the signal processing tool, along with a basic example which shows the performance of the statistical estimators which have been used in the computation of the cumulants. This example also motivates the use of HOS in time-series characterization.

III. HIGHER-ORDER STATISTICS

Higher-order cumulants are used to infer new properties about the data of non-Gaussian processes [5], [11]. In multiple-signal processing it is very common to define the combinational relationship among the cumulants of r stochastic signals, $\{x_i\}_{i \in [1,r]}$, and their moments of order p, $p \leq$, given by using the *Leonov-Shiryaev* formula [12], [13]

$$Cum(x_1, ..., x_r) = \sum (-1)^{p-1} \cdot (p-1)! \cdot E\{\prod_{i \in s_1} x_i\}$$
$$\cdot E\{\prod_{i \in s_2} x_j\} \cdots E\{\prod_{i \in s_p} x_k\}, \quad (1)$$

where the addition operator is extended over all the partitions, like one of the form $(s_1, s_2, \ldots, s_p)$, $p = 1, 2, \cdots, r$; and $(1 \leq i \leq p \leq r)$; being s_i a set belonging to a partition of order p, of the set of integers $1, \ldots, r$.

Let $\{x(t)\}$ be an rth-order stationary random real-valued process. The rth-order cumulant is defined as the joint rth-order cumulant of the random variables $x(t)$, $x(t+\tau_1), \ldots, x(t+\tau_{r-1})$,

$$C_{r,x}(\tau_1, \tau_2, \ldots, \tau_{r-1})$$
$$= Cum[x(t), x(t+\tau_1), \ldots, x(t+\tau_{r-1})] \quad (2)$$

The second-, third- and fourth-order cumulants of zero-mean $x(t)$ can be expressed via [14]:

$$C_{2,x}(\tau) = E\{x(t) \cdot x(t+\tau)\} \quad (3a)$$

$$C_{3,x}(\tau_1, \tau_2) = E\{x(t) \cdot x(t+\tau_1) \cdot x(t+\tau_2)\} \quad (3b)$$

$$\begin{aligned} C_{4,x}(\tau_1, \tau_2, \tau_3) \\ = E\{x(t) \cdot x(t+\tau_1) \cdot x(t+\tau_2) \cdot x(t+\tau_3)\} \\ - C_{2,x}(\tau_1)C_{2,x}(\tau_2 - \tau_3) \\ - C_{2,x}(\tau_2)C_{2,x}(\tau_3 - \tau_1) \\ - C_{2,x}(\tau_3)C_{2,x}(\tau_1 - \tau_2) \end{aligned} \quad (3c)$$

By putting $\tau_1 = \tau_2 = \tau_3 = 0$ in Eq. (3), we obtain

$$\gamma_{2,x} = E\{x^2(t)\} = C_{2,x}(0) \quad (4a)$$

$$\gamma_{3,x} = E\{x^3(t)\} = C_{3,x}(0,0) \quad (4b)$$

$$\gamma_{4,x} = E\{x^4(t)\} - 3(\gamma_{2,x})^2 = C_{4,x}(0,0,0) \quad (4c)$$

The expressions in Eq. (4) are measurements of the variance, skewness and kurtosis of the distribution in terms of cumulants at zero lags (the central cumulants).

Normalized kurtosis and skewness are defined as $\gamma_{4,x}/(\gamma_{2,x})^2$ and $\gamma_{3,x}/(\gamma_{2,x})^{3/2}$, respectively. We will use and refer to normalized quantities because they are shift and scale invariant. If $x(t)$ is symmetrically distributed, its skewness is necessarily zero (but not *vice versa*); if $x(t)$ is Gaussian distributed, its kurtosis is necessarily zero (but not *vice versa*). In the experimental section, results are obtained by using sliding cumulants, i.d. a moving window in the time domain over which to compute the each cumulant (3rd-order and 4th-order cumulants for zero time-lag).

IV. DATA ARRANGEMENT AND EXPERIMENTAL RESULTS

A. Data arrangement and feature extraction: input and output spaces

In this paper we have focussed in oscillatory transients, which is a very common anomaly, usually associated to parasitic discharges. The decaying law for the amplitude is exponential, and each transient vanishes within less than half a cycle (2 ms), corresponding to the power line. A sample frequency of 0.1 ms is chosen. Each time-series under test for computation performance contains eight cycles of the power-line, with one transient to be targeted.

Computation is explained over Fig. 2. For each register, the 3th and 4th-order cumulants are computed according to the following procedure. For a given statistical order (three or four) the feature extraction algorithm computes a cumulant over a sliding window, whose length is equal to two periods of the power line (20 ms), and then it jumps to the following starting point (overlapping 98 %). Besides, each nth-order cumulant, $Cum_{n,x}[i]$, associated to the ith computation window has been normalized by $(Cum_{2,x}[i])^2$, so that to give a real statistical characterization. When the computation swept along a signal finishes, the maximum and the minimum are calculated for all the sliding windows results. Consequently, each signal is characterized by its third and fourth-order extremes. The constant-value zone which has been remarked in Fig. 2, establishes another feature for the higher statistical orders. This fact is object for future works.

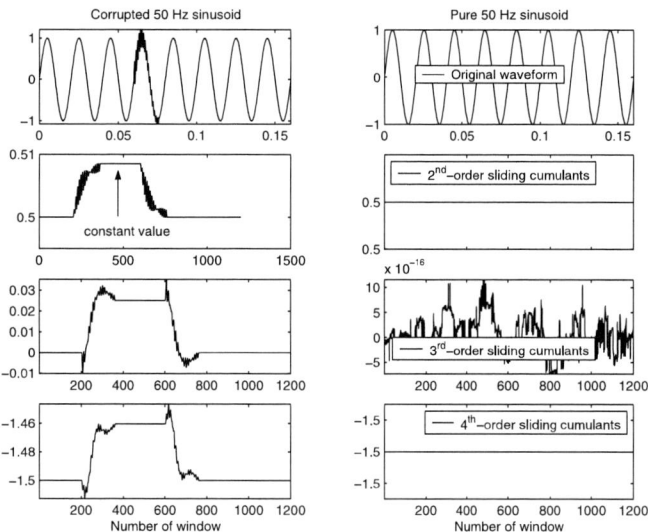

Fig. 2. Example of computation on a 1000 Hz transient, which shows the maxima and the minima associated to each cumulant time-series and the constant value zone, between the extremes, which will be used in future work.

As mentioned before, the computation results in the feature extraction stage consist of two couple extremes, associated to the 3rd and 4th-order cumulants. Since each pair can be interpreted as coordinates in a 2D space, we will consider that the mathematical procedure characterizes each evaluate signal as two points, in two 2D spaces. One of them projected into the space defined by 3rd-order cumulants' extremes (maxima and minima) and the other one into the 4th-order analogous space. These spaces are called output spaces hereinafter.

In order to test if transients' characteristics can be inferred-tracked from the output spaces, an experiment is proposed. Transient's amplitude and frequency will be modulated and their projections will be analyzed. Hence, families of transients are generated and organized in two groups attending to: the named constant-frequency-variable-amplitude (CFVA) and constant-amplitude-variable-frequency (CAVF) groups.

A CFVA family of transients comprises a battery of 50 transients with equal frequencies and randomly selected initial amplitudes, distributed according to a uniform probability distribution between 0.1 and 0.9. We have tested the CFVA families associated to seven frequency values (200, 300, 500, 1000, 1500, 2000, 2500 Hz). In the same way, nine CAVF families, with amplitudes (0.1, ... , 0.9 V) and random frequencies between 200 - 2500 Hz, have been also tested.

Both for the CFVA and CAVF groups, 3rd and 4th order statistics are computed. As amplitude and frequency define a 2D space, it will be interpreted as the input space in order to describe the whole proposed methodology as a spatial transformation from $\Re^2$ to $\{\Re^2, \Re^2\}$.

B. Results

The results are organized in four couples of graphs (each figure contains a couple of graphs). Each couple is composed of the 2D graph and its associated 3D. The 3D graph allows a better understanding of the data assemblies, since the z-axis represents the random variable in each case. CFVA graphs are depicted in Fig. 3 and Fig. 4, and the homologous CAVF are Fig. 5 and Fig. 6.

The random conception of the experiment allows to appreciate that data are naturally arranged along their corresponding curves. These curves may be interpreted as isolines, since they are traced using a fixed parameter. Additionally, this finding means that the frequency (or amplitude) of the transient confines it to a curve along which, its amplitude (or frequency) determines the higher-order 2D coordinates (maxima and minima of the skewness and the kurtosis). This fact conveys the idea of an automatic learning process.

A number of remarks have to be made for these graphs. Attending to the CFVA graphs both for 3rd and 4th order (3 and 4) it is seen that as the amplitude decreases (moving towards the origin) exists a major data concentration nearby the origin, it becomes more difficult to distinguish a transient by its frequency, due to the close separation of the curves. For the 3rd-order curves, as the frequency increases curves confuse each other, thereby making difficult frequency identification. This fact is also observed in the 4th-order family, but we have found an exception, in the curve corresponding to 2500 Hz. Higher frequencies curves would overlap the former high-frequency curves (below 2500 Hz).

If we pay attention to CAVF curves (Fig. 5 and Fig. 6), each effect observed in CFVA curves has a "mirror" or symmetrical phenomenon. Hence, as the amplitude decreases (moving

978-1-4244-8806-3/11 $26.00 © 2011 IEEE

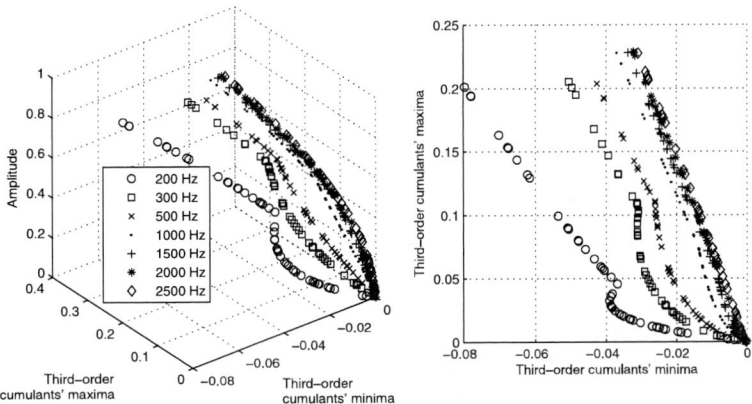

Fig. 3. Constant frequency curves corresponding to the third-order extremes. Each curve is associated to a frequency.

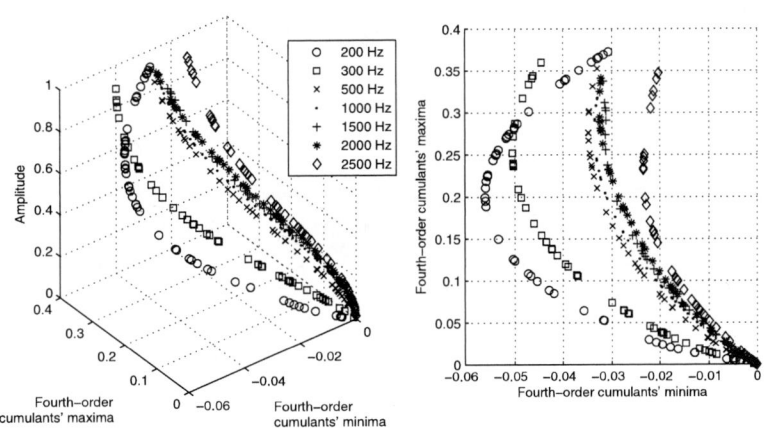

Fig. 4. Constant frequency curves corresponding to the fourth-order extremes. Each curve is associated to a frequency.

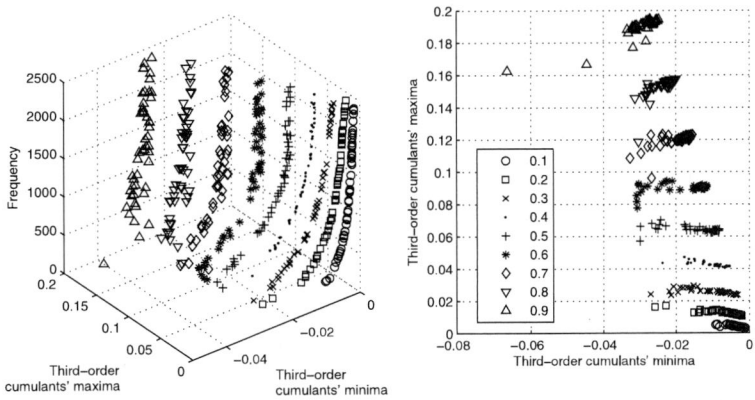

Fig. 5. Constant amplitude curves corresponding to the third-order extremes. Each curve is associated to an initial amplitude of the transient.

978-1-4244-8806-3/11 $26.00 © 2011 IEEE 47

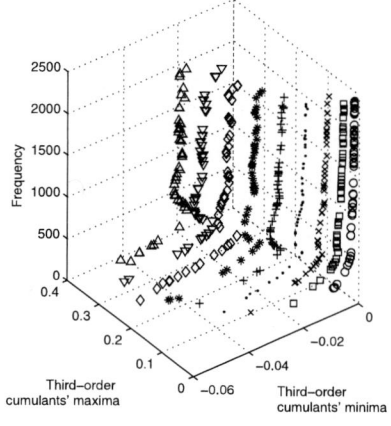 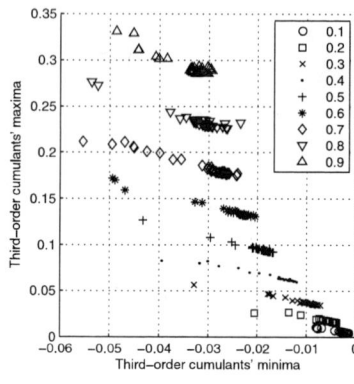

Fig. 6. Constant amplitude curves corresponding to the fourth-order extremes. Each curve is associated to an initial amplitude of the transient.

towards the origin) the random distributions are enclosed or confined to shorter ranges. This fact was expected because CFVA curves are closer nearby the origin, limiting the range of dispersion. In the same way, as the frequency increases (moving right within each curve), extremes tend to overlap each other and to concentrate along a definite trend line, coincident with the observed overlapping of high-frequency curves in CFVA. A particular effect of CAVF curves is that as the frequency decreases points are scattered from the principal alignment observed in the main trend.

In the former graphs it is observed that each transient's frequency or amplitude is associated to a specific curve. Therefore, a point result of the intersection between a CFVA and a CAVF curve should define a transient with a specific frequency and amplitude. In other words, an input-space point, defined by a particular frequency and amplitude, must have associated an image within both CFVA and CAVF specific curves. This fact conveys the idea of a mapping or space projection, in which coordinates of the input space (amplitude-frequency) have an specific associated point in the output spaces.

In order to visualize the transformation, a mesh (grid) is mapped from the input space to the output ones (Fig. 7). Each cross-point, with a known amplitude and frequency, is mapped and then connected to their consecutive neighbors, using iso-frequency lines (continuous line in Fig. 7) and iso-amplitude line (discontinuous line in Fig. 7). Obviously, the resulting structure reminds the graphs obtained from the random experiment, but assembling the amplitude and frequency information.

The direct mapping has been explained with the goal of showing that the higher-order features associated to a given transient adopt spatial concrete patterns. Collaterally, it is shown that the inverse mapping action is equally described, and it would allow the inference of the transient's characteristics (amplitude and frequency) from the basis of the measurement of the higher-order features (skewness andurtosis), i.e. from the output spaces, it would be possible to infer any

frequency and amplitude in the studied range. This is a future work. However, the objective of the ANNs in the present paper is to corroborate the subjacent geometry, thereby to show that these output spaces' geometries can be learned, in order to develop an automatic classification system in a future research stage.

V. CONCLUSION

The results of the present paper indicate that an automatic HOS-based detection and classification system for PQ transients can be implemented in future work. With a low computational cost, the system would process raw data to calculate the sliding skewness and the kurtosis and evaluate the extremes, which characterize each transient in a 2D space, either for the third and the fourth-order cumulants.

The most important finding is that the transients distribute themselves naturally along curves of constant frequency and constant amplitudes. Hence, once a signal with a coupled transient is analyzed, its image is located in a particular point of the higher-order spaces, determined by the transient's frequency and amplitude. It has been demonstrated that output spaces can be meshed, providing a background structure in which transients' characteristics can be tested, i. e., given the 2D coordinates in the higher-order space, the frequency and the amplitude of the transient can be bounded attending to the neighboring isolines. The worse measurement resolution is given by the cell's dimensions in the output space, being possible in practice a better inference's accuracy within each cell of the output space's grid. For high frequencies the cells are overlapped, making difficult to infer the frequencies, and consequently limiting the range of learning. On the other hand, but less important, the same phenomenon takes place for low amplitudes.

ACKNOWLEDGEMENT

The authors would like to thank the *Spanish Ministry of Science and Innovation* for funding the research project TEC2009-08988. Our unforgettable thanks to the trust we have

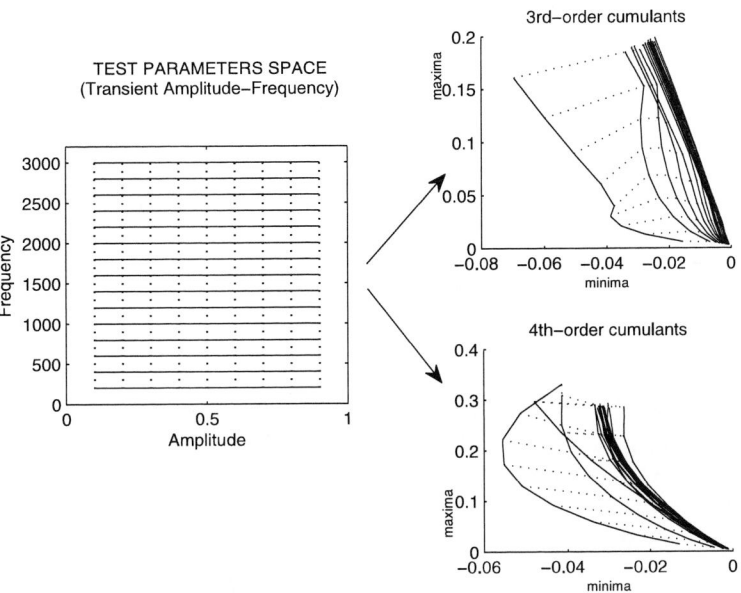

Fig. 7. Mapping process which shows the transformation of the input space into the output spaces, making the projections to the 3rd and 4th orders output spaces.

from the *Andalusian Government* for funding the Research Unit PAIDI-TIC-168 in *Computational Instrumentation and Industrial Electronics*.

REFERENCES

[1] A. Moreno and *et al, Mitigation Technologies in a Distributed Environment*, 1st ed., ser. Power Systems. Springer-Verlag, 2007.

[2] "IEEE Recommended practice for monitoring electric power quality," The Institute of Electrical and Electronics Engineers, Inc., Tech. Rep. IEEE Std. 1159-1995, 1995.

[3] J. J. G. De la Rosa, A. Moreno, and C. G. Puntonet, "A practical review on higher-order statistics interpretation. application to electrical transients characterization," *Dynamics of continous discrete and Impulsive Systems-Series B: Applications and Algorithms*, vol. 14, no. 4, pp. 1577–1582, August 2007.

[4] Ömer Nezih Gerek and D. G. Ece, "Power-quality event analysis using higher order cumulants and quadratic classifiers," *IEEE Transactions on Power Delivery*, vol. 21, no. 2, pp. 883–889, April 2006.

[5] J. J. G. De la Rosa, I. Lloret, C. G. Puntonet, and J. M. Górriz, "Higher-order statistics to detect and characterise termite emissions," *Electronics Letters*, vol. 40, no. 20, pp. 1316–1317, september 2004, Ultrasonics.

[6] J. J. G. De la Rosa, I. Lloret, C. G. Puntonet, R. Piotrkowski, and A. Moreno, "Higher-order spectra measurement techniques of termite emissions. a characterization framework," *Measurement (Ed. Elsevier)*, vol. 41, no. 1, pp. 105–118, January 2008, available online 13 October 2006.

[7] J. J. G. De la Rosa, R. Piotrkowski, and J. Ruzzante, "Third-order spectral characterization of acoustic emission signals in ring-type samples from steel pipes for the oil industry," *Mechanical Systems and Signal Processing (Ed. Elsevier)*, vol. 21, no. Issue 4, pp. 1917–1926, may 2007, available online 10 October 2006.

[8] J. J. G. De la Rosa, A. M. Muñoz, A. Gallego, R. Piotrkowski, and E. Castro, "Higher-order characterization of power quality transients and their classification using competitive layers," *Measurement (Ed. Elsevier)*, vol. 42, no. Issue 3, pp. 478–484, February 2009.

[9] J. J. G. De la Rosa and A. M. Muñoz, "Higher-order characterization of power quality transients and their classification using competitive layers," *Przegąd Elektrotechniczny-Electrical Review*, vol. 10, no. Issue 85, pp. 284–289, February 2009.

[10] M. H. J. Bollen, E. Styvaktakis, and I. Y.-H. Gu, "Categorization and analysis of power system transients," *IEEE Transactions on Power Delivery*, vol. 20, no. 3, pp. 105–118, July 2005.

[11] C. L. Nikias and A. P. Petropulu, *Higher-Order Spectra Analysis. A Non-Linear Signal Processing Framework*. Englewood Cliffs, NJ, Prentice-Hall, 1993.

[12] C. L. Nikias and J. M. Mendel, "Signal processing with higher-order spectra," *IEEE Signal Processing Magazine*, pp. 10–37, 1993.

[13] J. M. Mendel, "Tutorial on higher-order statistics (spectra) in signal processing and system theory: Theoretical results and some applications," *Proceedings of the IEEE*, vol. 79, no. 3, pp. 278–305, 1991.

[14] A. K. Nandi, *Blind Estimation using Higher-Order Statistics*, 1st ed., A. K. Nandi, Ed. Boston: Kluwer Academic Publichers, 1999, vol. 1, no. 1.

The through simulation of devices on the basis of the structural linguistic method

Andrey Nikiforov

Donetsk National Technical University, Ukraine

apniktel@yandex.ua

Abstract-**The through analysis, synthesis and design of multi-input devices of relay protection and automatics on the basis of the structural linguistic method are considered in the given article. The article shows the development of the steady-state devices in the conditions of variety of input meaningful situations. The article presents methods of further improvement of CAD through design.**

1. INTRODUCTION

The analysis, synthesis and design of multi-input devices of the relay protection and automatics (RP&A) operating with complex objects of control and protection (OCP) in the conditions of variety of input meaningful situations have caused the necessity to apply the structural linguistic (SL) method [1]. Preliminary studies have shown the effectiveness of SL method application to analyze the reasons of unstable operation of RP&A devices of the distribution networks with voltages 6-35 kV. The aim of this article is to investigate the problems of deterministic simulation in accordance with SL method during the through design of devices with specified characteristics. Another aim is to give some ways to improve RP&A devices designed on the basis of multi-core integrated circuits and suggest wider application of SL method in industry.

It is known that the effective development of the devices starts and finishes within the limits of CAD through design. The following stages are used when developing devices in CAD: from an original idea to structural diagrams of improved devices, well-functioning schematic diagrams and programs, then to printed circuit boards and construct and finally to the adjustment of the completed devices. The development of the devices is based on the simulation modeling of device components when signals of the real processes in OCP are sent to the device inputs. It is also based on combined simulation of devices including OCP.

2. THE DESCRIPTION OF SL METHOD

The method is based on the analogy between the structure of objects and the syntax of a language. Objects consist of linked sub-objects similar to phrases and sentences which are made up of words and words in their turn consist of the letters. The flow of real time meaningful information, which is transferred by means of the signal parameters of the input device coordinates, is divided by the information transducers with the threshold output, that is by terminal symbols (TS). The practical creation of TS is based on the methods, which have been justified technically and economically. When we assign the proper names to each elementary information component of TS and to the law how to process it (*P)*, we obtain the opportunity to define the characteristic features typical for any current meaningful situation with respect to permissible meaningful flow of information (to the transient process in the circuit that is the consequence of the emergency situation).

3. THE APPLICATION OF THE SL-METHOD TO CONSTRUCT THE STEADY-STATE DEVICES

The stages of device simulation, according to SL method, are the following: 1) decomposition of the signals of the device input coordinates into the structural information components of TS; 2) construction and analysis of the generalized tree to define the meaningful situations of the known devices; 3) construction of the optimal tree of the steady-state device; 4) verification of the optimal tree operation; 5) synthesis of the principal scheme, the construct of the device.

The simulation of the stages in CAD can be examined on the basis of any RP&A device. Let us take the device including all the levels of hierarchy in relation to the processing of the meaningful information. The example of such a device is the scanner-analyzer of the state of the loop of a circuit zero-sequence (LCZS) [2]. Let us briefly describe the purpose, principle of operation and the device characteristics.

To prove the expediency of SL method application in the theorem "*About the meaningful signal existing in the relay protection systems*" is introduced. According to the theorem, the RP systems operating as a result of perturbation may be transformed into the systems operating as a result of deviation. The feedback in such systems is closed with the help of the meaningful signal $S(t)$ (See Fig. 1). The change in signal $S(t)$ is proportional to the change in the meaningful states of OCP if the signal were imagined by the operator who would control OCP in manual operation. The following concepts can be referred to the characteristic properties of the meaningful signal $S(t)$: the presence of OCP in normal and other modes of operation, effective compression of the parametric information, graphic representation of any complicated meaningful situation, ease of application the signal for analysis, synthesis and design of simple and hierarchical RP&A devices.

The above properties of the meaningful signal $S(t)$ allowed us to unite the corresponding RP&A devices operating separately into the single system of the automatic stabilization of the normal mode of operation (ASNMO) LCZS. The aim of the system operation is to locate the faulty sub-circuit of LCZS in case of the one-phase grounding (OPG) and to provide the system performance in indefinable meaningful situations. The development of the system resulted in new design of selective RP devices as the devices

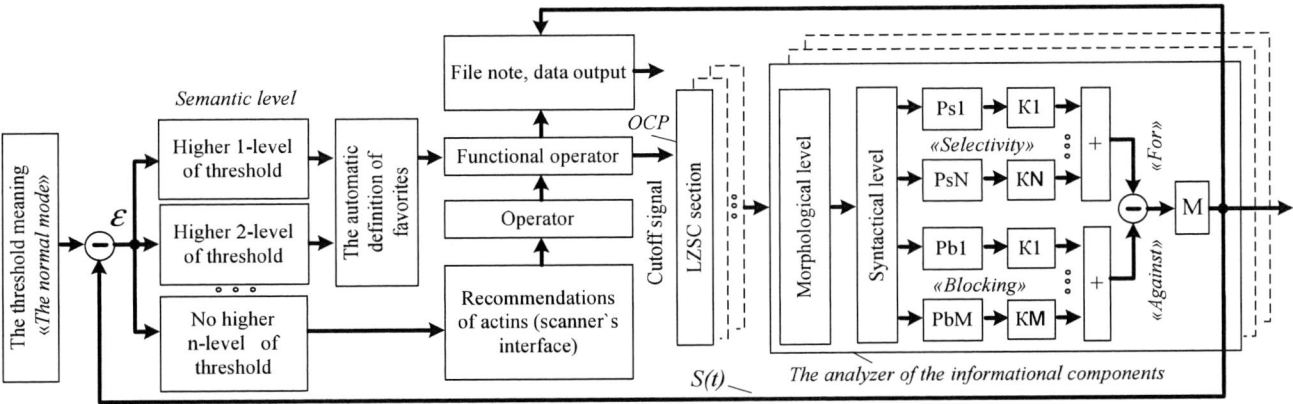

Fig. 1. Structural logical scheme of the "scanner-analyzer".

for the selective search (SS). It also resulted in the necessity to apply devices for self-compensation of OPG currents and alarm circuit recording to realize the search of the OCP faults. The main element of the ASNMO system is the "scanner-analyzer" which is the structural element of a new type of the RP &A "T-LCZS-1" terminal. The terminal can be installed in the cell of the voltage transformer and can substitute or complement the RP&A switchboard "The supply line insulation check". Each system device ASNMO forms the completion of the signal $S(t)$ by means of the universal algorithm: For–Against "Fig. 1". The group "FOR" includes TS, P that are responsible for forming the root symbol S corresponding to the classical OPG. The group "AGAINST" includes TS, P that are against forming S. The range of the meaningful signal $S(t)$ change is within the limits of [0–100%]. Having set the number of the threshold values P, for example $Pmax$=98%, we can control the presence and quality

of OPG detection, other processes in CZS as well as all high-voltage galvanic coupled machinery. In case of failure (or division) of any part of the ASNMO system, the threshold $Pmax$ can be controlled in every device separately. In case there is a lack of information or the latter is greatly distorted in the course of its formation, or given through long time intervals and so on, we can involve the operators. The expert system of the "scanner-analyzer" forms the diagnostic massage about OCP state, the possible cause of the transient process and ways to eliminate this cause.

4. SOLUTION OF THE SIMULATION PROBLEMS WHEN APPLYING SL METHOD

The purposeful transformation of the input meaningful information by means of cascade circuits and programs of device structural blocks imposes certain requirements to the simulation possibilities. In addition to the above-mentioned

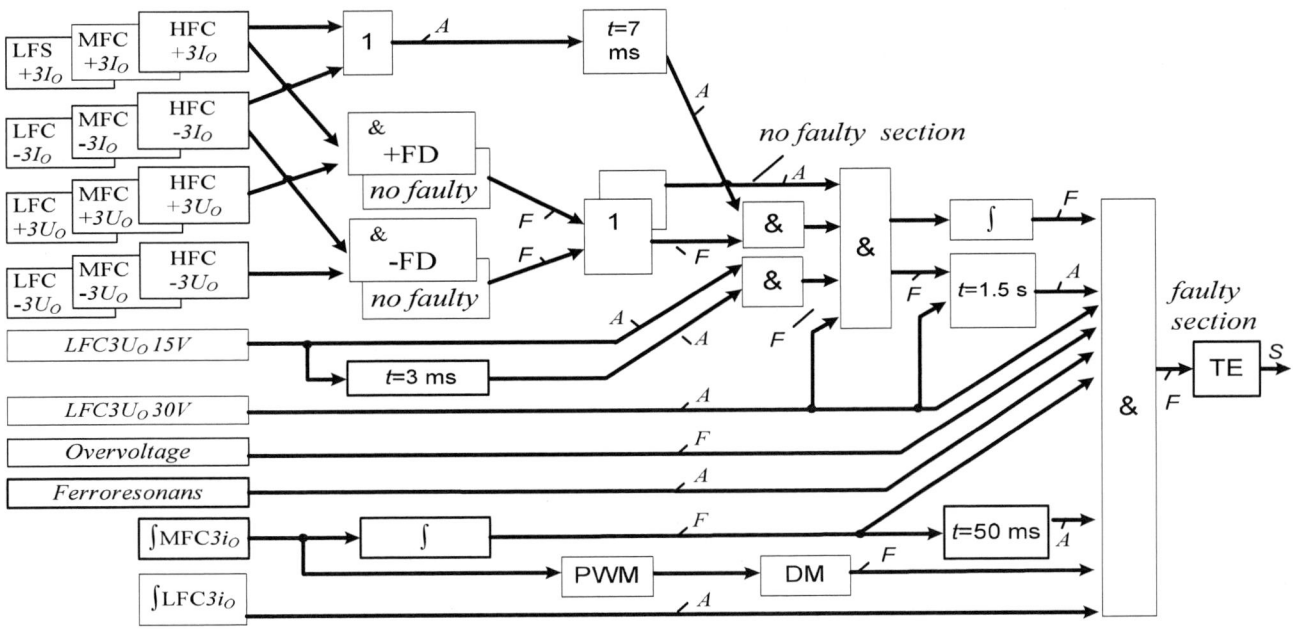

Fig. 2. Structural logical scheme of the "Syntactical level".

requirements we may add the possibility to simulate dynamically within one CAD information transformations by means of morphological, syntactical and semantic levels in consequence. The rapidity of calculation becomes very important.

Fig. 1 shows a multilevel hierarchy of the meaningful information processing, namely the morphological, syntactical and semantic levels according to SL method. Some simulation stages are easy to realize within CAD, but others are not easy to realize. The device simulation showed that CAD "OrCAD" is successfully suitable for the majority of SL method stages speaking about all the specified requirements. However, it was difficult to simulate the semantic level and we had to use some other CADs. "MatLab-Simulink" turned to be the most effective. Let us consider some problems when simulating by means of SL method.

5. LOW SIMULATION RATE WITHIN CAD

The combination of the hierarchy levels of SL method in one project is unacceptable concerning the simulation time. The reason is that every level operates at different frequencies. Thus, if the frequency of the input information change is 10 kHz (Fig. 2), then the change in the input signal $S(t)$ is chosen in the order of 10 Hz. The reduction of input information frequency by 1000 times is equivalent to the compression of the parametric information of signals, which is perfect to construct the system and devices, but not for combined simulation of the hierarchy levels.

The above problem can be solved by decomposition of the project in the following way. The system or the device is structurally divided into the sections realizing the hierarchy levels - the morphological, syntactical and semantic ones. It is usually possible to carry out the division because the number of connections between the levels in the transition points is

minimum. Each level is simulated separately at the corresponding frequencies. The transmission of the results from one level to another to complete the project is carried out by the file sources of the signals.

6. SIMULATION OF THE MORPHOLOGICAL LEVEL

The elements of the given level include information transducers (TS), operating at high-, medium-, low-frequency components (HFC, MFC, LFC). TSs include the elements of the galvanic isolation, filters, limiters, drivers, threshold elements (TE). TSs are realized by means of analogue, PWM circuits (perfectly within "OrCAD") and by means of microcontrollers (MC) on the basis of the given program. When constructing on the basis of the MC, during the analysis we manage to construct the structural schemes of digital filtration using available PSpice -models (ADC, adders and so on). During synthesis we have to use the additional CAD ("MatLab") to synthesize filters as there is no suitable tool in "OrCAD". The "OrCAD" development engineers should be aware of this problem.

The simulation of the principal TS schemes of the known devices within CAD is principally carried out by generating the real signals of the high frequency fault files of the training sample as there are no non-stationary signal components formed by using OCP models. At the given stage, we can find a great number of cases of improper operation of each of the TS component elements. In this case the typical morphological mistakes of TS formation may be divided into groups which must be properly checked - a) the quality and efficiency of operation of the same TS circuit when complex input signals are different; b) interaction of the natural transient oscillations of the filters with current impact effects in case of the arc OPG;

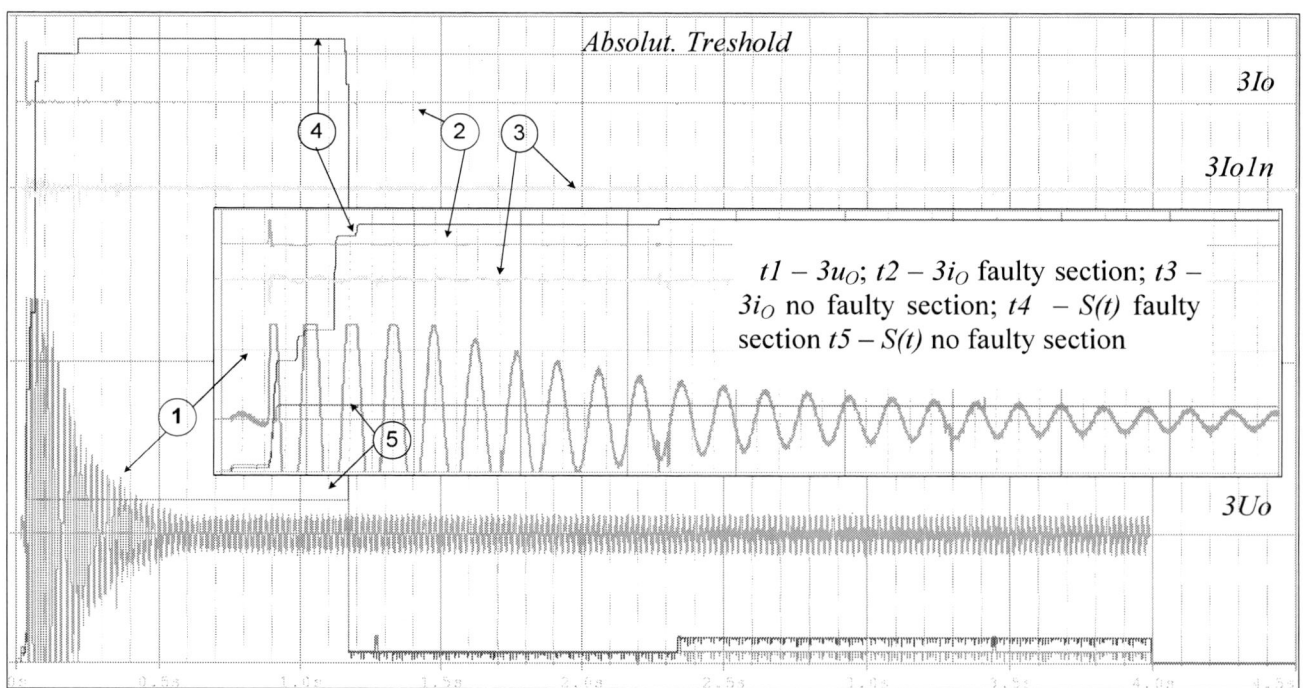

Fig. 3. Simulation of the meaningful situation "Burst" LCZS two sections.

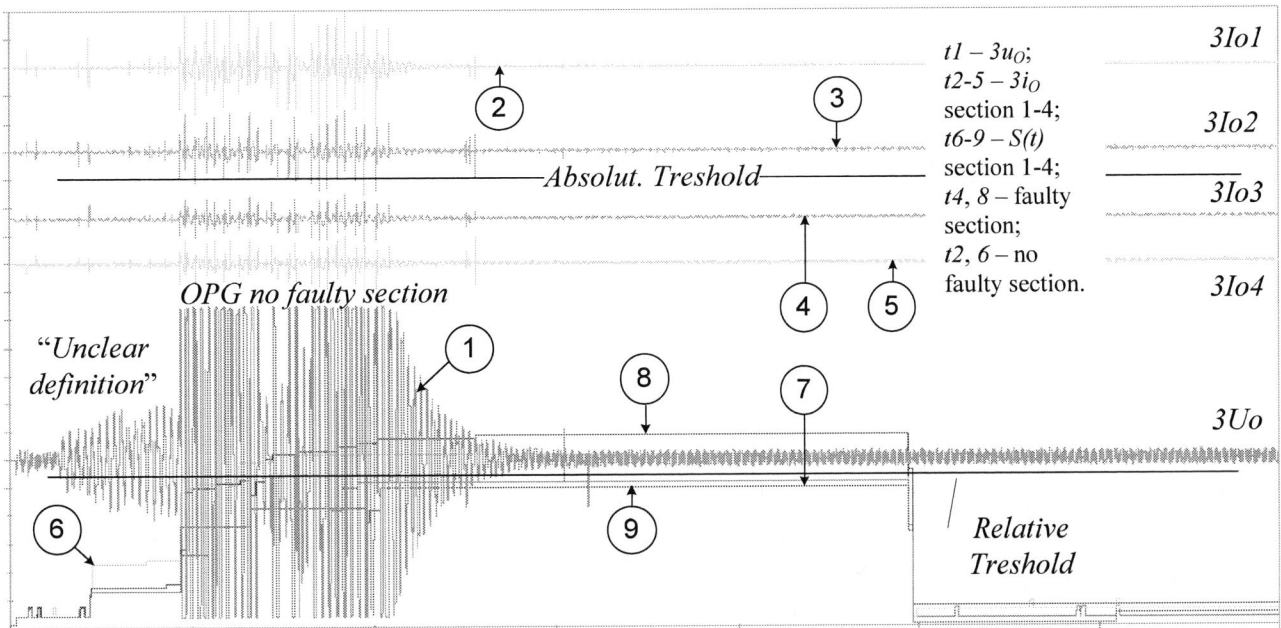

Fig. 4. – Simulation of the meaningful situation LCZS 4 sections. "*Indefinable situation*".

c) the shifting of HFS, MFS band for different OCP. Let us consider the above groups in detail.

As an example, let us analyze the operation of the connected components – an input filter (F)+TE of the certain SS devices (Fig.1, 2) –

a) From the very beginning the SS devices were designed to determine selectively the fault of LCZS section in case of the dead OPG in operating mode of circuit with the insulated neutral. Under existing conditions the regular OPG break down, as a rule, occurs during each half-wave of commercial frequency ω. From the point of view of the response of the connection FLFS+TE the signals of the input coordinates $3uo$, $3io$ are similar in case of arc and dead OPG (concerning repeatability, regularity are similar to frequency ω). According to the inventors of the SS devices, the similarity must increase especially when the arc OPG transforms into the dead one because of high breakdown current. The known statistics and the latest OPG monitoring using the fault recorders show that the number of steady-state processes in conditions of OPG at the frequency of ω (the dead OPG) doesn't exceed 1.5-2% in comparison with the number of all OPG but in some circuits the above-mentioned processes can occur very seldom. That is the similarity but not the determinacy of the device structure that helps to operate in case of the arc OPG. Simulation shows that the field of unstable mode of device operation is not limited.

b) The voltage recovery time of the fault phase in the circuit with the resistor-grounded neutral conductor is half cycle-a cycle ω. The interval between the OPG breakdowns is influenced by the individual characteristics of the circuit at the certain moment of its operation as well as by the change in the capacitive current when the modes of circuit operation and user operation changes. Each circuit has its own time intervals between the OPG breakdowns, which differ from the intervals in other circuits, but the parameters of the SS device circuit scheme remain the same.

Here we are faced with the problem of choice and adjustment of the SS device parameters according to the circuit characteristics.

c) TS "*LFC3Uo*" contains a lot of information about LCZS and OPG (See Fig. 2). However, it happens that this information is transmitted not to all SS devices. When the information is transferred to TS LFC the additional check of $3u_0$ presence doesn't play an important role.

d) As the arc interval is non-linear, OPG current stops flowing mainly during the first passage through the zero value of HFS of the recharging the circuit phase capacitances (discharge capacitance DHFC+charge capacitance CHFC). Duration of the OPG current flow in this case can be T_{OPG}=1-2 mS (See Fig. 2). When the time T_{OPG} elapses, the character of the transients in the circuit should be related not to OPG but to the stage of recovery of the normal operating conditions of the circuit and this stage does not have the selective phase shifts between LFC $3u_0$, $3i_0$ at ω frequency. How soon the next OPG breakdown occurs depends on the voltage recovery rate in the point of the breakdown (that is on the mode of the neutral grounding). The less the recovery time is, the greater is the probability of the next OPG breakdown. Thus, simulation shows there are practically no selective signals at the frequency ω in conditions of the arc OPG.

e) In different distribution networks DHFC, CHFC can be shifted along the frequency scale, there can be big or slight frequency diversity, DHFC can dominate over CHFC or vice versa. That is why P "*FD*" (phase detection) is determined in different ways, which is according to DHFC or according to CHFC or DHFC+CHFC in spite of stating the criteria of selectivity.

f) The qualitative formation of TS requires the reliable separation of LFC $3u_0$, $3i_0$ from the HFC signals. An intensive filtration improves its inertia. In the mode of the arc OPG the impact effect of the MFC $3u_0$, $3i_0$ with the peak amplitude and duration T_{OPG} (See Fig. 2) will excite the natural FLFC oscillations and result into TE operation during T_{FLFC}=(3-

5)ω^{-1}. Summation of natural FLFC oscillations caused by subsequent OPG breakdowns depending on the interval between breakdowns can lead to both TE operation and TE non-operation. When TE goes off at the output of FLFC+TE connection, the additional significant phase shift takes place because of by the width pulse modulation. Signals of such character can cause incorrect formation of P "HFC-FD".

g) HFC $3u_0$, $3i_0$ can be separated for "HFC-FD" by producing an intensive filter. APFC formation in known SS devices is carried out by means of either 2T–shaped notch-filters suppressing w frequency when the operating frequency band is 0.5-5 kHz or Bessel's filters of the third order (0.2-10 kHz) or additionally when differentiating the coordinate $3u_0$ or by using the ferroresonance transformers suppressing frequency w and separating (0.5-5 kHz). To reduce the phase distortion the input filters, as a rule, are produced in the same way. However, firstly, the HFC coordinate oscillation $3i_0$ takes place, mainly, with respect to the zero level, that is, the DHFC, CHFC amplitudes of non fault ZSC sections are much greater than those of LFC $3i_0$. Secondly, the amplitudes of the HFC $3u_0$ oscillation take place with respect to the front of LFC $3u_0$ rise, and they reach the phase value of the circuit which is by one order greater than the amplitudes of DHFC, CHFC $3u_0$. Also, as it is mentioned above, the first half-wave of $3u_0$, $3i_0$ coordinates change should be considered as the impact effect. Thus, for example, HFC-filter excited because of the interference (1 V, 0.2 mS) generates the transient process at the natural frequencies. This process actuates TE, TS and non-selective formation of P "HFC-FD" during two periods ω. The concept of forming the specified APFC is true mainly for steady-state signals. The specific character of the $3u_0$, $3i_0$ signals reduces sharply the efficiency of the high-quality filters to separate HRC $3u_0$, $3i_0$. Consequently, it is more advisable to take into account the inertia characteristics of the filters, but not the concept of the specified APFC. To synthesize the stable – state devices it is more effective to use the programmed FIR-filters having no feedback between input and output. Consequently, the latter have the linear FFC and don't store the previous information for a necessary period of time.

Having summarized the results of the morphological level simulation, we can say that the standard SS devices functioning at LFC $3u_0$, $3i_0$ in the conditions of the resistor grounded neutral should be applied to circuits where it is possible to guarantee the transition of the first OPG breakdown into the dead OPG if the active component of OPG current increases. It is natural that under existing conditions the total OPG current mustn't damage the phase-to-phase circuit insulation and the total technical and economic index of the circuit operating mode must meet requirements of the industrial engineering organization. It is clear that the above-mentioned restrictions are not easy to implement, consequently, the application of the resistor-grounded neutral itself is rather restricted. The same concerns the circuit with an insulated neutral the application of which is much more restricted.

7. SIMULATION OF THE SYNTACTICAL LEVEL

The analysis of the morphological level allowed us to show the number of essential disadvantages of the standard SS device modes. The number of these disadvantages is rather big to cause the unstable mode of device operation. The results of

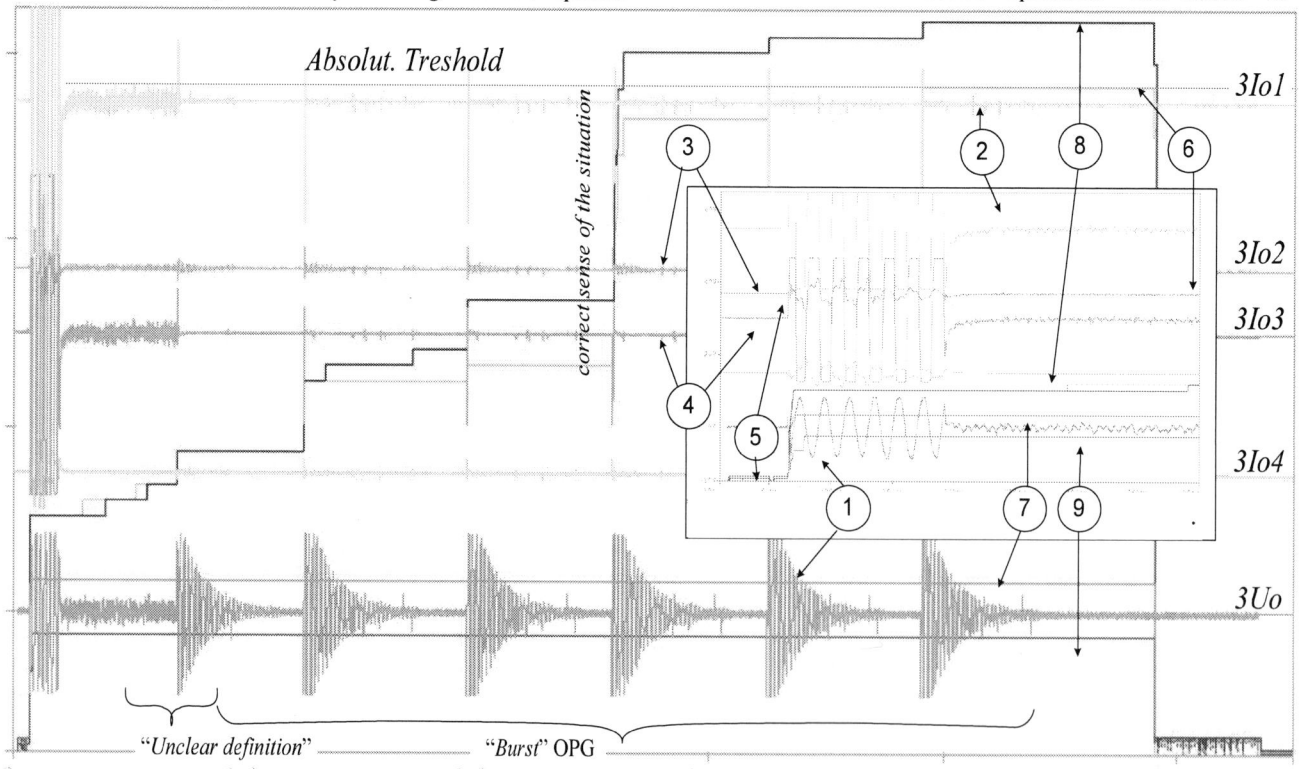

Fig. 5. – Simulation of the meaningful situation LCZS 4 sections. "Indefinable situation". t1 – $3u_O$; t2-5 – $3i_O$ section 1-4; t6-9 –S(t) section 1-4; t4, 8 – faulty section; t2, 6 – no faulty section.

the synthesis show that the simulation of the automatic circuit-breakers of the syntactical level within the "*OrCAD*" according to SL-method may be successfully carried out under conditions of the device realization on the basis of CPLD (logical elements, VHDL-language). Such realization is suitable and even preferable because the level mainly consists of the great number of the sections of the logical transformation of information functioning in parallel and this level requires a high speed of real time transformation which can't be realized properly by MC.

In case of level realization with the help of MC, the simulation within the "*OrCAD*" is impossible because there are no models of MC calculator and corresponding tools. The development engineers of "*OrCAD*" should be aware of this because the calculator of the latest MC is presented in VHDL-language. Thus, for example, in CAD the "*Proteus*" realizes the through simulation of analogous and constructive parts of the different MC-calculator. Within the CAD "*MatLab-Simulink-Stateflow*" the development of the elements of the program in C-language is supported and automatically formed. MC of the company "Texas Instruments" is simulated too. Unfortunately, other producers are not supported. In "*MatLab*" the development of the morphological level during the analogous and PWM realization is complicated; moreover, the device construction is not developed.

8. SIMULATION OF THE SEMANTIC LEVEL

The realization of the level within CAD "*OrCAD*" is very difficult. The reason is that there are no necessary models and corresponding tools and there arises the problems of optimization and speed of the calculation. For example, the quantity, efficiency and simplicity of realization when choosing TS, P, $S(t)$, diagnostic massages are all subject to optimization. Let us consider the tasks of simulation.

a) The first task is the choice of values of the weight factors to apply the rules P, which allows to define gradations of the meaningful signal $S(t)$ (See Fig.) distinguishing the current meaningful situation from the given one (100%). The task is solved within the "*OrCAD*" by means of the model construction formed into in the hierarchy blocks. However, when it is necessary to use additional TS, P on the previous levels - morphological and syntactical, it becomes necessary to carry out a new simulation for these levels and only then for the semantic level. Thus, in case of optimization, the decomposition of the general project is difficult to organize and is time-consuming.

b) The second task is to formulate the of P_{REL} rules of the relative criteria of comparison of the meaningful signals $S(t)$ from each faulty OCP section (See Fig. 1,2). To synthesize P_{REL} it is necessary to simulate SS devices not only of faulty and non faulty OCP sub-circuits as it was done during the synthesis of the known SS devises but to simulate three or four SS devises at least. The task of simulation of ASNMO LCZS sub-circuits is to define the order of preferable disconnections when there is the "*Indefinable situation*" and "*Unclear definition*" or several preferences. The formation of automatic control of P_{REL} thresholds, the automatic control of majority determination of the preferences and so on is carried out by means of construction using the elements available in "*OrCAD*". Consequently, the simulation is carried out within the limits of the structural scheme of automatic control

constructed by using not the programming language of realization of the level in a device but another language, that is, the (PWM, electrical) language.

c) The task is solved in practice. The simulation of several SS devises in one project complicates the organization of the working place when it is necessary to change the sources of the input signals preliminary prepared at the previous levels of the hierarchy. It is necessary to note that is case of simultaneous simulation of several devices, "*OrCAD*" supports the list of connections, but it takes a lot of time to calculate and the main point is that the calculation stability (the method of Newton-Rawson) significantly worsens.

d) The third task is the construction of the semantic automatic control to take a final decision to disconnect preferences, to choose and fill in the textual and diagnostic massages, to use the file of events and also to take decision in case of "*Indefinable situation*". The special feature of the task is not to take into account the connection of the semantic automats operation with the time axis. The automatic control is realized mainly by means of the programming. Its construction in "*OrCAD*" is rather ineffective. In this case it is necessary to transfer ("*MatLab-Simulink-Stateflow*") to CAD. When realizing the task on the basis of the MC and with the help of both the inserted operational system and the graphic shell, the simulation reduces to the debugging of the operational program of the MC on C-language for PC. The above mentioned solution of the simulation problem minimizes expenditures before starting the device realization and forms the model of the device to work with PC. However, the cost of the MC components is growing rapidly.

CONCLUSIONS

1. It is possible to realize effectively the purposeful analysis, improvement, device design and the organization of the working place of designers on the basis of SL-method. It is especially important for devices on the basis of multi-core integrated circuits.

2. The through design of the device on the basis of SL method shows the discontinuity of the simulation stages, preparation of the design specifications in spite of positioning of the certain CAD as the CAD of the trough design and it also shows the ways of further CAD improvement. In spite of the significant CAD development, the organization of working place of a designer is, unfortunately, carried out on the basis of CAD, maintaining the definite programmable integrated circuits or real electronic components.

3. The following devices of ASNMO LCZS system have been developed within the CAD "*OrCAD*", "*MatLab-Simulink-Stateflow*": the device of the selective search "*VCR-SS-1*", 8-channel high frequency breakdown recorder "*P-SS-1*", the terminal of the RP&A "*T-LZSC-1*".

REFERENCES

[1] G Tue, R.. Gonsalis. Principles of images recognition, M:"Mir", 1978.-411 p.

[2] Grebchenko N.V., Koval I.I., Sidorenko A.A., Smirnova M. A. Definition of complex admittance of electric isolation without disconnecting of electrical equipment // CPE2009. 6Th International Conference-Workshop 978-1-4244-2856-4/09/25.00. P. 61-66.

Electronic Synchronous Machine for Dynamic Power Conditioning in Wind Parks

Constantinos Sourkounis, Jan Wenske[*]

Ruhr-University Bochum, Bochum, Germany; sourkounis@eele.rub.de

[*]IWES, Fraunhofer-Gesellschaft, Bremerhaven Germany; jan.wenske@iwes.fraunhofer.de

Abstract- **The stochastic fluctuating offer of wind energy limits the capacity of supply in electrical grids by means of technical and economical reasons. To increase the supply, a concept of dynamic conditioning with the objective to increase the quality of supply was developed. This contribution describes the construction and the effect-principle of an electronic synchronous machine with active damper-circuit as well as its possible use to the energy respectively power conditioning and improvement of the power quality. The behaviour of synchronous machine is reproduced by static converters and short time storage. By the four-quadrant operation of the converters, short time fluctuation of the energy sources and changes of the load can be absorbed or smoothed directly on location. With the use of dynamic storage systems an intermediate storage of the stochastic offered primary energy is done. This low order energy can thereby be converted into high order recallable energy, which even can be used to satisfy peak demands. The local compensation of the fluctuation of the energy supply on the one hand and of the changes of the demand on the other hand, allows a more steady energy demand from the nationwide electrical grid.**

I. Introduction

When electrical grids are supplied by regenerative sources, like, e.g. wind energy converters (WEC) with stochastically fluctuating energy offer, unfavorable conditions may occur, particularly in grid parts with low short circuit capacity or in isolated grids. This situation limits the maximal possible connectable power of wind energy converters considerably.

In addition to non-periodically fluctuating effective power flows by wind gusts within a frequency range of <1Hz and periodic tower shadow effects in a range of around 0.3-3 Hz, further reasons for low or fluctuating energy qualities in coupled grid parts may be switching processes, harmonic component emissions [2] by converter-operated systems and plant control systems or plant managements which are not laid out in conformity with the requirements.

Heavy fluctuations of local power flows or non-sinusoidal currents [4] and the resulting drastic reduction of the grid quality at the point of common coupling (PCC) generally necessitate a power conditioning. The main objective of an universal energy conditioner [3] is therefore smoothing of effective and reactive power flows as well as a coherent grid management in isolated operation [1] with voltage and frequency control. For this purpose, a novel cascaded PI-state control has been developed and tested which allows for a load-dependent automatic adaptation/conditioning of the control structure.

II Concept of the Energy Conditioner

A simple radial distribution system topology is illustrated in Fig. 1. It features several producers (e.g. wind energy converters) which feed a bus bar via transformers which is connected via a superposed node PCC, to which further generating and consumer groups may be connected.

The requirements on an energy conditioning system can be determined, based on the major tasks of energy conditioning. First of all dynamic energy storage is needed along with a dynamic delivery and compensation of active and reactive power independent of each other. In other words a dynamic control element in four-quadrant operation is needed. A control structure for grid support and grid control is needed to establish such a system in an electrical grid, which can be realized as a dynamic detaching control structure. The whole system should also be short circuit stable and be usable for a wide range of rated power. All of these demands can be fulfilled by the concept of the electronic synchronous machine with an active damping circuit (ELSAD) (Fig. 1).

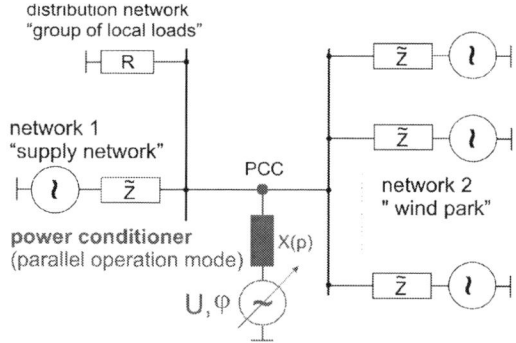

Fig. 1. Basic concept for energy conditioning in dezentralized electrical energy supply systems

The synchronous machine can basically be used for energy conditioning, because of the special sub transient and transient behavior as well as the possibility of adjusting the stator current versus the stator voltage by alternating the synchronous internal voltage over the excitation voltage and the load angle over the drive torque in all four electric quadrants during grid parallel operation. Therefore disadvantages for active power smoothing are the fixed rotation-torque characteristic, the poor control dynamic of the synchronous internal voltage, the rotor displacement angle at the range from 100ms to several seconds, and phase swinging.

The behavior of the synchronous machine is simulated

within the energy conditioner, which consists of a self-commutated inverter and an energy storage system (Fig. 2). To be able to do this, the inverter output voltage U_{WR} is selected in such a way, that a voltage difference between U_{WR} and the grid voltage U_{sN} occurs. This voltage difference specifies the current in amplitude and phase, which for the stabilization of the grid voltage is necessary, as also the delivered and received power respectively. Dissipation resp. absorption of effective and reactive powers correspond to those of a synchronous machine [3].

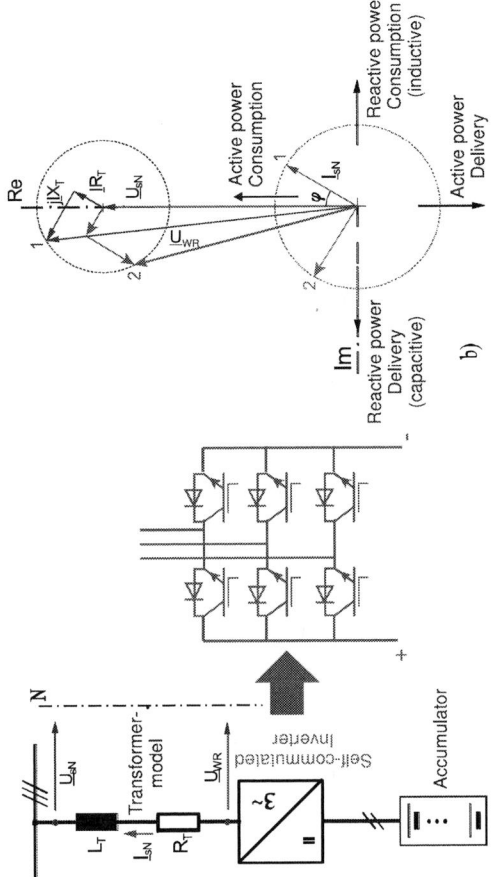

Fig. 2. Four quadrant operation of a self-commutated inverter at grid parallel operation

A constant frequency (in the case of an isolated grid) is assumed by the self commutated inverter over an internal frequency generator. Changes of active power in the grid have no impact on the frequency, in contrast to rotating generators, as long as the current is below a maximum tolerated value. The self-commutated inverter is, independent from the power flow in the grid, always operational, so that all other energy converters that go on line can synchronize on to the grid parameters given by the energy conditioner. Thereby phase shifts and short term frequency and voltage fluctuations respectively between the different energy converters can be avoided.

The design of ELSAD as shown in Fig. 3 is the result of the demand for a functional separation between energy conditoning

in the fundamental harmonic and the harmonic component range in order to increase the efficiency of the global system particularly when high powers are concerned.

The electronic synchronous machine with active damping circuit mainly consists of two self-commutated voltage converters [2] of different powers, a triple-wound power converter transformer of vector group Ynd5d5 and two combined passive LC filters.

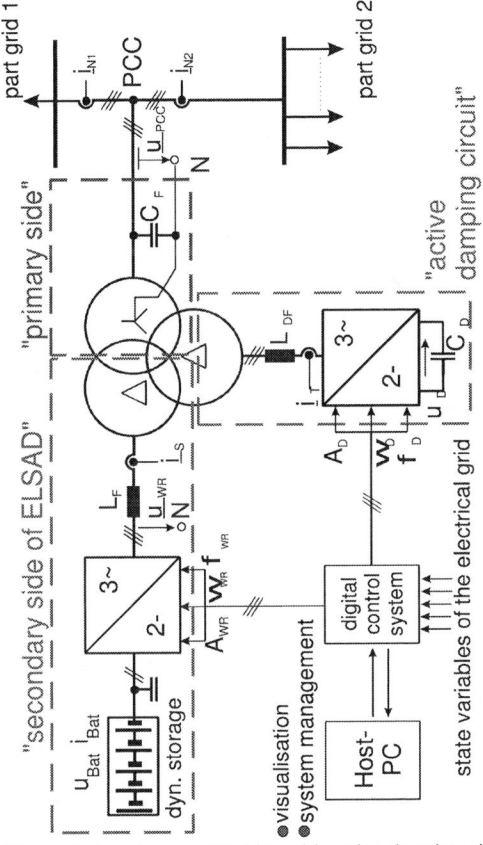

Fig. 3. Electronic Synchronous Machine with active damping circuit for energy conditioning

III. MODEL

The separation in circuitry between power flow control for the harmonic components by the active damping circuit and the fundamental harmonic compensation by means of high-power converters on the secondary side of the transformer allows for an almost independent layout of the controllers (so called dynamic decoupling) for both tasks. For the fundamental harmonic range a mathematic model of ELSAD can be established which allows for an analytical layout of the control.

The active damping circuit is fully neglected in this case, because its control strategy does not permit fundamental harmonic currents in the damping circuit and compensation procedures die out very quickly in the fundamental harmonic range because of a high interference suppression.

Furthermore , the transformer is assumed of being current-ideal, so that the quadrature-axis reactance of the main field inductivity must no more be taken into account. This simplification is admissible, since the general magnetising currents of transformers only vary within a single-digit percentage range of the nominal current. The relatively simple grid simulation according to Fig. 3 assumes a general load case with connected supply grids and variable short circuit power at the PCC. For both line taps, an ohmic-inductive internal grid resistance and a sinusoidal voltage source has been assumed. At suitable parameter selection, this grid model characterises the relevant proper motion of the system which are relevant for the controller layout.

At first, the equation system for the fundamental harmonic model at the hub is formulated according to Fig. 4 at a stationary coordinate system (index "S"). For the primary side (stator side) applies:

$$-\frac{\underline{u}_{PCC}^S}{\underline{ü}_P} = \underline{R}_1 \cdot \underline{ü}_P^* \cdot i_P^S + L_{1\sigma} \cdot \underline{ü}_P^* \cdot \frac{d\,i_P^S}{d\,t} - \underline{u}_{h1} \qquad (1)$$

and for the secondary side:

$$\frac{\underline{u}_S^S}{\underline{ü}_S} = \underline{R}_2 \cdot \underline{ü}_S^* \cdot i_S^S + L_{2\sigma} \cdot \underline{ü}_S^* \cdot \frac{d\,i_S^S}{d\,t} + \underline{u}_{h2} \qquad (2)$$

The complex electrical terminal parameters of the secondary side can be recalculated to match the primary side (transformed quantity). The terminal voltage can be recalculated for the primary side according to

$$\underline{u}_P^{S'} = \underline{u}_S^S \cdot \underline{ü}_{W12} \cdot \underline{ü}_P \cdot \underline{ü}_S^{-1} \qquad (3)$$

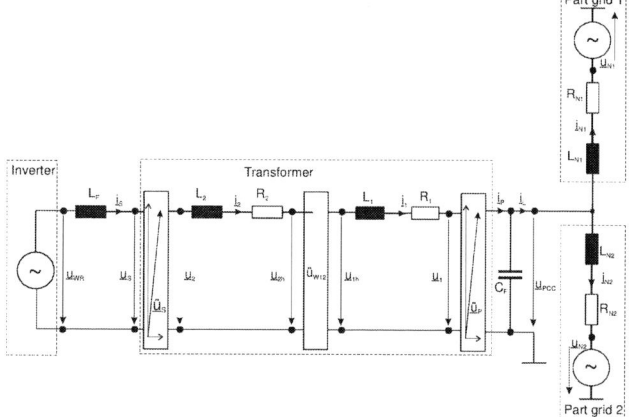

Fig. 4. Equivalent circuit diagram for the fundamental oscillation of the power conditioning system

and the current according to:

$$i_P^{S'} = i_S^S \cdot \left(\underline{ü}_{W12} \cdot \underline{ü}_P^*\right)^{-1} \cdot \underline{ü}_S^* \qquad (4)$$

Because of the recalculation of the quantities of the secondary side and under consideration of the upstream smoothing inductivity L_F on the secondary side of the transformers, of the transmission ratio and the switching group

$$\underline{u}_{WR}^S = L_F \cdot \frac{d\,i_S^S}{d\,t} + \underline{u}_S^S$$

$$\underline{u}_P^{S'} = \underline{u}_S^S \cdot \underline{ü}_{W12} \cdot \underline{ü}_P^* \cdot \underline{ü}_S^{-1} \qquad (5)$$

$$i_P^{S'} = i_S^S \cdot \left(\underline{ü}_{W12} \cdot \underline{ü}_P^*\right)^{-1} \cdot \underline{ü}_S^*$$

results from equation (2) for the voltage equation of the rotor side transformed for the primary side:

$$\underline{u}_{WR}^{S'} = \left(R_2 \cdot \underline{ü}_{W12}^2\right) \cdot \underline{ü}_P^* \cdot \underline{ü}_P \cdot i_P^{S'}$$
$$+ \left[L_{2\sigma} \cdot \underline{ü}_{W12}^2 + L_F \cdot \frac{\underline{ü}_{W12}^2}{\underline{ü}_S^* \cdot \underline{ü}_S}\right] \cdot \underline{ü}_P^* \cdot \underline{ü}_P \cdot \frac{d\,i_P^{S'}}{d\,t} + \underline{u}_{h1} \qquad (6)$$

The unknown voltage $\underline{u}_{h1}$ in Equation (6) of the ideal transformer can now be eliminitated by integration into equation (1). For the transformer switching group (Ynd5) of the test plant in this example, according to Fig. 3, at transmission ratios of

$$\underline{ü}_P = -1\,, \quad \underline{ü}_S = \frac{e^{-j\pi/6}}{\sqrt{3}}$$
$$\underline{ü}_{W12} = \frac{w_1}{w_2} = \frac{u_{1h}}{u_{2h}} = \frac{1}{\sqrt{3}} \qquad (7)$$

the following is then applicable:

$$\underline{u}_{WR}^S \cdot e^{j\pi/6} = \left[R_1 + \frac{R_2}{3}\right] \cdot i_P^S + \left[L_{1\sigma} + \frac{L_{2\sigma}}{3\cdot} + L_F\right] \cdot \frac{d\,i_P^S}{d\,t} + \underline{u}_{PCC}^S \qquad (8)$$

with $\quad i_P^S = i_P^{S'}$

A transformation into the rotating reference coordinate system leads to the following equation:

$$\underline{u}_{WR}^R \cdot e^{j\pi/6} = R \cdot i_P^R + L_\sigma \cdot \frac{d\,i_P^R}{d\,t} + j \cdot \omega_R \cdot L_\sigma \cdot i_P^R + \underline{u}_{PCC}^R \qquad (9)$$

with $R = R_1 + \dfrac{R_2}{3}, \quad L_\sigma = L_{1\sigma} + \dfrac{L_{2\sigma}}{3} + L_F$

The behaviour of the grid coupled to ELSAD with two line taps according to Fig. 3 is described by two space vector equations in a rotating reference coordinate system.

$$\underline{u}_{PCC}^R = R_{N1} \cdot i_{N1}^R + L_{N1} \cdot \frac{d\,i_{N1}^R}{d\,t} + j \cdot \omega_R \cdot L_{N1} \cdot i_{N1}^R + \underline{u}_{N1}^R \quad (10)$$

$$\underline{u}_{PCC}^R = R_{N2} \cdot i_{N2}^R + L_{N2} \cdot \frac{d\,i_{N2}^R}{d\,t} + j \cdot \omega_R \cdot L_{N2} \cdot i_{N2}^R + \underline{u}_{N2}^R \quad (11)$$

For the layout of a multivariable control is a mathematic formulation of the state-space equation system of 8th order in normalized form of advantage. Following references are used for the scaling:

-for the primary-side rated current of ELSAD

$\quad I_N$

978-1-4244-8806-3/11 $26.00 © 2011 IEEE

-the ELSAD rated voltage at PCC

$$\frac{U_N}{\sqrt{3}}$$

-and the rated series impedance of ELSAD

$$Z_N = \frac{U_N}{\sqrt{3} \cdot I_N}$$

To avoid a confusing symbolism, the mathematic equation system is shown in state-space representation form without changing the definitions of the respective scaled electrical quantities.

$$
\begin{bmatrix} i_{Pd}^R \\ i_{Pq}^R \\ u_{PCCd}^R \\ u_{PCCq}^R \\ i_{N1d}^R \\ i_{N1q}^R \\ i_{N2d}^R \\ i_{N2q}^R \end{bmatrix} =
\begin{bmatrix}
-\frac{R}{L_\sigma} & -\omega & -\frac{Z_N}{L_\sigma} & 0 & 0 & 0 & 0 & 0 \\
\omega & -\frac{R}{L_\sigma} & 0 & -\frac{Z_N}{L_\sigma} & 0 & 0 & 0 & 0 \\
\frac{1}{C_F} & 0 & 0 & -\omega & \frac{-1}{C_F Z_N} & 0 & \frac{-1}{C_F Z_N} & 0 \\
0 & \frac{1}{C_F} & \omega & 0 & 0 & \frac{-1}{C_F Z_N} & 0 & \frac{-1}{C_F Z_N} \\
0 & 0 & \frac{Z_N}{L_{N1}} & 0 & -\frac{R_{N1}}{L_{N1}} & -\omega & 0 & 0 \\
0 & 0 & 0 & \frac{Z_N}{L_{N1}} & \omega & -\frac{R_{N1}}{L_{N1}} & 0 & 0 \\
0 & 0 & \frac{Z_N}{L_{N1}} & 0 & 0 & 0 & -\frac{R_{N1}}{L_{N1}} & -\omega \\
0 & 0 & 0 & \frac{Z_N}{L_{N1}} & 0 & 0 & \omega & -\frac{R_{N1}}{L_{N1}}
\end{bmatrix}
\begin{bmatrix} i_{Pd}^R \\ i_{Pq}^R \\ u_{PCCd}^R \\ u_{PCCq}^R \\ i_{N1d}^R \\ i_{N1q}^R \\ i_{N2d}^R \\ i_{N2q}^R \end{bmatrix}
$$

$$
+ \begin{bmatrix} \frac{Z_N}{L_\sigma} & 0 \\ 0 & \frac{Z_N}{L_\sigma} \\ 0 & 0 \\ 0 & 0 \\ 0 & 0 \\ 0 & 0 \\ 0 & 0 \\ 0 & 0 \end{bmatrix} \cdot \ddot{U}_S \cdot \begin{bmatrix} u_{WRd}^R \\ u_{WRq}^R \end{bmatrix}
+ \begin{bmatrix}
0 & 0 & 0 & 0 \\
0 & 0 & 0 & 0 \\
0 & 0 & 0 & 0 \\
0 & 0 & 0 & 0 \\
\frac{Z_N}{L_{N1}} & 0 & 0 & 0 \\
0 & \frac{Z_N}{L_{N1}} & 0 & 0 \\
0 & 0 & \frac{Z_N}{L_{N2}} & 0 \\
0 & 0 & 0 & \frac{Z_N}{L_{N2}}
\end{bmatrix}
\begin{bmatrix} u_{N1d}^R \\ u_{N1q}^R \\ u_{N2d}^R \\ u_{N2q}^R \end{bmatrix}
\qquad (12)
$$

IV. CONTROL STRUCTURE

The determination of a suitable control strucure has the purpose to use ELSAD for the control of power flows at the PCC for random short-circuit powers. Limiting cases are thereby the operation in a solid grid and an isolated operation with solely passive loads. In the first case, an operation of ELSAD as a controlled current source is of advantage, in the second case its operation as a controlled voltage source. The operating behaviour shall adapt itself in the first-harmonic range, varying with the characteristics of the coupled grids, in order to influence the powers at the PCC with respect to a specific and stable operation. Besides the necessary adaptation of the control circuit structure, an adjustment of controller parameters is also vital. Additional requirements stipulate that secondary conditions like maintaining the tolerance range of the voltage at the PCC, a safe short-circuit protection and a transparent controller parameterization are considered.

The concept of cascaded PI-state control, pictured in this paper (Fig. 5), establishes among the known advantages of a PI-

state control the possibility of accomplishing detaching control circuits by cascading state control circuits with given limits for the set point values.

The traits of the controller change through structure modification, depending on the active limits, by seizing of determined control circuits through the limitation. Thereby the operating behavior is collected by the subordinated control circuits.

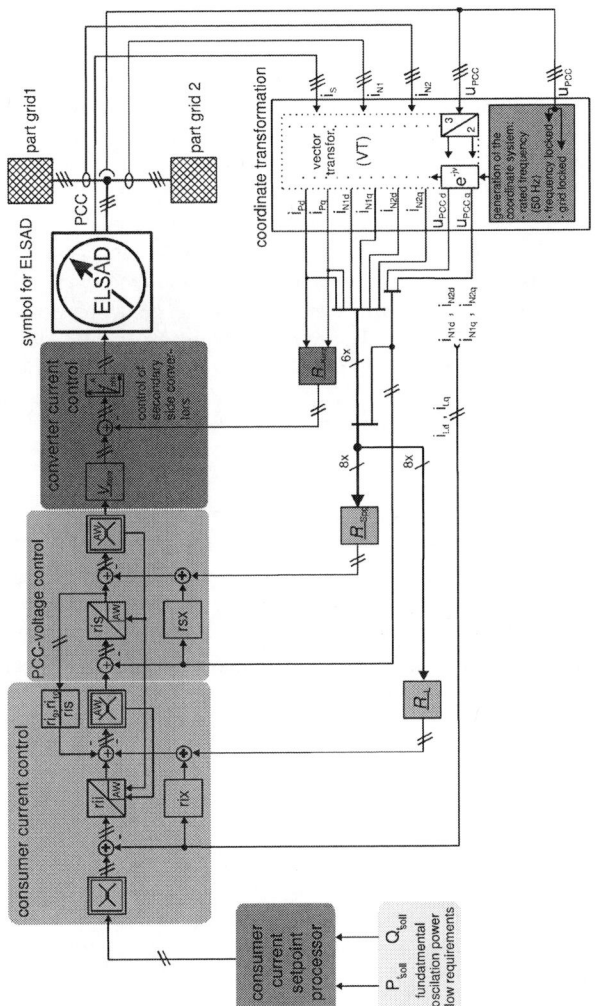

Fig. 5. Fundamental control structure

The inverter current control with limitation of the absolute value of the current establishes the innermost control loop of the dynamic compensator. It is used to limit the current in case of a short circuit, provided that it has an adequate design. The controller is based on a state control parameterized on the short circuit impedance of the inverter. This impedance, unlike the impedance of electrical grids coupled at the PCC, is time dependent. As a result of this, the control circuit parameters don't have to change with the charge states.

The actual voltage value at the PCC is kept in a small range of tolerance at an adjustable set point by a superior voltage control circuit. The voltage control circuit is only active in case of an actual voltage value outside of the range of tolerance due to the special control structure, and therefore only then impresses a voltage source like behavior on the system.

The consumer current control is superior to the PCC voltage control. When this control circuit is active, the dynamic conditioner behaves like a controlled current source at the PCC. Such a mode of operation has advantages at a coupling on a rigid net. This means the nominal power of the dynamic conditioner is slender compared to the coupled grid. In this case there can be no noteworthy impact on the working voltage at the PCC, neither by the load, nor the dynamic conditioner.

If the set point values of the voltage control are in their range of tolerance, the consumer current control should impress its behavior on the system and compensate the subordinated control circuits. This outer control circuit is also a PI-state control circuit to avoid stationary control deviation.

The calculation of the set point values of the current control circuit is based on orthogonal active and reactive power components. Hereunto a consumer current set point processor is heterodyned on the consumer current control circuit, which calculates the consumer current set point value from the power set point values forced by the operating control and from the actual voltage value at the PCC.

V. MEASUREMENTS AT THE TEST BAY

The ELSAD was put to test in a test bay (Fig. 6). Measurements were made under grid parallel and isolated operation conditions. With a computer controlled WEC-test bench, including an asynchronous generator, the characteristic of the active and reactive power curve, as shown in Fig. 7, were reproduced. A 60kVA ELSAD-test plant was then used to condition and smooth the energy that was supplied to the public power grid.

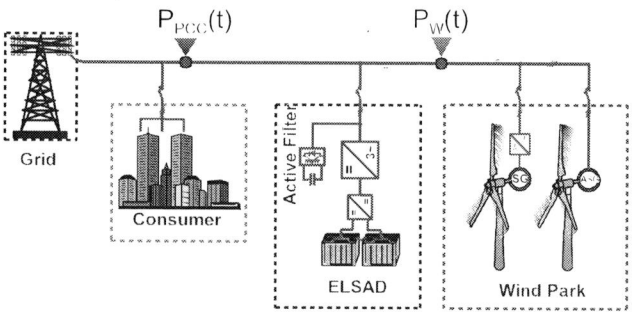

Fig. 6. Fundamental structure of test system

The influence of the simulated gusts and tower effects could be decreased by more than the factor of 10 and the reactive power consumption from the supply grid was completely compensated.

Figure 8 shows the voltage curves at the PCC when an asynchronous machine (ASM, 10 kVA) at a grid short-circuit power of 150 kVA is engaged (motor startup) at the time t_1 with and without the ELSAD system.

The measurement results show the effect of the dynamic increasing of the short-circuit power by means of the

subtransient and transient system characteristics of the ELSAD. The relative voltage drop at the PCC during engagement was decreased from 20% to 7% and was completely compensated after 50 ms.

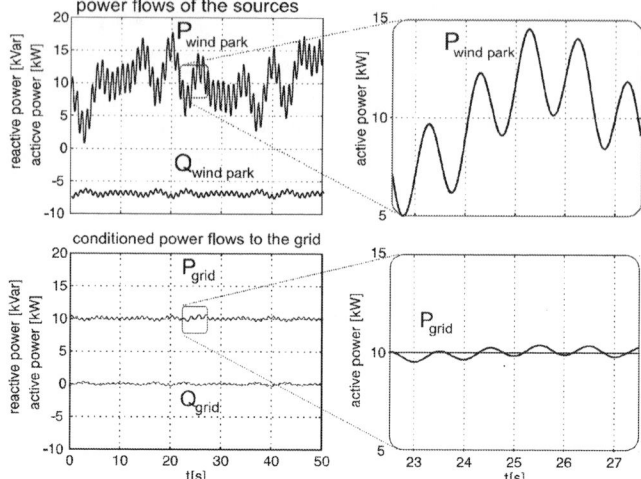

Fig. 7. Smoothing of wind caused power fluctuations and compensation of tower effects

In addition, the energy conditioning system with a limited power compared to the nominal power of the wind energy converter was applied for mid and long term smoothing of wind caused power peaks. The installed power of the energy conditioner represents 20% of the nominal power of the connected wind energy converters. The accumulation depth, defined by

$$T_{SP} = \frac{E_{SP}}{P_{WEc,N}} \qquad (16)$$

of the test facility was set up with 8.3 s in respect to the nominal power of the wind energy converter.

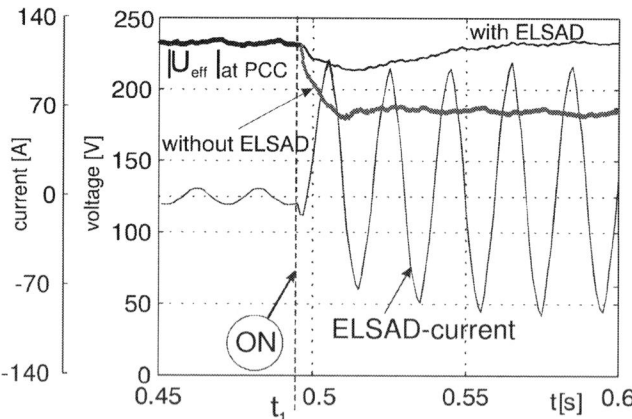

Fig. 8. Start-up process of a 10 kW asynchronous machine in a network spur with low short circuit power

The test results, as shown in Fig. 9, were evaluated using the classification method. It is clear, that the fluctuation of the supplied electrical power at the PCC could be limited to the designated tolerance band of $P_{mean} \pm 0.2 \cdot P_N$, using the energy conditioner. The distribution of the electrical power provided by the WEC shows values that noteworthily differ from the mean value.

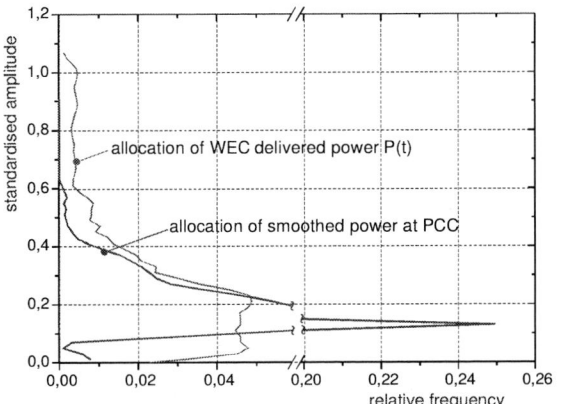

Fig. 9. Comparison of frequency distribution of the output power of the WEC with the smoothed power at the PCC

V. CONCLUSION

This paper describes an electronic synchronous machine with active damping circuit and cascaded PI state control as an instrument for energy conditioning. Because of its particular system properties in the fundamental harmonic and harmonic component ranges it is able to smoothen power flows at the PCC and control the effective power outputs and inputs from the grid to constant values.

In power ranges of up to 100 MVA, ELSAD can be built up economically by standardized components generally used in power converter engineering and simulates the function of a synchronous machine with damper cage very well.

The use of ELSAD for grid feedings with high proportions of regenerative energy sources into weak line taps has been tested in a complex test bay.

Within the scope of the presented study it was demonstrated, that it is possible, to smooth short term fluctuations in the energy supply of renewable energy sources (e.g. wind power) by the use of a dynamic energy conditioner in combination with dynamic storage.

Above that a limitation of the middle and long term changes in power output relating to the average value has been realized. This limitation of the middle and long term gradient of the output power allows long term predictions about the availability of energy from renewable energy sources in electrical public power supplies.

Thereby the share of electrical power generated by wind energy converters to cover the basic energy demand can be increased.

REFERENCES

[1] Wenske, J., "Elektronische Synchronmaschine mit aktivem Dämpferkreis zur Energiekonditionierung in elektrischen Versorgungssystemen," Dissertation 1999, TU Clausthal

[2] Beck, H.-P.; Sourkounis, C.; Wenske, J., "Statischer Synchronkonverter mit aktiver Dämpfungseinrichtung zur Energiekonditionierung," Archiv für Elektrotechnik, Volume 81, Number 6 April 1999

[3] Sourkounis, C.; Beck, H.-P.; Wenske, J., "Gesamtkonzept zur Konditionierung elektrischer Energie aus fluktuierenden Quellen in dezentralen Netzen am Beispiel der Windenergie," DEWEK '98 Tagungsband

[4] C. Sourkounis; "Windenergiekonverter mit maximaler Energieausbeute am leistungsschwachen Netz," Dissertation 1994, TU Clausthal

[5] H.-P. Beck, C. Sourkounis; "Autonomes, modulares Energieversorgungssystem für Inselnetze," Patent-Nr.: 4232516

[6] H.-P. Beck, C. Sourkounis, J.Wenske, Electronic synchronousmachine for the power conditioning in distributionnetworks with high wind energy share, 5th International Conference Electrical Power Quality and Utilisation, Sept. 15 – 17, 1999, Cracow, Poland, Pages 371-377

[7] G.-Myoung Lee, Dong-Chon Lee, Jul-Ki Seok; "Control of Series Active Power Filters Compensating for Source Voltage Unbalance and Current Harmonics," IEEE Transactions on Industrial Electronics, Vol. 51, No.1 February 2004, Pages 132 – 139

[8] A. Engler, C. Hardt, P. Strauss, M. Vandenbergh; "Parallel Operation of Generators for Stand-Alone Single-Phase Hybrid Systems - First Implementation of a new control Technology," 17th European Photovoltaic Solar Energy Conference and Exhibition, Munich 22.-26.10.2001

[9] Sourkounis, C.; Richter, F.: Increase of the Quality of Energy from Stochastic Fluctuation Sources by Dynamic Power Conditioning. International Journal of Energy Technology and Policy, Vol. 5, No. 3 (2007) pp. 271-279.

[10] Hardan, F.; Bleijs, J.A.M.; Jones, R.; Bromley, P.; Ruddell, A.J.; "Application of a power-controlled flywheel drive for wind power conditioning in a wind/diesel power system," Electrical Machines and Drives, 1999. Ninth International Conference on (Conf. Publ. No. 468), vol., no., pp.65-70, 1999

[11] Plotkin, J.; Hanitsch, R.; Schaefer, U.; "Power conditioning of a 132 MW wind farm," Power Electronics and Applications, 2007 European Conference on, vol., no., pp.1-9, 2-5 Sept. 2007

[12] Graovac, D.; Katic, V.; Rufer, A.; "Power quality compensation using universal power quality conditioning system," Power Engineering Review, IEEE, vol.20, no.12, pp. 58- 60, Dec 2000

[13] Zingel, R.; "Grid Compliance Conditioning of Renewable Power Sources by Means of Modern Power Electronics," Power Electronics and Applications, 2007 European Conference on, vol., no., pp.1-8, 2-5 Sept. 2007

[14] Akagi, H.; "The state-of-the-art of active filters for power conditioning," Power Electronics and Applications, 2005 European Conference on, vol., no., pp.15 pp.-P.15

[15] Nagy, I.; Jardan, R.K., "Power quality conditioning based on space vector transformation," Industrial Electronics, 2005. ISIE 2005. Proceedings of the IEEE International Symposium on, vol.2, no., pp. 789- 794 vol. 2, 20-23 June 2005

Electromagnetic Compatibility Test System

F. Domingo-Perez; J. M. Flores-Arias; A. Moreno-Munoz; J. J. G. De la Rosa; A. Gil-de-Castro; V. Pallares-Lopez
and I. Moreno-Garcia

Universidad de Córdoba. Departamento A.C., Electrónica y T.E. Escuela Politécnica Superior. Campus de Rabanales. E-14071
Córdoba. (Spain). Tel: +34-957-218373. Fax: +34-957-218316.
Email: p62dopef@uco.es
*Universidad de Cádiz. Área de Electrónica. Dpto. ISA, TE y Electrónica. Escuela Politécnica Superior
Avda. Ramón Puyol, s/n. E-11202-Algeciras-Cádiz, (Spain).
E-mail: juanjose.delarosa@uca.es

Abstract- **Voltage dips analysis is a complex stochastic issue, since it involves a large variety of random factors, such as: type of short circuits in the power system, location of faults, protective system performance and atmospheric discharges. Among all categories of electrical disturbances, the voltage dips (sags) and momentary interruptions are the nemeses of the automated industrial process. On the other hand, harmonic distortion is a steady state disturbance which is caused by the rectifier employed in adjustable speed drives, uninterruptible power supplies, electronic ballasts and other widely used energy-efficient technology. This paper describes a system for voltage dips testing according to IEC 61000-4-11 norm and it also test the supply current harmonic distortion according to the limits given in IEC 61000-3-2. The system equipment is described also with the test process. We use the created system to test the dip immunity of different topologies of PWM rectifiers and propose a new method of representing the results using power acceptability curves. The harmonic distortion is represented in a bar chart compared with the IEC 61000-3-2 limits.**

I. INTRODUCTION

As more and more electronic equipment enter the residential and business environment, the subjects related to Power Quality (PQ) and its relationship to vulnerability of installations is becoming an increasing concern to the users [1].

Power Quality is concerned with deviations of the voltage or current from the ideal single-frequency sine wave of constant amplitude and frequency. Poor PQ is a concern because it wastes energy, reduces electrical capacity, and can harm equipment and the electrical distribution system itself. Power quality deterioration is due to transient disturbances (voltage dips, voltage swells, impulses, etc.) and steady state disturbances (harmonic distortion, unbalance, flicker) [2].

Among all categories of electrical disturbances, voltage dips and momentary interruptions are the nemeses of the automated industrial processes. On the other hand, voltage swells (which are not so common) do not normally disrupt sensitive load, but can cause harm to equipment.

Voltage dip is commonly defined as any low voltage event between 10% and 90% of the nominal RMS voltage lasting between 0.5 and 60 cycles. Momentary voltage interruption is any low-voltage event of less than 10 percent of the nominal RMS voltage lasting between 0.5 cycles and 3 seconds.

Voltage dips can be caused by natural events (e.g., trees falling on power lines or lightning striking lines or transformers), utility activities (e.g., routine switching operations or human error) or customer activities. Voltage dips at a customer bus are different depending on his location in the electrical network. Because of the short duration of these PQ events, residential customers are rarely aware that a VQ (Voltage Quality) event has taken place. However, for many industrial customers, they pose a far more significant problem than outages because of their much greater frequency of occurrence and overall because of that their incidence can cause hours of manufacturing downtime.

In medium voltage distribution networks, voltage dips are mainly caused by power system faults. Even though the load current is small compared to the fault current, the changes in load current during and after the fault strongly influence the voltage at the equipment terminals. It has been discovered that the 85% of power supply malfunctions attributed to poor Power Quality are caused by voltage dips or interruptions of fewer than one second duration [3] [4].

The IEC norm defines a voltage dip as "a sudden reduction of the voltage at a particular point of an electricity supply system below a specified dip threshold follow by its recovery after a brief interval". The latest version of the IEC 61000-4-11 (second edition) dates from March 2004 and it is the only one which must be used since first of June 2007 [5].

This edition adds to the previous one a new dip level, whereas Edition 1 only used 0%, 40% and 70%, Edition 2 adds the dips level 80%. It also includes a definition of Equipment under Test (EUT) classes, specifying test level and durations. Both editions use the same criteria to classify the results, to summarize:

A) The EUT worked within the manufacturer's specifications.

B) The EUT suffered degradation in its performance but it recovered when the disturbance ceased without operator intervention.

C) The EUT suffered degradation in its performance and needed operator intervention to recover.

D) The EUT suffered degradation in its performance which was not recoverable.

A power acceptability curve is a kind of graph that plots the change of the supply voltage (usually in percent) versus its duration (in seconds or cycles). CBEMA and ITIC curves

978-1-4244-8806-3/11 $26.00 © 2011 IEEE

divides the graph in two regions, the acceptable power zone and the unacceptable zone, which is divided into overvoltage and undervoltage. Lower durations (below 8.33 ms, a 60 Hz half-cycle) are considered acceptable for voltage dips. From about five seconds (steady state) is considered unacceptable a variation below 87% (CBEMA) or 90% (ITIC). The construction of the CBEMA curve is discussed in [6] and reference [7] explains every aspect of the ITIC curve. Both curves suggest dip limits that the EUT should pass. In this work we use the power acceptability curve with the limits given by IEC 61000-4-11 edition 2.

With regard to harmonics, harmonic analysis is a primary matter of PQ assessment. With the widespread use of power electronics equipment and nonlinear loads in industrial, residential and commercial office buildings, the modeling of harmonic sources has become an essential part of harmonic analysis [8].

Apart from that, harmonic attenuation refers to the interaction of the load voltage and current distortion [9]. Various research works have shown that a nonlinear load supplied with distorted voltage will inject less harmonic currents than those generated when the load is supplied by undistorted voltage.

The IEC 61000-3-2 norm deals with the limitation of harmonic currents injected into the public supply system, this norm is applicable to electrical and electronic equipment having an input current up to and including 16 A per phase. The standard divides the equipment into four groups; the devices tested in this paper are analyzed as class C (lighting equipment).

II. METHODOLOGY

The electromagnetic compatibility (EMC) testing facility described next has the main purpose of being used for voltage dips testing in electronic ballasts.

The test procedure for voltage dips testing is very similar to that given in [10], except from this is a test system for single-phase equipment and the use of power acceptability curves at the end of every measurement.

A. Equipment

The following items were for the system development.

- PC: Intel P4 1.6 GHz, 512 MB RAM, 40 GB HDD, running Windows XP SP2, to run the test software and store the results.

- USB oscilloscope (min. 2 channels). We used the Elan Digital Systems USBScope50, with a maximum sampling frequency of 50 MHz; 8 bits resolution; full scale of 300 mV to 30 V with 1x probe, the scope is stackable up to 4 channels, they share the same reference terminal when they are stacked. In this application we only need 2 channels.

- Voltage dips generator controlled by USB serial adapter. The generator used by the authors was the Deneb Elettronica DNBGVD01, which incorporates predefined tests for de first edition of the voltage dips test norm, but it can be almost totally programmed using the serial port and run test with the

durations of the second edition. It complies with the requirements of dip raise and fall time of IEC norm, that is 1 to 5 microseconds.

- Current probe: to measure the input current, we used the Tektronix A622, which can measure AC/DC currents from 50 mA to 100 A peak over a frequency range of DC to 100 kHz.

- AC/DC PWM rectifier converters Evaluation Modules (EVM) with resistive loads. They will be described in the section "Test Example and Results".

B. Connection

The oscilloscope and the voltage dips generator are connected to the PC via USB. The generator input is plugged to the main line and the output is used to supply power to the EVM. Resistive loads are attached to the DC output of the EVMs.

One scope channel measure the generator signal that indicates the moment when the dip is running, the other channel measure the output voltage from the EVM.

In the harmonic test, the current probe is attached to the output of the dips generator.

C. Software

The control of the scope and the generator are separated in two virtual instruments. As this paper focuses in the monitorization of the tests it only shows the virtual instrument for the scope. The dips generator is run manual as it is not necessary to perform a test with other durations apart from the first edition of the norm (Table 1).

The test software of the voltage dips test consists of one window, which integrates the graph of the measured waveform; the graph of the portion of waveform that contains the dip; the CBEMA, ITIC or the curve designed with the IEC 61000-4-11 limits and the instruments' controls and indicators (Fig. 1). The figure shows only the undervoltage condition, as we are studying voltage dips it does not matter the overvoltage curves.

The main window allows configuring the most representative parameters of the scope and selecting the voltage dips test. Selecting a test number is only applied to the frequency of the scope, selecting a frequency to measure the dip with the maximum resolution. The scope will be measuring during a bit more time than the dip duration to show how the equipment recovers itself, usually the extinction of the disturbance is followed by an increase of the output voltage.

TABLE I
TEST LEVEL AND DURATIONS DESCRIBED BY IEC 61000-4-11 EDITION

Dip level (%)	Dip duration (periods)
0	0.5*
	1
40	5
	10
70	25
	50

*The half-cycle duration need to be tested for both positive and negative half-cycle

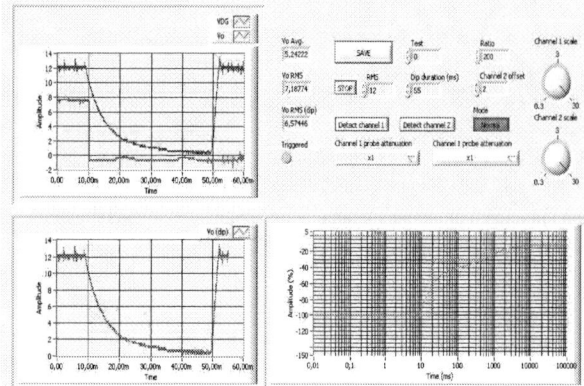

Fig. 1. Voltage dips test software main window.

During the test the RMS voltage is measured when the Evaluation Module (EVM) is working normally and at the end of the test, so the two parameters are taken to the power acceptability curve plotting the RMS voltage variation, activating an alarm signal when the new point falls into the unacceptable zone of the power acceptability curve.

The harmonic test software is simpler than the previous one. It only uses one channel instead of two; the controls are the same, the software processes the current waveform and obtains the harmonic spectrum, comparing with the limits in IEC 61000-4-11 and representing both in a bar graph.

D. Test process

First, the dip level must be selected by hand with the generator dial.

The scope is configure for triggering when channel 1 detects a rising slope, this signal come from the voltage dips generator and it is activated when the dip is running.

While the test is not running, the EVM is supplied by the main line, so it is working normally and the channel 2 is measuring the normal output voltage and storing it RMS and average value. Once the test starts channel 1 is triggered and the clock scope is configured covering the dip test duration so the entire voltage during the dip is measured and stored for later revisions. The RMS voltage previous to the first test is used to reckon the voltage variation with the RMS value obtained at the end of the test. Data are saved in order to be reviewed to consider the fail importance. The result is also shown in the power acceptability curve immediately. The process flowchart is shown in Fig. 2.

Note that the process described in Fig. 2 involves testing the EUT three times; this is a requirement of the IEC norm. The test must be done a minimum of three times with a gap between dips of ten seconds.

The harmonic test is simple, just plug the current clamp in the scope and measure the supply current, the harmonic orders are compared with 61000-3-2 limits.

III. TEST EXAMPLE AND RESULTS

The proposed test system was used to test the dip immunity level of different AC/DC PWM rectifiers checking the three main control strategies: Continuous Conduction Mode (CCM,

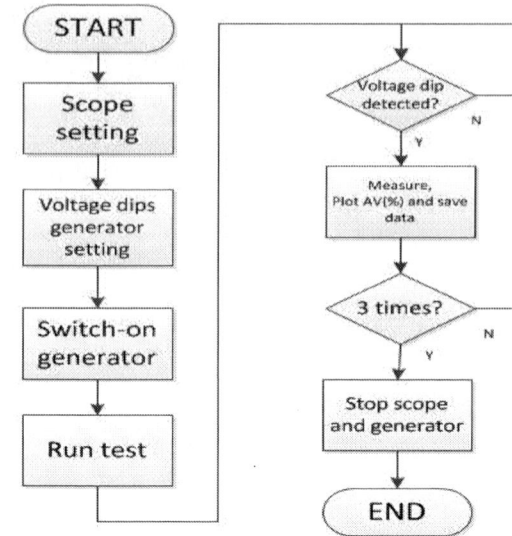

Fig. 2. Dips test process flowchart.

when the input current is always above zero), Discontinuous Conduction Mode (DCM, the input current is zero during part of the switching cycle) and Critical Conduction Mode (CRM, when the input current reaches zero the switch turns on).

All of the EVM tested were designed by Texas Instruments (TI) and they share as a common feature the existence of a universal input (from 85 to 265 V).

For each conduction mode we performed some tests to 2 or 3 different EVMs with different DC/DC stages which included boost, forward and 2-phase interleaved.

The dip levels from table 1 were tested with the following duration:

- Test 1: input voltage of 0% during 10 ms (positive half cycle)
- Test 2: input voltage of 0% during 10 ms (negative half cycle)
- Test 3: input voltage of 40% during 200 ms
- Test 4: input voltage of 70% during 1000 ms.

As previously mentioned, the input is universal, that makes the test 3 and 4 not very interesting, but the test 1 and 2, which tested short interruptions, presented some interesting results. We tried to approach much as we could to the maximum power the devices were able to deal with using serial and parallel connections of 50 W pure resistive loads.

Table 2 shows the results of the experiments, it shows the number code of the EVM given by TI, the conduction mode, the DC/DC stage topology, the experimental conditions of voltage and power and the test results according with the classification given in the introductory part. As we are testing rectifiers it is obvious that once the power is restored it will not be necessary the operator intervention unless the fuse is crashed, which never happened. Any of the modules suffered any damage so we only needed to use class A and B, we considered as a degradation of the equipment performance given by the manufacturer when the output voltage were out of the manufacturer tolerance.

978-1-4244-8806-3/11 $26.00 © 2011 IEEE

TABLE 2
TEST RESULTS

EVM	Control mode	DC/DC Topology	V_{out} (V) P_n (W) P_t (W)	Test (%V / Periods)	Result
28061	CCM	Boost	390 V±5 % 350 W 215.48 W	0 / 0.5(+)	A
				0 / 0.5(-)	A
				40 / 10	A
				70 / 50	A
28514	CCM	Boost	24 V±4 % 100 W 87.27 W	0 / 0.5(+)	A
				0 / 0.5(-)	A
				40 / 10	A
				70 / 50	A
28070	CCM	2-phase interleaved	390 V 300 W 215.48 W	0 / 0.5(+)	A
				0 / 0.5(-)	A
				40 / 10	A
				70 / 50	A
3817	DCM	Boost	385V±10% 250 W 209.9 W	0 / 0.5(+)	A
				0 / 0.5(-)	A
				40 / 10	A?
				70 / 50	A
38500	DCM	Boost + forward	12 V±5 % 100 W 87.27 W	0 / 0.5(+)	B
				0 / 0.5(-)	B
				40 / 10	A
				70 / 50	A
28061	CRM	2-phase interleaved	390 V 300 W 215.48 W	0 / 0.5(+)	A
				0 / 0.5(-)	A
				40 / 10	A
				70 / 50	A
38050	CRM	Boost	400V±6.25 % 100 W 80 W	0 / 0.5(+)	A
				0 / 0.5(-)	A
				40 / 10	A
				70 / 50	A

All the test were performed using an input voltage of 230 V. P_n and P_t show the nominal and tested power.

As it can be seen most of the tests were successfully passed and we only had some troubles with the DCM PWM rectifiers. The boost + forward topology could not pass the short interruption tests and the output voltage went to zero during almost two periods of the reestablished input voltage.

The DCM boost topology apparently passed all of the tests but we included a question mark in the 40% input voltage test, we will focus in these two tests for the DCM modules.

The scope has a resolution of 8 bits, when we tested modules with an output voltage above 300 V we needed a x100 probe that gave us a full scale of ±3000 V, i.e. 8 bits for 6000 V with a 23.5 V precision which makes the tests not to be very confident.

Next we tried was to replace the scope by another scope with a better resolution and differential inputs, which allowed us to measure both input and output voltage so it was not necessary to check the dip instant as we could see the real dip. The scope used was the Tie Pie Engineering HandyScope 4 DIFF (HS4D), with 12 to 16 bits of resolution and differential inputs. It also has a very long buffer (131071 samples), so we could store the previous and subsequent signal before and after the dip. Our objective was to verify the test using the USBScope50, which is a tiny, low cost PC scope, comparing with the results obtained with the HS4DIFF.

Checking all the tests again we only found some differences in the test 3 of the DCM boost module, the one with the question mark. The results for the 12 and 24 output voltage were almost the same since we have an acceptable resolution for a voltage below 30 V. Fig. 3 shows the result for both scopes for the DCM boost module, we can see how with the USBScope50 we obtained an almost constant signal (between 400 and 375 V, which is inside the manufacturer tolerance band) and with the HS4D we can see how the output voltage goes down to 300 V, which is out of the tolerance bands, so the test classified as A is actually a B.

In conclusion, the test system proposed with the actual scope is suitable for testing rectifiers which offer an output voltage below 300 V in order to get an acceptable resolution which is able to detect normal voltage variations.

Another improvement is the disappearances of those extreme peaks which reaches up to 550 V and are nonexistent, but they introduce errors in the average and RMS computation.

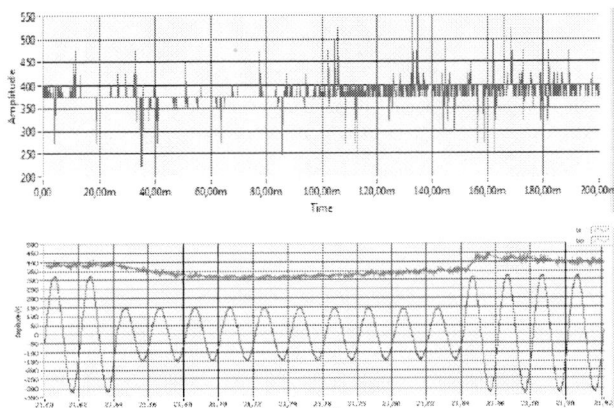

Fig. 3. DCM boost EVM result for test 3. Signal of USBScope50 is shown above and both the input (blue) and output (red) are shown in the bottom graph.

Finally, we want to suggest a new way of representing and reporting the test results. In the norm they only mention if the EUT is broken or not, but they do not go deeper in the performance during the test. We propose to use the power acceptability curve plotting along with the zero and lower voltage condition curves the tolerance bands given by the manufacturer or desired by the client. This new graph could be included in the comment section of the sample report given by the IEC 61000-4-11 norm. Fig. 4 shows the suggested representation, in red, dash lines is represented the tolerance bands, the black lines are the power acceptability curve, they show the limits of IEC 61000-4-11 class 3 (it is made for edition 2, which include the 80% limit) and also the zero line, which means no variation of the output voltage refers to the nominal value. The green points are the different results obtained with the tests, they must be inside the two red lines for a good dip immunity. With this graph we can see exactly how the EUT performance distances from the desired values.

The harmonic tests were a success for all of the EVM as they are designed for a full compliance with IEC 61000-3-2. Fig. 5 shows the results for the UCC28514EVM.

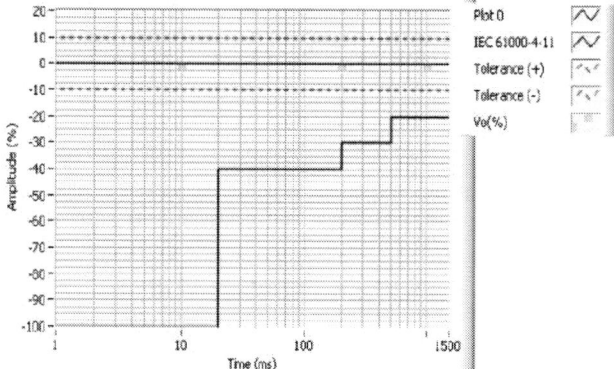

Fig. 4. Suggested representation of the results of dip tests.

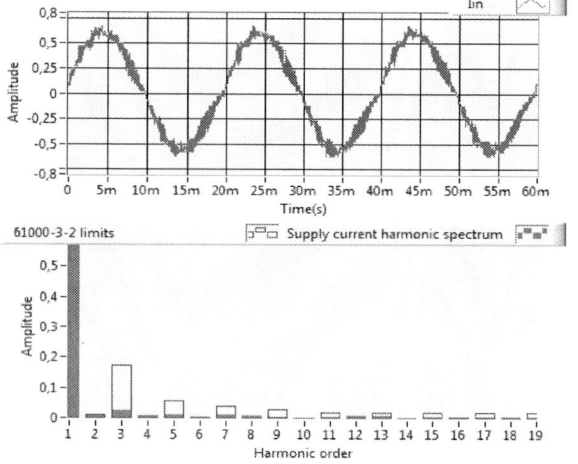

Fig. 5. Input current waveform (top) and harmonic distortion of UCC28514EVM input current (bottom) compared with the limits of IEC 61000-3-2.

IV. CONCLUSIONS

The system can be used to test equipment according to IEC 61000-4-11 and IEC 6100-3-2. The voltage dips generator can be programmed by software to define any dip duration between 10 ms and more than 1 s. Because of this reason there is no problem in adapting the software to new norm revisions.

The next step to improve the voltage dips test system should be the possibility of changing the dip level by computer instead of turning the dial by hand (e. g. with a stepper motor); this will allow the user not to be present during the test execution for changing the dip level. As all the other parameters can be changed by software, the dip level makes the software almost useless, as it is necessary to turn the dial manually.

That way, the system is only limited due to the hardware specification. The low oscilloscope resolution makes very inaccurate any measurement above 300V. On the other hand, the voltage dip generator configuration is limited to changing the dip level and duration and the pause between dips, but the norm also advices about critical phase angles which should also be tested.

Finally, we suggest a very intuitive representation of the dip test result based on power acceptability curves, testing the dip levels up to and below the undervoltage line. This representation method allows the user to quantify the compliance of the EUT with the IEC norm.

ACKNOWLEDGMENT

This research was supported partially by the Company Telvent Energy, Spain, through the project Malaga SmartCity under contract number 12009028. SmartCity's budget is partly financed by the European regional development fund (ERDF) with backing from the Junta de Andalucía and the Ministry of Science and Innovation's Centre for the Development of Industrial Technology. The authors would like to thank the Spanish Ministry of Industry, Tourism and Trade for funding the Project TSI-020100-2010-484 which partially supports this work. Our unforgettable thanks to the Spanish Ministry of Science and Innovation for funding the research project TEC2010-19242-C03-02.

REFERENCES

[1] De la Rosa, J. J. G.; Moreno Muñoz, A.; Gil de Castro, A.; Pallarés, V.; Sánchez Castillejo, J. A., "A web based distributed measurement system for electrical power quality assessment", Measurement Journal of the International Measurement Confederation, 43 (6), pp. 771-780, 2010.

[2] Gil de Castro, A.; Moreno Muñoz, A.; de la Rosa, J. J. G., "Characterizing the Harmonic Attenuation Effect of High Pressure Sodium Lamps", 14th IEEE International Conference on Harmonics and Quality of Power (ICHPQ), Italy, September 2010.

[3] Gil de Castro, A.; Moreno Muñoz, A.; Pallarés, V.; de la Rosa, J. J. G., "Improving Power Quality Immunity in Factory Automation", Power Electronics Technology, 35 (5), pp. 28-32, 2009.

[4] Moreno Muñoz, A.; de la Rosa, J. J. G., "Voltage sag in highly automated factories", IEEE Industry Application Society Annual Meeting, art. nº 4659120, 2008.

[5] Haefely EMC Technology, "Application Note: Dips and Interrupt Testings according to IEC 61000-4-11 Edition 2" [online], http://www.haefelyemc.com/.

[6] Kyei, J, Ayyanar, R, Heydt, G, Thallam, R, Blevins, J, "The design of power acceptability curves," IEEE Transaction on Power Delivery, vol. 7, n° 3, pp 828–833, July 2002

[7] Information Technology Industry Council (ITI), "ITI (CBEMA) curve application note, October 2000.

[8] Acarkan, B., Erkan, K. Harmonics modeling and harmonic activity analysis of equipments with switch mode power supply using MATLAB and simulink. 2007 Proceedings of IEEE International Electric Machines and Drives Conference, IEMDC 2007 1, art. no. 4270692, pp. 508-513.

[9] Mansoor, A., Grady, W.M., Chowdhury, A.H., Samotyj, M.J. An investigation of harmonic attenuation and diversity among distributed single-phase power electronic loads. 1995. IEEE Transaction Power Delivery, 10 (1), pp. 467-473.

[10] Keus, A. K.; Van Coller, J. M.; Koch, R. G, "A test facility for determining the response of industrial equipment to voltage dips (sags)." Electric Machines and Drives, 1999. International Conference IEMD, p. 210, May 1999.

Integration of Active Power Filters in a Harmonic Load Flow Algorithm for Optimizing Location and Strategy

Eva González-Romera, Enrique Romero-Cadaval, Sergio Ruiz Arranz and María Isabel Milanés-Montero
Research Group in Power Electrical and Electronic Systems, University of Extremadura
evagzlez@unex.es, eromero@unex.es; sruiz@coiiex.es; milanes@unex.es

Abstract- **This paper presents a new simple method for integrating the contribution of active power filters (APFs) in a harmonic power flow algorithm for distribution networks. A control strategy for the APFs inspired in the perfect harmonic cancellation (PHC) is proposed and evaluated. This proposed strategy is compared to other one inspired in droop control. Different locations of APFs are tested, evidencing the usefulness of the algorithm for analyzing and optimizing this parameter. The proposed strategy outperforms that based in droop control in terms of power quality and reduction of power losses, in the circumstances studied in this paper.**

I. INTRODUCTION

The presence of harmonics in power systems is nowadays an expected situation in most distribution networks, due to the proliferation of pollutant loads as computers or other electronic devices. Power quality and electromagnetic compatibility standards [1-3] limit their presence locally in generation and consumption facilities, but the integration of multiple pollutant loads and even pollutant generation systems can lead to a poor power quality in voltages of network buses, mainly in radially operated networks, the most commonly used structure for distribution systems.

Many devices, as passive, active or hybrid power filters, and operation strategies have been developed for locally correction of power quality problems [4-7]. However, the contribution of these devices to the power quality improvement of the whole grid is nowadays under study [8-10] and new tools must be developed to facilitate and automate this task.

On the other hand, harmonic power flow has been widely studied in time and frequency domain for power systems, with the aim of knowing the influence of pollutant loads in other buses of the network they are connected to [11-12].

This paper presents a simple method for integrating active power filters (APFs) in a load flow algorithm, with the aim of developing a tool for studying and optimizing their location and the strategy they must follow to reduce the Total Harmonic Distortion (THD) in the voltage of the network buses. Although the presented algorithm is prepared for radial networks, it can be easily extended to meshed networks.

The paper is structured as follows: Section II presents a simple method for analyzing the harmonic load flow in terms of frequency with the possibility of integration of active

power filters for power quality improvement; in Sections III and IV different models for lines, linear and nonlinear loads and active power filters with two different operation strategies are proposed; Section V deals with a simulation of a distribution network and shows results for different locations and strategies of the active power filters; finally, Section VI presents conclusions.

II. FREQUENCY DOMAIN HARMONIC LOAD FLOW

In this paper, a simple method for a harmonic load flow analysis is proposed for radial networks. In this situation, we can suppose that the voltage in the supplying bus is known and the current there injected has a good quality (without harmonics). This algorithm uses the Norton equivalent circuits of elements connected to the different nodes [11], and currents are considered positive when enter the node.

For considering the different harmonics, the algorithm is repeated for each one, modifying in each case the current injected by active elements and the values of passive elements, as it is explained in Sections III and IV.

Finally, once calculated the voltage harmonics in each bus, up to the maximum order of harmonic determined, the Total Harmonic Distortion (THD) of each bus and power losses of the whole grid are calculated. THD in the bus i is defined in (1). Power losses are calculated from the resistance of the lines and the current that pass through them, considering every harmonic.

Fig. 1 shows the proposed harmonic power flow algorithm.

$$THD_i(\%) = 100 \frac{\sqrt{\sum_{h=2}^{H} (V_{i,h})^2}}{V_{i,1}} \qquad (1)$$

In (1):

- $V_{i,h}$ is the h-harmonic voltage component in bus i
- $V_{i,1}$ is the fundamental voltage component in bus i
- H is the maximum order of harmonic considered.

III. MODELS FOR SOURCES, POWER LINES AND LOADS

Before applying the algorithm, the different elements of the network must be modeled to be considered in the analysis. As it can be seen from Fig. 1, the proposed harmonic load flow algorithm works with Norton equivalent circuits for active

978-1-4244-8806-3/11 $26.00 © 2011 IEEE

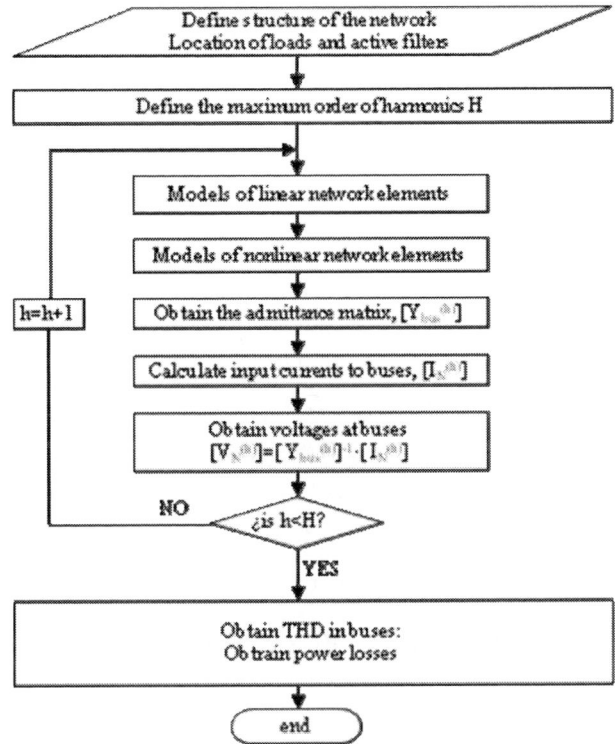

Fig. 1. General algorithm for harmonic load flow analysis in radial networks.

elements, as it calculates currents and then obtain resulting voltages.

For our simulation, power sources and loads have been considered active elements of the network and their Norton equivalent circuits have been obtained and considered in the algorithm.

Fig. 2 shows a radial three-phase balanced network used as an example to illustrate the usefulness of the proposed algorithm.

A. Model for the Source

As this work deals with a radial network, only one power source has been used for simulation. It is located in the node number 1 and it presents a pu impedance of 0.017j (every pu magnitudes in this paper are referred to 6 kVA and 220 V base). For calculating the Norton current, a voltage of 1 pu has been considered in this node. This element only acts as a

source for the first harmonic (fundamental component), as we consider that the power quality of the supply grid is good. For the other harmonics, this element is only considered as an passive shunt impedance.

B. Model for power lines

Power lines are usually modeled by a combination of a series impedance and a shunt capacitance. The length of the pieces of line in distribution network is not high enough to encourage to the use of distributed parameters for a steady state analysis, thus concentrated series resistance and reactance and shunt capacitance have been considered in this work. Capacitors are located in the network nodes for simulation.

Although a line resistance is only influenced for the frequency due to the skin effect, it can be considered negligible for the purpose of this paper. However, both reactance and capacitive reactance are proportional to the frequency. Therefore, values of these parameters must be adapted to the harmonic order. For each harmonic, the fundamental value of reactance and capacitive reactance of lines are multiplied by the order of the harmonic, as it determines the proportionality with frequency.

For the example presented in this paper, two different types of lines have been used (Fig.2). Their parameter values, for 50 Hz fundamental frequency, are shown in Table I.

TABLE I
DATA OF LINES (PU), REFERRED TO 6000 VA AND 220 V BASE

Type of line	line 1	line 2
Resistance (pu)	0.0062	0.0123
Reactante (pu)	0.0093	0.0186
Capacitive reactance (pu)	0.0913	0.0913

C. Model for linear loads

In linear loads, the current demanded can be calculated from the voltage in the node where it is connected and its impedance. For this work, one linear load has been used. It acts as a passive element with a shunt resistive impedance of 0.4 pu.

D. Model for nonlinear loads

A non-controlled rectifier is a source of harmonics in a power system. An idealized three-phase non-controlled 6-pulse rectifier, with constant dc current is used for simulation (Fig. 3), an acceptable approximation for rectifiers connected to R-L loads. For modelling these loads it is necessary to take into account the waveform distortion in order to achieve a

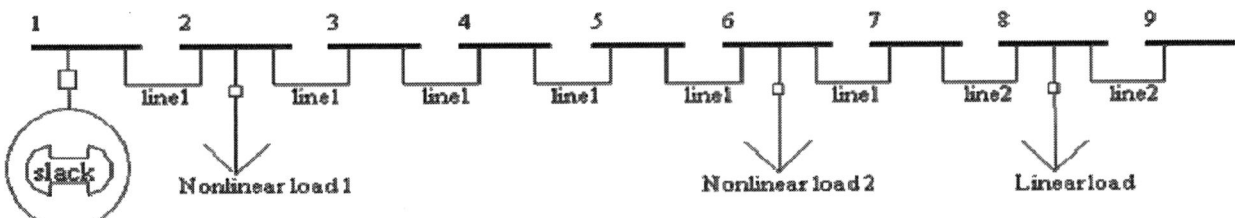

Fig. 2. Network used for simulation.

better description of the interaction with the network. The current wave can be decomposed in Fourier series and each harmonic component can be injected into the system as a power source; thus, it is possible to determine the system nodes harmonic voltages if a frequency sweep is made on the network.

Therefore, nonlinear loads can be modelled as constant current sources for each harmonic frequency and are calculated regarding to the fundamental frequency current. It is well known that an idealized three-phase 6-pulse rectifier, as describes above, demands a current composed by a fundamental component, $I_{R,1}$ and a combination of harmonics calculated as

$$I_{R,h} = \frac{I_{R,1}}{h} \qquad (3)$$

$I_{R,h}$ is the h-harmonic current component demanded by the rectifier. In this case even and triple harmonics are zero [13].

$I_{R,1}$ is the current drawn by the rectifier at the fundamental component, whose expression is obtained from (4).

$$I_{R,1} = \frac{4}{\pi} \int_{\pi/6}^{\pi/2} I_0 \cdot \sin(\theta) \cdot d\theta \qquad (4)$$

where I_0 is the dc current provided by the rectifier, assumed to be constant.

By integrating the previous expression, the next one is obtained:

$$I_{R,1} = \frac{4}{\pi} I_0 [\cos\theta]_{\pi/2}^{\pi/6} = \frac{4}{\pi} I_0 \frac{\sqrt{3}}{2} = \frac{2\sqrt{3}}{\pi} I_0 \qquad (5)$$

If V_0 is the average dc voltage provided by the rectifier, R is the dc equivalent resistance, provided by the known rated capacity of the rectifier ($S_{RECTIF.}$) and V_{LL} is the line-to line ac voltage:

$$I_0 = \frac{V_0}{R} = \frac{1.35 \cdot V_{LL}}{R} \qquad (6)$$

with

$$R = \frac{(1.35 \cdot V_{LL})^2}{S_{RECTIF.}} \qquad (7)$$

On the other hand, the equivalent star resistance of rectifier for the purposes of modelling its behaviour for the fundamental component can be expressed by:

$$R_{eq} = \frac{V_{LL}}{\sqrt{3} \cdot I_{R,1}} \qquad (8)$$

Consequently, the equivalent resistance results:

$$R_{eq} = \frac{1.35 \cdot \pi \cdot V_{LL}^{2}}{6 \cdot S_{RECTIF.}} \qquad (9)$$

For the first harmonic, rectifiers are considered to behave as passive resistances, with values shown in (9). For the remaining harmonics, the Norton equivalent circuit for the

rectifier is a current source calculated from (3). In this calculation, $I_{R,1}$ is obtained from the resistance in (9) and the obtained fundamental voltage in its node (8).

For the example presented in this paper, two ideal three-phase 6-pulse non-controlled rectifiers have been used (Fig.2). Their parameter values are shown in Table II.

TABLE II
DATA OF RECTIFIERS (PU), REFERRED TO 6000 VA AND 220 V BASE

Parameters:	Rectifier 1	Rectifier 2
Rated capacity, $S_{RECTIF.}$ (pu)	0.46	0.554
Equivalent resistance, R_{eq} (pu)	1.5356	1.2759

IV. MODELS AND STRATEGIES FOR ACTIVE POWER FILTERS

Active power filters (APFs) are being investigated and developed as a solution for power quality problems in nodes of networks where pollutant loads are connected. Several topologies and control strategies have been proposed for their performance. In [4], four control strategies for shunt active power filters are discussed and compared in different conditions of power quality of the current demanded by the pollutant load. In [4], the control strategy named perfect harmonic cancellation (PHC) is proved to be the most complete correction method, in presence of harmonics, unbalance and reactive power. The term "perfect harmonic cancellation" was introduced in [7] by the authors who proposed the strategy. Its philosophy consists on defining a reference current synchronized with the fundamental positive-sequence voltage in the connection bus, which is calculated with the aim of cancelling all the harmonic currents demanded by the load.

In the present paper, we introduce a distribution network with several linear and nonlinear loads connected in different buses. In this situation, the possibility of power quality correction in the case that location of pollutant loads is not exactly known must be considered. Therefore, the objective of the APFs must not be to correct power quality only in the bus where the pollutant load is connected, but in every nodes of the main grid.

In this section a performance strategy inspired in PHC [4] is proposed to reduce the voltage distortion in every node of a radial distribution network. Then, its performance is compared with a method based on droop control, and proposed by [8].

In both cases, the action of the APFs is introduced in the harmonic power flow algorithm described in Section II, and the Total Harmonic Distortion (THD) of the bus voltages, as defined in (1) is used to evaluate results.

A. Strategy inspired on perfect harmonic cancellation (PHC)

As it has been commented before, the objective of the APFs in this paper is to correct power quality in every nodes of the distribution grid, not only in the node where the pollutant loads are connected. Therefore, the strategy of the APFs must be different than the conventional ones, consisting

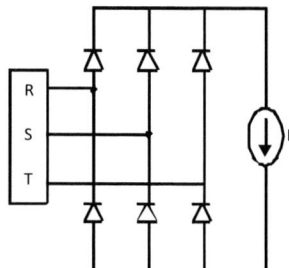

Fig. 3. Three-phase non-controlled 6-pulse rectifier.

978-1-4244-8806-3/11 $26.00 © 2011 IEEE

of cancelling harmonics due to the load connected to the same bus.

In this case, the control strategy of each APF is to cancel, if possible, every harmonic current present in the piece of the line located immediately upwards the APF node. As the line is radial, this strategy pursues to reduce and even to totally cancel every harmonic in nodes located upwards that of the APF.

With this strategy, THD is expected to be zero in every node if an APF is located in each node in which a pollutant load is connected. In other cases, the total cancellation of harmonics is not guaranteed. An obvious disadvantage of this strategy is that APFs located downwards the pollutant loads are not useful and cannot contribute to harmonic cancellation.

The integration of this strategy for APFs in the harmonic power flow algorithm is made in two steps:

1) First, the harmonic power flow is done without considering APFs. Thus, harmonic components of voltage in different nodes are calculated.

2) In the second step, the target current needed from the APFs to cancel each harmonic current in the upwards piece of line is calculated. For this purpose, the harmonic currents in that piece of line are calculated from the voltages of the node where APF is connected and the previous one and the known impedance if the piece of line, adapted to each harmonic. Once obtained the target current for the APF, the harmonic power flow is repeated adding the contribution of the APFs, calculated as described.

B. Strategy inspired on Droop Control [8]

For comparison purposes, the strategy proposed in [8] for the APF performance, based on droop control, is adapted and applied to the example under study in this paper.

In this case, each APF acts as a shunt conductance. For each harmonic, this conductance (G_{APF}) is multiplied by the calculated voltage in the node where the APF is connected ($V_{APF,h}$) to obtain the target current ($I_{APF,h}$) the APF must inject to the grid (10).

$$I_{APF,h} = G_{APF} \cdot V_{APF,h} \tag{10}$$

In [8], a droop relationship between the modelled conductance and the VA consumption of the APF is proposed (11).

$$G_{APF} = G_0 + b \cdot (S - S_0) \tag{11}$$

where:

- G_0= rated conductance of the APF (pu, original magnitude measured in Ω^{-1})
- b= slope of the droop equation (pu, original magnitude measured in V^{-2})

- S_0= rated capacity of the APF (pu, original magnitude measured in VA)
- S= consumption of the APF (pu, original magnitude measured in VA).

On the other hand, the APF consumption, for the harmonic h, can be determined multiplying RMS values of voltage and current in the node where APF is located (when pu magnitudes are used, and three-phase criterion is applied, this equation is valid for three-phase pu systems). Assuming that the APF suppress the harmonics significantly, thus the voltage at buses with APF is dominated by the fundamental voltage component and the equation for calculating the APF consumption can be approximated to (12).

$$S = \left| V_{APF,1} \right| \cdot \sqrt{\sum_{h=1}^{H} I_{APF,h}^2} \tag{12}$$

Combining (10) and (12) and considering that $I_{APF,1}$ is equal to zero, as the APF is not expected to inject any fundamental current component, we obtain (13).

$$S = \left| V_{APF,1} \right| \cdot G_{APF} \cdot \sqrt{\sum_{h=2}^{H} V_{APF,h}^2} \tag{13}$$

Finally, combining (11) and (13), we can obtain G_{APF}.

$$G_{APF} = \frac{G_0 - b \cdot S_0}{1 - b \cdot \left| V_{APF,1} \right| \cdot \sqrt{\sum_{h=2}^{H} V_{APF,h}^2}} \tag{14}$$

An iterative process is needed to obtain G_{APF} from the different calculated voltage components. Then, the APF is modeled as a current source, whose value is calculated from (10) and (14).

Thus, in this second strategy for the APF control, the harmonic power flow algorithm shown in Fig. 1 must be iteratively repeated towards convergence.

The parameters for two APFs applying this strategy in the example presented in this paper are shown in Table III.

TABLE III
DATA OF APFs (PU), REFERRED TO 6000 VA AND 220 V BASE

Parameters:	APF 1	APF 2
Rated capacity, S_0 (pu)	0.083	0.083
Rated conductance, G_0 (pu)	0	0
Slope, b (pu)	-38.72	-38.72

V. SIMULATION OF A RADIAL DISTRIBUTION NETWORK

Once described the models employed for the different elements of the network under study (Fig. 2), this section shows the results obtained applying the algorithm shown in Fig.1 with MATLAB™.

The proposed strategy for APFs, presented in Section IV-A is first applied. One APF is located in bus 2, where the first pollutant load is connected, with the aim of verifying that it can absolutely suppress the THD in previous nodes. Another APF is used, varying its location from bus 3 to 6, where the second pollutant load is connected. A location downwards both rectifiers is not tested, as APFs cannot be useful in this situation with the proposed strategy, as commented above.

Fig. 4 shows the resulting voltage THD in each node of the network when varying the location of the second APF. In every case, the original THD, obtained without APFs, is presented in red for comparison. It can be easily observed that the THD in nodes upwards both APFs are always equal to zero. On the other hand, those nodes located downwards both APFs present a reduced THD. This THD reduction is higher as closer to the second pollutant load the second APF is. The optimal situation is that in which both APFs are located in the same buses than both pollutant loads. In this case, THD is totally eliminated from every node of the network.

The second strategy presented in Section IV-B is applied only in the optimal situation, with the aim of comparison with the proposed first strategy. Fig. 5 shows the resulting THD in every buses of the network. It can be observe that THD reduction is less effective than that obtained with the strategy inspired in PHC. This strategy, however, presents one advantage compared with the first one: as it pursues reducing globally the THD in every node, its performance is not so dependant of the location of the pollutant loads.

Table IV shows reductions in power losses that APFs achieve in every simulated case. In this table, strategy A is referred to that described in Section IV-A and strategy B corresponds to that presented in Section IV-B.

TABLE IV
REDUCTION IN POWER LOSSES WITH DIFFERENT LOCATIONS AND
STRATEGIES FOR ACTIVE POWER FILTERS

APF Strategy	A	A	A	A	B
Location of APF (buses)	2,3	2,4	2,5	2,6	2,6
∇ losses by harmonics	60%	70%	90%	100%	30%
∇ total losses	0.47%	0.55%	0.71%	0.78%	0.24%

This table evidences another great advantage of the use of APFs in distribution networks. They not only improve power quality in network buses, but also contribute to the efficiency of the system, reducing losses produced by the harmonic currents circulating through the lines. The reduction of power

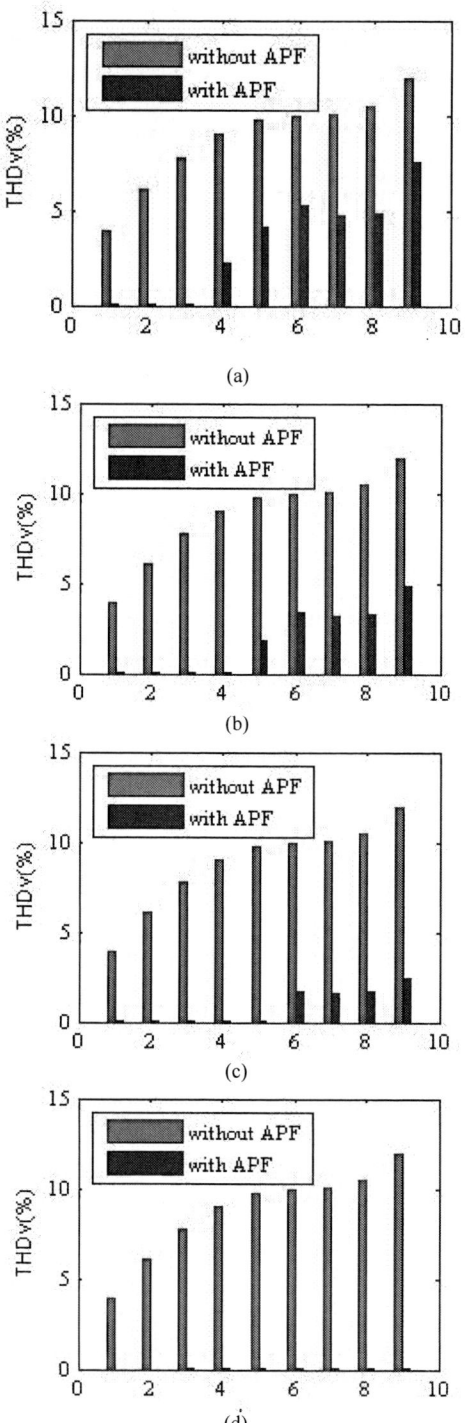

Fig. 4. THD (%) in bus voltages when varying the location of active power filters (APF) using the strategy inspired in PHC (Strategy A): (a) APF in buses 2 and 3; (b) APF in buses 2 and 4; (c) APF in buses 2 and 4; (d) APF in buses 2 and 6.

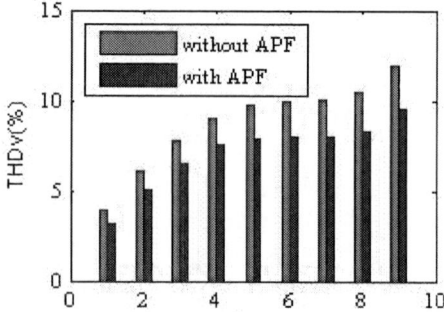

Fig. 5. THD (%) in bus voltages when droop control has been applied (Strategy B), locating APF in buses 2 and 6.

losses is very important when the proposed strategy (A) is used, even eliminating losses due to harmonics when APFs are located in the same buses than pollutant loads.

VI. CONCLUSIONS

This paper proposes an algorithm for harmonic load flow analysis of distribution networks with the presence of pollutant loads and active power filters (APFs). The strategy proposed for the APFs is inspired in perfect harmonic cancellation (PHC) presented in [4], and compared with other strategy adapted from that proposed in [8] and inspired in droop control.

The presented algorithm is simple to implement and permits the comparison of different strategies and locations of APFs with the aim of optimization.

A radial distribution network with linear and nonlinear loads and APFs is used as an example for simulation. The proposed strategy for APFs proves to be absolutely effective for voltage THD reduction in every nodes when the location of APFs is the same of the pollutant loads. In other cases, APFs significantly reduce THD in nodes and power losses due to current harmonics. The proposed strategy presents a better performance than that inspired in droop control in the cases simulated, although it presents a higher dependency of the location of the APFs.

REFERENCES

[1] EN 50160 "Voltage Characteristics in Public Distribution Systems", 1999.

[2] IEEE Std. 519-1992. "IEEE Recommended Practices and Requirements for Harmonic Control in Electrical Power Systems," 1992.

[3] EN 61000. "Electromagnetic Compatibility (EMC)"

[4] M. Milanés, E. Romero and F. Barrero, "Comparison of control strategies for shunt active power filters in three-phase four-wire systems," *IEEE Trans. Power Electronics,* vol. 2(1), pp. 229-236, January 2007.

[5] C.H. da Silva, R.R. Pereira, L.E.B. da Silva, G.L. Torres and J.O.P. Pinto, "Unified hybrid power quality conditioner (UHPQC)," *IEEE International Symposium on Industrial Electronics (ISIE),* pp. 1149 – 1153. 2010.

[6] A. Ghosh and G. Ledwich, "A unified power quality conditioner (UPQC) for simultaneous voltage and current compensation," *Electric Power Systems Research,* vol. 59, pp. 55–63, 2001.

[7] M.-R. Rafiei, H. A. Toliyat, R. Ghazi, and T. Gopalarathanam, "An optimal and flexible control strategy for active filtering and power factor correction under nonsinusoidal line voltages," *IEEE Trans. Power Delivery,* vol. 16, no. 2, pp. 297–305, Apr. 2001.

[8] P-T. Cheng and T-L. Lee "Distributed Active Filter Systems (DAFSs): A new approach to power system harmonics," *IEEE Trans. Industry Applications,* vol. 42(5), pp. 1301 – 1309, Sept.-Oct. 2006.

[9] S. Leng, II-Y. Chung and D.A. Cartes, "Distributed operation of multiple shunt active power filters considering power quality improvement capacity," *2nd IEEE International Symposium on Power Electronics for Distributed Generation Systems,* pp. 543 – 548, August 2010.

[10] W. Yan-song, S. Hua, L. Xue-min, L. Jun and G. Song-bo, "Optimal allocation of the active filters based on the tabu algorithm in distribution network," *International Conference on Electrical and Control Engineering (ICECE),* pp. 1418 – 1421. 2010.

[11] J. Arrilaga, A. Medina, M.L.V. Lisboa, M.A. Cavia and P. Sánchez, "The harmonic domain. A frame of reference for power system harmonic analysis," *IEEE Trans. Power Systems,* vol. 10(1), pp. 433-440, February 1995.

[12] H. J. Chuang, C. S. Chen, C. H. Lin, and H. M. Shiau, "Stochastic harmonic load flow analysis and distortion mitigation of mass rapid transit systems," *IEEE Bologna Power Tech Conference,* Bologna, Italy. 2003.

[13] N. Mohan, T.M. Undeland and W.P. Robbins, "Power electronics. Converters, applications and design," USA: John Wiley & Sons. 2003.

Generator Overloading owing to Neutral Wire Current

Kus Vaclav, Skala Bohumil
University of West Bohemia in Pilsen, Czech Republic
kus@kev.zcu.cz, skalab@kev.zcu.cz

Abstract - **The cogeneration unit uses high-quality synchronous generators. Its power is over 1 MVA and simultaneously the harmonic distortion is very low, usually less than 5%. If this generator works in the self-operated mode, its operation is without any problems. Problems arise when the synchronous generator works in parallel operation with the network and when the neutral point of the star connection has to be connected to a neutral wire of the network. In this case the synchronous generator is overloaded very often and the neutral wire current is up to 3 times higher than nominal phase current. As conclusions are presented the methods of how the neutral wire current can be decreased.**

I. INTRODUCTION

The present paper describes an example when this problem had to be solved as soon as possible, because the generator was already situated in the consumer's facility. The precise calculations and simulations confirmed the solution.

The first part of the paper describes the generator and the network. Using a calculation, the 3rd harmonic was discovered being very low. These values were confirmed by the measurement under the no-load test and under the loading. The load was a symmetrical 3-phase load of 200kW.

The next step was the mode of parallel operation with the network. When the generator had the same load, the neutral wire current increased; however, the value was still not dangerous. The 3rd harmonic was dominant. By the measurement it was discovered that the network distorsion was very low. Especially the 3rd harmonic was very low.

The backset is brought about by the situation where the network was assumed with the 5% distortion by the 3rd harmonic. This type of distortion is nothing special in the industrial areas. The calculation confirmed that in this case the neutral wire current is enormous and the overloading of the generator wing is significant.

In this paper are presented the results of the laboratory measurement as well as the results of the practical measurement performed directly on the cogeneration unit. The results of the zero sequence component of the reactance of the cogeneration unit are also presented. Some results are completed with the FFT analysis.

II. DESCRIPTION OF SYSTEM

Figure 1 is a basic circuit diagram of the situation. Due to the investigation of the third harmonic current is generated in the model of the third harmonic voltage at the same time are considered non-rotating reactance. If we assume symmetrical three-phase loads, so the neutral current will not flow through the fundamental harmonic.

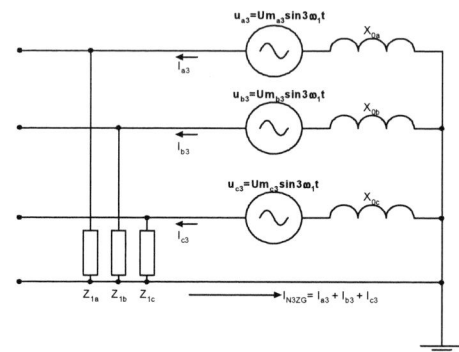

Fig. 1 Self operation mode – the 3rd harmonic is low

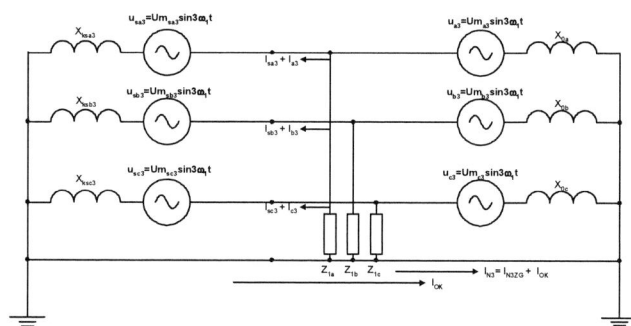

Fig. 2 Network parallel mode of operation – the 3rd harmonic is extremely high

A. Self operating mode

We assume the generator of a power 1 MW which is connected to a network 3x 400V (see Fig. 1). This generator works with very low voltage distortion. The third harmonics is a 3 percents only. Then

$$U_{a3} = 0{,}03 . U_{a1} = 0{,}03.230 = 6{,}9 \text{ V} \qquad (1)$$

The current of the third harmonics, which is flowing through the neutral wire, depends on

- load impedance
- zero sequence component of the generator
- magnitude of the third voltage harmonics.

We can assume the symmetrical load of 3x 200 kW in a star connection. Then we can derive for a impedance of a load (common motor with a power factor 0,5) the following:

978-1-4244-8806-3/11 $26.00 © 2011 IEEE

$$Z_{3L} = 3.Z_1 = 3.U_1^2 / P = 3.230^2 / 200.10^3 = 0,397 \qquad (2)$$

The zero sequence component of a generator was estimated to be $Z_0 = 0,02 \ \Omega$. Later the measurement confirmed this value. The current of the third harmonics is given by the voltage of the third harmonics and by the sum of both impedances. The third harmonics current is given by the sum of all three phases. This current is given by

$$I_{N3ZG} = 3.U_{a3}/(Z_{3L}+Z_0) = 3.6,9/(0,397+0,02) = 51,7 \ A \qquad (3)$$

From the relation (3) it follows that in the neutral wire there is a low current of a third harmonics. This current is low with respect to nominal current of the generator.

B.   Generator on the network operation mode

In this case is necessary to determine the network parameters (see fig. 2). This figure follows that the third harmonics current depends on many parameters. Current consist of the third harmonics current and of a load and circular current of the third harmonics.

The current of the third harmonics may be calculated similarly as in previous case. As follows from the relation (4), its value is still low with respect to the circular current.

We can assume that the voltage distortion of the network is 5 percents. If the short circuit power of a network is Sk=5 MVA we can calculate the reactance of a network for a third harmonic $X_{KS3}= 0,012 \ \Omega$. We assume the symmetrical network, the third harmonic currents are summed. Third harmonic current (I_{N3ZG}) caused by loads and flows through the generator neutral is not changed.

For a circular current calculation we assume the above determined voltage distortion of the generator (3%) and that of the network (5%). The total third harmonics for a voltage is 18,4 V. Circular current of the third harmonics is limited by the network impedance and the zero sequence component of the generator. Short-circuit power network at the parallel cooperation we assume $S_{KS} = 5$ MVA

We sum the current from all phases. The circular current in the neutral wire is

$$I_{OK} = 3.U_{a3}/(Z_0 + Z_{KS3}) = 3.18,4/(0,02+0,012) = 1725 \ A \quad (4)$$

The total current in the neutral wire is

$$I_{N3} = I_{OK} + I_{N3ZG} = 1725 + 51 = 1776 \ A \qquad (5)$$

This current is extremely high. This current flows through the neutral wire, loads the cables and increases the generator current. Due to this fact the temperature of the generator increases and its efficiency decreases. Mode of this operation is unsuitable. Therefore it must be considered whether such a design can be maintained.

If it is necessary to connect the neutral point of a generator to a neutral wire of a network, we must find the solution. In the following chapters is described the experience from the laboratory as well as from the factory which produces generator units.

III. MEASUREMENT IN THE LABORATORY

To take a rough estimate of the third harmonic component amplitude, a verification experiment was necessary. The measurements were performed for various loading of the synchronous generator. In each state the current in a neutral wire was measured and the FFT analysis was performed. In this way we determined the current (third harmonic) as dependent on the load of the synchronous generator. Please note that the loads were exclusively symmetrical.

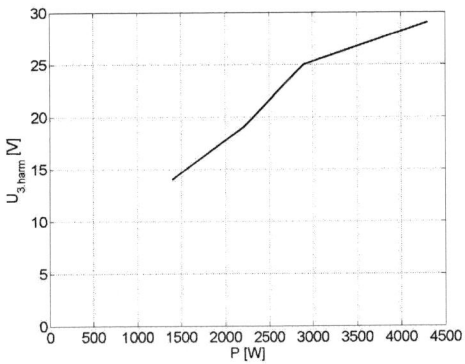

Fig. 3 Third harmonic component (voltage) as dependent on the load of the synchronous generator

In this situation it seemed that the only way is connect the additional reactance to a neutral star point of the generator winding to reduce the third harmonic current. As far as this consideration is right, the current will be significantly reduced. In the other hand, the neutral point voltage will be higher and its value cannot reach the safety voltage.

Because the generator can operate also separately (without the power network), the same situation have to be tested in this mode of operation as well.

Thank to the experiments we observed that the third harmonic current is almost independent on the additional reactance connection. This consideration is valid only under the condition that the load is balanced. In case of unbalance load the third harmonic component is marginal.

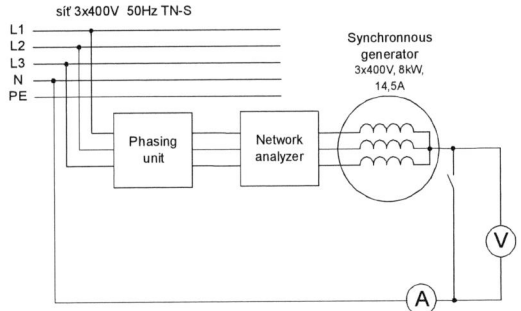

Fig. 4 – Measurement scheme of third harmonic component

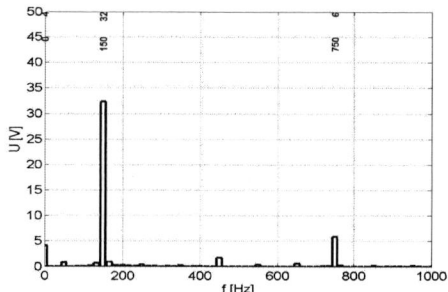

Fig. 5 – FFT analysis results for the third harmonic voltage (see Fig.4)

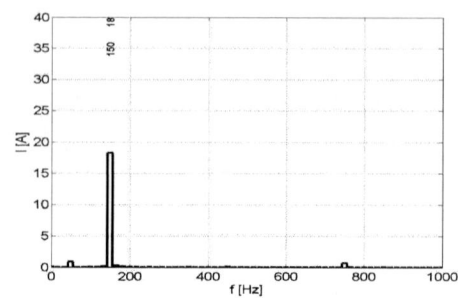

Fig. 9 – FFT analysis results for the third harmonic current (see Fig.7)

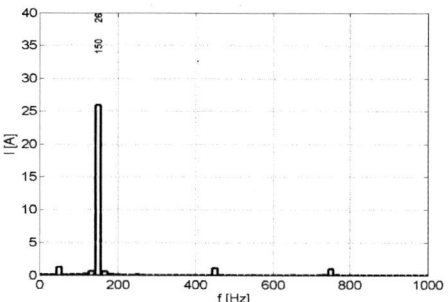

Fig. 6 – FFT analysis results for the third harmonic current (see Fig.4)

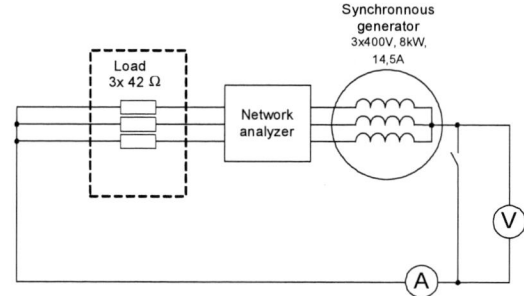

Fig. 10 – Measurement scheme of third harmonic component in the self-operation mode of the generator

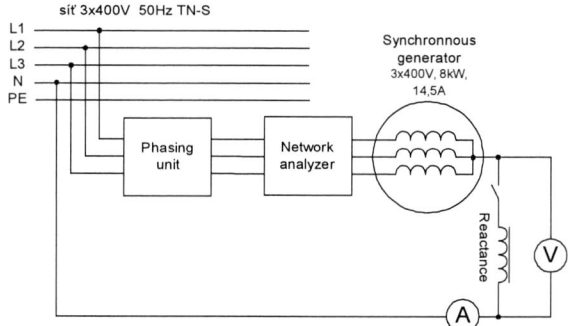

Fig.7 – Measurement scheme of third harmonic component with additional reactance (30%).

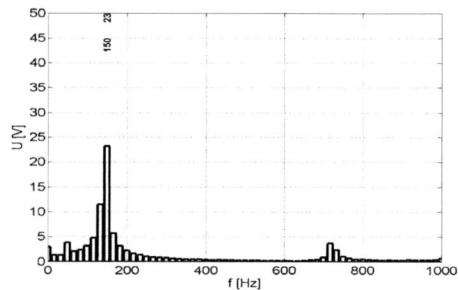

Fig.11 – FFT analysis results for the third harmonic voltage (see Fig.10)

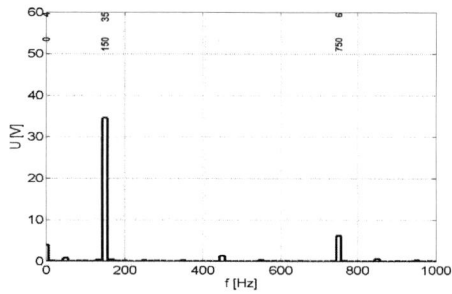

Fig. 8 – FFT analysis results for the third harmonic voltage (see Fig.7)

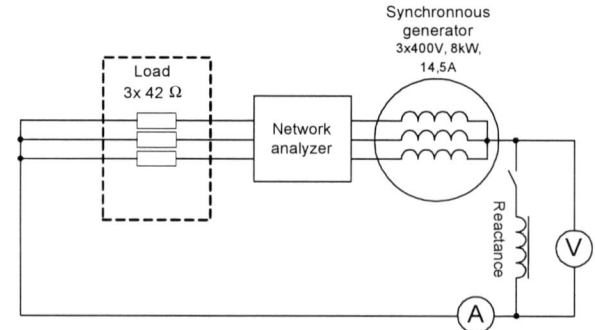

Fig.12 – Measurement scheme of third harmonic component in the self-operation mode of the generator with additional reactance (30%).

978-1-4244-8806-3/11 $26.00 © 2011 IEEE

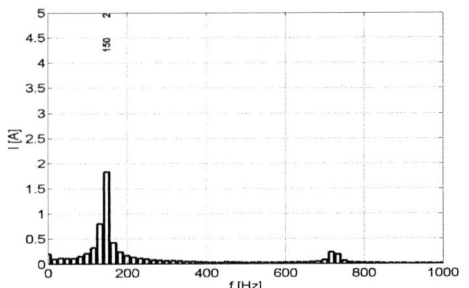

Fig. 13 – FFT analysis results for the third harmonic current (see Fig.12)

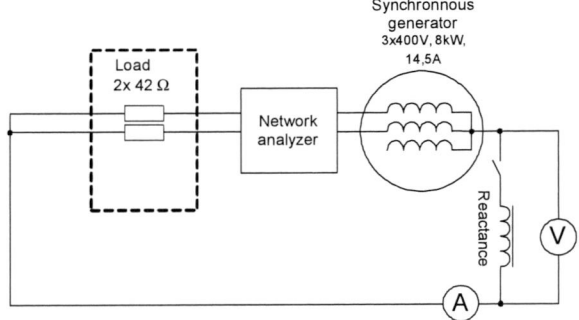

Fig.14 – Measurement scheme of third harmonic component in the self-operation mode of the generator, unbalanced load. The neutral point voltage has the dangerous value.

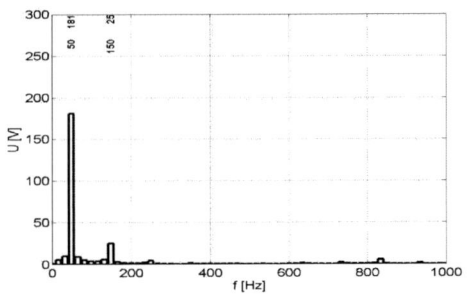

Fig.15 – FFT analysis results for the third harmonic voltage (see Fig.14)

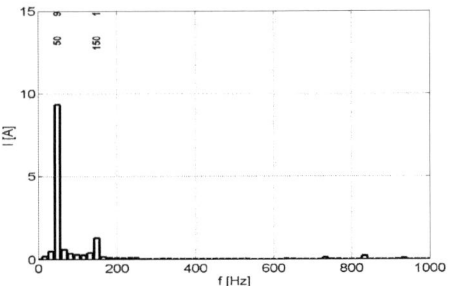

Fig. 16 – FFT analysis results for the third harmonic current (see Fig.14)

IV. PRACTICAL MEASUREMENT

At the manufacturing hall we measured the neutral wire currents. The generator was connected to the network and its active power was about 10% (i.e. 200kW) of a nominal power. The current was observed only 20A.

The next step was the zero-sequence reactance measurement. The brush-less exciter was disconnected from the rotating rectifier. Thanks to autotransformer and step down transformer in the cascade connection, we obtained the voltage source which was able to supply the winding with the current of up to 200A. The task of disconnecting the star connection of a winding and reconnecting it to a serial connection was too complicated. Consequently, the parallel connection was used. The current flowing into each phase is not the same; however the obtained results give us the rough estimate about the zero-sequence reactance value.

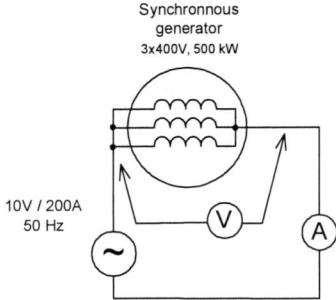

Fig.17 – The zero-sequence component measurement (scheme of connection).

V. CONCLUSION

As a conclusion, it was recommended to connect an additional reactor between the neutral point of a generator and power network. This reactor decreases the current of a third harmonic. The dimensioning of this reactor has to fulfill the following conditions:
-        Contact voltage must not reach the allowable value of voltage (safety rule)
-        Current dimensioning is on 1/3 of the nominal current of a 50 Hz

ACKNOWLEDGMENT

This paper was supported by the project No. GAČR 102/09/1164.

REFERENCES

[1]   Skala,B., Kůs,V.: Neutral wire current of the synchronous generator of the cogeneration unit. ISEM 2009 – XVII. International Symposium on Electric Machinery, Prague, 2009, ISBN: 978-80-01-04417-9
[2]   Dexters,A., Loix,T., Driesen,J., Belmans,R.: A comparison of grounding techniques for distributed generators implemented in four-wire distribution grids, UPS systems and mikrogrids. CIRED, 19th International Conference on Electricity Distribution, Vienna, -24 May 2007, paper 0638
[3]   Barrado,J,A., Grino,R.:Analysis of voltage control for a self-excited induction generator using a three-phase four-wire electronic converter. 11ª. Conferencia Hispano-Lusa de ingeneria electrica, Zaragoza, 2009.

Power Analysis of a Multimodular Wind Power System Including PMG, Active Rectifier and VSI

Sergey Kharitonov, Sergey Brovanov, Gennady Zinoviev
Novosibirsk State Technical University
kharit1@yandex.ru

Abstract - **This paper is concerned with the power characteristics of the wind power system based on modular principle. A mathematical model for analysis of energy characteristics of the electric power generation system consisting of a synchronous generator with excitation from permanent magnets, the active rectifier and the voltage inverter with PWM is considered. Various algorithms to control the active rectifier and inverter for variable speed wind turbine shaft are analyzed. Analytical relations for the calculation of currents, voltages and power generation in the system are obtained. Recommendations on the choice of control algorithms and structural circuits of the generation electrical energy at a variable speed of shaft of wind turbine (WT) are given.**

Index Terms -- **permanent magnet synchronous, active rectifier, voltage-source inverter, power characteristics.**

I. INTRODUCTION

A promising energy conversion system for the Wind Power Installations (WPI) with variable wind turbine speed is the system "Permanent Magnet Synchronous generator – Active Rectifier – Voltage Source Inverter" [1], Fig.1. Further the system is called Wind Power Generation System (WPGS).

WPGS-type systems can realize all the required options: generation mode with non-linear load, asymmetrical and non-stationary load, electric starter launch mode, in-phase and parallel work with electrical network and other WPI.

This paper evaluates general power characteristics of the system in S_G, S_R, S_I, S_T sections in condition of working for high power network. The main feature of this research is taking into account factors appearing due to variable speed of the wind turbine shaft. These factors are variable frequency and magnitude of the synchronous generator (G) output voltage and dependency of the generated power on wind turbine rotation frequency. Besides that, presence of the active rectifier (R) changes functional and energetic abilities of the WPGS.

II. MAIN POWER CHARACTERISTICS OF THE SYSTEM "SYNCHRONOUS GENERATOR – ACTIVE RECTIFIER"

Power characteristics of electromagnetic processes in WPGS define the efficiency of the mechanical energy conversion of shaft rotating with variable speed into electrical energy by synchronous generator and voltage source inverters. Also these characteristics evaluate degree of the WPGS influence on electrical network through its energy quality performance.

The main power characteristics are efficiency, power factor and THD. These characteristics are indirect because they require calculation of the RMS values for currents and voltages.

For the further analysis the following quantities have been chosen, as basic:

$$U_{dc} = \sqrt{3} U_{N'}; E_b = U_{dc}/\sqrt{3} = U_{N'}; \omega_b = E_b/\Psi_o;$$
$$X_b = \omega_b \left(L_f + L_d \right); I_b = I_{sc} = E_b/X_b; S_b = 3E_b I_b/2, \quad (1)$$

where $U_{N'}$ – is the magnitude of the electric network voltage reduced to the primary transformer winding, T, $\Psi_o = const$ – is the magnetic linkage generated by permanent magnets.

Let us denote $q = L_f/L_d$ and $k_L = L_q/L_d$ then taking into account (1) it can be written:

$$X_{fR}^* = \omega^* q/(1+q), X_d^* = \omega^*/(1+q),$$
$$X_q^* = \omega^* k_L/(1+q), X_{fR}^* = \omega^* q/(1+q),$$
$$X_d^* = \omega^*/(1+q), X_q^* = \omega^* k_L/(1+q),$$
$$X_{d\Sigma}^* = X_d^* + X_{fR}^* = \omega^*$$
$$X_{q\Sigma}^* = X_q^* + X_{fR}^* = \omega^* (k_L + q)/(1+q).$$

Let us assume, that the blank cycle EMF of the generator, $e_{Gm}^\bullet$, and active rectifier control voltage are conformed to the following equations:

$$\begin{cases} e_{Gm}^* = \omega^* \cos\left[\vartheta - (m-1) 2\pi/3 \right]; \\ u_{Rcm}^* = u_c/u_{sc} \sin(\theta_m) = M \sin(\theta_m); \end{cases}$$

where

$$\theta_m = \vartheta - (m-1) 2\pi/3 + \pi/2 - \phi_{Rc};$$
$$\vartheta = \omega t; m = 1, 2, 3 \ (a, b, c);$$

and u_c – is the amplitude of control input wave, u_{sc} – is the amplitude of sawtooth carrier wave, M – is the modulation index.

Neglecting by the active losses, equations of the "G-R" system in the rotating dq frame with angular location $\gamma(\vartheta) = \vartheta - \pi/2$ can be written as:

$$u_{Rd}^\bullet = -\omega^\bullet \frac{di_{Gd}^\bullet}{d\vartheta} + \omega^\bullet \frac{k_L + q}{1+q} \cdot i_{Gq}^\bullet;$$
$$u_{Rq}^\bullet = -\omega^\bullet \frac{k_L + q}{1+q} \cdot \frac{di_{Gq}^\bullet}{d\vartheta} - \omega^\bullet i_{Gd}^\bullet + \omega^\bullet. \quad (2)$$

Active (P), reactive (Q) and the total power (S) for the fundamental harmonic can be found through the following equations (here and further all fundamental components have index *0*):

978-1-4244-8806-3/11 $26.00 © 2011 IEEE

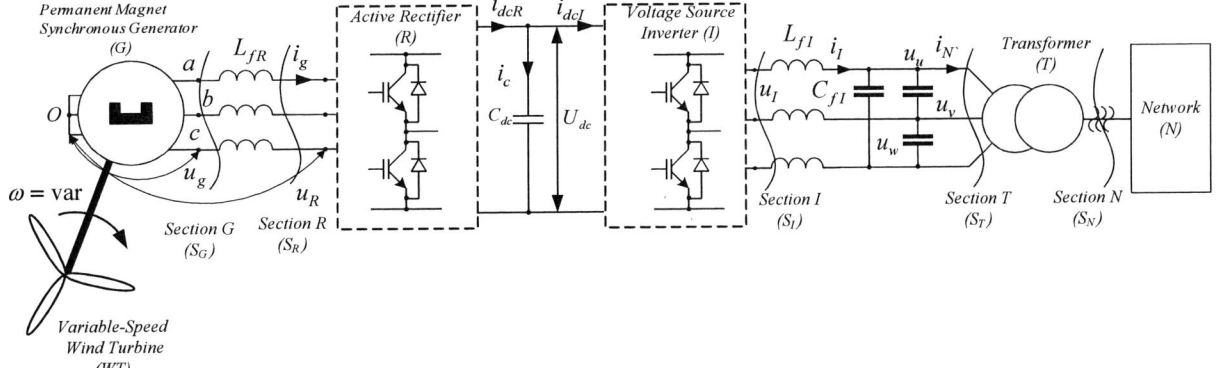

Fig. 1. Structure of the Wind Power Generation System.

$$\begin{cases} P^*_{SGo} = P^*_{SRo} = P^*_{SNo} = \omega^* i^*_{gqo}, \\ Q^*_{SGo} = u^*_{Gdo} i^*_{Gqo} - u^*_{Gqo} i^*_{Gdo}; \quad Q^*_{SRo} = u^*_{Rdo} i^*_{Gqo} - u^*_{Rqo} i^*_{Gdo}; \\ S^*_{SGo} = \sqrt{(P^*_{SNo})^2 + (Q^*_{SGo})^2}; \quad S^*_{SRo} = \sqrt{(P^*_{SNo})^2 + (Q^*_{SRo})^2}; \end{cases}$$

It has been taken into account that active power in each section is equal to the generated power for electric network.

Active power value is not changed with considering higher harmonics. Calculation of reactive and total power with considering higher harmonics is shown in [2]. Orthogonal components d and q are $M_d = M \sin\phi_{Rc}$ and $M_q = M \cos\phi_{Rc}$ respectively.

Taking into account high frequency ripples, equations for the active rectifier will be written as in (3) at the bottom of the page [2].

$$u^*_{Rd} = u^*_{Rdo} + \Delta u^*_{Rd}; \; u^*_{Rdo} = \sqrt{3} M \sin(\phi_{Rc})/2 = \sqrt{3} M_d /2; \\ u^*_{Rq} = u^*_{Rqo} + \Delta u^*_{Rq}; \; u^*_{Rqo} = \sqrt{3} M \cos(\phi_{Rc})/2 = \sqrt{3} M_q /2; \tag{3}$$

In the equation (3) u^*_{Rdo}, u^*_{Rqo} – are the fundamental harmonic orthogonal components in the active rectifier, Δu^*_{Rd}, Δu^*_{Rq} – are the higher harmonic orthogonal components in the active rectifier, $J_p(x)$ – is the first order Bessel function, $a_R = \omega_{cR}/\omega$ – is the angular frequency of PWM in the active rectifier.

Analytical equations for generator currents were defined in [2]. Generator current and voltage can be defined by the similar way.

$$i^*_{Gd} = i^*_{Gdo} + \Delta i^*_{Gd}; \quad i^*_{Gq} = i^*_{Gqo} + \Delta i^*_{Gq}$$
$$u^*_{Gd} = u^*_{Gdo} + \Delta u^*_{Gd}; \quad u^*_{Gq} = u^*_{Gqo} + \Delta u^*_{Gq}$$

For steady state condition it can be written:

$$i^*_{Gqo} = \frac{u^*_{Rdo}}{X^*_{q\Sigma}} = \frac{1+q}{\omega^*(k_L+q)} \cdot u^*_{Rdo} = \frac{\sqrt{3}}{2} \frac{1+q}{\omega^*(k_L+q)} M \sin(\phi_{Rc});$$

$$i^*_{Gdo} = 1 - \frac{u^*_{Rqo}}{\omega^*} = 1 - \frac{\sqrt{3}}{2} \frac{1}{\omega^*} M \cos(\phi_{Rc});$$

$$u^*_{Gdo} = u^*_{Rdo} - X^*_{Rf} i^*_{Gdo} = \frac{\sqrt{3}}{2} \frac{k_L}{k_L+q} M \sin\phi_{Rc};$$

$$u^*_{Gqo} = u^*_{Rqo} + X^*_{Rf} i^*_{Gqo} = \frac{1}{1+q} \left[\omega^* q + \frac{\sqrt{3}}{2} \frac{k_L}{k_L+q} M \cos\phi_{Rc} \right].$$

RMS values for voltage and current of the active rectifier will be defined using results of [2] and well known relations:

$$U^*_{R,rms} = \sqrt{M/2}; \quad \nu_{uR} - \Gamma^\top_{R(1),rms}/U^*_{R,rms} = \sqrt{3M}/2;$$
$$I^*_{G,rms} = \sqrt{(i^*_{Gdo})^2 + (i^*_{Gqo})^2 + (\Delta I^*_G)^2}/\sqrt{2},$$

where

$$\Delta I^*_G \approx \left(\frac{\sqrt{6}}{\pi \cdot \omega^*} \right) \cdot \left(J_1(\pi \cdot M)^2 \frac{a_R^2+1}{(a_R+1)^2(a_R-1)^2} \right)^{\frac{1}{2}};$$
$$\nu_{iG} = \sqrt{(i^*_{Gdo})^2 + (i^*_{Gqo})^2}/I^*_{G,rms}.$$

RMS value of the generator voltage will be defined assuming $k_L \to 1$:

$$U^*_{G,rms} = \frac{1}{\sqrt{2}} \sqrt{(\omega^* \cos\theta)^2 + \frac{1}{(1+q)^2} \cdot \left[M - \left(\frac{\sqrt{3}}{2} M \cos\phi_{Rc} \right)^2 \right]},$$

where θ – is the angle between the fundamental harmonics of the generator EMF and the generator voltage, Fig. 2, defining as $\theta = arctg\, u^*_{Gdo}/u^*_{Gqo}$.

The character of change of the generated power, voltages and currents will be considered further using previously derived equations. For WPI with variable speed of the wind turbine generated active power (P^*_{WTo}) in working speed range ($\omega^* \in \{\omega^*_{WT\min}, \omega^*_{WT\max}\}$) is defined by wind velocity and known turbine parameters.

Thus, generated active power can be found from equation (4)

$$P^*_{WTo} = \gamma \cdot (\omega^*)^3 \tag{4}$$

Where $\gamma = const$ – is a construction parameter. Without taking into account active losses $P^*_{Ro} \approx P^*_{WTo}$.

In the sections S_R and S_G q-components of the voltage with specified active power are defined by the following equations:

978-1-4244-8806-3/11 $26.00 © 2011 IEEE

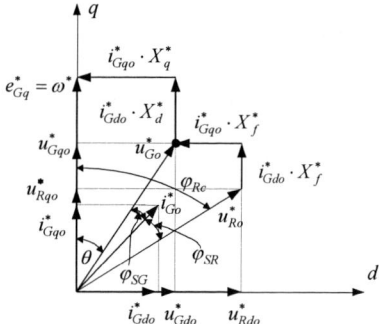

Fig. 2. Vector diagram of the Generator-Rectifier system in dq-frame.

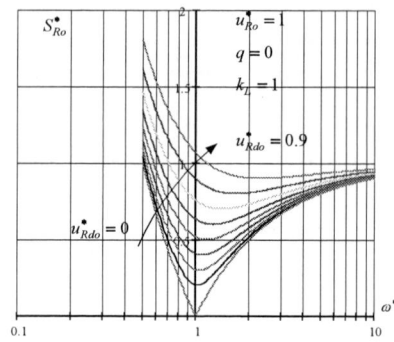

Fig. 3. Dependence S_R^* on ω^*

$$u_{Rqo}^* = \omega^* \cdot \left(1 - \frac{P_{Ro}^*}{u_{Rdo}^*}\right)\frac{k_L + q}{k_L - 1};$$

$$u_{Gqo}^* = \omega^* \cdot \left(1 - \frac{P_{Ro}^*}{u_{Gdo}^*(1+q)}\right)\frac{k_L}{k_L - 1}.$$

Generated active power of the system can be found through the d-components of this voltage:

$$P_{Ro}^* = e_{Gq}^* i_{Gqo}^* = \omega^* i_{Gqo}^* = \omega^* \cdot u_{Rdo}^* / X_{q\Sigma}^* =$$
$$= \omega^* \cdot u_{Gdo}^* / X_q^* = u_{Rdo}^* \cdot (1+q)/(k_L + q) =$$
$$= u_{Gdo}^* \cdot (1+q)/k_L ,$$

For $k_L \to 1$, $P_{Ro}^* \approx u_{Rdo}^* = (1+q)u_{Gdo}^*$. The total power and the displacement power factor in section S_R: $S_{Ro}^* = \left[(Q_{Go}^*)^2 + (P_{Go}^*)^2\right]^{\frac{1}{2}}$, $\cos\phi_R \qquad P_o^*/S_{Ro}^*$, where the reactive power Q_{Ro}^* is defined by:

$$Q_{Ro}^* = \frac{1+q}{\omega^*(k_L + q)}(u_{Rdo}^*)^2 + \frac{1}{\omega^*}(u_{Rqo}^*)^2 - u_{Rqo}^*.$$

As can be seen from the Fig. 3, there are minimums of the total power for certain values of the angular frequency, when the reactive power is zero. Let us denote the frequency when the minimum total power as ω_0^*. This frequency is defined by:

$$\omega_0^* = u_{Rqo}^* + \frac{(u_{Rdo}^*)^2}{u_{Rqo}^*}\frac{(1+q)}{(1+k_L)} = u_{Gqo}^* + \frac{(u_{Gdo}^*)^2}{u_{Gqo}^*}\frac{1}{k_L}$$

From the equation for Q_{Ro}^* an equation linking coordinates of u_{Rdo}^* and u_{Rqo}^* can be obtained:

$$\left(u_{Rdo}^*/\gamma_Q R_Q\right)^2 + \left[\left(u_{Rqo}^* - \omega^*/2\right)/R_Q\right]^2 = 1 \quad (5)$$

Where $\gamma_Q = \sqrt{(k_L + q)/(1+q)}$, $R_Q = \sqrt{\omega^* Q_{Ro}^* + (\omega^*/2)^2}$. Equation (5) is the ellipse equation with the big axis is $2a_d$ and the small axis is $2b_q$, where $a_d = \gamma_Q R_Q$; $b_q = R_Q$. The ellipse center is located in coordinates $(0, \omega^*/2)$. In the polar coordinates equation (5) will be following:

$$\rho(\varphi)^2 \cdot [1 - (\varepsilon\cos\varphi)^2] - 2\rho(\varphi)\rho_0\left[\left(\frac{b_q}{a_d}\right)^2 \cos\varphi\cos\varphi_0 + \sin\varphi\sin\varphi_0\right] +$$
$$+\rho_0^2 \cdot [1 - (\varepsilon\cdot\cos\varphi_0)^2] - b_q^2 = 0,$$

where $\varphi \in (0, 2\pi)$, ρ_0, φ_0 are the ellipse center coordinates and parameter ε is defined by the following equations:

$$\rho_0 = \omega^*/2; \quad \phi_0 = \pi/2; \quad \varepsilon^2 = 1 - (b_q/a_d)^2 = (k_L - 1)/(k_L + q).$$

Thus the locus equation $u_{Ro}^*(\varphi) = \sqrt{(u_{Rdo}^*)^2 + (u_{Rqo}^*)^2}$ in section S_R^* can be written:

$$u_{Ro}^*(\varphi) = \omega^* \sin\varphi + \frac{\sqrt{(\omega^*\sin\varphi)^2 + 4\omega^* Q_{Ro}^*\left(1 - \varepsilon^2\cos\varphi^2\right)}}{[2(1 - \varepsilon^2\cos\varphi^2)]}.$$

Locus $u_{Ro}^*(\varphi)$ for different values of ω^* and is shown on Fig. 5. Here and further circle with the radius $u_{Ro\,max}^* = \sqrt{3}/2$ limits mode with $M = 1$ i.e. out of this circle the modulation index is limited. Therefore all obtained equations are correct within this circle.

Varying values of u_{Rdo}^* and u_{Rqo}^* makes it possible to regulate generated active power and consumed reactive power by the fundamental harmonic. Since $P_{Ro}^* \approx u_{Rdo}^*$, it can be seen from Fig. 5 that the maximum active power, which is defined by the maximum d-component of the locus, also considerably depends on reactive power (Q_{Ro}^*) and ω^*. What is more, with the negative value of Q_{Ro}^* value of $P_{Ro\,max}^*$ decreases:

$$P_{Ro\,max}^* \approx \gamma_Q R_Q \approx \sqrt{\omega^* Q_{Ro}^* + (\omega^*/2)^2}$$

On conditions that $Q_{Ro}^* = tg\phi_{SR} P_{Ro}^*$ and $q = 0$ equation (5) can be rewritten:

$$\left(\frac{P_{Ro}^* - \frac{\omega^*}{2} tg\phi_{SR} \cdot k_L}{\sqrt{k_L} \cdot \frac{\omega^*}{2}\sqrt{1 + k_L (tg\phi_{SR})^2}}\right)^2 + \left(\frac{u_{Rqo}^* - \frac{\omega^*}{2}}{\frac{\omega^*}{2}\sqrt{1 + k_L (tg\phi_{SR})^2}}\right)^2 = 1 \quad (6)$$

$$a_d = \sqrt{k_L}\sqrt{1 + k_L (tg\phi_{SR})^2}\cdot\omega^*/2; b_q = \sqrt{1 + k_L (tg\phi_{SR})^2}\cdot\omega^*/2.$$

978-1-4244-8806-3/11 $26.00 © 2011 IEEE

$k_L = 1.05$, $q = 0$, $Q_{Ro}^\bullet = -0.05$

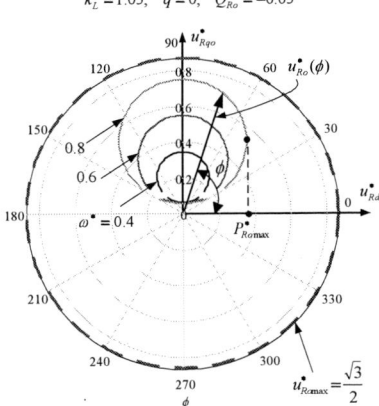

Fig. 4. Locus for $u_{Ro}^\bullet(\varphi)$

From this equation dependence $P_{Ro\,max}^\bullet$ on ϕ_{SR} can be obtained:

$$P_{Ro\,max}^\bullet = \sqrt{k_L} \cdot [\sqrt{k_L}\, tg\phi_{SR} + \sqrt{1 + k_L(tg\phi_{SR})^2}] \cdot \left(\frac{\omega^*}{2}\right)$$

Fig. 5 shows changing of $P_{Ro\,max}^\bullet$ ($P_{Ro\,max(1)}^\bullet$, $P_{Ro\,max(2)}^\bullet$, $P_{Ro\,max(3)}^\bullet$) with different values of ϕ_{SR}.

Thus, from the obtained results it can be concluded that there are operating modes with different values of $\cos(\phi_{SR})$ by varying $\omega^\bullet$. It became possible thanks to independent regulating of the orthogonal components of the reference vector u_{Ro} by the active rectifier. For instance, following modes are possible: fundamental harmonics of current and generator EMF are in-phase, Fig. 6, *a*; fundamental harmonics of current and generator voltage are in-phase, Fig. 6, *b*; fundamental harmonics of current and generator voltage are not in-phase and their phase difference varies.

$k_L = 1.05$, $q = 0$, $\omega^\bullet = 0.8$

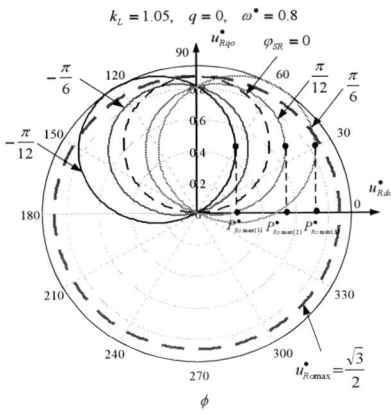

Fig. 5. Locus for $P_{Ro\,max}^\bullet$ ($P_{Ro\,max(1)}^\bullet$, $P_{Ro\,max(2)}^\bullet$, $P_{Ro\,max(3)}^\bullet$)

Let us consider the last mode assuming that in low frequency mode the current advances generator voltage and in high frequency mode generator voltage advances the current. Analysis shows that in this case the same power value can be obtained by two operating modes with different M_q.

Indeed, the active power depends on $u_{qo}^\bullet$ by the following relation:

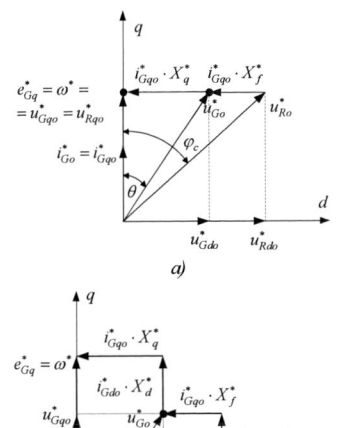

Fig. 6. Different operating modes in the system

$$\left[\left(P_{Ro}^\bullet - P_0\right)\big/\gamma_P R_{uP}\right]^2 + \left[\left(u_{qo}^\bullet - U_0\right)\big/R_{uP}\right]^2 = 1 \quad (7)$$

where

$$\gamma_P = \sqrt{\frac{k_L + q}{1 + q}}; \quad R_{uP} = \frac{\omega^*}{2}\sqrt{1 + \left(tg\phi_{SG}\Big/\sqrt{\frac{k_L + q}{1 + q}}\right)^2};$$

$$U_0 = \frac{\omega^*}{2}; \quad P_0 = \frac{\omega^*}{2} \cdot tg\phi_{SG}\Big/\left(\frac{k_L + q}{1 + q}\right).$$

From (7) we can obtain:

$$u_{qo1,2}^\bullet = U_0 \mp \sqrt{-\left(\frac{P_{Ro}^\bullet - P_0}{\gamma_P}\right)^2}; \quad u_{Ro}^\bullet = \frac{k_L + q}{1 + q} P_{Ro}^\bullet;$$

$$M_{q1,2} = \frac{2}{\sqrt{3}} u_{qo1,2}^\bullet; \quad M_d = \frac{2}{\sqrt{3}}\frac{k_L + q}{1 + q} P_{Ro}^\bullet. \quad (8)$$

It can be concluded from (8) and Fig. 6 that the same active power value can be obtain from two values of the voltage $u_{Rqo}^\bullet$ on the transverse axis, Fig. 7. In the relations (8) and further indexes *1* and *2* denote the operating modes *1* and *2* in accordance with Fig. 6.

Maximum power for certain frequency $\omega^\bullet$ is defined from the following relation:

$$P_{Ro\,max}^\bullet = P_0 + \gamma_P R_{uP} \equiv \omega^*/2 \quad (9)$$

Relations (8) make possible to obtain voltage dependences in the system as angular frequency functions with different values of ϕ_{SG} and parameters q and k_L.

The choice of an operating mode for the system as a WPI part will be considered further. Let us assume $q = 0$, $k_L = 1$ for the certainty. In this case equation (7) will be rewritten as:

$$\rho(\phi) = \omega^* \sec(\varphi_{SG})\sin(\phi + \varphi_{SG})$$

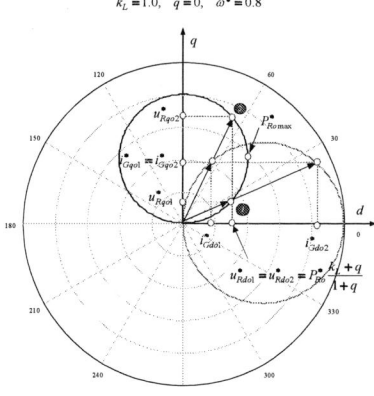

$$k_L = 1.0, \quad q = 0, \quad \omega^* = 0.8$$

Fig. 7. First and second operating modes of the system

Besides, d- and q-components will be defined as:

$$P_{Ro}^*(\varphi) = \omega^* \sec(\phi_{SG}) \sin(\phi + \phi_{SG}) \cos(\varphi);$$
$$u_{qo}^*(\varphi) = \omega^* \sec(\phi_{SG}) \sin(\varphi + \phi_{SG}) \sin(\varphi). \tag{10}$$

Fig. 8 shows proposed dependences of ϕ_{SG} and $\cos \phi_{SG}$ on the shaft angular frequency. This operating script makes possible to work with the second operating mode. Thereto match the system operating point for maximum WT power with maximum possible system power. Above and beyond it, maximum power $P_{Ro\max}^*$ must be conformed to maximum modulation index $M = 1$, Fig. 9.

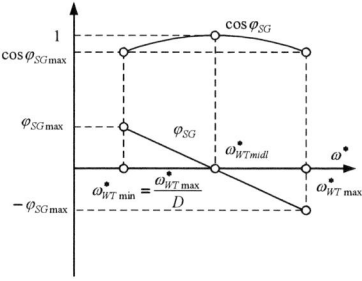

Fig. 8. Dependence of the angle between current and voltage in the generator

Let us denote $P_{Ro\max}^* = P_{Ro\max\lim}^*$. Angle φ_{max} is defined from condition $d\rho(\varphi)/d\varphi = 0$ i.e. $\varphi_{max} = \pi/4 - \phi_{SG}/2$. Rotating frequency for $P_{Ro\max}^* = P_{Ro\max\lim}^*$ can be found from the equation: $\rho(\varphi_{max}) = \sqrt{3}/2$ hence:

$$\omega_{\max}^* = \frac{\sqrt{3}}{2} \cdot 1 \bigg/ \sec(\varphi_{SG}) \sin\left(\frac{\pi}{4} + \frac{\varphi_{SG}}{2}\right)$$

As a result we obtain $P_{Ro\max\lim}^* = \frac{\sqrt{3}}{2} \cos\left(\frac{\pi}{4} - \frac{\phi_{SG}}{2}\right)$. Further, let $\omega_{WT\max}^* = \omega_{\max}^*$; $P_{WTo\max}^* = P_{Ro\max\lim}^*$. From equation (4) we can obtain:

$$\gamma = P_{Ro\max\lim}^* \big/ (\omega_{\max}^*)^3, \quad P_{WTo}^*(\omega^*) = P_{Ro\max\lim}^* \cdot (\omega^*)^3 / (\omega_{\max}^*)^3$$

In accordance with Fig. 9 let us assume $\phi_{SG} = -\phi_{SG\max}$ on condition $\omega^* = \omega_{\max}^*$. Also, the variation law of ϕ_{SG} within operating range $\omega^* \in \{\omega_{WT\min}^*, \omega_{WT\max}^*\}$ will be following:

$$\phi_{SG}(\omega^*) = \phi_{SG\max}[1 - 2 \cdot (\omega^* - \omega_{\min}^*)/(\omega_{\max}^* - \omega_{\min}^*)]$$

where $\omega_{\min}^* = \omega_{WT\min}^* = \dfrac{\omega_{WT\max}^*}{D_{WT}} = \dfrac{\omega_{\max}^*}{D_{WT}}$, $D_{WT} = \dfrac{\omega_{WT\max}^*}{\omega_{WT\min}^*}$.

Then

$$\omega_{\max}^* = \frac{\sqrt{3}}{2} \cdot 1 \bigg/ \sec(\varphi_{SG}) \cdot \sin\left(\frac{\pi}{4} - \frac{\varphi_{SG\max}}{2}\right);$$

$$P_{Ro\max\lim}^* = \frac{\sqrt{3}}{2} \cos\left(\frac{\pi}{4} + \frac{\varphi_{SG\max}}{2}\right).$$

Minimum power with $\omega^* = \omega_{WT\min}^*$ will be

$$P_{Ro\min}^* = P_{Ro\max\lim}^* \big/ (D_{WT})^3.$$

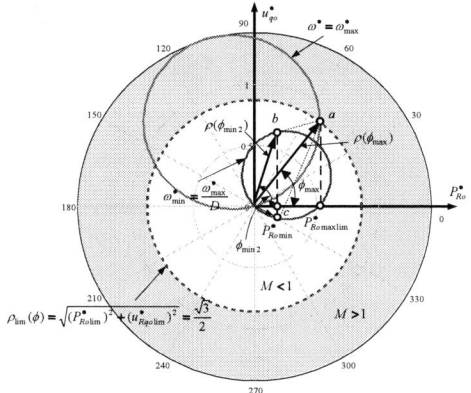

Fig. 9. Operating scripts of the system with varying ω^*

The locus for $\omega^* = \omega_{WT\min}^*$ is following:

$$\rho(\phi) = \omega_{WT\min}^* \sec(\varphi_{SG\max}) \sin(\phi + \varphi_{SG\max})$$

Angle $\varphi = \varphi_{\min}$ for that condition is found from the following equations:

$$\omega_{WT\min}^* \sec(\varphi_{SG\max}) \sin(\phi_{\min} + \varphi_{SG\max}) \cos\phi_{\min} = P_{Ro\max\lim}^* \big/ (D_{WT})^3$$

$$\varphi_{\min 1} = \frac{1}{2} \arcsin\left[2\frac{P_{Ro\max\lim}^*}{D_{WT}^2 \omega_{WT\max}^* \sec(\phi_{SG\max})} - \sin(\phi_{SG\max})\right] - \phi_{SG\max};$$

$$\varphi_{\min 2} = \frac{\pi}{2} - (\varphi_{\min 1} + \phi_{SG\max}).$$

In these equations angles $\varphi_{\min 1}$ and $\varphi_{\min 2}$ are conformed to the operating modes *1* and *2* respectively.

There are two possible trajectories with varying frequency $\omega_{WT\min}^* \leftrightarrow \omega_{WT\max}^*$: $a \leftrightarrow c$ (in the first operating mode) and $a \leftrightarrow b$ (in the second operating mode).

Low power factor and high generator current are typical for the first operating mode. Therefore the second mode is more reasonable. Then:

978-1-4244-8806-3/11 $26.00 © 2011 IEEE   82

$$M_d(\omega^*) = 2P^*_{WTo}(\omega^*)/\sqrt{3};$$

$$M_q(\omega^*) = \omega^*/2 + \sqrt{(\omega^*/2)^2 - [P^*_{WTo}(\omega^*)]^2} + \omega^* tg[\phi_{SG}(\omega^*)]P^*_{WTo}(\omega^*).$$

For this case when $\phi_{SG\,max} = \pi/12$ for $D_{WT} = 2$ on Fig. 11 calculation results of the current, voltage, power factor, active power and $\cos\phi_{SG}$ in dependence on $\omega^* \in \{\omega^*_{WT\,min}, \omega^*_{WT\,max}\}$ are presented.

It should be noticed that chosen linear variation law of $\phi_{SG}(\omega^*)$ is not only possible. In case of the prevalent wind velocity is known for given area, the shaft rotating frequency is calculated for this velocity on condition $\cos\phi_{SG} = 1$. In this case variation law of $\phi_{SG}(\omega^*)$ can be optimized an accordance with variation law of the wind velocity.

What is more, equality $\cos\phi_{SG} = 1$ in boundary values of operating range $\{\omega^*_{WT\,min}, \omega^*_{WT\,max}\}$ is not necessary.

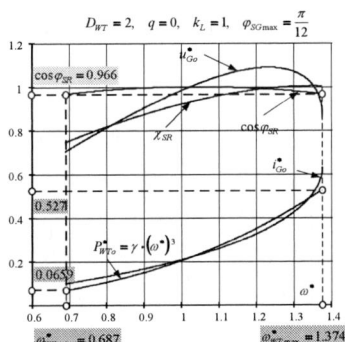

Fig. 10. System power characteristics

As can be seen from Fig. 11, this operating script choice makes possible to realize expanded operating range by increasing ω^*_{max} for specified value of $\cos\phi_{SG}$. Dependence of $\phi_{SG\,max}$ on specified angle value $\phi_{SG\,max}$ is shown by Fig. 11. Figure also shows that maximum possible frequency is ω^*_{max} for chosen operating script.

Thus, the operating script of the WPGS with varying $\cos\phi_{SG}$ by specified variation law in condition of varying ω^*_{WT} provides to increase maximum operating frequency and to save the second operating mode which has relatively high power factor.

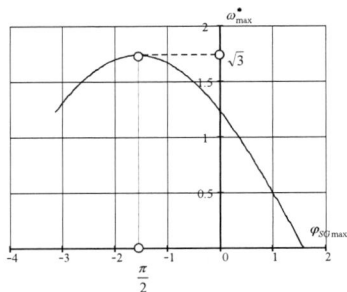

Fig. 11. Dependence of maximum operating frequency in the WPI on maximum angle between current and voltage in the generator.

III. Main Power Characteristics of the System "VSI – Electric Network"

Principle circuit of the system "Voltage Source Inverter – electric network" is given on Fig. 12.

Let us assume that variation law of the voltage in the electric network is following: $u_{N'm} = U_{N'} \cos[\upsilon - (m-1)2\pi/3]$, where $\upsilon = \Omega t$, $m = 1,2,3$ (u, v, w).

Besides, inverter control signals are defined by the equation: $u_{Icm} = u_c \cos(\theta_m)$, where $\theta_m = \upsilon - (m-1)2\pi/3 + \phi_{Ic}$. The following quantities have been chosen as basic frequency and resistance: $\omega_\delta = \Omega$, $X_\delta = \omega_\delta L_f$. Let us introduce the term of the frequency ratio: $a_I = \omega_{cI}/\Omega$, where ω_{cI} – is the angular PWM frequency and Ω – angular frequency of electric network. One more assumption is $U_{dc} = \sqrt{3} \cdot U_{N'} \cdot \delta_{Udc}$, where $\delta_{Udc} \geq 1$ is exceeding of the minimum possible dc-link voltage.

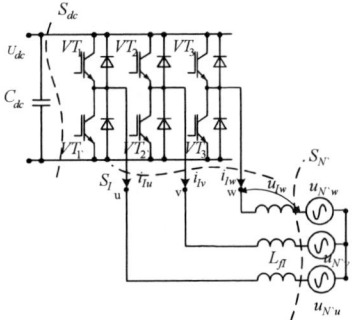

Fig. 12. Principle circuit of the system "Voltage Source Inverter – electric network"

Taking into account all above mentioned assumptions and denoted quantities a mathematical model of the circuit in the dq-frame can be presented as:

$$u^*_{Id} = di^*_{Id}/d\upsilon - i^*_{Iq}, \quad u^*_{Iq} - 1 = di^*_{Iq}/d\upsilon + i^*_{Id}, \tag{11}$$

where $\upsilon = \Omega t$. In (11) Electric network voltage is oriented by q axis.

Voltages u^*_{Id} and u^*_{Iq} can be defined by the following equations:

$$\begin{aligned}
u^*_{Id} &= u^*_{Ido} + \Delta u^*_{Id}; \\
u^*_{RIq} &= u^*_{Iqo} + \Delta u^*_{Iq}; \\
u^*_{Ido} &= \delta_{Udc} M \sqrt{3}/2 \cdot \sin(-\phi_{Rc}) = -\delta_{Udc} M_d \sqrt{3}/2; \\
u^*_{Iqo} &= \delta_{Udc} M \sqrt{3}/2 \cdot \cos(-\phi_{Rc}) = \delta_{Udc} M_q \sqrt{3}/2;
\end{aligned} \tag{12}$$

Here u^*_{Ido}, u^*_{Iqo} – are d and q components of the fundamental harmonic of the inverter output voltage; Δu^*_{Id}, Δu^*_{Iq} – are d and q components of the higher harmonic of the inverter output voltage.

Higher harmonics for SPWM are defined by the equation (3) with taking into account coefficient δ_{Udc}.

In steady state condition analytical equations for inverter currents can be defined through the equations (3), (11) and (12). In this

978-1-4244-8806-3/11 $26.00 © 2011 IEEE

case d and q components of the fundamental harmonic are defined as $i_{Iqo}^* = -u_{Ido}^*$; $i_{Id}^* = u_{Iqo}^* - 1$.

Let us consider the control algorithm of the inverter when WPI generates into the electric network active power only. With this algorithm $u_{Iqo}^* = 1$; $i_{Io}^* = i_{Iqo}^* = -u_{Ido}^*$; $i_{Id}^* = 0$ and generated active power is defined by the following equation:

$$P_{No}^* = i_{Iqo}^* = i_{Io}^* = -u_{Ido}^* \qquad (13)$$

D and q components of the inverter control signals and angle ϕ_{Ic} are defined as: $M_q = 2/(\sqrt{3} \cdot \delta_{Udc})$, $M_d = 2 \cdot P_{No}^* /(\sqrt{3} \cdot \delta_{Udc})$, $\phi_{Ic} = arctg\, M_d / M_q = arctg P_{No}^*$.

Linear working range of the inverter is limited by the following condition:

$$\begin{cases} \sqrt{(M_d)^2 + (M_q)^2} \le \{1 - SPWM, 2/\sqrt{3} - SVPWM; \\ \dfrac{2}{\sqrt{3}} \cdot \dfrac{1}{\delta_{Udc}} \sqrt{1 + (P_{No}^*)^2} \le \{1 - SPWM, 2/\sqrt{3} - SVPWM, \end{cases} \qquad (14)$$

From this equation (in case of equality) we can obtain equation for the maximum active power which can be transferred into the network without current distortion:

$$P_{No\,max}^* = \begin{cases} \sqrt{\left(\sqrt{3}/2 \cdot \delta_{Udc}\right)^2 - 1}, & -SPWM \\ \sqrt{\left(\delta_{Udc}\right)^2 - 1}, & -SVPWM. \end{cases} \qquad (15)$$

From (15) follows that the minimum value of $\delta_{Udc\,min}$ for generating active power is defined by the equation:

$$\delta_{Udc\,min} = \begin{cases} 2/\sqrt{3} - SPWM; \\ 1 - SVPWM. \end{cases}$$

Dependence of the active power P_{No}^* on $\delta_{Udc} = U_{dc}/\sqrt{3}U_N$. and on modulation index can be found through this equation: $P_{No}^* = \sqrt{\left(\sqrt{3}\delta_{Udc} M/2\right)^2 - 1}$. With taking into account active losses at the output of the WPI the fundamental harmonic of the current will be following:

$$i_{Io}^* = P_{No}^* = \frac{1}{1 + (\omega_R^*)^2} \cdot \left\{ \sqrt{[1 + (\omega_R^*)^2]\left(\frac{2}{\sqrt{3}\delta_{Udc} M}\right)^2 - 1} - \omega_R^* \right\};$$

where $\omega_R^* = R/X_\delta$, R — equivalent active inverter phase resistance.

The plot of $P_{No}^* (M)$ for SV PWM is shown on Fig. 13. It can be seen from Fig. 13 that regulating range decreases if δ_{Udc} decreases. It should be noticed that using of SV PWM provides significantly increasing generated active power. Fig. 14 also shows that for each value of δ_{Udc} there is a minimum modulation index M_{min}. If the modulation index less than $M_{min} = 2/(\sqrt{3}\delta_{Udc})$ the active power is zero.

Dependence of inverter current THD on the WT shaft frequency will be defined further.

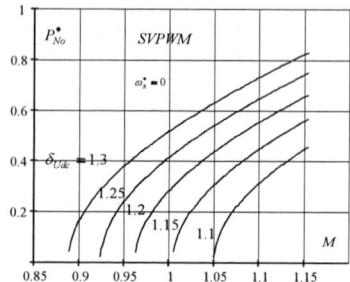

Fig. 13. Generated power dependence on the modulation index.

In accordance with (13) and results obtained in [2], RMS values of the fundamental harmonic of current $i_{Io,rms}^*$, ripple component $\Delta i_{I,rms}^*$ and total RMS value of the inverter current are defined by the following equations:

$$i_{Io,rms}^* = \frac{1}{\sqrt{2}} P_{No}^*,$$

$$\Delta i_{I,rms}^* = \left(\frac{\sqrt{3}\delta_{Udc}}{\pi \cdot X_\Sigma^*}\right) \cdot \left(J_1\left(\pi \cdot M\right)^2 \frac{a_I^2 + 1}{(a_I + 1)^2 (a_I - 1)^2}\right)^{\frac{1}{2}},$$

$$i_{I,rms}^* = \sqrt{(i_{Io,rms}^*)^2 + (\Delta i_{I,rms}^*)^2}.$$

Modulation index depends on the WT shaft frequency as follows (in accordance with the condition (15)):

$$M = 2 \cdot \sqrt{\left[P_{No\,max}^* \cdot \left(\omega/\omega_{WT\,max}\right)^3 \right]^2 + 1} \Big/ \sqrt{3}\delta_{Udc}.$$

Knowing of this dependence can be used to find power characteristics of the generated power as functions of WT shaft frequency. As an example Fig. 14 shows plots of $THD_{iI}\left(\omega/\omega_{WT\,max}\right)$ for SV PWM.

Since in high power WPI frequency ratio a_I is limited by the dynamic losses in switches, it can be concluded that it is impossible to meet the power quality demands by increasing PWM frequency without LC filter. Increasing of δ_{Udc} causes increasing of the dc-link voltage and, as a consequence, rated power of the inverter.

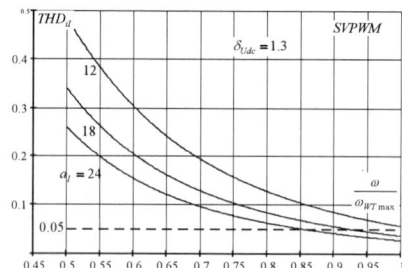

Fig. 14. THD of the generator current in dependence on WT angular frequency.

Changing of the inverter topology or decreasing of the regulating range of power with SVPWM can solve this problem. Therefore design of WPI with rated power 1MW or more must be oriented to multilevel inverter topologies.

Another solution is parallel connection of m inverters. All inverters are connected to the same dc voltage source and controlled by the same control signal in each phase, but switching times in every inverter are delayed by angle $2\pi/m$ in relative to each other. It can be realized, for example, by introducing m control signals with specified phase delay. This solution increases power of the system above and beyond energy quality performance. It also provides modular construction for such systems. Advantage of modular construction is saving the same efficiency of the system with low and high WT shaft frequency. This feature is provided by different numbers of modules. Number of the modules is a function of the shaft frequency.

Increasing of channels number causes excluding groups of mixed harmonics from the current spectrum with the following harmonic orders: $\nu = n \cdot \omega_{kl} \pm p \cdot \Omega; \quad n < m$. Calculation shows that decreasing of the current THD can be evaluated by this equation: $THD_{iI(m)} \approx THD_{iI(1)}/m^2$, where $THD_{iI(1)}$ – is the total harmonic distortion with $m=1$.

Besides that, number of mixed harmonics groups is also decreases in the amplitude-frequency spectrum of the current in dc-link i_{dc}. Power factor $\chi_{Sdc(m)}$ in section S_{dc}, Fig. 12, in the inverters' inputs can be evaluated by the following equation [4]:

$$\chi_{Sdc(m)} = \left[\sqrt{1 + \frac{1 - \left(\chi_{Sdc(1)}\right)^2}{\left(\chi_{Sdc(1)}\right)^2} \cdot \frac{1}{m^2}} \right]^{-1},$$

where $\chi_{Sdc(1)}$, $\chi_{Sdc(m)}$ – are power factors in section S_{dc} for single channel and m channels respectively.

It can be concluded that parallel inverter connection causes increasing of the generated power by m times, decreasing THD about m^2 times. What is more, for the same frequency ratio a_f, reactive power density of the dc-link capacitor C_{dc} decreases m times. On the other hand, for the same THD using of parallel inverter connection provides to decrease frequency ratio m^2 times.

Thus, performed analysis of the system "VSI – Electric network" makes possible to evaluate main power characteristics of the system analytically in different operating conditions and control algorithms. In conditions of low wind velocity THD of the system does not meet the demands of power quality when WPI works for high power networks. This problem is solved by using of multilevel converters, applying SVPWM control, modular inverter topology. Latest solution increases the efficiency of WPGS by disabling of several channels when low shaft frequency. Modular construction principle can be applied in WPGS. Fig. 15 shows an example of modular-based WPGS.

Fig. 16 shows dependence of the power in a generating module of WPGS in different intervals of the shaft frequency variation.

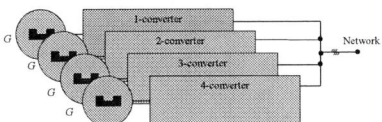

Fig. 15. Modular construction principle in WPGS

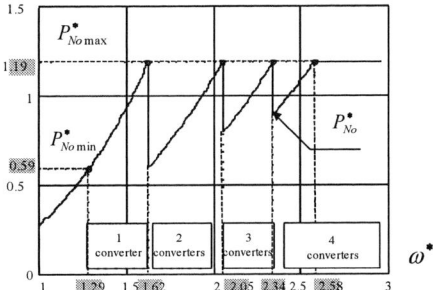

Fig. 16. Dependence of the power in a generating module of WPGS

It can be seen that first one module works, then two modules, after – three etc. Increasing number of the modules decreases rated range of each module and increases system efficiency in condition of low wind velocity. Besides that maximum power of the system increases and its power quality still high. Using this technology a WPI named "Raduga" was designed and built near Elista city. WPI has rated power 1MW and consists of four modules. Each module includes permanent magnet synchronous generator, frequency converter, control system and matching transformer. Operating wind velocity range: 4–25m/s.

Conclusions

A mathematical model for analysis of power characteristics of the system including permanent magnet synchronous generator, active rectifier and voltage source inverter has been considered.

A control algorithm for the active rectifier and inverter with taking into account variable wind turbine shaft has been considered.

Analytical equations for generated power and its different graphic dependences on different parameters and operation modes were presented.

Useful recommendations for choosing control algorithms and structures of WPI considering variable wind turbine were formulated.

Acknowledgment

The authors would like to thank the Ministry of Knowledge and Economy which partly supported this research through the Network-based Automation Research Center (NARC) at the University of Ulsan.

References

[1] Z. Chen, J. M. Guerrero, and F. Blaadjerg "A Review of State of the Art of Power Electronics for Wind Turbines" *IEEE Trans. on Power Electronics*, vol. 24, NO. 8, August 2009, pp. 1659 – 1875.

[2] Kharitonov S.A. "The power system "permanent magnet synchronous generator - active rectifier". (Mathematical model)", *Elektrotehnika. M. 2009, №12, p. 33-41.*

[3] Corn G., Corn T. "Mathematics handbook (for science officers and engineers)". - *M. - the Science, 1974. in Russian*

[4] Kharitonov S.A. "Integrated parameters and characteristics of voltage inverters in structure of generating systems of an alternating current such as "variable speed - constant frequency" for wind-energetic installations", *Scientific bulletin NSTU, Novosibirsk, 1999. pp. 92 – 120, in Russian.*

Operation of Photovoltaic Power Systems with Energy Storage

Bernard J. Szymanski
ELPOL
Electronics and Automation Center
www.elpol.biz
ul. Wrzosowa 10/1, 26-600 Radom
Email: bs@elpol.biz

Lukasz Roslaniec
Institute of Electrical Power Engineering
Warsaw University of Technology
Plac Politechniki 1, 00-661 Warsaw
Email: roslanil@ee.pw.edu.pl

Antoni Dmowski
Institute of Electrical Power Engineering
Warsaw University of Technology
Plac Politechniki 1, 00-661 Warsaw
Email: a.dmowski@ee.pw.edu.pl

Kamil Kompa
Institute of Information Technology
Warsaw University of Technology
Nowowiejska 15/19, 00-665 Warsaw
Email: kkompa@elka.pw.edu.pl

Jerzy Szymanski
Faculty of Transport and Electrical Engineering
Radom University of Technology
Malczewskiego 29, 26-600 Radom
Email: j.szymanski@pr.radom.pl

Abstract—Photovoltaic (PV) power systems are described in the article. Power generated from renewable energy sources is not stable because of the energy fluctuations caused mainly by atmospheric conditions. Therefore, in order to improve the parameters and stability of the power system, PV power plants should cooperate with electrical energy storage system such as electrochemical batteries. Moreover, because of the reason that PV power plants are often located on the terrain which is accessible by the man, galvanic isolation may be obligatory. Since the efficiency of the power conversion in PV power systems has to be maximized, soft switched resonant power converters, in which switching losses are minimized, are utilized in such PV power systems. Moreover, grid-connected PV inverter which is a crucial element of the PV power system is presented in the article.

Index Terms—Solar power generation, photovoltaic systems, resonant inverters, energy storage

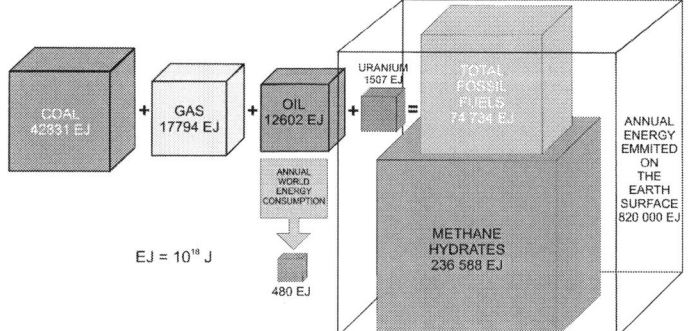

Fig. 1: World primary energy resources [2]

I. INTRODUCTION

Solar energy is a basic energy form among all energy sources and was processed by means of bioorganic processes to high concentration form of fossil fuels which are nowadays used as the main energy source. Figure 1 presents the comparison between annual renewable solar energy emitted on the surface of earth and total available primary energy resources [1] [2]. Figure 1 shows that annual energy emitted on the earth's surface is much higher than total energy which can be achieved from all conventional energy sources.

Currently we can convert the solar energy by means of renewable energy technologies such as solar thermal power plants, photovoltaic (PV) sun concentrators and photovoltaic power systems [3].

The energy conversion in a solar thermal plant starts with collecting the sunlight, converting it into heat which is then powering a thermodynamic engine [3]. In the last stage the engine drives a generator which produces the electricity. The heat conversion path is similar to any other conventional fossil

in Europe is a Seville's solar power tower located in Spain. However, the biggest solar thermal power plant is planned to be founded on the Sahara desert in Africa. The project has the name DESERTEC [4] and was officially started in July 2009 by consortium of European companies. Produced electricity is going be transmitted to European and African countries by means of high voltage DC lines.

Sun concentrators which use lenses or mirrors to concentrate the sunlight onto PV cells are considered as well. This, in turn, allows to reduce the cell area which is required for producing a given amount of power. However, it has turned out, that this is very difficult [5] in practice.

Power plants which use the sun light in order to produce electricity directly, by means of photovoltaic (PV) cells, are the next group. Here, sunlight is converted into electrical current in the process of excitation of electrons in the semiconductor junction. Standard PV cells have about 10% efficiency, which is decreasing over the time. Modern technologies such as the multi–junction PV cell or the organic PV cell are used to increase power conversion efficiency. In the PV systems, grid by means of power electronics converter. Nowadays, the rapid development of PV installations can be observed. At present, this manner of electrical energy production is used especially in residential areas as the building integrated PV power systems. However, large scale (>200 kWp) grid-connected PV power plants exist as well. The PV power plant in Braindis (Germany) with 40 MWp or PV power plant in Puertollano (Spain) with 47 MWp can serve here as an example [6].

The growth of new PV installations in Europe is mainly caused by introduction of feed-in tariff [7]. The introduction of feed-in tariff was the main cause of installation of additional 2500 MW and 1500 MW photovoltaic power in Spain and Germany respectively in 2008. Simultaneously in Czech Republic which also introduced the feed-in tariff, the increase of 50 MW of new installed photovoltaic power was observed in 2008. It can be expected that the introduction of the feed-in tariff in other countries will lead to the same results.

In 2008, approximately 3.8 GWp was gained from large scale (>200 kWp) photovoltaic power plants [6], whereas in 2006 it was approximately 500 MWp. Since the rated power of a single photovoltaic power converter has started to exceed 1 MVA, this technology should be taken into consideration as an alternative to fossil fuels.

II. PHOTOVOLTAIC POWER SYSTEMS

The photovoltaic power systems can be divided into two main groups:

- Stand-alone PV systems
- Grid-connected PV systems

Stand-alone PV systems are in the power range up to several kW. These systems have no connection to the electrical grid and are used to supply local loads. Such a solution is mainly used when the cost of connecting particular localization to the grid is larger than the cost of the PV power system.

On the other hand, grid-connected PV systems are connected to the electrical grid by means of suitable power electronics inverter which converts the DC power produced by the PV cells into alternating current (AC), which is synchronized with the utility grid. This allows to sell the produced energy to other users connected to the grid.

Since the PV array is the most expensive element of the whole PV system, the power extracted from PV array should be maximized. Therefore, the power converter which serves as an interface between a PV array and an utility grid has to track maximal power point (MPP) [8] of PV array.

III. ENERGY FLUCTUATIONS

The energy which comes from the PV power plants is not stable because it depends on the weather conditions and the time of the day. The generated current fluctuations in case of PV and wind power plants are depicted in Figure 2. Therefore, high power PV plants connected to the utility grid can have influence on the parameters and stability of the power system. It can be especially critical in case of high power PV plants

Instability of PV power plants can cause the flicker effect [9]. This disturbance is dangerous to the electrical motors supplied from the utility grid [10] [11]. Described problem can be minimized by means of proper control of maximum power output of the PV power plant. Unfortunately, this solution leads to decrease of efficiency of the PV power plant.

If the PV power plant is connected to the node with stiff voltage and frequency parameters, than the influence of the PV power plant on the power system can the minimized by means of proper amount of ready reserve. Nevertheless, this solution causes the decrease of the efficiency of power system because of the utilization of high value of ready reserve and transport losses.

Usually maintaining system stability (in case of large PV power plants) requires the continuous contact of PV installation's control system with the local electrical energy distributor.

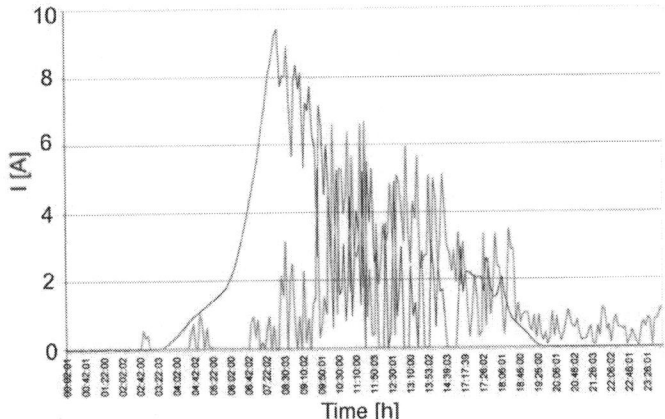

Fig. 2: Current generated from photovoltaic (blue) and wind (red) power plants [12]

IV. PHOTOVOLTAIC POWER SYSTEM WITH ENERGY STORAGE

The disadvantages which were described above can be limited to a large extent or even eliminated if the PV power plants installations are connected with energy storage system. There are many technical solutions which can be utilized as the storage elements (e.g. flywheels, supermagnetic coils, capactiors or electrochemical storage elements [13] [14]). In case of PV power systems, the best solution is an utilization of electrochemical storage element. Several types of batteries are currently available, i.e. lead-acid, nickel-cadium, zinc-bromide, zinc-chloride, sodium-sulphur , nickel-hydrogen, re-dox and vanadium batteries. Nowadays, the development of cost effective electrical energy storage element is one of the main challenges.

The PV power system with battery storage element should fulfill following operation modes:

- Provide the electrical energy to the utility grid from the

978-1-4244-8806-3/11 $26.00 © 2011 IEEE

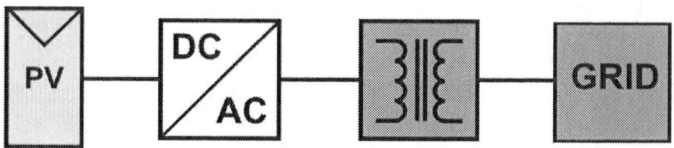

Fig. 3: Galvanic isolation realized by means of 50 Hz transformer

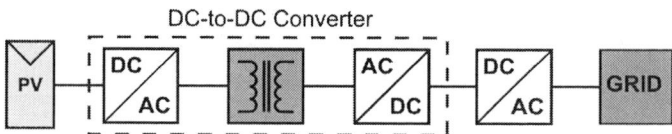

Fig. 4: Galvanic isolation realized by means of high-frequency transformer

- Provide the electrical energy to the utility grid from the PV generators and battery storage system.
- Provide the electrical energy to the utility grid only from the battery storage system.
- Charge the battery storage system from the utility grid with the excess of the electrical energy produced by other sources.
- Provide a battery backup operation in case when there is fault in the utility grid and important receivers need to have power supply.

V. GALVANIC SEPARATION

For the reason that high power PV power plants are mainly located on the terrain which is accessible for the man, it is advantageous when PV array is isolated from the utility grid. The galvanic isolation from the utility grid can be realized by means of 50 Hz isolating transformer (old solutions). Such a transformer is located on the grid side (Figure 3). It is also possible to isolate PV array by means of the modern high-frequency transformer, which is a part of the PV DC-to-AC converter (Figure 4). The solution with high-frequency transformer is advantageous in respect to the low-frequency transformer because of:

- much smaller dimensions
- higher power density
- higher efficiency

If the PV power plant with energy storage system is fully dispositional, then it helps the grid operator in control of the power system [15] [16]. In Figure 5, an example of such a PV power plant is shown. It can be seen that the system consists of several blocks, i.e: PV array, DC-to-DC converter where maximal power point tracking (MPPT) is implemented, DC-to-AC converter which feeds the energy to the utility grid, AC-to-DC and DC-to-DC converters which are responsible for management of battery storage system. The control unit, which allows to control the power converters according to the grid operator commands, is crucial. In such a system, the galvanic isolation is realized by means of high-frequency transformers. System presented in figure 5 utilizes separate active front

power storage units. This allows to maintain these units separately. In case of low power installation, it may be more cost effective to implement other power flow paths and in result to minimize power losses and number of converters. Power plant topology should be always selected according to particular application.

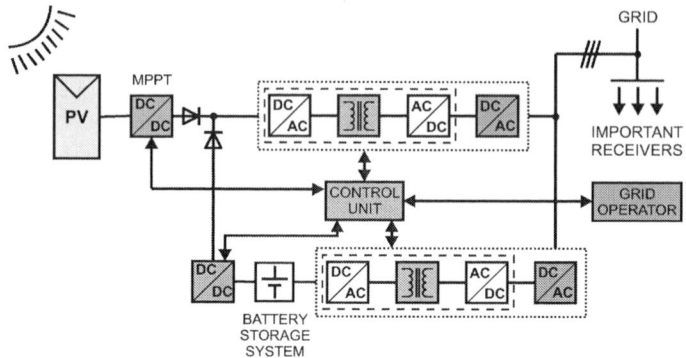

Fig. 5: Fully dispositional PV power plant

VI. RESONANT CONVERTERS IN PHOTOVOLTAIC POWER SYSTEMS

Nowadays, efficiency and density of processed power is a crucial factor in case of power converters. In order to decrease power losses, as well as the volume and weight of the power converter (increase the processed power density), high switching frequency of power transistors is used. High frequency operation results in reduced size and weight of high power magnetic components (e.g. separation transformers). In case of hard switching converter, high switching frequency would result in a very high switching losses. However, utilization of soft switching methods allows to reduce switching losses significantly. Thus, high-frequency soft-switching converter is much more efficient than typical low-frequency hard-switching converter. High power DC-to-DC converters are realized as the multi-phase resonant converters (e.g. three-phase).

The usage example of multiphase series resonant DC-to-DC converter [17] is presented in Figure 6. In this situation, each PV panel has its own integrated converter where Maximal Power Point (MPP) tracking is implemented. Thus, the power produced by the PV system is maximized. These DC-to-DC converters are connected in parallel to the single, three-phase resonant converter which in turn serves as DC-to-DC transformer providing galvanic isolation and proper DC voltage level to the DC-to-AC converter.

An example of the topology of the three-phase DC-to-DC resonant power converter is presented in Figure 7. The power converter has a high frequency isolating transformer and uses series resonannt circuit in order to convert the energy and maintain soft-switching of power transistors.

The presented topology, along with unique control algorithm, allows to charge batteries and maintain soft-switching in the entire operation range. The power converter is controlled by means of frequency and pulse density modulation

978-1-4244-8806-3/11 $26.00 © 2011 IEEE

Fig. 6: Multiphase series resonant converter in stand alone PV power system [17]

Bold red and green curves show the battery charging process (two different possibilities of current and voltage changes on the converter's output over the time). Resonant frequency of the converter's tank circuit is 100kHz. Curves tagged $F = 150\,kHz$ and $F = 200\,kHz$ show converter's output current and voltage dependences at the particular switching frequency. In the region below the 200 kHz curve, it is possible to choose any current level by changing operating frequency (frequency control mode). In this mode there soft-switching is maintained.

In the presented example, highest possible converter's switching frequency is 200 kHz. Thus, to get low value of the current on the converter's output (especially when the voltage is low) it is necessary to utilize pulse density control method. This method assumes transport of energy in pulses and therefore any mean current value may be achieved. Moreover, when the energy is transported in the described manner, value of current in the resonant tank circuit is always high enough to maintain soft-switching (ZVS) while power transistors are operating. Choke on the output of the converter allows to keep constant current in the battery during pulse density control.

The 8 kW prototype of described converter was built and tested. Figure 8 presents experimental results, i.e. the resonant tank voltage u_R (yellow) and current i_R (blue). In this case the resonant current and voltage waveforms have the frequency of:

$$f \approx 200\,kHz. \qquad (1)$$

Inductors in the resonant tank circuit are realized in form of leakage inductance of the transformer and small external chokes. Patent is pending for the described battery charging method.

VII. BIDIRECTIONAL CONVERTERS IN PHOTOVOLTAIC POWER SYSTEMS

Cost of the converter, as well as power losses, may be minimized by integration of power conversion's electronics of the PV power plant and the storage system.

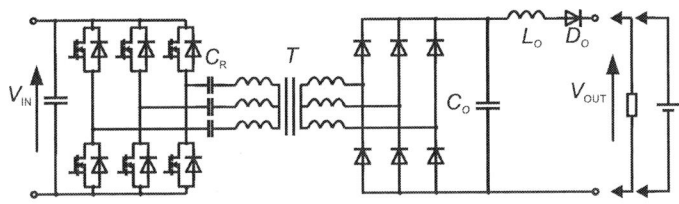

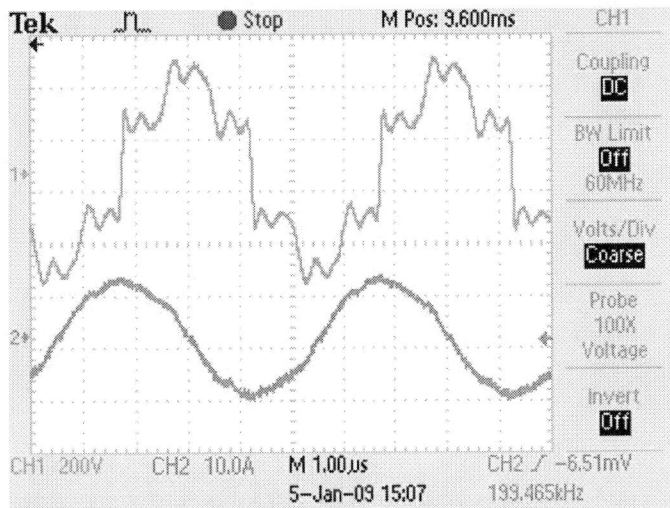

Fig. 8: Experimental results - frequency control of three-phase resonant DC-to-DC converter

Since bidirectional power flow between battery and power system is necessary, as well as low loss battery charging from the PV array, the possible usage of bidirectional resonant power converters in such systems is an issue which needs further investigation.

The example of the bidirectional DC-to-DC series resonant converter is presented in Figure 10 [19]. When bidirectional power converters are utilized, the topology of the PV power system depicted in Figure 11 arises. The presented power plant has two bidirectional converters i.e.: DC-to-DC converter which cooperates with battery storage system and DC-to-AC converter which couples PV power plant with utility grid.

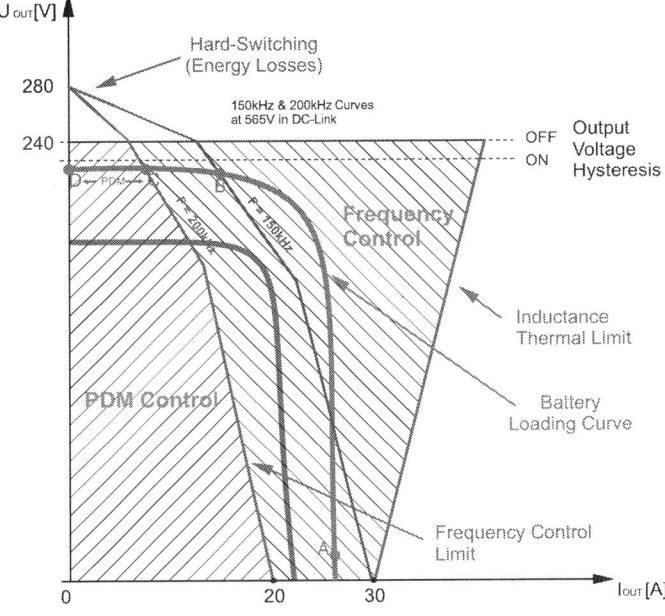

Fig. 9: Output characteristic of the three-phase resonant power

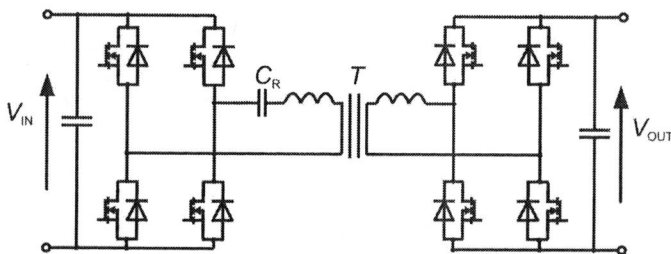

Fig. 10: Bidirectional DC-to-DC series resonant converter

VIII. GRID-CONNECTED PHOTOVOLATIC INVERTERS FOR ENERGY TRANSMISSION

Another crucial part of every grid-connected PV power plant is a grid-connected inverter which is transmitting energy from a DC link to the electrical power system. Typically voltage source inverters (VSI) are used for cooperation with the grid. Such an inverter has to be properly built and it has to use modern control techniques to maintain its current quality under versatile normalized disturbances on the grid side. Control and construction of such converters is a wide subject. Properly designed modern inverters are able to fulfill very restrict requirements.

The grid-connected converter is able to operate as an active compensator during energy transmission to the grid. This allows to maintain voltage level in proper range in the place where inverter is connected. Polish low–voltage network may have problems with rapidly growing capacity factor of PV systems. Production of reactive power by PV power plants can decrease losses related to reactive power transmission in the grid.

Typical construction of a single–phase system is shown in Figure 12 and its simulation model is depicted in Figure 13. It consists of DC-side capacitance, IGBT full bridge converter, grid-side LCL filter, surge arrester, as well as diverse current and voltage sensors and appropriate control system.

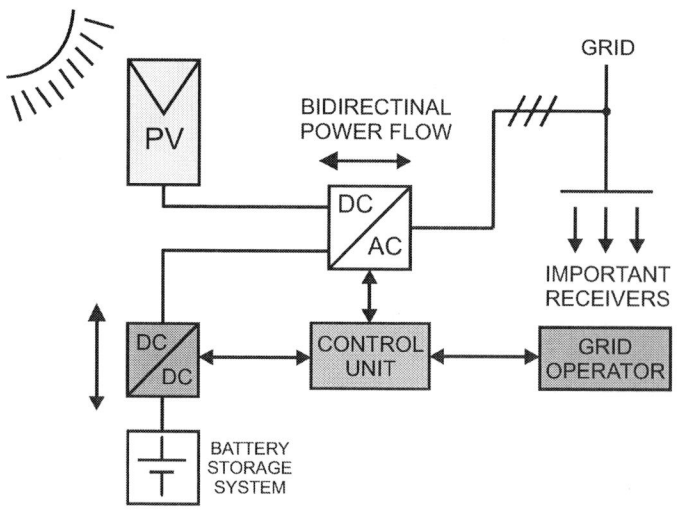

Very significant to the system performance are the parameters of the LCL filter and EMI filters. Electromagnetic emission, current disturbances and power losses in the system are minimized by proper design of those filters and the control algorithm. Control algorithm's behavior should be investigated for wide range of normalized grid disturbances. High power grid simulators are used in such experiments.

Most of modern grid-connected converters used in PV systems are hard-switching converters, which use fully controllable power switches such as MOSFETs and IGBTs and generally use pulse width modulation (PWM) in order to produce the AC output.

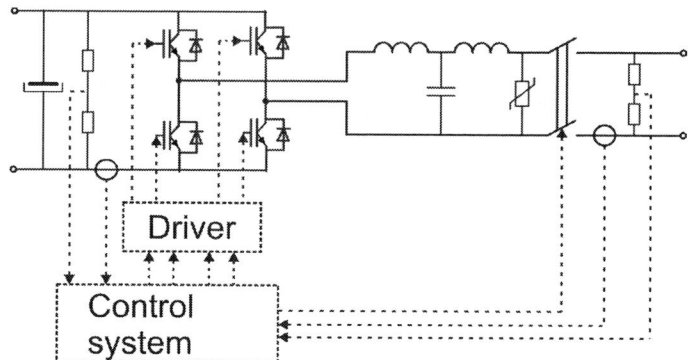

Fig. 12: Single-phase voltage source inverter

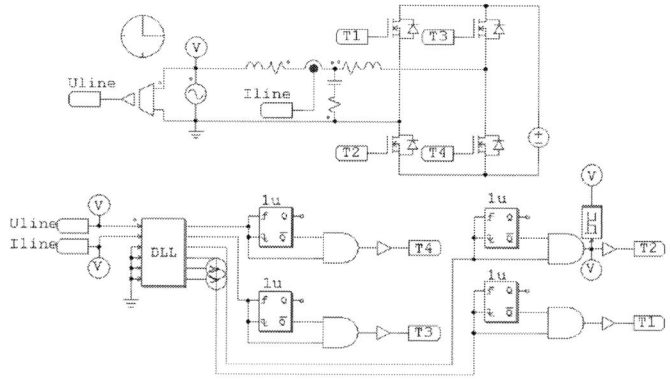

Fig. 13: Simulation model of single-phase voltage source inverter

IX. SUMMARY

In the article issues concerning generation of electrical energy from PV power plants are described. The difference between stand alone PV power system and grid connected power system is explained. Moreover, PV power systems with energy storage are described, along with the function which such a system has to fulfill. The issue of galvanic isolation and resonant power converters utilization in PV power system is introduced. Furthermore, grid-connected power plant concept

X. ACKNOWLEDGMENTS

This work has been supported from the Grant N N510 325537 of the polish Ministry of Science and Higher Education.

REFERENCES

[1] M. Bartosik, *Globalny kryzys energetyczny - mit czy rzeczywistość ?* Proceedings of 10th International Conference Nowoczesne Urzadzenia zasilajace w energetyce, 14-16 March, Zakopane, Poland, 2007, 2007.

[2] *Survey of energy resources.* World energy council, 2004.

[3] K. Heinloth, *Energy Technologies.* Springer Verlag, 2006.

[4] D. Foundation, *Red Paper - An overview of Desertec Concept - 2nd Edition.* Desertec Foundation, 2009.

[5] A. Luque and S. Hegedus, *Handbook of Photovoltaic Science and Engineering.* John Wiley & Sons Ltd, 2003.

[6] D. Lenardic, *Large scale PV power plants - Annual and cumulative installed power output capacity - Key statistical indicators.* Annual Review 2008, 2008.

[7] S. Pietruszko, *Taryfa stala (Feed-in Tariff) motorem rozwoju odnawialnych zródeł energii.* Centre for Photovoltaics of the Warsaw University of Technology, 2009.

[8] F. Blaabjerg, F. Iov, R. Teodorescu, and Z. Chen, *Power electronics in renewable energy systems.* Proceedings of 12th International Power Electronics and Motion Control Conference (EPE), 2006.

[9] R. Albarracin and H. Amaris, *Power Quality in distribution power networks with photovoltaic energy Sources.* Proceedings of International Conference on Environment and Electrical Engineering, 10-13 May, Karpacz, Poland, 2009.

[10] A. Bien and A. Rozkrut, *A measurement scale for the light flickering phenomenon.* 6th International Conference, Electrical Power Quality and Utilisation, 19-21 September, 2001.

[11] A. Bien and Z. Hanzelka, *Power Quality Application Guide. Voltage Disturbances.Flicker Measurement.* Copper Development Association, October, 2005.

[12] P. Biczel, *Optymalne wykorzystanie pierwotnych nośników energii na przykładzie hybrydowej elektrowni słonecznej z ogniwami paliwowymi.* PhD thesis, Warsaw University of Technology, 2003.

[13] M. Rashid, *Energy Technologies.* Elsevier Inc. 2nd Edition, 2007.

[14] D. Sauer, T. Blank, J. Kowal, and D. Magnor, *Energy Storage Technologies for Grids With High Penetration of Renewable Energies and for Grid Connected PV Systems.* 23rd European Photovoltaic Solar Energy Conference,1-5 September, Valecia, Spain, 2008.

[15] A. Dmowski, B. Szymański, and L. Rosłaniec, "Photovoltaic power plants as an alternative to conventional power generating systems," *Aktualne Problemy w Elektroenergetyce Conference Proceedings, Jurata 3-5 June,Poland,* 2009.

[16] B. Szymański and A. Dmowski, *Battery charging system in photovoltaic application.* X International PhD workshop OWD 2008, 18-21 October 2008, Wisła, Poland.

[17] J. Jacobs, *Multi-Phase Series Resonant DC-to-DC Converters.* Aachen,Germany, RWTH Aachen University, Aachener Beitraege des ISEA Band 42, PhD Thesis, 2006.

[18] J. Matysik, *Metody sterowania integracyjnego tranzystorowych falowników napięcia klasy D z szeregowym obwodem rezonansowym.* Prace naukowe elektryka, zeszyt 114, Oficyna wydawnicza Politechniki Warszawskiej, Warszawa 2001.

[19] F. Krismer, J. Biela, and K. J, "A comparative evaluation of isolated bi-directional dc-to-dc converters with wide input and output voltage range," *Fourtieth IAS Annual Meeting Industry Applications Conference Conference Record of the 2005, Vol.1, pp. 599-606.*

Generalized Proportional-Integral Control for Voltage-Sag Compensation in Dynamic Voltage Restorers

Alfonso Parreño Torres, Pedro Roncero-Sánchez, Xavier del Toro García and Vicente Feliu Batlle

School of Industrial Engineering, University of Castilla-La Mancha

alfonso.parreno@pcyta.com, pedro.roncero@uclm.es, xavier.deltoro@uclm.es, vicente.feliu@uclm.es

Abstract—**The Dynamic Voltage Restorer (DVR) is a custom power device used to protect sensitive loads in power distribution systems from the most frequent voltage disturbances, such as sags, swells, imbalances and harmonics. This paper presents a control strategy for voltage-sag compensation in DVR systems which offers an accurate tracking of the voltage compensation reference and a very fast transient response. The proposed control scheme is based on the technique known as Generalized Proportional-Integral (GPI) control. This type of control is closely related to the use of the classical compensation network controllers and consists in the estimation of certain state variables by means of integral reconstructors avoiding the use of derivative operators. Furthermore, the control scheme is completed by adding a feedforward term in which an approximation of the inverse dynamics of the system is used to improve and quicken the transient response. The paper includes a brief theoretical introduction to GPI control design and simulation results to corroborate the suitability of the GPI controller for voltage-sag compensation in DVR systems.**

I. INTRODUCTION

The presence of disturbances in the electrical grid, causing poor power quality, is becoming an important issue due to their financial impact and negative effects on the end-user satisfaction. Voltage sags, swells and harmonics are among the most frequent disturbances nowadays [1]. These disturbances are caused by short-circuits, the connection and disconnection of large loads and the presence of non-linear loads.

The negative impact of these disturbances has motivated the development of a variety of custom power devices to mitigate the effects they produce. Voltage sags are particularly frequent and can produce an important damage on sensitive loads. They are characterized as a short interruption, typically of several cycles of the fundamental frequency, in which the voltage amplitude is between 10% and 90% of its nominal value. These conditions are sufficient to cause tripping and an unaffordable damage on certain sensitive loads and industries such as semiconductor, paper and textile manufacturing. The compensation of voltage sags to protect sensitive loads requires a custom power device able to provide a very fast and stable response, being the DVR the current solution that better fulfils these requirements. Furthermore, DVRs can additionally compensate voltage imbalances in three-phase systems and

the voltage harmonic distortion caused by non-linear loads. The first DVR was built in the U.S. by Westinghouse for the Electric Power Research Institute (EPRI), and it was installed in 1996 on the Duke Power Company grid system to protect an automated yarn manufacturing and weaving factory [2].

Several topologies for the DVR implementation have been proposed, having all of them in common the use of a Voltage Source Converter (VSC) connected in series with the supply system. The series connection is usually done by means of a transformer, although transformerless configurations are also possible. A detailed comparison of several topologies depending on the energy storage employed is carried out in [3]. Furthermore DVR systems in which the VSC DC-Link is supplied from one distribution grid to compensate disturbances in a different distribution grid are also proposed in literature.

The control of the DVR has to be designed to quickly apply the voltage in series with the grid line that compensates the sag and other disturbances such as imbalances and harmonics. The simplest control solution consists in a feedforward action based on the error between the reference voltage that has to be supplied to the load and the voltage supplied by the grid during the sag. This solution however is not able to completely cancel errors while tracking the reference value. Moreover, it provides undesired voltage oscillations due to the poor damping of the resonant frequency introduced by the output LC filter, which is used to attenuate the switching frequency components of the VSC output voltage. These drawbacks have been addressed by introducing a feedback control loop to improve the tracking of the references and provide active damping of the filter resonant frequency. Several solutions can be found in literature regarding the feedback control implementation including: State-Feedback control [4], Proportional-Integral control in a synchronous reference frame [5], robust H∞ control [6], Proportional-Resonant control [7], Repetitive control [8] and Predictive control [9].

This paper presents a control strategy to design the feedforward compensation and the feedback control loop by means of an approximation of the dynamic inversion and a GPI control [10] respectively. The proposed method provides a very fast response and an accurate tracking of the voltage

978-1-4244-8806-3/11 $26.00 © 2011 IEEE

compensation reference. Moreover, the controller exhibits a good performance in terms of rejection of load disturbances and damping during transients.

The paper is organized as follows. A conventional model of a DVR system is presented in Section II. In Section III, the proposed control scheme is described in detail, including a brief introduction to the GPI control basic principles and the design methodology for the DVR application. Simulation results in PSCAD/EMTDC of the control response for different types of voltage sags are presented in Section IV. Finally, the main conclusions of the paper are given in Section V.

II. MODEL OF THE DYNAMIC VOLTAGE RESTORER (DVR)

A generic scheme of a power system including a DVR is shown in Fig. 1. Different loads are connected to the point of common coupling (PCC) such as induction motors, sensitive equipment, etc. The DVR is designed to compensate voltage sags in electrical grid lines feeding sensitive equipment and mainly includes the following four components [11]:

- The energy storage system used as a DC voltage source.
- The VSC, operating as a DC-AC converter.
- An output filter, which is required to remove the high order harmonic components generated by the VSC switching operation.
- A transformer to couple the VSC output and the grid line. The primary of the coupling transformer is connected in series with the protected load.

The DVR is a nonlinear device due to the VSC switching nature. Nevertheless, if the converter operates at a sufficiently high switching frequency, it can be considered as an ideal linear amplifier [12] and, therefore, the dynamic response of the DVR is mainly determined by the output filter. The state-space model of the DVR considering the output filter shown in Fig. 2 can be defined as follows:

$$\frac{d}{dt}\begin{bmatrix} v_c \\ i_{L_f} \end{bmatrix} = \begin{bmatrix} 0 & \frac{1}{C_f} \\ -\frac{1}{L_f} & -\frac{R_f}{L_f} \end{bmatrix}\begin{bmatrix} v_c \\ i_{L_f} \end{bmatrix} + \begin{bmatrix} -\frac{1}{C_f} & 0 \\ 0 & \frac{1}{L_f} \end{bmatrix}\begin{bmatrix} i_s \\ u \end{bmatrix} \quad (1)$$

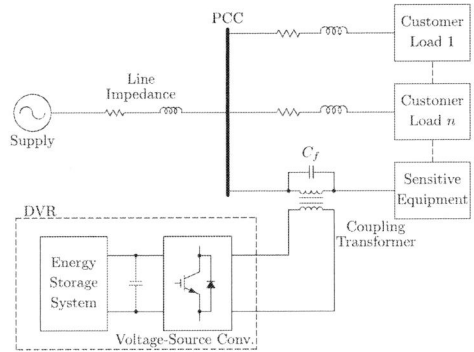

Fig. 1: Basic scheme of a power system with a DVR.

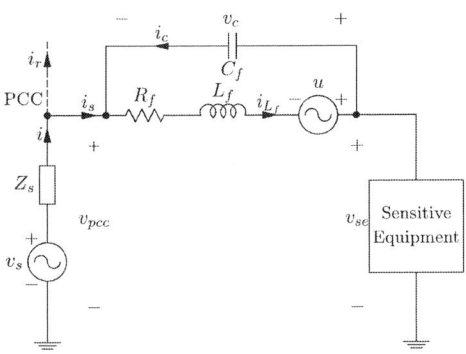

Fig. 2: Single-phase equivalent circuit of the output filter.

where L_f, R_f and C_f are the leakage inductance of the coupling transformer, the winding resistance and the filter capacitance, respectively; v_c is the capacitor voltage, i_{L_f} is the current through the leakage inductance, i_s is the load current and u is the output voltage generated by the VSC.

The state-space model in (1) shows that load current, i_s, can be considered as an input disturbance when controlling the capacitor voltage. Since this current can be easily measured, the resulting disturbance can be compensated by means of a feedforward compensation. In Fig. 3, the DVR control scheme is presented including the following transfer functions in the model of the plant:

$$P_1(s) = \frac{1}{R_f + L_f(s)}, \ P_2(s) = \frac{1}{C_f(s)} \quad (2)$$

The load current, i_s, can be compensated introducing the block $\hat{P}_1^{-1}(s)$ as shown in Fig. 3. This block is an estimation of the voltage drop in the $R_f - L_f$ branch of the filter. Using this current compensation loop, possible changes in the load current will not affect the capacitor voltage and, hence, the open-loop transfer function of the system can be expressed as the following second order system in the Laplace domain:

$$G(s) = \frac{V_c(s)}{U_c(s)} = \frac{\omega_n^2}{s^2 + s2\xi\omega_n + \omega_n^2} \quad (3)$$

$U_c(s)$ being the reference voltage before the load current, $\omega_n^2 = 1/R_f L_f$ and $\xi = R_f/2\sqrt{L_f C_f}$.

Finally, equation (3) can be easily expressed in the time domain as:

$$\ddot{v}_c(t) + \dot{v}_c(t)2\xi\omega_n + v_c(t)\omega_n^2 = u_c(t)\omega_n^2 \quad (4)$$

III. PROPOSED DVR CONTROL SCHEME

The proposed control for the DVR is depicted in Fig. 3. The scheme consists of a feedforward compensation to speed up the transient response and a feedback loop with a GPI controller to guarantee a highly accurate tracking of the reference in steady state. It should be noted that the proposed control is able to compensate balanced and imbalanced voltage sags with the same control scheme, and no coordinate transformations are used.

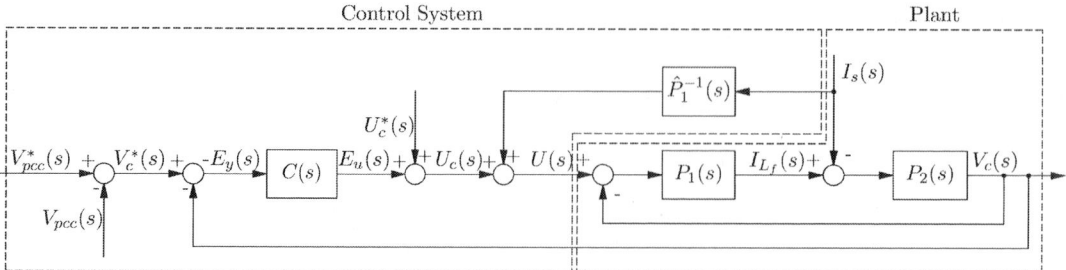

Fig. 3: Closed loop control scheme.

Following the scheme depicted in Fig. 3, the capacitor voltage reference, $V_c^*(s)$, which is the desired voltage to be applied by the DVR, is calculated from the difference between the ideal pre-sag voltage reference in the PCC, $V_{pcc}^*(s)$, and the measured voltage in the PCC, $V_{pcc}(s)$. The error between the capacitor voltage reference and the actual voltage measured in the capacitor is the input for the GPI controller represented by $C(s)$. The feedforward control action, $U_c^*(s)$, should ideally be the dynamic inversion of the plant that can be obtained from the differential equation of the system defined in (4). In order to obtain the inverse of the dynamics it is necessary to compute the first and second time derivative of $V_c^*(s)$. The time derivative of signals containing noise and sudden changes is generally a troublesome issue in terms of computation. In this paper, it is assumed that $V_c^*(s)$ is a pure sinusoidal (an infinitely differentiable function) represented in the time domain by:

$$v_c^*(t) = K \sin(\omega_1 t + \phi) \tag{5}$$

ω_1 being the fundamental angular frequency, K the amplitude of the signal and ϕ the phase. Then $u_c^*(t)$ can be obtained by means of the following equation:

$$u_c^*(t) = -\frac{\omega_1^2}{\omega_n^2} v_c^*(t) - \frac{\omega_1 2\xi}{\omega_n} \int_0^t v_c^*(\tau) d\tau + v_c^*(t) \tag{6}$$

Since $U_c^*(s)$ is only an approximation of the dynamic inversion of the plant, a feedback loop is also required to cancel errors and achieve the desired control of the DVR. This paper proposes the use of a GPI controller in the feedback loop.

A. GPI control principles

The control strategy known as GPI is closely related to the classical compensation network control. It consists in the design of a feedback controller using the structural model-based estimation of the unknown state variables [13]. This estimation is given in terms of a linear combination of the input and output variables, and a finite number of iterated integrals of them, the so-called integral reconstructors, avoiding the use of derivative operators. The estimation errors produced by the unknown initial conditions and the possible disturbances are compensated by adding successive integrals of the output error. The number of integrals added depends on the particular

problem and, more closely, on the type and order of the disturbance input to be rejected and/or the reference input to be tracked. A GPI controller is designed by developing a control law which incorporates:

1) The integral reconstructor obtained from the model of the system to estimate the unknown state variables of interest for the control.
2) A certain number of integrators added to compensate the unknown initial conditions and the possible presence of disturbances. Furthermore, an improvement regarding the tracking of variations in the reference is also achieved by means of these integrators.

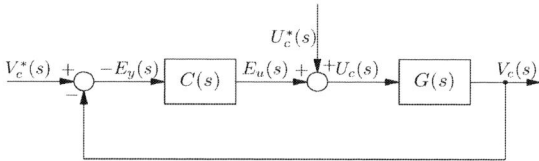

Fig. 4: Classical scheme for control design and terminology employed.

Fig. 4 shows a classical compensation scheme, where $G(s)$ is the transfer function of the system to be controlled and $C(s)$ is the transfer function of the controller. A generic case in which the order of $G(s)$ is n, $E_y(s) = V_c(s) - V_c^*(s)$ and $E_u(s) = U_c(s) - U_c^*(s)$ is considered. The closed-loop transfer function can be extracted from Fig. 4 and the resulting output tracking error $E_y(s)$ is governed by the following expression:

$$(1 + G(s)C(s)) E_y(s) = 0 \tag{7}$$

To guarantee the exponentially asymptotic convergence of $E_y(s)$ to zero, the denominator of the closed loop transfer function must be a Hurwitz polynomial (a polynomial in which the coefficients are positive real numbers and all the zeros are located in the left half of the complex plane), [14].

B. Design of the GPI controller

Considering the DVR model shown in equation (4) and defining the following errors in the time domain:

$$e_y(t) = v_c(t) - v_c^*(t) \tag{8}$$

$$e_u(t) = u_c(t) - u_c^*(t) \tag{9}$$

978-1-4244-8806-3/11 $26.00 © 2011 IEEE

The error $e_y(t)$ is governed by the following differential equation:

$$\ddot{e}_y(t) + \dot{e}_y(t)2\xi\omega_n + e_y(t)\omega_n^2 = e_u(t)\omega_n^2 \quad (10)$$

The integral reconstructor that estimates the error derivative [15], can be represented as:

$$\hat{\dot{e}}_y(t) = \omega_n^2 \int_0^t e_u(\tau)d\tau - 2\xi\omega_n e_y(t) - \omega_n^2 \int_0^t e_y(\tau_1)d\tau_1 \quad (11)$$

It should be noted that $\dot{e}_y = \hat{\dot{e}}_y(t) + e_y(0)$, where $e_y(0)$ is the unknown initial condition. A control law based on a classical compensation network for the system under control, which corresponds to a PID controller, can be formulated as follows:

$$e_u(t) = -K_d \dot{e}_y(t) - K_p e_y(t) - K_i \int e_y(\tau)d\tau \quad (12)$$

The derivative term of the control law has to be either measured or estimated. In practice, the estimation requires the sampling of the error signal at a high frequency rate to compute the derivative. This method is, however, very sensitive to noise and the derivative term is either filtered, introducing some delay, or simply not considered in the control law. Alternatively, the derivative term can also be considered replacing $\dot{e}_y(t)$ by the integral reconstructor, $\hat{\dot{e}}_y(t)$, following the GPI control design methodology that avoids the aforementioned problems. Furthermore, introducing several successive integrals of the output error, the unknown initial conditions and disturbances can be compensated, and additionally, the tracking of the reference is improved. The use of three successive integrals in the DVR control system offers a good tradeoff between tracking accuracy and controller complexity in terms of computation and it leads to the following expression:

$$e_u(t) = -K_5 \int e_u(\tau)d\tau - K_4 e_y(t) - K_3 \int e_y(\tau)d\tau -$$
$$- K_2 \int^{(2)} e_y(\lambda)d\lambda d\tau - K_1 \int^{(3)} e_y(\gamma)d\gamma d\lambda d\tau -$$
$$- K_0 \int^{(4)} e_y(\sigma)d\sigma d\gamma d\lambda d\tau \quad (13)$$

Operating and rearranging terms leads to the following expression in the Laplace domain for the control signal $U_c(s)$:

$$U_c(s) = U_c^*(s) - \underbrace{\left[\frac{s^4 K_4 + s^3 K_3 + s^2 K_2 + s K_1 + K_0}{s^3(s + K_5)}\right]}_{C(s)} E_y(s)$$
$$(14)$$

The total transfer function of the system can be obtained by combining equations (14) and (3). The sixth-order denominator of the resulting transfer function is:

$$p(s) = s^6 + (K_5 + 2\omega_n\xi)s^5 + (\omega_n^2 + K_4\omega_n^2 + K_5 2\omega_n\xi)s^4 +$$
$$+ \omega_n^2(K_5 + K_3)s^3 + K_2\omega_n^2 s^2 + K_1\omega_n^2 s + K_0\omega_n^2 =$$
$$= s^6 + \alpha_5 s^5 + \alpha_4 s^4 + \alpha_3 s^3 + \alpha_2 s^2 + \alpha_1 s + \alpha_0$$
$$(15)$$

The set of design coefficients $(\alpha_5, \alpha_4, \alpha_3, \alpha_2, \alpha_1, \alpha_0)$ can be selected to obtain a Hurwitz polynomial with the desired roots [16]. All six poles can be chosen to be the same ($s = -a$):

$$p(s) = (s + a)^6 \quad (16)$$

The design coefficients can be directly obtained combining the developed equation (16) with equation (15):

$$\alpha_5 = 6a, \ \alpha_4 = 15a^2, \ \alpha_3 = 20a^3$$
$$\alpha_2 = 15a^4, \ \alpha_1 = 6a^5, \ \alpha_0 = a^6 \quad (17)$$

It should be noted that one of the most important advantages of the GPI control strategy is the possibility to choose all the poles in the transfer function of the closed loop system due to the number of parameters that can be adjusted.

IV. SIMULATION RESULTS

The designed control scheme for voltage-sag compensation has been simulated in PSCAD/EMTDC. The simulated power system is based on the scheme shown in Fig. 1, in which the DVR protects a sensitive load and only a single customer load has been considered. Two different loads are connected to the PCC: a squirrel-cage induction machine (induction motor 1) as customer load 1 in Fig. 1, and a three-phase sensitive load made up of a resistance in series with an inductance in each star-connected phase. A squirrel-cage induction machine (induction motor 2) is connected in parallel with the sensitive load to test the correct voltage compensation when there is a sudden change in the load current demanded.

The DVR comprises a DC energy source, three H-bridge converters and the output capacitors. A coupling transformer has been used to connect the DVR between the PCC and the sensitive load. Table I summarizes the most relevant parameters of the simulated setup.

Table I: Main parameters of the PSCAD/EMTDC simulated system.

Electrical Grid	RMS line-to-line voltage: 400 V Frequency: 50 Hz
Line Impedance	Connection resistance: $R_s = 10$ mΩ Connection inductance: $L_s = 0.5$ mH
Customer Load 1: **Induction Motor 1**	Connection inductance: $L_1 = 0.5$ μH Mechanical power: $P_m = 11$ kW Rated voltage: $V_{nom} = 400$ V
Sensitive Load: **Resistive-Inductive Load** + **Induction Motor 2**	**Resistive-inductive load** Resistance: $R_{sl} = 3$ Ω Inductance: $L_{sl} = 50$ mH **Induction Motor 2** Mechanical power: $P_m = 5.5$ kW Rated voltage: $V_{nom} = 400$ V
Coupling Transformer	Rated complex power: $S = 20$ kVA Rated voltage windings: 400 V/400 V Copper loss resistance: $R = 0.5$ Ω Leakage inductance: $L = 3$ mH No-load losses have been considered
Output Filter	Capacitor: $C_f = 5$ μF Cutoff frequency: $f_c = 1300$ Hz
Voltage-Source-Converter	Swithching frequency: 7350 Hz DC voltage: $V_{dc} = 600$ V 3 H-Bridge converters

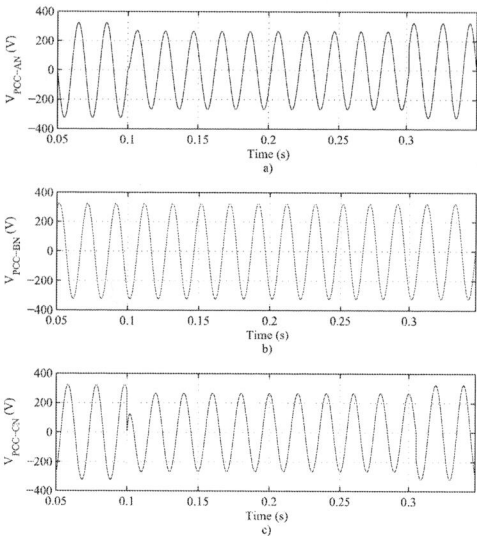

Fig. 5: Line-to-neutral voltages at the PCC during the imbalanced voltage-sag (from $t = 0.05$ s to $t = 0.35$ s), a) V_{PCC-AN}, b) V_{PCC-BN} and c) V_{PCC-CN}.

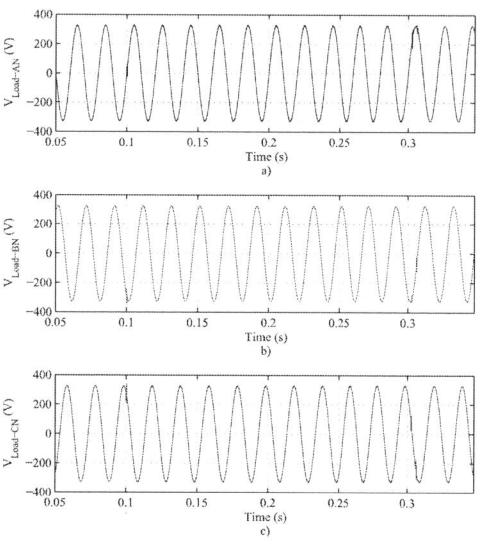

Fig. 6: Line-to-neutral voltages in the sensitive load during the imbalanced voltage-sag (from $t = 0.05$ s to $t = 0.35$ s), a) $V_{Load-AN}$, b) $V_{Load-BN}$ and c) $V_{Load-CN}$.

The GPI controller used in simulation has been designed considering six poles at $s = -4000$. In the simulated case, at $t = 0$ s the DVR and the sensitive load are connected. A short-circuit fault in two of the phases is applied at the PCC from $t = 0.1$ s to $t = 0.3$ s via a fault resistance of 0.25 Ω. This fault causes a voltage-sag in the two affected phases with an amplitude close to the 20 % of the voltage nominal value. The induction machine connected to the PCC (induction motor 1) is started at $t = 0.4$ s producing a balanced sag in the grid. Finally, at $t = 0.5$ s the induction machine in parallel with the sensitive load (induction motor 2) is started, producing an

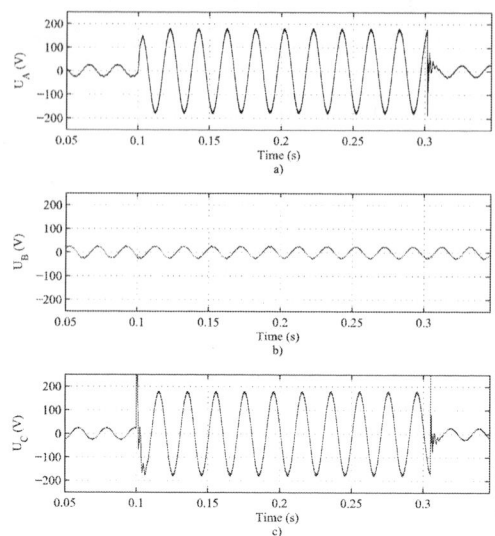

Fig. 7: VSC control voltage (from $t = 0.05$ s to $t = 0.35$ s), a) U_A, b) U_B and c) U_C.

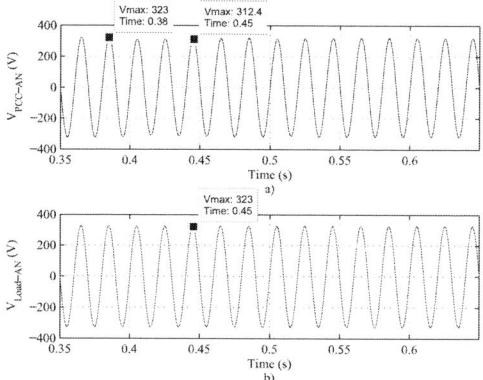

Fig. 8: Line-to-neutral voltages at the PCC and at the sensitive load during the balanced voltage-sag produced by induction motor 1 (from $t = 0.35$ s to $t = 0.65$ s), a) V_{PCC-AN} and b) $V_{Load-AN}$.

increase in the load current demanded by the sensitive load branch. The simulation finishes at 0.8 s.

Fig. 5 shows the three line-to-neutral voltages at the PCC. These voltages are plotted from $t = 0.05$ s to $t = 0.35$ s. The fault is cleared at $t = 0.3$ s and the three line-to-neutral voltages at the PCC return to their nominal values.

The three line-to-neutral voltages measured in the sensitive load are shown in Fig. 6. It can be observed that the DVR is able to quickly compensate the imbalanced sag and exhibits a very fast transient response (≈ 1 ms). Moreover, the voltages demanded at the output of the VSC, which are plotted in Fig. 7, do not exceed a reasonable voltage level (the peak voltage demanded is approximately 270 V). It is noteworthy that even in the absence of a voltage sag at the PCC, the control voltages are not zero because the DVR has to compensate the voltage drop in the transformer impedance due to the load current.

978-1-4244-8806-3/11 $26.00 © 2011 IEEE

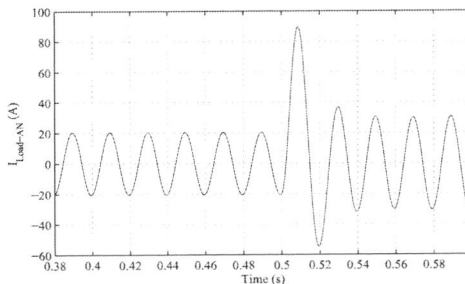

Fig. 9: Current demanded by phase A in the sensitive load (from $t = 0.38$ s to $t = 0.6$ s).

At $t = 0.4$ s the induction motor connected to the PCC starts, and produces a balanced voltage-sag at the PCC decreasing the amplitude of the phase voltages to 312.4 V. The line-to-neutral voltage in phase A at the PCC and the line-to-neutral voltage in phase A at the sensitive load are plotted in Fig. 8 from $t = 0.35$ s to $t = 0.65$ s. Note that the DVR is able to compensate the simulated balanced sag.

Finally, Fig. 9 shows the load current in phase A. At $t = 0.5$ s the induction motor connected in parallel with the sensitive load starts, demanding a high current peak during the transient as illustrated in the plot. The increase in the load current is compensated by the control, which exhibits a good performance in terms of disturbance rejection as illustrated in Fig. 8.

V. CONCLUSIONS

The negative impact of balanced and imbalanced sags in sensitive equipment has motivated the development of custom power devices such as the DVR to mitigate their effects. In this study, the modeling of a DVR in a generic power system has been developed and a control scheme is proposed for balanced and imbalanced voltage-sag compensation. The control is based on the combination of a feedforward action, which approximates the dynamic inversion of the system, and a feedback loop consisting in a GPI controller. The GPI control extends the use of classical compensation networks replacing the required derivative terms in the control law by integral reconstructors, avoiding the problems associated to the computation of the derivative operator. Furthermore, the control scheme does not require coordinate transformations.

Simulation results in PSCAD/EMTDC prove that the control scheme provides an accurate tracking of the voltage reference and a very fast transient response. Moreover, it exhibits good damping of voltage oscillations for the studied disturbances. Finally, it should be noted that a wide range of operating conditions has been tested to corroborate the robustness of the proposed control system.

ACKNOWLEDGEMENTS

This work has been partially supported by the Spanish Ministry of Science and Innovation in the framework of the National Program of Scientific Research, Development and Technological Innovation (2008-2011), by the European Regional Development Fund and by the Council of Castilla-La Mancha under the Research Project PSE-440000-2009-8.

Additionally, this work has been partially supported by the European Regional Development Fund and by the Council of Castilla-La Mancha under the Research Project PII2I09-0088-1366.

REFERENCES

[1] M. H. J. Bollen, "Understanding power quality problems: Voltage sags and interruptions," *Piscataway, NJ: IEEE Press*, 2000.

[2] N. H. Woodley, L. Morgan, and A. Sundaram, "Experience with an inverter-based dynamic voltage restorer," *IEEE Transactions on Power Delivery*, vol. 14, no. 3, pp. 1181–1186, July 1999.

[3] J. G. Nielsen and F. Blaabjerg, "A detailed comparison of system topologies for dynamic voltage restorers," *IEEE Transactions on Industry Applications*, vol. 41, no. 5, pp. 1272–1280, September/October 2005.

[4] H. Kim and S.-K. Sul, "Compensation voltage control in dynamic voltage restorers by use of feed forward and state feedback scheme," *IEEE Transactions on Power Electronics*, vol. 20, no. 5, pp. 1169–1177, September 2005.

[5] M. J. Newman, D. G. Holmes, J. G. Nielsen, and F. Blaabjerg, "A dynamic voltage restorer (dvr) with selective harmonic compensation at medium voltage level," *IEEE Transactions on Industry Applications*, vol. 41, no. 6, pp. 1744–1753, November/December 2005.

[6] Y. W. Li, D. M. Vilathgamuwa, F. Blaabjerg, and P. C. Loh, "A robust control scheme for medium-voltage-level dvr implementation," *IEEE Transactions on Industrial Electronics*, vol. 54, no. 4, pp. 2249–2261, August 2007.

[7] Y. W. Li, P. C. Loh, F. Blaabjerg, and D. M. Vilathgamuwa, "Investigation and improvement of transient response of dvr at medium voltage level," *IEEE Transactions on Industry Applications*, vol. 43, no. 5, pp. 1309–1319, September/October 2007.

[8] P. Roncero-Sánchez and E. Acha, "Dynamic voltage restorer based on flying capacitor multilevel converters operated by repetitive control," *IEEE Transactions on Power Delivery*, vol. 24, no. 2, pp. 951–960, April 2009.

[9] J. D. Barros and J. F. Silva, "Multilevel optimal predictive dynamic voltage restorer," *IEEE Transactions on Industrial Electronics*, vol. 57, no. 8, pp. 2747–2760, August 2010.

[10] M. Fliess, R. Marquez, E. Delaleau, and H. Sira-Ramírez, "Correcteurs proportionnels-intégraux généralisés," *ESAIM: COCV*, vol. 7, pp. 23–41, 2002.

[11] T. I. El-Shennawy, A. Moussa, M. A. El-Gammal, and A. Y. Abou-Ghazala, "A dynamic voltage restorer for voltage sag mitigation in a refinery with induction motors loads," *American Journal of Engineering and Applied Sciences*, vol. 3, no. 1, pp. 144–151, 2010.

[12] Y. Li, D. M. Vilathgamuwa, and P. C. Loh, "Design, analysis, and real-time testing of a controller for multibus microgrid system," *IEEE Transactions on Power Electronics*, vol. 19, no. 5, pp. 1195–1204, September 2004.

[13] H. Sira-Ramírez and R. Silva-Ortigoza, *Control Design Techniques in Power Electronics Devices*. Springer, 2006.

[14] M. Fliess and H. Sira-Ramírez, "Closed-loop parametric identification for continuous-time linear systems via new algebraic techniques," in *Identification of Continuous-time Models from Sampled Data*, ser. Advances in Industrial Control. Springer London, 2008, pp. 363–391.

[15] H. Sira-Ramírez, "Sliding modes, delta-modulators, and generalized proportional integral control of linear systems," *Asian Journal of Control*, vol. 5, no. 4, pp. 467–475, December 2003.

[16] R. Morales, V. Feliu, and H. Sira-Ramírez, "Nonlinear control for magnetic levitation systems based on fast online algebraic identification of the input gain," *IEEE Transactions on Control Systems Technology*, pp. 1–15, 2010.

Concept of Modular Uninterruptible Power Supply System with Alternative energy storages and sources

Andrew Stepanov , Ilya Galkin
Riga Technical University
astepanov@eef.rtu.lv, ilja.galkins@gmail.com

Abstract-**The paper describes uninterruptible power supply module that can be used to build most UPS topologies. It also gives possibilities to change such UPS parameters as: power, back-up time, storage device, power source, number of redundant modules. In this paper simulation and experimental results are given.**

I. INTRODUCTION

Different modern electrical equipment gives many advantages. For correct operation this equipment requires qualitative power supply. Good example is a computer which gives us endless opportunities: helps at work, studies, communication, entertainment, but this equipment is sensitive to input power faults and hence may cause operation failures. To prevent this a load must be connected through an uninterruptible power supply (UPS).

There are few topologies of UPS that give different class of protection and protect from different grid failures. To select a correct UPS type it is necessary to know what problems are typical to local grid and load requirements to power quality. Various events can occur in the electrical grid [1], most of them are characterized by a change in the magnitude of the voltage and they can have different time durations from milliseconds up to hours.

In theory, all of these grid events (voltage deviation outside the ±10% of nominal voltage) can cause electrical equipment fault, especially sensitive electronic equipment. Often overvoltages above +10% cause reduction of equipment lifetime, but high overvoltages can damage the equipment. Undervoltage can cause equipment immediate shutdown. Such faults are most common (up to 95% of all faults) and they are classified in power quality category as a voltage dips or voltage sags. Voltage dips are probably the most annoying of all power quality problems. US statistic shows that momentary interruptions (voltage dips and short interruptions of supply voltage) cause 2/3 of all damages (losses) [2]. UPS protect from short interruptions (up to 1h) and back up time depends on battery volume. To protect load from long and very long (up to several days) interruptions additional voltage sources like diesel-generator or fuel-cells are required. In this case, UPS task is to supply equipment while reserve voltage source reaches full power. For diesel-generator this time may be up to 1 minute (depending on diesel-generator construction, usually 8-15 seconds and the best ones up to 3-4 second).

Regarding EN 62040-3 standard there are 3 classes of UPS: Offline or Standby, Line interactive, Online or double conversion. Offline is voltage and frequency dependent and mainly protect against: mains failure, voltage dip (sag), voltage peaks; line interactive UPS is voltage independent and protect against: all previous plus better protection from undervoltage and overvoltage; Online is voltage and frequency independent and protect against all previous plus: surges, frequency variations, voltage distortions (burst), voltage harmonics.

As it was mentioned above back up time depends on storage capacity. Lead acid batteries are the most used type of rechargeable batteries in all types of UPS. However there are several disadvantages:

1) Most significant disadvantage of lead-acid batteries is described by Peukert's law [3]. Output of the given energy depends on the discharge current.

2) Batteries life also depends on discharge depth. The deeper discharge, the longer time the battery is in discharging state, the fewer are charge-discharge cycles.

3) Battery lifetime depends on ambient temperature. Each extra 10 ^{0}C cause 2 times decreasing of its lifetime.

4) One more factor which can affect battery lifetime is charging current. In order for battery to accumulate its maximum capacity, it is necessary to charge it correctly.

Nowadays the use of traditional batteries (lead-acid) and diesel-generator set is not always the best choice. Today various chemical, mechanical and electrical energy storage elements and different energy sources can be found in the market as alternative for diesel-generators and lead-acid batteries.

In last years more popular become UPS with energy storage based on flywheels and supercapacitors due to higher reliability. Usually flywheels and supercapacitors have higher power density, but smaller energy density than lead-acid battery.

Supercapacitors have some advantages over batteries:

- Rapid recharge capability within minutes
- Wider temperature range versus batteries
- Low degradation (up to 1 000 000 cycles)
- High efficiency (above 95%)
- No maintenance

But there are also some disadvantages:

- High price
- Energy density lower that has batteries
- Voltage depends on a degree of charge

Another alterative to lead-acid batteries is a flywheel. Advantages of a flywheel over lead-acid batteries are:
- Higher power density
- Fast recharge after use
- Wide temperature range
- Lifetime of more than 20 year

Disadvantages of flywheel over lead-acid battery:
- more complex system
- smaller energy density (but can be close to the lead acid battery)
- higher standby losses for low speed flywheels

As it can be seen supercapacitors and flywheels have many advantages over batteries. Main disadvantage of alternative energy storages is low relative energy density and higher price at longer backup times. Main advantage is high power density and long lifetime. Supercapacitors and flywheel can be the cheapest way to get high power for short time. Usually it is up to 20-30 seconds.

Small data centre UPS system (without diesel-generator) usually has back up time about 1 hour. By statistical data equipment is protected from 90-99% of all grid problems. It is assumed that grid reliability is near 99.9%. Simple UPS can improve power availability up to 99.99%. As it was mentioned above, usually UPS backup time does not exceed 1 hour. To ensure 1 hour or above backup time large amount of batteries is required, and it also increases costs of UPS. In such cases to increase the backup time use of additional voltage sources such as diesel-generators, fuel-cells, etc is proposed. This ensures almost unlimited backup time and can increase power availability up to 99.999%. It seams that five 9's is very reliable system, but this number means that system's downtime can be up to 5.256 minutes per year. For some users or services stop for 5 minutes can cause huge loses.

Modern data-centers and server-rooms need power availability for 99.9999% or even higher, to ensure it, it is necessary to use redundant (paralleled) systems.

Back to the question: "When a standby diesel-generator (or another alternative voltage source) is necessary for backup system?"

The higher data centre power the shorter reasonable supplemental battery time. For example, for 12kW after 15 minutes it is more reasonable to put a diesel-generator [4]. For higher powers (100kW to 1 MW) this time can be few minutes and below.

Another question: "What energy storage is better for reliable and cost effective system"?

As it was considered most UPS systems have diesel-generator or alternative voltage source. As the fact diesel-generator can start operation at full power in 8-15 seconds. Well (in time) maintained diesel-generator has start-up reliability near 99.5% [6]. It means that from 1000 start-up attempts only 5 can be unsuccessful.

From this it can be concluded that not always 5-15 minutes of battery backup time in system with diesel-generator is reasonable. 10-20 years ago such time was enough for system soft shutdown, but today such approach is intolerable. In practice if a diesel-generator does not start during first seconds there is a great probability that it will not start in 15 minutes also [6].

Due to this fact and taking in to account supercapacitor high reliability, supercapacitor is a good candidate for use in highly reliable UPS system.

In complex systems with availability above 99.999% each system link is important. The availability of all UPS parts (rectifier, inverter, energy storage charger) and availability switch mode power supply (SMPS) parts (rectifier, pfc, step-down converter) is also important.

There is a possibility to kill two birds with one stone: to increase efficiency and to improve reliability of system. Use of AC UPS must be abandoned; instead DC UPS system can be used. DC UPS system lacks TWO energy conversion steps: there are no inverters in UPS and no rectifiers or power factor correctors (PFC) in SMPS of computer. Due to smaller number of active switches efficiency and reliability of system increases [7].

Additional positive side effect of smaller number of parts in DC UPS system is its lower price system. Another advantage of such system is that it simplifies integration of alternative energy sources such as solar power, fuel cell and another DC-based distributed generation systems [8]. High DC-link voltage (300-400V) is considered as disadvantage due to pore safety. However, high DC-voltage will have lower power losses in conductors. An indirect advantage of higher efficiency of UPS system is lower power of cooling system.

II.   CONCEPT OF MODULE-BASED UPS SYSTEM

As mentioned before there are many different cases, grid problems, load requirements, and each require different solution of uninterruptible supply. In some cases more simple solution like standby UPS can be used, in other cases it is better to implement online UPS, or DC-UPS system. There are the cases when it is difficult to make a decision which approach is the best. In some cases it is hard to choose even rated power of UPS due to unknown power consumption in the future. For such case many manufacturers offer modular UPS [9,10]. Such system has several advantages over conventional system: 1) Lower mean time to repair; 2) more UPS modules can be added when load requirements increase; 3) modular system has higher efficiency and better input characteristics due to better power management between modules (in case of low load power it is possible to switch-off some modules and to increase other module loading).

Other manufacturers offer even more flexible UPS system like hybrid-UPS or online on demand [5]. It is modular scalable UPS and it can operate like line interactive UPS with high efficiency, but when grid event happens UPS automatically switches into online mode ensuring higher protection. If it is necessary it also can be forced to operate like online UPS all the time. Back up time also can be increased by connecting up to 4 extended batteries.

In this paper even more flexible system is described. It consists of one basic versatile module. This module can operate like rectifier, inverter or buck/boost converter [11]. Connection of few such modules can operate in parallel to ensure higher power or to operate like more complex converter: n-phase inverter/rectifier, frequency converter, active filter, uninterruptible power supply. Using such modules UPS with different combinations of 1-phase/2-phase/3-phase/n-pfase inputs and outputs can be made. Combination of modules can operate in standby or online UPS mode.

III.   UPS MODULE DESCRIPTION

It was decided to develop 4kW module, which is a complete device that can operate alone or in connection with other modules. Basic parts of module power board are: 4 IGBT transistors with counter parallel diodes that create full-bridge topology, one DC-link capacitor, contacts (for inductor, input/output, dc-link), switch S1 (Fig.1). As well module has measurement circuits, driver circuits, control devise. Using user interface it can be chosen one of mode in which will operate UPS module.

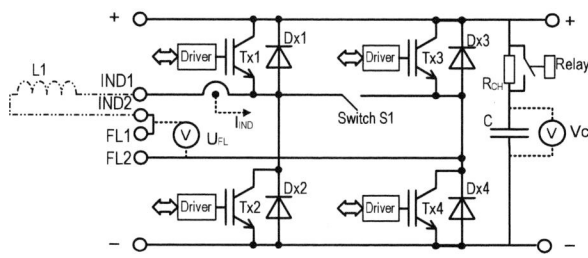

Fig.1. Power circuit of versatile module

S1 switch can be used for paralleling both half-bridge arms. Some converters have no need in full-bridge topology (like boost or buck converter) and due to that both arms can be paralleled to increase maximal current of the converter.

Rch is necessary for DC-link capacitor inrush current limitation in first seconds after module is connected to grid or to another module.

Measured signals (DC-link voltage, input/output voltage, inductor current) ensure feedback for control system.

To minimize size of heat sinks it was decided to use liquid cooling system (fig. 2). It was calculated that more effective is serial connection of heat sinks. In this case each following heat sink gets warmer coolant (in this case water), however calculated data (and later experiments) approve that the coolant temperature difference ΔT between heat sink input and output is near 1^0C. Due to that showed 6 to 8 modules can be connected in series. To connect module to cooling system quick connection fittings with valves are used and due to that there is no need in use of any tools to connect additional module.

For higher reliability cooling system with additional reserve water pump can be used.

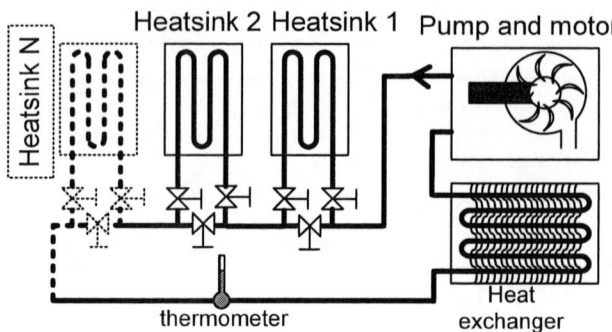

Fig. 2. Series connected liquid cooling system with possibility to add or remove heat sinks

IV.   MODULE IMPLEMENTATION

Such modules allow building of extremely flexible UPS systems. To show UPS system flexibility an example of 20kW server room is proposed. For basic protection simplest UPS topology like stand-by UPS (fig. 3) can be build using such modules. This UPS topology supplies load directly from grid and only connects an inverter in the case of grid fault. To build such UPS 6 modules are used. One module can charge battery, but other 5 modules can be paralleled to ensure 20kW.

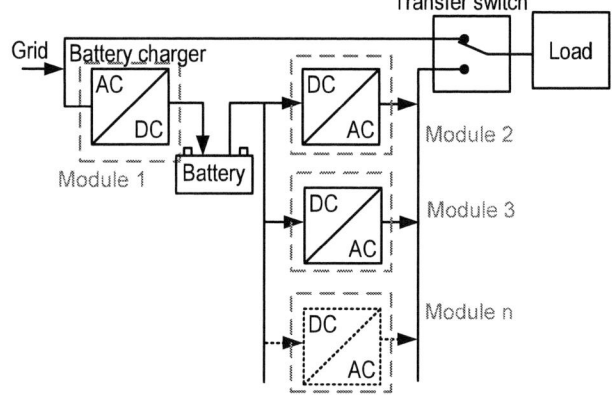

Fig.3. Stan-by UPS from versatile UPS modules

In case of load power increase up to 24kW an extra inverter can be added. In case of necessity of back-up time increase required amount of additional batteries could be added. With larger amount of batteries and existing charger battery charging time will increase due to the limited power of the existing charger. To increase charging speed one more charger module could be added.

In the case of repeated load power increase (up to 32kW) and also needs in better protection, modules can be reconfigured to another UPS topology like on-line UPS (fig.4). In this case it is not necessary to buy new UPS. 10 modules (18 modules in total) could be added: 8 modules for rectifier, plus 1 redundant module and 8 modules for inverter, plus 1 redundant.

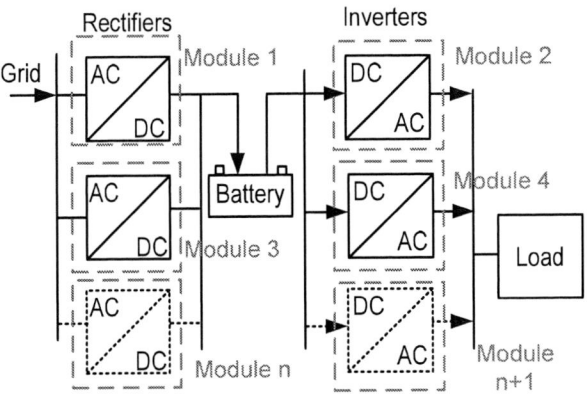

Fig. 4. On-line UPS topology from versatile UPS modules

In case of modernization of server room with increased efficiency of power distribution system, UPS can be reformed to DC-UPS system (fig.5). In this case 18 modules can create 60kW UPS system. 15 modules will operate as rectifiers plus 1 redundant module and 1 module can supply AC loads (if necessary) plus 1 redundant.

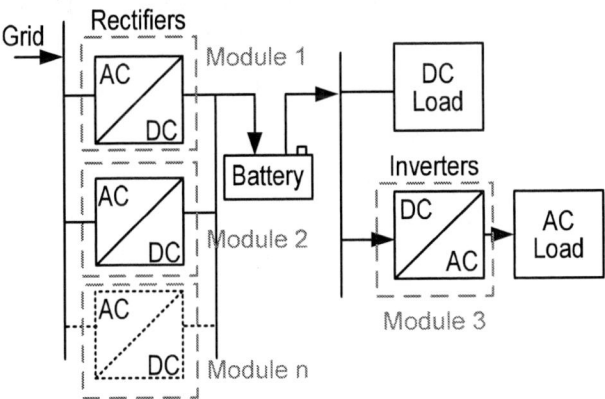

Fig. 5. DC-UPS topology from versatile UPS modules

V. MODULE SIMULATION AND EXPERIMENTAL TESTING

Preliminary simulations were done in Matlab Simulink. As an example an on-line UPS is chosen. Matlab Simulink model consists of full-bridge inverter, active full-bridge PFC rectifier and battery charger/discharger.

Simulated data shows that efficiency of PFC rectifier at full power (4kW) is near 96%, but THD of input current at full power is near 5%. At 25% of rated power efficiency is 97% and THD near 10%. It must be noted that model does not take in to account power consumption for power converter itself: like drivers, measurements and control system.

Full-bridge inverter with resistive load simulation data shows that at full-power efficiency is 95.8%, but at 25% of rated power it is near 94%. This means that on-line UPS (double conversion) at full power has efficiency near 92%. In Fig.6 UPS response at voltage disconnection is shown.

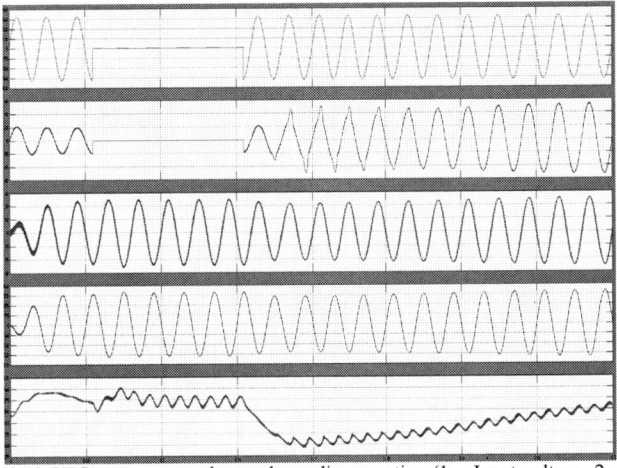

Fig.6. UPS response on short voltage disconnection (1 – Input voltage, 2 – input current, 3- output current, 4 –output voltage, 5 – DC-link voltage)

In Matlab model 125V, 63F supercapacitor is used as a back-up storage device that can supply 4kW load 60 seconds.

Experiments with developed UPS module were also done. In Fig.7 can be seen UPS voltages and currents in normal operation mode when load (3.5kW resistive load) is supplied from grid. As it can be seen the grid voltage (blue line) is distorted; and in order to keep PF close to 1 the rectifier should consume current of the same shape. In spite of distorted input voltage the output voltage is of pure sine wave.

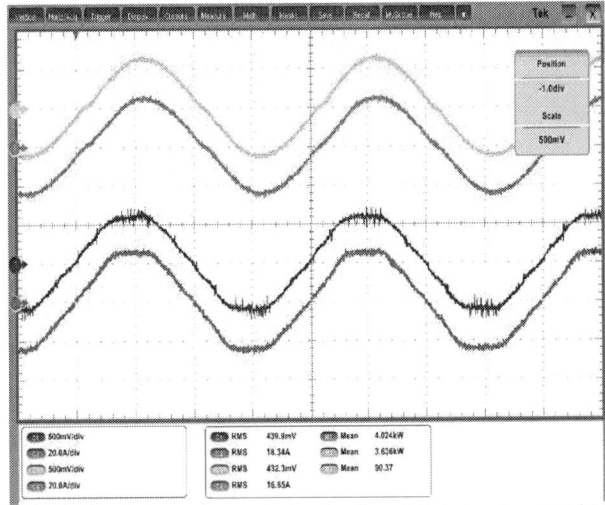

Fig. 7. UPS voltages and currents. input voltage (blue), input current (red), output voltage (green) and output current (purple).

Experimental efficiency of UPS is about 90% that is lower than simulated. Main reason to this is that the chosen Matlab-Simulink model of IGBT switch does not take some parameters into account. The most important of them is switch-on time. Fig. 8 depicts switching process of the transistor model.

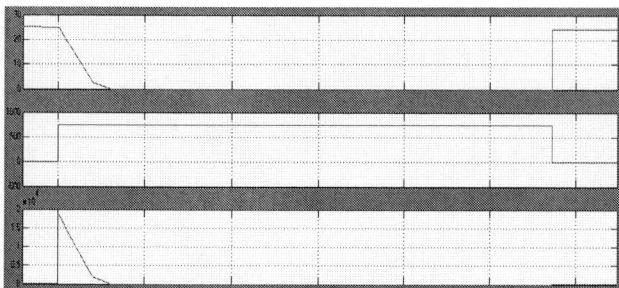

Fig.8. Switching process of the Matlab-simulink transistor model. Curves from above: switching current, switching voltage, momentary power losses

In future the attempts to develop more accurate IGBT model will be made and new simulation data will be presented. As well efficiency of power converter also will be improved by optimizing input and output inductors.

CONCLUSIONS

4kW versatile UPS module has been simulated, designed, assembled and experimentally tested. Modularity of the UPS has such advantages as: scalability, flexibility in UPS topology selection, efficiency and economic efficiency at modernization.

In case of necessity a higher output power can be reached by means of parallel connection of several modules. UPS topology can be changed without load disconnection.

Due to power management between modules it is possible to obtain higher efficiency. Using these UPS modules it is also possible to connect to UPS system alternative power sources like solar panels or wind turbines without disconnecting load from grid or UPS.

REFERENCES

[1] Standard EN 50160, http://www.leonardo-energy.org/drupal/node/3935
[2] Kristina Hamachi LaCommare, Joseph H. Eto, "Understanding the Cost of Power Interruptions to U.S. Electricity Consumers", Berkeley, California 94720, September 2004
[3] http://www.smartgauge.co.uk/peukert_depth.html
[4] Wendy Torell, Victor Avelar, "Four Steps to Determine When a Standby Generator is needed for Small Data Centers and Network Rooms", APC White Paper #52
[5] http://powerquality.eaton.com/Products-services/Backup-Power-UPS/BladeUPS.aspx?cx=3
[6] D. DeCoster, "15 Seconds Flywheel Reserve or 15 Minutes battery Reserve? The RELIABILITY difference", Mission Critical West Inc.
[7] .My Ton , Brian Fortenbery , William Tschudi, "DC Power for Improved Data Center Efficiency", March 2008
[8] Florin IOV, Frede BLAABJERG, Roger BASSETT, Jon CLARE, Alfred RUFER, Stefano SAVIO, Peter BILLER, Paul TAYLOR, Brigitte SNEYERS, "Advanced power converter for universal and flexible power management in future electricity network", 19th International Conference on Electricity Distribution, Vienna, 21-24 May 2007
[9] Joe Oreskovic, "The Modular UPS Responding to the Market's Need", Eaton Power Quality Company, INFOBATT, April 2010, Toronto
[10] http://www.upspower.co.uk
[11] Koen De Gusseme, David M. Van de Sype, Jeroen Van den Keybus, Alex P. Van den Bossche, Jan A. Melkebeek, "Fully Equipped Half Bridge Building Block for Fast Prototyping of Switching Power Converters", 35th Annual IEEE Power Electronics Specialists Conference Aachen, Germany, 2004

NordPoolSpot price pattern analysis for households energy management

Aivar Auväärt[1], Argo Rosin[1], Nadezhda Belonogova[2], Denis Lebedev[1]

[1] Tallinn University of Technology, [2] Lappeenranta University of Technology

aivar.auvaart@ttu.ee, vagur@cc.ttu.ee, Nadezhda.Belonogova@lut.fi, denis.lebedev@saksa-automaatika.ee

Abstract - This paper describes the analysis of price fluctuations in the Nord Pool Spot (NPS) and the possibilities to introduce consumption scheduling and energy storage equipment to reduce price fluctuations in households using the real-time open market electrical energy tariffs.

I. INTRODUCTION

The deregulation of electricity industry is giving way to global trends toward the commodization of electric energy. [1,2]. This trend has intensified in Europe and North America, where market forces have pushed legislators to begin removing artificial barriers that have shielded electric utilities from competition. The price of electricity is far more volatile than that of other commodities normally noted for extreme volatility. Relatively small changes in load or generation can cause large changes in price and all in a matter of hours. Unlike in the financial markets, electricity is traded every hour of the year - including nights, weekends and holidays. Unlike other commodities, electricity cannot be stored efficiently. Therefore, delicate balance must be maintained between generation and consumption 8760 hours a year.

There is, however, a great difference between electricity and the other energy (and commodity) markets in that the variable costs of production vary so greatly between different types of installation – Wind and Hydropower with a virtual nil cost at one extreme and Gas Turbines at the other end of the scale. In order to satisfy fluctuating consumer demand at the lowest cost, a broad variety of generating techniques are required. Some installations are capital intensive but can be run year round and are relatively fuel efficient (hydro, nuclear, coal-fired). Other units such as co-production of heat and power are used less frequently to cover winter heating demand at times of higher prices. Whilst energy intensive units such as Gas Fired Turbines are used for brief periods of very high price and demand.

Although the principle of generation electricity is simple, generating electricity for an area as large as Europe, means a complex balancing process. One of the biggest problems faced by the system operator is congestion. When congestion occurs, zonal prices supersede power exchange's market clearing price, which is based on the aggregated energy supply and demand curve intersection point for each hour [5].

In such a case, electricity prices can increase or decrease dramatically. The primary role of a market price is to establish equilibrium between supply and demand. This task is especially important in the power markets because of the inability to store electricity efficiently and the high costs associated with any supply failure. NPS runs the largest market for electrical energy in the world, offering both day-ahead and intraday markets to its participants. 330 companies from 20 countries trade on the Exchange. In 2009 the NPS group had a turnover of 288TWh [7]. The spot market at NPS is an auction based exchange for the trading of prompt physically delivered electricity. The spot market carries out the key task of balancing supply and demand in the power market with a certain scope for forward planning. In addition to this, there is a final balancing process for fine adjustments in the real time balancing market. The spot market receives bids and offers from producers and consumers alike and calculates an hourly price which balances these opposing sides. NPS publishes a spot price for each hour of the coming day in order to synthetically balance supply and demand. Every morning Nord Pool participants post their orders to the auction for the coming day. Each order specifies the volume in MWh/h that a participant is willing to buy or sell at specific price levels (€/MWh) for each individual hour in the following day. The SESAM (Elspot trading system) calculation equation (1) is based on an application of the social welfare criteria in combination with market rules. SESAM is maximizing the value of the objective function subject to physical constraints; like volume constraints, area balances, transmission and ramping constraints.

$$Max \sum_n \left\{ \int_0^{d^a} D^a(x)dx - \int_0^{s^a} S^a(y)dy \right. , \tag{1}$$

where a represents an area, d^a is the demand in the area a and D^a is the demand function in the area a, s^a is supply in the area a, S^a is the supply function in the area a, and n is the number of areas. The system price (SP1) for each hour is determined by the intersection of the aggregate supply and demand curves which are representing all bids and offers for the entire Nordic region [7]. In addition to area price there is also an annual fixed fee and a variable trading fee for all market participants.

978-1-4244-8806-3/11 $26.00 © 2011 IEEE

In the political debate surrounding energy, this type of price formation is labeled a marginal price setting. This gives a false impression that the establishment of prices in the electricity market is different from the price formation process in other commodity markets. The only difference lies in the significantly higher requirements for the secure delivery of electricity because it must be delivered at the precise moment it is needed by the consumer. The inelasticity caused by the inability to store electricity is the reason of this difference.

II.   AVERAGE DAILY PRICE

To find the possibilities to use the possible fluctuations we constructed an average day from actual data from the NPS trading system. We studied a period of seven months starting in April 2010. An average was calculated with the well-known formula of a generalized mean

$$\bar{x} = \sqrt[m]{\frac{\sum_{i=1}^{n} x_i^m}{n}} \,, \qquad (2)$$

where X_i is the price of electrical energy in the instance i

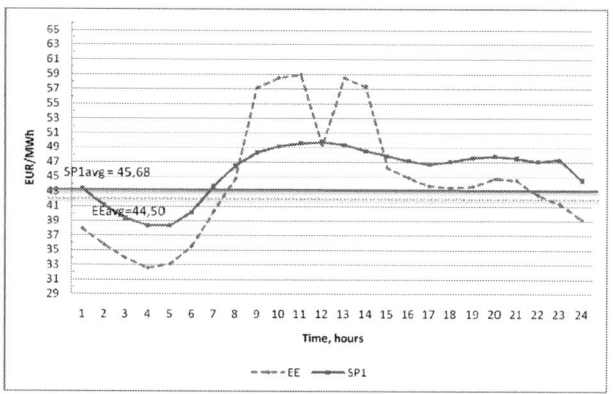

Fig. 1 Average daily price at the EE area and SP1 area

Equation (2) is used to calculate the average price during the period in the EE area. It is calculated as 44.500€/MWh. One hour is the smallest time interval when prices can change, because on spot electricity trading prices are set constant for delivery of power during a certain hour. The chart in Fig. 1 compares an arithmetic average price during the day in the NPS SP1 area and EE area. It shows very clearly that fluctuations in the system area are small - around 11.00€/MWh, but in the EE area the amplitude of the price during the day is much higher at 26.35€/MWh. The high price amplitude in the local market provides opportunities to use consumption scheduling models in residential areas to gain economy.

We can also observe differences on workdays and at weekends in figures 2 and 3

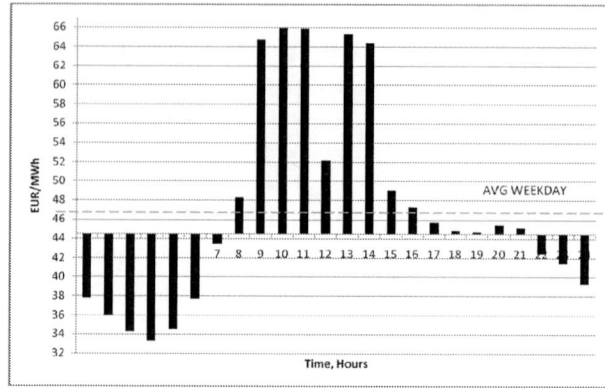

Fig. 2 Average daily price on workdays in the EE area

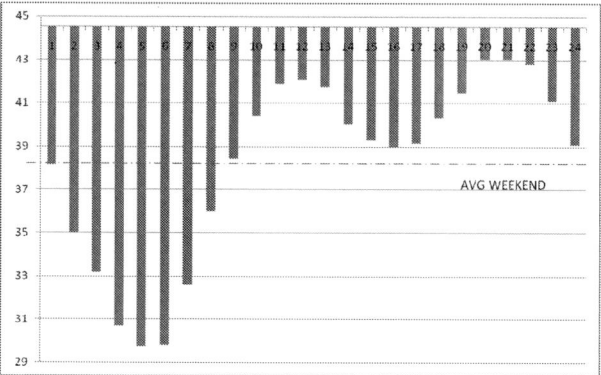

Fig. 3 Average daily price at weekends in the EE area

The price curve is not similar on workdays and at weekends. The maximum price on workdays is 65.93€/MWh and the minimum is 33.35€/MWh. At weekends the maximum and minimum prices are 43.05€/MWh and 29.76€/MWh,

Average price below the EE area average (44.500€/MWh) is 38.02€/MWh (-14.55%) on workdays and 38.256€/MWh (-14.03%) at weekends. Average price above the EE area average is 53.496 €/MWh (20.22%) on workdays and does not exceed the average at weekends.

III.   AVERAGE PRICE DEVIATION

Fig. 4 shows the deviation calculated by the simple formula (3) from the average price to analyze possibilities to use off-peak hours to store energy or shift the load to off-peak hours. We needed an assurance of off-peak hours available to recharge the batteries or other storage equipment. We found that the average duration of peaks that are higher than the average area price is 9.59 hours and the average duration of off-peaks is 13.48 hours. That means there is plenty of time to recharge storage equipment during the off-peak time.

Deviation from an average price is higher at peak hours, but peak hours last less than off-peak hours. It is most profitable to save energy between the 23...06 o'clock, then the price is lower than 10% compared to average. There is also a possibility to save energy between 16...19 o'clock when the price is about 2-3% lower than average.

$$S = \frac{\sum_{n=1}^{k} \frac{x_i}{n} - X_F}{X_F}, \qquad (3)$$

where X_i is the price of electrical energy in the instance i (from 0-24 hours) and X_F is the average area price.

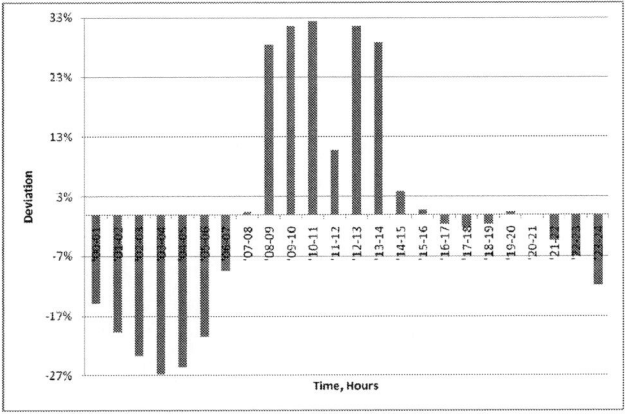

Fig. 4 Average EE area price deviation

Another feature of volatility of electricity price is its seasonal character. The daily and weekly seasonality can be illustrated by intra weekly plot of mean absolute hourly price changes (Fig. 5). The patterns of volatility are clearly correlated to the on-peak/off-peak specification of the market. The lowest volatility is observed at the weekends and during night. The huge increase of price within hours 33-39 is a result of emergency shutdown of the thermal power plant section in the EE area. For electricity spot price returns there is strong 7-day dependence. It is surprising that this dependence lasts almost forever [4].

Another seasonal phenomenon is observed on yearly basis as we compare NPS SP1 prices in 2009 the price is much higher on the winter season but remains nearly the same in other seasons as plotted in Fig. 6. The price curve in summertime for the EE area in 2010 is quite similar to trends observed in the SP1 area on summertime.

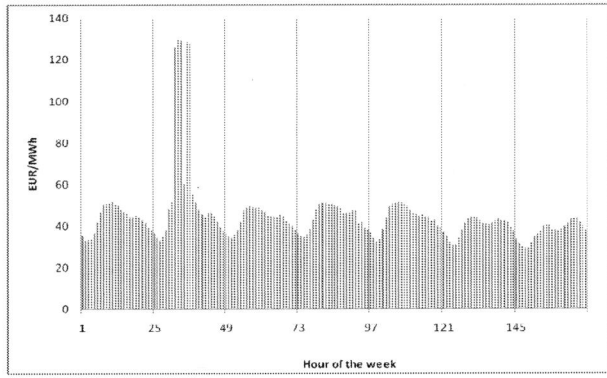

Fig. 5 Intra-weekly plot of mean absolute hourly EE area price changes for the NPS market. The statistical week is divided into 168 hours from Monday 0:00 to Sunday 24

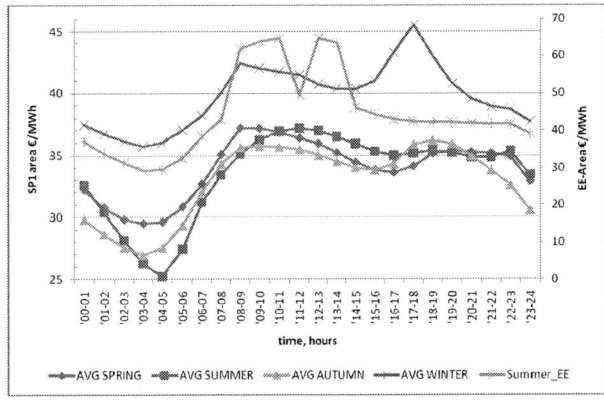

Fig. 6 Average daily price at the SP1 area on the seasonal scale compared to the EE price at summertime

IV. DISTRIBUTION OF PRICE RANGE

As seen in Fig. 7 the distribution of prices is symmetric and leptokurtic. With the leptokurtic distribution, the price will have a relatively low amount of variance, because return values are close to the mean. This could mean that energy producers will not try to invest to storage facilities as there could be quite small return on investment. This gives us an opportunity to continue our research on the profitability of using energy storing and shifting on the demand side.

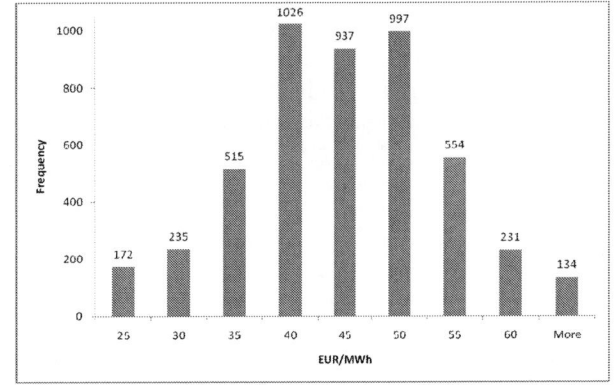

Fig. 7 Distribution of price range in the EE area

V. HOUSEHOLD ENERGY CONSUMPTION COMPARED TO AVERAGE PRICE DEVIATION

Energy consumption in households in the UK is reported in [8] and in Estonia in [3]. Peak hours for UK households form 06-08 and 13-18. Main peak hours for Estonian average households are at 7-8 and 19-21 on workdays and 12-14 and 19-21 at weekends. It is quite easy to see the possible use of energy storage to smoothen the loads at morning or midday use and even the evening use at weekends. However, some exact calculations are needed in terms of the possibilities to conserve energy at low price before evening peak hours on working days.

VI. POTENTIAL OF ENERGY STORAGES IN HOUSEHOLDS

NPS prices have been already implemented for large industrial customers in Finland. Spot market prices open opportunities also for demand response, i.e. load control for small customers. Moreover, price demand response has been already tested in Finland on residential customers and has revealed the benefits for energy costs optimization [9].

In future, energy storages will help reduce price fluctuations in spot market if used smartly. Storages can be classified into heating and electrical ones. In heating storage energy is charged and discharged in form of heating, while in electrical energy storage – in form of electricity. Electrical energy storage can be stationary (batteries) or mobile (electric vehicles). Heating energy storage represents water and/or space heater in residential houses with electrical heating loads. Although this kind of energy storage is already available in many households, as well as it has rather high energy density compared to other household appliances and therefore could have a significant impact on market prices if adjusted to them, it is less flexible in time because its usage is limited by customer's comfort requirements.

The usage of electrical energy storage is, on the contrary, less dependent on customer's requirements, but on market prices. It is in customer's interests to use (discharge) energy storage in high price hours and charge it in low price hours. The recent research shows that technical profitability of peak power reduction using energy storages is limited to 30% of network penetration [10]. Optimization of electrical energy storage charging/discharging cycles according to market prices is an important further research question. The main objective of customers is to minimize their energy costs. At the same time, they contribute to leveling peak powers in the network and to reducing the volatility of spot prices in the market. That way, residential customers can contribute to their own welfare and welfare of other market players, since the price risk of energy supplier will be minimized due to stable market prices as well as both distribution company and customer will benefit from good quality of electricity supply.

VII. CONCLUSION

This paper analyzes the fluctuations of electrical energy price in the NPS EE price area and to some extent in the SP1 area. The NPS area is currently the largest free electricity market today with a turnover of 277TWh/year. Our analysis shows that energy price in the open market is far more volatile than other commodities, but it does not behave like most financial instruments as it has a strong seasonal character.

There is a possibility to use renewable energy for local production of electricity during peak times. Solar and wind power generation would be most suitable for households.

The most perspective type could be solar power as it is usable at high peak times during the day, but the main problem in the NPS area is winter period when the effectiveness of photovoltaic panels decreases. Another possibility is to use wind power but this source is far less reliable than the sun and could not be generated without opposition from inhabitants in densely populated areas.

We observed the EE area price during 4802 hours starting from 1 April 2010 when Estonia entered the NPS market. During that time an average hourly price for the EE price area was 44.5€/MWh and it is slightly lower than the price in the system area. The price curve is similar on weekdays and at weekends. At weekends the average hourly price remains under an average area price during the observation time.

The price will always be lower than average at off-peak times. An average off-peak time lasts for 13.48 hours, which is long enough to store energy with cheaper storage equipment or shift the power usage to a less expensive time period without losing customer's comfort requirement.

Some economic impact on consumers who will buy their electricity from the open market could occur. As the prices for next day are known at least 12 hours in advance, the complex prediction models for scheduling or storing energy are not necessary. It is quite clear that until the electrical energy producers will not use energy saving technologies, the price will fluctuate almost in the same way as described in this paper. Additionally, some questions remain about changes in the behavior of prices when the households or other micro grids join the NPS market.

ACKNOWLEDGEMENTS

Authors thank Estonian Ministry of Education and Research (Tallinn University of Technology Grant No SF0140016s11 and B613A) and Archimedes Foundation (project Doctoral school of energy- and geotechnology II) for financial support of this study.

REFERENCES

[1] Strecker S, Weinhardt C.; Electronic OTC Trading in the German Wholesale Electricity Market, Lecture Notes in Computer Science, 2000, Volume 1875/2000, 280-290.
[2] S. Green, Power Eng. Int. 7 (4) (1999) 45.
[3] Rosin, A.; Hõimoja, H.; Moᴏller, T.; Lehtla, M.; , "Residential electricity consumption and loads pattern analysis," *Electric Power Quality and Supply Reliability Conference (PQ), 2010* , vol., no., pp.111-116, 16-18 June 2010.
[4] Weron, R., Przybylowicz, B., Hurst analysis of electricity price Dynamics, Physica A: Statistical Mechanics and its Applications Volume 283, Issues 3-4, 15 August 2000, Pages 462-468.
[5] Weron, R., Energy price risk management, Physica A285 (2000) 127-134.
[6] Kian, A.; Keyhani, A.; , "Stochastic price modeling of electricity in deregulated energy markets," *System Sciences, 2001. Proceedings of the 34th Annual Hawaii International Conference on* , vol., no., pp. 7 pp., 3-6 Jan. 2001.
[7] NPS homepage www.nordpoolspot.com (2010).
[8] Firth S., Lomas K., Wright A., Wall R., Identifying trends in the use of domestic appliances from household electricity consumption measurements, Energy and Buildings 40 (2008) 926–936.
[9] Koponen P., Kärkkänen S., Experiences from spot-market based price response of residential customers, CIRED Vienna, 21-24 May 2007.
[10] Belonogova N, Lassila J, Partanen J, „Effects of Demand response on the end-customer distribution fee", CIRED Workshop Lyon 7-8 June 2010, Paper 0081.

Power Electronics Interfaces for Low Voltage Distribution Generation – EMC Issues

Robert Smoleński, Marcin Jarnut, Grzegorz Benysek, Adam Kempski
Institute of Electrical Engineering, University of Zielona Góra, Poland
R.Smolenski@iee.uz.zgora.pl, M.Jarnut@iee.uz.zgora.pl, G.Benysek@iee.uz.zgora.pl, A.Kempski@iee.uz.zgora.pl

Abstract-In recent years there is observed an intensive growth of activities in matter of bidirectional electrical energy distribution systems. In situation of fast industrial progress the Electrical Power System (EPS) has to be developed to meet energy demand. This development in accordance to new environmental requirements [1] should be managed very cleaver. Because of these the participation of small renewable and clean energy sources in over all Energy Mix is still growing and is restricted by national and international legislations [2, 3]. These small electrical energy sources dispersed in EPS called as Distributed Generation (DG) are the one of most significant part of the future grids – Smart Grids [4]. Although many issues have been widely discussed and solved [5-6] the number of tasks for consideration has not been reduced. This paper introduces some investigation results of DG interfacing with Low Voltage Distribution System (LVDS) using Power Electronics Interfaces (PEI). The EMC phenomena generated by power electronics inverters has especially considered.

I. TECHNICAL AND LEGISLATIVE BACKGROUND

A. Needs of LVDS system improvement

Low Voltage Distribution System (LVDS) with its specific structure bring many problems to LVDS Operator. There is many difficulties joined with required voltage profile maintenance in all nodes of the system. This situation is especially observed in large rural network where the length of the line and the distance from substation to end user could be significant. In this network the nonlinear loads could cause voltage distortion, the volatile and repetitive – percussion loads could cause voltage mitigation. Moreover, the influence of those loads on system voltage grows with the distance from substation. In Poland, the rural network in bigger part is constructed from overhead lines what in situation of sudden and unpredictable weather phenomenon like strong wind, rime frost never noted before in such a scale often causes

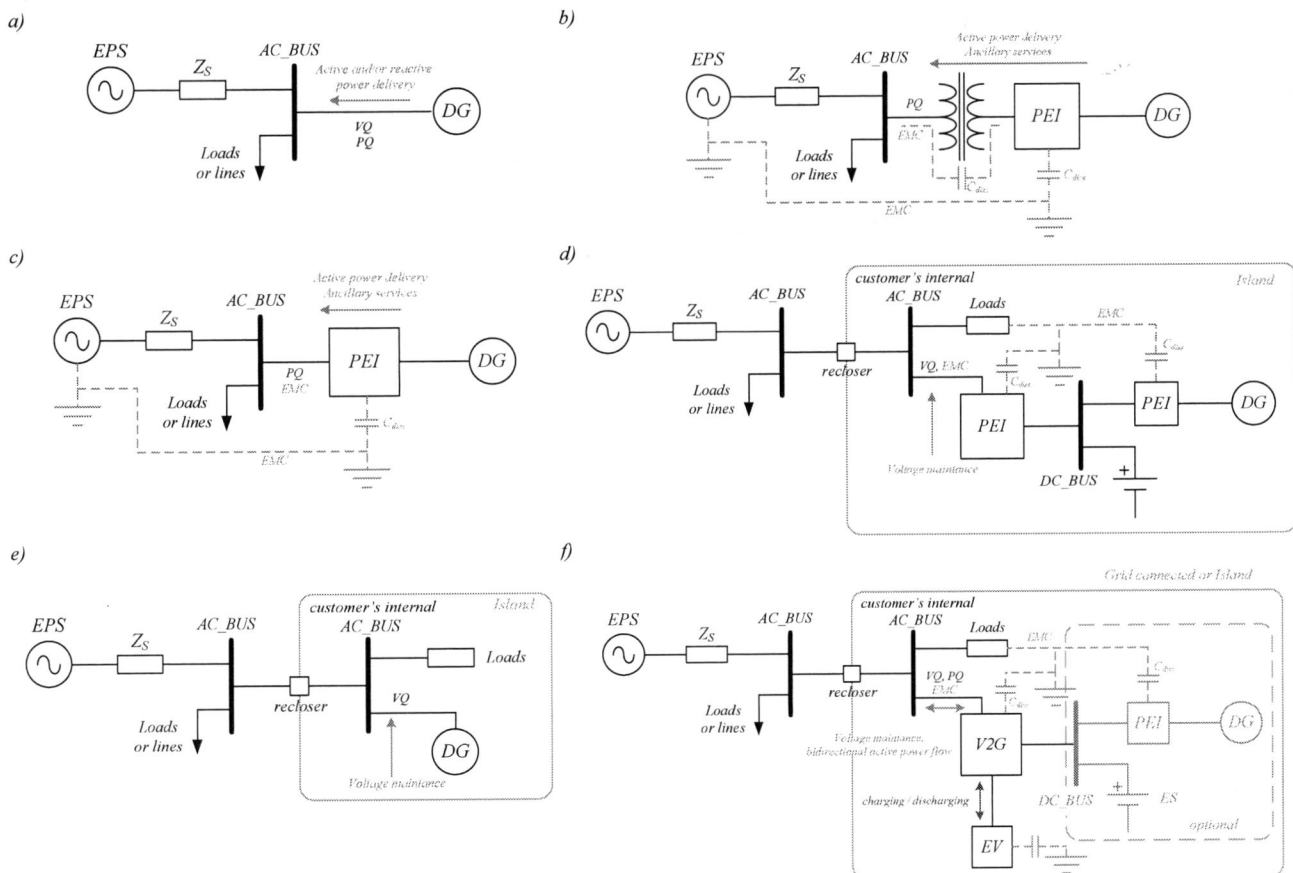

Fig.1. Low voltage distribution generation.

voltage outages. The average time of voltage outages in Poland is about ten times greater than in Germany.

The quite new type of problems is EMC disturbances penetration of LVDS caused by power electronics converters used in energy saving solution like fluorescent compact light, variable speed drives and etc.. These could be a reason of improper working of sensitive equipment connected to the same grid.

B. Traditional solutions for LVDS voltage quality improvement

Up to now the LVDS service has the limited number of solution for voltage and power quality improvement. The long – term voltage variations are dumped by setting of transformer tap changer inside range satisfying all customers connected to grid supported by this transformer. Unfortunately the most of substations are not equipped with automatic voltage controllers and on – load tap changers, so the voltage mitigation remain to be suppressed by different ways. One of them is connecting the loads which are the source of this disturbance as close as possible to energy sources. It requires some system reconfiguration and could be expensive. The voltage dips caused by line overloading with nonactive power are compensated using follow – up reactive power compensators. This type of equipment is sensitive for voltage distortion and should be used very prudently.

C. Present and future applications supported with energy sources

Many problems with voltage profiles in LVDS can be solved using small power sources (DG) dispersed in distribution system [4]. The Distribution System Operators (DSO) reluctantly treat these sources in their region of operation claiming that the system behavior with DG could be unpredictable. Partially it is the truth, but having enough information of system states and knowing DG characteristics the operator could have the additional instrument for system reliability improvement. It needs some activities like:
- development of the smart metering giving more information about system parameters;
- development of power management systems of DG including energy storage;
- development of standards and legislations.

Changes in LVDS operating structure means high cost which could be divided between Operators and DG units owners. In return for their investment they could have the benefits from ancillary services for LVDS. In the result all of local energy market participants could be content.

There are many types of DG sources in accordance to:
- generation type;
- interconnection type;
- mode of operation.

The major types of recent and future DG configuration have been introduced in Fig.1.

The DG's from Fig. 1 depending on power rate and interconnection type could provide different services for grid:
- active power delivery (local frequency control);
- voltage stabilization or maintenance;
- reactive power compensation.

D. EMC and PQ requirements

The DG sources together with grid interfaces have to fulfill compatibility requirements in field of voltage and power quality (VQ and PQ) [6, 7] or electromagnetic compatibility (EMC) according to 2004/108/EC. There are some ready standards or drafts of standards concerning DG equipment designing, earthing and safety [8]. Moreover compatibility standards have been supplemented by parts joined with low voltage dispersed generation [9]. Acquainting with those documents, unfortunately it have to be marked that there is a luck of taking to account some types of ancillary services. Omitting services joined with distortion power compensation, the DG systems should meet the requirements depicted in Table I. As can be observed, DG's with power electronics interfacing could be a source of EMC disturbances and have to meet more critical requirements.

II. LOW VOLTAGE DISPERSED GENERATION

A. Power generation and energy storage

The DG collected in Table I have the different primary energy sources. From this matter point of view they can be divided onto three groups:
- with ability for fuel flow control – primary regulation;
- with electrical regulation – secondary regulation;
- energy storage based.

The regulation type of DG output voltage and power parameters determinates the way of interconnection and the range of possible ancillary services [4]. As can be concluded from data series in Table I, the first one group mentioned above are based on AC synchronous generators (gas, hydro, biomass, diesel). They can be direct tied (DT) with LVDS and they haven't tendency for EMC problems generation. The ancillary services provided to the grid by these DG are focused on activities joined with fundamental harmonic of LVDS voltage parameters regulation (frequency control, RMS voltage control, in some permissible cases of islanding mode – loads supplying) [4].

The DG in the second one group are based on renewable energy sources (PV, wind) where the generated power is strongly depended on weather conditions. They need power electronics arrangements (PEI) for interconnection with LVDS. The PEI matches the DG output parameters with specified requirements in point of common coupling, moreover PEI can smooth the characteristic of injected power what can be used for damping of voltage mitigation. The PEI based DG can provide wider range of ancillary services to LVDS enriched in comparison to previous group with services joined with higher harmonic of system's voltages and currents compensation. Unfortunately, because of implementing high frequency converters in PEI structure, they are the source of EMC disturbances.

The energy storage (ES) based DGs (batteries, CAES, V2G, fuell cell) accumulate the advantage of two previous groups. Using stored energy and PEI based interconnection the ES can provide the widest range of ancillary services,

978-1-4244-8806-3/11 $26.00 © 2011 IEEE   108

TABLE I.
DISTRIBUTED GENERATION BENEFITS AND REQUIREMENTS

DG Type	Generation Type	Grid interfacing		Compatibility Requirements			Other Standards (connection, safety, etc.)	Mode of operation		Grid Services	
				VQ	PQ	EMC		Grid - connected	Island	Power Delivery	Ancillary Services
				EN 50160							
		DT	PEI			IEC 61000					
Gas (cogeneration)	AC	√		√	√		IEC60034 ISO3977	√		√	√
Wind	AC or DC		√	√	√	√	IEC61400	√	√ (customer -side)	√	√
Hydro	AC	√		√	√		IEC60034	√	√ (customer -side)	√	√
Biomass (bio-gas)	AC	√	√	√	√	√	IEC60034	√	√ (customer -side)	√	√
Diesel (bio-diesel)	AC	√		√			ISO8528	√		√	
PV	DC		√	√	√	√	IEC61730 VDE0126	√ (large scale)	√ (customer -side)	√	√
Fuel Cell	DC		√	√		√	IEC62282		√	√	
Batteries	DC		√	√	√	√	IEC60896	√ (large scale)	√	√	√
CAES	AC or DC		√	√	√	√	97/23/EC	√	√ (customer -side)	√	√
V2G	DC		√	√	√	√	SAE J2293	√	√ (customer -side)	√	√

DT – Direct Tied; PEI – Power Electronics Interface; VQ – Voltage Quality; PQ – Power Quality; EMC – Electro-Magnetic Compatibility; PV – Photo Voltaic; CAES – Compressed Air Energy Storage; V2G – Vehicle to Grid.

whereas have the big disadvantage – restricted capacity of energy. This energy could be supplemented from external energy sources for example DG or from grid. That is the reason of using ES for smoothing system's load characteristic and using in hybrid systems with DG. The ES systems for LVDS is saturated with power electronics arrangements for control of charging and discharging processes, so the EMC phenomena there will be the important problem.

B. Power electronics interfaces

There can be used both type of inverters in PEI [4]. Although current source inverter (CSI) is sometimes used for coupling ES devices like superconducting magnetic energy storage (SMES) with LVDS, the dominant number of PEI uses voltage source inverters (VSI). Controlling of amount of delivered power to the grid is realized in VSI using two modes of operation:

- voltage mode of operation, where the output voltage control loop is implemented to satisfy requirements of desired services;

- current mode of operation, where the output current control loop is implemented.

They have the common advantage of bidirectional power flow and each of mode has the unique features. For example there is not possible to realize loads supplying after system reclosing into islanding mode when the VSI current mode of operation is implemented and in contrast to this the current compensation is not possible in voltage mode of operation

when the maintenance of the voltage profile in point of common coupling is required.

All of VSI working as PEI have one the same feature which is EMC phenomena generation [9]. The strength and the depth of LVDS system penetration by these disturbances strictly depend on system configuration especially on inverter's construction and modulation technique as well as way of interconnection with distribution network. Disconnection to islanding mode (Fig. 1e, f) doesn't eliminate the problem but only reduces the area of EMC disturbances diffusion to the customer's installations or to the local grid inside isolated island.

Some results of investigation in DG system with PEI based on VSI have been introduced in the next paragraph. Although control strategies of PEI from ancillary services point of view have been widely considered, the dumping of EMC phenomena caused by PEI seems to be the more urgent problem to solve now.

III. CONDUCTED EMI CAUSED BY 4-QUADRANT FREQUENCY CONVERTERS

The four-quadrant frequency converter is currently commonly used in novel asynchronous drives and in distributed power generation systems as PEI for asynchronous and permanent magnet variable speed generators, Fig.1 [9]. The main circuit of this converter consists of two IGBT based three-phase bridges and an intermediate circuit allowing two-way energy flow and four-quadrant operation. The switching of IGBT devices with typically high *dv/dt* causes high level of electromagnetic interferences (EMI) especially in conducted electromagnetic emissions frequency range (9 kHz-30 MHz). The origin of a common mode (CM) interferences is the CM voltage source, which inevitably exists in the system as the result of temporary electrical asymmetry introduced by pulse width modulation strategy using a three-phase two-level converter bridge. From the EMC analysis point of view, the motoring and generating quadrants are virtually the same.

We have tested a two-pole, 10kW induction generator connected with mains by industrial four-quadrant frequency converter supplied via LISN (Line Impedance Stabilization Network). Fig.2 shows the results of measurements which have been carried out on the system consisting of LISN and EMI receiver ESCS-30, in the frequency range specified in EN 61800-3. We can observe spectra envelopes that indicate oscillation modes of EMI current waveforms at frequencies 2.5 MHz, 3.8 MHz and 4.7 MHz. The limits are slightly exceeded at the frequency 2.5 MHz, and significantly at the beginning of the CISPR B band (150 kHz). The shape of the EMI envelope [11] implies that its source could be located in a lower frequency range.

Additional measurement in the lower frequency part of the spectrum (CISPR A), not required by standards, have shown repeatable changes at frequencies 40 kHz, 80 kHz, ... We have identified them with the time of the synchronized impulse of transistors switching (25 μs) and its harmonics. The envelope of this spectrum is characteristic for a damped sine wave pulse at a frequency of about 70 kHz [10].

IV. PENETRATION OF EMI INTO LV ELECTRIC GRID

The main reason for presented investigations concerning deepness of interference penetration into mains were observed malfunctions of electronic equipment caused by 4-quadrant converter located in relatively distant circuits.

In order to evaluate the deepness of interference penetration into electric grid the measurements of CM currents in PE wires in different points of local mains had to be performed. The converter was connected straightly to mains without LISN. The result of CM current measurement in CISPR A frequency band is shown in Fig.3.

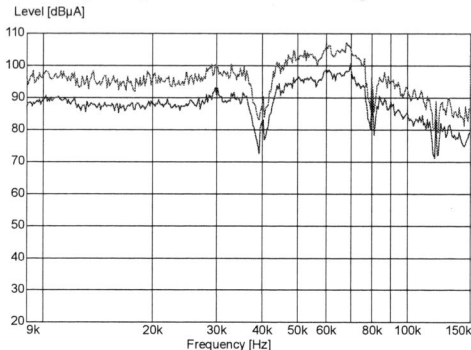

Fig. 3. Spectrum of CM current in converter PE wire.

Fig.4 shows spectrum of CM current in the PE wire of the power cable that supplies the laboratory. Measuring point was located in transformer station near common PE bus above 200 m away from interference source. Fig.4a shows spectrum of the background noises and Fig.4b shows spectrum measured during converter operation [12].

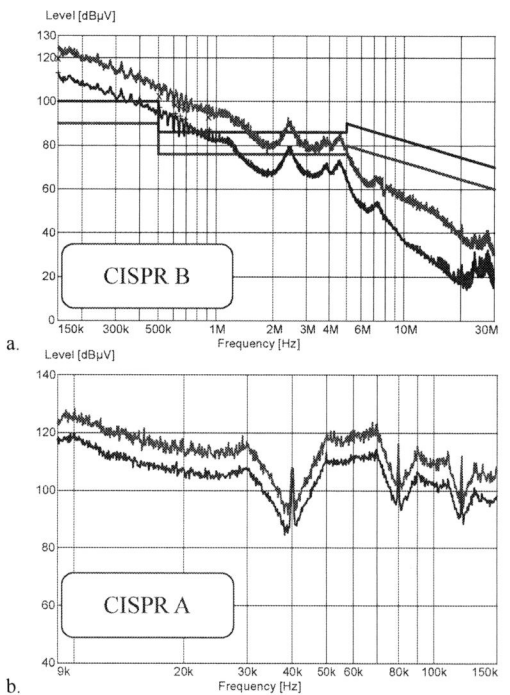

Fig. 2. Conducted EMI spectra (system without filters) for frequency ranges: a. CISPR B, b. CISPR A.

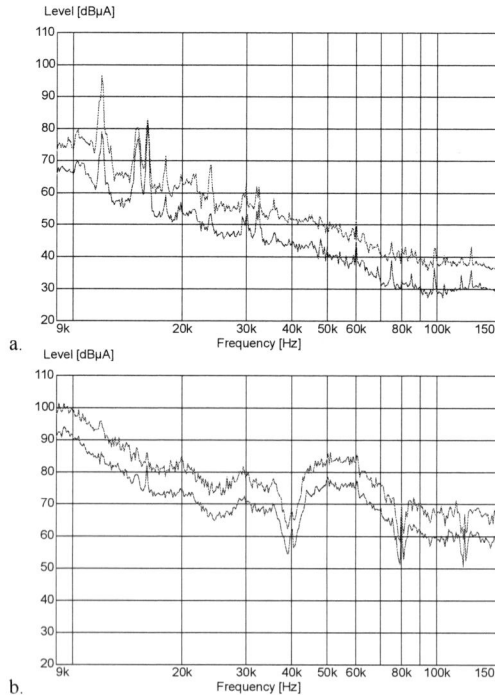

Fig. 4. Spectra of current in PE wire of power cable at transformer terminal for: a.) switched off converter, b.) switched on converter.

The level of the interference in Fig.4b increases significantly. At the frequency 60 kHz that constitutes main oscillatory mode of the current the level of interferences increased 100 times (40dB) compared to background interferences. The observed level is only 20dB lower in comparison with interferences measured in PE wire near converter in spite of the existence of many alternative paths for the interference flow in the laboratory hall.

V. PENETRATION OF EMI INTO MV ELECTRIC GRID

Fig.5 shows the results of magnetic field strength measurements in the power transformer station on both low and medium voltage sides.

The necesity of interference investigations in medium voltage (MV) grids forced the application of field measuring method for conducted electromagnetic interferences. The active loop antenna was used for measurements of interference penetration depth into MV grid. The investigations were carried out in an urban type transformer station. The generator and converter was connected to LV side of the 160 kVA transformer.

It is important to note that the LV and the MV side of the transformer were located in relatively distant points on opposite buildings' sides. Presented experimental results show that interferences introduced by the converter, in systems presented in Fig.1b, are transferred by parasitic capacitive couplings onto MV side of the transformer (not according to transformer ratio). In this case the transformer cannot be treated as an attenuating device for high frequency interferences.

Further investigations were performed under overhead MV lines. The first measurement was taken 20 m away from transformer station. The second measuring point was located under overhead MV line 1300 m away from the transformer station. In both cases the loop antenna was oriented along the lines in order to assure maximum level of interferences measured in the near field.

Fig.6 shows an increase of interferences caused by converter in comparison with background interferences under MV overhead lines 20 m and 1300 m away from transformer station.

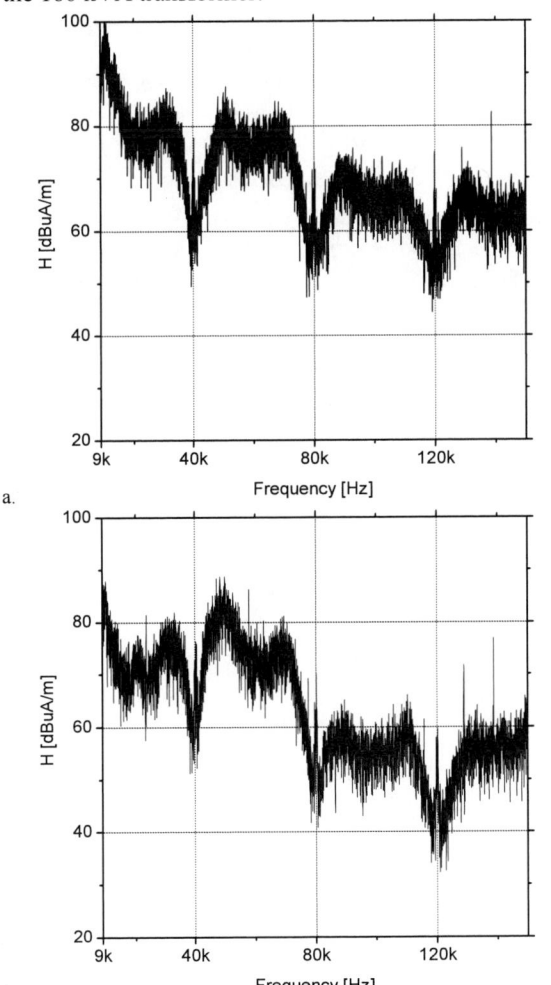

Fig. 5. Magnetic field strength on both sides of power transformer: a.) low voltage side, b.) medium voltage side.

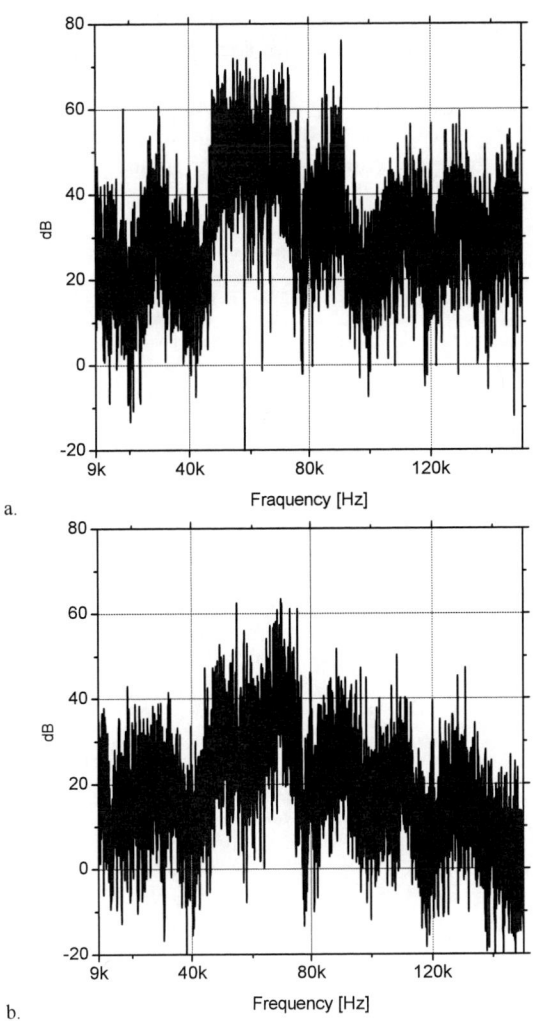

Fig. 6. Increase of interferences caused by converter under MV overhead lines: a) 20 m away from station, b) 1300 m away from station.

Presented results show that the four-quadrant converters, generating high level of conducted EMI, connected to LV grid may cause 40-60dB increase of interferences in distant points under MV lines. For characteristic, oscillatory mode frequencies, introduced by the converter, the observed attenuation came to 10-20dB for those two points. It is important to note that due to travelling wave, especially standing wave phenomena, the measured attenuations should be treated as approximate levels and might change along the line.

VI. EMI FILTRATION

The application of the dedicated EMI filter, designed on the basis of interference paths study, allow to decrease significantly the level of interferences introduced to electric grid. Typical filter design bases on the creation of alternative lower impedance path for interference flow inside of the converter with simultaneous increase of mains impedance. The application of proper CM filter usually assure high attenuation of interferences without decreasing of working currents of DM type. However, the most effective interference mitigating technique is passive or active compensation of interference sources at input and output of converters [10, 13].

Fig.7 shows the effect of application of CM filters at input and output of four quadrant converter [10]. The used CM filters have provided high interference attenuation, in comparison with those presented in Fig.2, and are cost-effective.

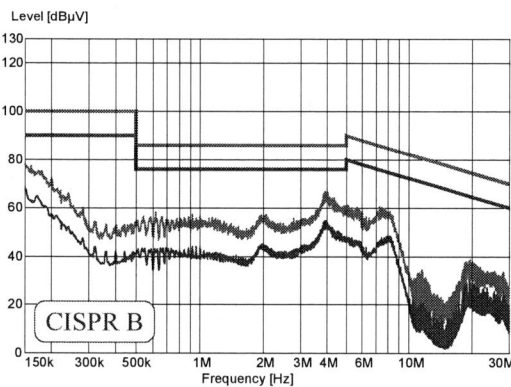

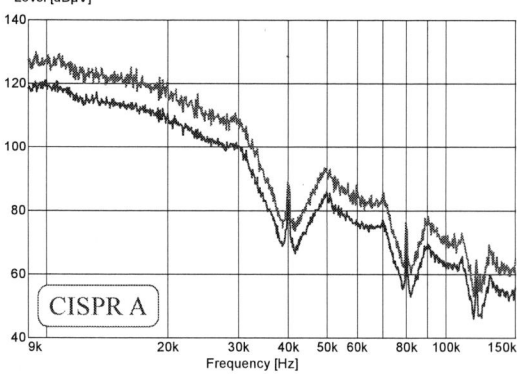

Fig. 7. Conducted EMI spectra (system with input and output filters).

VII. CONCLUSION

The application of the power electronic interfaces often requires the connection of the susceptible control and communications electronic equipment to high emission power electronic converters that implies the caution and in-depth EMC analysis to ensure system reliability. Thus, on the basic of legislation development and presented experimental results, the assurance of electromagnetic compatibility seems to be the crucial factor conditioning the development of modern power systems.

The presented measurement results have shown that interferences caused by a four quadrant converter, which is commonly used as a part of the power electronics interfaces for low voltage distribution system, can reach a distant point of the local low voltage grid. The interferences coupled by common impedance might cause immunity problems in the distorted mains. Moreover, the experimental results have shown that the interferences, introduced by power electronic converters into LV grid, can be transferred by means of parasitic couplings of a transformer (not according to transformer ratio) onto a MV side. The application of the field measuring method for conducted EMI frequency range allows to investigate the EMI flow in LV distributed installations and MV overhead lines. The performed researches have shown that the conducted EMI might spread over the wide circuits and MV lines can be a part of the paths of its passage.

ACKNOWLEDGMENT

Scientific work financed from the scientific research resources in years 2009–2011 as the research project no. N N510 333537.

REFERENCES

[1] *Directive 2008/1/EC of the European Parliament and of the Council of 15 January 2008 concerning integrated pollution prevention and control.*

[2] *Directive 2009/28/EC of the European Parliament and of the Council of 23 April 2009 on the promotion of the use of energy from renewable sources and amending and subsequently.*

[3] Polish Ministry of Economy, *Energy Policy of Poland until 2030*, Warsaw, 2009.

[4] G.Benysek, *Improvement in the quality of delivery of electrical energy using power electronics systems*, Springer-Verlag, London, 2007.

[5] EPRI Final Report, *Integration of Distributed Resources in Electric Utility Distribution Systems: Distribution System Behavior Analysis for Suburban Feeder*, 1998

[6] Dugan RC, Key TS, Ball GJ. *Distributed resources standards.* IEEE Ind Appl Mag 2006;12 (1):27–34

[7] *IEC 61000 Electromagnetic compatibility (EMC)*

[8] *IEEE P1547/D08. Draft Standard for Interconnecting Distributed Resources with Electric Power Systems*

[9] R. Strzelecki, G. Benysek, *Power electronics in smart electrical energy networks*, Springer-Verlag, London, 2008.

[10] R. Smolenski, M. Jarnut, A. Kempski, G. Benysek, *Compensation of CM Voltage in Interfaces for LV Distributed Generation*, IEEE International Symposium on EMC, 2011, in press.

[11] J. Luszcz, *Motor cable influence on the converter fed AC motor drive conducted EMI emission*, CPE'09, pp. 386 - 389.

[12] R. Smolenski, *Selected conducted electromagnetic interference issues in distributed power systems*, Bulletin of the Polish Academy of Sciences: Technical Sciences - 2009, Vol. 57, no 4, pp. 383-393.

[13] H. Akagi, H. Hasegawa, T. Doumoto, *Design and performance of a passive EMI filter for use with voltage source PWM inverter having sinusoidal output voltage and zero common-mode voltage*, IEEE Trans. on Power Electr. Vol. 19, 2004, pp. 1069-1076.

Voltage Sag Independent Operation of Induction Motor based on Z-Source Inverter

Uroš Flisar, Danjel Vončina, Peter Zajec
University of Ljubljana, Faculty of Electrical Engineering
uros.flisar@fe.uni-lj.si, voncina@fe.uni-lj.si, peter.zajec@fe.uni-lj.si

Abstract- **This paper describes an adjustable speed drive system for driving an induction motor beyond its nominal speed, even in the presence of input voltage sags. The system is based on the Z-source inverter, which offers several advantages over traditional current or voltage source inverters as it can operate in both buck or boost mode. The boost operation is achieved with controlled short-circuiting of the inverter phase legs that is otherwise forbidden in traditional inverters. These shoot through states are accomplished with the modification of the space vector modulation which is thoroughly explained. In order to assure the required output voltage and ride-through ability during voltage sags, the method for selecting the proper inverter voltage is introduced. The control of the induction motor is carried out with the field oriented control coupled with the field weakening regime of the induction motor. The experimental setup is based on a prototype with a DSP control system to verify the operation of the proposed system.**

I. INTRODUCTION

The traditional power converters used for the control of motor drives are voltage source inverter (VSI) and current source inverter (CSI). However, both have limitation in their operation. Because the ac output voltage of the VSI is limited below the dc bus, the VSI usually requires an additional boost converter. Similarly the buck converter is often added to the CSI. This additional converter stage increases cost and complexity and lowers the overall efficiency of the power conversion system.

The Z-source inverter (ZSI) overcomes the restrictions of the previously mentioned topologies [1]. Its structure is comprised of two capacitors and inductors connected in a unique impedance network that is usually coupled with a voltage source and Inverter Bridge. The operating principle of the ZSI includes a controlled short circuiting (also called shoot through) of the inverter phase legs. This enables the boost operation, whereas without it, the ZSI acts as a traditional voltage source inverter. The ZSI was originally intended for power systems with fuel cells [2-4], due to their distinctive operating curve. As their output voltage dramatically decreases with the increased current demand, the need for the boost operation becomes essential. A similar conclusion can be drawn in conjunction with solar cells, since their output voltage changes according to the change in temperature and sun radiation. Another demanding power conversion process takes place in wind turbines, where the wind energy is transformed into the electrical energy.

Because the output of the wind turbine is directly proportional to the change in wind speed, the uninterrupted power delivery to the electrical grid is essential.

Recently, the use of the ZSI is advancing into the motor drive applications [5], because it offers several advantages over traditional solutions. The nature of ZSI's operation makes it less sensitive to EMI which could short-circuit inverter phase legs that would normally destroy the switching devices. Because this kind of shoot through is allowed, the insertion of dead time is not necessary anymore. Consequently, this reduces the current and voltage harmonics. Controlling the short-circuiting of inverter phase legs can theoretically step up the input voltage to any value up to the infinity. The practical values are however limited with the device voltage ratings. Nevertheless, with controlled boost of the input voltage an important benefit of the ZSI is gained - the ability to provide ride-through during voltage sags.

The ac output of the VSI is usually controlled with sinusoidal pulse-width modulation (SPWM) or with a computational more intensive space vector modulation (SVM). Regarding the utilization of the inverter voltage, the SVM is preferable as it is capable to utilize the inverter voltage of about 15 percent more than the SPWM. Both strategies have to be modified in order to include the shoot through states, needed for the ZSI's boost operation [6-8]. Another factor that favors the use of the SVM is the compatibility with Field Oriented Control (FOC) of the induction motor (IM). This type of control is frequently used with the operation in the extended speed range, where the field weakening enables the motor operation with the constant power above base speed.

II. CONFIGURATION AND CONTROL OF THE ZSI FOR MOTOR DRIVES

The voltage source PWM inverter is the common choice for supplying the induction motor drives. The preferred PWM method for voltage source inverter is SVM, where the concept of space vectors (representing the states of the inverter switches) is used to control the ac output of the inverter bridge. Fig. 1 depicts how the output voltage vector $\mathbf{V}$ can be expressed as a linear combination of adjacent space vectors $\mathbf{V}_1$ and $\mathbf{V}_2$

978-1-4244-8806-3/11 $26.00 © 2011 IEEE

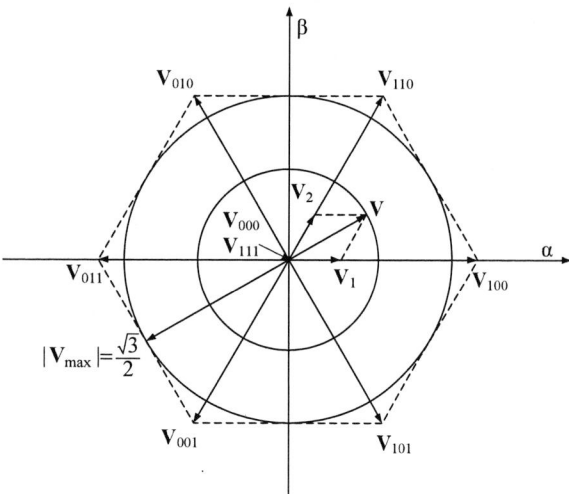

Fig. 1. Space vectors of three phase inverter bridge showing voltage trajectory and voltage vector limit

$$\mathbf{V} = \mathbf{V}_1 + \mathbf{V}_2 = \mathbf{V}_{100}\frac{T_1}{T_S} + \mathbf{V}_{110}\frac{T_2}{T_S} + \left(\mathbf{V}_{000} + \mathbf{V}_{111}\right)\frac{T_0}{T_S} \quad (1)$$

T_0 is time duration of zero state $\mathbf{V}_{000}$ and $\mathbf{V}_{111}$, T_1 and T_2 are time durations for any of the neighboring active states and T_S is the switching period. In the linear or undermodulation region the output voltage vector $\mathbf{V}$ always remains within the inscribed circle in the hexagon formed by the six space vectors.

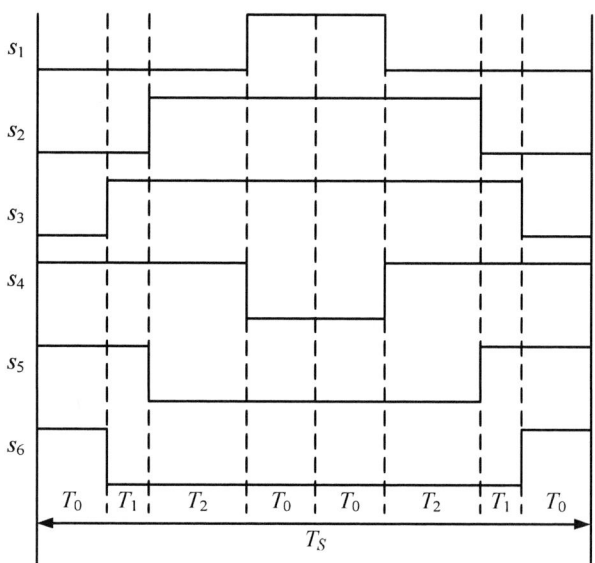

Fig. 2. Symmetrical pulse pattern of SVM for three phase inverter bridge

The pulse pattern of SVM for three phase inverter is illustrated in Fig. 2. The state sequence begins with the zero state where the $\mathbf{V}_{000}$ is impressed (all upper transistors are

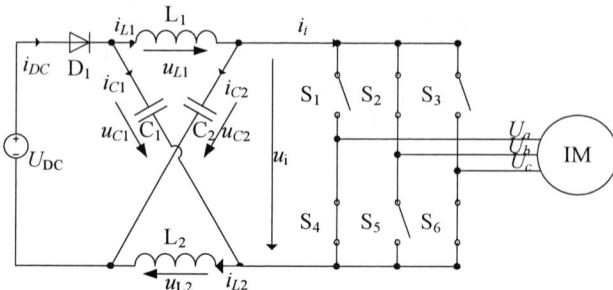

Fig. 3. ZSI with IM

open). This is followed by the two active states and ends with another zero state $\mathbf{V}_{111}$, where all upper transistors are closed. After half of the switching period T_S, this sequence repeats in reverse order. The SVM with the symmetrical pulse pattern has the two zero states distributed equally on both ends of the active states as illustrated in Fig. 2.

The maximum output line-to-line rms voltage (U_{ab}) of the VSI that can be utilized [9] with the SVM from the inverter voltage u_i is

$$U_{ab\_max} = \sqrt{\frac{3}{2}} \cdot \hat{U}_{an\_max} = \sqrt{\frac{3}{2}} \cdot \frac{2}{3} \cdot u_i \cdot |\mathbf{V}_{max}| = \sqrt{\frac{2}{3}} \cdot u_i \cdot \frac{\sqrt{3}}{2} = \frac{u_i}{\sqrt{2}}. \quad (2)$$

When using an IM with the nominal line-to-line rms voltage of 177 V as an example, it requires a minimum inverter voltage of 250 V. This value needs to be further increased if the compensation of voltage sags is required. Although the VSI usually includes input capacitor for such occurrences, it is a low energy storage element and is inefficient in case of severe voltage sags. Furthermore, increasing the input voltage is not always a viable solution. Often an additional converter stage is inevitable, which can be an additional boost converter or an addition of the Z source impedance network.

1. Z-Source Inverter Operation

If the impedance network is added to existing VSI, the later is transformed into a ZSI shown in Fig. 3. It can be separated into four major parts: the voltage source U_{DC} and diode D_1 to block the input voltage during shoot through; the symmetrical Z-source impedance network, with capacitors $C_1 = C_2$ and inductors $L_1 = L_2$; a three phase inverter bridge and the IM as the load.

The basic operating principle and control of the ZSI have been detailed in [1]. The summary of the ZSI main operating modes are:

Mode 1): The inverter bridge is operating in one of the six active states and the diode D_1 is conducting. From the load point of view, the inverter bridge behaves as a current source as depicted in Fig. 4. Because of the symmetrical impedance network the capacitor voltages are $u_{C1} = u_{C2} = u_C$, also the

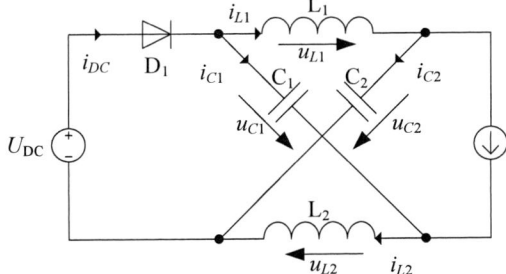

Fig. 4. ZSI operating in active mode

inductor currents $i_{L1} = i_{L2} = i_L$ and voltages are $u_{L1} = u_{L2} = u_L$. The voltage and current relationships are therefore

$$u_L = U_{DC} - u_C; \; u_i = 2u_C - U_{DC}. \tag{3}$$

$$i_{DC} = i_L + i_C; \; i_i = i_L - i_C \rightarrow i_{DC} = 2i_L - i_i. \tag{4}$$

Mode 2): The input diode is still conducting and the inverter bridge is operating in the zero state. The upper or the lower transistors are closed and the inverter bridge acts as an open circuit viewed from the Z-source network. The voltage relationships are the same as in (3) while the currents are

$$i_{L1} = i_{L2} = i_{C1} = i_{C2} = \frac{i_{DC}}{2}. \tag{5}$$

Mode 3): Fig. 5 illustrates the shoot through mode, where the input diode is reverse biased because the sum of the capacitor voltages is higher than the input voltage. The shoot through can be done in one phase leg, two or in all three phase legs. The voltage and current relationships in the shoot through mode are

$$u_L = u_C; \; u_i = 0 \tag{6}$$

$$i_{DC} = 0; \; i_{L1} = -i_{C1}; \; i_{L2} = -i_{C2}. \tag{7}$$

The duration of the shoot through (T_{sh}) depends on the required boost ratio (B) and can be calculated with

$$T_{sh} = \frac{B-1}{2 \cdot B} \cdot T_S, \tag{8}$$

where B is

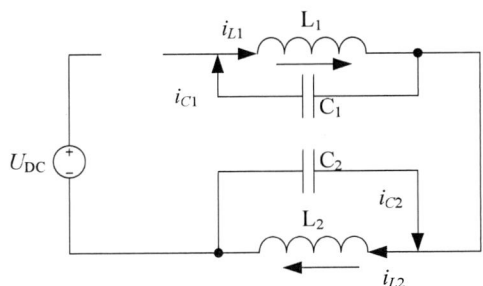

Fig. 5. ZSI operating in active mode

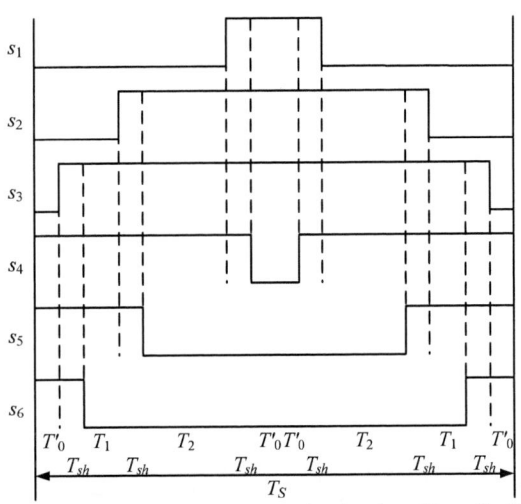

Fig. 6. Modified SVM switching pattern showing the uniform distribution of T_{sh}

$$B = \frac{u_i}{U_{DC}} = \frac{1}{1 - 2 \cdot \frac{T_{sh}}{T_S}}. \tag{9}$$

The maximum duration of the shoot through is limited to half of the switching period, where the resulting inverter voltage is theoretically boosted to infinity. However in practical application the maximal boost or the maximal inverter voltage is limited with the device voltage ratings.

To summarize, the ZSI behaves in a similar way as a traditional VSI, however it also introduces a new operating state, called shoot through state that boosts the inverter voltage. This enables the ZSI to produce almost any output voltage, provided the inverter voltage stays within device voltage ratings and the current capabilities of the input voltage source are sufficient.

2. Modified SVM

In order to include the shoot through states, we must modify the SVM (MSVM). We do this by positioning the shoot through states at the transients of the switching states so that upper and lower transistors on-time is overlapping as illustrated in Fig. 6. The shoot through states are distributed equally among all three phase legs and they utilize the zero states symmetrically. This kind of placement allows the shoot through states to leave the existing active states (T_1 and T_2) uncompromised and more importantly, the number of switchings remains unchanged, however, at the cost of reduced duration of zero state (T'_0).

It can be evident from (9) and Fig. 6 that the boost ratio depends on the available duration of the zero state. If the required T_{sh} is greater than the available T_0, the required voltage boost can not be achieved. Because the voltage boost

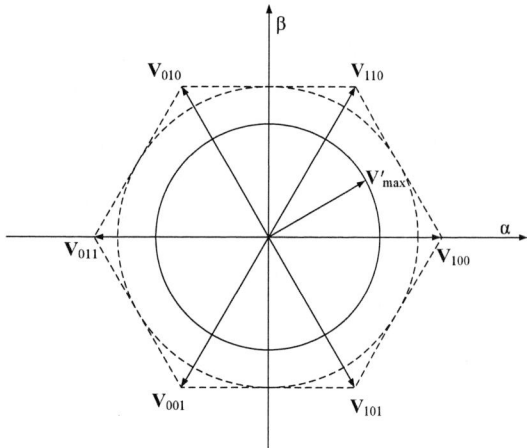

Fig. 7. Modified trajectory of the clamped voltage vector V'$_{max}$

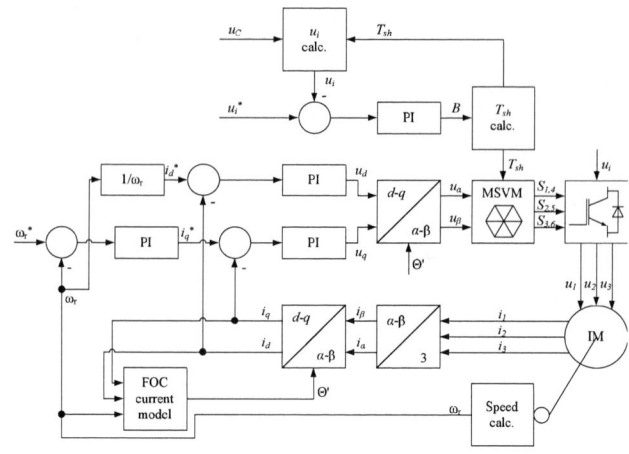

Fig. 9. Functional diagram for FOC with the MSVM and control of the inverter voltage

$$U_{ab} = \sqrt{\frac{3}{2}} \cdot \hat{U}_{an} = \sqrt{\frac{3}{2}} \cdot \frac{2}{3} \cdot u_i \cdot |\mathbf{V'}_{max}| = \sqrt{\frac{2}{3}} \cdot G_U \cdot U_{DC}. \quad (12)$$

has a higher priority, the duration of the active states has to be reduced. Fig. 7 illustrates the clamped maximum voltage vector ($\mathbf{V'}_{max}$), which is reduced according to the required T_{sh} in order to assure the required voltage boost of inverter voltage

$$|\mathbf{V'}_{max}| = \frac{\sqrt{3}}{2} \cdot (1 - \frac{T_{sh}}{T_S}). \quad (10)$$

The total voltage gain G_U is a combination of the output voltage vector and the boost ratio. The relationship between G_U and T_{sh} is illustrated in Fig. 8 and can be calculated with

$$G_U = B \cdot |\mathbf{V'}_{max}| \quad (11)$$

The choice of the optimal G_U depends on the requirements for the intended application. However in general, we should maximize $\mathbf{V}$ and minimize B to reduce the voltage stress of the switching devices. The line-to-line rms voltage of the ZSI with MSVM is

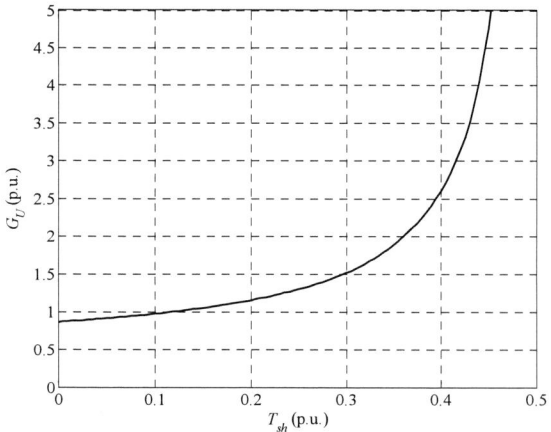

Fig. 8. Maximum voltage gain of ZSI with MSVM

Operating an IM with a ZSI requires careful control of the voltage gain to guarantee the proper operation during voltage sags. Usually the u_i is kept constant and the B is adapting according to the change of input voltage. The inverter voltage that will enable uninterrupted motor operation during voltage sag can be calculated by solving the (11) and (12) for u_i

$$u_i = 2\sqrt{2} \cdot U_{ab} - U_{DC\_min}. \quad (13)$$

The resulting u_i is valid for SVM based ZSI. If a different PWM technique is used, such as the SPWM, the (12) should be modified accordingly. It should also be noted, that the (13) considers the voltage vector is maximal and boost ratio is minimal. If a different voltage gain strategy is preferred, the (10) should be taken into consideration when choosing the desired inverter voltage.

III. EXPERIMENTAL RESULTS

The experimental verification has been carried out on the proposed system with the following parameters:

- Input voltage: 100 – 300 V$_{DC}$,
- Z-source network: $L_1 = L_2 = 165$ µH; $C_1 = C_2 = 1000$ µF
- Switching frequency: $f_S = 10$ kHz,
- Load: Induction motor ($U_n = 177$ V, $I_n = 14{,}8$ A, $N_n = 1456$ rpm, $M_n = 20$ Nm, $f_n = 50$ Hz, cos(φ = 0,785),
- Position sensor: 1024-lines incremental encoder,
- Control system: DSP TMS320F2808.

978-1-4244-8806-3/11 $26.00 © 2011 IEEE

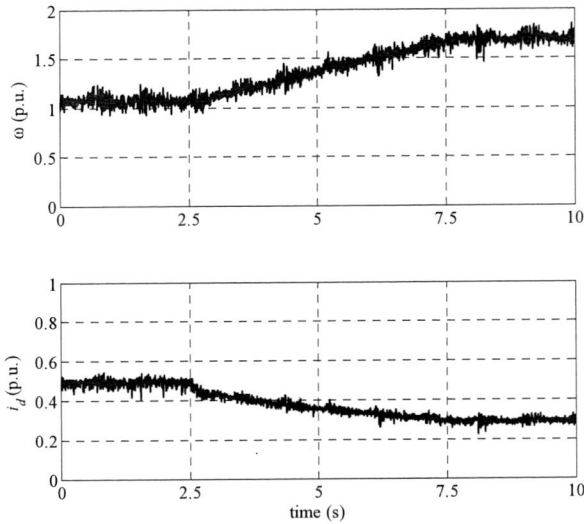

Fig. 10. Ramp up of the speed of IM with VSI (U_{DC} = 280 V)

The IM was controlled with the FOC which is often used for adjustable speed drive applications, where the field weakening is applied to extend the speed range. The principle of field weakening was achieved by decreasing the flux producing current i_d according to the "$1/\omega_r$" method, while increasing the torque producing current i_q.

A typical functional diagram of the indirect FOC for the IM can be seen on Fig. 9. The implementation of the existing

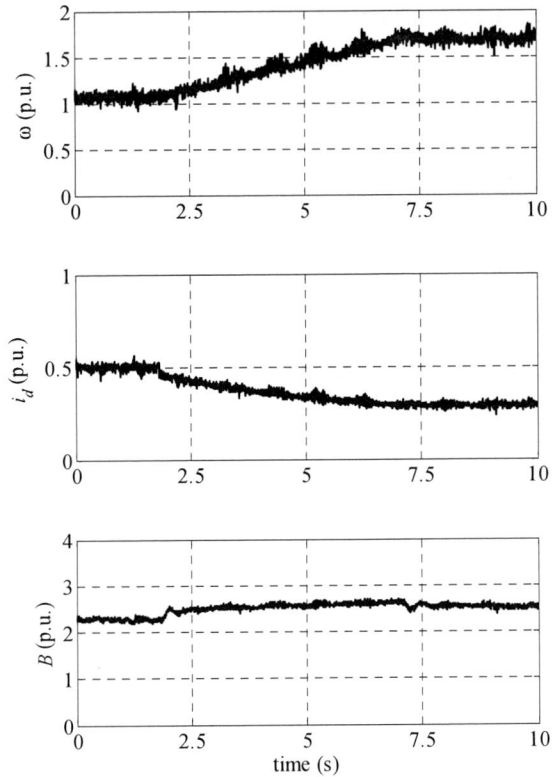

Fig. 11. Ramp up of the speed of IM with ZSI (U_{DC} = 180 V)

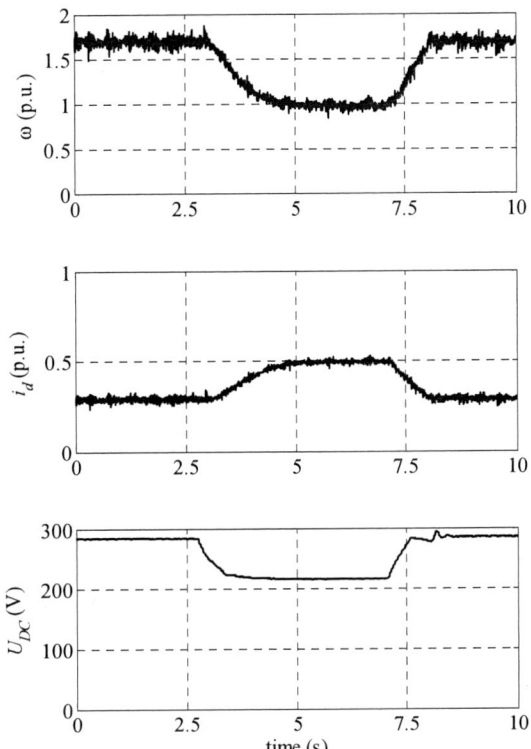

Fig. 12. Motor operation with VSI when exposed to voltage sag

FOC for VSI based IM requires minor modifications. These are mostly related to the chosen PWM method and to the control of the voltage boost. The measuring of the inverter voltage u_i is somewhat cumbersome because the u_i is zero at the time of the shoot through. A more elegant solution is measuring the capacitor voltage u_C and calculating the inverter voltage u_i from

$$u_i = \frac{u_C}{1 - \dfrac{T_{sh}}{T_s}}. \tag{14}$$

From here forth, the control of the u_i is performed with a PI regulator, which outputs the required B. The T_{sh} is calculated with (8).

The operation of the IM was first verified with the VSI without the Z-source network. For the VSI, the minimal inverter voltage according to (2) is 250 V. Including a safe margin of 10 %, the u_i was set to 280 V. Fig. 10 shows the ramp up of the rotor speed from nominal value up to 2400 rpm, together with the waveforms of i_d current. The motor was loaded with the 50 % of the nominal torque, because when the motor is operating at high speed, the required torque is normally low.

The same measurements were repeated with a ZSI and the results are shown in Fig. 11. The input voltage was decreased to 180 V to demonstrate the capabilities of the ZSI. At this u_i, the VSI would be unable to reproduce the

978-1-4244-8806-3/11 $26.00 © 2011 IEEE

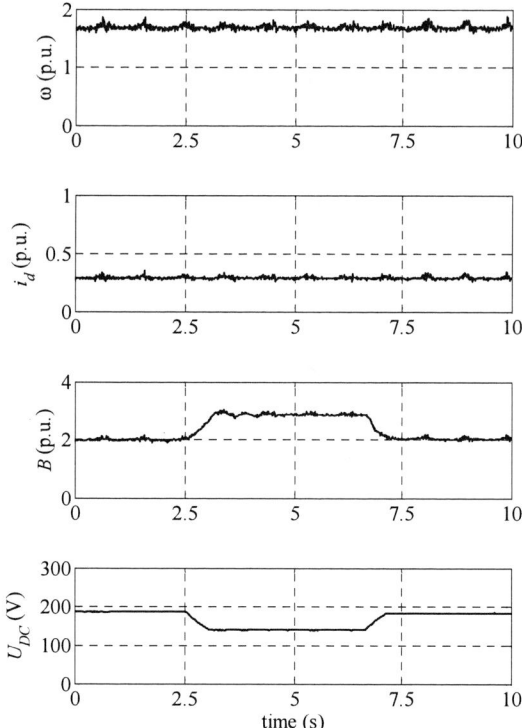

Fig. 13. Motor operation with ZSI when exposed to voltage sag

The waveforms clearly demonstrate that the inverter voltage is boosted without any negative influence on the flux current i_d, which is maintained at a desired level without any interruptions.

IV. CONCLUSIONS

This paper presents the motor drive system based on the ZSI and the IM. The FOC which is extensively used in high-performance drive applications was successfully adapted to the ZSI, where the modification of the SVM and control of the inverter voltage was explained. The operation of the system was verified with the practical use with the DSP control system together with the operation in the extended speed range. The use of the Z-source based inverter is beneficial if the input voltage is limited, or if the voltage source is subjected to frequent voltage sags. The ZSI can easily adapt to such disturbances and keep the operation of the motor system reliable. In general, the transition from VSI to ZSI is beneficial and requires additional control of the voltage boost and the modification of the existing PWM method. However, because of this modification, the overall efficiency at higher duty cycles would be lower and in that case the combination of VSI and separate boost converter would be more reasonable.

The findings of the performed experiments could also help when designing power converters with renewable energy sources in mind. In such cases, the changes in input voltage are seldom that sudden, nevertheless an insight into the behavior of the ZSI under comparable circumstances has been presented.

nominal output voltage needed for the IM. The waveforms in Fig. 11 verify the proper operation of the ZSI at a decreased input voltage, where the rotor speed is again increased to 2400 rpm at the 50 % of nominal torque. The cause of a minor increase of B can be explained with the increased back-EMF, because of the increased rotor speed.

When operating the motor in the field weakening regime, voltage sags at the DC input present an even bigger problem, which can lead to rapid reduction of the rotor speed. Voltage sags are common when using renewable energy sources, especially when using solar cells or wind turbines. Usually the duration of the output voltage drop is sustained over longer periods of time. For the purpose of this experiment the total duration of the measurements was limited to 10 s. Fig. 12 illustrates the effect of simulated voltage sag of about 25 % of the input voltage for the VSI. Because the voltage sag is too severe, the rotor speed decreases below its nominal value. On the contrary, the flux producing current i_d greatly increases, which indicates the disrupted operation of the IM in the field weakening regime.

For the ZSI to be able to endure the input voltage drop of 25 %, the minimal inverter voltage has to be set according to (13). The resulting u_i is 365 V. However the inverter voltage was set 10 % higher, at 400 V. Fig. 13 shows the IM operating at 2400 rpm and at 50 % of the nominal torque. The ZSI successfully adapts its boost ratio during the voltage sag and the operation of the IM remains uninterrupted.

REFERENCES

[1] Peng, F.Z., "Z-source inverter", IEEE Transactions on Industrial Applications, Vol. 39, No. 2, pp. 504-510, March/April 2003.

[2] Shen, M.; Joseph, A.; Wang, J.; Peng, F.Z.; Adams, D.J., "Comparison of traditional inverters and Z-source inverter for fuel cell vehicles", IEEE Transactions on Power Electronics, Vol. 22, No. 4, pp. 1453 – 1463, July 2007.

[3] Peng, F.Z; Shen, M.; Holland, K., "Application of Z-Source Inverter for Traction Drive of Fuel Cell-Battery Hybrid Electric Vehicles", IEEE Transactions on Power Electronics, Vol. 22, No. 3, pp. 1054 – 1061, May 2007.

[4] Holland, K.; Shen, M.; Peng, F.Z., "Z-source inverter control for traction drive of fuel cell-battery hybrid vehicles", Industry Applications Conference, Vol. 3, pp. 1651 – 1656, October 2005.

[5] Peng, F.Z., "Z-source inverter for motor drives", Power Electronics Specialists Conference, 2004, Vol. 1, pp. 249 – 254, November 2004.

[6] Loh P.L.; Vilathgamuwa, D.M.; Lai, Y.S.; Chua G. K.; Li, Y., "Pulse-width modulation of Z-source inverters", IEEE Transactions on Power Electronics, Vol. 20, No. 6, pp. 1346 – 1355, November 2005.

[7] Peng, F.Z.; Shen, M.; Qian, Z., "Maximum boost control of the Z-source inverter", IEEE Transactions on Power Electronics, Vol. 20, No. 4, pp. 833 – 838, July 2005.

[8] Shen, M.; Wang, J.; Joseph, A.; Peng, F.Z.; Tolbert, L.M.; Adams, D.J., "Constant boost control of the Z-source inverter to minimize current ripple and voltage stress", IEEE Transactions on Industrial Applications, Vol. 42, No. 3, pp. 770 – 778, June 2006.

[9] Bose, B. K., "Modern Power Electronics and AC Drives", Prentice – Hall, Upper Saddle River, 2002.

Voltage Drops Mitigations Using Flywheel Energy Storage System in Production Lines

Ahmad Al-Diab, Constantinos Sourkounis

Ruhr-University Bochum / Institute of Power System Technology and Power Mechatronics, Bochum, Germany.
aldiab@eele.rub.de, sourkounis@eele.rub.de

Abstract— **A novel flywheel energy storage system is proposed in this study. The energy storage system consists of a flywheel coupled to permanent magnets synchronous machine. The stored energy is used for drop correction for the critical load. In motoring mode (no drop on the grid), the grid connected converter lets the system deliver power and boosts the DC link voltage until the flywheel reaches the maximum speed. In generating mode, the grid connected converter supplies sinusoidal voltage to the grid. The permanent magnets synchronous machine is selected for the design of a three-phase machine for the flywheel energy storage system and Hysteresis current control is used for controlling both converters.**

Key Words: Flywheel Energy Storage System (FESS), Permanent Magnets Synchronous Machine (PMSM), Voltage Drop, Point of common Coupling (PCC).

I. INTRODUCTION

Modern production lines are moving to more electric based equipment, thus improving the reliability of the overall system. Electrical equipment on such systems will include critical loads that require high power for short periods of time which produces voltage drops on the grid. The current approach to mitigate voltage drops at these periods is to use energy storage systems (ESS) where the required energy to eliminate the drop is stored and then discharged as a pulsed power when it is occur. Now days, energy storage systems are necessary to eliminate the cost and weight of oversized generation equipment to support the desired pulsed power of these applications.

Energy storage can be with batteries, superconducting magnetic energy storage (SMES) coil, or a flywheel. Advantages in no existing memory effect, cost, ruggedness, greater number of charge–discharge cycles, environmental friendliness, and high power density make flywheel a viable alternative.

Among these innovative energy storage systems, flywheel energy storage system (FESS) has an important technical role on the applications of Grid Voltage Support, Grid Frequency Support and Power Quality Improvement and Unbalanced Load Compensation. FESSs have some unique finical features such as longer lifetime and lower requirements than battery systems which make it an important alternative as a replacement for battery in the UPS.

In this paper, a FESS is designed, the power of which can reach 140kVA. The FESS is composed of three main parts: 1) the Flywheel to store or to release the Energy. 2) Permanent Magnets Synchronous Machine. 3) Power Electronics Conditioning Unit.

The Power Electronics Conditioning Unit consists of two power converter working at modes (Inverting/Rectification) according to the flow of the power coupled to each other by a DC link. One converter is holding the deriving of the machine, to accelerate or to decelerate the flywheel to maintain a fixed DC voltage at the DC link. While, the other converter is holding the power quality at the grid by sensing the grid's voltage and phase in order to absorb or to supply power to the grid.

The detailed time domain modeling of each component of a flywheel-energy-storage-system (FESS) for voltage drop correction is presented in this paper. This includes the grid components, PMSM, controllers and flywheel energy storage. The simplified model is used to simulate the operation of pulse loads on a production line using MATLAB/Simulink. The impact of the pulse loads is studied in the presence of the FESS. Modeling was done. The corresponding results validate the proposed design.

II. MODEL SYSTEM

The FESS is used to inject a dynamically controlled current generated by an IGBT converter in series to a bus. The stored energy is used for drop correction for the critical loads. Fig.1. shows an overview of the proposed FESS.

It consists of a flywheel coupled to permanent magnets synchronous machine (PMSM) The flywheel releases the stored energy by reversing the power flow direction and using the motor as a generator. As the flywheel releases its stored energy it slows until it reaches the minimum designed speed.

A. Configuration of Power System

Fig.1. Shows the power system model used in this paper. A power station is modeled by a three-phase voltage source (20kV) with 20 km transmission line. A three phase transformer (20kV/400V phase-phase rms, 2MVA nominal power, 0.4% inductive voltage drop) is connected. A critical resistive load has the value of 10 Ohm and switches to 0.7 Ohm for 1 second each 7 second is connected at the end of the line.

978-1-4244-8806-3/11 $26.00 © 2011 IEEE

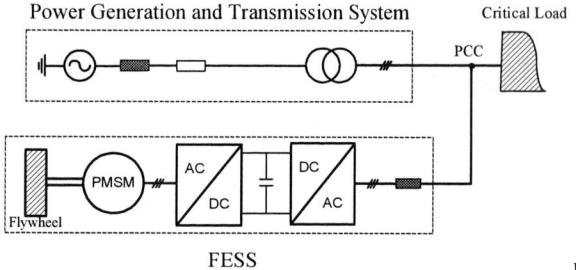

ig. 1. System Model.

B. Configuration of FESS

Fig.2. shows a model configuration of FESS. It consists of a PMSM, a flywheel for energy storage (Kinetic Energy) and two forced commutated converters coupled to each other by a DC link.

The Grid connected converter controls detects the grid's voltage and stabilizes it by supplying or absorbing active/reactive power to/from the grid.

C. Operation of FESS

The FESS has three modes of operation which are Motoring, Generating and stable modes.

In motoring mode (no drop voltage on the grid), the grid connected converter lets the grid deliver power and boosts the DC link voltage. On the other side the machine connected converter trying to maintain the DC link voltage fixed by converting the DC power into AC power supplied to the machine in order to transform it to mechanical power to speed up the flywheel until it reaches the maximum designed speed. A speed limiter is used to stop the power flow from the grid to the FESS. In generating mode (drop voltage on the grid occurred), the grid connected converter supplies active and reactive power to the grid by converting the DC power to AC power which cause a drop on the DC link voltage. This drop on the DC link forces the PMSM connected converter to compensate this drop by reversing the direction of the current on the machine. On other words, in order to compensate the drop on the DC link the flywheel will start decelerating and releases the stored energy as a kinetic energy, this energy will be converted through the machine into AC electrical power and the machine connected converter transforms this power into DC power stored on the DC link Capacitor in order to be transferred to the grid by the grid connected converter.. While the stable mode indicates the absence of voltage drop and the flywheel is running at maximum speed or the drop voltage took a place that the flywheel reaches the designed minimum speed where it is forbidden to the flywheel to run under it.

The development of robust controllers for flywheel energy storage systems has been extensively studied over the past decade [4,5]. Different types of controllers have been proposed for the control design phase. H∞ control, fuzzy logic controllers, nonlinear controllers, and PI-based

vector control systems were used for.

A novel PI-based vector controllers are implemented with the FESS to reach the desired performance in steady state and in transient conditions are introduced in section V. A control algorithm which acts to regulate the operation of the FESS in both charging (motoring) and discharging (generating) modes is developed. The controller of the PMSM connected converter regulates the operation of the FESS by charging the flywheel with constant current during charge mode based on regulating the DC bus voltage during discharge mode under constant power. The grid connected converter controller regulates and monitoring the grid's voltage in order to control the direction of the power flow.

The electromechanical part of a FESS system composed of two subsystems, electrical and mechanical subsystems. The electrical subsystem describes the electrical part on the machine. While the mathematical model of the mechanical subsystem describes the Machine's rotor, shaft, the friction and damping behavior of the system, all of these parameters will be considered on the design.

III. FLYWHEEL MODEL

Flywheels have many advantages as high efficiency, longer life, pollution free and flywheel stores energy in the form of kinetic energy. The amount of energy stored varies linearly with the moment of inertia of the flywheel and the square of its angular velocity. Hence, most of the FESS systems run at very high speeds and sophisticated high speed drives are required [2].

The mechanical subsystem can be modeled as the so-called two-mass model which similar to the drive train of a wind turbine. The model, in graphical form, is presented in Fig.2 and Fig.3. The subsystem can be described by the following set of equations [13,14]:

$$n_{sys} = \frac{1}{j_{machine}} \cdot \int (T_{ele} - T_{shaft}) \cdot dt \qquad (1)$$

$$T_{shaft} = C. \int (n_{sys} - n_{flywheel}) \cdot dt + K.(n_{sys} - n_{flywheel}) \qquad (2)$$

$$T_{flywheel} = T_{shaft} - T_{friction} \qquad (3)$$

$$n_{flywheel} = \frac{1}{j_{flywheel}} \cdot \int T_{flywheel} \cdot dt \qquad (4)$$

$$T_{friction} = F. n_{flywheel} \qquad (5)$$

Where:

n_{sys}: System Speed (rad/sec)

$n_{flywheel}$: flywheel Speed (rad/sec)

T_{ele}: machine Torque (N.m)

T_{shaft}: Shaft Torque (N.m)

$T_{friction}$: Torque generated from the friction (N.m)

K: Spring Constant (N/m)

C: Friction Coefficient

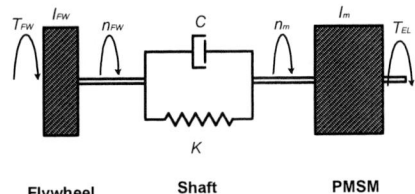

Fig. 2. FESS Mechanical Subsystem Model [2].

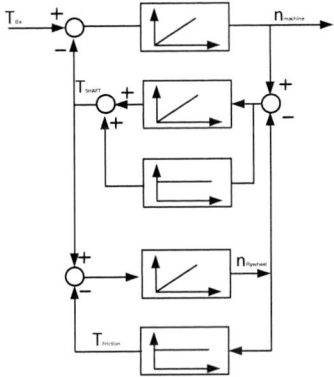

Fig. 3. The Mathematical Model of the Mechanical Subsystem.

IV. PMSM MODEL

The dynamic model of PMSM has been derived in the dq rotating reference frame. Fig.4 shows the equivalent electric circuit of PMSG in dq reference frame The mathematical model of the PMSM in the d-q reference frame, with the voltage and torque equations are given by:

$$V_d = i_d.R_s + L_d \frac{di_d}{dt} - \omega_r.L_q.i_q \qquad (6)$$

$$V_q = i_q.R_s + L_q \frac{di_q}{dt} - \omega_r.L_d.i_d \qquad (7)$$

$$T_e = \frac{3}{2}.p.i_q.(L_d - L_q).i_d \qquad (8)$$

Where vd and vq are voltages in the d-q axis, id and iq are the current in the d-q axis, Rs is the stator resistance, Ld and Lq are the d-q axis inductance, ωr is electrical rotational speed and p is the number of pole pairs.

V. CONTROL METHODS

The basic functions of a controller in a FESS are the detection of voltage drops in the grid, computation of the correcting current, generation of pulse width modulation and to shift the DC-AC inverter into rectifier and vice versa to charge or discharge the capacitor in the DC link in the Absence or occurrence of voltage drops, Fig.5. shows a

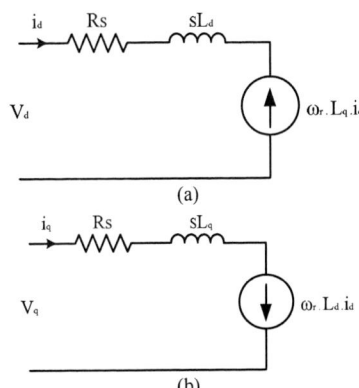

Fig. 4. Equivalent circuit of PMSG in d-q reference frame.

detailed block diagram of the control structure of the FESS.

Park's transformation is used to control the FESS. The dqo method gives the grid's voltage and phase shift information at the whole process time. The quantities are expressed as the instantaneous space vectors. It converts the voltage from abc reference frame to dqo reference. Because of the balanced loads, zero phase sequence component is ignored.

The control scheme for the grid connected converter is based on the comparison of a voltage reference and the measured terminal voltage (Va,Vb,Vc).The voltage drops is detected when the supply drops below the reference value. The error signal is used as a modulation signal that allows generating a commutation pattern for the IGBT constitutes the converter. The commutation pattern is generated by means of the current hysteresis method.

The control scheme for the machine connected converter, is shown in Fig.8, is based on the comparison of a reference DC link voltage and the measured voltage. When the a voltage drop occurs at the grid the DC voltage will be dropped, to maintain the DC voltage fixed a change on the direction of the machine to work as a generator and the speed of the flywheel will be continually decelerating until it reaches it's minimum speed or the drop vanished.

At the absence of a voltage drop at the grid, the voltage

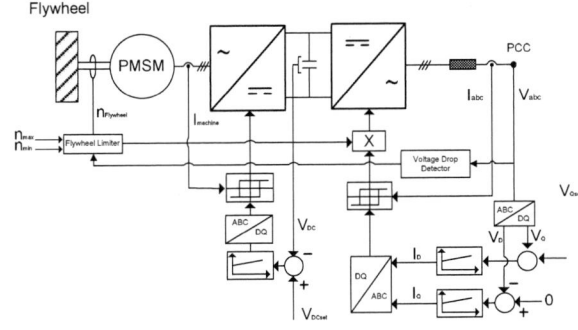

Fig. 5. Detailed Control System of the FESS Model.

978-1-4244-8806-3/11 $26.00 © 2011 IEEE

of the DC link rises which means, the machine will run as a motor and the speed of the flywheel rises until it reaches it's maximum where a speed limiter is used to stop the power flow when the maximum speed is reached.

The limiter based on the flywheel speed and the voltage drop detection information is used to disconnect the FESS from the grid by switching off the converter in the stable mode. In this paper the Monitoring of the D-Component of the voltage in Vector Controller has been used to detect an occurrence of voltage drop at the grid side. A brief description of this technique is held in [17].

PLL technique is used to extract the phase angle of the grid voltages. The SRF-PLL is implemented in dq synchronous reference frame, and its schematic is illustrated in Fig. 9. As it can be noticed, this structure needs the Park's transformation to get the positive

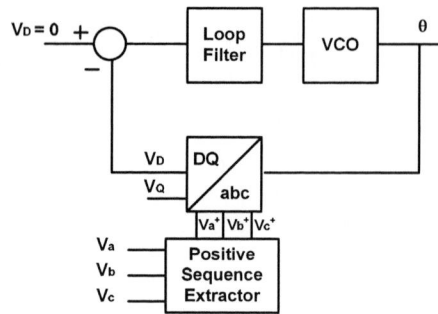

Fig. 9. SRF-PLL Block Diagram..

variables and the lock is realized by setting the reference Vd to zero. A PI controller is used to control this variable, and the output of this regulator is the grid frequency. After the integration of the grid frequency, the phase is obtained.

VI. SIMULATION EVALUATION

A detailed FES system has been modeled by MATLAB/SIMULINK to study the efficiency of suggested structure and control strategy.

A determination of the energy storage capacity is very important for designing energy storage system. The power rating of FESS is decided as 140kVA to supply the grid with a current equals 200 A rms when a three-phase voltage drop at the grid on 1sec and ended at 2sec. is created, resulting in a balanced three-phase voltage on the PCC.

Fig. 10 and Fig. 11. show the flywheel speed at stable, discharging and charging modes, DC link voltage, a single phase voltage and current at the PCC, the current through the FESS, the electromechanical torque of the PMSM and the active / reactive power generated and absorbed from the storage system at the operating modes.

From the simulation results, it can be concluded that the proposed FESS with 140kVA power rating can maintain the grid's voltage of the model system constant at 555V when the drop occurred with an efficiency reached 98.75%. The system start to recover after the absence of the drop and when it reaches the maximum speed the flow of power will be stopped except some pulses needed to the FESS on the stable mode.

VII. CONCLUSIONS

In this paper, a proposed flywheel energy storage system (FESS) for mitigating voltage drop on production lines has been evaluated by simulation. The power system components for voltage drop studies have been modeled and simulated by using a transients program. The model and the results can be used during the design task of the mitigation devices and ride-through alternatives. The main advantage of this FESS is low cost and high energy

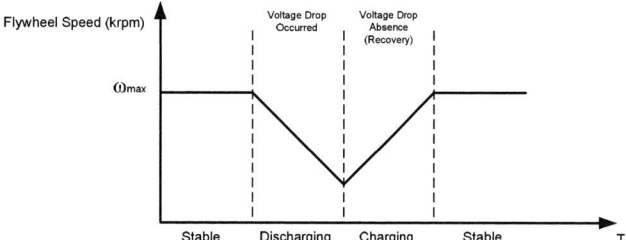

Fig. 6. Flywheel Operating Modes.

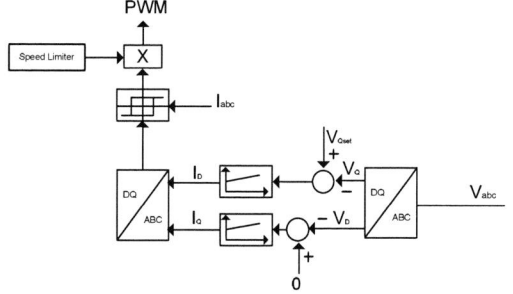

Fig. 7. Grid Connected Converter Control Scheme.

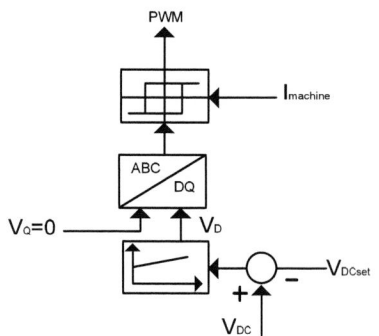

Fig. 8. PMSM Connected Converter Control Scheme.

density. It can mitigate long duration voltage drops efficiently. The future work will include more detailed analysis of the energy storage system for unbalanced drops and comparison with a laboratory flywheel interfaced to an analog model power system.

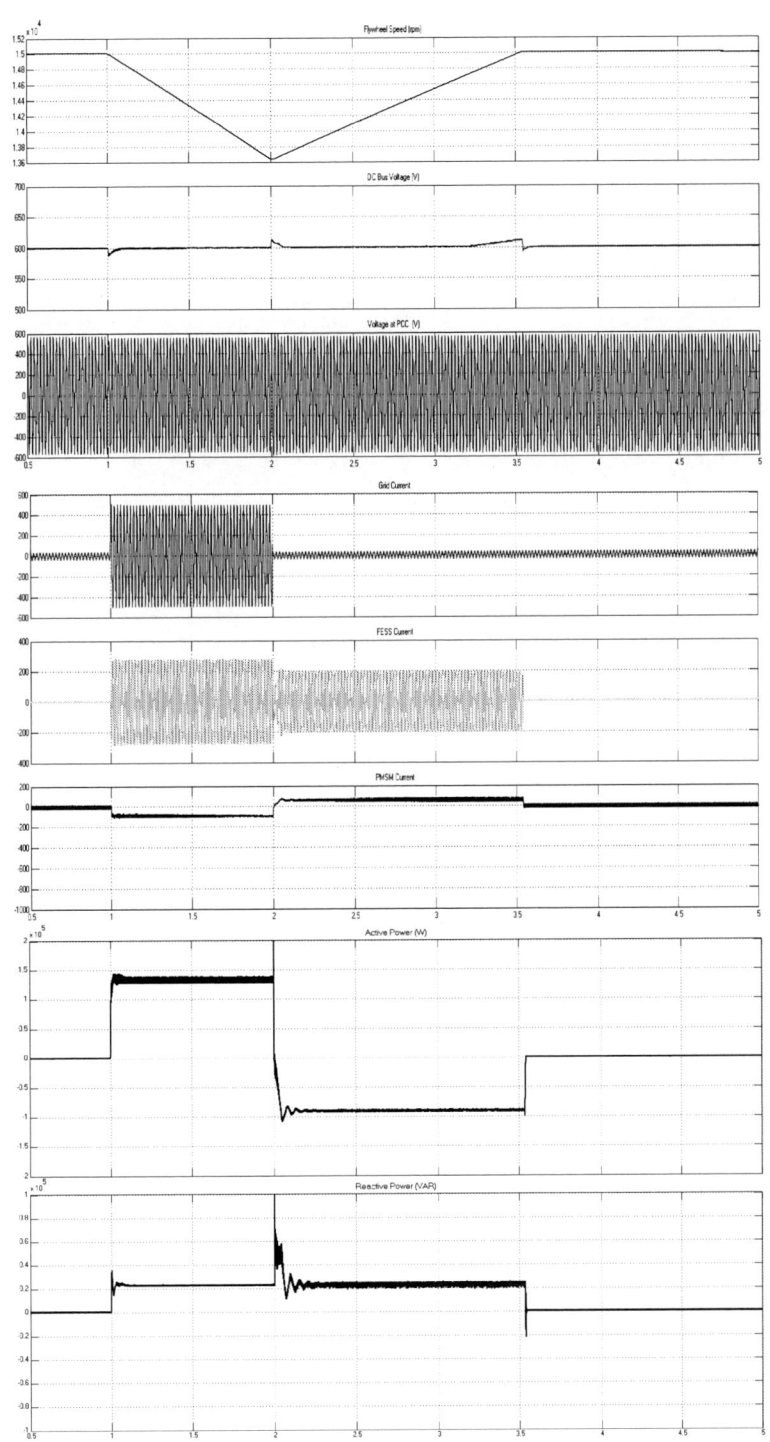

Fig. 10. Time Domain Simulations for FESS's Operating Modes.

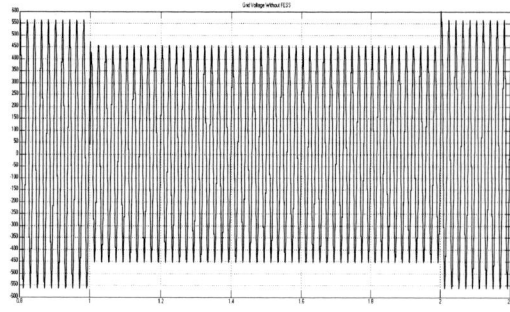

(a) . Grid Voltage on the drop occurrence without FESS.

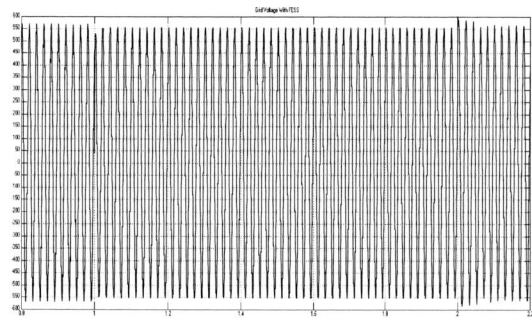

(b) . Grid Voltage on the drop occurrence with FESS.

Fig. 11. Grid Voltage on the drop occurrence with and without FESS.

REFERENCES

[1] M. F. McGranaghan, D. R. Mueller, M. J. Samotyj, "Voltage sags in industrial systems," IEEE Trans. Ind. Applicat., vol. 29, pp. 397–403, Mar./Apr. 1993.

[2] Daryoush Mehrzad, Javier Luque, Marc Capella, "Vector control of PMSG for wind turbine applications," Aalborg University: Electrical Energy Technology, Denmark.

[3] Alaa Mohd, Egon Ortjohann, Andreas Schmelter, Nedzad Hamsic, Danny Morton, "Challenges in Integrating Distributed Energy Storage Systems into Future Smart Grid," IEEE International Symposium on Industrial Electronics, 2008. ISIE 2008.

[4] Satish Samineni, Brian K. Johnson, Herbert L. Hess, Senior, Joseph D. Law, "Modeling and Analysis of a Flywheel Energy Storage System for Voltage Sag Correction," IEEE Transaction on Industry Applications, VOL. 42, NO. 1, Jan./Feb. 2006.

[5] Saurabh Kulkarni, Surya Santoso, "Impact of Pulse Loads on Electric Ship Power System: With and Without Flywheel Energy Storage Systems," Electric Ship Technologies Symposium, 2009. ESTS 2009. IEEE, April 2009.

[6] Ken Furusawa, Hideharu Sugihara,Kiichiro Tsuji,Yasunori Mitani, "A new operation framework of demand-side energy storage system cooperated with power system operation," international Conference on Power System Technology - POWERCON 2004.

[7] W. H. Kim, J. S. Kim, J. W. Baek, H. J. Ryoo, G. H. Rim, S. K. Choi "Improving Efficiency of Flywheel Energy Storage System with A New System Configuration," 29th Annual IEEE Power Electronics Specialists Conference, 1998. PESC 98 Record..

[8] James Larminie, John Lowry " Electric Vehicle Technology Explained," John Wiley & Sons, 2003ISBN 978-0-470-85163-0.

[9] Rion Takahashi, Junji Tamura, "Frequency Stabilization of Small Power System with Wind Farm by Using Flywheel Energy Storage System," IEEE International Symposium on Diagnostics for Electric Machines, Power Electronics and Drives, 2007. SDEMPED 2007.

[10] Hanmin Lee, Euijin Joung, Gildong Kim, Cheonheon An, "A Study on the Effects of Energy Storage System," International Conference on.Information and Multimedia Technology, Dec.2009. ICIMT '09.

[11] C. Sourkounis, "Active Dynamic Damping of Torsional Vibrations by H4-Control," 12th International Conference on Optimization of Electrical and Electronic Equipment, OPTIM 2010, 2010.

[12] Sourkounis, C.; Wenske, J.; Richter, F, "Dynamic Conditioning of Stochastic Fluctuating Energy from Wind Parks," 15th Power Systems Computation Conference (PSCC `05), 2005.

[13] Sourkounis, C.; Ni, B., "Optimal Control Structure to Reduce the Comulative Load in the Drive Train of Wind Energy Converters," European Conference on Power Electronics and Applications (EPE 2005), 2005.

[14] Sourkounis, C.; Ni, B., "Drive Train Control for Wind Energy Converter Based on Stochastic Dynamic Optimization," IEEE Industrial Electronics Conference (IECON 2006), Special Session: Self-optimizing systems and advanced control, 2006.

[15] Satoshi Uemura, Shinichi Nomura, Ryuichi Shimada., "Stabilization of Electric Power System using the Variable Speed Flywheel Generator," Proceedings of Power Conversion Conference - PCC '97.

[16] N.Hamsic, A.Schmelter, A.Mohd, E.Ortjohann.E.Schultze, A.Tuckey, J.Zimmermann, "Stabilising the Grid Voltage and Frequency in Isolated Power Systems Using a Flywheel Energy Storage System," The Great. Wall World Renewable Energy Forum, 2006.

[17] Chris Fitzer, Mike Barnes, Peter Green "Voltage Sag Detection Technique for a Dynamic Voltage Restorer," IEEE Transaction on Industry Applications, VOL. 40, NO. 1, Jan/Feb 2004.

[18] Changjiang Zhan, Vigna Kumaran Ramachandaramurthy, Atputharajah Arulampalam, Chris Fitzer, Stylianos Kromlidis, Mike Barnes, Nicholas Jenkins, "Dynamic Voltage Restorer Based on Voltage-Space-Vector PWM Control," IEEE Transaction on Industry Applications, VOL. 37, NO. 6, Nov./Dec. 2001.

Modular-Based PFC for Low Power Three-Phase Wind Generator

Freddy Flores-Bahamonde*, Hugo Valderrama-Blavi*, Josep M. Bosque*, Luís Martínez-Salamero*

DEEEA, Universitat Rovira i Virgili, Avinguda Països Catalans 26, Tarragona, CAT, Spain.

freddy.flores@urv.cat, hugo.valderrama@urv.cat

Abstract—**Hybrid Distributed systems include many sources, storage elements, and various local loads, connected to a common distribution bus. In this work, different matching methods to adapt a 3-phase wind alternator to a variable DC-bus are reviewed, and finally a modular solution based on single-phase Boost-based PFC's is proposed. Nevertheless, Boost modules must be adapted to operate with non-isolated sources, like 3-4 wire, 3-phase generators. Magnetic coupling is introduced here to enhance the isolation among phases, reducing also converter losses and size. Sliding mode approach has been applied to the whole converter as a single unit, to verify that independent regulation of all phase modules was possible, even sharing a common output capacitor. To verify the theoretical analysis, a 1.5 kW prototype has been built confirming independent phase control and good sinusoidal input waveforms.**

I. INTRODUCTION

GREENHOUSE emissions can be significantly reduced using small renewable generation sources. Energy production optimization requires improvements in generators, conversion stages, and transport networks. Distributed generation (DG) development can contribute to generate electricity with lower transport losses and pollution.

According to its Electricity Directive, the European Union (EU) defines distributed generation as the whole suite of power plants connected to the distribution network [1].

Nevertheless, main interest of DGS is to increase efficiency approaching generation sites to the consumption centers. In this context, big power plants far from consumers imply excessive transport costs, so DGS are oriented to low power production centers, frequently based on renewable energies.

Common renewable-based generation, very dependent on weather conditions, have a grid penetration limit around 25% because these systems consider the electrical grid an infinite sink, causing instability problems, especially at weak grids.

Small DGS include normally storage devices and different energy sources creating an energy reservoir. As a result, these systems can implement active grid-cooperation promoting their expansion. In few words, active policies intend that small production systems behave like conventional power plants, abandoning the infinite sink conception of the electrical grid.

In practical terms, this means that: primary frequency regulation, sag compensation, predictable/constant energy injection, and voltage regulation at connection point, will be mandatory soon [2].

A distributed generation system is under development in our laboratory. A variable voltage DC Bus, (270-370 V) is the system core. The plant is tied to the grid in a single connection point (PCC) by a 6 kW inverter. All the generators, storage elements, and loads are connected to the DC-bus through an adaptor circuit [3]. The main features of the DC microgrid elements are given in Table I.

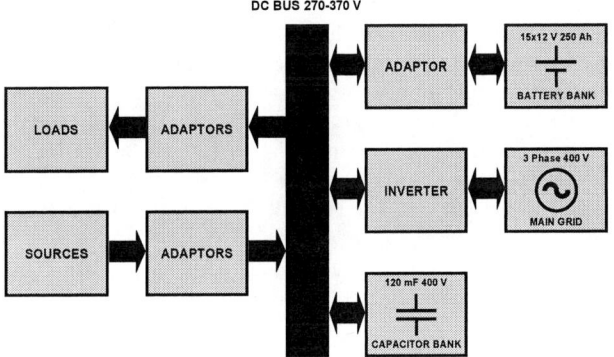

Fig. 1. GAEI Laboratory DC-Microgrid

As can be seen in Table I, user-programmable and weather-dependent sources have been considered, the last ones regulated by MPPT controllers. Controllable devices are required to investigate microgrid operation under diverse generation and consumption profiles. Besides, controllable generators behave as energy reservoir for active grid-policies.

TABLE I
MAIN DC-MICROGRID ELEMENTS

Microgrid Element	Size	Voltage	Power Control
PV-Field (Two)	2x1.62 kW	259 V_{oc}	MPPT
PV-Field (Two)	2x1.02 kW	252 V_{oc}	MPPT
Lead-Acid Battery Bank	250 Ah	180 V_{nom}	user
Ballard Nexa Fuel Cell	1.2 kW	26 V_{nom}	user
FS Wind Generator	1 kW	0-85 (3ϕ)	MPPT
Wind Gener. Work-Bench	1kW	0-100 (3ϕ)	user
Diesel Generator	5.5 kW	400 (3ϕ)	user
Controllable Loads (Three)	3x2 kW	80-400 V	user

978-1-4244-8806-3/11 $26.00 © 2011 IEEE

The Wind Work-Bench in Table I, is made coupling mechanically a wind permanent magnet alternator with an induction motor driven by a variable frequency inverter. The controllable loads, commercial MPPT inverters supplied from programmable sources, are non-dissipative.

Connecting a three-phase AC source to a DC-bus, is a quite complicated issue, motivating this work. The remaining part of this paper is organized as follows. At Section II a quick review of 3-phase PFC's is presented. Next, at Section III and IV, the proposed converter and its control are discussed, and finally at Section V some experimental results are given.

II. THREE-PHASE PFC APPROACHES

Design of high efficiency PFC converters with reduced harmonic content, low EMI, cost, and size continues to be an interesting research subject. Many different PFC approaches assuring a good sinusoidal input current and output voltage regulation can be found in the literature.

Boost converter and derived topologies are very common at PFC applications. Indeed, elementary boost is frequently seen at single-phase PFC's, mainly due to the input current source characteristic [4]. For 3-phase systems, several PFC solutions can be proposed. To reduce the work scope, we analyze only the PFC interface between a 3-phase generator and a higher voltage DC-Bus.

Fig. 2 depicts two common PFC topologies for 3-phase systems. The upper part of the figure shows the Six-Switch Boost Rectifier. The second circuit, derived also from the classical Boost converter, is the Vienna rectifier.

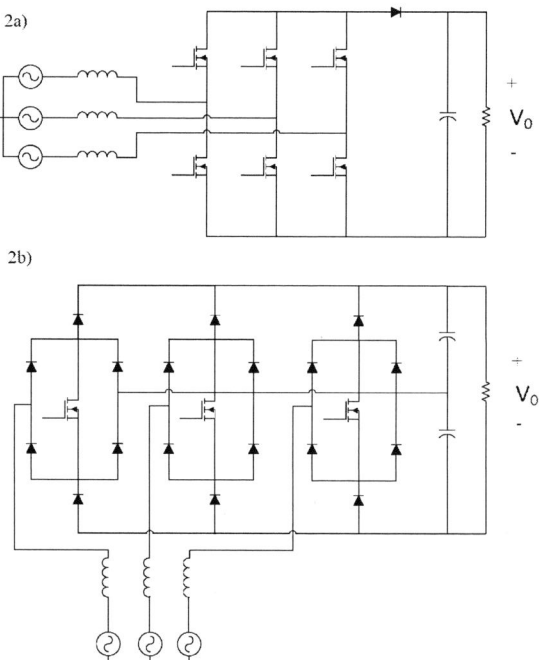

Fig.2. Examples of Boost-derived 3-phase rectifiers

Various versions of Vienna rectifier [5] can be found in the literature. The version shown in fig. 2b reduces the stress on switching devices using the neutral wire to split the output voltage. Despite of the stress reduction, this system can operate only with star connected 3-phase generators.

Six-Switch Rectifier [6] main advantage is a good input current shaping capability. Its main drawback is that requires double number of active switches than other approaches, namely Vienna Rectifier, and the circuit is not modular and has no redundancy. Therefore, if a failure occurs, all the generated power is lost. Conversely, redundancy is precisely the main advantage of Vienna rectifier and other proposals, like modular PFC appearing, in fig.3.

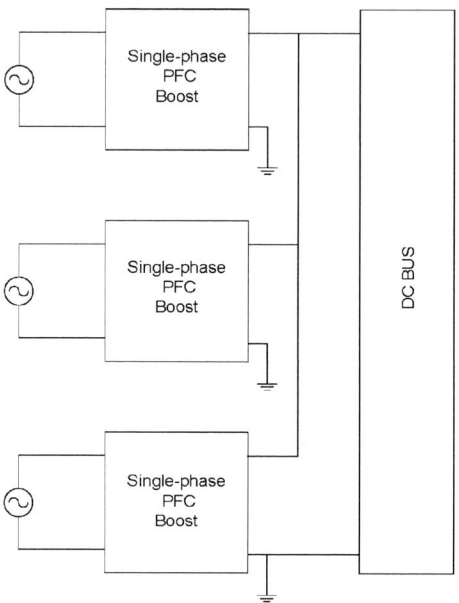

Fig.3. Modular 3-Phase PFC rectifier

The third PFC proposal, based on modularity, requires the same number of devices as the Vienna Rectifier, and even 3 diodes less, if the input sources are floating as in shown in fig.3. Besides, by means of a Scott transformer the system can be converted to a bi-phase system [7], reducing the number of single-phase modules and devices. The modular approach main problem appears when input sources are not floating, as in a conventional 3-4 wire generator. Then, the current into a phase module is different from its return current, causing control interference among the three phases.

To solve this problem, some researchers [8] have proposed to change the converter topology introducing a transformer into each PFC. This solution, that avoids stage-interaction keeping the input phases balanced, reduces the converter efficiency due to the transformer leakage losses, and increases the PFC cost and size.

References [9-10] propose a second solution to improve isolation among stages, although in this case, Boost converter topology is retained, the diode and the inductor, are both split, as explained in the next section.

978-1-4244-8806-3/11 $26.00 © 2011 IEEE   126

The work presented in this paper continues that research work, improving the isolation performance using coupled inductors.

III. CIRCUIT DEVELOPMENT

In the previously section, a modification of a common boost topology to keep the input phases balanced is mentioned. In this section, we explain the interaction problem appearing in the modular PFC (fig. 3) when the single-phase modules are connected to a 3-4 wire 3-phase generator, instead of being connected to three floating generator (six-wire connection). Simultaneously, the solutions found in the literature are discussed, and after, the improvements proposed in this work are presented. An experimental 1.5 kW prototype, presented in Section V, will prove the system performance.

A. Previous System.

The interaction problems can be easily understood analyzing the circuits shown in fig.4. Here, for three-phase four-wire input system, three single-phase PFC rectifiers are connected to the same output DC-link capacitor without isolation. We will explain progressively how can be partially solved, first splitting the diode and then splitting the inductor.

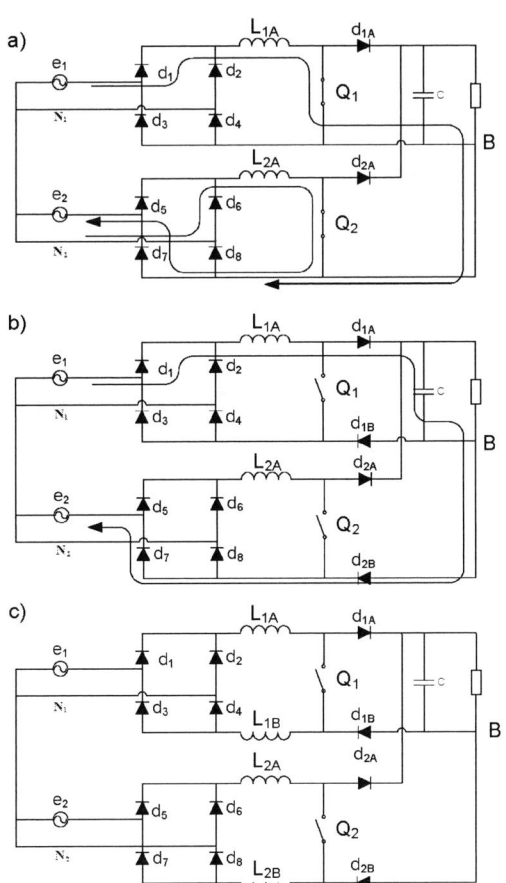

Fig.4.Single-Phase PFC converter, (a) Common Boost. (b) Boost with diode split. (c) Boost with diode and inductor split

To simplify the explanation, only two of the three single-phase PFC rectifiers are considered, see fig 4a. In this circuit, any switches combination can be made. To analyze the circuit, let's consider a case where the voltage phases have opposite sign ($e_1>0$ and $e_2<0$) and both switches are in the ON state. In this moment, node B (fig. 4) is more negative than e_1 neutral wire (N_1), because e_2 has a negative voltage. This blocks D_4 diode in the upper stage, and then, the upper converter current goes through the second converter.

This first interaction problem can be partially solved inserting a second freewheeling diode D_{1b} as can be seen in fig.4b. As this diode is reverse biased, the current cannot go through the second converter, and goes through the first one.

Until now, we have solved the interaction problem while the mosfets are both at ON-state. Next case to examine is when both switches are in the OFF-state. This case is shown at fig. 4b. In this moment, B node is still more negative than N_1 wire, diodes D_4 and D_{1b} are blocked, and the current of the upper stage returns through the second converter again. To solve that, a second inductor L_{1b} is added between the active switch and D_4, forcing the conduction of diode D_{1b}, see fig. 4c.

If these two corrections are applied to all boost inductors, and freewheeling diodes, each input current should return through its own stage, balancing the system [9-10].

B. Final Proposed Circuit.

To enhance the effect of the split inductor, we propose to couple the original boost inductor and the second one, as can be seen in the single-phase boost circuit of fig. 5.

With the coupled inductor, current ripple, cost, and weight reduction can be achieved. Besides, a transformer effect is reached, forcing the system to keep balanced.

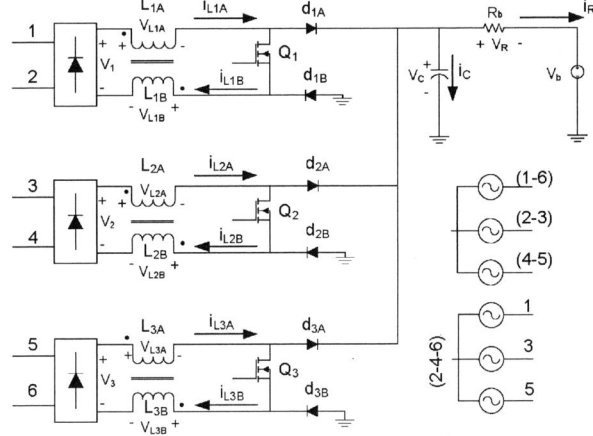

Fig.5.Single-Phase three-phase PFC converter proposed circuit

Coupled inductors have an important role in power converters. Many authors used coupled inductors to correct non-minimum phase response in boost, and similar converters [11-13]. Others [14-15] used them to get ideal zero current-ripple in some DC-DC converters. In fact, the induced voltage effect caused a significant ripple reduction, and the ripple slope was related to the coupling factor K.

978-1-4244-8806-3/11 $26.00 © 2011 IEEE

Ripple reduction has also other positive effects like EMI reduction. Higher efficiency can be achieved because the number of magnetic cores is halved. Besides, the RMS value of the current is slightly reduced, and therefore the system conduction losses.

Nevertheless, our system is not working at fixed frequency, because the control-loop is based on sliding mode approach. In that case, ripple is determined by the hysteresis band controller, so instead of ripple amplitude reduction, the main effect of coupling is switching frequency reduction.

As presented in the next section, if coupling is perfect and both inductors are equal, the equivalent input inductance is increased by a factor 4. Therefore, in our case, the switching losses are divided also by four.

On other hand, under the same hypothesis about perfect coupling and equal size inductors, the coupled inductors behave like a transformer with a turns ratio of 1. This means that both inductor currents i_{LA} and i_{Lb}, have to be equal, enhancing the balancing effect of the previously explained split or second inductor.

IV. CIRCUIT ANALYSIS AND CONTROL

Circuit in fig.5 has three boost cells connected in parallel, sharing the output capacitor. Interaction among stages has been apparently solved but, dynamic responses of each converter could be coupled through the output common capacitor. In this section we will prove that independent phase control is possible even sharing the output capacitor. Therefore single phase analysis could be sufficient to describe the circuit operation.

A. Single-Phase Circuit State-Space Model

Figure 6 shows the electrical model of two coils magnetically coupled. The boost converter is a variable structure system, where the switch action distinguishes two circuit topologies, as can be seen in fig. 7a to 7c. For control purposes, a binary signal $u(t)=\{0,1\}$ is used to indicate each topology, where ON topology is associated to $u(t)=1$

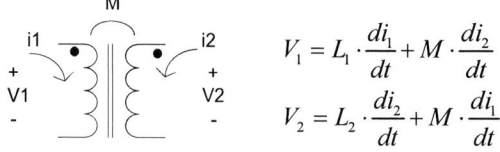

$$V_1 = L_1 \cdot \frac{di_1}{dt} + M \cdot \frac{di_2}{dt}$$

$$V_2 = L_2 \cdot \frac{di_2}{dt} + M \cdot \frac{di_1}{dt}$$

Fig.6. Electrical Model of Coupled inductor

In the circuits of fig. 7a to 7c, V_b is the DC-Bus voltage, the input generator and the full-bridge are replaced by a voltage source V_1, and finally R_b, L_1 and C are the remaining circuit parameters, where $L_1=L_{1A}=L_{1B}$. According to the circuit topologies, the state descriptions are given below.

$$\overset{o}{X}_{ON} = \begin{bmatrix} 0 & 0 \\ 0 & -1/R_bC \end{bmatrix} \cdot \begin{bmatrix} i_L \\ v_c \end{bmatrix} + \begin{bmatrix} V_1/4L_1 \\ V_b/R_bC \end{bmatrix} \qquad (1)$$

$$\overset{o}{X}_{OFF} = \begin{bmatrix} 0 & -1/4L_1 \\ 1/C & -1/R_bC \end{bmatrix} \cdot \begin{bmatrix} i_L \\ v_c \end{bmatrix} + \begin{bmatrix} V_1/4L_1 \\ V_b/R_bC \end{bmatrix} \qquad (2)$$

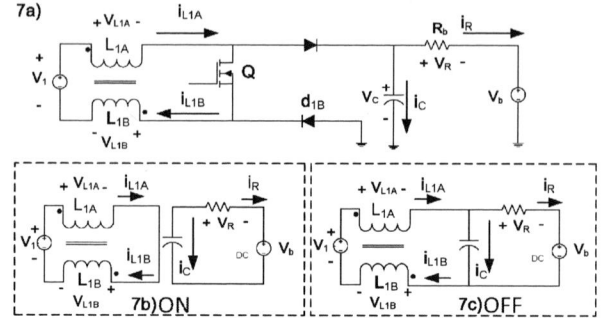

Fig.7. Circuital scheme of Single Phase Converter with coupled inductor. (a) On State (b) OFF State

B. Full Circuit Analysis

Considering the full circuit of fig. 5, the three switches are controlled independently, and eight possible switching states can appear. The switches Q_1, Q_2, Q_3, control signals are respectively u_a, u_b, u_c, where $\{u_a, u_b, u_c\}=\{0,1\}$.

After analyzing the eight circuit topologies, it has been found that the dynamic matrix A_{abc} depends on the circuit topology, but the excitation matrix is always the same. The state vector is given as (3), equation (4) describes each topology, and (5) gives the values of A_{abc} and B.

$$X^T = \begin{bmatrix} i_{L1}, & i_{L2}, & i_{L3}, & V_C \end{bmatrix}^T \qquad (3)$$

$$\overset{o}{X} = A_{abc} \cdot X + B \qquad (4)$$

$$A_{abc} = \begin{bmatrix} 0 & 0 & 0 & \frac{u_a-1}{4L} \\ 0 & 0 & 0 & \frac{u_b-1}{4L} \\ 0 & 0 & 0 & \frac{u_c-1}{4L} \\ \frac{1-u_a}{C} & \frac{1-u_b}{C} & \frac{1-u_c}{C} & \frac{-1}{R_bC} \end{bmatrix} \quad B = \begin{bmatrix} \frac{V_1}{4L} \\ \frac{V_2}{4L} \\ \frac{V_3}{4L} \\ \frac{V_b}{R_bC} \end{bmatrix} \qquad (5)$$

By multiplying each topology equation by its activation function, a converter full-time model can be found.

$$\begin{aligned}
&\left[\overset{o}{X} = (A_{000} \cdot x + B)\right] \cdot (1-u_a) \cdot (1-u_b) \cdot (1-u_c) \\
&+\left[\overset{o}{X} = (A_{001} \cdot x + B)\right] \cdot (1-u_a) \cdot (1-u_b) \cdot u_c \\
&+\left[\overset{o}{X} = (A_{010} \cdot x + B)\right] \cdot (1-u_a) \cdot u_b \cdot (1-u_c) \\
&+\left[\overset{o}{X} = (A_{011} \cdot x + B)\right] \cdot (1-u_a) \cdot u_b \cdot u_c \\
&+\left[\overset{o}{X} = (A_{100} \cdot x + B)\right] \cdot u_a \cdot (1-u_b) \cdot (1-u_c) \\
&+\left[\overset{o}{X} = (A_{101} \cdot x + B)\right] \cdot u_a \cdot (1-u_b) \cdot u_c \qquad (6) \\
&+\left[\overset{o}{X} = (A_{110} \cdot x + B)\right] \cdot u_a \cdot u_b \cdot (1-u_c) \\
&+\left[\overset{o}{X} = (A_{111} \cdot x + B)\right] \cdot u_a \cdot u_b \cdot u_c \\
\hline
&\overset{o}{X} = f(X) + g_a(X) \cdot u_a + g_b(X) \cdot u_b + g_c(X) \cdot u_c
\end{aligned}$$

where $f(X)$, $g_a(X)$, $g_b(X)$, and $g_c(X)$ are given by (7)

$$
\begin{aligned}
f(X) &= A_{000} X + B \\
g_a(X) &= \left[A_{100} - A_{000} \right] \cdot X \\
g_b(X) &= \left[A_{010} - A_{000} \right] \cdot X \\
g_c(X) &= \left[A_{001} - A_{000} \right] \cdot X
\end{aligned}
\qquad (7)
$$

After analyzing the model, it can be easily deduced that we have a bilinear system with three control inputs, where each current is affected only by the respective control input, whereas the capacitor voltage is affected by the three.

C. Sliding Mode Control Laws

To assure unity power factor, LFR concept is used. According to that, the generator will see the converter like three resistors, one per control input, and each input current will be in phase with the corresponding input voltage.

$$
S(X) = \left[i_{L1} - g_1 V_1, \quad i_{L2} - g_2 V_2, \quad i_{L3} - g_3 V_3, \quad 0 \right] = 0 \quad (8)
$$

The equivalent control is obtained when the first derivative of S(x) is equal to zero, as shown in (9). Considering the system matrices, the equivalent control is given by (10) where all the transversality conditions are satisfied.

$$
\overset{o}{S} = \nabla S(X) \cdot \overset{o}{X} = \nabla S \cdot \left(f(X) + g_a(X) \cdot u_a + \dots \right.
$$
$$
\left. \dots + g_b(X) \cdot u_b + g_c(X) \cdot u_c \right) = 0
\qquad (9)
$$

$$
u_{eq} = \left[1 - \frac{V_1}{V_C}, \quad 1 - \frac{V_2}{V_C}, \quad 1 - \frac{V_3}{V_C} \right]
\qquad (10)
$$

The ideal sliding dynamics (11) are obtained considering the surface constraints, and substituting the equivalent control in the bilinear expression (6). Realize that the capacitor voltage dynamics is non linear, and the system dynamics is reduced from four to first order. In case of negligible bus impedance, the converter would have no dynamics.

From the equilibrium point, at zero bus impedance $R_b=0$, the capacitor and the bus voltage are the same. If there is no DC-Bus ($V_b=0$), all the generated power is delivered directly to the "converter load" R_b, and the sliding motions only will exists while $V_C > \max(V_1, V_2, V_3)$.

$$
\begin{cases}
i_{L1} = g_1 V_1, \quad i_{L2} = g_2 V_2, \quad i_{L3} = g_3 V_3 \\
\overset{o}{V}_C = \dfrac{g_1 V_1^2 + g_2 V_2^2 + g_3 V_3^2}{C V_C} + \dfrac{V_b - V_C}{R_b C}
\end{cases}
\qquad (11)
$$

$$
\begin{aligned}
V_C^* (R_b \to 0) &= V_b \\
V_C^* (V_b \to 0) &= \sqrt{R_b \left(g_1 V_1^2 + g_2 V_2^2 + g_3 V_3^2 \right)}
\end{aligned}
\qquad (12)
$$

By linearizing capacitor dynamics around the equilibrium point, the local stability can be easily verified. There is a first order pole placed in the left-side of the s-plane.

$$
s = -\left[\frac{g_1 V_1^2 + g_2 V_2^2 + g_3 V_3^2}{C V_C^{*2}} + \frac{1}{R_b C} \right]
\qquad (13)
$$

V. Experimental Results

To match a 3-phase wind generator with the microgrid DC voltage, 270 V$<V_b<$370 V, a prototype has been developed. The passive components are L=150 μH, C=33 μF. The DC-Bus circuit can be appreciated in fig. 8. The bus is supplied from two power-supplies, connected in series, with the output current limited to 20 A. The first one gives 270 V, and the second, the remaining voltage. A blocking diode avoids inverse currents to the power-supplies that may cause their destruction. The bus has a load of R_L=60 Ω, this means 1.7 kW or 5.33 A at 320 V, an impedance of R_b=50 mΩ, and finally a capacitor bank of C_b=28,200 μF.

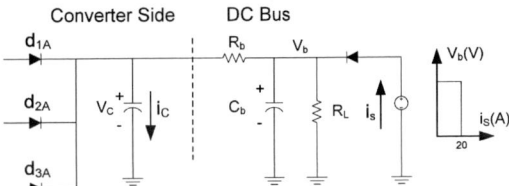

Fig. 8. Converter Connection with Test DC-Bus

Photograph in fig.9 depicts the prototype, the DC-Bus, and the Work-Bench made with a Whisper 200 alternator, an induction machine, and a variable frequency inverter.

Fig. 9. Prototype, DC-Bus and Wind-Bench

Two different experiments have been carried out. During first experiment, shown at fig 10a-10b, 1.4 kW are delivered to the DC-Bus with a steady voltage of V_b=320 V. During second experiment, depicted in Fig. 10c, the DC-Bus changes slowly, during 75ms, from V_b=285 V to V_b=355 V.

In the first experiment, when the wind generator is stopped, the power-supply feeds the load, delivering 5.32 A at 320 V. When the generator begins to turn, the current given by the power supplies progressively decreases, until a minimum value of 1.2 A is reached. At this point, the power injected to the microgrid is slightly over 1.3 kW.

As can be seen at fig 10a, the input currents have a good sinuosidal shape. There are slight differences in amplitude, due to small differences in conductance g_1, g_2, g_3 values.

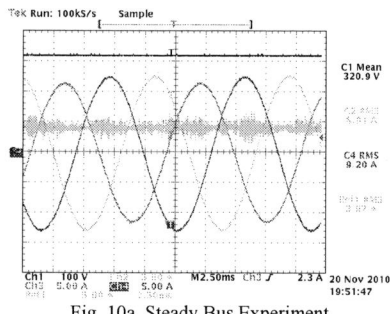

Fig. 10a. Steady Bus Experiment

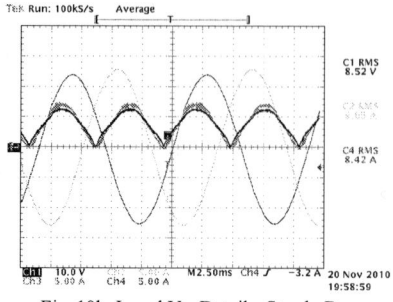

Fig. 10b. I_L and V_{ref} Details, Steady Bus

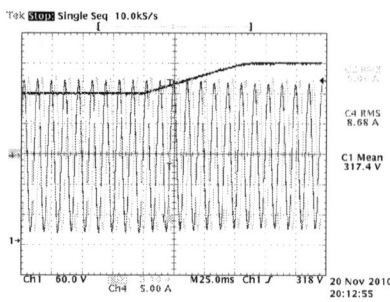

Fig 10c. DC-Bus Voltage Transient

In oscilloscope caption of fig. 10b, we can appreciate one of the inductor currents I_L, as well as its reference signal, that is proportional, to the rectified input voltage. Inductor ripple current is negligible because the oscilloscope is working in the "Average" mode.

The transient experiment is shown at fig. 10c. The bus voltage slope during the transient, is defined by the capacitor bank C_b and the maximum available current, 20 A.

The converter presented in this work should perform like a power source. Consequently, if the bus voltage increases, the current injected in the bus must decrease, and the converter power would remain unchanged. Therefore, the converter input currents should remain constant for any bus voltage. The caption at fig. 10c shows a good agreement between practical results and theoretical hypothesis, as the converter input currents remain practically constant.

VI. CONCLUSIONS

In this work, three-phase modular PFC's developed from single-phase boost PFC circuits are investigated.

When a single-phase boost PFC is connected in a poly-phase system without isolation either at the input, or at the output, balancing problems among input currents appear.

To avoid these undesired interactions, diverse techniques can be found in the literature. Magnetic coupling has been introduced to enhance the isolation among phases, mainly due to the transformer effect, forcing the input and output wire currents to be equal. Besides, magnetic coupling incorporates also other improvements like losses, cost, and weight reduction. Namely, the number of magnetic cores can be halved if compared to previous approaches.

To assure good sinusoidal waveforms, sliding mode control and Loss-Free Resistor have been considered. Good experimental waveforms have corroborated our analysis.

Development of three-phase PFC systems with a reduced number of storage elements, as well as new hybrid PFC architectures, will continue this work.

VII. ACKNOWLEDGEMENT

This work has been partially sponsored by the Spanish Ministry of Research and Science, under grants: DPI2009-14713-C03-02, and Consolider RUE CSD2009-00046.

VIII. REFERENCES

[1] M. Scheepers et al. "Regulatory Improvements for Effective Integration of Distributed Generetion into Electricity Distribution Networks", Summary of the DG-GRID project results. ECN-E-07-083; 2007.

[2] Corin Millais, Luisa Colasimone, "Large Scale Integration of Wind Energy in the European Power Supply: Analysis Issues and Recommendations", Report of The European Wind Energy Association (EWEA), 2005.

[3] Valderrama-Blavi, H., Bosque-Moncusi, J.M., Marroyo, L., Guinjoan, F., Barrado, J.A.; Martinez-Salamero, L., "Adapting a low voltage PEM fuel-cell to domestic grid-connected PV system," Industrial Electronics, 2009. IECON '09. 35th Annual Conference of IEEE, vol., no., pp.160 165, 3-5 Nov. 2009.

[4] Sebastian, J., Jaureguizar, M.; Uceda, J., "An overview of power factor correction in single-phase off-line power supply systems," Industrial Electronics, Control and Instrumentation, 1994. IECON '94., 20th International Conference on , vol.3, no., pp.1688-1693 vol.3, 5-9September 1994.

[5] Alahuhtala, J., Tuusa, H., "Four-Wire Unidirectional Three Phase /Level / Switch (VIENNA) Rectifier," IEEE Industrial Electronics, IECON 2006 - 32nd Annual Conference on , vol., no., pp.2420-2425, 6 10 Nov. 2006.

[6] Hengchun Mao, Boroyevich, D., Ravindra, A., Lee, F.C. , "Analysis and design of high frequency three-phase boost rectifiers ," Applied Power Electronics Conference and Exposition, 1996. APEC '96. Conference Proceedings 1996., Eleventh Annual , vol.2, no., pp.538-544 vol.2, 3-7 March 1996.

[7] Jaehong Hahn, Enjeti, P.N., Pitel, I.J. , "A new three-phase power-factor correction (PFC) scheme using two single-phase PFC modules," Industry Applications, IEEE Transactions on , vol.38, no.1, pp.123-130, Jan/Feb 2002.

[8] Chapman, D., James, D., Tuck, C.J., "A high density 48 V 200 A rectifier with power factor correction-an engineering overview," Telecom-munications Energy Conference, INTELEC '93. 15th International, vol.1, no., pp.118-125 vol.1, 27-30 Sep 1993.

[9] Spiazzi, G., Lee, F.C., "Implementation of single-phase boost power-factor-correction circuits in three-phase applications," Industrial Electronics, IEEE Transactions on , vol.44, no.3, pp.365-371, Jun 1997.

[10] Calvente, J., Martinez-Salamero, L., Valderrama, H., Vidal-Idiarte, E., "Using magnetic coupling to eliminate right half-plane zeros in boost converters," Power Electronics Letters, IEEE , vol.2, no.2, pp. 58- 62, June 2004

[11] Sanchis-Kilders, E., Ferreres, A., Maset, E., Ejea, J.B., Esteve, V.; Jordan, J., Calvente, J., Garrigos, A., "Bidirectional High-Power High-Efficiency non-isolated step-up DC-DC Converter," Power Electronics Specialists Conference, 2006. PESC '06. 37th IEEE , vol., no., pp.1-7, 18-22 June 2006

[12] Sable, D.M., Cho, B.H., Ridley, R.B., "Elimination of the positive zero in fixed frequency boost and flyback converters," Applied Power Electronics Conference and Exposition, 1990. APEC '90, Conference Proceedings 1990., Fifth Annual , vol., no., pp.205-211, 11-16 Mar 1990

[13] Milanovic, M., Mihalic, F., Jezernik, K., Milutinovic, U., "Single phase unity power factor correction circuits with coupled inductance," Power Electronics Specialists Conference, 1992. PESC '92 Record., 23rd Annual IEEE , vol., no., pp.1077-1082 vol.2, 29 Jun-3 Jul 1992

[14] Slobodan Cuk, "DC-to-DC switching converter with zero input and output current ripple and integrated magnetics circuits". United States Patent 4,257,087, from March 17, 1981.

[15] Sira-Ramirez, H., "Sliding motions in bilinear switched networks," Circuits and Systems, IEEE Transactions on , vol.34, no.8, pp. 919- 933, August 1987.

[16] Alves, R.L.; Barbi, I.; "Analysis and Implementation of a Hybrid High-Power-Factor Three-Phase Unidirectional Recitfier", Power Electronics, IEEE Transactions on, vol. 24, no. 3, pp. 632-640, March 2009 .

Frequency Dependent Grid-Impedance Determination with Pulse-Width-Modulation-Signals

Michael Jordan, Hauke Langkowski, Trung Do Thanh, Detlef Schulz

Helmut-Schmidt-University, Faculty of Electrical Power Systems, Holstenhofweg 85, 22043 Hamburg, Germany

michael.jordan@hsu-hh.de, hauke.langkowski@hsu-hh.de, trung.dothanh@hsu-hh.de, detlef.schulz@hsu-hh.de

Abstract- **The power quality in electrical grids is increasingly influenced by generators and loads with power electronic grid coupling. Even comparatively small harmonic currents caused by these units can lead to unacceptable high voltage harmonics at the point of common coupling (PCC) due to highly resistive resonance points of the grid impedance. Therefore, the effective grid perturbation is defined by the frequency dependent grid impedance at the PCC. In this contribution an approach is presented and analyzed to measure the frequency and time dependent grid impedance. The measured grid impedance is an important parameter for filter and controller design and can help to evaluate power quality problems of generators and loads in advance.**

I. INTRODUCTION

Today the number of power generation units and consumer loads are connected to the grid with electronic power converters is increasing. Especially the progressively rising share of renewable energies in the power production with a high number of smaller distributed power plants, like wind turbines or photovoltaic panels, contribute to this trend. Since more and more of these power units are integrated in the power supply system their impact on the power quality is growing [1],[2],[3]. Depending on the primary energy carrier (e.g. wind speed or global radiation) the generated power of these plants fluctuates strongly over time, which leads to flicker effects in the grid voltage [4]. Further on, most of these units feed the grid partly or fully through alternating current or voltage converters. As a consequence harmonic currents besides the fundamental frequency are generated and fed into the grid. Depending on the grid impedance at the particular PCC the harmonic currents can create high harmonic voltages, which may damage other operational equipment in the grid. Under certain circumstances unacceptable voltage harmonics can even be caused by small harmonic currents, if the grid impedance contains highly resistive resonance points at these specific frequencies [5].

To maintain the power quality and supply guarantee in the system standard grid codes have to be fulfilled in many countries by power generators in order to get grid access. Upon other terms these statutory limits restrict the maximum permitted voltage harmonics and flickers to a tolerable level. As the grid feedback of these disturbances is directly related to the frequency characteristic of the grid impedance, the information about it at a specific PCC can be essential for filter and controller dimensioning. In addition, the exact knowledge of the grid impedance at the fundamental frequency reflects the real short-circuit power and hence the possible capacity of a PCC.

In practice the grid impedance at a PCC is in most cases only estimated for the fundamental frequency on the base of the maximum short-circuit power of the corresponding voltage-level [6]. Sometimes grid-simulation-programs are used to determine the grid impedance. However, the complex and time varying grid structure makes an exact simulation of the spectral characteristic difficult. In contrast to simulation, measuring the grid impedance in situ reflects the real grid structure at a certain PCC. The measured grid impedance includes all operational equipment in the grid and combines a number of serial and parallel resonant circuits into a single parameter. Fig.1 illustrates this relationship. The complex impedance network of a few resonance circuits displayed as Pi-elements is merged into a single parameter. The measured parameter can be imported as look-up table in computer simulation programs to assess power quality problems and to optimize filter circuits and controller algorithms of power electronic converters.

To determine the spectral characteristic of the grid impedance different methods have been proposed. In [7] and [8] the transient response of voltage or current spikes injected into grid are analyzed. The procedures presented in [9], [10], [11] feed sinusoidal current signals with variable frequencies into the grid and measure the phase and magnitude of the corresponding grid voltage at the specific frequency in steady state. Further on, the harmonic distortion caused by switched capacities [12] banks and inrush currents of transformers [13] have been successfully used to determine the grid impedance in AC supply systems. In this paper an approach is presented

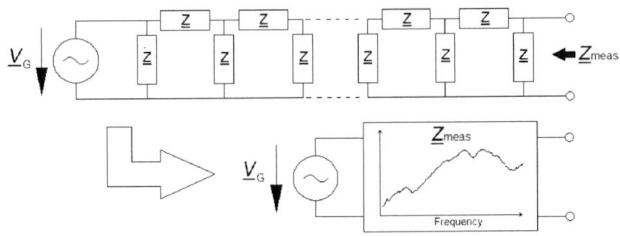

Fig. 1. Merging a complex impedance network of Pi-elements to a single measured parameter

978-1-4244-8806-3/11 $26.00 © 2011 IEEE

to measure the frequency dependent grid impedance by switching a resistive load with different pulse pattern at a PCC. In contrast to the methods stated in [7-13] the procedure offers a simple solution to identify even asymmetrical phase impedances in a three-phase three-wire system.

A. Measurement principle

A basic method for frequency dependent grid impedance identification is shown in Fig. 2 for a single phase impedance measurement. In the circuit the grid is simplified to a voltage source $\underline{V}_G$ with serial impedance $\underline{Z}_G$. In order to determine the grid impedance $\underline{Z}_G$ three basic steps are necessary. In the first step, the open circuit voltage $\underline{V}_1$ is measured, which generally includes also harmonic frequencies beside the fundamental. In the second step, a resistive load R_l is switched with a pulse pattern while the corresponding voltage $\underline{V}_2$ and the current $\underline{I}_2$ through the resistive load are recorded. Finally, the measured data is transformed into the frequency domain and the grid impedance is calculated with the following simple equation:

$$\underline{Z}_G = \frac{\Delta \underline{V}}{\Delta \underline{I}} = \frac{V_2 - V_1}{\underline{I}_2 - \underline{I}_1} = \frac{\Delta \underline{V}}{\underline{I}_2} \qquad (1)$$

In a three-phase four-wire system, normally present at the low-voltage level, this method can be adopted to all three phases. Fig. 3 shows the basic measurement setup. The grid impedance of every phase can be determined by switching a resistor to neutral line. With this method the calculated values of $\underline{Z}_a$, $\underline{Z}_b$ and $\underline{Z}_c$ also include the neutral line impedance. The measured grid impedance can be used as input parameter in grid simulation models to evaluate the harmonic perturbation of single phase consumer loads or generation units. In addition they can be transformed into symmetric components to analyze short-circuit behavior [6].

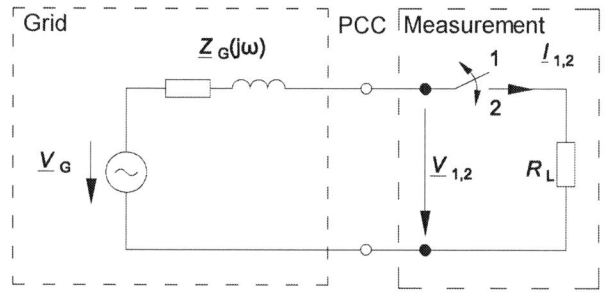

Fig. 2. Basic method for single phase grid impedance determination

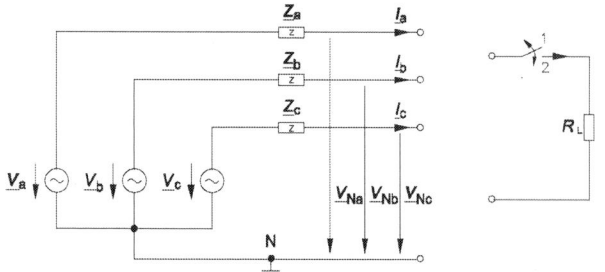

Fig. 3. Grid impedance determination in a three-phase four-wire system

B. Determination in a three-phase three-wire system

On the medium- and high-voltage level a three-phase three-wire system is installed most of the time. Due to the missing neutral line the measurement methods descripted above cannot be directly applied. In order to determine the grid impedance on these voltage levels a modified method, called "asymmetrical phase switching", can be used [14]. The principle is illustrated in Fig. 4. First, the three open circuit line to line voltages are measured. Afterwards a resistive load is switched between two phases while the corresponding voltage and current signals are recorded. This process is repeated three times for every phase combination, which lead to a set of three linearly independent equations:

$$\underline{V}_{1,ab} - \underline{I}_{2,ab} \cdot \left(\underline{Z}_a + \underline{Z}_b + R_L\right) = 0$$
$$\underline{V}_{1,bc} - \underline{I}_{2,bc} \cdot \left(\underline{Z}_b + \underline{Z}_c + R_L\right) = 0 \qquad (2)$$
$$\underline{V}_{1,ac} - \underline{I}_{2,ac} \cdot \left(\underline{Z}_a + \underline{Z}_c + R_L\right) = 0$$

The voltages measured during the switching process are equal to the currents times the resistive load:

$$\underline{V}_{2,x} = \underline{I}_{2,x} \cdot R_L \qquad (3)$$

Further on, the open circuit voltages $\underline{V}_{1,x}$ can be subtracted from $\underline{V}_{2,x}$ which leads to $\Delta \underline{V}_x$ like in (1). Combining (1), (2) and (3), the grid impedance of all three phases can be calculated with a matrix operation:

$$\begin{pmatrix} \underline{Z}_a \\ \underline{Z}_b \\ \underline{Z}_c \end{pmatrix} = \begin{pmatrix} \frac{1}{2 \cdot \underline{I}_{2,ab}} & -\frac{1}{2 \cdot \underline{I}_{2,bc}} & \frac{1}{2 \cdot \underline{I}_{2,ac}} \\ \frac{1}{2 \cdot \underline{I}_{2,ab}} & \frac{1}{2 \cdot \underline{I}_{2,bc}} & -\frac{1}{2 \cdot \underline{I}_{2,ac}} \\ -\frac{1}{2 \cdot \underline{I}_{2,ab}} & \frac{1}{2 \cdot \underline{I}_{2,bc}} & \frac{1}{2 \cdot \underline{I}_{2,ac}} \end{pmatrix} \cdot \begin{pmatrix} \Delta \underline{V}_a \\ \Delta \underline{V}_b \\ \Delta \underline{V}_c \end{pmatrix} \qquad (4)$$

In order to get correct results with this method, $\underline{Z}_a$, $\underline{Z}_b$ and $\underline{Z}_c$ as well as the RMS-value and frequency characteristic of the open circuit grid voltages must stay fairly constant during the identification process. To satisfy these conditions the measurement time should be as short as possible. Yet, the requirements are not always fulfilled in AC supply systems and will be discussed later in this paper. A detailed description of the mathematical derivations presented in this section and further theoretical and practical investigations of the impedance identification process in three and four wire systems can be found under [14][15][16].

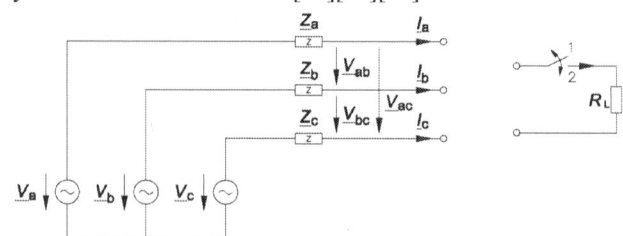

Fig. 4. Grid impedance determination in a three-phase three-wire system

II. SIMULATION OF THE IDENTIFICATION PROCESS

To determine the frequency dependent grid impedance the grid has to be excited by current pulses with spectral components in the desired frequency range. In order to find appropriate pulse signals a Matlab/Simulink model displayed in Fig. 5 is used to simulate different pulse sequences. The grid is modeled by an ideal voltage source with an impedance consisting of resistor and an inductive component. An excitation unit is connected to the grid realized by diode rectifier combined with a resistive load and a serial connected ideal IGBT (**I**nsulated **G**ate **B**ipolar **T**ransistor) with parallel snubber circuit.

A. Broad spectral excitation with RPWM signals

For a wide spectral excitation of the grid impedance random pulse width modulation signals (RPWM) can be applied to the gate unit of the IGBT. Fig. 6 shows such a sequence and the resulting voltage and current characteristic. The time between "on" and "off" state of the power switch changes stochastically. A switch-on of the IGBT leads to fast current rise and a sudden drop in the corresponding voltage characteristic. By the time the switch is opened the current through the ideal IGBT immediately stops causing a voltage peak partly absorbed by the snubber capacitor. In both cases also transients can be observed in the voltage and current curves. From a system theory point of view the shape of the transients in the voltage at the grid connection can be seen as system response from the grid impedance to a current signal applied to the grid. Fig. 7 shows the discrete Fourier transform of the current signal presented in Fig. 6. Besides the fundamental 50 Hz frequency the spectrum contains higher frequency components.

The spectral distribution of these components is determined by the switching characteristic of the IGBT (in this simulation ideal), the grid impedance, the dimensioning of the snubber circuit and strongly by the pulse sequence applied to the switch. The state changes of the pulse sequence can only be distributed within a defined time interval. By changing the limits of this time interval and/or the distribution-function it is possible to create RPWM signals, which stimulate the grid quite evenly in specific spectral ranges. The mathematical theory of these custom-made excitation signals and their application for frequency dependent grid impedance determination is presented in detail under [16].

Fig. 8 shows absolute value and phase of grid impedance calculated with the simulated data by using (1) as well as the calculated grid impedance as reference. The noise in the grid impedance characteristic is directly related to the current excitation at the corresponding frequencies. To reduce the noise at frequencies with low current excitation interpolation can be used. In addition, the slope can be straightened by digital filtering and averaging over several iteration loops. In practice a few iteration loops are required to reduce the system and measurement noise during the identification process. Further on, to realize a smooth broad spectral excitation and a good spectral resolution the measurement time interval has to be increased. These measures all lead to longer overall process time, which limits the possible time resolution. Despite of these disadvantages a sufficiently precise identification of the grid impedance over a wide spectral region and with high spectral resolution can be realized with RPWM signals.

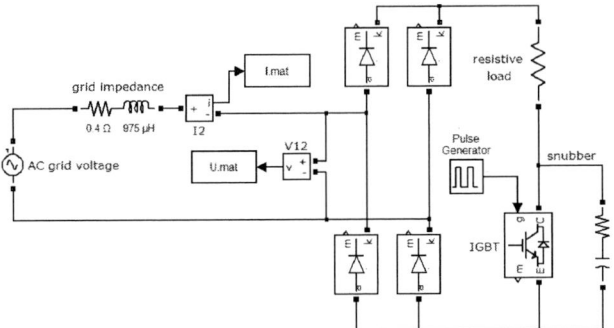

Fig. 5. Simulink model of the grid impedance identification process.

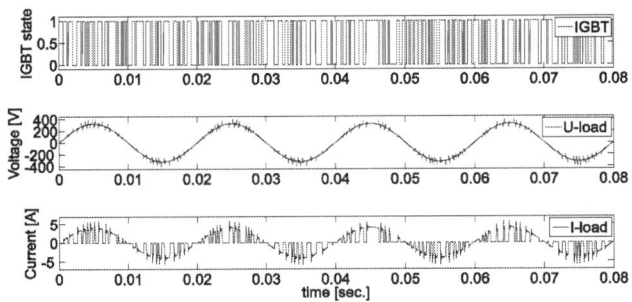

Fig. 6. RPWM sequence and the corresponding voltage and current characteristic.

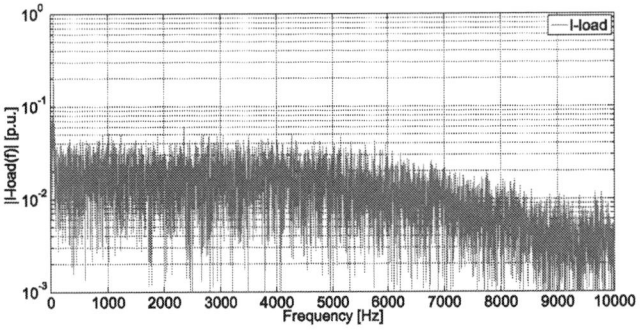

Fig. 7. Fourier transform of the current characteristic.

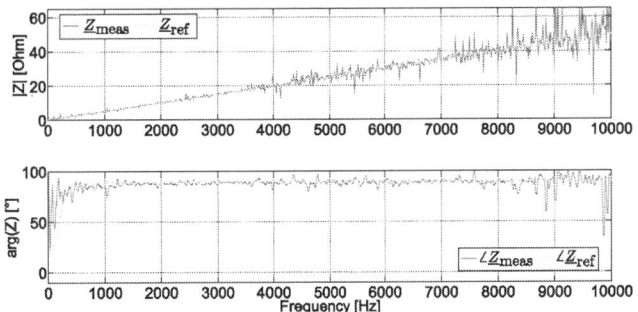

Fig. 8. Absolute value and phase angle of the grid impedance.

B. Discrete excitation with PWM signals

Another approach for grid impedance identification is to use pulse width modulation signals with a constant duty cycle to switch the IGBT. With these signals only certain discrete frequencies are strongly excited in the grid. Fig. 9 shows the corresponding current characteristic and it's Fourier-transformation of a 300 Hz pulse sequence with 50 % duty cycle applied to the IGBT. Besides the fundamental odd factors of the pulse frequency harmonics plus/minus the fundamental can be seen in the frequency domain. The spectral distribution of the current characteristic can be easily derived from system theory. The current signal is in time domain a result of a multiplication by the 50 Hz voltage source, the total admittance of the circuit, the pulse signal and a rectangle function for the measurement interval. In frequency domain the multiplication changes to a convolution of the corresponding Fourier-transformed functions, which leads to the observed spectrum. The strongly excited frequencies can be calculated with the following equation:

$$f_d(n) = f_{pwm} \cdot (1 + 2n) \pm 50 \ [Hz] \ with \ n = 0,1,2 \dots \quad (5)$$

At these frequencies the grid impedance can be measured more precisely than with RPWM signals. Because a high spectral resolution is not required data acquisition time can be small and the Fourier-transforms of the time signals have to be calculated just for a number of discrete frequencies, which makes the overall process time short. Fig. 10 shows the calculated grid impedance for a single simulation loop with a 300 Hz pulse sequence applied to the IGBT. The calculated grid impedance fits well to the reference at frequencies which satisfy (5). In practice the grid impedance at certain harmonics of the fundamental is of special interest in order to evaluate the grid feedback of power electronic converters. The method presented in this section offers a possibility to accurately measure the grid impedance at these harmonic frequencies in a short time. Through a variation of the pulse sequence a frequency sweep can also be realized, which can for example be used for a precise analysis of resonance points or other important spectral characteristics of the grid impedance.

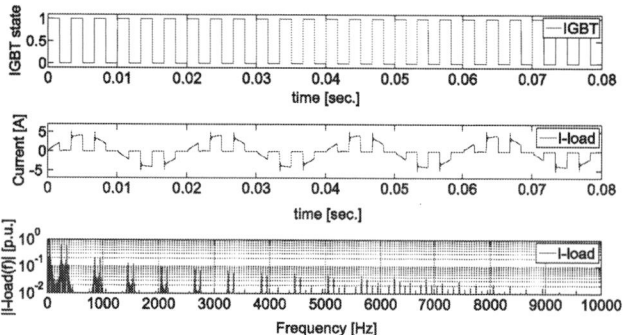

Fig. 9. Pulse sequence with 300 Hz and the corresponding current characteristic in time and frequency domain.

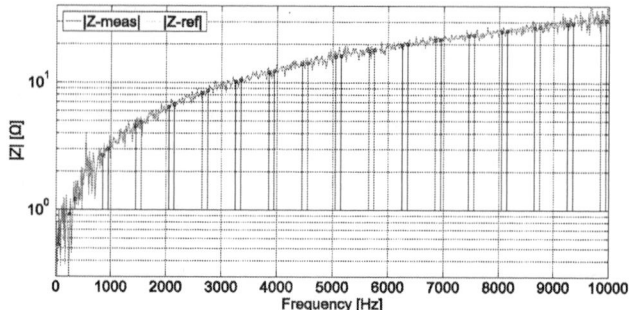

Fig. 10. Grid impedance measured with a 300 Hz pulse sequence.

III. MEASUREMENTS UNDER IDEAL GRID CONDITIONS

For verification and optimization of the grid impedance identification method a simple laboratory test circuit is setup similar to the simulation model. The schematic arrangement is illustrated in Fig. 11. The emulation of the grid is realized by a voltage source with high output power, which generates a pure 50 Hz voltage signal. The voltage source is connected to a norm-impedance consisting of a resistor and coil for every line. The measurement unit includes a diode rectifier bridge connected to serial circuit of a resistor and an IGBT power switch with a parallel snubber-capacitor. The PWM signals for the switch are generated by a microcontroller. The voltage and current signals are measured with precise broad band probes. The output signals of the probes are simultaneously sampled by a 16-bit transient recorder. Data acquisition and pulsing are synchronized on the positive zero crossing of the voltage fundamental.

In the first measurement a RPWM signal is used for switching. The switching is applied over eight fundamental oscillations while measuring current and voltage. Afterwards the open circuit voltage is recorded for eight periods leading to a spectral resolution of 12.5 Hz and a 320 ms data acquisition time. The sampling rate is set to 409.6 kHz, which results in 65536 data points for every measured current and voltage parameter. The data is transformed into the frequency domain with an FFT-algorithm. The frequency dependent grid impedance is calculated with (1). Fig. 12 shows the absolute value, the phase and the real and imaginary components of the measured grid impedance averaged over 20 iteration loops with a total data acquisition time of 6.4 s.

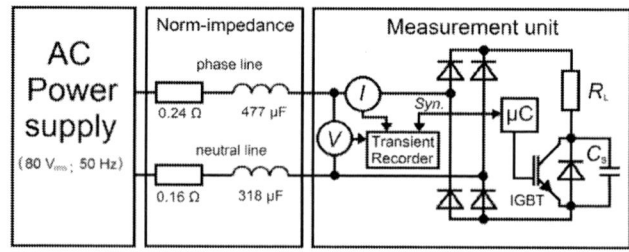

Fig. 11. Schematic setup for the single phase impedance measurement under ideal grid condition

978-1-4244-8806-3/11 $26.00 © 2011 IEEE

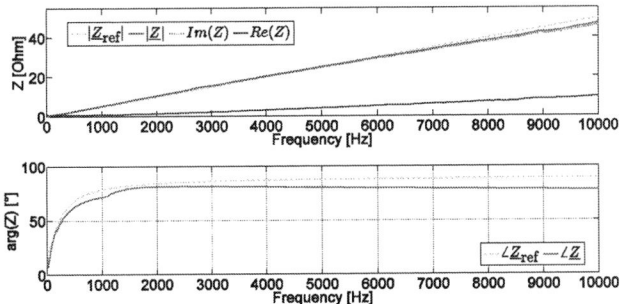

Fig. 12. Absolute value and phase of the measured grid impedance.

Due to the dominating inductive reactance the grid impedance increases with the frequency. The measured curve deviates more and more from the reference for higher frequencies. This can be explained by a capacitive reactance of the cables and the voltage source in the setup, which is not considered in the calculated reference. The capacitive reactance of the grid impedance leads to smaller measured phase angle and to a reduced inclination of the imaginary component for higher frequencies. This contributes to a slightly decrease of the phase angle at frequencies above 2 kHz and a small phase lag at about 1 kHz. It can also be observed, that the real component of the measured grid impedance rises for higher frequencies. This results from eddy current effects in the norm-impedance, which is also not considered in the reference. These effects contribute to the decline of the phase angle at higher frequencies as well.

IV. MEASUREMENT IN THREE-PHASE SYSTEMS

A. Variation of the system frequency

In contrast to the ideal laboratory grid setup shown in Fig. 11, the grid frequency varies over time. The frequency variation can lead to errors in the impedance identification process due to artificial phase shifts and aliasing effects caused by the Fourier-transformation [17]. This is generally the case if the recorded time interval does not accurately fit to multiples of the grid frequency. In order to reduce these effects window functions and digital filters applied in time domain as well as phase correction algorithm in the frequency domain can be used [18]. Another approach is a sampling rate adjustment to the present grid frequency [19]. For this purpose a phase-locked loop (PLL) circuit determines phase and frequency of the grid voltage shortly before every measurement cycle. The phase and frequency information is processed by a microcontroller, which generates matching clock and synchronisation signals for the analog to digital converters. Through several hardware-implemented counters and clock dividers/multipliers in the microcontroller (μC) the clock signal can be precisely adjusted in steps of 1 Hz up to 500 kHz. The ideal sampling rate to a corresponding grid frequency is set by the microcontroller as follow:

$$f_S(f_G, n) = f_G \cdot 2^n \; [Hz] \quad with \; n = 1, 2, 3 \dots \quad (6)$$

with sampling rate f_S and grid frequency f_G. The factor 2^n determines the number of sampling points in a fundamental period of the grid voltage. A power of two has been chosen to

transform the recorded data with a FFT-algorithm in the frequency range. In this investigation the parameter n is set to 13, which results in 8192 data points per fundamental period and a sampling rate of 409.6 kHz in case of an exact 50 Hz grid frequency.

B. Grid voltage and impedance fluctuations

For the grid impedance identification method the line impedances as well as the grid voltage must be persistent into certain limits during the measurement in order to get reliable results. However, these parameters can be strongly influenced by changing loads, generators and step transformers connected in close distance to the point of measurement. The impact of these fluctuations on the impedance identification process, especially the changes caused by nearby non-linear loads and generators are part of ongoing research and may be presented in future publications. To avoid significant variations in the grid voltage and impedance during the identification process the measurement time should be as short as possible. Further on, all open circuit voltages ($\underline{V}_{ab}$, $\underline{V}_{bc}$, and $\underline{V}_{ac}$) are recorded before and after switching the resistive load. In the data processing the RMS-values and spectral components are calculated for every voltage data set. Afterwards the variation of the RMS-values and the level of significant included harmonics are analyzed. If the variations of these parameters stay in defined limits, the line impedances are calculated with (4). If the limits are exceeded, the recorded data are discarded. The same evaluation is applied to data set of different iteration loops with slightly higher limits. This approach offers a possibility to detect substantial grid fluctuations during the measurement process.

C. Measurement on a Low-voltage level

The schematic arrangement for the three-phase impedance measurement is presented in fig. 13. The asymmetrical phase switching is realized by three excitations units, which are delta-connected to the grid. Every excitation unit consists of a diode rectifier and a resistive load in serial to an IGBT power switch. The Fig. 13 shows the absolute value and phase angle of three line impedances measured with this setup at a connection point on the low-voltage level.

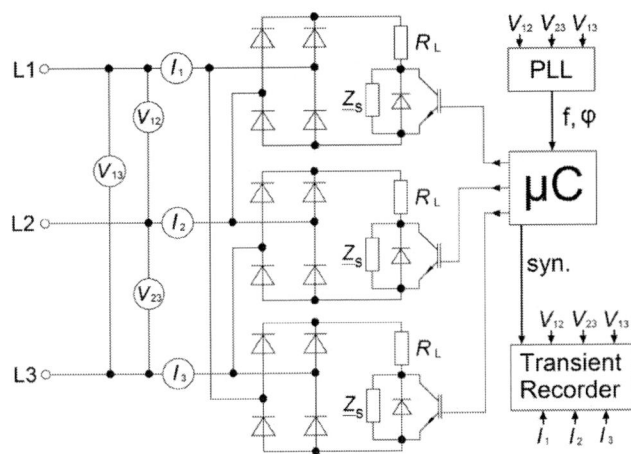

Fig. 13. Schematic setup for the three-phase impedance measurement.

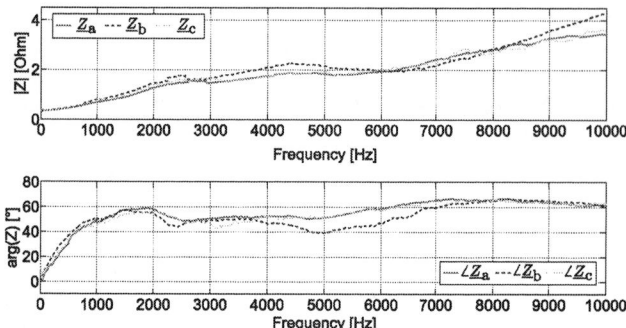

Fig. 14. Phase impedances measured on the Low-voltage level.

For this measurement every phase combination is sequential pulsed with a RPWM signal for 8 fundamental oscillations of the grid voltage. To reduce measurement and process noise digital filters and correlations techniques are applied to the reordered data in time domain [20]. The three grid impedances are calculated with (6) and averaged over 5 iteration loops. The characteristic of all three line impedances are fairly symmetrical up to 1 kHz and strongly increase with higher frequencies due to the dominant line and transformer inductance. At frequencies above 1 kHz resonance points can be observed mainly caused by distributed capacities in the grid. Due to the high number of asymmetrical loads on the low-voltage level these resonance points vary in amplitude and frequency.

V. CONCLUSION

Since more and more renewable energy sources with power electronic converters are connected to the grid, their influence on the power quality is growing. The grid perturbation of electronic converters is strongly determined by grid impedance at the grid connection point. In contrast to simulation, measuring the grid impedance reflects the real grid structure at a specific connection point and can be used as parameter for filter design and power quality evaluations.

This paper has introduced a basic method for frequency dependent grid impedance determination. It has been shown, that with random pulse width modulation signals for grid-excitation the grid impedance can be identified over a wide spectral range in an acceptable time and precision. Furthermore, pulse sequences with constant duty cycle can be used to determine the grid impedance at discrete frequency with higher precision and time resolution than with RPWM signals. Through a variation of these sequences a frequency sweep can also be realized, which can be used for more detailed determination of resonance points in the grid impedance.

ACKNOWLEDGMENT

This topic was investigated within the project "Development of a measurement device for the determination of the time and frequency dependent grid impedance on the medium-voltage-level". The project is funded by the German Ministry for the Environment, Nature Conservation and Nuclear Safety under the support code 0325049.

REFERENCES

[1] G.J. Wakileh, *Power Systems Harmonics - Fundamentals, Analysis and Filter Design.* Berlin: Springer, 2001.

[2] Z. Hanzelka, A. Kempski, and R. Smoleński, "Quality Problems in Smart Networks" in *Power System Electronics in Smart Electrical Energy Networks*, R Strzelecki and G. Benysek, Ed., 1st ed. London: Springer, 2008, pp. 327-374.

[3] D. Schulz "Grid Integration of Wind Energy Systems" in *Power System Electronics in Smart Electrical Energy Networks*, R Strzelecki and G. Benysek, Ed., 1st ed. London: Springer, 2008, pp. 107-146.

[4] D. Schulz, *Power Quality - Theory, Simulation, Measurement and Assessment.* (In German) Offenbach: VDE-Verlag, 2004.

[5] H. Langkowski, T. Do Thanh, K.-D. Dettmann, and D. Schulz, "Grid impedance determination - relevancy for grid integration of renewable energy systems," in *Conference of the IEEE Industrial Electronics Society (IECON'09)*, Porto, 3-5 November 2009.

[6] J. Schlabbach and W. Mombauer, *Power Quality – Origin and Assessment of Phase Effects; Grid Connection of Renewable Energies.* (In German) Offenbach: VDE-Verlag, 2008.

[7] M. Sumner, B. Palethorpe, and D. W. P. Thomas, "Impedance measurement for improved power quality-part 1: the measurement technique," *IEEE Transactions on Power Delivery*, vol. 19, no. 3, pp. 1442-1448, 2004

[8] M. Sumner, B. Palethorpe, D. W. P. Thomas, P. Zanchetta, and M. C. Di Piazza, "Intelligent protection for embedded generation using active impedance estimation," in *IEEE 2nd International Symposium on Power Electronics for Distributed Generation Systems (PEDG)*, HeFei, 16-18 June 2010.

[9] A. Knop and F. W. Fuchs, "High frequency grid impedance analysis with three-phase converter and FPGA based tolerance band Controller," *Compatibility and Power Electronics (CPE'09)*, Badajoz, 20-22 May 2009

[10] J. Xie, Y. X. Feng, and N. Krap, "Network impedance measurements for three-phase high-voltage power systems," *Asia-Pacific Power and Energy Engineering Conference (APPEEC)*, Chengdu, 28-31 March 2010.

[11] Y. L. Familiant, K. A. Corzine, J. Huang, and. M. Belkhayat, "AC impedance measurement techniques," *IEEE International Conference on Electric Machines and Drives*, pp. 1850-1857, 15 May 2005.

[12] M. Nagal, W. Xu, and J. Sawada, "Harmonic impedance measurement using three-phase transients," *IEEE Transactions on Power Delivery*, vol. 13, pp. 272-277, January 1998.

[13] C. Wie, S.B. Tennakoon, , R. Langella, D. Gallo, A. Testa, and A. Wixon. "Harmonic impedance measurement of a 25 KV single phase AC supply system," in *Proceedings of Ninth International Conference on Harmonics and Quality of Power*, Orlando, vol. 1, pp. 214-219, 1-4 October 2000.

[14] D. Schulz, T. Do Thanh, H. Langkowski and M. Jordan, *Device for measuring the impedance of the electrical power supply system.* German patent application DE 10 210 018 996.0, 2010.

[15] T. Do Thanh, T. Schostan, K.-D. Dettmann, and D. Schulz, "Nonsinusoidal power caused by measurements of grid impedances at unbalanced grid voltages," *Proceedings of the IEEE conference of the International School on Nonsinusoidal Currents and Compensation ISNCC*, Lagow, 10-13 June 2008.

[16] M. Jordan, T. Do Thanh, H. Langkowski, and D. Schulz, "Strategies for frequency dependent grid impedance measurement at the medium- and high-voltage level," *5th International Ege Energy Symposium and Exhibition (IEESE-5)*, Denizli, 27-30 June 2010.

[17] A. Phadke, M.A. Pai, A. Stankovic, J.S. Thorp, *Synchronized Phasor Measurements and Their Applications.* New York: Springer, 2008.

[18] R. N. Bracewell, *The Fourier Transform and Its Applications.* International Student Edition, New York: McGraw-Hill, 1986.

[19] A. Moreno-Muñoz, *Power Quality –Mitigation Technologies in a Distributed Environment.* London: Springer, 2007.

[20] H. Unbehauen and G. P. Rao, *Identification of Continuous Systems.* North-Holland Systems and Control Series Volume 10, Amsterdam: North-Holland, 1987.

Trans-Z-source-like Inverter with Built-in DC Current Blocking Capacitors

Marek Adamowicz[1,2], Jarosław Guzinski[2], Dmitri Vinnikov[3], Natalia Strzelecka[1]

[1]Gdynia Maritime University
Morska 81-87, 81-225
Gdynia, POLAND
madamowi@am.gdynia.pl

[2]Gdansk University of Technology
Narutowicza 11/12
Gdansk, POLAND
jarguz@ely.pg.gda.pl

[3]Tallin University of Technology
Ehitajate tee 5, 19086
Tallin, ESTONIA
dm.vin@mail.ee

Abstract - **Many renewable generation units (e.g. PVs, fuel cells, small wind turbines) require integrated DC-AC converters providing efficient high step-up power conversion. The recently proposed Z-source-type integrated buck-boost inverters characterize one stage energy processing and wide range of output voltage regulation. This paper presents the possibility of further extending Z-source-type inverters application area through the topology reconfiguration. The unique property of the proposed Z-source-type inverter with applied high frequency transformer and built-in DC current blocking capacitors is that no energy is stored in the transformer windings. That prevents core saturation and enables high voltage gain. Simulation and experimental results are shown to verify the proposed topology.**

I. INTRODUCTION

Integrated buck-boost voltage source inverters have received increasing attention in the last decade for renewable generation applications (PV cells, fuel cells, wind generators and batteries). The position of the Z-source type buck-boost inverters [1]-[10] among other topologies extracting energy from low voltage DC sources has been established in the literature. The advantages of the Z-source type buck-boost inverters are: one stage energy processing, wide range of output voltage regulation, robustness to incorrect turn on of transistors and improved immunity to the electromagnetic noise. The Z-source inverters (ZSI) [1]-[3] as well as its modifications, ie. embedded Z-source inverters (EZSI) [4] and quasi Z-source inverters (qZSI) [5] utilize passive input circuits enabling the shoot-through of the voltage source inverter (VSI) bridge legs to boost voltage in a DC link. The unique Z-source type network (EZ-source, qZ-source) is connected between a low voltage DC source and a transistor bridge to form the Z-source type buck-boost inverter topology. Z-source type impedance network serves as power storage and guarantees double filtration grade at the input of the inverter, and therefore dumping current ripples and voltage pulsation in the DC circuit of the VSI. Unfortunately the disadvantage of Z-source type inverters is that two control variables: shoot-through interval T_{shoot} during switching period T and modulation index M are interdependent. Longer shoot-through period provides higher DC link voltage but also decreases a period of active PWM states. That imposes limitation on boosting of output voltage as the increase in boost factor B resulting in lower modulation index M.

Very recently two new topologies have been developed to extend the voltage boost properties of Z-source type impedance source inverters. These are T-source inverters also named trans-Z-source inverters (TZSI) [8]-[10] and cascaded quasi-Z source inverters (CqZSI) [11]-[13].

The CqZSI characterizes higher output voltage gain compared to that of the traditional qZSIs with the same shoot-through ratio. This is possible thanks to cascade connection of two or more qZ-source input networks.

The TZSI provides larger voltage gain without cascading its input sub-circuits. Contrary to the CqZSI, the TZSI utilizes turns ratio of two transformer windings and therefore characterizes lower element count.

Preliminary results of the CqZSI and TZSI presented in literature [10]-[12] show that none of them is claimed to be the best and dominant over the other. The CqZSI utilizes the effect of an electronic boost pump obtained by increasing number of inductors, diodes and capacitors appearing in the input circuit and therefore involves higher element count and costs. Parasitic parameters of the CqZSI cascade input network, particularly its series resistances of inductors and voltage drops on diodes may negatively impact on CqZSI performance.

In turn the performance of TZSI is affected by the used transformer. The transformer leakage inductance causes large voltage spikes on the transistors and the input diode. The secondary transformer winding is not paralleled by any capacitor during non-shoot-through-state as distinguished from ZSI and qZSI. Therefore the transformer should be designed with a very low leakage inductance and consequently with a very small air gap. However the DC currents appearing in the transformer with negligible air gap will saturate the core and affect the TZSI performance.

The aim of the work presented in this paper is to obtain a modified Z-source inverter topology containing the high frequency transformer as in the case of T-source inverter and TZSI and combining the advantages of high voltage gain and high immunity to electromagnetic noise as in the case of the qZSI.

In chapter III a topology reconfiguration of quasi Z-source inverter is presented. A kind of Trans-Z-source inverter with built-in DC current blocking capacitors is obtained which characterizes alternating currents in the transformer. Analysis and preliminary results of proposed inverter in two configurations: with discontinuous input current and with continuous input current are presented.

978-1-4244-8806-3/11 $26.00 © 2011 IEEE

II. Z-SOURCE TYPE IMPEDANCE SOURCE INVERTERS

A. Z-source and quasi-Z-source inverters

The ZSI topology features a DC link consisting of a symmetrical lattice network consisting of two inductors and two capacitors. Different configurations of 3-phase impedance source inverter (ZSI, EZSI or qZSI) are shown in Fig. 1a-Fig. 1c. The Z-source inverter can assume nine states, that is one more than in conventional VSI system. Additional 'ninth' state is the third "null" state, occurring when the load is being shorted simultaneously by lower and upper group of transistors. This state, is defined as a "shoot-through" state and may be generated in seven different ways. All three topologies from Fig. 1 characterize DC current of both inductors.

During the shoot-through state of ZSI no energy flows from the source to the load. Instead, energy within the Z-source network is re-orientated with electrostatic energy from the DC electrolytic capacitors transferred to magnetic energy stored in the Z-source inductors, before the stored energy is released for output voltage boosting during the next non shoot-through interval [4].

Modifications of ZSI from Fig. 1b and Fig. 1c focus on changing the topology arrangement of the connections and did not focus on changing the basic structure. EZ-source and qZ-source impedance networks produce the same voltage transfer gain as Z-source network, but have the additional advantage of inherently filtering the currents drawn from the two DC sources in EZSI and one DC source in qZSI [4], [5]. Waveform of qZSI input current i_{IN} is shown in Fig. 1e.

The qZSI, when compared to the ZSI, features lower DC voltage on capacitor C_2. All three topologies from Fig. 1 can be controlled using any methods which can be used to control the traditional ZSIs [6],[7].

B. Trans-Z-source inverter

TZSI extends the ZSI output voltage regulation [8]. Three impedances: two inductances, and one capacitor form the passive input circuit of discontinuous input current (DIC) TZSI described in Fig. 2a. The equivalence of Z-source network from Fig. 1a and TZ-source network from Fig. 2a can be proven using circuit theory [9], [14], [15] also for configuration in which the ZSI inductors L_1 and L_2 are not coupled. TZSIs offer wide inverter output voltage manipulation and reduction of shoot-through coefficient thanks to windings turns ratio $n_1:n_2 > 1$ [8]-[10]. The extended voltage gain and reduced element count are main advantages of TZSI over ZSI and qZSI.

To produce the same ac output voltage TZSI requires a smaller shoot-through duty ratio $D = T_{shoot}/T$ (accordingly a larger modulation index M which depends on D) compared to the ZSI and qZSI.

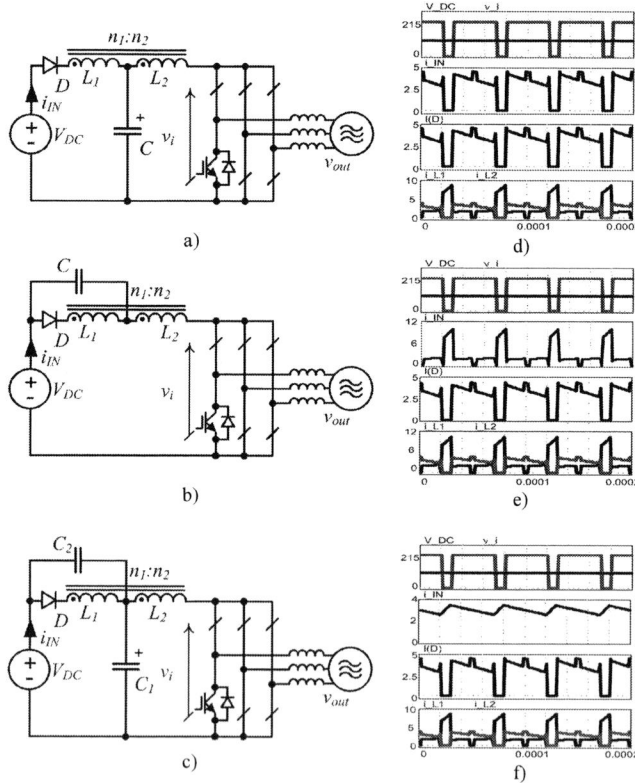

Fig. 2. Circuit schematics of T-source (trans-Z-source) inverters with turns ratio $n_1:n_2=2$; discontinuous input current (DIC) TZSI [8],[9] (a); DIC trans-quasi-ZSI [10] (b); continuous input current (CIC) TZSI (b); and typical waveforms ($D=0.19$) of DIC TZSI (d); DIC trans-quasi-ZSI (e); and CIC TZI (f).

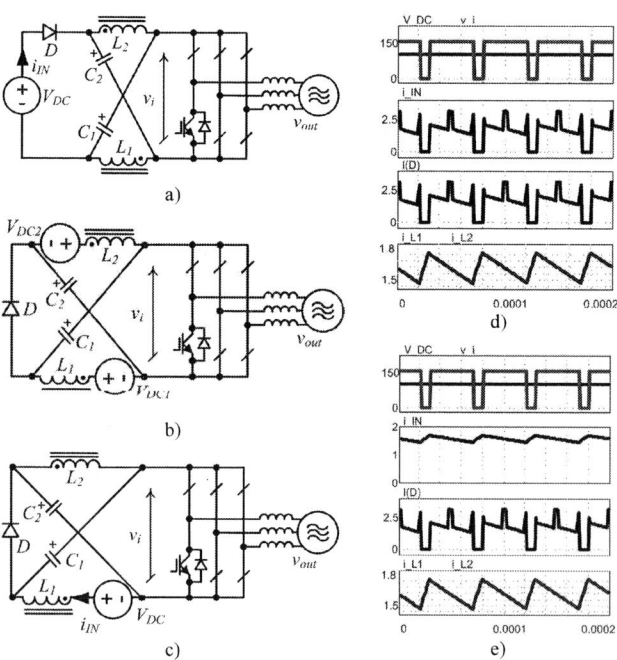

Fig. 1. Circuit schematics of Z-source type buck-boost inverters: ZSI (a); EZSI (b); qZSI (c) and typical waveforms ($D=0.19$) of ZSI (d) and qZSI (e).

$$v_i = \frac{1}{1-\left(1+\dfrac{n_1}{n_2}\right)D}V_{DC} = B \cdot V_{DC} \qquad (1)$$

where the boost factor B is [10]:

$$B = \frac{1}{1-\left(1+\dfrac{n_1}{n_2}\right)D} \qquad (2)$$

When the maximum constant boost control (MCBC) is applied to TZSI, the voltage gain ($G=MB$) is defined as [10]:

$$G = \frac{M}{1-\left(1+\dfrac{n_1}{n_2}\right)\left(1-\dfrac{\sqrt{3}}{2}M\right)} \qquad (3)$$

For TZSI transformer turns ratio $n_1:n_2$ over 1, the higher DC link voltage (v_i) boost gain (3) is obtained compared to ZSI. For larger turns ratio the D ratio beneficially decreases with the same voltage gain.

Contrary to qZSI, the voltage across L_2 inductance is not constrained by any capacitor and therefore can be imposed proportional to the voltage across L_1 by changing the turns ratio $n_1:n_2$. Hence two windings of winding is not constrained by capacitor can behave like a transformer, except for the stored energy.

From the equivalence of circuits the capacitance of DIC TZSI with $n_1:n_2=1$ is equal to sum of capacitances from basic ZSI topology. For larger turns ratios TZSI capacitance should by properly adjusted.

T-quasi-ZSI shown in Fig. 2b is a variant of TZSI [10]. The capacitance of T-quasi-ZSI with unity turns ratio can be two times reduced compared to TZSI. However after replacing the capacitance from TZSI to T-quasi-ZSI the shape of input current will worsen (Fig. 2e).

Continuous input current (CIC) operation can be achieved in another variant of TZSI shown in Fig. 2c by adding second capacitor C_2 to the TZSI impedance network. The additional capacitor plays the role of built-in input filter.

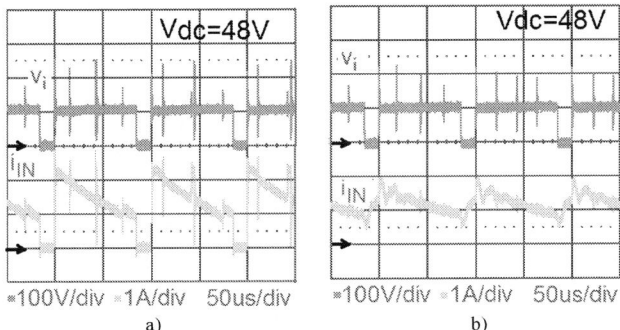

Fig. 3. Typical waveforms of DIC T-source inverter and CIC T-source inverter; V_{DC}=48V

The CIC T-source inverter (CIC ZSI) has not been published yet. Its performance and usability strongly depend on parasitic parameters of input sub-circuit V_{DC}-C_1-C_2. Particularly large parasitic inductance could impose unwanted resonance in input current i_{IN}. The capacitance ratio C_1:C_2 in CIC TZSI should be adjusted to windings turns ratio n_1:n_2.

It is convenient to use the power planar transformers with extremely low leakage inductance for realization of both the DIC T-source inverter (DIC TZSI) and CIC T-source inverter (CIC TZSI). Fig. 3. presents the experimental results for DIC TZSI (Fig. 3a) and CIC TZSI (Fig. 3b). The 1:4 planar transformer HIMAG149402 was used in experiments.

III. TRANS-Z-SOURCE-LIKE INVERTER WITH BUILT-IN DC CURRENT BLOCKING CAPACITORS

This chapter discusses the integration of LC qZ-network shown in Fig. 4a with a high frequency transformer (Fig. 4b) [16]. The proposed technique combines the pair of inductor L2 and capacitor C1 into unique two port network of high frequency (HF) transformer (a pair of coupled inductors LT1 and LT2) and a blocking capacitor CT. The HF transformer can be represented by an equivalent circuit with three uncoupled inductors. The T-equivalent for transformer from Fig. 4b is shown in Fig. 4c.

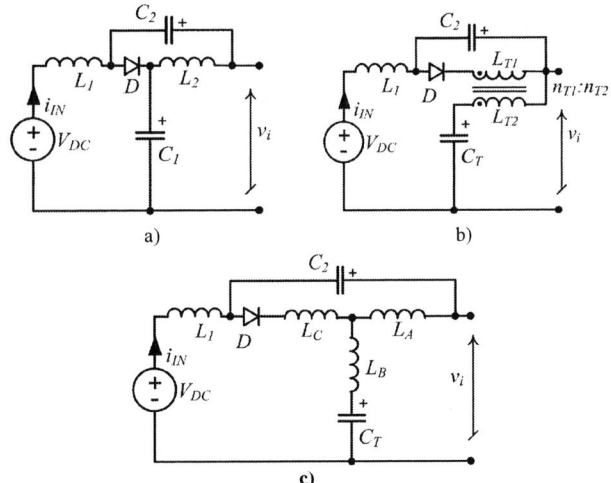

Fig. 4. Proposed technique of replacing LC qZ-network (a) with novel LCCT -Z-network (b) and a T-equivalent of transformer (c) [16].

From Fig. 4b and Fig. 4c it can be written [13]:

$$L_A = L_M = k\sqrt{L_{T1} \cdot L_{T2}} \qquad (4)$$

$$L_B = L_{T2} - L_M \qquad (5)$$

$$L_C = L_{T1} - L_M \qquad (6)$$

where L_M denotes mutual inductance of the pair of windings and $0<k<1$ denotes coupling coefficient defined by:

978-1-4244-8806-3/11 $26.00 © 2011 IEEE

$$k = \frac{L_M}{\sqrt{L_{T1} \cdot L_{T2}}} \qquad (7)$$

For the condition $(L_B \to 0,\ L_C \to 0)$ the impedance networks from Fig. 4a and Fig. 4c are equivalent.

Theoretical possibility of Z-source network integration with built-in HF transformer has been also illustrated in [8]. HF transformer integrated within Z-source circuit would extend voltage gain of ZSI utilizing turns ratio higher than one. In alternative passive networks presented in [8] no energy will be stored in windings of additional built-in transformer and average currents of integrated transformer will equal zero at the end of inverter output voltage period. The system could therefore display behavior that is contrary to TZSI, in which transformer windings store the energy carrying DC current during boost mode.

A. Discontinuous Input Current Configuration

The schematic of the proposed DIC trans-Z-source-like inverter with built-in DC current blocking capacitors is shown in Fig. 5.

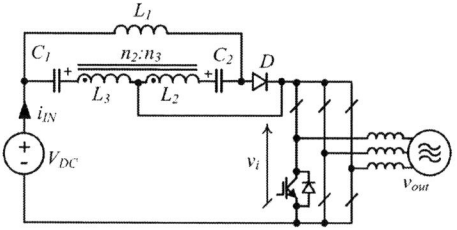

Fig. 5. Circuit schematic of a novel DIC TZS-like inverter with built-in DC current blocking capacitors.

The key waveforms of proposed inverter: DC link voltage, input current, diode voltage and current, inductor current and transformer currents for DIC configuration are shown in Fig. 6.

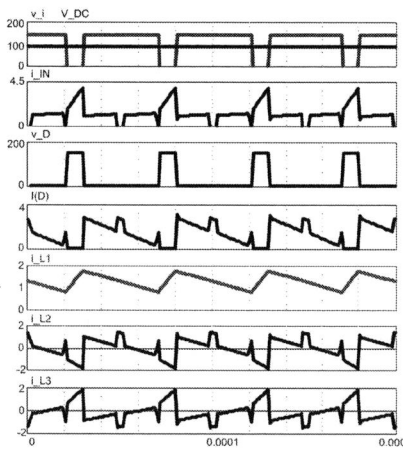

Fig. 6. Characteristic waveforms of the DIC TZS-like inverter with built-in DC current blocking capacitors for $n_2{:}n_3 = 1$.

The input circuit of inverter shown in Fig. 5 is created by four element LTCC topology (inductance + transformer + two capacitors) and contains five reactive components (three inductances and two capacitances) [16]. Although the number of inductances increases compared to previous T-source inverter (TZSI) topology, number of coils in which the energy is stored reduces. The unique property of proposed inverter is that no energy is stored in the transformer windings. This is thanks application of capacitors connected in series with transformer.

Reconfiguration of previous TZSI and arrangement of DC current blocking capacitors eliminates several disadvantages of Z-source type inverters providing:

(1) extended voltage boost capability and reduced shoot-through coefficient compared to ZSI and qZSI,

(2) improved immunity to parasitic parameters of transformer inductances compared to TZSI,

(3) alternating character of transformer currents preventing core saturation,

(4) reduced element count compared to CqZSI with the same voltage gain,

B. Continuous Input Current Configuration

To improve on topology from Fig. 5 a CIC trans-Z-source-like inverter with built-in DC current blocking capacitors (Fig. 7) is proposed. Key waveforms of CIC inverter system are presented in Fig. 8.

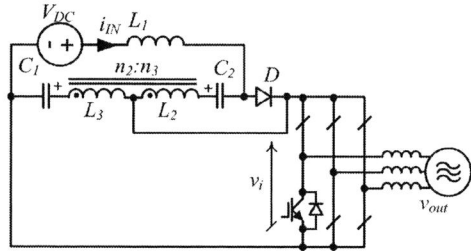

Fig. 7. Proposed CIC TZS-like inverter with built-in DC current blocking capacitors.

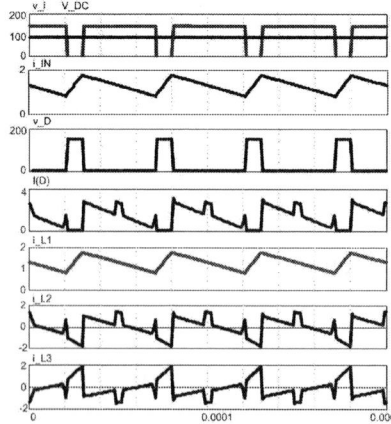

Fig. 8. Characteristic waveforms of CIC TZS-like inverter with built-in DC current blocking capacitors for $n_2{:}n_3 = 1$.

From the equivalent circuit of proposed CIC TZS-like inverter with built-in DC current blocking capacitors from Fig. 9 we have:

$$V_{DC} = V_{C1} - V_{C2} \qquad (8)$$

$$v_{L1} = -v_{L2} - v_{L3} \qquad (9)$$

$$v_i = V_{C1} + v_{L3} \qquad (10)$$

The equivalent circuit of CIC TZS-like inverter with built-in DC current blocking capacitors in shoot-through states is presented in Fig. 9a. In the shoot-through state ($v_i=0$) for an interval $D \cdot T$ a diode is reversely blocked. Mathematical expressions governing the inverter dc voltages can be written as:

$$v_{L3} = -V_{C1} \qquad (11)$$

$$v_{L2} = \frac{n_2}{n_3} v_{L3} = -\frac{n_2}{n_3} V_{C1} \qquad (12)$$

During non shoot-through state from the equivalent circuit, Fig. 9b, for an interval $(1-D) \cdot T$ it can be written:

$$v_{L3} = v_i - V_{C1} \qquad (13)$$

$$v_{L2} = V_{C2} = V_{C1} - V_{DC} \qquad (14)$$

The average voltages of the two inductors L_1, L_2 over one switching period should be zero in steady state. From (8) - (14) we obtain a system of two equations:

$$\langle v_{L3} \rangle = \frac{(-V_{C1}) \cdot D \cdot T + (v_i - V_{C1}) \cdot (1-D) \cdot T}{T} = 0 \qquad (15)$$

$$\langle v_{L2} \rangle = \frac{(-(n_2/n_3)V_{C1}) \cdot D \cdot T + (V_{C1} - V_{DC}) \cdot (1-D) \cdot T}{T} = 0 \qquad (16)$$

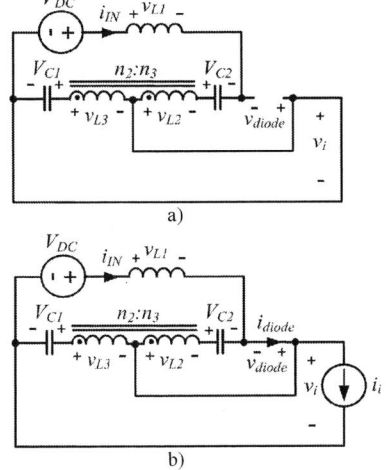

Fig. 9. Equivalent circuits of CIC TZS-like inverter with built-in DC current blocking capacitors: during shoot-through state (a), and during active (non shoot-through) state (b).

From the system (15) and (16) the capacitor voltage V_{C1} equals:

$$V_{C1} = \frac{1-D}{1 - \left(1 + \dfrac{n_2}{n_3}\right)D} V_{DC} \qquad (17)$$

and the inverter dc link voltage v_i in the non shoot-through states can be boosted to the value V_i as follows:

$$V_i = \frac{1}{1-D} V_{C1} = \frac{1}{1 - \left(1 + \dfrac{n_2}{n_3}\right)D} V_{DC} = B \cdot V_{DC} \qquad (18)$$

where $n_2/n_3 \geq 1$.

Hence the maximum dc-link voltage of proposed CIC TZS-like inverter with DC current blocking capacitors is the same as the previous TZSI topology (1) - (2).

When applying sinusoidal PWM algorithm to the proposed inverter the amplitude of output voltage V_{out} will equal

$$V_{out} = M \frac{V_i}{2} = \frac{M}{1 - \left(1 + \dfrac{n_2}{n_3}\right)D} V_{DC} . \qquad (19)$$

IV. SIMULATION RESULTS

Fig. 10 and Fig. 11 show preliminary simulation results of proposed CIC TZS-like inverter with built-in DC current blocking capacitors. Simulations of three phase (~400V, 50Hz) grid connected 2kW system supplied from 200V DC source were carried out using PSIM simulation software with the parameters as follows:

- impedance network: input inductor L_1= 2mH, HF planar transformer n_2:n_3= 4:1 with primary inductance L_{prim}=2.94mH, DC current blocking capacitors C_1= C_2= 48μF,

- output filter and load: L_f= 1.5mH, C_f= 6μF, R_{load}= 20Ω.

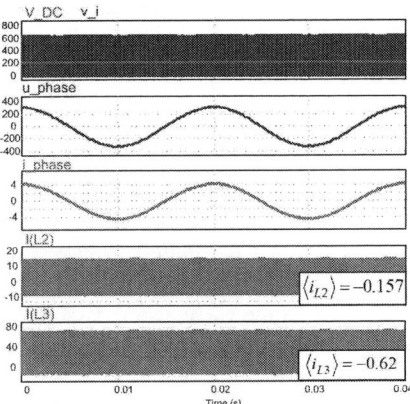

Fig. 10. Simulation results of grid connected three phase TZS-like inverter (n=4:1) with DC current blocking capacitors (D=0.15): DC source voltage V_{DC}=200V, inverter DC link voltage v_i, output phase voltage and current, transformer currents i_{L2}, i_{L3} with pointed out average values.

978-1-4244-8806-3/11 $26.00 © 2011 IEEE

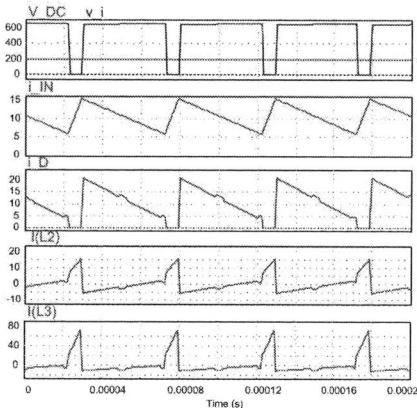

Fig. 11. Simulation results of grid connected three phase TZS-like inverter (n=4:1) with DC current blocking capacitors (D=0.15): DC source voltage V_{DC}=200V, inverter DC link voltage v_i, input current i_{IN}= i_{L1}, diode current i_D and transformer currents: i_{L2} and i_{L3}.

The switching frequency of simulated inverter was 20kHz. The modified MCBC control was applied in performed simulation with the shoot-through coefficient D=0.15. Simulation results confirm improved performance of proposed inverter over previous TZSI. Peak-to-peak values of transformer currents i_{L2}, i_{L3} (Fig. 10) are 17A and 68A while their average values are -0.157A and -0.62A respectively. Simulation results show that proposed configuration of TZS-like inverter could guarantee functionality and reliability in grid connected renewable source applications. Continuous is input current is extracted from low voltage DC source. DC link current during shoot-through state equals the sum of transformer currents i_{L2}, i_{L3}. During shoot-through mode it is distributed among three inverter legs. Therefore maximum transistor currents of grid connected inverter supplied from 200V DC source do not exceed 32A in simulated 2kW application.

V. EXPERIMENTAL EVALUATION

Preliminary experimental results were obtained from two laboratory models: 1kW and 3kW CIC TZS-like inverters with built-in DC current blocking capacitors operated at 10kHz. Two available on the market low leakage inductance (L_σ<5uH) planar transformers were used for realization of proposed LTCC-type TZSI input circuits: HIMAG134403, E64-2kW, 350V, n_2:n_3=14:12, with inductance measured from primary side L_{prim}=2.25 mH and HIMAG 149402 E64-6kW 650V, n_2:n_3=16:4 , with primary inductance L_{prim}=2.94 mH. DC current blocking capacitances were realized using triple 16μF high alternating current low parasitic inductance capacitors (total 48μF each). Two configurations of DC source were accessible in experimental setup: 50V and 100V. Fig. 12 and Fig. 13 present experimental results for proposed TZSI with 14:12 transformer and 100V DC source. It can be seen that for conventional transformer (without air gap) the diode current dramatically increases for large shoot-through

duty ratio (D>0.25). This is due to negative peaks of i_{L2} as diode current is a sum of input current and negative i_{L2} current. Modification of used transformer, particularly application of a small gap to the transformer core should help improve diode current waveform. Ringing, which can be also observed in diode current waveform could be due to parasitic parameters of test circuit.

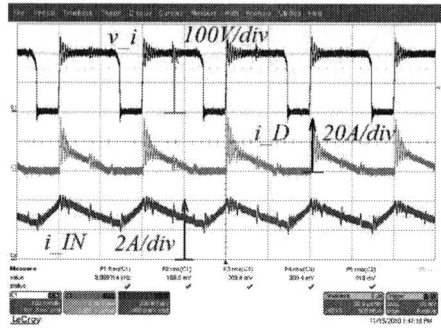

Fig. 12. Experimental waveforms of proposed CIC TZS-like inverter with DC current blocking capacitors utilizing 14:12 planar transformer (MCBC control, D=0.26, V_{DC}=100V): inverter dc link voltage v_i (100V/div), diode current i_D (20A/div) and input current extracted from DC source i_{IN} (2A/div).

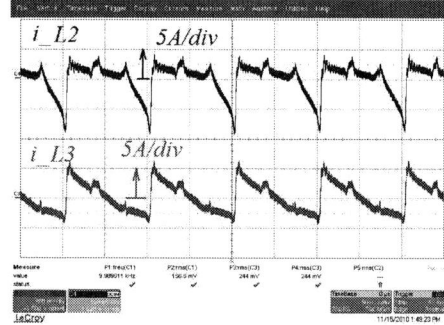

Fig. 13. Experimental waveforms of proposed CIC TZS-like inverter with DC current blocking capacitors utilizing 14:12 planar transformer (MCBC control, D=0.26, V_{DC}=100V, 50μs/div): transformer currents i_{L2}, i_{L3} (5A/div)

Fig. 14 to Fig. 16 present experimental results for proposed TZS-like inverter with built in DC current blocking capacitors utilizing 16:4 planar transformer and connected to 50V DC source. The advantage of investigated TZS-like inverter with DC current blocking capacitors over previous TZSI is that no voltage spikes can be observed in dc link voltage waveforms. The parasitic parameters are visible in test circuit however no negative impact of parasitic parameters on dc link voltage can be observed. Thanks to larger turns ratio of transformer (n_1:n_2=16:4) inverter dc link voltage obtained two times input voltage with significantly reduced shoot-through ratio D compared to system wit 14:12 transformer. Although the used planar transformer has no core gap the diode can carry large currents and the effect of saturation starts only if the value of diode current is over 10 amps. This is an other advantage of the proposed system over the previous TZSI.

978-1-4244-8806-3/11 $26.00 © 2011 IEEE

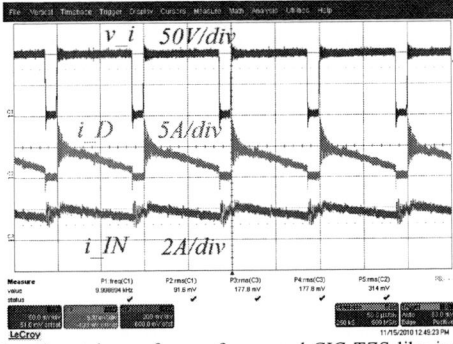

Fig. 14. Experimental waveforms of proposed CIC TZS-like inverter with DC current blocking capacitors utilizing 16:4 planar transformer (MCBC control, D=0.125, V_{DC}=50V, 50μs/div): inverter dc link voltage v_i (50V/div), diode current i_D (5A/div) and input current extracted from DC source i_{IN} (2A/div).

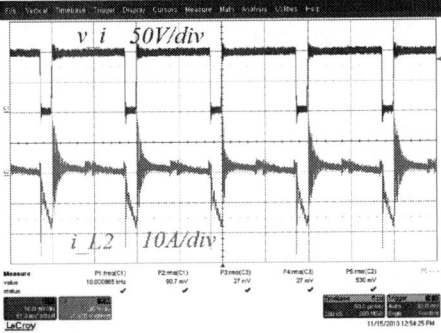

Fig. 15. Experimental waveforms of proposed CIC TZS-like inverter with built in DC current blocking capacitors utilizing 16:4 planar transformer (MCBC control, D=0.125, V_{DC}=50V, 50μs/div): inverter dc link voltage v_i (50V/div), transformer current i_{L2}(10A/div)

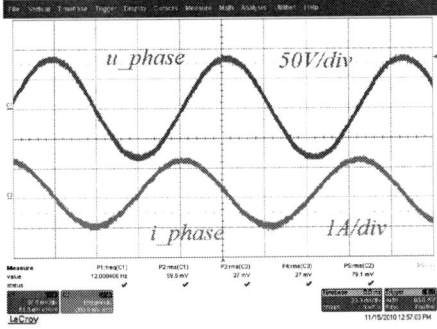

Fig. 16. Output voltage and current of proposed TZSI with built in DC current blocking capacitors applied in small induction motor drive (20ms/div)

The diode current starts ringing after the negative peak of i_{L2} current exceeds 15A.

VI. CONCLUSIONS AND FUTURE WORK

Investigated topology of novel TZS-like inverter with built in DC current blocking capacitors demonstrates improved performance over previous TZSI. Thanks to circuit reconfiguration the negative impact of parasitic parameters on inverter DC link voltage was significantly reduced. Thanks to series connection of built in capacitors with transformer windings two alternating currents flow through transformer coils and therefore proposed inverter performs higher immunity to core saturation than previous TZSI. Only one inductive element is used to story the energy during boost operation compared to previous TZSI. Proposed inverter is also interesting alternative to other Z-source type buck-boost inverters including qZSI and CqZSI as it extracts continuous current from input DC source contrary to previous TZSI. Proposed inverter characterizes reduced element count compared to recently developed CqZSI if similar voltage gain is. To obtain high performance of operation the application of small core air gap is recommended in used transformer. Proposed topology can be further developed using power planar magnetics.

ACKNOWLEDGMENT

The Polish Ministry of Science and Higher Education finances this work from the resources for the years 2009 – 2011.

REFERENCES

[1] Peng F.Z., Z-Source Inverter, *IEEE Trans. on Industry Applications*, vol. 39, n.2, 2003, pp. 504-510.

[2] Shen M., Joseph A., Wang J., Peng F.Z., Adams D.J., Comparison of Traditional Inverters and Z-Source Inverter, *Proc. of 36th IEEE Power Electronics Specialists Conference PESC'05*, 2005, pp.1692 – 1698.

[3] Strzelecki R., Adamowicz M., Wojciechowski D., Buck-Boost Inverters with Symmetrical Passive Four-terminal Networks, *Proc. of 5th IEEE Conference-Workshop CPE'07*, 2007, CD-ROM, pp. 1-9.

[4] Loh P.C., Gao F., Blaabjerg F., Goh A.L., Buck-Boost Impedance Networks, *Proc. European Conference on Power Electronics EPE'07*, 2007, CD-ROM, pp. 1-10.

[5] Anderson J., Peng F.Z., Four Quasi-Z-Source Inverters, *Proc. IEEE Conference PESC'08*, 2008, pp. 2743 – 2749.

[6] Rabkowski J., Barlik R., Nowak M., Pulse Width Modulation Methods for Bidirectional/High Performance Z-source Inverter, *Proc. of IEEE Power Electronics Specialists Conf. PESC'08* (2008), 2750-2756

[7] O. Ellabban, J. Van Mierlo, P. Lataire: Comparison between Different PWM Control Methods for Different Z-Source Inverter Topologies, *Proc. IEEE Conf. EPE 2009*, pp. 1-10.

[8] R. Strzelecki, M. Adamowicz, N. Strzelecka and W. Bury, New type T-Source inverter, *Proc. IEEE Compatibility and Power Electronics, CPE '09*, 2009, pp. 191 - 195.

[9] M. Adamowicz, N. Strzelecka: T-source inverter, *Electrical Review*, ISSN 0033-2097, vol. 85, n. 10, 2009, pp.1-6.

[10] W. Qian, F. Z. Peng; H. Cha, Trans-Z-source inverters, *Proc. IEEE International Power Electronics Conference IPEC*, 2010, pp. 1874 - 1881.

[11] M. Adamowicz, R. Strzelecki, Boost-buck inverters with cascaded qZ-type impedance networks, *Electrical Review*, ISSN 0033-2097, vol. 86 n. 2, 2010, pp.370-375.

[12] C.J. Gajanayake, F. L. Luo; H. B. Gooi, P. L. So, L. K. Siow, Extended-Boost Z-Source Inverters, *IEEE Transactions on Power Electronics*, vol. 25 , n. 10, 2010, pp. 2642 - 2652.

[13] Vinnikov D., Roasto I., Strzelecki R., Adamowicz M., Performance improvement method for the voltage-fed qZSI with continuous input current, *Proc. 15th IEEE Mediterranean Electrotechnical Conference MELECON*, 2010, pp. 1459 - 1464.

[14] Willoner G., Tihelka F., A Phase-Shift Oscillator with Wide-Range Tuning, *Proc. of the I.R.E.* (1948), 1096 – 1100.

[15] Wing O., Classical Circuit Theory, Springer 2008.

[16] Adamowicz M., "LCCT-Z-Source Inverters," Proc. International Conference on Environment and Electrical Engineering EEEiC 2011, Rome, 8-11 May, 2011, CD-ROM, p. 1-6.

Performance Comparison of SiC Schottky Diodes and Silicon Ultra Fast Recovery Diodes

Marek Adamowicz[1,2], Sebastian Giziewski[1], Jedrzej Pietryka[1], Zbigniew Krzeminski[1]

[1]Gdansk University of Technology
Narutowicza 11/12, Gdansk, POLAND
zkrzem@ely.pg.gda.pl, lider@ely.pg.gda.pl

[2]Gdynia Maritime University
Morska 81-87, 81-225, Gdynia, POLAND
madamowi@am.gdynia.pl

Abstract- **Advanced control systems combined with high speed gate driver circuits enable extremely high rate of change of power devices voltages, up to hundreds of kV/us. Short rise times of power devices could cause significant EMC problems, which are unacceptable in majority of power electronics applications. It is known that voltage variations during diode switch-off depend on how long it takes for the charge stored near the *p-n* junction to be recovered during voltage reversing. In fast switching applications good forward recovery characteristics are needed. The silicon carbide (SiC) diodes characterize almost zero reverse recovery charge. However the lossless operation in connection with extremely high dv/dt could cause the SiC diodes less effective in damping the voltage ringing. The compromise between high efficiency and low EMI emission is therefore the actual aim of the research. The paper compares the static and dynamic characteristics of ultra fast silicon (Si) and SiC Schottky diodes and presents the study of the mechanism of parasitic high frequency oscillations during turn-off transient.**

I. INTRODUCTION

The hard-switched insulated gate bipolar transistors (IGBTs) and metal-oxide semiconductor field-effect transistors (MOSFETs) integrated with free-wheeling diodes made in silicon (Si) technology have been commonly used in the power converters of electrical drives and renewable generation systems for last two decades. IGBTs, with breakdown voltages above 1kV, have been historically preferred in high-voltage, low-frequency applications (<20kHz) due to efficiency restrictions. MOSFETS have been mainly chosen for high-frequency (up to hundreds kHz), low-voltage and low output power applications due to higher on-state losses. However, the recently introduced low on-resistance MOSFETs characterizing high blocking voltage (up to 900V) and high dv/dt capability can also be used in high-current high power applications.

The significant part of the overall losses of Si-based power converter are the reverse-recovery switching losses of Si diodes [1]. The reverse recovery of Si diodes affects the transistor causing additional turn-on losses [2] and lead to a significant amount of noise (EMI) in the system. The recovery softness factor (*RSF*) is determined [3] for power diodes as the ratio of reverse recovery current fall time to reverse recovery charge Q_{RR} removal time. Low *RSF* may indicate that the diode will produce large amplitude

voltage spike due to snappy recovery [3]. The snap-off in the Si Ultrafast diode causes oscillations in the IGBT voltage, which generate EMI. Excessive voltage spikes during turn-off process even can destroy the diode. These phenomena can be attenuated by the use of ultra fast soft recovery Si diodes. The diode has a soft recovery characteristic for *RSF* equal 1 or greater.

At present the performance improvement in power converters is also accomplished through the change of semiconductor material [1]-[9]. The application of silicon carbide (SiC) to Schottky diodes reduces their reverse-recovery current almost to zero and greatly improves the efficiency of power converter [1], [9]. Power electronic devices made in SiC technology can operate at high temperatures, very high voltages and very high switching frequencies. However, the introduction of high frequency SiC devices involves also an area of scientific problems. The very high dv/dt in connection with parasitic parameters of the circuits also would cause possible ringing and noticeable radiated EMI in a high frequency range (30MHz - 2GHz).

The paper compares the static and dynamic characteristics of ultra fast Si and SiC Schottky diodes operated as the free-wheeling diodes with fast-switching *normally-off* SiC JFETs. The performance of 100kHz SiC JFET-based inverter is shown in the paper.

II. STATIC CHARACTERISTICS

Three diodes: two SiC Schottky diodes S1 and S2 and one ultra fast recovery Si diode S3 have been investigated for their potential use in high frequency SiC JFET based inverter. Table I shows the ratings of investigated diodes and two transistors used in laboratory test circuits.

The investigated SiC Schottky power diodes S1 and S2, capable of carrying currents of 20A, have been fabricated by parallel connection of two diode elements on a single die. The parallel operation results in slight decrease in the forward voltage. The SiC Schottky diodes characterize almost zero reverse recovery charge and has only a small capacitive charge Q_c of the junction. The first SiC Schottky diode (S1) has total capacitive charge Q_{c1}=122nC and active area of $2*0.0489cm^2$. The second investigated SiC diode (S2) has total capacitive charge Q_{c2}=129nC at typical conditions of V_R=1200V and di/dt=500A/μs.

978-1-4244-8806-3/11 $26.00 © 2011 IEEE

TABLE I
DIODES UNDER TEST AND TRANSISTORS USED IN TEST CIRCUITS

	Abbre-viation	Voltage Ratings	Current Ratings at 100°	Parameters
SiC Schottky Diode 1	S1	1200V	20A	Q_c=122nC at di/dt=500A/µs
SiC Schottky Diode 2	S2	1200V	20A	Q_c=129nC at di/dt=500A/µs
Si Ultra Fast Recovery Diode	S3	1200V	30A	Q_{rr}=2000nC at di/dt=500A/µs
SiC JFET	T1	1200V	30A	$R_{DS(on)}$=0.063Ω $E_{TS,typ}$=440µJ
Si MOSFET	T2	900V	25A	$R_{DS(on)}$=0.14Ω

The reverse recovery charge Q_{rr} of third investigated ultrafast Si diode (S3) varies depending on:
- rate of change of current through zero crossing di_F/dt,
- value of forward current i_F,
- temperature.

The Q_{rr} of S3 starts from 500nC for junction temperature of 100°C and i_F= 15A and reaches over 2000nC for i_F= 30A and high di_F/dt. Fig. 1 describes the test circuit for determination of diodes static characteristics. Unlike S3 the Q_c of SiC Schottky diodes is almost independent on these boundary conditions.

Measurements of the static characteristics were carried out in the test circuit shown in Fig. 1. The circuit was supplied from low voltage V_{DC}= 60V. A Si MOSFET transistor was used to discharge the capacitors through the 2.2Ω resistor and the investigated diode (*DUT -diode under test*) during the test.

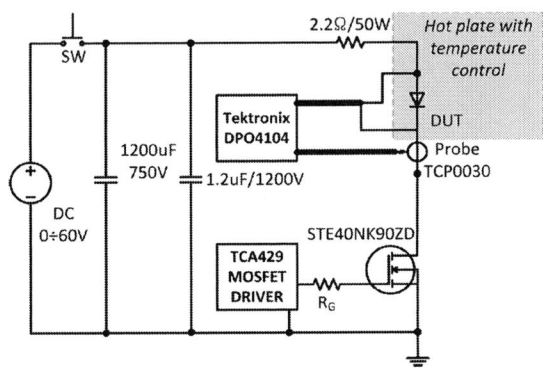

Fig. 1. Low voltage test circuit; DUT – device under test, R_G =10Ω; SW is used to charge the capacitors to the voltage set on the DC supply.

The *i-v* characteristics of three investigated diodes S1, S2 and S3 were obtained at different temperatures in the 25°C to 125°C ambient temperature range. These characteristics are shown in Fig. 2 - Fig. 4.

The features of SiC diodes and ultra fast silicon diode differ significantly. The most important difference is the positive temperature coefficient of the forward voltage of the SiC diodes. The positive temperature coefficient allows to operate SiC diodes in parallel.

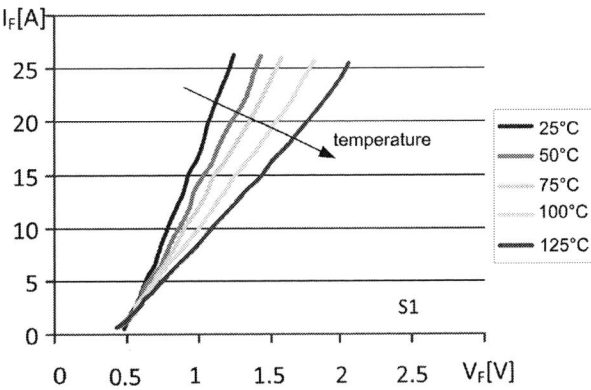

Fig. 2. *i-v* characteristics of first investigated SiC Schottky diode S1 at different operating temperatures.

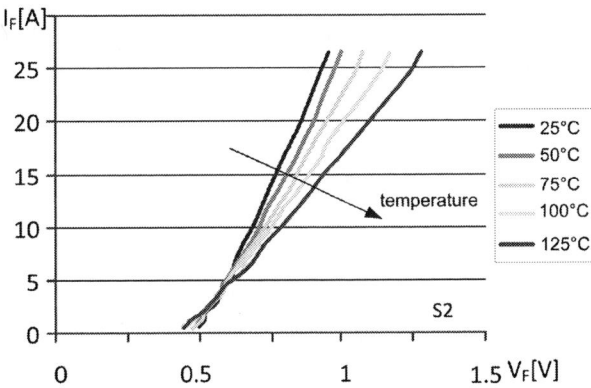

Fig. 3. *i-v* characteristics of second investigated SiC Schottky diode S2 at different operating temperatures.

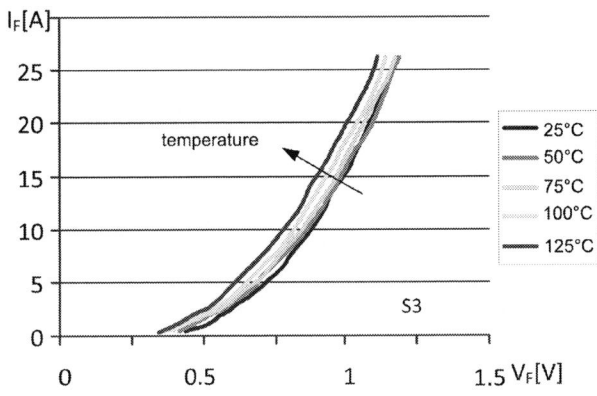

Fig. 4. *i-v* characteristics of investigated ultra fast recovery diode S3 at different operating temperatures.

That enables the increase of power rating of power converters using commercially available (<30A) SiC devices.

The *i-v* curves of S1 and S2 are almost linear and deviate from exponential behaviour seen for S3. S1 and S2 differ in forward voltage drop. S2 characterizes lower forward voltage than S1.

III. DYNAMIC CHARACTERIZATION

The introduction of high frequency SiC devices characterizing high dv/dt involves wide area of scientific problems connected with the impact of parasitic inductances and capacitances on the switching performance and EMI problems. The application of zero reverse-recovery current SiC Schottky diodes improves the efficiency of the power converter, however, the possible oscillations and resonances during high speed switching are the source of noticeable EMC/EMI emission in the converter system. The current oscillations of high magnitude, caused by high dv/dt and circuit parasitics decrease the converter performance and might cause the devices failure and breakdown. The locations of possible failure events and EMI problems connected witch applications of high speed SiC devices within the power converter are shown in Fig. 5.

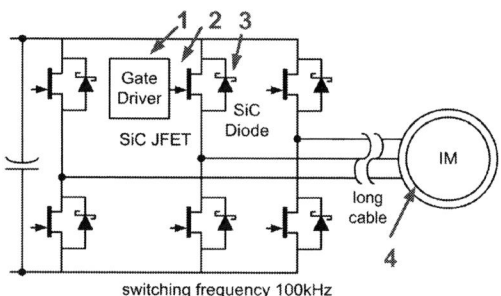

Fig. 5. The occurrence of failure events and EMI problems in SiC based power converter: (1)gate driver failures caused by oscillations/EMI problems in gate driver signals; (2), (3) device failures caused by oscillations/EMI problems during switching; (4) IM bearings failures caused high dv/dt and common mode voltages.

To predict the scale of possible failure events and EMI problems, the dynamic characteristics of different combinations of Si and SiC diodes with SiC JFET were measured in the high voltage test circuit described in Fig. 6. The current and voltage waveforms were observed and recorded using Tektronix DPO4104 oscilloscope equipped with TCP0030 120MHz current probe, P5205 high voltage differential and P6931A 500MHz voltage probe. The experimental test setup is shown in Fig. 7. The FR4 double layer PCB was used for realization of the test circuit.

From the experimental test setup the switching waveforms of S1 and S2 and T1 were obtained to understand the switching features of investigated SiC devices. The gate resistor R_1 was 2.7Ω for both investigated combinations: T1 - S1 and T1 - S2.

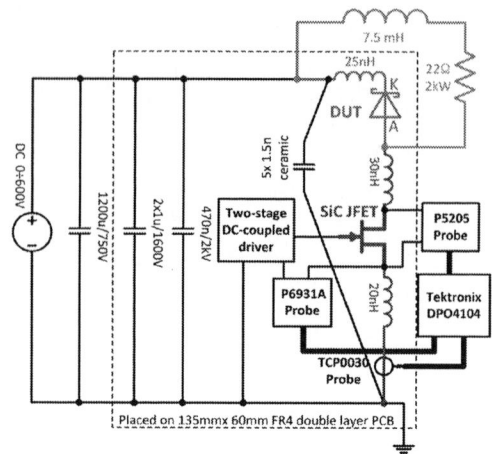

Fig. 6. Schematic of the high voltage test circuit.

Fig. 7. Experimental test setup

The measured switching characteristics of the investigated diodes S1-S3 are shown in Fig. 8 - Fig. 10.

The turn-off waveforms of Si diode S3 are shown in Fig. 8.

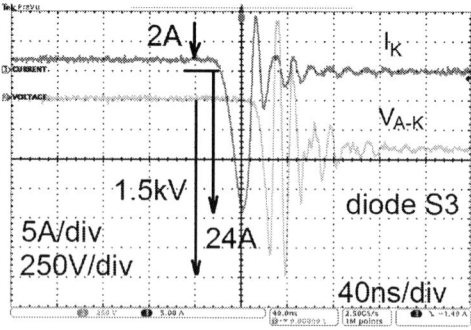

Fig. 8. Turn-off waveforms of combination III (SiC transistor T1 and Si diode S3): Diode S2 anode to cathode voltage V_{A-K} and diode current I_K.

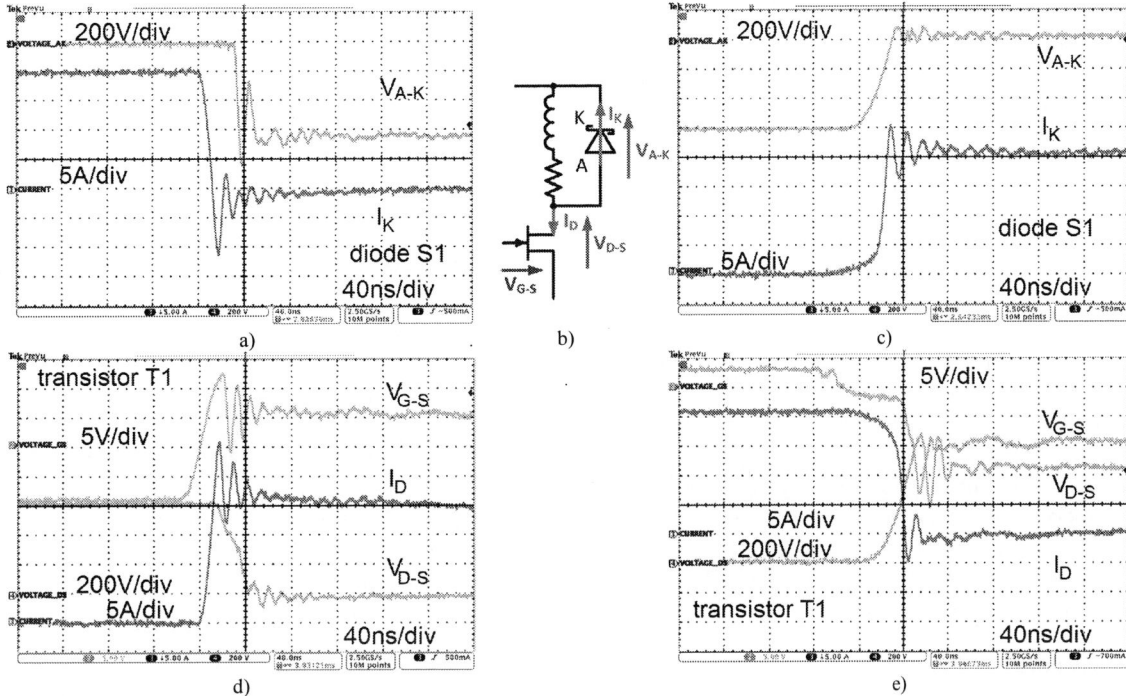

Fig. 9. Switching waveforms of combination I (SiC transistor T1 and SiC diode S1): Diode S1 anode to cathode voltage V_{A-K} and diode current I_F during turn-off (a) and turn-on (c); SiC JFET (T1) gate to source voltage V_{G-S}, drain to source voltage V_{D-S} and drain current I_D during turn-on (d) and turn-off (e); schematic of directions of measured voltage and currents.

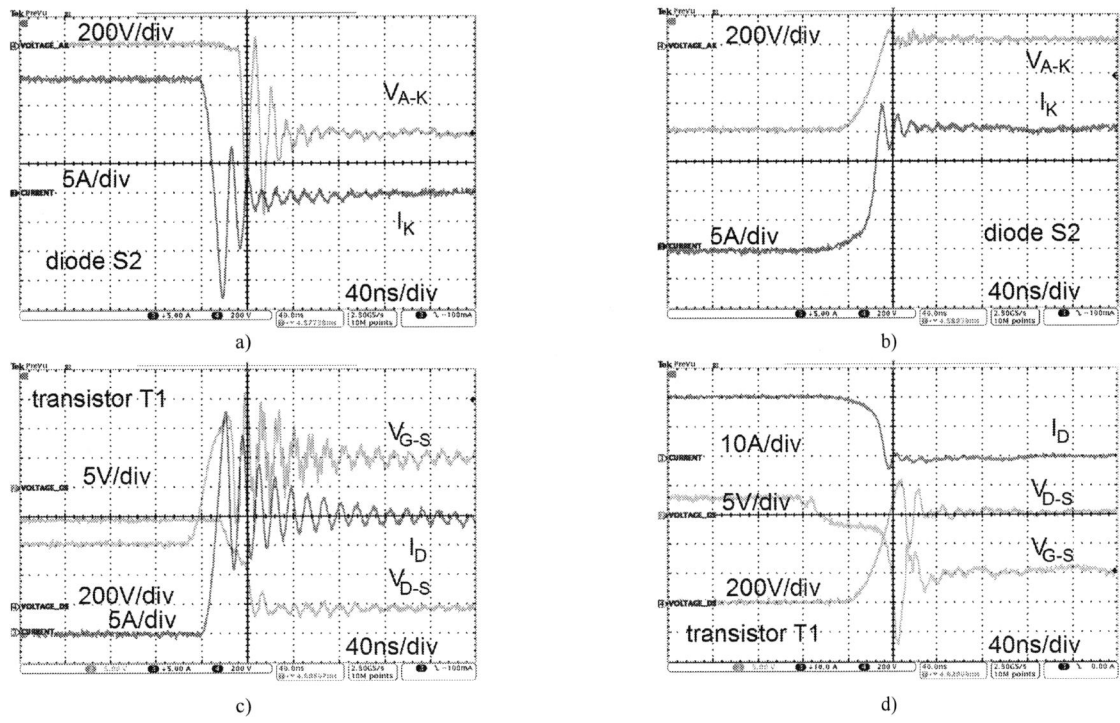

Fig. 10. Switching waveforms of combination II (SiC transistor T1 and SiC diode S2): Diode S2 anode to cathode voltage V_{A-K} and diode current I_K during turn-off (a) and turn-on (b); SiC JFET (T1) gate to source voltage V_{G-S}, drain to source voltage V_{D-S} and drain current I_D: during turn-on (c) and turn-off (d).

As it can be seen from Fig. 8, the combination of T1 and S3 diode was not able to carry currents higher than 2A due to high magnitude of reverse recovery current (>24A) during the S3 diode turn-off and considerable voltage oscillations (>1.5kV in peak) observed in the test circuit. The Q_{rr} of the ultra fast Si diode at di/dt >1.2kA/µs was unacceptable.

The high dv/dt (>100kV/µs) and di/dt (>1.6kA/µs for S1 and >1.9kA/µs) values can be observed during turn-off transients of both investigated SiC diodes. As it can be seen from Fig. 9c and Fig. 10b, the turn-on waveforms of two SiC diodes are almost identical. This is because the SiC JFET turn-off process is not so much influenced by diode performance and does not depend considerably on diode characteristics.

It can be seen from Fig. 10a and Fig. 10c that noticeable oscillations appear during transistor switch-on and simultaneous diode turn-off in the investigated system of T1 and diode S2 (combination II). S1 and S2 characterize similar current and voltage ratings and similar Q_C however differ in the damping of the ringing. The observed oscillations in main circuit influence gate voltage V_{G-S} waveforms and the gate circuit should be protected with the use of additional Zener diode. It can be deduced that V_{G-S} ringing is caused by high magnitude voltage variations on SiC JFET source lead parasitic inductance.

The insufficient damping of voltage and current oscillations might be explained by a very small diode losses dissipated during switching and very fast run of switching process. These undesirable phenomena can be attenuated by slowing down the switching process [10].

IV. PERFORMANCE OF SiC DIODES IN HIGH-FREQUENCY SiC-BASED INVERTER

The picture of designed and investigated three phase SiC JFET-based inverter with applied S2-type SiC Schottky diodes is shown in Fig. 11a [10]. Six pairs of transistors and free-wheeling diodes are mounted together with DC-link capacitors and current transducers on main round-shaped printed circuit board (PCB). A secondary round-shaped PCB containing six fast-switching two-stage DC-coupled gate drivers is fixed to the main PCB. The ADSP-21363 (333MHz) based universal control card containing programmable logic unit CYCLONE II EP2C8F256 (Fig. 11b), worked out at Gdansk University of Technology, is used to control SiC-based inverter. The control system hardware enables three independent interruptions: IRQ1, during which the analog to digital conversion is executed of voltage and currents measurements (every 5µs), IRQ2 executing PWM algorithm (every 10µs) and IRQ3 during which state observer differential equations are solved and power flow control with generator control procedures are executed (every 50µs).

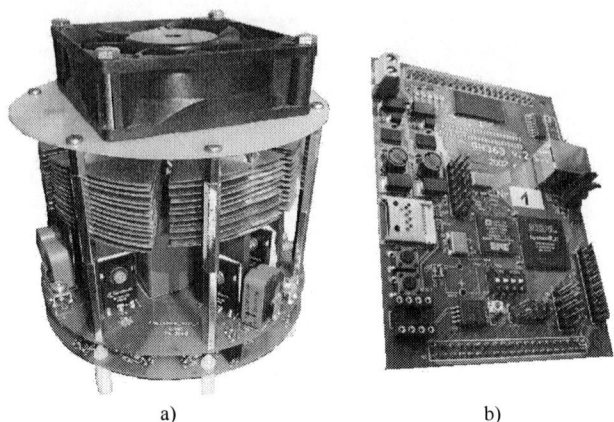

a)   b)

Fig. 11. Hardware prototype of *normally-off* SiC JFET-based DC-AC converter (a) and ADSP-21363 based control card (b).

Unlike IGBT-based inverters where switching frequencies of 2 to 10 kHz in combination with long generator leads (>5m) involve the application of large passive output filters, the increased switching frequency (>100kHz) in SiC JFET-based DC-AC converter considerably reduces the passive filter size.

Fig. 12 shows output currents and voltages of the investigated SiC-based inverter fed small induction machine and equipped with small size 60µH (15 turns) output inductors.

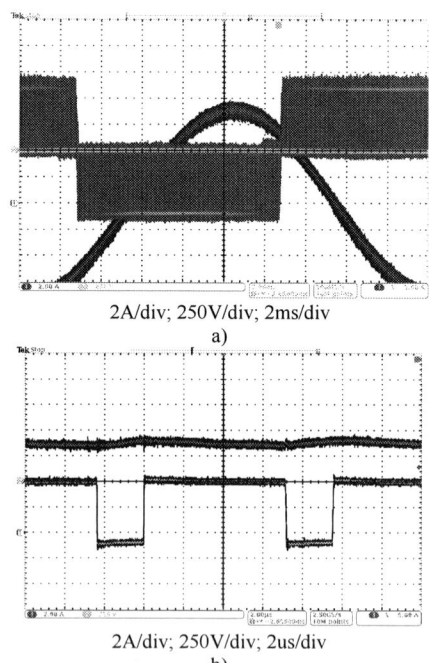

2A/div; 250V/div; 2ms/div

a)

2A/div; 250V/div; 2us/div

b)

Fig. 12. Phase output current and line-to-line output voltage of investigated high frequency inverter with applied S2-type SiC Schottky diodes.

V. CONCLUSIONS

Three 1.2kV fast diodes: two SiC Schottky diodes and one ultra fast recovery Si diode were investigated for the purpose of potential application in high speed SiC-JFET based inverter. The SiC diodes characterize better performance during transients. While the first investigated SiC Schottky diode has larger voltage drop during conduction which could cause higher on-state losses, the second SiC diode has worse damping coefficient during switching. Unlike SiC diodes the ultra fast recovery Si diode, when applied to SiC JFET as e free-wheeling diode, was not capable to carry currents over 2A. The preliminary tests of SiC JFET based inverter equipped with SiC Schottky diodes and small passive filter were completed successfully.

ACKNOWLEDGEMENT

This scientific work is financed from resources for science in 2010-2012 years as a LIDER research project of the National Centre for Research and Development.

REFERENCES

[1] Elasser A., Kheraluwala M.H., Ghezzo M., Steigerwald R.L., Evers N.A., Kretchmer J., Chow T.P., "A comparative evaluation of new silicon carbide diodes and state-of-the-art silicon diodes for power electronic applications" *IEEE Transactions on Industry Applications*, Vol. 39 , No 4, 2003, pp. 915 - 921.

[2] S. Bontemps, A. Calmels, S. D. Round, J. W. Kolar, Low profile power module combined with state of the art MOSFET switches and SiC diodes allows high frequency and very compact three-phase sinusoidal input rectifiers., Proc. of the Conference for Power Electronics, Intelligent Motion, Power Quality (PCIM'07), Nuremberg, Germany, May 22 - 24, CD-ROM (2007).

[3] Jovalusky J.," New low reverse recovery charge (QRR) high-voltage silicon diodes provide higher efficiency than presently available ultrafast rectifiers," Proc. Twenty-Third Annual IEEE Applied Power Electronics Conference and Exposition APEC 2008, pp.918-923.

[4] Johnson C.M., Rahimo M., Wright, N.G., Hinchley, D.A., Horsfall A.B., Morrison, D.J., Knights A., "Characterisation of 4H-SiC Schottky diodes for IGBT applications," *Proc. IEEE Industry Applications Conference, 2000*, Vol. 5, pp. 2941 - 2947.

[5] Lu B., Dong W., Zhao Q., Lee F.C., "Performance evaluation of CoolMOS™ and SiC diode for single-phase power factor correction applications," *Eighteenth Annual IEEE Applied Power Electronics Conference and Exposition APEC '03*, 2003, Vol. 2, pp. 651 - 657.

[6] Galigekere, V.P.; Kazimierczuk, M.K., " Performance of SiC Schottky diodes," *50th Symposium MWSCAS* 2007, pp. 682 - 685.

[7] Millan J., Banu V., Brosselard P., Jorda X., Perez-Tomas A., Godignon P., " Electrical performance at high temperature and surge current of 1.2 kV power rectifiers: Comparison between Si PiN, 4H-SiC Schottky and JBS diodes," *Proc. IEEE Semiconductor Conference, CAS 2008*, pp. 53 - 59.

[8] B. Ozpineci, M. S. Chinthavali, L. M. Tolbert, A. Kashyap, H. A. Mantooth, "A 55 kW Three-Phase Inverter with Si IGBTs and SiC Schottky Diodes," Proc. IEEE Applied Power Electronics Conference and Exposition APEC '06, 2006, pp. 448 - 454.

[9] J. Richmond, " Hard-Switched Silicon IGBTs? Cut Switching Losses in Half with Silicon Carbide Schottky Diodes," Application Note CPWR-AN03, 2006, www.cree.com/power.

[10] M. Adamowicz, S. Giziewski, J. Pietryka, M. Rutkowski, Z. Krzeminski, "Evaluation of SiC JFETs and SiC Schottky Diodes for Wind Generation Systems", *accepted for publication at* 20th IEEE International Symposium on Industrial Electronics ISIE'2011, 27-30 June, 2011, Gdansk, Poland.

Three-Phase Multilevel Inverter Based on LeBlanc Transformer

V. Fernão Pires[*][ᴨ], Manuel Guerreiro[*], J. F. Martins[*] and J. Fernando Silva[♦][ᴨ]

[*] Escola Superior Tecnologia Setúbal, Instituto Politécnico Setúbal, Setúbal, Portugal
[ᴨ] Center for Innovation in Electrical and Energy Engineering (CIEEE), Lisbon, Portugal
[❖] CTS, UNINOVA, Departamento de Engenharia Electrotécnica, Faculdade de Ciências e Tecnologia, Universidade Nova de Lisboa, Costa Caparica, Portugal
[♦] Instituto Superior Técnico / Universidade Técnica de Lisboa, Lisboa, Portugal

Abstract — **A new power conversion structure for a three-phase multilevel inverter is proposed in this paper. This structure uses a LeBlanc transformer and three single-phase voltage source inverters. The power converter topology has two single-phase outputs that connect to the LeBlanc transformer. This transformer allows deriving two-phase voltage to a three-phase source. This multilevel power conversion structure allows obtaining a five-level three-phase inverter. To control the proposed converter a sliding mode controller with a vectorial modulator was used. Simulation results of the proposed converter structure with their control system are presented. From the obtained results it possible to confirm the voltage multilevel waveforms and the effectiveness of the proposed control system technique.**

Index Terms—**Multilevel converter, LeBlanc transformer, cascade structure, sliding mode controller, vectorial modulator, power converter.**

I. INTRODUCTION

Many applications use a three-phase inverter. The classical three-phase inverter is very limited regarding the output voltage levels. This power converter only allows obtaining a phase-to-phase three output voltage levels. In this way, multilevel inverters have been proposed in order to overcome this problem. These kind of power converters are very interesting for the power industry due to attractive features like reduced current and voltage harmonics on the ac side, high voltage capability and low dv/dt's. The topology, modulation strategy and performance of these power converters have been extensively studied over the last decades. In this way several multilevel topologies have been proposed. The most used topologies are the neutral point clamped [1]-[4], the flying capacitor [5]-[8] and the cascaded H-bridge inverter [9]-[11]. A hybrid multilevel topology has also been proposed [12]. This topology is similar than the cascaded H-bridge inverter but uses only one DC source.

In addition to the above attractive features of the multilevel inverters, in medium-power and high-power applications safety and robustness are also important. In this way isolated systems in low frequency are used. So, multilevel converters associate to line-frequency transformer have been proposed. Three single or one three phase transformer are normally used. However, another kind of transformers have also been used. In this way, power converters with special line-frequency transformers such as Scott or LeBlanc have also been proposed [13]-[15]. These special transformers allow deriving two phase current from a three phase source or vice-versa.

Given these considerations a new three-phase power conversion structure is proposed. This structure is based on a multilevel inverter and on a LeBlanc transformer. The multilevel power converter uses a five-level cascade structure. The outputs of the power converters generate two voltages with a 90° phase angle between them. The use of the LeBlanc transformer provides a three-phase system.

There are several controllers for multilevel power converters, such as the sinusoidal PWM extended to multiple carrier arrangements [16]-[18]. However, other control techniques have been used. One of the fast and robust techniques is the sliding mode control [19]-[20]. The proposed power converter can also be controlled by a sliding mode approach. In this way, a vectorial sliding mode controller adapted to this power converter is also proposed. Simulation results are presented in order to illustrate the circuit performance.

This paper is organized in six sections. After this introduction the proposed multilevel inverter based on LeBlanc transformer is presented in Section II. In section III it is presented the developed theoretical expressions for the study of this multilevel converter. The control system for the proposed system is described in Section IV. Some simulation results obtained from the proposed converter are presented in Section V. Finally, in section VI the conclusions of the work are synthesized.

II. PROPOSED MULTILEVEL INVERTER BASED ON LEBLANC TRANSFORMER

Fig. 1 shows the new power conversion structure for a three-phase multilevel inverter. This topology uses two dc sources and three single-phase voltage source converters in order to maximize the output voltage. A 2-to-3 phase transformer is used at the output of the voltage source converters. Due to the connection of the three single-phase inverters, at the output of the power converter structure a five level voltage is obtained (2V,V,0,-V,-2V). One of the drawbacks of this topology, is that two of the single-phase inverters have to withstand a voltage of 2V.

The transformer uses a three-phase core with a winding system as presented in Fig. 2 (a). This transformer is known as Leblanc transformer. The primary windings are fed by two different voltages $V_{ab}(t)$ and $V_{cd}(t)$. These voltages are generated by the power converter with a phase angle 90° between them. The secondary windings are the three-phase system $V_R(t)$, $V_S(t)$ and $V_T(t)$. The phasor diagram is presented in Fig. 2 (b).

978-1-4244-8806-3/11 $26.00 © 2011 IEEE   150

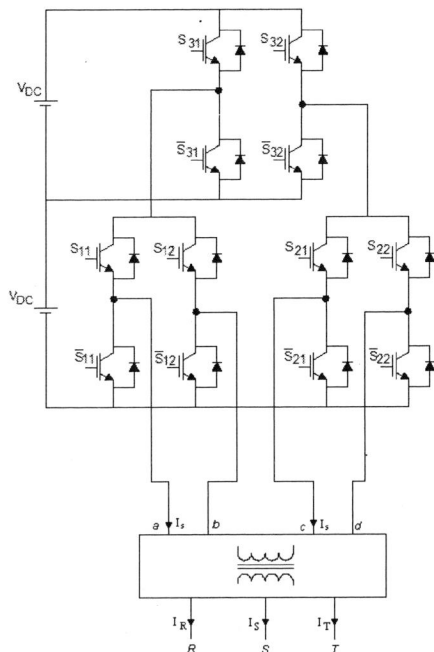

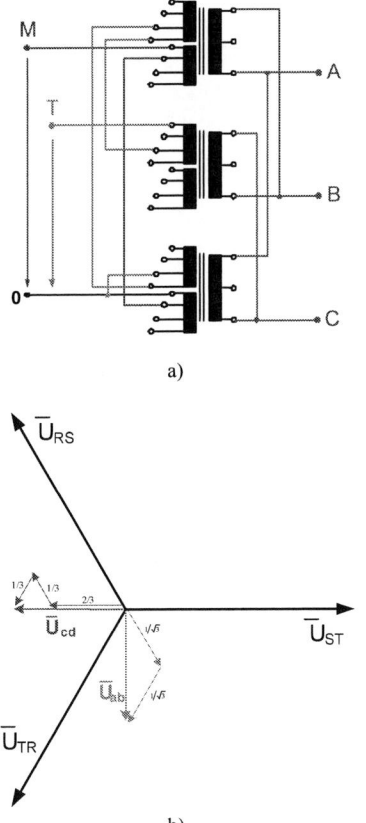

Figure 1. Proposed conversion structure for a three-phase multilevel inverter.

Figure 2. Connection a) and phasor diagram b) of the Leblanc transformer.

III. MULTILEVEL CONVERTER MODEL

In order to obtain suitable theoretical expressions for the study of the proposed multilevel converter with their control system, it will be assumed ideal power switches. In this way, the states of the switches of the kth, $k \in \{1,2,3,4,5,6\}$, power converter leg can be represented by the time dependent variable γ_{ij}, $i \in \{1,2,3\}$, $j \in \{1,2\}$, defined by (1).

The output multilevel voltages U_{ab} and U_{cd} are dependent of the switches states. So, according condition (1) these output voltages are obtained by (2).

$$\gamma_{ij} = \begin{cases} 1 & if \ S_{ij} \ is \ ON \ \wedge \ \bar{S}_{ij} \ is \ OFF \\ 0 & if \ S_{ij} \ is \ OFF \ \wedge \ \bar{S}_{ij} \ is \ ON \end{cases} \tag{1}$$

$$U_{ab} = \left(\gamma_{31} \gamma_{11} - \gamma_{31} \gamma_{12} + \gamma_{11} - \gamma_{12} \right) U_{DC}$$

$$U_{ab} = \left(\gamma_{32} \gamma_{21} - \gamma_{32} \gamma_{22} + \gamma_{21} - \gamma_{22} \right) U_{DC} \tag{2}$$

Fig. 3 shows the correspondent output voltage vectors. Table I shows the output voltage vectors according to the state of the switches.

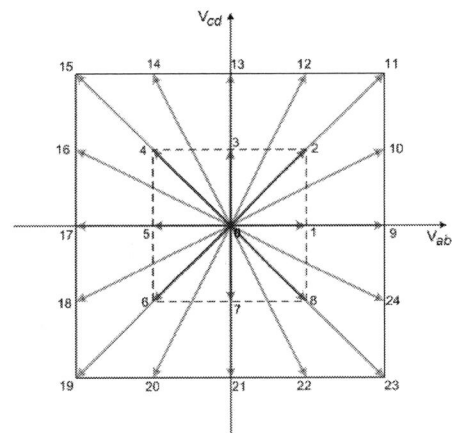

Figure 3. Output voltage vectors of the proposed power converter.

TABLE II
OUTPUT VOLTAGE VECTORS FOR THE PROPOSED TOPOLOGY

S_{31}	S_{32}	S_{21}	S_{22}	S_{11}	S_{12}	V_{ab}	V_{cd}	Nº
0	0	0	0	0	0	0	0	0
0	0	0	0	0	1	-V	0	5
0	0	0	0	1	0	+V	0	1
0	0	0	0	1	1	0	0	0
0	0	0	1	0	0	0	-V	7
0	0	0	1	0	1	-V	-V	6
0	0	0	1	1	0	+V	-V	8

978-1-4244-8806-3/11 $26.00 © 2011 IEEE

0	0	0	1	1	1	0	-V	7
0	0	1	0	0	0	0	+V	3
0	0	1	0	0	1	-V	+V	4
0	0	1	0	1	0	+V	+V	2
0	0	1	0	1	1	0	+V	3
0	0	1	1	0	0	0	0	0
0	0	1	1	0	1	-V	0	5
0	0	1	1	1	0	+V	0	1
0	0	1	1	1	1	0	0	0
0	1	0	0	0	0	0	0	0
0	1	0	0	0	1	-2V	0	17
0	1	0	0	1	0	+2V	0	9
0	1	0	0	1	1	0	0	0
0	1	0	1	0	0	0	-V	7
0	1	0	1	0	1	-2V	-V	18
0	1	0	1	1	0	+2V	-V	24
0	1	0	1	1	1	0	-V	7
0	1	1	0	0	0	0	+V	3
0	1	1	0	0	1	-2V	+V	16
0	1	1	0	1	0	+2V	+V	10
0	1	1	0	1	1	0	+V	3
0	1	1	1	0	0	0	0	0
0	1	1	1	0	1	-2V	0	17
0	1	1	1	1	0	+2V	0	9
0	1	1	1	1	1	0	0	0
1	0	0	0	0	0	0	0	0
1	0	0	0	0	1	-V	0	5
1	0	0	0	1	0	+V	0	1
1	0	0	0	1	1	0	0	0
1	0	0	1	0	0	0	-2V	21
1	0	0	1	0	1	-V	-2V	20
1	0	0	1	1	0	+V	-2V	22
1	0	0	1	1	1	0	-2V	21
1	0	1	0	0	0	0	+2V	13
1	0	1	0	0	1	-V	+2V	14
1	0	1	0	1	0	+V	+2V	12
1	0	1	0	1	1	0	+2V	13
1	0	1	1	0	0	0	0	0
1	0	1	1	0	1	-V	0	5
1	0	1	1	1	0	+V	0	1
1	0	1	1	1	1	0	0	0
1	1	0	0	0	0	0	0	0
1	1	0	0	0	1	-2V	0	17
1	1	0	0	1	0	+2V	0	9
1	1	0	0	1	1	0	0	0
1	1	0	1	0	0	0	-2V	21
1	1	0	1	0	1	-2V	-2V	19
1	1	0	1	1	0	+2V	-2V	23
1	1	0	1	1	1	0	-2V	21
1	1	1	0	0	0	0	+2V	13
1	1	1	0	0	1	-2V	+2V	15
1	1	1	0	1	0	+2V	+2V	11
1	1	1	0	1	1	0	+2V	13
1	1	1	1	0	0	0	0	0
1	1	1	1	0	1	-2V	0	17
1	1	1	1	1	0	+2V	0	9
1	1	1	1	1	1	0	0	0

period T, the output voltages U_{ab} and U_{cd} average values must be equal to their reference average values (3).

$$\begin{cases} \dfrac{1}{T}\int_0^T U_{ab\,ref}\,dt - \dfrac{1}{T}\int_0^T U_{ab}\,dt = e_{Uab} = 0 \\[2ex] \dfrac{1}{T}\int_0^T U_{cd\,ref}\,dt - \dfrac{1}{T}\int_0^T U_{cd}\,dt = e_{Ucd} = 0 \end{cases} \tag{3}$$

From the previous assumption, a sliding surface $S(e_{ab,cd},t)$ can be obtained. Equation (4) shows the sliding surface that allows to enforce the control goal $U_{ab,cd}=U_{ab,cd}ref$, where $k_{ab,cd}$ is used to impose the switching frequency.

$$S\big(e_{ab,cd},t\big) = \frac{k_{ab,cd}}{T}\int_0^T \big(U_{ab,cd\,ref} - U_{ab,cd}\big)dt = 0 \tag{4}$$

The control strategy must guarantee that the system trajectory moves towards and stays on the sliding surface from any initial condition. This can be obtained by the stability condition (5).

$$S\big(e_{ab},e_{cd},t\big)\,\dot{S}\big(e_{ab},e_{cd},t\big) < 0 \tag{5}$$

To ensure the sliding surface (4), the switching law must be select according to the conditions expressed in (6).

$$\begin{cases} S(e_{ab,cd},t) > 0 \Rightarrow \dot{S}(e_{ab,cd},t) < 0 \Rightarrow U_{ab,cd} > U_{ab,cd\,ref} \\[2ex] S(e_{ab,cd},t) < 0 \Rightarrow \dot{S}(e_{ab,cd},t) > 0 \Rightarrow U_{ab,cd} < U_{ab,cd\,ref} \end{cases} \tag{6}$$

In order to ensure this, the switching strategy must select the proper values of $U_{ab,cd}$ from the available outputs. From the analysis of the proposed topology it is possible to verify that, there are 25 different output voltage vectors. Fig. 3 and Table II show the output voltage vectors that can be used in this power converter topology.

TABLE I
OUTPUT VOLTAGE VECTORS FOR THE PROPOSED TOPOLOGY

+2V	15	14	13	12	11
+V	16	4	3	2	10
0	17	5	0	1	9
-V	18	6	7	8	24
-2V	19	20	21	22	23
$\begin{matrix} V_B \\ V_A \end{matrix}$	**-2V**	**-V**	**0**	**+V**	**+2V**

IV. POWER CONVERTER CONTROL

To control the multilevel inverter a vector sliding mode controller will be used. The sliding mode controller will be obtained from the controllable canonical form of the system model. The controller should impose that, in each switching

To implement this system, two five-level hysteretic comparators (with hysteresis ε in order to limit the maximum switching frequency) at the output of the sliding mode controller (4) have been used. Fig. 4 shows the five-level hysteretic comparator that has been used. The output of these comparators are the integer variables λ_{ab} and λ_{cd} (λ_{ab}, λ_{cd} $\in \{2;1;0;-1;-2\}$) corresponding to the five selectable levels. These integer variables are related with the desired V_{ab} and V_{cd} voltages. According variables λ_{ab}, λ_{cd} and table I the switches that must be on are selected.

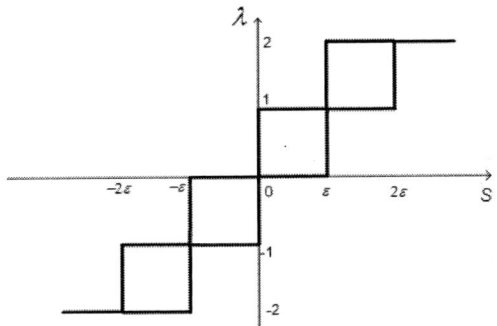

Figure 4. Output voltages of the power converters.

V. SIMULATION RESULTS

In order to investigate the behavior of the proposed power converter several simulations have been carried out. These results have been obtained using the program Matlab-Simulink/Power System Blockset. For the simulated system it was used the following parameters: V_{DC}=200 V, L=5 mH and R=100 Ω.

Fig. 5 shows the output voltages v_{ab} and v_{cd}. As can be seen there are five levels with a 2Vdc maximum output voltage. At the transformer output a RL load was connected. Fig. 6 shows the converter output currents I_{ab} and I_{cd}. As expected these currents are nearly sinusoidal and 90º phase-shifted from each other with equal amplitudes. The sinusoidal waveform is due to the RL load and LeBlanc transformer.

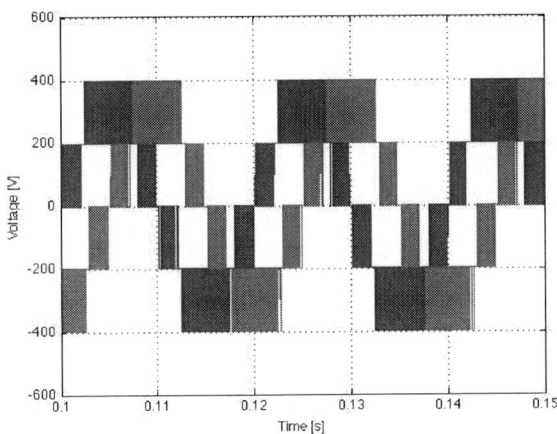

Figure 5. Output voltages of the power converters.

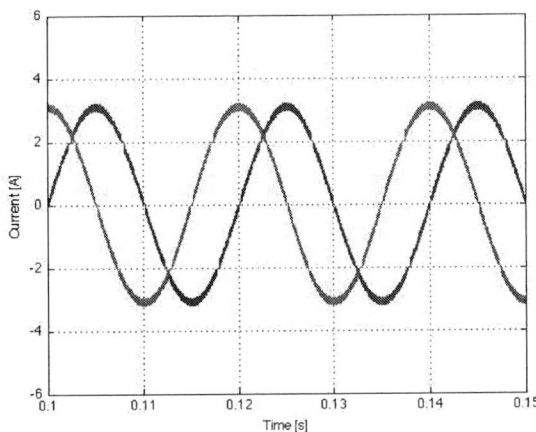

Figure 6. Multilevel inverter output currents.

Fig. 7 shows the three-phase load currents I_R, I_S and I_T. From this result it is possible to verify that the output currents are balanced and nearly sinusoidal. The obtained results also show the effectiveness of the proposed control system.

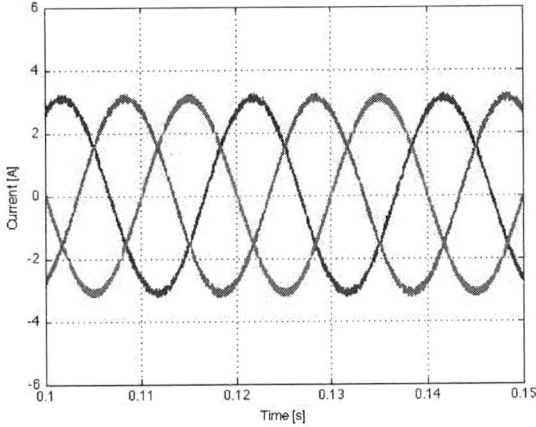

Figure 7. Three-phase load currents.

VI. CONCLUSIONS

A new configuration for a three-phase multilevel inverter has been proposed. In the proposed structure, a three-phase five-level inverter was obtained. This structure is accomplished with three single-phase voltage source inverters and a LeBlanc power transformer. In this way it is possible to obtain a three-phase system from a two-phase power converter using two power supplies. To control this power converter it was adopted a vectorial sliding mode controller. This fast and robust converter allows obtaining the desired five level output voltages. Since this topology allows obtaining 25 different output voltage vectors, two five-level hysteretic comparators were used to implement the vectorial modulator. The proposed power converter structure has been validated using computer simulations.

978-1-4244-8806-3/11 $26.00 © 2011 IEEE

REFERENCES

[1] A. Nabae, I. Takahashi, and H. Akagi, "A new neutral-point clamped PWM inverter," IEEE Trans. Ind. Appl., vol. IA-17, no. 5, pp. 518–523, Sep./Oct. 1981.

[2] X. Yuan, I. Barbi, "Fundamentals of a New Diode Clamping Multilevel Inverter", IEEE Transactions on Power Electronics, vol. 15, nº 4, pp. 711-718, July 2000.

[3] J. Dionísio Barros, J. Fernando Silva, "Optimal Predictive Control of Three-Phase NPC Multilevel Converter for Power Quality Applications, " IEEE Transactions on Industrial Electronics, vol. 55, nº 10, pp. 3670-3681, October 2008.

[4] J. Rodriguez, J.S. Lai, F.Z. Peng, "Multilevel inverters: a survey of topologies, controls, and applications," IEEE Transactions on Industrial Electronics, vol. 49, nº 4, pp. 724 – 738, Aug. 2002.

[5] T. A. Meynard, H. Foch, "Multi-level choppers for high voltage applications," Eur. Power Electron. J., vol. 2, no. 1, pp. 45–50, March 1992.

[6] X.M. Yuan, H. Stemmler, I. Barbi, "Self-balancing of the clamping-capacitor-voltages in the multilevel capacitor clamping inverter under sub-harmonic PWM modulation, " IEEE Trans. on Power Electronics, vol. 16, no. 2, pp. 256–263, March 2001.

[7] C. Feng, J. Liang, and V. G. Agelidis, "Modified phase-shifted PWM control for flying capacitor multilevel converters," IEEE Trans. Power Electron., vol. 22, no. 1, pp. 178–185, Jan. 2007.

[8] R. H. Wilkinson, H. de Mouton, and T. Meynard, "Natural balance of multicell converters: the two-cell case," IEEE Trans. Power Electronics, vol. 21, no. 6, pp. 1649-1657, November 2006.

[9] Peter W. Hammond, "A New Approach to Enhance Power Quality for Medium Voltage AC Drives", IEEE Transactions on Industry Applications, Vol. 33, No 1, pp. 202-208, January/February 1997.

[10] P. C. Loh, D. G. Holmes, and T. A. Lipo, "Implementation and control of distributed PWM cascaded multilevel inverters with minimal harmonic distortion and common-mode voltage," IEEE Trans. Power Electron., vol. 20, no. 1, pp. 90–99, Jan. 2005.

[11] V. M. E. Antunes, V. F. Pires, J. Fernando Silva, "Narrow Pulse Elimination PWM for Multilevel Digital Audio Power Amplifiers Using Two Cascaded H-Bridges as a Nine-Level Converter", IEEE Transactions on Power Electronics, vol. 22, no. 2, pp. 425-434, March 2007.

[12] J. N. Chiasson, B. Ozpineci, L. M. Tolbert, "A Five-Level Three-Phase Hybrid Cascade Multilevel Inverter Using a Single DC Source for a PM Synchronous Motor Drive," Applied Power Electronics Conference, APEC, pp.1504-1507, February-March 2007.

[13] A. Ruffer, Ch.-B. Andrianirina, "A symmetrical 3 phase 2-switch PFC-power supply for variable output voltage," EPE'95: European Conference on Power Electronics and Aplications, (1995).

[14] A. A. Badin, I. Barbi, "Unity Power Factor Isolated Three-Phase Rectifier With Split DC-Bus Based on the Scott Transformer, " IEEE Transactions on Power Electronics, vol. 23, no. 3, pp. 1278-1287, May 2008.

[15] V. F. Pires, Manuel Guerreiro, J. F. Martins, J. Fernando Silva, "Three-Phase PWM Rectifier Employing Two Single-Phase Buck-Boost PFC Modules and a Scott Transformer", Przeglad Elektrotechniczny, vol. 85, no. 10, pp 213-216, October, 2009.

[16] G. Carrara, S. Gardella, M. Marchesoni, R. Salutari, and G. Sciutto, "A new multilevel PWM method: A theoretical analysis," IEEE Trans. Power Electron., vol. 7, no. 3, pp. 497–505, Jul. 1992.

[17] B. P. McGrath and D. G. Holmes, "Multicarrier PWM strategies for multilevel inverters," IEEE Trans. Ind. Electron., vol. 49, no. 4, pp. 858–867, Aug. 2002.

[18] S. Kouro, P. Lezana, M. Angulo, and J. Rodriguez, "Multicarrier PWM with DC-link ripple feedforward compensation for multilevel inverters," IEEE Trans. Power Electron., vol. 23, no. 1, pp. 52–59, Jan. 2008.

[19] W. Gao, J. Hung, "Variable structure control: A Survey," IEEE Transactions on Industrial Electronics, vol. 40, no. 1, pp. 2-22, February 1993.

[20] J. Fernando Silva, "Sliding Mode Control Design of Drive and Regulation Electronics for Power Converters," Special Issue on Power Electronics of Journal on Circuits, Systems and Computers, vol. 5, no. 3, pp. 355-371, September 1995.

Alleviating Bottlenecks in Power Interconnectors by Means of FACTS

Rolf Grünbaum
ABB AB
rolf.grunbaum@se.abb.com

Talivaldis Podins
ABB SIA
talivaldis.podins@lv.abb.com

Abstract-**Power interconnections are becoming increasingly important in various parts of the world, as incentives for power exchange between countries are growing. A current example is the Baltic Energy Market Interconnection Plan launched by the European Council.**

For a variety of reasons, it is desirable to keep transmission corridors as slender as possible, i.e. keeping the number of lines as limited as possible, while still keeping adequate stability and power transmission capacity over the corridor. This is true, no matter whether it concerns a green-field project, or if it is a question of expanding an existing transmission corridor into higher power transmission capability.

To achieve this, FACTS (Flexible AC Transmission Systems), based on state of the art high power electronics, is a highly useful option, from technical, economical and environmental points of view, to increase the utilization and stability of a transmission system or intertie.

The paper presents salient design features as well as benefits of recently installed FACTS devices, more specifically SVC and series capacitors, for enabling or improving cross-border as well as interregional power transfer in a cost-effective and environmentally friendly way.

I. INTRODUCTION

Effective interconnection of the Baltic Sea region has been identified as a priority energy infrastructure project by the European Council. The main goal within the context of the EU's 20/20/20 objectives is the full integration of the three Baltic States into the European energy market through the strengthening of interconnections with their EU neighbouring countries [1].

Effective interconnection of the region requires HVAC as well as HVDC interties.

For an HVAC grid interconnector, several types of system stability need to be maintained:

- Angular stability, to keep the system in synchronism for all operating conditions;
- Synchronous stability, with sufficient electro-mechanical damping and no sustained power oscillations;
- Steady-state and dynamic voltage stability.

Furthermore, in cases of several parallel lines in a power transmission corridor, power has to be divided between circuits in such a way that optimum overall transfer capability is obtained, and bottlenecks due to uneven power sharing are avoided.

Minimizing the environmental impact is another key issue. A very efficient way of doing this is avoiding the building of more transmission lines, or, in other words, utilizing existing transmission lines more efficiently. And, of course, avoiding the need to build more transmission lines also leads to a minimizing of investment costs, as transmission lines are expensive to build, not to mention time-consuming, in comparison with available options. In green-field projects, with FACTS (Flexible AC Transmission Systems), the number of transmission lines needed to attain a specific amount of power transmission can be kept to an absolute minimum.

FACTS cover several power electronic systems used for AC power transmission [2]. FACTS solutions are particularly justifiable in applications requiring rapid dynamic response, ability for frequent variations in output, and/or smoothly adjustable output. Under such conditions, FACTS is a highly useful option for enabling or increasing the utilization of transmission grids and interties. In green-field projects, AC plus FACTS will likewise in many cases prove attractive and cost-effective in relation to other options. Applications treated in this paper involve SVC (Static Var Compensators), as well as Series Capacitors (SC).

II. DYNAMIC VOLTAGE STABILITY

Behind a voltage collapse, there is usually a deficit of reactive power, as typically, reactive power is needed to maintain proper voltage levels in a power system.

However, reactive power should not travel over long distances, because it is associated with voltage gradients as well as power losses. Therefore, reactive power should be provided where it is needed (load centres).

Reactive power is consumed by loaded lines. When a fault occurs in a power system, such as a short circuit, the faulted line is disconnected, whereupon remaining lines pick up flow from the disconnected line. Reactive power is then consumed to an increasing degree. If reactive power supply is limited, the increased line loading will cause a voltage drop over the system. Then, if reactive power is not provided at the receiving end, the voltage can fall precipitously. At which point, the transmission system can no longer transfer electrical energy. There is a system blackout.

It is obvious that providing reactive power at the right instant and at the right location(s) is a potent way of preventing, or at least limiting, blackouts. Providing an SVC at the load point will, within its range, maintain the load voltage within rated limits (Fig. 1). Theoretically, if the SVC

978-1-4244-8806-3/11 $26.00 © 2011 IEEE

had unlimited rating, as shown in the figure, it would be possible to hold constant voltage at the load busbar for any load condition. In reality, the SVC must be sufficiently rated to cope with the most severe case of loading in a given situation.

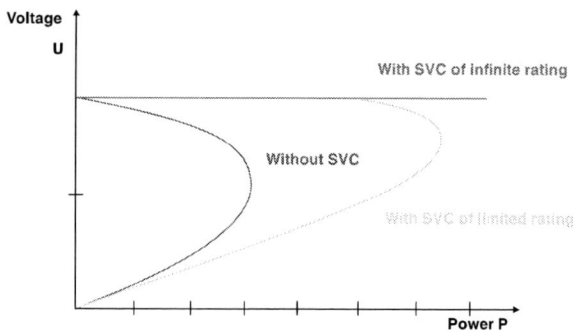

Fig. 1. Voltage at a load busbar as a function of loading, with and without SVC.

III. TRANSIENT STABILITY IMPROVEMENT

The voltage angle difference between the ends of a line, Ψ, is a key quantity for power transmission stability:

$$P = \frac{U_1 U_2}{X} \sin\Psi \quad (1)$$

We can see that for a certain power transmission level (P), there will be a certain angular separation, Ψ. This separation, for the sake of system stability, should never be allowed to exceed a certain limit. Series compensation is a tool to keep the angular separation to a minimum for any power transmission level, thereby increasing the system stability as well as power transmission capability.

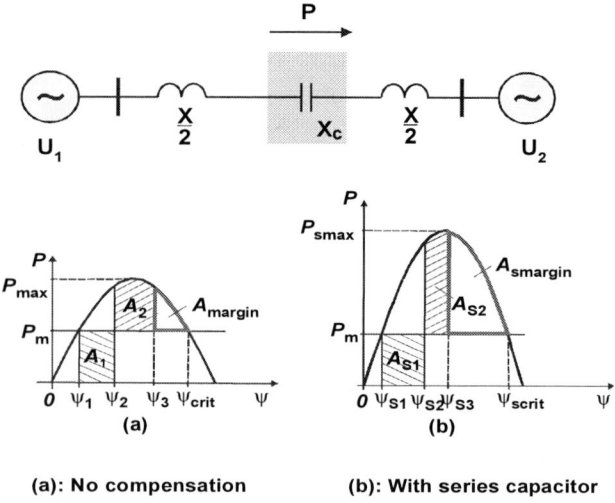

(a): No compensation   (b): With series capacitor

Fig. 2. Improvement of transient stability of a power system by means of series compensation.

In Fig. 2, the power transfer P as a function of angular displacement Ψ is shown for two cases: with no compensation (a), and with series compensation (b). At a certain moment, a fault is incepted in the system, during which the transmitted electrical power P becomes zero, while the mechanical input power to the generator remains the same, P_m.

After the fault has been cleared, the critical condition for system stability is that the angular displacement does not exceed Ψ_{crit}, for if it does, the system cannot get back to equilibrium, and synchronism is lost. We can find the location of the maximum angular displacement Ψ_3 in both cases shown in Fig. 2 by recognizing that the areas A_1, A_{s1} represent the accelerating energy and areas A_2, A_{s2} the decelerating energy during and after the fault, and letting these areas be equal ("Equal Area Criterion"). The "stability margin" in each case will then be given by the area locked in between P_m and $P(\Psi)$ in the interval to the right of Ψ_3, Ψ_{s3}. From the graph, it is obvious that the series capacitor improves the stability margin, by how much will depend of the degree of compensation in each case.

IV. INCREASED POWER TRANSMISSION CAPACITY

Series compensation, in its classical appearance, has been in commercial use since the 1960s. From its basic mechanism, i.e. a decrease of transfer reactance at power frequency, a number of benefits are attainable, all contributing to an increase of the power transmission capability of new or existing transmission circuits [3]:

- An improvement of system stability
- Improvement of voltage regulation and reactive power balance
- Improved load sharing between parallel circuits.

Also a valuable benefit, a decrease of transmission losses is enabled in many cases.

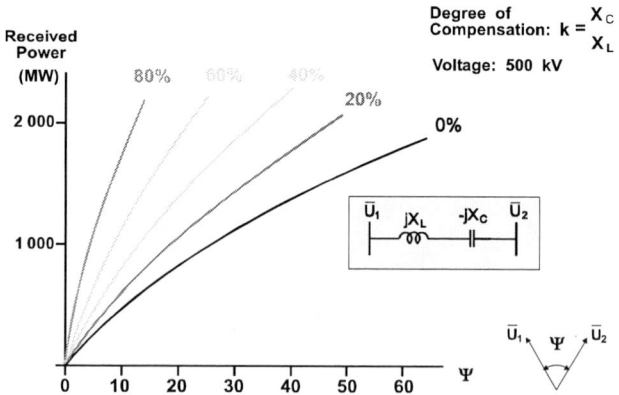

Fig. 3. Impact of series compensation on power transmission capability.

The impact of series compensation on power transmission capability can be illustrated as in Fig. 3. Here, the quantity k is the degree of compensation of the series compensated line, equal to the ratio between the capacitive reactance of the series capacitor (X_C) and the inductive reactance of the transmission line (X_L). Ψ is the angular difference between the end voltages of the line, as previously. In power transmission applications, the degree of compensation is usually chosen somewhere in the range $0{,}3 \leq k \leq 0{,}7$.

For a fixed angular difference, the active power transmission capability of the line increases as the degree of compensation increases. Vice versa, for a fixed amount of power transmission over the line, the angular difference decreases as k increases, which is a measure of increased dynamic stability of the transmission system.

V. STATIC VAR COMPENSATORS

An SVC is based on thyristor controlled reactors (TCR), thyristor switched capacitors (TSC), and harmonic filters. Two common configurations, both having their specific merits, are shown in Fig. 4a and 4b.

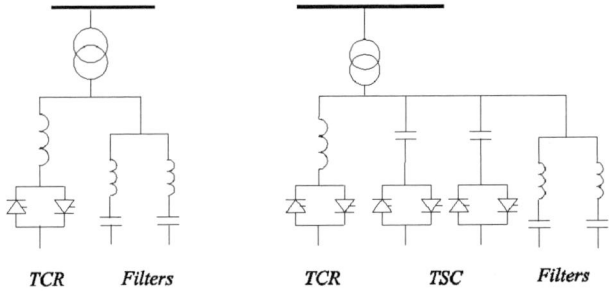

Fig. 4a. TCR / FC configuration. Fig. 4b. TCR / TSC configuration.

A TCR consists of a fixed reactor in series with a bi-directional thyristor valve. TCR reactors, as a rule, are of air core type, glass fibre insulated, epoxy resin impregnated.

A TSC consists of a capacitor bank in series with a bi-directional thyristor valve and a damping reactor which also serves to de-tune the circuit to avoid parallel resonance with the network. The thyristor switch acts to connect or disconnect the capacitor bank for an integral number of half-cycles of the applied voltage. The TSC is not phase controlled, which means it does not generate any harmonic distortion.

A complete SVC based on TCR and TSC may be designed in a variety of ways, to satisfy a number of criteria and requirements in its operation in the grid. In addition, slow vars by means of Mechanically Switched Capacitors (MSC) can be incorporated in the schemes, if desired.

The fast var capabilities of SVC make it highly suitable for performing the following functions:

- Steady-state as well as dynamic voltage control, yielding power transfer capability increases and reduced voltage variations.
- Synchronous stability improvements, yielding increased transient stability and improved power system damping.

- Power quality improvement in grids feeding heavy industrial loads.

A. SVC characteristics

An SVC has a voltage-current (VI) characteristic as shown in Fig. 5. The SVC current/susceptance is varied to regulate the voltage according to a droop characteristic, or slope. The slope setting is important in coordination with other voltage control equipment in the grid. It is also important in determining at what voltage the SVC will reach the limit of its control range. A large slope setting will extend the active control range to a lower voltage, but at the expense of voltage regulation accuracy.

In Fig. 5, the voltage improving effect of the SVC is demonstrated for three different load cases by letting the SVC characteristic intersect with system load lines for the following cases (Slope: X_s):

1: Nominal voltage & load
2: Under-voltage, e.g. due to generator outage
3: Over-voltage, e.g. due to load rejection.

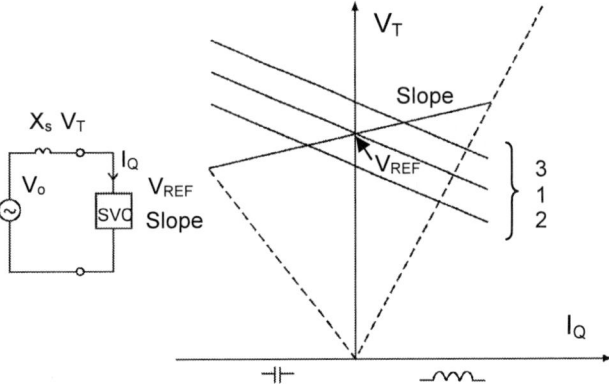

Fig. 5. System voltage correction by means of SVC.

B. SVC control

The main objective of the control system is to determine the SVC susceptance needed in the point of connection to the power system, in order to keep the system voltage close to some desired value. This function is realised by measuring the system voltage and comparing it with a set (reference) value. In case of a discrepancy between the two values, the controller orders changes in the susceptance until equilibrium is attained.

The controller operation results in a susceptance order from the voltage regulator which is converted into firing orders for each thyristor. The overall active SVC susceptance is given by the sum of susceptances of the harmonic filters, the continuously controllable TCR, and the TSC if switched into operation. The control system also includes supervision of currents and voltages in different branches. In case of need, protective actions are taken.

C. Thyristor valves

The The thyristor valves consist of single-phase assemblies (Fig. 6). The thyristors are electrically fired. The energy for firing is taken from snubber circuits, also being part of the valve assembly.

Between thyristors, heat sinks are located. The heat sinks are connected to a water piping system. The cooling media is a low conductivity mixture of water and glycol. In the most recent SVCs supplied, the thyristor valves are equipped with Bi-Directional Control Thyristors (BCT). In such devices, two thyristors are integrated into one wafer with separate gate contacts. Thus, the valves comprise only one thyristor stack in each phase instead of two, which enables considerable compacting of the valve design.

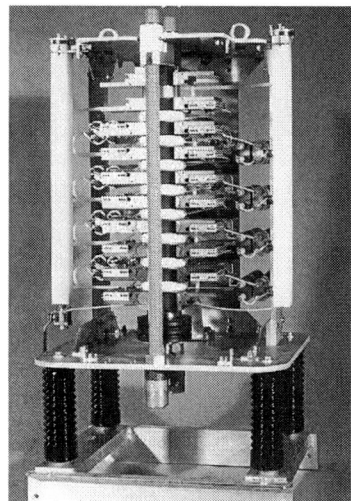

Fig. 6. Thyristor valve of BCT design.

VI.   SERIES CAPACITORS

The main circuit diagram of a state of the art series capacitor is shown in Fig. 7.

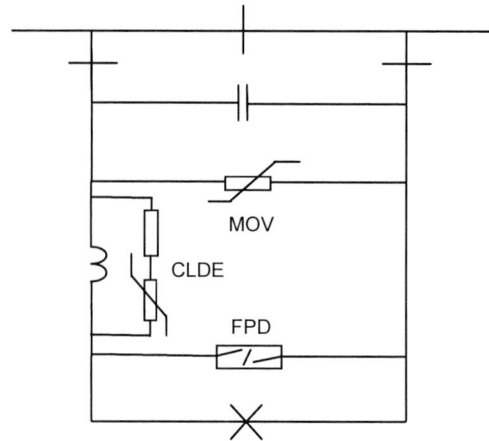

Fig. 7. Series capacitor scheme.

The series capacitor protective scheme consists of a Metal Oxide Varistor (MOV), Current Limiting Damping Equipment (CLDE), a Fast Protective Device (FPD), and a Bypass Switch. The CLDE consists of a current limiting reactor, plus a resistor and a varistor in parallel with the reactor. The purpose of the resistor is to add damping to the capacitor discharge current, and thus quickly reduce the voltage across the capacitor after a bypass operation. The purpose of the varistor is to avoid fundamental frequency losses in the damping resistor during steady state operation.

The FPD scheme is based on a hermetically sealed and very fast high power switch, CapThor™, which replaces conventional spark gaps (Fig. 8). The FPD works in combination with the MOV, and allows bypassing in a very controlled way in order to reduce the energy dissipation in the MOV. The FPD scheme has advantages over previous, conventional schemes with spark gaps such as [4]:

- More compact
- Unaffected by the environment
- Capacitor by-passing possible for a wide range of voltages
- Added flexibility for future series capacitor upgrading.

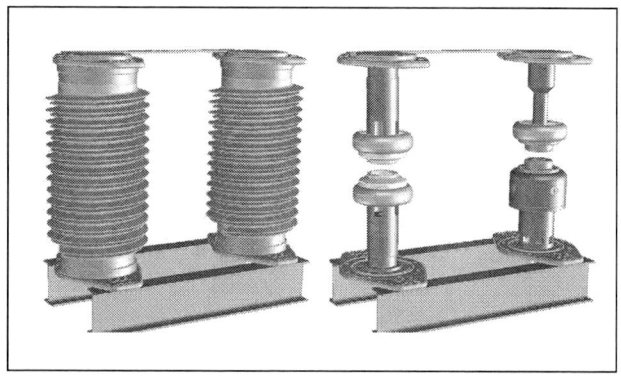

Fig. 8. External and internal view of CapThor. In the internal view, the plasma switch is seen to the left and the mechanical switch to the right.

A. Series capacitor layout

Series capacitors need to be laid out in an arrangement that allows full insulation to ground of all main circuit equipment. To achieve this, each phase of the series capacitor is located on a steel platform insulated from ground by means of support insulators. The by-pass switch alone is located on ground, to enable easy access of its operating mechanism. Fig. 9 shows the lay-out arrangement for one phase of a series capacitor.

The capacitor bank is located on the part of the platform closest to the beholder. The part deeper into the picture houses the protective devices (MOV, CLDE and CapThor).

978-1-4244-8806-3/11 $26.00 © 2011 IEEE   158

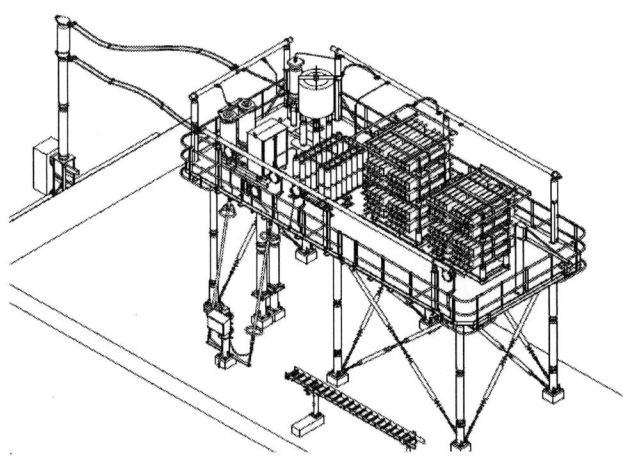

Fig. 9. Lay-out arrangement of one series capacitor phase.

VII. NORDIC SERIES CAPACITORS

In 2009, two series capacitors went on line in the 400 kV power transmission system in northern Finland. The series capacitors are located at Asmunti and Tuomela, close to the top end of the Gulf of Bothnia, as shown in Fig. 10. Their purpose is to increase power transmission capacity in northern Finland and to assure power system security.

The series capacitors add transmission capacity towards neighbouring Sweden by 200 MW and are thereby needed to meet the rising market demand on power transmission between northern Finland and Sweden and also internally in Finland. Series capacitors enable added transmission capacity to the existing power lines and help to maintain the grid stability in the Finnish power system.

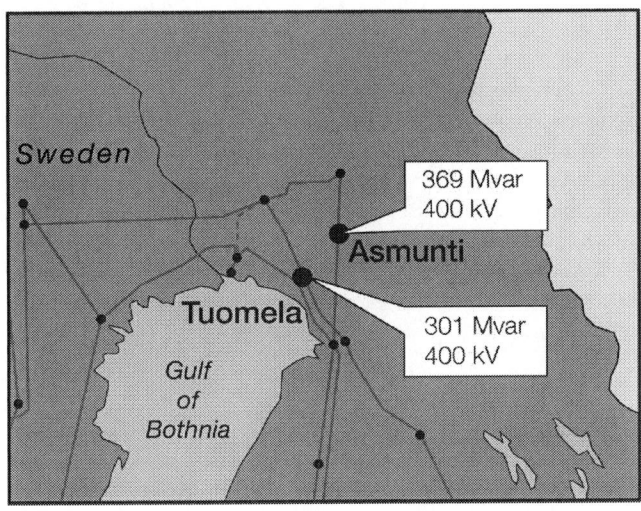

Fig. 10. 400 kV series compensated grid in Finland.

The main technical data of the series capacitors can be summarized as follows (Table I):

TABLE I
MAIN TECHNICAL DATA OF SERIES CAPACITORS

	Asmunti	Tuomela
Rated system voltage	400 kV	400 kV
Rated reactive power	369 Mvar	301 Mvar
Rated current per phase	1800 A	1800 A
Overload current, 30 min	2430 A	2430 A
Rated phase reactance	38 Ω	31 Ω
Degree of compensation	70%	70%
Rated MOV energy	130 MJ/3-phase	85 MJ/3-phase

The ambient conditions are typical northern Scandinavian, with long, cold winters and summer seasons that, at times, can reach quite high temperatures. Based on these conditions, the series capacitors are rated to withstand an ambient temperature span of -50 to +40 degrees C.

A site picture of one of the series capacitors (Asmunti) is shown in Fig. 11.

Fig. 11. 400 kV series capacitor in Finland.

VIII. AN SVC CASE

As a result of large power demanding industry development in central Norway, the demand in the region has increased dramatically and is expected to grow further. The power import capacity to the region was previously limited by the risk of voltage collapse. As a remedy, two SVCs were commissioned in 2008, as well as nine MSCs. With the installation of the SVCs, the power import capacity to the region has increased by 200-400 MW, depending on the operating conditions [5].

The SVCs, each rated at -/+ 250 Mvar, were installed at Viklandet and Tunnsjödal substations in the 420/300 kV power transmission network, Fig. 12. At Viklandet, the system voltage is 420 kV. At Tunnsjödal, the voltage is limited to 300 kV, but earmarked to be upgraded to 420 kV at a later stage. Also at Tunnsjödal, a 100 Mvar MSC is part of the scheme, for increased short term capacitive capability, utilizing the inherent short term overload capability of the SVC transformer.

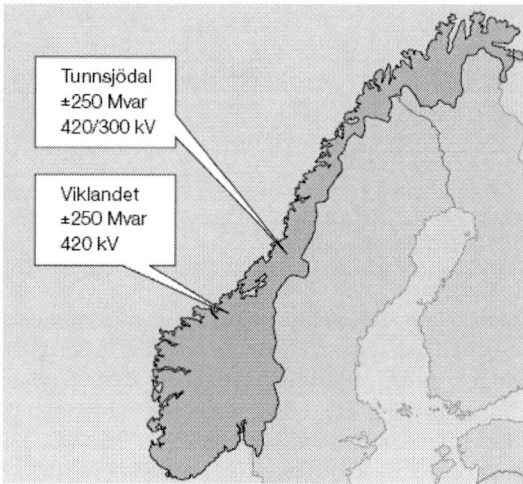

Fig. 12. 420/300 kV SVCs in Norway.

The dynamic range is achieved by means of three TCR and two TSC branches in parallel in each SVC. The build-up of the thyristor valves can be seen in Fig. 13.

Fig. 13. Thyristor valves for TCR and TSC branches.

The purpose of the SVCs is to perform system control tasks as follows:

- Steady state voltage control at Viklandet 420 kV bus as well as Tunnsjödal 420 (300) kV bus.
- Enhanced damping of system electro-mechanical oscillations by means of POD (Power Oscillation Damping) based on active power measurements.
- Control of an external MSC in the substations. This will reserve a dynamic SVC range to be utilized for system contingencies.

A. Power oscillation damping

There are several well-known local area as well as inter-area power oscillation modes in the Norwegian power system [5]. These modes include an inter-area 0.45 Hz oscillation between Norway and Finland, a 0.65 Hz oscillation between Norway and southern Sweden, and a local 0.85 Hz oscillation within the Norwegian grid. The Viklandet and Tunnsjödal SVCs are equipped with Power Oscillation Dampers to damp particularly the local 0.85 Hz oscillation mode.

A photo of one of the SVCs is displayed in Fig. 14.

Fig. 14. Norwegian -250/+250 Mvar SVC.

IX. CONCLUSION

Effective interconnection of the Baltic Sea region has been identified as a priority energy infrastructure project by the European Council. Effective interconnection of the region requires HVAC as well as HVDC interties. For an HVAC grid interconnector, several types of system stability need to be maintained.

At the same time it is desirable to keep transmission corridors as slender as possible, i.e. keep the number of lines as limited as possible, while still keeping adequate stability and power transmission capacity over the corridor.

To achieve this, FACTS is a highly useful option, from technical, economical and environmental points of view, to increase the utilization and stability of a transmission system or intertie.

In the paper, the usefulness of SVC and series capacitors for the purpose has been demonstrated. For each kind of device, a recent case of utilization has been treated, where increased power transmission capacity of interconnectors has been attained.

REFERENCES

[1] European Commission, "Energy Infrastructure, Baltic Energy Market Interconnection Plan (BEMIP)", 2008.
[2] L. Gyugyi, N. Hingorani, "Understanding FACTS", IEEE Press, 2000.
[3] R. Grünbaum, J. Samuelsson, "Series capacitors facilitate long distance AC power transmission," *IEEE Power Tech Conference*, 2005.
[4] J. Redlund et al, "A new fast protective device for high voltage series capacitors," *IEEE Conference*, 2005.
[5] M. Meisingset, O. Skogheim, B. Ekehov and K. Wikström,, "Viklandet and Tunnsjodal SVCs - Design, project execution and operating experience", *Cigré B4-106, Paris*, 2010.

Angle Estimation using High Frequency Injection for Sliding Mode Controlled Permanent Magnet Synchronous Machines Drives.

V.Repecho [1], D.Biel [1,2], A.Arias [2], F. Guinjoan [2]

[1] Institute of Industrial and Control Engineering, Universitat Politècnica de Catalunya, Barcelona, Spain
[2] Department of Electronics, Universitat Politècnica de Catalunya, Barcelona, Spain
victor.repecho.del@upc.edu, biel@eel.upc.edu, arias@eel.upc.edu, guinjoan@eel.upc.edu.

Abstract – **In this paper, high frequency injection (HF) technique for rotor position estimation using Sliding Mode Control (SMC) in Permanent Magnet Synchronous Machines (PMSMs) is presented. The novelty of this paper is the fact that the SMC is used instead of the conventional Field Oriented Control (FOC) and therefore the injected HF test signals for tracking the machine' saliency (difference between Ld and Lq) is naturally eliminated by the SMC. Such suppression is tackled when using the SMC and the position signals are calculated to properly estimate the permanent magnet (PM) position.**

Finally, some simulation results are presented, confirming the proper angle estimation at low speed reversal and load impact at zero speed.

I. INTRODUCTION

Permanent Magnet Synchronous Machines (PMSMs) have higher power density, higher efficiency and better dynamic performance than Induction Machines. The accurate control of these motors requires rotor position to implement the coordinate transformation of the vector control, in order to obtain speed and position feedback control signals. Significant research efforts have been conducted in order to achieve vector control of PMSMs without encoders or resolvers. These techniques can be broadly divided into model based techniques, where the back-*emf* of the machine is used for rotor magnet flux detection, and injection techniques, where a test signal, either high frequency (HF) AC voltage or voltage pulse, is used to detect the rotor saliency (difference between L_D and L_Q) position [1] [2].

Model based techniques e.g. [3] [4] successfully achieve sensorless control at medium and high rotor speed but fail at low excitation frequencies due to the reduction and eventual disappearance of the back-*emf* induced by the rotor magnets at low rotor speed.

Injection methods, on the other hand, detect the angular dependent saliency of the machine and rotor position estimation is fundamentally speed independent [1] [2].

There are mainly two different injection techniques. The first one consists on the superposition to the fundamental voltage vector of a HF injection either in the alpha/beta frame [5],[6],[7],[8] or in rotating d/q synchronous frame [9] [10]. The second ones are based on the modification of the fundamental PWM pattern to include a voltage pulse test [11],[12],[13].

For all injection methods to function, some level of machine saliency is necessary. This makes the technique straightforward for salient machines such as interior permanent magnet (PM) motors. Surface Mount PMSM, on the other hand, only have a saliency due to stator tooth saturation and it is generally of small magnitude.

Such mentioned injection techniques are mostly implemented using Field Oriented Control (FOC) schemes where the band width of the inner current control loops is typically of lower value than the injected frequency. Therefore, the system does not react against such a perturbation not eliminating the further position information. On the other hand, when Sliding Mode Control (SMC) is used the band width, (which strictly speaking is not defined since it is not a linear controller), is within the range of the injected test signal, therefore the system tends to cancel it treating it as a perturbation.

In this paper, the HF injection technique for tracking the rotor position in a Surface Mount PMSM Controlled by Sliding Mode is addressed. The novelty of the paper relies on the use of SMC instead of the well known FOC technique and therefore the issue is to tackle the natural suppression of the injected signal. The well-know benefits of the SMC [14] [15], like fast torque response, robustness, reduction system order or high immunity against parameters drift are reached, improving the FOC dynamics.

Position signals and the injection algorithm are discussed. Finally, some speed reversal and load impact results showing proper angle estimation for further Sensorless SMC are presented.

II. PMSM MODEL

The electrical equations that model the PMSM in the rotating reference frame (*dq*) are shown on (1) and (2).

$$\frac{di_D}{dt} = \frac{v_D}{L_D} - \frac{R}{L_D} \cdot i_D + \omega_e \frac{L_Q}{L_D} i_Q \tag{1}$$

978-1-4244-8806-3/11 $26.00 © 2011 IEEE

$$\frac{di_Q}{dt} = \frac{v_Q}{L_Q} - \frac{R}{L_Q} \cdot i_Q - \omega_e \cdot \frac{L_D}{L_Q} i_D - \omega_e \cdot \frac{\phi_M}{L_Q} \qquad (2)$$

Where:

v_D, v_Q = Stator dq Voltages
i_D, i_Q = Stator dq currents
L_D, L_Q = Stator dq Inductance
R = Stator Resistance
Φ_M = Permanent Magnet Flux
ω_e = Electrical angular speed
p = Pole pairs number
B = Friction coefficient
J = Moment of inertia
T_L = Load torque
ω_m = Mechanical angular speed

The related electromagnetic torque equation to dq current components is expressed at (3):

$$Te = \frac{3 \cdot p \cdot \phi_M}{2} \left[i_Q - (L_D - L_Q) \cdot i_D \cdot i_Q \right] \qquad (3)$$

The PMSM motion equation is:

$$\frac{d\omega_m}{dt} = \frac{1}{J} Te - \frac{B}{J} \omega_m - \frac{1}{J} T_L \qquad (4)$$

From the expressions (3) and (4) the third state space system equation is obtained:

$$\frac{d\omega_m}{dt} = \frac{3 \cdot p \cdot \phi_M}{2 \cdot J} \cdot i_Q + \frac{3 \cdot p}{2 \cdot J} (L_Q - L_D) \cdot i_D \cdot i_Q - \frac{B}{J} \omega_r - \frac{1}{J} T_L \qquad (5)$$

III. PMSM SLIDING MODE CONTROL

The SMC is applied to the system defined by equations (1) and (2). The SMC is executed above a control variable, which will take two discrete values at any time, choosing the proper one depending on the system state. The control forces the systems states trajectories towards the switching surface at any time. The switching surfaces are defined by the systems errors, where the errors are the difference between the desirable and the real values of the controllable variable.

In order to control PMSM, the chosen controllable variables are dq currents, and consequently the control variables are dq voltages.

The switching surfaces (S) for controllable variables are defined, using directly the error of the reference value (x^*) and real value (x) of a given variable x, then:

$$S_D = i_D^{\ *} - i_D \qquad (6)$$
$$S_Q = i_Q^{\ *} - i_Q \qquad (7)$$

Defining the state error as: $e_{ID} = i_D^{\ *} - i_D$ and $e_{IQ} = i_Q^{\ *} - i_Q$ and using the expression (1), (2), (6) and (7), and knowing that the equilibrium point of the system is forced by the SMC, the errors system dynamics is rewritten as follows:

$$\frac{de_{ID}}{dt} = \frac{v_D}{L_D} - \frac{R}{L_D} \cdot e_{ID} + \omega_e \frac{L_Q}{L_D} \cdot e_{IQ} \qquad (8)$$

$$\frac{de_{IQ}}{dt} = \frac{v_Q}{L_Q} - \frac{R}{L_Q} \cdot e_{IQ} - \omega_e \cdot \frac{L_D}{L_Q} \cdot e_{ID} \qquad (9)$$

The control will be defined by two discrete values, depending on the S sign. The discrete dq voltages values are defined as:

$$v_D \in \{v_{D0}, -v_{D0}\} \qquad (10)$$
$$v_Q \in \{v_{Q0}, -v_{Q0}\} \qquad (11)$$

The SMC has to ensure the system trajectories are always directed towards the sliding surface, and this is achieved when (12) and (13) are fulfilled:

$$S_D \cdot \dot{S}_D < 0 \qquad (12)$$

$$S_Q \cdot \dot{S}_Q < 0 \qquad (13)$$

Using the equivalent control definition (v_{Deq}, v_{Qeq}), which is defined as the input values of control variable that satisfy a given system condition, expression (16) and (17) are extracted from (14) and (15). The physical meaning of the equivalent control is that the system variables are on the switching surfaces and remain there.

$$S_D \cdot [v_D - v_{DEQ}] < 0 \qquad (14)$$
$$S_Q \cdot [v_Q - v_{QEQ}] < 0 \qquad (15)$$

That yields:

$$v_D = -v_{D0} \cdot sign(S_D) \qquad (16)$$
$$v_Q = -v_{Q0} \cdot sign(S_Q) \qquad (17)$$

Taking a look of the control law expression summarized in (16) and (17), it could be noted that the control does not depend on any system state variable or parameters value (R or L), which proves the high robustness against parameters drifts.

In order to ensure that the control can be successfully employed, the sliding domain must be guaranteed. Indeed, the

978-1-4244-8806-3/11 $26.00 © 2011 IEEE

two variable control discrete values have to be able to produce the equivalent control values, necessary to satisfy the references values for any given conditions.

$$-v_{D0} < v_{Deq} < v_{D0} \qquad (18)$$

$$-v_{Q0} < v_{Qeq} < v_{Q0} \qquad (19)$$

After the SMC has been deduced and developed, the PMSM can be controlled in terms of *dq* current with the control law deduced.

For setting the reference current values, it is necessary to understand clearly the physical meaning of the *dq* model. The *dq* motor model, as it was noted before, is a rotating reference frame system aligned with the motor PM. The *d* current component is aligned in the flux PM direction, and the *q* current component, its 90° phase-shifted from *d* component. From this point of view, the *dq* PMSM model equation is highly similar to a stator equation of a DC motor. Besides, from this deduction, is clear that the preferred desired current value on *d* component has to be set to 0, in order to avoid fluxes that could demagnetize the PM. Torque current component, that is completely perpendicular to PM flux, is created by i_Q, as it can be deduced from (3).

III. SLIDING MODE CONTROL AND HIGH FREQUENCY INJECTION

a. *Analyzing the Control Band Width*

The injection of a rotating *HF* voltages for saliency tracking when using FOC has been proposed by several authors [1] [3].

At this point, analyzing the SMC proposed on section II, a big difference with FOC control arises. Meanwhile the bandwidth of the FOC control can be easily known by the PI controllers, the SMC have, theoretically, an infinite bandwidth (BW).

Typically, HF carrier signal frequency for angle estimation is fixed around 1-2 kHz (between fundamental electrical frequency and PWM frequency). Fig. 1, shows how the HF injection is inside SMC BW. This means that the SMC will perform a determined action control against this HF injection, being treated as an external disturbance.

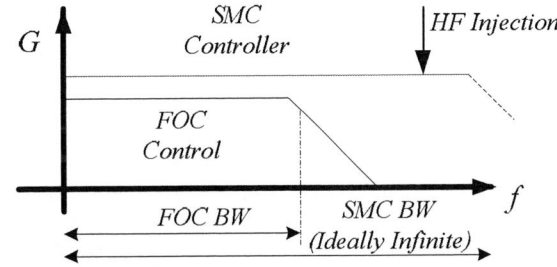

Figure 1. FOC and SMC Band Width

In FOC control, the HF carrier voltage is injected on alpha beta (*αβ*) plane, the PI controllers are not sensible at this frequency and the system does not perform any action against this disturbance. The HF signals pass through the PMSM, and the HF currents outputs have the desired angle information. If the same technique is employed to the present SMC, the higher BW of the control performs against the HF injection, and the HF voltage passing through the PMSM is ideally zero, and consequently, the HF output currents.

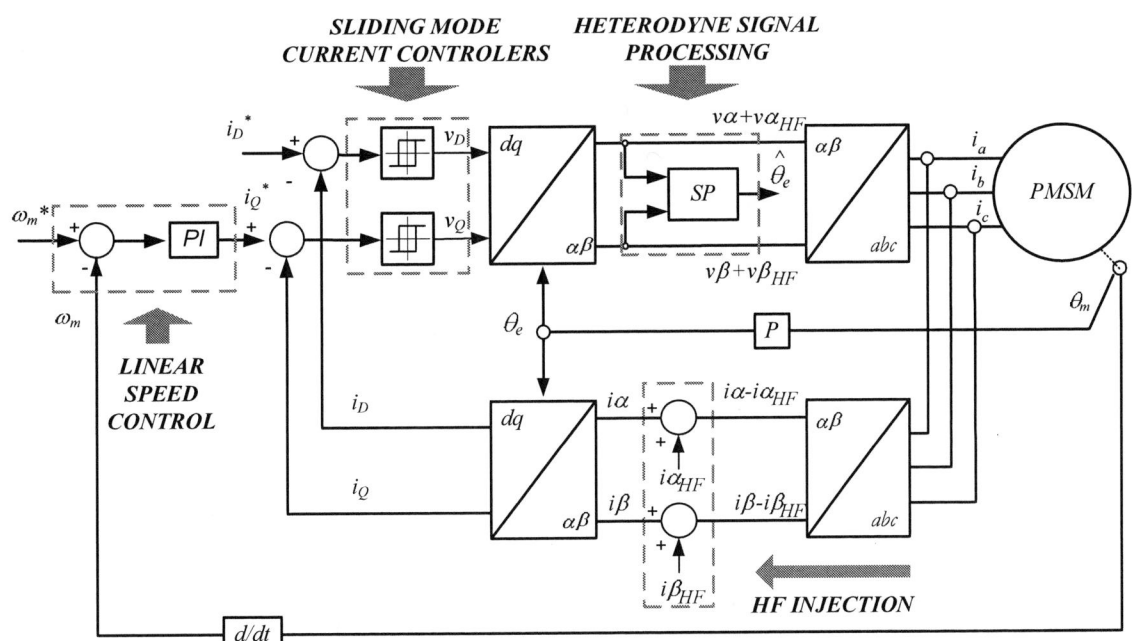

Figure 2. Full System Proposed

978-1-4244-8806-3/11 $26.00 © 2011 IEEE

Taking into account the idea explained before, is reasonable to choose a HF current carrier for the system. In the same way, the control will perform against the perturbation, but in this case, the current signal generated by the control, ideally equal to the disturbance with a opposite signal, are passing through the motor. In this injection method, the signal processing to obtain the angle estimation must be done above the $\alpha\beta$ voltages. On the Fig. 2, the proposed system with the HF current injected and HF voltage output processed is presented.

Note that a outer speed control loop has been implemented, based on the slower mechanical motion PMSM described on equation (4). A classical PI linear controller is used, which will give the q current reference to the SMC, keeping the d current component always equal to zero. If the sliding domain is ensured, we can extract the dynamics of the system related to the angular speed from equation (4).

b. HF PMSM System Model

In the PMSM with saliency, there are variations of the inductance values depending on the rotor position. These changes are caused by the different flux levels on the corresponding PMSM areas, depending on PM position. The inductance factor change in front of PM position is modelled (20):

$$\begin{bmatrix} L_\alpha \\ L_\beta \end{bmatrix} = \begin{bmatrix} Ls - \Delta Ls \cdot \cos(2\theta r) & -\Delta Ls \cdot \sin(2\theta r) \\ -\Delta Ls \cdot \sin(2\theta r) & Ls + \Delta Ls \cdot \cos(2\theta r) \end{bmatrix} \tag{20}$$

Where:

$$\Delta Ls = \left(L_Q - L_D\right)/2 \tag{21}$$
$$Ls = \left(L_Q + L_D\right)/2 \tag{22}$$

Note that in a non-saliency PMSM ($L_D = L_Q$), the inductance versus angle variation is not produced.

The $\alpha\beta$ PMSM model it is defined by (23).

$$\begin{bmatrix} v_\alpha \\ v_\beta \end{bmatrix} = r_s \cdot \begin{bmatrix} i_\alpha \\ i_\beta \end{bmatrix} + \frac{d}{dt} \begin{bmatrix} Ls - \Delta Ls \cdot \cos(2\theta r) & -\Delta Ls \cdot \sin(2\theta r) \\ -\Delta Ls \cdot \sin(2\theta r) & Ls + \Delta Ls \cdot \cos(2\theta r) \end{bmatrix} \begin{bmatrix} i_\alpha \\ i_\beta \end{bmatrix}$$
$$+ \psi_m \cdot \frac{d}{dt} \begin{bmatrix} \cos(\theta r) \\ \sin(\theta r) \end{bmatrix} \tag{23}$$

Where r_s and ψ_m are the stator resistances and the PM flux respectively.

If we assume the disturbance frequency bigger enough than the phenomena related to the motor rotation, the HF motor model can be simplified as stated in (24).

$$\begin{bmatrix} v_\alpha \\ v_\beta \end{bmatrix} = \frac{d}{dt} \begin{bmatrix} Ls - \Delta Ls \cdot \cos(2\theta r) & -\Delta Ls \cdot \sin(2\theta r) \\ -\Delta Ls \cdot \sin(2\theta r) & Ls + \Delta Ls \cdot \cos(2\theta r) \end{bmatrix} \cdot \begin{bmatrix} i_\alpha \\ i_\beta \end{bmatrix} \tag{24}$$

IV. ANGLE ESTIMATION IN SLIDING MODE CONTROL SCHEME

The current injection added on the $\alpha\beta$ currents is :

$$\begin{bmatrix} i_{\alpha P} \\ i_{\beta P} \end{bmatrix} = -A_P \cdot \begin{bmatrix} -\sin(\omega_I \cdot t) \\ \cos(\omega_I \cdot t) \end{bmatrix} \tag{25}$$

Under sliding motion the current vector cancelling the disturbances can be modelled as (26):

$$\begin{bmatrix} i_{\alpha H} \\ i_{\beta H} \end{bmatrix} = A_P \cdot \begin{bmatrix} -\sin(\omega_I \cdot t) \\ \cos(\omega_I \cdot t) \end{bmatrix} \tag{26}$$

The equation (24) is solved using as an input currents the ones showed on (26). Solving the proposed expression, (27) is reached.

$$v_{HF} = \begin{bmatrix} v_{\alpha HF} \\ v_{\beta HF} \end{bmatrix} \approx \begin{bmatrix} v_0 \cdot \cos(\omega_I \cdot t) + v_1 \cdot \cos(2 \cdot \theta e - \omega_I \cdot t) \\ v_0 \cdot \sin(\omega_I \cdot t) + v_1 \cdot \sin(2 \cdot \theta e - \omega_I \cdot t) \end{bmatrix} \tag{27}$$

Where $\quad v_0 = -\omega_I \cdot A_P \cdot Ls \quad$ and $\quad v_1 = \omega_I \cdot A_P \cdot \Delta Ls$

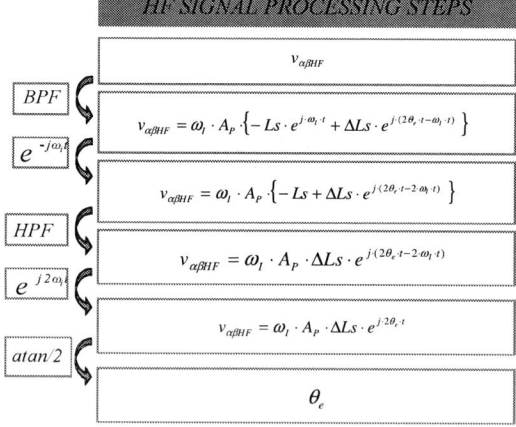

Figure 3. Signal Processing Used

The resultant output voltages have the same structure than the output current voltages resultant when a HF voltage signal is injected in FOC application. Therefore, the signal processing, shown in Fig. 3, for angle estimation (θe) is almost equal.

978-1-4244-8806-3/11 $26.00 © 2011 IEEE   164

V. RESULTS

The simulation results showed have been obtained using the PMSM characterized by the table II parameters. The PI control and SMC were adjusted for a good dynamic response and acceptable error on the angle estimation obtained.

TABLE II
PERMANENT MAGNET SYNCHRONOUS MACHINE

Surface Mount PMSM	
Rated power	3.8 kW
Poles number	6
Nominal speed / Rated torque	314.15 rad/s / 12.2 Nm
Rs / Ld / Lq	0.94 Ω / 7 mH / 8.3 mH
Magnetic flux linkage [Wb]	0.2515
Friction Coefficient [N·m·s]	0.03833
Moment of Inertia [kg·cm^2]	20.5

TABLE III
ANGLE ESTIMATION BLOCK PARAMETERS

Injection Frequency	1 kHz
Injection Signal Amplitude	250 mA
Band-Pass Filter	4th Order Butterworth Type LP = 800Hz HP= 1200Hz
High-Pass Filter	1st Order Butterworth Type HP= 60 Hz

A. Position Signals

Fig. 4 and Fig. 5 were obtained at steady state Control, at 1% of the nominal speed (3.14 rad/s) and full load torque (12.2 Nm), which implies that the frequency of the position signals is at 3 Hz, considering the number of pole pairs and the fact that the position signal frequency is twice the fundamental electrical frequency. The results showed on Fig. 4 are directly the signals obtained at the end of the SP, not applying any additional filter. In order to validate the quality of the position signals the FFT is performed as Fig. 5. The quality of the signal is remarkable since any sub harmonic appears which might cause an error in the estimated angle.

From the $\alpha\beta$ position signals the *arc tangent* function is executed and the estimated angle is obtained.

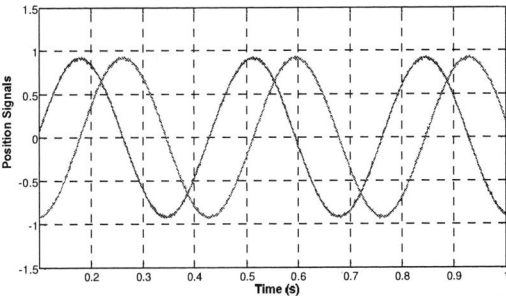

Fig. 4. Alpha and beta position signals.

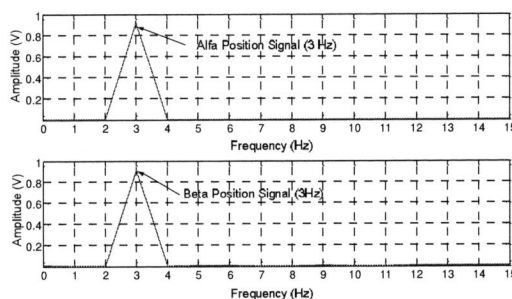

Fig. 5. Position signals' FFT.

B. Speed Reversal

Fig. 6 illustrates the performance under a step speed change (+/-1 rad/s) at full load torque (12.2Nm) .

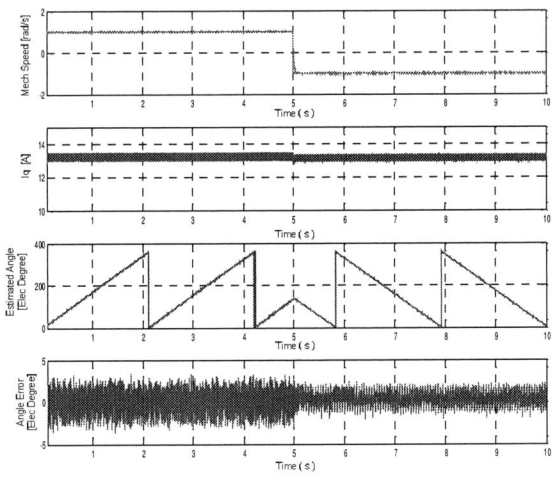

Fig. 6 Sensored Speed Reversal (±1 rad/s) at full load (12.2 Nm). From top: sensored and reference speed (rad/s); current torque demand (A); estimated angle; angle error. All angles in electrical degrees.

The angle error oscillates within ±3.0 electrical degrees (±1 mechanical degrees) even during the speed reversal and when the speed is set to zero. The angle estimation is good enough and the SMC used eliminates properly the perturbation included. The i_Q current has an acceptable ripple value caused by the SMC, and any undesired effect caused by the perturbation signal appears. The external PI controller speed loop works properly in speed reversal shown in Fig. 6.

C. Load Impact

The second experiment involves a Sensored load impact test with the speed reference set at zero. Fig. 7 shows the speed response for a load impact of 100% and back to 0 again. The angle error is kept at smaller values than ±4 electrical degrees (<±1.3 mechanical degrees) taking into account the worst situation under the transient response.

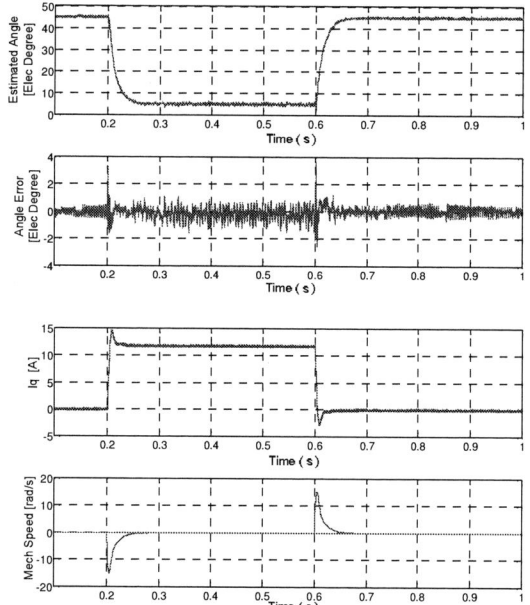

Fig. 7. Sensored at zero Speed response to 100% load impact
From top: rotor position (electrical degrees); angle error
(electrical degrees) ; i_Q current(A); mechanical speed (rad/s)

VI. CONCLUSIONS

This work has introduced an angle estimation algorithm for Permanent Magnet Synchronous Machines (PMSMs) at low and zero speed based on the high frequency injection technique to track the machine saliency. The novelty of the work relies on the fact that the PMSM drive is under Sliding Mode Control (SMC) for the inner current loops instead of the traditional Field Oriented Control (FOC), based on PI controllers.

The traditional voltage injection process used under FOC schemes does not work in SMC due to the ideally infinite band width of SMC controllers. Alternatively, the high frequency test signal has been injected in the feedback *alpha- beta* currents. Despite the elimination of the injected high frequency current, the angle information can be obtained from the equivalent control voltage values generated by the SMC.

Simulation results (speed reversal and load impact) shows the validity of this angle estimation algorithm since the angle error value is at all times less than 4 electrical degrees. Further research is focused on modelling the VSI, the full Sensorless Sliding Mode Control and a final implementation in order to obtain experimental results.

REFERENCES

[1] J. Holtz. "Sensorless Control of Induction Machines- With or Without Signal Injection?" IEEE Trans. on Industrial Electronics. Vol. 53, pp, 7-30, Feb. 06.

[2] P. P. Acarnley and J. F. Watson. "A Review of Position Sensorless Operation of Brushless Permanent-Magnet Machines" IEEE Trans. on Industrial Electronics. Vol. 53, pp. 352-362, April 06.

[3] R. Wu, G. R. Slemon, "A Permanent Magnet Motor Drive Without a Shaft Sensor," IEEE Transactions on Industry Applications, vol. 27, pp. 1005-1011, Sep. / Oct. 1991.

[4] R. Dhaouadi, N. Mohan, L. Norum, "Design and Implementation of an Extended Kalman Filter for the State Estimation of a Permanent Magnet Synchronous Motor," IEEE Transactions on Power Electronics, vol. 6, pp. 491-497, Jul. 1991.

[5] C. Silva, G. M. Asher, M. Sumner, K.J. Bradley, "Sensorless control in a surface mounted PM machine using rotating injection," EPE Journal. Vol. 13, No. 3, Aug. 03.

[6] A. Arias, C. Silva, G. M. Asher, J. C. Clare, P. W. Wheeler. "Use of a Matrix Converter to Enhance the Sensorless Control of a Surface-Mount Permanent-Magnet AC Motor at Zero and Low Frequency" IEEE Trans. on Industrial Electronics. Vol. 53, pp 440-449, April 06.

[7] J. Cilia, G.M. Asher, K.J. Bradley, M. Sumner, "Sensorless Position Detection for Vector-Controlled Induction Motor Drives using an Asymmetric Outer-Section Cage," IEEE Transactions on Industry Applications, Vol. 33, pp. 1162-1169, 1997.

[8] N. Teske, G.M. Asher, M. Sumner, K.J. Bradley, "Encoderless Position Estimation for Symmetric Cage Induction Machines under Loaded Conditions" IEEE Transactions on Industry Applications, Vol. 37, no. 6, pp. 1793-1800, Nov/Dec 2001.

[9] P. L. Jansen, R. D. Lorenz, "Transducerless Position and Velocity Estimation in Induction and Salient AC Machines," IEEE Trans. on Industry Applications. Vol. 31, pp. 240-247, March/April 1995.

[10] J.-S. Kim, S.-K. Sul, "High performance PMSM drives without rotational position sensors using reduced order observer" IEEE-IAS Annual Meeting, pp 75-82, 1995.

[11] C. S. Staines, C. Caruana, G. M. Asher, M. Sumner. "Sensorless Control of Induction Machines at Zero and Low Frequency Using Zero Sequence Currents" IEEE Trans. on Industrial Electronics. Vol. 53, pp. 195-206, Feb. 06.

[12] J. Holtz, J. Juliet. "Sensorless Acquisition of the Rotor Position Angle of Induction Motors with Arbitrary Stator Windings". IEEE Trans. on Industry Applications. Vol. 41, pp. 1675-1682, Nov./Dec. 05.

[13] P. L. Jansen, R. D. Lorenz, "Transducerless Position and Velocity Estimation in Induction and Salient AC Machines," IEEE Transactions on Industry Applications, vol. 31, pp. 240-247, 1995.

[14] V. Utkin, J. Guldner, J. Shi, "Sliding Mode Control in Electromechanical Systems" CRC PRESS 1999.

[15] Z. Yan, C. Jin, V.I. Utkin, "Sensorless Sliding-Mode Control of Induction Motors" IEEE Trans. on Industrial Electronics. Vol. 47, pp. 1286-1297, Dec. 00.

978-1-4244-8806-3/11 $26.00 © 2011 IEEE

A New Criterion for Selecting the Inductors of an Active Power Line Conditioner

P. González-Castrillo, E. Romero-Cadaval, M. I. Milanés-Montero, F. Barrero-González, M. A. Guerrero-Martínez

Power Electrical and Electronic Systems (PE&ES), School of Industrial Engineering
(University of Extremadura), (http://peandes.unex.es)

Abstract-This paper analyzes the inductances used at the terminals of an Active Power Line Conditioner (APLC) based on a voltage source inverter (VSI) in some current correction topologies. The criteria to select the optimal inductances values are determined and a novel design criterion, named "control criterion", is proposed. This novel method is applied to situations where there are sharp current variations as the case of a three phase rectifier feeding a resistive load.

A parametrical analysis of the expressions obtained for the inductors is performed, obtaining the intervals of possible values of inductance that can meet the initial specifications. Finally a simulation is carried out to verify how the method presented reaches these specifications.

I. INTRODUCTION

The presence of renewable energies in recent years in the utility has meant an increase in the number of power electronic devices connected to the grid, but also the existence of nonlinear loads, influence the quality and the reliability of the distribution network. In order to solve it, traditionally different topologies of APLC have been used [1 - 3].

The selection of filters is one of the main design parameters of a conditioner and must be calculated taking into account the topology, load current to be compensated, the selected correction strategy and voltage of DC link. The inductance of AC side affects the static and dynamic performance of the whole system when the grid-connected conditioner is working. [4].

The value of inductor must be selected resolving a compromise between two criteria. Firstly, to ensure that the ripple due to the switching frequency of the corrected current waveform is lower than a default value; that is the *ripple criteria* [5], [6]. Secondly, the current provided by the filter should be able to follow the slopes needed for a correct compensation, it is the *slope criteria* [5].

In some situations, the rate of change of current produced by AF to achieve the desired current at the point of common coupling (PCC), shows very high values that are impossible to reach in normal operation of the conditioner. This happens with the load demanded by a three phase rectifier with a resistive load, so the quickness criteria named before should be replaced by a novel criterion.

The criterion are focused to L filters, even LCL filters have been studied extensively and offer better harmonic attenuation and efficiency than L filters [7], [8], [9]; the purpose is to evaluate the criterion firstly with conventional L filters.

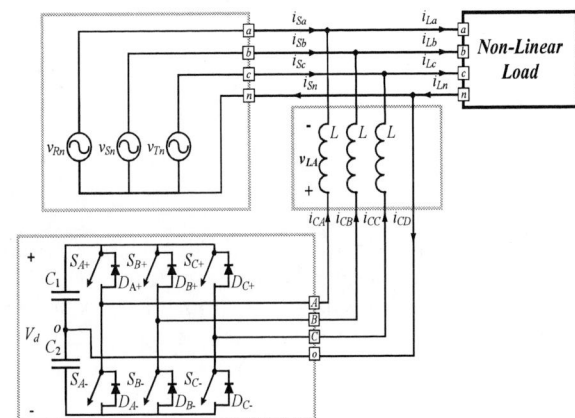

Fig. 1. Current conditioner topology: three-phase four-wire source with nonlinear load.

This paper analyzes the criteria for selecting the inductors values should be placed between the APLC-VSI and the PCC where a nonlinear load is connected and also a novel criterion, the **control criteria**, is presented to complement the previous ones.

II. INDUCTOR SELECTION STRATEGY

The objective of the inductors is mainly to filter the high order harmonics generated by VSI that correspond to the switching frequency and its multiples. Currently two criteria (ripple and slope) are used to select the inductors values. These criteria could be contradictories, the first provides a minimum value for the inductance, while the second provides a maximum value for the inductance [5].

1. Ripple criterion: minimum value.

This criterion ensures that the error between the reference current and current injected by the conditioner is within a margin. In this situation, the switching ripple of the fundamental current will be limited, and also the harmonics, since there is a high dependence between the two magnitudes: ripple and content of harmonics.

The topology used for this criterion is three phase four wire without inductor in neutral wire (Fig. 1).

Analyzing the Fig. 1, the following expression can be deduced for the first branch:

$$v_{LA} = v_{Ao} - v_{Rn} = q_A V_d - v_{Rn} \qquad (1)$$

where q_A is a switching function that can take the value 0.5 or -0.5, depending on whether branch switch is closed at the

978-1-4244-8806-3/11 $26.00 © 2011 IEEE

instant of time considered (t) is the upper or the lower respectively, they are the only two possible states of the branches.

The inductor current increase, assuming a constant voltage applied to it will be given by the expressions:

$$v_{LA} = L\frac{di_{CA}}{dt}$$

$$\Delta i_{CA}(t) = \frac{1}{L}\int_{t-\frac{T}{2}}^{t+\frac{T}{2}}(q_A V_d - v_{Rn})\,dt \qquad (2)$$

To ensure that the ripple is lower than a given value $I_{C,r}$ for any situation and instant of time considered, it is necessary to satisfy

$$|\Delta i_C(t)| < I_{C,r} \qquad (3)$$

The greatest positive current increase (it could get a similar reasoning to the most negative increase) appears when higher voltage is applied to the inductor for as long as possible. This maximum time is the switching period, when the duty cycle applied on the inverter branch is equal to unity, the variable T used in (2), is the switching period T_C.

The greatest increase occurs when the inductor voltage is maximum, this will happen when the top switch of the branch is closed and the phase voltage is minimum. In this situation, shown in Fig. 2, there will be a current increase given by:

$$\Delta i_{C,\max} = \frac{1}{L}\int_{\frac{3T_1}{4}-\frac{T_c}{2}}^{\frac{3T_1}{4}+\frac{T_c}{2}}\left(0,5V_d - \hat{V}_{Rn}\sin\omega_1 t\right)dt$$

$$\Delta i_{C,\max} = \frac{1}{L}\left[0,5V_d T_c + \frac{\hat{V}_{Rn}}{\omega_1}2\sin\frac{\omega_1 T_c}{2}\right] \qquad (4)$$

Applying the condition imposed on the ripple (3), the minimum value of the inductor is calculated:

$$I_{C,r} > \Delta i_{C,\max} = \frac{1}{L}\left[0,5V_d T_c + \frac{\hat{V}_{Rn}}{\omega_1}2\sin\frac{\omega_1 T_c}{2}\right]$$

$$L > \frac{1}{I_{C,r}}\left[0,5V_d T_c + \frac{\hat{V}_{Rn}}{\omega_1}2\sin\frac{\omega_1 T_c}{2}\right] = L_{\min} \qquad (5)$$

2. Slope criterion: maximum value.

In order to eliminate the distortion, the conditioner must provide a current slope at least equal to the maximum load current slope, in the worst conditions of the supply voltage. The maximum current slope provided by the conditioner sets a minimum value for the ratio between the inductor and its voltage.

Therefore, this criterion determines what values of inductance guarantees an enough current to follow the slopes variations that permits to correct the load current.

The load current is analyzed to determine the current that must be provided by ALPC and the maximum slope of the ALPC current is obtained, Si_C.

It must ensure that the slope of the inductor current is greater than that determined in the previous analysis, the next condition must be satisfied:

$$\frac{di_C(t)}{dt} > Si_{C,\max} \qquad (6)$$

The derivative of the current in the inductor can be calculated from the inductor voltage:

$$\frac{di_C(t)}{dt} = \frac{1}{L}(q_A V_d - v_{Rn}) > Si_{C,\max} \qquad (7)$$

From this expression the maximum value of the inductor is obtained to satisfy the previous condition. The worst condition for a positive slope is the case of top switch closed and maximum phase voltage (Fig. 3).

$$L < \frac{1}{Si_C}(q_A V_d - v_{Rn}) \le \frac{1}{Si_C}\left(0,5V_d - \left(+\hat{V}_{Rn}\right)\right)$$

$$L < \frac{0,5V_d - \hat{V}_{Rn}}{Si_C} = L_{\max} \qquad (8)$$

III. NOVEL INDUCTOR SELECTION STRATEGY

A new control criterion is proposed to replaces slope criterion, when the slope of current is very sharp. In these cases the application of the slope criterion is not possible, because the minimum inductors values are very high and there is not an interval of inductors values to make the APLC follows the slopes of the load current.

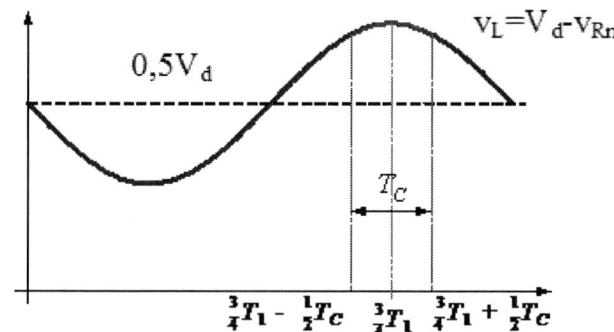

Fig. 2. Worst interval time to select the inductor value that guarantees a ripple in the current lower than a certain value.

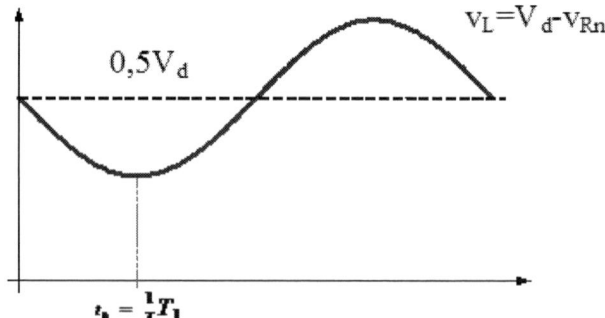

Fig. 3. Worst interval time to select the inductor value that guarantees a slope of conditioner current greater than a certain value PiC, max.

978-1-4244-8806-3/11 $26.00 © 2011 IEEE

This alternative criterion ensures the maximum time that conditioner current is out of control, until the conditioner follows again the reference current.

This case is shown in Fig.4, the conditioner can not generate the current to follow a great increase in the slope of the reference current.

The interval time, when the conditioner is out of control and can not follow the reference current, is named (t_{oc}).

The instantaneous increase of the reference current is determined, for phase A, by the following expression:

$$\Delta i_{CA} = \int_0^{t_{oc}} \left(q_A(t)V_d - v_{Rn} \right) dt \qquad (9)$$

The worst situation for the case described in (Fig. 4), in which the current increase is positive, and therefore the top switch of the branch is closed (so $q_A(t)=+0.5$) will occur when the phase voltage reaches its maximum. This situation is shown in Fig. 5.

In this case:

$$\Delta i_{CA} = \Delta i_{C,ref} = \frac{1}{L} \int_{t_0 - \frac{t_{oc}}{2}}^{t_0 + \frac{t_{oc}}{2}} \left(0,5V_d - \hat{V}_{Rn}\sin\omega_1 t \right) dt \qquad (10)$$

The phase voltage reaches its maximum value in t_0, considering the voltages references, ωt_0 is $\pi/2$ and the expression obtained:

$$L\Delta i_{C,ref} = 0,5V_d t_{oc} - \frac{2\hat{V}_{Rn}}{\omega_1}\sin\frac{\omega_1 t_{oc}}{2} \qquad (11)$$

A maximum inductor value is calculated to determine a certain current increase in the interval t_{oc}:

$$L < \frac{1}{\Delta i_C}\left(0,5V_d t_{oc} - \frac{2\hat{V}_{Rn}}{\omega_1}\sin\frac{\omega_1 t_{oc}}{2} \right) = L_{max} \qquad (12)$$

IV. PARAMETRICAL ANALYSIS

Three expressions to select the inductors values have been calculated in (5), (8) and (12). Different parameters take part of these expressions:

- Capacitors voltages (depending on the conditioner topology used).
- Frequency modulation (relation between the constant switching frequency used in the conditioner and fundamental frequency).
- Maximum current increase (function of the load).
- Maximum ripple permitted.
- Maximum interval time out of control permitted.

In order to develop a parametrical analysis to select the inductors values, all the criterion have been described using some constants which are shown in Table I.

The constants described in Table I, are used in the expressions (5), (8) and (12) to obtain:

$$L > \left(L_{min} \right) = \frac{\hat{V}_{Rn}}{\omega_1 \hat{I}_C}\left[\frac{\pi}{K_r}\left(\frac{0,5K_v}{K_f} + \frac{1}{\pi}\sin\frac{\pi}{K_f} \right) \right] \qquad (13)$$

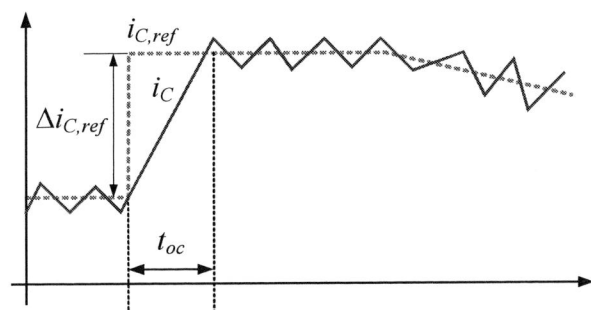

Fig. 4. Condition that leads conditioner out of control when the APLC is not able to follow the evolution of the reference current.

$$L < \left(L_{max,slope} \right) = \frac{0,5V_d - \hat{V}_{Rn}}{Si_C} = \left(\frac{\hat{V}_{Rn}}{\omega_1 \hat{I}_C} \right)\frac{1}{K_p}(0,5K_v - 1) \qquad (14)$$

$$L < \left(L_{max,control} \right) = \frac{\hat{V}_{Rn}}{K_i 2\hat{I}_C}\left(0,5K_v\left(K_c\frac{2\pi}{\omega_1} \right) - \frac{2}{\omega_1}\sin\left(K_c\pi \right) \right) \qquad (15)$$

The previous expressions have been normalized taking into account a normalization value (L_N), it corresponds with an inductance connected to a sinusoidal feeder with the nominal APLC current:

$$L_N = \frac{\hat{V}_{Rn}}{\omega_1 \hat{I}_C}, \qquad (16)$$

The normalized and parametrical expressions obtained are:

$$l > \left(\frac{L_{min}}{L_N} \right) = \left[\frac{\pi}{K_r}\left(\frac{0,5K_v}{K_f} + \frac{1}{\pi}\sin\frac{\pi}{K_f} \right) \right] \qquad (17)$$

$$l < \left(\frac{L_{max,slope}}{L_N} \right) = \frac{0,5V_d - \hat{V}_{Rn}}{Si_C} = \frac{1}{K_p}(0,5K_v - 1) \qquad (18)$$

$$l < \left(\frac{L_{max,control}}{L_N} \right) = \frac{1}{2K_i}\left(\pi K_v K_c - 2\sin\left(K_c\pi \right) \right) \qquad (19)$$

where l, is the normalized inductance equal to:

TABLE I
CONSTANTS USED IN THE PARAMETRICAL ANALYSIS

Constant	Relation	Description
K_f	$\frac{T_1}{T_c} = \frac{f_c}{f_1} = \frac{\omega_c}{\omega_1}$	**Switching constant.** Relation between switching frequency and 1th harmonic frequency
K_v	$\frac{V_{dc}}{\hat{V}_{Fn}}$	**Voltage constant.** Relation between DC link voltage and peak value of the phase-neutral voltage
K_r	$\frac{I_{C,r}}{I_{C,pp}} = \frac{I_{C,r}}{2\hat{I}_C}$	**Ripple constant.** Relation between peak to peak ripple value of the current and the peak to peak value of the current
K_p	$\frac{Si_C}{\hat{I}_C\omega_1}$	**Slope constant.** Relation between the maximum current slope and the maximum current slope at fundamental frequency
K_c	$\frac{t_{oc}}{T_1} = \frac{\omega_1 t_{oc}}{2\pi}$	**Control constant.** Relation between the interval time when the conditioner is out of control and the period of fundamental harmonic
K_i	$\frac{\Delta i_C}{I_{C,pp}} = \frac{\Delta i_C}{2\hat{I}_C}$	**Instant variation constant.** Relation between instant current increase and peak to peak current value

$$l = \frac{L}{L_N} \qquad (20)$$

Usually, the switching constant K_f has high values and it has been assumed in (17) that:

$$\sin \frac{\pi}{K_f} \approx \frac{\pi}{K_f} \qquad (21)$$

Also, when the interval time out of control t_{oc} is small, a simplification is applied in (19):

$$\sin\left(K_c\pi\right) \approx K_c\pi \qquad (22)$$

The final simplified expressions are:

$$l > \left(\frac{L_{min}}{L_N}\right) = \left(L_{min,N}\right) = \pi \frac{0,5K_v+1}{K_r K_f} \qquad (23)$$

$$l < \left(\frac{L_{max,slope}}{L_N}\right) = \left(L_{max,slope,N}\right) = \frac{1}{K_p}\left(0,5K_v-1\right) \quad (24)$$

$$l < \left(\frac{L_{max,control}}{L_N}\right) = \left(L_{max,control,N}\right) = \pi \frac{K_c}{K_i}\left(0,5K_v-1\right) (25)$$

V.   RESULTS

In this part the evolution of the possible values of the inductances are presented, for a predetermined design conditions, introducing changes on one of the parameters. The results shown were obtained by simulation using MATLAB-SIMULINK, in the case of a VSI with three-phase four-wire topology and neutral point clamped capacitor (Fig. 1), that corrects the current demanded by a three phase controlled rectifier, feeding a resistive load (R = 88 Ω) (Fig. 5).

In Fig. 6, the load current, the PCC and the APLC currents are represented for a firing angle of 30° of the rectifier.

Previously, an analysis of the evolution of more representatives factors of the APLC current are presented, when the firing angle of the rectifier changes its value from 0 to π/2. (Fig. 7 (a)).

The filter inductance value is conditioned by the other design parameters of the APLC and the characteristics of the current that the APLC must provide. The expressions to select the inductances value have been developed in (23) and (25), under the hypothesis of a high frequency modulation (parameter K_f) and according to the desired ripple. In this

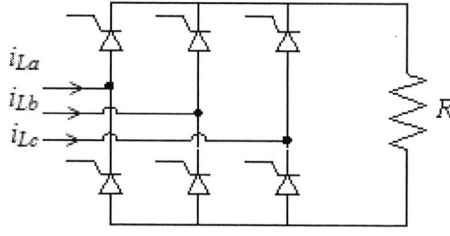

Fig. 5. Three phase controlled rectifier scheme.

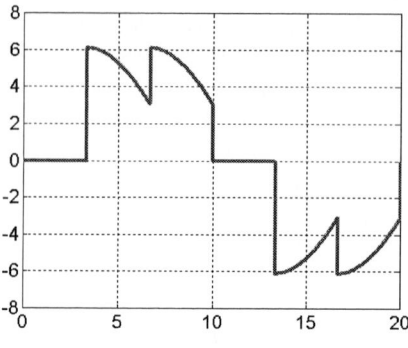

(a)

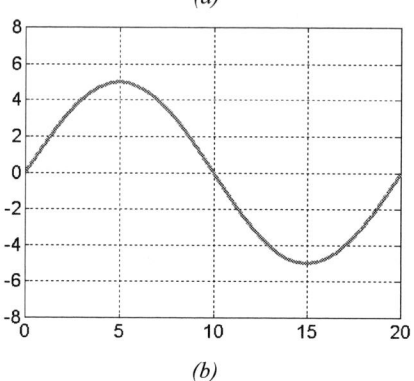

(b)

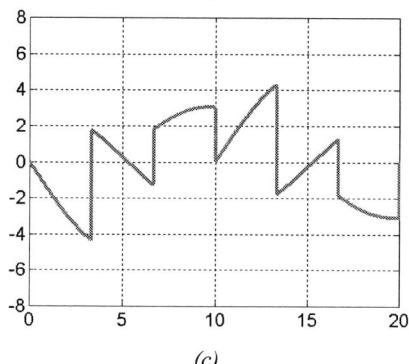

(c)

Fig. 6. Evolution of ideal currents (A), with a firing angle of 30° in the controlled rectifier for 20 ms: (a) load current (i_L) (b) PCC current (i_S) (c) APLC current (i_C)

case, fast current variations appear and therefore the slope criterion can not be applied.

The value used to normalize these expressions (23) and (25) is:

$$L_N = \frac{\hat{V}_{Rn}}{\omega_1 \hat{I}_C} = \frac{\sqrt{2}\cdot 220}{100\pi\cdot 4,5} = 220 \text{ mH} \qquad (26)$$

The maximum peak value of the active filter current and the maximum instantaneous increase of the active filter current are obtained from Fig. 7 (parameter K_i) Applying a switching frequency of 10 kHz (parameter K_f)and a maximum ripple of 10% (parameter K_r), the studied parameters values are calculated with the proposed load and resume in Table II.

$$K_i = \frac{\Delta i_C}{2\hat{I}_C} = \frac{6,075}{2\cdot 4,5} = 0,675 \qquad (27)$$

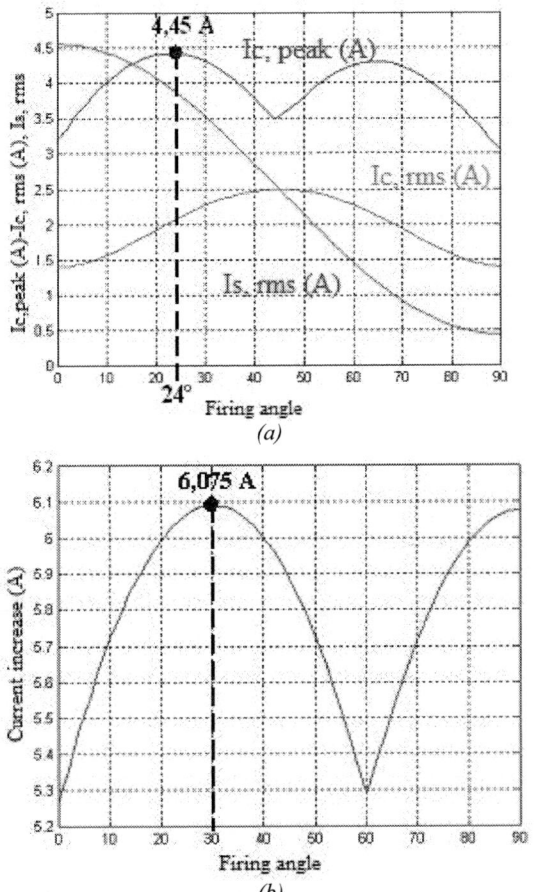

(a)

(b)

Fig. 7. For different values of the firing angle of the rectifier: (a) Peak value (IAF,pico) and rms value (IAF,rms) of the APLC current and rms value of the supply current (Is,rms), (b) Maximun APLC current increase.

$$K_f = \frac{f_c}{f_1} = \frac{10k}{50} = 200 \qquad (28)$$

$$K_r = \frac{I_{C,r}}{I_{C,pp}} = \frac{10\% \cdot I_{C,pp}}{I_{C,pp}} = 0,1 \qquad (29)$$

In Fig. 8, the evolution of maximum and minimum inductances values, given by the ripple (23) and control criteria (25) for different K_v and K_c values, is shown. The other parameters, K_i, K_f and K_r, take the values calculated before in the expressions (27-29).

TABLE II
SIMULATION PARAMETERS

$\hat{I}_{max}$	4,45 A (firing angle = 24°)
ΔI_{max}	6,075 A (firing angle = 30°)
$\hat{V}_{Fn}$	311 V
ω	314,16 rad/s
K_f	200
K_r	0,1
K_i	0,675
L_N	220 mH

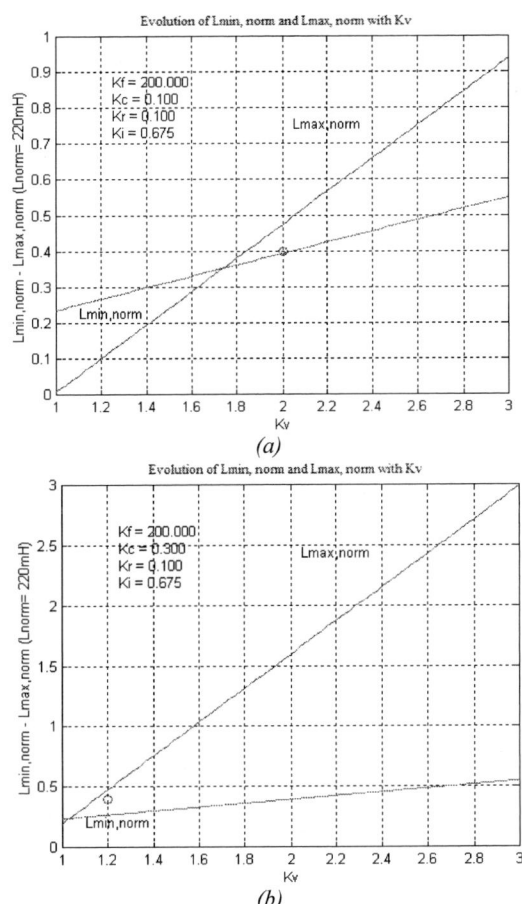

(a)

(b)

Fig. 8. Evolution of inductance values changing Kv from 1 to 3: (a) with Kc = 0.1 (maximum time out of control: 2 ms), (b) with Kc = 0.3 (maximum time out of control: 6 ms).

It can be seen in Fig. 8 (a) and Fig. 8 (b), how the interval of possible inductances values between $L_{min,N}$ and $L_{max,N}$ is greater when the DC voltage increases (with a greater K_v) and the same happens when the time out of control increases (with a greater K_c).

The corrected current, taking into account the values for K_v and K_c indicated in Fig. 8 (a) and (b), is shown in Fig. 9 and Fig. 10. The parameters used to obtain both figures are:

- a inductor value $\left(L = 0,4 \cdot L_N = 0,4 \cdot 200mH = 80mH\right)$
- a voltage constant: K_v=2 (assuming a V_{dc}=622 V) in Fig. 9, and K_v=1,2 (assuming a V_{dc}=373,2 V) in Fig. 10,
- a control constant: K_c=0,1 (assuming a t_{oc}=2 ms) in Fig. 9, and K_c=0,3 (assuming a t_{oc}=6 ms) in Fig. 10.

It can be seen in Fig. 9 and Fig.10 that for the same inductor value, the corrected current by the APLC presents a ripple greater when the DC voltage is greater and with decreasing the time out of control.

VI. EXPERIMENTAL RESULTS

The control criterion was tested with an APLC prototype in laboratory using a rapid prototyping control platform

DSPACE supported by MATLAB/SIMULINK. In Fig. 11, captures from oscilloscope are showed for a switching frequency of 20 kHz. The waveforms obtained confirm the simulation results showed in Fig. 9 and Fig. 10.

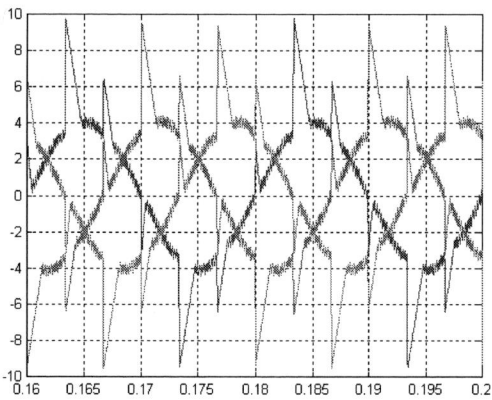

Fig. 9. Corrected current (A), with $Kc = 0,1$, $Kv = 2,0$ and $L = 0,4\ L_N$.

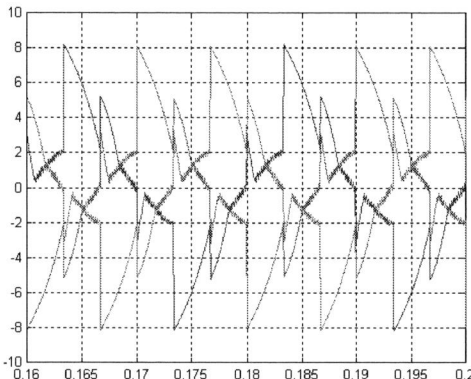

Fig. 10. Corrected current (A), with $Kc = 0,3$, $Kv = 1,2$ and $L = 0,4\ L_N$.

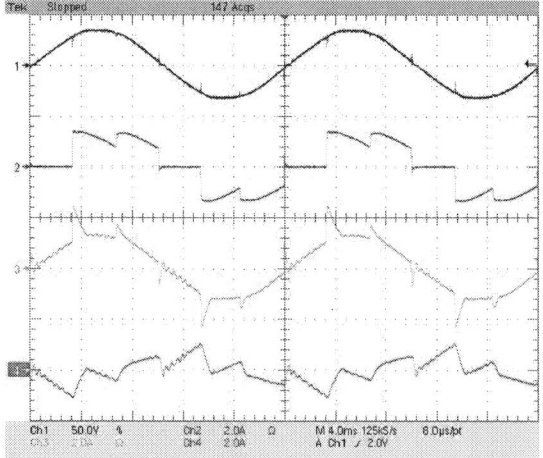

Fig. 11. Evolution of experimental waveforms. From top to bottom: PCC voltage, load current (i_L), PCC current (i_S) and APLC current (i_C).

VII. CONCLUSIONS

A new selection criterion, named the ***control criterion***, of the inductances of an APLC is presented in this paper.

A parametrical analysis of the inductances expressions obtained taking into account all the influence parameters, has been carried out in order to make easier the inductor selection for any current and voltage ranges.

This parametrical analysis has been verified by simulation proving that:

The interval of possible inductances values is greater (a greater working power range) when the DC voltage increases, but in the other hand, the ripple of the corrected current is greater too.

If the time out of control in the APLC current is decreased, the ripple of the corrected current is greater and also the interval of possible inductances values is smaller so the power range is limited.

A compromise, between DC voltage used in the APLC topology and the interval time that the APLC used to follow the changes in load current slope, must be reached.

To show a perfect current compensation is not the objective of the simulation results, rather to verify that the criteria proposed are useful to select the APLC inductors values reaching a given specifications. Some laboratory tests have been developed to contrast the simulation results.

This new criterion is analyzed in a first step for conventional L filters and will be applied for LCL filters in future work.

REFERENCES

[1] H. J. Azevedo, J. Ferreira, A. P. Martins, A. S. Carvalho. "Direct current control of an active power filter for harmonic elimination, power factor correction and load unbalancing compensation". Proc 10th Europ Conf Power Electron Appl, Toulouse, pp. 1–10, 2003.

[2] A. Chandra, B. Singh, B. N. Singh, K. Al-Haddad, "An improved control algorithm of shunt active filter for voltage regulation, harmonic elimination, power-factor correction, and balancing of nonlinear loads". IEEE Transactions on Power Electronics, vol.15, pp.495-507, May 2000.

[3] T.F. Wu, H.S. Nei, C.L. Shen, G.F. Li, "A single-phase two-wire grid-connection PV inverter with active power filtering and nonlinear inductance consideration". IEEE Conference on Applied Power Electronics, PP. 1566 - 1571, Vol.3, 2004.

[4] Yang Yong, Ruan Yi, Shen Huan-qing, Tan Yan-yan, Yang Ying, "The design of inductance in AC side of grid-connected inverter in wind power generation". Proc International Conference on Electrical Machines and Systems, ICEMS 2008, pp. 2464-2469, 2008.

[5] A. G. Pedder, A. D. Brown, J. N. Ross y A. C. Williams. "A Parallel-Connected Active Filter for the Reduction of Supply Current Distortion". IEEE Transactions on Industrial Electronics, vol.47, pp.1108-1117, October 2000.

[6] A. M. Al-Zamil, D. A. Torrey. "A Passive Series, Active Shunt Filter for High Power Applications". IEEE Transactions on Power Electronics. Vol. 16, pp. 101-109. January 2001.

[7] C. Parikshith, J. Vinod. "Filter Optimization for Grid Interactive Voltage Source Inverters". IEEE Transactions on Industrial Electronics, Vol. 57, pp. 4106-4114. December 2010.

[8] M. Liserre, F. Blaabjerg, S. Hansen. "Desing and Control of an LCL-Filter-Based Three-Phase Active Rectifier". IEEE Transactions on Industry Applications, Vol. 41, pp. 1281-1291. October 2005.

[9] R. Teodorescu, F. Blaabjerg, M. Liserre, A. Dell'Aquila. "A Stable Three-Phase LCL-filter Based Active Rectifier without Damping". 38th Industry applications Conference. Vol. 3, pp. 1552-1557. October 2003.

978-1-4244-8806-3/11 $26.00 © 2011 IEEE

Design and Implementation of a Digital Control and Monitoring System for an AC/DC UPS

Andriy Palamar[1], Mykola Karpinskyy[2], Valery Vodovozov[1]

[1]Tallinn University of Technology, Department of Electrical Drives and Power Electronics

[2]University of Bielsko-Biala, Department of Electrical Engineering and Automatic

andrij777@mail.ru, mkarpinski@ath.bielsko.pl, valery.vodovozov@ttu.ee

Abstract - **This paper presents the design and implementation of a digital control and monitoring module for an AC/DC uninterruptible power supply (UPS). A hardware and software of the system is developed. Moreover, a friendly graphical-user interface is provided. Management strategy of the UPS in different operation modes is described. The proposed digital control and monitoring system allows to realize the complex algorithms and this way increase the reliability and efficiency of the UPS.**

I. INTRODUCTION

Because of the increasing demand for the devices sensitive to energy quality, such as computers, medical equipment, telecommunication systems, ensuring continuity of power supply is very important. For this purpose, the installation of uninterruptible power supply (UPS) units with energy storage devices is necessary. UPS systems provide continuous, reliable, and high-quality power for important loads in order to prevent expensive hardware failure, data loss or corruption, cessation of the business process in cases of disturbance in the main electrical grid [1].

AC uninterruptible power supply systems are commonly used to provide energy backup to wide ranges of electronic loads. Nevertheless, DC UPS are receiving more and more attention with increasing demand for smaller and cost effective uninterruptible power supplies [2]. Such UPS systems are used in low voltage devices in the telecommunications and computer industry.

Uninterruptible power supply systems are still in their research stage. Although UPS has applications in many various fields, it still requires improvement of many technologies for customer needs. Availability of low cost, user friendly and compact control and monitoring system is one of them. Monitoring of input and output electrical parameters of an UPS, digital control of charging and discharging of a battery, and the archiving of events in real time are necessary parts of such control systems. Although numerous interesting applications of intelligent control for AC UPS are available [3] – [5], the digital control system for DC UPS with its peculiarities of topology still needs further research. In response to the requirements above, a digital control and monitoring system for AC/DC UPS is developed and implemented in this work. The purpose of the paper is to design a hardware and software of the system and realize the control algorithms for different operation modes of the UPS.

II. OVERALL STRUCTURE OF THE SYSTEM

Control block diagram of an AC/DC uninterruptible power supply system based on a microcontroller is shown in Fig. 1. It consists of an AC/DC converter, a lead-acid battery, a control and monitoring module, sensors and switches. The AC/DC converter rectifies the input AC voltage of the main grid into the output DC voltage, provides filtration and stabilization. The value of the output voltage varies depending on the input analog control signal level, which is generated by the control module. The output DC voltage of the converter is used to supply the local loads and to charge the battery storage. The switches are used for disconnection of the battery from the loads in order to avoid deep discharge of the battery storage in case of lack of power provided by the main grid during a long period. The technical specifications of the UPS are listed in Table I.

TABLE I
TECHNICAL SPECIFICATIONS OF THE DEVELOPED UPS

Item	Characteristic
Input voltage	185 V ÷ 255 V
Maximum output power	3.6 kVA
Output voltage	42 ÷ 56 V
Maximum output current	75 A
Back up time	2 hours

The control and monitoring module is the core of the management system of the proposed UPS. It comprises a hardware and software. The possibility of computer connection is also present in the system.

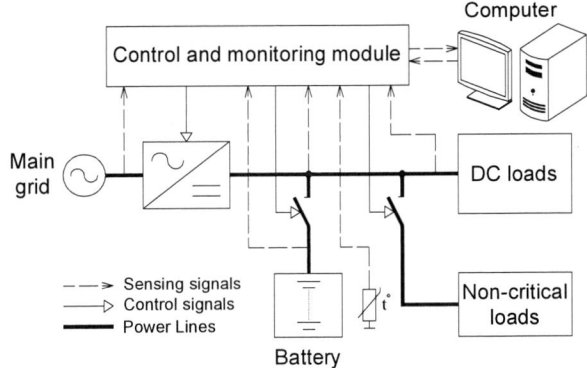

Fig. 1. The control block diagram of the AC/DC UPS.

978-1-4244-8806-3/11 $26.00 © 2011 IEEE

III.   DESIGN OF THE HARDWARE

The control and monitoring module is composed of a control board, liquid crystal display (LCD), and keyboard. The control board includes a microcontroller, circuits of convert and amplification of analog signals, input / output ports, a real time clock (RTC) module, a ferroelectric random access memory (FRAM), a voltage amplifier circuit, and a galvanic isolation scheme. The block diagram of the control module and its peripheral circuits is shown in Fig. 2. The main functions of the control and monitoring module include the following:

- collecting the data measured;
- providing visualization of the received information;
- generating control signals;
- managing power switches;
- archiving information;
- connecting with a computer.

A.   Microcontroller

The control board is based on the Analog Devices ADuC831 microcontroller. It is a cost-effective, programmable 8-bit MCU on a single chip which is executing the control algorithms in real time.

B.   Input Circuits

The outputs of the sensors are not always in a desirable voltage form. The purpose of the analog signals converting circuits is to transform the measured voltages into the clean voltage waveforms so that they can be converted into discrete values by an inner analog-to-digital converter (ADC). The circuits of signal amplification from the current sensors (shunts) are based on the operational amplifiers. They amplify the standard shunt output $0 \div 75$ mV to the level that is normal for the input range of the ADC ($0 \div 4.5$ V). After the conversion and amplification, voltage, currents, and temperature signals are measured by using inner 12-bit ADC with a resolution of ± 1.2 mV. Relative error of measurement equals 0.3 %.

Digital signals of the UPS status are being read periodically by using the input / output port of the microcontroller. Furthermore, the module provides an opportunity for audio alarm to inform of the emergency condition of the system.

C.   Scheme of Displaying and Keeping of Data

All deviations of the measured data of voltage, currents, temperature from the normal values, alarms as well as time of an event are stored in a non-volatile memory of the control module. The archived information can be subjected to off-line processing and analysis.

Furthermore, the received data are visualized on the four-line LCD. Electrical parameters of the system as well as displaying mode can be set or changed using a keyboard with four buttons on the front of the module.

D.   Real Time Clock

RTC module is connected to the microcontroller using I2C interface. It is used to keep track of the current time. The RTC circuit uses a lithium battery as an alternate source of power to guarantee keeping time while the primary source is off.

E.   Output Circuits

The microcontroller produces an analog signal in order to control the output voltage level of the AC/DC converter according to a preset algorithm. For this purpose an inner digital-to-analog converter (DAC) and output amplifier are used. Digital output signals are responsible for managing of the process of connection/reconnection of the battery storage and non-critical loads.

F.   RS-232 Interface

A galvanic isolation scheme and a signal converter are used to ensure reliable communication with a computer via standard RS-232 serial interface. The control module is connected to the computer in order to transfer the collected archive information, set or change the electrical parameters, and remote monitoring of the system.

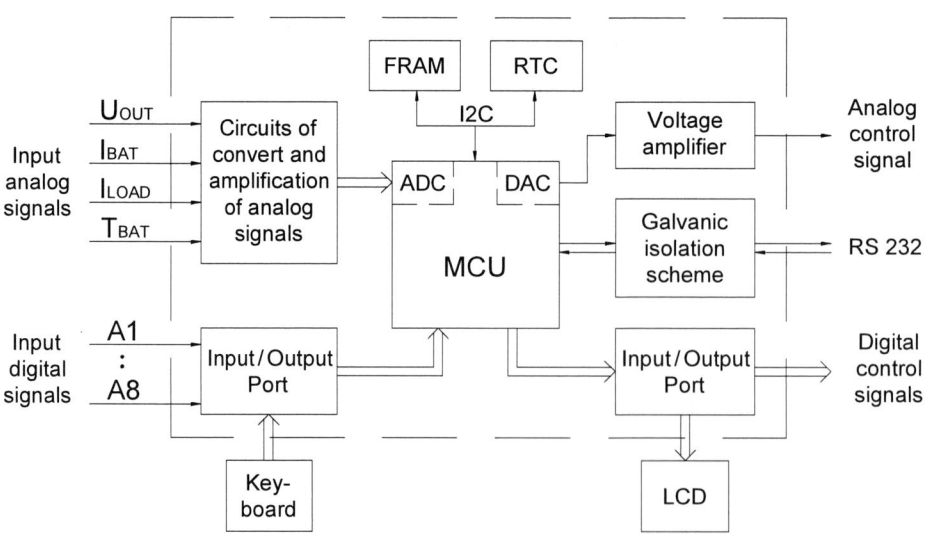

Fig. 2. Block diagram of the control module and its peripheral circuits.

IV. DEVELOPMENT OF THE SOFTWARE

The software of the control and monitoring system comprises of two parts. The first part is software burnt into the microcontroller of the control board. It is implemented in the Assembler programming language. The second part of the software is a computer monitoring application – graphical user interface. The main program of the microcontroller includes the following subroutines:

1. System initialization
2. Data collection and processing
3. Display of information
4. Control unit
5. Data transfer unit

All of these procedures are described in detail below.

A. System Initialization

This procedure is performed once immediately after the system startup, before the main program circle. It is necessary for setting of interrupts, assigning initial values of variables, initializing of LCD, ADC, DAC, serial port, etc.

B. Data Collection and Processing

The function of this subroutine is to provide digitizing of the collected data and their previous processing. Measured information required for the proposed monitoring and control system includes:

- Output DC voltage – U_{OUT};
- Battery current – I_{BAT};
- Load current – I_{LOAD};
- Battery temperature – T_{BAT}.

The inner ADC of the microcontroller is used for digitizing the analog signals. Measurement range of the currents for each channel may vary depending on the programmed coefficients in the flash-memory of the microcontroller.

Information received by the ADC module is processed in real-time. During the previous data processing, the software algorithms of accumulation and averaging the measured information have been used in order to improve a signal / noise ratio and provide pulse obstacles filtering.

The measured data are kept in the inner memory of the control module and transmitted to the computer in cases of connection with the monitoring program. This information is used for management purposes of the control system.

C. Display of Information

This subprogram is designed to display the measured information on the LCD in a real-time and to provide access to archived data. Furthermore, it allows one to set and to modify parameters of the system. The subroutine manages three levels of the menu: "Measurements", "Archive", "Parameters". The subprogram uses the keyboard with four buttons: "up", "down", "enter", and "esc" in order to execute viewing, setting, and changing the information.

D. Control Unit

The following modes of control are available: fully automatic control (with manually adjusted set-points) and local direct manual control. In the automatic mode, the control module implements optimal operation conditions of the battery storage during charge, recharge, and backup processes. Furthermore, it sets the electrical parameters of the system, such as voltage and current, taking into account the temperature of the battery. The manual control mode allows an operator to set and modify the required values of the electrical parameters of the system using the key-board or the graphical-user interface. This mode of control shall be password-protected.

The UPS system could be operated in three different modes: current limitation, charging, and backup. Control strategies of the management system of the UPS for all of these modes are described in detail in the next section of the paper.

E. Data Transfer Unit

This procedure is responsible for transferring the measured, data to the computer through the serial interface RS-232. The amount of data to be transmitted includes the measured data of the system and the archived information from the memory of the control module. Furthermore, an opportunity of using of Ethernet protocol in order to provide remote access to the UPS is also present.

F. Graphical User Interface

The graphical user interface of the software is illustrated in Fig. 3. It has the following functions: visualizing the waveforms in real-time; displaying alarms; providing possibility of viewing and saving the archive information in a convenient form; setting and modifying of the electrical parameters of the system. The graphical user interface includes: the menu, toolbar, control parameters setting section, data monitor, archiving table, alarming panel, status bar, and interface for connection with the control module.

With this user-friendly interface, system operators are able to easily track the overall system status and performance. Furthermore, this software enables its user to generate a report on the recorded electrical information of the system for a particular day as well as to generate a graph of the same based on the choice of the user in the interactive front end.

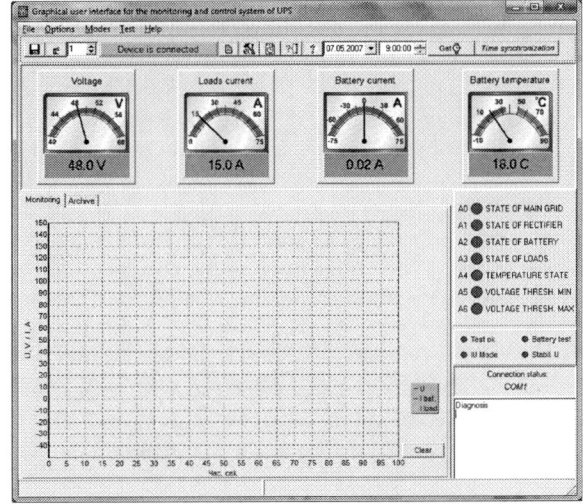

Fig. 3. Graphical user interface for the monitoring and control system.

V. CONTROL STRATEGIES OF THE UPS

A. Mode of Current Limitation

In this mode, the AC/DC converter is always able to provide the maximum load current and battery charge current. This mode always comes after the connection of the UPS to the main electrical grid. In that case the current rises smoothly from zero to the pre-established maximum value. The UPS enters into the mode of limitation of the output current if the total of load current and battery charge current exceeds the preset current-limit value of the device. In this mode, the control module reduces the output voltage to a level when the total current will be equal to the threshold value of the current limitation of the device. The microcontroller sends the command in order to switch off the battery storage when the output voltage decreases below the minimum threshold value. As a result, the total output current is decreased. Thereby, the system exits from the mode of output current limitation.

B. Charging Mode

The UPS enters into the charging mode when the main grid is in the normal state and the battery voltage is simultaneously lower than the desired value. The AC/DC converter supplies power to the load and charges the battery in this mode in order to provide backup power for the backup mode of operation. The charge process could be implemented using two methods: "IU" and "U". Fig. 4 and Fig. 5 shows curves of the voltage and current during charge process using "IU" and "U" methods respectively.

The "IU" method assumes that the charge process takes place in two stages. The first stage – the charge process is carried out using a stabilized current. It is necessary to continually increase the output voltage of the AC/DC converter in order to maintain the charge current at a constant level. Once the voltage reaches to the maximum value a transition to the second stage of the charge occurs. Then the charge process is carried out using a stabilized voltage. Hence, the charge current herewith decreases.

The "U" method is a partial case of the "IU" method. During this process the microcontroller measures continuously the battery charge current comparing it with a pre-established limit value. If the current exceeds this value, the microcontroller using the ADC and the amplifier reduces the analog control signal that is connected to the input of the AC/DC converter. As a result, charger voltage decreases to a level when the charge current becomes equal to or less than the pre-established value. The output voltage of the converter rises slowly in this period, maintaining the charge current at a constant level. This process continues until the instant at which the charge current becomes lower than the set value. After that the level of the charger voltage will depend on the temperature of the battery storage. The ideal battery operation temperature is the 20 °C [6]. If it deviates from this value the control system adjusts the charge voltage according to a preset algorithm.

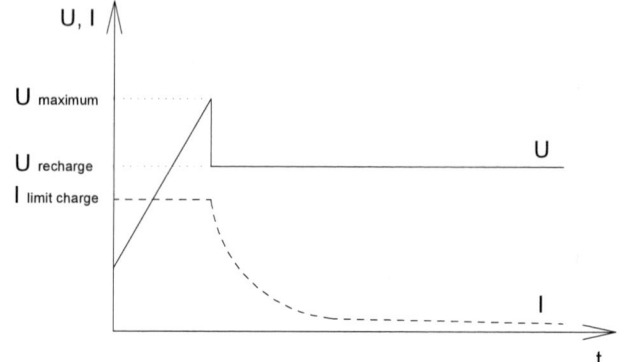

Fig. 4. IU method of battery charge.

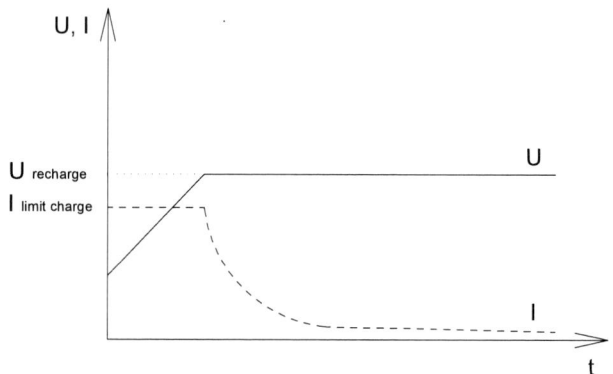

Fig. 5. U method of battery charge.

If the temperature exceeds the threshold value of 45 °C the charge process terminates in order to reduce the battery temperature, because the further rising of the temperature can cause enhanced self-discharge of the storage device.

Batteries are fully charged, if the residual charge current is not changed within two hours. The batteries have to receive recharge current continuously in order to compensate of the self-discharge process.

C. Backup Mode

When the main grid AC voltage is outside the preset tolerance, the UPS enters a backup mode which means that the loads are totally supplied from the battery storage. The duration of this mode is the duration of the preset UPS backup time or until the main grid is available again.

The battery voltage condition is continuously monitored. If the voltage decreases below the first preset threshold value, the control system sends the signal to detach non-critical loads. Once the output voltage falls below the second certain threshold level, the control module disconnects all the loads in order to avoid deep discharge of the battery storage.

When the main grid voltage fulfills the desired requirements, the battery storage automatically reconnects to the AC/DC converter, which starts to charge the battery. In the same way if non-critical loads are detached, they are also reconnected.

978-1-4244-8806-3/11 $26.00 © 2011 IEEE

VI. PRACTICAL IMPLEMENTATION

The designed control system has been applied to an AC/DC UPS that is used mainly for power supply for telecommunication equipment, such as cellular base station. A photograph of the prototype of the management module is illustrated in Fig. 6. Fig. 7 shows the implementation of the control and monitoring system into the UPS. Using an RS-232 serial connection, a laptop can monitor the UPS in real-time. The operating status of the UPS can be monitored by the computer software as shown in Fig. 3. This program was implemented for Windows platform with Delphi 7.0. The system can monitor the voltage, currents, and temperature of the UPS. The faults and alarming information can be automatically recorded and restored. In addition, the protective functions are implemented in the management system: preventing battery charge with unacceptably high current, and preventing discharge of the storage device to unacceptably low voltage. Such capabilities clearly add to the overall reliability of the UPS system.

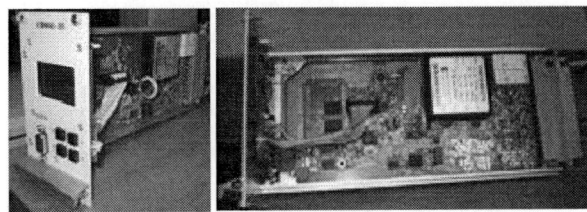

Fig. 6. Photograph of the control and monitoring module.

Fig. 7. The implementation of the control and monitoring system.

VII. CONCLUSIONS

In this paper a digital control and monitoring system for AC/DC UPS has been designed and implemented. The hardware and software of the management system was developed. A user-friendly graphical interface is provided. The software was designed for the monitoring of electrical parameters and for managing operation modes of the UPS. A control strategy of the UPS in different operation modes was described in detail. The proposed control and monitoring system can increase the reliability, safety, and efficiency of the AC/DC uninterruptible power supply.

ACKNOWLEDGEMENT

This research work has been supported by Estonian Ministry of Education and Research (Project SF0140016s11), Estonian Science Foundation (Grant ETF8020) and Estonian Archimedes Foundation (project „Doctoral School of Energy and Geotechnology II").

REFERENCES

[1] A. Emadi, A. Nasiri, and S. B. Bekiarov, "Uninterruptible power supplies and active filters," *CRC Press LLC,* Boca Raton, Florida, pp. 3-11, 2005.

[2] E. Rodriguez, H. Visairo, and J. Arau, "A High Efficiency DC-UPS with PFC," in *VII IEEE International Power Electronics Congress,* Mexico, 2000, pp. 150-155.

[3] F. Wataru and Y. Tomoki, "Construction of FPGA Based Hardware Controller for Autonomous Decentralized Control for UPS Application," in *12th International Power Electronics and Motion Control Conference,* Portoroz, Slovenia, 2006, pp. 846-851.

[4] C.H. Lai and Y.Y. Tzou, "DSP-Embedded UPS Controller for High-Performance Single-Phase On-Line UPS Systems," in *28th Annual Conference of the Industrial Electronics Society,* vol.1, Sevilla, Spain, 2002, pp. 268-273.

[5] Z. He, M. Li, and Y. Xing, "Core Techniques of Digital Control for UPS," in *International Conference on Industrial Technology,* Hong Kong, 2005, pp. 546-551.

[6] P. Bentley, and D. Bond, "The UPS Handbook," *Uninterruptible Power Supplies Ltd.,* Berkshire, UK, pp. 116-124, 2007.

Three-Phase Regenerative Electronic Load to Test Shunt Power Conditioners

C. Roncero-Clemente, *Student Member, IEEE*, M.I Milanés-Montero, *Member, IEEE*, V.M. Miñambres-Marcos, *Member, IEEE*, E. Romero-Cadaval, *Senior Member, IEEE*

Power Electrical & Electronic Systems (PE&ES), School of Industrial Engineering, University of Extremadura. Badajoz, Spain

Abstract- **A 3-phase regenerative electronic load to test shunt equipments such as active power filters and power conditioners at high power levels is presented in this paper. The electronic load is composed by two inverters with a common DC- bus. One inverter draws from the grid a configurable load current with the harmonic and imbalance desired levels, and the other one is used to inject the energy into the grid trying to minimize the losses, while the active filter performance is being tested. The control strategies and the current controller for both inverters are explained. Simulations and experimental results showing the proper operation of the electronic load to test shunt power conditioners are presented.**

I. INTRODUCTION

Some electronic devices, such as uninterruptible power systems (UPS), power supplies or energy storage systems need to be tested in a load bank to check the behavior of the equipment and to improve its reliability and stability [1].

Other devices which need a load bank to be tested are active power filters and power conditioners (PC). These equipments are connected in parallel with nonlinear loads in order to compensate reactive power and harmonic and unbalanced currents demanded by the load.

The load bank is usually designed with dissipative components, resistors, inductors and capacitors, which generate a lot of power losses, heat and few possibilities to carry out variable tests. For high power applications these power losses are unacceptable [2].

Electronic loads have been developed and commercialized [3] to extend the conditions of the test with constant power, constant current or constant voltage modes of operation, and even behaving as non lineal loads [4]. This development is very important to test active power filters and power conditioners. However, they all have the drawback of consuming a great amount of energy during the tests. Also, some authors [5-7] have developed control algorithms for electronic loads simulators, achieving better performance and effectiveness.

At high power levels another aspect to improve is the energy recycling during the tests. Whit this objective, simulators have been implemented considering the energy regeneration and trying to emulate a variety of load characteristics. Tests energy are fed back to the power grid with little loss and high power factor. These simulators have been designed with a topology made by two electronic converters, with a common DC bus [8, 9]. A lot of control strategies have also been implemented as repetitive control strategies to manage these simulated devices [10].

There are few research works about active loads with energy recycling and only experimental small-scale prototypes or simulators can be found. In high power applications it is necessary to develop regenerative type active electronic loads to return the power to the grid, with these characteristics:

- Highly energy-efficient system, since the energy can be returned back to the utility, so a large amount of energy can be saved during test procedures. It is especially advantageous in case of high-power equipments.
- Compact size, because the active regeneration scheme allows reduction on passive power components and associated cooling system. According to it, the size can be reduced as compared to the conventional scheme.
- Flexible dc control design: the system typically requires a front-end ac-dc converter and a regenerative dc-ac inverter. The controller can be flexibly designed with dc link voltage either controlled by the converter or by the inverter [11].

The electronic loads with energy recycling found in the literature are designed and controlled to test series equipments, as different types of power supplies. In order to test parallel equipments at high power levels, it is essential the performance of an energy feed-back electronic load with the capability of choosing the harmonic and imbalance levels of the current demanded from the grid. With this objective, this paper presents the design of a 3-phase regenerative electronic load to be able to conduct tests of shunt equipments for high power applications. Simulations and experimental results are presented to validate the operation of the load to test a shunt power conditioner to improve the power quality.

II. REGENERATIVE ELECTRONIC LOAD TOPOLOGY WITH AN ACTIVE FILTER CONNECTED IN PARALLEL

Fig. 1 shows the electronic load scheme, composed by two electronic converters with a common DC bus, with a power conditioner connected in parallel to be tested. One inverter demands from the grid a load current while the other inverter injects the energy into the grid. The topology of a 3-phase regenerative AC electronic load is displayed in Fig. 2. The values of the parameters of the electronic load are summarized in Table I.

978-1-4244-8806-3/11 $26.00 © 2011 IEEE

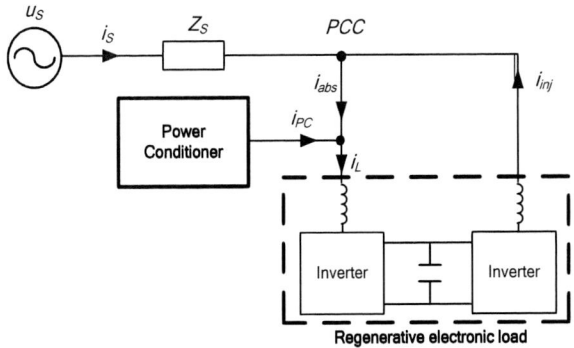

Fig. 1. Electric scheme of the regenerative electronic load and shunt power conditioner.

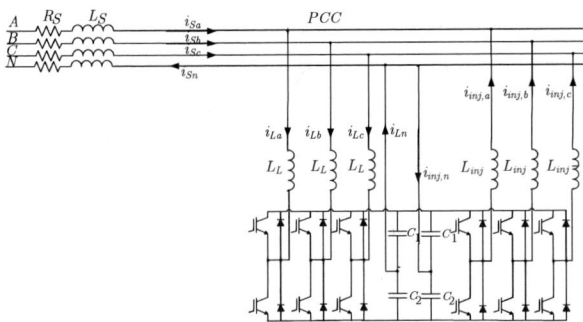

Fig. 2. Three-phase regenerative electronic load topology.

TABLE I
PARAMETERS OF THE REGENERATIVE ELECTRONIC LOAD

U_n(V)	I_n(A)	S_n (kVA)	C_1 (mF)	C_2 (mF)
400	200	100	28,2	28,2
$U_{dc,ref}$ (V)	$f_{s,\,max,load}$ (kHz)	$f_{s,max,inj}$ (kHz)	L_{load}(mH)	L_{inj} (mH)
800	10	20	3	1,75

Each electronic converter is a 3 branches inverter with midpoint DC bus. Each branch is connected to the grid with a filter inductor, L_L or L_{inj}, and the middle point of the DC bus is connected to the neutral conductor directly, with no impedance.

III. ELECTRONIC LOAD CONTROL STRATEGY

The first inverter is used as programmable load. To control this converter a friendly application in *Simulink* (Fig.3 (a) and Fig.3 (b)) is developed, which allows introducing the desired load current specifications. The parameters that can be introduced are the *RMS* value, the harmonic order, the phase angle and the sequence for each component. With all of these parameters one can compose the reference load current with the harmonic and/or imbalance levels needed.

The developed block needs the angle of the positive

sequence fundamental component of the voltage at the point of common coupling (PCC). It is attained with the Autoadjustable Synchronous Reference Frame (ASRF) block [12] shown in Fig. 3(a), which needs the measurements of the PCC voltages u_{PCC} *(a, b, c)*.

The second inverter is used to inject the energy into the grid. To generate the reference injection current it is necessary to analyze the circuit shown in Fig.2. Applying the First Kirchhoff law in the circuit, one has:

$$i_S + i_{inj} = i_L,\qquad(1)$$

where i_S is the current supply, i_{inj} is the reference injection current and i_L is theload current.

Despising the losses ($i_S \approx 0$), it is obtained:

$$i_{inj} = i_L.\qquad(2)$$

Also the control of DC bus is required. This control will be done by the injector device. For this reason a new term is included in (2):

$$i_{inj} = i_L - k_C(s)(U_{dc,ref} - U_{dc,meas})u_{pcc_1^+},\qquad(3)$$

where $U_{dc,ref}$ is the reference DC bus voltage , U_{dcmeas} is the measured DC bus voltage, $u_{pcc_1^+}$ is the positive-sequence fundamental component of the PCC voltage and $kc(s)$ is the transfer function of a proportional-integer (PI) controller.

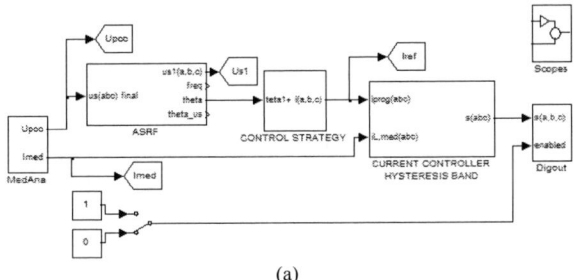

(a)

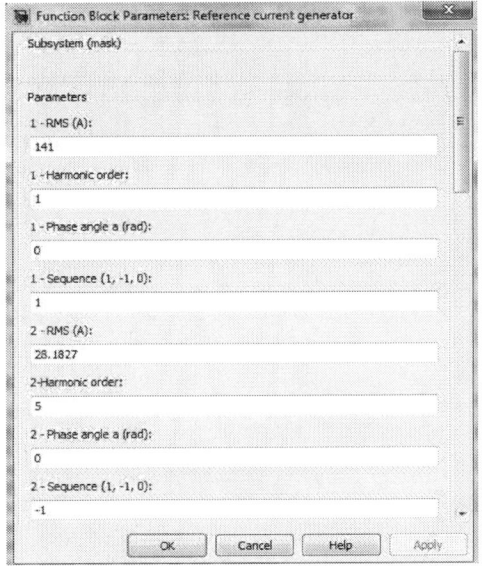

(b)

Fig. 3. Programmable load: (a) Control block. (b) Data entry screen.

IV. ELECTRONIC LOAD CURRENT CONTROLLER

As controller technique to track the reference current, a hysteresis band (see Fig.4) has been employed. This technique compares the reference current with the measured current, and depending on the error the switching signals of the IGBT's are generated. The same tracking technique has been implemented for the programmable load and for the injector.

The operation of the tracking technique for the injector is displayed in Fig. 5. If the error is positive, the injector must inject a greater value of current (positive slope), so the upper IGBT of the inverter branch must be closed and the lower IGBT must be open. The inverse operation is needed (negative slope) in case of a negative error.

V. ELECTRONIC LOAD SIMULATION RESULTS

The 100 kVA three-phase regenerative electronic load operation has been simulated using *Matlab/Simulink* with the topology shown in Fig. 2 and the parameters collected in table 1. To improve the injector reference current tracking the inductance of the injector is half that of the load and the injector switching frequency is twice that of the load.

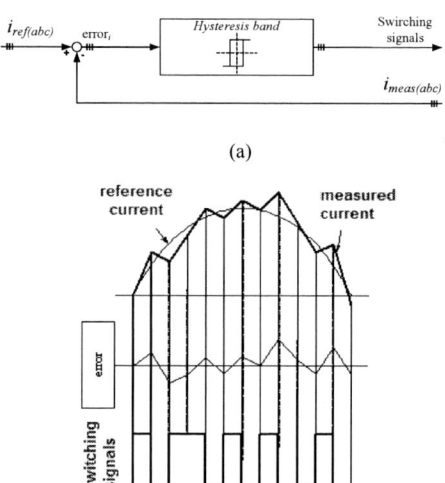

(a)

(b)

Fig. 4. Hysteresis band controller.

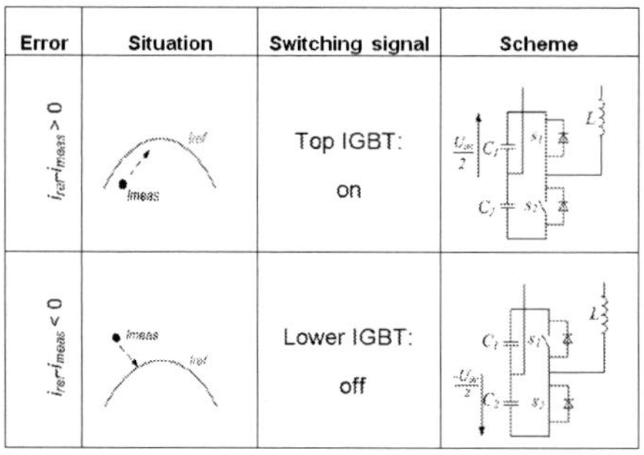

Fig. 5. Injector control modes.

The simulation tests of the electronic load in case of distorted and unbalanced conditions are shown in the following sections.

A. Current with harmonic content(h=1, h=5 and h=7)

The reference current of the load in this case is programmed with an individual harmonic distortion of IHD_5=20 % and IHD_7=14, 28 % respectively. The *RMS* value of the fundamental component is 141 A. The reference load current and the demanded load current are shown in Fig. 6 where it can be seen an appropriate tracking. Below, in Fig. 7 they are displayed the reference current and injected current by the injector to recycle the energy. Finally, the current supply *RMS* value is shown in Fig. 8. The energy recovery ratio is over the 90%.

B. Unbalanced current

In this case, the electronic load with energy recycling is tested with an unbalanced current. The unbalanced ratios are I^-/I^+ = 22% and I^0/I^+ = 28,8%. The *RMS* value of the positive-sequence component is 115 A. The reference load current and the demanded current are shown in Fig.9. In Fig. 10, they are displayed the reference current and injected current by the injector. The energy recover ratio is similar to the one obtained in the first test.

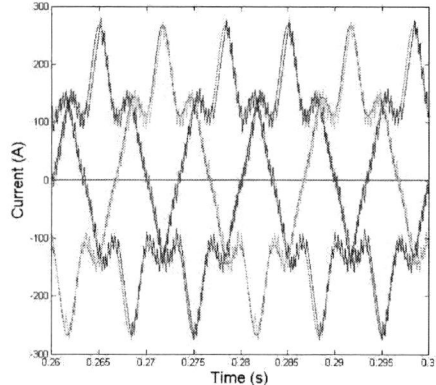

Fig. 6. Reference and demanded load current under distorted conditions (simulation case A).

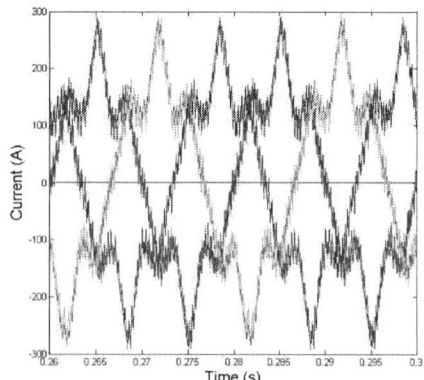

Fig. 7. Reference and injected injector current under distorted conditions (simulation case A).

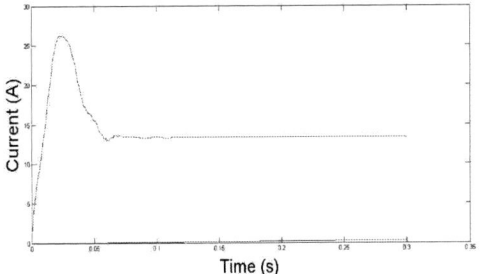

Fig. 8. Supply *RMS* current under distorted conditions (simulation case A).

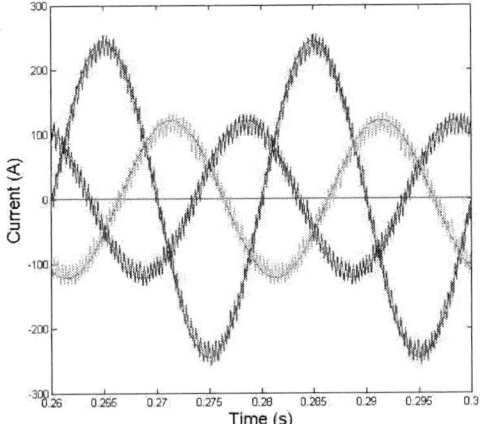

Fig. 9. Reference and demanded load current under unbalanced conditions (simulation case B).

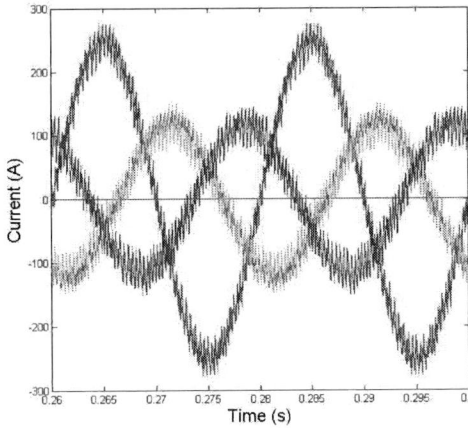

Fig. 10. Reference and injected injector current under unbalanced conditions (simulation case B).

VI. APPLICATION OF THE REGENERATIVE ELECTRONIC LOAD TO TEST POWER CONDITIONERS. SIMULATIONS RESULTS.

In this section an application of the electronic load to test shunt equipments is presented. A power conditioner connected in parallel with the regenerative load has been tested according to the scheme in Fig. 1.

The power conditioner topology is a 3 branches inverter with middle point DC bus (see Fig. 11). This device is used to correct the harmonics and unbalanced currents demanded from the grid by nonlinear loads and to compensate reactive power to improve the power factor.

In the power conditioner a Perfect Harmonic Cancelation (PHC) [13] control strategy has been implemented needing the positive sequence fundamental component of the PCC voltage. This value is provided by an ASRF.

The topologies and connection of the shunt power conditioner and regenerative load are presented in Fig. 11. Simulations have been conducted using *Matlab/Simulink*. The simulations parameters are summarized in Table II.

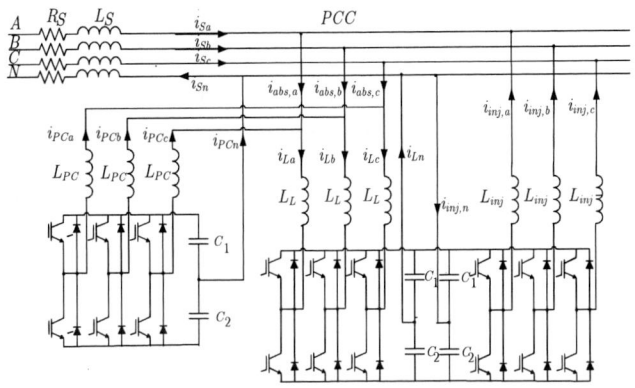

Fig. 11. Regenerative electronic load and shunt power conditioner topologies.

TABLE II
DEVICES PARAMETERS IN THE SIMULATION

U_n(V)	I_n(A)	S_n (kVA)	C_1 (mF)	C_2 (mF)
400	200	100	28,2	28,2
$U_{dc,ref\,(load)}$ (V)	$U_{dc,ref(PC)}$ (V)	$f_{s,\,max,load}$ (kHz)	$f_{s,max,inj}$ (kHz)	$f_{s,max,PC}$ (kHz)
800	800	10	20	20
L_L(mH)	L_{inj} (mH)	L_{PC} (mH)		
3	1,75	1,75		

The proper operation of the regenerative electronic load, when testing the performance of the power conditioner in case of a distorted and balanced nonlinear load is presented.

The harmonic content of the load current in this test is the same as that presented in section V, case A. The reference load and demanded load currents are shown in Fig. 12. In Fig. 13 the compensation current of the power conditioner is displayed. The current demanded from the grid, i_{abs} in Fig. 1 and Fig. 11, once the power conditioner is operating is presented in Fig. 14. As the power conditioner is responsible for the harmonic components of the load current, this current is sinusoidal and balanced, validating the good performance of the power conditioner. Finally, Fig. 15 shows the injector current in order to recycle the energy into the grid. This current is also sinusoidal and balanced due to the operation of the power conditioner.

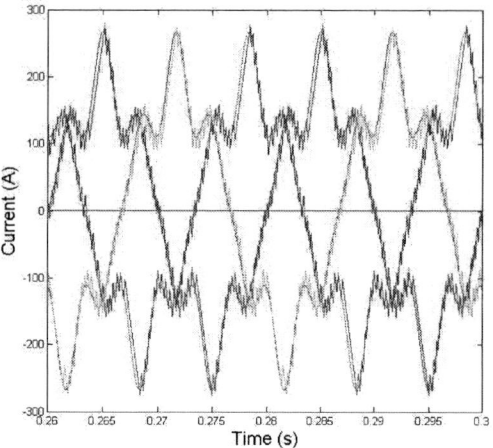

Fig. 12. Reference load current and demanded current to test the power conditioner operation (i_L).

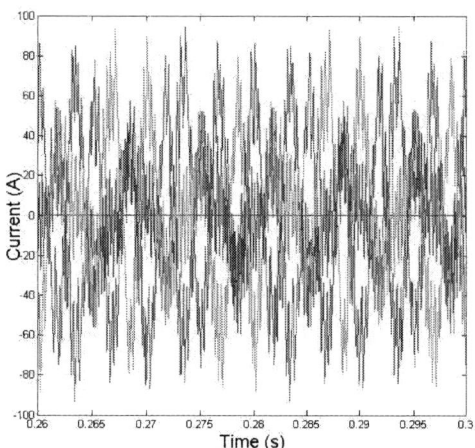

Fig. 13. Power conditioner compensation current (i_{PC}).

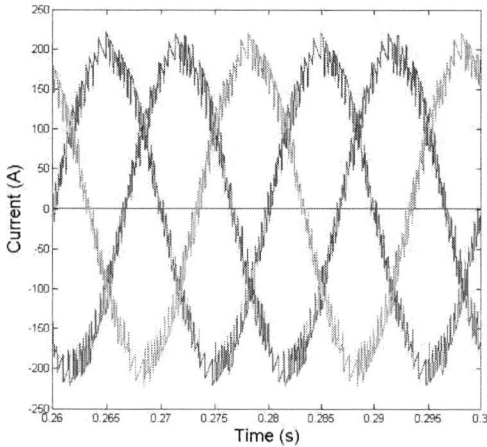

Fig. 14. Demanded current from the grid while the power conditioner operation test (i_{abs}).

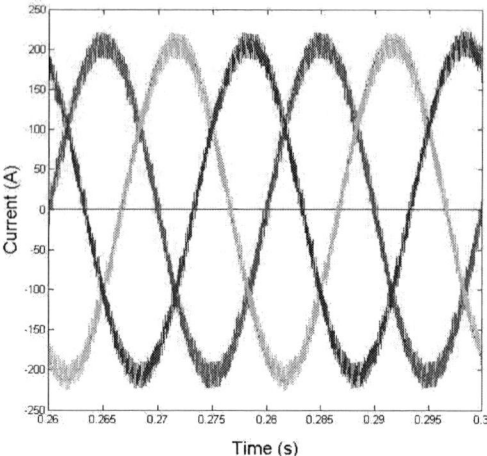

Fig. 15. Injected current into the grid by the regenerative load while the power conditioner operation test (i_{inj}).

VII. APPLICATION OF THE REGENERATIVE ELECTRONIC LOAD TO TEST POWER CONDITIONERS. EXPERIMENTAL RESULTS

The electronics converters used in the experimental test are 3 branches inverter by *Semikron* with hall effect sensors of current in each arm and voltage sensors to capture the middle point DC bus and PCC voltages.

Each inverter branch is connected to the grid with a filter inductor, L_L, L_{inj} and L_{PC} respectively, and the middle point of the DC bus is connected to the neutral conductor directly, following the topologies shown in Fig. 11.

XPC Target from *Matlab/Simulink* has been used as control platform. Three industrial PCs have been used as Target to run the presented models in real time, equipped with data acquisition cards (DAQ's), and one PC is used as Host to control the other PCs.

The experimental test has been conducted at low power level with the parameters shown in table 3. The programmed load current has a harmonic content equal as the presented in the simulation tests, but with a *RMS* fundamental component value reduced to 1,5 A.

Due to XPC Target control platform limitations, the switching frequency has been decreased to 4 kHz. In Fig. 16 the experimental prototypes used to carry out the test are shown and the experimental results are displayed in Fig. 17. In this figure they are presented, from top to bottom, the source current, i_{abs}, the injected current, i_{inj}, the programmable load current, i_L and the power conditioner current, i_{PC}. One can notice that the source current and injected current are sinusoidal without harmonics, since the harmonic components of the load current are delivered by the power conditioner. Also, the source current is in phase with the positive-sequence fundamental component of the source voltage, attaining an unity displacement power factor.

978-1-4244-8806-3/11 $26.00 © 2011 IEEE   182

TABLE III
DEVICES PARAMETERS IN THE EXPERIMENTAL TEST

U_n(V)	I_n(A)	S_n (VA)	C_1 (mF)	C_2 (mF)
30	1,5	135	28,2	28,2
$U_{dc,ref\,(load)}$ (V)	$U_{dc,ref\,(PC)}$ (V)	$f_{s,\,max,load}$ (kHz)	$f_{s,max,inj}$ (kHz)	$f_{s,max,PC}$ (kHz)
120	120	4	4	4
L_{load}(mH)	L_{inj} (mH)	L_{PC} (mH)		
3	1,75	1,75		

Fig. 16. Experimental prototypes: power conditioner (highlighted in red) and regenerative electronic load (highlighted in blue).

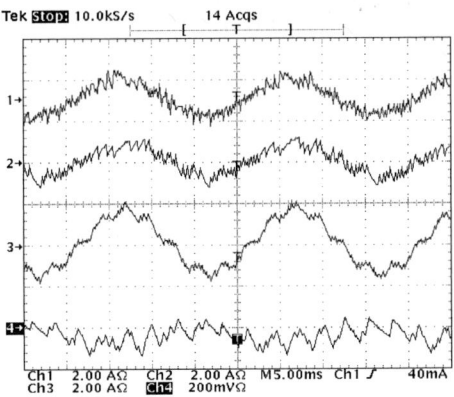

Fig. 17. Experimental results. From top to bottom: source current, i_{abs}, injected current, i_{inj}, programmable load current, i_L and power conditioner current, i_{PC}.

VIII. CONCLUSIONS

This paper has presented the design and control of a three-phase regenerative programmable electronic load to be used to test shunt electronic devices for high power applications. Simulation results of the electronic load under distorted and unbalanced conditions at high power levels are shown. The application of the regenerative load to test a power conditioner is verified by simulation at high power levels and experimentally with a small-scale laboratory prototype.

ACKNOLEDGMENTS

The authors gratefully acknowledge the financial support provided by the Junta de Extremadura (Regional Government), Spain, under project PDT08A046.

REFERENCES

[1] M.T. Tsai. C. Tsai. "Energy Recycling for Electrical AC Power Source Burn-In Test". IEEE Transactions on Industrial Electronics, Vol. 47, No. 4, pp 974-976. 2000.
[2] Ju- Won Baek; Myung- Hyo Ryoo; Jong Hyun Kim; Jih- Sheng Lai. "50 kVA Regenerative Active Load for power test system". European Conference on Power Electronics and Applications. pp 1-8. 2007.
[3] Z. Rong; C. Jian; "Repetitive control algorithms for a real-time dynamic electronic load simulator". Conference and Exposition on Applied Power Electronics APEC'06. Twenty-First Annual IEEE. 2006.
[4] M. Kazerani "A high-performance controllable AC load". 34th Annual Conference on Industrial electronics. IECON. pp 442.447. 2008.
[5] Y. Srinavasa Rao; M. Chandorkar, "Rapid Prototyping Tool for Electrical Load Emulaion using Power Electronic Converters". Conference on Thermal Phenomena in Electronic Systems, 1996. pp 352- 358. 1996.
[6] H. Akagi, Y. Kanazawa, A. Nabae. " Instantaneous reactive power compensators comprising switching devices without energy storage components". IEEE Transaction on Industrial Electronic. Vol. IA-20, n° 22. pp.625. 1984.
[7] M.Y. Chang, Jiann-Yow Lin, Y. Tzou. "DSP- based fully digital control of a AC/DC converter with a nonlinear digital current mode control". IEEE PESC Conference. pp 721-726. 1994.
[8] W. Shaokun, H. Zhenyi, P. Chuanbiao. " A repetitive control strategy of AC electronic load with energy recycling". International Technology and Innovation Conference ITIC. pp 1-4. 2009.
[9] Z. Kai, Kang Yong; X. Jian; et al. "Direct repetitive control of SPWM inverter for UPS purpose. IEEE Transactions on Power Electronics. pp 784-792. 2003.
[10] J. Zhao, S. Pan; X. Wang. " High Power Energy Feedback AC Electronic Load and its Application in Power System Dynamic Physical Simulation". Conference on Industry Application. 42 nd IAS Annual Meeting. pp 2303-2310. 2007.
[11] J. Baek, M. H. Ryoo, J. Hyun, K. Jih-Sheng Lai. " 50 kVA Regenerative Active load for power test system". European Conference on Power Electronics and Applications. pp 1-8. 2007.
[12] M.I. Milanes-Montero; E. Romero-Cadaval; A. Rico; V. Miñambres-Marcos; F. Barrero-González. "Novel method for synchronization to disturbed three-phase and single-phase systems". IEEE International Symposium on Industrial Electronics. Junio, 2007.
[13] M.I Milanés Montero, E. Romero Cadaval, F. Barrero Cadaval. "Comparison of Control Strategies for Shunt Active Power Filters in Three-Phase Four- Wire Systems". IEEE Transactions on Power Electronics. Vol.22, pp. 229-236. Enero 2007.

High Power, Medium Voltage, Modular Active Power Filtering System

Daniel Wojciechowski[1], Ryszard Strzelecki[2]

[1, 2] Gdynia Maritime University, Faculty of Marine Electrical Engineering, 83 Morska Str., 81-225 Gdynia, Poland

[1] dwojc@am.gdynia.pl, [2] rstrzele@am.gdynia.pl

Abstract-**The paper presents the concept of 2 MVA, medium voltage active power filtering system. The system is modular, thus provides full scalability. All the power modules of the system share a special, unified topology of 3rd order LCL passive circuit, which together with predictive control provides very high filtering effectiveness and, at the same time, zero current ripples. This paper contains preliminary simulations. The full power, MV system is currently being applied.**

I. INTRODUCTION

Shunt active power filters (APFs) are nowadays an industry acclaimed solution to limit current harmonics in a distribution networks. Among their strong points are high effectiveness of compensation, adaptability to changes on both the load side and the supply side, immunity to resonances and overloads, some capability to act toward stabilization of the network, and usually programmable, flexible modes of compensation. The week points are the price (although getting lower and lower), and in most solutions PWM related current ripples. The latter drawback is very severe, and limits possible applications of APF to networks with local loads which are very immune to high frequency disturbances of the grid voltage.

APF comprises voltage source inverter (VSI) connected to the grid via passive circuit. For given both DC voltage of VSI and switching frequency the level of current ripples depends only on characteristics of this passive circuit. Usually, simple inductor is used as a connecting passive circuit. Although its simplicity and ease to implement the control this solution provides very poor attenuation of current ripples. To ensure good attenuation of these ripples and, at the same time, low attenuation of harmonics within the APF compensation bandwidth it is necessary to apply 3rd order LCL connecting circuit [1].

The control of APF with LCL circuit is challenging, because of necessity to overcome the resonance problem and, at the same time, to provide high current control dynamics. In the literature there are presented control systems for VSI with LCL circuit mainly for active rectifier application [2]-[5]. These control systems most often utilize various resonance damping methods [2], [4] which, unfortunately, decrease current control dynamics or utilize a low dynamics current controllers [5]. There is also a group of control methods with passive or partly passive resonance damping, which deteriorate frequency response of LCL circuit and introduce additional losses [3], [6]. These methods are unstable or provide unacceptable low control dynamics for APF application. In [1] authors introduced experimentally tested algorithm of predictive control for APF with LCL circuit, which provides dead beat dynamics and full, active only suppression of resonance.

In LCL based APF it is very important to design properly the connecting circuit. Some design considerations are given in [8]. Among the most important parameters is the LCL resonance frequency. It should be set to provide high enough pass-band limit, usually not less than 2 kHz. In consequence, PWM switching frequency of VSI should be not less than 8 kHz. This requirement limits the maximum rated power of the single power module of APF due to semiconductor switching loses. Therefore, there are two possibilities to design a high power APF system:

- several power modules operating in parallel,
- utilization of multilevel VSIs.

It is also possible to use both solutions at the same time.

Parallel operation of the LCL based APF power modules can be realized by applying a separate LCL circuit for each module [1]. This method ensures fully independent operation of all the modules, but is not optimal from the economical point of view. This paper introduces the novel, joined topology of LCL circuit for APF system with power modules operating in parallel. This topology utilizes a single grid side inductor for the whole APF system, thus it substantially reduces the overall cost of the application.

This paper contains simulations of the 6.3 kV, 2 MVA system with two 1 MVA power modules operating in parallel with joined LCL topology. Power modules are based on NPC multilevel VSIs. The 2 MVA APF system, operating in the mine hoist drive application, is presently being applied.

II. DESCRIPTION OF THE SYSTEM

Fig. 1 shows general block diagram of the APF system with power modules operating in parallel. In Fig. 1a there is presented system with independent LCL circuit, whereas in Fig. 1b the system with proposed in the paper, joined LCL circuit. The joined LCL circuit is clearly advantageous from the economical point of view, and is especially advantageous for medium voltage applications, where the inductor L_2 can be replaced by a single MV/LV transformer with appropriate leakage inductance.

978-1-4244-8806-3/11 $26.00 © 2011 IEEE

a)

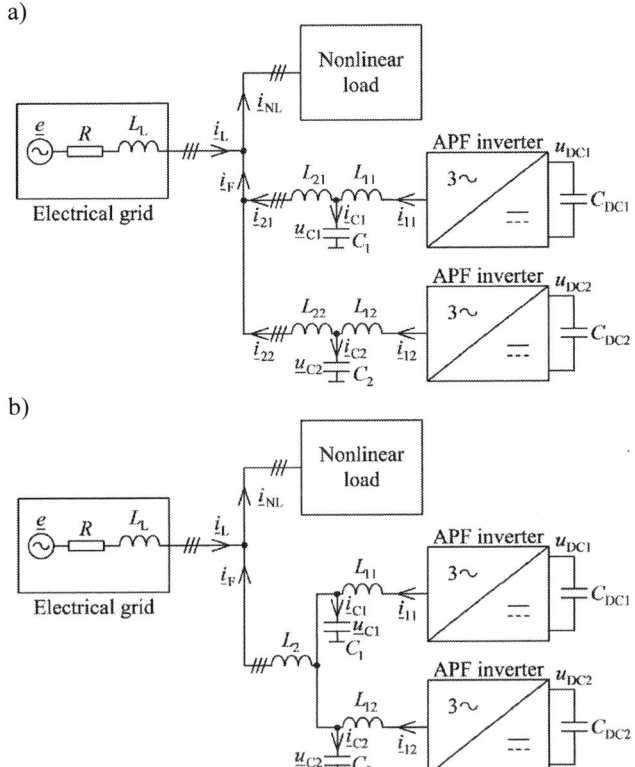

Fig. 1. APF system with power modules operating in parallel, with use of independent (a) and joined (b) LCL circuit.

A. LCL circuit design considerations

The most important factor influencing effectiveness of current compensation realized by APF system is the current control dynamics. There are two kinds of factors which define this dynamics: the hardware and the software (control). The hardware factors are:

- topology and parameters (characteristics) of the connecting circuit,

- voltage on the DC side of VSI u_{DC}.

Hardware factors define the maximum capability of current shaping dynamics, whereas the control task is to make the most of it. Moreover, the connecting circuit has to suppress current ripples and influences the level of ripples conducted by VSI's semiconductors. The above reasons justify a very high importance of APF connecting circuit.

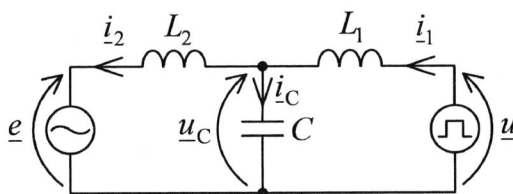

Fig. 2. Equivalent circuit of LCL-based APF.

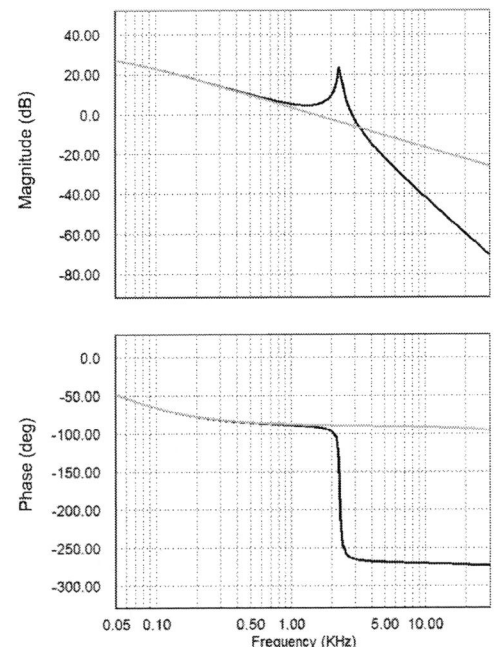

Fig. 3. Frequency response of first-order L circuit (grey line) and third-order LCL circuit (black line).

The joined LCL filter from Fig. 1b can be analyzed by using simplified model shown in Fig. 2, where:

$$C = \sum_n C_n , \quad L_1 = \left[\sum_n \left(L_{1n} \right)^{-1} \right]^{-1} , \qquad (1)$$

and:

$$\underline{i}_1 = \sum_n \underline{i}_{1n} , \quad \underline{u}_C = \underline{u}_{Cn} . \qquad (2)$$

The LCL circuit is described by the equations:

$$L_1 \frac{d\underline{i}_1}{dt} = \underline{u} - \underline{u}_C , \quad C \frac{d\underline{u}_C}{dt} = \underline{i}_1 - \underline{i}_2 , \quad L_2 \frac{d\underline{i}_2}{dt} = \underline{u}_C - \underline{e} . \qquad (3)$$

Fig. 3 compares characteristics of connecting circuits: LCL and L. Parameters of LCL has been chosen to equalize the inductances of both circuits, that is $L = L_1 + L_2$. It is clearly seen, that characteristics of both circuits are identical up to almost resonance frequency of LCL. In consequence, the influence of LCL connecting circuit on current control dynamics capability of the APF can be analyzed by using model with simple inductor, that is the model from Fig. 2, but without capacitor C. For such a model it can be easily derived, that the hardware limits (upper and lower) of the current dynamics region are defined by the following equations:

$$\left.\frac{\mathrm{d}i(t)}{\mathrm{d}t}\right|_{\mathrm{APF,upper}} = \frac{\sqrt{2}U_{\mathrm{f,RMS}}}{L_1+L_2}\sin\left(\omega t+\alpha\right)+\frac{\sqrt{3}}{3}\frac{U_{\mathrm{DC}}}{L_1+L_2} \qquad (4)$$

$$\left.\frac{\mathrm{d}i(t)}{\mathrm{d}t}\right|_{\mathrm{APF,lower}} = \frac{\sqrt{2}U_{\mathrm{f,RMS}}}{L_1+L_2}\sin\left(\omega t+\alpha\right)-\frac{\sqrt{3}}{3}\frac{U_{\mathrm{DC}}}{L_1+L_2} \qquad (5)$$

where α is the initial angle of grid voltage in particular phase: A, B, or C, defined as the value for $t=0$. The limit functions vary between two extremes, which absolute values can be derived as:

$$\left.\left|\frac{\mathrm{d}i}{\mathrm{d}t}\right|\right|_{\mathrm{limit,\,max}} = \left|\frac{1}{L_1+L_2}\left(\frac{\sqrt{3}}{3}U_{\mathrm{DC}}+\sqrt{2}U_{\mathrm{f,RMS}}\right)\right|, \qquad (6)$$

$$\left.\left|\frac{\mathrm{d}i}{\mathrm{d}t}\right|\right|_{\mathrm{limit,\,min}} = \left|\frac{1}{L_1+L_2}\left(\frac{\sqrt{3}}{3}U_{\mathrm{DC}}-\sqrt{2}U_{\mathrm{f,RMS}}\right)\right|. \qquad (7)$$

The hardware limits of APF current dynamics are shown in Fig. 4. The lower the value L_1+L_2 the broader the current dynamics limits. To achieve capability of APF to fully compensate the current of nonlinear load it is necessary to set the L_1+L_2 value of LCL circuit to achieve limits (defined by (4) and (5)) that exceed transients of derivative of desired compensating current (i.e.: unwanted components of nonlinear load current). On the other side, inductances have to be not too small, because it would lead (for given LCL resonance frequency) to high level of current ripples conducted by VSI, high reactive current supplying LCL's capacitor from the VSI side and, moreover, high sensitivity to both feedback errors and non-ideal accuracy of PWM.

The ratio between L_1 and L_2 has to give the compromise between the level of current ripples and reactive current supplied form VSI, and attenuation of current ripples at the grid side. Usually, L_1/L_2 ratio from the range $\langle 1, 2\rangle$ gives the best results.

For given inductors, capacitor C of the LCL defines a very important parameter – the resonance frequency. This resonance frequency is defined as follows:

$$f_{\mathrm{res,LCL}} = \frac{1}{2\pi}\sqrt{\frac{L_1+L_2}{L_1\cdot L_2\cdot C}}, \qquad (8)$$

and sets the upper limit of APF compensation bandwidth, by separating the pass-band and stop-band of the LCL circuit (Fig. 3). The PWM switching frequency has to be at least about 4 times greater than LCL resonance frequency because of stability reasons [8].

According to (1) and (8), for given parameters of joined LCL circuit its resultant resonance frequency depends on the number of power modules being in operation in a particular instant of time. Passive circuits of non operating APF power modules have to be disconnected from the grid, because they can cause an uncontrollable resonances. There are two possible methods of disconnection: including capacitor C_{n}, (see Fig. 1b) or leaving this capacitor connected. Fig. 5 shows the diagram with variations of resultant resonance frequency of connected parts of joined LCL circuit as a function of number of power modules being connected to the grid. There are presented relative frequencies for both methods mentioned above, and for the system with 8 power modules. For the case of disconnecting both L_{1n} and C_{n} the resonant frequency raises with decreasing the number of power modules being in operation, whereas for the case of disconnecting only L_{1n} the resonance frequency is getting lower. This means, the for the latter case the overall compensation bandwidth of APF system is getting narrower as the power modules are switched off and disconnected. This drawback can be eliminated by using the former method, but it requires to leave the sufficient margin between resonance frequency and PWM switching frequency. The best choice depends on the total number of power modules of APF system. For the two power modules it seems that better is the former method, because for half a modules in operation (1 of 2 in this case, 4/8 in Fig. 5)) there is a slight increase of resonant frequency (no pass-band deterioration, only slight decrease of stability margin). For the higher number of power

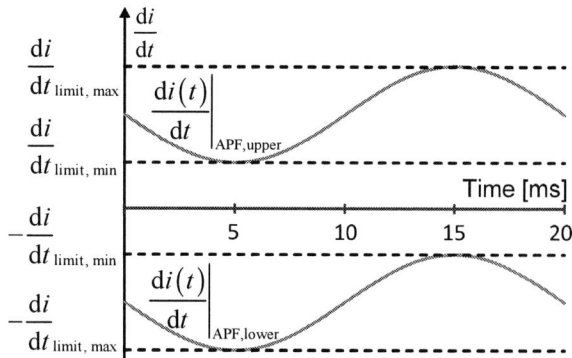

Fig. 4. Hardware limits of APF current control dynamics.

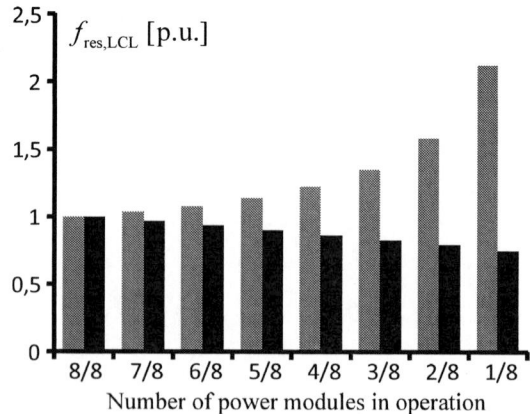

Fig. 5. Relative changes of resonant frequency of joined LCL circuit depending on the number of APF power in operation. Disconnection of both L_{1n} and C_{n} (grey bars), and L_{1n} only (black bars).

modules in the system, the better method is the latter one, because otherwise, the increase of resonance frequency would be too high. These rules are not strict, and may depend on requirements for the particular application of APF system.

B. The control system

The APF system has been designed to obtain maximum independence of the power modules. However, the modules share L_2 inductor of the joined LCL circuit, and there is necessity to exchange data between them. The control of the single module is in general the same as described in [1], and is presented in Fig. 6. Every power module uses independent DSP controller. There are several signals that has to be shared among the power modules: grid emf $\underline{e}$, nonlinear load current $\underline{i}_{\mathrm{NL}}$, current $\underline{i}_2$ of shared inductor L_2 and, moreover, PWM synchronization clock, and some configuration and diagnostics data. It is important, that all the real time signals which have to be shared are the feedback signals, thus there is no necessity to apply wide bandwidth real time transmission between controllers of each power modules. The only shared real time signal is a simple PWM synchronization clock.

The control system comprises predictive reference current calculator, model based predictive current controller, and grid voltage predictor to achieve zero-error current tracking within dynamics boundaries limited only by the APF hardware. The set voltage is realized by using Space Vector Pulse Width Modulation (SVPWM) for NPC multilevel inverter with voltage balancing and minimized IGBT switching. The indirect current control technique (in opposite to various direct control methods) with separate blocks of current controller and PWM is flexible in term of using any of the VSI topology, in this case the NPC multilevel topology for medium voltage (MV) application. Although the control is capable of grid voltage sensorless operation, in the high power application it is necessary to measure the voltage because of diagnostics requirements.

In the proposed control system the model based predictive current controller for LCL-based APF is implemented. The controller provides full utilization of APF operating region and performs a full control signals prediction, thus it ensures maximum dynamics of current control which can be achieved for particular circuits parameters of APF, that is: grid voltage e, DC voltage u_{DC}, inductances L_1, L_2 and capacitance C.

The equations of the controller are based on LCL filter state equations (3), and have the following form:

$$\underline{u}_{\mathrm{c}}\left(k+1\,|\,k+2\right)=\frac{L_2}{T_{\mathrm{S}}}\Big[\underline{i}_2^{\mathrm{ref}}\left(k+2\right)-\underline{i}_2^{\mathrm{ref}}\left(k+1\right)\Big]+\tilde{\underline{e}}\left(k+1\,|\,k+2\right), \quad (9)$$

$$\hat{\underline{i}}_1\left(k+1\right)=\frac{C}{T_{\mathrm{S}}}\Big[\underline{u}_{\mathrm{c}}\left(k+1\,|\,k+2\right)-\hat{\underline{u}}_{\mathrm{c}}\left(k\,|\,k+1\right)\Big]+\underline{i}_2^{\mathrm{ref}}\left(k+1\right), \quad (10)$$

$$\underline{u}^{\mathrm{set}}\left(k\,|\,k+1\right)=\frac{L_1}{T}\Big[\hat{\underline{i}}_1\left(k+1\right)-\hat{\underline{i}}_1\left(k\right)\Big]+\hat{\underline{u}}_{\mathrm{c}}\left(k\,|\,k+1\right). \quad (11)$$

The currents and voltages denoted with hat are the predicted values, and are calculated based on equations:

$$\hat{\underline{u}}_{\mathrm{c}}\left(k-1\,|\,k\right)=\frac{T_{\mathrm{S}}}{C}\Big[\underline{i}_1\left(k-1\right)-\underline{i}_2\left(k-1\right)\Big]+\underline{u}_{\mathrm{c}}\left(k-2\,|\,k-1\right), \quad (12)$$

$$\hat{\underline{i}}_1\left(k\right)=\frac{T_{\mathrm{S}}}{L_1}\Big[\underline{u}^{\mathrm{set}}\left(k-1\,|\,k\right)-\hat{\underline{u}}_{\mathrm{c}}\left(k-1\,|\,k\right)\Big]+\underline{i}_1\left(k-1\right), \quad (13)$$

$$\hat{\underline{i}}_2\left(k\right)=\frac{T_{\mathrm{S}}}{L_2}\Big[\hat{\underline{u}}_{\mathrm{c}}\left(k-1\,|\,k\right)-\tilde{\underline{e}}\left(k-1\,|\,k\right)\Big]+\underline{i}_2\left(k-1\right), \quad (14)$$

$$\hat{\underline{u}}_{\mathrm{c}}\left(k\,|\,k+1\right)=\frac{T_{\mathrm{S}}}{C}\Big[\hat{\underline{i}}_1\left(k\right)-\hat{\underline{i}}_2\left(k\right)\Big]+\hat{\underline{u}}_{\mathrm{c}}\left(k-1\,|\,k\right). \quad (15)$$

where $(k–1)$ denotes the instant of the last measurements.

The predicted values of grid emf which are present in (9) and (14), and denoted with tilde, are calculated by using predictor of distorted and unbalance voltage described in [1].

III. SIMULATIONS OF THE HIGH POWER APF SYSTEM

The APF system described in the paper is currently being applied for the hoist drive in the coal mine in Poland. Fig. 7 shows the simulation model of the system. Simulations have been realized in PSIM 8.0 software. The APF system is connected to the 6.3 kV medium voltage line via Yy0 transformer of ratio 6.3/1.1 kV. The system has the total rated power of 2MVA, and it is compound of two power modules, 1 MVA each, operating at 1.1kV terminal voltage. Power modules utilize 3-level NPC inverters with water cooling. Both modules share proposed in the paper, joined LCL connecting circuit. Transformer acts as the L_2 inductor by using its leakage inductance. The APF system operates together with SVC and doesn't compensate for the reactive power of the compensated thyristor rectifier.

The performance of the proposed control system has been investigated with simulations. Symbols are consistent with the ones used in the previous text and figures.

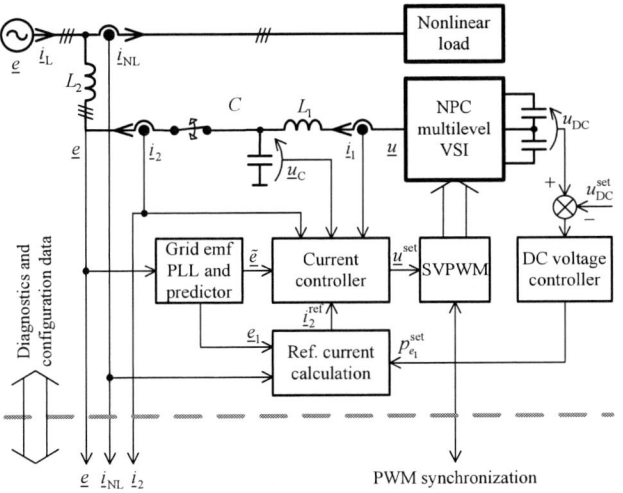

Fig. 6. The control of the single power module of a high power APF system.

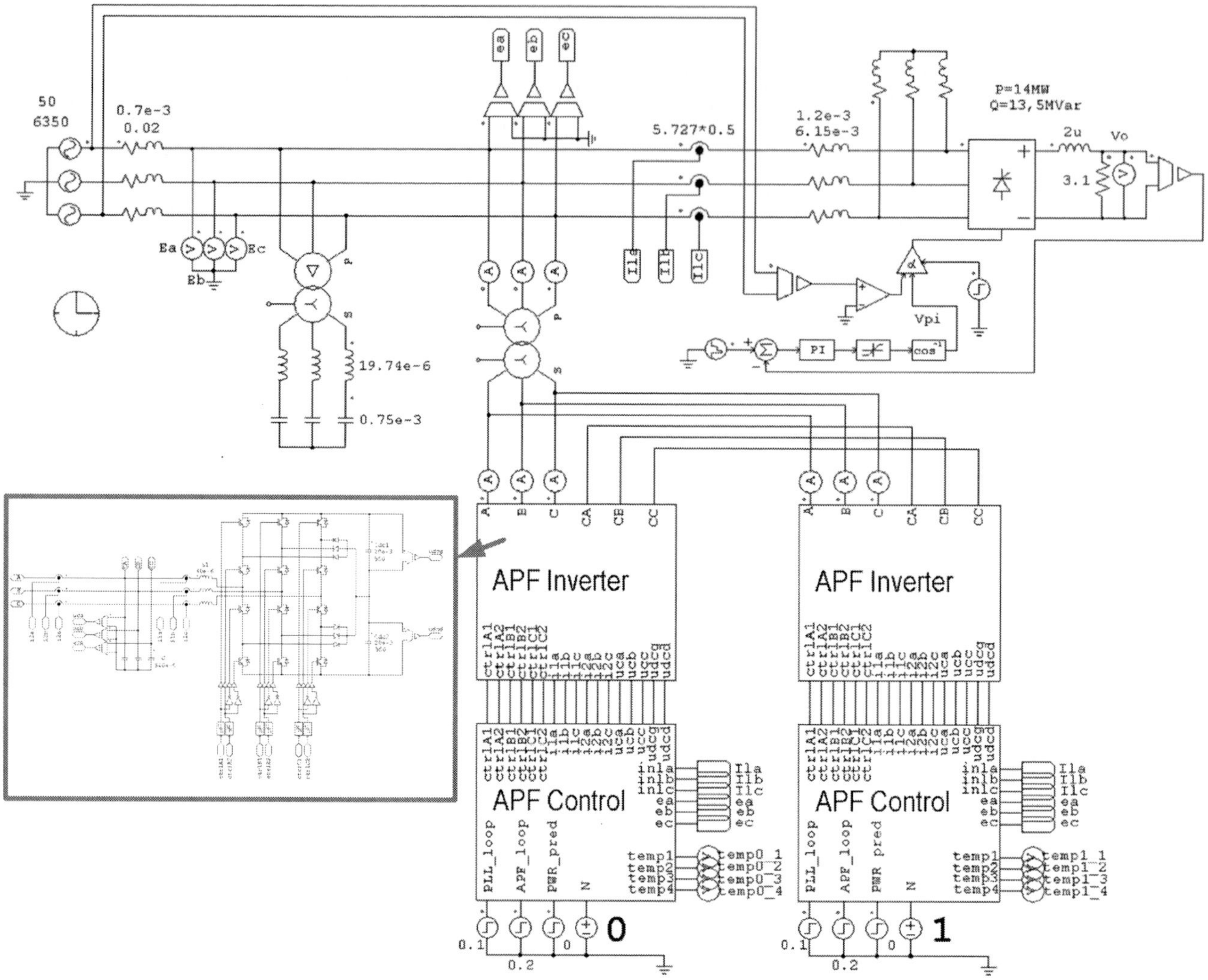

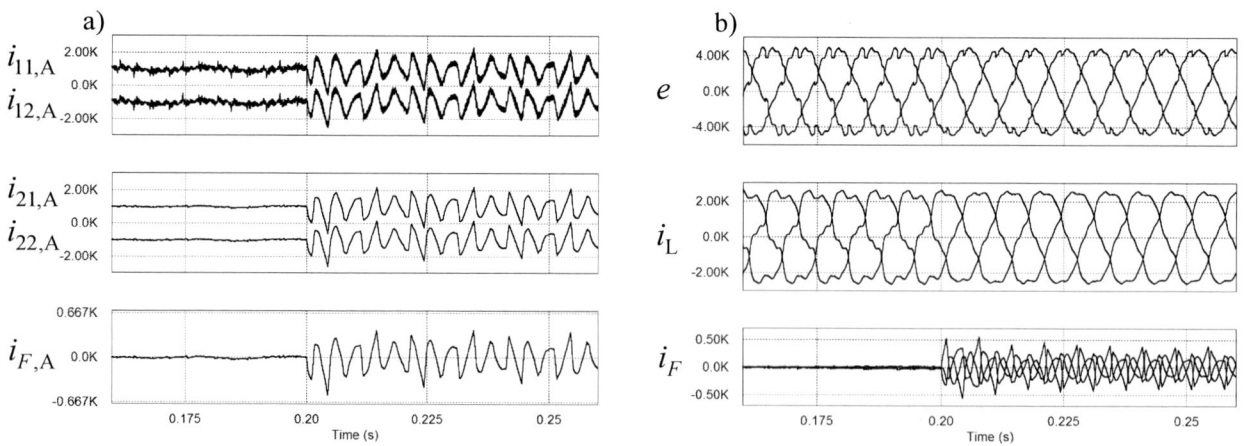

Fig. 7. Simulation model of the APF system with two modules operating in parallel, utilizing the joined LCL circuit with transformer instead of inductor L_2.

Fig. 8. Simulation; turning on the compensation. Power sharing among the modules; ripple cancelation by using LCL (a). Performance of the system (b).

Fig. 8a shows the phase currents i_1 of inductors L_1 for two power modules (with added offsets to demonstrate both), currents i_2 before the connecting point of modules, and current i_F of the whole APF system on the medium voltage side. Transients prove an excellent power sharing between the APF power modules in both steady and transient state. This figure also demonstrates the effectiveness of current ripple attenuation by using an LCL filter. PWM switching related harmonic currents are present in the currents i_1, but the output currents i_2 are practically free of these ripples.

Figure 8b presents grid voltage e, currents drained from the grid i_L, and the total output currents of the APF system i_F during activating of compensation. Because of the economical reasons, the rated power of the system is designed as a trade-off between the compensation effectiveness and the cost, thus it has not to compensate the harmonics completely, but only to the levels below the designed limits. The APF which compensate the harmonics completely would be of substantially higher power than 2 MVA. The system compensates the supply current from THD_i=20% to 8, and the grid voltage (indirectly) from THD_u=12% to 6%. It is important to note, that the APF system has to operate properly in presence of dip commutation notches and other voltage distortions caused by the compensated nonlinear load.

IV. CONCLUSION

Presented results of simulations confirm excellent active filtering performance of high power APF system with the proposed, predictive control, and joined LCL circuit. Compensating currents do not contain PWM-related current ripples. The resonance of LCL is completely suppressed by the current controller, and the power is shared equally between the power modules.

The 2 MVA APF system is currently being applied for the thyristor rectifier fed hoist drive in the coal mine in Poland.

APPENDIX

TABLE I
PARAMETERS OF ELECTRICAL CIRCUITS AND CONTROL SYSTEM

Quantity	Value
Total rated power	2 MVA
Line to line supply voltage	3 x 6.3 kV
Terminal voltage of the VSIs	3 x 1.1 kV
Line voltage frequency	50 Hz
DC voltage	1800 V
Inductances L_1 in each power module	40 µH
Leakage ind. of transformer - LV side	26 µH
Capacitances C in each power module	340 µF
Series capacitances on DC side	2 x 20mH
PWM carrier frequency	8 kHz
Sampling frequency	16 kHz

ACKNOWLEDGMENT

The research work has been partially realized within grant nr N N N510 326237.

REFERENCES

[1] D. Wojciechowski, R. Strzelecki, "Predictive Control of Active Filter System with LCL Coupling Circuit," *IPEC 2010*, Sapporo, Japan, 2010

[2] E. Wu, P.W. Lehn, "Digital Current Control of a Voltage Source Converter With Active Damping of LCL Resonance," *IEEE Trans. Power Electron.*, vol. 21, no. 5, pp. 1364-1373, September 2006.

[3] M. Liserre, F. Blaabjerg, and S. Hansen, "Design and Control of an LCL-Filter-Based Three-Phase Active Rectifier," *IEEE Trans. Ind. Appl.*, vol. 41, no. 5, pp. 1281-1291, September/October 2005.

[4] L.A. Serpa, S. Ponnaluri, P.M. Barbosa, J.W. Kolar, "A Modified Direct Power Control Strategy Allowing the Connection of Three-Phase Inverters to the Grid Through LCL Filters," *IEEE Trans. Ind. Appl.*, vol. 43, no. 5, pp. 1388-1400, September/October 2007.

[5] M. Malinowski and S. Bernet, "A Simple Voltage Sensorless Active Damping Scheme for Three-Phase PWM Converters With an LCL Filter," IEEE Trans. Ind. Electron., vol. 55, no. 4, pp. 1876-1880, April 2008.

[6] M. Routimo and H. Tuusa, "LCL Type Supply Filter for Active Power Filter – Comparison of an Active and a Passive Method for Resonance Damping," *PESC 2007*, pp. 2939-2945, Orlando, FL, 17-21 June 2007.

[7] Z. Qiu, J. Kong, and G. Chen, "A Novel Control Approach for LCL-based Shunt Active Power Filter with High Dynamic and Steady-State Performance," *PESC 2008*, pp. 3306-3310, Rhodes, 15-19 June 2008.

[8] G. Zeng, T. W. Rasmussen, L. Ma, R. Teodorescu, "Design and Control of LCL-Filter with Active Damping for Active Power Filter," *ISIE 2010*, Bari, Italy, 2010.

Novel and Simple Method for Power Electronics Compensator Design and Optimization

C. Martínez, V. Valdivia, J. Lourido, I. Quesada, A. Lázaro, A. Barrado

Power Electronics System Group (GSEP)

Universidad Carlos III de Madrid

Leganés, Madrid, Spain

cdmartin@ing.uc3m.es / http://gsep.uc3m.es

Abstract— **The development of power converters leads to design the control loop. In order to reduce developing times, it is very useful to have an initial design criterion that lead to a stable feedback loop. In this paper a new concept for power electronics compensator design and optimization has been developed. This concept provides the different combinations of crossover frequency and phase margin which lead to stable solutions.**

On the other hand, in order to optimize the performance of the control loop, other optimization features are also provided, such as the relationship between the phase margin and the settling time when the crossover frequency is placed below the power stage resonant frequency.

I. INTRODUCTION

A variety of synthesis techniques can be found in the scientific literature [1]-[5]. Most of these methods obtain a straightforward compensator synthesis from two numerical inputs, e.g. phase margin (PM) and crossover frequency (f_{cross}). Although, the zeroes and poles placement is direct, the selection of these two last parameters for the optimization of the loop overall performance is not always immediate.

One of the key issues when designing a control loop is the appropriate selection of the abovementioned f_{cross} and PM. In order to ease the first attempt when designing a control loop, the "solutions map" feature has been developed. It provides the PM that can be achieved for a determined f_{cross} assuming stability. The solutions map also provides the attenuation at the switching frequency to avoid high frequency instability. When designing a control loop there are two possibilities regarding the resonant frequency of a second order power stage; settling f_{cross} of the open loop transfer function (T) below or above this frequency. Therefore, it is important to remark that there are two relationship regarding the phase margin; PM vs. damping ratio and PM vs. settling time (T_S).

The definition of the solutions map is shown in section II. The designing zones provided by the solutions map are studied in section III. Another important feature that provides the solutions map is the capability of selection the most suitable compensator for a determined power stage. It will be shown in section IV. In section V some experimental results are shown in order to validate the correct design using the solutions map and the relationship exposed in section IV. Finally, section VI is dedicated to the implementation of the solutions map within a CAD tool to ease the designing of a control loop.

II. DEFINITION OF THE SOLUTIONS MAP

The solutions map provides a "safe operating area" for different combinations of f_{cross} and PM that lead to stable systems based on the converter power stage, sensor and type of compensator. Thus, this map minimizes the number of trial and error iterations associated to the optimization of the control loop.

The boundaries, which determine the valid area, represent the maximum and minimum phase margin that can be achieved for any kind of compensator. In terms of frequency, for a simple control loop, the range is delimited by the switching frequency (f_{sw}). In the case of the external loop within a double control loop design, this range is delimited by the crossover frequency of the inner loop. The simple integrator is a particular case of any compensator, therefore it provides the lower phase margin limit (PM_{min}) by adding 90 degrees to the phase of the open loop transfer function without compensator (power stage, sensor and modulator). PM_{min} is represented in expression (1).

$$PM_{min} = 90 + P(f) \qquad (1)$$

Where P(f) is the phase of the open loop transfer function without compensator.

The upper limit is given by the maximum phase boost provided by each kind of compensator. The solutions map explained along this work takes into account two possibilities of compensators; lead compensator (Type 2) and lead-lag compensator (Type 3).

Taking into account the K-factor method [1], for any f_{cross} and PM and fixing K to its maximum value (K_{max}), the upper limit of the solutions map for Type 3 compensator (PM_{max_T3}) and for a Type 2 compensator (PM_{max_T2}) can be obtained by the rearranging of the K-factor equations [1] as is shown in (2) and (3) respectively.

$$PM_{max\_T3} = 4 \cdot arctan\left(\sqrt{K_{max}}\right) - 90 + P(f) \qquad (2)$$

$$PM_{max\_T2} = 2 \cdot arctan\left(K_{max}\right) + P(f) \qquad (3)$$

Fig. 1 and Fig. 2 show the solutions map for a Type 3 and a Type 2 compensators respectively, for the case of a voltage mode control (VMC) Forward converter. In all cases, the lower boundary corresponds to a simple integrator. Within the valid area (white color between the two curves) each set of f_{cross} and PM corresponds to a stable solution.

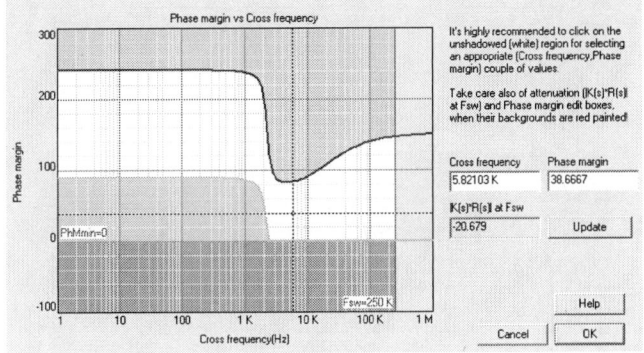

Fig. 1: Solutions map of a VMC Forward converter with a Type 3 compensator.

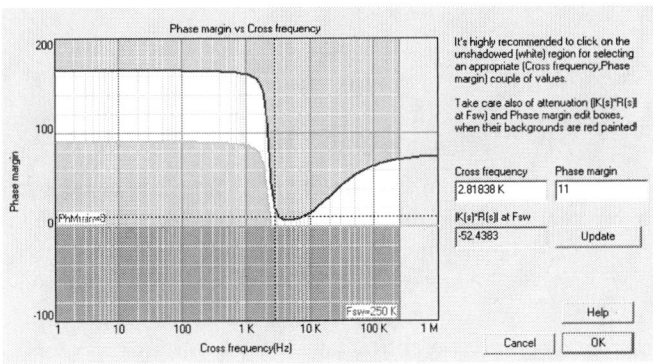

Fig. 2: Solutions map of a VMC Forward converter with a Type 2 compensator.

It is important to remark that the value of K_{max} needs to be delimited. If K_{max} is too large, poles and zeroes are too separated from each other, so the designed control loop will present less gain at low frequencies and less attenuation at f_{sw}. These two effects are undesirable when designing a control loop.

III. ZONES OF THE SOLUTIONS MAP

It has been considered a second order power stage in order to illustrate the two design possibilities mentioned in the introduction: f_{cross} set below the resonant frequency and f_{cross} set above it. In these conditions, over the solutions map can be distinguished two zones as it can be seen in Fig. 3:

- Zone 1: f_{cross} placed on the left of the power stage resonant frequency
- Zone 2: f_{cross} placed on the right of the power stage resonant frequency

A. Zone 2: f_{cross} set above resonant frequency

Zone 2 is the commonly designing zone used by designers and f_{cross} is set above the resonant frequency. Within this zone the design parameters are the damping factor as a function of the PM, the attenuation at the switching frequency and the crossover frequency. The crossover frequency should be as high as possible and the PM will be related to the damping factor.

The step response of the closed-loop operation is well known and it is related to the PM that can be achieved. In [5] the relationship between the damping factor and the PM is shown. Taking into account the abovementioned relationship and the attenuation at the switching frequency provided by the open loop transfer function, the designer is able to obtain an optimized feedback control loop.

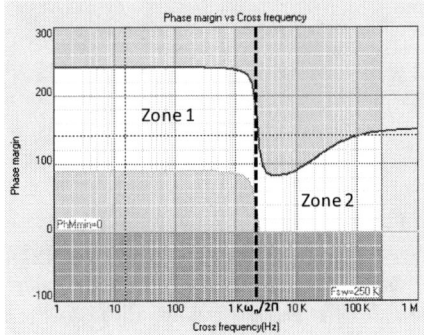

Fig. 3: Zones of the solutions map for a VMC Forward converter with a Type 3 compensator. Dotted line represents the undumped natural frequency of the power stage.

B. Zone 1: f_{cross} set below resonant frequency

Although dominant pole compensation (Zone 1) is not the proper designing solution, it is feasible from a stability point of view. In this zone the compensator behaves as an integrator and only PMs higher than 90° are achievable. Placing f_{cross} in this zone, synthesis methods tend to synthesize the compensator as an integrator. It should be noticed that there are two ways to implement the compensator as it can be observed in Fig. 4 [6], where both the analog implementation and the feedback block diagram are shown.

978-1-4244-8806-3/11 $26.00 © 2011 IEEE

The most common implementation of the compensator is shown in Fig. 4.b. This is the structure that is allowed to be implemented by most of commercial control ICs [7].

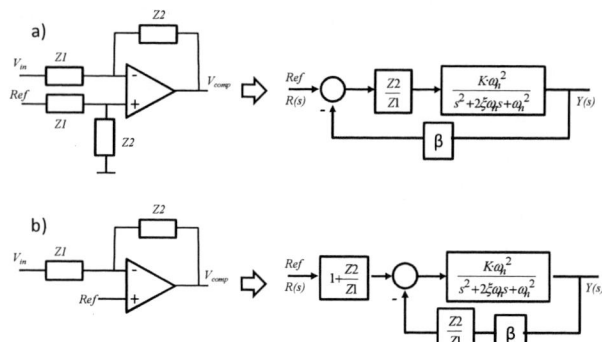

Fig. 4: Compensator implementation and feedback block diagrams.

Consider a dominant pole as the system compensator ($Z2/Z1=\omega i/s$). The closed loop operation of the system a) and b) in Fig. 4 are represented in equations (4) and (5) respectively.

$$\frac{Y(s)}{R(s)} = \frac{K\cdot\omega_n^2\cdot\omega_i}{(s^2+2\xi\omega_n s+\omega_n^2)\cdot(s+p)} \tag{4}$$

$$\frac{Y(s)}{R(s)} = \frac{K\cdot\omega_n^2\cdot(s+\omega_i)}{(s^2+2\xi\omega_n s+\omega_n^2)\cdot(s+p)} \tag{5}$$

Where K is the static gain of the power stage, ω_n is the natural frequency, ξ is the damping factor, ω_i is the dominant pole frequency and p is given in equation (6).

$$p = \beta\cdot K\cdot\omega_i \tag{6}$$

In Fig. 5 both closed loop frequency responses for equations (4) and (5) are shown for a generic second-order transfer function. As it can be seen, the frequency response for the closed loop system for eq. (5) is very close to the power stage transfer function (dashed trace in Fig. 5).

Taking into account the two ways to implement the compensator, there are two different reference step responses (small signal). The first one is for the closed loop system shown in (4) where the transient response corresponds to a third-order system (dashed trace in Fig. 6) reduced to a first order system due to the effect of the dominant pole (s+p).

On the other hand, the transient response for the equation (5) is formed by a first-order system plus a second-order system (solid trace in Fig. 6). In this case the effect of the zero is significant and it provokes a huge overshoot.

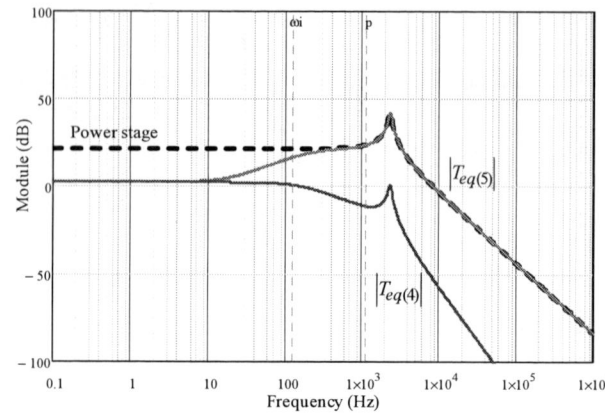

Fig. 5: Closed loop frequency responses for VMC Forward converter with dominant pole compensation. Red trace: closed loop transfer function from eq. (5). Blue trace: closed loop transfer function from eq. (4). Dashed trace: Power stage transfer function.

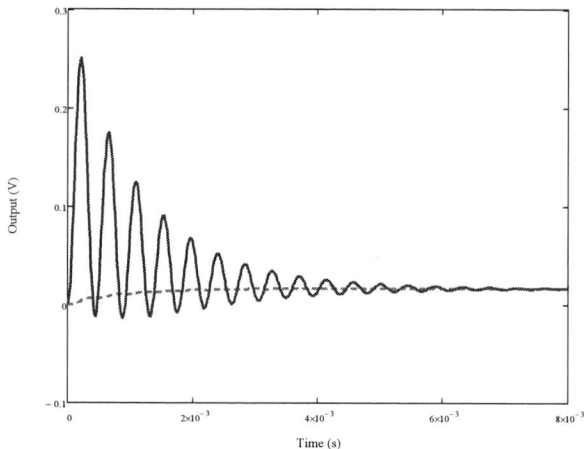

Fig. 6: Closed loop transient responses. Dashed trace: system that corresponds to eq. (4). Solid trace: system that corresponds to eq. (5).

IV. RELATIONSHIP BETWEEN TS AND PM

When designing in Zone 1 and for the compensator implementation shown in Fig. 4.b), it is not correct to take into account the damping factor due to the fact that this factor will be provided by the complex conjugate poles of the power stage. Therefore, under a reference voltage step, the system will be underdumped and it will oscillate at the natural frequency (ω_n) as it can be seen in Fig. 7. So, when designing in Zone 1 the settling time (T_S) is the factor to be considered (see Fig. 7) because it is higher than in Zone 2.

The theoretical development that relates the settling time and the phase margin for the closed loop system shown in equation (5) is very complex and so, this relationship has been carried out trough simulation.

978-1-4244-8806-3/11 $26.00 © 2011 IEEE

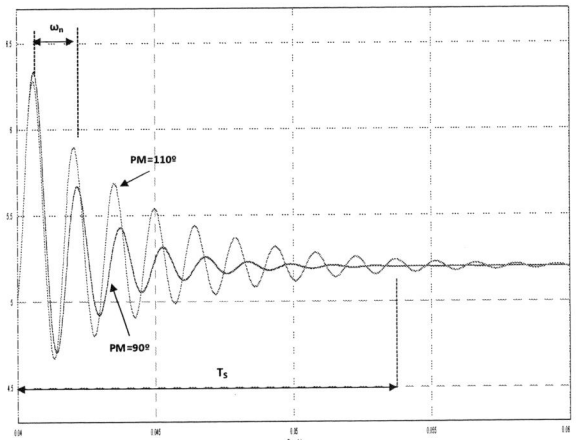

Fig. 7: Closed loop transient responses for VMC Forward converter with Type 3 compensator under different PM designs in Zone 1.

Fig. 8 shows the relationship between Ts and PM for the closed loop operation of a 2nd order power stage (i.e. VMC Forward converter with Type 3 compensator) under a reference voltage step for a Zone 1 design. As it can be seen in Fig. 8 this relationship is linear dependent with the PM (7).

$$\frac{\Delta T_s}{\Delta PM} = \frac{T_{s1} - T_{s2}}{PM1 - PM2} = K \tag{7}$$

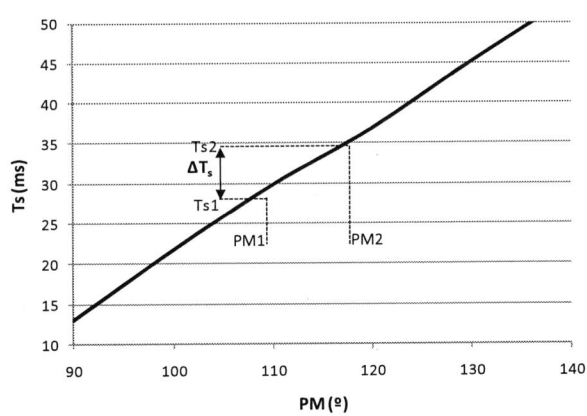

Fig. 8: Simulated relationship between Ts and PM for a 2nd order power stage controlled with a Type 3 compensator in Zone 1 (i.e VMC Forward converter).

V. EXPERIMENTAL RESULTS

This section aims to validate experimentally the proposed work along the paper. For this purpose a VMC Forward converter has been developed which transfer function is represented in Fig. 9. In order to implement the different compensators, only it is necessary to change resistors and capacitors of the control loop, so this task is very simple since the compensator has been synthesized.

This section has two purposes. The first one involves comparing the reference step transient response of two designs,

taking into account different PMs and zones of the solutions map in order to check the stability of both designs.

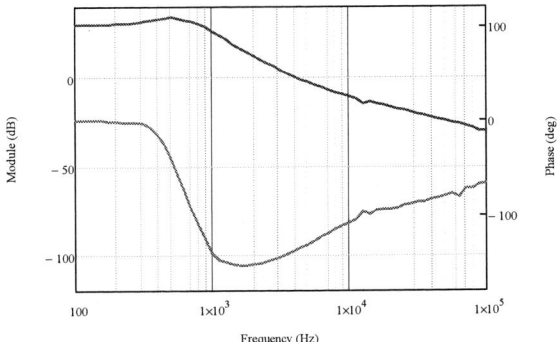

Fig. 9: Power stage transfer function bode plots.

Fig. 10 and Fig. 11 show the transient response under a reference voltage step for designs in Zone 2 and 1 respectively. The open loop characteristics for Zone 2 are $f_{cross}=1.15$ kHz with PM=56° and the characteristics for Zone 1 are $f_{cross}=100$ Hz with PM=90°.

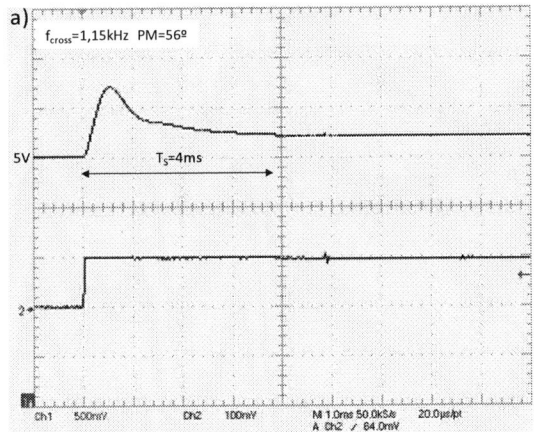

Fig. 10: Experimental results for Zone 2 design: $f_{cross}=1.15$kHZ, PM=56°.

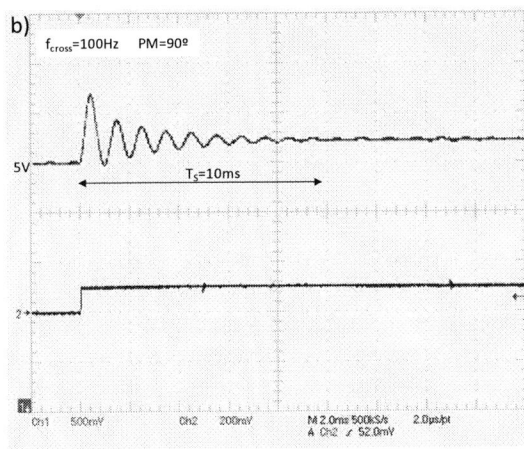

Fig. 11: Experimental result for Zone 1 design: $f_{cross}=100$Hz, PM=90°

The second objective to be achieved with the experimental results is to validate the relationship between PM and T_s. In this aspect, a sweep, respect the PM, has been carried out in order to trace the curve on Fig. 12, where the abovementioned relationship is shown. It should be noted that Fig. 11 only shows one result for one of the points in Fig. 12. However, it is important to remark that all the points in this curve has been obtained experimentally.

As it can be seen in Fig. 12, the relationship between Ts and PM obtained experimentally is linear dependent with the PM as it has been shown in section IV. Therefore, when designing below the resonant frequency and taking into account the attenuation of T at ω_n the designer is able to predict ΔT_s as a function of the selected PM.

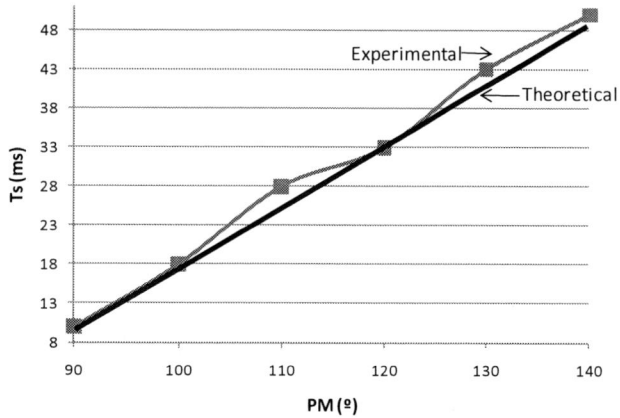

Fig. 12: Experimental relationship between Ts and PM for the VMC Forward converter with Type 3 compensator at Zone 1.

VI. IMPLEMENTATION WITHIN A CAD TOOL

In order to have a complete design tool, the solutions map has been introduced into a CAD tool for power electronics compensator design and optimization. In this tool all plots are real updated under any change made in the solutions map. Therefore, the designer is able to observe every parameter effect at real time, and so to select the optimized compensator design. The main characteristics included within the CAD tool are the following:

A. Avalaible Topologies and Operating Point

The CAD tool includes some conventional topologies such as Buck, Boost, Buck-Boost, Flyback and Forward converters. All of them present both voltage and current control loops. Additionally, a powerful characteristic is provided: the CAD tool is able to easily import any transfer function from another platforms. In this way, the designer is able to design the control loop of power stages with complex dynamics without deriving its small signal model. Thus, the developing time can be reduced significantly.

When designing a DC/DC converter, the static operating point of the converter is a key aspect to be taken into account by the designer. Thus, in the data entry window itself, the conduction mode and the inductance current values (maximum, minimum and average) are calculated from the input data. Hence, the operating point of the converter, which is a key aspect in control loop design, is calculated and displayed.

B. Sinthesys Methods

In the CAD tool have introduced three different synthesis methods such as the well known K-factor, the Kplus method and the manual method.

The Kplus method is an optimization method and it is based on the K-factor method [1]. Unlike K-factor method [1], f_{cross} is no longer the geometric mean of the zeroes and the poles frequencies. The Kplus method provides an additional design degree of freedom with respect to the conventional K-factor [1], since the new method places the zeroes frequency f_z a factor "α" below f_{cross} ($f_z = f_{cross}/\alpha$) and the poles a factor "β" above f_{cross} ($f_p = f_{cross} \cdot \beta$) (see Fig. 13).

The additional design degree of freedom that can be achieved with the Kplus method can be used as follows: if "α" is set to be lower than K, higher gain at low frequencies but less attenuation at f_{sw} are achieved (see Fig. 14). On the contrary, if "α" is set higher than K, the control loop has less gain at low frequency but more attenuation at f_{sw} (see Fig. 14). It should be remarked that the phase margin is the same in all cases.

The manual method places poles and zeroes independently from each other. It is used to optimize the results obtained from the K-factor [1] and Kplus methods or when these automatic methods do not provide a valid solution.

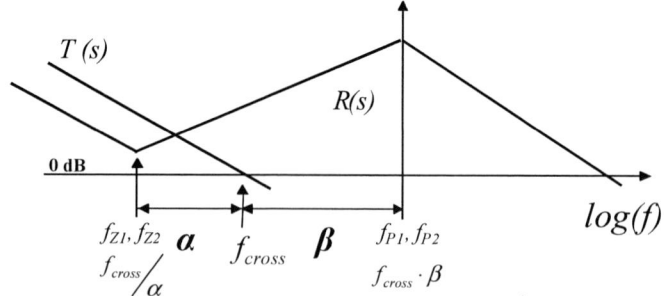

Fig. 13: Basis of the Kplus method: example of a Type 3 compensator design. R(s) is the regulator transfer function. T(s) is the open loop transfer function. Both f_{Z1} and f_{Z2} are the designed zeroes frequency and f_{P1}, f_{P2} are the designed poles frequency once zeroes have been fixed.

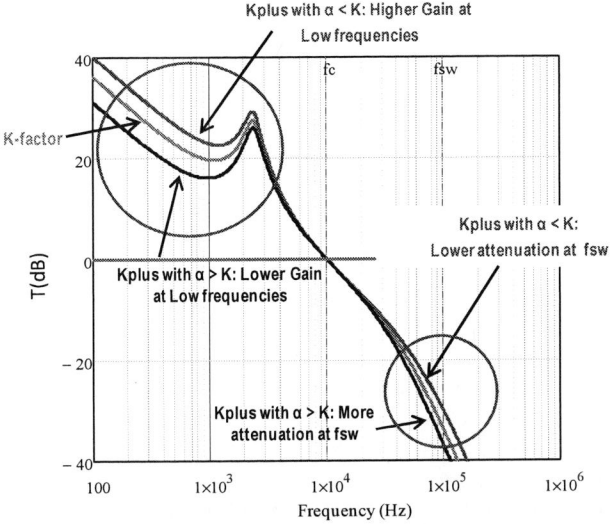

Fig. 14: Comparative Bode plots of K-factor vs Kplus.

C. Analog and Digital Compensator

The CAD tool allows obtaining both analog and digital compensators. The digital compensator has been obtained from the continuous-time compensator by means of discretization through either the bilinear transform or the Tustin method.

The effects of the time delays are included once the compensator has been obtained. If the time delay implies a phase delay too high, the continuous compensator should be redesigned with more phase margin. On the other hand, the rounding of the coefficients which appears in fixed-point systems has been taken into account. From the number of available bits to represent data in digital system, the coefficients are rounded and the new discrete compensator with rounding coefficients is analyzed.

D. Transient response

Transient response specifications, such as settling time and overshoot are usually critical specifications when designing the compensator of a power converter. Therefore, providing a real time view of the converter transient response may greatly help the designer while designing the control loop. This feature reduces dramatically the designing time, because it is not necessary to simulate the transient response of the system each time any parameter of the control loop is modified.

E. Parametric Sweep

An important feature that presents the proposed new CAD tool is the parametric sweep. By means of this characteristic, once the compensator has been obtained, the power stage and the compensator values can be changed in order to observe every effect in the control loop. In this way, a sensitivity analysis can be made by the designer.

Conclusions

In this paper a new concept for power electronics compensator design has been proposed. The most important features may be summarized in a correct selection of f_{cross} and PM taking into account the attenuation at the switching frequency and the most suitable compensator for the power stage to be compensated.

When designing below the resonant frequency and under a reference step, the transient response of the feedback converter presents more settling time for increasing PM. The relationship between these two magnitudes is linear and it has been validated by means of experimental measurements in a VMC Forward converter with a Type 3 compensator.

The new concept proposed along this paper has been introduced into a CAD tool in order to have a complete design solution for power converters. All features presented within the CAD tool allows to reduce the designing process of the converter control loop.

Acknowledgment

This work has been supported by Ministerio de Educación y Ciencia (Spain), by means of the research projects FLAME (DPI: 2010:21110-C02-02) and SAUCE (DPI: 2009-12501).

References

[1] VENABLE, H. D. "The k-factor: A New mathematical Tool for Stability, Analysis, and Synthesis". Proceeding of. Powercon 10, San Diego, CA, March 22-24, 1983.

[2] O'DWYER, A. "Handbook of PI and PID controller tuning rules". Imperial College Press, 2006. ISBN 978-1-84816-242-6.

[3] Voltage Mode Boost Converter Small Signal Control Loop Analysis Using the TPS61030. TI. application report SLVA274–May 2007.

[4] ROSSETTO, L., SPIAZZI, G. "Design Considerations on Current-Mode and Voltage-Mode Control Methods for Half-Bridge Converters" in Proc. of IEEE App. Power Elect. Conf. (APEC'97). ISBN 0-7803-3704-2.

[5] ERICKSON, R. W. "Fundamentals of Power Electronics 2nd edition", Springer, 2001. ISBN 0-7923-7270-0.

[6] FERNANDEZ, A., TONICELLO, F., AROCA, J., MOURRA, O. "Battery Discharge Regulator for Space Applications based in the Boost Converter" in Proc. of IEEE Applied Power Electronics Conference (APEC 2010). ISBN 978-1-4244-4783-1.

[7] Texas instruments UC3823 datasheet.

Influence of a High Precision Current Sensor for Improving the Efficiency of PV Power Systems

Abdoulkarim Bouabana, Ahmad Al-Diab, Constantinos Sourkounis

Ruhr-University Bochum / Institute of Power System Technology and Power Mechatronics, Bochum, Germany.

Abstract—**Maximum power point tracking (MPPT) techniques are employed in photovoltaic (PV) systems to maximize the PV array output power which depends on solar irradiance and temperature. Most of these techniques depend on the measurement of the current sensor which needs to be simulated on the design phase. Simulation tools like Matlab/SIMULINK, PSpice, Multisim..etc, don't simulate the practical behaviour of measurement devices and only deal with ideal characteristics which make it difficult to understand the exact response of such system on the operation mode. Current sensor model is implemented to enhance the design and the implementation of the proposed modified P&O algorithm. The algorithm is designed in order to overcome the drawbacks in traditional P&O MPPT algorithm. Varying the step size value and the sampling time reduces the oscillations around the MPP and leads to a faster response. When a step change in the solar irradiance occurs, the step size and sampling time are automatically tuned according to the operating point. If the operating point is far from the MPP, it increases the step size and decreases the sampling time which enables a fast tracking ability with smaller oscillation steps around the MPP.**

I. INTRODUCTION

RENEWABLE energy sources are expected to play an essential role in the field of power production due to the increase of the world's power demand, of their independence from limited power sources and their low impact on the environment. Photovoltaic generation systems are currently considered to convert one of the most useful natural energy sources due to the continuous cost reduction, stable system, fast technological progress, being maintenance and pollution-free. They are used today in many applications such as battery charging, water pumping, home power supply, satellite power systems etc. The minimum element in PV systems is the PV module. Typical module is composed of series-connected solar cells, because of the insufficient power of one PV module is not enough. It is necessary to connect number of PV modules either in parallel or in series until the desired current and voltage levels are achieved.

Two existing drawbacks encountered while generating power from PV systems the first one that the efficiency of electric power generation is very low, especially under low radiation states, the other drawback is the amount of electric power generated by solar arrays is influenced by with weather conditions, i.e., irradiation and temperature.

Therefore a Maximum Power Point Tracking (MPPT) is an essential part to ensure that PV systems operate at the maximum power. The P&O Maximum Power Point Tracking (MPPT) algorithm is mostly used, due to its ease of implementation. It is based on the calculation of the PV array output power and the power change by sensing both the PV current and voltage. The tracker operates periodically by comparing the actual value of the power and current with the previous value to determine the change (incrementing or decrementing) on the solar array current. If a given perturbation leads to an increase (decrease) in the output power of the PV, then the subsequent perturbation is generated in the same (opposite) direction. When the MPP is reached, the operating point oscillates around the MPP giving rise to the waste of some amount of available energy. In order to minimize the oscillation, the step size should be reduced. However, a smaller step size slows down the MPPT to overcome this, the sampling time is reduced.

Simulation tools like Matlab/SIMULINK, PSpice, Multisim..etc, don't simulate the practical behavior of measurement devices and only deal with ideal characteristics. The problem of ideal current sensor is the accuracy of the calculated value. The simulated value has a high accuracy, which depends on of the processor. For example, a 32bit processor achieves an accuracy of 1/32bit =2,328.10-10. This will not be realizable in metrology, which will produce a mismatch between simulation results and experimental results. Therefore, a more realistic simulation model of a current sensor is required to express the significant meaning of the simulation results. Different disturbances such as, the offset (depends on temperature), the non linearity and the noise influence on the accuracy of a practical current sensor. The implemented simulation model includes these typical influences that gives the simulated value a more realistic view and provides more accurate information about the simulation results.

This current sensor model is implemented to enhance the design and the implementation of the proposed modified P&O algorithm. The algorithm is designed in order to overcome the drawbacks in traditional P&O MPPT algorithm. Varying the step size value and the sampling time reduces the oscillations around the MPP and leads to a faster response. When a step change in the solar irradiance occurs, the step size and sampling time are automatically tuned according to the operating point. If the operating point is far from the MPP, it increases the step size and decreases the sampling time which enables a fast tracking ability with smaller oscillation steps around the MPP.

978-1-4244-8806-3/11 $26.00 © 2011 IEEE

II. PV SYSTEM

Photovoltaic generates (DC) electrical power from semiconductors when they are illuminated by photons [14]. As long as light is shining on the solar cell, it generates electrical power. The most common solar cells technologies nowadays are the mono-crystalline, the poly-crystalline-silicon and Amorphous cells which are based on traditional, and expensive, semiconductor manufacturing processes and differs on the physical characteristics. PV array consists of number of PV modules connected either in series or in parallel according to the required power.

On the other side, a PV module consists of a number of connected solar cells in series, which are the basic element of any PV-System.

Practical PV cell

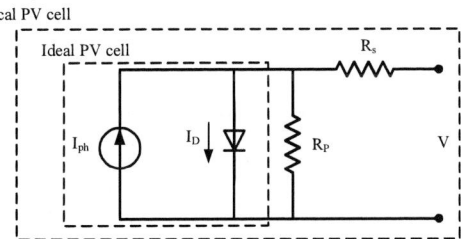

Fig. 1. PV cell.

A. The Solar Cell

A typical solar cell consists of a p-n junction formed in a semiconductor material similar to a diode. It acts either as a voltage source or as a current source. The equivalent circuit of a solar cell is shown in Fig. 1.

The photo-generated Current (IPh) is proportional to solar radiation. The maximum value of the solar cell output current (I) is the Short Circuit Current (ISC) which produced under short circuit conditions (V=0). While the maximum voltage at zero current, whereas the short circuit current is the maximum current at zero voltage, is called the Open Circuit Voltage (VOC).

The relationship between current and voltage may be determined from the diode characteristic equation:

$$I = I_{ph} - I_0 \left\{ \exp\left[\frac{e(V + IR_s)}{K.T_{cell}} \right] - 1 \right\} \qquad (1)$$

$$V = V_t \ln\left(\frac{I_{ph}}{I_0} \right) \qquad (2)$$

where

I	Cell Output Current (A)
I_{ph}	Photogenerated Current (A)
I_0	Diode Saturation Current (A)
V	Cell Output Voltage (V)
R_S	Series Resistor (Ω)
e	Electron Charge 1.6×10^{-19} ($coul$)
K	Boltzman Constant (j/K)
T_{Cell}	Cell Temperature (K)
V_{OC}	Open Circuit Voltage (V)
V_t	Thermal Voltage (V)

Power (P) against current (I) characteristic curve of a PV module, Fig. 2, shows a unique point on the curve which the solar cell will generate maximum power. This is known as the maximum power point (Vmp, Imp), which depends on the solar irradiance and on the temperature. Therefore; a MPPT is an essential part of PV system to ensure that inverters operate at the maximum power of the PV array.

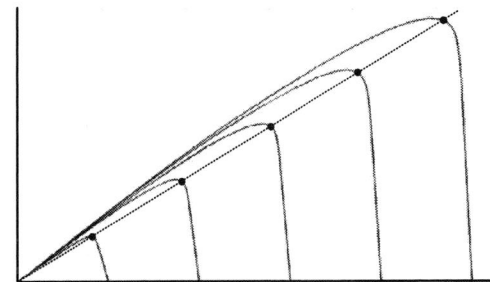

Fig. 2. Change in power-current characteristics due to change of irradiance.

B. The Perturb & Observe (P&O) Method

The most commonly used MPPT algorithm is the (P&O), due to its simplicity of implementation in its basic form. However, it has some drawbacks, like oscillations around the MPP in steady state operation, slow response speed due to the change of the solar irradiance which makes it less favorable for rapidly changing operating conditions.

P&O algorithm is based on the calculation of the PV array output power and the power change by sensing both the PV current and voltage. The tracker operates periodically by comparing the actual value of the power with the previous value to determine the change (incrementing or decrementing) on the solar array voltage or current (depending on the control strategy). If a given perturbation leads to an increase (decrease) in the output power of the PV, then the subsequent perturbation is generated in the same (opposite) direction [7, 11, 13, 20]. The algorithm is summarized in Fig. 3. When the MPP is reached, the system then oscillates around the MPP. In order to minimize the oscillation, the perturbation step size should be reduced. However, a smaller step size slows down the MPPT.

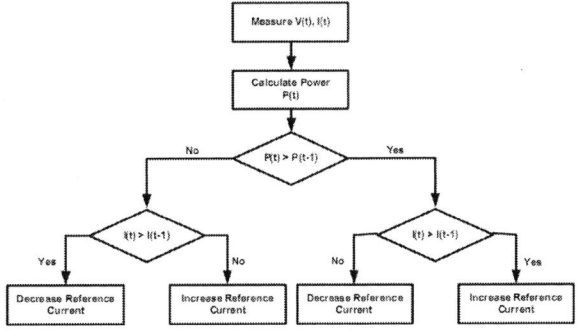

Fig. 3. Traditional Perturb & Observe (P&O) Method.

A modified P&O algorithm is introduced in this paper in order to overcome this drawback by varying the step size and the sampling time.

C. Modified Perturb & Observe (P&O) Method

In general, MPPT algorithms use a fixed step size, which is determined by the accuracy and tracking speed requirements. However, if the step size is increased for tracking speed-up, the accuracy is decreased, resulting in a comparatively low efficiency and vice versa. In this paper, a modified variable step size variable sampling time algorithm is proposed for the P&O MPPT method and is dedicated to find a simple, effective way to improve tracking accuracy and to overcome the drawbacks in traditional P&O MPPT algorithm.

The flowchart of the modified variable step size variable sampling time P&O MPPT algorithm is shown in Fig. 4. When a step change in the solar irradiance occurs, the step size and sampling time are automatically tuned according to the operating point. If the operating point is far from the MPP, it increases the step size and reduces the sampling

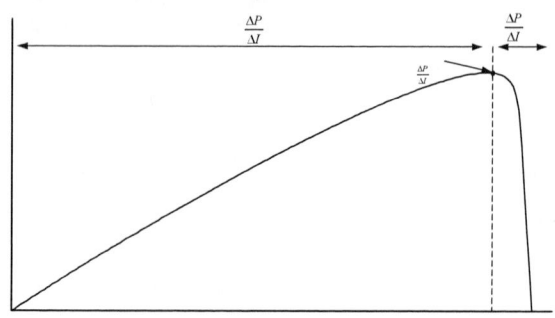

Fig. 5. (ΔP/ΔI) Range on the PI-Curve.

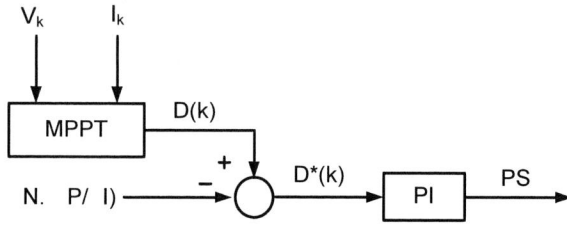

Fig. 6. Phase Shift controller.

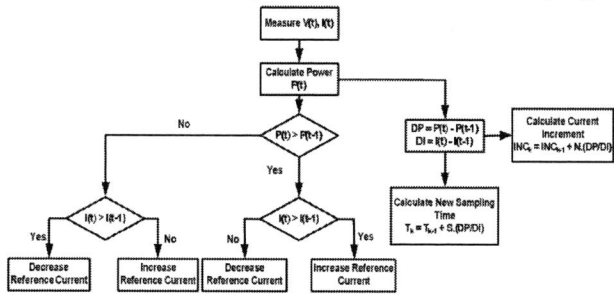

Fig. 4. Proposed Perturb & Observe (P&O) Method.

time which enables a high accuracy fast tracking ability.

In most applications, the MPPT is achieved by connecting a power conditioner (dc/dc or dc/ac converter) between the PV array and load. To simplify the control system the PV array operation point is employed to directly control the converter duty cycle based on a Phase Shift Pulse Width Modulation (PSPWM).

The variable step size adopted to reduce the problem mentioned above is shown as follows:

$$D^*(k) = D(k) - N . \frac{\Delta P}{\Delta I} \qquad (3)$$

Where N is the scaling factor determines the performance of the MPPT system and is tuned at the design process. The MPPT control systems for phase shift PWM and for the variable sampling time are shown in Fig.6 and Fig.7. which has an inherent characteristic as follows:

$\dfrac{\Delta P}{\Delta I} > 0$ PV operating point at the left of the MPP.

$\dfrac{\Delta P}{\Delta I} = 0$ PV operating point at MPP .

$\dfrac{\Delta P}{\Delta I} < 0$ PV operating point at the right of the MPP.

On the first case when the value of $\frac{\Delta P}{\Delta I} > 0$, the step size should be increased which makes the magnitude of the phase shift to be produced will be decreased, therefore; the PV current flow through the high frequency inverter will be increased, causing an increase on the generated power. However, it will be the opposite for the third case. If the $\frac{\Delta P}{\Delta I} = 0$

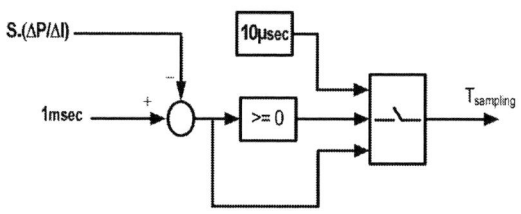

Fig. 7. Sampling Time Calculator.

Array is running at the MPP, the value of   , this keeps the value phase shift without any change.

III.   CURRENT SENSORS

Nowadays, the technical world like control systems depends on different measurement devices, which can measure the current.

The current measurement is an important part. There are different (physical) options to measure the current.

For example, there are the Hall-Effect, the Magneto-Restive Effect or the shunt method. Each one of these methods has advantages and disadvantages.

All current sensors will be described based on the accuracy. A real sensor has some influences such as, the offset, the non-linearity and the noise. Because of that it has not an accuracy of 100%. Current sensors which work with the Hall-Effect, can measure the current with the open loop method or with the closed loop method. The output signal of the open loop method is the measured voltage (after amplifier) of the Hall device. Thus, it will be measured by the magnetic flux in the air gap. The closed loop method is related, but the measured voltage generates a feedback current in the control cycle. This secondary current should compensate the magnetic flux of the primary current.

The disadvantage of the open loop method is the offset and the linearity (hysteresis). The reason of the drawback is the saturated ferrite core. In addition, the offset is dependent on temperature. The advantage of both methods is the electrical isolation.

The measurement with a shunt has no offset and non-linearity (advantage), but the absence of the electrical isolation causes problems (disadvantage).

The following part describes a Matlab/SIMULINK model of a current sensor. The current sensor model deals with realistic characteristics and will achieve a more realistic simulation.

A. Construction of a current sensor model

The block diagram of current sensor model is represented in Fig. 8. It shows the part with the influenced values of the accuracy (gain, offset, the non-linearity and the noise). This part will be added to ideal value.

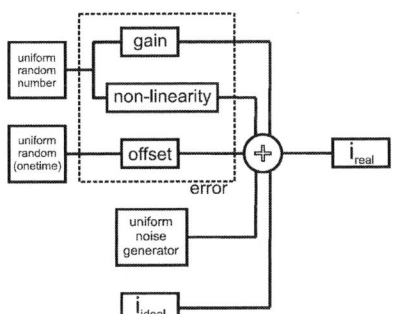

Fig. 8. Current Sensor Model.

The "error block" is subdivided to three several blocks (gain, non-linearity and offset). Each of these blocks are described by a mathematical equation. The equations (4-6) calculate the absolute error of the gain, the non-linearity and the offset.

The three error types (gain, non-linearity and offset) are shown in Fig. 9.

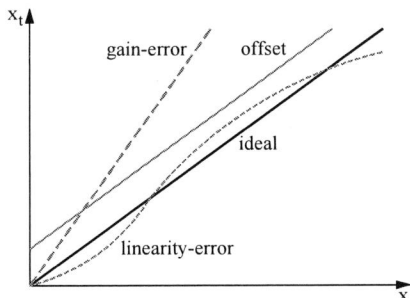

Fig. 9. Error Types (real vs. ideal).

$$fa_{gain} = \left(\frac{fr_{total}}{100} - \frac{offset}{x_N} \right) \cdot x_N \cdot random(-1,1) \qquad (4)$$

$$fa_{lin} = \frac{fr_{lin}}{100} \cdot x_N \cdot \sin\left(2\pi \cdot random(-1,1) \right) \cdot random(-1,1) \qquad (5)$$

$$fa_{offset} = \left(T_c \cdot \left(\frac{(T_o - T_a)}{100} \right) + \frac{offset}{x_N} \right) \cdot x_N \cdot unifrnd(-1,1) \qquad (6)$$

where:

fa_{gain}	*absolute gain error* $[A]$
fa_{lin}	*absolute linearity error* $[A]$
fa_{offset}	*absolute offset* $[A]$
fr_{total}	*accuracy* $[\%]$
fr_{lin}	*relative linearity error* $[\%]$
offset	*offset @25°C*
x_N	*nominal value* $[A]$
T_c	*temperature drift* $[\% / K]$
T_o	*operating temperature* $[°C]$
T_a	*ambient temperature* $[°C]$
$random(-1,1)$	*random function* $[]$
$unifrnd(-1,1)$	*uniform random number* $[]$

The first part in (4) is the relative error of the gain. The variable fr_{total} (in %) is the accuracy value (datasheet) of a commercial current sensor. This variable will be divided by 100 and subtracted with the term $offset/x_N$. Afterwards, the relative gain error will be multiplied by nominal value x_N in order to receive the absolute gain error. The random function generates values between -1 and 1, because the error is not fix, but rather stochastic.

Equation (5) of the absolute linearity-error has three terms: the random function (see above), the relative linearity-error (fr_{nl}) with x_N and a "stochastic sinus function". The "stochastic sinus function" should approximately describe the non-linearity characteristics of a current sensor.

The offset of a current sensor depends on the temperature. In this case, the equation (6) has the temperature drift (T_c). The drift will be multiplied by the temperature difference and added with the offset at 25°C. The last term is an uniform random, which returns only onetime a number. The offset is a stochastic value, but it is "constant" for a measurement period.

Another distribution, which does not appear in the simulation tools with the ideal characteristics, is the noise.

Therefore, in the model is a selectable uniform noise generator, which generates a factor for the noise (% of nominal value).

The "real" simulated value is now the addition of the "error block", the noise value and the ideal value. Fig.10 shows an example of a parameter mask. In this mask the required values of the current sensor model can be enter.

In the mask, there is also the possibility to choose the measurement mode. The current can be measured with or without a shunt. Current sensors, which are based on a shunt, have a voltage drop. In this case, it is useful to regard this resultant voltage drop.

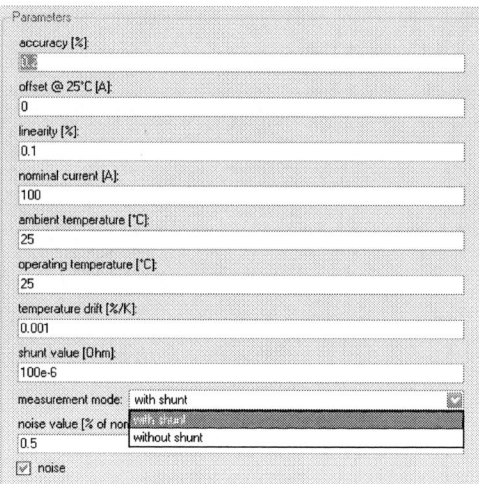

Fig. 10. Parameter Mask.

IV. SIMULATION EVALUATION

To verify the performance of the proposed modified P&O MPPT algorithm, a Matlab/SIMULINK model of the PV system is initially developed. Three PV string Each string contains 12 modules, one module contains of 60 solar cells. The PV modules have the following manufacturing specification: Voc= 36.9V, Isc= 8.07A, voltage at maximum power= 30.2V, current at maximum power= 7.42A, taking in concern the deviations on the specifications due to the semiconductors manufacturing tolerances. A comparison is held to verify the modified P&O algorithm with the common fixed step size P&O MPPT method. To compare the performance of the modified P&O MPPT under in different current sensor's specifications under a variable irradiance conditions with step size 0.2 seconds for one string and the other two strings have an implemented irradiance equals 1000 W/m². The simulations are configured under exactly the same conditions to compare the performances. The sampling period used for MPPT algorithm is chosen as 1ms as a maximum value and 10μs as a minimum value.

The current sensor has the following specifications: 0.1% linearity factor, nominal current 26A, temperature drift 0.001[%/K] and a shunt resistor equals 0.5 mOhm.

The Tolerance of the current sensor is varied (1% or 0.1%) to verify the proposed P&O MPPT algorithm.

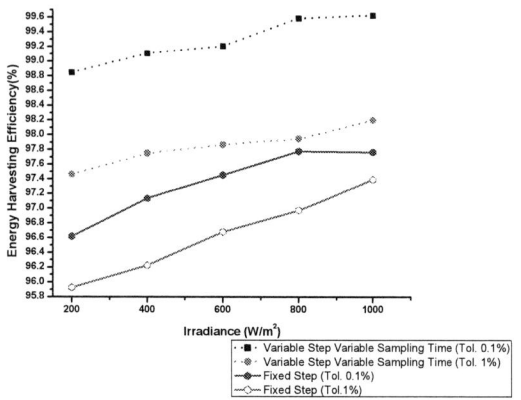

Fig. 11. Energy Yield for Fixed and Variable P&O MPPT Methods for Different Current Sensor's Tolerances.

The Energy yield for both fixed step size and the modified P&O with 0.1% and 1% tolerance of the current sensor is presented in Fig.11.

The Energy yield for modified P&O algorithm under

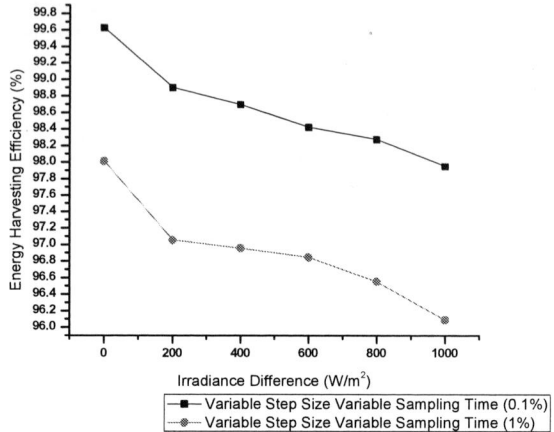

Fig. 12. Energy Yield for the Proposed P&O MPPT Methods.

different current sensor tolerance is presented in Fig.12.

The Power Response for the fixed step and the proposed P&O MPPT algorithms under partial shading are shown on Figs. 13-14.

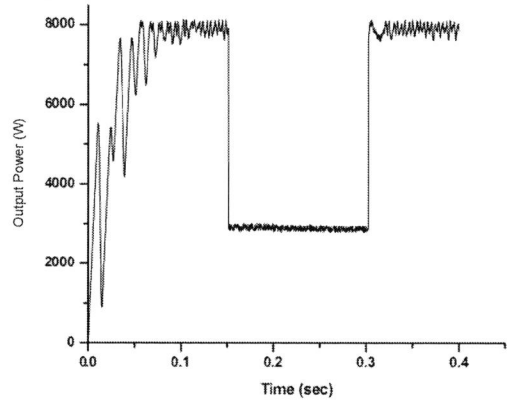

Fig. 13. Power Response for Fixed Step P&O MPPT Method Under Partial Shading.

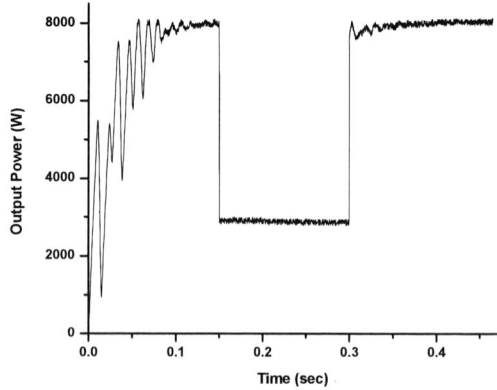

Fig. 14. Power Response for the Proposed P&O MPPT Method Under Partial Shading.

From the simulation results, the proposed algorithm has higher energy harvesting efficiency (2%) than the fixed algorithm. Regarding to the response, the proposed algorithm has fast response with small oscillations which yields to a small energy waste.

978-1-4244-8806-3/11 $26.00 © 2011 IEEE   200

A. Conclusions

In this paper, a modified P&O MPPT algorithm has been presented, which is able to improve the dynamic and steady state performance of the PV system simultaneously. The algorithm is designed in order to overcome the drawbacks in traditional MPPT algorithms. Varying the step size value and the sampling time reduces the oscillations around the MPP and leads to a faster response to reach it. When a step change in the solar irradiance occurs, the step size and sampling time are automatically tuned according to the operating point. If the operating point is far from the MPP, it increases the step size and decreases the sampling time which enables a fast tracking ability. Moreover, at the start process of the MPPT, the PV system may exhibit comparable large step change in the output voltage and current due to the large step size. Both fixed step size and the proposed variable size P&O MPPT methods are implemented with Matlab/SIMULINK for simulation. The simulation results verify the feasibility and effectiveness of the proposed method, the developed current sensor model gives extra enhancements to the proposed P&O MPPT algorithm due to it's parameters that simulates practical current sensor which makes the simulation results more realistic.

References

[1] Soeren Baekhoej Kjaer, Frede Blaabjerg, "A Review of Single-Phase Grid-Connected Inverters for Photovoltaic Modules," IEEE Transaction on Industrial Electronics, VOL. 41, NO. 5, February 2005.

[2] Eftichios Koutroulis, Kostas Kalaitzakis, Nicholas C. Voulgaris, "Development of a Microcontroller-Based, Photovoltaic Maximum Power Point Tracking Control System," Transactions on Power Electronics, VOL. 16, NO. 1, January 2001.

[3] Fangrui Liu, Shanxu Duan, Fei Liu, Bangyin Liu, and Yong Kang, "A Variable Step Size INC MPPT Method for PV Systems," Transactions on Industrial Electronics, 55, NO. 7, July 2008.

[4] Jae Ho Lee , HyunSu Bae and Bo Hyung Cho, "Advanced Incremental Conductance MPPT Algorithm with a Variable Step Size," 12th International Power Electronics and Motion Control Conference EPE-PEMC, pp. 603-607, Aug. 2006.

[5] Ahmad Al-Diab , Constantinos Sourkounis, "Multi-Tracking Single-Fed PV Inverter," The 15th IEEE Mediterranean Electrotechnical Conference MeleCon2010, April 2010.

[6] Ahmad Al-Diab , Constantinos Sourkounis, "Variable Step Size P&O MPPT Algorithm for PV Systems," 12th International Conference on Optimization of Electrical and Electronic Equipment OPTIM 2010, May 2010.

[7] Eduardo Román, Ricardo Alonso, Pedro Ibañez, "Intelligent PV Module for Grid-Connected PV Systems,"IEEE Transaction on Industrial Electronics, VOL. 53, NO. 4, August 2006.

[8] A. O. Zue, A. Chandra, "Simulation and Stability Analysis of a 100 kW Grid Connected LCL Photovoltaic Inverter for Industry," IEEE Power Engineering Society General Meeting, June. 2006.

[9] M. Ciobotaru, T. Kerekes, R. Teodorescu,A. Bouscayrol, "PV inverter simulation using MATLAB/Simulink graphical environment and PLECS blockset ," IECON 2006 - 32nd Annual Conference on IEEE Industrial Electronics, Pages 5313 – 5318, November 2006.

[10] Trishan Esram, Patrick L. Chapman, "Comparison of Photovoltaic Array Maximum Power Point Tracking Techniques," IEEE Transactions on Energy Conversion, Vol. 22, No.2, June 2007.

[11] N. Khaehintung, P. Sirisuk, "Implementation of maximum power point tracking using fuzzy logic controller for solar-powered light-flasher applications ," The 47th Midwest Symposium on Circuits and Systems, Vol. 3, Pages 171-4, July 2004.

[12] Roberto González, Jesús López, Pablo Sanchis, Luis Marroyo, "Transformerless Inverter for Single-Phase Photovoltaic Systems," IEEE Transactions on Power Electronics, Vol. 22, No.2, March 2007.

[13] Philip T. Krein, Robert S. Balog, Xin Geng, "High-Frequency Link Inverter for Fuel Cells Based on Multiple-Carrier PWM," IEEE Transactions on Power Electronics, Vol. 19, No.5, September 2004.

[14] J. M. A. Myrzik, M. Calais, "String and Module Integrated Inverters for Single-Phase Grid Connected Photovoltaic Systems - A Review ," IEEE Bologna Power Tech Conference Proceedings, June 2003.

[15] D Xu, C Zhao, H Fan, "A PWM plus phase-shift control bidirectional DC-DC converter ," IEEE Transactions on Power Electronics, Vol. 19, No.3, May 2004.

[16] W. Xiao, W. G. Dunford ,"A modified adaptive hill climbing MPPT method for photovoltaic power systems," IEEE Power Electronics Specialists Conference (PESC), 2004, pp1957–1963.

[17] Bimal K. Bose, "Power Electronics And Motor Drives: Advances and Trends," Elsevier Inc, 2006, ISBN: 978-0-12-088405-6.

[18] Muhammad H. Rashid, "Power Electronics Handbook: Devices, Circuits and Applications," Prentice-Hall Inc., 3rd edition, 2003, ISBN: 0-13-101140-5.

[19] Hiren Patel, Vivek Agarwel, , "Maximum Power Point Tracking Scheme for PV Systems Operating Under Paertially Shaded Conditions," IEEE Transaction on Industrial Electronics, VOL. 55, NO. 4, April 2008.

[20] Nobuyoshi Mutoh, Masahiro Ohno, Takayoshi Inoue, "A Method for MPPT Control While Searching for Parameters Corresponding to Weather Conditions for PV Generation Systems," ," IEEE Transaction on Industrial Electronics, VOL. 53, NO. 4, August 2006.

[21] Mohammad A. S. Masoum, Hooman Dehbonei, and Ewald F. Fuchs, "Theoretical and Experimental Analyses of Photovoltaic Systems With Voltage- and Current-Based Maximum Power-Point Tracking," ," IEEE Transaction on Energy Conversion, VOL. 17, NO. 4, December 2002.

[22] K. K. Tse, Billy M. T. Ho, Henry Shu-Hung Chung, S. Y. Ron Hui, "A Comparative Study of Maximum-Power-Point Trackers for Photovoltaic Panels Using Switching-Frequency Modulation Scheme," IEEE Transaction on Industrial Electronics, VOL. 51, NO. 2, April 2004.

[23] Sourkounis,C., "Stromsensor für die Anwendung in derAntriebstechnik und Automotive-Appliktionen," Ruhr-University Bochum, 2007.

Modeling and Analysis of Three-Phase Hybrid Transformer Using Buck-Boost MRC

Jacek Kaniewski
Institute of Electrical Engineering, University of Zielona Góra
J.Kaniewski@iee.uz.zgora.pl

Abstract- **This paper deals with the modelling and analysis of the three-phase AC transformer with electromagnetic and electric (hybrid) coupling. The electromagnetic coupling is realized by means of the conventional transformer. The electrical coupling is realized by means of a buck-boost matrix-reactance chopper (MRC), which is supplied from an auxiliary secondary winding of the transformer. In this paper there are descriptions of the proposed solution, with a presentation of their modeling and analysis of their properties.**

I. INTRODUCTION

The dynamic states in AC electric systems, such as faults, fast load changes, switching effects, generate for the consumer undesirable effects like voltage sags, swell and interrupt [1]. In the case of sensitive devices such as computers, transceivers devices, medical systems, erratic supply parameters cause failure or defective devices [2]. In the case of big plants and factories, voltage sags and swells may cause very large financial damage. The application of an AC-AC converter using Pulse Width Modulation (PWM) control strategy to build secondary supply sources (voltage sag and swell compensators and voltage regulators) mitigate the unwanted effects of supply [3]–[5]. The conception of a single–phase new generation distribution transformer is presented in [6], where a conventional transformer works together with a unipolar matrix converter. Circuit with conventional transformer and matrix (MC) or matrix-reactance (MRC) chopper has two couplings (hybrid coupling). Electromagnetic coupling is realized by means of the conventional transformer and electric one realized by means of a matrix converter. From this reason it is called a hybrid transformer (HT). Single phase and three-phase topologies of HT was described in [7]–[12]. This paper presents three-phase hybrid transformer using buck-boost type matrix-reactance chopper. Presented in this paper there is a description of the proposed solution and modeling by using averaged state space equations, space vector and d-q transformation method.

II. DESCRIPTION OF THE PRESENTED HT

Schematic diagram of the presented three-phase hybrid transformer (HT) with ideal switches is shown in Fig. 1. As is visible in Fig. 1 the circuit of the HT contains two main units. The first one is three-phase conventional transformer (TR), with two secondary windings in each phase. The second one is a three-phase buck-boost type matrix-reactance chopper (MRC b-b). Considered HT is assumed as symmetrical and balanced circuit

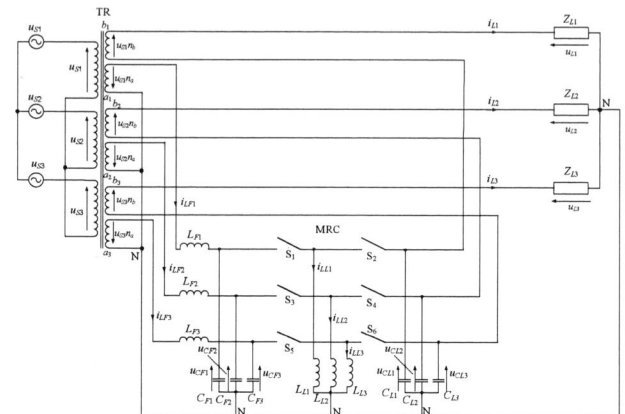

Fig. 1. Schematic diagram of the three-phase hybrid transformer with matrix-reactance chopper with ideal switches

Primary windings are in Y-connections. The main secondary windings (a_1, a_2, a_3) of TR also have Y configurations and, by input filter LC, are connected with the MRC b-b. Secondary phase windings (b_1, b_2, b_3) are connected in series with the required phase output connectors of MRC. Output voltages of HT (u_{L1}, u_{L2}, u_{L3}) are the sum of secondary voltages ($p_b u_{S1}$, $p_b u_{S2}$, $p_b u_{S3}$) and phase output voltages of the MRC. The voltages of the transformer secondary windings a_1, a_2, a_3 and b_1, b_2, b_3 are equal $p_a = 4/3$ and $p_b = 2/3$ respectively. More detail description of the considered HT was described in [12]. In the considered circuit there are two operating states. In the on state (Fig. 2) switches S1, S3, S5 are turned on, while in the off-state (Fig. 3) switches S2, S4, S6 are turned on

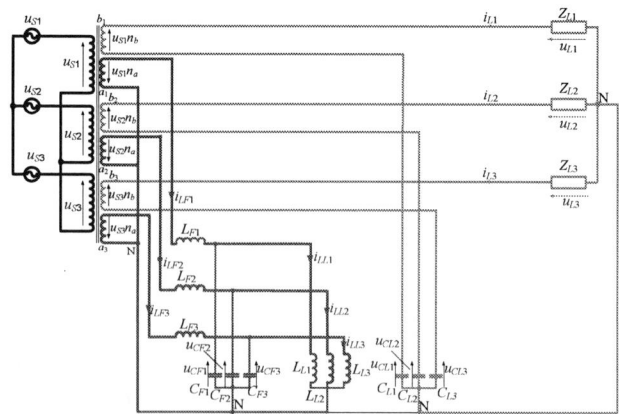

Fig. 2. Schematic diagram of HT in on-state

978-1-4244-8806-3/11 $26.00 © 2011 IEEE

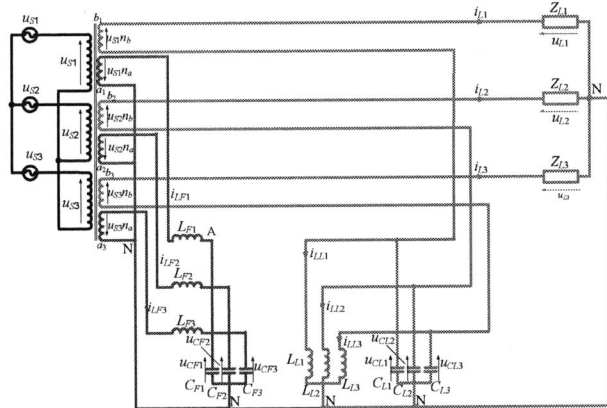

Fig. 3. Schematic diagram of HT in off-state

Exemplary voltage phasors of voltages of the presented HT are shown in Fig. 4 and are described by (1) – (5).

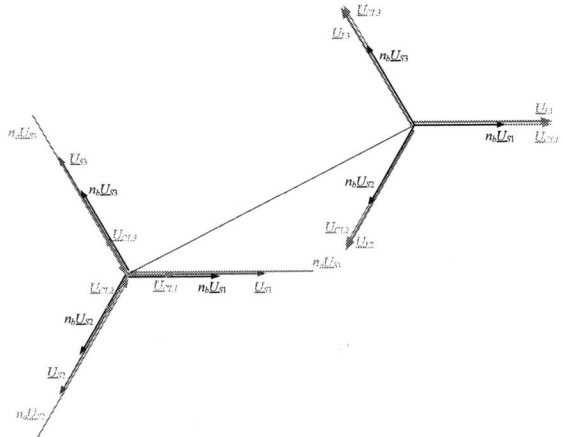

Fig. 4. Voltage phasors of presented HT

$$\underline{U}_{S1} = U_S \cdot e^{j0} \qquad\qquad n_a\underline{U}_{S1} = -n_aU_S \cdot e^{j0}$$

$$\underline{U}_{S2} = U_S \cdot e^{-j\frac{2\pi}{3}}, \quad (1) \quad n_a\underline{U}_{S2} = -n_aU_S \cdot e^{-j\frac{2\pi}{3}}, \quad (2)$$

$$\underline{U}_{S3} = U_S \cdot e^{j\frac{2\pi}{3}} \qquad\qquad n_a\underline{U}_{S3} = -n_aU_S \cdot e^{j\frac{2\pi}{3}}$$

$$n_b\underline{U}_{S1} = n_bU_S \cdot e^{j0} \qquad\qquad \underline{U}_{CL1} = n_aU_SH_U^{b-b} \cdot e^{j0}$$

$$n_b\underline{U}_{S2} = n_bU_S \cdot e^{-j\frac{2\pi}{3}}, \quad (3) \quad \underline{U}_{CL2} = n_aU_SH_U^{b-b} \cdot e^{-j\frac{2\pi}{3}}, \quad (4)$$

$$n_b\underline{U}_{S3} = n_bU_S \cdot e^{j\frac{2\pi}{3}} \qquad\qquad \underline{U}_{CL3} = n_aU_SH_U^{b-b} \cdot e^{j\frac{2\pi}{3}}$$

$$\underline{U}_{L1} = n_a\underline{U}_{S1} + \underline{U}_{CL1}$$
$$\underline{U}_{L2} = n_a\underline{U}_{S2} + \underline{U}_{CL2} . \qquad (5)$$
$$\underline{U}_{L3} = n_a\underline{U}_{S3} + \underline{U}_{CL3}$$

III. THEORETICAL ANALYSIS

Theoretical analysis is based on averaged state space method [13] (6), space vector (7) and d-q transformation method (8) [14], [15].

$$\dot{\overline{\mathbf{x}}} \cong \mathbf{A}(D)\overline{\mathbf{x}} + \mathbf{B}(D)\mathbf{u}_S,$$
$$\overline{\mathbf{y}} \cong \mathbf{C}(D)\overline{\mathbf{x}} + \mathbf{D}(D)\mathbf{u}_S, \qquad (6)$$

where: $\overline{\mathbf{x}}$ - vector of the averaged variables, $\mathbf{A}(D)$ – averaged state matrix, $\mathbf{B}(D)$ – averaged input matrix, $\mathbf{C}(D)$ – averaged output matrix, $\mathbf{D}(D)$ – averaged input-output matrix.

$$\mathbf{x}_{SV} = \mathbf{K}\mathbf{x}_{ABC}, \quad (7) \qquad \mathbf{x}_{DQ} = \mathbf{x}_{SV}e^{j\theta(t)}, \quad (8)$$

where: $\mathbf{x}_{ABC}$ – vector of averaged variables in three-phase electrical system (9), $\mathbf{x}_{SV}$ – space vector (10), $\mathbf{x}_{DQ}$ – time invariant vector in d-q coordinate system

$$\mathbf{x}_{ABC} = \begin{bmatrix} x_a \\ x_b \\ x_c \end{bmatrix}, \quad (9) \qquad \mathbf{x}_{SV} = \begin{bmatrix} x_0 \\ x_Z \\ x_P \end{bmatrix} \begin{matrix} \Leftarrow zero \\ \Leftarrow forward \\ \Leftarrow beckward \end{matrix}, \quad (10)$$

$$\mathbf{K} = \frac{2}{3}\begin{bmatrix} 1 & 1 & 1 \\ 1 & a^2 & a \\ 1 & a & a^2 \end{bmatrix}, \quad (11) \qquad \underline{a} = e^{j\frac{2\pi}{3}}, \quad (12)$$

$$\theta(t) = \pm\omega t + \theta_0 . \qquad (13)$$

According to (7) averaged state space equation of the considered circuit of HT (Fig. 1) in matrix form can be described as (14).

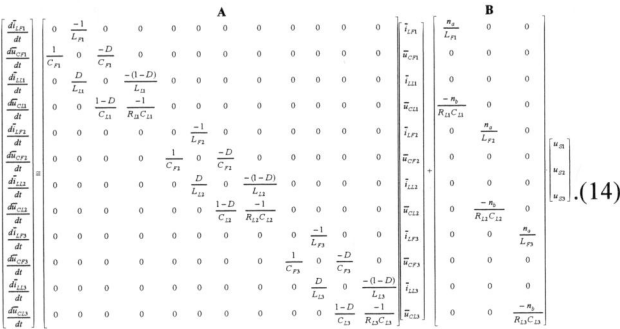

Take into account (14) averaged circuit model of HT is constructed (Fig. 5).

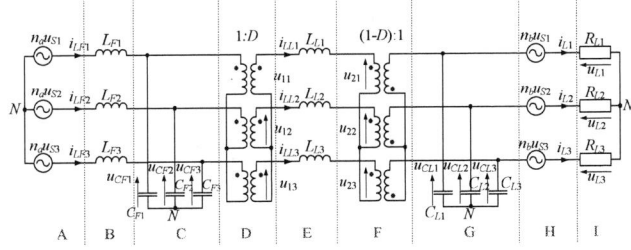

Fig. 5. Three-phase averaged circuit model of considered HT

Assumed parameters of considered HT $U_{S1}=U_{S2}=U_{S3}$, $L_{F1}=L_{F2}=L_{F3}$, $C_{F1}=C_{F2}=C_{F3}$, $L_{L1}=L_{L2}=L_{L3}$, $C_{L1}=C_{L2}=C_{L3}$,

978-1-4244-8806-3/11 $26.00 © 2011 IEEE   203

$R_{L1}=R_{L2}=R_{L3}$ the source voltages $\mathbf{u}_S$ in ABC coordinate system (9) are defined as:

$$\mathbf{u}_S = \begin{bmatrix} u_{S1} \\ u_{S2} \\ u_{S3} \end{bmatrix} = U_S \begin{bmatrix} \cos(\omega t + \varphi_S) \\ \cos\left(\omega t - \dfrac{2\pi}{3} + \varphi_S\right) \\ \cos\left(\omega t + \dfrac{2\pi}{3} + \varphi_S\right) \end{bmatrix}, \quad (15)$$

were φ_S – initial phase.

Take into account (7) space vector of source voltages is described as (16)

$$\underline{\mathbf{u}}_{S-SV} = \frac{2}{3}\left(U_{S\max}\cos(\omega t + \varphi_S) + U_{S\max}\cos\left(\omega t - \frac{2\pi}{3} + \varphi_S\right)e^{j\frac{2\pi}{3}} + U_{S\max}\cos\left(\omega t + \frac{2\pi}{3} + \varphi_S\right)e^{-j\frac{2\pi}{3}}\right) =$$

$$= \frac{2}{3}U_{S\max}\left(\frac{3}{2}(\cos(\omega t + \varphi_S) + j\sin(\omega t + \varphi_S))\right) = U_{S\max}e^{j(\omega t + \varphi_S)} \quad (16)$$

According to (8) the space vector of source voltages (16) is transformed to d-q coordinate system and can be described as (17) and (19).

$$\underline{\mathbf{u}}_{S-DQ} = \underline{\mathbf{u}}_{S-SV}e^{j\theta(t)}, \quad (17)$$

where

$$\theta(t) = (\pm\omega t + \varphi_0). \quad (18)$$

Take into account voltage ratios n_a i n_b and assumed $\varphi_0 = 0$ (18), the source voltages (part A and H) (Fig. 5) are described by (20) and (21).

$$\underline{\mathbf{u}}_{S-DQ} = U_{S\max}e^{\varphi_S} = U_{S\max}(\cos\varphi_S + j\sin\varphi_S), \quad (19)$$

$$n_a\underline{\mathbf{u}}_{S-DQ} = n_aU_{S\max}e^{\varphi_S} = n_aU_{S\max}(\cos\varphi_S + j\sin\varphi_S), \quad (20)$$

$$n_b\underline{\mathbf{u}}_{S-DQ} = n_bU_{S\max}e^{\varphi_S} = n_bU_{S\max}(\cos\varphi_S + j\sin\varphi_S). \quad (21)$$

Schematic diagrams of the source voltages $n_a\mathbf{u}_S$, $n_b\mathbf{u}_S$ before and after transformation are shown in Fig. 6a, b. Assumed conditions $L_{F1} = L_{F2} = L_{F3} = L_F$ the voltages $\mathbf{u}_{LF}$ (Fig. 5, part B) are given by (22).

$$L_F\frac{d}{dt}\mathbf{i}_{LF} = \mathbf{u}_{LF}, \quad (22)$$

where

$\mathbf{i}_{LF} = \begin{bmatrix} i_{LF1} & i_{LF2} & i_{LF3} \end{bmatrix}^T$ - currents vector,

$\mathbf{u}_{LF} = \begin{bmatrix} u_{LF1} & u_{LF2} & u_{LF3} \end{bmatrix}^T$ - line voltages vector.

According to (7) The space vector of voltages $\mathbf{u}_{LF}$ is given:

$$\underline{\mathbf{u}}_{LF-SV} = \frac{2}{3}(u_{LF1} + u_{LF2}e^{j\frac{2\pi}{3}} + u_{LF3}e^{-j\frac{2\pi}{3}}) =$$

$$= \frac{2}{3}L_F\frac{d}{dt}(i_{LF1} + i_{LF2}e^{j\frac{2\pi}{3}} + i_{LF3}e^{-j\frac{2\pi}{3}}) = L_F\frac{d}{dt}\underline{\mathbf{i}}_{LF-SV}. \quad (23)$$

According to (8) space vector (23) is transformed to the d-q coordinate sysem (24) (Fig. 6c).

$$\underline{\mathbf{u}}_{LF-DQ} = \underline{\mathbf{u}}_{LF-VS}e^{j\omega t}. \quad (24)$$

Take into account (23) and (24) we obtain:

$$\underline{\mathbf{u}}_{LF-DQ} = L_F\frac{d}{dt}\underline{\mathbf{i}}_{LF-VS}e^{j\omega t} = L_F(\frac{d}{dt}\underline{\mathbf{i}}_{LF-VS}e^{j\omega t} + j\omega\underline{\mathbf{i}}_{LF-VS}e^{j\omega t}) =$$

$$= L_F\frac{d}{dt}\underline{\mathbf{i}}_{LF-DQ} + j\omega L_F\underline{\mathbf{i}}_{LF-DQ}. \quad (25)$$

The same method is used to describe voltages $\mathbf{u}_{LL}$ (Fig. 5, part E). The voltages $\mathbf{u}_{LL}$ in d-q coordinate system (Fig. 6d) are given as:

$$\underline{\mathbf{u}}_{LL-DQ} = L_L\frac{d}{dt}\underline{\mathbf{i}}_{LL-DQ} + j\omega L_L\underline{\mathbf{i}}_{LL-DQ}. \quad (26)$$

Assumed conditions $C_{F1} = C_{F2} = C_{F3} = C_F$, vector of currents $\mathbf{i}_{CF}$ (Fig. 5, part C) is given by (27).

$$C_F\frac{d}{dt}\mathbf{u}_{CF} = \mathbf{i}_{CF}, \quad (27)$$

where:

$\mathbf{u}_{CF} = \begin{bmatrix} u_{CF1} & u_{CF2} & u_{CF3} \end{bmatrix}^T$ - line voltage vector,

$\mathbf{i}_{CF} = \begin{bmatrix} i_{CF1} & i_{CF2} & i_{CF3} \end{bmatrix}^T$ - line current vector.

Take into account (7) and (27) the space vector of the currents $\mathbf{i}_{CF}$ is described by (28).

$$\underline{\mathbf{i}}_{CF-SV} = \frac{2}{3}(i_{CF1} + i_{CF2}e^{j\frac{2\pi}{3}} + i_{CF3}e^{-j\frac{2\pi}{3}}) =$$

$$= \frac{2}{3}C_F\frac{d}{dt}(u_{CF1} + u_{CF2}e^{j\frac{2\pi}{3}} + u_{CF3}e^{-j\frac{2\pi}{3}}) = \frac{2}{3}C_F\frac{d}{dt}\frac{3}{2}\underline{\mathbf{u}}_{CF-SV} = C_F\frac{d}{dt}\underline{\mathbf{u}}_{CF-SV} \quad (28)$$

According to (8) the space vector (28) is transformed to the d-q coordinate system (29) (Fig. 7a).

$$\underline{\mathbf{i}}_{CF-DQ} = \underline{\mathbf{i}}_{CF-SV}e^{j\omega t}. \quad (29)$$

Take into account (28) and (29) we obtain (30):

$$\underline{\mathbf{i}}_{CF-DQ} = C_F\frac{d}{dt}\underline{\mathbf{u}}_{CF-SV}e^{j\omega t} = C_F(\frac{d}{dt}\underline{\mathbf{u}}_{CF-SV}e^{j\omega t} + j\omega\underline{\mathbf{u}}_{CF-SV}e^{j\omega t}) =$$

$$= C_F\frac{d}{dt}\underline{\mathbf{u}}_{CF\_DQ} + j\omega C_F\underline{\mathbf{u}}_{CF\_DQ}. \quad (30)$$

The same method is used to describe currents $\mathbf{i}_{CL}$ (Fig. 5, part G). The currents $\mathbf{i}_{CL}$ in d-q coordinate system (Fig. 7b) are given as:

$$\underline{\mathbf{i}}_{CL-DQ} = C_L\frac{d}{dt}\underline{\mathbf{u}}_{CL\_DQ} + j\omega C_L\underline{\mathbf{u}}_{CL\_DQ}. \quad (31)$$

Assumed conditions $R_{L1} = R_{L2} = R_{L3} = R_L$ the load voltages vector $\mathbf{u}_L$ of considered HT can be described as:

$$\mathbf{u}_L = R\mathbf{i}_L, \quad (32)$$

where: $\mathbf{u}_L = \begin{bmatrix} u_{L1} & u_{L2} & u_{L3} \end{bmatrix}^T$ - line load voltages vector,

$\mathbf{i}_L = \begin{bmatrix} i_{L1} & i_{L2} & i_{L3} \end{bmatrix}^T$ - line load current vector.

Load voltage space vector is given as:

$$\underline{\mathbf{u}}_{L\_SV} = \frac{2}{3}R(i_{L1} + i_{L2}e^{j\frac{2\pi}{3}} + i_{L3}e^{-j\frac{2\pi}{3}}) = R\underline{\mathbf{i}}_{L\_SV}. \quad (33)$$

Equation (34) is described the load voltage in d-q coordinate system (Fig. 7e).

$$\underline{\mathbf{u}}_{L\_DQ} = \underline{\mathbf{u}}_{L\_SV}e^{j\omega}. \quad (34)$$

Take into account (33) and (34) we obtain (35):

$$\underline{\mathbf{u}}_{L\_DQ} = R\underline{\mathbf{i}}_{L\_SV}e^{j\omega} = R\underline{\mathbf{i}}_{L\_DQ}. \quad (35)$$

The voltages vector of part D and F (Fig.5) are described as (36) and (37) respectively.

$$\mathbf{u}_1 = \mathbf{u}_{CF}D, \quad (36)$$

$$\mathbf{u}_{CL} = \mathbf{u}_2(1-D), \quad (37)$$

where: $\mathbf{u}_{CF} = \begin{bmatrix} u_{CF1} & u_{CF2} & u_{CF3} \end{bmatrix}^T$, $\mathbf{u}_1 = \begin{bmatrix} u_{11} & u_{12} & u_{13} \end{bmatrix}^T$, $\mathbf{u}_{CL} = \begin{bmatrix} u_{CL1} & u_{CL2} & u_{CL3} \end{bmatrix}^T$, $\mathbf{u}_2 = \begin{bmatrix} u_{21} & u_{22} & u_{23} \end{bmatrix}^T$.

According to (7) and (36) the space vector of voltage $\mathbf{u}_1$ is given by equation (38).

$$\underline{\mathbf{u}}_{1\_SV} = D\frac{2}{3}(u_{CF1} + u_{CF2}e^{j\frac{2\pi}{3}} + u_{CF3}e^{j\frac{2\pi}{3}}) = D\underline{\mathbf{u}}_{CF\_SV}, \quad (38)$$

According to (8) and (38) we obtain vector voltage $\mathbf{u}_1$ in d-q coordinate system (39) (Fig. 7c).

$$\underline{\mathbf{u}}_{1\_DQ} = D\underline{\mathbf{u}}_{CF\_SV}e^{j\omega} = D\underline{\mathbf{u}}_{CF\_DQ}. \quad (39)$$

By the same method, we obtain voltage $\mathbf{u}_{CL}$ (37) as space vector (40) and vector in d-q coordinate system (41) (Fig. 7d).

$$\underline{\mathbf{u}}_{CL\_SV} = (1-D)\frac{2}{3}(u_{21} + u_{22}e^{j\frac{2\pi}{3}} + u_{23}e^{j\frac{2\pi}{3}}) = (1-D)\underline{\mathbf{u}}_{2\_SV}, \quad (40)$$

$$\underline{\mathbf{u}}_{CL\_DQ} = (1-D)\underline{\mathbf{u}}_{2\_SV}e^{j\omega} = (1-D)\underline{\mathbf{u}}_{2\_DQ}. \quad (41)$$

Schematic diagrams of each elements of HT before and after d-q transformation are shown in Figs. 7 and 8.

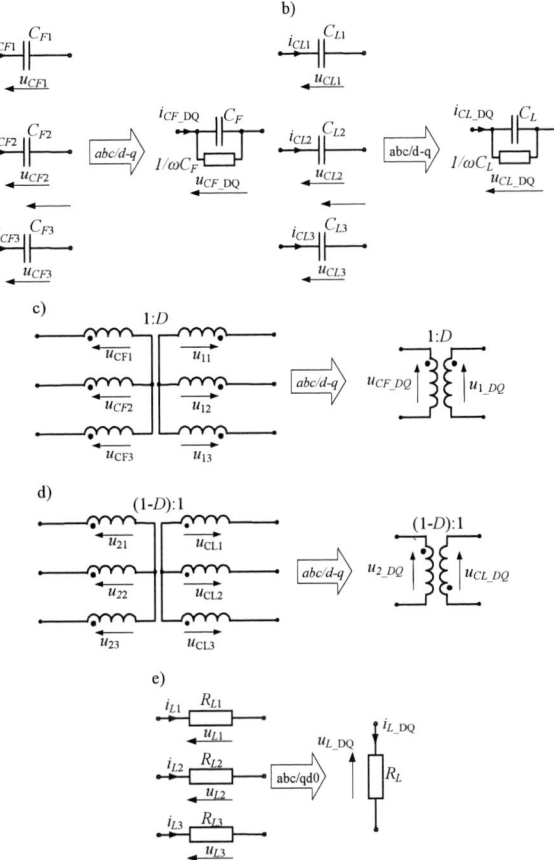

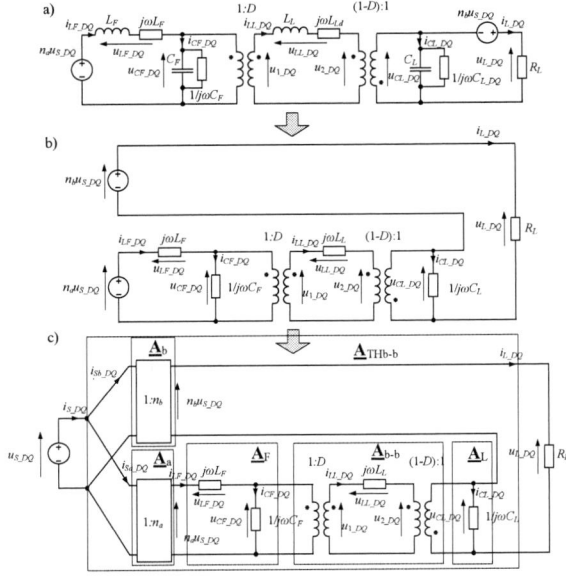

Fig. 7. Schematic diagrams of each parts of the considered HT (Fig. 5) before and after d-q transformation, 7 a), b), capacitances (part. C, G), c), d), averaged models of switches (part. D, F), e) load resistances (part. I)

The steady state characteristics can be obtained by considering the d-q circuit model of considered HT (Fig.8a).

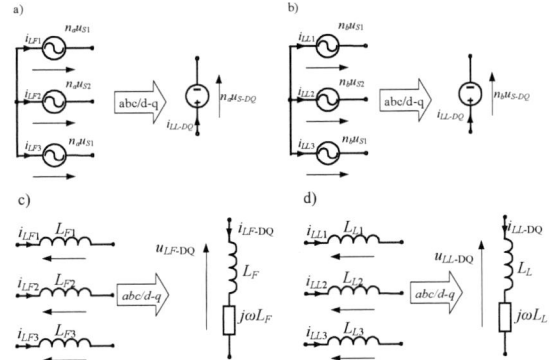

Fig. 6. Schematic diagrams of each parts of the considered HT (Fig. 5) before and after d-q transformation, a), b) source voltages (part A and H), c), d), inductances (part B, E)

Fig. 8. Circuit model of considered HT in d-q coordinate system, a) general form, b) with imaginary resistances, c) with four terminal description

978-1-4244-8806-3/11 $26.00 © 2011 IEEE

For steady state analysis a single-phase circuit model is divided into four terminal networks (Fig. 8c) [7]-[12]. Referring to Fig. 8 four-terminal chain equations in complex form can be written as (42) – (46).

$$
\begin{bmatrix} \underline{U}_{S\_DQ} \\ \underline{I}_{Sa\_DQ} \end{bmatrix} = \begin{bmatrix} \dfrac{1}{n_a} & 0 \\ 0 & n_a \end{bmatrix} \begin{bmatrix} n_a \underline{U}_{S\_DQ} \\ \underline{I}_{L\_DQ} \end{bmatrix} = \underline{\mathbf{A}}_a \begin{bmatrix} n_a \underline{U}_{S\_DQ} \\ \underline{I}_{L\_DQ} \end{bmatrix}, \quad (42)
$$

$$
\begin{bmatrix} \underline{U}_{S\_DQ} \\ \underline{I}_{Sb\_DQ} \end{bmatrix} = \begin{bmatrix} \dfrac{1}{n_b} & 0 \\ 0 & n_b \end{bmatrix} \begin{bmatrix} n_b \underline{U}_{S\_DQ} \\ \underline{I}_{LF\_DQ} \end{bmatrix} = \underline{\mathbf{A}}_b \begin{bmatrix} n_b \underline{U}_{S\_DQ} \\ \underline{I}_{LF\_DQ} \end{bmatrix}, \quad (43)
$$

$$
\begin{bmatrix} \underline{U}_{CL\_DQ} \\ (1-D)\underline{I}_{LL\_DQ} \end{bmatrix} = \begin{bmatrix} 1 & 0 \\ j\omega C_L & 1 \end{bmatrix} \begin{bmatrix} \underline{U}_{CL\_DQ} \\ \underline{I}_{L\_DQ} \end{bmatrix} = \underline{\mathbf{A}}_L \begin{bmatrix} \underline{U}_{CL\_DQ} \\ \underline{I}_{L\_DQ} \end{bmatrix}, \quad (44)
$$

$$
\begin{bmatrix} n_b \underline{U}_{S\_DQ} \\ \underline{I}_{LF\_DQ} \end{bmatrix} = \begin{bmatrix} 1-j\omega^2 L_F C_F & j\omega L_F \\ j\omega C_F & 1 \end{bmatrix} \begin{bmatrix} \underline{U}_{CF\_DQ} \\ \underline{I}_{LL\_DQ} \end{bmatrix} = \underline{\mathbf{A}}_F \begin{bmatrix} \underline{U}_{CF\_DQ} \\ D\underline{I}_{LL\_DQ} \end{bmatrix}, \quad (45)
$$

$$
\begin{bmatrix} \underline{U}_{CF\_DQ} \\ D\underline{I}_{LL\_DQ} \end{bmatrix} = \begin{bmatrix} \dfrac{1-D}{D} & \dfrac{j\omega L_L}{D(1-D)} \\ 0 & \dfrac{D}{1-D} \end{bmatrix} \begin{bmatrix} \underline{U}_{CL\_DQ} \\ (1-D)\underline{I}_{LL\_DQ} \end{bmatrix} = \underline{\mathbf{A}}_{b-b} \begin{bmatrix} \underline{U}_{CL\_DQ} \\ (1-D)\underline{I}_{LL\_DQ} \end{bmatrix}, \quad (46)
$$

Take into account connection between four terminal networks (Fig. 8c) we can easily obtain chain parameters of considered HT described by (47) – (54).

$$
\begin{bmatrix} \underline{U}_{S\_DQ} \\ \underline{I}_{S\_DQ} \end{bmatrix} = \underline{\mathbf{A}}_{THb-b} \begin{bmatrix} \underline{U}_{L\_DQ} \\ \underline{I}_{L\_DQ} \end{bmatrix} = \begin{bmatrix} \underline{A}_{THb-b11} & \underline{A}_{THb-b12} \\ \underline{A}_{THb-b21} & \underline{A}_{THb-b22} \end{bmatrix} \begin{bmatrix} \underline{U}_{L\_DQ} \\ \underline{I}_{L\_DQ} \end{bmatrix} \quad (47)
$$

where:

$$
\underline{A}_{THb-b11} = \frac{1}{p_b + \dfrac{(-1+D)Dp_a}{-(-1+D)^2 + \omega^2 M1}}, \quad (48)
$$

$$
\underline{A}_{THb-b12} = -\frac{i\omega(-D^2 L_F + (-1+\omega^2 C_F L_F)L_L)}{M2}, \quad (49)
$$

$$
A_{THb-b21} = \frac{i\omega p_a^2(-D^2 C_L + C_F((-1+D)^2 - \omega^2 C_L L_L))}{M2}, \quad (50)
$$

$$
A_{THb-b22} = p_b + p_a\left(\frac{(-1+D)D(1+\det \underline{\mathbf{G}})}{M1} + \frac{D(D(p_a - p_b) + p_b) - p_a \omega^2 C_F L_L}{M2}\right), (51)
$$

$$
M1 = -(-1+D)^2 + \omega^2((C_L(D^2 L_F + L_L) + C_F L_F((-1+D)^2 - \omega^2 C_L L_L)), \quad (52)
$$

$$
M2 = (-1+D)(Dp_a - p_b) + p_b \omega^2(C_L(D^2 L_F + L_L) + C_L L_L((-1+D)^2 - \omega^2 C_L L_L)), \quad (53)
$$

$$
\det \underline{\mathbf{G}} = \frac{p_a^2 \omega^2(-D^2 L_F + (-1+\omega^2 C_F L_F)L_L)(-D^2 C_L + C_F((-1+D)^2 - \omega^2 C_L L_L))}{M1^2} - \left(p_b + \frac{(-1+D)Dp_a}{M1}\right)^2. \quad (54)
$$

In accordance with four terminal theories, we obtain:

$$
|\underline{H}_U| = \left|\frac{\underline{U}_L}{\underline{U}_S}\right| = \left|\frac{1}{\underline{A}_{HTb-b11} + \underline{A}_{HTb-b12}/R_L}\right|, \quad (55)
$$

$$
\arg \underline{H}_U = \arg\left(\frac{1}{\underline{A}_{HTb-b11} + \underline{A}_{HTb-b12}/R_L}\right), \quad (56)
$$

$$
\lambda_S = \frac{P_S}{S_S} = \cos\left[\arg\left(\frac{\underline{A}_{HTb-b11}R_L + \underline{A}_{HTb-b12}}{\underline{A}_{HTb-b21}R_L + \underline{A}_{HTb-b22}}\right)\right]. \quad (57)
$$

Static characteristics as a function duty pulse factor D of magnitude (55), phase (56) and input power factor (57) are shown in next section in Figure 9a, b and 10 respectively.

IV.   SIMULATION AND EXPERIMENTAL TEST RESULTS

The parameters of the circuit shown in Fig 1, which has been investigated, are collected in the appendix in table I. The presented results have been obtained for matching conditions described by (58).

$$
\sqrt{\frac{L_F}{C_F}} \approx \sqrt{\frac{L_L}{C_L}} \approx R_L. \quad (58)
$$

Static characteristic magnitude (55), phase (56) and input power factor (57) as a function pulse duty factor D are shown in Figs. 9 and 10. The results of theoretical analysis are collected with simulation and experimental test results.

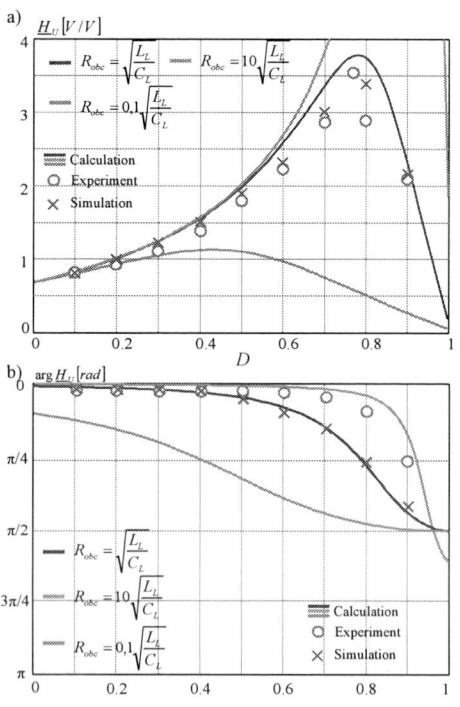

Fig. 9. Static characteristic HT as a function of the pulse duty factor D, a) magnitude of voltage transmittance, b) phase of voltage transmittance,

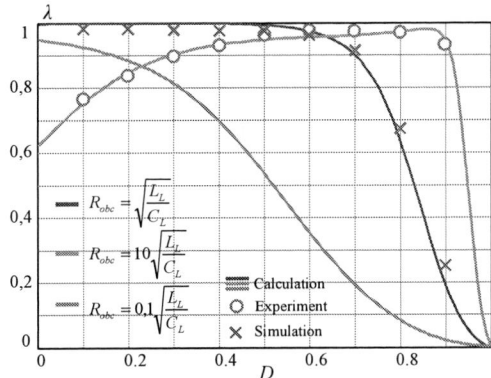

Fig. 10. Static characteristic HT as a function of the pulse duty factor D of input power factor

Experimental source and load line voltage and current time waveforms for a various value pulse duty factor D are shown in Fig. 11.

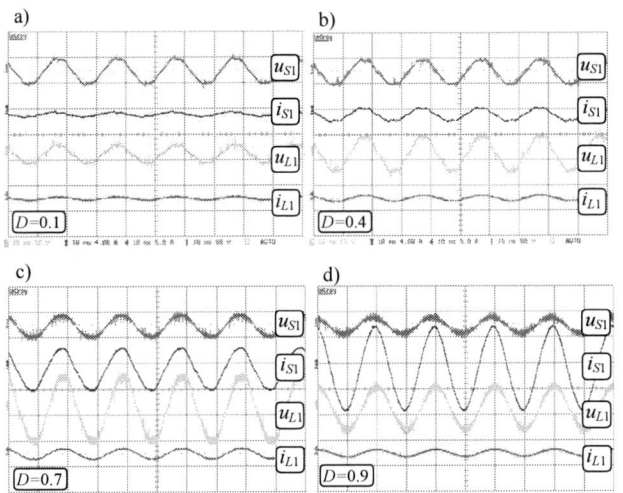

Fig. 11. Experimental source and load line voltage and current time waveforms for, a) $D = 0.1$, b) $D = 0.4$, c) $D = .7$, d) $D = 0.9$

As is visible from Fig. 9a and Fig. 11 the output voltage of considered HT is less, approximately equal or greater than source voltage for $D < 0.2$, for $D = 0.2$ and for $D > 0.2$ respectively. The range of change of output voltage is from $0.66u_S$ to more than $3u_S$.

V.   CONCLUSIONS

In this paper the results of modeling of a three–phase hybrid transformer using buck-boost type matrix-reactance chopper has been presented. Modeling of HT is based on averaging method and four–terminal description method in d-q coordinate system. Generally the results of the simulation investigation confirm the results of theoretical study. Differences between analytic and simulation / experimental results are caused by higher harmonics taking into account during simulation / experimental investigation (non-stationary circuit).The range of change of output voltage gives the possibility of using proposed HT for compensation of 50%

sags and over 50% swells of source voltage. Further research will be focused on detailed analysis of non-balanced circuit.

VI.   APPENDIX

Table I. Circuit parameters

Parameter	name	value	unit
p_a	voltage ratio	4/3	-
p_b	voltage ratio	2/3	-
L_F / L_L	MRC inductance	1	mH
C_F / C_L	MRC capacitance	10	μF
R_L	Load resistance	10 *	Ω
f_s	switching frequency	5	kHz
U_S	supply voltage	3x400 *	V
* experiment - R_L=60 Ω, U_S = 50V			

REFERENCES

[1] J. Milanowić, I. Hiskansen: "Effect of load dynamics on power system damping," IEEE Trans. on Power System Vol. 10 No. 2, pp. 1022 - 1028   May 1995

[2] Z. Djokic, J. Desment, G. Vanalme, J. Milanovic, K. Stockman: "Sensitivity of personal computer to voltage sags and short interruptions," IEEE Trans. on Power Delivery, vol. 20, No.1, pp. 375 - 383   Jan. 2005

[3] E. Aeloiza, P. Prased, P. Enjeti, L. Moran, O. Montero-Hernandez, S. Kim.: Analysis and design of new voltage sag compensator for critical loads in electric power dystrybution system. IEEE Trans. on Ind. Applications, vol. 39 No.4, pp.1143 - 1150 July / Aug. 2003

[4] O. Montero-Hernandez, P. Enjeti: "Application of a boost ac-ac converter to compensate for voltage sags in electric power distribution system," in Proc. 2000 IEEE 31TH Power Engineering Society Transmission and Distribution Conf. pp. 470 – 475

[5] S. Subramanian, M. K. Mishra, "Interphase AC-AC topology vor sag supporter", IEEE Tran. on Power Elec, vol. 25, No. 2, Feb. 2010

[6] E. Aeloiza, P. Enjeti, L. Moran, I. Pite: Next generation distribution transformer: to address power quality for critical loads .PESC'03 IEEE vol. 3 pp.1266 – 1271 June 2003.

[7] J. Kaniewski: "Single phase hybrid transformer using matrix converter" Wiadomości Elektrotech. 03.2006, pp: 46-48. (In Polish)

[8] Z. Fedyczak, J. Kaniewski: "Single phase hybrid transformer using bipolar matrix – reactance chopper," Przegląd Elektrotechniczny 07-08.2006, pp: 80-85. (In Polish)

[9] Fedyczak Z, Kaniewski J., "Modeling and analysis of three-phase hybrid transformer using matrix converter", Compatibility in Power Electronics - CPE 2007, 5th International Conference-Workshop. Gdańsk, Polska, 2007, Gdynia, 2007

[10] Kaniewski J, Fedyczak Z., "Modelling and analysis of three-phase hybrid transformer using matrix-reactance chopper", Przegląd Elektrotechniczny, 2009, nr 2, s. 100-105

[11] Kaniewski J., Fedyczak Z., Klytta M., Łukiewski M., Szcześniak P.,"Implementation of a three-phase hybrid transformer using a matrix chopper", 13th European Conference on Power Electronics and Applications, EPE 2009. Barcelona, Hiszpania, 2009

[12] Kaniewski J., Fedyczak Z., "Modeling and Analysis of Dynamic Properties of the Hybrid Transformer with MRC", "International school of Nonsinusoidal Current and Compensation", ISNCC 2010, Łagów, Poland

[13] Middlebrock R. D., Ćuk S.: "General unified approach to modeling switching converter power stages". Rec. IEEE PESC,76, pp. 18 – 34, 1976

[14] Soo-Bin Han, Gyu-Hyeong Cho, Bong-Man Jung, Soo-Hyun Choi: "Vector-Transformed Circuit Theory and Application to Converter Modeling/Analysis" 1998 IEEE, pp. 538-544

[15] C. T. Rim, D. Y. Hu, G. H. Cho: "Transformers as Equivalent Circuits for Switches: General Proofs and D-Q Transformation – Based Analyses," IEEE Trans on Ind Apl. vol.26 No. 4 July/Aug 1990

Dynamical Operational Behavior of the Power Drain of Wind Energy Converters with PMSM Considering Different Current Control Methods

Constantinos Sourkounis, Alexander Broy
Ruhr-University Bochum, Power Systems Technologie and Power Mechatronics
sourkounis@eele.rub.de, broy@eele.rub.de

Abstract-- **With increasing nominal wind park power, interest in the operating behavior of the plants and the wind park is developing. Main criteria for the quality of the operating behavior are based on mains pollution (e.g. harmonics) and the power fluctuation through variations of the wind speed. The dynamic and stationary operating behavior of wind energy converters (WEC) are defined by the drive train structure and the control algorithm. The essence of this paper is the investigation on the operating behaviors of different WECs generator systems in relation to their influence on the mechanical drive train's operating performance. Because of practical experiences so far, only a few concepts are suitable for the design of the electrical part. Therefore, in this paper, a theoretical and simulation based comparison will be carried out. Five different concepts (some state-of-the-art and some new creations especially for off-shore use) therefore, are compared in relation to necessary requirements.**

I. INTRODUCTION

The development of wind energy utilisation at the last twenty years contributes from technical and economical point of view to a high acceptance. These facts cause a rapid growth of the wind energy branch. The rapid trend has been and is still involved with a continuous high demand of wind energy capable areas. Now nearly all of this areas are already in use for wind energy utilisation, so there is a big interest, besides the utilisation of areas with low wind potential, in developing off-shore areas for the installation of wind energy plants. Because of the increased installation costs of off-shore wind parks there is a trend to wind parks with installation power up to 1 GW to decrease the specific costs.

The main parts of the investment are caused on the one hand by the necessary connection to the on-shore supply system, which is realised by a three-phase transmission line (up to 50 km) or by a High-Voltage-DC-Transmission (50 km and above) and on the other hand by the fixing of the wind energy converters to the sea ground [6]. Because of this there is, besides the high installed power per wind park, a demand for higher plant power. The current generation of wind power plants for off-shore wind parks reaches a nominal plant power of 5 MW and more.

With the increasing nominal power of the plants and the cumulated wind park power the interest in the operation behaviour of the plants and finally of the wind park is rising. One of the main criteria for the quality of the operating behaviour is based on the mains pollution (e.g. harmonics) caused by the wind energy converter itself and by the power fluctuation through the fluctuation of the wind speed. Furthermore the availability and the life time oft the subsystems as well as of the components constitute economical benefits. The dynamic and stationary operating behaviour of wind energy converter is on the one hand defined by the structure of the drive train and on the other hand by the control algorithm [8], [9]. The essence of this paper is the investigation on the operating behaviour of different concepts of wind energy converters with different concepts of rectifier control algorithms.

II. VALID SYSTEM TECHNICAL REQUIREMENTS

Based upon general valid requirements, which apply to every utilization method and power class for the operation of wind energy converters, are the system engineering requirements, which gain importance due to extreme requirements of off-shore utilization and the 5MW power class, and which are to be specified. The valid general requirements are as follows:

- operation with maximum possible power coefficient of the wind rotor, and
- least possible fluctuation in the power grid.

Operation with the maximum possible power coefficient of the wind rotor, which leads to maximum possible energy yield and the system's improved economic feasibility, is realized by adaptation of the rotation speed to the current wind speed. For partial load operation, this means a variable rotational speed. Simultaneously a correspondingly designed controlling process of the variable speed rotation system can be used to smooth temporary wind caused power fluctuations and reduce the resulting flicker effect. For this the rotating masses (i.e. the wind rotor) are used as a kinetic energy storage. During a gust of wind the rotor speed must be increased, so that the rotor operates at a maximum power coefficient. Thereby a portion of the energy drawn from the wind during the process of acceleration is saved in the rotating mass. During the following reduction in wind speed, energy gained during the matching of the rotor speed is converted into electrical energy and fed into the power grid.

The system engineering requirements, which complete the requirements defined by the process, are those of:

- a high availability,
- a long service life and

• long life time respectively.

The importance of these requirements of high uptime and a long service life for off-shore implementation is the result of high costs associated with the inaccessibility of the installation. Both of the system's technical requirements are, amongst other things, coupled with the requirement of smoothing fluctuations caused by the wind. Smoothing the temporary power fluctuations caused by the wind by using the mass's rotating reserve means a similar time lapse between the wind power and the rotational speed in the mechanical-electrical drive train. This leads to a reduction of the cumulated load in the drive train, and therefore also to reduced long-term damage of the components in the drive train. Furthermore to the mentioned high uptime and long life time, the interaction between the generator system and the mechanical power train are of essential importance.

Based upon the explained relationships between the process flow and the operating behaviour of the wind energy converter, the speed variable system is seen as a basic condition for fulfillment of the mentioned requirements. A technical solution for speed variable wind energy converter can be realized based on a permanent excited synchronous machine as generator in combination with a frequency converter. Because of the system's low maintenance requirements, the technically interesting variation of the synchronous generator for its off-shore use is the permanent magnet synchronous system, whereby slip-rings and brushes are avoided. The influence of generator systems with different rectifier systems and firing control algorithms on the behaviour of electro-mechanical drive train will be investigated.

III. CONCEPTS OF GENERATOR SYSTEMS WITH PM-SYNCHRONOUS MACHINE

A range of technical concepts exists for the realization of a variable speed system for wind energy converters based on a permanent magnet synchronous machine. By the basic concept with the synchronous machine as generator, the necessary decoupling of the generator's rotation speed is realized through the use of a DC link converter. Building upon the basic concept, several reasonable modifications from modern technology can be conceived. These deal with combinations between the functional principal of the generator side rectifier and the structure of mechanical drive train (see Fig. 1).

Concepts relevant for off-shore use of the mechanical electrical energy conversion system for WECs are shown in Fig. 1. The first system shows the concept of PMSM with an uncontrolled rectifier. This concept has a speed dependent generator voltage, therefore to hold the dc-link voltage constant a step up chopper is necessary. The second generator system consists of a PMSM and dc-link inverter with a self-commutated rectifier. The self-commutated rectifier allows an operation with cos φ = 1. Also an underexcited operation of the synchronous generator is possible, whereby the technical cost in relation to the height of the rotor's permanent flux can

be reduced. The technical gain from realisation of the permanent-field synchronous generator is relativised by the high cost of the necessary self-commutated rectifier.

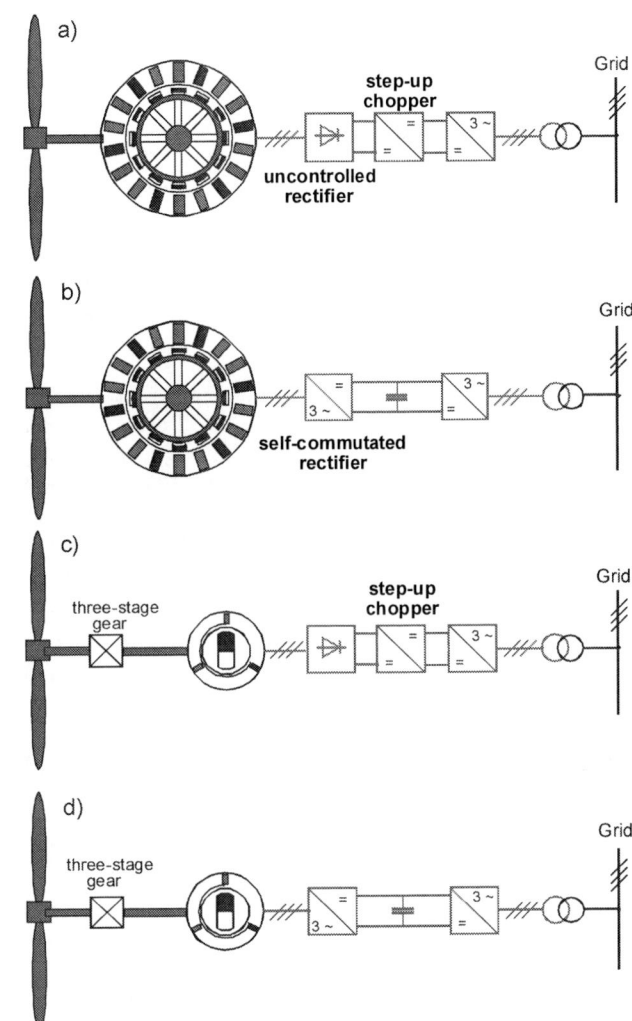

Fig. 1. Comparison of different power drain for speed variable WECs; a) gearless PMSM with uncontrolled rectifier, b) gearless PMSM with self-controlled rectifier, c) PMSM with uncontrolled rectifier, d) PMSM with self-controlled rectifier

The operating behaviour of the self-commutated rectifier depends on the control method. Therefore the operating behaviour of the generator system with self-commutated rectifier with different control methods were studied and compared to each other. Following control methods were implemented:

- the current hysteresis method,
- die sine-delta-modulation.

The interaction between the mechanical part of the power train and the generator system (electrical part of the power train) defines amongst others the dynamical operating behaviour of the mechanical-electrical power train. Their direct influence on the cumulated load in the power drain also influences the availability and life time of the components in the power drain. Two different structures of the mechanical part of the power train were considered in this respect. By the use of multi-pole synchronous machines, the gear drive can be avoided, whereby a highly integrated mechanical layout of the power train can be realised. The tight construction leads to a high torsional stiffness and a relatively high first natural resonance. In contrast, power trains with discrete layout (Fig. 1c, 1d) have a low torsional stiffness and therefore a low first natural resonance, 4 Hz to 12 Hz. Because of the different torsional stiffness and natural resonance frequency, it is expected that the influence of the generator system on the operating behaviour of the mechanical part of the power train are noteworthy different from each other, depending on the layout.

IV. COMPARATIVE INVESTIGATIONS

The goal of the investigation is to generate a proof if the posed requirements are fulfilled by the listed concepts. The emphasis of the investigation is to take into account, on the one hand, the effect of the interaction between generator and rectifier on the energy converter's degree of utilization, and its effect on the dynamic operational behaviour of the mechanical drive train. The dynamics, as well as the selection of the deployed actuator (rectifier) effect should, on the other hand, be portrayed by the process flow of the wind energy converter, as well as on the installation's dampening capability.

A. Analytical Investigations

Investigations of the interaction between generator and rectifier can be reduced to two basic rectifier topologies. The externally commutated rectifier in its two varieties (controlled or uncontrolled) portrays one basic type. The other is portrayed by the self-commutated higher switched IGBT-converter. With an externally commutated rectifier, the generator is run overexcited and delivers the reactive power for the commutation, and in the case of a controlled bridge, the phases control the reactive power. That leads to a higher necessary synchronous generated voltage in comparison to a system with a self-commutated rectifier (power factor of 1), and therefore to a low utilization of the generator. The following investigations in the scope of this contribution are carried out with the uncontrolled rectifier only, because this topology is still being used due to its low technical complexity. The externally commutated rectifier based on thyristors lost practical relevancy for pulse generation due to its high technical complexity. With uncontrolled rectifiers the generator has to provide the reactive power for commutation. That especially leads to, with permanent synchronous-field generators, a higher technical expenditure and a greater weight. The commutation time span determines the needs of reactive power. The so called overlapping, which expresses the time span of the commutation, is calculated according to [1], [3]

$$\cos(\alpha + u) = \cos(\alpha) \cdot \frac{2\omega L_k \cdot I_{dc}}{\sqrt{2} \cdot U_k} \tag{1}$$

and depends on the effective individual L_k in the commutation circuit. In this case it deals mainly with the reactance $x_k = x_d = x_q$ of the generator, which lies at approximately 30% - 50% p.u. by the permanent-field synchronous machine. The equation in p.u. calls for a trigger delay angle α of $\alpha = 0$ (uncontrolled):

$$\cos(u_0) = 1 - \frac{\pi}{3} \cdot x_k \cdot i_{dc} \tag{2}$$

Separately excited synchronous generators show a slightly active commutation impedance through the mentioned damper cage with:

$$x_k = \frac{x_d'' + x_q''}{2} \tag{3}$$

The overlapping of normal commutation impedances from $x_k = 8 \ldots 10\%$ can be significantly reduced, and thereby an improvement of the generator's utilization level can be achieved. The rectifier necessary for the excitation is used as the final controlling element in the power control structure. The DC link voltage will be supplied, as needed, over the excitation circuit. The otherwise implemented step-up converter is thereby no longer needed.

Besides the need for reactive power, the uncontrolled rectifier generates harmonics in the generator's phase currents, due to its functional principle. The harmonics have the ordinal numbers of

$$v = k \cdot p \pm 1 \tag{4}$$

with $k = 1, 2, 3, \ldots$and p being the pulse number (for B6-topology $p = 6$). Because the fundamental oscillation frequency of the generator current directly depends on the rotational speed n_G and the number of pole pairs p_z,

$$\omega_1 = p_z 2\pi n_G \quad \text{and} \quad f_1 = p_z n_G, \tag{5}$$

the analysed variable speed WECs have varying fundamental oscillation frequencies and therefore also varying current harmonics frequencies. For low operation speeds the frequencies of the current harmonics shift close to the mechanical fundamental frequencies, whereby torsional oscillations can be induced. The harmonics caused by the rectifier superimpose with the harmonics of the dc-link current, which are caused by the step-up chopper. It has a frequency spectrum where the first frequency component corresponds to the pulse frequency of the actuator. By the superimposition of the harmonics with different frequencies results a periodical amplification and attenuation of the harmonics amplitude. This results in a harmonic with a frequency lower than $5f_1$.

The amplitude of the specific harmonics depends on the order of the harmonics and the duration of the commutation, in particular on the effective inductance in the commutation circuit in relation to the voltage available for commutation. A

978-1-4244-8806-3/11 $26.00 © 2011 IEEE

longer commutation generally leads to lower amplitudes of the harmonics. For the permanent synchronous machine the high inner resistance ($x_d = x_q = 0.3 \ldots 0.6$ p.u.) has positive effects on the generator's current ripple. Utilization of a self-commutating rectifier is another approach towards increasing the generator's level of utilization. The advantage thereby is that the permanent-field synchronous generator can be designed for underexcited operation (i.e. $\cos \varphi = 0.9$ inductive). The reactive power necessary for that is delivered over the stator winding by the rectifier. In spite of an over-dimensioning of the rectifier, a reduction in the specific weight by the mechanical-electrical converters (generator and DC link converter) is achieved. Besides the possibility of a high utilisation rate of the PMSM by operation with $\cos \varphi = 1$ or even an inductive $\cos \varphi < 1$ (e.g. 0.9), lower generator current ripple are the result of high pulse frequency of self commutated rectifiers. Therewith results less load from the distortional reactive power on the synchronous generator.

For rectifier with sine-delta control the first harmonic of the generator current is equal to the pulse frequency [3]. Further harmonics occur at multiples of the pulse frequency. The high frequencies lead to low amplitudes of the harmonics, because inner reactances of the generator (L_d und L_q) have stronger effect at higher frequencies. The current hysteresis control directly limits the deviation of the actual current from the given sine wave by putting a tolerance band around the reference sine wave. Depending on the width of the tolerance band Δi_B and with a fixed voltage ratio between the generator and dc-link, a mean pulse frequency $f_{s,av}$ results. During one period the pulse frequency varies depending on the slope of the sine wave. A simple approach is given by:

$$u_{dc} - \hat{u}_{P,L} \sin \omega t \approx L_d \, \Delta i_B / \Delta t . \qquad (6)$$

Therewith the actual pulse frequency can be determined:

$$
\begin{aligned}
f_s &\approx \frac{u_{dc} - \hat{u}_{P,L} \sin \omega t}{2 \cdot L_d \Delta i_B} = \frac{u_{dc} \left(1 - \hat{u}_{P,L}/u_{dc} \cdot \sin \omega t\right)}{2 \cdot L_d \Delta i_B} \\
&\approx \frac{u_{dc} \left(1 - \sqrt{2}\, \Psi_{PM} \, \omega / u_{dc} \cdot \sin \omega t\right)}{2 \cdot L_d \Delta i_B} ,
\end{aligned}
\qquad (7)
$$

with Ψ_{PM} : permanent magnetic flux,
p_z : number of pole pairs,
n_G : generator's rotational speed,
L_d : generators's synchronous inductance,
u_{dc} : dc-link voltage,
$u_{P,L}$: generator' pulse wheel voltage.

From these equations, it is obvious to see that a spectrum of harmonics results and not specific harmonics with dedicated frequencies. The frequencies of the harmonics, for a constant dc-link voltage, are directly depending on the rotational speed of the synchronous generator. With increasing rotational speed the frequency spectrum drops and with it also the current harmonics. Caused by the current

harmonics and according to:

$$M_i = \frac{3}{2} p_z \cdot \Psi_{PM} I_{sq}$$

with $i_{sq}(t) =$

$$\frac{2}{3}\left(i_a(t) \cos \omega_n t + i_b(t) \cos\left(\omega_n t - \frac{2\pi}{3}\right) + i_c(t) \cos\left(\omega_n t + \frac{2\pi}{3}\right) \right) \qquad (8)$$

and $\omega_n = 2\pi p_z n_G$

torque ripples results, which can have noteworthy load on the mechanical parts of the drive train by inducing torsional oscillations. For the consideration of the influence of the generator system on the operating behaviour of the drive train's mechanical parts, a simplified mathematical model of the drive train is taken into account. The basic structure of the power train has been used for the conceptional design and layout of the torque control. As shown in Fig. 2 it can be divided into a mechanical and electrical subsystem.

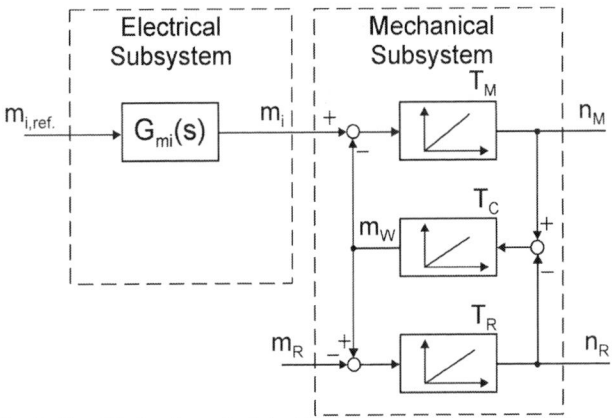

Fig. 2. Simplified mathematical electric-mechanical power train model

Based on the general differential equation of movement, the dynamic behavior of the mechanical subsystem is represented by the mathematical model of a two-mass-spring-system (see Fig. 2), simplified by the assumption that the mass of the shaft is negligibly low. This simplification also means that the mechanical subsystem of the power train and its operating behavior respectively can be defined substantially by the only dominant self-motion [2], [4]. From the simplified mathematical model derives the transfer function of the shaft torque m_w. After conduction of the Laplace transformation and scaling, the following transfer function is derived:

$$m_w(s) = \frac{1}{v+1} \cdot \frac{1}{s^2 T_{ef}^2 + 1} \cdot m_i(s) + \frac{v}{v+1} \cdot \frac{1}{s^2 T_{ef}^2 + 1} \cdot m_L(s)$$

$$\text{with} \quad T_{ef} = \frac{1}{\omega_{ef}} = \sqrt{\frac{T_C T_M}{(1+v)}}, \quad v = \frac{J_M}{J_R} = \frac{T_M}{T_R} . \qquad (9)$$

Input variables are here the air gap moment m_i and the load torque m_L. The mechanical values (mass inertia and spring constant) have been taken into account by introducing the time constants

T_M: start-up time constant of the motor,
T_R: start-up time constant of the load mass
T_C: time constant of the shaft spring constant
and
ν: mass inertia ratio.

From the transfer function a direct correlation occurs between generator torque M_i propagation in the power train and the mass inertia ratio ν. At decreasing mass inertia ratio ν but constant total mass inertia, higher impact of generator torque on the time behavior of the shaft torque, as well as higher torque oscillations have to be expected in the mechanical drive train.

B. Simulation studies

For the simulation studies following machine parameters were modeled, generator power $P_{Gen} = 5MW$, $p_z = 72$, nominal rotor speed $n_{nenn,R} = 14.4$ rpm, $U_{P,L} = 3.3$ kV and $x_d = x_q = 0.45$. The simulation results of the different generator systems were analysed and compared to each other. To improve the comparability of the simulation results, relevant machine parameters are normalised to their mean

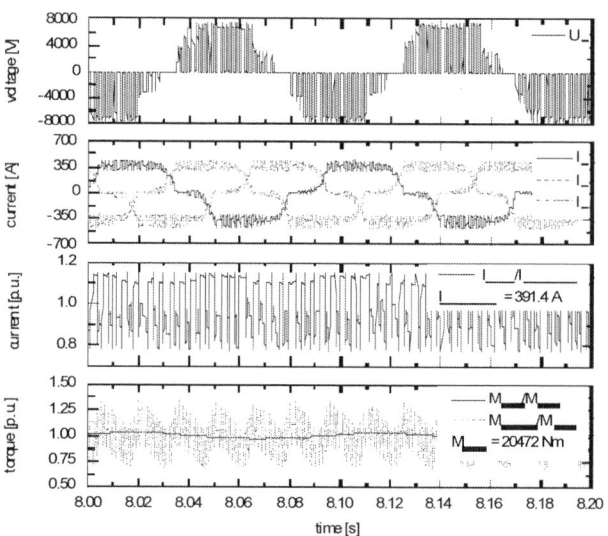

Fig. 3 Time diagram for uncontrolled rectifier

value. In Fig. 3, 4, 5 the time diagram of the generator voltage, phase current, dc-link current and the electrical generator torque are shown for each generator system separately. In the time diagram of the voltage and phase current the influence of the rectifier's functional principle can be seen clearly. With the uncontrolled rectifier the current shows a nearly rectangular behaviour for 120° electrical. The simulated synchronous machine does not have damper windings, the relatively high effective inner transient reactance leads to a relatively long commutation duration of about u = 30° elec., which has positive effects on the current.

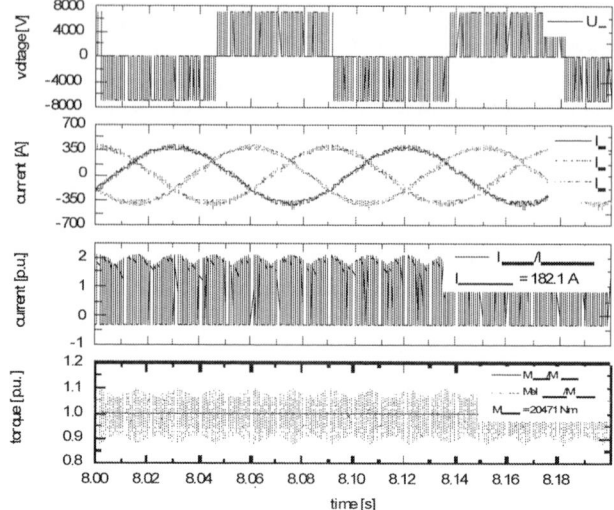

Fig. 4 Time diagram for sine-delta-modulation

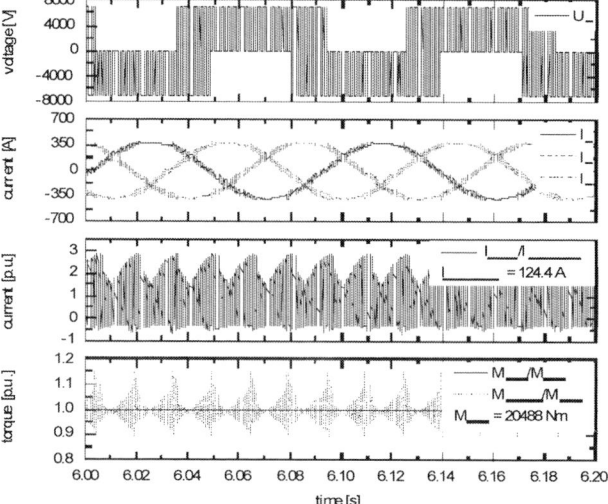

Fig. 5 Time diagram for current hysteresis control.

Though the low utilisation ratio of the synchronous generator, cos φ = 0.96, is disadvantageous. The current and voltage diagram shows pulsations which are caused by the step-up chopper in the dc-link. The thereby resulting high frequency harmonics in the phase current also cause, on their part, ripples which superimpose the ripples in the diagram of the electrical generator torque caused by the uncontrolled rectifier. The superimposition causes a beat frequency. The phase current diagram of the generator system with self-commutated rectifier shows a sinusoidal bahaviour with superimposed high frequency oscillations. The spectrum of the harmonics is directly depending on the control method of the rectifier. Therefore the diagrams of the generator torque show ripples with amplitudes of 20 % for the current hysteresis control and 10 % for the sine-delta modulation, in relation to the mean torque. Furthermore the torque diagram shows a beat frequency of approx 65 Hz, which is about six times the generator's natural frequency. The beat frequency is more obvious with the current hysteresis method. To judge the influence of the generator system on the mechanical parts of the drive train, the shaft torque and the generator's counter torque were analyzed in the frequency domain. The results are shown in Fig. 6 and 7.

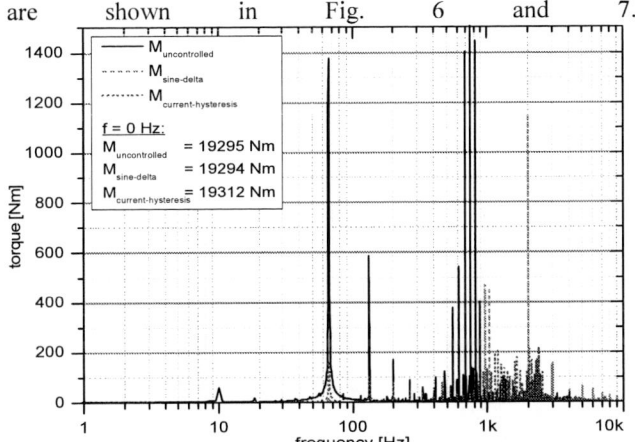

Fig. 6 Frequency analysis of the electrical generator torque

The frequency spectrum of the electrical generator torque from a system with uncontrolled rectifier has harmonics at 65 Hz, which is six times the generator's natural frequency. The first harmonic has an amplitude of about 7.1 % of the mean value, with decreasing amplitudes for higher harmonics. The step-up chopper, which works with a current hysteresis control (bang-bang control), causes harmonics in the range from 400 Hz to 1.1 kHz, with amplitudes of up to 7.5 % of the mean value. Because of the mentioned ripple in the elec. generator torque, especially the harmonics up to 65 Hz, torsional oscillations with an amplitude of 3.7 % from the mean value are induced. The frequency corresponds to the first natural frequency of the mechanical drive train. The first harmonic of the electrical generator torque propagates with a low amplitude through the mechanical part of the drive train. The electrical generator torque for the current hysteresis control has a harmonics spectrum from 500 Hz to 2.5 kHz, with amplitude of maximal 1 % of the mean value. Therefore no influence on the mechanical system can be identified.

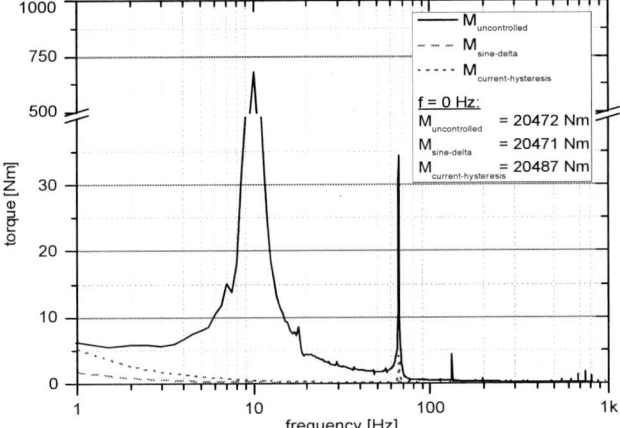

Fig. 7 Frequency analysis of the shaft torque

The sine-delta-modulation causes harmonics in the electrical torque close to the pulse frequency (1 kHz), so called sidebands, and multiples of the pulse frequency. The highest amplitude lies at 2 kHz with an amplitude of 5.8 % of the mean value. The frequency spectrum is rotation speed dependent, because the induced voltage varies with the speed and therefore the duty cycle changes [5]. Because of the high frequency separation to the first natural frequency, no torsional oscillations are induced in the mechanical part of the drive train.

V. CONCLUSION

Based on the permanent magnet synchronous machine, different generator systems for variable speed WEC with a nominal power of 5 MW were analysed. Generator concepts with low maintenance demand, long life time and high utilisation rate were considered. The interactions between generator system and mechanical drive train have strong influences on life time and operational availability. The required rectifiers cause harmonics in the generator current, which lead to generator torque ripples. They might excite torsional oscillations in the mechanical power train. This is the case for PMSM with uncontrolled rectifier, because the generator's ripple frequency is similar to the mechanical natural frequency. The generator systems with self-commutated rectifier do not have noteworthy influence on the mechanical power train.

REFERENCES

A. Sourkounis, C.; "Windenergiekonverter mit maximaler Energieaus-beute am leistungsschwachen Netz," (Wind Energy Converter with Maximum Energy Efficiency in Low Power Networks), Ph.D. Dissertation, TU Clausthal 1994, Papierflieger Clausthal 1995

B. Sourkounis C.; "Drehzahlelastische Antriebe unter stochastischer Belastungen," (Speed Flexible Power Drives under Stochastic Load); Habilitationsschrift, TU Clausthal 2004 (Book of Habilitation Research Work); Papierflieger, Clausthal 2004   ISBN ISBN 3-89720-737-0

C. Meyer, M.; "Leistungselektronik: Einführung, Grundlagen, Überblick," Springer Verlag, Berlin-Heidelberg-New York, 1990

D. Sourkounis, C.; "Stochastische Optimierung von Regelstrukturen zur Lastkollektivminimierung im Antriebsstrang von Windkraftanlagen," (Stochastic Optimization of Control Algorithms for Reducing of Cumulatitve Load in the Power Train of Wind Energy Converters), Antriebstechnisches Kolloquium, ATK 2001, Aachen; Tagungsband

E. Beck, H.-P.; Sourkounis C.; "Comparison of 5-MW-Offshore Wind Energy Converters," World Wind Energy Conference 2002, Berlin Germany, Proceedings (CD-ROM)

F. Häusler, Michael; "   Netzanbindung von offshore-Windparks mit Gleichstrom," (Connecting offshore-windfarms to the grid via DC), Elektrizitätswirtschaft, Jg 99 (2000), Heft 24

G. J. F. Fuller, E. F. Fuchs, and K. J. Roesler, "Influence of harmonics on power distribution system protection," *IEEE Trans. Power Delivery*, vol. 3, pp. 549-557, Apr. 1988.

H. Ni, B.; Sourkounis, C.; "Influence of Wind-Energy-Converter Control Methods on the Output Frequency Components," IEEE Transactions on Industry Application, vol. 45, pp. 2116-2122, Nov.-Dec. 2009

I. Beltran, B.; Ahmed-Ali, T.; El-Hachemi Benbouzid, M.; "High-Order Sliding-Mode Control of Variable Speed Wind Turbines," IEEE Transactions on Industrial Electronics, vol. 56, pp. 3314-3321, Sep. 2009

Influence of control parameters of SRM for output characteristics

M. Parchomiuk, K. Tomczuk. Electrotechnical Institute, Power Converters Department.
Pozaryskiego Street 28, 04-703 Warsaw, POLAND. Email: k.tomczuk@iel.waw.pl

Abstract – **The paper describes the influence of parameters controlling the motor operations on values of the electromagnetic torque obtained. A mathematical model, a method of parameters identification in equations and some selected results of laboratory measurements of drive system are presented. The main purpose of the paper is to present the realization characteristics of traction algorithm for a switched reluctance motor, required for drives of mechanical vehicles.**

I. INTRODUCTION

Switched reluctance motors (SRM) belong to a group of motors with electronic commutation, where the phenomenon of variable reluctance is used in their operations. Because of the absence of rotor windings and of a relatively simple machine phase winding system they belong to motors with the lowest production costs and they are characterized by high durability, similar to that of induction machines. The SRM have several advantages which cause that they are ideally suitable for use in certain applications for drives. These machines do not have mechanical commutator, that is there are not any problems related to commutator operation. They can be operated in explosive environments and in conditions of high dustiness, without any special casing. In case of any damage to one winding , their operation can be continued at a reduced load which constitutes an essential advantage for fans and pumps. The SRM do not have permanent magnets or rotor windings which permits the machines to operated at much higher temperatures, and permits the machine to be designed with reduced dimensions. A drive with a reluctance motor (SRM and converter) has higher efficiency [10] than one realized using an induction motor with converter. It is of basic importance to minimize the amount of energy consumption by vehicles with electric drive.

II. MEASUREMENTS OF SRM PARAMETERS

A 6/4 switched reluctance motor with a power of 1kW was used for laboratory tests. Fig. 1 presents a schematic diagram of a switched reluctance motor with indicated method of determining position of the rotor and of the windings in the motor. The torque was measured by means of a suitable lever,

mounted on the rotor shaft. This lever permitted a precise control of the rotor position with respect to the stator by means of the worm gear used.

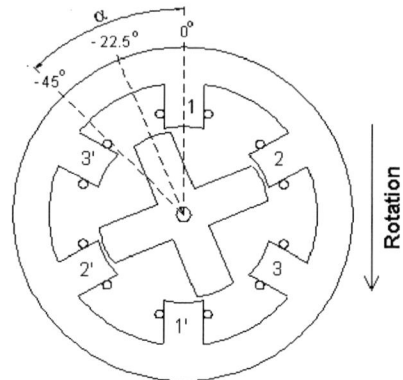

Fig. 1. Designation of position angles

Fig. 2 presents variations of the static electromagnetic torque generated by one phase winding of the motor with a blocked rotor. If the machine has to operate in the range of motor operation the control angles must be set in the range of $\alpha = -45° \div 0°$, while in order to work in generator mode the angles must be in the range of $\alpha = 0° \div 45°$.

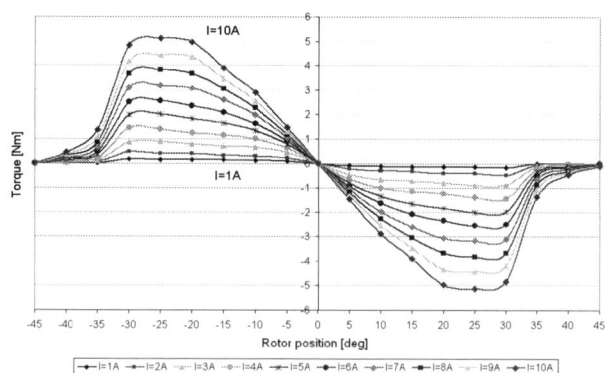

Fig. 2. Variations of the static electromagnetic torque generated by the SRM winding

Measurements of electromagnetic flux values flowing in the magnetic circuit of the machine is presented In Fig. 3. Calculations of the flux were made basing on current time

978-1-4244-8806-3/11 $26.00 © 2011 IEEE

variations recorded (with rotor blocked in known position) flowing in the winding, and on voltages across its terminals.

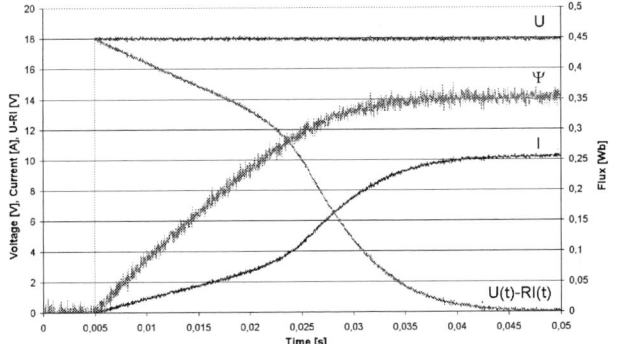

Fig. 3. Time variations of U(t), I(t) and additional curve U(t)-RI(t)

Measurement of the electromagnetic flux value generated by the motor phase winding was connected with determining an additional curve U(t)-RI(t), and by integrating it then according to Formula (1).

$$\psi(t) = \int_0^t [U(t) - R \cdot I(t)] \cdot dt \qquad (1)$$

where: U – voltage [V], I - current [A], R – winding resistance [Ω], Ψ – flux [Wb]

Measurements of the flux were repeated for other rotor positions. The value of the voltage applied across winding terminals was selected in such a way that it forced the flow of rated current in the winding in stationary state.

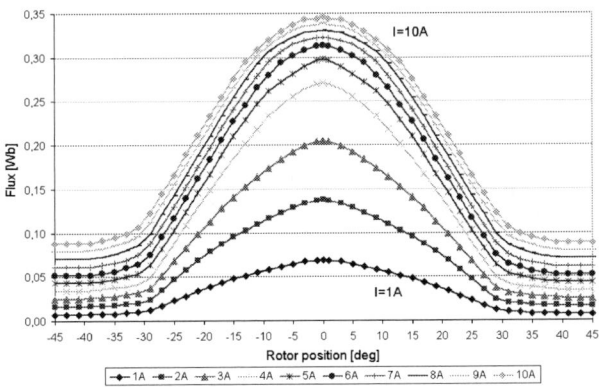

Fig. 4. Variations of the electromagnetic flux

Fig. 4 shows variations of the electromagnetic flux as a function of rotor position for various values of the current flowing in the motor phase winding, while Fig. 5 presents electromagnetic flux variations as a function of the current in the motor phase winding for different positions of the rotor.

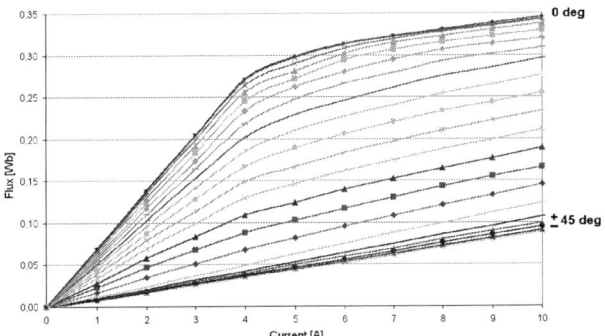

Fig. 5. Variations of the electromagnetic flux

The results were introduced into the simulation model (Formulas 3 and 4).

III. VERIFICATION OF THE MOTOR PARAMETERS

A field model of the motor was elaborated in the programme FEMAG-DC in order to verify the values of torque and flux.

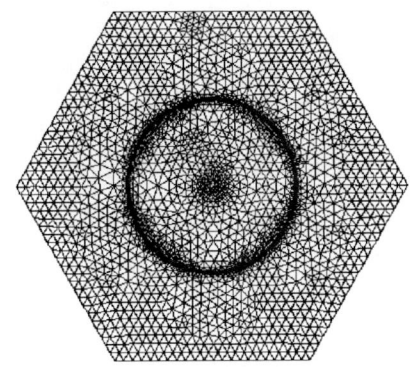

Fig. 6. Grid in the whole cross-section of the motor

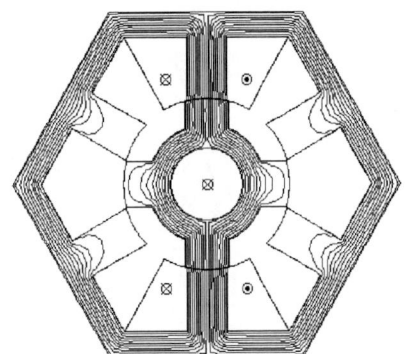

Fig. 7. Magnetic flux lines for the position $\alpha=0°$ and current $I=10A$

978-1-4244-8806-3/11 $26.00 © 2011 IEEE   215

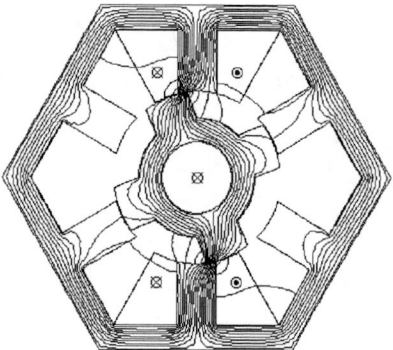

Fig. 8. Magnetic flux lines for the position
$\alpha=-22.5°$ and current $I=10A$

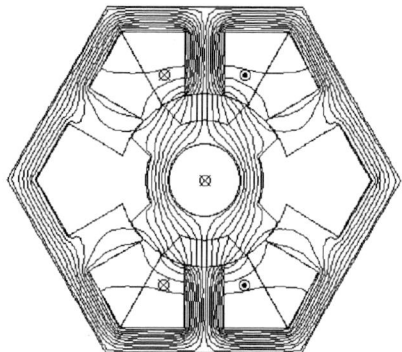

Fig. 9. Magnetic flux lines for the position
$\alpha=-45°$ and current $I=10A$

The field model of the reluctance motor was made introducing into the programme the real machine dimensions and the characteristic of magnetic circuit material.

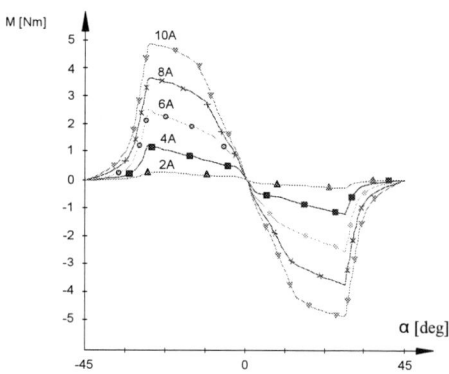

Fig. 10. Variations of the electromagnetic flux determined on the basis of field model tests In the FEMAG-DC programme

On the basis of field model tests there were determined magnetic flux values and variations of electromagnetic torque generated in the motor for different rotor positions. The results are shown in Figs. 11 and 12.

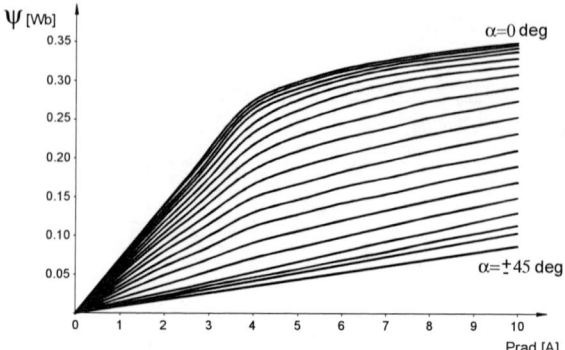

Fig. 11. Variations of electromagnetic flux determined on the basis of field model tests In the FEMAG-DC programme

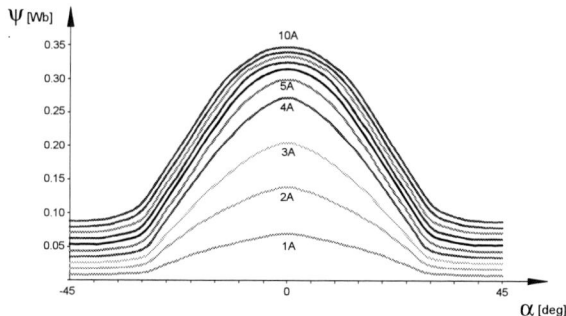

Fig. 12. Variations of electromagnetic flux determined on the basis of field model tests In the FEMAG-DC programme

TABLE I
COMPARISON BETWEEN LABORATORY AND FIELD MODEL RESULTS

Current [A]	Max. values of the torque [Nm]		Max. values of the flux [Wb]	
	Laboratory model (Fig. 1)	Field model (Fig. 10)	Laboratory model (Fig. 4)	Field model (Fig. 12)
2	0.5	0.2	0.138	0.142
4	1.5	1.2	0.271	0.276
6	2.8	2.5	0.314	0.320
8	3.9	3.6	0.331	0.335
10	5.1	5.0	0.346	0.348

Satisfactory convergence of measurement results was obtained from both models. The obtained results were used as input data for the simulation model realized in the Matlab-Simulink programme. The results are shown in Table I.

IV. THE MATHEMATICAL MODEL OF SRM

Formula (2) show the voltage equation for the SRM phase winding.

$$U_j = R_j \cdot i_j + \frac{d\Psi_j(\alpha, i_j)}{dt} \qquad (2)$$

where: U – supply voltage, R – winding resistance, Ψ – flux, j – winding number, α - rotor position, i – winding current

The value of electromagnetic flux depends on the rotor position "α" and on the value of current "i" in the winding, thus partial derivatives of the variables appear in Formula (3).

$$\frac{\partial \Psi_j(\alpha, i_j)}{dt} = \frac{\partial \Psi_j(\alpha, i_j)}{\partial \alpha} \cdot \frac{d\alpha}{dt} + \frac{\partial \Psi_j(\alpha, i_j)}{\partial i_j} \cdot \frac{di_j}{dt} \quad (3)$$

The voltage equation for one phase winding with partial derivatives is shown by Equation (4) which was used in the simulation programme.

$$U_j = R_j \cdot i_j + \frac{\partial \Psi_j(\alpha, i_j)}{\partial \alpha} \cdot \frac{d\alpha}{dt} + \frac{\partial \Psi_j(\alpha, i_j)}{\partial i_j} \cdot \frac{di_j}{dt} \quad (4)$$

The rotation speed of rotor was calculated according to Formulas (5) and (6).

$$M = J\frac{d^2\alpha}{dt^2} + D\frac{d\alpha}{dt} + T_f \, \mathrm{sgn}\left\{\frac{d\alpha}{dt}\right\} + TL \quad (5)$$

where: M – sum of torques generated by motor phases, J – rotor inertia, D – friction, T_f – friction, TL – external load

$$\alpha = \int_0^t \omega \cdot dt + \alpha_0 \quad (6)$$

where: α – rotor position, ω – rotation speed, α_0 – starting position

V. SELECTED LABORATORY TEST RESULTS AND VERIFICATION OF THE MATHEMATICAL MODEL

Fig. 13 presents a family of M=f(n) characteristics for various current switching angles α_{on} and for a constant current switching angle α_{off} = -1.4°.

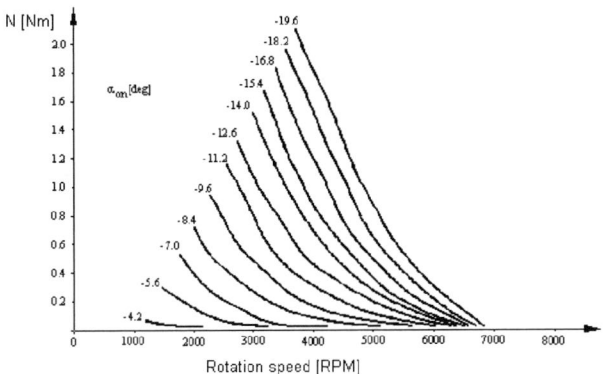

Fig. 13. Characteristics M=f(n) for Uz=320V ; UG=400V ; α_{on}=var ; α_{off}=-1.4°

The measurements presented were performed on a laboratory model of a drive with a reluctance motor, in order to verify the mathematical model. Fig. 10 presents selected measurements results of the M=f(n) characteristics obtained from the laboratory and simulation model. A satisfactory convergence of measurements was obtained from both models. Thus it can be assumed that the mathematical model is correct.

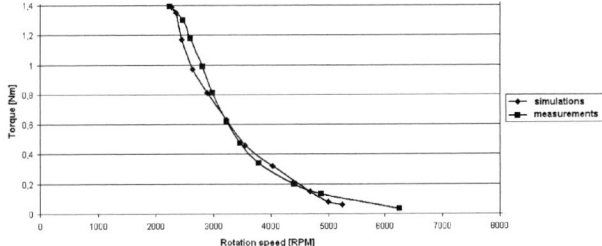

Fig. 14. Comparison of results obtained from the simulation model with the measurement for Uz=320V ; UG=400V; α_{on}= -11.2°; α_{off}=-1.4°

Figs. 15a, 15b, 15c presents exemplary test results of the simulation model of the variables of rotational speed (rys. 15a), torque ripples (rys. 15b) and of the efficiency (rys. 15c), at changes of the firing angle α_{on} and of the current conduction angle (Δ_{on}= |α_{on} - α_{off}|). The tests were performed for the converter supply voltage Uz=320V, UG=640V, motor load torque TL=4Nm.

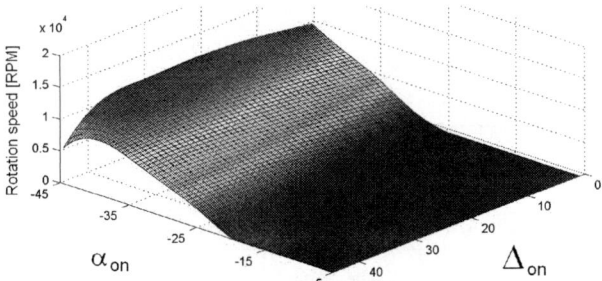

Fig. 15a. Variability of the rotation speed

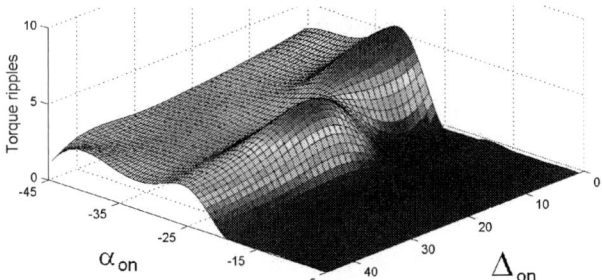

Fig. 15b. Variability of the torque pulsation

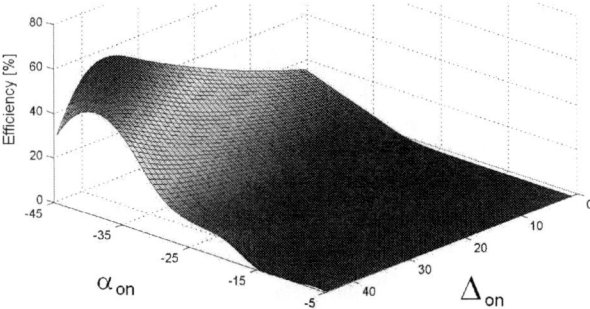

Fig. 15c. Variability of the efficiency of the motor with converter

According to the test results it is possible to obtain simultaneously a minimum of torque pulsation and maximum of SRM and converter efficiency. It is possible to obtain these parameters for α_{on}= -40° ÷ -35° and Δ_{on}= 40° ÷ 35°.

VI. TRACTION CHARACTERISTIC REALIZATION ALGORITHM

Two control methods can be distinguished during realization of the traction characteristic: at a constant torque and a constant power. During the controlling with a constant torque the force applied to the vehicle wheels is constant and it does not depend on speed. The power consumed by the motor increases as well as the temperature of windings. The moment of going over from the first method to the second one is determined by the thermal safety limit which should not be exceed, as it can lead to overheating the motor windings. If a further increase in the vehicle speed is required, then the control algorithm shall be changed to another one, in which the force applied to the vehicle wheels decreases with rise in the rotation speed. The power supplied to the motor is then constant and consequently, so is also the windings temperature.

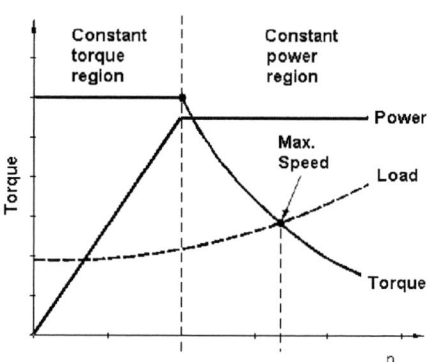

Fig. 16. Description of the traction characteristic

The realization algorithm of traction characteristic was defined on the basis of tests of the simulation model. The results are presented in Table II.

TABLE II
TRACTION CHARACTERISTIC REALIZATION ALGORITHM

Speed range [RPM]	α_{on} [°]	Δ_{on} [°]	I_{max} [A]	Torque ripples
0 – 500	-45	45	10	1,4
500 – 1100	-45	40	10	1,5
1100 – 2000	-45	35	10	1,6
2000 – 3500	-45	30	10	1,7
> 3500	-45	25	10	1,5

where: α_{on} – firing angle, Δ_{on} – conduction angle, I_{max} – current limitations in motor windings

The highest values of the torque generated were acheived when the phase currents were fired at α_{on} = -45° and

conducted by the angle Δ_{on} = 45°. The current conduction angle shall undergo reduction when the rotation increases. The variation of the torque shown in Fig. 17 is not mild, there are two local decreases of the torque value for n = 2000 RPM and n = 4000 RPM. It is related to the fact that the conduction angle Δ_{on} in the algorithm was reduced in steps of 5° (see Table II).

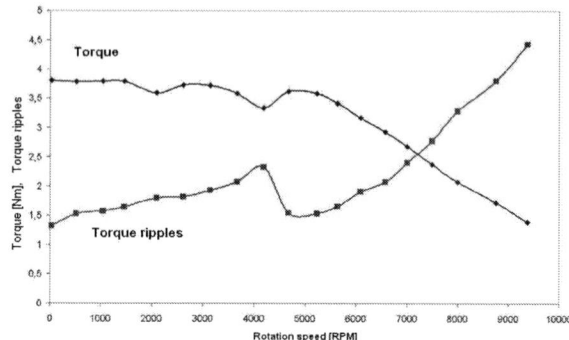

Fig. 17. Graphical interpretation of Table II

If simulation is performed by steps of 1°, then the variation of torque becomes smooth. At the same time the number of rotation speed ranges described In Table II, must be increased.

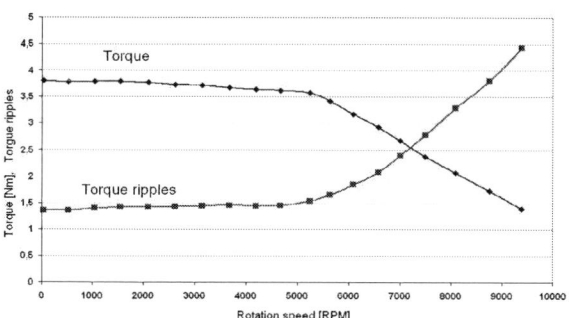

Fig. 18. Limit-wise interpretation of Table II in case of the number of control intervals increased by five times with a reduction step Δ_{on} every 1°.

VII. C-DUMP CONVERTER

A C-DUMP converter was used to supply the laboratory model of a switched reluctance motor with power supplied from mains of 230V, 50Hz. It permitted the currents in motor windings to be switched on and off in precisely defined positions, determined by the α_{on} and α_{off} angles. During switching off the currents in windings, the energy collected in the magnetic fields L1, L2, L3 charges the CG capacitor through the D1, D2, D3 diodes. The system has the possibility of controlling the UG voltage which influences the speed of the current decay in motor windings. Such a capability of the system is particularly useful in the range of high rotation speeds (high voltage across UG). The value of the UG voltage is controlled by the TG transistor. The control system of the converter had the possibility to control the α_{on} and α_{off} in the range from -45° to 45°. The value of UG could be set in the range from 325 Vdc to 1200 Vdc.

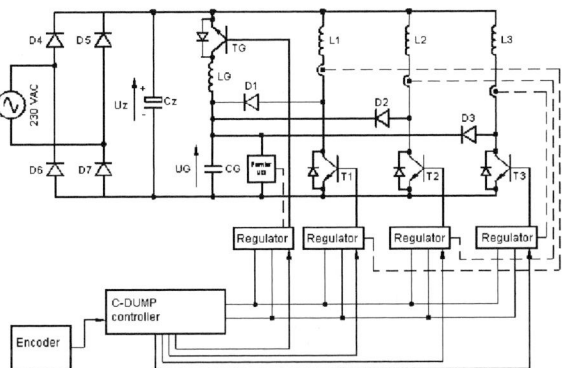

Fig. 19. Connection topology of the C-DUMP converter

Fig. 20. View of enkoder and the SRM

Fig. 21. View of the motor with an electric dynamometer

Fig. 22. View of the converter laboratory model

CONCLUSIONS

The present paper describes a method of measuring the torque and the electromagnetic flux in a switched reluctance motor. The investigations carried out determine the possibility of applying SRM in electric and hybrid vehicles. An algorithm of realization of a traction characteristic required for the drive of mechanical vehicles is presented. Characteristics of M=f(n) for constant α_{on} and α_{off} angles used in simple control algorithms are shown.

REFERENCES

[1] X. D. Xue, K.W.E.Cheng, J. K. Lin, Z. Zhang, K. F. Luk, T. W. Ng, N. C. Cheung. "Optimal Control Method of motoring Operation for SRM Drives in Electric Vehicles". IEEE TRANSACTIONS ON VEHICULAR TECHNOLOGY, VOL. 59, NO. 3, MARCH 2010.

[2] V. L. Do, M. C. Ta. "Modeling, Simulation and Control of Reluctance Motor Drives for High Speed Operation". Energy Conversion Congress and Exposition. September 12-16, 2010. IEEE.

[3] P. Krishnamurthy, W. Lu, F. Khorrami, A. Keyhani. "Robust Force Control of an SRM-Based Electromechanical Brake and Experimental Results". IEEE TRANSACTIONS ON CONTROL SYSTEMS TECHNOLOGY, VOL. 17, NO. 6, NOVEMBER 2009

[4]. W. Jazdzynski, M. Majchrowicz. "An Approach to Find an Optimum Designed SRM for Electric Vehicle Drive". Proceedings of the 2008 International Conference on Electrical Machines

[5] Ch. Mademlis, I. Kioskeridis. "Smooth Transition between Optimal Control Modes in Switched Reluctance Motoring and Generating Operation". International Conference on Power Systems Transients (IPST'07) in Lyon, France on June 4-7, 2007

[6] K. Ha, Ch. Lee, J. Kim, R. Krishnan, S. G. Oh. "Design and Development of Low-Cost and High-Efficiency Variable-Speed Drive System With Switched Reluctance Motor". IEEE TRANSACTIONS ON INDUSTRY APPLICATIONS, VOL. 43, NO. 3, MAY/JUNE 2007.

[7] Jin-Woo Ahn, Sung-Jun Park, Dong-Hee Lee. "Novel Encoder for Switching Angle Control of SRM". IEEE TRANSACTIONS ON INDUSTRIAL ELECTRONICS, VOL. 53, NO. 3, JUNE 2006.

[8] M. Zeraoulia, M. E. H. Benbouzid, D. Diallo. "Electric Motor Drive Selection Issues for HEVPropulsion Systems: A Comparative Study". IEEE TRANSACTIONS ON VEHICULAR TECHNOLOGY, VOL. 55, NO. 6, NOVEMBER 2006.

[9] Chih-Hong Lin S. J. Chiang. "Torque-Ripple Reduction in Switched Reluctance Motor Drive Using SHRFNN Control". Power Electronics Specialists Conference, 2006. PESC '06. 37th IEEE.

[10] P. Andrada, B. Blanqué, J.I. Perat, M. Torrent, E. Martínez, J. A. Sánchez. "Comparative efficiency of switched reluctance and induction motor drives for slowly varying loads". INTERNATIONAL CONFERENCE ON RENEWABLE ENERGY AND POWER QUALITY (ICREPQ'06), Palma de Mallorca, 5, 6, 7 April, 2006

Design Guidelines using Selective Harmonic Elimination Advanced Method for DC-AC PWM with the Walsh Transform

Jesus Vicente, Rafael Pindado, Inmaculada Martinez
Technical University of Catalonia (UPC)
jesus.vicente@upc.edu, rafael.pindado@upc.edu, inmaculada.martinez@upc.edu

Abstract – **The use of the Walsh transform in DC-AC PWM waveform generation allows the calculation of the switching angles by means of linear equations which depend on the fundamental amplitude. However, when it is needed a wide regulation of the fundamental amplitude, conventional application of Walsh transform forces to switch among different equation sets, due to the range limitation of each one of them. In this paper it is described an advanced method to obtain the switching angles that permits full regulation of the fundamental amplitude, with only a switching interval vector, in single phase systems. The paper shows Matlab simulation results that proof the efficiency of the advanced method algorithm.**

Index Terms –**Algorithm optimization, PWM inverters, Selective harmonic elimination, DC-AC conversion, Walsh transform.**

I. INTRODUCTION

In DC-AC PWM waveforms generation with selective harmonic elimination the real time control of the fundamental amplitude takes a great amount of time to compute the nonlinear equations. This question can be solved by the off-line calculation of the switching angles, which establishes a compromise between the desired precision and the memory storage required.

On the other hand, it is possible the on-line calculation of the switching angles as a linear function of the fundamental amplitude using the Walsh functions [1]. These functions permit the linearization of the equation set that leads to the harmonic cancellation [2-4].

The Walsh functions, defined by J.L.Walsh in 1923, are made of a set of orthogonal and orthonormal functions that only have two values, {+1 −1}. In this paper the functions are normalized in time (unity period). As an example, Fig. 1 depicts the 17 first Walsh functions subset.

Every periodic function f(t) may be expressed as a sum of Walsh functions as:

$$f(t) = W_0 \, wal(0,t) + \sum_{k=1}^{\infty} W_k \, wal(k,t) \; , \; t \in [0,1] \quad (1)$$

where:

$$W_0 = \int_0^1 f(t) \, wal(0,t) \, dt = \int_0^1 f(t) \, dt \quad (1a)$$

$$W_k = \int_0^1 f(t) \, wal(k,t) \, dt \quad (1b)$$

For evident reasons Walsh series expansion is especially useful for the representation of PWM waveforms. In particular, as PWM waveforms generated in DC-AC converters have half-wave symmetry its series expansion will only have Walsh functions with 4n-3 order, where n is an integer {n=1,2,3,4,...}.

$$f(t) = \sum_{n=1}^{\infty} W_{(4n-3)} \, wal(4n-3,t) \quad (2)$$

With this technique the use of M angles per quarter of period permits the cancellation of M-1 harmonics and the regulation of the fundamental amplitude.

However, this technique has the drawback of obtaining a great number of solutions that difficult the selection process of the better cases and also increments the computation time, especially when a big number of switching angles is used.

On the other hand, there are different solutions for each value of fundamental amplitude depending on the position of the switching angles in the intervals of the first quarter period (switching interval vector). Obviously, these solutions cause different harmonic distributions.

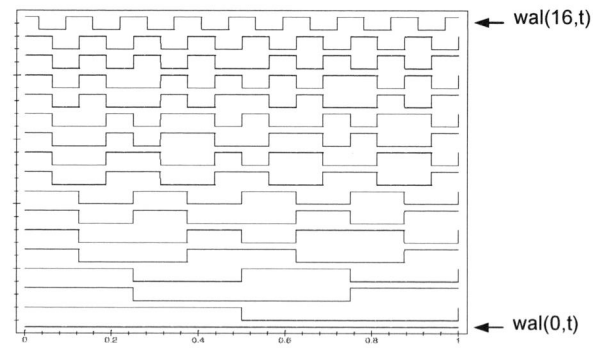

Fig.1. Walsh functions: wal(i,t) , i ∈ {0,..,16}

In order to compare the differences in these harmonic distributions, the distortion factor defined in (3) has been chosen as the parameter to be considered. This factor is preferred to the THD, because the harmonic order, k, diminishes the influence of the harmonic amplitude. This behavior is analogous to the filtering obtained when an inductive load is connected to the inverter. In this paper, the maximum order selected to do the computing is k=39.

978-1-4244-8806-3/11 $26.00 © 2011 IEEE

$$DF = \frac{100}{A_1}\left(\sum_{k=3}^{k\to\infty}\left(\frac{A_k}{k}\right)^2\right)^{1/2} \tag{3}$$

The efficiency of the control algorithm of the inverter will be better if it achieves a wide fundamental range, a low distortion factor and a small processing time.

In the following sections there is a comparison between the conventional Walsh transform method and the advanced method, proposed by the authors [5].

II. CONVENTIONAL METHOD ANALYSIS

Since the problem of harmonic elimination is inherent to the frequency domain, the analysis requires the availability of reciprocal conversion tools between Walsh and Fourier transforms. The relationship between Fourier and Walsh coefficients is the starting point to compute the switching angles of the PWM waveform and can be expressed by the transformation (4) developed in references [2], [3], [6], where G_F and G_W are the coefficients of the PWM signal by Fourier and Walsh expansions respectively, and B is the Fourier-Walsh conversion matrix .

$$G_F = B\ G_W \tag{4}$$

The Walsh coefficients are obtained as follows:

$$G_W = C\ \phi + D \tag{5}$$

where $\phi = [\phi_1\ \phi_2\ ...\ \phi_M]^T$ is the vector of switching angle fractions, each one referred to the end of its switching interval, $\phi_i \in (0,1)$, shown in figure 2, while C and D are the set of coefficients that relate the angles with the Walsh expansion coefficients.

Each quarter period is subdivided in N intervals, from 0 to N-1, but only M of which include one, and only one, switching angle. Those intervals $m(\alpha_i)$ form the elements of the switching interval vector (6).

$$m = [m(\alpha_1)\ m(\alpha_2)\ ...\ m(\alpha_M)\ ] \tag{6}$$

To simplify the notation the elements of vector m would be referred as follows:

$$m = [m(1)\ m(2)\ ...\ m(M)\ ] \tag{7}$$

The relationship between the fractions (ϕ_i) and the switching angles (α_i) is given by (8) and can be observed in figure 2.

$$\alpha_i = \frac{\pi}{2N}(m(i) + 1 - \phi_i) \tag{8}$$

Fixing N as the power of 2 greater or equal to 4 times the number of angles, it can be made that all Walsh functions used in the expansion of the PWM signal have a constant value in each subinterval, reducing the complexity of the algorithm. Another simplification can be obtained by fixing the end of the switching (β_i) in the next interval if the value of m(i) is less than (N/2)-1, and in the same interval otherwise [2].

Fig. 2 shows the first quarter of a PWM signal with two switching angles (α_1, α_2) at intervals 2 and 6. In this case: M=2, N=8, m(1)=2 and m(2)=6.

Taking this into account, matrix C and column vector D are easily derived from the Walsh matrix (WAL) whose elements are obtained from (9).

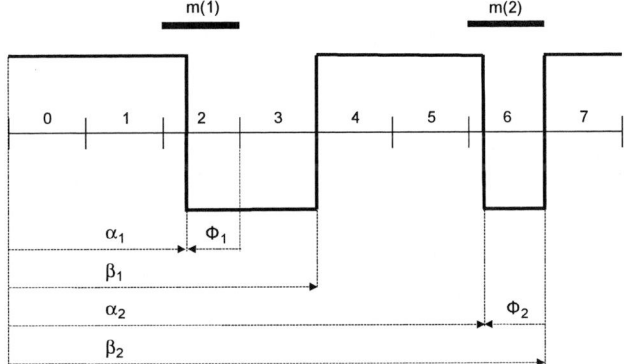

Fig. 2. PWM signal with two switching angles (conventional method).

$$\mathrm{WAL}_{i,j} = wal(4i-3, \frac{j-1}{4N})\quad i,j = 1...N \tag{9}$$

Matrix C is formed with the M columns of WAL corresponding to vector m, multiplied by the factor 2/N.

$$C_{i,j} = -\frac{2}{N}\mathrm{WAL}_{i,m(j)+1}\quad i=1...N\ , j=1...M \tag{10}$$

The N-dimensional column vector D is obtained by (11).

$$D_i = \frac{1}{N}\left[\sum_{j=1}^{N}\mathrm{WAL}_{i,j} - 2\sum_{j=1}^{M}\mathrm{PW}_{i,m(j)+1}\right]\quad i=1...N \tag{11}$$

Matrix PW is obtained by shifting left one column matrix WAL and making zero the columns whose index is greater than (N/2)-1.

Combining equations (4) and (5) we obtain (12):

$$G_F = B(\ C\ \phi + D) = E\phi + F \tag{12}$$

We can obtain the linear system for the switching angles from (13):

$$\phi = E^{-1}\ (G_F - F) \tag{13}$$

Now, leaving the fundamental amplitude (A_1) variable, and making zero the rest of the harmonics (A_3, A_5, ...), the linear system is represented by (14):

$$\phi = P A_1 + K \qquad (14)$$

This linear system has the restrictions that A_1 and $\phi_i \in [0,1]$. Also, in order to validate the equations, the range of fundamental amplitude should be positive.

III. ADVANCED METHOD ANALYSIS

With the advanced method, the PWM signal has notches placed in adjacent intervals and symmetry properties that are clearly shown in Fig. 3. In this case the end of the switching (β_i) is variable.

As it will be shown this way of generating the PWM signal makes it possible to achieve ranges of variation of the fundamental amplitude that are near to 100%, value that is impossible to reach with the conventional method.

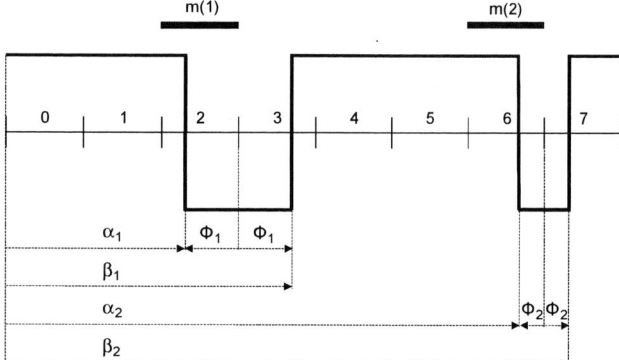

Fig. 3. PWM signal with two switching angles (advanced method).

Again, the Walsh coefficients are obtained from (15) as it has been shown in (5) but the matrices **C** and **D** will differ.

$$G_W = C \phi + D \qquad (15)$$

where $\phi = [\phi_1 \ \phi_2 \ ... \ \phi_M]^T$ is the vector of switching angle fractions, referred to the center of the notches. $\phi_i \in (0,1)$. The relationship between these fractions and the switching angles (α_i) is given by (16).

$$\alpha_i = \frac{\pi}{2N}(m(i)+1-\phi_i) \qquad (16)$$

The end of the switching (β_i) is given by (17).

$$\beta_i = \frac{\pi}{2N}(m(i)+1+\phi_i) \qquad (17)$$

These changes produce a notable simplification of the expressions needed to find the switching vectors which have solution.

$$C_{i,j} = -\frac{2}{N}\left(WAL_{i,m(j)} + WAL_{i,m(j)+1}\right)$$
$$i = 1...N \ , j = 1...M \qquad (18)$$

$$D_i = \frac{1}{N} \sum_{j=1}^{N} WAL_{i,j} \qquad (19)$$

Notice the simplification achieved in the computation of the elements of matrix **D**.

Following the same reasoning done with the conventional method the linear system that links the angle fractions with the fundamental amplitude will be:

$$\phi = P A_1 + K \qquad (20)$$

IV. RESULTS COMPARISON

In order to compare the methods described above, an exhaustive process has been developed. Several questions have to be considered.

First of all, figure 4 shows the number of interval combinations tested and the computing time needed to find which of them have a positive range of variation of the fundamental amplitude, as a function of the angles number. It can be seen that the processing times are similar in both methods. The processor used was an Intel T5270 CPU working at a frequency of 1.4 GHz.

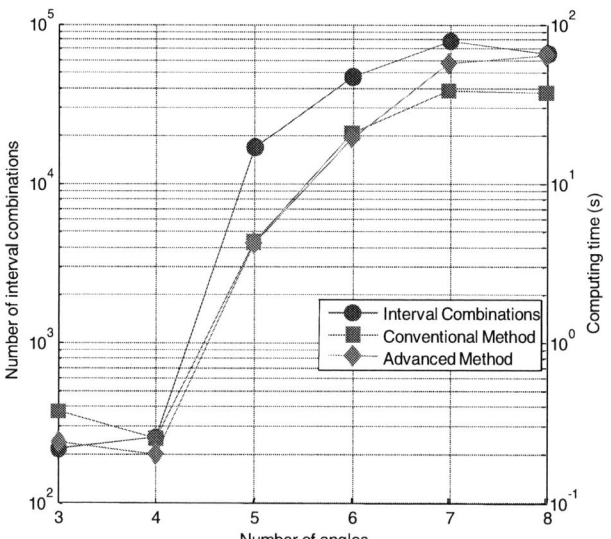

Fig. 4. Number of interval combinations and computing time.

Secondly, figure 5 shows the results obtained using the conventional and the advanced Walsh transform methods, respectively. This figure shows the solutions for each number of angles. The columns grouped to the left are for the conventional method (C) and the columns to the right are for the advanced method (A). It can be observed that the advanced method achieves greater ranges of variation of the fundamental amplitude.

978-1-4244-8806-3/11 $26.00 © 2011 IEEE

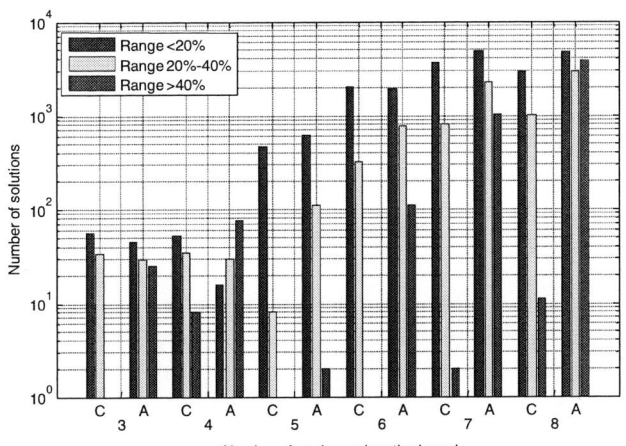

Fig. 5. Number of solutions obtained.

In tables I and II it can be seen the minimum and maximum values of fundamental amplitude (per unit), considering all the solutions. They also show the maximum fundamental range available with a single switching interval. Finally the values of distortion factor are shown, making the calculation when the fundamental amplitude is the maximum in the switching interval.

V. SIMULATION EXAMPLES

In this section some simulation examples for 4 and 8 angles are described, for the two methods discussed above. The switching intervals chosen are those which have the greatest range of variation for the fundamental amplitude.

These switching intervals are **m** = [1 6 11 14] and **m**=[1 5 8 13 19 21 26 29] with the conventional method and **m**=[1 5 9 13] and **m**=[2 6 10 14 18 22 26 30] using the advanced method. For this switching intervals, in tables III to VI there are represented the linear equations for the angles (21)-(24) and the minimum and maximum values that can be obtained for A_1.

TABLE I
AMPLITUDE RANGE AND DISTORTION FACTOR FOR CONVENTIONAL METHOD

Number of angles	A1 min	A1 max	A1 range (%)	DF (%) min	DF (%) max
8	0.53	1.01	43.9	1.8	6.6
7	0.56	1.02	40.1	2.3	7.5
6	0.65	1.02	32.9	2.5	6.5
5	0.77	1.03	24.2	2.9	7.3
4	0.46	1.04	51.0	3.8	16.2
3	0.60	1.06	43.4	4.7	15.6

TABLE II
AMPLITUDE RANGE AND DISTORTION FACTOR FOR ADVANCED METHOD

Number of angles	A1 min	A1 max	A1 range (%)	DF (%) min	DF (%) max
8	0.06	1.01	94.3	1.5	3.8
7	0.23	1.02	78.7	1.8	5.3
6	0.39	1.02	62.4	2.3	8.1
5	0.57	1.03	45.0	2.8	8.1
4	0.03	1.04	98.6	3.7	18.7
3	0.32	1.07	74.2	4.8	19.8

It should be remarked that the values shown are not for the same switching interval vector. This is the reason why, for 8 angles in the conventional method, although a fundamental minimum value of 0.53 and a maximum of 1.01 can be obtained, the maximum fundamental range is, only, 43.9%.

It can be seen that the advanced method achieves a wider range of amplitudes, with similar values of distortion factor. The best results in amplitude range, for the advanced method, are obtained for 8 and 4 angles, being the corresponding switching interval vectors: **m**=[2 6 10 14 18 22 26 30] and **m**=[1 5 9 13]. In both cases the number of angles is a power of two and the intervals are equally distributed in the first quarter of the period.

TABLE III
EQUATIONS AND AMPLITUDE RANGE CONVENTIONAL METHOD

m = [1 6 11 14]	
$\Phi_1 = -0.3590\ A_1 + 1.1490$	
$\Phi_2 = -1.5803\ A_1 + 1.5483$	
$\Phi_3 = -1.8984\ A_1 + 1.8922$	(21)
$\Phi_4 = -1.3055\ A_1 + 1.3941$	
A_1 min = 47.00% , A_1 max= 97.97%	

TABLE IV
EQUATIONS AND AMPLITUDE RANGE ADVANCED METHOD

m=[1 5 9 13]	
$\Phi_1 = -0.1954\ A_1 + 0.9018$	
$\Phi_2 = -0.5565\ A_1 + 1.0187$	
$\Phi_3 = -0.8328\ A_1 + 0.9938$	(22)
$\Phi_4 = -0.9824\ A_1 + 1.0016$	
A_1 min = 3.37% , A_1 max= 101.96%	

TABLE V
EQUATIONS AND AMPLITUDE RANGE CONVENTIONAL METHOD

m=[1 5 8 13 19 21 26 29]	
$\Phi_1 = 0.7935\ A_1 + 0.1434$	
$\Phi_2 = -1.2594\ A_1 + 1.4115$	
$\Phi_3 = -0.1541\ A_1 + 0.2644$	
$\Phi_4 = -2.1312\ A_1 + 2.1001$	
$\Phi_5 = -1.3492\ A_1 + 1.6192$	(23)
$\Phi_2 = -1.7046\ A_1 + 1.9318$	
$\Phi_7 = -1.7874\ A_1 + 1.8833$	
$\Phi_8 = -2.0562\ A_1 + 2.0576$	
A_1 min = 54.66%, A_1 max = 98.4%	

TABLE VI
EQUATIONS AND AMPLITUDE RANGE ADVANCED METHOD

m=[2 6 10 14 18 22 26 30]	
$\Phi_1 = -0.1418\ A_1 + 1.0034$	
$\Phi_2 = -0.3350\ A_1 + 0.9945$	
$\Phi_3 = -0.5201\ A_1 + 1.0077$	
$\Phi_4 = -0.6621\ A_1 + 0.9891$	
$\Phi_5 = -0.8191\ A_1 + 1.0164$	(24)
$\Phi_2 = -0.8771\ A_1 + 0.9723$	
$\Phi_7 = -1.0310\ A_1 + 1.0607$	
$\Phi_8 = -0.7059\ A_1 + 0.7071$	
A_1 min = 5.88%, A_1 max = 100.18%	

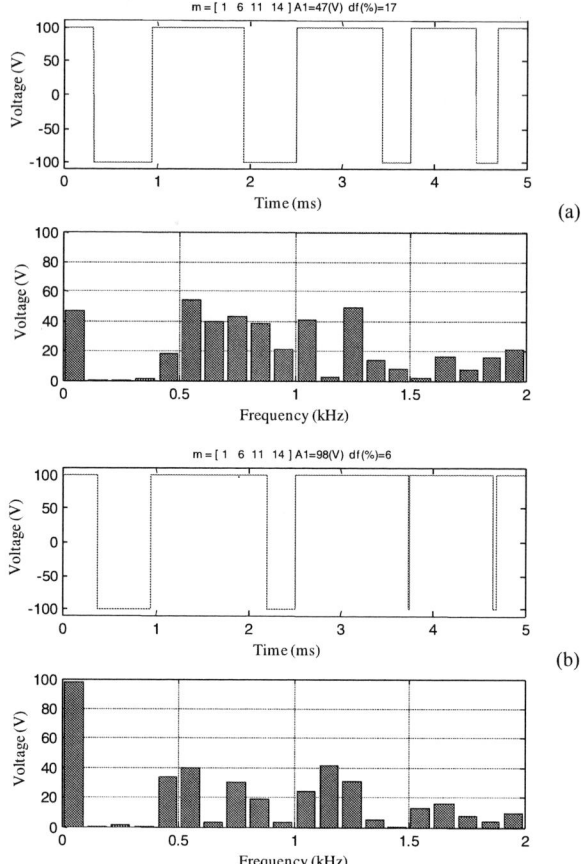

Fig. 6. 4 angles, PWM signal and harmonic distribution Conventional Method. (a) A_1 =47%, (b) A_1 =98%

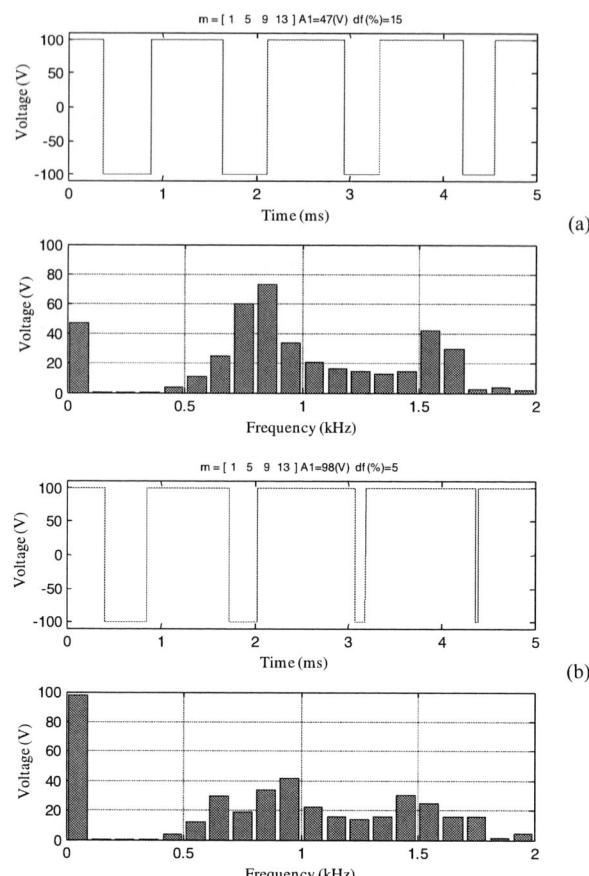

Fig. 7. 4 angles, PWM signal and harmonic distribution. Advanced Method. (a) A_1 =47%, (b) A_1 =98%

Simulation results for 4 angles are shown in figures 6 (Conventional Method) and 7 (Advanced Method).

In order to compare more faithfully the harmonic distributions and the distortion factors, there have been taken the values 47.00% and 97.97% that are the minimum and maximum A1 values for the conventional method.

The obtained values of distortion factors using the advanced method are 15% and 5%, for the lower and higher fundamental amplitudes, respectively. Using the conventional method the obtained values are 17% and 6%, respectively.

Figures 8 (Conventional Method) and 9 (Advanced Method) show the simulation results for 8 angles. As before, there have been taken the fundamental amplitude values of 54.66% and 98.40%, to compare the distortion factor and the harmonic distribution.

The obtained values of distortion factors using the advanced method are 6% and 2%, for the lower and higher fundamental amplitudes, respectively. Using the conventional method the obtained values are 7% and 3%, respectively.

It can be seen that the PWM waveforms obtained from the advanced method have a regular shape: the notches are thinner as they approach the quarter of period. Also the distortion factors for the same A_1 values are smaller when the advanced method solution is taken.

Also, figure 9, shows that the orders of the two harmonics with greatest peak amplitudes are 31 and 33, that complies with equation (25). This fact is also verified using advanced method, with 4, 16, 32 and 64 angles when the switching interval vectors are [2:4:4·M].

$$k_{\text{peak}} = 4 \cdot M \pm 1 \qquad (25)$$

VI. DESIGN GUIDELINES

The following paragraphs give the design guidelines.

First, select the distortion factor tolerable in the particular application in agreement with the required fundamental amplitude. This leads to choose a number of angles that cancels enough harmonics as seen in tables I and II.

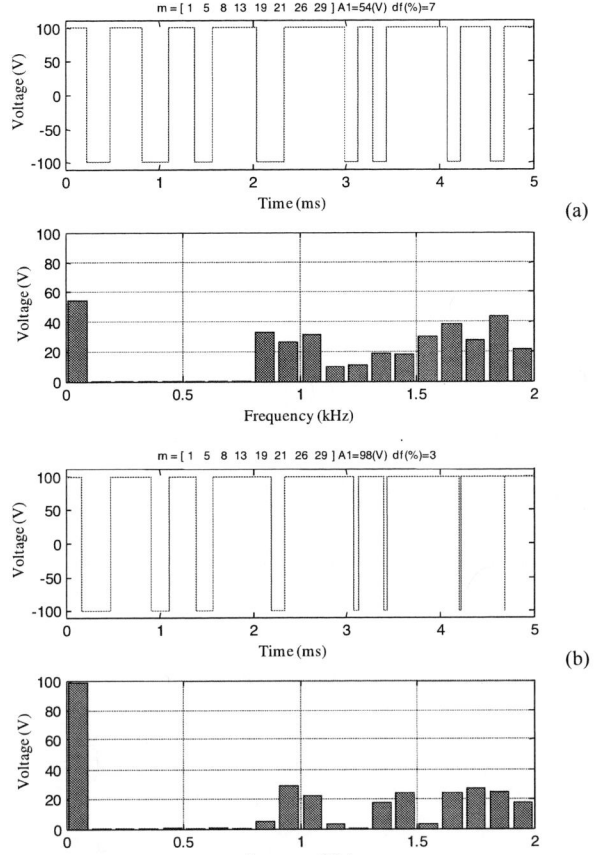

Fig. 8. 8 angles. PWM signal and harmonic distribution. Conventional Method. (a) A_1=54%, (b) A_1=98%

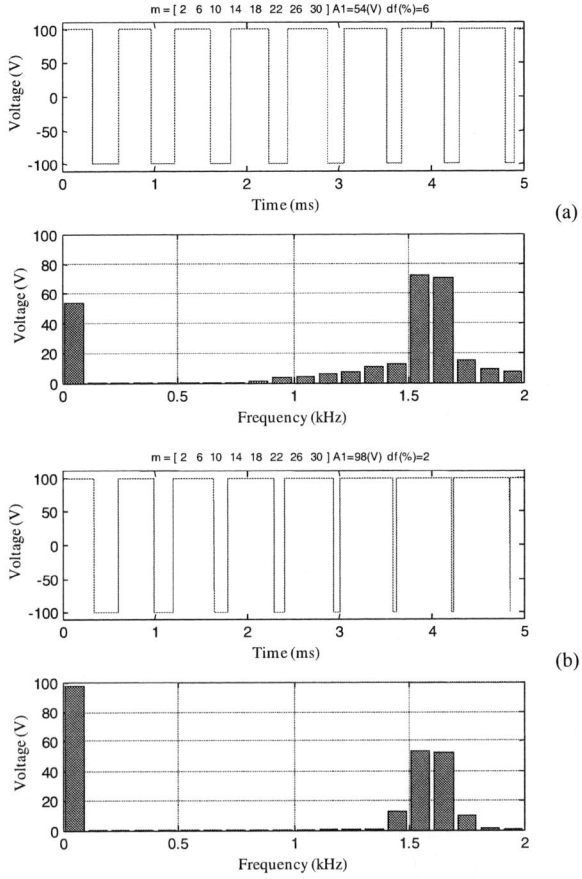

Fig. 9. 8 angles. PWM signal and harmonic distribution. Advanced Method. (a) A_1=54%, (b) A_1=98%

Another consideration is the range of fundamental amplitude needed. When a wide range is necessary, it should be chosen the advanced method.

Finally, when a big amount of harmonics should be cancelled, the advanced method gives the best solutions for the switching interval vectors whose number of angles is a power of two and the intervals are equally distributed in the first quarter of the period.

VII. CONCLUSION

An advanced algorithm has been presented, based on the Walsh functions. This algorithm produces a set of linear equations that allows a wide range of fundamental amplitude regulation, near 100 per cent, with a single switching vector, reducing the processing time needed for the computation of the switching angles for the required output voltage.

Another significant contribution of this work is the design considerations to guide in the selection of the switching vector value for the optimal performance of the converter, obtaining the maximum regulation range of the fundamental component.

REFERENCES

[1] T.J. Rivlin and E.B. Saff, *Joseph L. Walsh: selected papers*, New York: Springer, 2000.

[2] F. Swift and A. Kamberis, "A New Walsh Domain Technique of Harmonic Elimination and Voltage Control in Pulse Width Modulated Inverters". *IEEE Trans. Power Electron.*, vol. 8, no. 2, pp.170-185, April 1993.

[3] T.J. Liang, R.M. O'Connell and R.G Hoft, "Inverter Harmonic Reduction Using Walsh Function Harmonic Elimination Method", *IEEE Trans. Power Electron.*, vol. 12, no. 6, pp. 971-982, Nov. 1997.

[4] J. Vicente, R. Pindado and I. Martinez. "Algorithm optimization for PWM signal generation with selective harmonic elimination using the Walsh transform". International Conference on Renewable Energy and Power Quality, proceedings. ICREPQ'03. ISBN 84-607-7173-3, 9-12 April 2003.

[5] J. Vicente, R. Pindado, I. Martinez and J. Pou, "A new efficient algorithm for DC-AC PWM waveform generation with full fundamental regulation on a single linear equation set," Industrial Electronics Society, 2003. IECON '03. The 29th Annual Conference of the IEEE , vol.2, no., pp. 1835- 1839 Vol.2, 2-6 Nov. 2003.

[6] K.H. Siemens and R. Kitai, "A Nonrecursive Equation for the Fourier Transform of Walsh Function", *IEEE Trans. Electromagn. Compat.*, vol. 15, no. 2, pp. 81-83, May 1973.

Energy Saving Possibilities in the Industrial Robot IRB 1600 Control

Anton Rassõlkin, Hardi Hõimoja, Raivo Teemets

Tallinn University of Technology/Department of Electrical Drives and Power Electronics, Tallinn (Estonia)
anton.rassolkin@tptlive.ee, hardi.hoimoja@ttu.ee, raivo.teemets@ttu.ee

Abstract- **The paper presents the approaches for electric energy saving possibilities and electricity consumption characteristics in the modern industrial robots together with practical examples concerning robot programming and positioning. The paper is based on measurements, made in the laboratory of Tallinn University of Technology with an industrial robot IRB 1600.**

I. INTRODUCTION

The development of modern robotic technologies is a matter of different subdisciplines with constantly augmenting application in various manufacturing processes. An industrial robot might be observed as an actuator mechanism, requiring energy for motion. As the environmental assets are limited, more and more attention is paid to energy saving opportunities and robotics is not an exception. The very first energy saving option is based on the advantage of robots before humans, i.e. on the fact that robots can operate in dark and cold environments, which means fewer expenses on lighting and heating. The current paper casts additional light to some energy saving possibilities in industrial robots with improved control methods. Modern industrial robots are essentially intelligent assemblies, able to choose the optimal operation mode, motion trajectory and other parameters on their own [1].

The manufacturing companies often do not disclose the full data about their products, as this may affect their competitive market potential, therefore this paper is based on the independent measurements carried out on a conventional industrial robot without prior detailed information about its construction. The presented measurements were made on the ABB robot system, located in the laboratory of Department of the Electrical Drives and Power Electronics at the Tallinn University of Technology. The central part of the studied robotic system is the welding robot *IRB 1600*, manufactured by ABB. To determine the effect of different control possibilities on industrial robot energy consumption, four experiments were made:

1) determination of optimal motion trajectory;
2) determination of optimal tool weight;
3) determination of optimal workpiece position;
4) determination of optimal operation speed.

In the next sections these experiments are described in more detail.

II. ENERGY SAVING BASICS IN ROBOTIC SYSTEMS

A. The Essence of Energy Saving

Energy saving means reducing both energy consumption and losses in manufacturing processes. Frequent change of temperature, caused by the inner losses, bring along rapid aging and deterioration of devices. Therefore, fighting the losses can extend the life cycle of devices and minimise the repair costs.

Energy-efficient operation is important also from the viewpoint of the market economy, because it reduces energy transmission costs and losses; increases duty time of the energy storage units and provides an opportunity to reduce the capacity and costs of such units, reduces costs of energy per product, thus increasing their competitiveness.

Another aspect of energy savings is related to mobile robots, operating on batteries [2][3]. It is self-evident, that increasing the operational efficiency of such robots yields increased running time and autonomy.

B. Motion Characteristics of a Manipulator

Industrial robot *IRB 1600*, used in described measurements, has 6 degrees of freedom (DOFs). Each DOF has its own synchronous motor. The motor data, unfortunately undisclosed by the manufacturer, can give a basis to investigate the energy consumption of an industrial robot. Moreover, the use of electric motors itself is related to hardly noticeable and measurable losses concerning friction, heat and magnetic leakage [4]. So it is more reasonable to carry out practical measurements and assess the energy consumption of the robot on a concrete example.

Industrial robot energy consumption depends on the characteristics of its movement. Different trajectories mean the involvement of different DOFs, which in turn means operating different motors. That is why it was necessary to select such a path that engages all DOFs, represented by the ∞-sign like closed trajectory, lopsided in one plane.

The active power exerted by the robot's mechanics is expressed by the equation [5]

$$P_{robot} = \sum_{i=1}^{n} T_i \cdot \omega_i \cdot \frac{1}{\prod_{i=1}^{n} \eta_{mec,i} \cdot \eta_{el,i}}, \quad (1)$$

where n is the number of DOFs, T_i is the torque applied to the ith DOF, ω_i the angular velocity of the ith DOF, $\eta_{mec,i}$ and $\eta_{el,i}$

978-1-4244-8806-3/11 $26.00 © 2011 IEEE

mechanic and electric efficiencies of the ith DOF drives, respectively. The active energy consumed by a robot is the integral of active power over time 0 ... t_f:

$$W_{act} = \int_0^{t_f} P_{robot} \cdot dt . \qquad (2)$$

In ac circuits also reactive power flows exist, though the use of a common diode rectifier (Fig. 1) must yield unity power factor. The differing results, discussed below, can only be explained by undisclosed data about the research object.

III. MEASUREMENTS ON THE IRB 1600 ROBOT

A. Measurement Conditions

Studies were made with 3-phase power quality analyzer *Fluke 434*. Measurement points were chosen with provision of losses and power consumption of other functional units (Fig. 1), for example the controller itself poses almost the same load (0.3 kW) than the manipulator in low duty (0.43 kW).

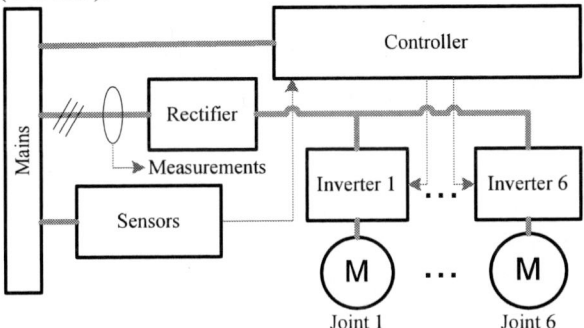

Fig. 1. Generalized power diagram of the IRB 1600 robot.

During the experiments robot was moving along the pre-programmed path. In each measurement the robot repeated the path 50 times, each measurement repeated three times in order to reduce the random error. Additional conditions, such as speed, movement character, weight of the tool etc are explained separately for each experiment.

B. Determination of the optimal motion trajectories

The load of robots drives depends on the movement direction. When the manipulator moves almost vertically, then the gravity has the opposite direction with upward movements and the same direction with downward movements. If the manipulator moves almost horizontally, the gravity force has the same influence in both directions.

In the first part of experiment the points P1, P2 and P3 were parallel to robot y-axis on the xy-plane, with the sketch shown in Fig. 2.

In the second part of experiment the points P1, P2 and P3 were chosen parallel to robot z-axis on the yz-plane, as shown in Fig. 3. Additional conditions of measurements were as follows: number of cycles – 50, the tool weight - 2.5 kg; velocity - 500 mm/s.

Both experiments lasted 305 s. Fig. 4 illustrates the results of measurements: the real power as well as the apparent

power in case of nearly horizontal movement is insignificantly higher than in case of nearly vertical movement. This can be explained by the fact that during one half of the trajectory, the gravity has the same sign with the motion [6], thus motors in generating quadrant supply other motors in motoring quadrant over the common dc bus.

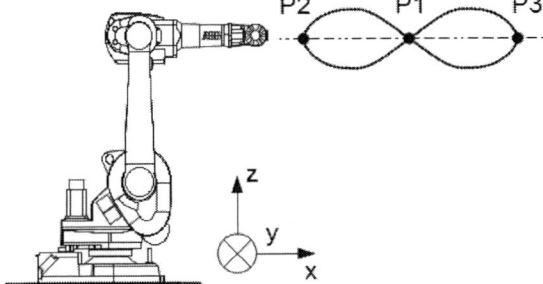

Fig. 2. Measured trajectory in the horizontal xy-axis plane.

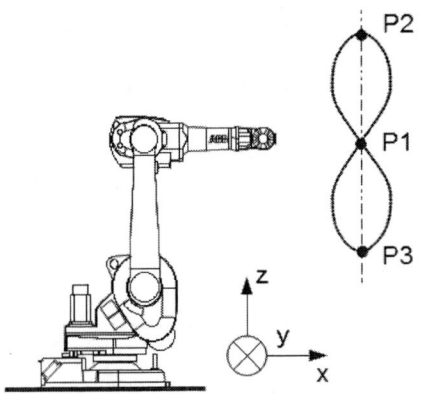

Fig. 3. Measured trajectory in the vertical yz-axis plane.

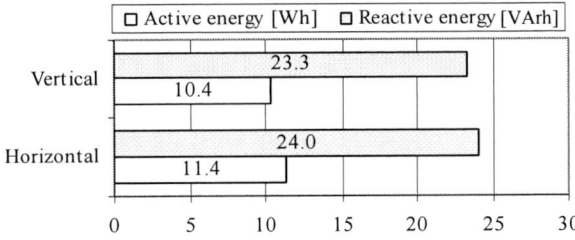

Fig. 4. Robot's energy consumption at different motion directions.

In case of the horizontal motion the x-axis is leaned forward, thus increasing the effective radius affecting the moment of inertia. From the classical equation of motion

$$T_i = T_{L,i} + J_i \frac{d\omega_i}{dt} , \qquad (3)$$

where $T_{L,i}$ is the static load and J_i the moment of inertia, it might be concluded that increased J_i yields additional energy need, as theoretically explained by Eq. (1) and (2) as well as the conducted experiment.

C. Effect of the Tool Weight on the Energy Consumption

Industrial robots have different payloads, depending on the robot's weight and application. The weight of a robot tool can

978-1-4244-8806-3/11 $26.00 © 2011 IEEE

be permanent (e.g. a welding robot) or variable (e.g. a pick-and-place robot). During this test robot was loaded with three different weights:
1) 0 kg – without payload;
2) 2.5 kg – the weight of the tool used in studying process;
3) 5 kg – the maximum possible payload of IRB 1600.

Additional conditions of measurements were as follows: the plane of the movements - horizontal; velocity - 500 mm/s. Experiments with three possible payloads lasted 305 s like during the previous measurements.

The results, shown on Fig. 5, can be explained by the robot motor characteristics. In the permanent magnet synchronous motors, the current is proportional to the torque, which depends on the tool's weight. The small differences are due to the fact, that the payloads are relatively small compared to the weight of the robot's links. In larger robots where payloads are heavier, the differences are even more remarkable.

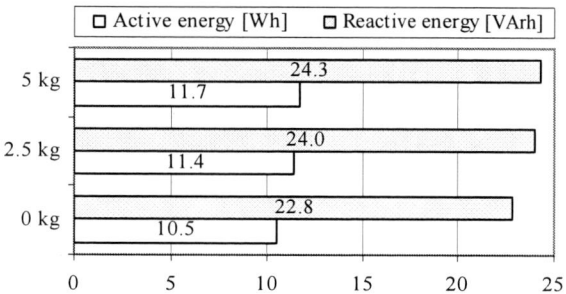

Fig. 5. Robot's energy consumption at different tool weights.

D. Determination of Optimal Workpiece Position

The main objective of this test was to get to know how the energy consumption of the robot depends on the workpiece position. The reference point of the IRB 1600 robot was defined by its home position, as shown in Fig. 6. During the measurements robot was moving alongside the pre-programmed path on 10 different heights. One was 150 mm over the reference position and other ones were below, decreasing by 150 mm increments. The lowest working plane was on the same level with the manipulator's base, determined by the possible working range. All the experiments lasted 305 s.

Test results are presented on Fig. 7. The most energy-efficient position of workpiece is 600 mm above the manipulator's base plane, the highest energy consumption is above the reference position.

Usually the workpiece is on a conveyor line or a positioner. As follows, the positioner IRBP 250 L used in the robotic system with IRB 1600 is preferably located 600 mm above the base plane. The positioner's location is selected taking into consideration the kinematical characteristics, so that operation are is the broadest.

A. Determination of Optimal Operating Speed

Working speed of the robot depends on the actual operation mode. For example, a typical welding speed is in the range of 100 mm/min - 500 mm/min, depending from welding current,

material, detail thickness etc [7]. Typical speed for pick-and-place robots is around 3000 mm/min - 15000 mm/min. During the experiments the manipulator was moving along the pre-programmed path 50 times with different speeds. The speed was increased from 100 mm/s to the maximum, the latter depending on the load. Additional conditions of measurements were as follows: the plane of the movements - horizontal, the tool weight - 2.5 kg.

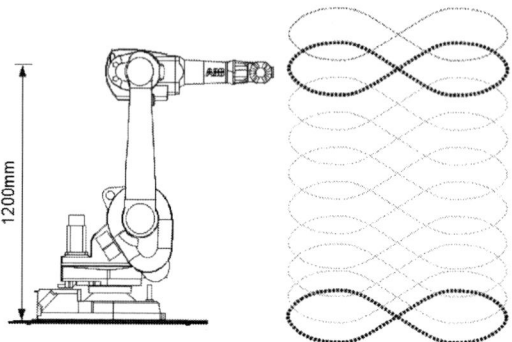

Fig. 6. Determination of the optimal workpiece position.

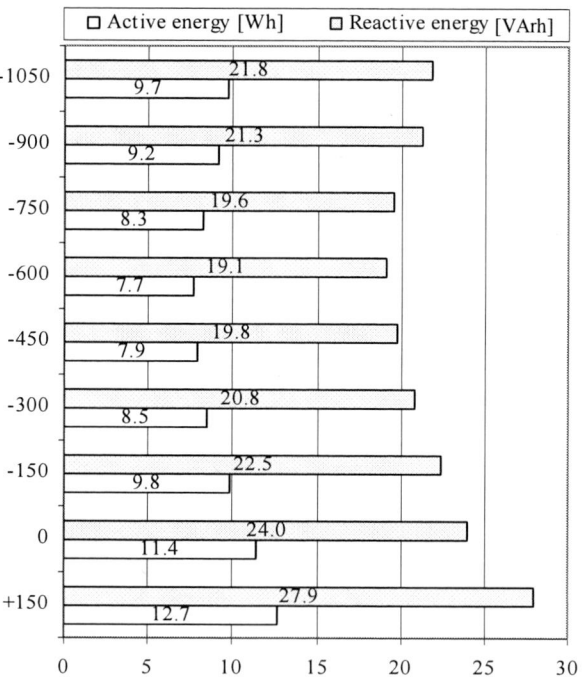

Fig. 7. Results of the optimal workpiece position measurements.

The results of the optimal operational speed determination are shown in Fig. 8. Under the given condition the lowest energy consumption was at 600 mm/s. Although reducing the motion speed can minimise the energy consumed by the robot, the increase in the time needed to carry out the operations counteract to the set objectives by additional energy consumption. Energy savings in terms of speed reduction is not always thinkable, especially when it comes to mass production, where the duration of a cycle is crucial [8].

978-1-4244-8806-3/11 $26.00 © 2011 IEEE   228

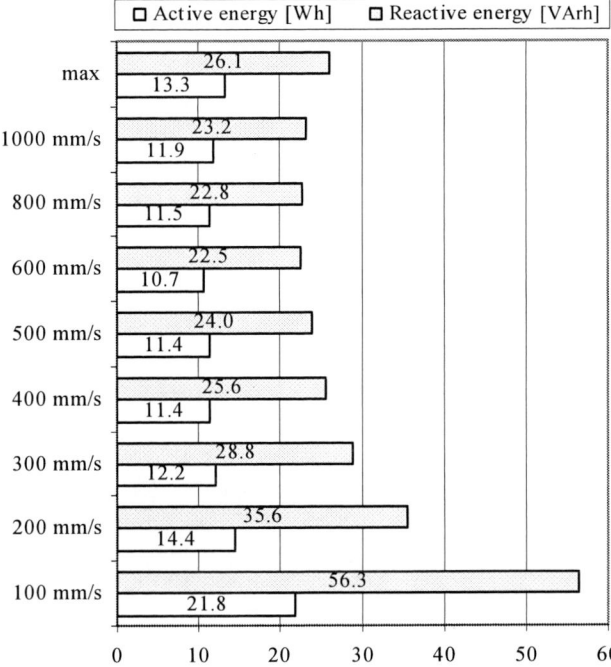

Fig. 8. Results of the optimal operation speed measurements.

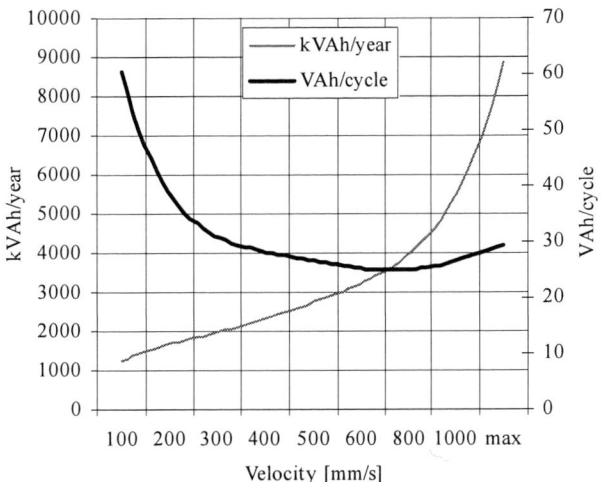

Fig. 9. Optimal operational speed vs. productivity.

IV. CONCLUSIONS AND FUTURE WORK

A. Results of the Measurements

The results of the measurements can be divided into two parts: the results of two first tests – determination of optimal trajectory (Fig. 4) and energy consumption with different tools weight (Fig. 5) – do not yield enough energy savings, explained by the low capacity of IRB 1600, where the bulk of the weight is constituted by the mass of the joints themselves; the results of two other tests – determination of optimal workpiece position (Fig. 7) and optimal operating speed (Fig. 8) – give already some hints for more essential energy savings. In that case one can conclude that a properly installed and correctly tuned robot can operate with improved energy efficiency.

B. Economic Benefits

Though the differences in consumed energy, determined during performed experiments might seem insignificant, one must remember that an industrial robot is often running continuously. Thus, when multiple robots are applied in an industrial process, the yearly savings would be remarkable [9]. In terms of economic benefits, finding a relationship between optimal operational speed and yearly energy consumption might be interesting. In Fig. 9, this optimum is defined as the intersection point between the two curves, in current case approximately 700 mm/s.

C. Future Prospects

To improve the results it would be useful to repeat the tests with a more powerful robot. Carrying out additional test like investigating the performance of a robotic system as a part of a smart grid would be a topic.

Distributed systems are also one possibility to use the robots more rationally [10]. The main point of distributed system is to use multiple robots in the system with a upstream main controller, which coordinates and forecasts the actions of the individual robots on the basis of minimal energy and maximal productivity ratio [11].

ACKNOWLEDGMENT

This research work has been supported by Estonian Ministry of Education and Research (Project SF0140016s11) and Estonian Archimedes Foundation (project „Doctoral School of Energy and Geotechnology-II").

REFERENCES

[1] W. Khalil, *Modeling, identification & control of robots.* London, Kogan Page Science, 2004.
[2] Y. Mei et al., "Energy-efficient motion planning for mobile robots". *IEEE International Conference on Robotics and Automation*, vol. 5 pp. 4344-4349, 2004.
[3] Y. Mei et al., "Energy-efficient mobile robot exploration". *IEEE International Conference on Robotics and Automation*, pp. 505-511, 2006.
[4] E.S. Sergaki, G.S. Stavrakis, A.D. Pouliezos, "Optimal Robot Speed Trajectory by Minimization of the Actuator Motor Electromechanical Losses", *Journal of Intelligent and Robotic Systems*, vol. 33, pp. 187–207, 2002.
[5] H. Choset et al., *Principles of robot motion: theory, algorithms, and implementation.* London : MIT Press, 2005.
[6] D. Verscheure et al., "Time-Energy Optimal Path Tracking for Robots: a Numerically Efficient Optimization Approach", *IEEE 10th Annual Workshop on Advanced Motion Control*, pp. 727-732, 2008.
[7] J.N. Pires, A. Loureiro and G. Bölmsjo, *Welding Robots Technology. System Issues and Applications.* London, Springer, 2006.
[8] S.A. Alshahrani, H. Diken, A.A.N. Aljawi, "Optimum trajectory function for minimum energy requirements of a spherical robot", *The 6th Saudi Engineering Conference*, vol. 4, pp. 613-625, 2002.
[9] Y. Li and G.M. Bone, "Are Parallel Manipulators More Energy Efficient?" *IEEE International Symposium on Computational Intelligence in Robotics and Automation*, pp. 41-46, 2001.
[10] T.D. Ngo, H. Raposo, H. Schioler, "Potentially Distributable Energy: Towards Energy Autonomy in Large Population of Mobile Robots", *IEEE International Symposium on Computational Intelligence in Robotics and Automation*, pp. 206-211, 2007.
[11] A. Vergnano et al., "Embedding detailed robot energy optimization into high-level scheduling". *IEEE 6th Annual Conference on Automation Science and Engineering*, pp. 386-392, 2010.

Analysis and Optimization of Sinusoidal Voltage Source Inverter Losses for Variable Output Power Applications

Enrique Romero-Cadaval, Víctor Miñambres-Marcos, María-Isabel Milanés-Montero
University of Extremadura (Badajoz, Spain)
eromero@unex.es, vminmar@unex.es, milanes@unex.es

Abstract-In this paper, a general analysis of converter losses is done. The aim of this analysis is to determine the parameters that have great influence in these losses. The losses study presented in this paper is focused in a Two-level Three-branch Voltage Source Inverter (VSI) and it is particularized for the case in which this inverter has to generate sinusoidal currents, showing the losses and efficiencies of the converter for different output power. Finally, the opportunities of using a topology based in a DC/DC converter connected in cascade with the VSI for decreasing the losses and achieving greater converter efficiencies are shown.

I. INTRODUCTION

Society is demanding a bigger quantity of energy as its level of development is growing. Conventional energy resources are limited, so authorities and governments are promoting energy savings and energetic efficiency. Also, renewable energies have been sustained and promoted by these authorities and governments as an alternative to limited conventional energy resources. To manage in an efficient way electrical devices and generation plants usually electronic power converters are needed.

One of the most used converter topology is the two-level voltage source inverter (Fig.1), that are used mainly for renewable energy integration (photovoltaic and wind generation plants) are for controlling conventional electrical load (most of them AC motors). An attempt to optimize the losses in VSI could have a great impact in energy savings due to the big quantity of converters used in real applications.

These converters present high efficiency ratios when they are operated at its nominal power, but these ratios decrease quickly when the output power is a low fraction of the nominal value. Most applications where VSI are used, present an out power that can vary from nearly zero to the nominal power. This is the case, for example, of photovoltaic plants when operates at low value of sun irradiance (cloudy days) or motor drives when load torque is low.

In this paper, a general analysis of VSI losses is carried out in the first section. In next section this analysis is focused when the VSI is generating sinusoidal currents. Finally, the advantages of using a cascade DC/DC converter to accommodate the DC voltage that supplies the VSI are evaluated.

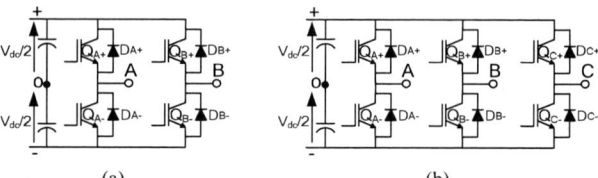

(a)   (b)

Fig. 1. Classical Two-levels VSI: (a) Full-bridge, Single-phase, (b) Three-phase, Three branches.

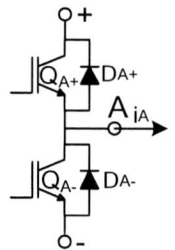

Fig. 2. Two-level VSI branch.

II. TWO-LEVEL VOLTAGE SOURCE INVERTER LOSSES

The losses produced in the semiconductor devices that compose a conventional Two-levels VSI can be calculated in a per branch analysis. If a single branch is considered (Fig.2) the current that flows from the branch to the load (terminal named A in Fig.1 and 2) will only pass thought one of the four semiconductors of the branch depending on the transistor state and the current direction:

$$s_{A+} \begin{cases} \text{state } 1:1 \begin{cases} i_A > 0 \Rightarrow Q_{A+} \\ i_A < 0 \Rightarrow D_{A+} \end{cases}, \\ \text{state } 2:0 \begin{cases} i_A > 0 \Rightarrow D_{A-} \\ i_A < 0 \Rightarrow Q_{A-} \end{cases} \end{cases} \quad (1)$$

Table I. Possible branch states according to the status of the two transistor of it.

Branch state	Transistor Q_{A+}	Transistor Q_{A-}	Comments
0	OFF	OFF	Only applied in branch dead times for avoiding state 3
1	ON	OFF	
2	OFF	ON	
3	ON	ON	Not allowed. Never applied

978-1-4244-8806-3/11 $26.00 © 2011 IEEE   230

where s_{A+} is the switching signal for the transistor Q_{A+}, that will be complementary with the switching signal for transistor Q_{A-}, because only the states 1 and 2 of the possible states of the inverter branch (Table I) are considered.

If d_{A+} is defined as the duty cycle of transistor Q_{A+}, $(1-d_{A+})$ will be the duty cycle of the complementary transistor Q_{A-} and the average voltage V_{Ao} in a switching period can be expressed as:

$$V_{Ao} = \frac{V_{dc}}{2}(2d_{A+}-1) \quad (2)$$

where o is the DC bus mid-point that can actual exist or can be only a virtual voltage reference point.

From expression (2), the duty cycle of transistor Q_{A+} will be:

$$d_{A+} = \left(\frac{1}{2} + \frac{V_{Ao}}{V_{dc}}\right). \quad (3)$$

If it is considered that, when conducting, an electronic switch (ES) made with a semiconductor has a voltage drop given by a first order approximation

$$v_{ES} = V_{ES,on} + r_{ES,on}I_A, \quad (4)$$

then the ES conduction losses will be

$$P_{on} = v_{ES}I_A = V_{ES,on}I_A + r_{ES,on}I_A^2. \quad (5)$$

From the previous analysis, the conduction losses in a switching period for the different semiconductors of the branches (assuming that both transistors and both diodes are identical) will be function of the branch current and the duty cycle as follow:

$$I_A \begin{cases} > 0 \begin{cases} P_{QA+} = \left(V_{onQ}I_A + r_{onQ}I_A^2\right)d_{A+} \\ P_{DA+} = 0 \\ P_{QA-} = 0 \\ P_{DA-} = \left(V_{onD}I_A + r_{onD}I_A^2\right)(1-d_{A+}), \end{cases} \\ < 0 \begin{cases} P_{QA+} = 0 \\ P_{DA+} = \left(V_{onD}I_A + r_{onD}I_A^2\right)d_{A+} \\ P_{QA-} = \left(V_{onQ}I_A + r_{onQ}I_A^2\right)(1-d_{A+}) \\ P_{DA+} = 0 \end{cases} \end{cases} \quad (6)$$

where V_{onQ} and V_{onD} are the ON voltage values for a transistor and a diode respectively, and r_{onQ} and r_{onD} are the ON resistance values for them.

Assuming that the current exiting the branch is positive, $I_A>0$, the conduction losses in a switching period, kT_S, can be determined using (6):

$$\begin{aligned} P_{on}(kT_S) &= P_{QA+} + P_{DA-} = \\ &= \left(V_{onQ}I_A + r_{onQ}I_A^2\right)d_{A+} + \left(V_{onD}I_A + r_{onD}I_A^2\right)(1-d_{A+}) \\ &= I_A\left(V_{onQ}d_{A+} + V_{onD}(1-d_{A+})\right) + I_A^2\left(r_{onQ}d_{A+} + r_{onD}(1-d_{A+})\right) \end{aligned} \quad (7)$$

and substituting the value of the duty cycle given by (3) and simplifying could be obtained that

$$\begin{aligned} P_{on}(kT_s) &= I_A(kT_s)\left(\frac{V_{onQ} + V_{onD}}{2} + \frac{V_{onQ} - V_{onD}}{V_{dc}}V_{Ao}(kT_s)\right) + \\ &+ I_A^2(kT_s)\left(\frac{r_{onQ} + r_{onD}}{2} + \frac{r_{onQ} - r_{onD}}{V_{dc}}V_{Ao}(kT_s)\right) \end{aligned} \quad (8)$$

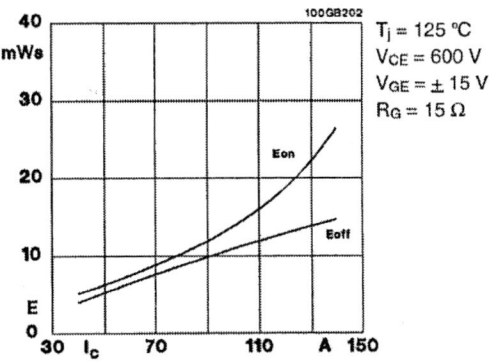

Fig. 3. Turn-on/-off energy dissipation as function of I_C [1].

A similar result can be obtained if the current exiting the branch is negative, $I_A<0$, and finally, a general expression can be obtained

$$\begin{aligned} P_{on}(kT_s) &= |I_A(kT_s)|\left(\frac{V_{onQ} + V_{onD}}{2} + \frac{V_{onQ} - V_{onD}}{V_{dc}}V_{Ao}(kT_s)\right) + \\ &+ I_A^2(kT_s)\left(\frac{r_{onQ} + r_{onD}}{2} + \frac{r_{onQ} - V_{onD}}{V_{dc}}V_{Ao}(kT_s)\right) \end{aligned} \quad (9)$$

Switching losses can be determined using the energy needed for turning on and off semiconductors that are given by manufacturers on the semiconductor datasheets [2]. These energies are function of the current driven by the semiconductor (Fig. 3) and usually they are determined for a reference value of voltage and current. Under the assumption of these energies present a linear dependence with the semiconductor voltage and current at the switching instant, and it is considered that one transistor is turning on and off and one diode is also turning on and off in any switching period, then switching losses in a switching period can be calculated as

$$P_{sw}(kT_s) = \left(E_{on} + E_{off} + E_{rr}\right)f_s \frac{|I_A(kT_s)|V_{dc}}{I_{ref}V_{ref}} \quad (10)$$

where E_{on}, E_{off} and E_{rr} are respectively the energy needed to turn on the transistor, to turn off the transistor and to turn off the diode, and I_{ref} and V_{ref} are the reference current and voltage for which this energies has been determined.

Total average losses of the converter could be determined taking into account the sign of the current exiting the branch by

$$P = \frac{1}{NT}\sum_{k=1}^{N}\left(P_{on}(kT_S) + P_{sw}(kT_S)\right) \quad (11)$$

where T is the fundamental period and N are the number of switching periods in a fundamental period.

If the switching period, T_S, is much smaller than the fundamental period, T, the previous expression could be substitute by the next integral expression

$$P = \frac{1}{T}\int_0^T \left(P_{on}(t) + P_{sw}(t)\right)dt \quad (12)$$

III. SINUSOIDAL VSI LOSSES

When the VSI has to produce a sinusoidal current that, in general, will not be in phase with the voltage (Fig. 4), the current flows thought the four different semiconductors that compose the branch according to Fig. 5.

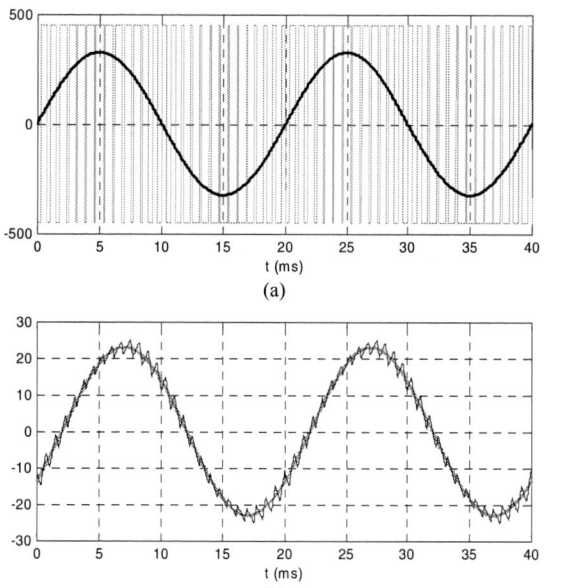

(a)

(b)

Fig. 4. (a) Voltage at VSI terminals (V_{Ao}) and its fundamental component ($V_{Ao,1}$), (b) Current thought VSI terminal A and its fundamental component.

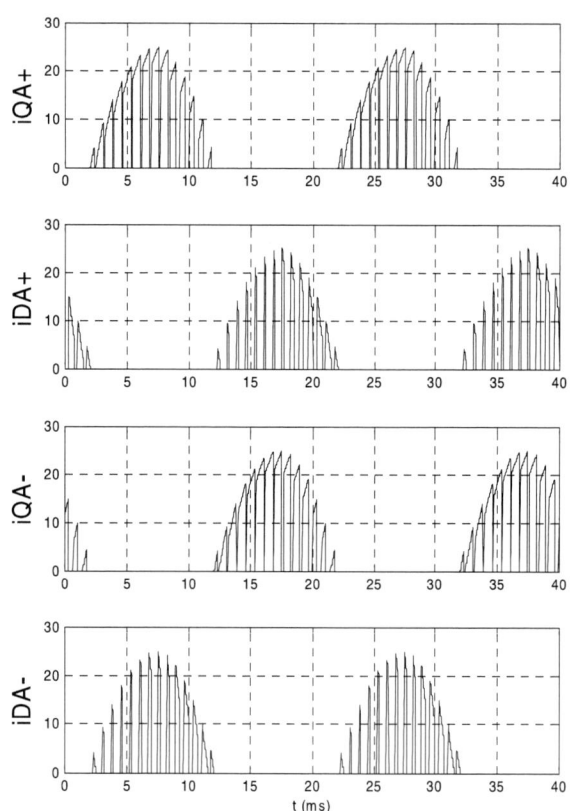

Fig. 5. Waveform of the currents that flow by each of the four VSI branch semiconductors.

Table II. Conditions used for simulation.

Nominal frequency (fundamental)	50 Hz
Nominal voltage (phase-neutral)	230 V
Nominal impedance	11.5 Ω - fp 0.8
DC bus voltage	700 V
Switching frequency	10,5 kHz

Table III. Parameters of Electronic Switches used in simulation.

Transistor ON voltage	3.7 V
Transistor ON resistance	76 mΩ
Diode ON voltage	1.8 V
Diode ON resistance	32 mΩ
Transistor ON energy	10.5 mJ
Transistor OFF energy	7.5 mJ
Diode Reverse Recovery energy	3 mJ
Reference current for energies	25 A
Reference voltage for energies	1200 V

If one applies the expression calculated in the previous section to this case, where the VSI has to generate sinusoidal currents, then it will be obtain the losses shown in Fig. 6 when the system operates with the parameters and conditions summarized in Table II and Table III.

Expressions (9) and (10) could de applied to this case assuming that

$$I_{A,1}(\theta_s) = \sqrt{2}I_A \sin\theta$$
$$V_{An,1}(\theta_s) = V_{Ao,1}(\theta_s) = \sqrt{2}V_{An} \sin(\theta + \varphi) \qquad (13)$$

where V_{An} and I_A are the fundamental component RMS values of branch voltage and current respectively, and φ is the displacement angle between fundamental component of voltage and current.

Next expression can be obtained by integrating expression (12) in a fundamental semi-period (considering the symmetry the waveforms have):

$$P_{on} = I_A \left\{ \left(\frac{V_{onQ} + V_{onD}}{2} \right) 0.9 + \left(V_{onQ} - V_{onD} \right) \frac{V_{An}}{V_{dc}} \cos\varphi \right\} +$$
$$+ I_A^2 \left\{ \left(\frac{r_{onQ} + r_{onD}}{2} \right) + \left(r_{onQ} - r_{onD} \right) \frac{8\sqrt{2}}{3\pi} \frac{V_{An}}{V_{dc}} \cos\varphi \right\} \qquad (14)$$

$$P_{sw} = \left(E_{on} + E_{on} + E_{rr} \right) 0.9 f_s \frac{I_A}{I_{ref}} \frac{V_{dc}}{V_{ref}} \qquad (15)$$

The output voltage of a VSI is related with the DC voltage by:

$$V_{An} = V_{An,1,RMS} = V_{Ao,1} = m_a \frac{\sqrt{2}V_{dc}}{4} \Rightarrow m_a = \frac{2\sqrt{2}V_{An}}{V_{dc}} \qquad (16)$$

where m_a is the amplitude modulation index used for generating the switching signal by a pulse width modulation technique.

For the proposed case, Fig.6(a) shows how the losses associated to the mean value of the ON voltage of the transistor and the diode are much greater that the losses due to the second term in expression (14) associated to the difference between the ON voltages of both semiconductors. The losses of the first term do not depend on the power factor directly but the average absolute output current.

From Fig.6(c), one can note that losses produced by ON voltage are more significant than losses associated to the ON resistance. Also, one can observe in this figure that these last losses are not negligible.

978-1-4244-8806-3/11 $26.00 © 2011 IEEE

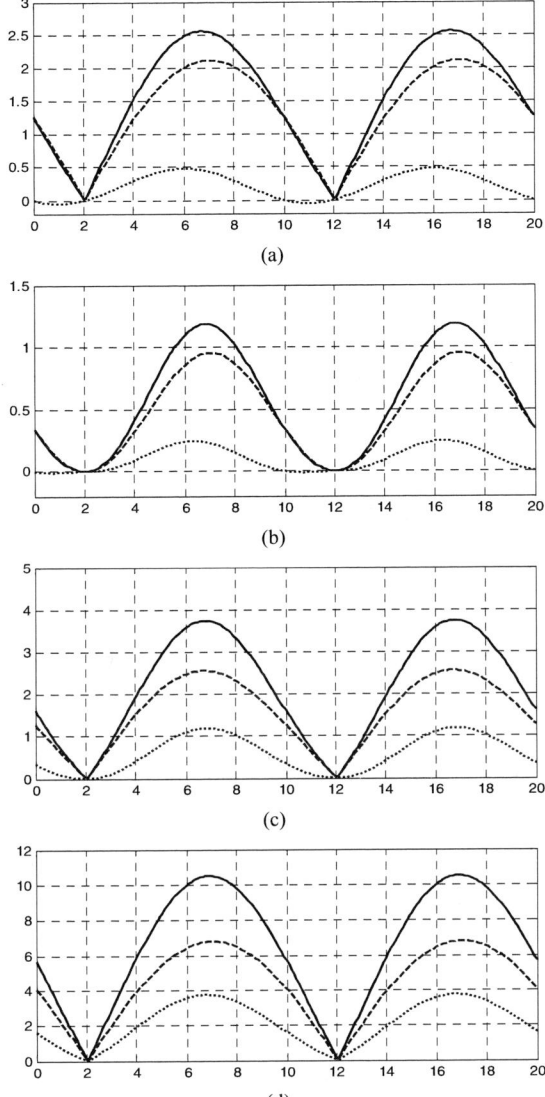

Fig. 6. VSI losses when generating a sinusoidal current expressed as percentage of the output power in a fundamental period (20ms): (a) Losses produced by ON voltage (solid: total losses, dashed: due to average ON voltage, dotted: due to the different values of ON voltage), (b) Losses produced by ON resistance (solid: total losses, dashed: due to average ON resistance, dotted: due to the different values of ON resistance), (c) Conduction losses (solid: total, dashed: due to ON voltage, dotted: due to ON resistance), (d) VSI losses (solid: total, dashed: switching losses, dotted: conduction losses).

Finally, one can see in Fig.6d that, for the proposed system, switching losses are greater than conduction losses.

The losses determined in Fig.6 correspond to the nominal operation point (defined in Table II). But, when the VSI is used for controlling a variable electrical load, the output power of this VSI could change from nearly zero to the nominal power. The losses change and the converter efficiency varies when the output power changes, and becomes very important to know the losses dependence on output power.

Fig.7 shows the converter losses variation when the output power ($P_{3\phi}$) varies for two types of sinusoidal loads:

- the first one (on the top, Fig.7a and Fig.7b) corresponds to a constant impedance load that needs a voltage from the VSI given by

$$V_{An} = \sqrt{\frac{P_{3\Phi}}{3} z_N} \qquad (17)$$

- and the second one (on the bottom, Fig.7d and Fig.7e) corresponds to a constant current load, needing a VSI voltage given by

$$V_{An} = \frac{P_{3\Phi}}{3} \frac{1}{I_N} \qquad (18)$$

versus the amplitude modulation index, determined by (16).

The converter efficiency for both cases, considering the output power of the VSI, is shown is shown in Fig.7c and Fig.7f.

From Fig.7 one can observe that the efficiency decrease heavily when the output power is below 30% for the constant impedance case, and bellow 50% for the constant current case.

IV. CASCADE DC/DC CONVERTER

In this section it will be analyzed how a cascade DC/DC converter (Fig.8) could improve the converter efficiency when the VSI is applied to a load with a power consumption that varies from nearly zero to its nominal value.

In this case we have to consider the additional losses produce by the DC/DC converter. These losses, following and analysis similar to the one done in previous section and assuming that transistor and diode of the DC/DC converter have the same parameters than the VSI semiconductors, could be determined by the next expressions:

$$P_{on} = I_o \left(V_{onQ} d_{dc} + V_{onD} (1 - d_{dc}) \right) + \\ + I_o^2 \left(r_{onQ} d_{dc} + r_{onD} (1 - d_{dc}) \right) \qquad (19)$$

$$P_{dc,sw} = (E_{on} + E_{on} + E_{rr}) f_s \frac{I_o}{I_{ref}} \frac{V_{dc}}{V_{ref}} \qquad (20)$$

where d_{dc} is the duty cycle of transistor Q_{dc} that is calculated to control the intermediate voltage V_{oi} using:

$$d_{dc} = \frac{V_{oi}}{V_{dc}} \qquad (19)$$

The intermediate voltage is controlled to guarantee that the VSI operated with an amplitude modulation index near unity ($m_a \approx 1$) for any output power supplying from VSI to the load. The voltage at its DC terminals, obtained using (16), is:

$$V_{oi} = \frac{2\sqrt{2} V_{An}}{m_a} = 2\sqrt{2} V_{An} \qquad (19)$$

The losses of the VSI can be determined by expressions (14) and (15), substituting V_{dc} by V_{oi}, and applying (17) and (18).

The results of the proposed system when supplying a load in the same cases than in previous section (constant impedance load and constant current load) are shown in Fig 9. In this figure, one can see (Fig.9a, 9b, 9d and 9e) how the VSI switching losses stay constant for any output power, and how a new losses, produced by the DC/DC converter, appear in the proposed topology. But the final result is that the reduction on the VSI losses is bigger than the new DC/DC converter losses, and so a global losses reduction is achieved, that is important when the VSI output power ratio is small.

LOAD BEHAVES AS A CONSTANT IMPDANCE

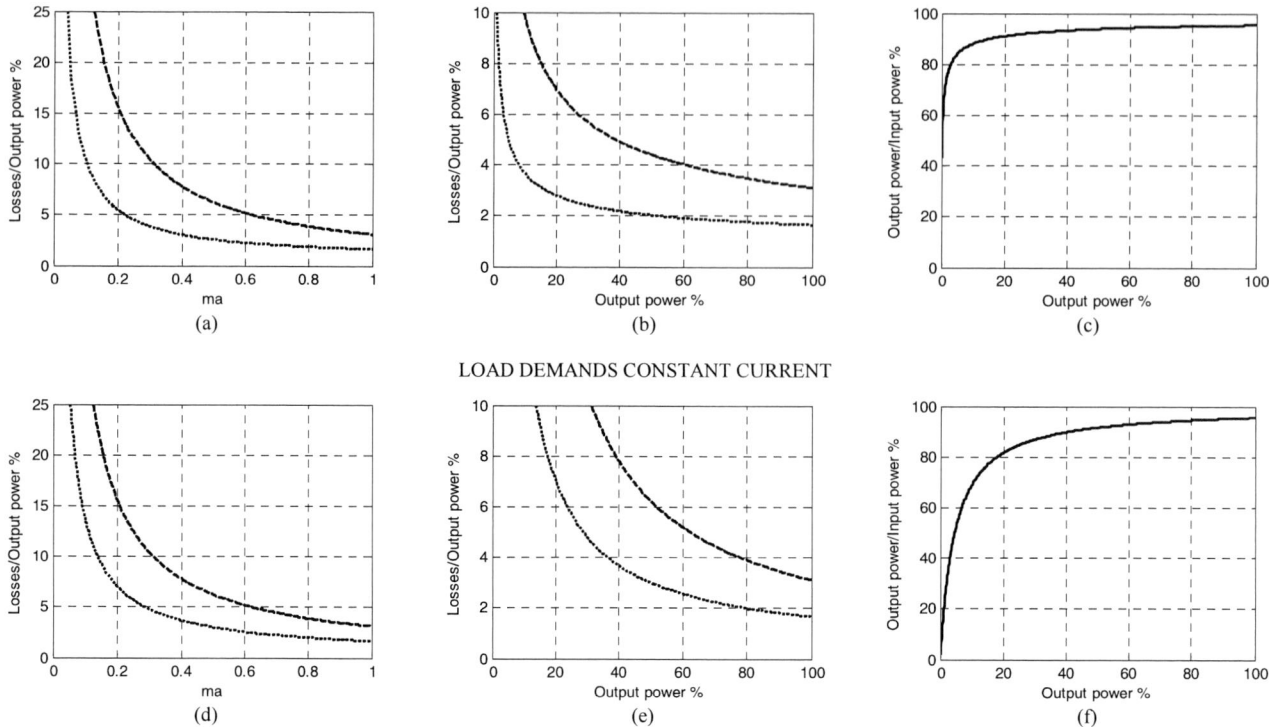

Fig. 7. Analysis of VSI losses: Variation of conduction losses (dotted) and switching losses (dashed): (a) when load behaves as a constant impedance versus the amplitude modulation index; (b) when load behaves as a constant impedance versus the output power ratio; (d) when load demands constant current versus the amplitude modulation index; (e) when load demands constant current versus the output power ratio.
Global efficiency versus the output power ratio: (c) when load behaves as constant impedance; (f) when load demands a constant current.

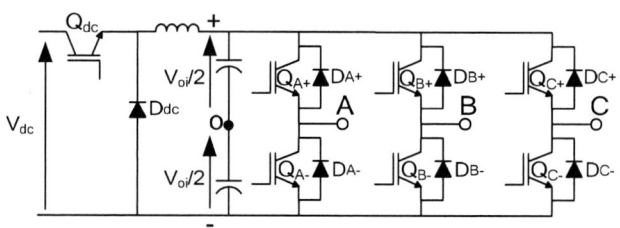

Fig. 8. Proposed topology variation, with a DC/DC converter for conditioning the VSI DC voltage.

The resulting efficiency ratio when varying the output power (Fig.9c and Fig 9f) is compared with the ones obtained in the previous section for a conventional VSI.
It could be observed how the proposed topology achieves that the reduction of efficiency at low output power ratios is smaller than if there is no a previous DC/DC converter for conditioning the initial DC voltage.

V. CONCLUSION

In this paper, the losses produced in a VSI (a commonly used converter in power electronic systems) have been analyzed, determining the parameters that are most significant in these losses.
It has been observed that the conduction losses produced by the ON resistance of the electronic switches are not negligible compared with the losses produced by the ON voltage, so both parameters must be considered for every electronic switches (transistor or diode).

The conduction losses are due to two terms, one depends on the average value of the parameters of the transistor and diode and the second one, that is less significant, depends on the difference between the values presented for these parameters (ON voltage and resistance) in both kinds of semiconductors. If this second terms is neglected, the conduction losses depend mainly on the average value of the absolute current the VSI branch is supplying, and do not depend neither the DC voltage nor $\cos\varphi$ of the load

A key parameter that has a direct influence in switching losses (note that switching losses are more significant than conduction losses in the cases studied in this paper due to the switching frequency selected to operate the VSI) is the DC voltage.

It has been proposed the inclusion of a DC/DC converter for conditioning the DC voltage that supplies the VSI (that in this case operates with a fixed m_a), reducing the initial DC voltage when the VSI operates with low output power ratios presenting low efficiency ratios.

The proposed topology allows to obtain better efficiency ratios, although additional losses (due to the DC/DC converter) occurs, the reduction on the losses of the three branches that compose the VSI compensate this increase, achieving even a reduction in total losses.

LOAD BEHAVES AS A CONSTANT IMPDANCE

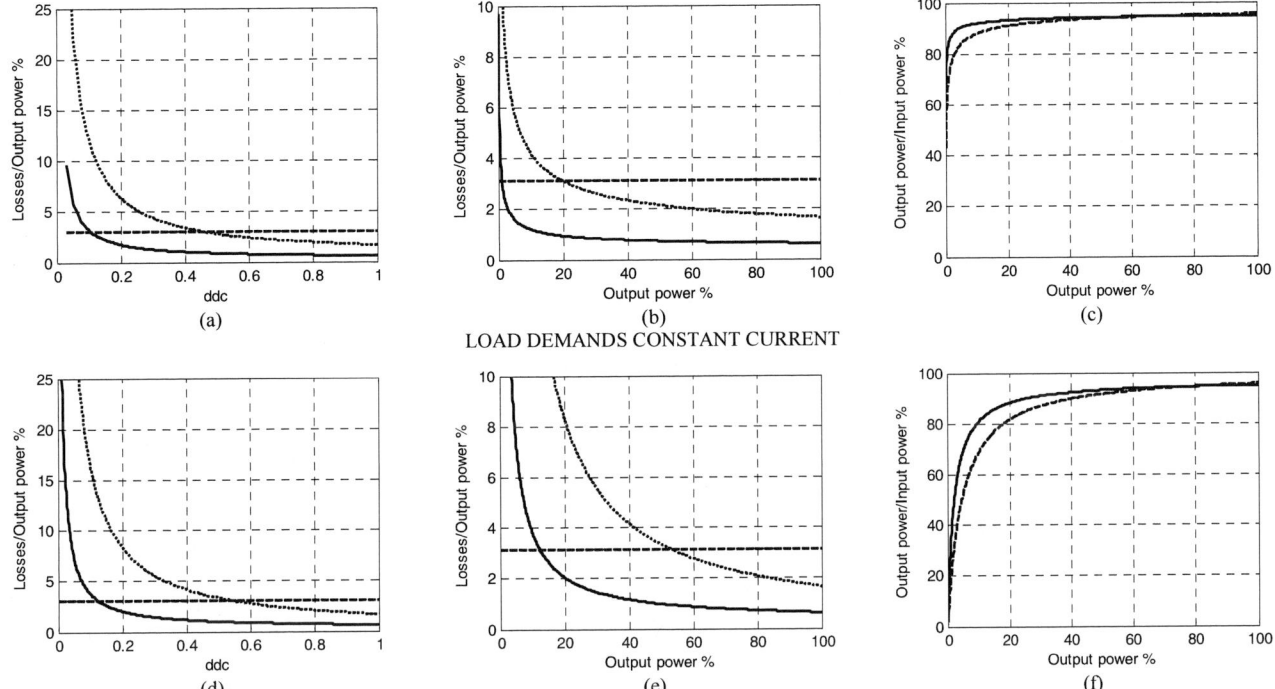

Fig. 9. Analysis of Fig.8 system losses: Variation of VSI conduction losses (dotted), VSI switching losses (dashed), and DC/DC converter losses (solid): (a) when load behaves as a constant impedance versus the amplitude modulation index; (b) when load behaves as a constant impedance versus the output power ratio; (d) when load demands constant current versus the amplitude modulation index; (e) when load demands constant current versus the output power ratio.
Global efficiency (solid) and global efficiency without DC/DC converter (dashed) versus the output power ratio: (c) when load behaves as constant impedance; (f) when load demands a constant current.

It could be say that it is included a branch that produces additional losses, but it is achieved a reduction of switching losses in three branches. One in front of three, this is the key. For a single phase VSI the same reduction could not be achieved.

Further research could be done for determining if there exists an optimum value for the fixed m_a used in the VSI when it is associated with a cascade DC/DC converter.

In the topology analyzed in the last section of this paper, the DC/DC converter is operated with the same switching frequency than the one used for the VSI. But a small switching frequency could be used with the aim of reducing even more the conduction losses, and taking advantage of the fact that the DC/DC converter has to condition the DC voltage with a less exigent dynamic specifications than the ones the VSI has to generate the sinusoidal currents.

Authors are working at the present in a prototype for validating these analytical and simulation results through the corresponding experimental tests.

REFERENCES

[1] SEMIKRON, "Datasheet parameters for MOSFET, IGBT, MiniSKiiP- and SKiiPPACK modules"
http://www.semikron.com/skcompub/en/SID-FEB6ECF9-7554810B/eng_2_3_3.pdf

[2] Semikron application manual. Web-document. (Available in http://www.semikron.com, accessed in April 2010).

[3] R. Ramin, M. Di-Lella, "SEMITOP® 3-level inverter for UPS". Presentation. (Available in http://www.semikron.com, accessed in April 2010).

[4] M. Di-Lella, R. Ramin, "IGBTs for 3-level inverters. Improved efficiency in DC/AC conversion". Bodo's Power Electronics in Motion and Conversion. September 2008, pp. 22-24. (Available in http://www.semikron.com, accessed in April 2010).

[5] Wu, Y.; Shafi, A.; Knight, M.; McMahon, A, "Comparison of the Effects of Continuous and Discontinuous PWM Schemes on Power Losses of Voltage Sourced Inverters for Induction Motor Drives". Power Electronics, IEEE Transactions on, Volume: PP , Issue: 99, 2010 , Page(s): 1 – 1.

[6] Green, T.C.; Hernandez-Aramburo, C.A.; Smith, A.C.; "Losses in grid and inverter supplied induction machine drives". Electric Power Applications, IEE Proceedings. Volume: 150 , Issue: 6. 2003 , Page(s): 712 - 724

[7] Mapelli, F.L.; Tarsitano, D.; Mauri, M.; "Plug-In Hybrid Electric Vehicle: Modeling, Prototype Realization, and Inverter Losses Reduction Analysis". Industrial Electronics, IEEE Transactions on. Volume: 57 , Issue: 2. 2010 , Page(s): 598 - 607

[8] Yan-jun Yu; Feng Chai; Shu-kang Cheng; "Analysis of modulation pattern and losses in inverter for PMSM drives". Vehicle Power and Propulsion Conference, 2008. VPPC '08. IEEE. 2008 , Page(s): 1 - 4

Determination of defects of electric motors without their disconnection

Grebchenko N.V., Smirnova M. A., Sidorenko A.A., Ragulina N.V., Bielchev I.V.
Donetsk National Technical University
Artema str., 58, Donetsk-00, 83000, Ukraine
tel./fax +38(062) 301-03-72, E-mail: gvn@dgtu.donetsk.ua

Annotation – The article shows the results of the pilot researches of an electric motor behavior with initiation of insulation damages in various points of the supply cable and the electric motor winding. The methods to calculate the insulation defect parameters taking into account the results of electric motor behavior are suggested.

Key words – pilot researches, electric motor, local insulation defect

INTRODUCTION

The widespread use of electric motors (EM) in all branches of industry and production processes imposes heavy demands on trouble-free operation of motors. EM design perfection, technical characteristics of the material used to manufacture EM, running efficiency of EM effect EM performance reliability. To provide the necessary level of service, it is necessary to monitor continuously the state of all components of EM, that is to do on-line diagnoses.

It is obvious that the performance of electric motors 6-10 kV depends not only on the operability of EM itself, but also on the operability of the cable supplying EM. The operation experience shows that the percentage of such kinds of EM damages can reach 40% of all damages of electrical equipment, and in case of cables, it is up to 10% [1]. Thus, almost half of the damages are related to the connection cable - electric motor. The most damageable component of EM is the stator (that is 60% of failures from the total number of EM damages). The percent of damages of the bearing section is about 20%, and as for the rotor, it is about 10% [1].

The main disadvantage of the known methods of EM diagnostics while the motor is running is that the place of a local insulation defect is not defined, and, therefore, it takes a lot of time to define its location.

To prevent the development of insulation defects it is necessary to carry out preventive measures in due time to restore the dielectric properties of the insulation. If we provide the ways to identify the local defect occurrence and its place when the motor is running, then the time for carrying out the above-mentioned kind of work can be significantly reduced.

Currently, many organizations are continuing to improve EM relay protection and develop new methods and means of diagnosing electrical equipment [2-8]. The particular attention is given to finding the defect location. For example, in [7] an original method to localize a single phase –to - ground defect is proposed. It focuses on identifying defects in cables lines.

However, there are no methods to detect local defects in insulation of electric motor windings and to determine the parameters of these defects (the distance from the beginning of the winding b and the value of insulation resistance in the point of the defect Z_{def} of a stator winding) in the running conditions.

In this connection, we set the task to develop more modern techniques to determine insulation defects in the supplying cable and in a stator winding when a motor is running and to detect the broken rods of the rotor short-circuited winding. The parameters of EM in the normal running conditions are taken as the basic information.

The aim of the work is to develop techniques to identify the defects in electrical motors without cutting motors off for realization on the basis of microprocessors. These techniques will allow us to prevent a lot of electric motor failures and accidents related to them.

DEVELOPMENT OF THE DIAGNOSING METHOD

Local insulation defects or damages of asynchronous EM components (for example, damages of the short-circuited rotor winding), which do not require an immediate shutdown, break the electrical or mechanical symmetry of EM. In the EM equivalent circuit such phenomena are reflected in the change of the longitudinal or lateral resistances. We can determine whether there is the asymmetry of the longitudinal or lateral resistances by measuring the running conditions parameters (currents and voltages of the phase with respect to earth).

Thus, the sequence of operations to identify the main defects when EM is running can be represented in the form of the algorithm (Fig. 1). The mathematical description of the connection cable-motor in the running conditions is done with the help of the equivalent circuit (Fig. 2). There are three complex longitudinal resistances of the phases in the connections $\underline{Z}_A$, $\underline{Z}_B$, $\underline{Z}_C$, which depend on the operating conditions of EM. If there are no insulation defects, these resistances can be determined according to the expressions:

$$\underline{Z}_A = \frac{\dot{U}_{Ag}}{\dot{I}_A}; \quad \underline{Z}_B = \frac{\dot{U}_{Bg}}{\dot{I}_B}; \quad \underline{Z}_C = \frac{\dot{U}_{Cg}}{\dot{I}_C}$$

The longitudinal resistance in the phase A $\underline{Z}_A$ is divided into two parts. This allows us to simulate a local defect of insulation $\underline{Z}_{def}$ in different points of the phase A by changing the numerical value of the distance b.

978-1-4244-8806-3/11 $26.00 © 2011 IEEE

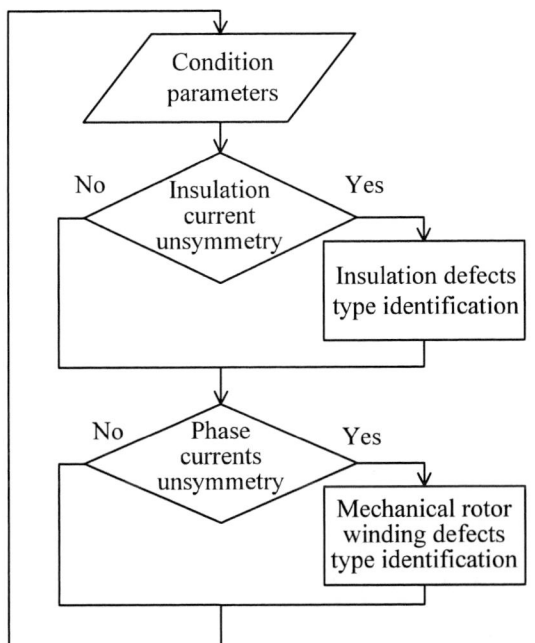

Fig. 1. Main algorithm to identify the defects in EM when it is running.

The parameters of the phase insulation of connection elements with respect to ground are considered to be symmetrical. They are not taken into account in the scheme because the capacitive admittances of the connection under control are significantly less than the capacitive admittances of the circuit phases.

To determine the parameters of the insulation defect (b, $\underline{Z}_{def}$) during operation we applied the algorithms based on the solution of the equation system of the current state. The system is made up in accordance with the equivalent circuit (Fig. 2).

The system of the equations describing the operating conditions of the connection (current state) is the following:

$$\left.\begin{array}{l} -\dot{U}_{Ag} = \dot{I}_A \cdot \underline{Z}_A \cdot b + \dot{I}_d \cdot \underline{Z}_{def} \\ -\dot{U}_{Og} = -(\dot{I}_A - \dot{I}_d) \cdot \underline{Z}_A \cdot (1-b) + \dot{I}_d \cdot \underline{Z}_{def} \end{array}\right\} \quad (1)$$

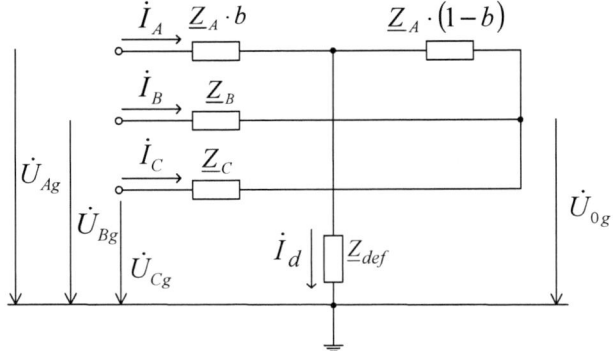

Fig. 2. Equivalent circuit of the connection cable - electric motor.

Summarizing the equations in (1) and performing some transformations, we obtain:

$$b = \frac{-\dot{U}_{Ag} - \dot{U}_{Og} + \dot{I}_A \cdot \underline{Z}_A - \dot{I}_d \cdot \underline{Z}_A - 2 \cdot \dot{I}_d \cdot \underline{Z}_{def}}{\underline{Z}_A \cdot (2 \cdot \dot{I}_A - \dot{I}_d)} \quad (2)$$

To determine the longitudinal resistance of the phase of connection Z_A let's write the equation for the other loop:

$$-\dot{U}_{Ag} + \dot{U}_{Og} = \dot{I}_A \cdot \underline{Z}_A \cdot b + (\dot{I}_A - \dot{I}_d) \cdot \underline{Z}_A \cdot (1-b) \quad (3)$$

On the basis of the equation (3) we can define phase impedance of a connection Z_A:

$$\underline{Z}_A = \frac{-\dot{U}_{Ag} + \dot{U}_{Og}}{\dot{I}_A + \dot{I}_d \cdot (b-1)} \quad (4)$$

In the expressions (2) and (4) we can take the zero-sequence current $3\dot{I}_0$ as the current $\dot{I}d$ in the point of the defect without taking into account the self-capacitance of the connection itself.

By solving the equation system of the current state of the connection [8] we define $\underline{Z}_{def}$ and then on the basis of the expression (2) we can determine the distance to the point of the defect occurrence b. Sometimes it is difficult to do.

As it is difficult to solve the equivalent system, we propose the second method of determining the defects parameters when there is no need to pre-determine the resistance of the insulation defect.

To determine the distance b when the resistance of the defect $\underline{Z}_{def}$ is unknown we compose a new system of the equations according to the circuit (Fig. 2):

$$\left.\begin{array}{l} -\dot{U}_{Bg} + \dot{U}_{Cg} = \dot{I}_B \cdot \underline{Z}_B - \dot{I}_C \cdot \underline{Z}_C \\ -\dot{U}_{Ag} + \dot{U}_{Bg} = \dot{I}_A \cdot \underline{Z}_A \cdot b + (\dot{I}_A - \dot{I}_d) \cdot \underline{Z}_A \cdot (1-b) + \dot{I}_B \cdot \underline{Z}_B \end{array}\right\} \quad (5)$$

If we assume, that the longitudinal phase resistances are equal to $\underline{Z}_A = \underline{Z}_B = \underline{Z}_C = \underline{Z}_N$ and $\dot{I}_d = I_A + I_B + I_C$, then on the basis of (5) we find

$$b = \frac{\dot{U}_{Bg} - \dot{U}_{Ag} + \underline{Z}_N \cdot (2 \cdot \dot{I}_B + \dot{I}_C)}{\dot{I}_d \cdot \underline{Z}_N}, \quad (6)$$

where the longitudinal resistance of the load phase is determined according to the first equation of the system (5):

$$\underline{Z}_H = \frac{\dot{U}_{Cg} - \dot{U}_{Bg}}{\dot{I}_B - \dot{I}_C} \quad (7)$$

According to the equivalent circuit (Fig. 2) we can write down the equation for the other loop:

$$-\dot{U}_{Ag}+\dot{I}_A\cdot\underline{Z}_A\cdot b-\dot{I}_d\cdot 3\underline{Z}_{def}=0 \qquad (8)$$

On the basis of (8) we obtain the expression to determine the resistance of the defect:

$$\underline{Z}_{def}=\frac{-\dot{U}_{A0}+\dot{I}_A\cdot\underline{Z}_A\cdot b}{3\cdot\dot{I}_d} \qquad (9)$$

The suggested methods for determining the parameters of the insulation defect were checked on the physical analogue of the connection cable - electric motor.

The tests were carried out with the help of the physical analogue of the connection (Fig. 3). The model includes the electric motor 0.22 kV voltage, which is connected to the isolating transformer 0.38 / 0.22 kV by means of 6 kV cable. In phase A of the winding of the EM stator there are taps in different points of the winding, these taps are connected to a special terminal block. The scheme of the above-mentioned physical analogue is shown in Fig. 3.

Physical modeling of the circuit was performed by means of three capacitances connected by star. It was assumed that the zero point was taken as ground. Oscillographic testing of the operating parameters (phase currents and phase voltages with respect to earth) was carried out by means of the device, which input the information to a personal computer. The basis of the above-mentioned device is the board with the 12-bit analog-digital converter L-154 (the maximum integral nonlinearity of the conversion is 0,06%). In each current and voltage channel the modules of galvanic insulation and signal conditioning are installed. The conversion accuracy of these modules is 0.05%.

The research program included experimental tests on the modeling of local insulation defects in the different points of the cable and the stator winding of the motor. As a result of the experiments concerning the connection of different stator winding taps through a resistor 200 Ohm to ground we obtained time dependences of digital sequences,

which correspond to the phase current vectors $\dot{I}_A, \dot{I}_B, \dot{I}_C$ and phase voltage vectors $\dot{U}_{Ag}, \dot{U}_{Bg}, \dot{U}_{Cg}$. Sampling time was 0.1 ms.

These sequences were approximated by polynomial regression using the least - squares method. 1000 points were considered in each operating parameter. The sinusoid of 50 Hz was adopted as a function of approximation.

The sinusoidal values found as the result of approximation were transformed into vector quantities. The values obtained were used as the input data to the developed algorithms shown in Fig. 4 and Fig. 5.

The presence of EM in the connection under consideration leads to the fact that the three-phase methods of insulation defect detection suitable for other types of loads, are of low accuracy in the given case [8]. This results from the static and dynamic asymmetry of EM, which expresses itself in the fact that the values of the longitudinal resistances of EM phases are not equal to each other and are changeable in time. It may be the consequence of turn-to-turn defect in one of the stator winding phases, breaks of the short-circuited rotor bar, etc.

To improve the accuracy of determining the distance b to the point of a local defect it is proposed to perform the calculation of the longitudinal resistance $\underline{Z}_A$ of the phase with a insulation defect at the moment t_b. To calculate b we use the operating parameter values at the same moment of time t_b. If we apply a preliminary approximation of current values to calculate the vectors of the operating parameters, we have to take the parameters in a digital form of 1-2 periods of mains frequency. When we calculate $\underline{Z}_A$, it is necessary to take into account the voltage drop caused by zero-sequence current in the part of the circuit from the beginning of the cable (at the buses) to the point of the defect, i.e. in the part $b\cdot\underline{Z}_A$.

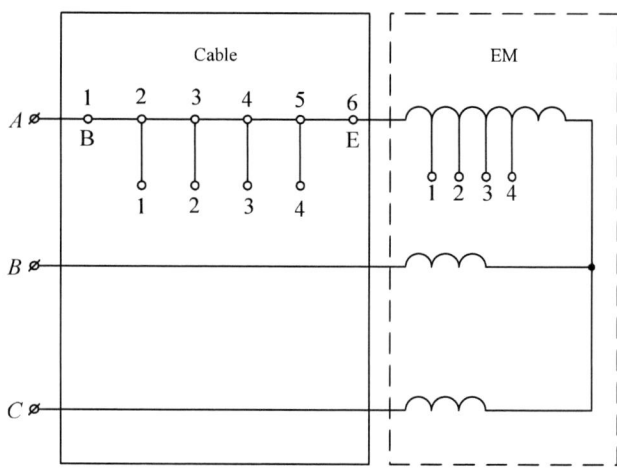

Fig. 3. Scheme of the physical analogue of the connection cable-electric motor.

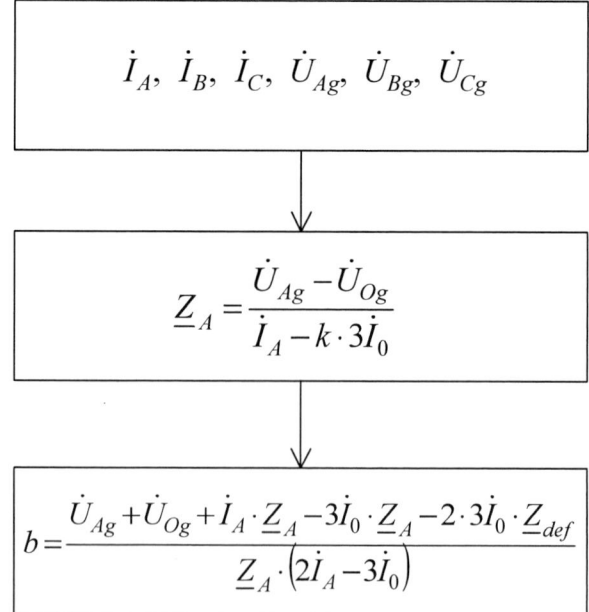

Fig. 4. Algorithm for determining the distance b when $\underline{Z}_{def}$ is known.

978-1-4244-8806-3/11 $26.00 © 2011 IEEE   238

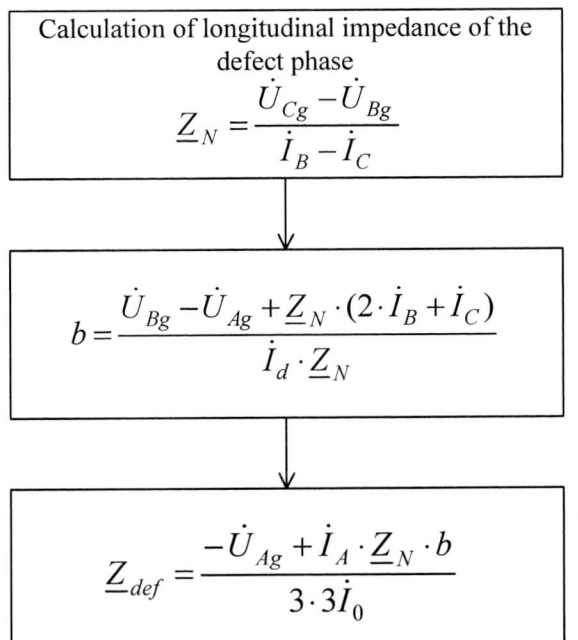

Fig. 5. Algorithm for determining the distance b when $\underline{Z}_{def}$ is unknown.

The table shows the results of determination the distance using the algorithm shown in Fig. 5.

The asterisk indicates the experiments where 5% of turns of EM stator winding were short-circuited. The turn-to-turn defect in the stator winding was created deliberately to evaluate the accuracy of the proposed algorithms.

On the basis of the data shown in the table we can conclude that the error of the proposed method does not exceed 7,4%. The simultaneous application of both developed methods can improve the accuracy of determining the parameters of insulation defects in the whole range of their possible changes.

TABLE
ERROR IN DETERMINING THE DISTANCE B TO THE POINT OF A LOCAL DEFECT
(R_{def}= 200 OHM) IN THE STATOR WINDING OF ELECTRIC MOTOR, %.

Number of an experiment	1	2*	3	4*	5	6
Real value of b, p.u.	0,513	0,513	0,565	0,565	0,687	0,761
Results of calculation of b, p.u.	0,551	0,545	0,5995	0,603	0,720	0,785
Error of definition of b, %	- 7,32	- 6,24	- 6,11	- 6,64	- 4,86	- 3,15

DEVELOPMENT OF FREQUENCY SCANNING METHOD

Frequency scanning is the third method proposed to detect local defects in the insulation of the connection cable – motor with the running motor. The essence of the method is that the test alternating voltage whose frequency varies linearly from f_{MIN} to f_{MAX} is applied to the connection cable – motor, and the amplitude of the output voltage is taken at the same time. If the insulation defect appears, time dependence U_{OUT} will be different from the same dependence obtained when there is no defect in the connection.

The analysis of the amplitude-frequency characteristics obtained in both cases allows us to determine the location and seriousness of the defect.

Technical implementation of scanning is based on the use of measuring current and voltage transformers to supply the test signals to the equipment under control.

To measure the controlled diagnostic parameter it is more convenient to use the phase current transformers or zero sequence current transformers.

The main advantage of application of a measuring current transformer to send the test signal and receive the diagnostic parameter is that it provides the galvanic insulation and makes it unnecessary to connect and disconnect the corresponding test and diagnostic units.

Frequency scanning can be used for both running and temporarily de-energized connections.

Frequency scanning allows us :
- to identify local and distributed defects of insulation and to determine the parameters of local defects;
- to diagnose the separate connections independently of other connections of a bus section due to the fact that the windings of the transformer supplying the bus section have a shunting effect at certain frequency.

The scheme for the research of the method of frequency scanning of the connection cable-motor is shown in Fig. 6.

At the first stage the method research was carried out when the connection cable-motor was de-energized. In addition, the motor stator winding was short-circuited by a plug wire that is the motor was excluded from the scanning area.

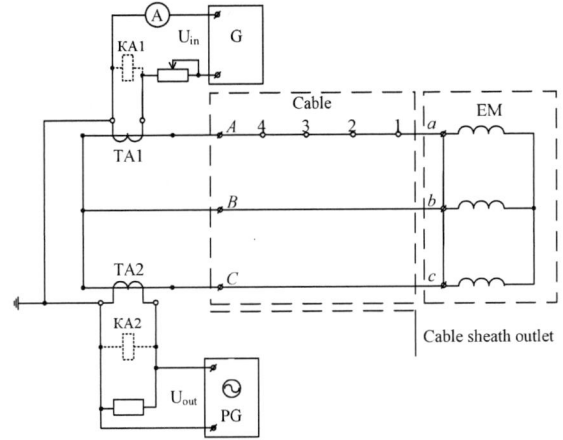

Fig. 6. The scheme of an experimental installation for research the method of frequency scanning (G - sound frequency generator, PG - light-beam oscillograph).

The physical analogue of the connection consists of an asynchronous motor (0,4 kW) and a cable 50 m long with aluminum cable conductors with cross-section 70 mm^2, and there are taps simulating defects in insulation every 10 meters (points 1,2,3,4 in Fig. 6).

To apply the test signal to the connection we used a standard current transformer in the phase A (TA1), which was also the part of the over-current protection of the connection. With the help of a sound-frequency generator G the test signal with the frequency 200×10^3 Hz and the peak voltage 4.7 V was generated (current of the generator Ig = 5 mA). To measure the output voltage a current transformer installed in the phase C (TA2) was used, and a light-beam oscillograph PG was connected to the second winding of the transformer.

In the course of the experimental tests, the defects were simulated by short-circuiting various taps to the cable sheath. Three cases were studied: firstly, when both current relays KA1, KA2 were disconnected, secondly, when only the relay KA1 was connected and thirdly, when both relays were connected.

In the first case (KA1 and KA2 were disconnected) when point 1 was short-circuited, the output voltage of the output signal changed insignificantly and these changes were not practically reflected on the oscilloscope. When the short circuit took place in points 3 and 4, the amplitude of the output voltage increased by about 7%.

When the amplitude of the test signal reached 10 V (Ig = 12 mA), the amplitude of the output voltage was 2,6-2,7 V.

When points 2, 3, 4 were short-circuited to the cable sheath, the amplitude of the output voltage increased by 8-10%, and when point 1 was short-circuited, it decreased by 2-4%.

The connection of the current relay KA1 did not change any previous results. At the same time, the connection of the relay KA2 led to a significant decrease of the output voltage and distortion of its curve shape. To eliminate the distortion the generator current was reduced to 8 mA.

Short-circuiting of points 3 and 4 led to increase of amplitude by 10-12%.

To give mathematical descriptions of the processes under consideration the equivalent circuit shown in Fig. 7 was used.

Unlike the pilot circuit where the electric motor was shunted between the points a, b and c, in the mathematical model the electric motor was presented in the circuit with the corresponding parameters of the equivalent circuit.

The current transformers were presented in the equivalent circuit by the equivalent complex resistances $\underline{Z}_{in}$ and $\underline{Z}_{out}$.

The resistances of the insulation defects were also added to the circuit in each phase of the connection. Under normal condition, the value of each of the resistances was taken 1MOhm. Decrease in resistance to 1 Ohm simulated the insulation defect in the certain part of the circuit. To obtain the analytic dependence of the output voltage on the input voltage $U_{out}(f) = f(U_{in}(f))$ the integrated equivalent circuit (Fig. 7) was transformed into an equivalent one (Fig. 8).

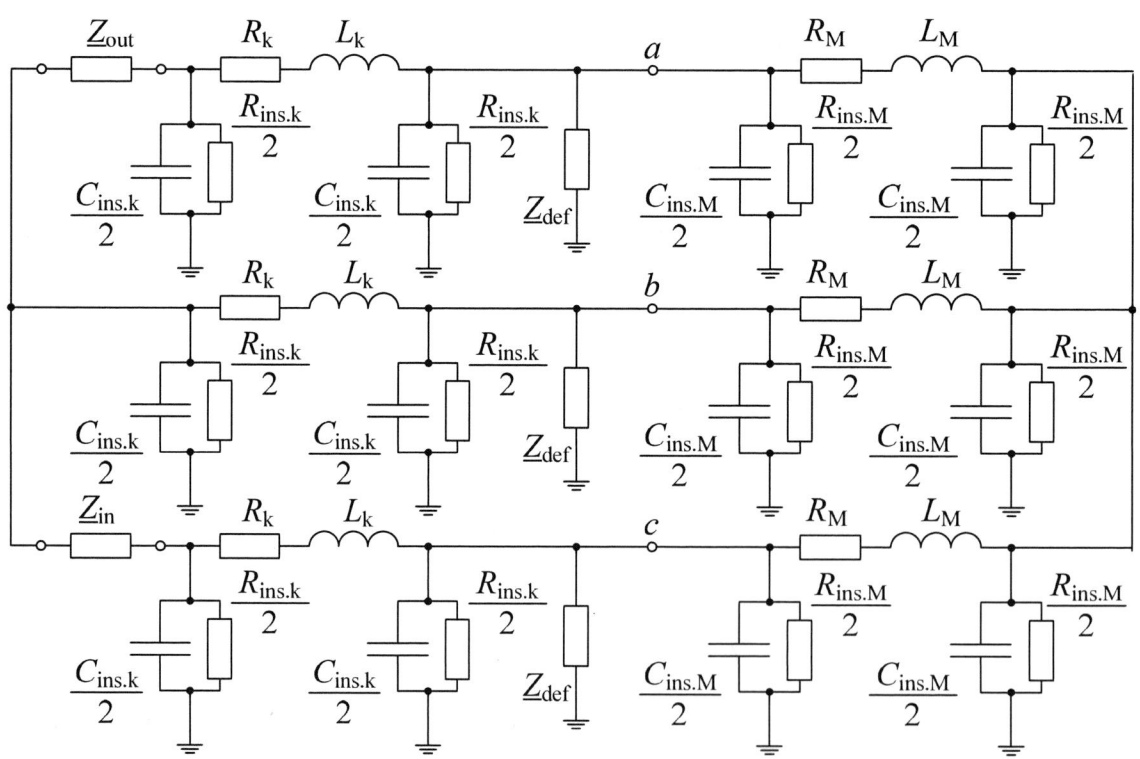

Fig. 7. Equivalent circuit of the connection cable - electric motor in the mode of frequency scanning.

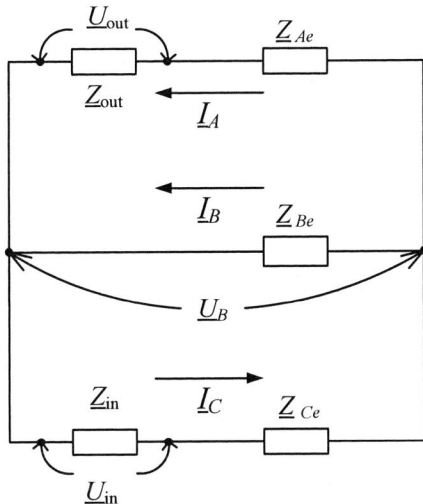

Fig. 8. Equivalent circuit of the connection cable - electric motor.

The resulting dependence is as follows:

$$U_{out}(f) = \underline{I}_A \cdot \underline{Z}_{out} = \frac{U_{in}(f) \cdot \underline{Z}_{out}(\underline{Z}_{in} + \underline{Z}_{Ce})}{\underline{Z}_{in}(\underline{Z}_{out} + \underline{Z}_{Ae})} \qquad (10)$$

The amplitude of the output voltage increased from 5 V to 5.046 V when the mathematical modeling of a defect in the cable insulation of phase A was performed at the frequency 200×10^3 Hz, which completely confirmed the results of the experiment.

The amplitude of the output signal decreased by 0.046 V when the insulation defect was simulated in the cable of phase C.

The changes in the output signal were insignificant when the insulation defect was simulated in the cable of phase B. This can be explained by the symmetry of the equivalent circuit with respect to that phase.

It was stated that there were simultaneous changes in phase alongside with the changes in output signal amplitude. This fact may also be used as a diagnostic parameter.

Mathematical modeling of the electric motor insulation defects gave the results similar to physical modeling, and in the given case, the values of amplitude and phase changes were less than in the case of the cable defect simulation.

Based on the obtained results, the decision was made that it was possible to use a vector of the output voltage as a diagnostic parameter. It allowed us to define the element of the connection and the phase in which the insulation defect took place.

CONCLUSIONS

1. The principle to identify the most common defects of high voltage electric motors without cutting motors off is proposed. The methods proposed to define the place of the local defect and the value of the resistance of the given defect in the connection cable-electric motor are intended for microprocessor realization.

2. The results of physical modeling of local insulation defects in various points of the power cable and the stator windings of the motor are given. The results of the calculations of the defect place on the basis of experimental data show that the error of the developed algorithms does not exceed 7,4%.

3. The method of frequency scanning the connection cable-motor, which allows us to identify the occurrence of insulation defects in the running conditions and to determine their parameters, is proposed.

4. The laboratory studies and the mathematical modeling results confirm the technical efficiency of the scanning method and the algorithm of its realization.

LITERATURE

[1] V.A. Saveliev, V.A. Martinov, «Metodi i sredstva upravlenija tekhnitcheskim sostojaniem elektrooborudovanija elektrostancij», 15 rossijskaja nautchno-tekhnitcheskaja konferencija «Nerazruschayuschij control i diagnostika», Moscow, p. 56-59, June 1999.

[2] F.P. Schkrabets, E.P. Mesyats, M.S. Kiritchenko, «Struktura sistemi nepriririvnogo kontrolya parametrov isolatsii v setyakh napriaghenijem 6-10 kV hornikh predpriyatij», IV mighnarodna naukovo-technitchna konferencija «Efektivnist ta yakist elektropostatchannja promislovikh pidprijemstv», Mariupol, Ukraine, p. 401-403, May. 2008.

[3] Bulitchev A.V., Nudelman G.S. «Upreghdayuschije funktsii relejnoj zaschiti», Sbornik dokladov meghdunarodnoj NTK «Sovremennije napravlenija razvitija system relejnoj zaschiti i avtomatiki energosistem», M., p. 72-78, 2009.

[4] Figurnov E.P., Bodrov P.A. «Opredelenije mesta odnofaznogo zamikanija na zemlyu v visokovoltnikh linijakh elektrosnabghenija avtoblokirovki gheleznikh dorog», Relejnaya zaschita I avtomatika energosistem 2004. Sbornik dokladov. – ARC, Moskow, c. 88-93, 2004.

[5] Welfonder T., Leitloff V., Feuillet R., «Vitet S. Location Strategies and Evaluation of Detection Algorithms for Earth Faults in Compensated MV Distribution Systems», IEEE Transactions on Power Delivery, vol. 15, No. 4, Oct. 2000.

[6] Katchesov V.E. «Metod opredelenija zoni odnofaznogo zamikanija v raspredelitelnikh setyakh pod rabotchim napryaghenijem», Elektritchestvo. – 2005. - № 6. – p. 9-19.

[7] Stogniy B.S., Rogosa V.V., Sopel M.F., Golubov O.Y. «Opredelenije mesta odnofaznogo zamikanija na zemlyu», Tekhn. Elektrodinamika, p.60-63, № 2, 2007.

[8] Grebchenko N.V., Koval I.I., Sidorenko A.A., Smirnova M. A. «Definition of complex admittance of electric isolation without disconnecting of electrical equipment», Compatibility and Power Electronics CPE2009. 6[Th] International Conference-Workshop 978-1-4244-2856-4/09, pp. 61-66, Apr. 2009.

Computer aided investigation system of the switched reluctance motors

Vasyl Tkachuk
Institute of Power Engineering and Control Systems
National Univ. "Lvivska Politechnika", Lviv / Ukraine
Tkachuk@polynet.lviv.ua

Marius Klytta
Department of Electrotechnics and Information Technology
Univ. of Applied Sciences Mittelhessen, Giessen / Germany
Marius.Klytta@ei.th-mittelhessen.de

Abstract - **The paper describes a system for automated, i.e. computer aided investigation of electric drives with switched reluctance motors in the design phase. The system usage and the obtained results are presented.**

I. INTRODUCTION

The computer engineering became essential not only for accumulation of knowledge. With reference to the electrical machines it enables tests and simulation of processes in them. The next step can be computer aided investigation already in the design stage, to obtain optimized machine structure and parameters in this faster and cheaper way.

The switched reluctance motor (SRM) with energy buffer belongs to insufficiently known electrical machines [5].

The creation of perfect schemes and their effective analysis depends on availability of adequate mathematical models of electromechanical processes in such systems.

The Fig.1 shows the basic electrical scheme of switched reluctance motor with series energy buffers and pulse-width speed regulation.

This scheme has an increased reliability and ensures recuperation of magnetic field energy of commutated section [1].

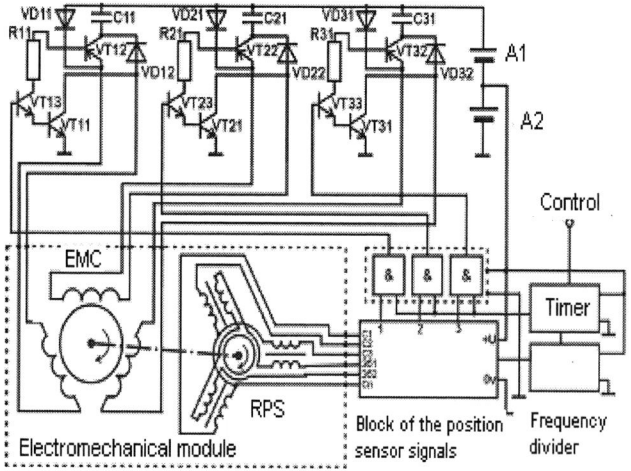

Fig.1. The common electrical scheme of the three-section SRM with capacity storage and pulse-width regulator (EMC - electromechanical converter, RPS – rotor position sensor)

II. MATHEMATICAL MODEL

For the creation of mathematical model of a SRM had been accepted assumptions, which allow obtaining relative simple mathematical dependences without distorting the real physical processes, i.e. still adequately showing energy conversion in this motor:

- the voltage source supplying the inverter part has internal resistance equal to zero
- there are no magnetic influences between sections
- the power switches of the electronic commutator have no inertia
- the parameters of the windings are concentrated

The mathematical analysis of electromechanical processes in all states or operation modes is based on the description of SRM by a system of differential equations of electrical and mechanical balances with their following solution with regard to currents.

In the general case the motor torque is determined by a derivative magnetic co-energy by a transition angle, which in turn depends on flux linkage of the excited outline. Therefore, for calculation of the electromagnetic torque of SRM with magnetic insulated sections fundamental is the current dependence and mutual rotor-stator angle position dependence of section flux linkage. Flux linkage of a magnetic insulated section of SRM is an unambiguous nonlinear function, which can be approximated by analytical expression. In [2] is given such approximated equation:

$$\psi(\theta, i) = \left(\psi_{10} - \psi_{1t} \cdot Sin\frac{\theta}{2} \right) \cdot i + \psi_y \cdot Sin\frac{\theta}{2} \cdot \left(1 - e^{-a \cdot i \cdot Sin\frac{\theta}{2}} \right) \quad (1)$$

where:

Θ - electrical angle between accesses of a rotor groove and a stator tooth; i - current of the activated section; $\psi_{10}, \psi_{1t}, \psi_y, a$ - factors, which are determined by performances of SRM magnetic core in two extreme positions: in the case of coincidence of rotor groove and stator tooth accesses ($\Theta = 0$) and in the case of coincidence of rotor and stator teeth ($\Theta = \pi$) (Fig. 2).

The factors: $\psi_{10} = \psi_y / I_s$, $\psi_{1t} = \psi_{10} \dfrac{\psi_s}{I_s} \dfrac{\psi_y}{I_s}$, $a = \dfrac{(3 \div 6)}{I_s}$

As shown, in order to obtain the factor Ψ_y and so other factors it is necessary to extrapolate the saturated part of the magnetization curve for the case $\theta = \pi$. According to the theory of electromechanical energy transformation from the equation (1) results equation (2) for calculation of an electromagnetic torque:

978-1-4244-8806-3/11 $26.00 © 2011 IEEE

$$M(\theta, i) = 0.5 \cdot z_r \cdot i \cdot Cos\frac{\theta}{2} \cdot \psi_y \cdot \left(1 - e^{-a \cdot i \cdot Sin\frac{\theta}{2}} \cdot \frac{\psi_{1t} \cdot i}{2 \cdot \psi_y}\right) \qquad (2)$$

where z_r - number of rotor teeth.

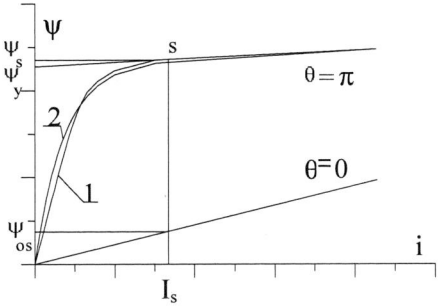

Fig.2. The magnetization curves (1) and their approximations (2) for extreme electrical angles Θ

The changing of magnetic flux causes eddy currents in the iron core and losses in it. The flux has meanwhile one polarity and limited variations. The hysteresis losses can be therefore neglected. The iron resistance in connection with eddy currents can be calculated as [4]:

$$R_s = 4.9 \cdot 10^4 \cdot \frac{w_z^2 \cdot S \cdot q}{p_0 \cdot \gamma_s \cdot l_m}$$

where: $p_0, \gamma_s, S, l_m, w_z, q$ - specific losses in iron, magnetic core material density, magnetic force line square and length, coils number on one tooth, teeth number of one stator section accordingly. The leakage inductance of eddy current circuit is [4]:

$$L_\sigma \cong \frac{\Delta t}{12} \cdot R_s$$

where: Δt - integration step of the differential equations.

Corresponding to the assumptions the electrical relation in each section of the m-sectional brushless motor can be examined separately. They are linked only in the creating of the common electromagnetic motor torque.

Simulation is carried out under additional assumptions that the diodes volt-ampere performance can be approximated by equation: $\Delta U_D = \ln\frac{i_D + I_0}{I_0}/b$ (where I_0 and b - the diode inverse current and coefficient of a temperature potential), that the switch-on and switch-off processes of EC transistors occur immediately, that the resistance of transistor in off-state is infinite and that the volt-ampere performance of the saturated switch is described by equation

$$\Delta U_T = U_{CE} = U_{T0} + R_d \cdot i,$$

where: U_{T0} and R_d - the appropriate transistor parameters.

Conditions of EC transistors are stipulated by a mutual rotor – stator position. They are guided by signals of rotor position sensor (RPS). Let's define a useful formal coefficients K_j, which acquires significance "1" (if the switch "j" is open) or "0" (if this switch is closed).

The values of the formal coefficients K_j depend on the rotor position and can be set according to the following switching function:

$$K_j = 1 \quad \text{if}$$

$$\beta + 2 \cdot (N_j - 1) \cdot \pi + (j-1) \cdot 2 \cdot \frac{\pi}{m} \langle \theta \le \beta + 2 \cdot (N_j - 1) \cdot \pi + (j-1) \cdot 2 \cdot \frac{\pi}{m} + \gamma$$

$$(3)$$

and $K_j = 0$ for all other significances, where in (3):

$$N_j = \frac{\theta + \pi + (j-1) \cdot 2 \cdot \dfrac{\pi}{m}}{2 \cdot \pi} + 1 \quad \text{- number of periods for appro-}$$

priate section. Taking into account the above mentioned, the nonlinear system of differential equations (n.s.d.e.) that describes electromechanical processes in SRM with capacity storage has the form:

$$\frac{di_j}{dt} = \left[u_j + R \cdot i_j + \frac{A_j}{L_\sigma} \cdot (u_j - R \cdot i_j + R_s \cdot i_{sj}) - B_j \cdot \omega \right] / A_j;$$

$$\frac{di_{sj}}{dt} = -(R_s \cdot i_{sj} + u_j - R \cdot i_j)/L_\sigma;$$

$$\frac{du_{cj}}{dt} = (1 - K_j - K_{j+m}) \cdot \frac{i_j}{C};$$

$$\frac{d\omega}{dt} = \left\{ \sum_{j=1}^{m} \left[\begin{array}{c} \dfrac{z_r}{2} \cdot (i_j + i_{sj}) \cdot Cos\theta_{pj} \cdot \psi_y \times \\ \times \left(1 - e^{-a(i_j + i_{sj})Sin\theta_{pj}} - \dfrac{\psi_{1t} \cdot (i_j + i_{sj})}{2 \cdot \psi_y} \right) \end{array} \right] - M_L \right\} \cdot \frac{z_r}{J};$$

$$\frac{d\theta}{dt} = \omega,$$

where $u_j = U \cdot K_j + u_c \cdot (K_j + K_{j+m} - 1) - \Delta U_T \cdot (K_j + K_{j+m}) - \Delta U_D \cdot (2 - K_j - K_{j+m});$

$$A_j = \psi_{10} - \psi_{1t} \cdot Sin\theta_{pj} + \psi_y \cdot a \cdot Sin^2\theta_{pj} \cdot e^{-a(i_j + i_{sj})Sin\theta_{pj}};$$

$$B_j = \frac{Cos\theta_{pj}}{2} \cdot \left[\psi_y - \psi_{1t} \cdot (i_j + i_{sj}) - \psi_y \cdot e^{-a(i_j + i_{sj})Sin\theta_{pj}} \times \right.$$
$$\left. \times (1 - a \cdot (i_j + i_{sj}) \cdot Sin\theta_{pj}) \right];$$

$$\theta_{pj} = \frac{\theta}{2} - (j-1) \cdot \frac{2 \cdot \pi}{m};$$

(4)

$j = 1, 2, \ldots m$; J – rotor moment of inertia; M_L - load torque. For solution of the above n.s.d.e. system the fourth order Runge-Kutta method with constant integration step was applied.

The described mathematical model is the basis for creation of an automated research system of salient pole switched reluctance motors with buffers of energy. It allows complex investigation of SRM as a part of a whole mechatronic system in different operation modes.

III. INVESTIGATION SYSTEM

As an automated investigation system of SRM we understand a system of programs suitable for computing of the instantaneous values of currents, electromagnetic torque, rotor speed and angle position, voltages and other quantities in dynamic and quasi-steady state operation modes of the simulated electric drive with a switched reluctance motor. On the basis of these quantities the system generates further quantities (e.g. input and output power, losses) as well as other values (average values or amplitudes).

The system was developed using principle of modular programming. It consists of the main module (main program), four further sub-modules (sub-programs), the file of input data and the files for result preservation. The block diagram of the whole system is given in the Fig. 3.

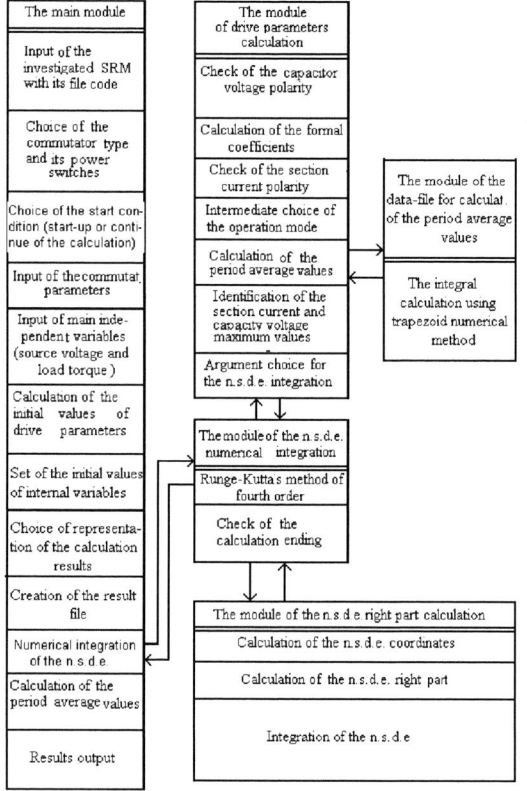

Fig. 3. The structure of an automated investigation system of switched reluctance motors

The system can research the salient poles switched reluctance motors in different configuration (with series capacity storages in each section, with a common capacity and separate feeding switches, with a common capacity and a common feeding switch as well as with a parallel capacity storage). The number of SRM sections can be chosen from three to six.

The system gives the user a broad auxiliary possibility to performance investigation from input data to the results calculation and the result visualization in form of "oscillograms" of

drive quantities (currents, electromagnetic torque, rotor speed, and capacitor voltage curves).

The data provision of necessary parameters is carried out automatically from the input file. The data are: approximation factors of EMC magnetization curves, number and resistances of sections, active resistance representing losses in steel and rotor moment of inertia. Besides it, from this file the system gets information about the chosen machine rating values (power, speed, voltage and current of the used motor).

The configuration choice (type of electronic commutator, buffer number, boost scheme of section currents) is carried out in conversational work mode. In the same way the user can set the exterior quantities: supply voltage and load torque as well as the integration time step for the differential equations system.

The setting of the initial conditions for these equations to zero allows the investigating of transitional process during the SRM start-up. If desirable or necessary the user can set any other values of the initial condition.

Thanks to the assumed relative exact transistors and diodes volt-ampere characteristics the mathematical model of SRM takes into account the losses in power elements of electronic commutator.

The system can calculate static (load, mechanical or control) characteristics of SRM. For this purpose there are separate inputs of static load torque and supply voltage and the possibility of computing prolongation from any of previously reached points.

For purposeful usage of the investigation system it is necessary to supervise the reaching of steady state of operation. This feature gives the user the possibility to interrupt the further computing within few integration periods and to save the reached results in the intended file. All this is carried out with the help of sub-programs.

The steady state operation can be identified thanks to the evaluation of differences between the values of the same section current in neighbouring operation periods. The residual value should be smaller than ε, which was estimated as:

$$\varepsilon = 0.001 \cdot \frac{P_N \cdot \gamma \cdot m}{\eta \cdot U_N \cdot 2 \cdot \pi} \quad (5)$$

where P_N, γ, m, η and U_N - rating value of an initial power, switching interval of section, sections amount, efficiency factor and rating value of supply voltage accordingly.

Input values of supply voltage, static load torque, n.s.d.e. integration step, average value of static performances of SRM and primary values of the n.s.d.e. coordinates are noted in the file, which name sets the user of the SRM designing system and where also the information about a researched SRM (input file) are kept.

The instantaneous values of all quantities in dynamic and in quasi-steady state modes of operation are noted in appropriate value files, which have the status "old". The file contents are changed due to the wished investigation mode.

As already marked, the system calculates in a foreground processing the instantaneous values of quantities, but also average values of several important quantities. In the file with the parameters of the simulated motor are stored during the

978-1-4244-8806-3/11 $26.00 © 2011 IEEE

computing also data of the chosen electronic commutator switches and the capacity buffer.

The system can be used also for construction of static performances. In this case there is no necessity of storing of the instantaneous values of all variables, as it would take significant memory capacity of the computer.

At the beginning of each session the system is set up on a desirable mode of results saving: to note transient, to note quasi-steady state mode, or to keep only average values of static drive performances. For investigation of the static modes it is necessary to set all relevant input data (which are for the analysis of static mode stored in an input file), to execute the computing with certain load till reaching quasi-steady state mode of operation. After changing of one independent parameter (for example magnitude of a static load on the motor shaft) the investigation can be prolonged.

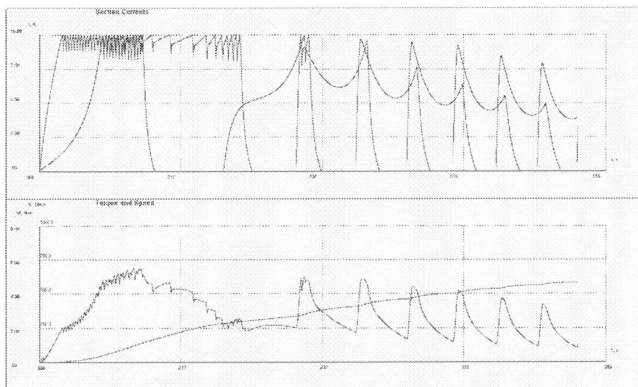

Fig. 4. Performances during the start-up of three-phase SRM with the series capacity storages

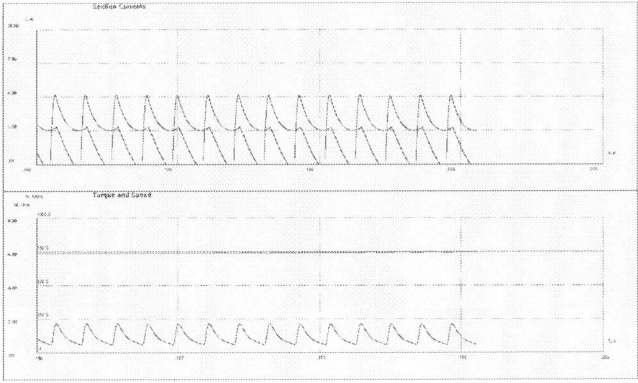

Fig. 5. The quasi-steady state operation of three-phase SRM with the series capacity storages

The Fig.4 and 5 show the example of instantaneous values calculated for starting and quasi-steady state mode (section currents, torque and speed) in the electric drive using a three-phase SRM with the series capacity storages of energy.

CONCLUSION

The paper describes the computer aided simulation and investigation system of SRM. It allows examinations of salient poles switched reluctance motors with various schemes of transistor commutator and with series or parallel capacity energy buffers.

The system can be used for investigation of the influence of various factors on the SRM static and dynamic performances. There can be estimated: the instantaneous section current, eddy currents of EMC magnetic core, electromagnetic torque, rotor speed, angle of a mutual rotor – stator position and voltages on the capacity storages. The comparison of the computed results with the values in the real system showed differences which didn't exceed 5%.

The system enables to select the structure and the power elements of commutator as well as to evaluate and to minimize pulsation of motor electromagnetic torque already in the designing phase. It can be done correctly without the necessity of the previous realization, i.e. faster and by considerably lower costs.

There is also simplifying of research process in case of simulation and investigation of already existing drive systems in order to optimize them.

The results obtained and stored in the described system have the convenient form. They can be used for any further analysis using other appropriate software tools.

REFERENCES

[1] Tkachuk V.I., Osidach Y. *"Транзисторні комутатори з ємнісними накопичувачами енергії"* Visnyk NU "Lvivska Politechnika" № 301, *Електроенергетичні та елек-тромеханічні системи.* Lviv, DULP, 1997, p. 115 – 122.

[2] Tkachuk V.I., Kasha L.V. *"Вентильний реактивний двигун з широтно-фазовим керуванням та його математична модель"* Visnyk SNU №1. Lugansk, 2002

[3] Tkachuk V.I., Klytta M. „*Switched reluctance motor and its mathematical model"*, International Conference and Workshop CPE. Gdansk, Mai 2007

[4] Tkachuk V.I., Klytta M. „*Electronic Commutator with Parallel Capacity Storage for Switched Reluctance Motor"*, International Conference and Workshop CPE. Badajoz, Mai 2009

[5] Krishnan Ramu „*Switched reluctance Motor Drives"*, Industrial Electronics Series, January 2001

Estimation of Parameters of Induction Motor with Broken Rotor Bars

Danilo Makuc, Klemen Drobnič, Vanja Ambrožič, Damijan Miljavec, Rastko Fišer, and Mitja Nemec
University of Ljubljana, Faculty of Electrical Engineering, Tržaška cesta 25, 1000 Ljubljana, Slovenia
E-mail: Klemen.Drobnic@fe.uni-lj.si

Abstract-**This paper describes the evaluation of parameters of a three-phase induction motor with broken rotor bars. Such fault causes electric and magnetic asymmetry in the machine. Using standard d-q model of induction motor this asymmetry can be expressed through differently modified rotor parameters in both axes. In pursuing the goal to establish accurate motor parameters, a non-standard, one-phase measuring method was employed. In this way angle dependence of rotor parameters following a sinusoidal shape was determined. Additionally to measurements an extensive finite element analysis has been performed for two purposes. Firstly, to assess the measurement results of the method and, secondly, to predict various situations which can not be easily achieved by measurements due to hardware limitations (e.g. different numbers of broken bars). Finite element analysis evidences the correctness of the machine parameters obtained from proposed measurement method.**

I INTRODUCTION

Equivalent circuit model constitutes a basic, yet sufficiently general mathematical description of induction motor and is derived directly from a corresponding transformer model. The rotor side quantities, which include rotor current, leakage reactance and resistance, are referred to stator by a coil ratio. A few variants are known, of which basic T-model is the most usual choice [1]. Nowadays, a number of complex models are also available, often combining magnetic field theory with classic circuit approach which can now easily deal with once "hard-to-calculate" effects of induction machine, such as magnetic saturation, iron losses, etc [2].

Appearance of broken rotor bars in induction machine gives rise to some well-known phenomena discernible in electric and mechanic quantities [3]. Typical approach to embrace the effects of this specific fault is to model internal state of the machine with sufficient precision. Therefore, finite element analysis (FEA) [4] or models with high-number of differential equations [5, 6] have been used for simulations yielding very accurate results and taking into account even less tangible, 2nd order effects. For example, in [7] it is shown how the redistribution of the current in rotor bars amplifies local saturations which in turn decrease the expression of the fault on stator current. Obvious disadvantage of these advanced approaches is rather high computational burden and often machine-specific model.

Recently introduced two-axis model of induction machine with broken rotor bars [8] equals the former approaches in purpose but differs in fundamental approach. Model is based on the assumption that broken bar can be described as some sort of salience of a squirrel cage resulting in different rotor parameters in *d*-, and *q*-axes, respectively. Consequently, measured from the stator side, these parameters will vary with the rotor angle [9].

Accuracy of this model clearly depends on the exactness of parameters, which must be determined beforehand. Standard induction motor tests as no-load, rated load and short-circuit, which are designed for symmetrical rotor, cannot take into account rotor saliency and must therefore be replaced with measurements capable of establishing the proper relationship between rotor angle and parameters. In this paper an evolution of a method for determining angle dependent rotor parameters, is presented. Induction motor is supplied only with one phase voltage while keeping the rotor locked. By changing alignment of rotor in respect to stator magnetic axis a dependence of the machine parameters on rotor angle can be obtained.

Independently from measurements, the machine parameters were computed using FEA of the very same induction motor with faulty rotor. On one hand, FEA provides a verifying tool for the measurements, but it also enables us to study many different cases and combinations of rotor faults for which measurements are difficult to perform due to the limited test hardware. Additionally, dependence of rotor parameters on frequency was also dealt with, taking into account skin-effect and local saturations of the iron core.

II INDUCTION MOTOR MODEL

The model of induction motor with broken bars derives directly from a classic two-phase *d-q* model depicted in Fig.1. Its obvious advantage is that solely two voltage equations are necessary to describe electromagnetic phenomena in the machine. One of the prerequisites that concede the use of this model is machine's internal electrical symmetry, which translates itself to symmetrical stator windings and rotor cage [10, 11].

With the onset of rotor fault in form of broken or cracked bars internal symmetry becomes distorted. At this point the classic *d-q* model needs to be reevaluated and adjusted to the new conditions. Question whether Park transformation, essential to derivation of *d-q* model, is still valid despite rotor asymmetry has been dealt with in [8]. Namely, the occurrence of fault modifies rotor parameters in a certain way. It was shown in [8] that rotor resistance and inductance change depending on rotor position (i.e. its electrical angle) following sinusoidal shape. In this way, the transformation remains

978-1-4244-8806-3/11 $26.00 © 2011 IEEE

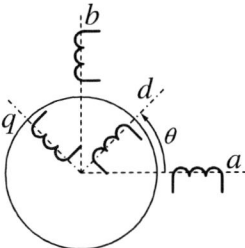

Fig. 1. Two-phase model of induction motor.

valid with electrical asymmetry taken into account.

In order to benefit from this simple yet satisfactory accurate model, its parameters must be ascertained. A starting point for the analysis is an induction machine single phase equivalent circuit (Fig. 2). Following established test procedure for induction motor using no-load and short-circuit test, the parameters of the machine can be determined. At no-load, when a slip value s is practically zero, all of the stator current flows through the magnetizing inductance thus substitute inductance equals $L = \sigma_s L_s + L_m$ and substitute resistance $R = R_s$. In a short-circuit state (locked rotor) the slip equals one and most of the current flows through the rotor branch thus $L = \sigma_s L_s + \sigma_r L_r$ and $R = R_s + R_r$.

Since the rotor with broken rotor bars exhibits electrical asymmetry, measurement in a rotating magnetic field or with a rotating rotor in a non-rotating filed cannot yield the dependence of machine parameters on rotor angle. The only way to obtain the parameters of the asymmetrical machine is with locked rotor and a stationary sinusoidal pulsating magnetic field [9]. The stator windings are supplied with one phase voltage only, where rotor angle θ is defined as the angle between the axis of stator field and the axis of rotor fault. Measurement of the parameters at different orientations of the rotor regarding the stator field should give the dependence of the parameters on the rotor position (Fig. 3).

However, machine parameters cannot be considered as constant. As it was already pointed out, the rotor resistance and inductance change duly with respect to rotor position. Additionally, machine is influenced by some other more or less pronounced effects. For example, with growing supply frequency rotor bars are subject to "skin-effect", which can seriously hamper accurate determination of rotor resistance [12]. In order to take account of this particular phenomenon, frequency dependence of machine parameters has to be evaluated as well.

Being the machine short-circuited, the supply frequency corresponds to the slip $s = 1$. However, the maximum reasonable frequency (at pull-out slip) is to low to presume the absence of a magnetizing inductance as stated before.

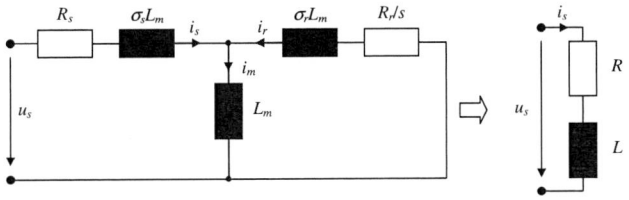

Fig. 2. Single phase equivalent circuit of induction machine.

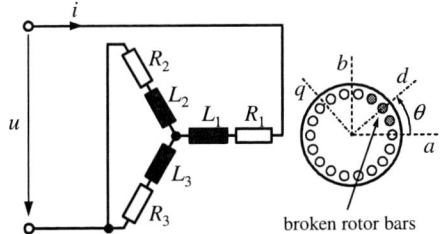

Fig. 3. Single phase measurement of machine parameters at standstill (assuming 2-pole machine).

Therefore, parameters have to be measured at various frequencies thus emulating real no-load and short-circuit condition. Which frequency should be considered to correspond to the virtual short-circuit (enabling to presume $R = R_s + R_r$) is hard to say, especially considering the above mentioned parasite effects and non-linearities. To demonstrate these phenomena, let us observe substitute resistance and inductance just for a healthy rotor obtained from measurements and FEA at different frequencies (Fig. 4). Obviously, at high frequency, as the substitute resistance in simulation tends to stabilize, the actual measurement show different behavior. It is the goal of this paper to analyze and explain these parasitic effects influencing parameter estimation, thus helping to obtain more accurate models for parameter estimation as demanded in [9].

In presented analysis two rotors were considered. Apart from healthy one (no broken bars), a rotor with 7 broken bars out of 44 (as being the only available), was used and is subsequently referred as "faulty".

III CALCULATIONS USING FEA

Using a 2D FEA, a case study of different rotor

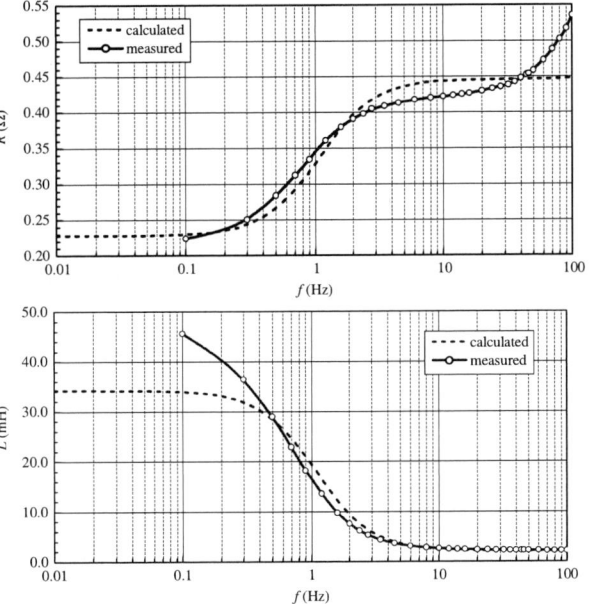

Fig. 4. Frequency dependence of machine parameters for single phase measurement and FEA for a healthy rotor at standstill.

asymmetries due to broken rotor bars was considered and machine parameters were calculated at different rotor angles. Results of FEA were used for comparison with measured data and to compute parameters or investigate the operating conditions that cannot be realized or measured on a real induction machine. Broken rotor bars were modeled as non-conducting bars in a squirrel cage where all bars are connected in parallel.

Due to a 2D model the stray inductances of windings' heads were neglected. A frequency domain solver was used to analyze this configuration, while stator windings were supplied by the same current values as in measurements.

A. Computations of motor parameters at lower frequencies

At low supply frequencies (below 1 Hz) the induced currents in rotor winding (bars) are also low and stator current consists mainly of magnetizing current. The influence of short circuited rotor winding on the shape and magnitude of the magnetic field in the machine is very weak. Flux lines at different rotor (fault) positions, while the supply frequency is 0.1 Hz, are shown in Fig. 5. As expected, the shape of the magnetic field is not influenced by rotor position. Furthermore, the field would almost be the same if there were no conducting bars in rotor slots at all. Although in this case the rotor cage has seven broken bars (in order to make FEA more pronounced), there is just a negligible difference when the position of the fault is rotated from direct (Fig. 5a) to quadrature axis (Fig. 5b). Even though the absence of seven rotor bars produces current asymmetry in the rotor, the magnetizing current (through L_m) in such case is much higher.

To get a complete dependence of a substitute resistance R and inductance L on healthy and faulty rotor due to rotor angle (position) θ, calculations were performed for a span of 180 electrical degrees. The results are shown in Fig. 6. One can observe a slight variation of resistance depending on the position of the fault axis, while the substitute inductance $(\sigma_s L_m + L_m)$ is practically constant irrespectively to rotor angle θ. It can be deduced that the total stator inductance (consisting of stator stray and magnetizing inductance) practically does not depend on the damage level of the rotor.

B. Computations of motor parameters at higher frequencies

At frequencies above 1 Hz the impact of broken rotor bars on substitute parameters is a lot more accentuated, since the rotor currents are now much higher and have considerable

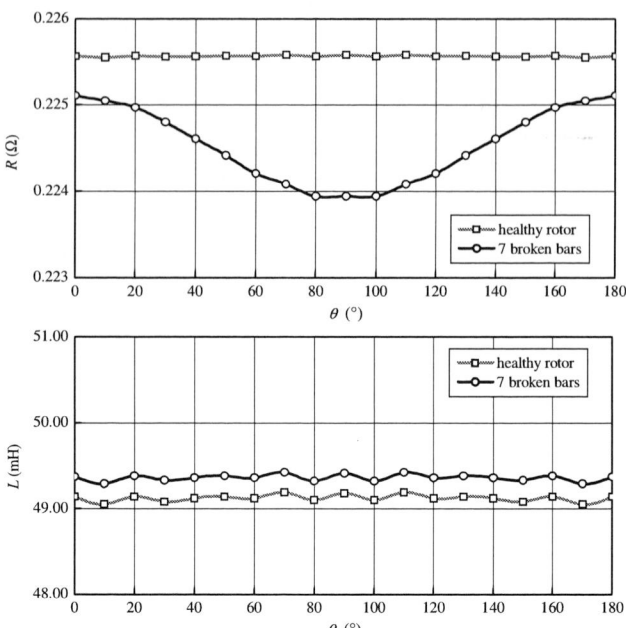

Fig. 6. Calculated resistance and inductance at 0.1 Hz versus rotor position for healthy rotor and rotor with 7 broken rotor bars.

influence on magnetic field in the machine.

Taking it to extreme, values of substitute resistance and inductance at supply frequency of 100 Hz for different rotor positions θ are shown in Fig. 7. As with a low frequency supply, the healthy rotor show that both the resistance and inductance are independent on θ, while with (seven) broken rotor bars both quantities are highly dependent on rotor position. Deviations from healthy values are maximal when the fault position (rotor axis d) is aligned with a stator quadrature axis b. When the fault is aligned with a stator

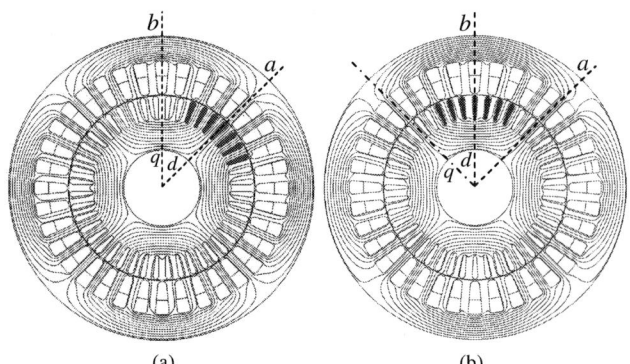

Fig. 5. Magnetic field in induction machine with 7 broken rotor bars supplied by a single phase voltage at 0.1 Hz, when broken rotor bars are in (a) direct axis and (b) quadrature axis.

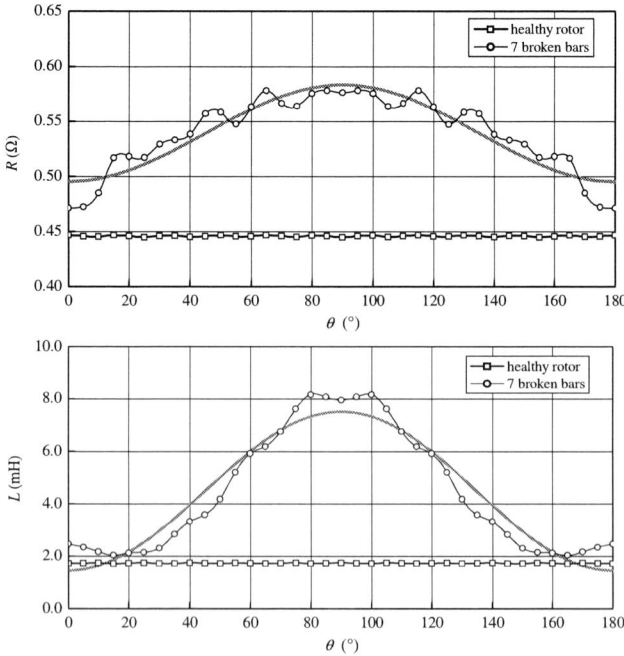

Fig. 7. Calculated resistance and inductance at 100 Hz versus rotor position for healthy rotor and rotor with 7 broken rotor bars.

direct axis *a* the values almost match the healthy ones.

In this case the frequency is so high that large induced rotor currents almost compensate the stator excitation, therefore only stray magnetic field of stator and rotor windings are present in the machine. The magnetizing current in a parallel branch of equivalent circuit can easily be neglected in such circumstances (Fig. 2).

In Fig. 8 the flux lines in both utmost rotor positions are shown, where almost no flux line encloses both the stator and the rotor winding. In both figures (Fig. 8a and Fig. 8b) directions of rotor currents are indicated. It is obvious that broken bars prevent currents to flow, but it is also evident how position of rotor fault actually influences the current distribution. When the fault is aligned with a direct axis (Fig. 8a) the currents are in fact not very disturbed, since also healthy rotor has practically no current in rotor bars which are positioned close to the axis of magnetic field.

We encounter a quite opposite situation when faulty bars lie in a quadrature axis. In a healthy rotor these are exactly the bars which conduct the majority of the rotor current. In the presented case with seven broken rotor bars just two remaining bars have to conduct the whole current (Fig. 8b). That influences the raise of the substitute resistance since the cross section of two bars is considerably smaller than that of nine, which would normally conduct the current.

The variation of the resistance has a sinusoidal profile with a period being half of a pole-pair period (Fig. 7). Also the minimal substitute resistance of a motor with broken rotor bars is somewhat higher than with healthy rotor, since it is evident that even in alignment of direct axis such an extended fault (7 broken bars) effects the current distribution very much.

The substitute inductance is also expectedly constant for a healthy machine and has similar sinusoidal profile as the resistance for the machine with broken rotor bars. The inductance rises due to the fact that part of the stator winding has no counter-ampere-turns and therefore the magnetic field spreads almost undisturbed into the rotor. Thus the magnetizing component of the current slightly increases which also affects the value of already small substitute inductance. Since the number of broken bars in the model is very high, this effect has also a relatively big influence on a value of a substitute inductance. In spite of that, the substitute inductance at high frequencies consists mainly of stator and rotor stray inductances ($\sigma_s L_m + \sigma_r L_m$).

The reason for the calculated resistance and inductance profile not being completely sinusoidal is the fact that, due to simplicity, a 2D FEA model was used and the effect of a skewed rotor was not taken into account. Rotor teeth and slots consequently form magnetic saliency, which influences the resultant profiles. Since the main goal was to show how and how much the rotor faults effect the motor parameters, the obtained results entirely confirm our expectations.

C. Number of broken rotor bars

The magnitude of parameter change due to rotor faults is practically defined by the number ob broken rotor bars. In this

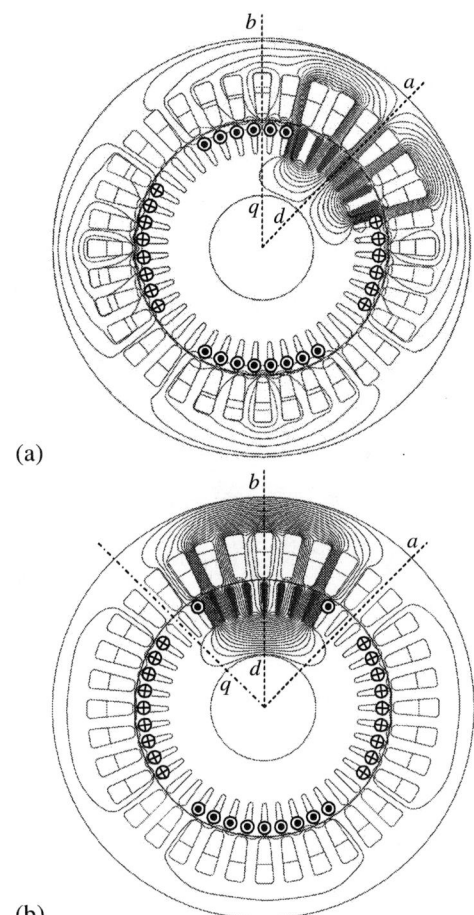

(a)

(b)

Fig. 8. Magnetic field in induction machine with 7 broken rotor bars supplied by a single phase voltage at 100 Hz when broken rotor bars are in (a) direct axis and (b) quadrature axis. In both cases directions of rotor currents are indicated.

case only faults where broken bars appear in a group were analyzed by means of FEA. Results of substitute parameters for 1, 3, 5, and 7 broken bars are shown in Fig. 9. A sinusoidal approximation was used for all obtained profiles. As expected the saliency diminishes with a smaller number of broken rotor bars.

D. Frequency response

To verify the assumption that the values of substitute quantities at very low and very high frequencies remain almost constant, a frequency response of the machine was calculated and measured in two positions – rotor fault aligned with stator axis *a* and axis *b*, respectively. Simulated response for different fault degrees (increasing number of broken bars) is presented in Fig. 10, while the measurement results are shown in Fig. 11. The latter could have been performed only for seven broken bars due to the available damaged rotor. In Fig. 10 only quadrature resistance and inductance are shown, since direct components do not show mentionable deviations from values of a healthy rotor. It can be seen that the substitute inductances start from the same value and

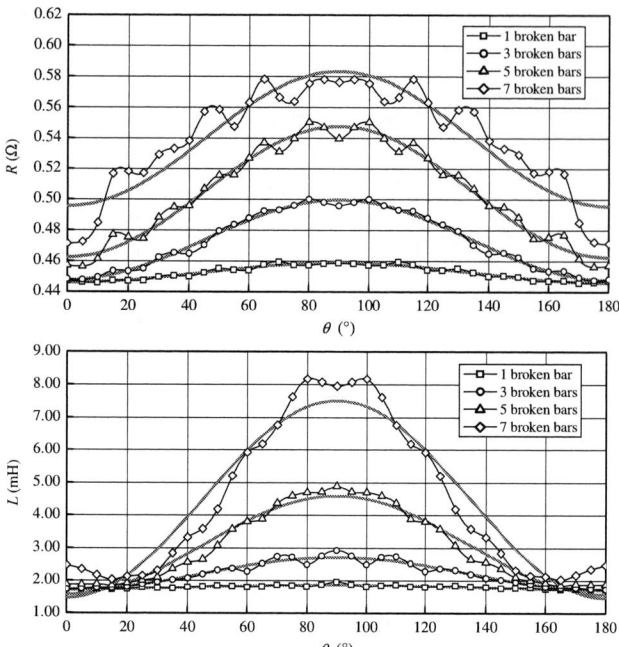

Fig. 9. Calculated resistance and inductance at 100 Hz versus rotor position for motor with 1, 3, 5 and 7 broken rotor bars.

asymptotically approach finite values at high frequency (depending on the level of fault. This is not the case with a substitute resistance. When the supply frequency rises above 20 Hz the resistance tends to steeply increase. Since the rotor bars are massive, skin effect is a main cause for such behaviour. The effect is evident with a healthy rotor as well as with broken rotor bars. This fact aggravates the simple

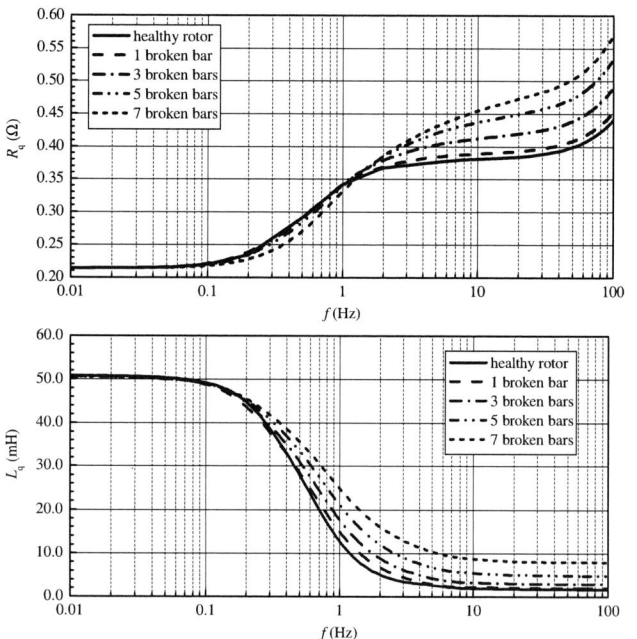

Fig. 10. Calculated frequency response of quadrature resistance and inductance for different numbers of broken rotor bars.

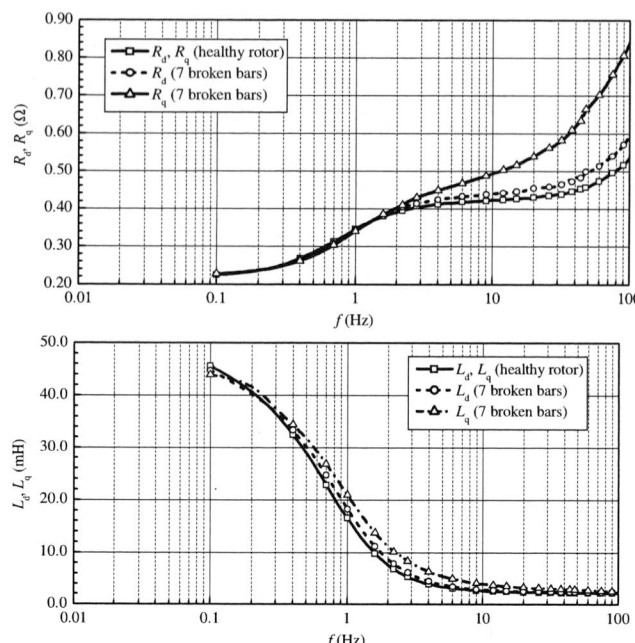

Fig. 11. Measured frequency response of direct and quadrature resistance and inductance for healthy and faulty rotor.

estimation of equivalent circuit parameters.

Since there is no special need for monitoring the parameters at so high (above rated) frequency, the resistance at frequencies above 20 Hz can be approximated with a nonlinear function [12] and then extrapolated to 0 Hz, which gives a value of substitute resistance without increase caused by a skin effect. This value can be then easily distributed among stator and rotor resistance ($R_s + R_r$). In Fig. 12 the calculated frequency response of quadrature resistance with the approximation function are shown.

IV EXPERIMENTAL RESULT

Experimental measurements were made with a variable frequency voltage/current source and an oscilloscope. Measurement principle was already described in Fig. 3. Substitute resistances and inductances were calculated from phase angle between impressed phase voltage and phase current, and their amplitudes. As expected from simulation

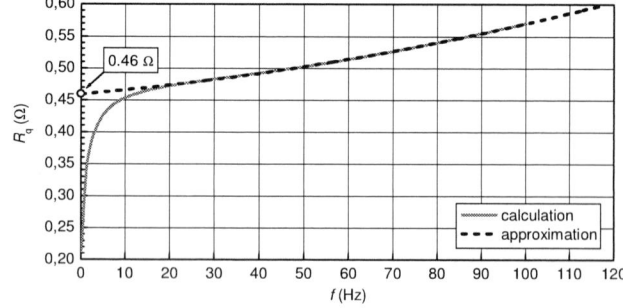

Fig. 12. Calculated frequency response of quadrature resistance for 7 broken bars and the approximation function which is extrapolated to 0 Hz.

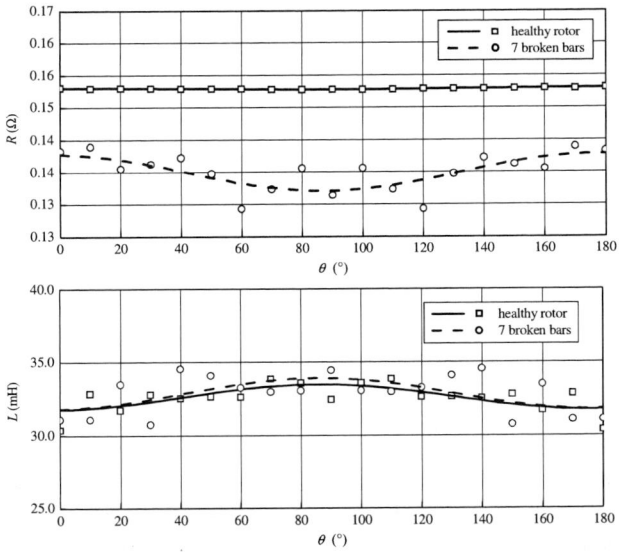

Fig. 13. Measured resistance and inductance at 0.1 Hz versus rotor position for healthy rotor and rotor with 7 broken rotor bars.

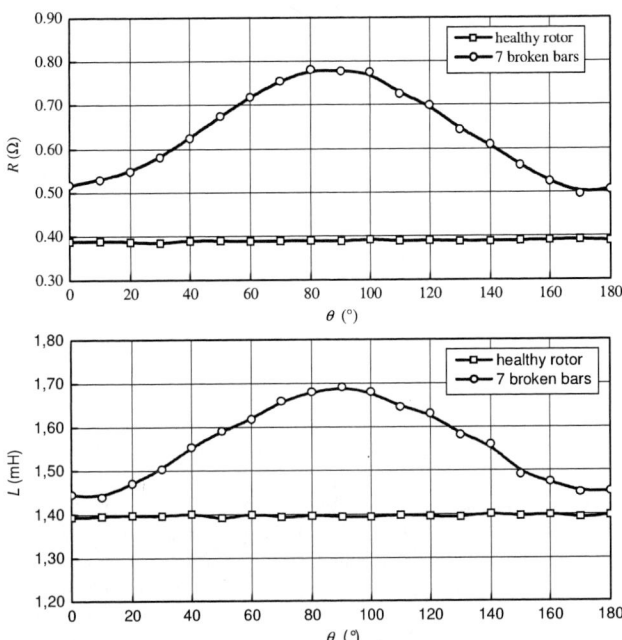

Fig. 14. Measured resistance and inductance at 100 Hz versus rotor position for healthy rotor and rotor with 7 broken rotor bars.

results (Fig. 6), broken rotor bars do not have significant impact at low supply frequencies (Fig. 13). Note that phase angle needed for impedance determination is hard to obtain precisely at extremely low frequencies.

At higher frequencies we can observe in Fig. 14 the same sinusoidal profile that was evident from simulations (Fig. 7). While there are some discrepancies in the offsets as well as in magnitudes between simulations and measurements they could be attributed to simplified simulation model (only 2D simulation, no skewing of rotor bars, neglected stray inductance, ...).

V CONCLUSION

In this paper, parameter variations in an induction machine with a healthy and gradually damaged rotor as well as their intrinsic nonlinearities have been investigated and explained through both FEA and measurement. In order to determine the presumed angular distribution of parameters, an already developed method using single-phase supply at different frequencies had to be adapted. By simple extrapolation of the substitute resistance frequency response the influence of skin effect present at higher supply frequencies can be mitigated, thus giving correct parameters at low frequencies. The future work will be focused on developing various dynamical models of faulty rotors based on measured parameters.

APPENDIX

TABLE I
RATED PARAMETERS OF INDUCTION MACHINE (HEALTHY ROTOR).

P	3 kW	R_s	0.214 Ω
U_s	177 V	L_s	35.4 mH
I_s	14.8 A	R_r	0.231 Ω
n	1456 rpm	L_r	35.0 mH
T	20 Nm	L_m	34.1 mH

Manufacturer's data of the applied induction machine is presented in Table 1.

REFERENCES

[1] M. Stieber, "Determining induction machine parameters using equivalent circuit models," ICEM 2004, Krakow, Poland, 5-8. Sept. 2004.

[2] M. Dems, K. Komeza, "Comparison of different models for the calculation of transient processes of large power induction motors on soft and frequency start-up", ICEM 2006, Chania, Greece, 2-5 Sept. 2006.

[3] M.E.H. Benbouzid, G.B. Kliman, "What stator current processing-based technique to use for induction motor rotor faults diagnosis?," IEEE Transactions on Energy Conversion, vol.18, no.2, pp. 238- 244, June 2003.

[4] R. Fišer, D. Makuc and V. Ambrožič, "Evaluation of the induction motor cage fault stage using finite element method," SDEMPED 2001, Grado, Italy, Sept. 1-3 2001.

[5] J. Jung and B. Kwon, "Corrosion model of a rotor-bar-under-fault progress in induction motors," IEEE Transactions on Industrial Electronics, vol. 53, no. 6, pp. 1829 -1841, Dec. 2006.

[6] H. A. Toliyat and T. A. Lipo, "Transient analysis of cage induction machines under stator, rotor bar and end ring faults," IEEE Transactions on Energy Conversion, vol. 10, no. 2, pp. 241 -247, June 1995.

[7] J. Sprooten, J.C. Maun, "Influence of Saturation Level on the Effect of Broken Bars in Induction Motors Using Fundamental Electromagnetic Laws and Finite Element Simulations," IEEE Transactions on Energy Conversion, vol.24, no.3, pp.557-564, Sept. 2009.

[8] M. Nemec, D. Makuc, V. Ambrožič, and R. Fišer, "Simplified model of induction machine with electrical rotor asymmetry," ICEM 2010, Rome, Italy, 6-8 Sept. 2010.

[9] M. Nemec, D. Makuc, "Parameters estimation using single phase measurement of three phase induction machine", SAEM 2010, Ptuj, Slovenia, 30 May-2 June 2010.

[10] C. V. Jones. Unified theory of electrical machines, London: Butterworths, 1967.

[11] P. C. Krause, S. D. Sudhoff, O. Wasynczuk, "Analysis of electric machinery and drive systems", New York: Wiley Interscience, 2002.

[12] A. Boglietti, A. Cavagnino, L. Ferraris, M. Lazzari, "Skin effect experimental validations of induction motor squirrel cage parameters," ICEM 2008, 6-9 Sept. 2008.

PWM-VSI Overmodulation Technique with Carrier Harmonics Reduction for Auxiliary Railway Power Supply

C. Martinez[1], R. Vazquez[2], I. Quesada[1], C. Lucena[1], A. Lazaro[1], A. Barrado[1]

[1]Universidad Carlos III de Madrid - GSEP
Leganés, Madrid, Spain
cdmartin@ing.uc3m.es / http://gsep.uc3m.es

[2]Sistemas Electrónicos de Potencia S. A., SEPSA
Pinto, Madrid, Spain
rvazquez@sepsa.es

Abstract -- **Auxiliary railway power supplies are needed onboard rolling stock to provide medium or low voltage to different systems from the catenary. In this kind of applications, the THD reduction of the filtered output voltage and a great regulation capability associated to the wide input catenary voltage range are the main goals to be reached. In order to achieve these objectives a novel overmodulation strategy for voltage-source inverters is presented. The new overmodulation technique has been tested and validated trough an auxiliary railway power supply prototype. Experimental results demonstrate the feasibility and the effectiveness of the proposed technique.**

Index Terms— **Overmodulation, rail transportation power systems, inverters, low THD.**

I. Introduction

Nowadays the railway transportation is widely used in cities, because is a low cost and reliable solution for the users. New trends regarding energy efficiency and safety lead to new designs in new vehicles, however it is not always economically feasible. Therefore it is very important to optimize the systems that are currently shipped on railways.

European catenary standard voltages can be either AC or DC depending on the country and on the application [1]. High speed or heavy duty new lines are provided with AC electrification system while tramways and underground railways usually work on DC supply voltages.

Auxiliary railway power supplies (ARPS) are needed onboard rolling stock to provide medium or low voltage to different systems and equipment from the overhead power cable (see Fig. 1). All of these systems can be included into a group called auxiliary services of the train.

This work is focused on auxiliary railway power supplies for DC catenaries. It is important to remark that according to EN-50163, catenary voltages can vary from 70% of nominal value up to 120%. This wide input voltage range requires a great regulation capability for the ARPS.

Another constraint associated to the input voltage is when the railway is fed by means of the garage's catenary. In this case the catenary voltage is 600V and it is required the whole power from the converter. Due to this fact, high currents are demanded, so the overall losses will be increased.

Due to specifications, the galvanic isolation is needed in this kind of systems. The conventional and widely used topology for ARPS use a low frequency transformer (50/60Hz) (see Fig.

1.a). The weight of this element is a third of the whole ARPS [2]. Therefore a new topology has been developed in recent years in order to reduce the size and weight of the transformer (see Fig. 1.b). In this case the transformer operates in a mid frequency range (4÷10 kHz).

In either of the abovementioned applications, the power supply output voltage is usually under very restrictive constraints regarding its output voltage Total Harmonic Distortion (THD) ([3]). In fact, the IEEE recommended practice guide (IEEE standard 519) establishes that it must be kept below 10% in order to avoid further damage of the loads.

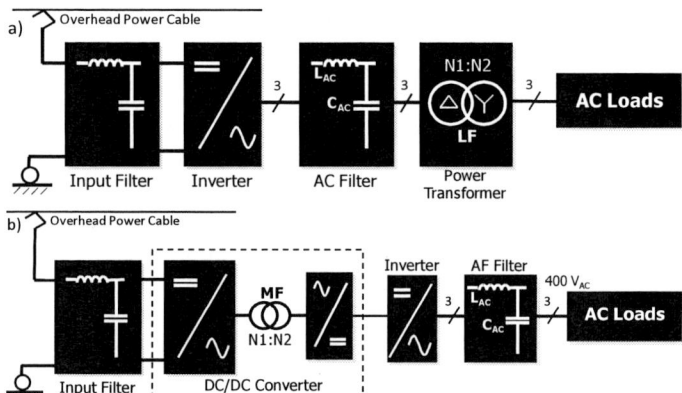

Fig. 1: Block diagram of a typical auxiliary railway power supply for a metro vehicle. a) Low frequency transformer. b) Medium frequency transformer

In Fig. 1, a basic AC voltage generation system has been included. On it, it can be observed that the inverter is fed by means of the overhead power cable through an LC filter intended for protection purposes. Due to the slow dynamic imposed by the LC input filter, the use of fast control loops or active filters are prevented because of the stability constraints imposed by the input filter [4].

Thus, the converter regulation can be considered quasi-static and so a control strategy based on the selection of the appropriate modulation index is enough to achieve a successful performance. The modulation index variation can be implemented by means of a feedforward of the input voltage and a feedback loop which controls the AC output voltage in order to compensate the effect of the resistive voltage drops.

978-1-4244-8806-3/11 $26.00 © 2011 IEEE

Along the last decades, PWM modulation techniques have been object of intensive research. From the original analog SPWM to the digital PWM techniques developed as the microprocessors technology was improved, a huge number of modulation techniques have been studied and carried out.

Due to the large number of existing modulation techniques, different authors have developed interesting surveys and comparisons in order to ease the selection of the most proper modulation technique for different applications.

However, most of these studies are focused on AC drives and their particular characteristics [5]-[7], while the present work aims to study the performance of the different modulation techniques applied to AC bus voltage generation in onboard systems.

As discussed in detail in [7]-[10] and [14], the performance of the different PWM modulation techniques is modulation-index dependant. Therefore, the combination of at least two different PWM techniques and on-line selection of the proper modulation technique as a function of the modulation index, may improve the overall performance of the system

Overmodulation issues have attracted the attention of many researches and overmodulation strategies have been developed [5], [11]-[13]. In all of these studies the overmodulation presents pulse dropping and so a high output THD. Is should be noted that the abovementioned strategies are focused in inverters motor drives.

Taking into account both the wide input voltage variation and the THD of the output voltage, this work aims to increase the regulation capability of the converter and at the same time to reduce the THD of the output voltage by means of the inverter modulation. It is important to notice that the proposed modulation technique can be applied to both systems represented in Fig. 1.

Along section II, the main idea of the new modulation technique is proposed. Right after this description, the sections III and IV describe the operating conditions of the proposed technique.

Both simulation and experimental results that validate the previous sections are shown in section V. And finally, the conclusions of this work are summarized.

II. PROPOSED MODULATION TECHNIQUE

The main characteristic of the new modulator presented along this epigraph is based on increasing the inverter modulation index, which is defined in (1), beyond the maximum 1.15 provided by the PWM with third-harmonic reference injection (hereinafter THIPWM) or space vector modulation [14], moving into overmodulation and reducing the output voltage THD. Therefore the converter regulation capability can be increased.

$$M = \frac{2 \cdot T_r \cdot V_1}{V_{dc}} \quad (1)$$

Where V_{dc} is the DC link voltage, T_r is the transformer conversion ratio and V_1 is the RMS line to neutral inverter output voltage.

The problem associated to the overmodulation lies in the pulse dropping in the central zone of the half period, so large amounts of subcarrier frequency harmonics (5th, 7th, etc.) are generated, and as the six-step mode is approached, these harmonics become increasingly dominant in determining the waveform quality (see Fig. 3). Thus, in order to achieve a good waveform quality, it is necessary to increase the carrier waveform frequency (f_c) in order to keep constant the number of switching events per quarter period, hereinafter "chops", and to concentrate them in the sidebands of the half period of the reference waveform (see Fig. 2). The increasing of the carrier waveform frequency will be shown in detail in section III.

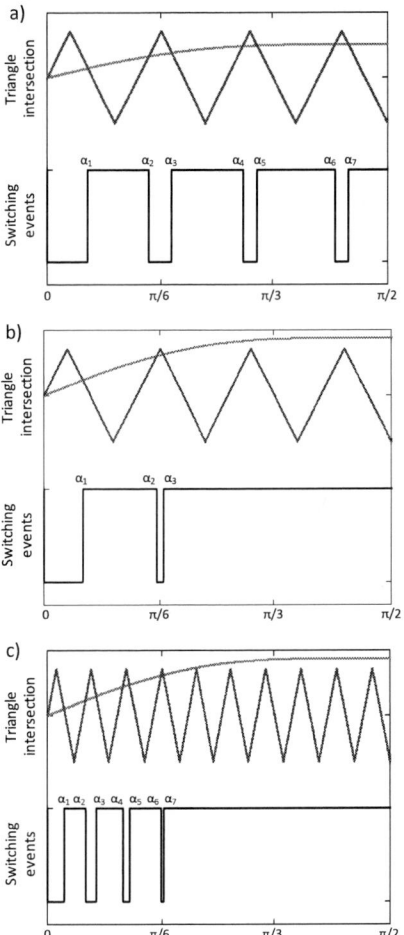

Fig. 2: Pulse generation for a 50Hz fundamental frequency. a) Linear modulation. b) Overmodulation with pulse dropping. c) Overmodulation with fixed number of "chops".

Due to the increased carrier frequency, the sidebands harmonics are moved to a higher frequency, so with the same filter designed for a linear modulation, the attenuation achieved is higher (see Fig. 3). In this way the sidebands harmonics introduced by the carrier waveform are reduced, so the THD is also reduced.

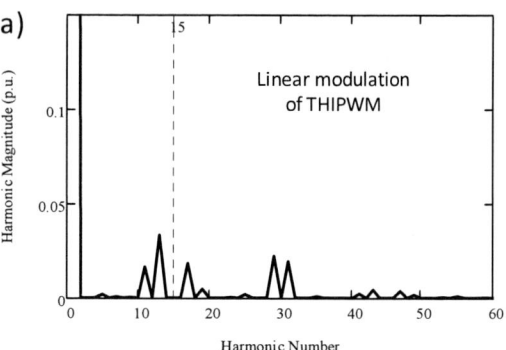

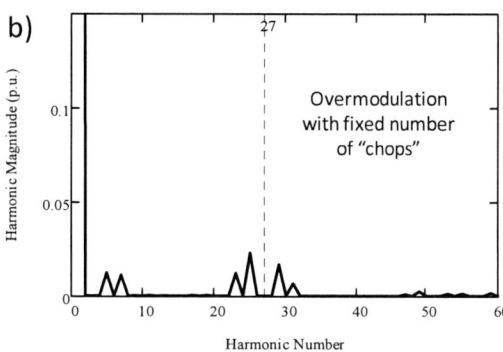

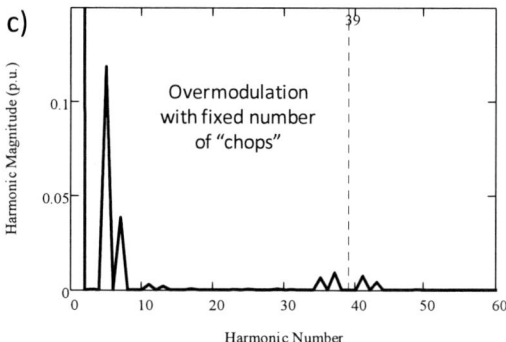

Fig. 3 Harmonic spectrum of the line to line filtered output voltage. a) Linear modulation for M=1. b) Overmodulation for M=1.16. c) Overmodulation for M=1.20.

III. LOOK-UP TABLE CONTROL

As it has been mentioned before, the PWM strategies are modulation index dependant. On the other hand, the converter regulation can be considered quasi-static and so a control strategy based on the selection of the appropriate modulation index within a look-up table is enough to achieve a successful performance.

A) Carrier ratio

Consider the carrier ratio (C.R.=f_c/f_o) where f_c is the carrier frequency and f_o the fundamental frequency. The particular carrier ratio values are not obvious. These values must be odd, but there are two possibilities regarding the multiple of three.

i. Carrier ratio multiple of three.

If the carrier ratio is set odd and multiple of three, for the three phase system shown in Fig.1, in the line to line output voltage will not appear inter-harmonics. Therefore, the carrier ratio is set as follows: it must be odd and triplen and such that the number of chops is kept constant. So the carrier frequency will be adapted to achieve the aforementioned characteristics.

The carrier frequency should be increased such that the range 0-β (see Fig. 4) has the desired number of chops, i.e. the carrier waveform should have a determined number of periods (n_periods). The value of β is determined where the value of the reference wave is equal to one, as seen in the expression (2).

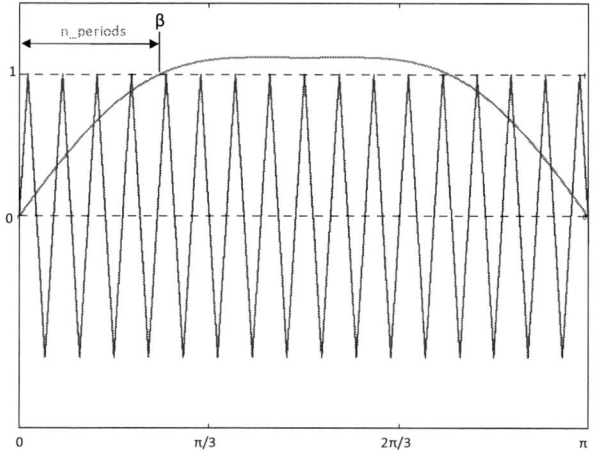

Fig. 4 Carrier ratio multiple of three vs. modulation index for 7 "chops".

$$M \cdot Sin(\beta) + M \cdot M_3 \cdot Sin(3\beta) = 1 \qquad (2)$$

Where M is defined in (1) and M_3 is the amount of third harmonic injection. The value of β has been obtained by means of numerical methods.

Once β has been defined, the carrier ratio value is given by expression (3).

$$C.R. = \frac{2 \cdot \pi \cdot n\_periods}{\beta} \qquad (3)$$

It is important to remark that the value of the carrier ratio obtained through the expression (3) is not an integer result. Therefore, in the case of a triplen carrier ratio, it should be rounded up to the next odd number taking into account the multiplicity with three.

In Fig. 5 has been represented the carrier ratio against the modulation depth for a constant number of chops (7 per quarter period). As it can be observed, the relationship between both of them has a staggered form. The step width, δ, is given by the number of "chops", which must remain constant for the whole range of modulation index.

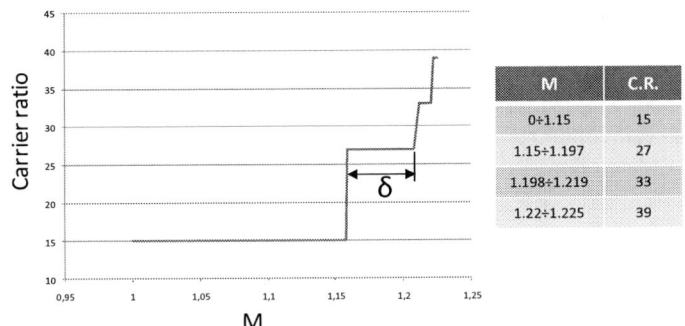

M	C.R.
0÷1.15	15
1.15÷1.197	27
1.198÷1.219	33
1.22÷1.225	39

Fig. 5 Carrier ratio multiple of three vs. modulation index for 7 "chops".

Since the carrier ratio is set triplen and the switching pattern presents quarter wave symmetry, it is only necessary to store in the look-up table the switching events of a quarter period for one inverter leg for each modulation index. Both symmetry and the other legs are implemented by software.

It is important to notice that for each modulation index contained in an interval δ, the carrier ratio does not change, but in the look-up table will be stored the switching events for each modulation index.

ii. Carrier ratio non-multiple of three.

The fact that the carrier ratio is not a multiple of three eliminates one of the constraints, but the control strategy must be precise in order to avoid inter-harmonics. The look-up table should store three switching patterns, one for each inverter leg.

Taking into account all possible odd values for the carrier ratio, the expression (3) should be rounded up to the next odd number. The relationship between it and the modulation index, for a number of chops of seven, is shown in Fig. 6.

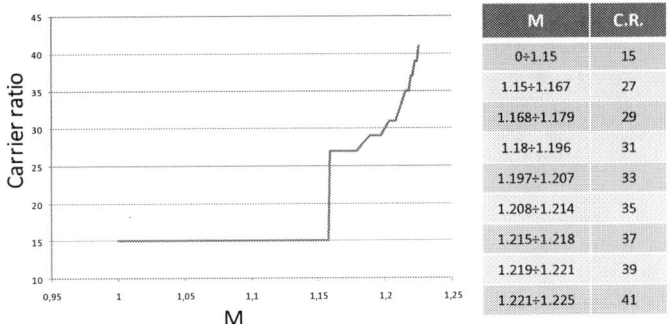

M	C.R.
0÷1.15	15
1.15÷1.167	27
1.168÷1.179	29
1.18÷1.196	31
1.197÷1.207	33
1.208÷1.214	35
1.215÷1.218	37
1.219÷1.221	39
1.221÷1.225	41

Fig. 6 Carrier ratio non-multiple of three vs. modulation index for 7 "chops".

IV. THEORETICAL RESULTS

The theoretical results shown along this epigraph have been obtained for the system shown in Fig. 1.a, in which the load power considered is 1280W, the line to line output voltage is 400V_{RMS} and the fundamental frequency is 50Hz.

The performance criteria selected in order to carry out a comparison between the proposed new modulation technique and other techniques such as THIPWM or Harmonic Elimination (HE) are the Total Harmonic Distortion (THD) of the filtered output voltage and the Weighted Total Harmonic Distortion (WTHD) of the inverter output voltage which is defined in (4). The WTHD is related to the copper losses [14].

$$WTHD = \frac{1}{V_{AN\_rms\_1}} \cdot \sqrt{\sum_{n=2}^{\infty}\left(\frac{V_{AN\_rms\_n}}{n}\right)^2} \qquad (4)$$

So far, the best solution in terms of modulation for this kind of systems is explained in [10]. On it, an improved modulator is proposed that is based on combining different modulation strategies (THIPWM and HE).

In Fig. 7 the new modulation technique for both carrier ratios, multiple and non-multiple of three, has been compared in terms of output voltage THD against THIPWM and HE. As it can be seen, the HE presents a limitation regarding the modulation index for a value of M=1.15 due to the fact that the equations system has no solution above this value.

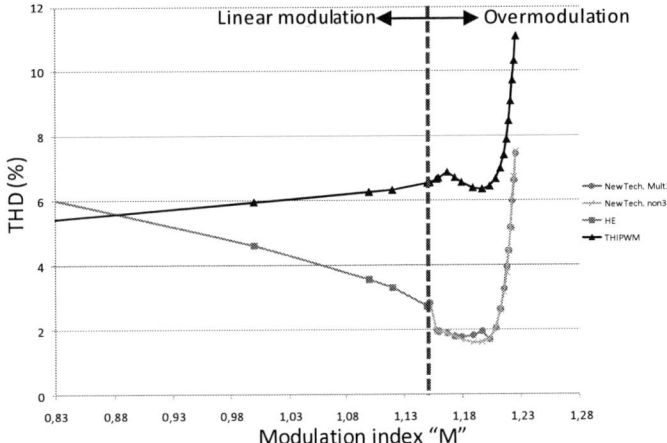

Fig. 7: Output voltage THD vs. modulation index. Red trace: HE. Black trace: THIPWM. Blue trace: New modulation technique for a carrier ratio multiple of three. Green Trace: New modulation technique for a carrier ratio non-multiple of three.

When the modulation index moves into the overmodulation zone, the new proposed technique provides lower THD than the obtained with the THIPWM and HE. If it is considered the carrier ratio non-multiple of three, it can be obtained better results in terms of THD for the interval between M=1.15 and M=1.25.

Consider the best solution provided by the modulator shown in [10]. The modulation method is determined by the value where both THD curves, THIPWM and HE, intersect each other. In Fig. 7, this value is around M=0.9 and below it, the modulation strategy that achieves better results regarding THD is THIPWM and above it, the better results are provided by HE. If the maximum THD, that is achieved for M=0.9, is taken into account, in order to obtain the same THD in the overmodulation zone, the modulation index could be increased up to M=1.22. Therefore, with the proposed overmodulation strategy, the modulation index can be increased up to 6%.

On the other hand and considering the HE technique, its minimum THD value is achieved for the maximum modulation index (M=1.15). Taken into account this technique, the modulation index can be increased by 3%.

In Fig. 8 the comparison between all techniques mentioned above has been carried out in terms of WTHD. It can be observed that when the modulation index moves into overmodulation, the WTHD is lower for the new technique than the obtained with the classic techniques.

As for the THD, when the carrier ratio is non-multiple of three, better results can be achieved than the obtained for a carrier ratio multiple of three.

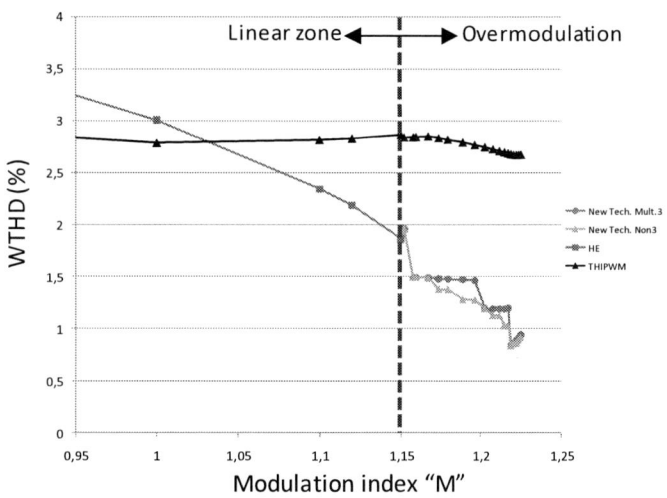

Fig. 8: Output voltage WTHD vs. M. Red trace: HE. Black trace: THIPWM. Blue trace: New technique with carrier ratio multiple of three. Green trace: New technique with carrier ratio non-multiple of three

V. EXPERIMENTAL RESULTS

The nominal output voltage (line to line voltage on the secondary side of the transformer) is 400 Vrms (50Hz) and the nominal power consumed by the load is 1280W (on a resistive load, $L_{LOAD}=0$). The rest of the prototype physical characteristics are summarized in Table I.

The experimental results shown below have been carried out for a carrier ratio multiple of three and keeping constant the number of chops in seven per quarter period. It should be noticed that the number of chops can be increased taking into account the minimum pulse width that can be achieved for a IGBT technology.

Table I. Prototype characteristics

L_{AC}	R_{LAC}	C_{AC}	CUT-OFF FREQUENCY	N1	N2
22mH	1.61Ω	4.5µF	270Hz	255	292
Vrms=400V			P=1280W		

In Fig. 9 have been compared, in terms of THD, the experimental results against the theoretical ones in order to validate them.

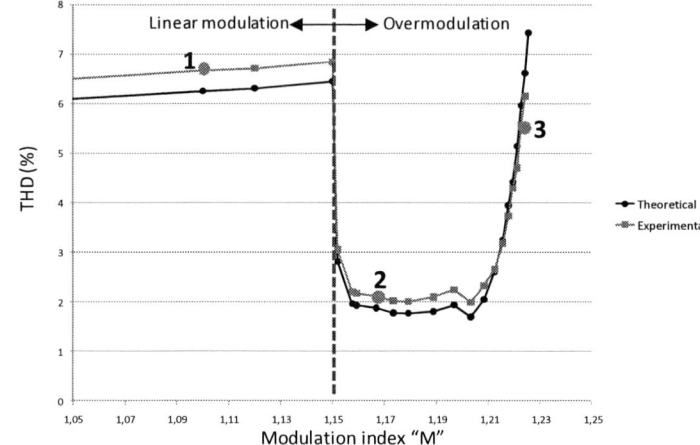

Fig. 9: Output voltage THD vs. M. Red trace: Experimental results. Black trace: Theoretical results.

As it can be observed in Fig. 9, both the experimental and the theoretical results are very close. Therefore, the theoretical results have been validated.

In Fig. 9 three different zones can be distinguished (dots 1, 2 and 3). The first one is the linear zone where the THD increases for increasing modulation index. In Fig. 10.a is shown the gate voltage of the top IGBT of the leg A and the output voltage applied to the load. The output voltage harmonic spectra is shown in Fig. 10.b.

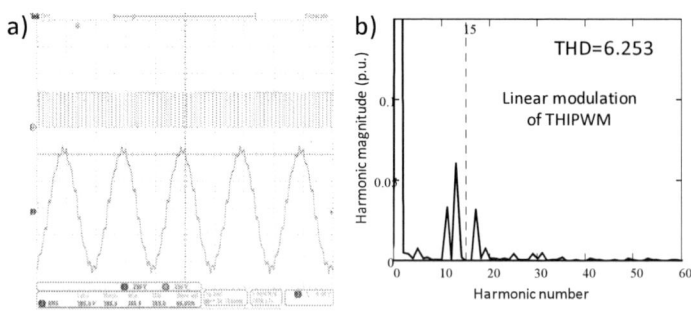

Fig. 10: M=1.1 (Dot 1 in Fig. 9) and 7 "chops" per quarter-period: a) From top to bottom: IGBT gate voltage, filtered output voltage. b) Output voltage harmonic spectrum.

978-1-4244-8806-3/11 $26.00 © 2011 IEEE

For the linear modulation, the THD is sidebands harmonics dependant, due to the fact that these harmonics are placed at low frequencies (around odd multiples of the carrier fundamental). For increasing modulation index, the THD will be high because the magnitude of the first carrier harmonic dominates the THD calculation.

The second zone shown in Fig .9 is for a modulation index between 1.15 and 1.21 where the THD remains constant. In Fig. 11 the output voltage and its harmonic spectrum for an experimental result in this zone (M=1.179) are shown. When the modulation index moves into overmodulation, the carrier ratio increases to keep constant the number of chops. Therefore, the sidebands harmonics are moved to higher frequencies and in this zone, the subcarrier frequency harmonics (5th, 7th, etc.) are low, so the THD is reduced.

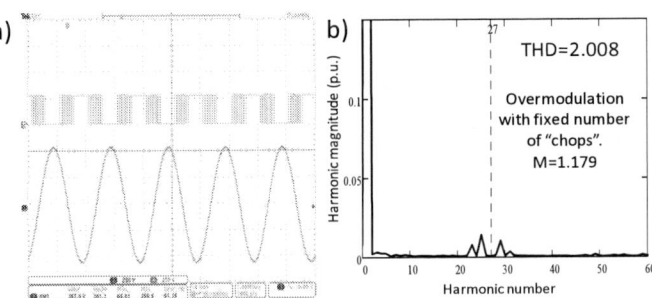

Fig. 11: M=1.179 (Dot 2 in Fig. 9) and 7 "chops" per quarter-period: a) From top to bottom: IGBT gate voltage, filtered output voltage. b) Output voltage harmonic spectrum.

The third zone that can be observed en Fig. 9, is where the THD has an exponential increase. The experimental result for this zone is shown in Fig. 12.

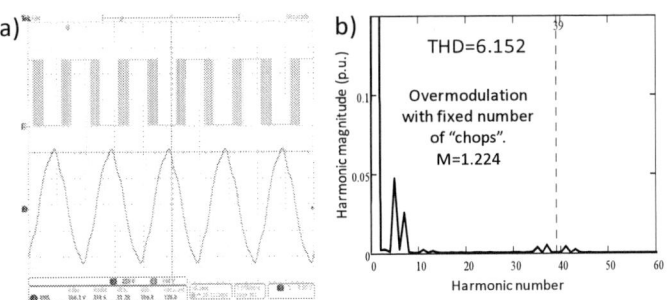

Fig. 12 M=1.224 (Dot 3 in Fig. 9) and 7 "chops" per quarter-period: a) From top to bottom: IGBT gate voltage, filtered output voltage. b) Output voltage spectrum.

For this zone, although the sidebands harmonics are placed at higher frequencies, the subcarrier frequency harmonics contribution to the THD has been increased. Therefore the output voltage is distorted with a high THD.

CONCLUSIONS

In this paper a new overmodulation strategy (patent pending) for voltage source inverters for railway applications has been proposed and validated. The strategy is based on increasing the carrier frequency in order to keep constant the

number of chops in linear zone, so the sidebands harmonics are moved to higher frequencies.

If the maximum allowable THD is established as the maximum THD achieved by the best modulator found in the literature, then, the proposed overmodulation technique is able to increase the modulation index by a 6% (from 1.15 to 1.22).

Additionally, within the overmodulated region, there is a minimum in the THD which improves the obtained THD by a 75% if THIPWM is considered for comparison or by a 40% if HE is considered.

Regarding WTHD, it is also reduced when the modulation index moves into the overmodulation region. So, the converter losses are not penalized.

ACKNOWLEDGMENT

This work has been supported by Ministerio de Educación y Ciencia (Spain), by means of the research projects FLAME (DPI: 2010:21110-C02-02) and SAUCE (DPI: 2009-12501).

REFERENCES

[1] A. Steimel, "Electric Traction-Motion Power and Energy Supply: Basics and Practical Experience", München: Oldenbourg Industrieverlag, 2007.

[2] R. Petrov, "Design issues of the high-power high-frequency transformer", Proceedings of International Conference on Power Electronics and Drive Systems, 1995, pp.401-410, vol.1.

[3] "IEEE Guide for harmonic control and reactive compensation of static power converters". IEEE standard 519.

[4] R. D. Middlebrook, "Input filter considerations in design and application of switching regulators," IEEE Ind. App. Society Annual Meeting, October 11-14, 1976, Chicago, IL. IAS Republished in Advances in Switch-Mode Power Conversion, Vol. I and II, 2nd edition, TESLAco, 1983, pp. 91-107.

[5] A. M. Hava, R. J, Kerkman, T. A. Lipo, "Carrier-based PWM-VSI Overmodulation Strategies: Analysis, Comparison and Design". *IEEE Transactions on Power Electronics*, Volume 13, No 4, July 1998.

[6] J. Holtz, "Pulsewidth modulation, A survey" *IEEE Transactions on Industrial Electronics.* Volume 39, Issue 5, October 1992, pp: 410-420

[7] M. A. Boost, P. D. Ziogas, "State of the Art Carrier PWM Techniques: A critical Evaluation". *IEEE Trans. on Ind. App.*, Vol. 24, No 2, April 1988.

[8] H. W. Van Der Broeck, "Analysis of the harmonics in voltage fed inverter drives caused by PWM schemes with discontinuous switching operation" Conf. Rec. European Power Electronics Conf., 1991, pp. 261-266.

[9] J. W. Kolar, H. Ertl, F. C. Zach, "Influence of the Modulation Method on the Conduction and Switching Losses of a PWM Converter System", IEEE Trans. on Industry App., Vol. 27, No 6, Nov./Dec. 1991, pp:1063-1075.

[10] I. Quesada, et al " Improved modulator for THD reduction in onboard generation of three phase AC voltage ". IEEE Compatibility and Power Electronics, 2009. CPE '09, pp. 450-456.

[11] R. J. Kerkman, T. M. Rowan, D. Leggate and B. J. Seibel "Control of PWM voltage inverters in the pulse dropping region". IEEE Applied Power Electronics Conference, 1994. APEC'94, pp. 521-528.

[12] T. M. Rowan, R. J. Kerkman and T. A. Lipo, "Operation of naturally sampled current regulators in the transition mode". *IEEE Transactions on Industry Applications*, volume 23, July/Aug. 1987.

[13] Kai Sun, Qing Wei, Lipei Huang and Kouki Matsuse, "An overmodulation method for PWM-Inverter-Fed IPMSM Drive with Single Current Sensor". *IEEE Trans. on Industrial Electronics*, volume 57, Oct. 2010.

[14] D. G. Holmes, T. A. Lipo, " Pulsewidth modulation for power converters. Principles and Practice", USA, John Wiley & Sons, Inc. 2003.

Feasible Solutions Space for the Harmonic Cancellation Technique

I. Quesada[1], C. Martinez[1], C. Raga[1], A. Lazaro[1], A. Barrado[1], R. Vazquez[2], I. Gonzalez[2]

[1]Universidad Carlos III de Madrid - GSEP
Leganes, Madrid, Spain
iquesada@ing.uc3m.es / http://gsep.uc3m.es

[2]Sistemas Electrónicos de Potencia S. A., SEPSA
Pinto, Madrid, Spain
rvazquez@sepsa.es

Abstract -- **AC power supplies generated by means of an inverter are under very restrictive specifications regarding their AC output voltage THD, even when they are supplying both linear and nonlinear loads. The presence of the low frequency harmonic content due to nonlinear loads operation, greatly increase the THD of the AC power supply output voltage.**

Traditional state of the art techniques approach this problem from two main sides: filtering and fast control loops. However, the proposed technique aims to solve the aforementioned problem from a different point of view: the inverter modulation technique. Therefore, the Harmonic Cancellation Technique (HCT) has been developed. It is a modulation technique which is able to cancel out the low frequency harmonics due to the nonlinear load operation through the proper pre-distortion of the inverter output voltage.

The present paper presents the limitations of the available solutions space for the proposed modulation technique. It also provides an insight of the converter design parameters which may help to enlarge the ranges of operating conditions for which HCT can be applied.

Index Terms—Rail transportation power systems, inverters, low THD, Inverters modulation

I INTRODUCTION

Auxiliary Railway Power Supplies (ARPS) implemented by means of an inverter is a widespread application of the power electronics. The interface characteristics that must be satisfied by an ARPS are detailed in [1]. Regarding the ARPS output voltage characteristics, it is said that the maximum allowable THD at its output voltage shall be established as a trade-off between the car builder and the converter manufacturer; since it is a compromise between the cost of the inverter filtter, maintainability and the harmonic content that the loads are able to accommodate.

Therefore, depending on the loads connected to the inverter output, the maximum allowable THD may be different. As an example, three commonly used maximum THD values are: 4%, 7% and 14%.

Keeping the output voltage THD within the limits is an important challenge, especially when the inverter must feed simultaneously linear and nonlinear loads, since the nonlinear loads operation provokes additional low frequency harmonic content which distorts the power supply output voltage and so increases significantly the THD.

Along the work presented in this paper, a typical auxiliary railway power supply (ARPS) has been considered in order to develop and check the proposed modulation technique. The auxiliary power supply is used to feed low and medium voltage equipment from the catenary (lightning, air conditioning, battery charger, etc.).

The typical ARPS is formed by an input filter which provides overvoltage protection to the converter placed downstream, a Voltage Source Inverter (VSI), an AC output filter (L_F, C_F) which attenuates the switching ripple and finally the auxiliary services of the train (modeled as an AC linear load) and a battery charger. See Fig. 1.

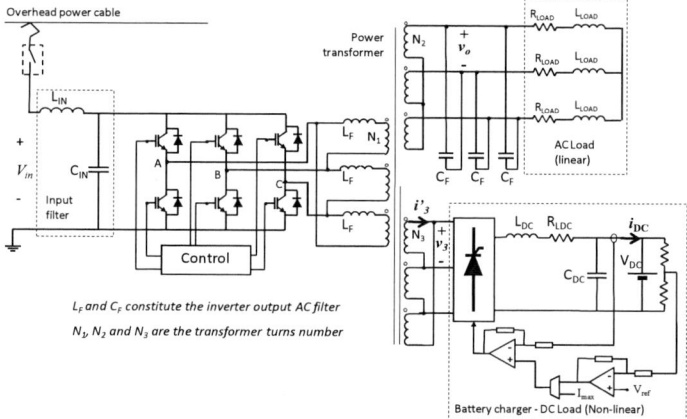

Fig. 1: Simplified scheme of a typical railway converter. VSI is used to feed simultaneously the auxiliary services of the train (AC linear load) and to charge some back-up batteries (nonlinear load).

Although some other options can be considered for the battery charger (PWM rectifiers [2], [3]), it is usually implemented by means of a three phase full bridge thyristor rectifier due to its robustness and cost effective characteristics, despite its non linear nature.

The presence of this nonlinear load provokes additional low frequency harmonics content which flow back through the impedances of the system and distort the AC filter output voltage. Thus the output voltage THD increases and doesn't comply with the specifications.

State of the art solutions reduce the output voltage THD by means of either multi-loop strategies ([4]-[6]) or active filtering

978-1-4244-8806-3/11 $26.00 © 2011 IEEE

([8]-[10]). However neither multi-loop strategies nor active filtering are applicable to the case of an ARPS due to stability constraints or unaffordable increase of cost, size and complexity, respectively [11].

Therefore, the pre-distortion of the inverter output voltage is a convenient strategy which is able to cancel out the low frequency content due to the non linear load without any significant increment of the size and cost of the system. The switching events are calculated analytically as it is summarized in Section II.

The present paper deals with the boundaries that limit the feasible solutions space of the Harmonic Cancellation Technique. The boundaries for a given design have been explored and the critical operating point, in terms of existence of solution, is identified. It should be remarked that the feasible solutions space is not only determined by the operating conditions but it is also influenced by the converter design parameters. So, an insight of the converter design parameters which may help to enlarge the ranges of operating conditions for which HCT can be applied are also considered in this paper.

Along Sections IV and V the limitations of the feasible solutions space and the design parameters which might help to enlarge the solutions space are described. In section VI the experimental performance of the Harmonic Cancelation Technique is shown. Finally some conclusions are extracted.

II OPERATION PRINCIPLE

The Harmonic Cancellation Technique (hereinafter HCT) aims to reduce the output voltage THD through the proper pre-distortion of the inverter output voltage. Since the pre-distortion of the inverter output voltage is calculated analytically for each given set of operating conditions, the HCT is a technique suitable in those cases in which the non linear load is concentrated and well known.

Let's consider the per phase equivalent circuit (Fig. 2) of the AC power supply in Fig. 1. The current drawn by the non linear load, referred to the primary side of the transformer is represented by means of a current source. The low frequency current harmonics drawn by the nonlinear load flow through the filter inductance and provoke a voltage drop across it (v_F) which distorts the filter output voltage (v_o).

The proposed modulation technique is able to generate an inverter output voltage formed by a controlled fundamental component as well as a set of low frequency harmonics which added to the ones across the filter inductance, they cancel each other out. Thus, the inverter output voltage has been represented qualitatively through a sinusoidal voltage source (V_{AB_1}) which represents the fundamental voltage and a non sinusoidal voltage source (v_F) that represents the low frequency content intended to cancel out the harmonics due to the non linear load operation.

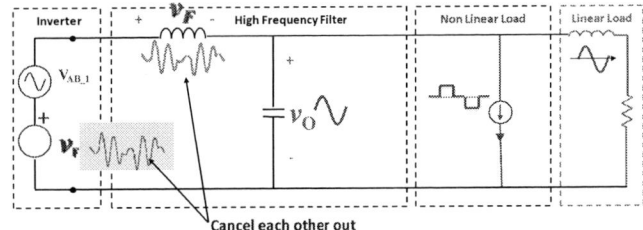

Fig. 2: Principle of operation of the Harmonic Cancellation Technique (HCT) on the per phase equivalent circuit of a Three phase AC power supply by means of an inverter, which supplies both a linear and a nonlinear load.

The implementation of the proposed modulation technique is based on the analytical determination of the switching events for a given set of operating conditions, in order to cancel out the harmonic content due to the non linear load. Those switching events are stored within a look-up table.

Applying the superimposing principle on the circuit represented in Fig. 2, the mathematical expression of the filter output voltage (v_o) can be obtained (1).

$$v_o = v_{AB} \cdot G_F - i_3 \cdot G_F \cdot X_F \qquad (1)$$

Where:

$G_F = \dfrac{Z_o}{Z_o + X_F}$ Filter Gain

v_{AB} Complex n^{th} harmonic components of the line to line inverter output voltage

i_3 Non linear load current referred to the primary side of the transformer

v_o Line to line output voltage of the group inverter-filter

X_F Complex impedance of the filter inductance at the nth harmonic frequency.

Z_o Complex output impedance at the nth harmonic frequency. It is formed by the parallel of the linear load and the filter capacitor.

Additionally, it should be taken into account the following considerations in order to determine the harmonic content of v_{AB}:

✓ The triple order harmonic components are cancelled naturally because of the composition of two voltages phased out 120°.

✓ The required fundamental component is given as an specification. So, applying this condition on (1), equation (2) is obtained.

$$v_{o\_spec} = v_{AB\_1} \cdot G_F(j \cdot \omega_1) - i_{3\_1} \cdot G_F(j \cdot \omega_1) \cdot X_F(j \cdot \omega_1) \qquad (2)$$

✓ In order to cancel out the low frequency harmonics, the voltage at the filter output for each of the low frequency harmonics considered must be zero. So, to achieve a successful cancelation, each harmonic component must satisfy (3).

978-1-4244-8806-3/11 $26.00 © 2011 IEEE

$$0 = v_{AB\_n} - i_{3\_n} \cdot X_F(jn \cdot \omega_1) \quad \text{for n>1} \tag{3}$$

III ANALYTICAL DETERMINATION OF THE SWITCHING EVENTS

Although the development of the HCT has been already published ([11]), in Section III a brief summary has been included in order to improve the understanding of the following sections.

The key idea of the modulation technique is grounded on the generation at the inverter output of a set of low frequency harmonics which cancel out those across the filter inductance due to the current drawn by the non linear load. Therefore, control over both the phase and module of each harmonic is needed. Additionally, in order to prevent the presence of even harmonics in the system, half wave symmetry is required. Thus, the inverter output voltage (v_A) can be described by a Fourier series formed by odd terms in sine and cosine.

The Fourier coefficients of v_A as a function of the switching angles (α_p) are known [12] and provided in (4) and (5).

$$A_{n\_vA} = \frac{4}{\pi} \cdot \frac{V_{in}}{2} \cdot \frac{1}{n} \sum_{p=1}^{N} (-1)^p \cdot \sin(n \cdot \alpha_p) \tag{4}$$

$$B_{n\_vA} = \frac{4}{\pi} \cdot \frac{V_{in}}{2} \cdot \frac{1}{n} \left[-1 + \sum_{p=1}^{N} (-1)^{p-1} \cdot \cos(n \cdot \alpha_p) \right] \tag{5}$$

Where:

N	Number of switching angles to be determined
V_{in}	Inverter DC input voltage
α_p	Switching angles
n	Harmonic order

As in any other inverter, the voltage of the middle point of any leg reproduces the switching pattern of its upper switch. So, if the desired pre-distorted voltage at the middle point of a leg A (v_A) is determined, the switching patterns of the six switches can be derived.

Therefore, the first step to determine the switching events of the inverter must be the characterization of the desired inverter output voltage (v_A). In order to determine the inverter output voltage, it has been used the basic equations which define the system operation: (2) and (3).

After the characterization of the current drawn by the non linear load and some algebraic manipulation of the equations, the Fourier coefficients of v_A are obtained as a function of the operating point of the converter and its physical parameters. The complete theoretical development is already published in [11].

Finally, replacing the obtained Fourier coefficients in (4) and (5), the equation system in (6) and (7) is obtained. Solving this system of equations for a given set of operating conditions will lead to a switching pattern able to modulate the inverter output and cancel out the low frequency content due to the non

linear load operation.

The switching patterns calculated for different sets of operating conditions are stored within a look-up table (control table) in order to provide harmonic cancellation at any time during the regular operation of the converter. The control table will be calculated taking into account the ranges of variation of the operating conditions involved.

$$0 - \sin(n\alpha_1) + \sin(n\alpha_2) - \ldots + \sin(n\alpha_N) = \frac{A_{n\_vA}}{\frac{4}{\pi} \cdot \frac{V_{in}}{2}} \tag{6}$$

$$1 - \cos(n\alpha_1) + \cos(n\alpha_2) - \ldots + \cos(n\alpha_N) = \frac{-B_{n\_vA}}{\frac{4}{\pi} \cdot \frac{V_{in}}{2}} \tag{7}$$

IV LIMITATIONS OF THE FEASIBLE SOLUTIONS SPACE

Let's consider a fully designed converter, since all its physical parameters must be known a priori. Then, it is important to determine the feasible solutions space in order to be able to provide a successful harmonic cancellation for the whole range of operation of the converter.

There are three different operating conditions which might vary within a range and must be taken into account to determine the feasible solutions space:

✓ V_{in} *Inverter supply voltage from the catenary.*

The supply voltage of the inverter is not constant but it may vary within a pretty wide range. For instance, considering a nominal input voltage of 1500V the actual input voltage might vary within 1100V to 1800V.

✓ V_{DC} *Regulated output voltage of the battery charger.*

The battery charger is a regulated device which output will be regulated to the nominal voltage of the batteries connected to its output. Some usual values might be: $24V_{dc}$, $72V_{dc}$ or $100V_{dc}$.

Additionally, for a given nominal voltage, variation is allowed within a range as it can be seen in [1].

✓ P_{DC} *Output power delivered by the battery charger.*

It is defined in (8) as a percentage of the nominal power delivered to the linear load by the inverter (%P_{AC}) and it can vary up to 50% of the nominal power (P_{AC}).

$$P_{DC} = \%P_{AC} \cdot P_{AC} \tag{8}$$

Since there are three separate ranges of variation involved, along the following paragraphs the influence of each of them on the feasible solution space is going to be illustrated. In order to achieve this aim, an actual converter has been considered and its characteristics are shown in Table I.

Table I. Converter characteristics

L_{AC}	C_{AC}	L_{DC}	C_{DC}	N_1	N_2	N_3
3.5mH	220µF	405uH	45mF	102	40	29
V_{RMS_nom}=400V				P_{AC_nom}=170kW		

The HCT is implemented in 7 notches. This is, 14 switching events per half period must be determined in order to control 7 harmonic components (fundamental component and 6 additional non triplen odd harmonics: 5th, 7th, 11th, 13th, 17th and 19th). In Fig. 3 the evolution of the 14 switching angles vs. the inverter input voltage for some different conditions at the battery charger output (output power and output voltage) can be observed.

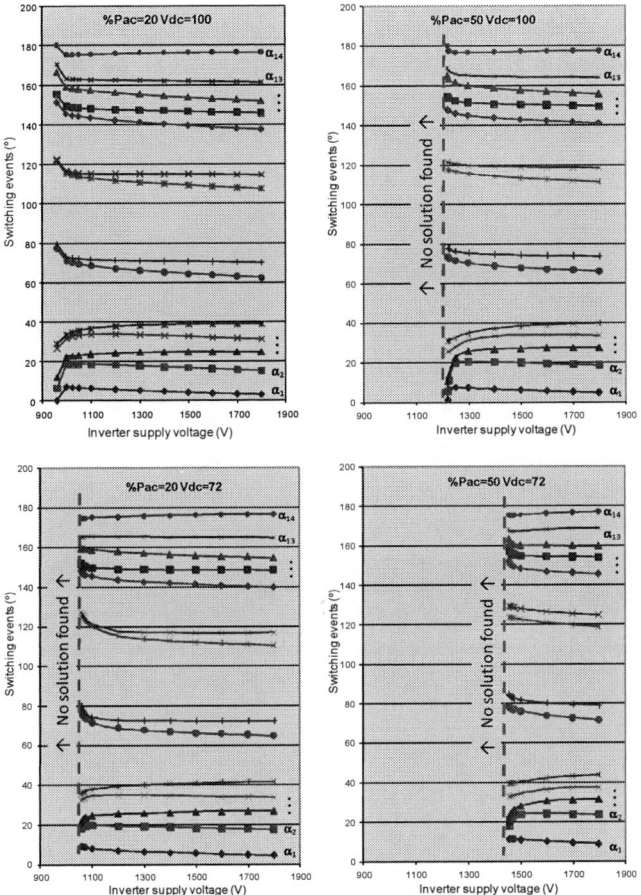

Fig. 3 Switching events evolution vs. inverter supply voltage for four different conditions at the battery charger output

It can be observed that the switching angles tend to get closer (the switching pulses gets narrower) as the input voltage decreases. Furthermore, below a given value of the inverter supply voltage, no feasible solution is found. Thus there is a lower limit for the feasible solutions space imposed by the inverter supply voltage.

Additionally, it is also noticeable that the minimum required input voltage is different depending on the non linear load operating conditions (%P_{AC} and V_{DC}):

✓ For a given non linear load output voltage (V_{DC}), the minimum inverter input voltage increases as the DC power delivered increases.

✓ For a given DC power, the minimum input voltage

increases as the DC output voltage decreases.

Thus, it can be said that the limiting factor is the voltage drop across the filter inductance due to the current drawn by the non linear load. As the current drawn by the battery charger increases, the voltage drop across the filter inductance increases and the minimum required input voltage to achieve a feasible solution increases.

Then, part of the available energy at the inverter input will be used to generate those low frequency harmonics. And so, as the low frequency harmonic content of the current drawn by the battery charger increases, the voltage across the filter inductance increases, and the low frequency harmonics at the inverter output will also be bigger. So a bigger amount of the available energy at the input will be injected at low frequencies.

Therefore, the worst case for the generation of the switching events will be the following combination of operating conditions:

✓ Minimum input voltage

✓ Maximum power delivered by the non linear load

✓ Minimum voltage at the non linear load

However, the feasible solutions space does not depend exclusively on the operating conditions, but it can be modified if the design of the converter is taken into account. This is, some parameters of the converter may help to enlarge the feasible solutions space. They are explained along the following section.

V   DESIGN PARAMETERS WHICH ENLARGE THE FEASIBLE SOLUTIONS SPACE

Take into account the simplified per-phase equivalent of the converter shown in Fig. 4). The current drawn by the non linear load is represented by means of a current source (i'_3).

Where, i'_3 is the current drawn by the battery charger seen at the primary side of the transformer. This current flows through the filter inductance provoking a voltage drop across that inductance (v_F). Therefore, it is easily observed that the size of the filter inductance directly affects the size of the low frequency harmonics which the inverter must generate to cancel out those due to the non linear load. Then, the effect of the size of this inductance is shown in Fig. 5.

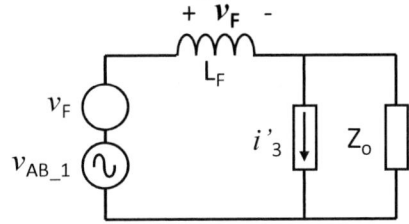

Fig. 4 Simplified per-phase equivalent of the Auxiliary Railway Supply which supplies a linear and a non linear load

On Fig. 5 the minimum input voltage required to achieve a

feasible solutions is represented against the power delivered by the battery charger. Additionally, in order to illustrate the influence of the filter inductance on the feasible solutions space, three different values of filter inductance have been considered, as well as two different DC output voltages. It can be observed that the use of lower DC output voltages penalizes the feasible solutions space, as it has been previously established.

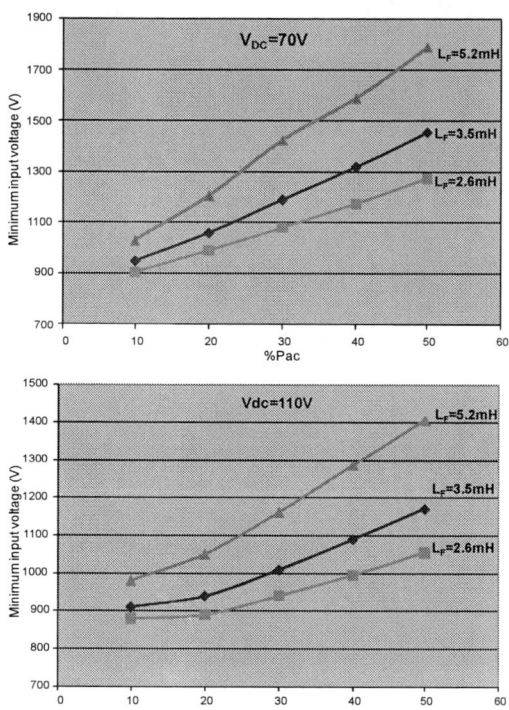

Fig. 5 Minimum input voltage required to achieve a feasible solution vs power delivered by the battery charger. Three different filter inductances have been considered as well as two different DC output voltages.

Regarding the influence of the filter inductance, the tendency is the same regardless the DC output voltage. This is, as the filter inductance value decreases the feasible solutions space is increased; since the minimum required voltage is lower.

Another design parameter involved, is the transformer ratio between its primary and tertiary sides (n_{13}). Since the modules of the harmonic components which flow through the filter inductance are affected by this transformer ratio.

As it has been illustrated previously, the larger the non linear load current, the bigger the minimum required input voltage. Then, if the number of turns of the transformer tertiary side is reduced, then the feasible solutions space will be enlarged.

Actual converters are subject to parasitic voltage drops such as parasitic resistances of the filter inductances, etc. Even though they must be taken into account within the calculation

of the switching events; their effect on the feasible solutions space is negligible compared to the effect of the operating conditions and the aforementioned design parameters.

VI HCT EXPERIMENTAL PERFORMANCE

The experimental performance of the Harmonic Cancellation Technique regarding the output voltage THD (9) has been carried out on a scaled-down prototype which characteristics are summarized in Table II. Considering an inverter input voltage range from 325V to 475V and a DC output power from 5% to 40%, the HCT is able to provide a successful performance in the whole range.

$$THD = \sqrt{\frac{\sum_{n=2}^{40} V_{o\_n}^2}{V_{o\_1}^2}} \tag{9}$$

Table II. Prototype characteristics

L_{AC}	R_{LAC}	C_{AC}	L_{DC}	R_{LDC}	C_{DC}	V_{DC}	P_{DC}	N_1	N_2	N_3
22 mH	1.61 Ω	4.5μF	5.4 mH	0.3Ω	16.4 mF	75V	513 W	255	286	90
V_{RMS_nom}=400V					P_{AC_nom}=1280W					

In order to illustrate the HCT performance, in Fig. 6 a comparison between the well known Selective Harmonic Elimination (SHE) versus HCT has been carried out. As it can be seen, the THD is reduced when the inverter is modulated with the HCT.

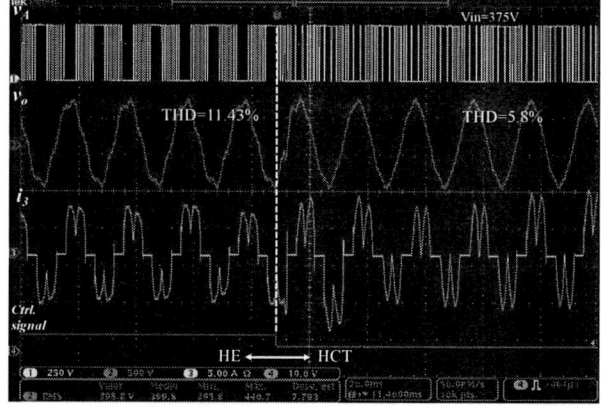

Fig. 6: Comparison of SHE vs. HCT for I=5.2A (CCM). v_A voltage in the middle point of the branch A of the inverter referred to ground, v_o line to line voltage on the secondary side of the transformer, i_3 current drawn by the nonlinear load

Fig. 6 only shows a trial for an inverter input voltage of 375V, but it has been carried out a sweep regarding the input voltage and the DC output power. The obtained results are shown in Fig. 7 and Fig. 8 for an inverter input voltage of 325V and 475V, respectively. Three different DC output powers have been considered for each voltage.

It can be observed that the THD is reduced in all cases. Since the high frequency THD is higher for higher input voltages, the THD reduction is slightly better for low input voltages than for high input voltages.

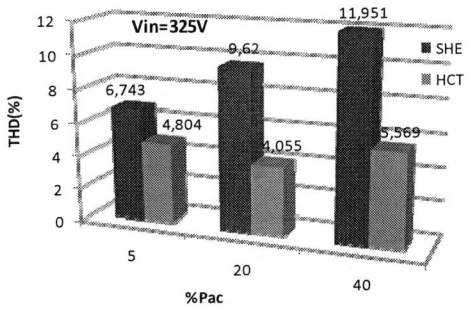

Fig. 7: Output voltage THD vs. %P_{ac} for an inverter input voltage of 325V

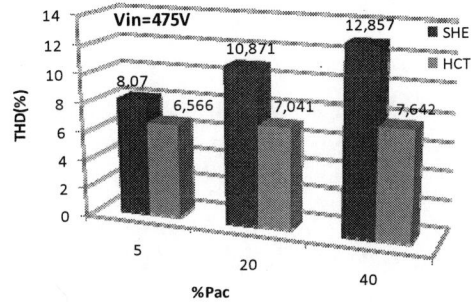

Fig. 8: Output voltage THD vs. %P_{ac} for an inverter input voltage of 475V

CONCLUSIONS

Along the present paper, the feasible solutions space for the Harmonic Cancellation Technique (HCT) has been studied. The HCT is able to pre-distort the inverter output in order to cancel out the harmonic content due to a non linear load performance. And so reduce the THD at the output voltage of the auxiliary railway power supply. For each set of operating conditions (V_{in}, V_{DC}, P_{DC}) the proper switching pattern is determined analytically.

However, not all the combinations of the aforementioned variables lead to a feasible solution. Thus, the determination of the feasible solutions space is needed to guarantee a successful cancellation of the low frequency content along the whole range of variation of the converter operating conditions.

For any converter the most restrictive operating conditions in terms of achieving a feasible solution, are: minimum input voltage, maximum power delivered by the non linear load and minimum voltage at the non linear load. So, if a solution is found in these conditions then the HCT can be used for the whole converter range of operation.

However the feasible solutions space does not depend only on the operating conditions, some physical parameters of the converter must be taken also into account. In order to enlarge the feasible solutions space, the filter inductance and the primary-tertiary transformer ratio should be designed as small as possible.

ACKNOWLEDGMENT

The authors would like to thank SEPSA for they support along the development of this work.

This work has been also supported by Ministerio de Educación y Ciencia (Spain), by means of the research projects FLAME (DPI: 2010:21110-C02-02) and SAUCE (DPI: 2009-12501).

REFERENCES

[1] IEEE Std. 1476-2000 "IEEE Standard for passenger train auxiliary power systems and interfaces" ISBN 0-7381-1964-4

[2] S. Choi "A three phase unity power factor diode rectifier with active input current shaping" IEEE Transactions on Industrial Electronics, Volume 52, Issue 6, pp: 1711-1714, December 2005

[3] M. Baumann, J. W. Kolar, "Parallel Connection of Two Three-Phase Three-Switch Buck-Type Unity-Power-Factor Rectifier Systems With DC-Link Current Balancing" IEEE Transactions on Industrial Electronics, Volume: 54 Issue: 6, *pp*: 3042 - 3053, Dec. 2007

[4] H. Deng, R. Orungati, D. Srinivasan "Modeling and control of single phase UPS inverters: a survey" International Conference on Power Electronics and Drives systems (PEDS) 2005, pp. 848-853

[5] S. Buso, S. Fasolo, P. Matavelli "Uninterruptible power supply multiloop control employing digital predictive voltage and current regulators" IEEE Transactions in Industry Applications, vol 37, 2001, pp. 1846-1854

[6] S. Sae-Sue, V. Kinnares, C. Tangsiriworakul, S. Potivejkul "Comparative performance evaluation of fixed and adaptive hysteresis band delta modulation techniques for UPS" International Conference on Power Electronics and Drive Systems 1999 pp. 956-960.

[7] F. Z. Peng "Harmonic sources and filtering approaches" IEEE Industry Applicatoins Magazine, Volume 7, Issue 4, Jul/Aug 2001 pp. 18-25

[8] K. Nishida, Y. Konishi, M. Nakaoka "Novel current control scheme with deadbeat algorithm and adaptive line enhancer for three phase current source active power filter" IEEE Industry Applications Conference Volume 1, Issue 30, Sept-Oct 2001, pp. 194-201

[9] H. Akagi "Modern active filters and traditional passive filters" Bulletin of the Polish Academy of Science, Vol 45, no 3, 2006.

[10] S. Bhim, K. Al Haddad, A. Chandra "A review of active filters for power quality improvement", IEEE Transactions on Industrial Electronics, vol 46, no 5,1999, pp. 960 – 971.

[11] I. Quesada, A. Lazaro; C. Martinez; A. Barrado; M. Sanz, C. Fernandez, R. Vazquez, I. Gonzalez, "Modulation technique for low frequency harmonic cancellation in Auxiliary Railway Power Supplies", Accepted for publication in a future issue of IEEE Transactions on Industrial Electronics

[12] D. Grahame Holmes, Thomas A. Lipo, "Pulse Width Modulation for Power Converters. Principles and Practice" ISBN 0-471-20814-0

Fundamental PMSM Model for Estimation of Cogging Torque Harmonic Components

Lovrenc Gašparin[1], Rastko Fišer[2]

[1]Iskra Avtoelektrika d.d., Polje 15, 5290 Šempeter pri Gorici, Slovenia, E-mail: Lovrenc.Gasparin@iskra-ae.com
[2]University of Ljubljana, Faculty of Electrical Engineering, Tržaška 25, 1000 Ljubljana, Slovenia, E-mail: rastof@fe.uni-lj.si

Abstract-A detailed analysis of cogging torque components reveals that besides well-know native cogging torque components exist also additional cogging torque components which are provoked by assembly tolerances in mass–production. Since these two groups of cogging torque components are within reference to common applied design techniques mostly in contradiction, a minimization of the total cogging torque becomes a challenging task for motor designers. A finite element method and Fast Fourier transformation were used to study and analyse a simple model of a PMSM which enables simulation of manufacturing tolerances and assembly imperfections. Calculations confirmed that imprecisions in mass–production cause the phenomena of additional cogging torque harmonic components, which are not present in the case of a perfectly assembled motor.

I. INTRODUCTION

One of the main requirements in modern drive systems, which nowadays mostly apply permanent magnet synchronous motors (PMSM), is an extremely low level of cogging torque [1]. A variety of methods for cogging torque reduction are known and applicable in practice [2], [3] but they are often inefficient, which is particularly noticeable in the case of mass-produced PMSMs [4]. Frequently observed substantial differences between measured and calculated (simulated) values of cogging torque are the result of manufacturing imperfections and tolerances, which are unavoidable facts in mass-production [5], [6].

II. STRUCTURE OF COGGING TORQUE HARMONIC COMPONENTS

Total cogging torque T_{cog} of PMSM is composed of several harmonic components (*HC*), which can be classified into two groups [7], [8]. The first group represents an array of native harmonic components (*NHC*) which originate in the combination of design parameters, mostly the number of stator slots Q and the number of rotor magnetic poles P. The second group is an array of additional harmonic components (*AHC*), which are the consequence of material and assembly imperfections and are not present in the case of a perfectly manufactured motor.

$$HC = NHC + AHC \qquad (1)$$

$$NHC(A_{NHCi}, N_{NHCi}, \varphi_{NHCi}) \qquad (2)$$

$$AHC(A_{AHCi}, N_{AHCi}, \varphi_{AHCi}) \qquad (3)$$

Each harmonic component could be described by the following parameters: an amplitude A, an order (number of repetitions in mechanical revolution) N, φ stands for a phase shift, and i is an integer. Cogging torque T_{cog} is a periodical function, which depends on rotor angular position α and is composed of two components

$$T_{cog}(\alpha) = T_{NHC}(\alpha) + T_{AHC}(\alpha) \qquad (4)$$

where T_{NHC} is a composed component of *NHC* contributions and T_{AHC} is a composed component of *AHC* contributions. For a clearer interpretation, T_{NHC} and T_{AHC} can be represented as a sum of sine functions

$$T_{NHC}(\alpha) = \sum_{i=1}^{\infty} A_{NHCi} \cdot \sin(N_{NHCi} \cdot \alpha + \varphi_{NHCi}) \qquad (5)$$

$$T_{AHC}(\alpha) = \sum_{i=1}^{\infty} A_{AHCi} \cdot \sin(N_{AHCi} \cdot \alpha + \varphi_{AHCi}) \qquad (6)$$

Native harmonic components *NHC* of cogging torque have orders defined by the following expression

$$N_{NHCi} = \mathrm{LCM}(Q,P) \cdot i \qquad (7)$$

where LCM stands for the least common multiple for a motor with ideally distributed PM on the rotor and equally spaced teeth on the stator. While *NHC*s of orders defined by expression (7) are well-known and often easily minimized, designers of PMSMs are generally not familiar with *AHC*s, therefore motor producers are frequently surprised and disappointed once theoretically optimized PM motors are manufactured and tested at output control [5], [6].

*AHC*s can be further classified as contributions of imperfections which originate either on the stator (AHC_S) or on the rotor side (AHC_R). The orders of *AHC*s are defined by expressions (8)-(10), where components $N_{AHC\gamma i}$ are caused by stator teeth misplacements, N_{AHCEi} are the consequence of interlocks in the stator back-iron, E is the number of symmetrically distributed interlocks, N_{AHCRi} arise due to rotor permanent magnet (PM) misplacements δ, thickness t, and width w variations, and i is an integer, respectively. Fig. 1 shows a detailed structure of harmonic components *HC* in the total cogging torque T_{cog} which is easily obtained when a Fast Fourier transformation analysis of the T_{cog} signal is performed.

$$N_{AHC\gamma i} = P \cdot i \qquad (8)$$

$$N_{AHCEi} = \mathrm{LCM}(E,P) \cdot i \qquad (9)$$

$$N_{AHCRi} = N_{AHC\delta i} = N_{AHCti} = N_{AHCwi} = Q \cdot i \qquad (10)$$

978-1-4244-8806-3/11 $26.00 © 2011 IEEE

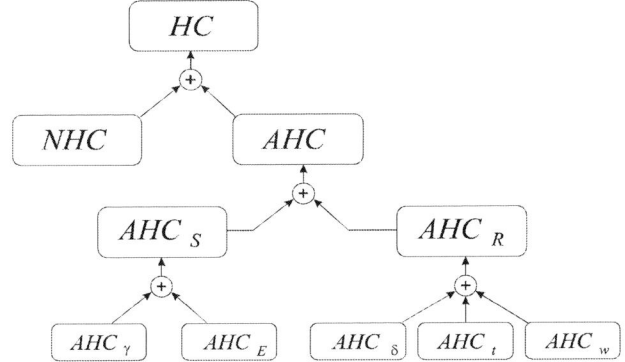

Fig. 1. Detailed structure of cogging torque harmonic components.

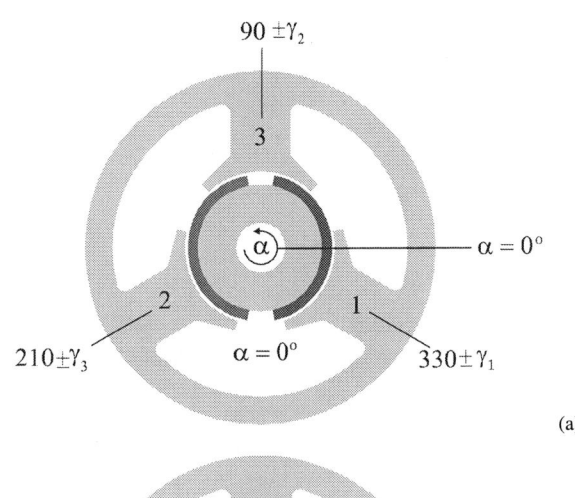

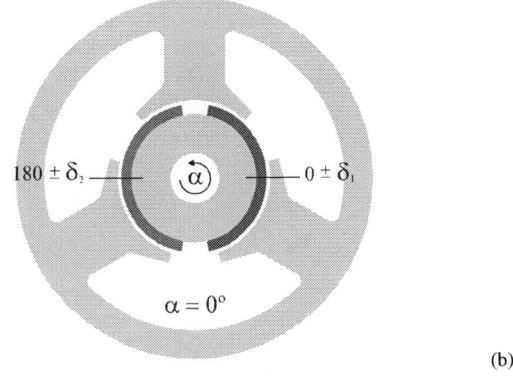

III. FUNDAMENTAL FEM MODEL

A simple parametric FEM motor model was used to study basic relationships among cogging torque harmonic components and to evaluate *AHC*s due to several possible imperfections on the stator and rotor side. For a three-phase motor, the simplest model has three stator teeth $Q = 3$ and two rotor magnetic poles $P = 2$ (Fig. 2) [9]. Such combination of Q and P results in a $F = 2$, where F represents a parameter of sensitivity to the *AHC*s [10]. Such value is preferred for the presented research as the model is sensitive to the influence of anomalies in construction and/or material properties [11]. It enables straightforward and distinctive study of changes in T_{cog} signal due to *AHC*s defined by (3), (6), (8) and (10) when applying asymmetries like stator teeth misplacement γ, rotor magnets' misplacements δ, magnets' thickness t and width w variations. So gained theoretical considerations can be advantageously applied on more complex geometries of real PMSMs [5]-[8]. Fig. 2 shows basic geometry of a PMSM under study with denoted possible stator and rotor imperfections.

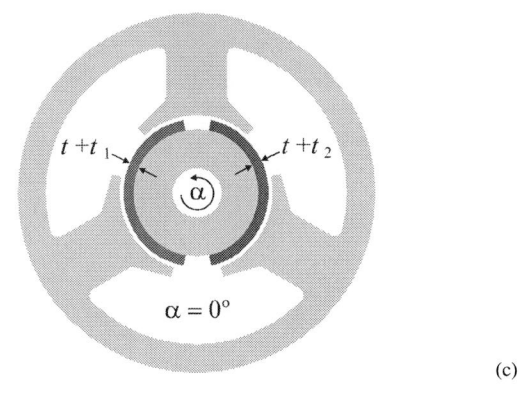

A. Modeling of Stator Teeth Misplacements

Fig. 3 shows results of cogging torque calculations for the case of stator asymmetry, when the stator tooth #1 is misplaced from ideal position for $γ_1 = -1°$. Because stator teeth are no longer symmetrically distributed, the angles among them differ from 120°. During rotation for one revolution (360°), both edges of each PM pass 4-times over the misplaced tooth #1 and 8-times over correctly positioned tooth #2 and #3 (Fig. 4). The crossing of PM under the misplaced tooth are in Fig. 3 and Fig. 4 marked with circles clearly indicating that T_{cog} amplitudes have diverse values at these positions. Amplitude value of T_{cog} has its minimum at rotor positions corresponding to α =15° in α =195° (Fig. 3), since both edges of PMs are at locations where the air-gap is the widest due to the misplaced tooth #1. On the contrary, T_{cog} has its maximum at positions related to α =105° and α =285° where the air-gap is the thinnest.

Results of FFT analysis of the T_{cog} for the case of symmetrical motor where only *NHC*s are present (emphasized in green) are gathered in Table I, while for

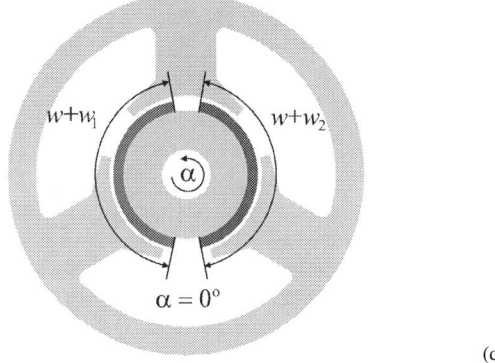

Fig. 2. Basic geometry of a PMSM with $Q = 3$, $P = 2$, which enables modeling of stator teeth misplacements γ (a), PM misplacements δ (b), PM thickness variations t (c), and PM width variations w (d).

978-1-4244-8806-3/11 $26.00 © 2011 IEEE   265

misplaced tooth #1 *AHC*s arise (emphasized in yellow) which also increases the maximal values of T_{cog}. The orders of *NHC*s correspond to (7), whereas the orders of *AHC*s match with (8) and all values of $A_{NHC\,i}$ and $A_{AHC\gamma\,i}$ are in mNm units. Some combinations of *NHC* and *AHC* components have the same orders $N_{NHC} = N_{AHC}$ which means that calculated amplitude values equal the sum of $A_{NHC\,i}$ and $A_{AHC\,i}$, marked in Table I as $A_{NHC+AHC}$.

B. Modeling of PM misplacements

Fig. 5 shows results of cogging torque calculations for the case of rotor asymmetry, when the red PM is misplaced from ideal position for $\delta_1 = -3°$. During the rotation (Fig. 6) for one revolution (360°) half of all PM crossings over the stator teeth happen on the side where both PMs are closer together (red circle) and the other half crossings occur where both PMs are shifted apart (blue circle). When observing simultaneously the instantaneous value of T_{cog} (Fig. 5) and the rotor position (Fig. 6) one can easily notice that T_{cog} reaches the highest values when PMs are crossing over the stator tooth's edge at the side where they are closer together due to the initial red PM's misplacement. The calculated T_{cog} has a geometrical

period 120° (Fig. 5), which indicates that besides *NHC*s there are also existing *AHC*s. Table II presents FFT results of the T_{cog} for the case of symmetrical motor where only *NHC*s are present (emphasized in green), while for misplaced red PM *AHC*s arise (emphasized in yellow) which also increases the maximal values of T_{cog} (from 390.8 to 399.5 mNm). The orders of *NHC*s correspond to (7), whereas the orders of *AHC*s match with (10). All other ascertainments are the same as for the Table I.

C. Modeling of PM thickness variations

Fig. 7 shows results of cogging torque calculations for the case of rotor asymmetry, when the red PM is thinner from initial thickness for $t_2 = -0.3$ mm. Let us observe crossings of red and blue PM over the stator teeth's edges (Fig. 8). As the red PM is thinner than the blue one, the air-gap region is locally wider, which causes lower amplitude value of T_{cog} in comparison to the crossings of the blue PM. The total number of adjacent PMs' crossings over all edges of stator teeth is 12 (Fig. 7). The period of T_{cog} repeats after four such crossings (120°), which indicates the existence of *AHC* having the order $N_{DHK\,t\,1} = 3$ and its multiples.

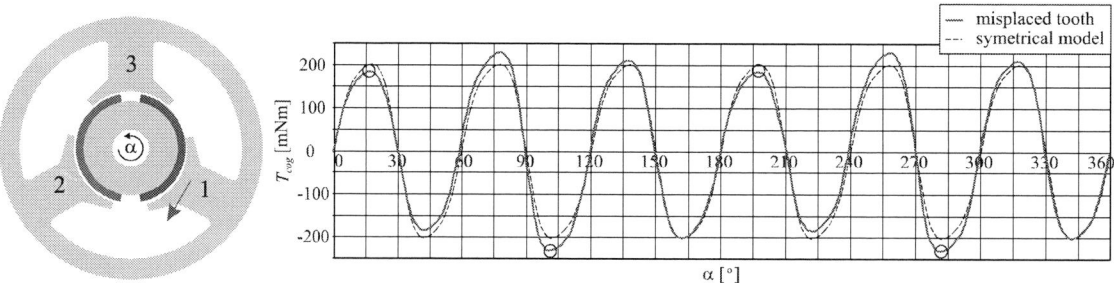

Fig. 3. FEM calculation of T_{cog} for the symmetrical model and for the model having stator tooth #1 misplaced for $\gamma_1 = -1°$, which causes additional harmonic components $AHC_\gamma(A_{AHC\gamma i}, N_{AHC\gamma i}, \varphi_{AHC\gamma i})$.

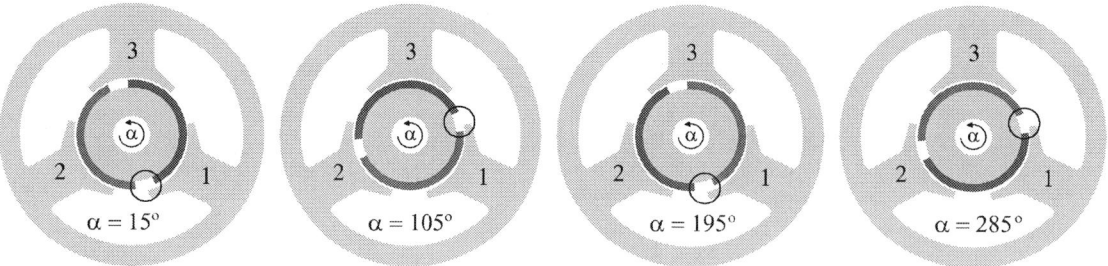

Fig. 4. Basic PMSM model with marked positions of PM's edges crossing the edges of misplaced tooth #1.

TABLE I
FFT ANALYSIS OF CALCULATED T_{cog} FOR THE CASE OF STATOR TOOTH MISPLACEMENT.

Imperfection			NHC			AHC						NHC + AHC			T_{cog} [mNm]
γ_1	γ_2	γ_3	i	1	2	i	1	2	4	5	7	$N_{NHC} = N_{AHC}$			
			$N_{NHC\,i}$	6	12	$N_{AHC\gamma\,i}$	2	4	8	10	14		6	12	
0	0	0	$A_{NHC\,i}$	207.1	15.4	$A_{AHC\gamma\,i}$	0.0	0.0	0.0	0.0	0.0				390.8
			$\varphi_{NHC\,i}$	179.9	−0.5	$\varphi_{AHC\gamma\,i}$	0.0	0.0	0.0	0.0	0.0				
−1	0	0				$A_{AHC\gamma\,i}$	8.0	15.2	12.3	2.9	2.5	$A_{NHC+AHC}$	208.5	16.3	438.8
						$\varphi_{AHC\gamma\,i}$	−35.9	207.9	144.9	44.7	−38.8	$\varphi_{NHC+AHC}$	180.4	7.6	

978-1-4244-8806-3/11 $26.00 © 2011 IEEE

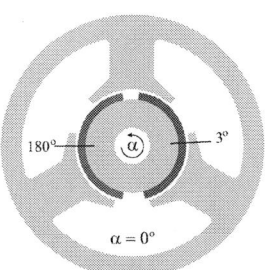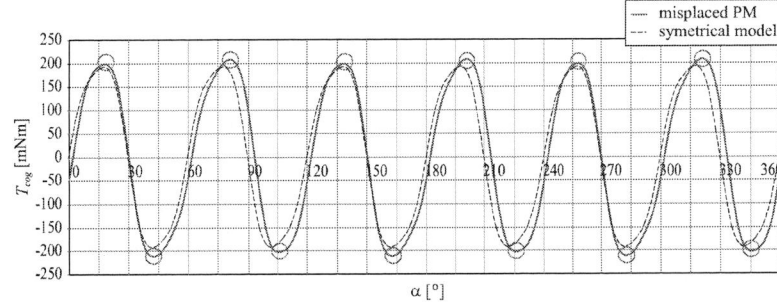

Fig. 5. FEM calculation of T_{cog} for the symmetrical model and for the model having red PM misplaced for $\delta_1 = -3°$, which causes additional harmonic components $AHC_\delta (A_{AHC\delta i}, N_{AHC\delta i}, \varphi_{AHC\delta i})$.

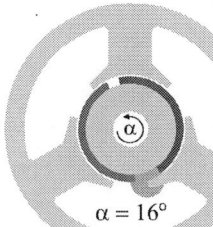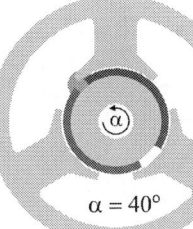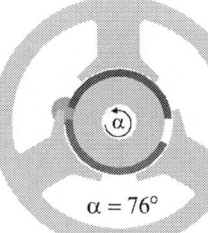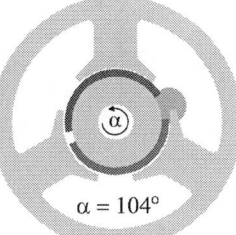

Fig. 6. Basic PMSM model with marked positions of misplaced PM's edges crossing the edges of stator teeth.

TABLE II
FFT ANALYSIS OF CALCULATED T_{cog} FOR THE CASE OF PM MISPLACEMENTS.

Imperfection		NHC			AHC				NHC + AHC			T_{cog} [mNm]
δ_1	δ_2	i	1	2	i	1	3	5	$N_{NHK} = N_{DHK}$			
		N_{NHCi}	6	12	$N_{AHC\delta i}$	3	9	15		6	12	
0	0	A_{NHCi}	207.1	15.4	$A_{AHC\delta i}$	0.0	0.0	0.0				390.8
		φ_{NHCi}	179.9	−0.5	$\varphi_{AHC\delta i}$	0.0	0.0	0.0				
−3	0				$A_{AHC\delta i}$	20.4	21.1	5.1	$A_{NHC+AHC}$	205.2	14.2	399.5
					$\varphi_{AHC\delta i}$	−4.0	166.4	−22.3	$\varphi_{NHC+AHC}$	170.9	−18.8	

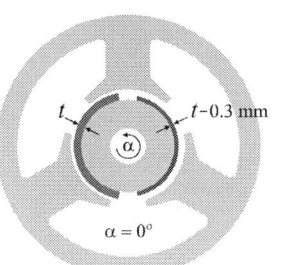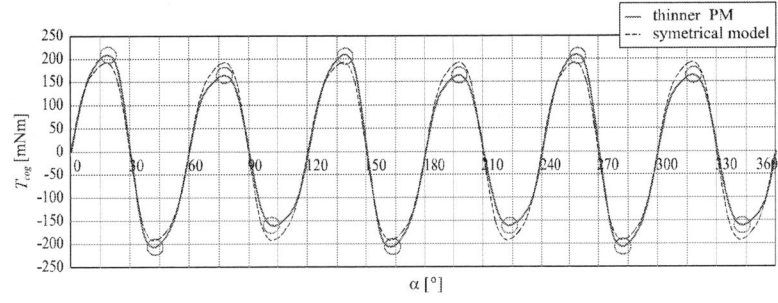

Fig. 7. FEM calculation of T_{cog} for the symmetrical model and for the model having the red PM thinner for 0.3 mm, which causes additional harmonic components $AHC_t (A_{AHCti}, N_{AHCti}, \varphi_{AHCti})$.

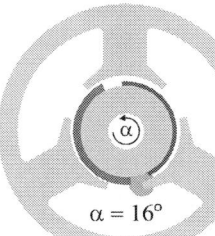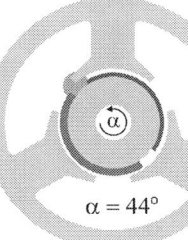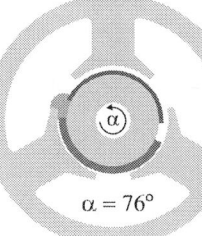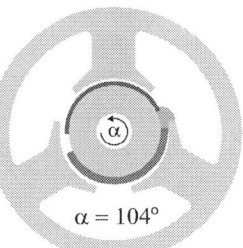

Fig. 8. Basic PMSM model with marked positions of thinner PM's edges crossing the edges of stator teeth.

978-1-4244-8806-3/11 $26.00 © 2011 IEEE

Table III presents FFT results of the T_{cog} for the case of symmetrical motor where only *NHC*s are present (emphasized in green), while for the thinner red PM *AHC*s arise (emphasized in yellow) which change the maximal values of T_{cog}. In this particular case of the thinner PM and consequently lower local magnetization, the T_{cog} decreases (from 390.8 to 369.5 mNm). The orders of *NHC*s correspond to (7), whereas the orders of *AHC*s match with (10).

D. Modeling of PM width variations

Fig. 9 shows results of cogging torque calculations for the case of rotor asymmetry, when the red PM is narrower from initial width for $w_2 = -3$ mm. Because of this difference (Fig. 9), the magnetizations of the red and blue PM do not have the same amplitude and phase angle. The period of the T_{cog} signal is 120°, which indicates the presence of *AHC*s having the order $N_{DHK\,w1} = 3$ and its multiples (Fig. 9). Maximal and minimal values of T_{cog} are due to the changed

TABLE III
FFT ANALYSIS OF CALCULATED T_{cog} FOR THE CASE OF PM THICKNESS VARIATIONS.

Imperfection		NHC			AHC				NHC + AHC			T_{cog} [mNm]
Δt_1	Δt_2	i	1	2	i	1	3	5	$N_{NHC} = N_{AHC}$			
		N_{NHCi}	6	12	N_{AHCti}	3	9	15		6	12	
0.0	0.0	A_{NHCi}	207.1	15.4	A_{AHCti}	0.0	0.0	0.0				390.8
		φ_{NHCi}	179.9	− 0.5	φ_{AHCti}	0.0	0.0	0.0				
0.0	− 0.3				A_{AHCti}	6.2	8.6	0.7	$A_{NHC+AHC}$	185.8	12.3	369.5
					φ_{AHCti}	91.0	− 89.6	93.5	$\varphi_{NHC+AHC}$	179.9	− 0.4	

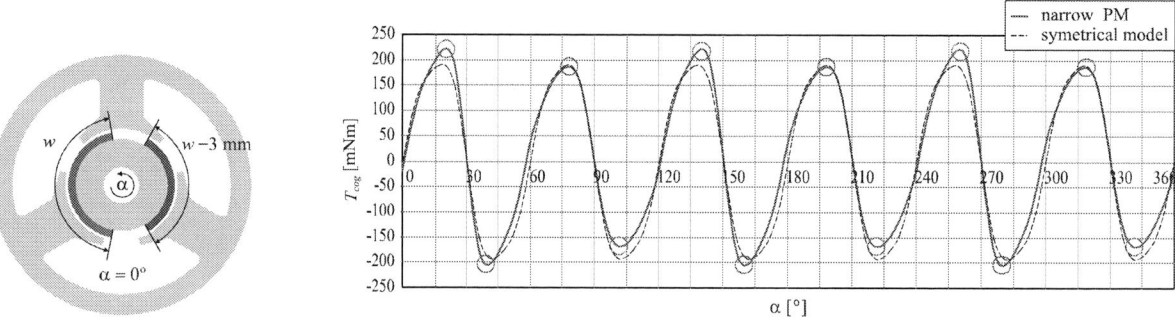

Fig. 9. FEM calculation of T_{cog} for the symmetrical model and for the model having red PM narrower for 3 mm, which causes additional harmonic components $AHC_w (A_{AHC\,wi}, N_{AHC\,wi}, \varphi_{AHC\,wi})$.

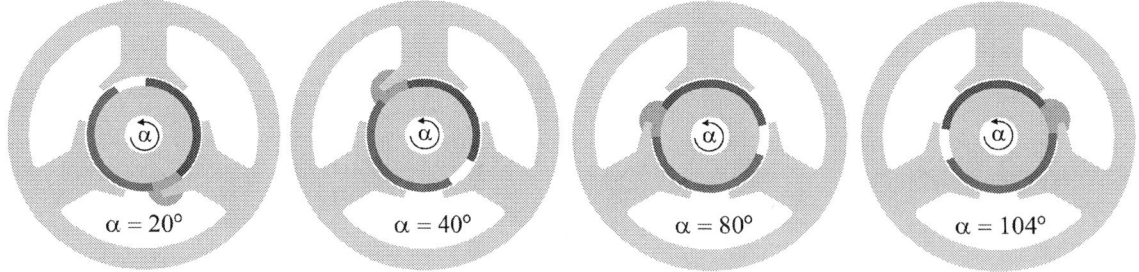

Fig. 10. Basic PMSM model with marked positions of misplaced PM's edges crossing the edges of stator teeth.

TABLE IV
FFT ANALYSIS OF CALCULATED T_{cog} FOR THE CASE OF PM WIDTH VARIATIONS.

Imperfection		NHC			AHC				NHC + AHC			T_{cog} [mNm]
Δw_1	Δw_2	i	1	2	i	1	3	5	$N_{NHC} = N_{AHC}$			
		N_{NHCi}	6	12	N_{AHCwi}	3	9	15		6	12	
0	0	A_{NHCi}	207.1	15.4	A_{AHCwi}	0.0	0.0	0.0				390.8
		φ_{NHCi}	179.9	− 0.5	φ_{AHCwi}	0.0	0.0	0.0				
0	− 3				A_{AHCwi}	1.0	21.4	1.8	$A_{NHC+AHC}$	206.9	32.0	449.0
					φ_{AHCwi}	269.9	89.9	269.4	$\varphi_{NHC+AHC}$	180.1	0.0	

width of the red PM at slightly different angles than in Fig. 7. Table IV presents FFT results of the T_{cog} for the case of symmetrical motor where only *NHC*s are present (emphasized in green), while for narrower red PM *AHC*s arise (emphasized in yellow) which also increases the maximal values of T_{cog} (from 390.8 to 449.0 mNm). The orders of *NHC*s correspond to (7), whereas the orders of *AHC*s match with (10).

IV. CONCLUSIONS

Manufacturing assembly imperfections like stator teeth and rotor PMs misplacements, PMs thickness and width variations cause the phenomena of *AHC*, which are not present in the case of an absolutely symmetrical motor. The paper presents only a few characteristic results of an extensive study based on the simplest possible PMSM design. Such model offers easily understood insight into the changes of T_{cog} and its harmonic components *NHC*s and *AHC*s. Similar analyses can be performed for all other possible imperfections (e.g. asymmetries in stator back-iron due to interlocks, notches and magnetic material inhomogeneity), as well as for any of their mutual combinations. In most cases the presence of such asymmetries increase the T_{cog} but it is also possible that the influence of an individual imprecision compensates with others due to different phase shifts which can result even in the decrease of total T_{cog}. The knowledge gained from a presented simulation PMSM model has proven its applicability when applying such research on real PMSM motor designs [5]-[11]. Considering the introduced theory motor designers are able to predict the entire cogging torque harmonic spectrum for any PMSM in advance, thus predetermine required manufacturing tolerances to minimize cogging torque and fulfil the stringent market demands.

REFERENCES

[1] J. Gieras, M. Wing, *Permanent–Magnet Motor Technology – Design and Applications*, New York, Marcel Dekker, 2002.

[2] N. Bianchi, S. Bolognani, "Design Techniques for Reducing the Cogging Torque in Surface–Mounted PM Motors," *IEEE Transactions on Industry Applications*, vol. 38, no. 5, pp. 1259–1265, Sept/Oct. 2002.

[3] L. Dosiek, P. Pillay, "Cogging Torque Reduction in Permanent Magnet Machines," *IEEE Transactions on Industry Applications*, vol. 43, no. 6, pp. 1565–1571, Nov./Dec. 2007.

[4] S. Islam, S. Mir, T. Sebastian, "Issues in Reducing the Cogging Torque of Mass–Produced Permanent–Magnet Brushless DC Motor," *IEEE Transactions on Industry Applications*, vol. 40, no. 3, pp. 813–820, May/June 2004.

[5] L. Gašparin, A. Černigoj, S. Markič, R. Fišer, "FEM Calculation of Additional Cogging Torque Components in Permanent Magnet Motors," *Proceedings of ISEF 2007 on CD*, Prague, Czech Republic, Sept. 13-15, 2007, pp. 1–4.

[6] L. Gašparin, R. Fišer, "Detection and analysis of additional cogging torque harmonic components in permanent–magnet synchronous motors," *Electrotechnical Review*, vol. 75, no. 3, pp. 129–135, 2008, (in Slovene).

[7] L. Gašparin, R. Fišer, "Detection and FEM Analysis of Additional Cogging Torque Components in PM Motors", *Przegląd Elektrotechniczny*, 84 (2008), No. 12, 163-166.

[8] L. Gašparin, A. Černigoj, S. Markič, R. Fišer, "Additional cogging torque components in permanent magnet motors due to manufacturing imperfections," *IEEE Transactions on Magnetics*, vol. 45, no. 3, pp. 1210–1213, March 2009.

[9] L. Gašparin, "Detection and Analysis of Additional Cogging Torque Harmonic Components in Permanent-magnet Synchronous Motors", *Ph.D dissertation*, (2009), Fac. Electr. Eng., Univ. Ljubljana, (in Slovene).

[10] L. Gašparin, A. Černigoj, S. Markič, R. Fišer, "A Sensitivity of Permanent-Magnet Motors to the Phenomena of Additional Cogging Torque Components due to Assembly Tolerances," *Proceedings of CEFC 2008*, Athens, Greece, May 12-15, 2007, pp. 325.

[11] L. Gašparin, R. Fišer, "Intensity of the Native and Additional Harmonic Components in Cogging Torque Due to Design Parameters of Permanent-Magnet Motors," *Proceedings of PEDS 2009 on CD*, Taipei, Taiwan, Nov. 2-5, 2009, pp. 1–6.

Comparison study of full-bridge and reduced switch count three-phase voltage source inverters

Roman Grinberg, Francisco Canales
ABB Corporate Research
Segelhofstrasse 1K
Baden-Dättwil, Switzerland
roman.grinberg@ch.abb.com, francisco.canales@ch.abb.com

Mikko Paakkinen
ABB Oy Drives
Hiomotie 13
Helsinki, Finland
mikko.paakkinen@fi.abb.com

Abstract – **Low voltage variable speed drives with 1-phase input are very much driven by cost considerations. In this respect, 4-switch 3-phase inverter topology was put under consideration. In this paper, review of pros and cons of 4-switch 3-phase inverter for motor drive is presented and practical case is shown. Results show that significantly higher DC link capacitance, higher semiconductor losses and active front-end requirement challenge application of 4-switch inverter.**

I. INTRODUCTION

Standard topology with diode rectifier, DC-link capacitors and three-phase 6-switch (B6) voltage-source inverter (VSI) for single-phase input used in low power drives is presented in Fig.1. In Fig.2 three-phase AC motor drive based on reduced switch count inverter is presented [1-4]. The inverter has 4 switches only (B4 inverter). It was shown [1-3] that in comparison with B6 two times higher DC-link voltage is required in B4 to power the same motor. Considering the same net and motor voltage, voltage-doubler is a passive front-end solution for B4 to achieve required DC link voltage as shown in Fig.2. Otherwise, active front-end is required.

In the literature [1-3], the following pros and cons of B4 were identified:

+ 2 switches&2gate drives out, so 'less' cost and corresponding losses;

- passive voltage-doubler rectifier or active front-end to obtain required DC-link voltage;

- higher voltage rating of the switches (twice);

- capacitor series connection is needed to have neutral point;

- higher capacitor current rating (fundamental motor current component flows through the DC-link);

- higher switching losses in B4 switch than in B6 one due to the increased switch voltage rating ;

- inverter output is more sensitive to modulation non-linearities (blanking time, etc.);

- in case of active front-end, neutral point control is required (two DC-link voltage sensors instead of one in B6);

- peak common mode voltage for B4 is 2Udc/3, whereas for B6 it is Udc/2.

In this study, comparison of drives based on B4 and B6 is presented for the same power rating considering practical case. The idea is to quantify advantages and disadvantages of B4, including ones that are specific for power rating considered.

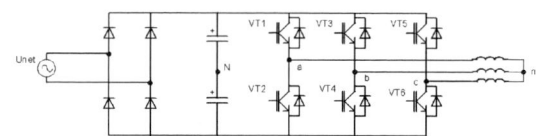

Fig.1. Standard topology for 3phase low voltage low power drive with 1-phase input

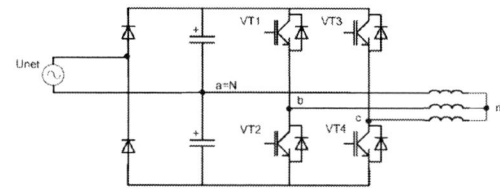

Fig.2. 4-switch topology with passive front-end [1-3]

II. COMPARISON SPECIFICATION & TOOLS

A. Specification
Input: Single-phase, 200-240V, +10%, -15%
Output: 1.4kVA
Switching frequency range: 4-16kHz
Output frequency range: <120Hz

B. Tools

In order to compute losses, harmonic distortion, etc. models of B6 and B4-based drives were developed in MatLab. In the model carrier-based space vector modulation is used for B6-inverter. For B4 – carrier based symmetrical modulation is used [5]. Inverter control is open-loop, where output voltage and frequency references are given before simulation. Modulator is carrier-based. Deadtime is not included in the model, since, typically, deadtime compensation techniques are used in commercial drives. Gate drives are not considered and their losses are neglected.

978-1-4244-8806-3/11 $26.00 © 2011 IEEE

To estimate the impact of the inverter type on the motor, induction motor model (M2QA 80M4B, 750W, 400V, 50Hz) is considered. Simple RL-model is used for input power source impedance corresponding to 100kVA transformer.

III. COMPARISON RESULTS

A. Dc link capacitance limitations – motor flux and torque ripple

First, B6 and B4 inverters operation with stiff DC link was analyzed. Continious time modelling was used in MatLab. Operating point: output voltage – 200V, output frequency 40Hz, nominal torque. Switching frequency – 4KHz. Constant voltage sources were used in DC link model. Full Dc link voltage: 325V – B6, 750V – B4. Simulation results showed that peak to peak high-frequency stator flux and torque ripple is about twice lower for B6. It is due to twice higher DC link voltage and leg-specific modulation pattern of output voltage. One can see that for line-line voltage BA and CA (see Fig. 3) modulation is 'bipolar', switching between +Udc/2 to –Udc/2. For line-line voltage BC, voltage is switching between +Udc, 0 and –Udc. However, due to machine inertia it is low frequency flux and torque ripple that is of interest. Therefore, in the following, B6 and B4-based drive operation with passive front-end is presented.

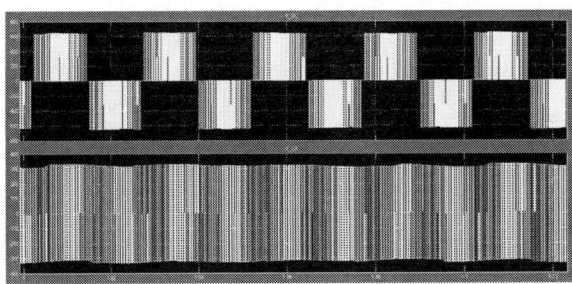

Figure 3. B4 inverter output voltages line-line BC and CA (non-stiff DC link). Axises X/Y – s/V

Operating point is the same as in the case above. No ripple compensation is used. Half DC link capacitance is chosen 2.3mF to reduce voltage ripple. In Fig.4 output currents for B4-based drive are shown. One can see that output currents in the phases have different amplitudes (phase C). It is due to uncompensated DC link voltage ripple effect. DC link voltage ripple causes low frequency ripple in stator flux and machine torque. DC link voltage ripple reduction can be achieved further increasing DC link capacitance. It is half DC link voltage ripple that is of concern for B4. It affects two line-line voltages that are between phase and neutral point. Therefore, DC link capacitance limitations for B4 were analyzed. From [5], dc-link voltage ripple amplitude that is due to the inverter operation is:

$$U_{C\_r\_inv} = \frac{1}{C} \frac{\sqrt{2}I_{rms}}{2\omega_o}, \quad (1)$$

where ω_0- output frequency, I_{rms} – rms current through the capacitor.

In Fig.5 relationship between inverter output frequency, DC link capacitance C and DC link voltage ripple amplitude $U_{C\_r\_inv}$ is presented. Output current rms value is 4A. Inverter output frequency range is 5-125Hz.

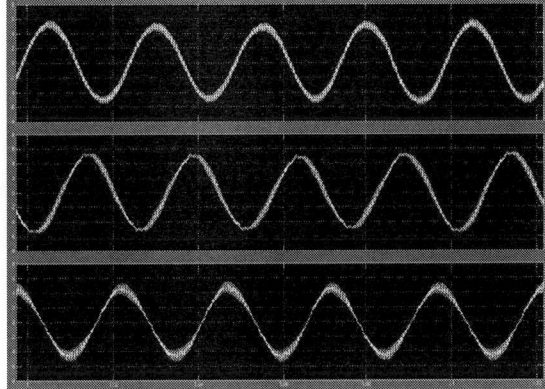

Figure 4. B4 inverter output currents: from top to bottom – phase A, B,C. Axises X/Y – s/A

For example, from Fig.5, half DC-link capacitance of 2mF is needed to have half DC link peak-peak ripple <100V in the frequency range 5-125Hz.

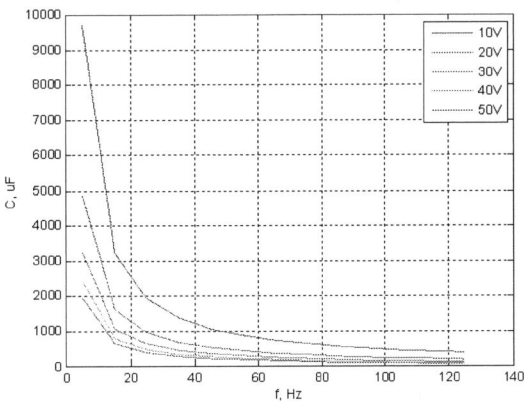

Figure 5. DC link capacitance C vs inverter output frequency for 4 A output rms current

DC link voltage ripple causes low frequency ripple in stator flux. Deviations in stator flux amplitude mainly cause radial forces between stator and rotor, which in turn may result in mechanical vibration and increase in acoustic noise [6]. Deviations in stator flux angle cause severe ripple in the electromagnetic torque [6]. In B6, DC link voltage ripple affects all three phases proportionally, whereas in B4 there is no such symmetry. Therefore, one can expect higher stator flux and torque ripple in case of B4 utilization. Compensation of DC link ripple in B6 is easy and can be done in carrier-based

978-1-4244-8806-3/11 $26.00 © 2011 IEEE

implementation. In B4, DC link ripple compensation technique requires on-line calculation of the switching times using measured half DC link voltages as shown in 0. Carrier-based implementation with DC link ripple compensation is not possible in B4. On-line calculation of the switching times requires DSP processor which leads to the higher cost of the control hardware in this very cost-sensitive drive segment.

Compensation of the DC link ripple can also be done, if the active front-end is used. Frequency of the DC link ripple imposed by B4 varies in AC drive. Input power delivered to the DC link has a fundamental frequency of 100Hz (50Hz input current and voltage). For effective ripple compensation with an active front-end, frequency of the compensated ripple should be well below 100Hz. This means a limitation on output frequency range of the drive.

In B6-based drive with passive front-end DC link voltage ripple depends on the load power. For the considered power rating for typical design, maximum peak-peak ripple is <50V.

In the following, it is assumed that DC link capacitor voltage ripple compensation in the inverter (adaptive modulation index) have the same performance in B4 and B6. Another assumption is made that DC link capacitor voltage peak-to-peak ripple in B6 and B4 produces the same effect on the stator flux and electromagnetic torque of the motor. Two cases are considered: constant torque loads (conveyor) and variable-torque loads (pump, fan). These loads present different requirements for B4 capacitance. Only inverter ripple imposed is considered.

1) Constant torque:

- With the constant torque load, drive is specified to deliver nominal current in the whole output frequency range. Since maximum ripple exists at the lowest frequency, realistic frequency of 5Hz is taken. Then, according to (1) 4mF capacitance (half DC link) is required to have <50V peak-peak ripple for inverter output frequencies >5Hz. Since two of these capacitors required for series connection, the total capacitance is as much as order of magnitude higher comparing to typical capacitance values used in B6 drives.

2) Variable torque:

- Pump and fans are variable torque loads with quadratic relation between torque and motor speed. So, inverter output current is proportional to the square of the output frequency. Then, according to (1), maximum ripple is at the nominal output frequency (nominal output current). From (3.1.5), around 400µF capacitance (half DC link) is required to have maximum <50V peak-peak ripple. As a result total capacitance is comparable to typical values used in the drives of that power rating.

Considering front-end, worst-case capacitor ripple is a sum of the peak-peak ripples imposed by front-end and B4. As demonstrated above, passive front-end utilization, results in prohibitively large DC link capacitance. If active front-end is utilized, DC link voltage ripple value can be controlled. In this case DC link capacitance reduction depends on compensation bandwidth. For effective ripple compensation with an active front-end, frequency of the compensated ripple should be well below input power frequency of 100Hz [8]. Based on that, 20Hz compensation limit is assumed. Then, applying (1) for constant torque loads around 1mF capacitance (half DC link) limit is imposed by B4.

Summarizing, in B4 inverter DC link capacitors are loaded with inverter output current fundamental component. Resulting DC link voltage ripple produces undesirable low frequency stator flux and torque harmonics. Variable speed operation requires significant DC link capacitance to limit the ripple. DC link ripple compensation on the inverter side requires on-line calculation of switching times. The latter means faster and more expensive control hardware. Analysis and simulation show that active front-end is necessary to control DC link voltage ripple, having realistic DC link capacitance. Considering voltage ripple imposed by B4 only, 2*1mF DC link capacitors required to meet DC link ripple criteria. In this case, drive can handle both constant torque and variable torque loads. Considering only variable torque loads, B4 needs at least 2*480 µF capacitors. As a result, significant DC link capacitance and active front-end requirement challenge B4 feasibility for cost-sensitive application.

B. Semiconductor losses

Comparison of semiconductor losses was done for B4 and B6 inverters. Devices used: B6 – 600V IGBT IRGB10B60KD (IRF), B4 – 1.2kV IGBT SKW15N120 (Infineon). Switching frequency range: 4 to 16KHz with 4kHz step. DC link voltage: B6: $U_{dc} = 230\sqrt{2}$ V, B4: $U_{dc} = 2*230\sqrt{2}$ V. DC link voltage ripple influence on inverter losses was neglected. Theoretical maximum modulation index values: B6 - 1.154, B4 -1. Representative operating point considered for loss calculation: B6 mod index – 1 ($U_{out l-l\, rms} = 200$ V); B4 mod index – 0.87 ($U_{out l-l\, rms} = 200$ V). Inverter output frequency – 40Hz.

For conduction and switching losses calculation was done using standard techniques based on IGBT datasheet parameters [9]. It should be noted that in B4 inverter, losses in the legs are not the same. The legs in B4 have different current distribution between IGBT and FWD. Therefore, losses were calculated separately for each phase leg and then summed up. Calculation results are shown in Fig.6. From Fig.6 total device losses in B6 are the same as in B4 at 4kHz. However, at higher switching frequencies device losses in B6 are lower than in B4. For

978-1-4244-8806-3/11 $26.00 © 2011 IEEE

16kHz, the difference in the total device losses is around 9W.

Additionally, calculation of losses at light load and low speed was also done. Operating point was as follows: 20% torque, inverter output frequency – 5Hz; mod.index – 0.17 (B6) and 0.15(B4); output speed n – 135rpm; output current, rms – 3.2A; power factor – 0.62.

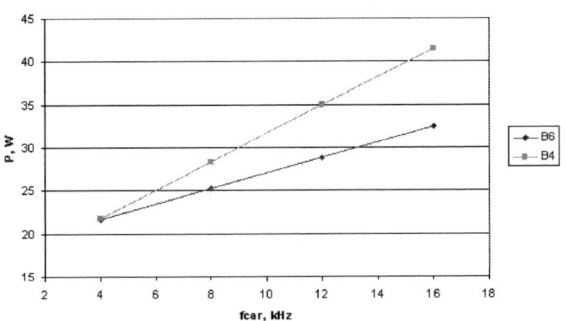

Total device losses vs switching frequency

Figure 6. Inverter losses vs switching frequency for B4 and B6 ($U_{out1-1rms} = 200$ V, $f_{inv} = 40$ Hz)

Operating point was calculated using inductor motor steady-state equations and given motor parameters. Switching frequency – 4kHz. Loss calculation results shows 12.4W losses for B6 and 13.1W for B4. One can see that for the light load and low speed conditions losses in B4 are higher than in B6.

Concluding, having less number of devices, B4 has the same total device losses as B6 at $f_{sw} = 4$kHz at the operating point close to maximum modulation index. However, at higher switching frequencies B6 inverter has lower total device losses. The difference increases with switching frequency to significant 9W at 16kHz. At light load and low speed conditions at low switching frequency, losses of B6 inverter lower than for B4 inverter.

C. DC link capacitor selection issues

DC link capacitor selection is based on voltage and current rating (losses). If capacitor current rating is not enough, larger capacitance value is typically taken.

Capacitor current defines capacitor lifetime. Capacitor current rating is rather complex matter since ESR is dependent on frequency and temperature. B6 inverter contribution in the passive front-end drive is negligible. It is rectifier DC link current component that defines capacitor losses and, consequently, lifetime in VSI. In B4 inverter output current at fundamental frequency flows through DC link resulting in additional capacitor losses. At light load low speed operation, DC link losses in B4 are still present due to motor magnetizing current.

Since capacitor ESR decreases with frequency, the highest capacitor losses are at the lowest inverter output frequency. In Fig.7 typical ESR vs frequency for EPCOS B43504 are presented [10]. Capacitor manufacturers provide methods to calculate capacitor lifetime which are based on empirical data [10]. There are analytical models of the capacitor impedance and ESR based on experimental data fit [11].

However, capacitor manufacturers typically do not provide ESR and ripple current data at frequencies lower than 50Hz-100Hz as shown in Fig.7.

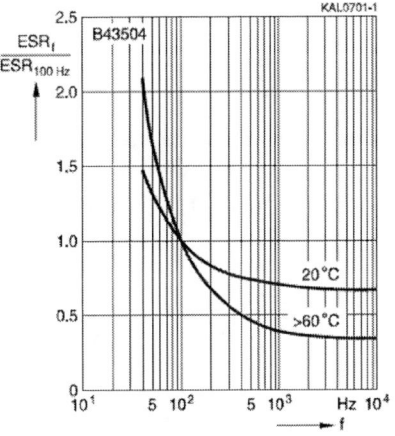

Figure 7. Typical ESR vs frequency for EPCOS B43504 0

Example estimation of B4 inverter contribution was done assuming ESR<40Hz is constant and equal ESR$_{40Hz}$. For B6 – two 330uF B43504 capacitors in parallel; B4 – same capacitors in series. Contribution of front-end was neglected. Capacitor current spectrum was obtained and losses for each harmonic were individually calculated (ambient temperature - 60 degrees). For full load conditions part A, resulting total capacitor losses were 1.2W for B4 and negligible for B6. For light load conditions part A, B4 – 0.5W, B6 – negligible. Capacitor losses were not dependent on switching frequency being dominated by fundamental output current component. From the calculations, significant contribution to capacitor losses in B4 topology can be seen.

For the capacitor operation in B4 another consideration can be important. At very low inverter output frequencies (close to zero), DC link capacitor provide high impedance. As a result, net current may contain harmonic at inverter output frequency. The magnitude of this harmonic depends on the relationship between DC link capacitance and net/rectifier equivalent impedance.

D. Output voltage harmonic distortion

In this comparison output voltage harmonic distortion is assessed. Unlike B6, B4 output voltage

978-1-4244-8806-3/11 $26.00 © 2011 IEEE   273

harmonic distortion is not the same for all the phases. Harmonic distortion of output line-line voltages that are between phase legs and the neutral point (U_{ab}, U_{ac}) is expected to be higher than the one between the phase legs (U_{bc}).

Comparison conditions are the same as in Part B. The criteria used for harmonic distortion analysis are as follows:

- THD$_U$ (total harmonic distortion)

$$THD_U = \sqrt{\sum_{i=2}^{n} U_i^2} / U_1 \qquad (2)$$

- WTHD$_U$ (weighted total harmonic distortion).

$$WTHD_U = \sqrt{\sum_{i=2}^{n} \left(\frac{U_i}{i}\right)^2} / U_1 \qquad (3)$$

WTHD$_U$ takes into account inductive nature of the motor load, where higher frequency voltage harmonics are better damped than low frequency ones. Voltage spectrums required to calculate THD$_U$ and WTHD$_U$ are obtained using Fourier transformation (5us resolution, f_{max}=100kHz). Results are presented in Fig.8.

Switching frequency, kHz		4	8	12	16
THD$_U$, %	B6	68	68	68	69
	B4 (U_{ab}/U_{bc})	128/140	128/140	128/140	128/140
WTHD$_U$,%	B6	0.44	0.46	0.23	0.43
	B4	0.95	1.14	0.53	0.95

Figure 8. B4 and B6 harmonic distortion indices vs switching frequency ($U_{out\,l-l\,rms}$ = 200 V, f_{inv} = 40 Hz)

In B6 inverter, THD$_U$ is independent on switching frequency and equal to around 68%. In B4 inverter, THD$_U$ is also independent on switching frequency and equal to around 128% for U_{ab} and 140% U_{bc}. THD$_U$ is different for U_{bc} then for U_{ab} (U_{ca}) since only two phases are modulated. According to the simulation results, WTHD$_U$ in B6 is not always lower as the switching frequency increases. It is due to the fact that, in reality, switching frequencies are not multiples of three or odd multiples of inverter output frequency. Simulation shows the same pattern for WTHD$_U$. It should be noted that unlike THD$_U$, WTHD$_U$ is the same for U_{ab} and U_{bc} in B4.

At lower modulation index, THD$_U$ increases as shown in Fig.8. It is because fundamental component amplitude decreases. However, relationship between THD$_U$ for B4 and B6 in principal stays the same. From Fig.8 B4 WTHD$_U$ also maintains the same pattern in comparison with B6. In Fig.9 harmonic distortion for field weakening case is presented. The results for f_{inv}=120Hz indicate that B6 has twice lower THD$_U$ and WTHD$_U$ then B4.

Summarizing, comparing harmonic distortion indices, one can see that B6 has significantly lower THD$_U$ than B4 in the whole switching frequency range and modulation index range. WTHD$_U$ of B6 is also significantly lower in the switching frequency range for modulation index value close to maximum. At lower modulation index WTHD$_U$ of B4 is also higher or at least the same as B6 (the results are obtained up to f_{inv}=20Hz). In field weakening range (f_{inv}=120Hz) B6 again shows significant advantage. Increased harmonic distortion in B4-based drive will result in higher losses in the motor increased output filter requirements in comparison with B6.

$U_{out\,l-l\,rms}$, V		200 f_{inv}=40Hz	160 f_{inv}=30Hz	120 f_{inv}=20Hz	200 (field weakening f_{inv}=120Hz)
THD$_U$, %	B6	68	92	125	68
	B4 (U_{ab}/U_{bc})	128/140	176/164	249/199	128/139
WTHD$_U$,%	B6	0.44	0.47	0.81	1.2
	B4	0.95	0.84	0.8	2.6

Figure 9. B4 and B6 harmonic distortion indices vs inverter output voltage (f_{sw} = 4 kHz)

E. Common-mode voltage

In Fig.10 and Fig.11 simulated waveforms of common mode voltage and their spectrums are presented. In the simulation model constant voltage sources are used in the DC link, switching frequency - 4kHz, $U_{out\,l-l\,rms} = 200$ V, $f_{inv} = 40$ Hz. B6 DC link voltage (B4 half DC link voltage) – $U_{dc} = 325$V.

In Fig.10, B6 common mode voltage is shown (DC link negative rail to load neutral). One can clearly see $3f_{inv}$ component in the spectrum due to the nature of space vector modulation (zero sequence injection). Dominant modulation harmonic component is at switching frequency. It equals around 60Vrms.

As discussed in introduction, peak common mode voltage for B4 is $2U_{dc}/3$, whereas for B6 it is only $U_{dc}/2$. In fact, common mode voltage in B4 is output voltage of phase A. That is why, spectrum in Fig.11 contains harmonic component at output fundamental frequency. Dominant modulation harmonic component is at switching frequency. It equals around 110Vrms which is almost twice higher than in B6.

In B6 inverter, common mode rms voltage increases as with decreasing fundamental output voltage amplitude. For V/f control theoretically it means that the largest common mode rms voltage is at zero output frequency. At zero output fundamental frequency, B6 common mode voltage is a rectangular waveform at f_{sw} with duty ratio of 0.5 and $U_{dc}/2$ amplitude. Since B4 common mode voltage is an output phase voltage, its pulses length can be up to switching period value in linear modulation range (as shown for example in Fig.11).

In case of overmodulation, common mode voltage in B4 is unipolar with pulse length that is longer than switching period. Therefore, common mode voltage filter inductor size will be bigger in B4. Also, high dV/dt in B-C line-to-line voltage should be pointed out. It is twice higher than in B6. High dV/dt elevates high-frequency ringing problems when long-cabling is used.

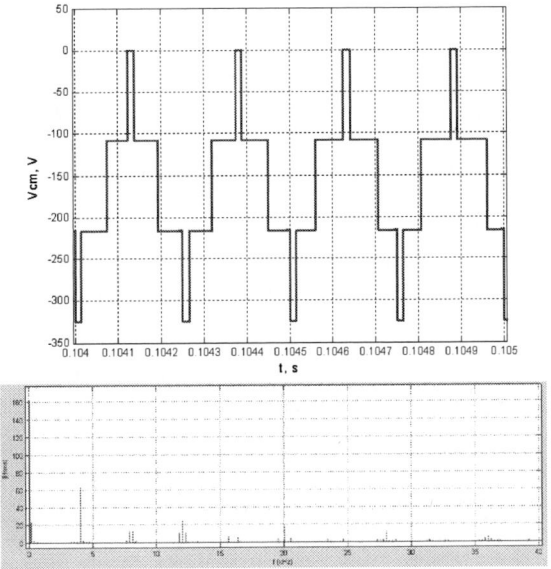

Fig.10. B6 common mode voltage (negative rail to load neutral) and its spectrum

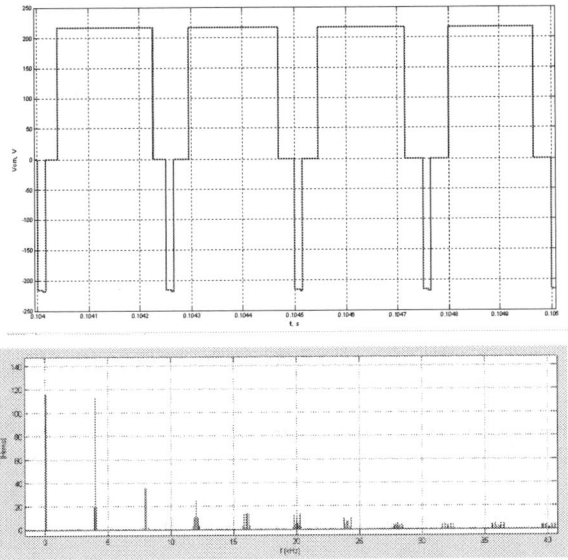

Fig.11. B4 common mode voltage and its spectrum

CONCLUSION:

In this paper comparative analysis of 4-switch voltage-source inverter (B4) advantages/disadvantages for low voltage drive application with single phase input was presented. Analysis and simulation showed that low frequency current loading of the capacitor in B4 results in significant DC link capacitance required. It was also shown that B4 DC link capacitance requirement depends on the type of motor load (inverter output current/frequency profile). Variable torque load results in significantly lower DC link capacitance required for B4 than constant torque load. Finally, to achieve realistic DC link capacitance active front-end is needed. In addition, calculation showed that reduced switch count does not result in reduced inverter device losses comparing. Harmonic distortion analysis showed significantly higher output voltage distortion for B4. Weighted harmonic distortion analysis indicate higher motor losses for B4 inverter operating at modulation index >0.5. Also, elevation of EMI problems as well as EMI inductor size and design issues is predicted due to the nature of common-mode voltage and different voltage spectrums in 4-switch 3phase inverter. Finally, all above mentioned shortcomings challenge application of 4-switch inverter concept for low voltage drives.

REFERENCES

[1] P.Enjeti, A.Rahman. A New Single-Phase to Three-Phase Converter with Active Input Current Shaping for Low Cost AC Motor Drives. IEEE transactions on IA, vol.29, N.4, July/August, 1993.pp.806-813

[2] G.L. Covic, G.L. Peters, J.T. Boys. An Improved Single Phase to Three Phase Converter for Low Cost AC Motor Drives. Power Electronics and Drives Systems Conference, 1995.pp.549-554

[3] C.B. Jacobina, M.B.R. Correa, A.M.N. Lima, E.R.C. da Silva AC Motor Drive Systems with a Reduced-Switch-Count Converter . IEEE Industry Applications Transactions, vol.39, N.5, Sep/Oct 2003.pp.1333-1342

[4] C.B. Jacobina, I.S. de Freitas, C.R. da Silva, M.B.R. Correa, E.R.C. da Silva. Reduced Switch-Count Six-Phase AC Motor Drive Systems Without Input Reactor. IEEE Industrial Electronics Transactions, vol.55, N.5, May 2008.pp.2024-2032

[5] S. Hansen, H.H. Hansen, S. Freysson, F. Blaabjerg. A New Optimized Space-Vector Modulation Strategy for a Component-Minimized Voltage Source Inverter. IEEE Power Electronics Transactions, vol.12, N.4, July 1997.pp.704-714.

[6] P. Thogersen, J.K. Pedersen. Stator Flux Oriented Asynchronous Vector Modulation For AC Drives. PESC'90, 1990.pp.641-648

[7] F.Blaabjerg, D. Neascu, J.K.Pedersen. Adaptive SVM to Compensate DC link Voltage Ripple for Four-Switch Three-Phase Voltage-Source Inverters. IEEE transactions on Power Electronics, vol.14, N.4, July, 1999.pp.743-752.

[8] J.Sander-Larsen, K.Jespersen,M.R.Pedersen, F.Blaabjerg, J.K.Pedersen. Control of a Complete Digital-based Component-minimized Single-phase to Three-phase AC/DC/AC converter. IECON, 1998.pp.618-625

[9] D. Graovac, M. Pürschel "IGBT Power Losses Calculation Using the Data-Sheet Parameters", *Infineon Apllication Note,* January 2009

[10]Aluminum electrolytic capacitors catalogue. Epcos. Edition 01/2007.pp.583

[11] M.L.Gasperi. A Method for Predicting the Expected Life of Bus Capacitors. IEEE IAS Conference'1997 proceedings, 1997

978-1-4244-8806-3/11 $26.00 © 2011 IEEE

Advances in CSI-fed Induction Motor Drives

Marek Adamowicz[1,2], Marcin Morawiec[2]

[1] Gdynia Maritime University
Morska Str. 81-87
Gdynia, POLAND
madamowi@am.gdynia.pl

[2] Gdansk University of Technology
Narutowicza Str. 11/12
Gdansk, POLAND
m.morawiec@ely.pg.gda.pl

Abstract- The application of current source inverters (CSI) in induction motor (IM) drives offers a number of advantages, including: voltage boosting capability, natural shoot-through short-circuit protection and generation of sinusoidal voltages. The exact generation of sinusoidal voltages is possible thanks to thanks to the effect of the output ac filter capacitors and may improve the accuracy of sensorless IM control. The elimination of high *dv/dt* in CSI fed IM drives provides also a noticeable electromagnetic interference (EMI) reduction. The paper presents recent achievements of current source inverters (CSI) which are the use of new genertion silicon carbide (SiC) devices (*normally-off* SiC JFETS and SiC Schottky diodes) and the application of modern nonlinear control which may considerably improve the performance of CSI fed IM drives.

I. INTRODUCTION

Among all types of ac machines the squirrel cage induction motors are most commonly used in the low and medium power (<160kW) variable-speed drives [1].

The features of induction motor drives that have established their position in different industrial sectors during last decades are:

- wide variable speed range operation,
- excellent dynamics and robustness,
- speed and flux sensorless operation including zero-speed and field weakening,
- decoupled speed - torque and field control,
- efficiency-optimized control,
- low cost.

Squirrel cage induction motors are simple in construction, economical, rugged, reliable, and are available in a wide power range [1]. Besides traditional industrial drives, i.e. pumps, fans, cranes, automation drives, at present the induction motors are used in hybrid electric vehicles (HEV) and pure electric vehicle (EV) [2] providing high performance of torque control. In the field of wind energy systems, for the last decade, the squirrel cage induction machines are increasingly employed in multi-stage geared small wind generation systems (<100kW) with back-to-back converters connected to grid.

Generally, the two groups of power converter topologies can be applied to variable speed IM drives: the voltage source inverters (VSI) and the current source inverters (CSI). For many years the VSIs were more often used than CSIs for the sake of better properties. Comparing to VSI, the CSI requires a DC current source and devices with symmetric blocking capability. The low switching frequency CSIs using thyristors and bulky DC link inductors were widespread mainly in the 80's of the last century. Large content of harmonics in stator currents as well as oscillations of torque on the motor shaft caused disappearance of development of these inverters. In recent decades the use of CSIs was limited to high power applications with characterizing switching frequencies below 1kHz [2], [3].

In the low and medium power applications the VSIs have been well established thanks to the use of fast-switching IGBTs with simple driver circuits controlled from the low cost DSP based control cards.

However, recently an attention has been paid to the VSIs drawbacks, particularly to their high *dv/dt*, resulting from the voltage pulse width modulation (PWM). High *dv/dt* has negative impact on the motor insulation an causes high frequency (> 1MHz) high magnitude current oscillations on parasitic capacitances and electromagnetic noise of the drive system. The IM drives with the long cable connections need therefore large and costly sinusoidal output filters. Similarly, sinusoidal output filters are needed in wind generation systems characterizing high towers and therefore also long cable connections. The generation of the excessive electromagnetic interference (EMI) is unacceptable in some kinds of sensitive systems. The impact of EMI effects coming from VSI voltage modulation on human being is expected not to be neutral and can not be ignored in applications like electric vehicles (EVs) and hybrid electric vehicles (HEVs).

Another disadvantage of the VSIs applied in the high power density, high temperature applications (like HEV and EV) is deteriorated performance of the bulky DC link capacitor with the temperature.

The application of high switching frequency (>20 kHz) silicon (Si) power transistors (IGBTs and RB-IGBTs) [4] - [7] and, recently, the application of high switching frequency(>100kHz) silicon carbide (SiC) devices (SiC JFETs and SiC diodes) [8]-[10] has an enormous influence on applications of systems based on CSI and creates new possibilities.

With the increase of switching frequency the volume of DC link inductors significantly decreases. Moreover the volume of the output filter capacitors decreases taking the fractional value of the VSI DC link capacitor. The SiC devices offer the possibility of a high blocking voltage and a low *on-state* resistance, which is the key parameter of the CSI.

978-1-4244-8806-3/11 $26.00 © 2011 IEEE

The recently proposed control methods of the CSI fed induction motor drives, e.g. the improved voltage control technique using the bang-bang controllers [5] and nonlinear voltage control taking advantages from linearization of multiscalar model of IM and decoupled control of DC-link input voltage e_d and stator current angular frequency ω_i [6], [7], [11], may make the CSI fed IM drives the potential replacement for the VSI fed IM drives. The CSI fed IM drives take also the advantages of sensorless control similarly to VSI fed variable speed drives.

The aim of the paper is to present and discuss the recent achievements of current source inverters (CSI) concerning the topics mentioned above.

II. HIGH SWITCHING FREQUENCY RB-IGBT-BASED CSI

The general circuit schematic of current source converter (CSC), in this case the RB-IGBT-based topology, is presented in Fig. 1. Such configuration provides bidirectional transmission of electrical energy between the grid and the IM drive with the possibility of an unity power factor on the grid side.

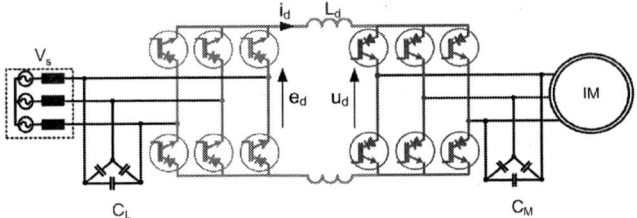

Fig. 1. The general circuit schematic of the current source converter fed IM drive.

Overall CSC topology comprises machine side CSI (on the right) generating the output current vector and connected to the IM terminals and a grid side Current Source Rectifier (CSR) which generates a DC voltage e_d to supply a DC-link circuit. The DC-link voltage e_d causes the DC-link current i_d flow:

$$e_d = i_d R_d + L_d \frac{di_d}{d\tau} + u_d, \qquad (1)$$

where u_d is the CSI input voltage (Fig. 1), R_d is the inductor resistance, L_d is the DC-link inductance.

The control strategies of high switching frequency CSIs (>3kHz) mainly utilize the changes of DC-link current magnitude, while the modulation index is maintained constant at its maximum value. Although the control of modulation index could guarantee higher dynamics of IM torque the operation with constant high DC-link current magnitude would be a reason of the high CSI losses.

The DC-link inductance L_d may be calculated from (1). Two criteria are taken into account in the L_d calculation algorithm:

- the minimization of currents ripples in the DC-link,
- the minimization of size and weight.

The ripple factor w_i detrmined for DC-link current can be written as [6]:

$$w_i = \frac{\Delta i_{d\max}}{i_d}, \qquad (2)$$

where:

$$\Delta i_{d\max} = \max(i_d(t_1) - i_d(t_2)), \qquad (3)$$

$i_d(t)$ is an average value of a DC-link current during one switching period.

The current ripples has an influence on the CSI output currents and on the whole commutation process. Therefore the stator current total harmonic distortion THD_i and stator voltage total harmonic distortion THD_u must be also taken into account in the numerical process of the L_d calculation. The numerical algorithm for selection of the inductor L_d and capacitors C_M has been demonstrated in [6]. The proper selection of of the DC-link inductance L_d and output capacitors C_M ensures the high performance of the CSI-based drive system and guarantee both sinusoidal output currents and small THD factor.

The six active vectors I_1-I_6 and three passive vectors I_7-I_9 can be distinguish in the CSI space vector modulation algorithm. These vectors are shown in Fig. 2.

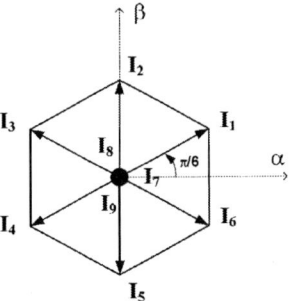

Fig. 2. CSI PWM space vectors in $\alpha\beta$ reference frame.

The typical waveform of PWM CSI current is shown in Fig. 3. The sinusoidal current at IM terminals is obtained thanks to the filtering action of the capacitor bank C_M connected at the CSI output.

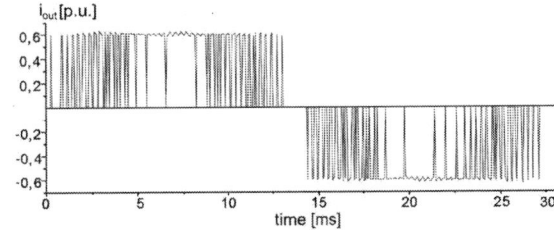

Fig. 3. Typical waveform of CSI output current

Fig. 4 presents DC-link current i_d and the output voltage measured in laboratory set-up with RB-GBT-based CSI, in this case with IXRH40N120 RB-IGBTs.

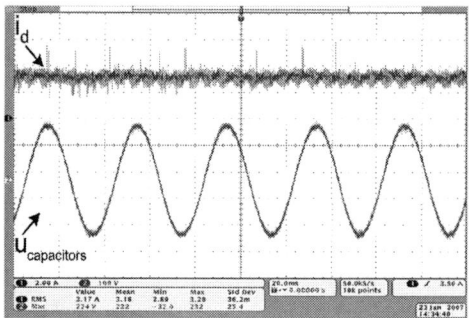

Fig. 4. DC-link current i_d and the output voltage in the experimatal setup with RB-IGBT-based CSI.

III. ALL-SiC JFET CURRENT SOURCE INVERTER

A. Suitability of CSI Topology for SiC JFETs

Silicon carbide (SiC) technology offers an opportunity to replace the RB-IGBTs with SiC JFETs and SiC Schottky diodes and thus to increase of CSI switching frequency of the CSI. SiC devices characterize switching frequencies above 100 kHz and enable the miniaturization of power electronics hardware, particularly the DC-link inductor L_d and output capacitors C_M. The high temperature functionality of SiC devices (>150°C) also allows the power density of the power converter to increase through the size reduction of the cooling system. High switching frequency SiC JFETs may make the SiC-based CSIs a potential replacement for the VSI also due to reduction in conduction losses.

Two types of SiC-JFETs can be distinguished: the *normally-on* and the *normally-off* devices. The *normally-on* SiC JFET-based current source inverters have been recently proposed in [8]-[10]. The usage of *normally-on* SiC JFETs in PWM CSIs guarantees an uninterrupted path for the DC-link inductor current. The *normally-on* JFET conducts the current when zero voltage or a slightly positive voltage is applied to the gate. The necessity of the use of zero gate-source voltage during switch-on and the negative voltage during switch off is the drawback of the *normally-on* SiC JFETs as it makes the design of the gate driver and protection circuits more complex than with standard normally-off devices [10]. Moreover some problems could arise with the generation of passive vectors involving shorting of all three phase legs and therefore all legs of output capacitors.

However the *normally-off* SiC JFETs can be also used in the CSI. The difference will mainly connected with the application of different gate drivers circuits and different space vector modulation strategy. The general scheme of the all *normally-off* SiC JFET CSI is shown in Fig. 5.

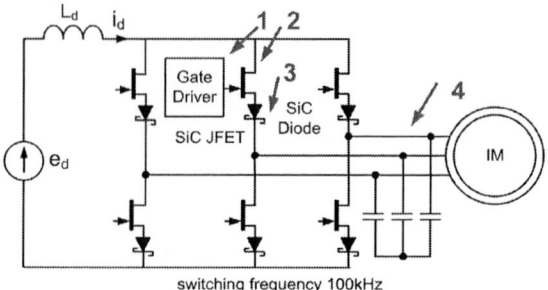

Fig. 5. Improvements in converter performance obtained in all *normally-off* SiC JFET-based CSI when compared to the SiC JFET-based VSI: reduction of dv/dt and EMI problems in the gate circuit (1) and the drain-source circuit (2) of the SiC JFET and SiC Schottky diode circuit (3), more compact output filter (4).

As it can be seen from Fig. 5 the following improvements can be obtained in all *normally-off* SiC JFET based CSI when compared to the all *normally-off* SiC JFET based VSI:

- the noticeable reduction in electromagnetic radiation and electromagnetic interference (EMI);
- reduced *dv/dt* ;
- sinusoidal output voltages and currents;
- more compact form of the inverter output filter.

B. Switching characteristics of normally-off SiC JFETs

Fig. 6 shows the dynamic characteristics of the *normally-off* SiC JFET measured in the experimental setup from Fig. 7.

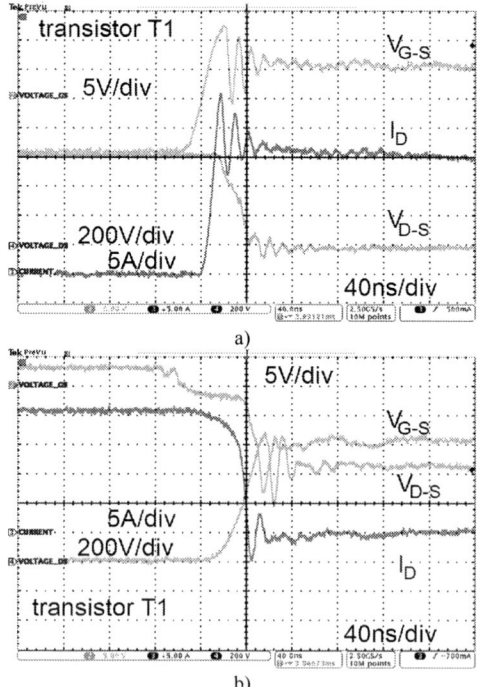

Fig. 6. Switching waveforms of the investigated *normally-off* SiC JFET: gate to source voltage V_{G-S}, drain to source voltage V_{D-S} and drain current I_D during turn-on (a) and turn-off (b).

978-1-4244-8806-3/11 $26.00 © 2011 IEEE

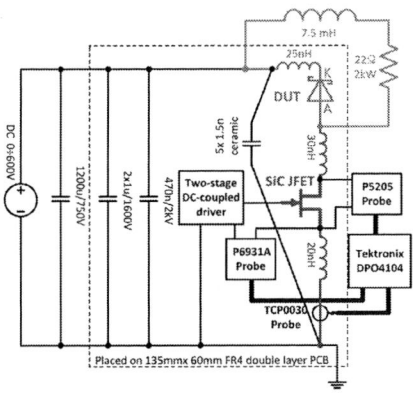

Fig. 7. Schematic of the test circuit with the investigated *normally-off* SiC JFET with the two-stage DC-coupled driver [12].

Table I shows the ratings of the investigated *normally-of* SiC JFET and SiC Schottky diode.

TABLE I
SiC DIODE AND SiC JFET UNDER TEST

	Abbre-viation	Voltage Ratings	Current Ratings at 100°	Parameters
SiC Schottky Diode	S1	1200V	20A	Q_c=122nC at di/dt=500A/µs
SiC JFET	T1	1200V	30A	$R_{DS(on)}$=0.063Ω $E_{TS,typ}$=440µJ

C. Gate Driver Design

The basic role of the gate driver is to deliver or remove the necessary gate charge required by internal gate-source and Miller capacitance in order to transit the device between on and off states [13], [14]. Fig. 8 shows the simulation scheme of the two-stage DC-coupled gate driver for *normally-off* SiC JFETs proposed in [12]. The two-stage DC-coupled driver circuit comprises three resistors R_1, R_2, R_3 coupled to the gate of SiC JFET [12]-[14]. The first resistor R_1 is coupled to the first turn-on circuit of the first sub-circuit A [12].

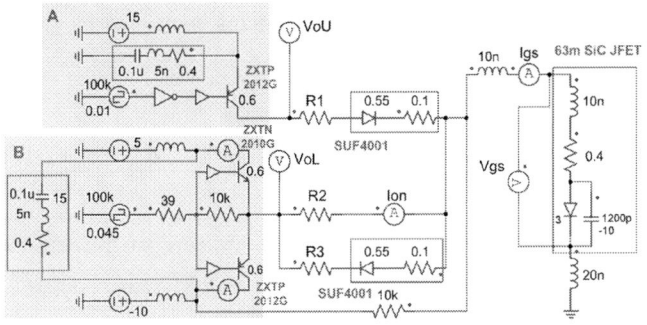

Fig. 8. Circuit diagram of the two-stage DC-coupled gate driver used for simulation in PSIM software: R_1=2.7, R_2=14, R_3=2.7 [12].

The switching performance of the SiC JFET strongly depends on the amplitude of forcing gate current and therefore on the R_1. Two resistors R_2 and R_3 are coupled to the sub-circuit B [12], whereby R_2 is connected to second turn-on circuit and R_3 is connected to the pull-down circuit. The resistance R_2 impacts the amplitude of the on-state gate current which should attain 100mA for investigated SiC JFET [12]-[14]. Fig. 3 shows simulated gate current at 100kHz switching frequency and duty cycle D =0.045.

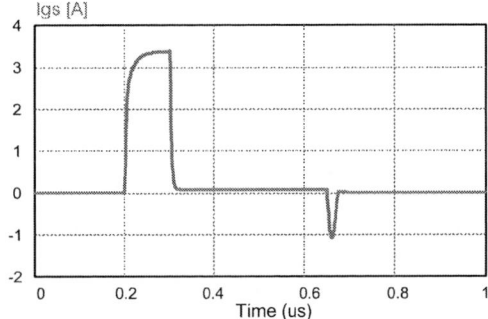

Fig. 9. Simulated gate current in the circuit with a basic model of the 0.063Ω SiC JFET gate.

Fig. 10 presents a picture of bottom layer of investigated two-stage DC-coupled gate driver circuit [12] and Fig. 11 shows current and voltage gate signals for R_1=2.7Ω, R_2=25Ω, R_3=2.7Ω, obtained in the investigated two-stage SiC JFET gate driver.

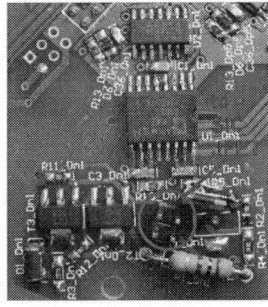

Fig. 10. Bottom layer of investigated two-stage DC-coupled driver circuit for SiC JFET [12].

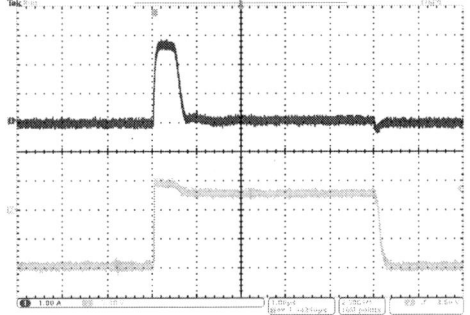

Fig. 11. Experimental gate signals: current (1A/div) and voltage (5V/div) at 100kHz and D =0.2.

IV. NONLINEAR CONTROL OF CSI-FED IM DRIVE

Nonlinear control methods rely on determining the nonlinear compensation which cancels the nonlinearities included in the IM model. This compensation is implemented as an inner feedback loop. For the linearized IM model the linear control system can be implemented as an outer feedback loop based on conventional linear theory [15]. Renewed attention has been recently paid to the analysis and design of nonlinear control methods due to the new possibilities offered by digital signal processors. Unlike the field oriented control (FOC) the nonlinear control using feedback linearization guarantees full decoupling between speed and flux amplitude of the induction motor [15].

The exact input-output linearization from stator voltages to torque, speed, and square of rotor flux amplitude has been proposed in [16] for fourth order IM model. The coordinate change proposed in [16] transforms the space vector induction motor model represented in the stationary frame (fixed with the IM stator) by the stator current vector $\mathbf{i_s}$ and rotor flux vector $\boldsymbol{\psi_r}$ into the multiscalar model independent of the coordinate system. Four multiscalar variables arbitrary selected in [16] for describing IM were: mechanical speed x_{11}, electromagnetic torque x_{12}, square of rotor flux x_{21} and reactive torque x_{22}.

Unlike [16] for the CSI fed IM another pair of vectors: the rotor flux vector $\boldsymbol{\psi_r}$ and the inverter output current vector $\mathbf{i_f}$ can be chosen to derive mathematical IM model as it is shown in Fig. 12.

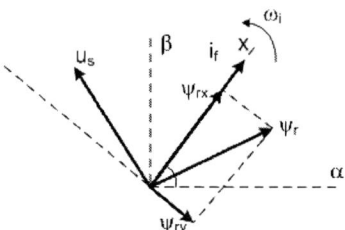

Fig. 12. Space vectors in rotating x-y frame fixed with inverter output current vector $\mathbf{i_f}$.

For inverter output current vector $\mathbf{i_f}$ and x-y frame rotating with the same angular frequency ω_i from Fig. 12 it can be written:

$$i_{fx} = i_f \ , \quad (4) \qquad i_{fy} = 0 \ . \quad (5)$$

In the case of CSI-fed induction motor the conventional IM space vector model can be extended to include the information about DC-link current i_d and output filter capacitors C_M voltages u_{cx}, u_{cy}:

$$\frac{di_d}{d\tau} = \frac{e_d}{L_d} - \frac{R_d}{L_d} i_d - \frac{u_{sx}}{L_d}, \quad (6)$$

$$\frac{du_{cx}}{d\tau} = \frac{1}{C_M} i_{cx}, \quad (7)$$

$$\frac{du_{cy}}{d\tau} = -\frac{1}{C_M} i_{cy}, \quad (8)$$

For output filter capacitors voltages and currents and IM stator voltages and currents it can be written:

$$u_{sx} = u_{cx} + R_c \left(i_f - i_{sx} \right), \quad (9)$$

$$u_{sy} = u_{cy} - R_c i_{sy}, \quad (10)$$

$$i_{cx} = i_f - i_{sx} , \quad (11)$$

$$i_{cy} = i_{sy} . \quad (12)$$

where i_{cx}, i_{cy} are the capacitors currents and R_c is a series resistance of the output capacitors C_M. Assuming the unitary commutation function $K=1$ of the CSI and high switching frequency (>100kHz) of the SiC JFET based CSI and taking into account (4) and (5), the inverter output current i_f and balance of active powers p_{DC} and p_{AC} at both CSI sides can be written as:

$$i_f = K \cdot i_d \approx i_d \quad (17)$$

$$u_d i_d = u_{sx} i_f \quad (18)$$

Moreover for high switching frequency of CSI transistors the impact of output filter C_M on CSI fed IM dynamics can be omitted and four multiscalar variables $x_{11}, x_{12}, x_{21}, x_{22}$ describing CSI fed IM can be chosen as [11]:

$$x_{11} = \omega_r, \quad (19) \qquad x_{12} = -i_d \psi_{ry}, \quad (20)$$

$$x_{21} = \psi_{rx}^2 + \psi_{ry}^2, \quad (21) \qquad x_{22} = i_d \psi_{rx}, \quad (22)$$

After differentiation of (19) - (20) the fourth order nonlinear multiscalar model of CSI fed IM is obtained [11]. The nonlinear compensation v_1, v_2 which cancels the nonlinearities included in the IM multiscalar model can be determined in the form:

$$v_1 = \frac{R_r L_m}{L_r} i_{sy} i_d - \frac{(u_{cx} - R_c i_{cx})}{L_d} \cdot \psi_{ry} + \frac{R_c}{L_d} i_{sx} \psi_{ry} + \frac{1}{T_i} m_1, \quad (23)$$

$$v_2 = -\frac{R_r L_m}{L_r} i_{sx} i_d + \frac{(u_{cx} - R_c i_{cx})}{L_d} \cdot \psi_{rx} - \frac{R_c}{L_d} i_{sx} \psi_{rx} + \frac{1}{T_i} m_2. \quad (24)$$

The linearized CSI fed IM multiscalar model can be written in the form of two decoupled subsystems controlled by two control variables m_1 and m_2:

- the electromechanical subsystem

$$\frac{dx_{11}}{d\tau} = \frac{L_m}{JL_r} x_{12} - \frac{1}{J} m_0, \quad (25)$$

$$\frac{dx_{12}}{d\tau} = \frac{1}{T_i} (-x_{12} + m_1), \quad (26)$$

where m_o denotes load torque, J - moment of inertia, T_i -time constant.

- the electromagnetic subsystem

$$\frac{dx_{21}}{d\tau} = -2\frac{R_r}{L_r}x_{21} + 2\frac{R_r L_m}{L_r}x_{22}, \tag{27}$$

$$\frac{dx_{22}}{d\tau} = \frac{1}{T_i}(-x_{22} + m_2). \tag{28}$$

Both subsystems: electromechanical (25) - (26) and electromagnetic (27) - (28) can be controlled using conventional proportional-integral (PI) controllers.

The reference value of the input DC-link voltage e_d for current source rectifier from Fig. 1 and command frequency ω_i for CSI modulator are calculated from:

$$e_d = -L_d \cdot \frac{\psi_{ry}v_1 - \psi_{rx}v_2}{x_{21}}, \tag{29}$$

$$\omega_i = \frac{\psi_{rx}v_1 + \psi_{ry}v_2}{i_d \cdot x_{21}} + \omega_r. \tag{30}$$

Startup of 5.5kW CSI, -fed induction motor (V_n=400V, I_n=10.7A, stator and rotor resistances: R_s=2.151Ω, R_r=2.059Ω, leakage inductances: $L_{\sigma s}$=$L_{\sigma r}$=0.0088H, L_m=0.254H) with proposed nonlinear control in experimental system, in this case with IXRH40N120 IGBTs, is shown in Fig. 13. Speed reversal from 0.8 p.u. to -0.8 p.u. of 5.5kW CSI-fed induction motor in experimental system with IXRH40N120 IGBTs based CSI and multiscalar control is shown in Fig. 14.

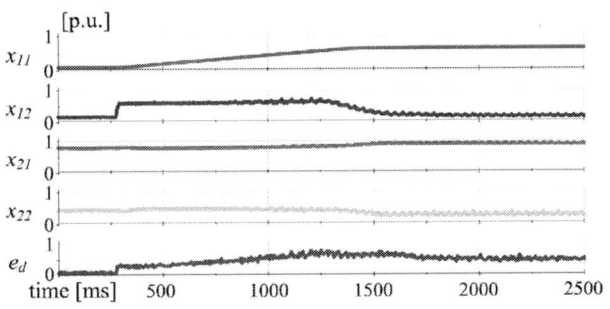

Fig. 13. Experimental waveforms during CSI-fed IM speed start-up to 0.7 p.u.

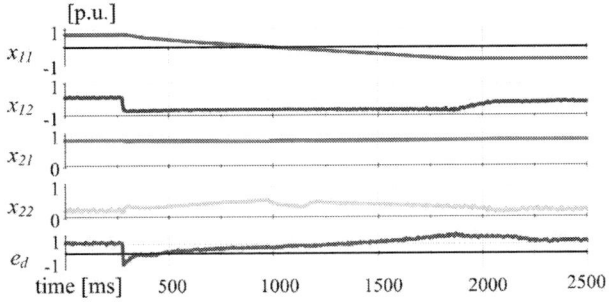

Fig. 14. Experimental waveforms during CSI-fed IM speed reversal from 0.8 p.u. to -0.8 p.u.

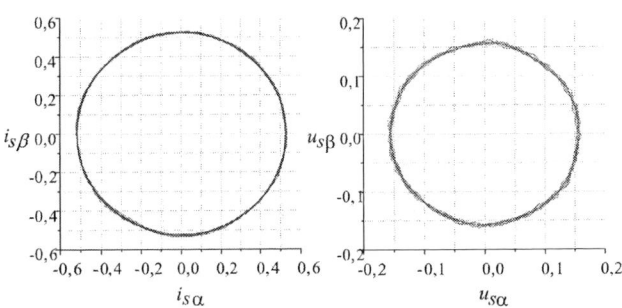

Fig. 15. The hodographs of currents and voltage in stationary $\alpha\beta$ coordinates taken from experimental setup of 5.5kW CSI-fed induction motor.

Fig. 15. shows the hodographs of currents and voltage in $\alpha\beta$ coordinates taken from experimental setup.

V. CONCLUSIONS

The recent advances in CSI-fed induction motor drives were presented in the paper which are:

- replacement of RB IGBTs with high switching frequency, high temperature and low on-state resistance SiC JFETs,
- application of high frequency (>100kHz) PWM which allows the reduction of DC-link inductor and passive output filters and therefore the minimization of the impact of output filter capacitors on IM dynamics, and inverter output current vector angular frequency ω_i,
- application of fourth order multiscalar IM model with state variables independent of reference frame to CSI-fed IM nonlinear control,
- application of the nonlinear compensation that linearizes IM multiscalar model and allows determination of DC-link voltage e_d reference value for current source rectifier and stator current frequency ω_i reference value for CSI modulator.

Measured transients of state variables and steady state waveforms of the DC-link current and output voltage confirm the performance of proposed multiscalar model based nonlinear control applied in experimental system of RB IGBT-based CSI fed IM drive. Moreover the obtained dynamic characteristics of *normally-off* SiC JFETs with two stage DC-coupled gate drivers confirm that they can beneficially replace the RB IGBTs in CSI topology increasing the operating frequency and causing further size reduction of the passive components.

ACKNOWLEDGMENT

This scientific work is financed from resources for science in 2010-2012 years as a LIDER research project of the National Centre for Research and Development.

978-1-4244-8806-3/11 $26.00 © 2011 IEEE

REFERENCES

[1] Bimal K. Bose: Power Electronics And Motor Drives: Advances and Trends, Elsevier, 2006.

[2] Yun Wei Li, Manish Pande, Navid Reza Zargari, Bin Wu: An Input Power Factor Control Strategy for High-Power Current-Source Induction Motor Drive With Active Front-End, *IEEE Transactions On Power Electronics,* vol. 25, No. 2, Feb. 2010, pp. 352-359.

[3] Eduardo P. Wiechmann, Pablo Aqueveque, Rolando Burgos, José Rodríguez: On the Efficiency of Voltage Source and Current Source Inverters for High-Power Drives, IEEE Transactions On Industrial Electronics, Vol. 55, No. 4, April 2008, pp. 1771-182.

[4] Fuchs F., Kloenne A.: DC link and Dynamic Performance Features of PWM IGBT Current Source Converter Induction Machine Drives with Respect to Industrial Requirements, *Proc. IEEE Conference IPEMC 2004,* Vol. 3, 14-16.

[5] Colli V.D., Cancelliere P., Marignetti F., Di Stefano R.: Voltage control of current source inverters, *IEEE Transactions on Energy Conversion* Vol. 21, Issue 2, June 2006 Pages: 451 – 458.

[6] M. Glab, Z. Krzeminski, A. Lewicki, " Multiscalar Control of Induction Motor Supplied by Current Source Inverter, " *Proc. Power Conversion Intelligent Motion Conference PCIM'07,* 2007, Nurnberg, p. 1-6.

[7] Glab M., *at. all*: The PWM CSI with IGBT transistors and multiscalar model control system, *Proc. Conference EPE 2005,* 2005, p. 1-6.

[8] T. Friedli, S. Round, D. Hassler, J. W. Kolar, "Design and Performance of a 200-kHz All-SiC JFET Current DC-Link Back-to-Back Converter," *IEEE Transactions on Industry Applications,* vol. 45, no. 5, pp 1868-1878, Sept/Oct 2009.

[9] Cass C. J., Wang Y., Burgos R., Chow T. P., Wang F., Boroyevich D., "Evaluation of SiC JFETs for a Three-Phase Current-Source Rectifier with High Switching Frequency," Proc. Twenty Second Annual IEEE Applied Power Electronics Conference APEC 2007, pp. 345 - 351.

[10] Rabkowski J., Barlik R., "Three-phase Grid Inverter with SiC JFETs and Schottky Diodes ," Proc. 16th International Conference "Mixed Design of Integrated Circuits and Systems" MIXDES 2009, June 25-27, 2009, pp. 181-184.

[11] Morawiec M, Krzeminski Z., Lewicki A., "Voltage multiscalar control of induction machine supplied by current source converter," *Proc. IEEE Conference ISIE2010,* 2010, Bari, Italy, pp. 1-7.

[12] Adamowicz M., Giziewski S., Pietryka J., Rutkowski M., Krzeminski Z., " Evaluation of SiC JFETs and SiC Schottky Diodes for Wind Generation Systems," Proc. IEEE International Symposium on Industrial Electronics, ISIE 2011, Gdansk (Poland) June 27-30, 2011, CD-Rom, *(accepted for publication).*

[13] R.L. Kelley, F. Rees, "Gate Driver For Enhancement-Mode And Depletion-Mode Wide Bandgap Semiconductor JFETs," United States Patent Application US 2010/0283515, from May 11, 2009.

[14] SemiSouth, "The SJEP120R063 – Reduced Switching Losses in the New Low-Inductance Test Fixture," White Paper WP-SS1, www.semisouth.com

[15] Kazmierkowski M. P., Blaabjerg F., Krishnan R. : Control in Power Electronics – selected Problems. Academic Press, USA, 2002.

[16] Z. Krzeminski, "Nonlinear control of induction motors," Proc. 10th IFAC World Congress, Munich, pp. 349-354, 1987.

Radiated Magnetic Field of a Low-Frequency Ferrite Rod Antenna

Alexander Stadler

STS Spezial-Transformatoren Stockach GmbH & Co. KG, 78333 Stockach, Germany

stadler@sts-trafo.de

Abstract-The frequency range below 135kHz is preferred for inductively coupled radio frequency identification (RFID) systems. At low frequency, the magnetic field coupling between reader and transponder can be increased by using a ferrite core transponder antenna. In this paper, the magnetic field pattern of such a device is investigated. A semi-numerical procedure is presented, how to calculate the magnetic fields resulting from coil and core with avoidance of virtual boundaries. Consequently, calculations at short as well as at large distances can be conducted with high accuracy and minimum computational expense. The model is finally verified by measurements and further simulation based on finite element method (FEM).

I. INTRODUCTION

In recent years, radio frequency identification (RFID) systems have become very popular and substituted barcode labeling technology in many industrial applications [1]. Electromagnetic fields are used for the data exchange between the reader and the data-carrying device (transponder) instead of galvanic contacts. Consequently, electromagnetic compatibility (EMC) is one of the key competences to ensure functionality and to improve system reliability. Fig. 1 depicts the currently available and practical RFID frequencies.

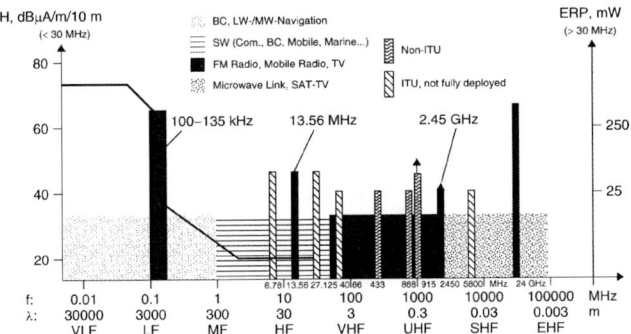

Fig. 1. Available and practical RFID frequencies [1].

This paper focuses on the low frequency (LF) range below 135kHz, where the distance r between reader and transponder is usually much shorter than the free-space wavelength λ:

$$r << \lambda = \frac{c_0}{f} \quad \text{with} \quad c_0 \approx 2.99792458 \cdot 10^8 \, \text{ms}^{-1} \,. \quad (1)$$

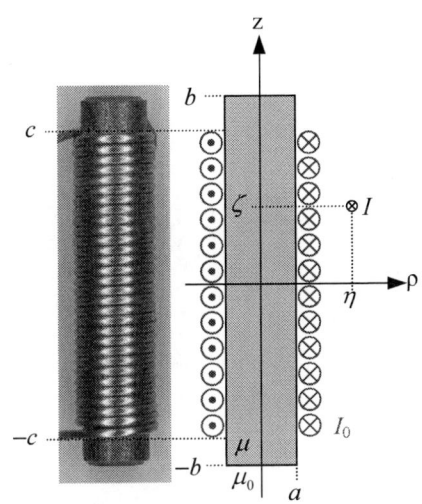

Fig. 2. Ferrite rod antenna conducting current I_0.

At LF, the magnetic coupling between the reader's and the transponder's antenna is the dominant mode of operation. In general, the coupling factor can be improved using a ferrite core transponder antenna [2][3].

Fig. 2 shows the basic configuration of such a device. All physical quantities are supposed to be independent of coordinate φ of the underlying axial symmetric coordinate system. The length of the rod is $2b$ and its diameter is $2a$. The width of the shown (one-layer) winding is given by $2c$. The winding conducts current I_0 that magnetizes the ferrite material of permeability $\mu = \mu_r \mu_0$. To investigate the magnetic coupling between reader and transponder as well as parasitic coupling to other adjacent electronic devices, the magnetic fields of coil and core have to be calculated.

II. CALCULATION

A. Problem Description

A solution of the investigated problem is presented in [4]. The method is based on magnetic charges, but the assumptions given by the authors are not true in praxis. It is described in the paper, that the accuracy of the model depends on the physical permeability of the core. Unfortunately, large errors are obtained for high permeability, as it is typical for practical ferrite materials. The finite differences method shown in [5] leads to time and memory consuming huge matrix systems, if the electromagnetic fields have to be evaluated at large distances from the rod. A similar

978-1-4244-8806-3/11 $26.00 © 2011 IEEE

problem also occurs in [6], [7], [8] and [9]. Periodicity has to be assumed there for solution, thus leading to worse results far from the ferrite rod.

Here, the solution is based on the basic configuration shown in Fig. 2. The axial symmetric problem with the magnetic flux density $\vec{B} = \vec{e}_\rho B_\rho(\rho, z) + \vec{e}_z B_z(\rho, z) = \nabla \times \vec{A}$ is represented by the magnetic vector potential $\vec{A} = \vec{e}_\varphi A_\varphi(\rho, z)$ that has to fulfill the partial differential equation

$$\frac{\partial^2 A_\varphi}{\partial \rho^2} + \frac{1}{\rho} \frac{\partial A_\varphi}{\partial \rho} - \frac{1}{\rho^2} A_\varphi + \frac{\partial^2 A_\varphi}{\partial z^2} = 0 \ . \qquad (2)$$

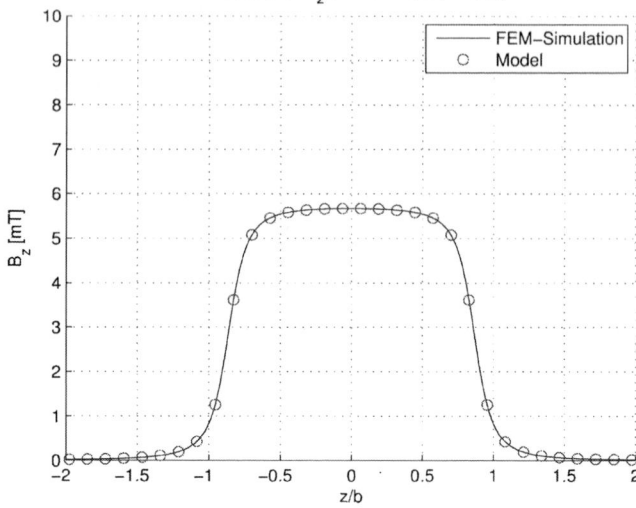

Fig. 3. Circular line currents as substitute for the magnetic fields of coil and core.

A well known practice [10] is to use an equivalent current distribution as substitute for the magnetization

$$\vec{M} = (\mu_r - 1)\vec{H} \ . \qquad (3)$$

Hence, $m = 1, 2, \dots M$ (unknown) circular line currents $I_m^{(mag)}$ are introduced on the surface of the ferrite rod as depicted in Fig. 3. A system of equations has to be found to calculate these currents with respect to the surface boundary conditions and the exciting currents $I_n^{(coil)} = I_0$ of the copper coil. After solution, the total magnetic field is given by the superposition of all contributions from $I_n^{(coil)}$ and $I_m^{(mag)}$.

B. Magnetic Field of the Air Coil

The magnetic vector potential of a circular line current I located at (η, ξ) e.g. can be found in [10]:

$$A_{\varphi,e}(\rho, z, \eta, \xi, I) = \frac{\mu_0 I}{\pi k} \sqrt{\frac{\eta}{\rho}} \left[\left(1 - \frac{1}{2}k^2\right)K(k) - E(k) \right]$$

$$\text{with} \quad k = \sqrt{\frac{4\eta\rho}{(\eta + \rho)^2 + (z - \xi)^2}} \ . \qquad (4)$$

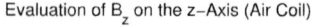

Evaluation of B_z on the z–Axis (Air Coil)

Fig. 4. Evaluation of B_z of the air coil on the z-axis ($2a = 10$mm, $2c = 70$mm, $N = 80$, $I_0 = 4$A).

Equation (4) contains the complete elliptic integrals of the first and second kind K and E. Carrying out the curl in cylindrical coordinates

$$\nabla \times \vec{A} = -\vec{e}_\rho \frac{\partial A_\varphi}{\partial z} + \vec{e}_z \frac{1}{\rho} \frac{\partial}{\partial \rho}(\rho A_\varphi) \ , \qquad (5)$$

the two components of the magnetic flux density are obtained by a collection of terms:

$$B_{\rho,e}(\rho, z, \eta, \xi, I) = \frac{\mu_0 I}{2\pi} \frac{z - \xi}{\rho\sqrt{(\eta + \rho)^2 + (z - \xi)^2}} \cdot$$

$$\cdot \left[-K(k) + \frac{\eta^2 + \rho^2 + (z - \xi)^2}{(\eta - \rho)^2 + (z - \xi)^2} E(k) \right] \qquad (6)$$

and

$$B_{z,e}(\rho, z, \eta, \xi, I) = \frac{\mu_0 I}{2\pi} \frac{1}{\sqrt{(\eta + \rho)^2 + (z - \xi)^2}} \cdot$$

$$\cdot \left[K(k) + \frac{\eta^2 - \rho^2 - (z - \xi)^2}{(\eta - \rho)^2 + (z - \xi)^2} E(k) \right] . \qquad (7)$$

The total magnetic vector potential of the air coil that conducts current I_0 is given by the superposition of all $n = 1, 2, \dots N$ windings with position $(\eta_n^{(coil)}, \xi_n^{(coil)})$:

$$A_\varphi^{(coil)}(\rho, z) = \sum_{n=1}^{N} A_{\varphi,e}(\rho, z, \eta_n^{(coil)}, \xi_n^{(coil)}, I_0) \ . \qquad (8)$$

The ρ-component of the magnetic flux density can be expressed as

$$B_\rho^{(coil)}(\rho, z) = \sum_{n=1}^{N} B_{\rho,e}(\rho, z, \eta_n^{(coil)}, \xi_n^{(coil)}, I_0) \ . \qquad (9)$$

The z-component is given by

$$B_z{}^{(coil)}(\rho, z) = \sum_{n=1}^{N} B_{z,e}\left(\rho, z, \eta_n^{(coil)}, \xi_n^{(coil)}, I_0\right) . \quad (10)$$

Fig. 4 shows a comparison of (10) and simulation data drawn from a commercial FEM software.

C. Influence of the Ferrite Rod

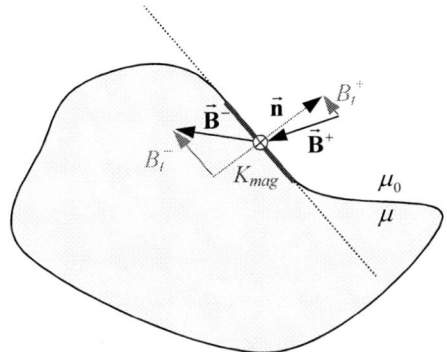

Fig. 5. Boundary conditions on the surface of a permeable body.

Fig. 5 depicts the boundary conditions on the surface of a permeable body. If $\vec{n}$ designates the surface normal vector and no "free" surface current density is assumed, the tangential components of the magnetic field strengths $\vec{H}^+ = \mu_0^{-1}\vec{B}^+$ and $\vec{H}^- = \mu^{-1}\vec{B}^-$ have to be continuous:

$$\vec{n} \times \left(\vec{H}^+ - \vec{H}^-\right) = \vec{0} . \quad (11)$$

Now, the magnetization is substituted by the "bound" surface current density

$$\vec{K}_{mag} = \vec{M} \times \vec{n} . \quad (12)$$

Consequently, the related tangential components of the magnetic flux density are connected by

$$\vec{n} \times \left(\vec{B}^+ - \vec{B}^-\right) = \mu_0 \vec{K}_{mag} . \quad (13)$$

Equations (11) and (13) form the equation system

$$\begin{bmatrix} \dfrac{1}{\mu} & -\dfrac{1}{\mu_0} \\ 1 & -1 \end{bmatrix}\begin{bmatrix} B_t^- \\ B_t^+ \end{bmatrix} = \begin{bmatrix} 0 \\ \mu_0 K_{mag} \end{bmatrix} \quad (14)$$

that can be solved for the tangential components of the magnetic flux density on the inner and on the outer surface of the permeable body B_t^- and B_t^+ (s. Fig. 5):

$$\begin{bmatrix} B_t^- \\ B_t^+ \end{bmatrix} = \frac{K_{mag}}{\mu_r - 1} \cdot \begin{bmatrix} \mu \\ \mu_0 \end{bmatrix} . \quad (15)$$

For the investigated problem, the "bound" surface current density has the form

$$\vec{K}_{mag} = \vec{e}_\varphi K_{mag} \quad (16)$$

and the tangential component B_t is defined in sections:

$$B_t = \begin{cases} -B_\rho & & \rho < a & & z = b \\ B_z & \text{if} & \rho = a & \text{and} & -b < z < b \\ B_\rho & & \rho < a & & z = -b \end{cases} . \quad (17)$$

If K_{mag} is represented by a step-function and each line current $I_m^{(mag)}$ is assumed to be equally distributed over the according step-width Δs_m, the approximate "bound" surface current density at the line current's position $\left(\eta_m^{(mag)}, \xi_m^{(mag)}\right)$ is given by

$$K_{mag}\left(\eta_m^{(mag)}, \xi_m^{(mag)}\right) \approx \frac{I_m^{(mag)}}{\Delta s_m} . \quad (18)$$

The two components of the magnetic flux density resulting from the magnetized ferrite rod are obtained analogously (9) and (10):

$$B_\rho{}^{(mag)}(\rho, z) = \sum_{m=1}^{M} B_{\rho,e}\left(\rho, z, \eta_m^{(mag)}, \xi_m^{(mag)}, I_m^{(mag)}\right) \quad (19)$$

and

$$B_z{}^{(mag)}(\rho, z) = \sum_{m=1}^{M} B_{z,e}\left(\rho, z, \eta_m^{(mag)}, \xi_m^{(mag)}, I_m^{(mag)}\right) . \quad (20)$$

D. Solution System

Considering (16), (17) and (18), the (outer) tangential component of the magnetic flux density given by (15)

$$B_t^+ = B_t{}^{(coil)} + B_t{}^{(mag)} = \frac{\mu_0 K_{mag}}{\mu_r - 1} \quad (21)$$

can be used to set up a linear system of equations. For this purpose,

$$B_t{}^{(mag)} = \frac{\mu_0 K_{mag}}{\mu_r - 1} - B_t{}^{(coil)} \quad (22)$$

is evaluated sequentially over all positions of the M circular line currents $I_m^{(mag)}$. With aid of (19) and (20), a unique $M \times M$ linear system of equations is defined that can directly be solved for the desired line currents $I_m^{(mag)}$ using the algebra of matrices.

Of course, the addends

$$B_{\rho,e}\left(\rho = \eta_m^{(mag)}, z = \xi_m^{(mag)}, \eta_m^{(mag)}, \xi_m^{(mag)}, I_m^{(mag)}\right)$$

and

$$B_{z,e}\left(\rho = \eta_m^{(mag)}, z = \xi_m^{(mag)}, \eta_m^{(mag)}, \xi_m^{(mag)}, I_m^{(mag)}\right)$$

in (19) and (20) that appear on the main diagonal of the coefficient matrix would diverge and therefore have to be replaced by adequate integral formulations:

$$B_{\rho,e}\left(\eta_m^{(mag)}, \xi_m^{(mag)}, \eta_m^{(mag)}, \xi_m^{(mag)}, I_m^{(mag)}\right) \approx$$

$$\approx \lim_{\varepsilon \to 0} \int_{\eta_m^{(mag)} - \Delta s_m/2}^{\eta_m^{(mag)} + \Delta s_m/2} B_{\rho,e}\left(\eta_m^{(mag)}, b+\varepsilon, \eta, b, \frac{I_m^{(mag)}}{\Delta s_m}\right) d\eta \qquad (23)$$

for evaluation on top side $(\rho < a, z = b)$ of the rod,

$$B_{\rho,e}\left(\eta_m^{(mag)}, \xi_m^{(mag)}, \eta_m^{(mag)}, \xi_m^{(mag)}, I_m^{(mag)}\right) \approx$$

$$\approx \lim_{\varepsilon \to 0} \int_{\eta_m^{(mag)} - \Delta s_m/2}^{\eta_m^{(mag)} + \Delta s_m/2} B_{\rho,e}\left(\eta_m^{(mag)}, -b-\varepsilon, \eta, -b, \frac{I_m^{(mag)}}{\Delta s_m}\right) d\eta \qquad (24)$$

for evaluation on bottom side $(\rho < a, z = -b)$ and finally

$$B_{z,e}\left(\eta_m^{(mag)}, \xi_m^{(mag)}, \eta_m^{(mag)}, \xi_m^{(mag)}, I_m^{(mag)}\right) \approx$$

$$\approx \lim_{\varepsilon \to 0} \int_{\xi_m^{(mag)} - \Delta s_m/2}^{\xi_m^{(mag)} + \Delta s_m/2} B_{z,e}\left(a+\varepsilon, \xi_m^{(mag)}, a, \xi, \frac{I_m^{(mag)}}{\Delta s_m}\right) d\xi \qquad (25)$$

for evaluation on the rod shell $(\rho = a, -b < z < b)$. The integrals (23), (24) and (25) have finite values even if their

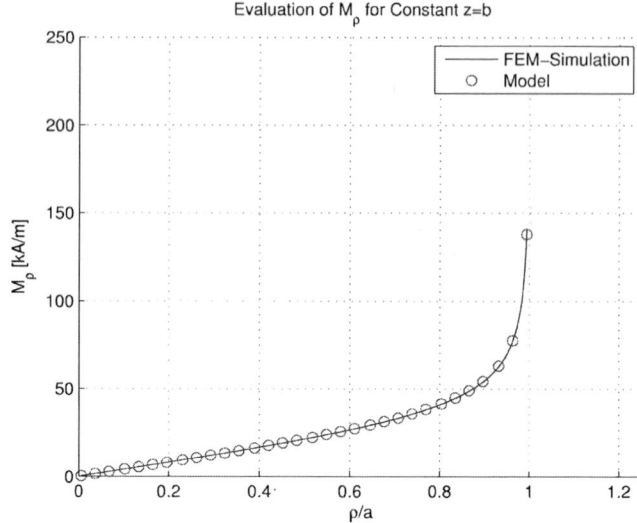

Fig. 7. Evaluation of M_ρ for constant $z = b$ ($2a = 10\text{mm}$, $2b = 81\text{mm}$, $2c = 70\text{mm}$, $\mu_r = 2000$, $N = 80$, $I_0 = 4\text{A}$, $M = 4096$).

integrands have not. Numerical integration of (23), (24) and (25) e.g. following adaptive Simpson quadrature leads to good results here.

Fig. 6 shows an evaluation of the z-component M_z of the magnetization (3) on the rod shell. It is found that the proposed model leads to good results in comparison with commercial FEM software – even the sharp increase of the magnetization at the edges $(\rho = a, z \to \pm b)$ is replicated with high accuracy. As depicted by Fig. 7, comparable results are obtained for the ρ-component M_ρ as well. Therefore, it is verified that the "bound" surface current density (12) is perfectly described by the proposed model. Fig. 8 shows the z-component B_z of the magnetic flux density on the z-axis. As expected, high precision is also obtained for the superposed magnetic fields of coil and core.

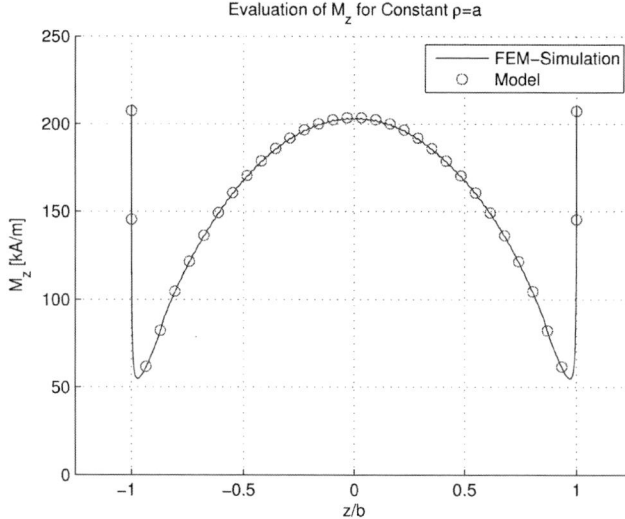

Fig. 6. Evaluation of M_z for constant $\rho = a$ ($2a = 10\text{mm}$, $2b = 81\text{mm}$, $2c = 70\text{mm}$, $\mu_r = 2000$, $N = 80$, $I_0 = 4\text{A}$, $M = 4096$).

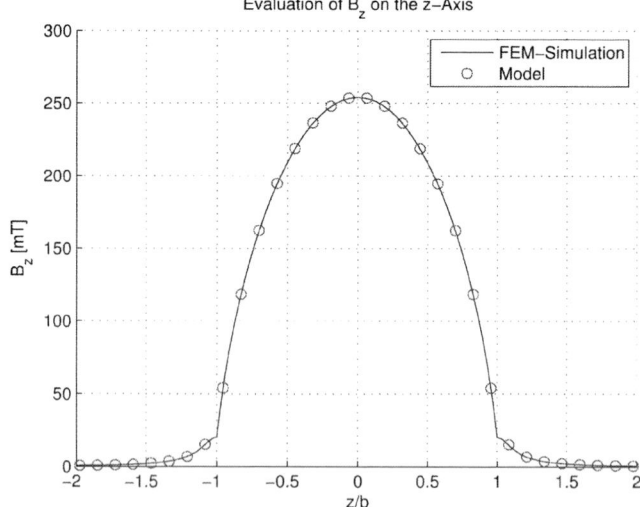

Fig. 8. Evaluation of B_z on the z-axis ($2a = 10\text{mm}$, $2b = 81\text{mm}$, $2c = 70\text{mm}$, $\mu_r = 2000$, $N = 80$, $I_0 = 4\text{A}$, $M = 4096$).

III. RESULTS

Fig. 9 shows the radiation patterns for three different values of the winding width $2c$. Therein, the absolute value of the magnetic flux density is plotted along a circular path around the ferrite rod antenna as a function of the angle. The distance is kept constant $\left(\rho^2 + z^2 = 4b^2\right)$ during evaluation. All results are compared on the condition of equal inductance L. Consequently, the number of turns is adjusted depending on the varying parameter $2c$. It is found that a slight attenuation of the radiated stray field is achieved, if the turns are more concentrated in the middle of the rod. A lower number of turns is needed to create a certain inductance in this case. This is due to the fact that a higher amount of flux is concentrated

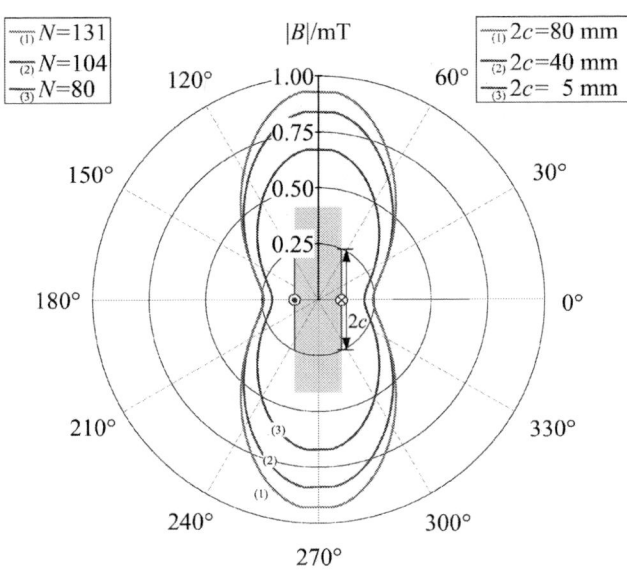

Fig. 9. Radiation patterns (constant distance $\rho^2 + z^2 = 4b^2$) for different values of the winding width $2c$, number of windings N adjusted for constant inductance $L = 600\mu H$ ($2a = 10mm$, $2b = 81mm$, $\mu_r = 2000$, $I_0 = 4A$, $M = 4096$).

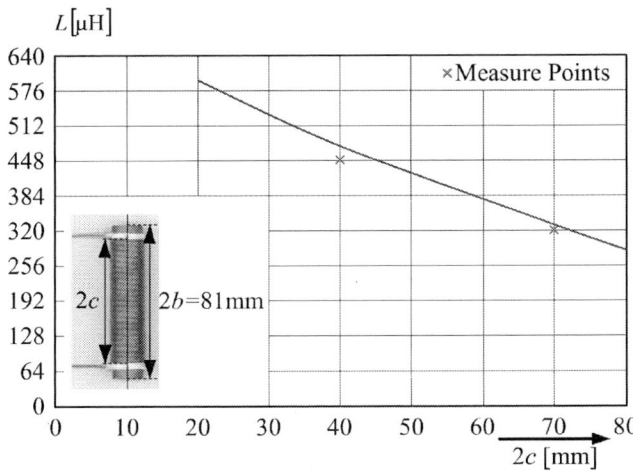

Fig. 10. Inductance L as a function of the winding width $2c$, measured points (small signal, low frequency) obtained from impedance analyzer ($2a = 10mm$, $2b = 81mm$, $\mu_r = 2000$, $M = 4096$).

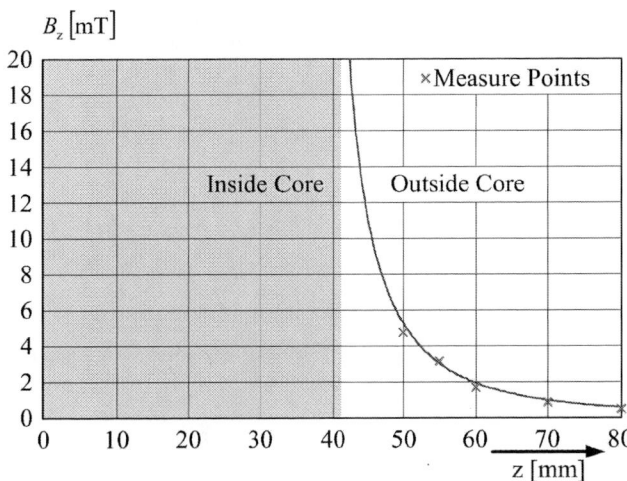

Fig. 11. Comparison of calculated and measured results of the z-component of the magnetic flux density B_z, test set-up shown in Figs. 12a,b. ($2a = 10mm$, $2b = 81mm$, $2c = 70mm$, $\mu_r = 2000$, $I_0 = 4A$, $N = 80$, $M = 4096$).

inside the windings and the magnetic coupling between the windings is increased.

This dependence is investigated more in detail in Fig. 10. The inductance L is calculated there as a function of the winding width $2c$. This time, the number of turns is fixed. The plotted measure points are obtained in the lower frequency range $f < 1kHz$ using an impedance analyzer. Considering the tolerance of information about magnetic material provided by the manufacturers and the observational accuracy, the values seem to be acceptable.

IV. VERIFICATION

Fig. 11 illustrates accurate measurements conducted with aid of the test set-up shown in Figs. 12a,b. A Hall sensor was used to compare the calculated and measured results for the z-component B_z of the magnetic flux density. An average error of 5-10 per cent validates the developed model.

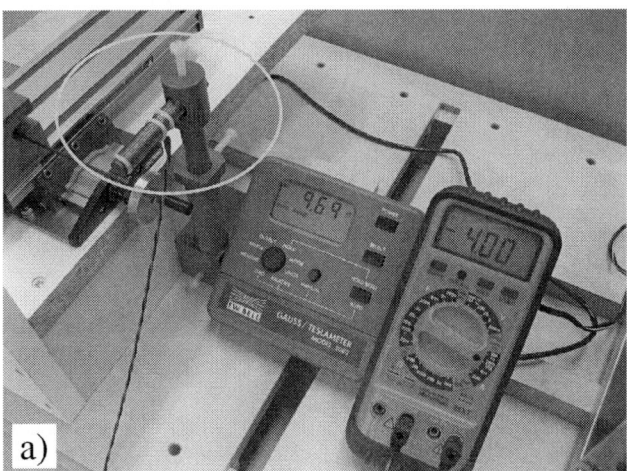

Fig. 12a. Test set-up with inductor and Hall sensor.

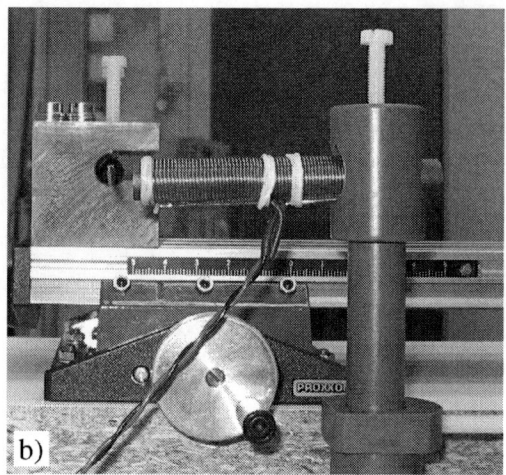

Fig. 12b. Rod core coil in Fig. 12a drawn to a larger scale.

V. CONCLUSION

With the method described above, the radiated magnetic field of a ferrite rod antenna can be calculated. The presented semi-numerical procedure allows a precise determination of the equivalent current distribution introduced as substitute for the magnetization. Consequently, accurate results are obtained for the magnetic fields of coil and core at short as well as at large distances. The results are verified by measurements and compared to further simulation data obtained from a commercial FEM software.

REFERENCES

[1] K. Finkenzeller, *RFID handbook: fundamentals and applications in contactless smart cards and identification, 2ⁿᵈ ed.*, John Wiley & Sons Ltd., May 2003.

[2] A. Tran, M. Bolic, and M. C. E. Yagoub, "Magnetic-field coupling characteristics of ferrite-coil antennas for low-frequency RFID applications," *International Journal of Computer Science Issues IJCSI*, vol. 7, issue 4, no. 1, July 2010, pp. 7-11.

[3] Philips Components, *Application note: ferrite rod antennas for RF-identification transponders, www.philips.com*

[4] O. Chadebec, L.-L. Rouve, and J.-L. Coulomb, "New methods for a fast and easy computation of stray fields created by wound rods," *IEEE Trans. on Mag.*, vol. 38, no. 1, Mar. 2002, pp. 517-520.

[5] H. Watanabe and K. Kato, "Field distribution and power loss assessment in conductive rod cores exhibiting hysteresis," *IEEE Trans. on Mag.*, vol. 32, no. 2, Sep. 1996, pp. 4293-4295.

[6] A. Stadler and M. Albach, "Analytical calculation of stray fields generated by ferrite rods in EMI suppression applications," *PCIM Eur. Conf.*, May 2006.

[7] M. Albach, A. Stadler, and M. Spang, "The influence of ferrite characteristics on the inductance of coils with rod cores," *10ᵗʰ Joint MMM/Intermag Conf.*, Jan. 2007, BF-03.

[8] M. Albach, A. Stadler, and M. Spang, "The influence of ferrite characteristics on the inductance of coils with rod cores," *IEEE Trans. on Mag.*, vol. 43, no. 6, June 2007, pp. 2618-2620.

[9] M. Spang and M. Albach, "Optimized winding layout for minimized proximity losses in coils with rod cores," *IEEE Trans. on Mag.*, vol. 44, no. 7, July 2008, pp. 1815-1821.

[10] W. R. Smythe, *Static and dynamic electricity, 3ʳᵈ ed.*, Hemisphere Publ. Corp., 1989.

Common Mode Model of Matrix Converter

Jordi Espina, Josep Balcells, Carlos Ortega, Antoni Arias, Liliana de Lillo and Lee Empringham.

Dep. d'Enginyeria Electrònica. Universitat Politècnica de Catalunya. C. Colom 1. 08222 Terrassa. Catalunya. Catalunya. (Spain). Jordi.espina@upc.edu

Abstract— **This paper deals with the validation of an EMI model to predict electromagnetic interferences (EMI) produced by direct AC/AC matrix converters (MC). The method is based on obtaining a high frequency equivalent circuit, using a combined time and frequency domain approach based on: "EMI source identification→ propagation path impedance→ derived disturbance". The advantages of the proposed procedure are the computational time reduction and the lack of convergence problems, which may arise when using pure time domain procedures. The paper is focused on the prediction of common mode (CM) EMI of a matrix converter. The simulation results of this case will permit the calculation of currents which leak through the ground connections.**

Index Terms— **Matrix Converters; EMI modeling; Common Mode Current.**

I. INTRODUCTION

NOWADAYS the electromagnetic interference (EMI) plays an important role in power electronics, to the extent that it can not be designed any electronic board without taking into account EMI emissions. A large number of contributions exist regarding the EMI propagation, emissions or susceptibility. Among them, a vast majority is focused on modeling the system in order to predict the EMI behavior. Such modeling techniques can be divided into three groups depending on the scale of study. In the first group, Maxwell's equations are used and therefore the order of magnitude for the models are micrometers [1]. Equivalent circuit methods are used when the order of magnitude is increased up to centimeters [2, 3]. In the third group, the scale increases up to meters and the dipole technique is employed [4]. Considering that all power converters are composed by different parts such as input and output stages, power devices and control boards, the use of Maxwell's equations is almost unfeasible. Also, such converters usually work in the kilohertz range and therefore the conducted emissions are the most significant and consequently the dipole method is not suitable either.

For all the previous reasons, in order to model a power converter, the equivalent circuit method is the most appropriated one. Such a modeling techniques are based on "Source – Path – Victim" identification and according to the reference system used they can be split in two sub groups: temporal methods [2] and frequency ones [3].

Temporal methods are essentially computed or solved by simulation program with integrated circuit emphasis (SPICE),

and their advantages are the high resolution and accurate results. However, their drawbacks are firstly the need of long simulation periods together with small simulation steps and secondly convergence problems. In [5], a complete drive formed by a voltage source inverter (VSI) and an electrical motor is modeled. Also in [6], a similar study is addressed, which models every part of the drive (VSI, cable, motor, IGBTs) and studies different output filters to reduce the EMI perturbations.

Alternatively, frequency methods bring a remarkable simplification in terms of modeling, reducing drastically the computational effort needed. On the other hand, their drawbacks are the limited frequency range, being valid only for EMI conducted emissions and the difficulty of accurately simulate the EMI sources. A boost converter in [7] has been analyzed by the frequency method, using the planar capacitor formula to calculate the leakage paths. Another boost converter with power factor correction (PFC) has been modelled in [8] using a simple approximation (simple circuits and theoretical sources). In [9], a matrix of impedances together with analytical (and therefore simplified) sources models a drive based on VSI to analyze common mode (CM) perturbations. Such impedances firstly have been experimentally obtained using a frequency sweep and secondly they have been approximated to *RLC* circuit equations. Finally, simplified EMI sources feed the previously obtained matrix of impedances to predict the CM currents which leak through the parasitic paths. In [10], an EMI study, which is based on [9], of an H bridge with PFC has been addressed. Reducing a three phase VSI to one phase equivalent CM circuit, composed by one EMI source and a few impedances, is another technique to estimate the leakage CM paths [11].

Another method that simplifies the three phase drive is presented in [12], where after the simplification it creates modules of each part (cable, motor, bulk capacitor, bridge, etc) and works with such parts using the "*quadripolar theory*". Such a modulation process permits to easily plug or unplug in cascade different parts of the system depending on the drive's configuration. Another frequency method based on "*Modular-terminal-behavioral*" obtains a Norton equivalent circuit per VSI leg [13]. This procedure allows the study of all VSI configurations by only firstly adapting and secondly adding/removing Norton circuits depending on the topology. In [14] the EMI suppression efficiency when connecting the Y capacitors at different places is addressed and a comparison using the model presented in [9] is done. Finally in [15], CM

Universitat Politècnica de Catalunya (e-mail: jordi.espina@upc.edu).

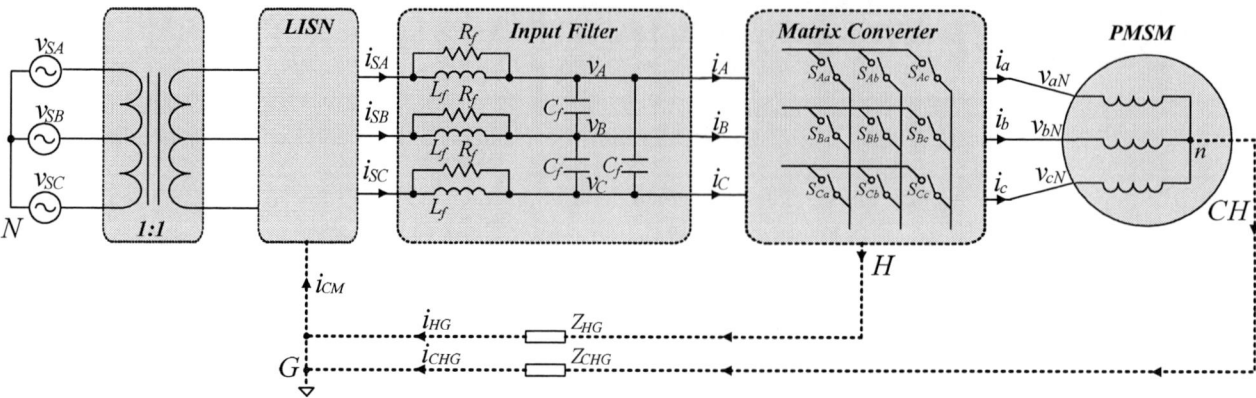

Figure 1. Matrix Converter's set-up and CM paths.

EMI disturbances are studied when using matrix converter (MC) fed wind turbine. The MC model, which is just a mono-phase model, is based on the indirect inverter-rectifier and virtual DC link approach.

MCs are direct AC-AC power converters without intermediary reactive elements, which are considered one of best advantages. Other advantages are less weight, compact size, low cost, power density and natural bidirectional energy transfer [16-18]. All these advantages make MCs really interesting and attractive in different fields of the state of the art industry [19, 20], such as wind turbines, aeronautic applications (replacing the hydraulic system that controls the flaps and as external sources when the plane is refueling [21]), and hostile environments and military applications.

This paper presents a complete EMI study of an adjustable speed drive (ASD), which is formed by a direct MC and a permanent magnet synchronous machine; as shown in Fig. 1. This study belongs to the frequency domain methods, more specifically is essentially based on circuit equivalent philosophy that was presented in [9].

This paper obtains an EMI model of a direct MC [17]. A better accuracy of the presented EMI model is achieved characterizing the matrix impedances with a non simplified array of values (for both magnitude and phase). Moreover, further resolution is achieved because the EMI sources not only contain the PWM information (for the A+ band) but also it might contain the ringing together with actual falling and rising times (for the B band). Finally, the model is corroborated comparing simulation and experimental waveforms spectrum.

II. COMBINED TIME-FREQUENCY METHOD

In order to properly model the MC drive shown in Fig. 1, this paper proposes a combined time-frequency domain method to identify "*sources, paths and victims*".

The proposed modeling method for a conducted EMI is based on a solution over the whole frequency range of interest for a CM equivalent circuit, as illustrated in Fig. 2. Therefore, identification of each part is a crucial step to achieve good model approximation [9]. The identification procedure can be divided in the following steps:

A. The sources

EMI sources are located at points with large dv/dt or di/dt generating CM or DM EMI perturbations respectively. In power electronics systems these types of EMI perturbations are normally produced by the power IGBTs or power diodes.

B. The paths

Mainly, the propagation paths are determined by either the susceptibility of any part of an electronic circuit to change its characteristic impedance along the frequency or the impedance relation between them. As mentioned above, CM EMI perturbations are produced by large dv/dt, therefore, the main EMI paths will be due to parasitic impedances appearing between tracks or between tracks and ground, for example: between semiconductor's terminals and heat-sink or between tracks and the ground plane.

C. The victims

Although any system, or system component, can be susceptible to be affected by EMI perturbations, in ASD systems the electrical grid and the electrical machine are considered victims of special concern. The electrical grid is affected by commutation harmonics deteriorating the distribution quality, whereas, the electrical machine is mainly affected by leakage current producing early winding failure and bearing deterioration, hence, drastically reducing the machine's operational life.

After proper identification of the sources, the paths and victims are identified so an equivalent circuit can be drawn as shown in Fig. 2. A mathematical model can be written in a compact matrix form by applying Kirchhoff laws to the equivalent circuit. The mathematical model can only be solved for a specific frequency; hence, a frequency sweep must be carried out in the frequency range of interest as shown in Fig. 2, where n is the number of harmonics, ω_b is the base frequency and $n \cdot \omega_b$ is the frequency range. The CM EMI frequency range from 9 kHz to 50 MHz, corresponding to conducted emissions, has been considered.

Despite the fact that the overall process described requires a significant number of iterations, it is still more effective than the time domain methods because it avoids long simulation times with small step calculus

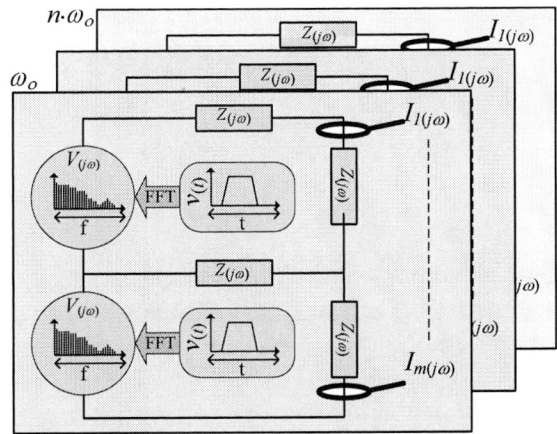

Figure 2. Compact time-Frequency model. Perturbation sources and impedances EMI circuit.

III. MATRIX EMI MODEL

This paper presents a conducted EMI model for a direct MC which allows the analysis of the whole conducted EMI frequency range. Fig. 1 shows the whole drive plus leakage current paths and the additional elements to isolate the drive from external perturbations. The equivalent circuit of the scheme of Fig. 1 is shown in Fig. 3. The EMI model has been created following the procedure explained in section II. The different parts of the model (sources, paths and victims) have been obtained as follows:

A. The sources

From Fig. 1, it can be observed that the large dv/dt appear between each output phase of the MC. An intuitive solution would be to take the EMI sources between each output phase (v_{ab}, v_{bc}). However, the internal impedances of the MC should be known for all the 27 possible combinations, making the approach unviable. This paper proposes a change in the sources reference by taking the EMI sources between each input and output phases of the MC (v_{aA}, v_{bB} and v_{cC}) , avoiding its internal impedances and, hence, simplifying the model as shown in Fig. 3.

B. The paths

In order to obtain a valid EMI model, proper current path identification is mandatory. Moreover, the parasitic impedances which will affect the model behaviour must be thoroughly determined, especially when CM perturbations analysis is of major concern. For example in Fig. 1 shows Z_{HG} and Z_{CHG} which are the main leakage path impedances corresponding to connection cables between the LISN and the MC that have been used to measure the leakage current.

All the impedances in the model have been characterized in frequency domain using two different instruments: an impedance analyzer (Fluke PM6303) for the range of 9 kHz to 1 MHz and a vector network analyzer (R&S ZVRE) for the range of 1 MHz to 50 MHz.

The LISN has been modeled by means of its impedances between the output and ground Z_{SAG}, Z_{SBG} and Z_{SCG} as shown in Fig. 3. Since the input side of the LISN is form by a low

pass filter which behaves like an isolated circuit in the frequency range of interest, only the impedances between the outputs and ground are taken into account

The MC has a low pass input filter consisting of an *LC* circuit with a parallel damping resistor as show in Fig. 1. The filter has been modeled as two lumped impedances per phase:

1. The serial impedances (Z_{SAA}, Z_{SBB} and Z_{SCC}), which account for the set R_f, L_f and the respective high frequency (HF) parasitic impedances.
2. The parallel impedances that appear between phases due to the capacitors C_f (Z_{AB}, Z_{BC} and Z_{AC}).

The MC employed in this work is an all silicon solution which packs all the bidirectional switches in a single power module. Because wiring is avoided, the solution significantly reduces the parasitic inductances. The module's characteristic impedances have been measured in each phase independently and are shown in Fig. 4. Notice that the dashed lines and impedances represent the parasitic elements that appear between terminals and ground (mainly heat-sink). For example, Z_{AH} in Fig. 4 represents the impedance measured between input phase *A* and heat-sink. Z_{BH} and Z_{CH} play the same role with regard to the other phases, *B* and *C* respectively. Using the same criteria, three parasitic impedances have been measured Z_{aH}, Z_{bH} and Z_{cH} between each out phase and heat-sink.

The load has been modeled taking into account the connection cables and the machine and measuring the impedances between each phase cable and the shielded ground cable. In Fig. 5, Z_{wireA} and Z_{parA} are the parasitic impedances representing the cable, Z_A is the impedance of the motor's phase *a* and Z_{CM} is the coupling impedance between motor's winding and ground motor frame [22].

It should be noted that in contrast to [11], where an equivalent circuit based on *R*, *L* and *C* elements emulates the real frequency response of each impedance, the approach presented in this paper works with an array of measured values thoroughly describing the impedance frequency response. The benefits of the proposed approach are more relevant in the high frequency range where the method adopted in [11] drastically reduces its accuracy.

C. The victims

As pointed out above, the victims in power electronics are mainly the electrical grid and the machine. Therefore, the leakage currents considered are I_{HG} and I_{CHG}. The leakage current I_{CHG} will affect the motor whereas the sum $I_{HG}+I_{CHG}$ will affect the electrical grid.

Finally, from the equivalent circuit in Fig. 3, the mathematical model has been obtained in the matrix form shown in (1) which can be easily solved by any mathematical software.

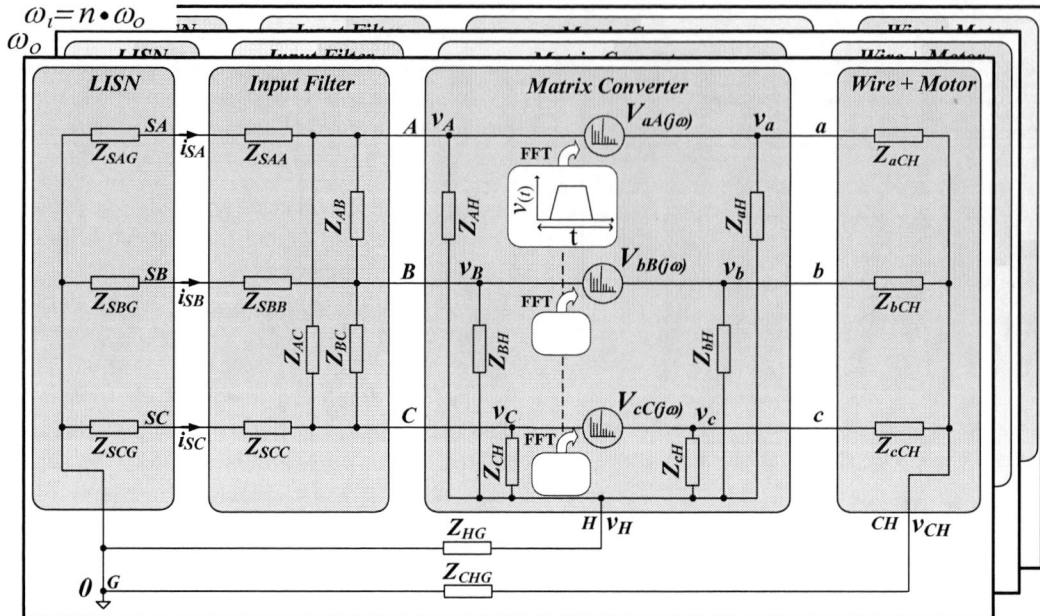

Figure 3. Whole EMI Model set-up; MC, load and LISN.

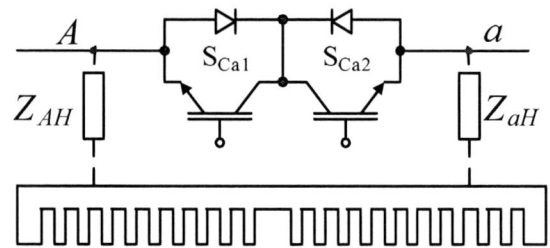

Figure 4. Input and output impedance to heat sink.

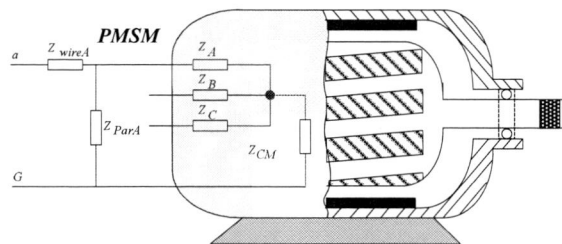

Figure 5. Load impedance model.

IV. SIMULATION

The EMI perturbations of the MC have been simulated using mathematical software MATLAB®. Due to limitations in the number of samples acquired, the simulations have been divided in two parts: Low and high frequency range. For the low frequency range, one electrical machine cycle containing a huge number of commutations corresponding to Space vector Modulation (SVM) patterns has been employed as the EMI sources. A single SVM commutation which includes rise and fall time and ringing effects has been considered as EMI sources as illustrated in Fig. 6.

As shown in Fig 6 the time domain signals are transformed to the frequency domain by applying the fast Fourier transform (FFT). These signals can be obtained in two different ways: From a simulation model which emulates the MC drive behavior or from an experimental setup capturing and recording the signals with and oscilloscope. A Tektronics® TDS510A has been used in this work.

Usually, when conducted EMI perturbations are analyzed, the frequency range is split into two bands; A band and B band. A band covers the frequency range from 9 kHz to 150 kHz whereas B band goes from 150 kHz up to 30 MHz.

However, in this work the low frequency band has been enlarged until 1 MHz because harmonics of power electronic converters easily go beyond the A band limit (150 kHz) and

therefore, the output spectra cannot be properly displayed. The enlarged band has been renamed in this work as A⁺ band.

In order to simulate the EMI perturbations, waveforms obtained from an MC dynamic simulation model driving a PMSM have been employed as EMI sources. The obtained waveforms are shown in Fig. 7 (a). For the HF range a single commutation of the SVM pattern has been applied. The waveforms have been obtained from the same dynamic model increasing the resolution accounting for the HF effects as Fig. 7 (b) illustrates.

All waveforms have been recorded with the motor running at 100 rad/s under no load conditions

Results of the simulations corresponding to A⁺ band and B band are shown in Fig. 8 (a) and Fig. 8 (b) respectively. Notice that the current shown in both figures is $I_{CM(j\omega)}=(I_{HG(j\omega)}+I_{CHG(j\omega)})$.

V. EXPERIMENTAL RESULTS

The experimental setup employed in this work corresponds to the one shown in Fig. 1. The characteristics of each part of the setup are listed below:

- *MC prototype*: Nottingham University, 7.5 kW, 380 V. It is constituted by an IGBT module FM35R12KE3ENG (Eupec).

$$
\begin{bmatrix}
\sum_A i = 0 \\
\sum_B i = 0 \\
\sum_H i = 0 \\
\sum_C i = 0 \\
\sum_{CH} i = 0 \\
v_{aA} \\
v_{bB} \\
v_{cC}
\end{bmatrix}
=
\begin{bmatrix}
\left(\frac{1}{(Z_{SAG}+Z_{SAA})}+\frac{1}{Z_{AC}}+\frac{1}{Z_{AB}}+\frac{1}{Z_{AH}}\right) & \left(\frac{1}{Z_{aH}}+\frac{1}{Z_{aCH}}\right) & \frac{-1}{Z_{AB}} & 0 \\
\frac{-1}{Z_{AB}} & 0 & \left(\frac{1}{(Z_{SBG}+Z_{SBB})}+\frac{1}{Z_{BC}}+\frac{1}{Z_{BH}}+\frac{1}{Z_{AB}}\right) & \left(\frac{1}{Z_{bH}}+\frac{1}{Z_{bCH}}\right) \\
\frac{1}{Z_{AH}} & \frac{1}{Z_{bH}} & \frac{1}{Z_{BH}} & \frac{1}{Z_{bH}} \\
\frac{1}{Z_{AC}} & 0 & \frac{1}{Z_{BC}} & 0 \\
0 & \frac{1}{Z_{aCH}} & 0 & \frac{1}{Z_{bCH}} \\
-1 & 1 & 0 & 0 \\
0 & 0 & -1 & 1 \\
0 & 0 & 0 & 0
\end{bmatrix}
$$

$$
\begin{bmatrix}
-\left(\frac{1}{Z_{aH}}+\frac{1}{Z_{AH}}\right) & \frac{-1}{Z_{AC}} & 0 & \frac{-1}{Z_{aCH}} \\
-\left(\frac{1}{Z_{AH}}+\frac{1}{Z_{bH}}\right) & \frac{-1}{Z_{BC}} & 0 & \frac{-1}{Z_{bCH}} \\
-\left(\frac{1}{Z_{HG}}+\frac{1}{Z_{AH}}+\frac{1}{Z_{aH}}+\frac{1}{Z_{BH}}+\frac{1}{Z_{bH}}+\frac{1}{Z_{cH}}+\frac{1}{Z_{CH}}\right) & \frac{1}{Z_{CH}} & \frac{1}{Z_{cH}} & 0 \\
\left(\frac{1}{Z_{CH}}+\frac{1}{Z_{cH}}\right) & -\left(\frac{1}{(Z_{SCG}+Z_{SCC})}+\frac{1}{Z_{AC}}+\frac{1}{Z_{BC}}+\frac{1}{Z_{CH}}\right) & -\left(\frac{1}{Z_{cH}}+\frac{1}{Z_{cCH}}\right) & \frac{1}{Z_{cCH}} \\
0 & 0 & \frac{1}{Z_{cCH}} & -\left(\frac{1}{Z_{aCH}}+\frac{1}{Z_{bCH}}+\frac{1}{Z_{cCH}}+\frac{1}{Z_{CHG}}\right) \\
0 & 0 & 0 & 0 \\
0 & 0 & 0 & 0 \\
0 & -1 & 1 & 0
\end{bmatrix}
\cdot
\begin{bmatrix}
v_A \\ v_a \\ v_B \\ v_b \\ v_H \\ v_C \\ v_c \\ v_{CH}
\end{bmatrix}
\qquad (1)
$$

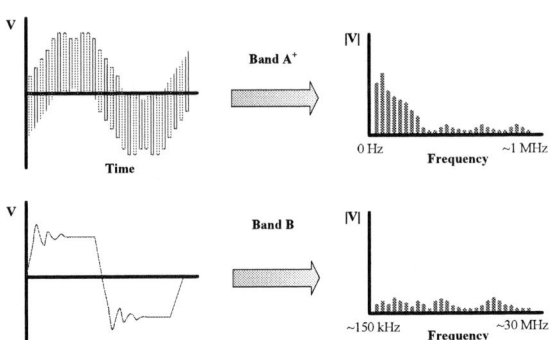

Figure 6. Temporal waveform conversion to frequency domain.

- *Permanent magnet synchronous motor*: Control Techniques, 3x220 V / 0.94 kW, 3000 rpm.

- *LISN*: Scwarzbeck Meßelektronik, NNLA8120, 50 Ω/5 Ω + 50 µH, 4x25 A.

- *Shielded cables*: Five conductors, section= 2.5 mm² + shielding braid.

In order to analyze the CM perturbations, the leakage currents flowing through the ground path have been measured employing a HF current probe and a spectrum analyzer (R&S ESPI).

The spectrums of I_{CM} corresponding to A$^+$ band and B band are shown in Fig. 8 (a) and Fig. 8 (b) respectively. A frequency reference of 9 kHz to 1 MHz with a Resolution Bandwidth (RBW) of 300 Hz have been set up for measurements corresponding to A$^+$ band. For B band measurements, the frequency reference was set to 150 kHz to 50 MHz with RBW of 9 kHz.

Comparing the experimental results with those obtained in

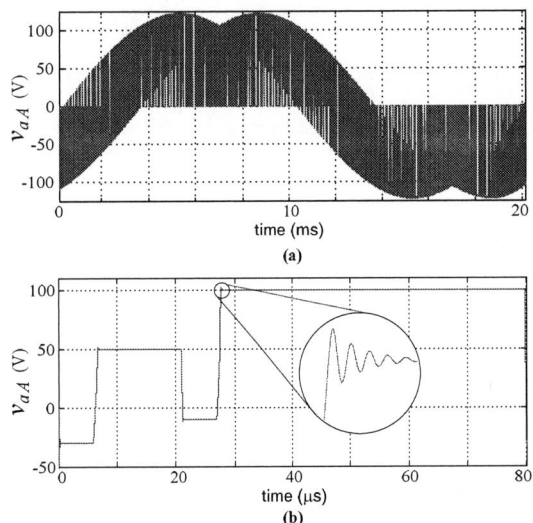

Figure 7. Simulation signals. a) EMI source used in A$^+$ band. b) EMI source used in B band.

section IV, it can be noted the similarity of both spectrums. However, when A$^+$ band is analyzed, small divergences between 8 dB and 14 dB can be observed. These divergences can be attributed to non-ideal effects that have not been taken into consideration in the MC dynamic model. On the other hand, comparing the results corresponding to B band, the only difference arises in the frequency range from 15 MHz to 50 MHz due to a spread spectrum clock that appears when the measurements are carried out on the experimental setup. After a tedious measurement work, the origin of this spread spectrum clock was located on the FPGA board that handles the SVM algorithm.

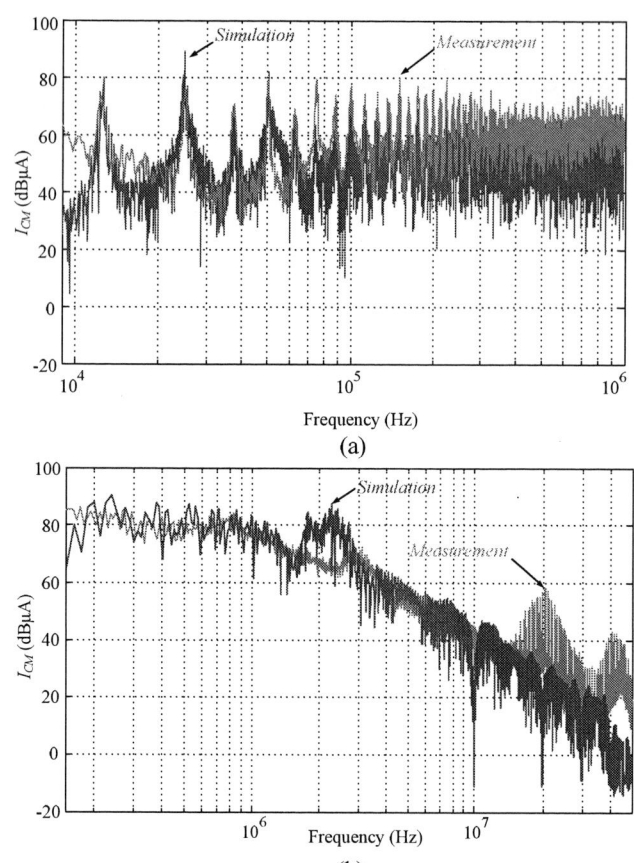

Figure 8. Simulation and experimental results. a) A$^+$ Band. b) B Band.

VI. CONCLUSIONS

A new EMI model of an MC has been developed in this work. The model allows the analysis of the conducted CM perturbations produced by an MC. Moreover, the paper also proposes a combined time-frequency simulation, reproducing the CM perturbations produced by an MC, with low-computational requirements.

The advantages of such simple model are

- It allows easy simulation of EMI when changing the switching pattern or the modulation strategy.

- It provides identification of the different propagation paths, and gives separate figures for CM, which eases the EMI filters design.

- No detailed layout description is needed. The path impedances can be measured or derived from previous simulations.

ACKNOWLEDGMENT

The authors acknowledge the financial support received from "Ministerio de Ciencia e Innovación de España" for realizing this work under "TEC2007-61582" research project.

REFERENCES

[1] V. Jithesh and D. C. Pande, "A review on computational EMI modelling techniques," in *Electromagnetic Interference and Compatibility, 2003. INCEMIC 2003. 8th International Conference on*, 2003, pp. 159-166.

[2] L. Ran, *et al.*, "Conducted electromagnetic emissions in induction motor drive systems. I. Time domain analysis and identification of dominant modes," *Power Electronics, IEEE Transactions on*, vol. 13, pp. 757-767, 1998.

[3] L. Ran, *et al.*, "Conducted electromagnetic emissions in induction motor drive systems. II. Frequency domain models," *Power Electronics, IEEE Transactions on*, vol. 13, pp. 768-776, 1998.

[4] J. R. Regue, *et al.*, "A genetic algorithm based method for source identification and far-field radiated emissions prediction from near-field measurements for PCB characterization," *Electromagnetic Compatibility, IEEE Transactions on*, vol. 43, pp. 520-530, 2001.

[5] M. Moreau, *et al.*, "Modeling of Conducted EMI in Adjustable Speed Drives," *Electromagnetic Compatibility, IEEE Transactions on*, vol. 51, pp. 665-672, 2009.

[6] C. Purcarea and P. Mutschler, "Investigation of EMI reduction techniques using time domain simulation of drives," in *Industrial Electronics, 2009. IECON '09. 35th Annual Conference of IEEE*, 2009, pp. 1174-1179.

[7] V. Tarateeraseth, *et al.*, "Assessment of Equivalent Noise Source Approach for EMI Simulations of Boost Converter," in *Electromagnetic Compatibility, 2009 20th International Zurich Symposium on*, 2009, pp. 353-356.

[8] K. Mainali and R. Oruganti, "Simple Analytical Models to Predict Conducted EMI Noise in a Power Electronic Converter," in *Industrial Electronics Society, 2007. IECON 2007. 33rd Annual Conference of the IEEE*, 2007, pp. 1930-1936.

[9] D. Gonzalez, *et al.*, "New simplified method for the simulation of conducted EMI generated by switched power converters," *Industrial Electronics, IEEE Transactions on*, vol. 50, pp. 1078-1084, 2003.

[10] J. C. Crebier and J. P. Ferrieux, "PFC full bridge rectifiers EMI modeling and analysis-common mode disturbance reduction," *Power Electronics, IEEE Transactions on*, vol. 19, pp. 378-387, 2004.

[11] E. Gubia, *et al.*, "Frequency domain model of conducted EMI in electrical drives," *Power Electronics Letters, IEEE*, vol. 3, pp. 45-49, 2005.

[12] C. Jettanasen, *et al.*, "Common-Mode Emissions Measurements and Simulation in Variable-Speed Drive Systems," *Power Electronics, IEEE Transactions on*, vol. 24, pp. 2456-2464, 2009.

[13] L. Qian, "Modular Approach for Characterizing and Modeling Conducted EMI Emissions in Power Converters," Electrical and Computer Engineering, Virginia Tech, Virginia, 2005.

[14] Z. Dongsheng, *et al.*, "Common-Mode DC-Bus Filter Design for Variable-Speed Drive System via Transfer Ratio Measurements," *Power Electronics, IEEE Transactions on*, vol. 24, pp. 518-524, 2009.

[15] S. Zhang and K. J. Tseng, "Modeling, simulation and analysis of conducted common-mode EMI in matrix converters for wind turbine generators," in *Power Electronics and Motion Control Conference, 2008. EPE-PEMC 2008. 13th*, 2008, pp. 2516-2523.

[16] C. Ortega, *et al.*, "Improved Waveform Quality in the Direct Torque Control of Matrix-Converter-Fed PMSM Drives," *Industrial Electronics, IEEE Transactions on*, vol. 57, pp. 2101-2110, 2010.

[17] J. F. Kolar, T. Krismer, F. Round, S., "The essence of three-phase AC/AC converter systems," *Przeglad Elektrotechniczny*, pp. 14-29, July 2008

[18] C. Ortega, *et al.*, "Reduction of the common mode voltage of a matrix converter fed direct torque control," *IEICE Electronics Express*, vol. 7, pp. 1044-1050, 2010.

[19] R. Vargas, *et al.*, "Predictive Torque Control of an Induction Machine fed by a Matrix Converter with Reactive Input Power Control," *Power Electronics, IEEE Transactions on*, vol. PP, pp. 1-1, 2010.

[20] R. K. Gupta, *et al.*, "Direct-Matrix-Converter-Based Drive for a Three-Phase Open-End-Winding AC Machine With Advanced Features," *Industrial Electronics, IEEE Transactions on*, vol. 57, pp. 4032-4042, 2010.

[21] S. Lopez Arevalo, *et al.*, "Control and Implementation of a Matrix-Converter-Based AC Ground Power-Supply Unit for Aircraft Servicing," *Industrial Electronics, IEEE Transactions on*, vol. 57, pp. 2076-2084, 2010.

[22] U. T. Shami and H. Akagi, "Experimental Discussions on a Shaft End-to-End Voltage Appearing in an Inverter-Driven Motor," *Power Electronics, IEEE Transactions on*, vol. 24, pp. 1532-1540, 2009.

Industrial DC/DC Converters in Terms of EMC

Zdenek Kubik, Jiri Skala
Department of Applied Electronics and Telecommunications
University of West Bohemia in Pilsen, Czech Republic
zdekubik@kae.zcu.cz, skalaj@kae.zcu.cz

Abstract- **This paper deals with industrial DC/DC converters in terms of EMC. The first part of paper describes basic information from standard EN 55022:2006, which establishes EMI limits for DC/DC converters. The second part shows the most interesting measured values of samples industrial DC/DC converters.**

INTRODUCTION

This paper deals with electromagnetic interference measurement of DC/DC converters. Disturbance limits of DC/DC converters are given by European standard EN 55022:2006 – Information technology equipment - Radio disturbance characteristics - Limits and methods of measurement [1]. The section 3.1 of this standard describes information technology equipment (ITE) – any equipment:

a) which has a primary function of either (or a combination of) entry, storage, display, retrieval, transmission, processing, switching, or control of data and telecommunications messages and which may be equipped with one ore more terminal ports typically for information transfer;

b) with a rated supply voltage not exceeding 600V [2].

The section 3.4 defines Module:

- Module is a part of an ITE which provides a function and may contain radio-frequency sources [2]. DC/DC converters are used like module, when are implemented into ITE.

The standard EN 55022 divides ITE into two classes. Class B ITE is intended primarily for use in the domestic environment and class A ITE are all others ITE [2].

For this reason, there exist two groups of limits. The first is being used for class A ITE and the second is being used for class B ITE. These limits are defined in the frequency range between 150kHz and 6GHz [3]. The frequency range is divided according to measurement procedures:

- conducted emission in the frequency range from 150kHz to 30MHz;

- radiated emission from 30MHz to 6GHz.

For limits measuring it is necessary to use measurement receiver with average detector, quasi-peak and peak detector in accordance with CISPR 16 – Specification for radio disturbance and immunity measuring apparatus and methods [4].

EN 55022 differentiated conducted disturbance limits on mains ports and telecommunication ports. For DC/DC converters are applied limits for mains ports. There are two possibilities for measurement of conducted disturbance from equipment under test (EUT): the first is a voltage method and the second is a current method. Test circuits for these methods represent Fig. 1. (voltage methods) and Fig. 2. (current method).

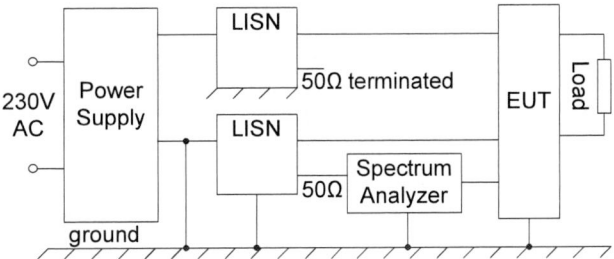

Fig. 1. Test circuit for measurement of conducted disturbance, voltage method.

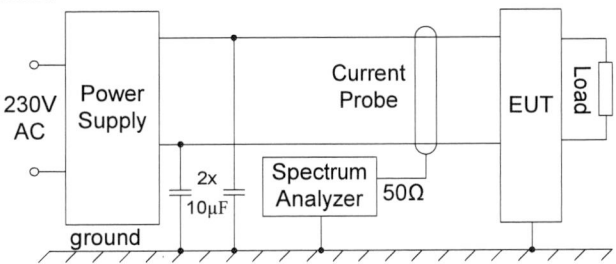

Fig. 2. Test circuit for measurement of conducted disturbance, current method.

The normalized work location with measuring antenna is used for measuring of radiated disturbance. Fig. 3. shows an example of the free space work location.

Measuring distance R between EUT and antenna depend on frequency range:

- R=10m in the frequency range from 30MHz to 1GHz;

- R=3m in the frequency range from 1GHz to 6GHz.

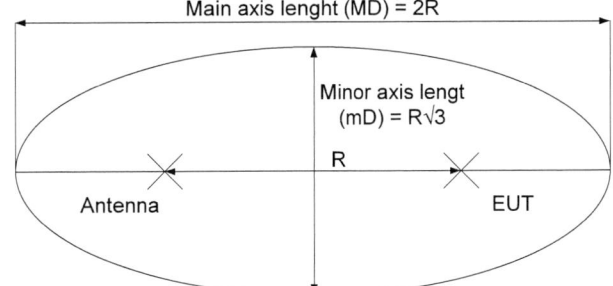

Fig. 3. Example of radiated disturbance test site.

Limits for conducted and radiated disturbance at mains ports and telecommunication ports are given in the Tables I – VII [2], [3]. The limits for telecommunications ports are lower than for mains ports.

TABLE I
LIMITS FOR CONDUCTED DISTURBANCE AT THE MAINS PORTS, A ITE

Frequency range [MHz]	Levels A ITE [dBuV]	
	Quasi-peak	Average
0.15 to 0.5	79	66
0.5 to 5	73	60
5 to 30	73	60

TABLE II
LIMITS FOR CONDUCTED DISTURBANCE AT THE MAINS PORTS, B ITE

Frequency range [MHz]	Levels B ITE [dBuV]	
	Quasi-peak	Average
0.15 to 0.5	66 to 56	56 to 46
0.5 to 5	56	46
5 to 30	60	50

TABLE III
LIMITS FOR CONDUCTED COMMON MODE DISTURBANCE AT THE TELECOMMUNICATION PORTS, A ITE

Frequency range [MHz]	Levels A ITE [dBuV]	
	Quasi-peak	Average
0.15 to 0.5	97 to 87	84 to 74
5 to 30	87	74

TABLE IV
LIMITS FOR CONDUCTED COMMON MODE DISTURBANCE AT THE TELECOMMUNICATION PORTS, B ITE

Frequency range [MHz]	Levels B ITE [dBuV]	
	Quasi-peak	Average
0.15 to 0.5	84 to 74	74 to 64
5 to 30	74	64

TABLE V
LIMITS FOR RADIATED DISTURBANCE , R=10M

Frequency range [MHz]	Levels A ITE [dBuV]	Levels B ITE [dBuV]
	Quasi-peak	Average
30 to 230	40	30
230 to 1000	47	37

TABLE VI
LIMITS FOR RADIATED DISTURBANCE, A ITE, R=3M

Frequency range [MHz]	Levels A ITE [dBuV]	
	Peak	Average
1000 to 3000	76	56
3000 to 6000	80	60

TABLE VII
LIMITS FOR RADIATED DISTURBANCE, B ITE, R=3M

Frequency range [MHz]	Levels B ITE [dBuV]	
	Peak	Average
1000 to 3000	70	50
3000 to 6000	74	54

DC/DC CONVERTERS MEASUREMENT

Two groups of converters were used for measurement. The first group consists of two 30W converters with similar parameters:
- Converter 1: 30W, output 15V/2A, input 36V-75V;
- Converter 2: 30W, output 15V/2A, input 36V-75V.

The second group consists of two 50W converters with similar parameters:
- Converter 1: 50W, output 5V/10A, input 36V-75V;
- Converter 2: 50W, output 5V/10A, input 36V-75V.

The voltage method was used for measurement of conducted disturbance. Block diagram of the test circuit is shown in Fig. 1. The type of Line impedance stabilization network (LISN) is 50Ω/50μH+5Ω, L2 produced by PMM, spectrum analyzer PMM8000+.

The power supply voltage was chosen with reference to application of converters at 48V. This value is standard for supplying of industrial power lines. For this reason the class A ITE limits was used.

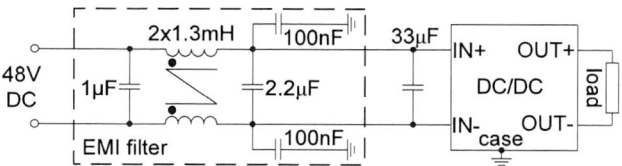

Fig. 4. EUT schematic diagram.

A. Conducted disturbance measurement
1. 30W converters

The Fig. 5. shows the quasi-peak conducted disturbance of loaded 30W converters in the frequency range between 150kHz and 5MHz. The class A limits for measurements at mains ports are also shown.

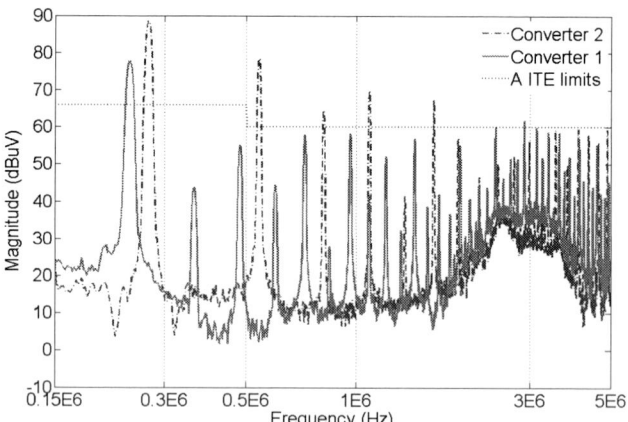

Fig. 5. Conducted disturbance in the frequency range 150kHz – 5MHz from 30W converters without input EMI filter. Quasi-peak magnitude.

The magnitude in this range exceeds limits for A ITE. The conducted disturbance at frequency higher than 5MHz is in compliance with the standard EN55022. Operating frequency is 240kHz (converter 1) and 272kHz (converter 2). At these frequencies magnitude exceeds limits by 11.8dBμV and

978-1-4244-8806-3/11 $26.00 © 2011 IEEE

22.6dBμV. For compliance with the standard the input EMI filter must be used. Schematic diagram of simple EMI filter is shown in Fig. 4.

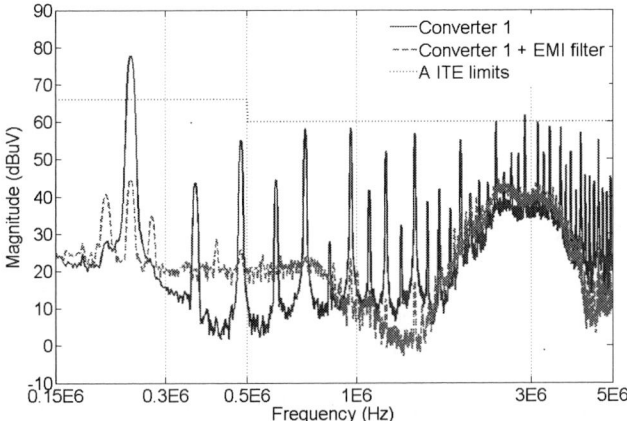

Fig. 6. Converter 1 (30W) - Conducted disturbance in the frequency range 150kHz – 5MHz, quasi-peak magnitude.

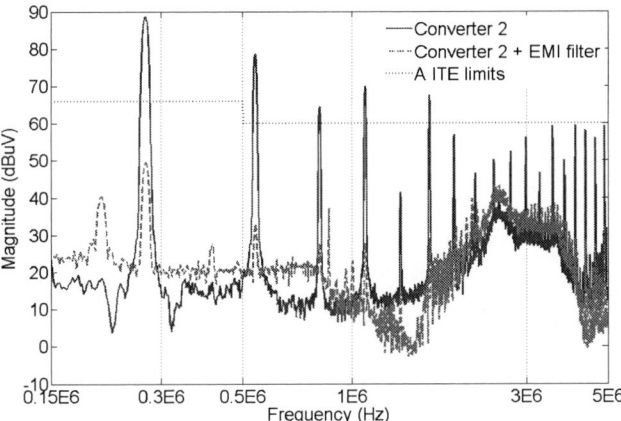

Fig. 7. Converter 2 (30W) - Conducted disturbance in the frequency range 150kHz – 5MHz, quasi-peak magnitude.

Results comparison of measurement 30W converters without or with input EMI filter is shown in the Fig. 6. and in the Fig. 7. Converters with the EMI filter are in compliance with the standard EN55022.

2. 50W converters

The Fig. 8. displays the quasi-peak conducted disturbance of loaded 50W converters in the frequency range between 150kHz and 30MHz. The class A limits for measurements at mains ports are also shown.

The magnitude in this range exceeds limits for A ITE. Operating frequency is 480kHz (converter 1) and 490kHz (converter 2). At these frequencies magnitude exceeds limits by 32.5dBμV and 27.7dBμV. For compliance with the standard the input EMI filter must be used. Schematic diagram of EMI filter is shown in Fig. 4, it is the same as EMI filter for 30W converters.

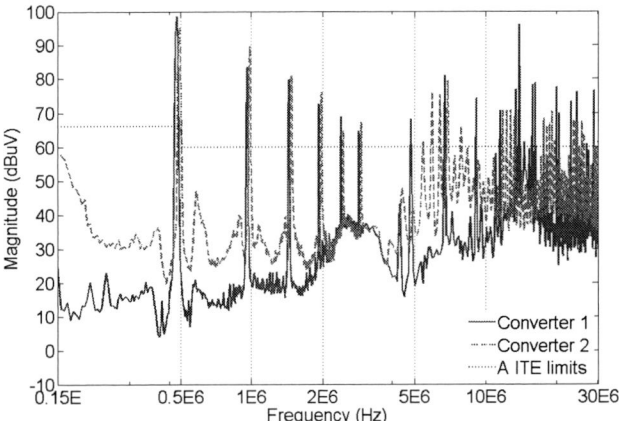

Fig. 8. 50W converters - Conducted disturbance in the frequency range 150kHz – 30MHz, without input EMI filter, quasi-peak magnitude.

Results comparison of measurement 50W converters without or with input EMI filter is shown in the Fig. 9. and in the Fig. 10. In these cases, the EMI filter didn't work correctly; at frequencies above 10MHz conducted disturbance magnitude exceeds limits of standard EN55022.

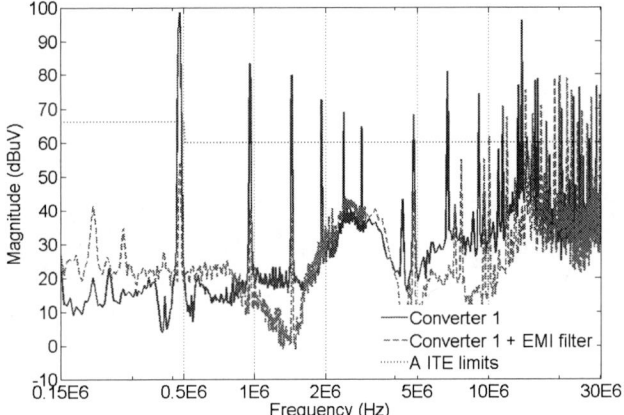

Fig. 9. Converter 1 (50W): Conducted disturbance in the frequency range 150kHz – 30MHz. Quasi-peak magnitude.

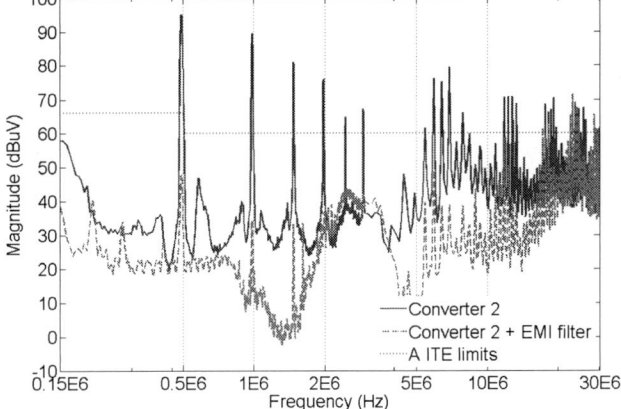

Fig. 10. Converter 2 (50W): Conducted disturbance in the frequency range 150kHz – 30MHz. Quasi-peak magnitude.

B. Radiated disturbance measurement

The radiated disturbance test was executed for all converters. The test site responds with Fig. 3., but the distance between EUT and antenna in the frequency range 30MHz and 200MHz was only 3m, because the size of test place forbids placing the antenna in 10m distance. Measurement was executed in the frequency range from 30MHz to 200MHz, at higher frequency with respect to operating frequencies of converters there is no occasion to measure. The biconical antenna was used for measuring in this frequency range. The Fig. 11. shows three trace, the first presents quasi-peak radiated disturbance of 50W converter 1 (the worst case of radiated disturbance), the second presents background of measuring place and the third shows the class A limit for measurements of radiated disturbance (for 10m distance between EUT and antenna).

Limits were exceeded in the frequency range from 88MHz to 106MHz, which is a range in which there are distribute signals of FM radio stations. In comparison with the background is likely that converters are in compliance with standard EN 55022.

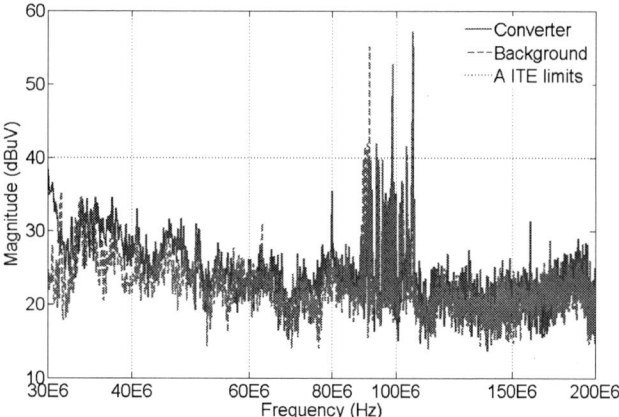

Fig.11 . Converter 1 (50W) - Radiated disturbance in the frequency range 30MHz – 200MHz, quasi-peak magnitude.

CONCLUSIONS

This paper presents the standard EN55022. The standard describes limits of conducted and radiated disturbance from information technology equipments.

The next part of paper deals with measurement of samples industrial DC/DC converters. All converters were measured in the frequency range from 150kHz to 30MHz (conducted disturbance) and from 30MHz to 200MHz (radiated disturbance). Only the most interesting results are displayed. Conducted disturbance magnitude of all converters without input EMI filter exceeds the limits for measurements at mains ports, it is necessary to use input EMI filter for all industrial DC/DC converters with similar parameters as there measured converters. 30W converters with simple input EMI filter meet all requirements of the EN55022. 50W converters with simple input EMI filter doesn't meet limits in the frequency range from 20MHz to 30MHz, in this case it is more suitable to use different multi-stage filter, which has higher attenuations in this frequency range. Radiated disturbance magnitude of all converters in the frequency range from 30MHz to 200MHz meets all requirements of the EN55022.

ACKNOWLEDGMENT

Research which is described in the paper was financially supported by the Czech Grant Agency under grant No. GA102/09/1164.

REFERENCES

[1] SLTA015A: EMI Considerations for DC to DC converters and Integrated Switching Regulators, Rev. 6/30/2000. <http://focus.ti.com>
[2] ČSN EN 55022 ed.2 (33 4290): *Zařízení informační techniky – Charakteristiky vysokofrekvenčního rušení – Meze a metody měření.* Praha: ČNI, 2007.
[3] ČSN EN 55022 ed.2 Změna A1 (33 4290): *Zařízení informační techniky – Charakteristiky vysoko-frekvenčního rušení – Meze a metody měření.* Praha: ČNI, 2008.
[4] ČSN EN 55016 (33 4210): *Specifikace přístrojů a metod pro měření vysokofrekvenčního ručení a odolnosti.* Praha: ČNI, 2005.

Application of Databases for Compatibility of Electrical Installation Design

M. Lehtla[1], M. Müür[1], I. Kõivastik[1], K. Müül
[1] Department of Electrical Drives and Power Electronics, Tallinn University of Technology
Ehitajate tee 5, 19086 Tallinn (Estonia)
E-mail: mlehtla@ttu.ee

Abstract - **Information systems and databases have significantly improved the design process of electrical systems. In recent years, several Internet-based information systems have become accessible directly from design software. Unfortunately integration of closed systems requires extra expenses from design companies. This paper gives a state of the art overview of information systems and databases used for design.**

I. INTRODUCTION

Wide application of complex building of automation systems with power electronics, converters and sensitive low-current equipment sets new requirements for electrical installation design, such as electromagnetic compatibility, control design and energy efficiency. New DC lighting systems, electronic transformers and lighting controllers cause changes in electrical installation design and wiring. New and uncommon systems should be well documented to avoid problems in installation and operation.

The 3D design is already in wide use in many different design fields. This brings new opportunities for energy efficient design of complex electrical heating, lighting, and ventilation systems. Re-use of component information and 3D models can integrate effort of different designers. Approaches that have been used separately in specific electronics or electrical design fields [6, 7] for more than ten years can now be integrated. Also, control circuits of electric drives for ventilation, heating, water supply and other application can be collected into databases.

Advanced CAD packages include databases with large amounts of component information. These databases have to be continuously updated. Symbol attributes on drawings are used to store links or names to databases. If a database software such as Microsoft Access or OpenOffice Base is not available, limited database engines such as Microsoft Access Runtime have the compact interfaces to enable CAD software to access database files. Common databases contain data for all CAD designers and projects for an entire company. These databases contain information about technical properties of devices, cables, lighting, heaters, etc. Product databases are stored in separate files that enable flexible updating of product data from different companies. The project-specific data is included into a project database, which is common for all drawings of the project. Typical projects usually include multiple types of drawings, such as floor-plans, installation drawings, layout drawings of electrical centers, circuit diagrams of electrical centers, etc. The project database gathers information from all these drawings together, and is updated according to project drawing files. Links between different files and databases are shown in Fig. 1.

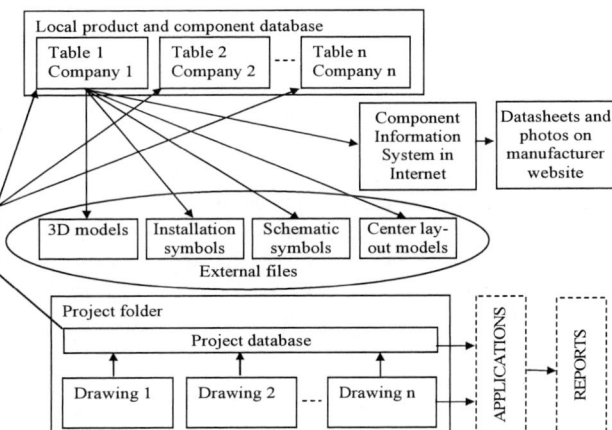

Fig. 1. Database structure in some CAD systems.

Typical problems and limitations of traditional 2D designs:

1. Height information about installing of cables and devices is shown by arrows on drawings. Fitting of additional information as arrows and texts is difficult on complex drawings. Mostly only relative height information is specified, only some pipes (sewage and cable pipes) and cable routes require absolute height information for compatibility with outside systems.

2. Designers and field-workers sometimes acquire insufficient information from 2D drawing when drawings describe objects with complicated geometry. Understanding and drawing objects with complicated geometry require high-skilled design staff. This information is very important in pipes, routes and construction to avoid collisions.

3. Clearance requirements between different systems are set in standards and compulsory normative but it is often complicated to fulfill these requirements with insufficient information.

4. Designs often do not have enough information about the amount of cables which leads to problems when too narrow cable routes are selected in the design or construction. Specifying clearances and cable types is additional work. Insufficient attention to the above could cause significant problems in large designs.

5. The cable routes and wall vias should have reserve for expanding and replacement of cables. This is especially important if new communication systems, such as fibre-optic computer networks, could be used in future. Drilling new vias should be avoided and cable route vias should often be sound, humidity and fire proof.

6. Implementation can differ from initial design because of misunderstandings in field work or changes by the owner or builder. This leads to changes in the initial design and thus to additional cost.

7. More complex designs could require special paper formats larger than A0, such as 900x1600mm.

II. MODELS

CAD drawings usually contain additional information inside drawings as symbol attributes. This information is also used for locating the given component for detailed visualization or modeling. The height data enables automatic conversion from 2D models to 3D models and correcting placement of component locations in 3D models. These drawing symbol attributes can also include product specific information for calculations. Modeling and simulation and calculation packages include libraries and databases with model parameters of products, including those for electrical, heating, lighting and ventilation design. Depending on the type of design it is possible to insert these models to the model of building or to the 3D layout model of an electrical centre.

Standardized model and drawing file formats, such as IFC 2x3, can enable file transfer between different application and software packages [1], including different analysis software. Validation software enables checking for compliance between different installation systems designs, such as electrical, water supply, heating and ventilation. Co-operation possibilities between different designers and users via standardized file formats and common databases are shown in Fig. 2.

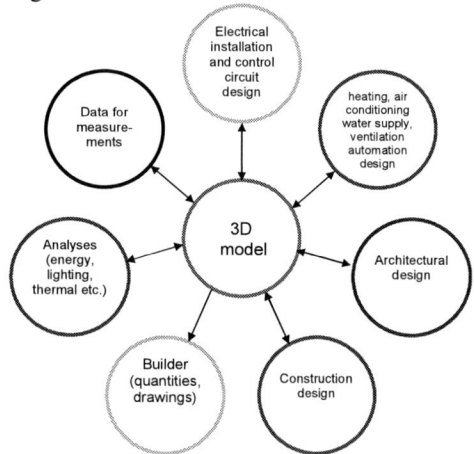

Fig. 2. Cross-usage of information for CAD models on building design.

Project-database based design enables linking elements between different diagram types and models. Project database is used for storing information which cannot be stored or which is not needed inside the diagram or model files. For example, manufacturers give 3D and electrical behavior models of their products as separate CAD drawings or model files and designers can associate these files via databases directly to their projects. Files are located in local hard disks or in network servers. The file that is located in the network can be accessed by multiple users. To avoid simultaneous unintended modification by other users, the mutual exclusion locking files are used. It is also possible to collect measurement data into such databases.

III. VISIBILITY FILTERS, LAYERS AND CLEARANCES

Filters and layers enable selecting which systems are visible and editable in the model design and which systems are printed. Electrical design packages can use standardized layering systems for different high and low current systems.

Drawings that include information of high-current systems have layers for installation, main electrical distribution, connections, such as cabling, lighting and electric heating. Drawings that include information of low-current systems can have separate layers for information systems, communication systems, signaling systems, security systems, computer networking systems, and automation systems.

Designers of particular systems can set other system models invisible but they cannot ignore other systems because of placement issues, clearances between different systems for fulfilling EMC requirements for weak current and communication cables, fire safety issues and sensor placement issues. Thus, systems have interactions and constraints according to specific building normatives.

Automatic component placement and cable routing programs can use all available information found in visible or invisible models, project database and constraints of each specific system. For manual placement of components into an installation model, these layers of other systems can be switched on and off on reference drawings during design or can be made partially transparent.

IV. CIRCUIT DIAGRAMS AND SCHEMAS

Circuit diagrams are a common way to describe electrical circuits and their functional operation. Block diagrams offer a simplified overview of entire systems or part of the systems such as an electrical center. Simplified block schemas are often needed besides a detailed layout used in 2D/3D models. These block schemas can be used as an interactive table of contents with Internet links for locating elements in models and layout drawings. The project database is a key for linking of different types of drawings and drawing elements in the project or across projects.

As an example, the electrical center supply groups for electrical protection that are specified on the installation drawing can be later used on the electrical circuit diagram of the centre. The detailed information about cables set in installation models such as types, lengths and cross sections [1, 2] enables correct calculation of currents of short-circuits. All this information is transferred via database and thus is accessible for other tools or applications.

V. FITTING AND ROUTING

Correct creation of a 3D model requires detailed height information about component locations, which are different for different components. Correct calculation of voltage drops and short-circuit currents requires correct information about

978-1-4244-8806-3/11 $26.00 © 2011 IEEE

conductors and their length. This information can be easily acquired from a correct 3D model.

Views or traditional 2D floor plans can be generated from a 3D model. This requires specification of component models for floor plan symbols. The links between 2D symbols and 3D model blocks can be specified via a component database or directly in a drawing file as symbol attributes.

Recent software packages allow automatic fitting and routing of cables. Automatic routing of floor-heating cables to given floor sections according to given constraints is shown in Fig. 3. A designer can select a suitable conductor or a cable with a given power rating and length from the database.

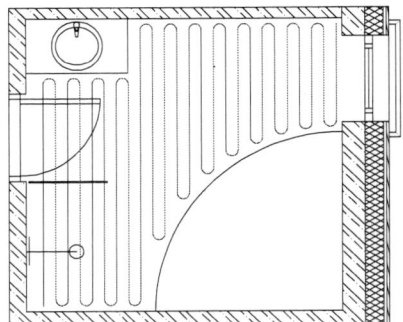

Fig. 3. Sample of automatically-routed layout of floor-heating cables.

Cables have different constraints for installations, such as minimum allowed installation radius, minimum via diameter, etc. All this data can be included into the product database and used directly from CAD systems. Recent CAD systems enable counting quantities [2] of cable trays, rails, raceways or tubes. The lengths of cables can be fitted according to cable trays, rails, raceways or tubes shown on drawings. Multiple cables with different outer diameters can be fitted into a single cable tray, rail, raceway or tube.

Besides automatic routing of conductors and cables, the software packages simplify the placement of different automation sensors by setting their operating sectors and areas on floor plans.

VI. INFORMATION SYSTEMS

Good cooperation between different designers is the key to benefit from the 3D design. Main problems in the cooperation are:

1. Different companies are designing different parts, such as electrical supply, weak current systems, water supply systems, ventilation and heating automation etc. using different and partially compatible software packages.
2. Different parts of the project are often made in different software packages with different style setting for linotypes and text styles (AutoCAD, MagicCAD, CADS, Microstation, etc.).
3. Project server problems can lead to problems in project database management and technical complications in cooperation. Project changes are often documented in separate databases.
4. Large files cannot be simultaneously edited from multiple computers which leads to the use of many reference

drawings and many drawing files in a single project server inside a company. The common practice is using separate files and separate drawing layers for separate systems, such as power supply, lighting, heating, etc. One file is locked and edited by a single designer.

To solve these problems Internet-based component and project information systems are used in many design companies.

Component database entries include links to manufacturer pages that often include product photos and datasheets. The same product can be described with multiple names for search.

The database access form contains several elements, such as product code, device groups that differ from company to company according to their stock needs, names, types, manufacturers, specifications, purchase methods, unit prices, and also CAD specific data, such as schematic symbols, layout symbols, 3D models and size.

In addition to local databases where product data is in different languages, the databases in design packages can include direct links to united Internet databases.

Local component databases include main installation system components divided into subgroups. High current component subgroups include electric cooking, heating, refrigerators, washing and post processing machines, home-hygiene and cosmetics, fans and pumps, motors and small-size housework machines, and batteries including battery based lighting systems. Low-current component subgroups include communication systems, alarm and monitoring systems, data communication devices, such as modems, phone system components, antenna systems and accessories, and audio system components. A set of components can be chosen from the database. This allows placing a set of components to the same position in drawings, such as lighting equipment with a lamp, fastenings, wiring etc.

Data flows of an Internet database, such as Finnish Electrical Wholesaling Industry (Sähkönumerot.fi, SSTL) online database, are shown in Fig. 4.

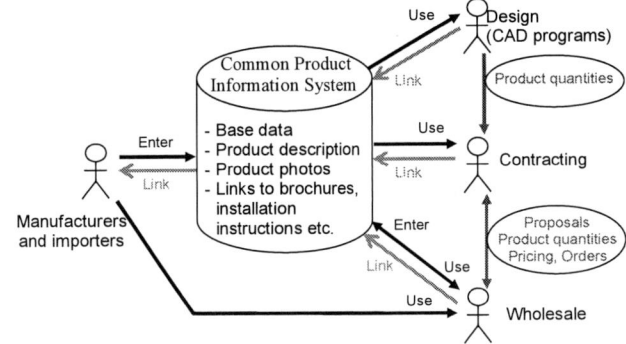

Fig. 4. Use-case diagram of an Internet-based product information system.

Product information includes a number [3, 4, and 5] with 7 digits, product names, reseller name, EAN-barcode, search names, product type, manufacturer product code, database entry editor, entry date, removal time, product group, and links to Internet pages. Internet database links are often accessible directly from the CAD program when the designer has selected a product from its local database.

All products in the Internet database are grouped into different subgroups, such as wires and cables including overhead line wires, communication cables, installation cables and power cables, accessories for wires and cables, switches, protection apparatus, lighting, electrical network components, tools, and measurement instruments.

Component databases updates should be fast and flexible, which can lead to additional work and expenses to component resellers.

CONCLUSIONS

There is a large variety of different design packages on the market; some of these already have add-ons for access of databases including Internet-based information systems. Many companies report that they will support open-standard based data exchange mechanisms in their products. Also, new modeling languages, such as SysML used in mechatronic and mechanical engineering [7], can be further used in electrical installation and building automation modeling together with different online databases. The future of building automation systems gives new challenges for modeling of control systems in electrical installation system design. Standardized and open format such as IFC [1] besides DWG enables interoperability between different software packages for modeling of different building systems, such as lighting design, short circuit current calculations, thermal calculations, etc. Besides the IFC 2x3 drawing format, other file formats such as DXF, DWF, XML, ODBC (Open Database Connectivity), gbXML (Green Building XML schema), and LandXML are used for compatibility of information interexchange.

Common problems of application of three-dimensional designs are:
1. Nowadays separate 2D drawings are anyway needed because field workers still need 2D drawings on the paper. This requires additional work effort from a designer. Solutions have to be developed for direct electronic use of 3D models and barcode readers on the field work. Maybe electronic drawings similar to electronic books and 3D glasses will solve this problem in future.
2. Re-education is necessary for using complicated 3D models and making printouts from these. 3D models offer broader range of layout and printout opportunities and also a broader range of error can be made when using or making 2D layouts and views for printouts.
3. High software upgrade costs. Simpler cost-effective CAD programs do not support 3D. Simpler 2D data formats such as 2D DXF cannot be used.
4. Conversion and generating 3D models from 2D models works only partially because the amount of available 3D component models is limited.

Three-dimensional designs of systems enable checking of different crossing and joint locations to avoid conflicts and collisions between designs of different systems and to assure required clearances between these systems. A 3D model can be better validated by companies or end-users who have limited skills and knowledge on a specific design field. This could help to assure correct component orders, better compatibility and thus could save time and money. New developments in design software packages allow avoiding conflicts already in the design phase and improve 2D drawing readability with better fitting of drawing elements.

The overall design cost and energy efficiency are increasingly important which leads to increasing complexity of design and conjunction of different systems. Better cooperation between different designers [8] is the key for benefiting from the 3D design. Databases and information systems provide support for cooperation of designers in different fields.

ACKNOWLEDGEMENTS

This research has been supported by Estonian Ministry of Education and Research (Project SF0140016s11) and Tallinn University of Technology (baseline funding grant BF123).

REFERENCES

[1] M. Haikka;, "Producing bills of quantities using CAD-programs" (in finnish), "*Massaluetteloiden tuottaminen CAD-ohjelmista: Sähkösuunnittelijan näkökulma*", Master's thesis of Helsinki University of Technology, 2006, http://lib.tkk.fi/Dipl/2006/urn006220.pdf, 04.2011

[2] K. Mäkeläinen;, "Benefits of electrical quantity calculation applications at the present moment" (in finnish), "*Selvitys sähköisen määrälaskennan nykytilasta ja käyttömahdollisuuksista tarjous-laskennan apuna*"; thesis of Tampere University of Applied Sciences, 2007, https://publications.theseus.fi/bitstream/handle/10024/9851/M%C3%A4kel%C3%A4inen.Keijo.pdf, 04.2011

[3] Sähkönumerot.fi service, http://www.sahkonumerot.fi/en/BrieflyInEnglish, 04.2011

[4] Guidelines for making graphical 3D product models for electrical design of buildings, http://www.sahkonumerot.fi/static_pages/NSS_STK_Guidelines_3D-product%20models.pdf, 04.2011

[5] O. Santana; B. T. Chia; J. L. Coulomb; J. P. Iafrate;, "Data bases for CAD applications", *IEEE Transactions on Magnetics*, Vol.25, No.4, pp.2956-2958, Jul. 1989

[6] M. L. Maher; J. H. Rutherford;, "A Model for Synchronous Collaborative Design Using CAD and Database Management", Architectural and Design Science, University of Sydney, Australia; *Research in Engineering Design* (1997) 9:85-98, Springer-Verlag London Limited, 1997

[7] R. Sell; M. Tamre; M. Lehtla; A. Rosin;, "A Conceptual design method for the general electric vehicle"; *Estonian Journal of Engineering*, Vol. 14, No. 1, March 2008; 16p.; Estonian Academy of Sciences, 2008

[8] H. Wenfa; H. Xinhua;, "A Case Study of Collaborative Education Based on Building Information Model", *International Conference on Computer Science and Software Engineering*, 2008 , vol.5, no., pp.198-201, 12-14, Dec. 2008

978-1-4244-8806-3/11 $26.00 © 2011 IEEE

Impact of Component Losses on the Voltage Boost Properties and Efficiency of the qZS-Converter Family

Dmitri Vinnikov, Indrek Roasto

Department of Electrical Drives and Power Electronics, Tallinn University of Technology (Tallinn, Estonia)
dmitri.vinnikov@ieee.org

Abstract- **This paper is devoted to the quasi-Z-source (qZS) converter family. Recently, the qZS-converters have attracted high attention because of their specific properties of voltage boost and buck functions with a single switching stage, which could be especially advantageous in renewable energy applications. As main representatives of the qZS-converter family, the traditional quasi-Z-source inverter as well as two novel extended boost quasi-Z-source inverters are discussed. Steady state analysis of three main topologies operating in the continuous conduction mode is presented. Input voltage boost properties of converters are compared for an ideal case. Mathematical models of converters considering losses in components are derived. Practical boost properties of converters are compared to idealized ones and the impact of losses on the voltage boost properties of each topology is justified. Finally, the impact of losses in the components on the boost conversion efficiency is analyzed.**

I. INTRODUCTION

Recently, the quasi-Z-source inverter (qZSI) topology (Fig. 1*a*) has attracted high attention because of its specific properties of voltage boost and buck functions with a single switching stage [1-3]. The qZSI also features such advantages as continuous input current, low or no inrush current during start-up, high immunity against EMI noise and misgating. Because of that the qZSI became a very attractive choice for renewable and alternative energy applications, where the

reliability and reduced number of energy conversion stages could play a vital role.

The conception of extending the qZSI boost capability without increasing the number of active switches has been recently proposed by several authors [4-9]. These new converters are known as cascaded (or extended boost) qZSIs and are generally classified as capacitor assisted and diode assisted. The topology of a capacitor assisted extended boost qZSI (CAEB qZSI, Fig. 1*b*) was derived by the adding of one diode (D_2), one inductor (L_3) and two capacitors (C_3 and C_4) to the traditional qZSI [7, 8]. The topology of a diode assisted extended boost qZSI (DAEB qZSI, Fig. 1*c*) was derived by the adding of one capacitor (C_3), one inductor (L_3) and two diodes (D_2 and D_3) to the traditional qZSI.

These three topologies (Fig. 1) have a common property - the input inductor L_1 that buffers the source current. It means that during the continuous conduction mode (CCM) the input current of the converter never drops to zero, thus featuring the reduced stress of the input voltage source.

Since all these three topologies have a different number of passive components in the qZS-networks, a detailed analysis of component losses and their impact on a converter's operating properties are especially topical issues to be addressed in order to obtain higher energy efficiencies.

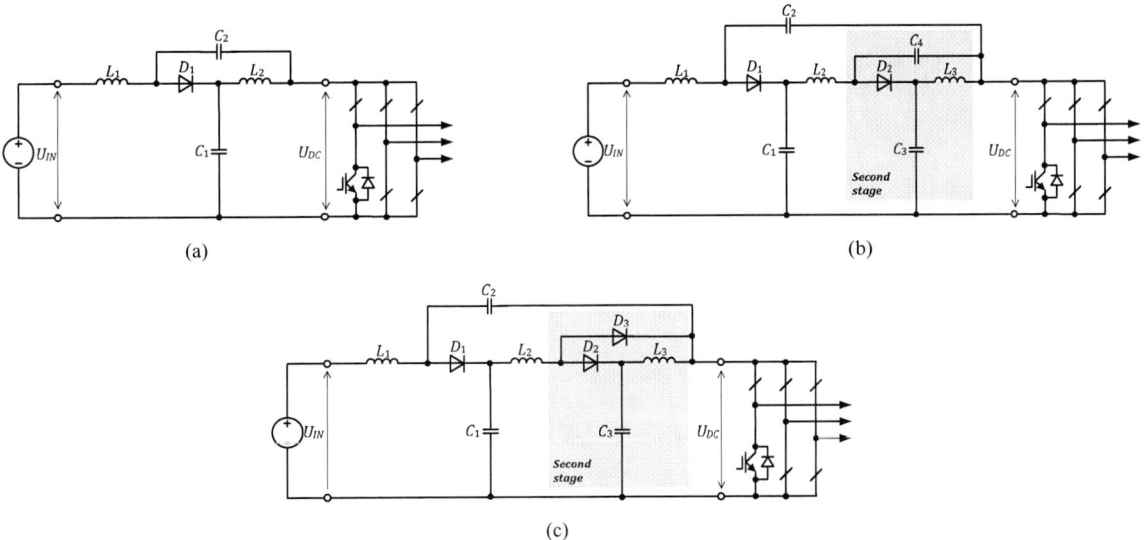

Fig. 1. Representatives of the qZS-converter family: traditional qZSI (a), capacitor assisted cascaded qZSI (b) and diode assisted cascaded qZSI (c).

978-1-4244-8806-3/11 $26.00 © 2011 IEEE

II. STEADY STATE ANALYSIS OF INVESTIGATED TOPOLOGIES

The topologies shown in Fig. 1 could be simply represented by the PWM inverter coupled with the appropriate qZS-network. Similarly to the traditional qZSI, the extended boost qZSI has two main types of operational states at the dc side: non-shoot-through states (i.e. the six active states and two conventional zero states of the traditional three-phase voltage source inverter (VSI)) and the shoot-through state (i.e. both switches in at least one phase leg conduct simultaneously). To simplify our analysis the inverter bridge was replaced by a switch S (Fig. 2). When the switch S is closed, the shoot-through state occurs and the converter performs the voltage boost action. When the switch S is open, the active (non-shoot-through) state emerges and previously stored magnetic energy in turn provides the boost of voltage seen on the load terminals.

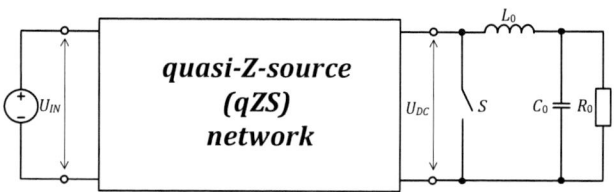

Fig. 2. Simplified power circuit of the qZS-converters used in the analysis.

The operating period of the qZS-converter in the CCM basically consists of a shoot-through state t_S and an active state t_A:

$$T = t_A + t_S . \tag{1}$$

Equation (1) could also be represented as

$$\frac{t_A}{T} + \frac{t_S}{T} = D_A + D_S = 1 , \tag{2}$$

where D_A and D_S are the duty cycles of an active and shoot-through states, correspondingly.

In order to simplify the analysis it was assumed that all the capacitors, inductors, and diodes of the qZS-networks of the investigated topologies are identical and lossless.

A. qZS-converter

Fig. 3 shows the equivalent circuits of the traditional qZS-converter operating in the CCM for the shoot-through (a) and active (b) states. At the steady state the average voltage of the inductors over one operating period is zero:

$$U_{L1} = \int_t^{t+T} u_{L1} dt = 0 ; \quad U_{L2} = \int_t^{t+T} u_{L2} dt = 0 . \tag{3}$$

Considering the above and defining the shoot-through duty cycle as D_S and the non-shoot-through duty cycle as $(1-D_S)$, the inductors' voltages over one operating period could be represented as

$$\begin{cases} U_{L1} = \overline{u}_{L1} = D_S(U_{IN} + U_{C2}) + (1-D_S)(U_{IN} - U_{C1}) = 0 \\ U_{L2} = \overline{u}_{L2} = D_S(U_{C1}) - (1-D_S)(U_{C2}) = 0 \end{cases} . \tag{4}$$

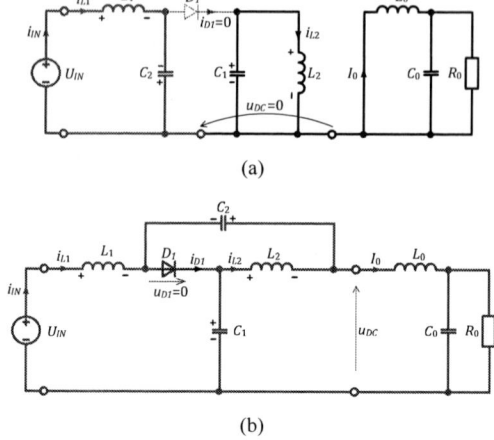

(a)

(b)

Fig. 3. Equivalent circuits of the traditional qZS-converter: during the shoot-through state (a) and during the active (non-shoot-through) state (b).

The peak DC-link voltage is

$$\hat{u}_{DC} = U_{C1} + U_{C2} = U_{IN} \frac{1}{1-2D_S} , \tag{5}$$

where $U_{C1} = U_{IN} \dfrac{1-D_S}{1-2D_S}$ and $U_{C2} = U_{IN} \dfrac{D_S}{1-2D_S}$.

B. CAEB qZS-converter

Fig. 4 shows the equivalent circuits of the CAEB qZS-converter operating in the CCM for the shoot-through (a) and active (b) states.

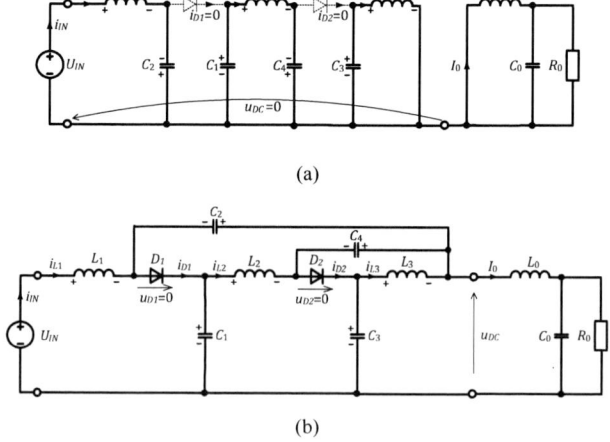

(a)

(b)

Fig. 4. Equivalent circuits of the CAEB qZS-converter: during the shoot-through state (a) and during the active (non-shoot-through) state (b).

At the steady state the average voltage of the inductors over one operating period is zero:

978-1-4244-8806-3/11 $26.00 © 2011 IEEE

$$U_{L1} = \int\limits_{t}^{t+T} u_{L1} dt = 0 \; ; \; U_{L2} = \int\limits_{t}^{t+T} u_{L2} dt = 0 \; ; \; U_{L3} = \int\limits_{t}^{t+T} u_{L3} dt = 0 \; . \; (6)$$

Based on that fact and defining the shoot-through duty cycle as D_S and the non-shoot-through duty cycle as $(1-D_S)$, the inductors' voltages over one operating period could be represented as

$$\begin{cases} U_{L1} = \overline{u}_{L1} = D_S \left(U_{IN} + U_{C2} \right) + \left(1 - D_S \right)\left(U_{IN} - U_{C1} \right) = 0 \\ U_{L2} = \overline{u}_{L2} = D_S \left(U_{C4} + U_{C1} \right) + \left(1 - D_S \right)\left(U_{C4} - U_{C2} \right) = 0 \\ U_{L2} = \overline{u}_{L2} = D_S \left(U_{C4} + U_{C1} \right) + \left(1 - D_S \right)\left(U_{C1} - U_{C3} \right) = 0 \\ U_{L3} = \overline{u}_{L3} = D_S \left(U_{C3} \right) - \left(1 - D_S \right)\left(U_{C4} \right) = 0 \end{cases} \; (7)$$

The peak DC-link voltage is

$$\hat{u}_{DC} = U_{C1} + U_{C2} = U_{IN} \frac{1}{1 - 3D_S} , \qquad (8)$$

where $U_{C1} = U_{IN} \dfrac{1 - 2D_S}{1 - 3D_S}$ and $U_{C2} = U_{IN} \dfrac{2D_S}{1 - 3D_S}$.

C. DAEB qZS-converter

Fig. 5 shows the equivalent circuits of the DAEB qZS-converter operating in the CCM for the shoot-through (a) and active (b) states.

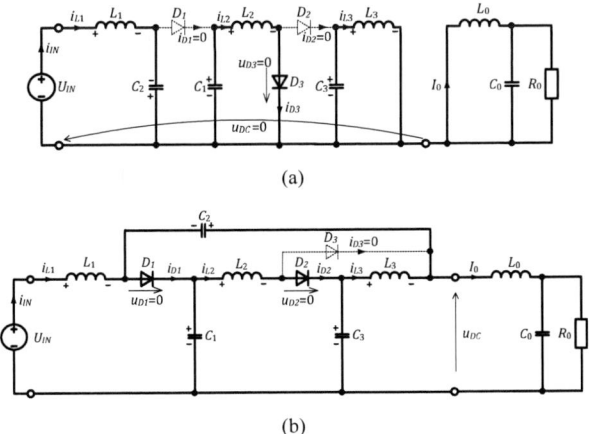

(a)

(b)

Fig. 5. Equivalent circuits of the DAEB qZS-converter: during shoot-through state (a) and during active (non-shoot-through) state (b).

Considering the fact that the average voltage of the inductors over one operating period is zero (6) and defining the shoot-through duty cycle as D_S and the non-shoot-through duty cycle as $(1-D_S)$, the inductors' voltages could be represented as:

$$\begin{cases} U_{L1} = \overline{u}_{L1} = D_S \left(U_{IN} + U_{C2} \right) + \left(1 - D_S \right)\left(U_{IN} - U_{C1} \right) = 0 \\ U_{L2} = \overline{u}_{L2} = D_S \left(U_{C1} \right) + \left(1 - D_S \right)\left(U_{C1} - U_{C3} \right) = 0 \\ U_{L3} = \overline{u}_{L3} = D_S \left(U_{C3} \right) + \left(1 - D_S \right)\left(U_{C3} - U_{C1} - U_{C2} \right) = 0 \end{cases} \; (9)$$

The peak DC-link voltage is

$$\hat{u}_{DC} = U_{C1} + U_{C2} = U_{IN} \frac{1}{D_S^2 - 3D_S + 1} . \qquad (10)$$

where $U_{C1} = U_{IN} \dfrac{D_S^2 - 2D_S + 1}{D_S^2 - 3D_S + 1}$ and $U_{C2} = U_{IN} \dfrac{2D_S - D_S^2}{D_S^2 - 3D_S + 1}$.

D. Comparison of idealized boost properties of the qZS-converter family

To evaluate the boost properties of the discussed qZS-converter topologies the boost ratio of the input voltage B was introduced:

$$B = \frac{\hat{u}_{DC}}{U_{IN}} ,$$

where $\hat{u}_{DC}$ is the amplitude value of the dc-link voltage U_{DC} and U_{IN} is the input voltage of the converter. Fig. 5 shows the boost ratio comparison chart of the three discussed topologies. It is seen that the CAEB and DAEB qZS-converters could provide up to 2 and 1.6 times higher input voltage gain, respectively, than the traditional qZS-converter.

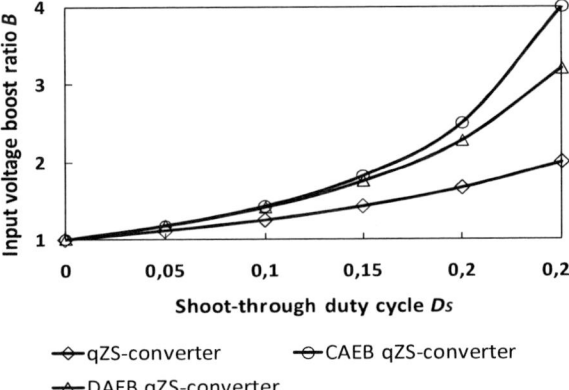

Fig. 6. Comparison of idealized boost ratios of the discussed topologies.

III. IMPACT OF COMPONENT LOSSES ON THE VOLTAGE BOOST PROPERTIES OF THE QZS-CONVERTER FAMILY

In the previous section the lossless models of the qZS-converter family were discussed. In real practice the voltage boost properties of the qZS-converters could be seriously affected by the losses in components. For more careful estimation of the operating characteristics of the converter the loss elements, such as winding resistances of inductors and voltage drops in semiconductors, should be added to the model. In our analysis the losses in inductors $L_1...L_3$ as well as in L_O are represented by the resistances r_L and losses in diodes during the conduction state are controlled by the voltage drop U_D. Moreover, it was assumed that the IGBT with the saturation voltage of U_S was used for switch S. To simplify the analysis it was stated that the capacitors and inductors of all the compared topologies are identical.

978-1-4244-8806-3/11 $26.00 © 2011 IEEE

A. qZS-converter

An extended equivalent circuit of the traditional qZS-converter considering losses in the components is presented in Fig. 7.

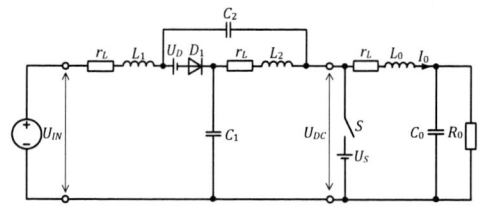

Fig. 7. Equivalent circuit of the traditional qZS-converter considering losses in the components.

Equation (4) written for the lossless system could be extended to (12). Considering losses in the components the peak dc-link voltage of the traditional qZS-converter could be expressed by (13).

B. CAEB qZS-converter

An extended equivalent circuit of the CAEB qZS-converter considering component losses is presented in Fig. 8.

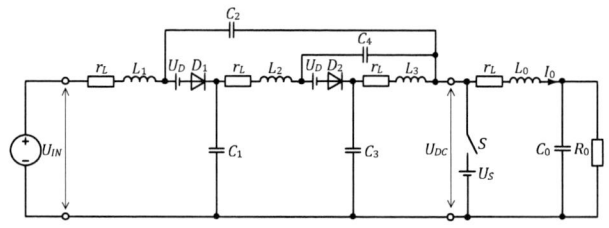

Fig. 8. Equivalent circuit of the CAEB qZS-converter considering losses in the components.

Equation (7) written for the lossless system could be extended to (14). Considering losses in the components the peak dc-link voltage of the CAEB qZS-converter could be expressed by (15).

C. DAEB qZS-converter

An extended equivalent circuit of the DAEB qZS-converter considering component losses is presented in Fig. 9.

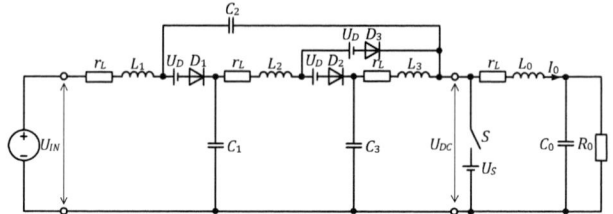

Fig. 9. Equivalent circuit of the DAEB qZS-converter considering losses in the components.

Equation (9) written for the lossless system could be extended to (16). Considering losses in the components the peak dc-link voltage of the CAEB, qZS-converter could be expressed by (17).

D. Comparison of idealized and practical boost properties of investigated topologies

In order to demonstrate the impact of losses in the components on the input voltage boost properties of the investigated topologies the lossless and lossy models were compared mathematically with the following parameters:

$$U_{IN} = 40 \text{ V}; \quad U_S = 1.7 \text{ V}; \quad U_D = 1.6 \text{ V};$$

$$R_O = 4.5 \text{ } \Omega; \quad r_L = 4 \text{ m}\Omega; \quad D_S = 0...0.25.$$

$$
\begin{cases}
U_{L1} = \overline{u}_{L1} = D_S\left(U_{IN} + U_{C2} - I_L r_L - U_S\right) + \left(1 - D_S\right)\left(U_{IN} - U_{C1} - I_L r_L - U_D\right) = 0 \\
U_{L2} = \overline{u}_{L2} = D_S\left(U_{C1} - I_L r_L - U_S\right) - \left(1 - D_S\right)\left(U_{C2} + I_L r_L + U_D\right) = 0 \\
U_{L0} = \overline{u}_{L0} = D_S\left(U_S - I_0 r_L - U_{C0}\right) + \left(1 - D_S\right)\left(U_{C1} - I_0 r_L + U_{C2} - U_{C0} + U_D\right) = 0 \\
I_{C1} = \overline{i}_{C1} = D_S I_L + \left(1 - D_S\right)\left(I_0 - I_L\right) = 0
\end{cases} \tag{12}
$$

$$U_{DC} = \frac{\left(4R_0 U_D - 4R_0 U_s + 2U_D r_L - 6U_s r_L\right)D_s^2 + \left(2R_0 U_s - 6R_0 U_D + 2R_0 U_{IN} - 2U_D r_L + 4U_s r_L + 2U_{IN} r_L\right)D_s + 2R_0 U_D - R_0 U_{IN} - U_{IN} r_L}{\left(-4R_0 - 6r_L\right)D_s^2 + \left(4R_0 + 8r_L\right)D_s - R_0 - 3r_L}. \tag{13}$$

$$
\begin{cases}
U_{L1} = \overline{u}_{L1} = D_S\left(U_{IN} + U_{C2} - I_L r_L - U_S\right) + \left(1 - D_S\right)\left(U_{IN} - U_{C1} - I_L r_L - U_D\right) = 0 \\
U_{L2} = \overline{u}_{L2} = D_S\left(U_{C1} + U_{C4} - I_L r_L - U_S\right) + \left(1 - D_S\right)\left(U_{C1} - U_{C3} - I_L r_L - U_D\right) = 0 \\
U_{L3} = \overline{u}_{L3} = D_S\left(U_{C3} - I_L r_L - U_S\right) - \left(1 - D_S\right)\left(I_L r_L + U_{C4} + U_D\right) = 0 \\
U_{L3} = \overline{u}_{L3} = D_S\left(U_{C3} - I_L r_L - U_S\right) + \left(1 - D_S\right)\left(U_{C3} - I_L r_L - U_{C1} - U_{C2} - U_D\right) = 0 \\
U_{L0} = \overline{u}_{L0} = D_S\left(U_S - I_0 r_L - U_{C0}\right) + \left(1 - D_S\right)\left(U_{C4} - I_0 r_L + U_{C3} - U_{C0} + U_D\right) = 0 \\
I_{C3} = \overline{i}_{C3} = D_S I_L + \left(1 - D_S\right)\left(\frac{I_0 - I_L}{2}\right) = 0
\end{cases} \tag{14}
$$

$$U_{DC} = \frac{\left(9R_0 U_D - 9R_0 U_s + 12U_D r_L - 12U_s r_L\right)D_s^2 + \left(3R_0 U_{in} - 12R_0 U_D + 3R_0 U_s - 18U_D r_L + 3U_{in} r_L + 6U_s r_L\right)D_s + 3R_0 U_D - R_0 U_{IN} + 6U_D r_L - U_{IN} r_L}{\left(-9R_0 - 12r_L\right)D_s^2 + \left(6R_0 + 12r_L\right)D_s - R_0 - 4r_L}. \tag{15}$$

$$\begin{cases} U_{L1} = \overline{u}_{L1} = D_S\big(U_{IN} + U_{C2} - I_{L1}r_L - U_S\big) + \big(1 - D_S\big)\big(U_{IN} - U_{C1} - I_{L1}r_L - U_D\big) = 0 \\ U_{L2} = \overline{u}_{L2} = D_S\big(U_{C1} - U_D - I_{L2}r_L - U_S\big) + \big(1 - D_S\big)\big(U_{C1} - U_{C3} - I_{L2}r_L - U_D\big) = 0 \\ U_{L3} = \overline{u}_{L3} = D_S\big(U_{C3} - I_{L3}r_L - U_S\big) + \big(1 - D_S\big)\big(U_{C3} - I_{L3}r_L - U_{C1} - U_{C2} - U_D\big) = 0 \\ U_{L0} = \overline{u}_{L0} = D_S\big(U_S - I_0 r_L - U_{C0}\big) + \big(1 - D_S\big)\big(U_{C2} - I_0 r_L + U_{C1} - U_{C0} + U_D\big) = 0 \\ I_{C1} = \overline{i}_{C1} = -D_S I_{L2} + \big(1 - D_S\big)\big(I_{L1} + I_{L3} - I_0 - I_{L2}\big) = 0 \\ I_{C2} = \overline{i}_{C2} = -D_S I_{L1} + \big(1 - D_S\big)\big(I_{L3} - I_0\big) = 0 \\ I_{C3} = \overline{i}_{C3} = -D_S I_{L3} + \big(1 - D_S\big)\big(I_{L2} - I_{L3}\big) = 0 \end{cases} \tag{16}$$

$$U_{DC} = \frac{\alpha D_s^{\,4} + \beta D_s^{\,3} + \gamma D_s^{\,2} + \lambda D_s + 3R_0 U_D - R_0 U_{IN} + 6U_D r_L - U_{IN} r_L}{\big(-R_0 - 2r_L\big)D_s^{\,4} + \big(6R_0 + 10r_L\big)D_s^{\,3} + \big(-11R_0 - 19r_L\big)D_s^{\,2} + \big(6R_0 + 14r_L\big)D_s - R_0 - 4r_L}, \tag{17}$$

where $\quad \alpha = R_0 U_D - R_0 U_s + 2U_D r_L - 2U_s r_L$;

$\quad\quad\quad \beta = 6R_0 U_s - 6R_0 U_D - 10U_D r_L + 9U_s r_L$;

$\quad\quad\quad \gamma = 13R_0 U_D - R_0 U_{IN} - 10R_0 U_s + 21U_D r_L - U_{IN} r_L - 15U_s r_L$;

$\quad\quad\quad \lambda = 3R_0 U_{IN} - 12R_0 U_D + 3R_0 U_s - 20U_D r_L + 3U_{IN} r_L + 6U_s r_L$.

Figs. 10, 11 and 12 show the impact of losses on the input voltage boost factor of the traditional, CAEB and DAEB qZS-converters, respectively. The diagrams show that at the maximal studied shoot-through duty cycle (D_S=0.25) the traditional, CAEB and DAEB qZS-converters have demonstrated the 9%, 14% and 13% reduction of the targeted DC-link voltage amplitude, respectively.

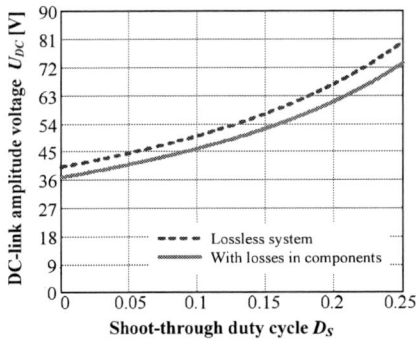

Fig. 10. Comparison of idealized and practical boost properties of the traditional qZS-converter.

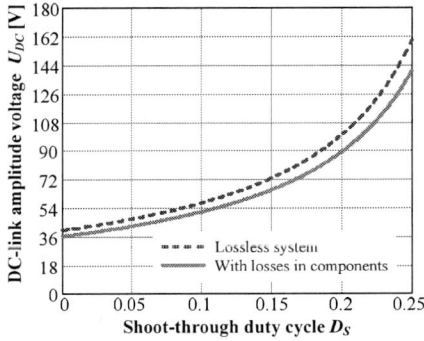

Fig. 11. Comparison of idealized and practical boost properties of the CAEB qZS-converter.

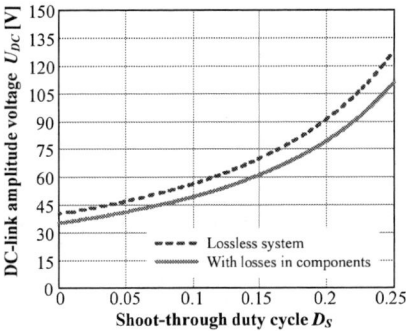

Fig. 12. Comparison of idealized and practical boost properties of the DAEB qZS-converter.

IV. IMPACT OF COMPONENT LOSSES ON THE EFFICIENCY OF THE QZS-CONVERTER FAMILY

In order to demonstrate the impact of component losses on the overall efficiency of the qZS-converter a number of experiments were performed. First, the impact of inductor winding resistance was studied. Figs. 13, 14 and 15 show that at maximal shoot-through duty cycle (D_S=0.25) and selected inductor's resistance (r_L = 4 mΩ), the efficiency variation of the investigated topologies lies in the range of 87...94%.

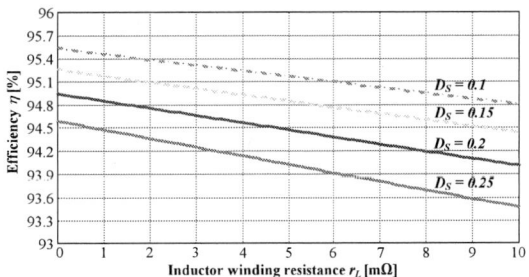

Fig. 13. Impact of the inductor's winding resistance on the efficiency of the traditional qZS-converter.

978-1-4244-8806-3/11 $26.00 © 2011 IEEE   307

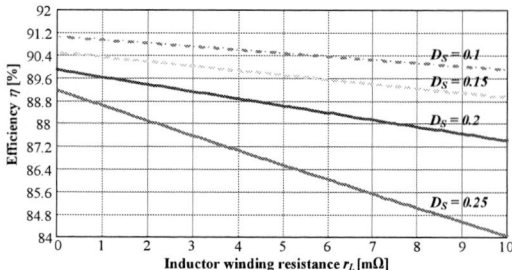

Fig. 14. Impact of the inductor's winding resistance on the efficiency of the CAEB qZS-converter.

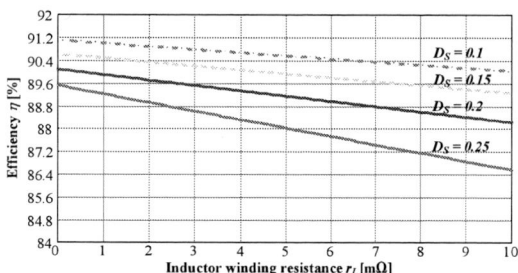

Fig. 15. Impact of the inductor's winding resistance on the efficiency of the DAEB qZS-converter.

In the second experiment, the impact of the forward voltage drop of diodes $D_1...D_3$ on the converter's efficiency was studied. Figs. 16, 17 and 18 show that at maximal shoot-through duty cycle (D_S=0.25) and selected fast recovery epitaxial diodes (U_D=1.6 V) the maximal efficiency that could be obtained with traditional, CAEB and DAEB qZS-converters is 94%, 87% and 88%, respectively. If the diodes are replaced with high-power Schottky rectifiers with a low forward voltage drop (U_D=0.6 V), the effective efficiency rise by at least 5% could be expected for all three topologies.

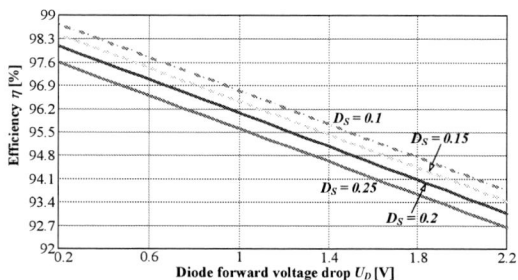

Fig. 16. Impact of diode forward voltage drop on the efficiency of the traditional qZS-converter.

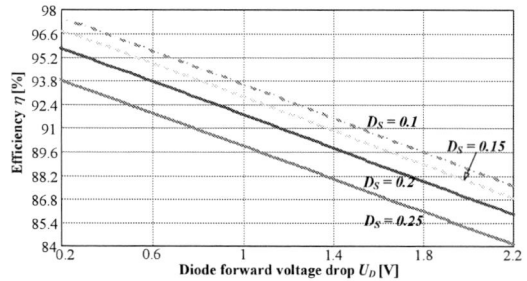

Fig. 17. Impact of diode forward voltage drop on the efficiency of the CAEB qZS-converter.

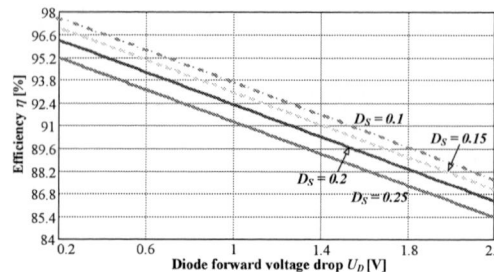

Fig. 18. Impact of diode forward voltage drop on the efficiency of the DAEB qZS-converter.

V. CONCLUSIONS

This paper is devoted to the analysis of the voltage boost capability of thee different quasi-Z-source converters. Special attention is paid to the impact of losses in passive components on the boost properties and boost conversion efficiency of the investigated topologies. It was found that for the same operating parameters (U_{IN} = 40 V; R_O = 4.5 Ω) and component values (U_S = 1.7 V; U_D = 1.6 V; r_L = 4 mΩ) the twofold boost of the input voltage could be realized at the efficiency of 93.9%, 89.2% and 89.3% for the traditional, CAEB and DAEB qZS-converters, respectively.

ACKNOWLEDGMENT

This research work has been supported by Estonian Ministry of Education and Research (Project SF0140016s11) and Estonian Science Foundation (Grants ETF8538 and ETF8687).

REFERENCES

[1] J. Anderson, F.Z. Peng, "Four Quasi-Z-Source Inverters", in Proc. *IEEE Conf. PESC'08*, pp. 2743 – 2749, June 2008.

[2] Yuan Li; Anderson, J.; Peng, F.Z.; Dichen Liu, "Quasi-Z-Source Inverter for Photovoltaic Power Generation Systems", in Proc. of IEEE Applied Power Electronics Conference and Exposition APEC'2009, pp. 918-924, 15-19 Feb. 2009.

[3] Jong-Hyoung Park; Heung-Geun Kim; Eui-Cheol Nho; Tae-Won Chun; Jaeho Choi, "Grid-connected PV System Using a Quasi-Z-source Inverter", in Proc. of IEEE Applied Power Electronics Conference and Exposition APEC'2009, pp. 925-929, 15-19 Feb. 2009.

[4] C. J. Gajanayake, F. L. Luo, H. B. Gooi, P. L. So, L. K. Siow, "Extended boost Z-source inverters", in Proc. *IEEE Conf. ECCE'09*, pp. 3845-385, Sept. 2009.

[5] Gajanayake, C. J., Gooi, H. B., Luo, F. L., So, P. L., Siow L. K., Vo, Q. N., "Simple modulation and control method for new extended boost quasi Z-source," in Proc. of IEEE Conference TENCON'2009.

[6] Strzelecki, R., Adamowicz, M., "Boost-buck inverters with cascaded qZ-type impedance networks", *Electrical Review*, vol. 86, no. 2, pp. 370-375, 2010.

[7] M. Adamowicz, R. Strzelecki, D. Vinnikov, "Cascaded Quasi-Z-Source Inverters for Renewable Energy Generation Systems", in Proc. *Ecologic Vehicles and Renewable Energies Conference EVER'10*, March 2010.

[8] D. Vinnikov, I. Roasto, R. Strzelecki, M. Adamowicz, "Performance Improvement Method for the Voltage-Fed qZSI with Continuous Input Current", in Proc. *IEEE Mediterranean Electrotechn. Conf. MELECON'10*, April 2010.

[9] Vinnikov, D.; Roasto, I.; Jalakas, T., "Comparative study of capacitor-assisted extended boost qZSIs operating in continuous conduction mode", in Proc. *12th Biennial Baltic Electronics Conference (BEC'2010)*, pp.297-300, 4-6 Oct. 2010.

978-1-4244-8806-3/11 $26.00 © 2011 IEEE

Multicell-type Transistor Converter with Combined Control for Micro Resistance Welding

Yuriy E. Paerand, Oleksandr F. Bondarenko, Iuliia V. Bondarenko

Electronic Systems Department, Donbas State Technical University

Lenin ave., 16, Alchevs'k, 94204, Luhans'k reg., Ukraine

paerand@mail.ru, bondarenkoaf@gmail.com, bondarenko.julie@gmail.com

Abstract – **Micro resistance welding is effective way to reliably join small-scale parts in electronics industry and instrument-making. High quality welded joints are obtained by accurate regulation of welding current with transistor converter module which is one of the most important sub-systems in micro resistance welding machine. Multicell-type transistor converter with combined control, represented in the paper, provides high accuracy of current regulation and high energy efficiency therefore is proposed for use in micro resistance welding.**

Index Terms – **Transistor converter, multicell-type structure, combined control, linear mode of transistor operation, switch mode of transistor operation, micro resistance welding.**

I. INTRODUCTION

Transistor converters (TC) are widely used in many different electronic and electro-technical devices and systems. In micro resistance welding machines the transistor converter module regulates welding current and is considered to be one of the most important sub-systems. Micro resistance welding is an effective way to join small-scale parts and is actively used in electronics industry and instrument-making. Joining is performed by squeezing parts between the electrodes, and then conducting electrical current through their interface. Generated heat in the welding area rises and reaches the melting temperature, thus creating a welded joint. To obtain high quality weld joints, the electrical current through parts' interface must be accurately adjusted and then precisely controlled in accordance with a predetermined profile [1-3].

It is known that operating TC transistors in their linear region results the most accurate current regulation [4-9]. The main disadvantage of using transistors in linear operation mode is substantial power losses. Losses could amount to kilowatts due to extremely large welding currents typically exceeding hundreds or sometimes thousands of amperes. Operating TC transistors in a switching mode reduces power losses, however does not provide the necessary current regulation accuracy for producing high quality micro resistance welding. Combining both, linear and switch transistor operation modes, benefits the process of producing reliable high quality welding while reducing power losses. These two modes could be used simultaneously or sequentially during current regulation process. The overview of known implementations of combined transistor operation modes is represented at [10].

DC transistor converter for micro resistance welding machines which provides improved energy efficiency and high current regulation accuracy due to multicell structure of con-

verter and utilization of combined transistor operation mode is described in the following section.

II. CONVERTER STRUCTURE AND PRINCIPLE OF OPERATION

Block diagram in Fig. 1 represents a welding current pulse generator which includes a multicell transistor converter. Energy Source provides the energy that is required for the welding process. Transistor Converter, which consists of n unified cells, regulates welding current in load (welded contact) in accordance with previously specified profile. Control System manages cells' switching and generates control signals $U_{CONT\ 1}$ ... $U_{CONT\ n}$ that must be tracked by converter cells' currents. Each converter's cell is composed of Power Block and Control Block. Power Block includes two simple converters: converter in linear mode and converter in switch mode. These simple converters of Power Block are built utilizing power transistors which accomplish cell's current regulation. Control Block generates continuous and pulse control signals (U_C and U_P) required by Power Block converters and switches them depending on the phase of current forming. Control System utilizes signals from current sensors in cells and in load for current monitoring.

In some designs of welding pulse generators not only current profile but also profiles of voltage and power in the welding area may be predetermined and controlled [5, 11, 12]. The delivered power is calculated by multiplying signals from current and voltage sensors (Fig. 1).

Fig. 2 illustrates how the multi-cell converter regulates current. The common welding current i_W in the load is formed as a sum of n currents provided by cells. Each of these n currents is regulated separately by its cell from 0 to the specified maximum current I_{MAX}. To increase resulting welding current additional cells are gradually activated until the desired current is reached. Algorithm of converter cells' switching is described below.

Initially, from t_1 the current in the load is generated by Cell 1 only. Until the t_2 is reached Cell 1 precisely regulates its current in linear mode in accordance with continuous control signal U_C. When its current reaches the predefined maximum value I_{MAX} (time moment t_2) Cell 1 maintains achieved current value I_{MAX} in switch mode under the effect of pulse control signal U_P until the end of welding. At the moment t_2 Cell 2 joins the welding process and starts generating current to the load.

978-1-4244-8806-3/11 $26.00 © 2011 IEEE

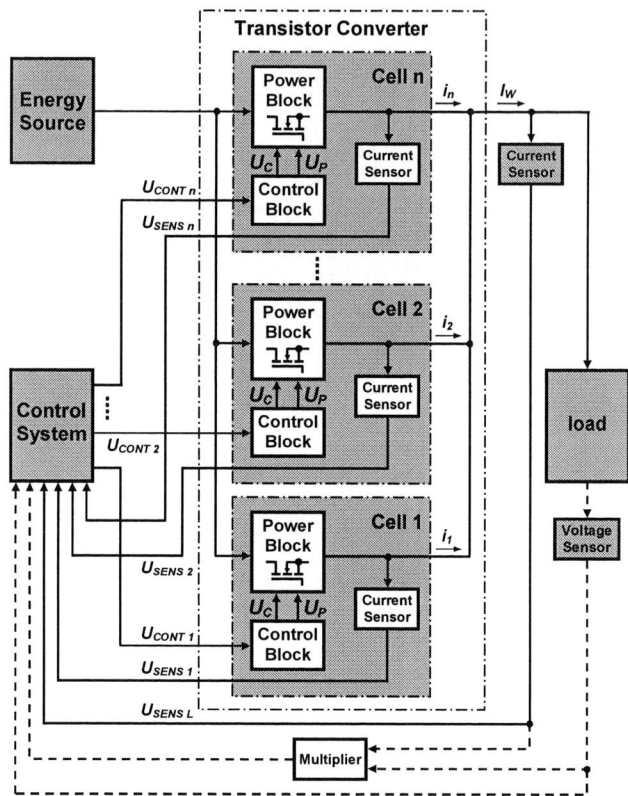

Fig. 1 Welding current pulse generator based on
multicell-type transistor converter with combined control

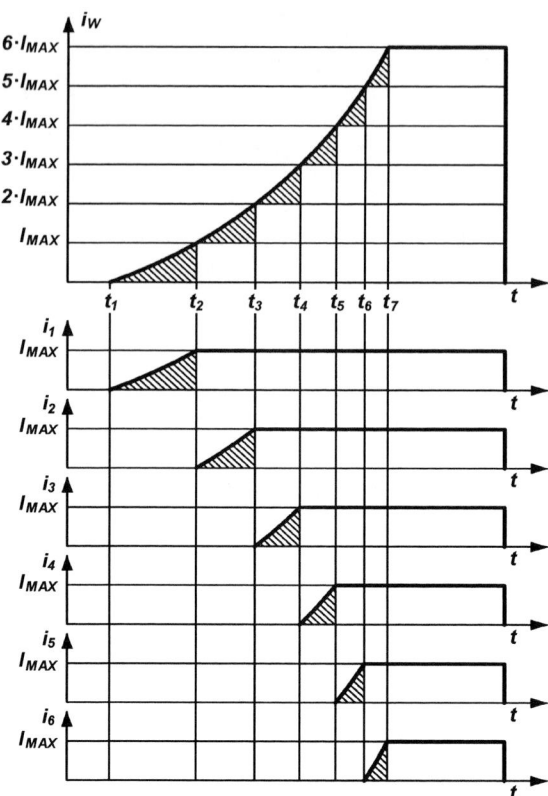

Fig. 2 Welding current diagrams
of multicell-type transistor converter with combined control

Cell 2 operates in the linear mode until t_3. After the t_3 mark Cell 2 retains constant current I_{MAX} operating in the switch mode. If additional current is required, next cells join its forming process and repeat the above sequence first operating in the linear mode and then switching to the pulse mode until the desired current is achieved. Fig. 2 demonstrates an example of how the common welding current was generated via six (6) cells activation. Shaded areas on Fig. 2 demonstrate time intervals where current regulation was carried out in linear mode of transistor operation.

It is obvious that power losses in the multicell-type converter with combined control are significantly smaller than in converter regulating current in linear mode for the full duration of welding process. Furthermore, the combined control allows for higher accuracy of current profile tracking as it operates in linear region of transistor output during crucial time intervals of current forming.

Fig. 2 illustrates a typical current waveform for micro resistance welding. It includes two stages:

– initial gradual current increase according to specified profile that is determined by strict welding conditions (materials, parts' thickness etc.);

– current stabilization at maximum level until the welding process is completed.

Special characteristics of the welding current profile depend on the particulars of the electro-physical processes in the welding contact [1-3, 5]. The initial stage of gradual current increase is the crucial stage in achieving high quality weld joints, therefore welding current changes during that stage must be accurately controlled.

III. CONVERTER CELL STRUCTURE

The converter cell, as it is shown in Fig. 3, is composed of two main blocks: Power Block and Control Block. The Power Block consists of two types of converters adjoined in parallel: Converter in Linear Mode and Converter in Switch Mode. Control Block generates the required control signals for both converters: continuous control signal and pulse control signal. Control signals' switching is performed by Switch Block.

Continuous control signal U_C is generated by Continuous Control Block and is equal to the difference between signal from Control System U_{CONT}, which specifies cell's current waveform, and the signal U_{SENS}, which is proportional to real cell's current value measured by current sensor. Obtained signal corrected by Continuous Regulator and adjusted by Amplifier then enters Converter in Linear Mode.

Pulse control signal U_P is generated by Pulse Control Block. First, the signal of difference between the signal U_{MAX}, which is proportional to specified maximum current value, and the signal U_{SENS}, which is proportional to real cell's current, is formed. The signal U_{SENS} initially passes through Low Pass Filter to reduce its high frequency component.

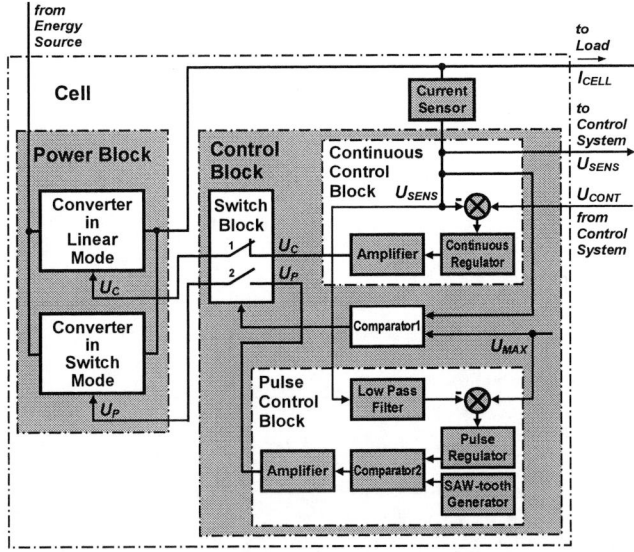

Fig. 3 Converter cell structure

Then Pulse Regulator corrects the obtained signal of the difference for further comparing it with SAW-tooth Generator signal by Comparator 2. The resulting output signal of Comparator 2 is formed as a sequence of rectangular pulses and is defined as pulse control signal U_P. The amplitude of the earlier obtained differential signal controls the Pulse Width Modulation (PWM).

Comparator 1 compares U_{SENS} with U_{MAX} and registers the moment when cell's current achieves specified maximum level. The output signal of Comparator 1 is used as the control signal for Switch Block. During the initial stage of cell's operation the output signal from Comparator 1 sets Switch Block to allow continuous control signal U_C pass to Converter in Linear Mode. When cell's current achieves specified maximum level, the output signal from Comparator 1 changes and sets Switch Block to allow pulse control signal U_P pass to Converter in Switch Mode.

IV. CONTROL SIGNAL FORMING FOR CONVERTER CELLS

Fig. 4 illustrates basic principle of control signal forming for converter cells. The reference signal U_{REF}, which defines the load current profile, is generates by Reference Signal Generator of Control System. The control signal U_{CONT1} for Cell 1 is produced by Subtracting Block 1 as a difference between the reference signal U_{REF} and the feedback signal from the load current sensor $U_{SENS\,L}$.

At the start of the welding process the control signal U_{CONT1} is directed to Cell 1. The current sensor in Cell 1 generates the signal U_{SENS1}, which is proportional to the real current. U_{SENS1} is compared with U_{CONT1} by Subtracting Block 2. While the current of Cell 1 is below its maximum value I_{MAX}, both signals (U_{SENS1} and U_{CONT1}) are equal to each other, therefore the output of Subtracting Block 2 is zero. Upon the Cell 1 current reaching its maximum value, the signal U_{SENS1} stabilizes at the value U_{MAX}, which is proportional to I_{MAX}, while U_{CONT1} continues to increase, resulting in a positive

differential signal from Subtracting Block 2. The differential signal $U_{CONT1} - U_{MAX}$ is used as the control signal U_{CONT2} for Cell 2. Upon the Cell 2 current reaching I_{MAX}, the signal U_{SENS2} stabilizes at the value U_{MAX}, while U_{CONT2} continues to increase, resulting in a positive differential signal from Subtracting Block 3. The differential signal $U_{CONT2} - U_{MAX}$ is used as the control signal U_{CONT3} for Cell 3. This sequence repeats itself for each successive cell.

The successive formation of control signals $U_{CONT2} \ldots U_{CONTn}$ is described by the formula:

$$U_{CONT\,k} = U_{CONT\,k-1} - U_{MAX} = U_{CONT1} - (k-1) \cdot U_{MAX}, \quad (1)$$

where k – the index of converter's cell (from 2 to n).

As soon as the last Cell n begins to operate, the signal $U_{SENS\,n}$ enters Limit Current Sensor which generates a signal for Display Block indicating that possible current level was reached.

When it is required to decrease the load current, cells are sequentially switched off as a result of negative control signal, which is formed when the reference signal is decreasing.

In cases when one or more cells become inoperative the current forming process is not interrupted. The zero signal from a non-operating cell current sensor at one input of subtracting block and the control signal at the other input result in a differential signal which is equal to the control signal, therefore it is used without alteration for the control of the next cell.

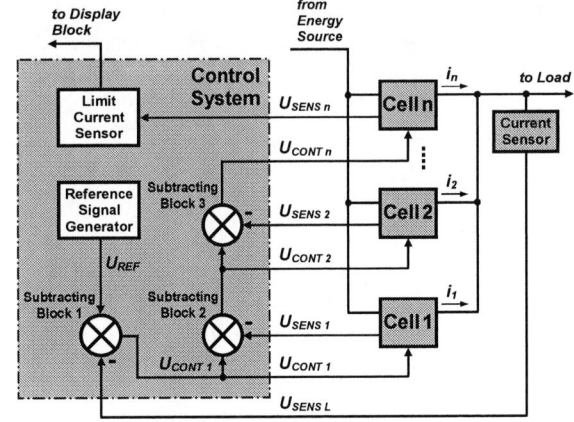

Fig. 4 The forming of control signals for converter cells

V. CONVERTER SIMULATION

Converter simulation was carried out to confirm the validity of the above stated current regulation. Fig. 5 illustrates the simulation model designed with MATLAB software. The model includes two cells with each cell consisting of Converter in Linear Mode and Converter in Switch Mode (in Fig. 5 they are labelled as Linear Converter and Switch Converter). Reference Signal Generator generates the signal profile for the load current. The load is imitated by resistance R and inductance L, together they represent main properties of welding circuit and welded contact.

Simulation parameters:
- maximum cell current: 100 A;
- energy source voltage: 10 V (DC);
- current waveform in the initial stage: square function;
- load resistance: 3 mOhm;
- load inductance: 300 nH;
- pulse control frequency: 100 kHz.

Fig. 6 illustrates currents obtained from the above simulation.

Diagrams (a) and (b) show cells' currents. Diagram (c) shows resulting load current.

Diagrams of currents graphically illustrate the switching process of cells. Initially, Linear Converter 1 is switched on and regulates the load current in accordance with the signal from Reference Signal Generator. When Cell 1 current reaches its maximum value (100 A), Current Detector & Switch 1 switches off Linear Converter 1 and at the same time switches on Switch Converter 1 together with Linear Converter 2. During the switching interval the current of Cell 1 initially sharply falls to "0" and then gradually increases as illustrated in Fig. 6 – diagram (a). The current increase rate depends on smoothing inductance value in Switch Converter 1 (the inductance is inside block Switch Converter). The profile of resulting current in the load is not distorted during the switching of converters, Fig. 6 – diagram (c), as the current gain in Cell 2 compensates current loss in Cell 1 clearly shown in Fig. 6 – diagram (b). Switch Converter 1 keeps the current at 100 A level while Linear Converter 2 coordinates current increase based on the reference signal and compensates the pulsating current of Switch Converter 1.

Upon the current of Cell 2 reaching 100 A, Current Detector & Switch 2 switches off Linear Converter 2 and switches on Switch Converter 2. Both Switch Converters 1 and Switch Converters 2 continue to supply the resulting current.

VI. ENERGY EFFICIENCY ESTIMATION OF MULTICELL-TYPE TRANSISTOR CONVERTER WITH COMBINED CONTROL

Energy efficiency estimation of the above represented multicell-type transistor converter with combined control was carried out by calculating the power losses in its transistors and comparing them with the losses in transistors of converter prototype, which operates in linear mode all the time during welding process. The converter prototype architecture and its principles of operation are described in [1].

Welding current forming by converter prototype in linear mode is illustrated in Fig. 7 – (a) and the current forming by proposed multicell-type transistor converter with combined control is illustrated in Fig. 7 – (b). The abbreviations for the newly introduced variables are:
- i_W – welding current through welded contact;
- $I_{W\,lim}$ – welding current limit value;
- I_{MAX} – maximum level of converter's cell current.

Shaded areas in Fig. 7 mark the intervals of current regulation in linear mode in order to visually compare power losses.

Due to the fact that initial gradual increase of current is the most important stage in achieving proper welding, energy efficiency was estimated for that stage only.

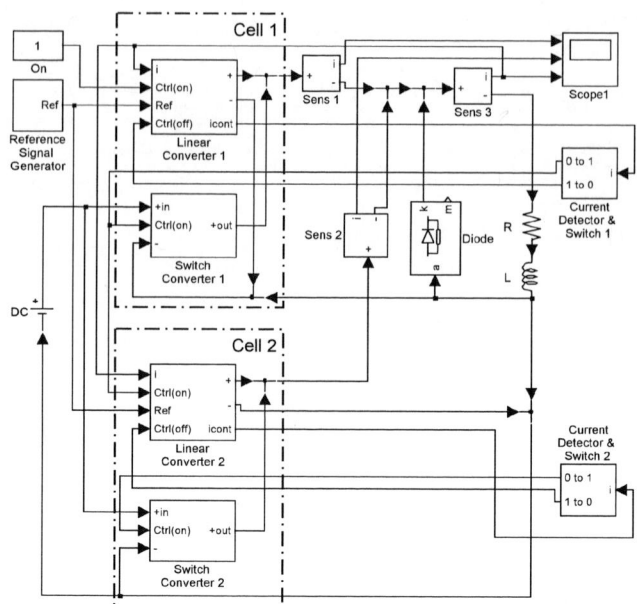

Fig. 5 Simulation model of multicell-type converter

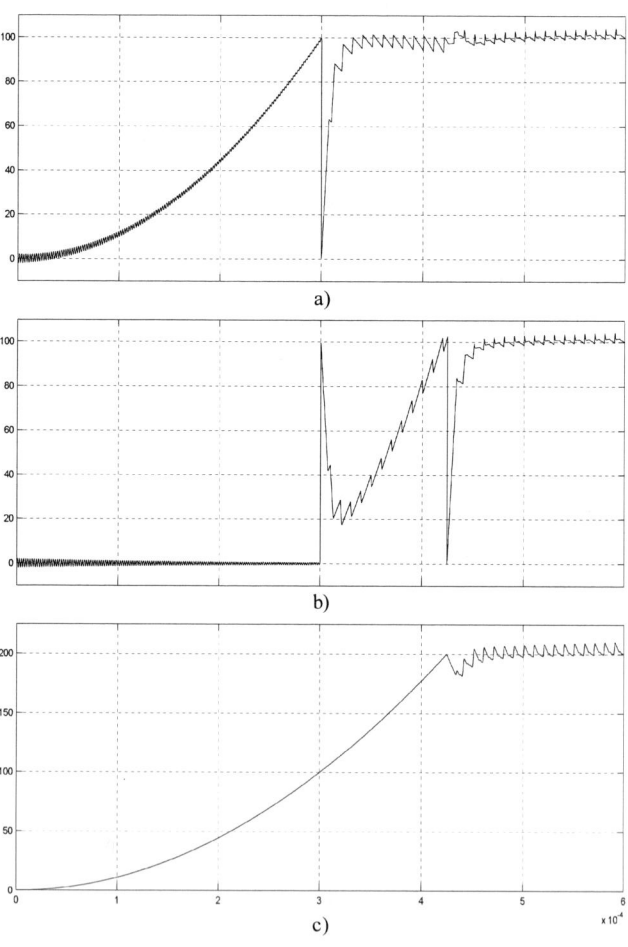

Fig. 6 Diagrams of currents
obtained with multicell-type converter simulation

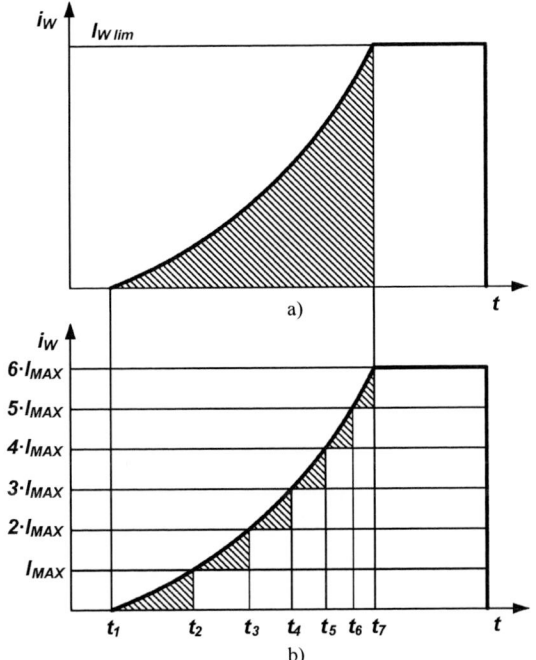

Fig. 7 The diagram of converter prototype welding current (a) and the diagram of proposed multicell-type converter welding current (b)

Averaged power losses in transistors of the converter prototype are represented as:

$$P_{LOSS\,1} = \frac{\int_{t_{start}}^{t_{end}} i_W(t) \cdot (U_S(t) - U_W(t))dt}{t_{end} - t_{start}}, \quad (2)$$

where $P_{LOSS\,1}$ – averaged power losses in transistors of converter prototype;

$i_W(t)$ – welding current through welded contact;

$U_W(t)$ – voltage across welded contact;

$U_S(t)$ – energy source voltage;

t_{start} – welding start time;

t_{end} – end time of initial gradual increase stage.

Averaged power losses in transistors of the multicell-type converter with combined control are calculated as:

$$P_{LOSS\,2} = \frac{\sum_{1}^{n}\left(\int_{t_k}^{t_{k+1}}(i_W(t) - I_{MAX}(k-1))(U_S(t) - U_W(t))dt\right)}{n \cdot (t_{end} - t_{start})}, \quad (3)$$

$P_{LOSS\,2}$ – averaged power losses in transistors of the multicell-type converter with combined control;

I_{MAX} – maximum value of converter cell current;

n – number of cells in the converter;

k – cell index (from 1 to n);

t_k – start time when cell k begins to operate;

t_{k+1} – stop time when cell k suspends linear mode and cell $k+1$ begins to operate.

The mathematical relationship between power losses in transistors of multicell-type converter with combined control and power losses in transistors of converter prototype may be represented as the relation between welding currents of converters and calculated as:

$$\frac{P_{LOSS\,2}}{P_{LOSS\,1}} = \frac{\sum_{1}^{n}\left(\int_{t_k}^{t_{k+1}}(i_W(t) - I_{MAX}(k-1))\right)}{n \cdot \left(\int_{t_{start}}^{t_{end}} i_W(t)dt\right)}, \quad (4)$$

The calculation was carried out with MathCAD software with the following parameters:

- maximum value of cell current: 100 A;
- energy source voltage: 10 V;
- current waveform on the initial stage: square function;
- number of converter cells: 6;
- welding current limit: 600 A;
- duration of the initial gradual current increase stage ($t_{end} - t_{start}$): 1 ms.

The power losses in converter transistors in switch mode are very small comparing to the losses in linear mode, therefore were ignored in present calculations.

The calculation using (4) generated the following ratio:

$$\frac{P_{LOSS\,2}}{P_{LOSS\,1}} = \frac{1}{4,314}.$$

Using the multicell-type converter with combined control resulted in reduction of power losses to more than 4 times with the above specified parameters.

CONCLUSION

Carried out simulations and calculations confirmed converter's ability to provide improved efficiency of energy consumption during micro resistance welding.

The combined control results in higher accuracy of current profile tracking than pulse control, due to use of the linear mode of transistor operation during the most crucial intervals of welding current forming.

The proposed method of control signal forming for converter cells as a differential signal is simple and effective and allows to keep the converter operating even when some of its cells are damaged.

Introduced multicell-type transistor converter with combined control is recommended for use in micro resistance welding machines to obtain high quality welded joints.

ACKNOWLEDGMENT

Authors thank Eli Marianovsky (AOS, Inc. U.S.A.) and Valentina Dyuzheva (DonSTU, Ukraine) for their help with the article editing.

REFERENCES

[1] Paerand Y.E., Bondarenko O.F. and Bondarenko Y.V. *"The former of special form current pulses for micro resistance welding"* // Compatibility and Power Electronics (CPE '09). – Badajoz, Spain. – 2009. – pp. 396-401. [Online]. Available: http://ieeexplore.ieee.org/stamp/stamp.jsp?arnumber=5156067&isnumber=5155997

[2] Паэранд Ю.Э. и Бондаренко А.Ф. *"Источник питания для контактной микросварки с программируемой формой сварочного импульса"*, Технология и конструирование в электронной аппаратуре, № 4(46). pp. 51-54. July – Aug. 2006. [Online]. Available: www.nbuv.gov.ua/portal/natural/tkea/texts/2006_4/51-54.pdf

[3] V.P. Leonov, V.E. Ataush, E.V. Bumbieris and M.A. Kalejs *"Device for Controlling Resistance Spot Welding Process"*, SU1214368, B23K11/24, Feb. 28, 1986. [Online]. Available: http://v3.espacenet.com/textdoc?DB=EPODOC&IDX=SU1214368&F=0

[4] Паэранд Ю.Э., Бондаренко Ю.В. и Бондаренко А.Ф. *"Формирователи импульсов тока для контактной сварки"* // Технология и конструирование в электронной аппаратуре, № 3(75). pp. 25-30. May – June. 2008. [Online]. Available: http://www.nbuv.gov.ua/portal/natural/tkea/texts/2008_3/25-30___.pdf

[5] Атауш В.Е., Леонов В.П. и Москвин Э.Г. *Микросварка в приборостроении.* Рига: РТУ, 1996, 332 с.

[6] Nippon Avionics Co., Ltd, *Welding Power Supply.* [Online]. Available: http://www.avio.co.jp/english/products/assem/lineup/welding/power/index.html

[7] MacGregor Systems, *Resistance Welding.* [Online]. Available: http://www.macgregorsystems.com/go/resistance_welding.php

[8] Miyachi Corp., *Fine Spot Welders.* [Online]. Available: http://www.miyachi.com/e/product/category/fine_spot/index.html

[9] Miyachi Unitek Corporation, *Resistance Welding Power Supplies.* [Online]. Available: http://www.miyachiunitek.com/Products_ResistanceWelding

[10] Паэранд Ю.Э., Бондаренко А.Ф. и Бондаренко Ю.В. *"Многоячейковый транзисторный регулятор тока с комбинированным управлением для установок контактной микросварки"* // Технічна електродинаміка. Тем. вип. Силова електроніка та енергоефективність. – 2009. – ч. 3. – С. 10 – 15. [Online]. Available: http://es.dmmi.edu.ua/news/files/alush_2009.pdf

[11] L.J. Brown and J. Lin *"Power Supply Designed for Small-Scale Resistance Spot Welding"*, Welding Journal, vol. 84, No. 7, 2005. [Online]. Available: http://www.aws.org/wj/2005/07/032/

[12] Леонов В.П. и Атауш В.Е. *"Малоинерционный источник питания для микросварки и пайки с обратной связью по электроэнергетическим параметрам"*, В кн.: Припои для пайки современных материалов. Киев: ИЭС им. Е.О. Патона, 1985, С. 133-139.

High Efficiency Adaptive Boost Converter for LED Drivers

Yang Lu and Dariusz Czarkowski

Polytechnic Institute of NYU

6 Metrotech Center

Brooklyn, New York 11201

Luyang2007@gmail.com

dcz@pl.poly.edu

USA

Wieslaw E. Bury

DeVry University

630 U.S. Highway One

North Brunswick, New Jersey 08902

wbury@devry.edu

USA

Abstract – **This paper presents the design and simulation results of an 89% power efficient CMOS adaptive controlled boost (step-up) LED driver implemented in 0.25 μm BCD (Bipolar CMOS DMOS) technology. A novel adaptive minimum frequency control provides up to 32 V from a single battery (3.6 V to 5.5 V) input supply to 10 series-connected LEDs at the output. The proposed control scheme provides an accurate load current (3--50 mA) while achieving 6% - 12% higher power efficiency than conventional fixed on-time schemes. The novel control circuitry is able to switch between PWM (Pulse width modulation) and PFM (Pulse frequency modulation) automatically by calculating the feedbacks from the inductor and LED's currents. This controller is functional from light to heavy loading situations which is critical in improvement of power efficiency and battery life-time for high-boost ratio applications in order as provide accurate LED current.**

Index Terms – PWM, PFM, dc-dc converter, boost, step-up, LED driver, high efficiency

I. INTRODUCTION

The demand for power in color displays is constantly increasing. Because of lower power consumption and higher reliability than other backlight options, white LEDs are an excellent choice for backlights of color displays, including those used in large screens such as notebook computers, tablet computers, and automotive displays. Today, LEDs are available for various colors, and they are also suitable for white illumination. However, an economical method of driving large arrays of white LEDs needs to be found to enable commercial applications. Approximately 32 V supply voltage is needed to forward-bias ten series-connected LEDs, whereas the nominal operating voltage of a single-cell lithium-ion battery is 3.6 V to

4.2 V [1]. Boost regulators usually provide such a supply voltage [2]. The switching losses and fixed losses of fixed-frequency PWM converter are independent of loading

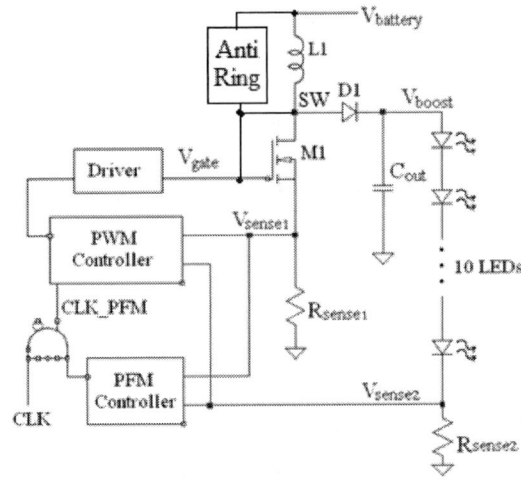

Fig. 1. Schematic of a boost converter with PWM and PFM control

conditions. The combination of PWM and PFM controlled schemes forms a better control option for light to heavy loading conditions because of lower switching frequency and lower switching loss [3], [4], [5]. Pulse-level-modulation (PLM) control method incorporated with conventional SMPS power stage realizes power efficient current regulator for driving high brightness LEDs [7]. LEDs and LED strings can not only be supplied by pure DC currents, but also by pulsating currents [8].

LEDs can be driven by almost all converter topologies like the buck, boost or fly-back converter. Resonant operating drivers are also suitable for LED supplies without a smoothing capacitor. A capacitor in series with a set of anti-parallel LEDs,

978-1-4244-8806-3/11 $26.00 © 2011 IEEE

in conjunction with application of a high frequency ac voltage, was found to help keep currents balanced [9].

This paper proposes a novel pulse-frequency modulation (PFM) technique which is compatible with fixed-frequency pulse-width modulation (PWM) to realize high power efficiency (86%) under full-to-light loading conditions (50 mA - 3 mA).

II.   ADAPTIVE CONTROL SYSTEM DESIGN

A.   Adaptive Control Concept

An IC driver circuit described in [6] supplies 6 LEDs from a 6 V battery to 29 V output at 20 mA load condition. The boost converter proposed in this paper, shown in Fig. 1, can drive 10 LEDs from battery voltage (3.6 V - 5.5 V) to boosted output voltage of 32 V at the current range of 3 mA to 50 mA.

The system works in the PWM mode when LED current is above 12 mA. By judging the signal from V_{sense1} and V_{sense2}, when the system works at light load, the output of the PFM controller is able to turn off the system clock by an AND gate which will finally turn off the PWM controller. When the output capacitor cannot support load requirements, the output of PFM controller is used to turn on the system clock which will subsequently turn on the PWM controller. The 10 LEDs, which are powered by output capacitor C_{out}, are in series with the sensing resistor R_{sense2}. Two feedback signals, V_{sense1} and V_{sense2}, are used by controllers to determine if the boost system works in the PWM or the PFM mode.

B.   PWM Operation

The PWM controller is shown in Fig. 2. Error Amp amplifies the voltage difference between V_{sense2} and V_{ref2}. V_{sense2} is the voltage over LED current sensing resistor

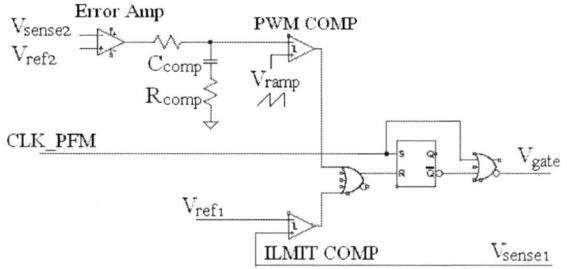

Fig. 2.   Schematic of PWM controller

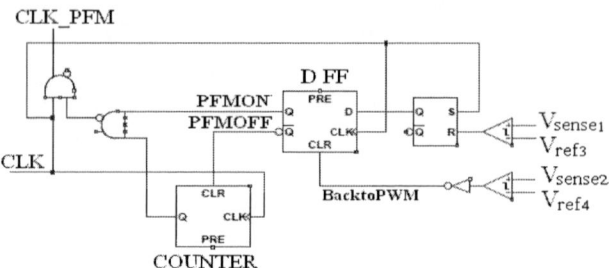

Fig. 3. Schematic of PFM controller

R_{sense2}. The output of the error amp is compared with V_{ramp} (ramp generator output signal) by a PWM comparator. C_{comp} is a compensation capacitor. R_{comp} is a compensation resistor. A high frequency zero is generated to cancel the high frequency pole.

The output of Error AMP is compared with a ramp voltage by a PWM COMP. PWM COMP is a hysteretic comparator which controls the current limit of the inductor. When $V_{sense1} > V_{ref1}$, PWM COMP will turn on the POWER NMOS M1 to avoid the over-charging of the inductor. When $V_{sense1} < V_{ref1}$, PWM COMP will do nothing. CLK_PFM is a system clock which is a constant 1MHz in the PWM mode.

C.   PFM Operation

The PFM controller schematic is shown in Fig. 3.

Vsense1 is the voltage over the sense resistor of POWER NMOS which is shown in Fig. 1. When $V_{sense1} < V_{ref3}$, the RS FF will be reset and PFMON is 0. So CLK_PFM equals CLK which is 1 MHz all the times. Boost converter works in the PWM mode. When $V_{sense1} > V_{ref3}$, the RS FF will be set. And PFMON will be 1 for the condition of $V_{sense2} > V_{ref4}$. When $V_{sense2} < V_{ref4}$, the D FF will be cleared by the signal of Back to PWM which is 1. That means that the LED current is lower than the threshold, and the system needs to switch back to PWM mode. The two sense threshold voltages can be determined by equations (1) and (2):

$$V_{ref3} = I_{NMOS} \times R_{sense1} \qquad (1)$$

$$V_{ref4} = I_{LED} \times R_{sense2} \qquad (2)$$

The counter, which is used to calculate the PFM operating time, provides an adaptive minimum frequency control. The minimum PFM operating frequency has been set to 60 kHz for lightest load conditions.

978-1-4244-8806-3/11 $26.00 © 2011 IEEE

D. Circuitry Design

A comparator with hysteresis is needed for fast comparison in a noisy environment. A complete comparator with internal hysteresis, including an output stage, is shown in Fig. 4. Biasing resistance is 25 kΩ.

The positive trip point and negative trip point can be calculated by equations (3) and (4):

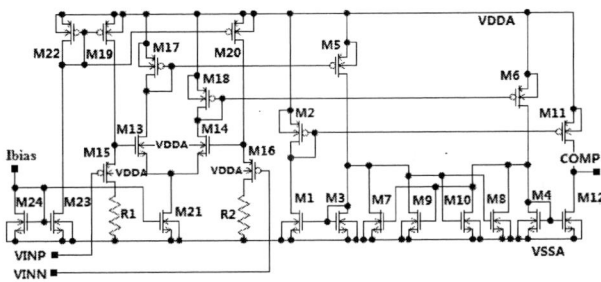

Fig. 4. Complete comparator with internal hysteresis

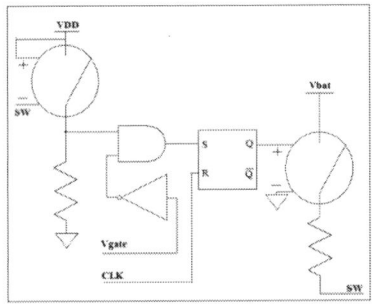

Fig. 5. Anti-ring topology

$$V_{TRP}^{+} = \sqrt{\frac{2i_1}{\beta_6[1+\frac{(\frac{W}{L})_3}{(\frac{W}{L})_9}]}} - \sqrt{\frac{2i_1}{\beta_5[1+\frac{(\frac{W}{L})_9}{(\frac{W}{L})_3}]}} + V_{T6} - V_{T5} \quad (3)$$

$$V_{TRP}^{-} = \sqrt{\frac{2i_1}{\beta_6[1+\frac{(\frac{W}{L})_{10}}{(\frac{W}{L})_4}]}} - \sqrt{\frac{2i_1}{\beta_5[1+\frac{(\frac{W}{L})_4}{(\frac{W}{L})_{10}}]}} + V_{T6} - V_{T5} \quad (4)$$

Anti-Ring is a critical design eliminating ringing of SW node as switch is turned off. Fig. 5 shows the topology of anti-ring circuits.

III. SIMULATION RESULTS

The fully integrated LED driver circuit is functional for LED currents up to 50 mA at low input supply voltages (3.0 V to 5.5 V). The LED driver with PWM & PFM controllers is fully functional in the 40 V process of 0.25 μm BCD technology, as shown by the simulated gate-drive, inductor current, LED current and output voltage waveforms presented in Fig. 6 and Fig. 7.

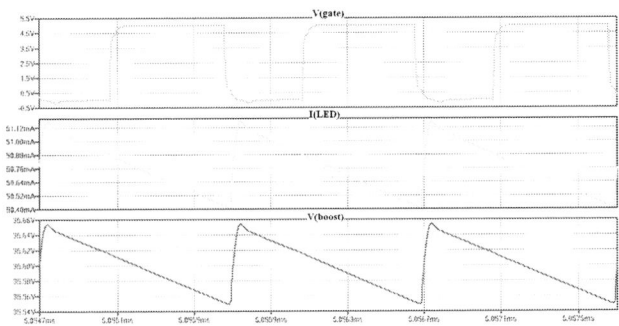

Fig. 6. Simulated gate-drive, inductor current, led current (50 mA) and output voltage waveforms of the proposed PWM mode

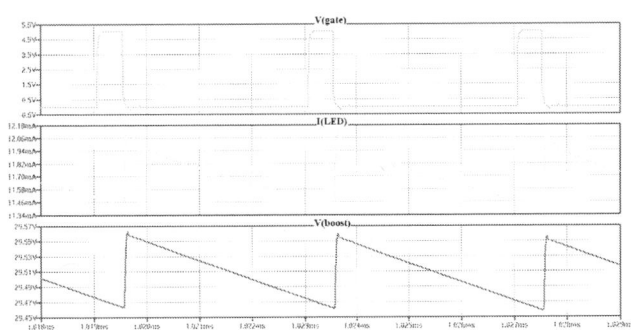

Fig. 7. Simulated gate-drive, inductor current, led current (16 mA) and output voltage waveforms of the proposed PFM mode

V(gate) is the gate drive signal of power MOSFET M1. I(L) and I(LED) are the inductor current and the LED current, respectively. V(boost) is the boost converter output voltage. In Fig. 6, the converter is operating at constant 1 MHz PWM mode. Ripplepk-pk of the output LED current is 1.34%. In Fig. 7, the controller operating frequency is not constant and is around 250 kHz. Ripplepk-pk of the output LED current in PFM mode is 2.75%.

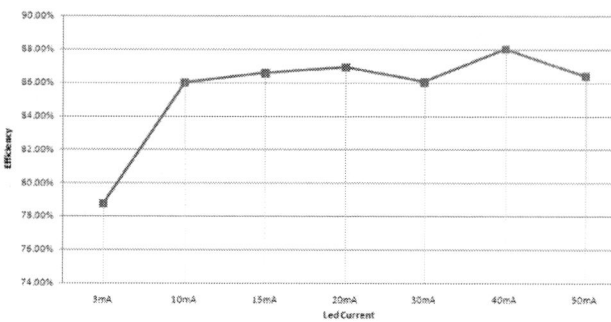

Fig. 8. Efficiency of 10 LED string

The efficiency of the proposed PWM/PFM controller has been measured when the LED current is changing from 3mA to 50 mA. The controller switches between PWM and PFM modes automatically with the changes of the LED current. The averaged efficiency of the converter is 86%. This boost LED driver provides up to 32 V from a single battery (3.6 V to 5.5 V) input supply to 10 series-connected LEDs at the output, 30mA is the switchover point between PFM mode and PWM mode. PFM acquires high efficiency when the LED current is less than 30mA. PWM acquires high efficiency when the LED current is larger than 30mA. The combination modulation of PWM and PFM realizes constant high power efficiency in a wide range of LED currents.

IV.   CONCLUSIONS

A novel combination of PWM and PFM controlled boost converter is proposed, designed, and simulated in 0.25 μm BCD technology. The converter provides up to 32 V from a 3.6 V to 5.5 V input supply for 10 serial LEDs at the output. To obtain the best power management efficiency, the controller switches between PWM mode and PFM mode automatically as the LED current changes. The averaged efficiency of the converter is 86% when the LED current is on the range of 3

mA - 50 mA. This controller is functional from light to heavy loading situations which is critical in improvement of power efficiency and battery life-time for high-boost ratio applications such as providing accurate LED current.

REFERENCES

[1] C. Richardson, Driving high-power LEDs in series-parallel arrays, National Semiconductor, Santa Clara, CA, 2008 [Online]. Available: http://www.edn.com/contents/images/6615611.pdf

[2] W. Ly, Circuit delivers dimming control for white-LED driver, National Semiconductor, Santa Clara, CA, 2004 [Online]. Available: http://www.edn.com/article/CA472839.html

[3] R. Erickson and D. Maksimovic, High efficiency, dc-dc converters for battery operated systems with energy management, Department of Electrical, Computer, and Energy Engineering, University of Colorado, Boulder [Online]. Available: http://ecee.colorado.edu/~rwe/papers/EnergyMangmt.pdf

[4] L. Zhu and F. Quanyuan, "Design of PWM controller for monolithic boost converter," *7th International Conference on ASIC*, pp. 660-663, 2007

[5] B. Sahu and G.A. Rincon-Mora, "An Accurate, Low-Voltage, CMOS Switching Power Supply with Adaptive On-Time Pulse-Frequency Modulation (PFM) Control," *IEEE Transactions on Circuits and Systems,* Volume 54, Issue 2, Feb. 2007 pp: 312 - 321

[6] C.H. Chang, H. M. Chen, and R.C. Chang, "A 2.3 V CMOS Monolithic, 84% Efficiency PFM Control DC-DC Boost Converter for White LEDs Driver IC," *PEDS 2005*, pp. 833 – 837

[7] Y. Fang, Siu-Hong Wong and L. Hok-Sun Ling, "A Power converter with pulse-level-modulation control for driving high brightness LEDs", Proc. Twenty-Fourth Annual IEEE Applied Power Electronics Conference and Exposition APEC, 2009, pp. 577-581

[8] H.V.D. Broeck, G. Sauerlander, and M. Wendt, "Power driver topologies and control schemes for LEDs", Applied Power Electronics Conf. (APEC2007), Mar. 2007, pp. 1319-1325

[9] S. M. Baddela, D. S. Zinger, "Parallel connected LEDs operated at high frequency to improve current sharing" IAS Conference, 2004. 39th IAS Annual Meeting. Conf. Rec. of the 2004 IEEE Vol. 3, 3-7 Oct. 2004 Pages: 1677-1681

978-1-4244-8806-3/11 $26.00 © 2011 IEEE

Staff Training for Servicing Special Power Electronic Applications

Zoja Raud and Valery Vodovozov
Department of Electrical Drives and Power Electronics
Tallinn University of Technology (TUT)
Tallinn, Estonia
e-mail: zoja@cc.ttu.ee

Abstract − **Based on a study of promoted innovative set-ups and practices of European enterprises, this paper describes a novel approach to curricula organisation that aims to facilitate and improve the quality and efficiency of in-service training, on-the-job training, and undergoing training provided in a workplace environment of power electronic engineering. The purpose is to ground an educational model suitable for planning optimal training trajectories in the industry context. Effective instruments are given to build the most appropriate professional thesauri for the company staff and to find an institution capable of providing training in such thesauri. Unlike the traditional environment, the proposed flexible curriculum has an exclusively dynamic nature. Any time when the professional level is raised, the curriculum may be changed simultaneously along with its background conceptual matrix. Thus, new disciplines are introduced, the contents of the corresponding disciplines refreshed, and the borders between the disciplines shifted fluently. This promotes designing the teaching modules in highly interdisciplinary areas and in the areas with specific needs.**

Keywords-e-learning; power electronics; service; education; electronic engineering

I. Introduction

In many sectors, electronics specialists are generally responsible for driving innovation and competition. Contemporary changes in the business sector, responses of enterprises to these changes, as well as available information and communication technologies pose a number of challenges to the electronics staff [1]:

- continuous learning of new technologies and methods
- fast promotion of projects in the frame of time scarcity
- maintenance of manufacturing systems based on the online tools and resources
- active personal development and competency improvement

Companies require an aptitude for collaborative work, team and task management, concept synthesis, and decision-making from their engineers, thus stimulating progress in the new environment for learning, which gives all of these job-related skills.

Over the past few years, there has been increasing interest in effective learning technologies for power electronics. To promote staff development, employees usually attend training courses, workshops, seminars, and conferences. With advancements in technology, organizations are engaging the use of different learning media to enhance staff skills and knowledge. In addition, training on the job and situation-based learning are increasingly considered in the modern fast changing knowledge society. However, most research and surveys indicate that enterprises have a limited capacity and participation in continuous education and training. It is confirmed also by our own findings that companies are slow to implement new educational approaches and the staff do not benefit much from training because of their context of work, productivity and time.

To recognise the optimal learning pathways, the definition of the professional profiles has been offered in [2]. Using such profiles, company representatives may choose between different courses and educational institutions to find the most appropriate for their profile specifics and targets. In [3], an overview of educational strategies used in companies is also given and a number of attempts to increase their effectiveness are listed. With that end in view, learning in industry is divided between formal training and vocational training. Formal training is classified as training arranged and packaged to cover a given subject, with clearly defined topics, eventually leading to the delivery of a certification whereas vocational training relates to training, the costs of which are supported by the company and the topics of which are related to individual jobs. Moreover, it is presented in [4] that apart from the normal working environment, staff training also occurs in social events and in everyday activities. This means that work activity is carried out in various social settings where employees collaborate and interact on specific subjects. Therefore, enterprises have to integrate learning from economic, human and social perspectives.

Following an analysis of the promoted innovative set-ups and practices in the European industrial companies, this paper proposes a novel approach to curricula organisation. The aim is to facilitate and improve the quality and efficiency of in-service training, on-the-job training, and undergoing training provided in a workplace environment. Further, an appropriate educational model is grounded enabling development of optimal training trajectories in the framework of the curricula organisation, particularly in the field of power electronics. The focus concerns reevaluation of the syllabi design along with the strengthening of e-learning role using the conceptual approach. Effective instruments are given to build the appropriate professional thesauri for the company staff and to find an institution capable of providing training in such thesauri.

II. CURRICULA ARRANGEMENTS

An educational curriculum may be viewed as a table or as a flowchart where the disciplines are linked by the knowledge flow from one course to another. Using the terminology of [5][6], particular disciplines in such models serve as the nodes whereas their links are the branches of the curriculum tree. The disciplines of the first semester are the roots of the tree as they have no input branches whereas the disciplines of the last semester are the leaves, as they have no output branches within a professional curriculum.

An analysis of multiple bachelor and master degree curricula of Tallinn University of Technology (TUT) and the partner institutions [7], [8] shows the following drawbacks of the curricula discussed:

- contents of different courses are poorly intercommunicated, therefore the interdisciplinary links are used or disregarded depending upon the course syllabi
- discipline contents often repeat one another to a great extent
- hard structure of the curricula is unsuitable for transformations due to the fixed volume of the courses and their invariable positions in the curricula

Thus, the traditional curricula management approach is inappropriate both for educators and students whose needs are in continuous enhancing of their professional knowledge.

To overcome these restrictions, a "curriculum container system" was proposed in [9] as a tree consisting of five levels − curriculum, unit, task, episode, and element. Clearly, this environment is sufficiently complex for teachers who have to follow it along with the learning process. The teacher's activity is restricted here by the curriculum "aggregates", thus any time when somebody wants to change a learning trajectory, a modified curriculum is to be developed.

The next widespread direction is known as a "concept mapping" [10]. The concept maps are used in mechanical, chemical, and computer engineering across a wide range of educational applications − from simple creating a conceptual overview of course objectives for students to implementing them for the curricula improvement [11]. Unfortunately, they concern the only one possible venue in clarifying the student's perception of the field of study venturing through a simple visual presentation of the curriculum components.

Another tool to design and improve curricular development is a knowledge representative "conceptual graph" [12]. This approach elaborates the concept mapping technique having stronger influence on the syllabi content rather than on the curriculum topology.

Therefore, redefining the structure and the content of a curriculum is not a simple task given the fact that all the traditional formats present the outcomes of the evolutionary developed education. Actually, in all the described systems the changes in curricula are conducted by addition, deletion, and sometimes content modification of the courses. The relationship between courses is often tenuous with the exception of what is termed as a pre-requisite [13]. Any modification of these types of curricula tends to crowd the learning period and places heavy demands on the student and faculty resources. This prioritisation usually results in slow, but definitely compressed content, narrow focus, and, in certain cases, discourages creativity.

III. EDUCATIONAL THESAURUS

In education, the skills and ideas are usually formed by means of the concepts to provide solutions of professional problems within an appropriate knowledge field. It is a way to describe the learning outcomes in terms of qualifications. It is reasonable to consolidate all the professional concepts into a *thesaurus* that is a kind of dictionary giving semantically controlled relationship of the terms and concepts in an application area.

As distinct from the traditional thesauri like [14], an educational thesaurus must meet the following important peculiarities.

First, it should define and display a concept by the dozens of other predetermined concepts of the thesaurus as well as through the standard terms using the minimum number of incorporated units. The concepts are introduced within the course flow whereas the terms are the elementary knowledge units given outside the academic curriculum that should be preliminarily explained during the college-school education. In other words, a student must use the previous knowledge to execute the subsequent learning activities.

Second, the educational thesaurus is a high dynamic system renewed in step with the development of the professional level. The new concepts enrich the thesaurus whereas the existing definitions are subjected to clarifying, and obsolete ones are erased. In contrast to the informational and technical systems, the educational thesaurus is based on the only concepts, which

978-1-4244-8806-3/11 $26.00 © 2011 IEEE

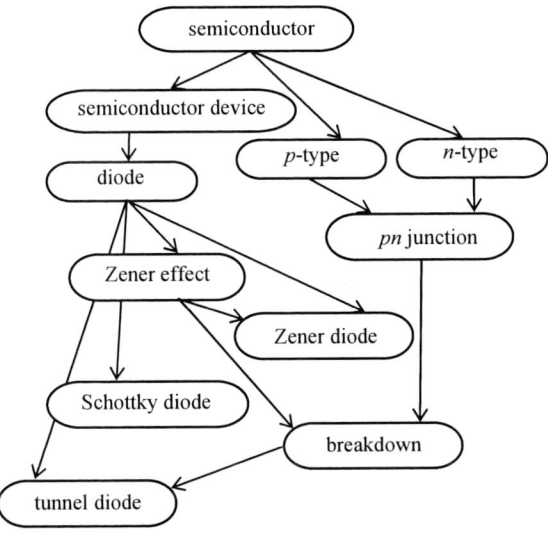

Figure 1. Fragment of a concept tree

TABLE 1. FRAGMENT OF THE CONCEPT TABLE

i	Name	C1	C2
1	breakdown	Zener effect	pn junction
2	diode	semiconductor device	
3	pn junction	p-type	n-type
4	Schottky diode	diode	
5	semiconductor		
6	Zener diode	diode	Zener effect
7	Zener effect	diode	
8	p-type	semiconductor	
9	n-type	semiconductor	
10	tunnel diode	diode	breakdown
11	semiconductor device	semiconductor	

are in active use in the professional activity defined earlier through other professional concepts.

Third, the conceptual apparatus of an educational system should be upgraded regularly along with the studies. It may well be so that the concepts which are similar in names but different in contents be repeated in the thesaurus. In learning, an elementary concept is introduced firstly, which is specified, enlarged, and illustrated with examples and exceptions. As a result, the particular studies stipulated by the syllabus help the students to build their own conceptual systems, thus approaching the system required by the curriculum. An efficiency of one or another educational environment is defined by the speed of mastering developed by the learners as well as by their knowledge completeness.

This is the reason why the thesaurus should be viewed as a concept tree. Fig. 1 gives a tiny fragment of the thesaurus, which concerns some concepts introduced in the course *Electronics and semiconductors*. Using such a bottom-up structure students and teachers can capture the concepts and evaluate the learning progress.

Assume that the thesaurus comprises m entries, each of which describes a particular node of the concept tree. An i-th concept within the thesaurus is written by an i-th entry, which is given by a finite sequence of n components, $n \geq 3$:

$$CON_i = \{i, Name_i, C_{i1}...C_{ip}\} = \{i, Name, C_{i\Sigma}\}.$$

Here, the component i ($i \in [1, m]$) designates an index of a concept identified by its *Name*. The concept group $C_{i\Sigma} = C_{i1}...C_{ip}$ comprises the entries introduced earlier, called *predetermined concepts* and used to define the current entry CON_i called a *defined concept*, $C_j \in [CON_1, CON_{i-1}]$, $p < m$. Table 1 represents a fragment of the thesaurus devoted to the concepts given in Fig. 1.

Conceptual approach proposes an effective instrument for organisations to build the most appropriate thesaurus for their staff and to find an institution capable of providing training in such a thesaurus.

Let us call the number of concepts m a student should acquire within the learning period as the length of an educational trajectory. As a defined concept CON_i depends on the predetermined concepts C_i and is independent of the rest concepts from m, the more is i the later the concept CON_i may be introduced into the learner's knowledge area. Therefore, the discipline thesaurus may be represented by the quadrant matrix occupied the columns $C1...Cm$ of Table 2, where $w_{j,j} \in \{1, 0\}$ are the binary connection weights.

TABLE 2. THESAURUS MATRIX

i	DIS	NAME	C1	C2	...	Cj	Cj+1	...	Cm−1	Cm
1		$NAME_1$	$w_{1,1}$	$w_{1,2}$	...	$w_{1,j}$	$w_{1,j+1}$	...	$w_{1,m-1}$	$w_{1,m}$
2	DIS_1	$NAME_2$	$w_{2,1}$	$w_{2,2}$	...	$w_{2,j}$	$w_{2,j+1}$	...	$w_{2,m-1}$	$w_{2,m}$
:		:	:	:	: :	:	:	: :	:	:
i		$NAME_i$	$w_{i,1}$	$w_{i,2}$	...	$w_{i,j}$	$w_{i,j+1}$	...	$w_{i,m-1}$	$w_{i,m}$
i+1	DIS_i	$NAME_{i+1}$	$w_{i+1,1}$	$w_{i+1,2}$	...	$w_{i+1,j}$	$w_{i+1,j+1}$	...	$w_{i+1,m-1}$	$w_{i+1,m}$
:		:	:	:	: :	:	:	: :	:	:
m−1	DIS_{m-1}	$NAME_{m-1}$	$w_{m-1,1}$	$w_{m-1,2}$	...	$w_{m-1,j}$	$w_{m-1,j+1}$	...	$w_{m-1,m-1}$	$w_{k-1,m}$
m		$NAME_m$	$w_{m,1}$	$w_{m,2}$	...	$w_{m,j}$	$w_{m,j+1}$	...	$w_{m,m-1}$	$w_{m,m}$

Consequently, an i-th predetermined concept group may be presented as follows:

$$C_{i\Sigma} = m(C_j \cdot w_{i,j}) = C_1 \cdot w_{i,1}, C_2 \cdot w_{i,2}, ... , C_m \cdot w_{i,m},$$

where the universal qualifier $m(C_j \cdot x_{i,j})$ comprises all the predetermined concepts that define CON_i and j as an index of the matrix column.

As the predetermined concepts are introduced into the curriculum before the defined concept, a properly designed thesaurus should be ordered and described by the left-triangular matrix as follows:

978-1-4244-8806-3/11 $26.00 © 2011 IEEE

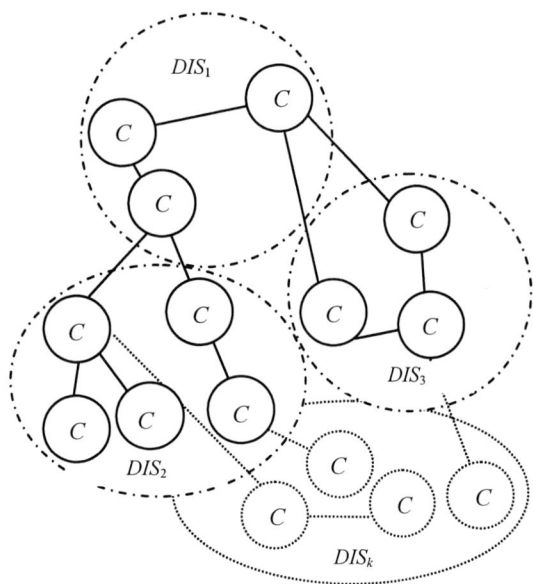

Figure 2. Concept aggregating with disciplines

$w_{i,j} = 0$ when $j \geq i$, $w_{i,j} \in \{0, 1\}$ when $j < i$.

An effective algorithm of the thesaurus ranging on the key C_i is proposed in [8].

IV. FLEXIBLE CURRICULUM

Since the thesaurus is ranged, the neighbour concepts may be involved into the groups outlined in the column DIS of Table 2 and in Fig. 2. Each group is properly associated with a discipline of the curriculum. From now on, a curriculum may be represented by a model having a couple of levels:

- top curriculum level as an ordered system of disciplines $DIS_1 \ldots DIS_k$, $k < m$, which shows an educational trajectory of a learner
- concept matrix, which serves as a background for the upper level

From this view, curriculum arrangement poses the process whereby the components are interpreted through the learning experiences. To integrate the disciplines into the concept model, modify the above given concept description as follows:

$$CON_i = \{i, Name_i, DIS_i, C_i\},$$

where DIS_i is a discipline whereby the concept CON_i and the following concepts are introduced.

Let i_{DIS} be a given starting instant where a discipline DIS_i begins. In the simplest case, when $p = 0$ (no predetermined concepts), $i = i_{DIS}$, which means that all such concepts CON_i may be introduced starting from the first study of DIS_i. The same concerns the defined concepts described by the components of the earlier passed disciplines. For instance, if $i_{DIS1} < i_{DIS2} < i_{DIS}$ and CON_i is defined by the predetermined concepts of i_{DIS1} or i_{DIS2}, then $i \geq i_{DIS}$.

To support the basic concepts of the application area and to reflect the challenges being an instrument for solving the practical problems of companies, a thesaurus should possess sufficient redundancy. Really, different professional branches of knowledge require the sets of defined concepts, thus an excess capacity of concepts and disciplines is the normal feature of any thesaurus. Particularly, in *Power Electronics* the known examples of such branches are: *Industrial electronics, On-board electronics, Aircraft electronics, Automotive electronics, Military electronics.*

As the number of concepts m a student should acquire within the learning period is less than the total number of the professional concepts m_{Σ}, the personal educational trajectories may differ depending upon the staff goals and the future degree that a learner approaches. To enhance the engineers' knowledge level for different enterprises, specific educational trajectories are needed, thus some groups of concepts and particular disciplines of the full thesaurus may be selected.

Hence, the proposed system of the consolidated disciplines aggregated into a curriculum is an enormously right tool to generate the required educational trajectories. Being connected by means of concepts, the disciplines successfully support the total plan for learning. In Fig. 2, the solid nodes and branches outline the appropriate educational trajectory whereas the rest of the concepts deleted from the total thesaurus are given by the dotted lines.

In addition to the concepts and disciplines, a curriculum may include everything that promotes the learners' intellectual, personal, social and physical preferences [15]. It may also involve the studies, extracurricular activities, approaches to teaching, learning and assessment systems, the quality of relationships within an institute, and the values embodied in the way the institute operates.

Unlike the traditional environment, the proposed model of learning has an exclusively dynamic nature; therefore, it may be called a *flexible curriculum*. Any time when the professional level is raised, the curriculum may be changed simultaneously along with its background conceptual matrix. Thus, the new disciplines are introduced, the contents of the corresponding disciplines refreshed, and the borders between the disciplines shifted fluently. This promotes designing the teaching modules in highly interdisciplinary areas and in the areas with specific needs.

V. DEFINING THE PERSONAL EDUCATIONAL TRAJECTORIES

Defining the optimum educational trajectories helps to provide the following procedures:

- evaluation of the complexity of the particular disciplines and the full speciality curriculum
- optimisation of the order of the disciplines in the curriculum

978-1-4244-8806-3/11 $26.00 © 2011 IEEE

- building the curriculum for additional education and second speciality acquisition

The common information problem may be formulated as follows. Previously studied discipline (speciality) is based on a system of concepts A. The new discipline (speciality) is based on a system of concepts B, some of which may be the concepts of A. Find an educational trajectory between the new and previously studied disciplines (specialities). If among the concepts B there is at least one concept defined by a term being a concept of A, then the two systems can be described together by a graph C and the solution of the problem is reduced to finding a path between these concepts in C using the known algorithms of the theory of information, for example, Dijkstra's algorithm [16]. Thus, the challenge is to find the terms of B that correspond to the concepts of A. If there are no concepts of A among the terms in B, but there is a concept from some other discipline K, which in turn has a concept from A, then K acts as an intermediary discipline. In this case we should find a path between the appropriate concepts in K using the same algorithm again.

An example from TTU learning system is given below.

Assume the previously studied discipline *Power electronic converters (PEC)* includes some concepts like these ones:

1. **power electronic converter (PEC)** − *electronic converter* that converts *energy* in *power electronic system*
2. **dc/dc converter** − **PEC** converting *dc* to *dc* of another level
3. **load** − object connected to the **PEC** output
4. **supply** − *power* line feeding the **PEC**
5. **switching dc converter** − **dc/dc converter** built on a *switching* principle of operation
6. **boosting** − production of **load** *voltage* higher than **supply** *voltage*
7. **booster** − **PEC** with **boosting** possibilities
8. **buck converter** − **switching dc converter** which output *voltage* is less than input *voltage*
9. **boost converter** − **booster**
10. **buck-boost converter** − **buck converter** united with **boost converter**

Here, the concept names are given by the bold type and an italic font is used for the terms coming from prior disciplines, such as *Electronics* and *Electrical Engineering*. The defined concept names occupy the left side of each line whereas the definition functions of the predetermined concepts are to the right. Figure 3 shows the graph of these concepts to the right, where the dotted arrows mark the incoming terms.

Let the new discipline *Energy engineering* include the concepts given below:

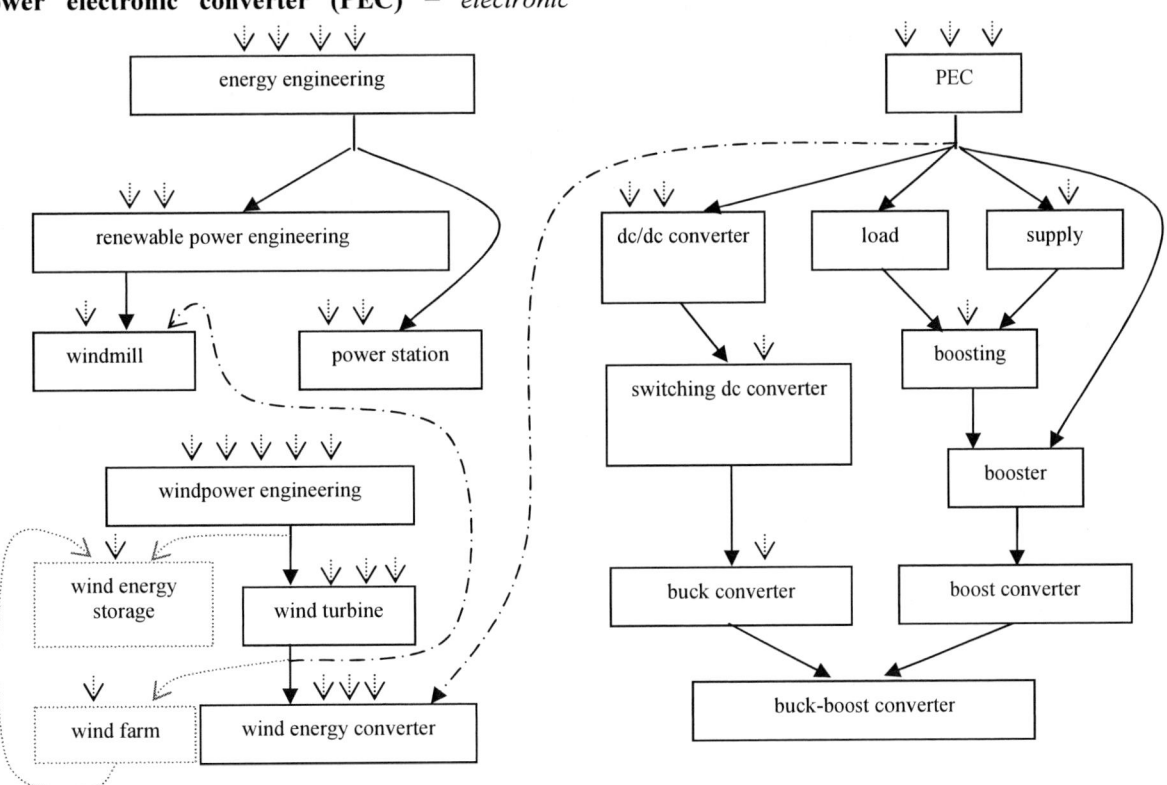

Figure 3. Semantic graph of a fragment of an educational thesaurus

1. **energy engineering** – field of *engineering* dealing with *energy management*, *plant engineering*, and *environmental compliance*
2. **power station** - **energy engineering** *system* for the generation of *electric power*
3. **renewable power engineering** – field of **energy engineering** dealing with *energy* which comes from *natural resources*, which are renewable
4. **windmill** – *machine* used in **renewable power engineering** to apply *mechanical energy* directly from *wind turbine*

An intermediary discipline *Wind power engineering* includes the following concepts:

1. **windpower engineering** - discipline focused on the *design engineering*, *maintenance*, *installation*, and *projects* related to the *wind power*.
2. **wind turbine** - *rotary device* used in **windpower engineering** to extracts *energy* from the *wind*
3. **wind energy converter** – *machine* to convert *mechanical energy* of **wind turbine** to *electricity* using *PEC*
4. **wind farm** - group of **wind turbines** in the same location used for production of *electric power*
5. **wind energy storage** - means used in **windpower engineering** to store *electricity* of **wind farm**

The graph fragments of *Energy engineering* and *Windpower engineering* are given in Fig. 3 to the left. Clearly, there are no concepts of *PEC* among the terms in *Energy engineering*. At the same time, the concept *windmill* is defined by the term *wind turbine*, which serves as a concept of the discipline *Windpower engineering* and there is a concept *wind energy converter* defined by *PEC* in the discipline *Windpower engineering*. Therefore, *Windpower engineering* acts as an intermediary discipline, in which we should find a path for a wind energy converter. Thus, the length of the full educational trajectory is equal to the length of the educational trajectory of the new discipline *Energy engineering* plus the length of the educational trajectory in the intermediary discipline *Windpower engineering* for the concept of the wind energy converter that is 3+2=5. The rest of the concepts of a wind farm and the wind energy storage of this discipline are not mandatory for study.

VI.   CONCLUSION

A novel model of the curriculum built on the professional thesaurus supports the overall set of concepts actual for definite enterprises and applies them to solve the practical problems. Unlike the traditional curricula, the approach proposes a flexible and easily upgradeable educational system. Therefore, this tool is suitable for numerous educational trajectories to be developed for different groups of learners. In this way, the learning outcomes are described in terms of qualifications, making institutions more responsive to the needs of the labour market and reinforcing the link between studies and employment needs.

ACKNOWLEDGMENT

This research work has been supported by Estonian Ministry of Education and Research (Project SF0140016s11), Estonian Science Foundation (Grant ETF8020) and Estonian Archimedes Foundation (project „Doctoral School of Energy and Geotechnology II").

REFERENCES

[1] V. Giannikopoulou, I. Hatzakis and A. Zafeiropoulos, "A technology enhanced flexible learning approach for SMEs," *IEEE 18th International Symposium on Personal, Indoor and Mobile Radio Communications (PIMRC' 07)*, Athens, Greece, 2007, pp. 1–5.

[2] L. Barreto, A. Vilaça, and C. Viana, "NetStart – Achieving new abilities with ICT," *IEEE Multidisciplinary Engineering Education Magazine*, vol. 4, no. 1/2, 2009, pp. 13–18

[3] V. Chang and C. Guetl, "E-Learning ecosystem (ELES) – A holistic approach for the development of more effective learning environment for small and medium sized enterprises," *Digital EcoSystems and Technologies Conference (DEST '07)*, Cairns, Australia, 2007, pp. 420–425.

[4] A. Cristea, A. Wentzler, E. Heuvelman, P. De Bra, "Adapting SME Learning Environments for Adaptivity," *6th International Conference on Advanced Learning Technologies (ICALT'06)*, Kerkrade, The Netherlands, 2006, pp. 130–132.

[5] N. Wirth, *Algorithms and Data Structures*, Prentice Hall, 1986, 288 p.

[6] D. Knuth, *The Art of Programming*, Addison Wesley, 1997, 650 p.

[7] V. Vodovozov and Z. Raud, "An object-oriented approach to curriculum design," *7th International Conference on Education and Information Systems, Technologies and Applications (EISTA'09)*, Orlando, FL, pp. 225–230.

[8] Z. Raud and V. Vodovozov, "Professional thesaurus of engineering educational system," *2nd International Multi-Conference on Engineering and Technological Innovation (IMETI'09)*, Orlando, FL, 2009, pp. 212–217.

[9] Y.-T. Wu, Y-M. Ku, Y.-H. Chen, J.-F. Wu, T.-W. Chan and J.-H. Wang, "Curriculum container system: a system to support curriculum and learning activity management," *5th IEEE International Conference on Advanced Learning Technologies (ICALT'05)*, Kao-Shiung, Taiwan, 2005, pp. 287–289.

[10] J. D. Novak and A. J. Canas, *The Theory Underlying Concept Maps and How to Construct and Use Them. Technical Report IHMC CmapTools 2006-01 Rev 01-2008*, Florida Institute for Human and Machine Cognition, 2008, Available at: http://cmap.ihmc.us/ Publications/ResearchPapers/TheoryUnderlyingConceptMaps.pdf.

[11] R. Morsi, W. Ibrahim and F. Williams, "Concept maps: Development and validation of engineering curricula," *37th Annual Frontiers in Education Conference – Global Engineering: Knowledge Without Borders, Opportunities Without Passports (FIE '07)*, Anaheim, CA, 2007, pp. T3H18–T3H23.

[12] Y. L. Wu, "A pilot study of applying hierarchical curriculum structure graph for remedial learning," *7th IEEE International Conference on Advanced Learning Technologies (ICALT'07)*, Niigata, Japan, 2007, pp. 103–107

[13] M. K. McCuddy, M. Pinar and E. F. R. Gingerich, "Using student feedback in designing student-focused curricula," *The International Journal of Educational Management*, vol. 22, no. 7, 2008, pp. 611–637.

[14] *Electropedia: The World's Online Electrotechnical Vocabulary*, Available at: http://www.electropedia.org/

[15] U. R. Nejib, "CAM: A tool for evaluating and adjusting engineering curriculum," *29th Annual Frontiers in Education Conference (FIE '99)*, 1999, San Juan, Puerto Rico, vol. 3, pp. 13B1/20–13B1/25

[16] E. W. Dijkstra, *Selected writings on computing: A personal perspective*, New York, Springer-Verlag, 1982, 362 p

DC-DC converter with isolation transformer for traction vehicles application – simulation studies

M.Parchomiuk, K.Tomczuk, Electrotechnical Institute, Power Converters Department,
Pozaryskiego 28, 04-703 Warsaw, POLAND. Email: m.parchomiuk@iel.waw.pl, k.tomczuk@iel.waw.pl.

Abstract - Single phase voltage converter structure with isolation transformer for applications in traction vehicles is presented. This paper shows simulation model of this type of converter and transformer model with modern magnetic material in OrCAD-PSpice simulation program. Two PWM control techniques are compared in order to eliminate voltage and current oscillations in primary winding of the transformer, decrease large peaks of current, decrease large system loss and improve system efficiency. Author shows a comparison of the PSpice simulation results with the real results.

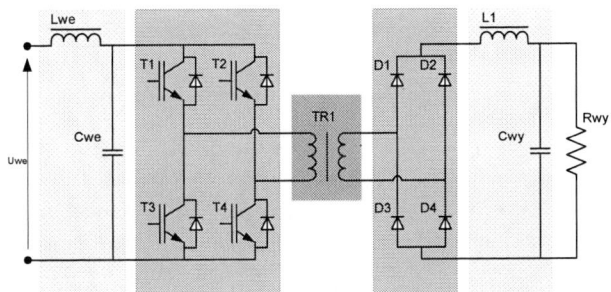

Fig.1. Model of single phase full bridge voltage inverter– selected structure for simulation studies.

I. INTRODUCTION

A progress in modern semiconductors influences on dynamic power converters development. These devices operate at higher frequencies and have to supply increasingly growing power needs. In addition the latest magnetic materials allow the construction of magnetic elements with high efficiency and small size. But at the stage of building and running power converter there are weaknesses in proper operation of such systems despite the fact we use technologically advanced components. For this reason the paper discusses the structure of DC-DC converter with isolation transformer for supplying the needs of traction vehicles.

II. DESCRIPTION OF THE CONVERTER

Selected simulation aspects were performed on a single phase full bridge voltage inverter with isolation transformer (Fig. 1) [1]. This is the structure, which has become an essential solution for high power traction converters in recent years. According to the standards traction devices are supplied with voltage of 600V (+30%-25%) [2]. The task of such a converter is to convert traction voltage to galvanically separated voltage of a certain value. According to the needs of traction vehicles, the most popular voltage source is 24V DC. In this respect simulation model presents this level of output voltage with nominal output power 10kW (output current approximately I_{out} = 400A).

For the assumed configuration there was a power converter used. It was operating at frequency of 16kHz, which allows to reduce the size of passive components, and thus reduce the dimensions of the device.

High frequency converter brings certain difficulties. With the increase of the frequency it results additional power losses in the windings of the transformer (the presence of the skin effect), transistors and core loss (hysteresis and eddy currents). In addition in this type of converter there are more important oscillations of the voltage and high current pulses in primary winding of transformer during the transistor switching.

In order to understand the phenomena occurring in this type of converters, a simulation model was created in OrCAD-Pspice program.

III. SIMULATION MODEL OF THE CONVERTER

Simulation model of DC-DC full bridge converter with isolation transformer was designed in Orcad-PSpice program (Fig. 2). It consists of: regulated voltage source, input low pass filter, voltage inverter, isolation transformer, rectifier, output low pass filter and load. To control the transistors epoly function was used. This function can operate as voltage source control. Voltage inverter was controlled using pulse width modulation PWM techniques. Two methods were used and compared: (traditional) symmetrical PWM and Phase Shifted PWM PS-PWM (see results). Control pulses of the transistor were generated alternately with a dead time taken into account.

To obtain simulation model parameters as close as possible to the real system, producers models of transistors and diodes were used. Very important issue is to use real transformer model with the appropriate identification of the core. This allows to observe and understand phenomena in the work of this type of converter.

IV. SIMULATION MODEL OF THE TRANSFORMER

An important issue to research and analyze a selected structure of the converter is to prepare core model of non-linear isolation transformer, with real magnetizing characteristic of material. OrCAD-PSpice is equipped with

978-1-4244-8806-3/11 $26.00 © 2011 IEEE

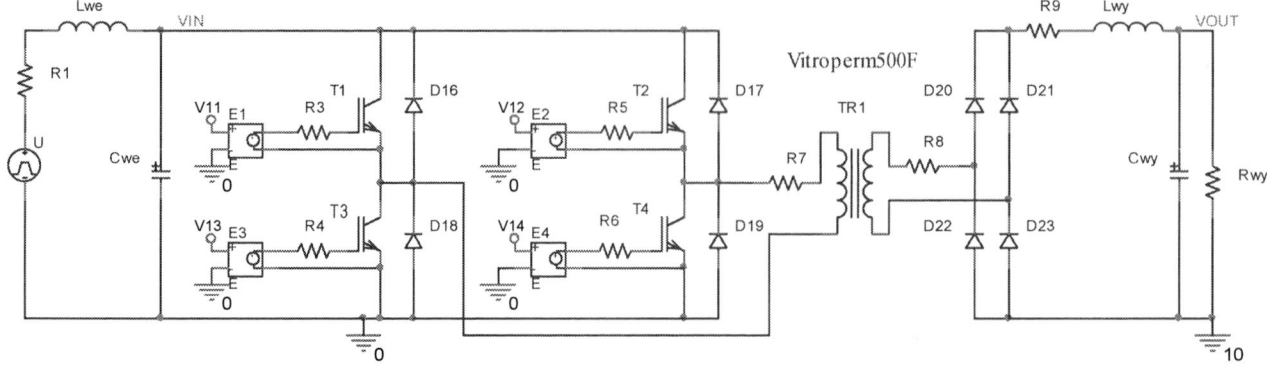

Fig. 2. Simulation model of a single phase full bridge voltage inverter in OrCad-PSpice program.

special tool to create this kind of model based on self-defined hysteresis loop. This program allows to introduce some parameters of the core, e.g. dimensions, initial permeability, hysteresis loop points. This is particularly important for modeling nonlinear magnetization characteristics of magnetic circuits. Currently there are many different magnetic materials available on the market (e.g. nanocrystalline or amorphous materials). These materials are suitable for many applications, have different shape, parameters and magnetic characteristics [3][4].

Nanocrystalline core model of Vitroperm 500F (Vacuumschmeltze product) was created (material parameters in Table I). Fig. 3 shows the view of modeled magnetization characteristics of this material in the Orcad-PSpice model editor. In order to achieve the best accuracy of the material, hysteresis loop was first measured with an oscilloscope using Dirac impulse response technique and then introduced to a model editor. To create a library of a core, the following parameters were used: A_e = 8,22 cm^2 (cross section of the core), l_e = 36,1cm (length of a path in the magnetic core), μ_0 = 30000 – initial permeability, toroidal shape of the core.

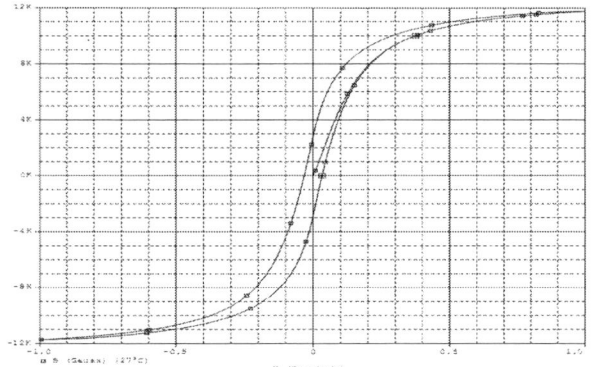

Fig. 3. Magnetization characteristics of nanocrystalline core of Vitroperm 500f. Function $B=f(H)$ were modeled in the Orcad-PSpice model editor (10kGauss=1T, 1Oe≈80A/m).

winding resistance), N_1 = 16 (number of turns of primary winding), N_2 = 1 (number of turns of secondary winding). Transformer model with actual magnetization characteristics of magnetic nanocrystalline material was created. It was used in the studies on simulation model of DC-DC converter.

V. THE RESULTS OF THE CONVERTER OPERATION

Fig. 4 presents the results of the operation of the system simulation model using real nonlinear model of the transformer core. Nonlinearity of the transformer was presented depending on supply voltage. Upper chart of this figure shows function $B=f(H)$ where transformer flux density reaches B = 0,7 T at input voltage U = 600V. It means that the operating point is placed on the linear part of the magnetization characteristics. This affects on stability of the circuit and prevents from saturation of the core. Proper value of the flux density was additionally confirmed according to the following formula (1).

$$B = \frac{U}{4 f N_1 A_e} \qquad (1)$$

where: U – input voltage of the converter, f – converter frequency, N_1 – number of primary winding turns of the transformer, A_e – a cross section of the core.

TABLE I
MAGNETIC CORE PARAMETERS OF VITROPERM 500F MATERIAL

Nr	Parameters	
	Type	Values
1	Tape width	25 μm
2	Flux density	1,2 T
3	Permeability	10000-150000
4	Resistivity	115 μΩcm
5	Magnetostriction	1e-8 – 1e-6
6	Core loss	80[W/kg] for 100kHz, B=0,3 T.
7	Curie temperature	600°C
8	Operation temperature	120-150°C

Core model was introduced to the nonlinear transformer model in Orcad-PSpice program. The nonlinear model is based on a theory which has been developed by D. Jiles and D Atherton in the eighties of the last century [5] (JA Model). The following transformer parameters were taken: $Rp1$ = 150 mΩ (primary winding resistance), $Rp2$ = 50 mΩ (secondary

978-1-4244-8806-3/11 $26.00 © 2011 IEEE   326

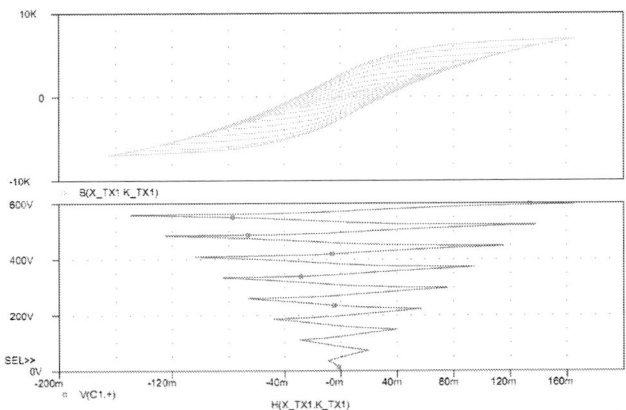

Fig. 4. Transformer magnetizing characteristic *B=f(H)* (upper chart) in respect to input voltage (lower chart) of converter simulation model. Input voltage range *U* = 0 – 600V. (10kGauss=1T, 1Oe≈80A/m).

Figure 5 shows results of the converter operation at input voltage range from 0V to 1200V. Deep saturation of the core is visible on the chart above voltage 600V. Created model allows the selection of the proper parameter of designed converter and transformer. In addition it gives opportunity to check its operation by quick changes of parameters.

This article highlights the operation of the converter at idle and under load, for various pulse width control. The following figures show the moment of transistor switching, where dangerous oscillation of the primary voltage of the transformer appear. It causes the occurrence of instantaneous high current peaks.

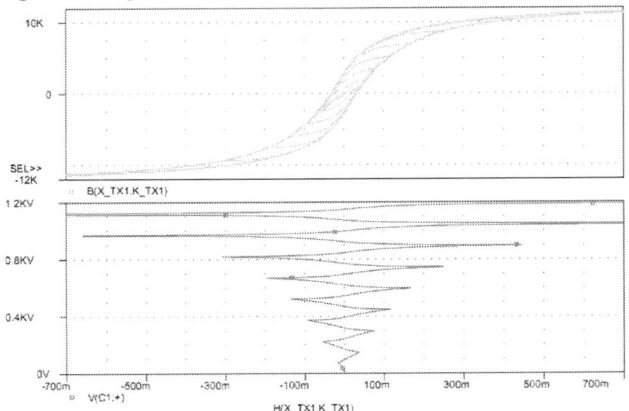

Fig. 5. Transformer magnetizing characteristic *B=f(H)* (upper chart) in respect to input voltage (lower chart) of converter simulation model. Input voltage range *U* = 0 – 1200V. (10kGauss=1T, 1Oe≈80A/m).

In real working conditions of the converter it is difficult to determine, whether the measured waveforms contain such a shape or whether it is a result of electromagnetic signal distortion derived from rapid changes in voltage and current in the circuit. Even highly specialized measuring equipment are not always able to determine this. Measurement should be made at least with two different measurement methods or systems. By measuring and comparing the results with the results of simulation there can be carried out a comparative analyses of the phenomenon and delineated.

By comparing the simulation results with those of the real system (Fig. 7 and 10) the cause of the oscillation voltage on the primary winding of the transformer was found. Leakage inductance of the transformer together with the transistor capacitance of the inverter causes signal oscillations. According to the Thomson formula (2) describing the resonant frequency of LC circuit, it appeared in simulation model as well as in real one, natural frequency of the circuit *f* = 850 kHz.

$$f = \frac{\omega}{2\pi} = \frac{1}{2\pi\sqrt{LC}} \qquad (2)$$

Oscillations (Fig.7, 10) in the system arise from the reactive energy collected in the electromagnetic circuit [1] [6]. The amplitude of those oscillations depend on current flowing through the magnetic circuit (the higher current values the more significant oscillations became) and squareness ratio of magnetic material hysteresis loop. On the one hand, the application of magnetic material with a rectangular shaped hysteresis loop, have favorable effect on the work of the converter, but result in the presence of small oscillations of voltage and current. On the other hand, the surface area of the hysteresis loop is proportional to the energy losses in the core. In addition, materials with high squareness ratio when used in pulse converters are more

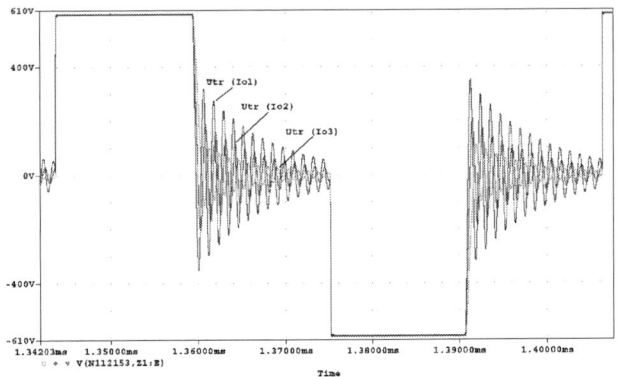

Fig. 7. Voltage oscillations in primary winding of the transformer for different current values Io1>Io2>Io3.

likely to core saturation than those with tilted magnetization characteristics. Designers of power electronics devices often use air gaps in order to tilt the magnetization characteristics in transformers and chokes.

The occurrence of these phenomena indicates that at the design stage of the power converter designer should apply a compromise in choosing the right magnetic material, as well as to adopt an adequate reserve in the selection of semiconductor elements, because of high current. This approach have effects in better efficiency of the system and higher capability.

978-1-4244-8806-3/11 $26.00 © 2011 IEEE

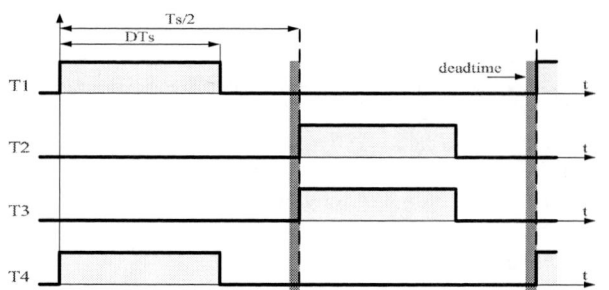

Fig. 8. Symmetrical Pulse Width Modulation – PWM technique. Where: DTs – pulse width, Ts – inverter operation period.

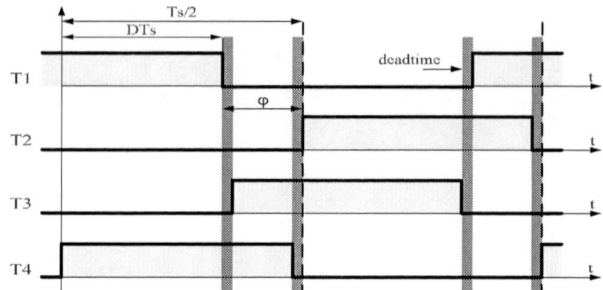

Fig. 9. Phase shifted Pulse Width Modulation PS-PWM technique. DTs – pulse width, Ts – inverter operation period, φ – phase shift.

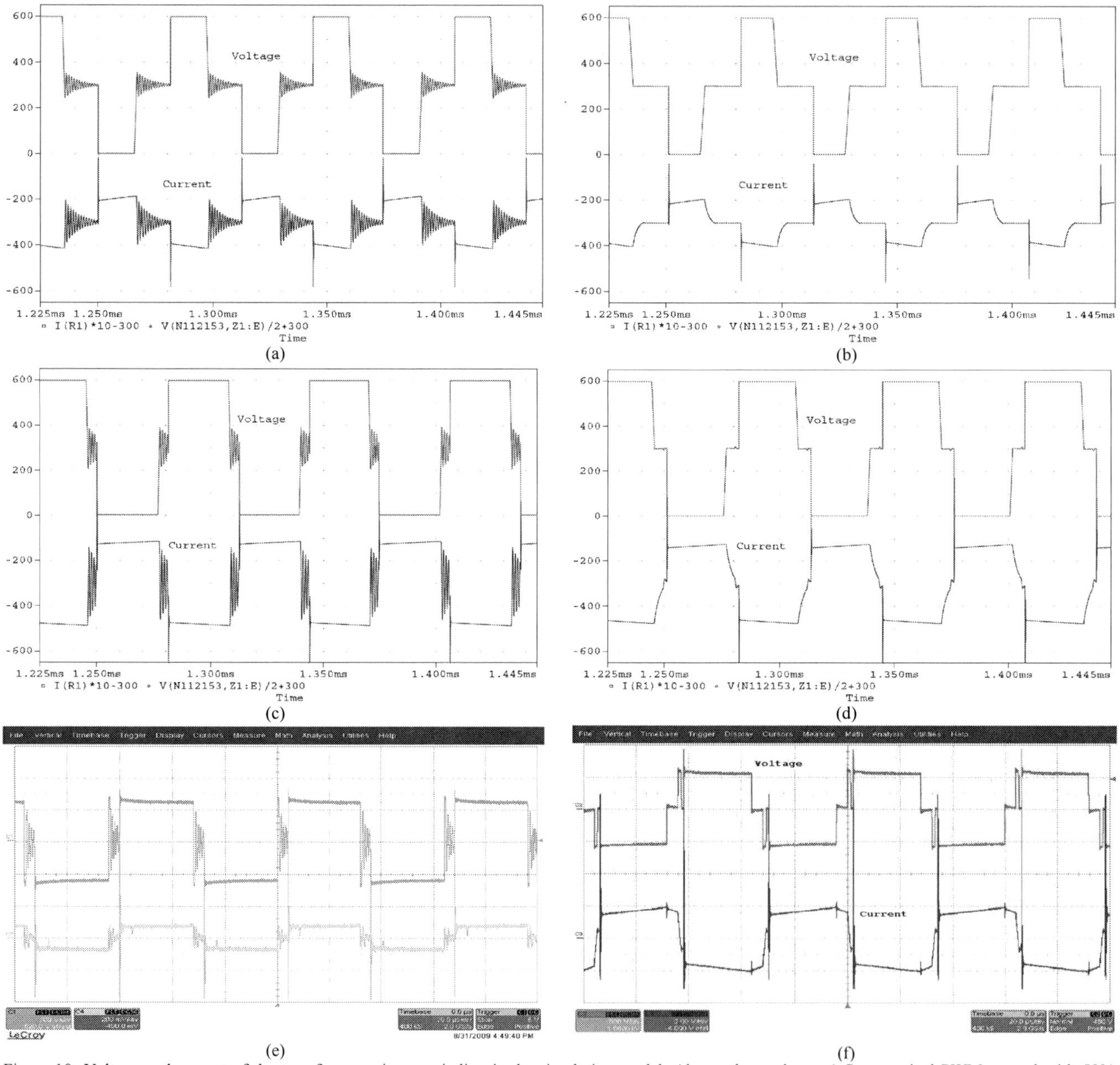

(a)

(b)

(c)

(d)

(e)

(f)

Figure 10. Voltage and current of the transformer primary winding in the simulation model. Above charts show: a) Symmetrical PWM control with 50% pulse width, b) Phase Shifted PWM (PS-PWM) control with 50% pulse width, c) symmetrical PWM control with 85% pulse width, d) Phase Shifted PWM (PS-PWM) control with 85% pulse width, e) Symmetrical PWM control with 85% pulse width in a real converter (C1 – 500V/div, C4 – 40A/div), f) Symmetrical PWM control with 85% pulse width in a real converter (C2 – 500V/div, C3 – 20A/div).

978-1-4244-8806-3/11 $26.00 © 2011 IEEE   328

One of the methods to eliminate oscillations in the primary winding circuit of the transformer is phase shifted pulse width modulation PS–PWM (Fig. 9), which were investigated in literature [6][7]. This method was compared with traditional symmetrical pulse width modulation (Fig. 8). The main difference is that gate signals of the transistors are phase shifted in diagonal of H-Bridge (Fig. 9) e.g. gate signals of T1 and T3 are shifted in respect to T2 and T4.This allows to discharge energy stored in the magnetic circuit and prevent from oscillations of voltage and current. Comparison results of two control techniques in the simulation model are shown (Fig.10). There are voltages and currents of the transformer primary winding presented. Charts are placed in two column.

On the left column there are simulation results with classical PWM technique. The right column presents results with PS-PWM control. These two control techniques are shown and compared in respect to a different pulse width. There are oscillations on voltage and current visible when classical PWM control is used due to energy stored in magnetic circuit. When PS-PWM control is used these oscillations disappear because energy in the circuit is discharged. It is possible due to the fact that in this control technique two transistors are always switched on. On the last charts of Fig.10 the real converter (Fig.11) results are placed. They are close to those from simulation model.

Fig. 11. Laboratory model of the real converter. Input voltage 3x400V, nominal output power 10kW, nominal output voltage 24V, up to 20kHz operation.

Very important issue is compensation of constant component, which is worth mentioning, but due to comprehensiveness of issues not developed in this work. Theoretically, magnetizing current and flux of the transformer should not include a fixed component, which may occur as a result of the lack of symmetry of the control transistors, transistor driver work asymmetry or irregular work of the load. A common cause of the constant component is the difference in switching times of transistors or different voltage drops on transistors. This leads to the occurrence of random faults in the device and its exemptions. One of the methods of constant component compensation is capacitor inclusion to primary winding of the transformer. In many applications, this solution is effective, which is described in the literature. Another way to avoid transformer core saturation is proper design of this magnetic element,

consisting of the working set point with large reserve, so as to have ability to work with a small asymmetry.

VI. CONCLUSION

Created simulation models reflecting the best working conditions and the results obtained from research of DC-DC converter allow to explain and understand some of the phenomena. Also allow to carry out impact studies on the parameters of individual elements of the converter operation, which are particularly important when constructing a model of the inverter. Compared to the real system there are no possibilities to change e.g. transistor capacitance, which depends on the structure and technical production process.

Currently available simulation tools such as OrCAD-PSpice or Matlab, allow designers to look carefully at the operation of the converters before they are built. These tools are complemented by programs to simulate and analyze the distribution of the magnetic field or heat, using the finite element calculations. This is critical for the proper design of magnetic circuits, which are important elements of converter.

Another reason for creating simulation models discussed in this paper is also growing demand for this type of devices in the industry. They are applied not only in urban transport and rail, but also by many companies requiring accurate energy supply with growing demand for power.

REFERENCES

[1] J.A.Sabate, V.Vlatkovic, R.B.Ridley, F.C.Lee, B.H.Cho, *"Design considerations for high power full-bridge ZVS-PWM converter"*, Proc. IEEE, APEC '90, 1990, pp.275-284.

[2] EN50163: 2004 – *Railway Applications Standards. Supply voltages of traction systems.*

[3] W.Shen, F.Wang, D.Boroyevich, C.Wesley, *High-Density Nanocrystalline Core Transformer for High-Frequency Resonant Converter*, IEEE Transactions on Industry Applications, vol.40, 2008.

[4] M.Parchomiuk, K.Tomczuk, *Projektowanie transformatora impulsowego w programie Matlab–Simulink,* Wiadomości Elektrotechnicze, 03/2010

[5] D.C. Jiles, D.L. Atherton, *Theory of ferromagnetic hysteresis*, 1986.

[6] H.Bai, Ch.Mi, *Eliminate reactive power and increase system efficiency of isolated bidirectional Dual-Active Bridge DC-DC Converters Using Novel Dual-Phase-Sift Control,* IEEE Transactions on Power Electronics, *VOL.23, 2008.*

[7] J.E.Baggio, H.L Hey, H.A.Grundling, H.Pinheiro, J.R.Pinheiro, *Isolated Interleaved-Phase-Shift-PWM DC-DC ZVS Converter,* IEEE Transactions on Industry Applications, vol. 39, 2003.

Power Electric Aiding Controller for Automated Bus Stopping

Jorge Godoy, Vicente Milanés, Joshue Pérez, Jorge Villagrá and Carlos González
AUTOPIA program, Centre for Automation and Robotics, UPM-CSIC
{jorge.godoy, vicente.milanes, joshue.perez, jorge.villagra, carlos.gonzalez}@car.upm-csic.es

Abstract— **Day by day the number of vehicles on roads is growing, increasing also the number of accidents, traffic jams and carbon dioxide emissions. For this reason, the European Union has adopted an action plan for improving urban mobility where the public transport is the main key. To present day, solutions as special bus lanes, electrical and hydrogen buses and rail-guided buses have been tested and implemented in several cities as London and Madrid. In this article an automated stopping system for electrical buses is presented. Stopping system have been implemented by the AUTOPIA program on an electric minibus and tested on a private circuit with satisfactory results.**

I. INTRODUCTION

During last decade, the number of vehicles on roads is continuously growing over the entire world. Only between years 2000 and 2006, the motorisation rate on the European Union (EU27) has increased from 422 to 466 passengers cars per 1000 habitants; moreover, in the same time interval the EU27 population has growth 2.12 %, which finally represents an increase of 12.77% - around 26 millions - in the number of passenger cars in the EU [1].

This motorisation growth also represents an increase of traffic jams and carbon dioxide emissions. According to the EU green paper for urban mobility [2], the EU economy lost around 100 billion Euros for delays on traffic jams. In a parallel line, and regarding the CO2 emissions, the urban traffic is responsible for 40% of them. Additionally, according to the European study "Attitudes on issues related to EU Transport Policy" [3] published on 2007, only 21% of the EU27 citizens use public transport as main mode of transportation while other 51% use passenger cars, being the schedule regularity the main reason for this election. Based on these factors, the EU has decided on 2009 to adopt an action plan for urban mobility where the use of public transport is considered as the main key to improve the urban mobility [4].

From the roads point of view, the Public Transport (PT), specifically buses, are more efficient than passenger cars due to the number of persons being transported per occupied area and fuel consumption. Moreover, the use of greener vehicles for PT would further improve this efficiency, reducing the carbon dioxide emissions. Between 2001 and 2006, the Clean Urban Transport for Europe project (CUTE) enclosed on the 5th Framework Program, tested electric bus fleets on 10 European cities with great results [5]. On the 6th

Framework Program, a one-year extension for CUTE project was included as the HyFLEET:CUTE project. Results for both projects showed the great functionality of these vehicles. As a matter of fact, London will include 10 hydrogen buses to their permanent fleet [6].

Furthermore efficiency, safety and accessibility are the pillars for PT in urban environments. According to [7], 21% of the EU citizens aged more than 55 years use PT as main mode of transportation; if other transit users like the vision impaired, children or people in wheelchairs are considered, the importance of this factor results more obvious. The major advance in improving bus accessibility is the implementation of Low-floor buses, pioneered by the German manufacturer NEOPLAN. The low-floor buses permit passengers to board and alight without stepping up or down from the sidewalk at the bus stop [8]; however, as a study conducted in Caen, France in 90's [9], even a small gap could be a hazard for the above mentioned users.

Up to now, several solutions have been proposed in response to this problem. On [10], the GIBUS and VISÉE systems' field test results are presented. The GIBUS - *Guidance des autobus en station* - project, supported by the French National Institute for Transportation and Safety Research (INRETS), installed a display on the bus board showing the lateral distance from the bus to the docking point to the driver. On the other hand, the VISÉE is a vision-based control system designed to guide the bus to the docking point by controlling only the steering wheel, while driver controls the longitudinal displacement - throttle and brake. More recently, the PATH project, from California University at Berkeley, developed a docking system based on magnetic marks installed on the roadways. By detecting the magnetic marks, the on board control system define the vehicle trajectory and controls the bus steering wheel, throttle and pneumatic brakes. PATH system was tested at Washington DC - United States - on 2003; results showed that the system was able to stop at the docking point with a longitudinal error lower than 50 cm [11]. As main drawback, this system needs to modify the infrastructure - in this case the road - so as to install the magnetic marks for vehicle's positioning.

In this paper, a longitudinal fuzzy-control differential global positioning system (DGPS)-based system for bus docking is presented. The main idea of this design is to

978-1-4244-8806-3/11 $26.00 © 2011 IEEE

develop an aiding system capable of stopping the electric minibus with high precision at the bus stop, making easier the access for children, disabled and old people. The system was tested using an AUTOPIA program's electric minibus at the Centre for Automation and Robotics (CAR, UPM-CSIC). CAR's private driving circuit was used in the experimental phase to validate the proposed system. A screen indicates driver when the system can be activated. When it is activated, the minibus longitudinal displacement is controlled by a main computer connected to the electric motor speed controller and an electro-hydraulic braking system specially designed for the application. In order to test the system in traffic situations as close as possible to the real traffic, the steering wheel is controlled by a human driver whether the system is activated or not.

The remainder of this paper is structured as follows: on section I a minibus description is presented, including the design for throttle and brake pedals automation. Section II introduces the trajectory estimation algorithm implemented and describes in detail the main fuzzy-controller designed for stopping the bus with precision. Tests and results data and analysis for both trajectory estimation and stopping performance are presented on section III. Finally, at the paper's end several conclusions and future work are presented.

II. ELECTRIC BUS DESCRIPTION

Molinero is the fifth vehicle automated by the AUTOPIA program. It is an electric minibus model EGK6152K (see fig. 1) with capacity for 14 passengers manufactured by Con+Auto [12]. This vehicle counts with four arrays of four 12V batteries each and a 5KW electric motor that allows it to run up to 35 Km/h with 100 Km autonomy. Molinero was chosen by the AUTOPIA program as a green option for automated public transport in urban environments.

As in previous vehicles, the main sensor on the Molinero's control architecture is a DGPS installed at vehicle rear-centre, which allows the vehicle to sense its position and velocity at 100 ms rate. The GPS correction is obtained through the WAAS/EGNOS network. It is well-know that, as velocity sensor, the GPS does not represent and optimal option; however, since Molinero does not have an electronic

calculator on board, the velocity estimation through DGPS is the simplest option with best results. As future work, it is being considered to install an electronic velocity sensor as the one described on [13].

The architecture's backbone is an On-board Control Unit which consists on a solid-state industrial computer for automotive applications. The solid state components allow the computer to work normally under driving conditions, avoiding the common hard disk failures due to vibrations.

III. LONGITUDINAL CONTROL

Previously, two electric Citroën Berlingo vans and two gas-propelled Citroën C3 were fully-automated. Although longitudinal control - i.e. throttle and brake pedals automation - has been previously implemented by the AUTOPIA group [14] [15], some modifications had to be applied for a PT vehicle. Specifically, number of passengers, road layout and battery energy were considered so as to design an appropriate longitudinal controller. On next subsections the throttle and brake pedals automation is described.

A. Throttle

On Molinero, the electric motor is controlled by a speed controller model 1204 manufactured by Curtis. Under normal operation, an electronic throttle pedal provides the throttle reference to the motor speed controller as a 0-5-volts signal and also incorporates a micro-switch that indicates the controller to leave the motor on freewheel when no throttle is applied. In order to emulate the pedal behaviour on the automatic mode, a digital-analogue I/O card was installed on the main computer to send the throttle reference; additionally an electronic switch controlled by Controller Area Network (CAN) bus connected on parallel to the pedal micro-switch allows the computer to control the freewheel mode. In fig. 2 a schematic outline of throttle is presented.

B. Electro-Hydraulic Brake

A bus is considered as a heavy-duty vehicle because it could carry so more people than a familiar vehicle, requiring more power on the brakes to stop the vehicle. However, since Molinero is an electric vehicle, the energy consumed by the automatic braking system is a very critical design factor. Additionally, it must be considered that adding elements to a commercial vehicle is usually a difficult task because

Fig. 1.   Electric minibus Molinero

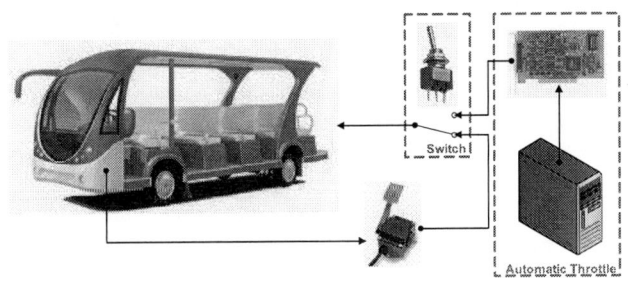

Fig. 2.   Throttle outline

manufactures do not leave much free space for external components. For these reasons, the braking system design should be balanced between power consumption and available space. Based on the AUTOPIA program experience, an electro-hydraulic braking system was adapted from a gas-propelled familiar car [16] to our electric minibus.

The hydraulic power system for the autonomous braking is provided by a gear pump coupled to a 12-volt 350-watts dc-motor. The pump has a one litre fluid tank which guarantees the needed displacement on the drum brakes. For safety considerations, a limiter tube fixed to 50 bars is added to protect the car elements from excessive pressure. As for throttle, in automated mode it is necessary to control the braking system from the main computer; for this, an electro-proportional pilot is installed after the safety loop. It is in charge of regulating the pressure to be applied from the electro-hydraulic braking power system. This pilot is controlled by a 0-10 V signal using the same digital-analogue I/O card used for the throttle. As the minimal pressure of the electronic pilot is not zero, a spool directional valve is added at the end of the electro-hydraulic braking system to solve this problem. This valve is activated at the same time that the electronic pilot. It is normally close when the braking system is activated. A power relay is in charge of activating this valve in order to permit brake fluid circulation.

The last stage consists on installing an adequate device to permit choosing either the original braking circuit or the automated hydraulic braking power system. Since throttle pedal only manages voltage signals, a two-way switch performs the commutation between original and automated systems. In the brake case, a device to permit introducing the brake fluid coming from the automated system into the minibus brake circuit was installed. To this end, two shuttle valves were installed to interconnect both systems just before the drum actuators. Each valve permits the flow from either of two inlet ports - electro-hydraulic or original braking system - to a common outlet - drums actuators - by means of a free-floating metal ball that shuttles back-and-forth according to the relative pressure at the two inlets. Since the systems must cooperate, the driver could stop the vehicle even when the automated mode is active, improving the safety in case of system failure. A schematic outline of the electro-hydraulic system design is shown in fig. 3.

IV. CONTROL ALGORITHM

Our goal is to develop a system capable of stopping the electric minibus with high precision with respect to a point in the road. This point is located in the bus stop. The automated system can be used as aiding system for disabled or old people in order to make easy the access for the kind of users to the PT.

Although all PTs are guided by a human driver, an aiding system capable of stopping the bus with enough precision can help the bus driver in his task making more comfortable and safe his work. So, two main requirements have to be set before proceeding with the design of the control system:

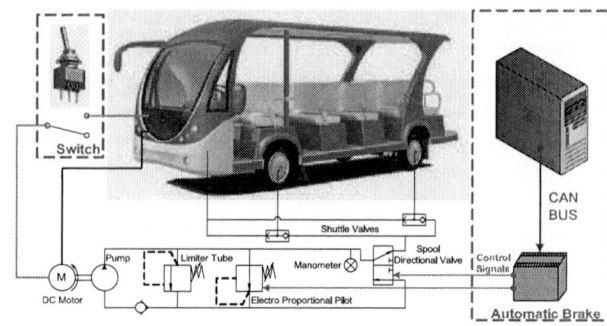

Fig. 3. Electrohydraulic brake outline

- The system has to be capable of detecting the moment to be activated.
- Once the system is activated, it has to be capable of stopping the bus with enough accuracy in the bus stop.

A. Trajectory estimation

Since the developed application works as an assistance system, one of the main aims is to avoid the need of complex trajectory maps. For this reason, an online simple trajectory estimation algorithm has been implemented to determinate the vehicle route and distance to the stop point. Based on the position data of the last seconds - obtained via the DGPS, the program calculates two regression lines: one for the current trajectory - i.e. the tangent with respect to the current position of the bus - and the other one based on the "accumulate" trajectory - i.e. the tangent with respect to the stretch the bus is covering. The difference between both line slopes permits the program to determinate if the vehicle is either in a curve or in a straight stretch. When the vehicle is in a straight stretch - difference between slopes is lower than 0.02 - the control program calculates the lateral and over-line distance to the stop point and finally, when the lateral distance is lower than 3 metres - vehicle is on a straight trajectory over the bus lane - allows the driver to activate the automated stop system.

B. Fuzzy controller

The vehicle longitudinal control is performed by a fuzzy controller. The controller is based on a fuzzy coprocessor named ORBEX (Spanish acronym for Fuzzy Experimental Computer) which had been developed previously at the CAR [17]. The ORBEX coprocessor allows to define and implement control rules in a quasi natural language by means of IF... THEN... sentences; for instance:

IF *time to arrive* big **THEN** *brake* null

where the words in italic are fuzzy variables, the words in bold are ORBEX language keywords, and the words in plain script are linguistic variable values [15]. For this application, two input variables, one output variable and a 9-rules set were defined.

As inputs to the fuzzy controller, the control program uses two variables: the Time-To-Arrival (TTA) and the Needed-Acceleration (NA). The time-to-arrival is defined as the over-line distance to stop point divided by the current vehicle velocity 1, while the Needed-Acceleration is defined as the constant deceleration needed to stop the vehicle at the stop point based on the current velocity 2. For both input variables, three membership trapezoidal functions were defined, as can be seen in fig. 4.

$$TTA = \frac{Distance}{CurrentVelocity} \quad (1)$$

$$NA = \frac{CurrentVelocity^2}{2Distance} \quad (2)$$

The controller output is defined as the normalized pressure applied by the Electro-Hydraulic circuit to brake. As can be seen in fig. 5, five singletons were defined for the output defuzzification. Negative values are a software consideration to indicate to the control program that brake is applied and any throttle input should be dismissed. Despite the throttle is automated, it was not implemented as controller output in order to guarantee an efficient brake, stopping the vehicle uniformly and making unnecessary increasing velocity in any moment. The fuzzy rule set is defined as below:

IF *TTA* Big **AND** *NA* Neg **THEN** *Pedal* Brake_Little
IF *TTA* Big **AND** *NA* Zero **THEN** *Pedal* Nothing
IF *TTA* Big **AND** *NA* Pos **THEN** *Pedal* Nothing
IF *TTA* Med **AND** *NA* Neg **THEN** *Pedal* Brake_Medium
IF *TTA* Med **AND** *NA* Zero **THEN** *Pedal* Brake_Little
IF *TTA* Med **AND** *NA* Pos **THEN** *Pedal* Nothing
IF *TTA* Small **AND** *NA* Neg **THEN** *Pedal* Brake_Much
IF *TTA* Small **AND** *NA* Zero **THEN** *Pedal* Brake
IF *TTA* Small **AND** *NA* Pos **THEN** *Pedal* Brake_Much

V. IMPLEMENTATION

A. Experimental driving circuit

All the experimental tests have been performed at the CAR's private driving circuit. In fig. 6 an aerial image of

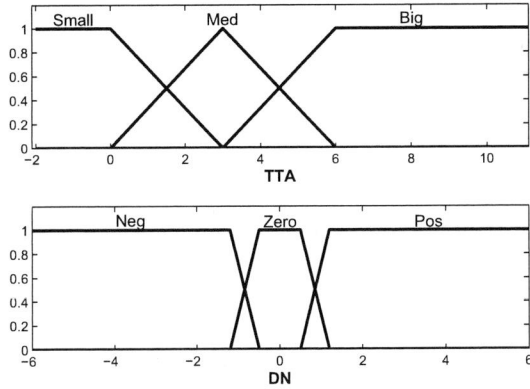

Fig. 4.   Fuzzy controller inputs membership functions

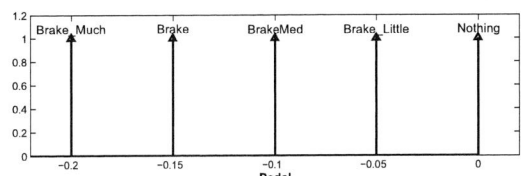

Fig. 5.   Fuzzy controller output membership functions

Fig. 6.   Experimental test cricuit

the circuit can be appreciated, including the start point, the stop point - located at the bus stop - and the approximate bus trajectory for all tests. Both start and stop points have been selected to demonstrate the trajectory adjustment algorithm and bus stopping performance.

B. Trajectory estimation

In order to validate the trajectory adjustment algorithm performance, several tests were performed before implementing the stop controller. After the first data analysis, it was found that the regression line adjustment was not precise enough at low speed - lower than 7 kilometres per hour - in some cases, because of high frequency changes due to the proximity of the position points measured. For this reason, the regression algorithm was modified to keep the last iteration line slope at velocities lower than 7 kilometre per hour and only calculate the mass centre of points where the line should cross.

In fig. 7, a test of the trajectory estimation is presented. The first graph represents the value of each calculated slope - current and "accumulate" trajectory - and the bottom graph shows the vehicle speed. Both graphs are presented over the same time axis. In the slope graph, one can see that, from the start point, the vehicle is over a curve trajectory; however, after the 10th second the difference between both values is almost null, indicating that the vehicle is over a straight stretch. Also it is possible to appreciate that when the speed decreases under the minimal value - near second 18.5 - the slope values are constant. So, these test results prove the correct algorithm performance for the trajectory estimation.

C. Stopping Performance

After the trajectory estimation algorithm was validated, the next step was calibration and validation of the stopping

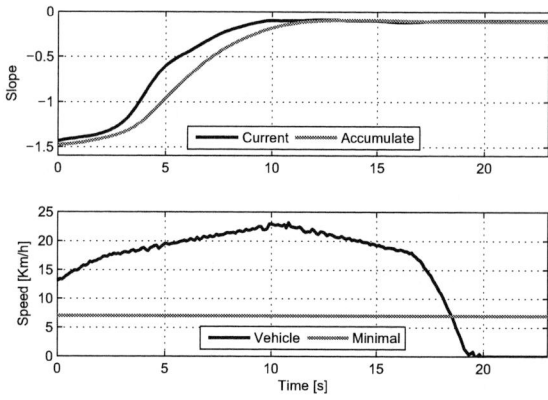

Fig. 7. Trajectory adjustment

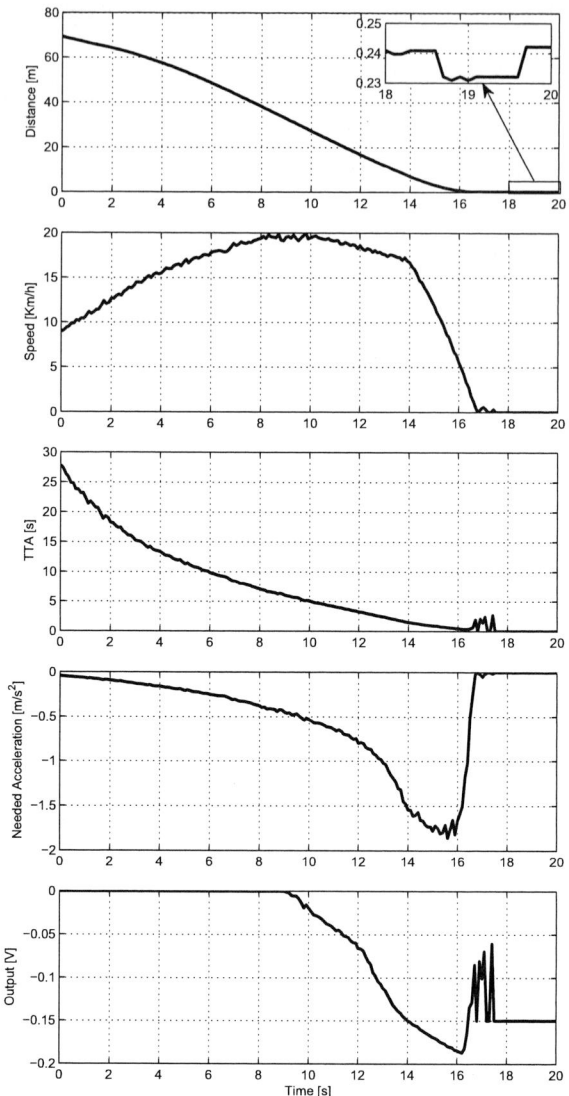

Fig. 8. Stopping controller test

controller. As it was mentioned before, the same route presented in fig. 6 was used for the stop test. The stop point position is fixed and it was determinated before tests using the same DGPS installed on vehicle.

In fig. 8 the evolution of the distance to the stop point, vehicle speed, and fuzzy controller inputs and output evolution over time for one case are presented. For this test, the stopping system was activated at around the 8th second, when the vehicle was about 40 metres from the stop point and at a speed of around 20 kilometres per hour. Finally, around the 17th second the vehicle was completely stopped.

From the needed acceleration graph, one can appreciate that the absolute value does not exceed 2 m/s^2, considered as the threshold for maximum acceleration maintaining the passengers comfort [18]. Regarding the controller output, it evolves continuously without reaching the maximum value or saturate. Peaks shown on the graph occur after the vehicle is stopped and they are caused by the front-back movement of the vehicle at the moment it is stopped. However, this effect does not considerably affect the system performance.

When the system is activated, the speed graph shows two deceleration phases around the 14th second, the first one between the 10th and 14th second, when the controller output is under 0.15; and other stage after 14th second when the controller output is bigger than 0.15. Despite this difference, at no given moment the deceleration of the vehicle is bigger than the comfort threshold and the vehicle decelerates uniformly on each stage. For this test the bus stopped at 24 cm from the stopping point.

Since the number of passengers on board is a factor that could affect the stopping performance due to the vehicle weight changes, several braking tests were made in order to determinate the influence of this factor over the system developed. Data analysis showed that, as supposed, the vehicle deceleration under constant brake pressure decreases when the number of passengers is bigger. Fig. 9 shows the speed evolution for 1 and 9 passengers and fixed values for the controller output. As can be appreciated on graph, with 9 passengers on board, the stop time increases 5, 4

and 0.5 seconds respect to the 1-passenger stop time for a 0, -0.1 and -0.2 pedal value respectively. Despite this stop time increments, automatic stopping tests performed with different number of passenger showed that the implemented fuzzy controller is able to manage this situation without any modification; meaning that the passenger number increase just causes a faster change on the output value without passenger comfort being affected.

Finally, the system was tested using different people as drivers. A screen was installed in the windshield of the bus so as to advise when the automated stopping system can be used. The system was available to be activated by the drivers while the *needed-acceleration* value was to be lower than 2 m/s^2. For all tests performed using different drivers and different number of passengers in the bus, the vehicle stopped between 5 and 35 cm from the stopping point. A

978-1-4244-8806-3/11 $26.00 © 2011 IEEE   334

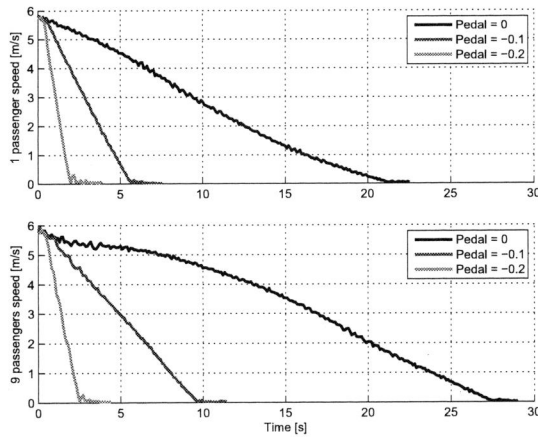

Fig. 9. Brake tests

video file of the stopping system could be retrieved from http://www.iai.csic.es/users/autopia/Videos/BusStopping.wmv

VI. CONCLUSIONS

The automated bus stopping control for an electric minibus using fuzzy logic has shown promising results. The required information for the control loop comes from a Differential GPS, notwithstanding the modularity of the control scheme proposed allows to add others sensors, which would improve the input data.

Furthermore, the simple trajectory estimation algorithm proposed in this work has shown good performance. Two regression lines allow to determine when the vehicle is in straight and curve segments. Moreover, the stopping system considers fast and smooth responses. The deceleration of the minibus never exceeds 2 m/s^2, providing a comfort sensation for all the passengers.

The control scheme, based on fuzzy logic, allows emulating the human behavior in the driving process. Final tests, considering an empty and a full minibus, show that the fuzzy logic is a good alternative to control system with different dynamics, as well as in the platform used in this work.

Finally, a new green and intelligent public transport system, which controls the brake and throttle in a commercial minibus, has been developed. This is the main contribution of this paper.

Other control strategies, such as neuro-fuzzy and genetic algorithm, will be considered to improve the tuning of the fuzzy controllers proposed in this work. An entire autonomous driving will allow a better use of energy on the minibus, helping to reduce the CO2 emissions in urban environments.

ACKNOWLEDGMENT

This work was supported by the Spanish Ministry of Science and Innovation by means of Research Grant TRANSITO TRA2008-06602-C03 and Spanish Ministry of Development by means of Research Grant GUIADE P9/08.

J. Godoy wants to specially thank to the JAE program (Consejo Superior de Investigaciones Científicas) for its support in the development of this work.

REFERENCES

[1] http://epp.eurostat.ec.europa.eu/.
[2] *Green Paper: Towards a new culture for urban mobility - COM(2007) 551.* European Commission, 2007.
[3] T. G. Organization, *Attitudes on issues related to EU Transport Policy - Analitical Report.* Eurobarometer - European Commission, 2007.
[4] *Action Plan on Urban Mobility - COM(2009) 490.* European Commission, 2009.
[5] http://www.global-hydrogen-bus-platform.com/.
[6] http://www.tfl.gov.uk/corporate/6585.aspx.
[7] *Informe 2006 OMM.* Observatorio Movilidad Metropolitana, 2008.
[8] J. B. Michael, "Safety analysis of concept systems for guidance and control of transit buses final report for mou 327," 1999.
[9] P. Lesauvage, M. Dejeammes, and C. J., "Experience of street-level access to buses in the french city of caen," in *Proceedings of the 22nd European Transport Forum: Public Tranport Planning and Operations,* 1994.
[10] M. Dejeammes, F. Coffin, T. Ladreyt, M. F. Dessaigne, V. Fouet, C. Dolivet, and R. Zac, "Bus stop design and automated guidance for low-floor buses: Evaluation of prototypes with investigation of human factors," *Journal of the Transportation Research Board,* vol. 1666, pp. 85–91, 1999.
[11] H.-S. Tan, "Develop precision docking function for bus operation final report for mou 397," 2003.
[12] http://www.conmasauto.com/.
[13] J. Perez, F. Seco, V. Milanes, A. Jiménez, J. Díaz, and T. de Pedro, "Rfid-based intelligent vehicle speed controller using active traffic signals," *SENSORS,* 2010.
[14] V. Milanés, J. Pérez, E. Onieva, C. González, and T. de Pedro, "Electric power controller for steering wheel management in electric cars," in *Proc. CPE '09. Compatibility and Power Electronics,* 2009, pp. 444–449.
[15] E. Onieva, V. Milanés, C. González, T. de Pedro, J. Perez, and J. Alonso, "Throttle and brake pedals automation for populated areas," *ROBOTICA,* vol. 28, no. 4, pp. 509–516, 2010.
[16] V. Milanes, C. Gonzalez, J. Naranjo, E. Onieva, and T. De Pedro, "Electro-hydraulic braking system for autonomous vehicles," *International Journal of Automotive Technology,* vol. 11, no. 1, pp. 89–95, Feb 2010.
[17] R. García and T. de Pedro, "First application of the orbex coprocessor: Control of unmanned vehicles," *EUSFLAT-ESTYLF Joint Conference. Mathware and Soft Computing,* vol. 7(2-3), pp. 265–273, 2000.
[18] V. Mílanés, "Sistema de control de tráfico para la coexistencia entre vehículos autónomos y manuales mediante comunicaciones inalámbricas," Ph.D. dissertation, Universidad de Alcalá, 2010.

Analysis of Current Doubler Rectifier Based High Frequency Isolation Stage for Intelligent Transformer

Viktor Beldjajev, Indrek Roasto, Dmitri Vinnikov
Tallinn University of Technology, Department of Electrical Drives and Power Electronics
vbeldjajev@gmail.com, indrek.roasto@ieee.org, dmitri.vinnikov@ieee.org

Abstract — **This paper proposes new topology for high frequency isolation stage of intelligent transformer based on bi-directional current doubler rectifier (CDR). Paper provides complete steady-state analysis of the proposed bi-directional CDR in boost mode. To facilitate the understanding of operation, the converter equivalent circuits for each operating mode with corresponding waveforms are presented. Also, mathematical analysis is done and compared with simulation results.**

I. INTRODUCTION

Modern trends in electrical energy technology are characterized by steadily growing need for renewable energy sources, energy storage, and smart grid technologies. Centralized generating facilities are giving way to smaller, more distributed generation. Thus, it is likely that future power generation and distribution will involve a lot of distributed renewable energy sources and micro grids. In order to effectively interconnect power generation and energy storage into a grid or micro grid intelligent energy management (IEM) is needed. IEM substations should have bi-directional energy flow control capability, intelligent control and communication interface. It becomes clear that traditional low frequency (50 Hz) distribution transformers are not any more suitable for such demanding applications.

Intelligent transformer (also known as solid state or power electronic transformer) is a good candidate for IEM subsystems. Intelligent transformer (ITR) is a new type of transformer that realizes voltage transformation, galvanic isolation and power quality enhancements in a single device.

The basic idea behind the ITR is to use a high frequency (f > 1 kHz) transformer instead of traditional low frequency (50 Hz) distribution transformer. Increasing the frequency allows higher utilization of the magnetic core and reduction in the size of the transformer. According to [1] size and weight reduction up to three times is possible. Moreover, ITR uses much less copper that decreases overall costs of the device.

For realization of the ITR, different topologies exist. In [2] an AC/AC buck topology has been proposed. Such converter has variable transformation ratio and the lowest component count. However, it provides no galvanic isolation and causes high voltage (HV) stress for semiconductors. Two stage AC/AC topology with matrix converters is presented in [3]. The benefits of this topology are low number of switches and components, and galvanic isolation. Limited switching

frequency that does not allow implementing of high frequency transformer is major drawback of such topology.

A recent approach is the ITR design that has modular structure consisting of one or several base cells, as shown in Fig. 1. The base cell consists of three stages: input, isolation, and output stage [4]-[6]. Input stage is a bi-directional controllable rectifier, which regulates the DC- link voltage and can also be used to shape the input current (reactive power compensation feature). The second stage provides galvanic isolation between the primary and the secondary side and reduces the DC-link voltage. First, the DC voltage is converted to a high frequency (HF) square-wave signal, then transferred through the HF transformer and finally rectified to form the reduced DC-link voltage on the secondary side. The output stage is a sinewave inverter, which converts reduced DC voltage back to low frequency (50 Hz) alternating grid voltage. It becomes clear that high frequency isolation stage is the key for building an optimal and efficient ITR. By selecting the proper inverter and rectifier topologies the efficiency, energy density and volumes of ITR can be improved.

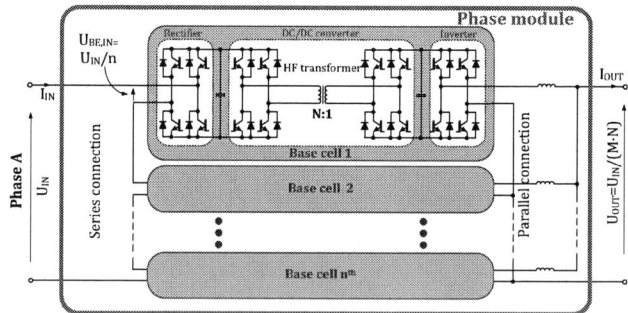

Fig. 1. One phase structure of ITR

II. SURVEY OF ISOLATION STAGE

Two basic topologies are generally suitable for the input side: full-bridge (FB) and half-bridge. Half-bridge topology has a simpler construction and lower number of switches. However, the drawbacks are higher current stress of switches and volt-second unbalance risk. Full-bridge allows to realize phase shift modulation control and zero-voltage switching without additional components, which can reduce switching losses. Due to higher number of switches the volt-second unbalance effect can be prevented and switches have lower

978-1-4244-8806-3/11 $26.00 © 2011 IEEE   336

current stress when compared to half bridge. Thus, FB seems to be more feasible solution for the input side.

In general, three basic topologies are suitable for output side: full-bridge, centre-tapped rectifier, and current doubler rectifier (CDR). With FB no centre tapped transformer is needed and it has a compact design. However, the transformer secondary and rectifier diodes carry full output current, which increases conduction losses. Centre-tapped rectifier has twice smaller number of rectifying diodes and only one diode is conducting at a time. Thus it has low switching losses. Main drawbacks are complicated design and rectifier diodes have to carry full output current, which increases conduction losses. Use of CDR offers many advantages in comparison with FB and centre-tapped rectifier. Transformers secondary current is half the output current allowing reduction of conduction losses especially in the case of higher currents. Using two transformers in the output provides better heat dissipation, and ripple current cancellation effect which allows to use smaller inductances. Operation of such topology in buck mode has been already presented in [16] - [19], where a detailed analysis can be found. However, using CDR for voltage elevation has been proposed only by few authors [11]-[15], thus needs more detailed analysis.

Several researches [11]-[13] propose to add auxiliary and HF transformer on one magnetic core in series with inductors to eliminate inductor currents at loads close to zero . A push-pull type converter in the input with CDR in the output is proposed in [14]. In addition, each coupled inductor is wound on the same core as input-to output windings, forming this way additional path for excessive energy in inductors. Research in [15] proposes to use two inductors in current-fed push-pull converter, to overcome the drawbacks of single inductor push-pull converter e.g. high output voltage ripple, high voltage across switches, high volt-ampere rating for transformer. However, none of these topologies is suitable for bi-directional energy flow control.

Problems also occur when CDR is working in bi-directional mode. It turns out that traditional CDR cannot provide bi-directional energy flow when duty cycle is smaller than 50 % [7]. A solution was proposed in [8] to overcome this drawback. It uses two auxiliary transformers with two diodes to transfer the leakage energy to the high voltage side. However it has high voltage stress on semiconductors. In the current high voltage application the voltage difference between input and output can be more than 5 times, which would require bulky transformers. Since small weight and volumes are important factors in the ITR design, those auxiliary transformers could be a serious drawback. In this paper a new topology is proposed where the auxiliary transformers can be designed much smaller. Also, the reverse voltage of diodes is lower due to smaller voltage difference.

III. NEW BI-DIRECTIONAL DC/DC CONVERTER

The proposed topology of the high frequency isolation stage is shown in Fig. 2. The proposed bi-directional dc-dc converter consists of a FB inverter, a high frequency transformer and a CDR. The energy can be transferred in both directions. Therefore, the diodes in traditional CDR are replaced with controllable switches T_5 and T_6. The high-frequency step-up isolation transformer TR provides required voltage gain as well as the galvanic isolation between the input and output sides of the converter.

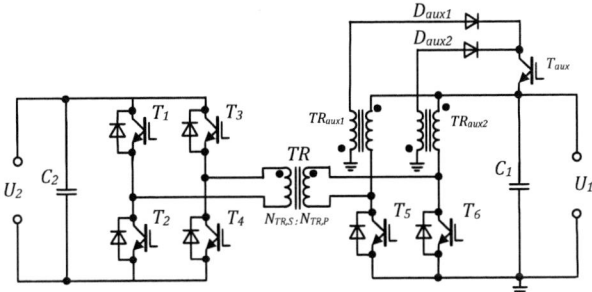

Fig. 2. Power circuit diagram of proposed HF isolation stage

Main benefits of the proposed high frequency isolation stage based on the FB inverter and bi-directional CDR are as follows:

- Zero voltage switching for the primary switches $T_1...T_4$ in a wide load range, achievable to utilize the energy stored in the output filter inductances [9].
- Bi-directional energy control.
- Use of HF transformer reduces size of the converter.
- Reduced current ripples due to transformers.
- Better thermal performance due to current doubling effect.
- Reduced transient response [10].

The proposed CDR can transfer energy in both directions, operating this way either in buck or boost mode in a wide range of duty cycles. In the buck mode the energy is transferred from high voltage side to low voltage side as shown in Fig. 2. In the boost mode the energy is transferred from low voltage side to high voltage side. Boost mode is also divided into three different states: normal operation with duty cycle D_A value from 50 % to 100 %, start-up operation in continuous conduction mode (with D_A value between 25 % and 50 %) and start-up operation in discontinuous conduction mode (with D_A value between 0 % and 25 %).

However, during start-up operation the freewheeling path of the leakage energy of the inductors is interrupted, which results in dangerous overvoltage. To provide an alternative freewheeling path the inductors are replaced with two auxiliary transformers TR_{aux1}, TR_{aux2}, which redirect leakage energy back to the low voltage side through diodes D_{aux1}, D_{aux2} and controllable switch T_{aux}, which is switched ON only during the start-up operation

IV. STEADY STATE ANALYSIS OF PROPOSED CONVERTER

In steady state analysis, all three operating modes are examined. Mathematical analysis and equivalent circuits for different states are provided.

978-1-4244-8806-3/11 $26.00 © 2011 IEEE

The analysis of proposed topology is based on the following assumptions: 1) all transistors and switches are ideal, 2) all inductances and capacitors are ideal, 3) auxiliary transformers TR_{aux1} and TR_{aux2} are identical.

A. Operation with duty cycle 50 % $\leq D_A \leq$ 100 %

When duty cycle is between 50% and 100 %, the converter is operating in normal operating mode. Fig. 3 shows the equivalent circuit of proposed converter in the normal mode.

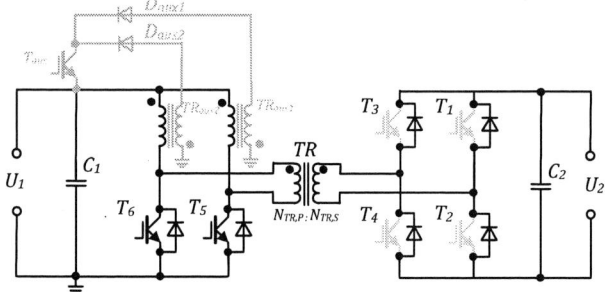

Fig. 3. Equivalent circuit in the boost mode for operation with $D_A \geq 50\%$

The operating period of one switch consists of the active and zero states. During the active state the switch on each branch is switched ON and energy is transferred to the isolation transformer. During the zero state the switch of the branch is switched OFF. The operating period can be written as

$$T = t_A + t_Z , \qquad (1)$$

where t_A and t_Z are durations of an active and zero states, correspondingly. For better appearance of equations the state intervals are represented with corresponding duty cycles as

$$\frac{t_A}{T} + \frac{t_Z}{T} = D_A + D_Z = 1 . \qquad (2)$$

The turns ratio of transformer is defined as follows

$$N = \frac{N_S}{N_P} , \qquad (3)$$

where N_S is the number of turns on the secondary side, and N_P is the number of turns on the primary side of the transformer.

Fig. 4 shows equivalent circuits for both, active and zero state of the switch in the normal operation mode. Typical waveforms for this operation can be found in [8].

Energy can be transferred to the output only during one transistor is switched ON. At the beginning of the active state of transistor T_6 the transistor T_5 is also switched ON and no energy is transferred to the HF transformer. When transistor T_6 switches ON (Fig. 4a), the energy is transferred to the output via the HF transformer. Due to the short zero state duration, transistor T_5 switches ON again during transistor T_6 is still ON and the whole process repeats. During the zero state of transistor T_6 and active state of transistor T_5, the energy is transferred to the output.

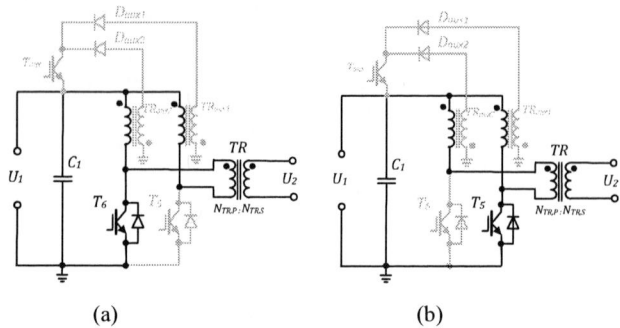

Fig. 4. Equivalent circuits of the transistor's T_6 active state (a) and zero state (b) during normal operation.

For the active state (Fig. 4a), it can be obtained that

$$u_{TRaux1} = u_1 . \qquad (4)$$

In the zero state (Fig. 4b) of transistor T_6 the transformer voltage can be obtained as

$$u_{TRaux1} = u_1 - u_{TR} = u_1 - \frac{u_2}{N_{TR}} , \qquad (5)$$

where N_{TR} is turns ratio of the HF transformer.

In the steady state operation, the average voltage over the inductor during one operating period is zero.

$$\frac{u_1 \cdot t_A + \left(u_1 - \dfrac{u_2}{N_{TR}} \right) \cdot t_Z}{T} = 0 . \qquad (6)$$

Equation (6) can be rewritten using duty cycles

$$u_1 \cdot D_A + \left(u_1 - \frac{u_2}{N_{TR}} \right) \cdot D_Z = 0 , \qquad (7)$$

where D_A represents the active state of the transistor, and D_Z represents the zero state. From (7) the output voltage can be found as:

$$u_2 = \frac{u_1 \cdot (D_A + D_Z)}{D_Z} \cdot N_{TR} = \frac{u_1 \cdot N_{TR}}{1 - D_A} . \qquad (8)$$

Accordingly, the boost factor is:

$$k = \frac{u_2}{u_1} = \frac{1}{1 - D_A} \cdot N_{TR} . \qquad (9)$$

B. Operation with 25 % $\leq D_A <$ 50 %

During the start-up operation the energy stored in transformer TR_{aux1} and TR_{aux2} primaries, is transferred out through secondary windings back into the input. Equivalent circuit for this operation is shown in Fig. 5 and typical waveforms in Fig. 6.

Operating period of the converter consists of one active and three zero states. During the active state the corresponding switch is conducting. The operating period can be defined as:

$$t_A + t_{Z1} + t_{Z2} + t_{Z3} = T . \qquad (10)$$

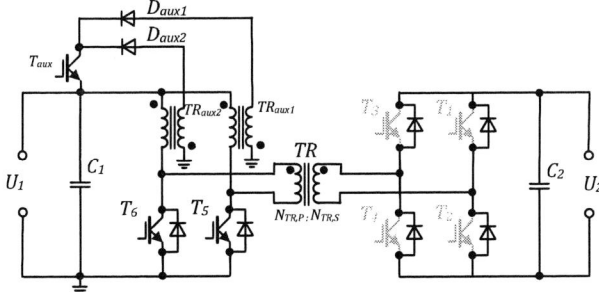

Fig. 5. Equivalent circuit for operation with $D_A < 50\%$.

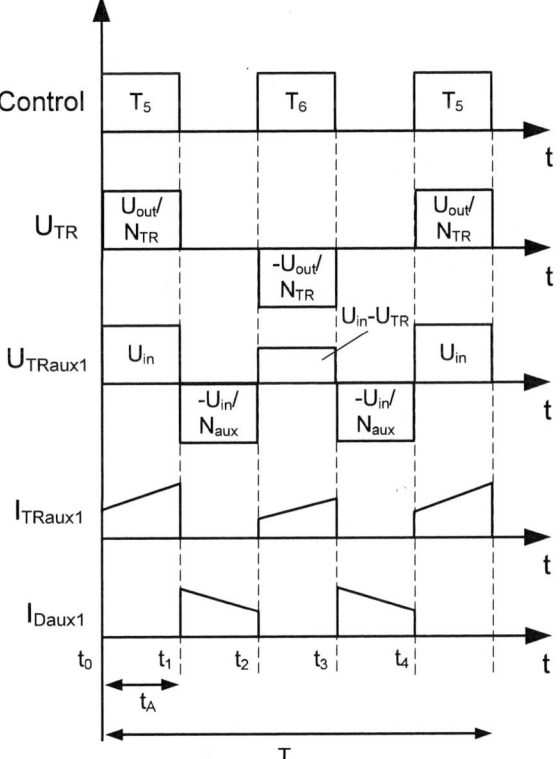

Fig. 6. Waveforms for operation with $D_A < 50\%$.

During active state t_A (**t₀ – t₁**) transistor T_6 is switched ON and energy is transferred to the output via the HF transformer. Current in the auxiliary transformer TR_{aux1} primary increases linearly with input voltage u_1 and stores energy into the transformer. Equivalent circuit for this state is shown in Fig. 4a. Voltage over the transformer TR_{aux1} primary is equal to (4).

During zero state t_{Z1} (**t₁ – t₂**), when both transistors are switched OFF, the currents i_{TRaux1} and i_{TRaux2} circulate over auxiliary transformer's secondary back to the input. Magnetizing current of transformer TR circulates on the secondary side [8]. Equivalent circuit for this mode is shown in Fig. 7.

Voltage over the transformer TR_{aux1} primary is equal to input voltage and N_{aux1} ratio.

$$u_{TRaux1} = \frac{u_1}{N_{aux}} . \qquad (11)$$

During zero state t_{Z2} (**t₂ – t₃**), transistor T_6 is switched ON and energy is transferred to the output via the HF transformer. At the same time current in the transformer TR_{aux2} primary increases linearly with the input voltage u_1. The voltage applied to TR_{aux1} primary is equal to (5).

Zero state t_{Z3} (**t₃ – t₄**), is identical to the interval t_{Z1}.

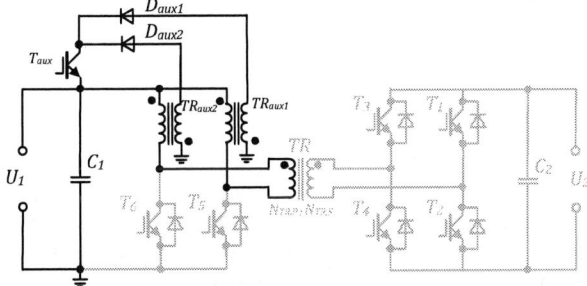

Fig. 7. Equivalent power circuit for intermediate state during start-up.

Equations for steady state analysis in CCM can be written as follows

$$\frac{u_1 \cdot t_A + \left(-\frac{u_1}{N_{aux}}\right) \cdot (t_{Z1} + t_{Z3}) + \left(u_1 - \frac{u_2}{N_{TR}}\right) \cdot t_{Z2}}{T} = 0 . \quad (12)$$

Equation (14) can be rewritten using duty cycles

$$u_1 \cdot D_A + \left(-\frac{u_1}{N_{aux}}\right) \cdot (D_{Z1} + D_{Z3}) + \left(u_1 - \frac{u_2}{N_{TR}}\right) \cdot D_{Z2} = 0 . \quad (13)$$

Taking into account that in the CCM the zero state duty cycle D_{Z2} is equal to the active state D_A and the zero state D_{Z1} is equal to D_{Z3}. Thus, we can simplify (10)

$$2D_A + 2D_Z = 1 , \qquad (14)$$

accordingly

$$D_Z = \frac{1 - 2D_A}{2} \qquad (15)$$

Equation (13) can be simplified as follows

$$u_2 = \frac{N_{TR}}{N_{aux}} \left(2N_{aux} - \frac{1 - 2D_A}{D_A} \right) \cdot u_1 , \qquad (16)$$

where boost factor for start-up operation is

$$k = \frac{u_2}{u_1} = \frac{N_{TR}}{N_{aux}} \left(2N_{aux} - \frac{1 - 2D_A}{D_A} \right) . \qquad (17)$$

C. Operation with $0\% \leq D_A < 25\%$

In case of small loads, relatively low switching frequency and small duty cycle the converter can start to operate in discontinuous conduction mode (DCM), when current

978-1-4244-8806-3/11 $26.00 © 2011 IEEE   339

through inductor goes to zero at the end of every operating cycle. As can be seen from (16). In DCM the discontinuous duty cycle D_D must also be taken into account, so that

$$k = \frac{u_2}{u_1} = \frac{N_{TR}}{N_{aux}}\left(2N_{aux} - \frac{1-2D_A-2D_D}{D_A}\right) \quad (18)$$

Boundary condition for DCM can be derived from (17)

$$D_A \geq \frac{1}{2(N_{aux}+1)}. \quad (19)$$

V. SIMULATION RESULTS

To verify the mathematical model of the bi-directional dc-dc converter, the electric circuit of this topology was simulated in PSIM environment. Moreover, the first experimental prototype was built and tested. Since conventional inductors were used instead of auxiliary transformers the prototype could be only tested in operation mode $D_A \geq 50$ %. The simulation circuit parameters correspond to experimental prototype's parameters, which are set as follows:

TABLE I
CIRCUIT SPECIFICATIONS

Parameter	Value
Low side voltage (U_1)	10 V
High side voltage range (U_2)	0…135 V
Power (P)	500 W
Switching frequency (f_s)	15 kHz
HF transformer turns ratio (N_{TR})	3.3
Auxiliary transformer's turns ratio (N_{aux})	1

Next, the simulation results in steady state operation in three different operating modes are presented. Fig. 8 presents the simulated voltage waveforms on low and high voltage sides in the operating mode with $D_A = 0.6$ in continuous conduction mode. It can be seen that both voltages are pure DC voltages with different layer. According to (8), in case of $U_1 = 10$ V, the $U_2 = 82.5$ V. As can be seen from Fig. 8, simulation results match the mathematical model of the topology.

Fig. 9 presents the experimental results for low and high side voltages, tested with the same parameters as simulation. The values of the voltages correspond to simulation results. Small peaks seen in the experimental results are caused by transistor switching.

Fig. 10 shows the simulated voltage waveforms on low and high voltage sides in the operating mode with $D_A = 0.3$ in continuous conduction mode. Voltage U_2 can be calculated accordingly to (16). In case of $U_1 = 10$ V, the $U_2 = 22$ V. As can be seen from Fig. 10 the simulation results correspond to mathematical model of the topology.

Fig. 11 shows the simulated voltage waveforms on low and high side voltages in the operating mode with $D_A = 0.15$. In this case the converter operates in discontinuous conduction

mode and additional discontinuous duty cycle value D_D in (18) must be considered.

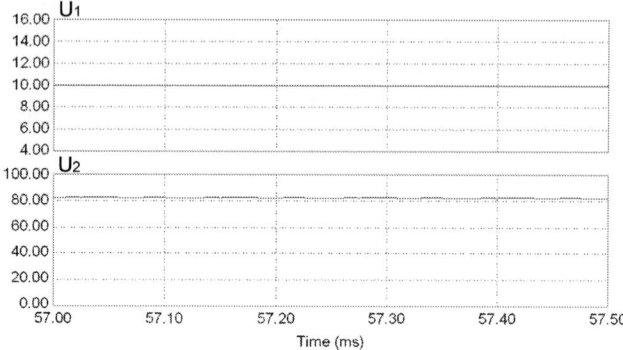

Fig. 8. Simulation results for low voltage side (U_1) as well as for high voltage side (U_2) during operation mode with $D_A = 0.6$.

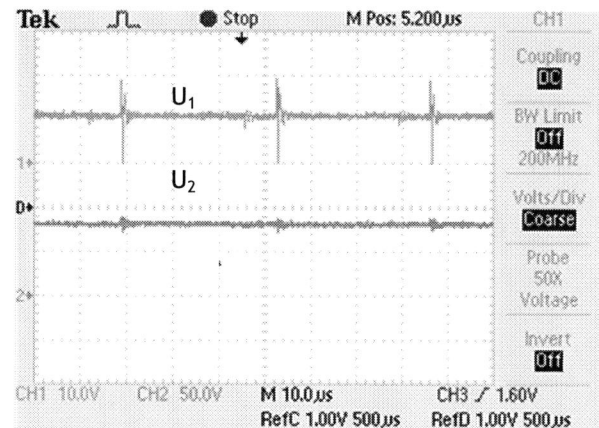

Fig. 9. Experimental results for low voltage side (U_1) as well as for high voltage side (U_2) during operation mode with $D_A = 0.6$.

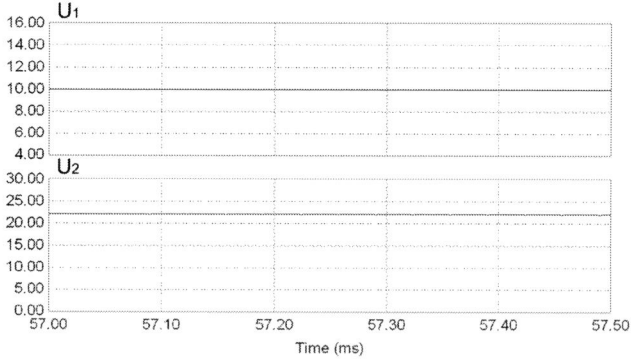

Fig. 10. Simulation results for low voltage side (U_1) as well as for high voltage side (U_2) during operation mode with $D_A = 0.3$.

TABLE II
CALCULATED VS SIMULATED RESULTS

	$D_A = 0.6$	$D_A = 0.3$	$D_A = 0.15$
Calculated U_2 [V]	82.5	22	0.2
Simulated U_2 [V]	82.5	22	0.016

Simulation results are compared with mathematically achieved results in Table II. All calculated voltages correspond to the simulation results. Slight difference in operating mode with $D_A = 0.15$ is caused by uncertainty of D_D.

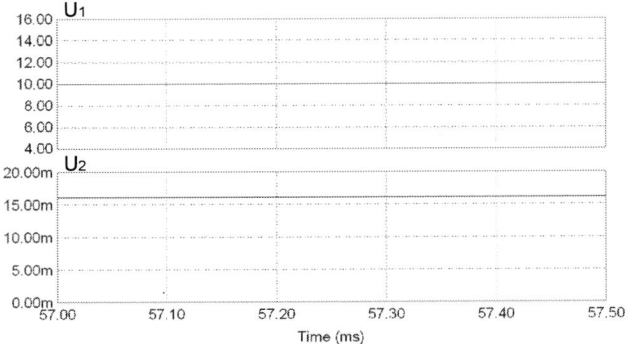

Fig. 11. Simulation results for low voltage side (U_1) as well as for high voltage side (U_2) during operation mode with $D_A = 0.15$.

CONCLUSION

This paper proposed new topology of HF isolation stage for ITR based on bi-directional CDR. CDR is not widely used for voltage elevation, however this topology is appropriate to ITR because of its bi-directional energy flow capability and operation in buck and boost mode, in a wide range of duty cycles. It has also a compact size, good heat dissipation, reduced danger of overvoltage and low current ripples. The converter is also able to operate in a duty cycle range lower than 50 % in CCM and DCM. Mathematical analysis of converter was done and verified with simulations. One operating mode, with $D_A \geq 0.5$ was also verified experimentally. Simulation results showed that the converter can operate in continuous conduction mode as well as in discontinuous conduction mode. The boundary condition for DCM was also derived. Mathematical models were confirmed by the simulation results, which gives a solid base to move on with real tests. As a next step the experimental prototype should be equipped with auxiliary transformers so that it could be tested in all three operating modes.

ACKNOWLEDGEMENT

This research and work has been supported by Estonian Ministry of Education and Research (Project SF0140016s11), Estonian Science Foundation (Grant ETF8687) and Estonian Archimedes Foundation (Project "Doctoral school of energy and geotechnology II").

REFERENCES

[1] T. Zhao, J. Zeng, S. Bhattacharya, M. E. Baran, A.Q. Huang "An Average Model of Solid State Transformer for Dynamic System Simulation". PES 2009, pp. 1-8.

[2] Srinivasan, S.; Venkataramanan, G.; "Comparative evaluation of PWM AC-AC converters", Power Electronics Specialists Conference, 1995, Atlanta, GA.

[3] Mirmousa, H.; Zolghadri, M.R.; "A Novel Circuit Topology for Three-Phase Four-Wire Distribution Electronic Power Transformer", 7th International Conference on Power Electronics and Drive Systems 2007. PEDS '07.

[4] Ronan, E.R.; Sudhoff, S.D.; Glover, S.F.; Galloway, D.L.; "A power electronic-based distribution transformer", Power Delivery, IEEE Transactions on, 2002.

[5] H. Iman-eini and Sh. Farhangi, "Analysis and design of power electronic transformer for medium voltage levels", in proc. IEEE Power Electronic Specialist Conference, PESC 2006, pp. 1- 5, June 2006.

[6] Jih-Sheng Lai; Maitra, A.; Mansoor, A.; Goodman, F.; "Multilevel intelligent universal transformer for medium voltage applications", in Proc. IEEE Industry Applications Conf., Vol. 3, October 2005.

[7] Torrico-Bascope, R.P; Torrico-Bascope, G.V.; Branco, C.G.C.; Oliviera Jr, D.S. "A New Current-Doubler Rectifier Based on Three-State Switching Cell for Buck Derived DC-DC Converters". PESC 2008, pp. 2492 – 2497.

[8] Flores, L.A.; Garcia, O.; Oliver, J.A.; Cobos, J.A. "High Frequency Bi-Directional DC/DC Converter Using Two Inductor Rectifier". IECON 2006, pp. 2793 – 2798.

[9] Ruan, X.; Li, B.; Wang, J.; Li, J. "Zero-Voltage-Switching PWM Three-Level Converter With Current-Doubler-Rectifier." IEEE Transactions on Power Electronics, vol. 19, No. 6, November 2004, pp 1523-1532.

[10] Mao, H.; Yao, L.; Deng, S.; Abdel-Rahman, O.; Liu, J.; Batarseh, I. "Inductor Current Sharing of Current Doubler Rectifier in Isolated DC-DC Converters." IEEE 2006.

[11] Liang, Y.; Lehman, B. Isolated Two-Inductor Boost Converter with One Magnetic Core. APEC '03. IEEE, pp. 879 – 885.

[12] Yungtaek, Y.; Jovanovic, M. M. New Two-Inductor Boost Converter with Auxiliary Transformer. APEC '03. IEEE, pp. 654 – 660, vol. 2.

[13] Yungtaek, Y.; Jovanovic, M. M. New Two-Inductor Boost Converter with Auxiliary Transformer. Power Electronics, IEEE Transactions on vol. 19, Issue 1, Jan. 2004, pp. 169 – 175.

[14] Chung-Wook, R.; Seung-Hoon, H.; Myung-Joong, Y. Dual Coupled Inductor Fed Isolated Boost Converter for Low Input Voltage Applications. Electronics Letters, vol. 35, Issue 21. 14 Oct. 1999, pp. 1791 – 1792.

[15] De Arago Filho, W.C.P.; Barbi, I. A Comparison Between Two Current-Fed Push-Pull DC-DC Converter-Analysis, Design and Experimentation. INTELEC '96. 18th International, 6-10 Oct. 1996, pp. 313 – 320.

[16] Kutkut, N.H.; Luckjiff, G.; "Current mode control of a full bridge DC-DC converter with a two inductor rectifier" PESC '97   Page(s): 203 -209 vol.1

[17] Kutkut, N.H.; "A full bridge soft switched telecom power supply with a current doubler rectifier" INTELEC 97., 19-23 Oct 1997 Page(s):344 -351

[18] Kutkut, N.H.; Divan, D.M.; Gascoigne, R.W.; "An improved full bridge zero-voltage switching PWM DC-DC converter using a two inductor rectifier "Industry Applications Society Annual Meeting, 1993., Conference Record of the1993 IEEE,2-8 Oct1993,Page(s):1065- 1072vol.2

[19] Kutkut, N.H.; Divan, D.M.; Gascoigne, R.W.; "An improved fullbridge zero-voltage switching PWM converter using a two-inductor rectifier", Industry Applications, IEEE Transactions on , Volume: 31 Issue: 1 , Jan/Feb 1995 Page(s): 119 -126.

High Power Li-Ion Battery Charger for Electric Vehicle

A. Kuperman*[1], U. Levy[2], J. Goren[2], A. Zafranski[2] and A. Savernin[2]

[1]Ariel University Center of Samaria, Israel [2]Gamatronic Electronic Industries LTD, Israel

*Corresponding Author, E-mail: alonku@ariel.ac.il

Abstract- **The manuscript presents a 50KW vehicle battery fast charger prototype design. The charger is basically a two-stage controlled rectifier with power factor correction. The input stage consists of a three phase full bridge rectifier combined with a shunt active power filter. The input stage creates an uncontrolled pulsating DC bus while complying with the grid codes by regulating the THD and power factor according to the permissible limits. The output stage is formed by six interleaved parallel groups of two DC-DC converters, fed by the uncontrolled DC bus and performing the charging process. Two independent control boards are employed: active filter control circuitry and the DC-DC control circuitry. The former is operated according to the predetermined grid interfacing behavior, while the operation of the latter is dictated by the requests from the Battery Management System (BMS). The charger is capable of operating in any of the two typical charging modes: Constant Current and Constant Voltage. Control loops are briefly explained throughout the paper and extended simulation/experimental results are presented.**

I. INTRODUCTION

The traction battery (typically lithium based chemistry) is undoubtedly the most critical component of an electric vehicle (EV), since cost and weight as well as the reliability and driving range of the vehicle are strongly influenced by the battery characteristics [1], [2]. Moreover, the battery must be properly managed, and in particular properly recharged in order to utilize its full capacity and preserve its nominal lifetime [3] – [6].

There are two types of vehicle battery chargers: the on-board (sometimes called slow or low power) charger, through which the battery is recharged when parking and plugged into a charging spot [7] – [14], and the off-board (so-called fast or high power) charger, located at the Battery Switch Station (BSS) [15], [16]. The slow charger usually operates at 0.1-0.2C rates, while the fast charger rate typically reaches 1-2C rates, i.e. while charging a 25KWh battery; the slow charger supplies 3-4KW while the fast charger peak power is around 30-50KW.

The typical concept of EV includes urban driving only, where the full battery charge is sufficient for short-range routes. Recharging is accomplished by the slow charger after plugging the car into charge spots placed at different city locations throughout the day and at driver's home during the night. The battery is recharged from empty to full state in about eight hours via the slow charger.

Recently, a paradigm shift towards closing the gap between EV and conventional vehicles has occurred, forcing the infrastructure to support EV intercity driving as well. The following concept of BSS was developed: when out of charge, the EV battery can be replaced at a BSS, allowing nearly uninterrupted long range driving. The near-empty battery, removed from a vehicle at the BSS, is charged by a fast charger (FC), and is available as quickly as possible for the next customer. The charging time should be as short as possible to reduce the battery stock at the BSS; however charging rates higher than 2C are seldom used because of potential instabilities of battery chemistry.

The FC is basically an off-grid supply, drawing the power from the three phase AC utility grid and injecting it into the traction battery. In order to turn out to be a feasible solution, the FC must satisfy both the grid code in terms of THD and power factor from the utility side and lithium-ion charging modes from the battery side. Since the BSS usually contains multiple FCs, its impact on the distribution grid is very significant, as was shown by previous research [17]-[23]. Therefore, the input stage of the FC usually performs rectification and power factor correction (PFC) according to the regulation requirements. It can be accomplished either by employing a bridgeless active rectifier [24]-[26], or a diode rectifier combined with a PFC circuit. The well-known single phase PFC approach, where a full-rating boost DC-DC converter is connected at the diode rectifier output [27] is unsuitable for the three-phase case. However, it can either be modified by splitting the three-phase rectifier into single-phase legs followed by boost PFC converters [28]. Alternatively, a more elegant approach employs a shunt connected active power filter (APF) at the rectifier input, supplying the reactive and harmonic current to the diode rectifier and thus achieving both near unity power factor and near zero THD by letting the utility to supply the active current, which is in phase with the utility voltage and of the same shape [29]–[35]. The use of either one three phase [29] – [32] or three single phase [33]-[35] APF configurations are potentially feasible for implementing a three phase PFC circuit. The additional advantage of the approach is the fact that because of the shunt connection, the APF rating is approximately 40% of the series-connected PFC circuit rating, since the APF supplies the reactive power only to the diode rectifier (which is around 40% of the active power, flowing through the rectifier), while the series connected PFC converter transfers the active power, demanded by the load.

978-1-4244-8806-3/11 $26.00 © 2011 IEEE

A lithium-ion battery charging is typically characterized by two main phases: constant current (CC) and constant voltage (CV), as shown in Fig. 1. The battery is charged by a constant current until the voltage reaches a predetermined level. From this point the voltage is kept constant while the current reduces as the capacity approaches 100%. When charging low capacitance home appliances batteries, the charger is performing CC-CV transitions while in traction batteries the BMS is responsible for the charging sequence by giving the appropriate power request to the charger. The charger measures the battery voltage and determines the current command. Nevertheless, the charger must be able to limit the battery voltage in case of BMS malfunction. Hence, the charger output stage must be capable of operating either as a current source (normal operation) or as a voltage source (BMS fault). Alternatively, it can be operated as a voltage supply with dynamic current limitation.

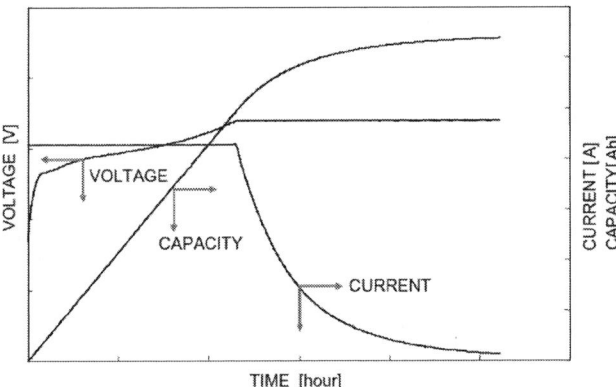

Fig. 1: Lithium-ion battery charging modes

In addition, the charger output current ripple should be kept as low as possible in order to prevent undesired influence on the battery chemistry and losses. The well-known approach, allowing splitting the load power between multiple modules in order to reduce the conduction losses and current ripple is interleaving [36], [37]. Interleaving employs parallel operation of converters, whose output current is shifted with respect to others such that when summed, the current ripples partially cancel each other to create a low ripple total output current.

It also reduces the implementation challenge of designing a single full rating converter by using several lower rating converters instead. The paper describes modeling and control of a 50KW FC, employing a three phase diode rectifier combined with three single phase APFs as the input stage and twelve buck DC-DC converters, divided into six interleaved groups as the output stage. The charger operates as a voltage supply with controllable dynamic current limitation. The charger operates from the 380V three phase utility grid and is able to charge lithium-ion batteries within the voltage range of 230 - 430V by supplying 0 − 125A current.

The rest of the paper is organized as follows. Section II presents the FC overview. The input stage operation is presented in Section III, followed by the output stage description, given in Section IV. Extended simulation and experimental results are shown in Section V. The paper is concluded in Section VI.

II. SYSTEM OVERVIEW

The block diagram of the 50KW FC is shown in Fig. 2. The charger is connected to a three phase four wire 380V AC distribution grid. The utility voltages are given by

$$v_{kN}(t) = V_m \sin(\omega t + \theta k) \qquad (1)$$

with $V_m = 230\sqrt{2}$ V and $k = R,S,T$. The input stage comprises a three-phase diode rectifier, enforced by three single phase APF as a PFC circuit. Each APF is connected between a power phase and the neutral line. The current, drawn by the rectifier

$$i_{Dk}(t) = \sum_{n=1}^{\infty} I_n \sin(n\omega t + \theta_{nk}), \qquad (2)$$

is created by a summation of the grid and APF currents. The THD of the current, drawn by a three-phase diode rectifier with a constant power load (which is the case in the FC) is around 32%, which is unacceptable by any grid code. Hence the APF should supply the rectifier harmonic content and the reactive part of the first harmonic,

$$i_{Fk}(t) = I_1 \sin\theta_{1k}\cos(\omega t) + \sum_{n=2}^{\infty} I_n \sin(n\omega t + \theta_{nk}), \quad (3)$$

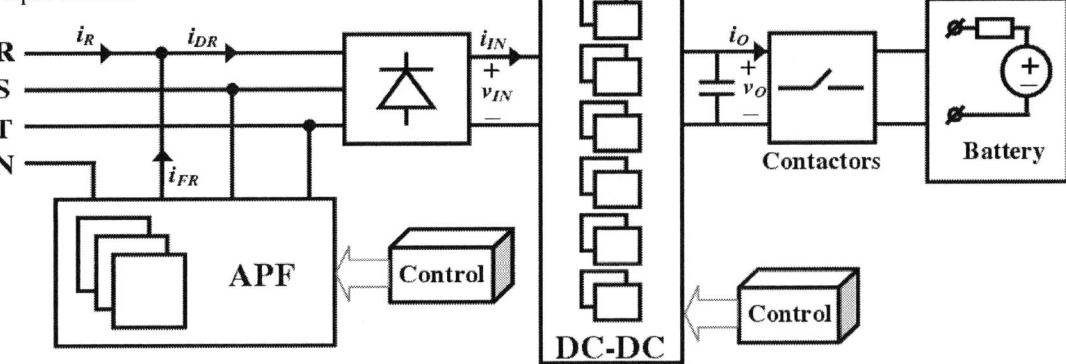

Fig. 2: Fast charger block diagram

letting the utility to supply the active part of the first harmonic only,

$$i_k(t) = I_1 \cos\theta_{1k} \sin(\omega t) \qquad (4)$$

The active power filters employs a separate control board, executing a control related to the grid requirements only, independent on the charging mode and power.

The uncontrolled pulsating rectified voltage v_{IN} given by

$$v_{in}(t) = \max(v_{RS}, v_{ST}, v_{TR}) \qquad (5)$$

is directly supplied to the output stage (note that there is no DC capacitor between the stages). The output stage comprises twelve 4.5KW buck DC-DC converters, connected in parallel. Thus the rectifier load possesses constant power characteristics. Denote the battery charging power P_O, hence the rectifier output current is given by

$$i_{in}(t) = \frac{P_O}{\eta \max(v_{RS}, v_{ST}, v_{TR})} \qquad (6)$$

where η is the efficiency of the output stage.

The DC-DC converters are operated in an interleaving mode, divided into six groups of two converters, allowing significant reduction of both input and output current ripples of the DC-DC stage. A single output capacitor is connected at the charger output to further smooth the output current and allow output voltage adjustment prior to battery charging. In order to prevent an uncontrolled current flow, the charger and the battery are separated by contactors. The DC-DC converters control board executes a dual level control algorithm. The high level algorithm creates voltage/current reference commands according to the desired charging mode, while the low level algorithm operates the DC-DC converters according to the high level control commands in an interleaved fashion. The detailed description of the hardware and the low level control is presented in the following sessions.

III. ACTIVE POWER FILTER

A single phase APF, employed in the input stage of the FC, is shown in Fig. 3. The APF is implemented as a four quadrant DC-AC buck converter, connected to the utility via an inductor. Since the APF supplies reactive power only, single capacitor at the DC side is sufficient. However, since there are some power losses in the APF, the power balance of the capacitor must be ensured by a respective control loop in order to compensate the losses in the APF by drawing a small amount of active power from the utility. The outer control loop operates to maintain the power balance by keeping a constant DC link voltage at a level above the line voltage. The inner loop implements an indirect current control, i.e. instead of the filter current the utility current is sampled and controlled.

Recall that the utility reference current should be in phase with the utility voltage and have the same shape, while its magnitude should be equal to the first harmonic magnitude. The output of the power balancing loop sets the utility current magnitude while the sinusoidal shape is achieved by sampling

the utility voltage and forcing the current loop reference current to follow it.

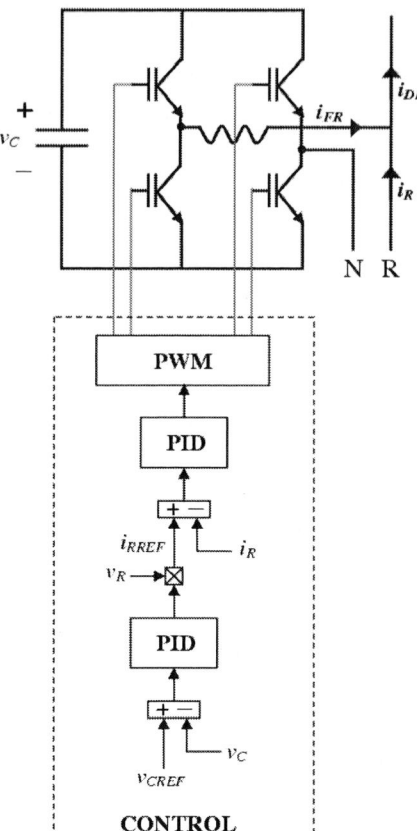

Fig. 3: Active power filter: hardware and control

IV. CHARGER OUTPUT STAGE

The output stage of the FC is shown in Fig. 4. As mentioned, it comprises six parallel buck cells, operated in an interleaved manner. Each cell is composed of two in-phase operated parallel buck converters, as shown in Fig. 5. The main loop of the output stage controls both the output current and voltage of the FC. The control command to each buck cell is the same and chosen according to the charging mode CC/CV. In order to implement an interleaved operation, the PWM clock to each buck cell is shifted by one-sixth of the PWM switching period D. Each DC-DC converter contains a parallel MOSFET-IGBT pair as the switching device. Such a configuration allows benefiting from each transistor advantage: fast MOSFET turn-on time and low conduction losses of the IGBT. The output current of each buck cell is controller utilizing a peak current mode control strategy, allowing both high-bandwidth current control and cycle-by-cycle current limit protection. All the buck cells employ similar control circuitry, since the interleaving is achieved by shifting the clock signals, supplied to the current mode control (CMC) blocks. An addition sub- circuit controls the on/off timing of the MOSFET-IGBT pair.

978-1-4244-8806-3/11 $26.00 © 2011 IEEE   344

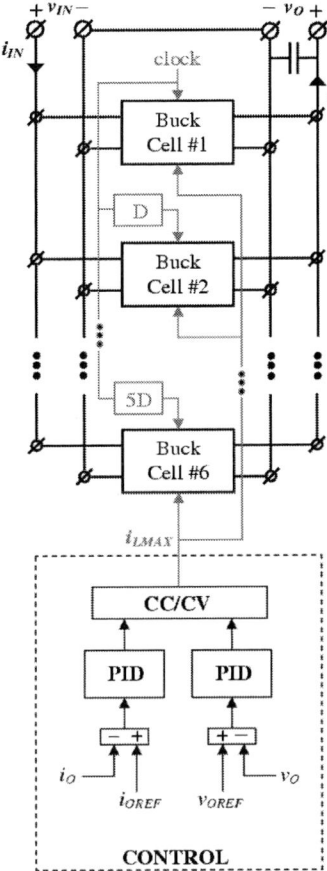

Fig. 4: Output stage overview and control

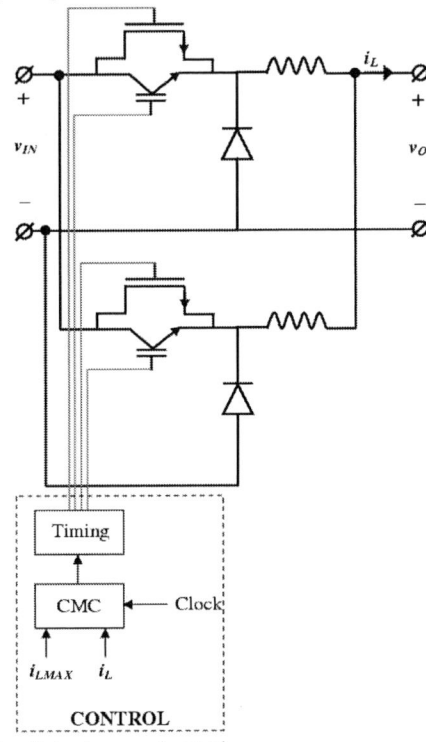

Fig. 5: Buck cell: hardware and control

V. RESULTS AND DISCUSSION

In order to validate the proposed design, the system was modeled and simulated using PSIM software. The input stage performance is shown in Fig. 6. Despite the highly nonlinear currents i_D drawn by the diode rectifier, the utility currents i_S are nearly sinusoidal and in phase with the utility voltages v_S, since the APF supplies the harmonic content i_F of the rectifier currents i_D. The rectifier output voltage and current are shown in Fig. 7. In order to emphasize the filtering function of the APF, the R phase spectra of the utility, rectifier and APF currents are shown in Fig. 8. Note the harmonic content of the diode rectifier currents, concentrated at $6n \pm 1$ harmonic, supplied by the APF, leaving the utility current spectra to contain energy mainly at first harmonic. Experimental results (35KW charging) are shown in Figs. 9 (phase R performance) and 10.

The output stage performance simulation is shown in Fig. 11 for a 45KW charging of at CC mode of 119A. Note the highly smooth transient performance and a stable steady-state operation. Despite the current ripple of nearly 5A of each buck cell, the interleaved operation causes the output current ripple to be as low as 0.8A, as shown in Fig. 12. In addition, even though the input current of each buck cell is highly discontinuous, ramping from zero at each switching cycle, the interleaved operation smoothes the total input current i_{IN} to an acceptable ripple level of 20%. The experimental results are shown in Fig. 13.

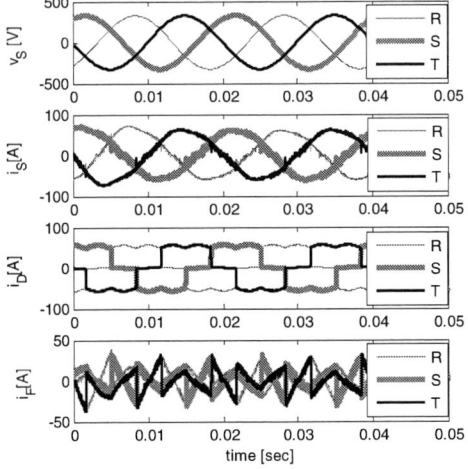

Fig. 6: Input stage performance, simulation.

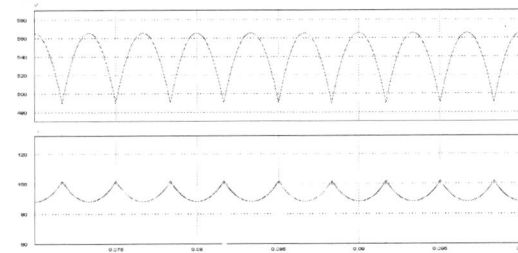

Fig. 7: Rectifier output voltage and current, simulation.

978-1-4244-8806-3/11 $26.00 © 2011 IEEE

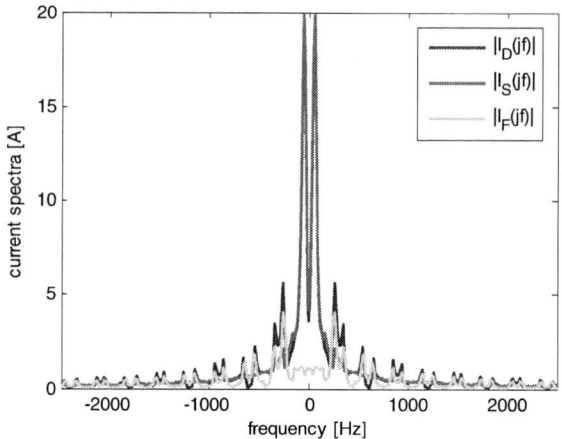

Fig. 8: The R phase spectra of the utility, rectifier and APF currents

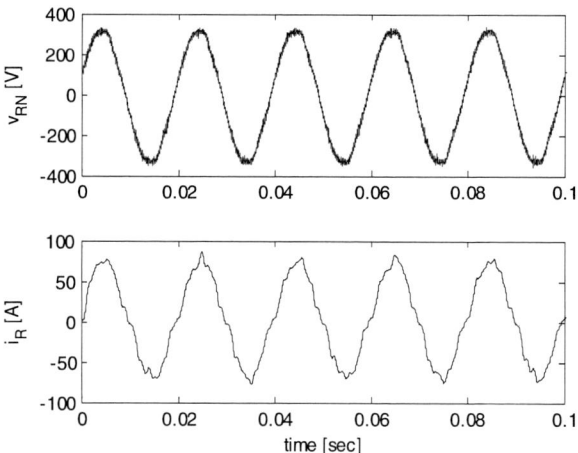

Fig. 9: Input stage performance, experiment (phase R).

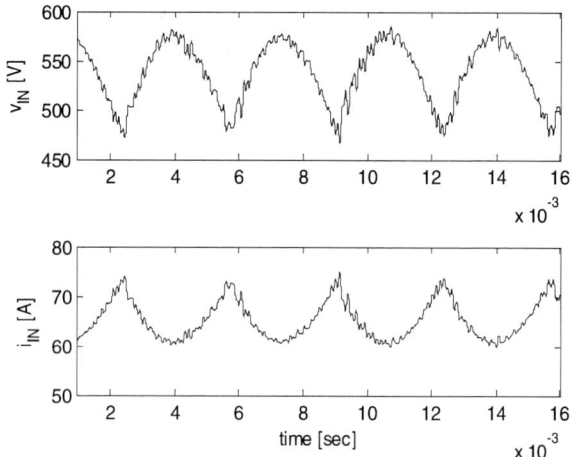

Fig. 10: Rectifier output voltage and current, experiment.

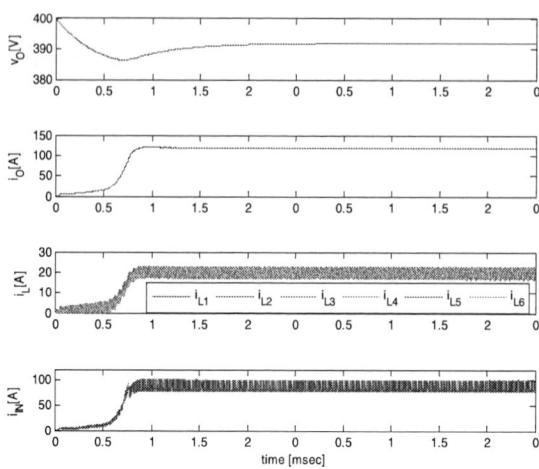

Fig. 11: Output stage performance, simulation

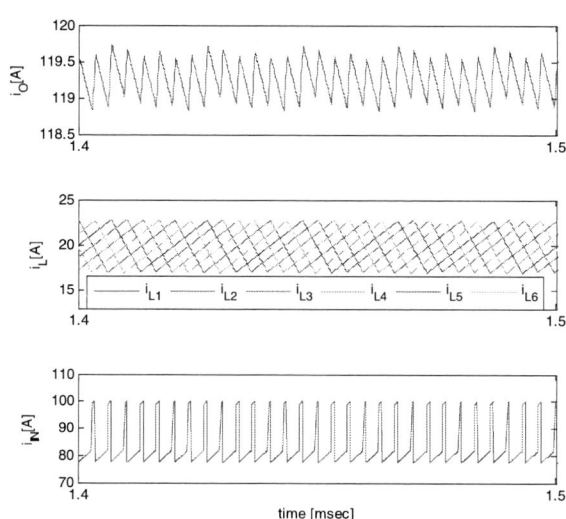

Fig. 12: Output stage zoomed currents, simulation

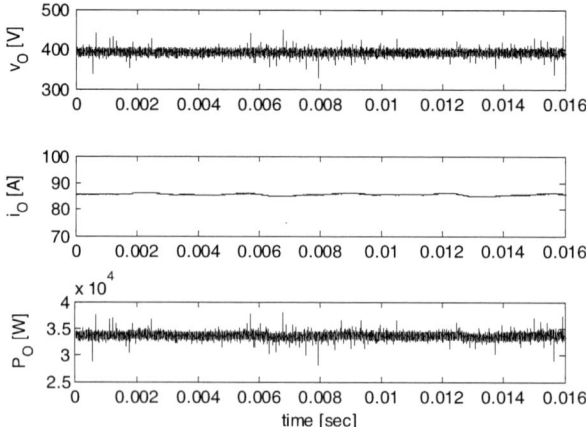

Fig. 13: Output stage performance, experiment.

978-1-4244-8806-3/11 $26.00 © 2011 IEEE

VI. CONCLUSION

A 50KW Li-Ion traction battery fast charger design was presented in the paper. The charger is a two-stage voltage power supply with dynamically controlled current limit function. The input stage includes a diode rectifier with a shunt connected active filter, achieving excellent results in terms of THD and power factor. The output stage is formed by six interleaved groups of two parallel connected buck DC-DC converters, employing current and voltage control according to the charging mode. Simulation and experimental results enforce the proposed design by showing similar results.

REFERENCES

[1] B. Kennedy, D. Patterson and S. Camilleri, "Use of lithium-ion batteries in electric vehicles," *J. Pow. Sources*, vol. 90, pp. 156 – 162, 2000.

[2] R. Gitzendanner, F. Puglia, C. Martin, D. Carmen, E. Jones and S. Eaves, "High power and high energy lithium-ion batteries for underwater vehicles," *J. Pow. Sources*, vol. 136, pp. 416 – 418, 2004.

[3] G-C. Hsieh, L-R. Chen and K-S. Huang, "Fuzzy-controlled Li-Ion battery charge system with active state-of-charge controller," *IEEE Trans. Ind. Elec.*, vol. 48(3), pp. 585 – 593, 2001.

[4] M. Chen and G. Rincon-Mora, "Accurate, compact, and power-efficient Li-Ion battery charger circuit," *IEEE Trans. Circ. Sys. II: Expr. Briefs*, vol. 53(11), pp. 1180 – 1184, 2006.

[5] K. Tsang and W. Chan, "A simple and low-cost charger for lithium-ion batteries," *J. Pow. Sources*, vol. 191, pp. 633 – 635, 2009.

[6] J-J Chen, F-C. Yang, C-C. Lai, Y-S. Hwang and R-G. Lee, "A high-efficiency multimode Li-Ion battery charger with variable current source and controlling previous stage supply voltage," *IEEE Trans. Ind. Elec.*, vol. 56(7), pp. 2469 – 2478, 2009.

[7] D. Thimmesch, "An SCR inverter with an integral battery charger for electric vehicles," *IEEE Trans. Ind. Appl.*, vol. IA-21(4), pp. 1023 – 1029, 1985.

[8] J. Bendien, G. Fregien and J. van Wyk, "High-efficiency on-board battery charger with transformer isolation, sinusoidal input current and maximum power factor," *IEE Proc. B*, vol. 133(4), pp. 197 – 204, 1986.

[9] S-K. Sul and S-J. Lee, "An integral battery charger for four-wheel drive electric vehicle," *IEEE Trans. Ind. Appl.*, vol. IA-21(4), pp. 1023 – 1029, 1995.

[10] B. Masserant and T. Stuart, "A maximum power transfer battery charger for electric vehicles," *IEEE Trans. Aerosp. Elec. Sys.*, vol. 33(3), pp. 930 – 938, 1997.

[11] D. Jackson, A. Schultz, S. Leeb, A. Mitwalli, G. Verghese and S. Shaw, "A multirate digital controller for a 1.5KW electric vehicle battery charger," *IEEE Trans. Pow. Elec.*, vol. 12(6), pp. 1000 – 1006, 1997.

[12] L. Solero, "Nonconventional on-board charger for electric vehicle propulsion batteries," *IEEE Trans. Veh. Tech.*, vol. 50(1), pp. 144 – 149, 2001.

[13] M. Egan, D. O'Sullivan, J. Hayes, M. Willers and C. Henze, "Power-factor-corrected single-stage inductive charger for electric vehicle batteries," *IEEE Trans. Ind. Elec.*, vol. 54(2), pp. 1217 – 1226, 2007.

[14] G. Pellegrino, E. Armando and P. Guglielmi, "An integral battery charger with power factor correction for electric scooter," *IEEE Trans. Pow. Elec.*, vol. 25(3), pp. 751 – 759, 2010.

[15] N. Kutkut, D. Divan, D. Novotny and R. Marion, "Design considerations and topology selection for a 120KW IGBT converter for EV fast charging," *IEEE Trans. Pow. Elec.*, vol. 13(1), pp. 169 – 178, 1998.

[16] C-H. Lin, C-Y. Hsieh and K-H. Chen, "A Li-Ion battery charger with smooth control circuit and built-in resistance compensator for achieving stable and fast charging," *IEEE Trans. Circ. Sys. I: Reg. Pap.*, vol. 57(2), pp. 506 – 517, 2010.

[17] J. Orr, A. Emanuel and K. Oberg, "Current harmonics generated by a cluster of electric vehicle battery chargers," *IEEE Trans. PAS*, vol. PAS-101(3), pp. 691 – 700, 1982.

[18] J. Orr, A. Emanuel and D. Pileggi, "Current harmonics, voltage distortion and powers associated with battery chargers," *IEEE Trans. PAS*, vol. PAS-101(8), pp. 2703 – 2710, 1982.

[19] J. Orr, A. Emanuel and D. Pileggi, "Current harmonics, voltage distortion and powers associated with battery chargers distributed on the residential power system," *IEEE Trans. Ind. Appl.*, vol. IA-20(4), pp. 727 – 734, 1984.

[20] S. Rahman and G. Shrestha, "An investigation into the impact of electric vehicle load on the electric utility distribution system," *IEEE Trans. Pow. Del.*, vol. 8(2), pp. 591 – 597, 1993.

[21] P. Staats, W. Grady, A. Arapostathis and R. Thallam, "A statistical method for predicting the net harmonic currents generated by a concentration of electric vehicle battery chargers," *IEEE Trans. Pow. Del.*, vol. 12(3), pp. 1258 – 1266, 1997.

[22] P. Staats, W. Grady, A. Arapostathis and R. Thallam, "A statistical analysis of the effect of electric vehicle battery charging on distribution system harmonic voltages," *IEEE Trans. Pow. Del.*, vol. 13(2), pp. 640 – 646, 1998.

[23] J. Gomez and M. Morcos, "Impact of EV battery chargers on the power quality of distribution systems," *IEEE Trans. Pow. Del.*, vol. 18(3), pp. 975 – 981, 2003.

[24] G. Gong, M. Heldwein, U. Drofenik, J. Minibock, K. Mino and J. Kolar, "Comparative evaluation of three-phase high-power-factor AC-DC converter concepts for application in future more electric aircraft," *IEEE Trans. Ind. Elec.*, vol. 52(3), pp. 727 – 737, 2005.

[25] B-R. Lin and T-Y. Yang, "Three-phase AC/DC converter with high power factor," *IEE Proc. Electr. Power Appl.*, vol. 152(3), pp. 757 – 764, 2005.

[26] M. Hartmann, S. Round, H. Ertl and J. Kolar, "Digital current controller for a 1MHz, 10KW three-phase VIENNA rectifier," *IEEE Trans. Pow. Elec.*, vol. 24(11), pp. 2496 – 2508, 2009.

[27] K. Raggl, T. Nussbaumer, G. Doerig, J. Biela and J. Kolar, "Comprehensive design and optimization of a high-power-density single-phase boost PFC," *IEEE Trans. Ind. Elec.*, vol. 56(7), pp. 2574 – 2587, 2009.

[28] J. Hahn, P. Enjeti and I. Pitel, "A new three phase power factor correction (PFC) scheme using two single phase PFC modules," *IEEE Trans. Ind. Appl.*, vol. 38(1), pp. 123 – 130, 2002.

[29] H. Akagi, A. Nabae and S. Atoh, "Control strategy of active power filters using multiple voltage source PWM converters," *IEEE Trans. Ind. Appl.*, vol. IA-22(3), pp. 460 – 465, 1986.

[30] J. Nastran, R. Cajhen, M. Seliger and P. Jereb, "Active power filter for nonlinear AC loads," *IEEE Trans. Pow. Elec.*, vol. 9(1), pp. 92 – 96, 1994.

[31] D. Pedder, A. Brown, J. Ross and A. Williams, "A parallel-connected active filter for the reduction of supply current distortion," *IEEE Trans. Ind. Elec.*, vol. 47(5), pp. 1108 – 1117, 2000.

[32] S. Kim, M. Todorovic and P. Enjeti, "Three-phase active harmonic rectifier (AHR) to improve utility input current THD in telecommunication power distribution system," *IEEE Trans. Ind. Appl.*, vol. 39(5), pp. 1414 – 1421, 2003.

[33] H-L. Jou, J-C. Wu and H-Y. Chu, "New single-phase active power filter," *IEE Proc. Electr. Pow. Appl.*, vol. 141(3), pp.129 – 134, 1994.

[34] C. Hsu and H. Wu, "A new single-phase active power filter with reduced energy-storage capacity," *IEE Proc. Electr. Pow. Appl.*, vol. 143(1), pp. 25 – 30, 1996.

[35] H. Komurcugil and O. Kukrer, "A new control strategy for single phase shunt active power filters using a Lyapunov function," *IEEE Trans. Ind. Elec.*, vol. 53(1), pp. 305 – 312, 2006.

[36] C. Chang and M. Knights, "Interleaving technique in distributed power conversion systems," *IEEE Trans. Circ. Sys. I: Fund. Theory Appl.*, vol. 42(5), pp. 245 – 251, 1995.

[37] D. Perreault and J. Kassakian, "Distributed interleaving of paralleled power converters," *IEEE Trans. Circ. Sys. I: Fund. Theory Appl.*, vol. 44(8), pp. 728 – 734, 1997.

978-1-4244-8806-3/11 $26.00 © 2011 IEEE

Neutral Point Clamped Quasi-Impedance-Source Inverter

Silver Ott, Indrek Roasto, Dmitri Vinnikov
Tallinn University of Technology, Department of Electrical Drives and Power Electronics
ott.silver@gmail.com, indrek.roasto@ieee.org, dmitri.vinnikov@ieee.org

Abstract- **In this paper a new modification of a three-level neutral-point-clamped inverter is presented. The proposed topology combines the advantages of the three-level neutral-point-clamped full-bridge inverter with those of the quasi-impedance-source inverter. The neutral-point-clamped quasi-impedance-source inverter is especially suitable for renewable energy sources.**

The steady-state analysis of a neutral-point-clamped quasi-impedance-source inverter in the case of continuous conduction mode is presented. The presented models as well as deduced equations are verified in a simulation.

I. INTRODUCTION

A three-level neutral-point-clamped full-bridge (NPCFB) inverter (Fig. 1*a*) has a lot of advantages, such as lower semiconductor voltage stress, lower required blocking voltage capability, decreased *dv/dt*, better harmonic performance, soft switching possibilities without additional components, higher switching frequency due to lower switching losses and balanced neutral-point voltage, in comparison with the two-level voltage source inverter. As a drawback it has two additional clamping diodes per phase-leg and more controlled semiconductor switches per phase-leg in comparison with the two-level voltage source inverter. The three-level NPCFB can normally perform only the voltage buck operation. In order to ensure voltage boost operation an additional DC/DC boost converter should be used in the input stage [1-2].

To obtain buck and boost performance the focus is turned into a quasi-impedance-source (qZS) inverter (Fig. 1*b*). The qZS inverter was first introduced in [3]. The qZS inverter

consists of two inductors (L_1, L_2), two capacitors (C_1, C_2), a diode (D_1) and a full-bridge (T_1, T_2, T_3, T_4), as shown in Fig. 1*b*. The qZS inverter can buck and boost DC-link voltage in a single stage without additional switches.

The qZS inverter can boost the input voltage by introducing a special shoot-through switching state, which is the simultaneous conduction (cross conduction) of both switches of the same phase leg of the inverter. This switching state is forbidden for traditional voltage source inverters because it causes a short circuit of the DC-link capacitors. Thus, the qZS inverter has excellent immunity against the cross conduction of top and bottom-side inverter switches. The possibility of using shoot-through eliminates the need for dead-times without having the risk of damaging the inverter circuit. The input voltage is regulated only by adjusting the shoot-through duty cycle. In addition the qZS inverter has a continuous mode input current (input current never drops to zero), which makes it especially suitable for renewable energy sources (e.g. fuel cells, solar energy, wind energy etc.). The main drawback of the qZS inverter is its poor performance in the case of small loads and relatively low switching frequency. In these conditions the qZS inverter starts to work in discontinuous conduction mode, which causes an over-boost effect and leads to instabilities [3-7].

In this paper a new inverter topology is proposed: a three-level neutral-point-clamped (NPC) qZS inverter Fig. 2). The proposed inverter combines the advantages of the two topologies described above. The static models of the proposed topology in the case of continuous conduction mode will be analysed and verified.

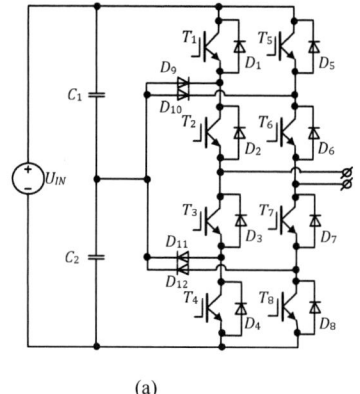

(a)

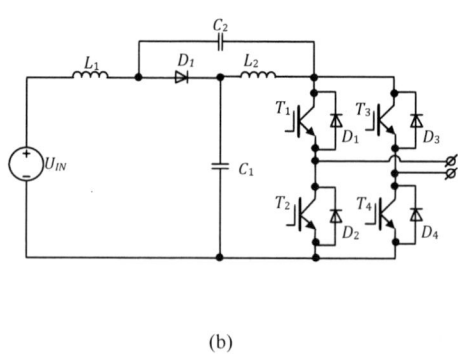

(b)

Fig. 1. Three-level neutral-point-clamped full-bridge inverter (a), quasi-impedance-source inverter (b)

978-1-4244-8806-3/11 $26.00 © 2011 IEEE   348

II. Neutral-Point-Clamped qZS Inverter

Fig. 2 illustrates the proposed topology of a single-phase three-level NPC qZS inverter. Each leg of the three-level NPC qZS inverter consists of two complementary switching pairs and four anti-parallel diodes. As an advantage this topology has continuous input current, the possibility to use shoot-through, lower switching losses and balanced neutral-point voltage in comparison with the traditional two-level voltage source inverter.

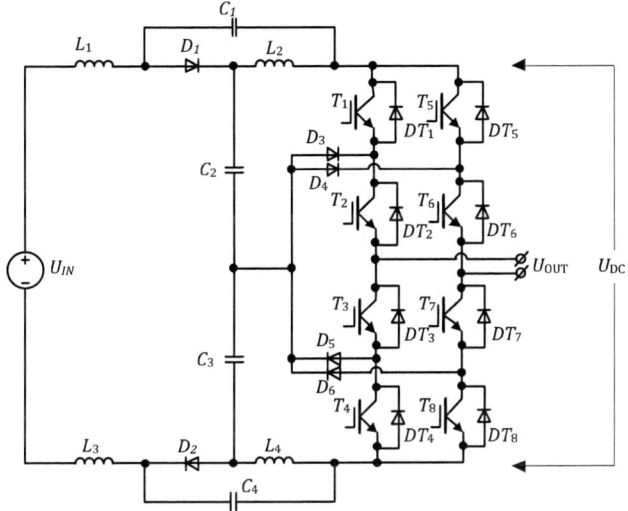

Fig. 2. Three-level NPC quasi-impedance-source inverter

Pulse width modulation with simple boost control has been used, as shown in Fig. 3.

Fig. 3 shows inverter switching states during one period. Shoot-through is generated during zero states. The zero and shoot-through states are spread over the switching period so the number of higher harmonics can be reduced. In order to reduce the switching losses of the transistors, the number of shoot-through states per period was limited by two. Moreover, in order to decrease the conduction losses of the transistors, the shoot-through current is distributed between both inverter legs, i.e. all switches conducting. As can be seen in Table I, the operating frequencies, thus switching losses of transistors are different. The switching frequency of T_1, T_4, T_5 and T_8 is three times higher and the switching frequency of T_2, T_3, T_6 and T_7, is twice higher than the frequency of the output voltage.

TABLE I.
INVERTER SWITCHING STATES DURING ONE PERIOD

	T_1	T_2	T_3	T_4	T_5	T_6	T_7	T_8
Zero-state	0	1	0	0	0	0	1	0
Shoot-through	1	1	1	1	1	1	1	1
Zero-state	0	1	0	0	0	0	1	0
Active state	1	1	0	0	0	0	1	0
Active state	1	1	0	0	0	0	1	1
Active state	0	1	0	0	0	0	1	1
Zero-state	0	0	1	0	0	1	0	0
Shoot-through	1	1	1	1	1	1	1	1
Zero-state	0	0	1	0	0	1	0	0
Active state	0	0	1	0	1	1	0	0
Active state	0	0	1	1	1	1	0	0
Active state	0	0	1	1	0	1	0	0

The inverter output voltage has three different levels: 0, $B \cdot (U_{IN}/2)$ and $B \cdot U_{IN}$ in positive and negative directions, where B is the inverter boost factor.

Four reference signals (REF1, REF2, REF3 and REF4) and two compare values (CMPR1 and CMPR2) are used to generate pulse width control signals for the switches ($T_1...T_8$), as shown in Fig. 3. The shoot-through vector is generated separately using reference signal REF5 and two compare values, V_p and V_n. Finally the shoot-though vector is mixed together with other control signals using OR-gates.

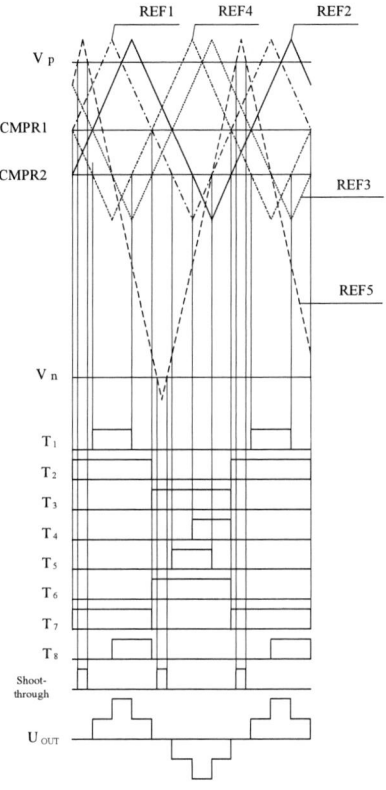

Fig. 3. Sketch of pulse width modulation with simple boost control ($D_S = 0.25$)

III. Steady-State Analysis of Three Level NPC qZS Inverter

In general, the operating period of the three-level NPC qZS inverter in continuous conduction mode may be divided into two states: non-shoot-through (t_N) and shoot-through state (t_S):

$$T = t_N + t_S. \tag{1}$$

Equation (1) could be represented in a form:

$$\frac{t_N}{T} + \frac{t_S}{T} = D_N + D_S = 1, \tag{2}$$

where D_N is the duty cycle of non-shoot-through and D_S is the shoot-through duty cycle. The equivalent schemes of the qZS inverter during the shoot-through and non-shoot-through modes are presented in Fig. 4.

978-1-4244-8806-3/11 $26.00 © 2011 IEEE

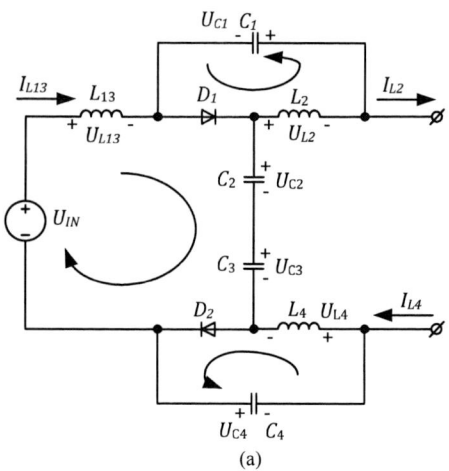
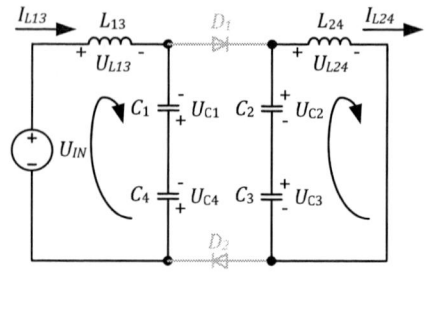

(a)   (b)

Fig. 4. Equivalent circuit of three-level NPC qZS inverter during non-shoot-through state (a) and shoot-through state (b).

Input capacitors and inductors are identical to each other, thus:

$$L_1 = L_3, \qquad L_2 = L_4; \qquad (3)$$

$$C_1 = C_4, \qquad C_2 = C_3. \qquad (4)$$

Considering that during non-shoot-through state inductors L_1 and L_3 are connected in a series they can be replaced by an equivalent inductor, L_{13}. From Fig. 4a inductor voltages during non-shoot-through mode can be found as follows:

$$u_{L2} = -U_{C1}; \qquad (5)$$

$$u_{L4} = -U_{C4}; \qquad (6)$$

$$u_{L13} = U_{IN} - U_{C2} - U_{C3}, \qquad (7)$$

where u_{L13} is the voltage of the equivalent inductance, L_{13}. Inductor voltages u_{L1} and u_{L3} can be found as:

$$u_{L1} = u_{L3} = \frac{u_{L13}}{2}. \qquad (8)$$

On the base of an equivalent circuit of a three-level NPC qZS inverter during shoot-through state (Fig. 4b) the inductor voltages could be presented as:

$$u_{L13} = U_{IN} + U_{C1} + U_{C4}; \qquad (9)$$

$$u_{L24} = U_{C2} + U_{C3}, \qquad (10)$$

where u_{L24} is the voltage of the equivalent inductance, L_{24}. Inductor voltages u_{L2} and u_{L4} can be found as:

$$u_{L2} = u_{L4} = \frac{u_{L24}}{2}. \qquad (11)$$

In steady state the average voltage of the inductor over one switching period is zero. Thus, from (5) to (11) we can obtain:

$$
\begin{cases}
U_{L1} = D_S \cdot \left(\dfrac{U_{IN} + U_{C1} + U_{C4}}{2} \right) + (1 - D_S) \cdot \left(\dfrac{U_{IN} - U_{C2} - U_{C3}}{2} \right) = 0 \\[2mm]
U_{L2} = D_S \cdot \left(\dfrac{U_{C2} + U_{C3}}{2} \right) + (1 - D_S) \cdot (-U_{C1}) = 0 \\[2mm]
U_{L3} = D_S \cdot \left(\dfrac{U_{IN} + U_{C1} + U_{C4}}{2} \right) + (1 - D_S) \cdot \left(\dfrac{U_{IN} - U_{C2} - U_{C3}}{2} \right) = 0 \\[2mm]
U_{L4} = D_S \cdot \left(\dfrac{U_{C2} + U_{C3}}{2} \right) + (1 - D_S) \cdot (-U_{C4}) = 0
\end{cases}
\qquad (12)
$$

By solving the equation system (12) the voltages across capacitors can be found:

$$U_{C1} = U_{C4} = \frac{D_S \cdot U_{IN}}{2 - 4 \cdot D_S}; \qquad (13)$$

$$U_{C2} = U_{C3} = \frac{U_{IN} \cdot (D_S - 1)}{4 \cdot D_S - 2}. \qquad (14)$$

The peak DC-link voltage U_{DC_p} is the sum of all capacitor voltages:

$$U_{DC\_p} = U_{C1} + U_{C2} + U_{C3} + U_{C4} = \frac{U_{IN}}{1 - 2 \cdot D_S}. \qquad (15)$$

The boost factor (B) of the three-level NPC qZS inverter equals:

$$B = \frac{U_{DC\_p}}{U_{IN}} = \frac{1}{1 - 2 \cdot D_S}. \qquad (16)$$

The boost factor of the three-level NPC qZS inverter is as big as that of the traditional qZS inverter.

Fig. 5 shows dependence between boost factor and shoot-through duty cycle.

978-1-4244-8806-3/11 $26.00 © 2011 IEEE   350

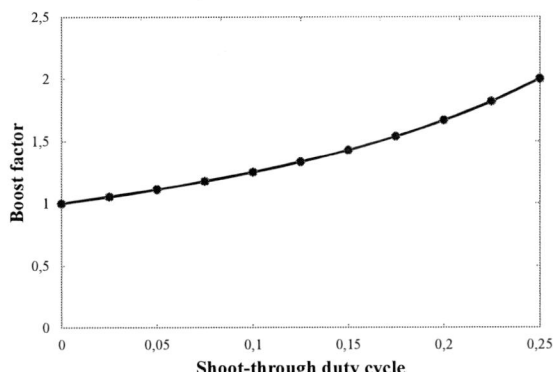

Fig. 5. Dependence between boost factor and shoot-through duty cycle.

IV. ANALYSIS OF SIMULATION RESULTS

For the verification of mathematical analysis a model in PSIM simulation software was developed. First, a general analysis was carried out with the duty cycle D_s=0.20. Then, to confirm mathematical models, simulations were performed at three operating points: D_S=0, D_S=0.1 and D_S=0.25. All simulations were carried out with ideal lossless components with parameters that are shown in Table II. To eliminate oscillating capacitor voltage and inductor current, a small series resistance (r_L) was added to inductors.

TABLE II.
SIMULATION PARAMETERS

Input voltage U_{IN}	220 V
Inductors L_1, L_2, L_3, L_4	50 µH
Inductor resistance r_L	4 mΩ
Capacitors C_1, C_2, C_3, C_4	240 µF
Equivalent resistor R_{load}	25 Ω
Switching frequency of T_1, T_4, T_6, T_7	60 kHz
Switching frequency of T_2, T_3, T_6, T_7	40 kHz
Output frequency f_{out}	20 kHz

Fig. 7 and Fig. 7 are verifying the simplifications which were presented in Fig. 4. Inductor L_1 and L_3 currents are similar as well as the inductor L_2 and L_4 ones.

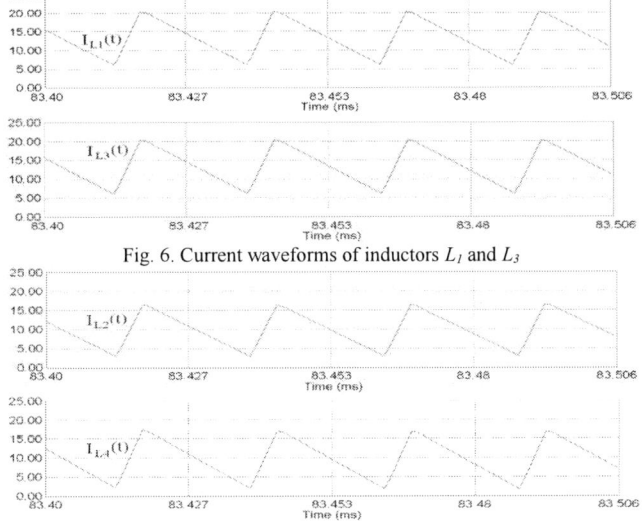

Fig. 6. Current waveforms of inductors L_1 and L_3

Fig. 7. Current waveforms of inductors L_2 and L_4

In the Fig. 8 the simulated waveforms of U_{L1} and U_{L3}, in Fig. 9 the simulated waveforms of U_{L2} and U_{L4} are shown. The simulated waveforms verify the equations (8) and (11). Inductor L_1 and L_3 voltages are similar as well as the inductor L_2 and L_4 ones.

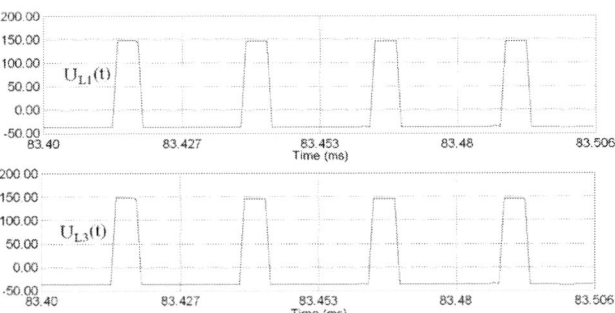

Fig. 8. Voltage waveforms of inductors L_1 and L_3

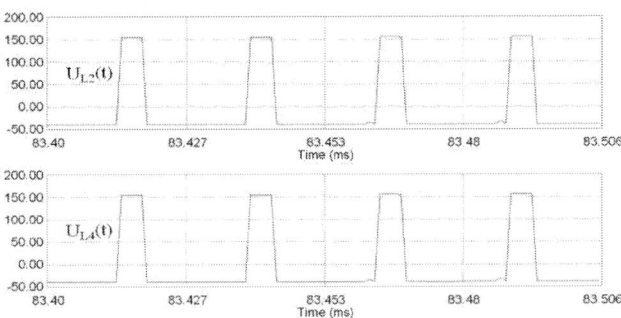

Fig. 9. Voltage waveforms of inductors L_2 and L_4

This increases the energy density of the inductors and allows using two coupled inductors, to minimize the number of needed inductors as well as the size of the prototype. One inductor will be needed for L_1 and L_3, the other one for L_2 and L_4.

Fig. 10 presents operating current and –voltage waveforms of qZS diodes D_1 and D_2. The average voltage value across the qZS diode D_1 is 32 V and the voltage peak value reaches 172 V. The peak value of current I_{D1} reaches 16 A. The average voltage value across the qZS diode D_2 is 36 V and the voltage peak value reaches 193 V. The peak value of current I_{D2} reaches 36 A. Since the diodes are unequally loaded it is advisable to avoid diodes in a single housing and to use two separate clamping diodes.

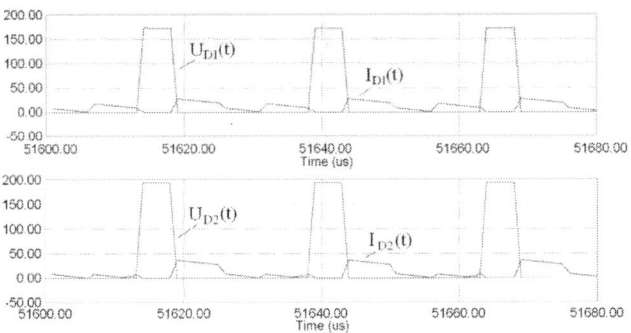

Fig. 10. Operating current and -voltage waveforms of D_1 and D_2

Fig. 11 shows operating current and –voltage waveforms of clamping diodes D_3 and D_4. The average voltage value across the clamping diode D_3 is 54 V and the voltage peak value reaches 173 V. The peak value of current I_{D3} reaches 27 A. The average voltage value across the clamping diode D_4 is 60 V and the voltage peak value 173 V. The peak value of current I_{D2} reaches 37 A. Also here are two separate clamping diodes preferable.

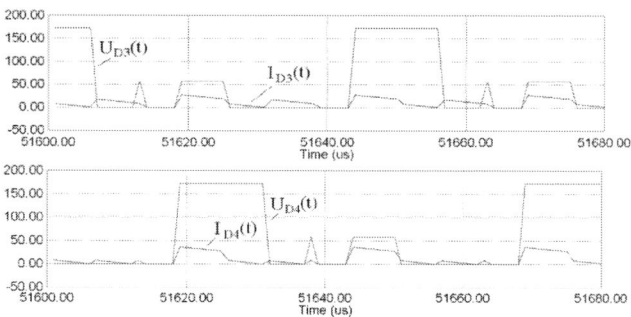

Fig. 11. Operating current and –voltage waveforms of D_3 and D_4

The collector-emitter currents and -voltages of switches T_1 and T_2 are presented in Fig. 12. The collector-emitter voltages of switches illustrate clearly the different switching frequency of the switches. Both switches are working in hard switching conditions.

The average collector-emitter voltage across T_1 is 84 V and the voltage peak value reaches 174 V. The peak value of current I_{T1CE} reaches 19 V. The average collector-emitter voltage across T_2 is 55 V and the voltage peak value reaches 193 V. The peak value of current I_{T2CE} reaches 19 V.

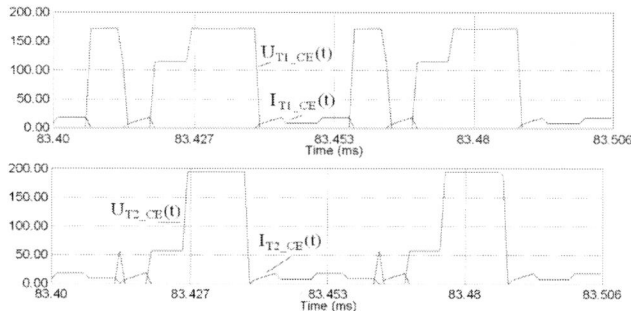

Fig. 12. Collector-emitter currents and –voltages of switches T_1 and T_2 (D_S=0.20)

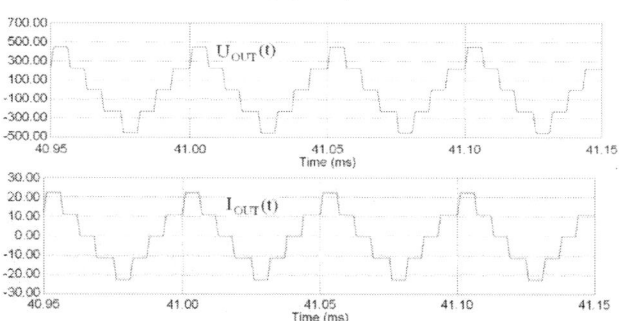

Fig. 13. Output voltage and –current of single-phase three-level NPC qZS inverter

By increasing the shoot-through duty cycle the amplitude value of the output voltage will increase, but the voltage shape will remain unchanged. Fig. 13 shows the output voltage and current of a single phase three-level NPC qZS inverter, when the D_S=0.20.

1. Operation with no shoot-through: $D_S = 0$

The first simulation was performed with no shoot-through, in other words $D_S = 0$. The inverter was controlled as a traditional three-level NPC.

Fig. 14 presents the waveform of continuous input current and DC-link voltage.

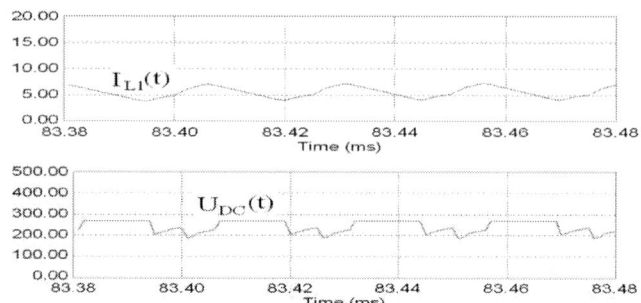

Fig. 14. Simulated waveforms of input current (a), -DC-link voltage (b) in case $D_S = 0$

Table III compares calculated and simulated results. As may be seen the results were predicted from the theoretical analysis.

TABLE III.
COMPARISON OF CALCULATED AND MEASURED RESULTS ($D_s = 0$)

Signal	Calculated result	Simulated result
U_{C1}	0 V	0 V
U_{C2}	110 V	109 V
U_{C3}	110 V	109 V
U_{C4}	0 V	0 V
U_{DCmax}	240 V	237 V

2. Operation with shoot-through: $D_S = 0.1$

The second simulation was performed with shoot-through duty cycle $D_S = 0.1$. Fig. 15 illustrates the simulation results under the named circumstances.

The converter is supplied with continuous mode input current. The DC-link voltage has been boosted according to (16).

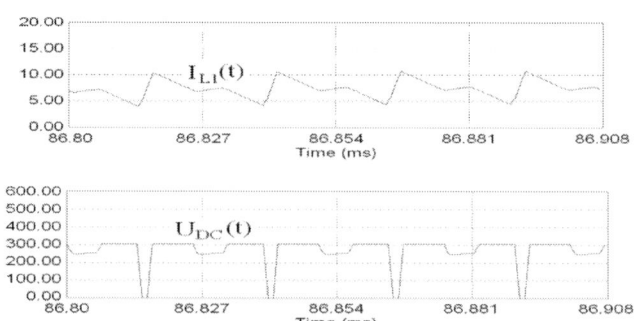

Fig. 15. Simulated waveforms of input current and DC-link voltage in the case of $D_S = 0.1$

Table IV compares calculated and simulated results. Minor differences between calculated and simulated results are caused by the slight distortion of the input current which was excluded in the theoretical calculations. The cause of named effect needs to be studied in more detail in the future.

TABLE IV.
COMPARISON OF CALCULATED AND MEASURED RESULTS ($D_s = 0.1$)

Signal	Calculated result	Simulated result
U_{C1}	13.75 V	14 V
U_{C2}	123.75 V	124 V
U_{C3}	123.75 V	124 V
U_{C4}	13.75 V	14 V
U_{DCmax}	275 V	276 V

3. Operation with shoot-through: $D_S = 0.25$

The third simulation was performed with the shoot-through duty cycle ($D_S = 0.25$). Fig. 16 illustrates the simulation results under the named circumstances.

The converter is supplied with continuous mode input current. The DC-link voltage has been boosted according to (16).

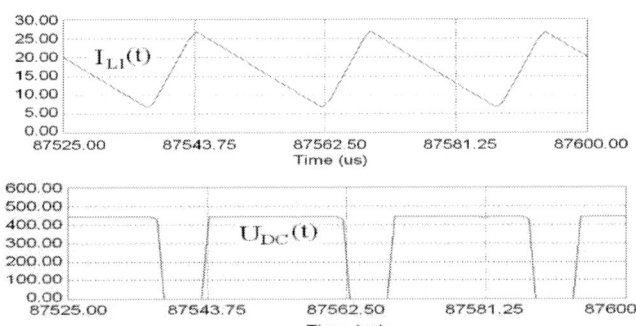

Fig. 16. Simulated waveforms of input current and DC-link voltage in the case of $D_S = 0.25$

Table V compares calculated and simulated results. As can be seen the results were predicted from the theoretical analysis.

TABLE V.
COMPARISON OF CALCULATED AND MEASURED RESULTS ($D_s = 0.25$)

Signal	Calculated result	Simulated result
U_{C1}	55 V	56 V
U_{C2}	165 V	166 V
U_{C3}	165 V	166 V
U_{C4}	55 V	56 V
U_{DCmax}	440 V	444V

V. CONCLUSION

A new topology three-level NPC qZS inverter was proposed and described. The NPC qZS inverter is a combination of the quasi-impedance-source inverter and three-level NPC full-bridge. The three-level NPC qZS comprises advantages from both topologies: it can buck and boost the input voltage, it has excellent short circuit immunity, due to the multilevel topology great energy density is achievable, etc. The steady-state analysis in the case of continuous conduction mode was carried out as well as the equation of boost factor was derived. The theoretical results were confirmed by the simulation results.

As future research a prototype with the presented topology will be built to confirm the theoretical analysis experimentally. Other control methods will be analysed experimentally and the analysis of the losses will be carried out.

ACKNOWLEDGEMENT

This research work has been supported by Estonian Ministry of Education and Research (Project SF0140016s11), Estonian Science Foundation (Grant ETF8687, Grant ETF8538) and Estonian Archimedes Foundation (project „Doctoral school of energy and geotechnology II").

REFERENCES

[1] F. Gao, P. C. Loh, F. Blaabjerg, D. M. Vilathgamuwa, "Dual Z-source inverter with three-level reduced common-mode switching", IEEE Transactions on industry applications, vol.43, no. 6, pp.1597-1608, 2007.

[2] P. C. Loh, S. W. Lim, F. Gao, F. Blaabjerg, "Three-level Z-source inverters using a single LC impedance network", IEEE Transactions on power electronics, vol. 22, no. 2, pp. 706-711, 2007.

[3] Anderson, J.; Peng, F.Z., "Four quasi-Z-Source inverters", in Proc. of IEEE Power Electronics Specialists Conference PESC'2008, pp. 2743-2749, June 15-19, 2008.

[4] F. Z. Peng, "Z-Source inverter", IEEE Transactions of Industry Applications, vol. 39, no. 2, pp.504-510, 2003.

[5] D. Vinnikov, I. Roasto, J. Zakis, "Mathematical models of cascaded quasi-impedance source converter", Технічна електродинаміка, 59 – 64, 2010.

[6] P. C. Loh, D. M. Vilathgamuwa, Y. S. Lai, G. T. Chua, Y. W. Li, "Pulse width modulation of Z-Source inverter", IEEE Transactions of Power Electronics, vol. 20, pp. 1346-1355, 2005.

[7] H. Rostami, D. A. Khaburi, "Voltage Gain Comparison of Different Control Methods of the Z-Source Inverter", International Conference on Electrical and Electronics Engineering, pp. 268-272, 2009.

[8] D. Li, F. Gao, P. C. Loh, F. Blaabjerg, K. K. Tan, "Hybrid-source impedance network and its generalized cascading concepts", International conference on power electronics and drive systems, pp. 1233-1238, 2009.

[9] R. Strzelecki, D. Vinnikov, "Models of the qZ-converters", Przeglad Elektrotechniczny, 86(6), 80 – 84, 2010.

[10] D. Li, F. Gao, P. C. Loh, F. Blaabjerg, K. K. Tan, "Hybrid-source impedance network and its generalized cascading concepts", International conference on power electronics and drive systems, pp. 1233-1238, 2009.

[11] J. Zakis, D. Vinnikov, I. Roasto, T. Jalakas, "Practical design guidelines of qZSI based step-up DC/DC converter", unpublished.

New Converter for Interfacing PMSG based Small-Scale Wind Turbine with Residential Power Network

Lauris Bisenieks[1,2], Dmitri Vinnikov[1], Ilya Galkin[2]

[1]Department of Electrical Drives and Power Electronics, Tallinn University of Technology, Tallinn, Estonia
[2]Institute of Industrial Electronics and Electrical Engineering, Riga Technical University, Riga, Latvia
lauris.bisenieks@rtu.lv, dmitri.vinnikov@ieee.org, ilja.galkins@gmail.com

Abstract-**This paper presents a new converter topology for interfacing a permanent magnet synchronous generator based variable speed wind turbine with a residential power network. The theory of wind energy conversion is analyzed first. Then an example of wind velocity distribution and normalized energy yield is discussed in order to formulate requirements for choosing a power converter. A new topology of an interfacing converter is analyzed and simulation results of a lossless model are presented. Simulation results of the proposed converter prove that its utilization in wind power applications is beneficial.**

I. INTRODUCTION

Sustainability is the main aspect that forces the renewable energy sources to be implemented for electric energy generation instead of fossil ones. Wind energy is quite attractive among other sources because of its commercial potential [72 TW] that is five times higher than world energy demand in all forms. However, the installed capacity in 2009 was only 159GW [1] and [2]. Large turbines play a main role on the market, but there is also demand for small turbines in the power range up to 11 kW as the power source for micro generators.

Micro generator is an electrical energy source that includes all interface units and operates in parallel with the distribution network. Current rating of such devices is limited up to 16 A per phase [3]. Some energy sources can be connected directly to the distribution network, but in the case of DC power sources or variable speed wind turbine (VSWT) systems it is necessary to use a power converter that interfaces the source and the grid.

VSWT based micro generators consist of a wind turbine, a generator and an inverter. Wind turbines capture wind energy and convert it to rotational mechanical energy. Variable speed operation of the wind turbine allows extraction of higher energy from wind than constant speed systems [4]. The generator converts mechanical energy into electricity. Different types of generators can be used in wind energy conversion systems (WECS), but permanent magnet synchronous generators (PMSG) play a main role on the market [5]. The main advantage of PMSG is the possibility of multipole design that offers slow speed operation and the possibility of gearless WECS construction. Another advantage is maintenance free operation since there are no brushes. The main drawback of PMSG is the dependence of

its output voltage on the rotation speed. The difference between the minimum and the maximum voltage can reach four times in VSWT applications [3]. This drawback can be easily overcome with the help of an appropriate interfacing converter.

The interfacing converter rectifies the input AC with variable voltage and frequency, adjusts voltage levels and inverts DC voltage into AC with grid voltage and frequency. Additionally, it should have maximum power point tracking (MPPT) functionality to extract more power from wind. Different topologies of the interfacing converter are discussed in the literature [4-19]. Basically they can be divided into two groups: topologies without galvanic isolation [16] (Fig. 1a) and those with isolation. Line frequency (LF) transformers (Fig. 1b) were widely used for galvanic isolation in last decades [9]. Main drawbacks of LF transformer are high weight and high price. For these reasons topologies with HF isolation (Fig. 1c) have became popular especially for photovoltaic applications [19], but there are only few topologies for wind applications studied in the literature [10,15].

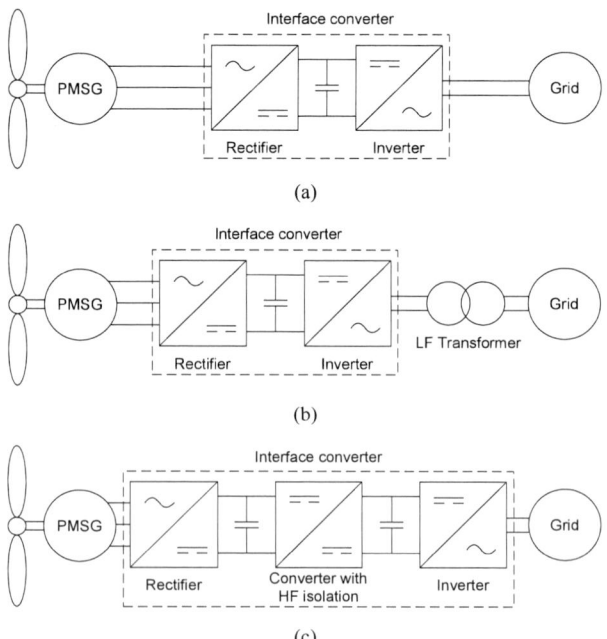

Fig. 1. Block diagrams of interface converters.

978-1-4244-8806-3/11 $26.00 © 2011 IEEE

All the topologies mentioned have a distinct DC link. It means that on the generator side there is a rectifier, but an inverter is placed on the grid side. Some low pass filter can be added to the inverter to fulfill standard requirements. Inverters have full bridge configuration in most cases because of lower DC link voltage. Since the inverter is well-known, this part of the interfacing converter would not be investigated in this paper.

The basic function of a rectifier is input voltage rectification, but it can ensure additional functions: power factor correction (PFC), DC link voltage stabilization and MPPT. Three-phase diode rectifiers are used in simple systems [11]. This configuration with diode rectifier is not suitable for VSWT since it cannot ensure necessary DC voltage level in all wind conditions. A boost converter can be added to this rectifier. This combination will improve the power factor and ensure appropriate DC voltage level, but extracted power is lower than with the controlled rectifier (inverter) [11].

Since the early presented converter topologies for PMSG based VSWT have several drawbacks, such us high complexity of dual LCL DC/AC converter [19] or insufficient voltage regulation capability of the isolated buck type converter with the uncontrolled rectifier [10], the new topology of the interfacing converter with the HF isolation transformer for PMSG based VSWT systems is presented in this paper. The topology presented has good voltage regulation capabilities at a relatively simple power circuit.

II. CHALLENGES OF PMSG BASED VSWT

This section introduces the properties of wind energy, emphasizing wind energy extraction by means of PMSG based VSWT. Operation modes of VSWT with fixed blades are analyzed and generator characteristics are given.

A. Wind turbine characteristics

Equation (1) gives the total power available in the wind, where A is the rotor area, ρ is the air density and v is the wind velocity.

$$P_{wind} = 0.5 A \rho v^3 \qquad (1)$$

Only a part of the total wind energy can be extracted. The available energy part in wind is described by the power coefficient Cp. The theoretical maximum value of this coefficient is 0.59 and it is called the Betz limit [7].

$$P_{turbine} = 0.5 C_p A \rho v^3 \qquad (2)$$

The practical values of C_p lie between 0.4 and 0.5 for industrial wind turbines [7]. This power coefficient is a function of the tip-speed ratio λ. An example of this function is shown in Fig. 2. The tip-speed ratio shows the relation between the circumferential velocity of the blade tips and the wind velocity:

$$\lambda = \frac{r\Omega}{v} \qquad (3)$$

where r is the rotor radius and Ω is the angular rotor speed.

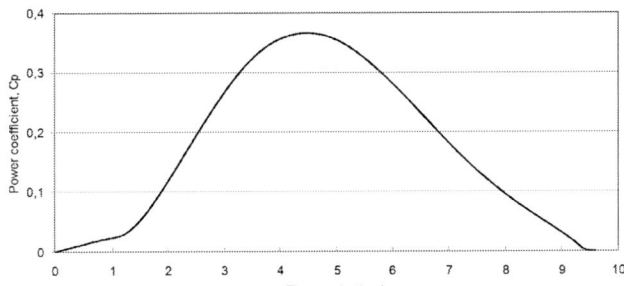
Fig. 2. Power coefficient Cp vs. tip-speed ratio.

Rotors are usually designed so that power coefficient C_p has the maximum values at the-speed ratio in the range from 4 to 8.

Since the coefficient C_p is the function of the tip-speed ratio the power extracted by the wind turbine depends on the wind velocity and the rotational speed. Power curves at different wind velocities for turbines with fixed blade position are shown in Fig. 3.

Fig. 3 indicates that the maximal power can be captured from wind turbines only if they are of a variable speed type. This figure illustrates also another feature of variable speed turbines: generator's speed is four times lower at the cut-in wind speed than at the rated velocity.

The wind velocity determines the rotational speed of the wind turbine and the generator. Since it has direct impact on power converter operation modes, an example of wind velocity distribution at 10 meter height is shown in Fig. 4, but the corresponding energy yield in Fig. 5.

Three distinct operating modes of the variable speed wind turbine generator can be emphasized: slow speed, rated speed and high speed. Slow speed occurs when the wind velocity lies in the range from 3m/s till 7m/s, rated speed - 7...8m/s and high speed mode is at higher velocities. This division of modes is made according to the normalized energy yield (Fig. 5). The rated speed corresponds to wind velocity with maximum energy.

These distributions are characteristic of Baltic coastal regions and so can be used as reference for interface converter design. The distribution of wind velocity per wind turbine modes is presented in Table I. It shows that the wind turbine is silent one quarter of the time and half of the time works at low speed.

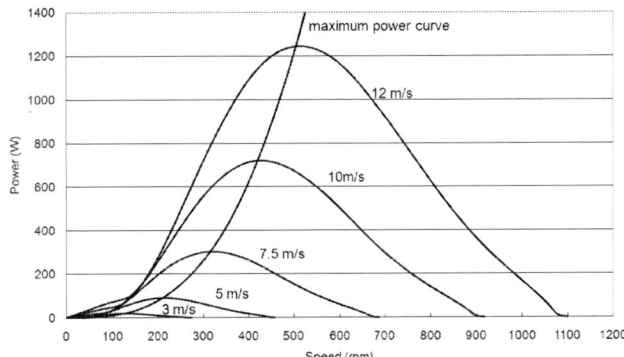
Fig. 3. VSWT power vs. rotation speed of turbine at different wind velocities.

978-1-4244-8806-3/11 $26.00 © 2011 IEEE

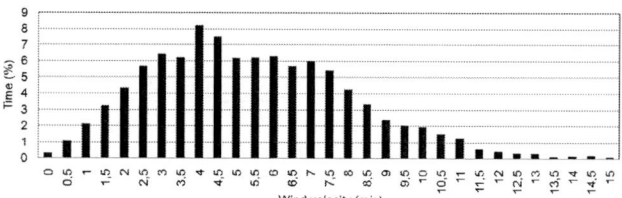

Fig. 4. An example of wind velocity distribution.

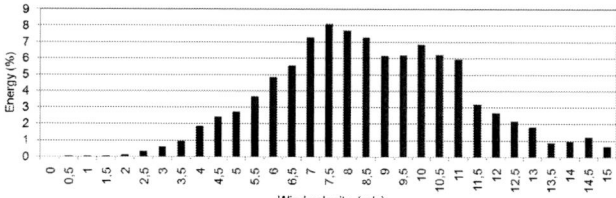

Fig. 5. An example of normalized energy yield.

TABLE I
WIND TURBINE MODES

	No speed	Slow speed	Rated speed	High speed
Wind velocity range	0-3	3.5-6.5	7-8	8.5-25
Time	23%	46%	16%	15%
Energy	0%	22%	24%	54%

B. PMSG based wind turbine

PMSG based VSWTs have three distinct operation modes: silent mode, variable speed operation mode and constant speed mode. A turbine is silent in two cases: wind speed is below a cut-in level or above the cut-off speed. If the speed is below its cut-in level it produces insufficient torque to move the turbine. At the same time winds above the cut-off level may damage the turbine that must be stopped at such conditions. A turbine usually starts to operate at 3 m/s and it should be stopped at the wind speed above 25 m/s [7]. Turbines operate at variable speed in the wind velocity range from cut-in to rated wind speed. Rated wind speed differs by turbine types, but often has the value of 12 meters per second. Constant speed mode takes palace above the rated wind speed. Turbine output power remains constant at this mode (Fig. 6).

PMSGs with 8 pole pairs are considered as a power source in this research. Its line voltage is 140V at 375 rpm, but it can operate up to 510 rpm. This speed is considered as the maximum power operational point for the turbine and the generator. Generator power reaches 1250W at this point, but the output voltage is 183V.

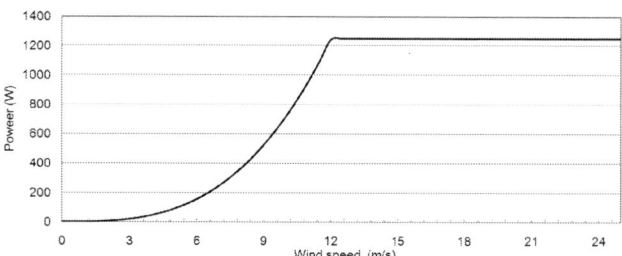

Fig. 6. Output power of VSWT.

Cut-in speed for a turbine is 125 rpm and it can produce 20W, but the generator voltage is only 48V at this point. So this is the lowest input voltage for a converter. Generator speed and voltage characteristics are shown in Fig. 7.

III. TOPOLOGIES OF INTERFACE CONVERTERS FOR PMSG BASED VSWT

Interface converter for a PMSG based VSWT can have different topologies. Topologies with HF isolation for VSWT application already known are studied in this section to evaluate their pros and cons in the context of the described PMSG utilization in such systems and a new interface converter topology with qZS DC/DC converter based HF isolation is offered.

A. Traditional topologies

There are only few converter topologies with high frequency isolation for small wind applications studied in the literature [10,15,19]. In [10] the authors have proposed a buck type isolated DC/DC converter (Fig. 8a). Variable generator voltage is rectified with a three-phase diode bridge firstly into proportional DC voltage. Stabilization of the second DC link voltage is obtained with the buck type isolated DC/DC converter by means of duty cycle variation. The main drawback of this solution is high currents in the transformer's primary winding at rated wind speeds that will reduce converter efficiency at this operating point. In [19] the authors have proposed a one-phase soft-switched dual LCL DC/AC converter (Fig. 8b). This converter utilizes a controlled rectifier for generator voltage rectification and DC link voltage stabilization. The soft-switched dual LCL DC/AC converter needs stable DC link voltage, so generator voltage should be boost up to its maximum voltage amplitude value in the whole input voltage range that will reduce the efficiency of the controlled rectifier at low generator speed.

B. New proposed topology

To improve the efficiency of the PMSG based VSWT system a new converter topology is introduced, presented in Fig. 8c. It consists of a three-phase full bridge controlled rectifier (generator side inverter) with PFC functionality and a quasi-Z-source (qZS) DC/DC converter with an HF transformer for galvanic isolation. Grid side inverters with output filters are not discussed here.

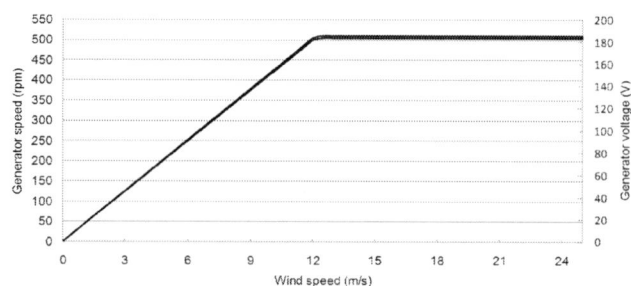

Fig. 7. Generator speed and voltage vs. wind speed.

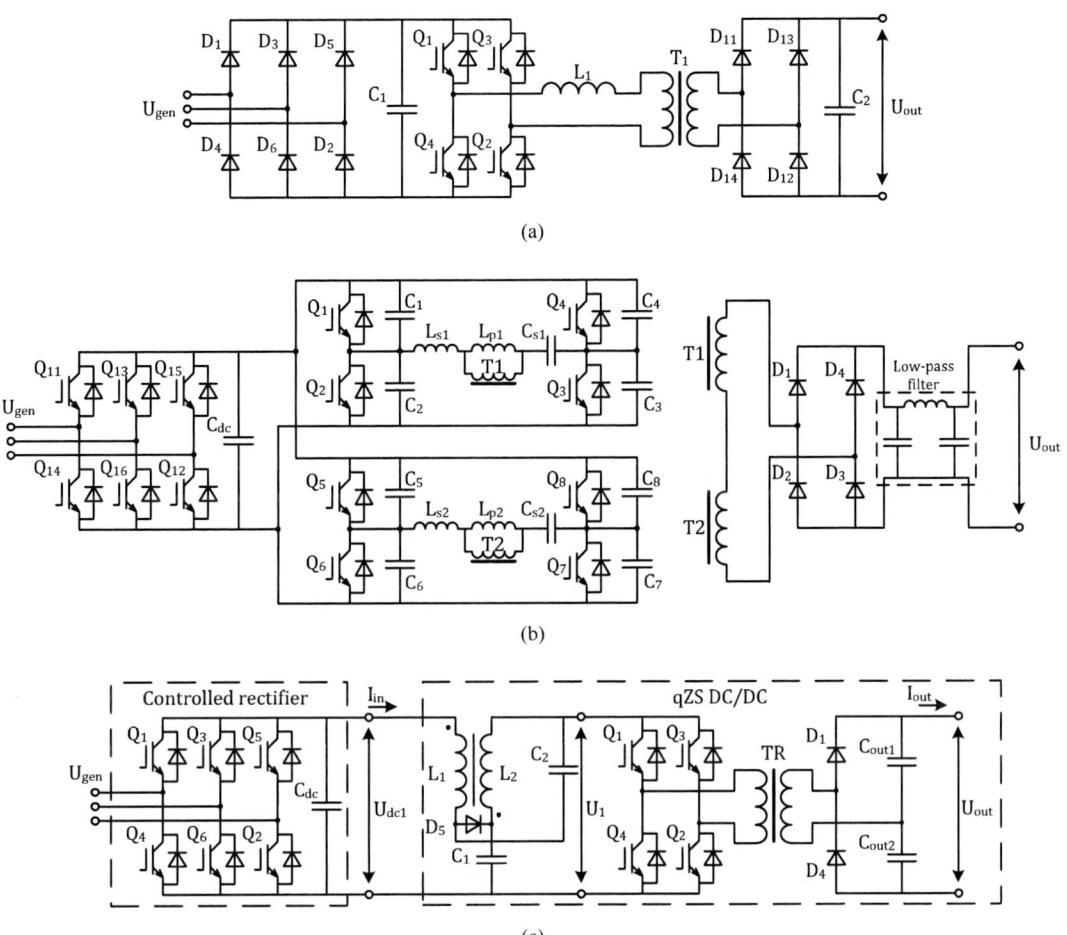

(a)

(b)

(c)

Fig. 8. Interfacing converter topologies: a) with a buck type isolated DC/DC converter, b) one-phase LCL DC/AC converter, c) proposed interface converter with qZS DC/DC.

A PFC inverter converts the variable voltage with variable frequency U_{gen} from the PMSG into a stabilized DC voltage U_{dc1}. The qZS DC/DC converter offers galvanic isolation and voltage level adjustment by means of the transformation coefficient. The unique qZS impedance network and appropriate control offer an additional voltage regulation capability at high efficiency. Stabilized DC link voltage U_{out} can be inverted into the grid current by an appropriate inverter.

IV. OPERATING PRINCIPLE OF THE NEW INTERFACE CONVERTER

This research attempts to prove the ability of the proposed topology to ensure stable grid side DC link voltage U_{out}. For this reason only the operating principles of the controlled rectifier and the qZS DC/DC converter will be studied in more detail. The power circuit of the qZS based DC/DC converter is shown in Fig. 8c, but operation modes in Fig. 9.

The voltage boost necessary is obtained by two steps. The PFC rectifier stabilizes the first DC link voltage U_{dc1} to a 150V level when the generator voltage is below 112V. The

DC link voltage U_{dc1} has the following relation with the generator line voltage U_{genf} in the PFC mode:

$$U_{dc1} = \frac{U_{zp-p} \cdot U_{genf} \cdot \sin \omega t}{U_r \cdot \sin \omega t}, \qquad (4)$$

where U_{zp-p} is the sawtooth voltage but U_r is the value of the reference signal amplitude.

The controlled rectifier works as a usual rectifier when the generator voltage U_{gen} is above 112V. In this mode the DC link voltage is changed proportionally to the generator voltage, at the range from 150V in rated speed conditions up to 250V at the maximal speed.

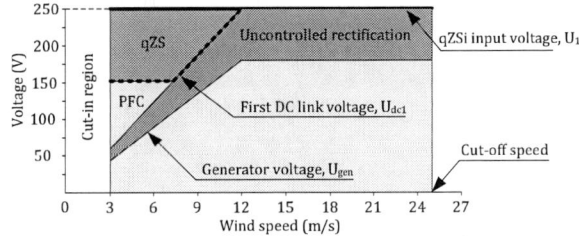

Fig. 9. Operation modes of a converter.

The qZS based HF isolation converter is stabilizing the HF inverter input voltage U_1 to 250V despite the voltage variations on the first DC link. The stabilized input voltage U_1 ensures inverter operation with the fixed duty cycle, thus ensuring constant volt second balance of the isolation transformer. The input voltage U_1 regulation is obtained by changing the shoot-through duty cycle [20]. The input voltage U_1 and the first DC link voltage U_{dc1} have the following relation:

$$U_1 = \frac{U_{dc1}}{(1 - 2 \cdot D_s)}, \qquad (5)$$

where D_s is the shoot-through duty cycle.

The shoot-through duty cycle is zero when the first DC link voltage U_{dc1} is equal to the necessary input voltage U_1. In such cases the HF inverter operates without shoot-through states.

V.   ANALYSIS OF SIMULATION RESULTS

Simulations of the controlled rectifier and the qZS DC/DC converter at different operation modes were performed to prove the functionality of the proposed topology. PSIM software was used for these simulations. Lossless models of the controlled rectifier and the qZS DC/DC converter were developed for the required simulations. Parameters of the developed models are summarized in Table II.

TABLE II
PARAMETERS OF PASSIVE COMPONENTS OF THE INTERFACE CONVERTER

PMSG phase resistance	1Ω
PMSG inductance	5mH
First DC link capacitance	470uF
Inductors of qZS network	600uH
Capacitors of qZS network	40uF
HF transformer turns ratio	1.25
Capacitors of voltage doubler rectifier	20uF

Simulation of the controlled rectifier was performed at three different operating modes: minimum generator voltage, edge of PFC operation and maximum speed conditions, taking into account available power. Fig. 9 illustrates controlled rectifier's ability to boost up generator voltage in cut-in speed conditions.

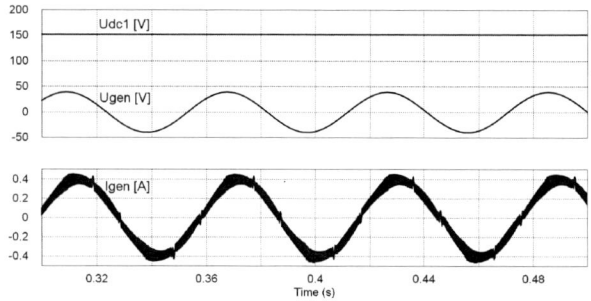

Fig. 9. Generator current I$_{gen}$ and rectifier output voltage U$_{dc1}$ at U$_{gen}$ = 48V.

Generator current is no more sinusoidal when the generator voltage U_{gen} reaches 112V (Fig. 10), but is not so distorted as in the case of pure rectifier operation (Fig. 11). Rectifier

simulation results shows that the voltage of the first DC link is kept in predefined limits in all generator operation modes.

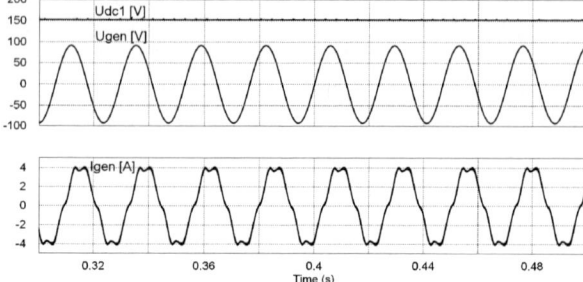

Fig. 10. Generator current I_{gen} and rectifier output voltage U_{dc1} at U_{gen} = 112V

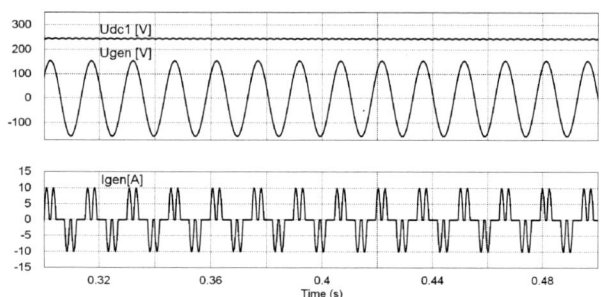

Fig. 11. Generator current I_{gen} and rectifier output voltage U_{dc1} at U_{gen} = 183V

Simulations of the qZS DC/DC converter with the voltage doubler rectifier were performed at two different modes. The first simulation was made in the rated conditions – the first DC link voltage $Udc1$ was 150V and the load of the voltage doubler rectifier was 300W. The first DC link and the HF inverter input voltage are shown in Fig. 12. It can be seen that the HF inverter input voltage has pulsations caused by shoot-through states, but amplitude is higher than first DC link voltage. Since active state is shifted away from shoot-through state, an amplitude voltage is attached to transformer's primary winding (Fig.13).

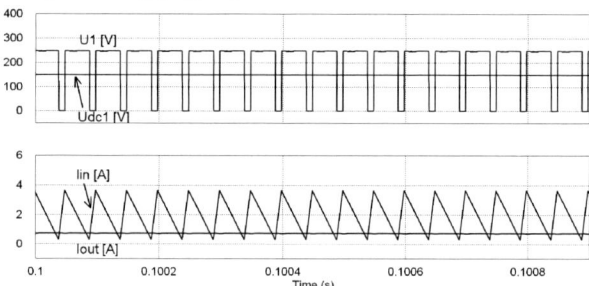

Fig. 12. Voltage and currents of qZS DC/DC converter in rated conditions

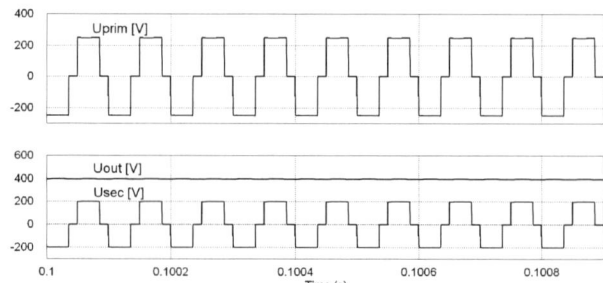

Fig. 13. Primary and secondary voltages of the HF transformer.

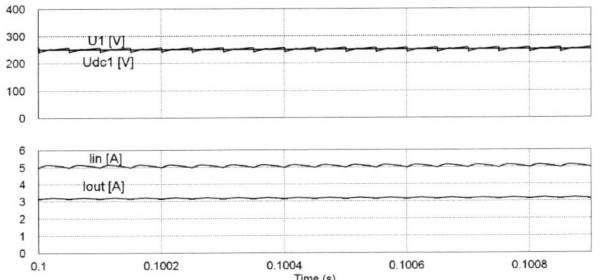

Fig. 14. Voltages and currents of the qZS DC/DC converter in maximum power conditions.

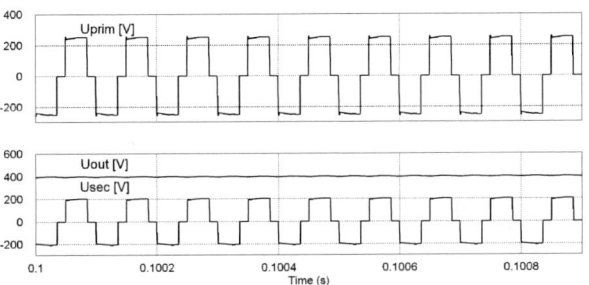

Fig. 15. Primary and secondary voltages of the HF transformer.

The second simulation was performed at the maximum generator voltage and 1250W load at the voltage doubler output. The qZS network input voltage U_{dc1} matches the HF inverter voltage U_1 in these conditions and there is no need for voltage boost. The HF inverter input voltage U_1 has small ripple due to the high transformer current (Fig.15). The ripple of the output voltage U_{out} is higher due to increased load.

VI. CONCLUSIONS AND FUTURE WORK

The study of the wind theory shows that a VSWT allows maximum power from air flow to be extracted. PMSG based VSWT characteristics and wind properties on the Baltic coastal regions were analyzed to define the converter operation modes. Analysis of the converter topologies studied earlier shows that they are not well suited for VSWT applications due to the high complexity of the parallel LCL DC/AC converter or low efficiency in the case of the isolated buck type converter.

For these reasons the new topology of PMSG based VSWT and grid interfacing converter is presented in this paper. The converter consists of a fully controlled rectifier, a quasi-Z-source DC/DC converter, a high frequency isolation transformer and a voltage doubler rectifier. Such topology offers good input voltage boost properties and a relatively simple power circuit.

Simulations of the converter were performed to verify its ability to ensure necessary output voltage at all modes.

The prototype of the proposed converter should be developed to obtain the experimental results and compare them with preliminary simulation results

ACKNOWLEDGMENT

This research work has been supported by Estonian Ministry of Education and Research (Project SF0140016s11), Estonian Science Foundation (Grant ETF8538) and Estonian

Archimedes Foundation (project „Doctoral School of Energy and Geotechnology II" and DORA5).

REFERENCES

[1] Cristina L. Archer and Mark Z. Jacobson. Evaluation of global wind power, 2005. Journal Of Geophysical Research, Vol. 110.

[2] World wind energy association. World Wind Energy Report 2009. [online].available:http://www.wwindea.org/home/images/stories/world windenergyreport2009_s.pdf.

[3] Requirements for the connection of micro-generators in parallel with public low-voltage distribution networks, EN 50438:2007.

[4] Tan, K.; Islam, S. "Optimum control strategies in energy conversion of PMSG wind turbine system without mechanical sensors" IEEE Transactions on Energy Conversion, vol.19, no.2, pp. 392- 399, June 2004.

[5] Arifujjaman, M.; Iqbal, M.T.; Quaicoe, J.E.; "A comparative study of the reliability of the power electronics in grid connected small wind turbine systems," Canadian Conference on Electrical and Computer Engineering, 2009. CCECE '09, pp.394-397, 3-6 May 2009.

[6] Yogesh M.; Kawale Mtech; Subroto Dutt; "Comparative study of Converter Topologies used for PMSG Based Wind Generation", Second International Conference on Computer and Electrical Engineering, Dubai, UAE, 2009, pp. 367-371.

[7] Manfred Stiebler. Wind Energy Systems for Electric Power Generation. Berlin: Springer-Verlag Berlin Heidelberg, 2008.

[8] Anderson, J., Peng, F.Z., "Four Quasi-Z-Source Inverters", 2008 IEEE Power Electronics Specialist Conference, PESC 2008, Rhodes, Greece, 2008, pp. 2743-2749.

[9] Yanto, H.A.; Chun-Ta Lin; Jonq-Chin Hwang; Sheam-Chyun Lin; "Modeling and control of household-size vertical axis wind turbine and electric power generation system," International Conference on Power Electronics and Drive Systems, 2009. PEDS 2009., pp.1301-1307, 2-5 Nov. 2009.

[10] Darbyshire, J.; Nayar, C.V. "Modelling, simulation and testing of grid connected small scale wind systems," Australasian Universities Power Engineering Conference, 2007. AUPEC 2007, pp.1-6, 9-12 Dec. 2007.

[11] Pathmanathan, M.; Tang, C.; Soong, W.L.; Ertugrul, N.; "Comparison of power converters for small-scale wind turbine operation," Australasian Universities Power Engineering Conference, 2008. AUPEC '08, pp.1-6, 14-17 Dec. 2008.

[12] Yanto, H.A.; Chun-Ta Lin; Jonq-Chin Hwang; Sheam-Chyun Lin; , "Modeling and control of household-size vertical axis wind turbine and electric power generation system," International Conference on Power Electronics and Drive Systems, 2009. PEDS 2009. pp.1301-1307, 2-5 Nov. 2009.

[13] Fujin Deng; Zhe Chen; , "Power control of permanent magnet generator based variable speed wind turbines," International Conference on Electrical Machines and Systems, 2009. ICEMS 2009, vol.pp.1-6, 15-18 Nov. 2009.

[14] Zhang, S; Tseng, K; Vilathgamuwa, M.; Nguyen, D.; "Design of a Robust Grid Interface System for PMSG-based Wind Turbine Generators," IEEE Transactions on Industrial Electronics, vol.PP, no.99, pp.1-1.

[15] Li, X.; Bhat, A.; "Multi-cell operation of a high-frequency isolated DC/AC converter for grid-connected wind generation applications," International Conference on Industrial and Information Systems (ICIIS), 2009, pp.169-174, 28-31 Dec. 2009.

[16] Dehghan, S.M.; Mohamadian, M.; Varjani, A.Y.; "A New Variable-Speed Wind Energy Conversion System Using Permanent-Magnet Synchronous Generator and Z-Source Inverter," IEEE Transactions on Energy Conversion , vol.24, no.3, pp.714-724, Sept. 2009.

[17] Arifujjaman, M.; Iqbal, M.T.; Quaicoe, J.E.; "A comparative study of the reliability of the power electronics in grid connected small wind turbine systems," Canadian Conference on Electrical and Computer Engineering, 2009. CCECE '09, pp.394-397, 3-6 May 2009.

[18] Shchur, I.; "Impact of nonsinusoidalness on Efficiency of alternative electricity generation systems," International School on Nonsinusoidal Currents and Compensation (ISNCC), 2010, pp.218-223, June 2010.

[19] Xiaodong Li; Bhat, A.; "A phase-modulated high-frequency isolated dual LCL DC/AC converter," Energy Conversion Congress and Exposition, 2009. ECCE 2009 IEEE , pp.350-357, 20-24 Sept. 2009.

[20] Vinnikov, D.; Roasto, I.; Zakis, J.; Strzelecki, R.; "New Step-up DC/DC Converter for Fuel Cell Powered Distributed Generation: Some Design Guidelines", Przeglad Elektrotechniczny, 86 245-252.

Energy-Efficient High-Voltage Switch Based on Parallel Connection of IGBT and IGCT

Andrei Blinov[1], Dmitri Vinnikov[1], Volodymyr Ivakhno[2]

[1]Tallinn University of Technology (Estonia), [2]Kharkiv Polytechnical Institute (Ukraine)

andrei.blinov@ieee.org

Abstract- **This paper presents an analysis of a hybrid high-voltage switch based on the parallel connection of IGBT and IGCT. The proposed configuration allows combining the advantages of both semiconductors, resulting in substantially reduced power losses. Such energy efficient switches could be used in high-power systems where decreased cooling system requirements are a major concern. The operation principle of the switch is described and simulated and power dissipation is estimated at different operation conditions.**

I. INTRODUCTION

High power densities together with a high functionality are the key aspects of modern power electronics. Further requirements are decreased volume and weight of the power systems as well as low cost. In order to fulfil these demands high switching frequencies of the semiconductors are necessary. Insulated gate bipolar transistors (IGBTs) are the major representatives in present day's medium- and high voltage electronics. In terms of blocking voltages (up to 6.5 kV) these devices have reached a level which can satisfy the majority of needs. The major advantages of IGBTs are easy driving and snubberless operation [1]. On the other hand, the switching behaviour of low voltage class IGBTs (<1 kV) is generally slower in comparison to MOSFETs, and high voltage class IGBTs (>3.3 kV) generally have higher conduction losses than GTO and IGCTs. In order to improve the performance of IGBTs, different approaches and methods were introduced and developed. For instance, at lower voltages, increased performance was achieved by a parallel IGBT-MOSFET-combination as shown in [2]. The hybrid integration of a unipolar and a bipolar power semiconductor in parallel allowed combining of their advantages whilst avoiding their disadvantages [3]. However, these positive results were observed only for certain applications and operation parameters.

Similarly, for high power applications the performance of high-power switches could be increased by a parallel connection of IGBT and IGCT switches [4-6]. This paper will focus on 4.5 kV class switches, since both IGBT and IGCT type semiconductors in press-pack type housings are commercially available, allowing easy connection of these devices in series by special cooling systems. The rated permanent DC voltage for both semiconductor devices is generally 2.8 kV. Using two- or three-level topologies, if necessary, this is sufficient to cope with the requirements of many traction and industrial applications with voltage ratings of 2.0-5.6 kV without the need of series connection of several semiconductors. Comparing parameters of two 4.5 kV class press-pack semiconductors: T0900EA45A-Westcode (Table 1 [7]) and 5SHY35L4512-ABB (Table 2 [8]), it could be observed that the on-state voltage U_T of IGCT is lower than the corresponding parameter $U_{CE(sat)}$ of IGBT. The turn-on behaviour is similar for both devices, while turn-off behaviour of IGCT is distinctly slower, which results in greatly increased losses during turn-off (Fig. 1).

The idea is based on the integration of positive properties of gate-commutated thyristors in terms of low turn-on and on-state power losses as well as high surge current capability and IGBTs with their relatively low losses during turn-off. This may allow creating high-voltage and high-current energy-efficient switches with increased switching frequency, which could be advantageous in high-power (>500 KVA) industrial and railway traction systems.

TABLE I
CHARACTERISTIC VALUES OF 900 A 4500 V IGBT (T0900EA45A)

Parameter	Symbol	Value
Collector-emitter voltage	U_{CE}	4500 V
Permanent DC voltage	U_{DC}	2800 V
Collector-emitter saturation voltage (I_C=900 A)	$U_{CE(sat)}$	4.7 V
Turn-on delay time	$t_{d(on)}$	1.6 µs
Rise time	t_r	2.3 µs
Critical rate of rise of diode current	dI_r/dt_{cr}	2000 A/µs
Turn-off delay time	$t_{d(off)}$	1.2 µs
Fall time	t_f	1.2 µs
Turn-off energy (I_C=900 A)	E_{off}	2.6 J

TABLE II
CHARACTERISTIC VALUES OF 4000 A 4500 V IGCT (5SHY35L4512)

Parameter	Symbol	Value
Peak off-state voltage	U_{DRM}	4500 V
Permanent DC voltage	U_{DC}	2800 V
On-state voltage (I_T=900 A)	U_T	1.15 V
Turn-on delay time	$t_{d(on)}$	3.5µs
Rise time	t_r	1 µs
Critical rate of rise of current	dI_r/dt_{cr}	1000 A/µs
Turn-off delay time	$t_{d(off)}$	11 µs
Turn-off energy (I_T=900 A)	E_{off}	6-8 J

II. OPERATION PRINCIPLE

The structure of the proposed hybrid switch (HS) configuration is presented in Fig. 2. The HS consists of a parallel connected asymmetrical press-pack IGCT and press-pack IGBT with an integrated freewheeling diode (FWD).

In the following analysis the HS is assumed to be operated in voltage - source inverter (VSI) circuits. The test circuit shown in Fig. 3 represents the main events that could occur in

978-1-4244-8806-3/11 $26.00 © 2011 IEEE

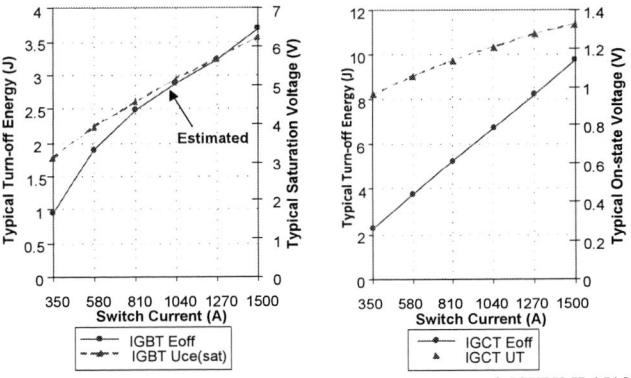

Fig. 1. Side-by side comparison ofT0900EA45A IGBT and 5SHY35L4512 IGCT on-state voltages and turn-off energies vs. current

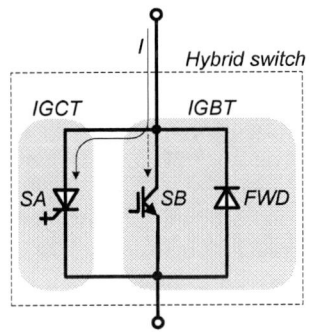

Fig. 2. Proposed hybrid switch configuration

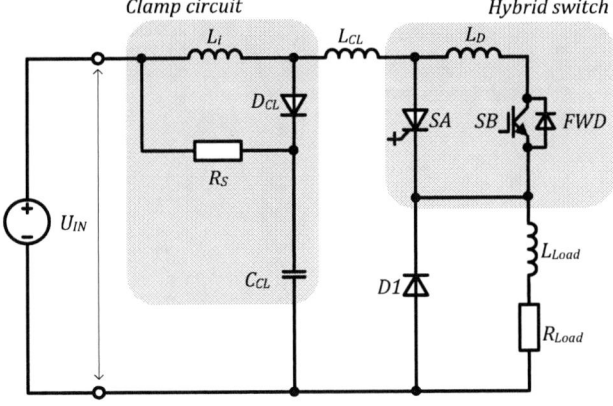

Fig. 3. Configuration of the commutation circuit

VSI topologies and includes the clamp circuit, hybrid switch, D1 (representing FWD of the opposite HS) and inductive load. The inductances L_{CL} and L_D represent the stray inductance of the clamp and the stray inductance between the IGCT and IGBT housings, respectively. The values of these inductances should be minimized in order to meet the specified SOA of the devices. The clamp circuit typically used in IGCT applications limits the surge reverse-recovery current of the turning-off freewheeling diodes and generally consists of a dI/dt limiting inductor L_i, a clamp capacitor C_{CL}, a clamping diode D_{CL} and a resistor R_S. In the case of a failure, the clamp inductance limits the short circuit current as well.

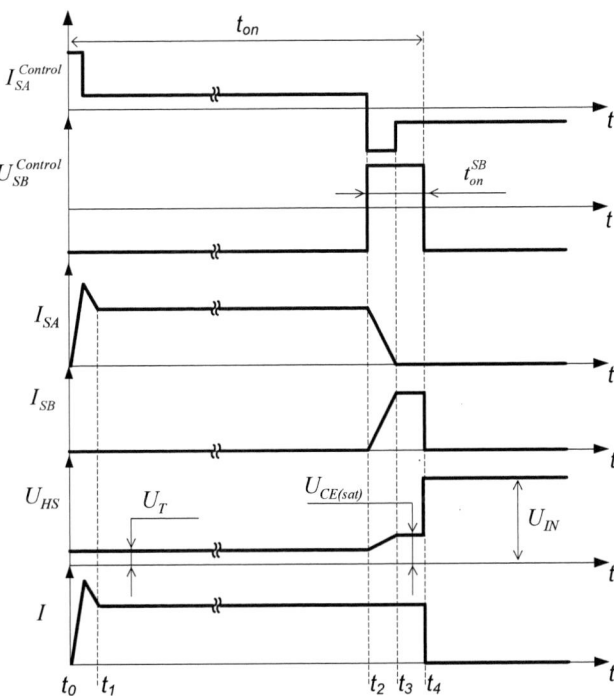

Fig. 4. Generalised operation principle and switching waveforms of the proposed HS

The generalised HS operation principle is shown in Fig. 4 and the following time intervals during the operation period can be distinguished:

t_0 – the beginning of each switching period of PWM. The thyristor SA of the HS is turned on by the control signal, applying full load current. During this time the transistor of the HS is turned off.

t_0-t_1 – freewheeling diode reverse-recovery process, duration and behaviour are dependent on the diode type and dI/dt.

t_1-t_2 – thyristor is conducting with low losses. The voltage across the HS determined by the voltage drop across the thyristor U_T.

t_2 – the turn-off control impulse is applied to the thyristor and simultaneously the turn-on impulse is applied to the transistor SB of the HS.

t_2-t_3 – as the turn-on behaviour of the IGBT is faster than the turn-off transient of the IGCT, the thyristor turn-off process occurs when the transistor is already in the on-state. The load current is distributed between both semiconductors.

t_3 – the SA returns to the blocking state, the full load current is applied to the transistor SB. Hence, the turn-off transient of the thyristor occurs when the voltage is limited to the voltage drop $U_{CE(sat)}$ across the conducting transistor SB of the HS. Moreover, during the current transfer to the transistor the voltage across its terminals is limited to the voltage drop across the SA during the on-state. The required duration of the

978-1-4244-8806-3/11 $26.00 © 2011 IEEE

transistor on-state should not be shorter than the turn-off transient of the thyristor.

t_4 – the turn-off of the HS occurs by applying negative gate voltage to the transistor after the thyristor returns to the blocking state. The turn off transient of the HV IGBTs is generally 2…7 µs. After the transistor is switched off, the voltage across HS and all its components become equal to the supply voltage.

III. SIMULATION MODEL

A. Simulation circuit

To simulate the HS operation the commutation circuit shown in Fig. 3 was modelled in PSpice software using idealised switch models. The same diode model was used in the topology for simplicity and the following simulation parameters were assumed: the input voltage is 2800 V, the maximum load current is 750. The values of the circuit's passive components are determined according to [9].

B. Control algorithm

Active states are generated using two-phase shifted triangle waveform generators operating with constant frequency and duty cycle and two comparators. Additional logic elements ensure that the SB is turned on right at the instant the turn-off of SA occurs (Fig. 5).

In real conditions the minimal and maximal duty cycle of the HS could be limited by a number of factors, such as IGCT gate driver limitations or behaviour of the clamp circuit and a turn-off snubber (if used). The minimal off-state time $t_{off(min)}$ should be maintained to stay within the safe operating area (SOA) of the circuit's components. The minimal on-state time $t_{on(min)}$ of the HS is generally not limited since during an operation with duty cycles near zero, only the transistor of the HS could be used. The example of SOA control flowchart is shown in Fig. 6.

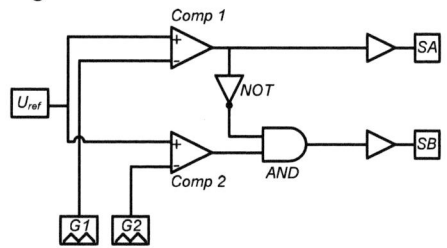

Fig. 5. Control algorithm of the HS

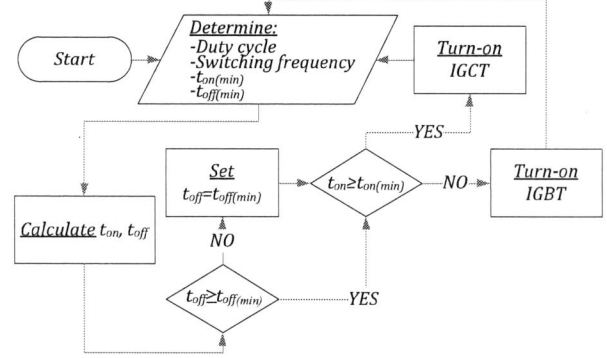

Fig. 6. Generalised SOA control flowchart of HS

The simulations confirm the estimated behaviour of the proposed switch configuration. At turn-on the HS operates like an IGCT with the *dI/dt* clamp (Fig. 7). The on-state voltage of the HS is equal to the voltage drop across the thyristor during its conducting period (Fig. 8). During turn-off of the HS the transistor is turned on for a short period; the turning-off thyristor current is then transferred to the transistor, which is closed right after the thyristor current becomes zero. The turn-off dynamics of the HS are greatly increased, while the excellent on-state characteristics of IGCT remain (Fig. 9) and all the elements are operated within the SOA.

IV. GENERALISED LOSS EVALUATION

The power loss estimation is one of the key points, crucial for the circuit mechanical structure and the cooling system design. The aim is to compare the power losses and performance of the proposed switch configuration with transistor- and thyristor-only counterparts at similar operation parameters. In the comparison, a variable switched current of $350A < I < 900A$ and $U_{DC} = 2800$ V is assumed. The total losses P_{tot} are calculated as the sum of conduction, turn-on and turn-off losses at maximum junction temperature (125°C) using the datasheet values of the devices.

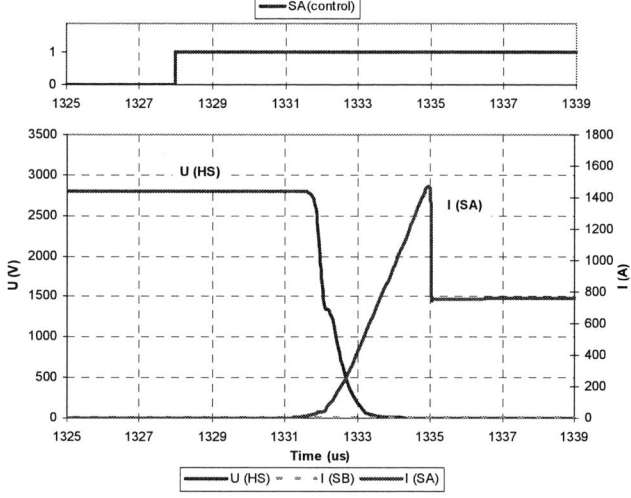

Fig. 7. Simulated turn-on behaviour of HS at I=750 A, U_{DC}=2800 V

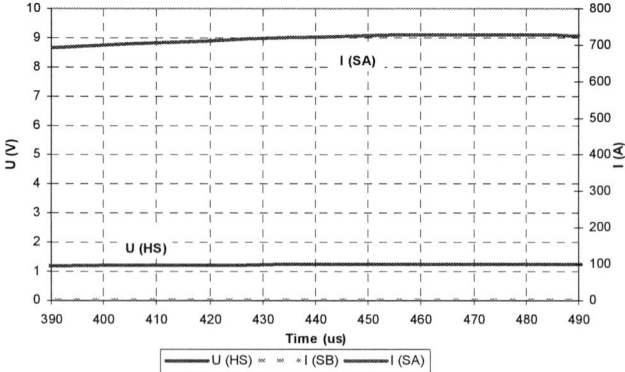

Fig. 8. Simulated conduction behaviour of HS at I=750 A, U_{DC}=2800 V

In the simulations of losses, the minimum IGBT switching losses with a very small gate resistances of R_{Gon}=4 Ω and R_{Goff}=2.5 Ω are assumed. In real industrial converters the IGBT gate units are adjusted to generate the desired dI/dt and dU/dt to avoid large voltage and current spikes during transients. However, the use of the gate resistor to control the dI/dt results in substantially higher switching losses in IGBT [10]. If a dI/dt limiting turn-on snubber is used with both IGBT and IGCT devices, the turn-on losses would be similar [11]. On the other hand, the turn-off losses of the device may increase slightly [12].

The turn-off losses of the IGCT were excluded in the simulations; however, according to the test results presented in the previous papers [13], the turn-off losses may not be completely removed due to several factors. Firstly, for a large area device, such as the IGCT, a significant output capacitance must be charged in order to establish the depletion region to support voltage. Another factor is the free carriers which had not recombined being swept from the junction. Nevertheless, an 89% reduction in turn-off losses was reported in [14]. In real conditions, the power losses of industrial applications could be distinctly higher than the simulated values.

After the turn-on of the IGBT, the current distribution between conducting IGBT and IGCT is mainly influenced by different characteristics of the semiconductors, temperature differences and asymmetrically distributed stray inductances in the circuit [15][16] Assuming both semiconductors in conducting state, the current sharing inside the HS neglecting cell resistances and inductances can be calculated by

$$k_I = \frac{I_{SA}}{I_{SB}} = \frac{U_{CE(sat)}(I)}{U_T(I)} \qquad (1)$$

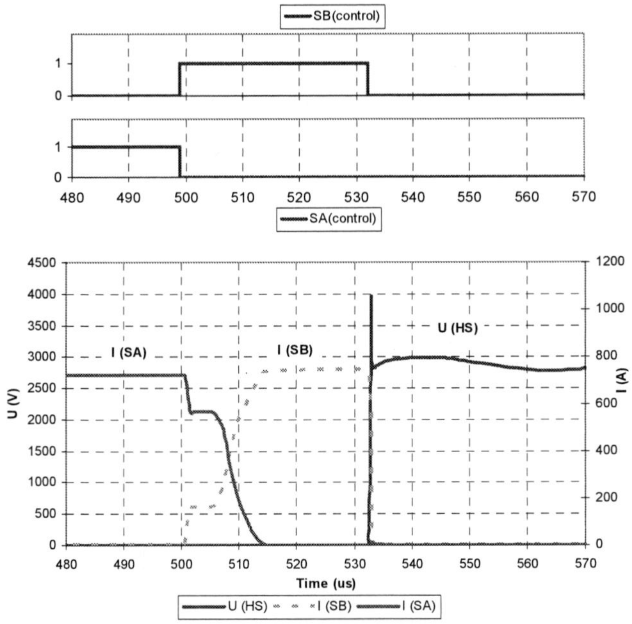

Using Eq. (1) the IGBT and IGCT currents could be obtained by

$$I_{SB} = I \cdot \frac{1}{1+k_I} \qquad (2)$$

$$I_{SA} = I \cdot \frac{k_I}{1+k_I} \qquad (3)$$

According to simulations, the considered IGCT is showing better dynamics for currents above 650 A, whereas the IGBT is performing better at lower currents (Fig. 10). The proposed switch configuration is estimated to provide 2.3...2.8 times increased switching frequency in comparison to single hard switched IGBT or IGCT with dI/dt clamp circuit exhibiting the same power dissipation of 3 kW. Assuming the same switching frequency in the range of 250...1050 Hz and switch current of 750 A the IGBT performs better than IGCT at frequencies above 450 Hz, whereas the HS provides substantial (1.9...2 times) decrease in power losses in comparison to single semiconductors (Fig. 11). Fig. 12 shows average losses of all considered switch solutions operating in the studied circuit with the wide range of duty cycles. The IGBT performs better than IGCT up to D=0.85. Again, the HS shows substantially (1.8...2.2 times) reduced power dissipation in comparison to single semiconductors.

Unlike in the case of the typical parallel connection of identical semiconductors, in the proposed HS both switches are conducting full input current during the operation, thus the current rating of both semiconductors must be sufficient. On the other hand, the overall power dissipation is decreased in comparison with single switches allowing one to increase the switching frequency or reduce cooling system requirements. Moreover, if one of the semiconductors fails, the other one can still continue to operate independently unless sufficient cooling is applied.

The economical feasibility of the HS implementation greatly depends on the application and its operation conditions. The comparison of semiconductor prices of discussed switch configurations is shown in Fig. 13. It should be mentioned, that semiconductor price is only a part of the overall power electronic system. The prices of the passive components greatly vary for different applications and are not considered in this paper.

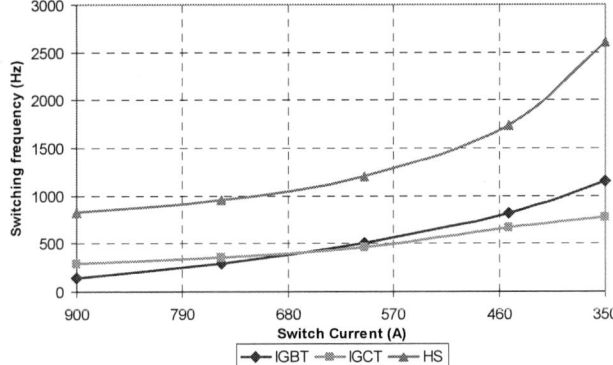

Fig. 10. Switch switching frequency vs. current for different semiconductor configurations corresponding to 3 kW total power dissipation at U_{DC}=2800 V, D=0.5

Fig. 9. Simulated turn-off behaviour of HS at I=750 A, U_{DC}=2800 V

978-1-4244-8806-3/11 $26.00 © 2011 IEEE   363

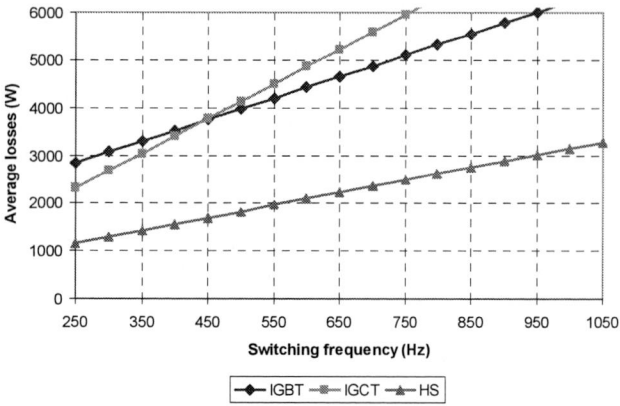

Fig. 11. Switch power dissipation vs. switching frequency for different semiconductor configurations at I=750 A, U_{DC}=2800 V, D=0.5

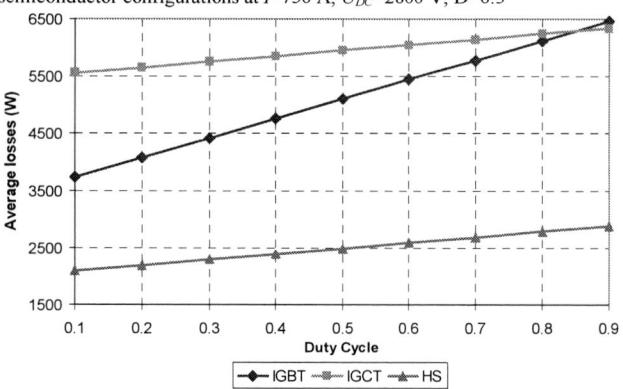

Fig. 12. Switch power dissipation vs. duty cycle for different semiconductor configurations at I=750 A, U_{DC}=2800 V, f_{sw}=750 Hz

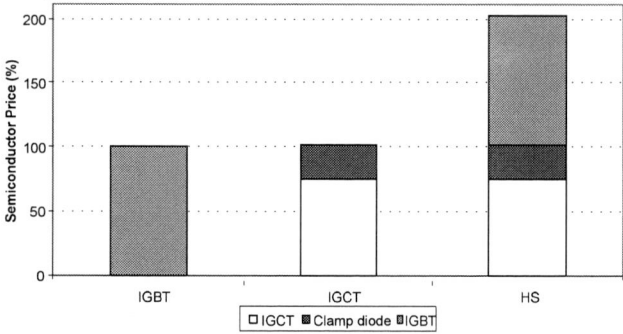

Fig. 13. Comparison of semiconductor prices of studied switch configurations

V.   CONCLUSION

Using commercially available 4.5 kV class IGBTs and IGCTs in press-pack housings it is possible to create energy efficient switches with essentially decreased power losses. Despite having decreased maximum current capabilities in comparison with parallel connected identical transistors or thyristors and a higher price than single semiconductor switches, the proposed switch configuration could be beneficial in rolling stock converters or other applications where higher switching frequencies are required or decreased cooling system requirements are essential.

The future research will include construction of the hybrid switch prototype, improvement of the simulation model according to test results as well as investigation of the benefits it could provide in modern converter topologies.

ACKNOWLEDGEMENT

This research work has been supported by Estonian Ministry of Education and Research (Project SF0140016s11), Estonian Science Foundation (Grants 7425, ETF8020) and Estonian Archimedes Foundation (project „Doctoral school of energy and geotechnology II").

REFERENCES

[1] Jaecklin, A.A.; "Advanced power bipolar devices," *Bipolar/BiCMOS Circuits and Technology Meeting, 1998. Proceedings of the 1998* , vol., no., pp.61-66, 27-29 Sep 1998

[2] Hoffmann, K.F.; Karst, J.P.; , "High frequency power switch - improved performance by MOSFETs and IGBTs connected in parallel," *Power Electronics and Applications, 2005 European Conference*, pp.11

[3] Kaerst, J.P.; Hoffmann, K.F.; , "High speed complementary drive of a hybrid MOSFET and IGBT power switch," *Power Electronics and Applications, 2005 European Conference*, pp.9

[4] Goncharov, J.; Ivakhno, V.; Nikulochkin, S.; Kipenski, A.; Pedan, E.; "Thyritor-Transistor Switch for Traction Converters Supplying Permanent Current Morors," *Implementation of converters in electro energetics, traction and electro technological applications, proceedings of international scientific conference pp. 34-36 1984.* /in Russian/

[5] Panasenko, M.; Panasenko, N.; Hvorost, V.; "Energy-Efficient High-Current High-Voltage Switches and Phase Modules Based on Them", *Electro technics and Electro mechanics*, pp.24-29 no.5, 2007 /in Ukrainian/

[6] Kozachok, V.; Nikulin, V; Panasenko, N; "Two-Quadrant Power Switches for Reserve Traction PWM-Converters," The collection of proceedings of the Ukrainian state academy of railway traction. pp.159-168, 2009 /in Ukrainian/

[7] Datasheet Insulated Gate Bi-Polar Transistor Type T0900EA45A, Westcode; 20 Sep. 2006

[8] Datasheet Asymmetric Integrated Gate-Commutated Thyristor 5SHY35L4512, ABB; 3 May 2008

[9] Setz, T.; Lüscher, M.; "Applying IGCTs," ABB Switzerland Ltd Application note no. 5SYA2032-03, Oct 2007

[10] Bernet, S.; "Recent developments of high power converters for industry and traction applications," *Power Electronics, IEEE Transactions on* , vol.15, no.6, pp.1102-11 17, Nov 2000

[11] Motto, E.; Yamamoto, M.; "New High Power Semiconductors: High Voltage IGBTs and GCTs," *Powerex Inc., Youngwood, Pennsylvania, USA; Mitsubishi Electric, Power Device Division, Fukuoka, Japan*

[12] Alvarez, R.; Bernet, S.; Lindenmueller, L.; Filsecker, F.; , "Characterization of a new 4.5 kV press pack SPT+ IGBT in Voltage Source Converters with clamp circuit," Industrial Technology (ICIT), 2010 IEEE International Conference, pp.702-709, 14-17 March 2010

[13] Bruckner, T.; Bernet, S.; "Investigation of a high power three-level quasi-resonant DC-link voltage source inverter," *Applied Power Electronics Conference and Exposition, 2000. APEC 2000. Fifteenth Annual IEEE* , vol.2, pp.1015-1022 vol.2, 2000

[14] Motto, K.; Zhang, B.; Huang, A.Q.; "Characterization of IGCT under zero-current-transition condition" *Industry Applications Conference, 2001. Thirty-Sixth IAS Annual Meeting. Conference Record of the 2001 IEEE* , vol.3, pp.1490-1496 vol.3, 30 Sep-4 Oct 2001

[15] Hermann, R.; Bernet, S.; Yongsug Suh; Steimer, P.K.; , "Parallel Connection of Integrated Gate Commutated Thyristors (IGCTs) and Diodes," *Power Electronics, IEEE Transactions on* , vol.24, no.9, pp.2159-2170, Sept. 2009

[16] Wang, X.; Caiafa, A.; Hudgins, J.; Santi, E.; "Temperature effects on IGCT performance," *Industry Applications Conference, 2003. 38th IAS Annual Meeting. Conference Record of the* , vol.2, pp. 1006- 1011 vol.2, 12-16 Oct. 2003

Some Aspects of Blended Learning for Tallinn University of Technology and Tallinn Center of Industrial Education

E. Brindfeldt, A. Grinko, M. Müür

Department of Electrical Drives and Power Electronics, Tallinn University of Technology
Ehitajate tee 5, 19086 Tallinn (Estonia)
E-mail: Eduard.Brindfeldt@tthk.ee

Abstract - **This article describes new directions in the modernization of higher and vocational education courses of automation and mechatronics on the basis of the blended learning. The authors review methodological approaches for the blended technology design of training, their realization in the higher and vocational education.**

Training courses are being implemented in the regular automation and mechatronics programmes to university and vocational students. In the future we plan to adapt the work results in the industry. The structure and main components of the module for practical training in the laboratory are described.

Such an approach to learning can be successfully used in the following courses: sensors and actuators in mechatronics and automation, application of PLC in industrial and building automation.

I. INTRODUCTION

Today's Estonian labor market is very flexible and we cannot predict with a high degree of accuracy the structure of the labor force that is needed in economy. New trends in the meaning of a professional career emerging - predictability of results, continuity and confidence in the future, are gradually replaced by more mobile forms of employment. Professional boundaries are changed or disappear altogether, and an increasing number of workplaces are temporary [1].

Below are some situations that the higher education system is interacting at present. The changing nature of economic relations, the evolution of forms of organization and the use of labor leads to the need for structural changes in priorities and the content of training students.

The educational system lacks flexibility and adaptation to the changing conditions of the environment is still low. There is a need to build entrepreneurial skills - the ability and willingness to support themselves and others through the creation of new workplaces [2].

However, students respond to changing economic conditions actively and increasingly combine work and school, at the same time agreeing to unattractive working conditions. This has become a significant convergence model of activity in European countries [3].

In recent years, the educational system in Estonia has stepped into a period of stable development. This situation has been achieved through the adoption of important laws and deep integration with educational structures of the EU:

1. Estonian Parliament passed laws which ensure the functioning of all levels of the educational system;

2. The Estonian Qualification Authority (trademark - Kutsekoda) has developed a system of professional standards for the majority of specialties;

3. The National Examination and Qualification Centre has developed national curricula for specialties: Mechatronics, Automatics and Electronics.

According to the requirements the professional qualifications system [4] designed the Vocational Education Standard [5] that clearly defines the following:

1. Requirements for professional skills arise from the professional standards approved by professional councils.

2. The content of studies is determined by the requirements for professional skills set forth by the professional standards. In the absence of a professional standard, the content of study shall be coordinated with the relevant professional associations.

The content of the national and school curricula in the Vocational Education Standard is defined as follows:

1. The national curriculum is a document which determines the purposes and functions of vocational training, the requirements for starting and graduating from the studies, the modules of curricula and the volumes thereof together with short descriptions, the possibilities of and conditions for electing modules and possibilities of specialization.

2. The national curriculum shall be reviewed and if necessary, a new version shall be approved if the professional standard which constitutes the basis for the national curriculum is repealed, a new professional standard is established, or the name of the professional standard or the requirements for professional skills provided in the standard are amended.

3. The school curriculum is the source document of vocational training. Schools shall prepare a curriculum for each profession being taught and for every type of vocational training, basing such curricula on the Vocational Education Standard and the national curriculum, and taking account of different forms of study.

978-1-4244-8806-3/11 $26.00 © 2011 IEEE

In the transition to the new school curriculum it is in line and in strict compliance with the national curriculum. The school curriculum may include elective (of your choice) study modules.

This approach to vocational education has a clear focus on the constant interaction with the labor market. All changes in the standards of the professional qualifications system are the reason for the adjustment of the national and the school curriculum that provides flexibility for vocational education.

The main aim of Estonian education is to prepare human resources for innovation and to develop of creative abilities of the individual. To achieve this goal it is necessary to apply the innovative educational technologies.

Today there is a social order for such professionals who are able to innovate for all spheres of activity. New technologies in the educational system aimed at fulfilling the important tasks are:

1. development of cognitive activity of students;
2. teaching of innovators;
3. improving communication skills, etc.

II. IDEOLOGY

The modern approach involves mastering separate knowledge and skills in the complex. In this regard, competences of students are developed on the basis of fundamental principles, universality and practical orientation.

Important characteristics of professional competence for the expert in the field of automation and mechatronics are:

1. innovation - research, formulate and solve fundamentally new problems and tasks of automation and design;
2. efficiency - successful professionally activity and analysis of technical information;
3. mobility - a willingness to upgrade existing experience and knowledge to adapt to changing conditions;
4. perspective - a willingness to continue their education, self-improvement, professional and personal growth.

To implement these objectives it is most appropriate to use a mixed mode of education – blended learning. In modern terms blended learning is based on the use of effective "mix" of traditional and distance learning technologies and innovative pedagogical methods of teaching [4]. The meaning of blended learning is that distance learning technologies are actively used in support of full-time basic education. They are means of storing educational resources and their delivery systems, as well as a platform for communication and collaboration [5].

To design the structure and algorithm representation of an educational material it is very helpful to use the general structure chart of automated manufacturing systems applications (Fig. 2.1.)[6]. The developers are using information, energy and physical manufacturing technology together with modern development methods in order to produce an automated manufacturing system. People, automatic control systems and driven energetic (mechanical) equipment are integrated during the system development. New manufacturing equipment enables us to produce more complicated products. The technology used in manufacturing products is becoming more complex. In the information age the production workers involved in making products regularly participate in further training.

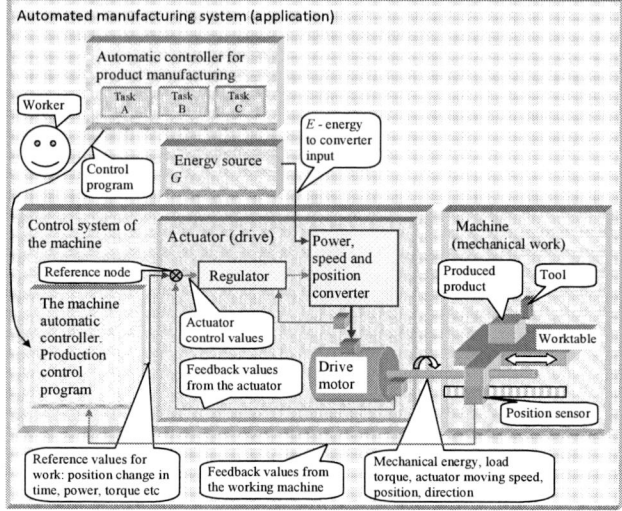

Fig. 2.1. Structure of an automated manufacturing system [6].

Blended learning allows teachers to use a web service, which includes the students into the processes of active learning and creative activities.

To create a course for blended learning it is necessary to:

1. create the structure of the module;
2. describe the content of the module structure;
3. upload and edit files on the site.

This method allows us to maintain constant contact with students and have an effective feedback.

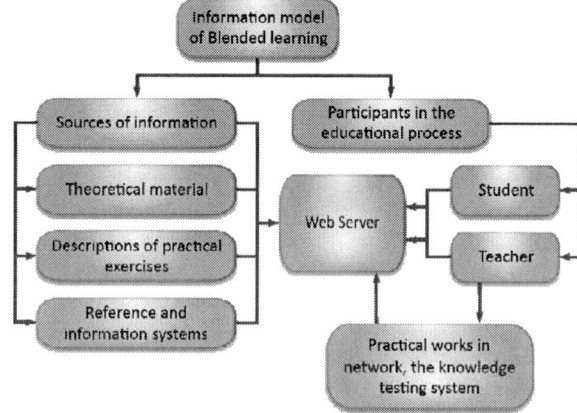

Fig. 2.2. The structure of the information model for blended learning.

Currently, there are quite large selections of various information models for blended learning [7]. As a result of their research and adaptation to the educational system of Estonia, resources currently available and most suitable are shown in Fig.2.2.

The course was created based on the following variants of combined elements in blended learning:

1. Parallel model - multichannel representation of educational material (in which students can choose the most convenient option for them);

2. Serial model - modular representation of educational material (when the content of discipline is divided into relatively independent parts and presented in different ways).

During testing of some elements of blended learning it was found that:

1. the highest efficiency is achieved online on discussion boards, and the maximum efficacy is achieved in the face-to-face sessions;

2. greater efficiency can be achieved through encouraging students to support each other through discussion forums;

3. more challenging issues were solved in face-to-face sessions with teachers.

Channels of communication used in the testing elements of blended learning in the form of a hierarchical pyramid of ascending effectiveness are shown in Fig. 2.3.

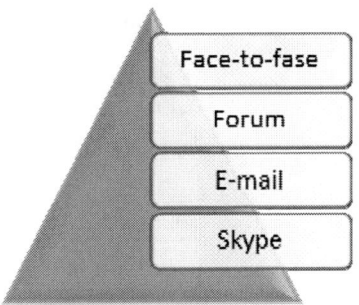

Fig.2.3. Communication channels on the course.

Development of new teaching methods is generally forward-looking and always involves a significant transformation of the existing method and technology of education. During the transformation commonly many new, unforeseen ways of solutions emerge. During the creation of courses with blended learning it is convenient to use the following sequence: "Analyzing product – Production process – Realization" [8].

At the beginning we must accurately define the requirements for a new way of learning. This is done using information obtained from the analysis of «the monitoring area» and resolving arising «mini-problems». After what we get the desired technology - a set of operations that must be done for the final product. After that during the operations it is fairly easy to use a large amount of information obtained from the analysis of «the monitoring area» and arising «mini-problems» resolving.

A feature of this approach is - upon receipt of each individual concept of mini-problem solution it is very useful to capture its main idea. In this case all situations are analyzed and after that we can choose the most appropriate solution, which is then used to build the new model of learning.

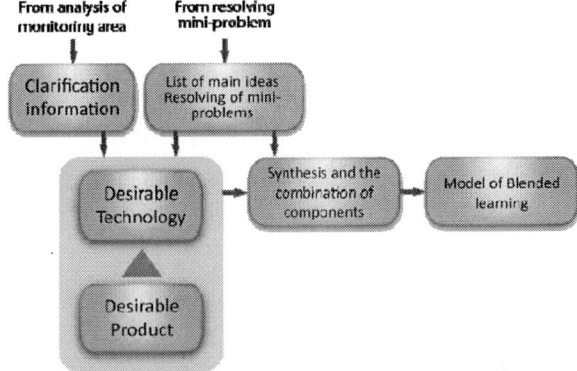

Fig. 2.4. Algorithm of blended learning course construction.

By performing the above actions, we receive the maximum efficient transformations that solve the initial problem and point to the most efficient ways of blended learning development (Fig.2.4).

III.   BASIC CONCEPTS OF BLENDED LEARNING FOR THE COURSE OF AUTOMATION

This work is partly based on some developments of the ongoing project which is supported by the European Commission under the "Leonardo da Vinci" Education and Culture Lifelong Learning Programme. Project participants are five educational and training establishments and enterprises from four different countries:

• Tallinn University of Technology (Estonia),
• Technical University of Gabrovo (Bulgaria),
• LUISS Guido Carli University (Italy),
• Multidisciplinary European Research Institute Graz (Austria),
• European Center for Quality Ltd. (Bulgaria).

The main goal of the project named "AutoMatic" is to develop innovative teaching and learning materials in the field of industrial automation. The materials will be used for teaching and training of employees in small and medium size enterprises, working people who want to find a job in this field, students, young graduates and unemployed people. Five main areas have been established for which the materials are to be created. These are:

1. Sensors in industrial automation,
2. Application of PLC in industrial automation,
3. Computer based means for automation,
4. Industrial networks and interfaces in industrial automation,
5. Actuators in industrial automation.

978-1-4244-8806-3/11 $26.00 © 2011 IEEE   367

As is well known, blended learning experience has the speed, scalability and efficiency, and many other benefits of e-learning. During the course and modules creation the right combination of methods to achieve the maximum effect of learning were very carefully defined. A feature of blended learning is its friendliness to students, but its application requires special treatment – it should not be used in cases where it is absolutely unnecessary.

We have considered the "Blended Learning" as the evolution of e-learning in the direction of integrated programmes with different methods of content delivery for solutions of educational problems. When creating an electronic and interactive content we used the method of "environment" and have achieved responsible e-learning programmes with realistic results (Fig. 3.1.).

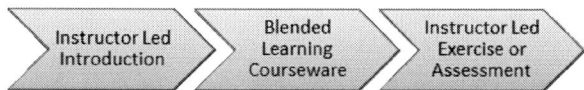

Fig. 3.1. The simplest approach.

During the course creation with the technology of blended learning, shortcomings of the traditional teaching of the scheme of "teacher-student-tutorial" and of e-learning courses were analyzed. The main stages of the development of blended learning courses were:

1.   collecting and preparation of materials;
2.   familiarization with the materials;
3.   training methods planning, to agree on the main elements of the course;
4.   methodological handling of the training material, preparation of texts and tests for the course;
5.   error correction;
6.   testing with a small group of students;
7.   maintenance and correction of errors discovered by the end user.

As a result, we received the following sequence of implementation of the course of Blended Learning (Fig. 3.2.).

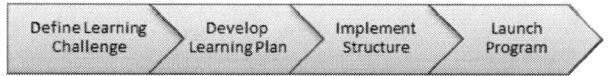

Fig. 3.2. The blended learning process.

During the development of the course the following factors interfering with successful results were taken into account:

1.   Lack of transparent and clear objectives of the course;
2.   Lack of a common concept of the course;
3.   Poor or absence of feedback;

4.   Boring exercises (repetitive testing tasks);
5.   Lack of emotions.

Among the major problems that were solved during the creation of the course are the following:

I.   development of prototype-based e-learning management software Moodle as a component of the information space of the school;
II.   working out and developing the functional facilities of the modules in the course;
III.   development of methods and educational methodical materials that support the process of blended learning.

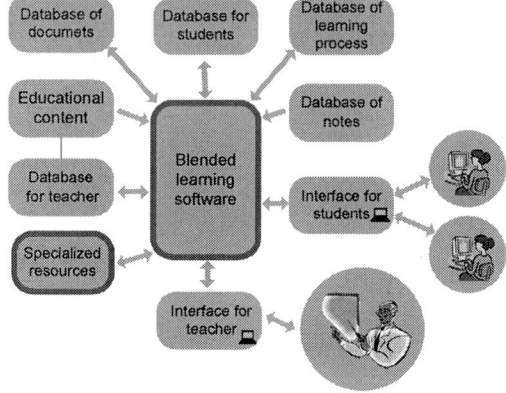

Fig. 3.3. General structure of a blended learning course.

The general structure of an integrated educational environment is shown in Fig. 3.3.

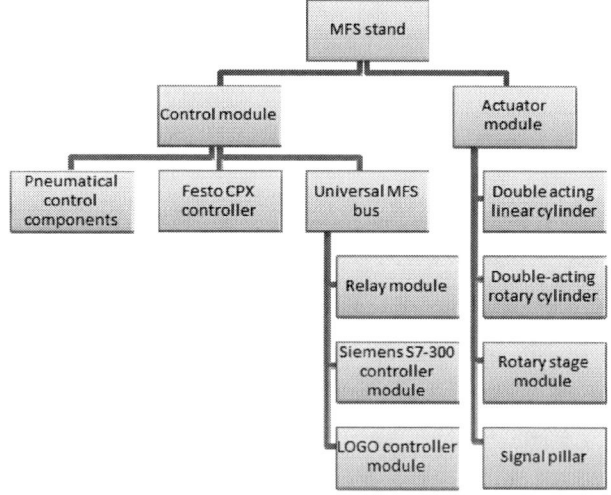

Fig. 4.1. MFS structure chart.

IV.   MODULE FOR PRACTICAL TRAINING IN A LABORATORY

This study material is prepared for students who learn the programmable automation controllers and actuators in an automation course. The multifunctional production automation stand (MFS) is one part of a blended learning system (Fig. 4.1.). It consists of two parts: the control module and the actuator modules. The control module itself is divided to different integrated control units: relay unit (Fig. 4.3.), LOGO controller unit and Siemens S7-300 controller unit (Fig. 4.4.). The actuator module consists of a pneumatically controllable input module, a placing module and an electrically controllable rotary stage module. Work piece feeding, repositioning and placing it to the rotary stage module take place in this module.

Control module enables the learner to get to know logic modules: installation of modules (construction and finding faults); programming (learning to use the software and programming instructions within it; doing programming exercises); debugging programs (Fig. 4.2).

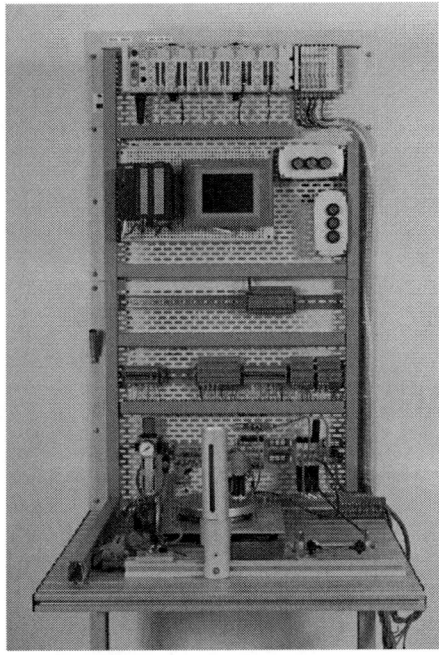

Fig. 4.2. The multifunctional stand.

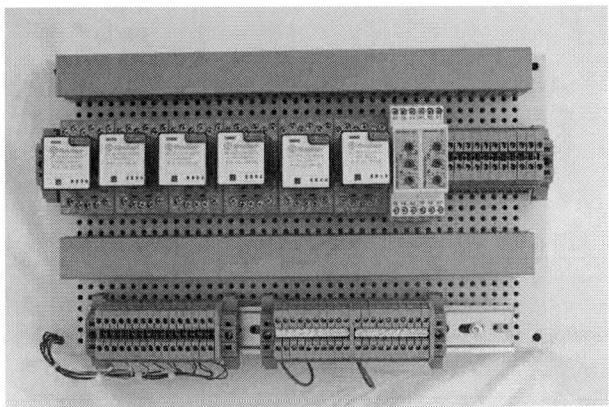

Fig. 4.3. Relay unit.

The following is learned about the programmable controllers:

1.   finding the build and installation errors;
2.   getting to know the software, setting hardware in the software,
3.   solving simple programming tasks (learning the programming instructions);
4.   program debugging, diagnostics and testing.

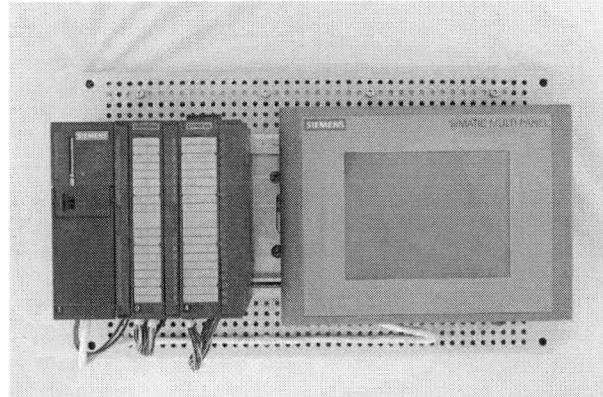

Fig. 4.4. Siemens S7-300 controller unit.

The following is learned about the text- and operator panels: text- and operator panels; installation (finding the building and installation errors); software (getting to know the software; object oriented programming; displaying variables, trends and moving pictures; solving simple tasks; testing).

Actuator module (Fig. 4.5.) enables learner to get to know the pneumatic control components and the actuators, the production automation systems and the installation of the automation system components.

Fig 4.5. Actuator module.

The control and the actuator modules together provide an excellent opportunity to get to know the automation systems: installation and troubleshooting; programming and tuning of flexible manufacturing modules; experimenting with flexible manufacturing.

978-1-4244-8806-3/11 $26.00 © 2011 IEEE

Installation of automation system components includes the following: installation of electrical, magnetic, optical, inductive and capacitive sensors, connecting them with measuring transducers or computers measuring interface; processing the measurement data with computers; software needed for processing of measurement data.

CONCLUSIONS

Blended learning is an effective training ground for research and testing of various new pedagogical solutions. The complex of activities done during the creation of blended learning has resulted in the following:

1. development of prototype-based e-learning management system Moodle as a component of the information space;

2. working out and developing the functionality of the modules of the course;

3. development of methods and teaching materials that support the process of blended learning;

4. strengthening of the role of independent cognitive activity of students;

5. introduction of blended learning technologies has resulted in the formation of the basis of funds for dynamically updated e-learning courses.

The final version of blended learning will be developed during the project "AutoMatic" (started on 01.10.2009 and to be finished on 30.10.2011) and will better organize the students' independent work.

REFERENCES

[1]   D. Stern , T. Bailey, and D. Merritt, "School-to-Work Policy Insights from Recent International Developments," *National Center for Research in Vocational Education, University of California at Berkeley*, January 1997.

[2]   "Helping to create an entrepreneurial culture - A guide on good practices in promoting entrepreneurial attitudes and skills through education," *European Commission*, 2006.

[3]   L. Chisholm, "Initial Transitions Between Education, Training and Employment in a Learning Society," *International Bulletin of Youth Research*, vol. 15 (1997), pp. 6 -16.

[4]   "The professional qualifications system," (in Estonian), [Online], Available: http://www.kutsekoda.ee/et/index.

[5]   "Vocational Education Standard," The National Examinations and Qualifications Center, *Regulation No. 90 of the Government of the Republic of 6 April 2006*, [Online], Available: http://www.ekk.edu.ee/vocational-education-standard.

[6]   E. Pettai, "Industrial automation," (in Estonian), *Technical University of Tallinn, Department of Electrical Drives and Power Electronics*, Tallinn, 2005, 336 p.

[7]   C. J. Bonk, and C. R. Graham, "The Handbook of blended learning: Global perspectives, Local designs," *John Wiley & Sons, Inc.*, 2006, 624 p.

[8]   N. Shpakovsky, H. J. Kim, E. Novitskaya, and V. Lenyashin, "Structural Scheme For Solving a Problem Using TRIZ," September 29, 2002, [Online], Available: http://www.gnrtr.com/Generator.html.

Three-phase Bidirectional Battery Charger for Smart Electric Vehicles

J. Gallardo-Lozano, M. I. Milanés-Montero, M. A. Guerrero-Martínez, E. Romero-Cadaval

Power Electrical and Electronic Systems (PE&ES), School of Industrial Engineering
(University of Extremadura), (http://peandes.unex.es)

Abstract- **A three-phase battery charger for electric vehicles is proposed in this paper. The charger is bidirectional, allowing the Charging and Vehicle to Grid operation modes. A novel Balanced Sinusoidal Source Current control strategy is proposed so that the charger demands or injects into the grid a perfect sinusoidal and balanced source current in phase with the positive sequence fundamental component of the phase-to-neutral grid voltage, achieving a unity displacement power factor. In this way, the charger turns the car into a smart vehicle, reducing the existing problem of harmonic current demand by electric and hybrid vehicles and improving the power quality of the electric power system. The topology and control stage of the charger are shown. Simulation tests are conducted to validate the proper operation of the charger under sinusoidal, distorted and unbalanced source voltages.**

I. INTRODUCTION

Hybrid Electric Vehicles (HEV) and Electric Vehicles (EV) are becoming more and more attractive due to the higher oil prices and the development of new battery technologies, such as Lithium-Ion, which have higher power and energy density. The battery charger converts the alternating current distributed by electric utilities into the direct current needed to recharge the battery. This mode of operation is known as Grid to Vehicle (G2V). However, a Vehicle to Grid (V2G) concept has arisen in the last years, allowing returning back to the grid the energy stored in the battery packs that has not been used by the vehicle. This operation mode will cause a great impact on the power grid. Specifications for V2G technology are still being developed by the Society of Automotive Engineers as part of the ZigBee Alliance [1]-[3].

At present there are commercial EVs with single-phase unidirectional G2V chargers, which demand a highly distorted current form the grid. As an example, the measured current demanded by a commercial car (Reva-i) while charging the battery is shown in Fig. 1(a) and the harmonic spectrum is displayed in Fig. 1(b). The Total Harmonic Distortion (THD) of the demanded current is over 20%, far exceeding the value permitted by the standard IEEE-519. If a proliferation of EVs is expected in the next years, it is very important to investigate in the control strategies of chargers in order to reduce the harmonic distortion demanded by the batteries or the power grid will suffer a power quality degradation.

Besides, the time required to recharge electric vehicle batteries depends on the total amount of energy that can be

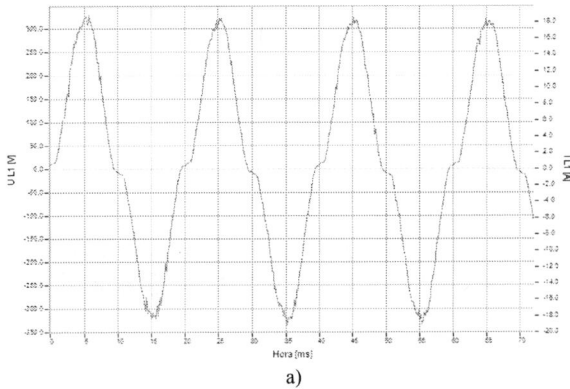

a)

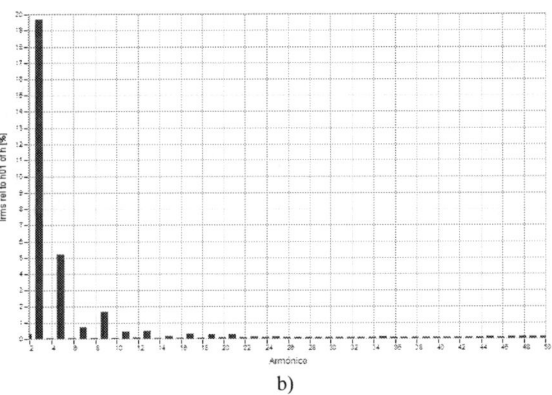

b)

Fig. 1. Demanded current to the grid by the electric vehicle Reva-I when charging the battery. (a) Waveform of the demanded current; (b) Spectrum of the demanded current (THDI = 20,475%).

stored in the battery pack, and the power available from the battery charger. According to it, the design of three-phase chargers will allow, on the one hand, managing a higher power flow between the battery and the grid and, on the other hand, providing or demanding power to help balance loads. Nowadays, a great study of charger models is being carried out [4]-[11].

In this paper a three-phase bidirectional (G2V and V2G operation modes) battery charger is presented. The novel control strategy proposed for the charger turns the car into a smart vehicle, demanding or injecting a sinusoidal and balanced source current, with unity displacement power factor, contributing to the improvement of the power quality. Simulation models of the battery pack, including the self-discharge effect, and the charger have been implemented to validate by simulation the proper operation of the charger under sinusoidal, distorted and unbalanced source voltages.

II. BATTERY MODEL

Some battery models have been implemented to be able to simulate the battery behavior in HEVs and EVs [12]. In general, existing battery models can be classified in three main types [13]: experimental, analytical and electric circuit-based models. The first type uses differential equations to simulate the complex electrical-chemical process in a battery [14] so, it requires intensive computations to solve the interdependent partial differential equations and they are difficult to be configured and used [15]. Analytical models simulate an equivalent mathematical representation to approximate the battery performance [16]-[18], but they ignore circuit features (voltage and internal resistance). Electric circuit-based models are useful to represent electrical characteristics of batteries [19]-[21].

According to it, an electric circuit-based model has been used for the battery model. A novelty in this model is the inclusion of the self-discharge effect, as it is an important parameter since cars can be parked for very long time. A controlled voltage source (modelled with a non-linear equation) in series with a constant resistance [19] is used to simulate the behavior of Lead-Acid, Lithium-Ion (Li-Ion) and Nickel-Metal-Hydride (NiMH) batteries, as shown in Fig. 2.

A Simulink block is created to implement the battery model with one input (the current provided by the BMS) and two outputs: State-of-Charge (SOC) and voltage between the electrodes.

The parameters of the model are obtained from the manufacturer's discharge curve and electrical characteristics from the datasheet of the battery (Fig. 3). The battery used in the different simulations is implemented with eight modules of "MP 176065 Integration" of Saft Batteries. It is important to mention that it is assumed the same characteristics for the charge and the discharge cycles [19].

In Fig. 4, the input parameters mask in the battery model is shown.

In addition to represent the battery behavior during driving, cars can be parked and left for long time, so the battery model must have into account the self-discharge and a new block has been designed and inserted into the battery model to include this parameter. The self-discharge is measured as the percentage per month of reduced stored charge of the battery without any connection between the electrodes, so the proposed block adds to the input current an additional amount

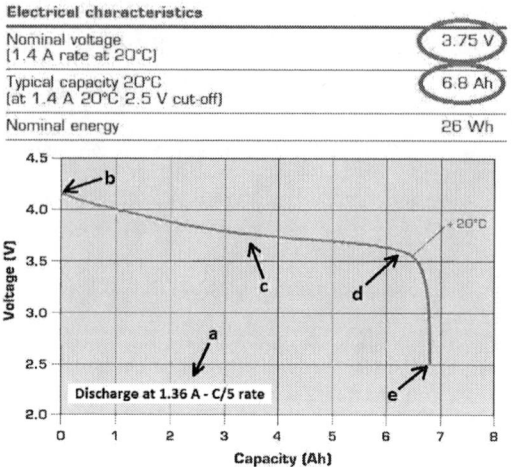

Fig. 3. Datasheet of the battery "MP 176065" of Saft Batteries.

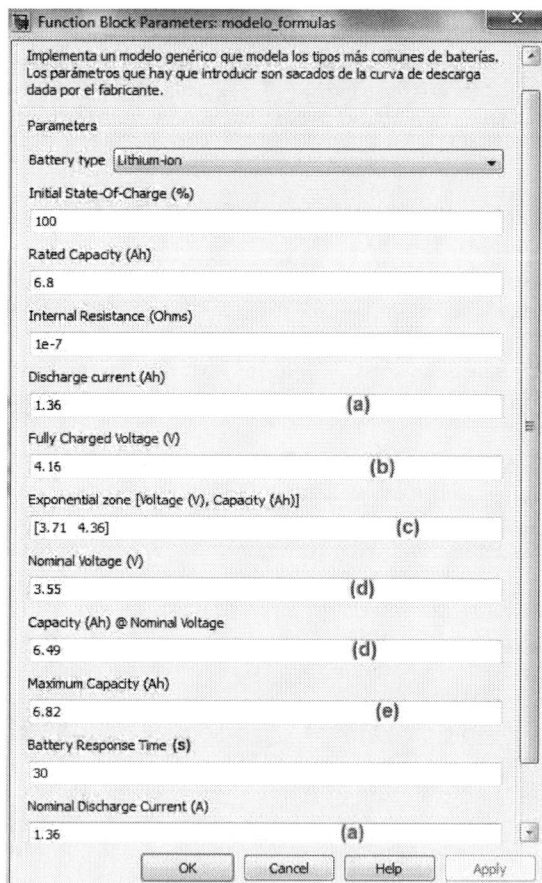

Fig. 4. Input parameters in the model for "MP 176065 Integration" of Saft.

that causes a battery discharge equivalent to the self-discharge and it is implemented as shown in Fig. 5.

From Fig. 5, it can be seen that no connection between the electrodes is detected when the current is zero and no variation occurs, what implies that the derivative of the battery current is zero. This detection keeps in memory the value that the SOC had when the disconnection occurred and

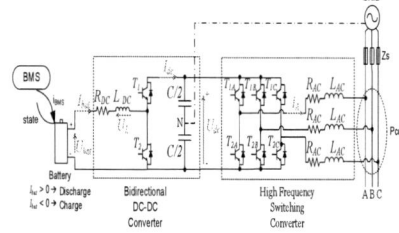

Fig. 2. Non-linear battery model.

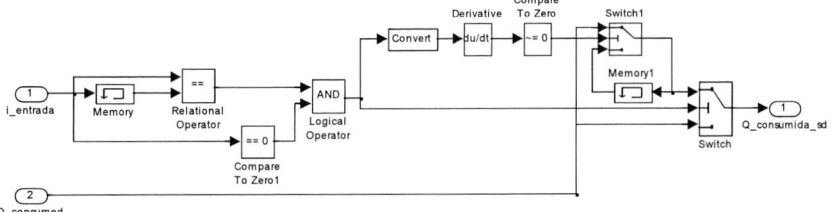

Fig. 5. Calculation of the battery self-discharge current.

calculates the amount of current needed to simulate the correct self-discharge.

A simulation has been implemented to check whether the self-discharge function works correctly (Fig. 6).

In Fig. 6, the battery model is shown with zero Amperes of reference input current, which simulates no connection between the electrodes. The instantaneous SOC and output voltage can be measured from the outputs of the model.

The initial SOC has been set to 100% and the self-discharge have been set to 2.5%. Attending to these values, in a period of a month, a drop of 2.5% (from 100% to 97.5%) of the SOC is expected. The result is shown in Fig. 7.

III. BIDIRECTIONAL BATTERY CHARGER

A. Topology

A three-phase bidirectional battery charger is studied in this paper. This bidirectional charger has two power stages: stage 1 is a DC/DC bidirectional converter; stage 2 is a bidirectional inverter.

In Fig. 8, the topology of the proposed charger is shown. The aim of this topology is to charge the battery demanding energy from the grid (in case the vehicle is pretended to be driven in a near future) or to discharge it returning the energy into de grid (for instance, when the car is parked at peaks of

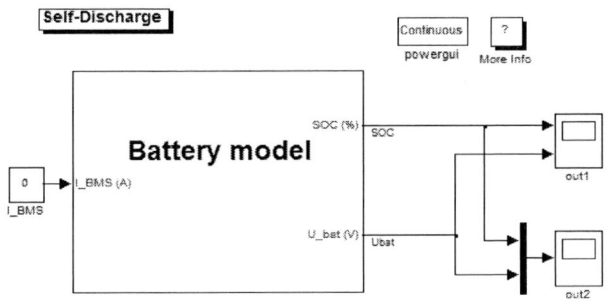

Fig. 6. Simulation to test the self-discharge function.

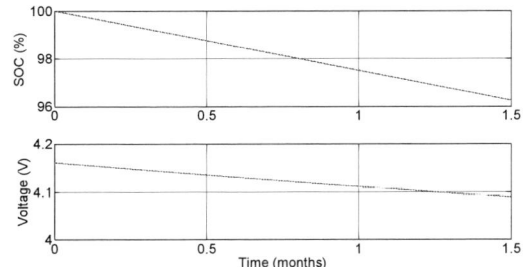

Fig. 7. Effect of the battery self-discharge.

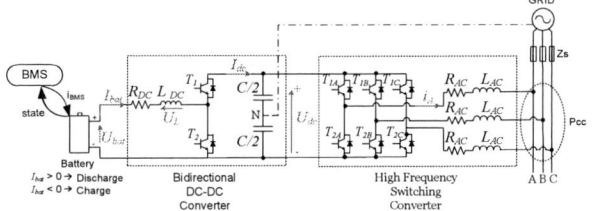

Fig. 8. General electric diagram of the bidirectional charger.

consumer energy demand to the grid).

The DC/DC converter is in charge of increasing the battery output voltage to the suitable inverter input voltage. In addition, the voltage in the inverter input is fixed to a constant level, independently of the voltage variation in the battery output due to its State-Of-Charge.

On one hand, the bidirectional inverter performs the function of converting the DC values (from the DC/DC converter) to the suitable AC values in order to inject (or demand) sinusoidal currents in phase with the voltage into the grid. On the other hand, in case the battery is being charged, the inverter converts the AC values (from the grid) to the suitable constant DC values with which the DC/DC converter works.

The criterion to determine the charge or discharge of the battery is the direction of the current. Positive battery current corresponds to a discharge process (flow from battery towards the grid). Therefore, negative battery current corresponds to a charge process (flow from the grid towards the battery).

B. DC/DC Bidirectional Converter

It has two operating modes: a first charging mode, in which the current flow is from the battery towards the grid, and a second mode, the discharging one, where the current flows from the grid towards the battery [22].

C. Bidirectional Inverter

A three-phase three-leg topology with mid-point DC bus is used for the bidirectional inverter, as it is shown in Fig. 9. The mid-point of each leg is connected to the grid by a filter inductor with an internal resistance R_{AC} and inductance L_{AC}, while the mid-point of the DC bus is connected with no impedance to the neutral conductor.

To carry out an analysis it is important to highlight the following points:

- It is needed a signal of control for each leg, that is to say, three signals of control, as its activation causes the activation of $S1$ switches in each leg and the

978-1-4244-8806-3/11 $26.00 © 2011 IEEE   373

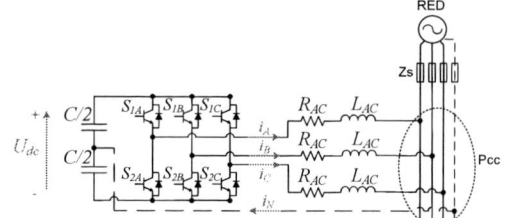

Fig. 9. Electric diagram of the three-phase inverter.

deactivation of *S2* switches (and vice versa). Each leg works with its own grid phase independently.

- It is necessary to satisfy

$$U_{dc} > 2 \cdot \hat{u}_{AN} , \qquad (1)$$

where U_{dc} is the DC inverter input voltage and $\hat{u}_{AN}$ is the peak value of the phase-neutral voltage (which is the maximum instantaneous value it can reach).

IV. CONTROL STRATEGIES

The control strategies generate the reference currents for each converter.

A. DC/DC Converter

The DC/DC converter is controlled by the battery current, I_{bat}. The reference current value comes from the BMS:

$$I_{bat,ref} = I_{BMS} . \qquad (2)$$

B. Bidirectional Inverter

The output instantaneous power is calculated by

$$p(t) = u_{AN} i_A + u_{BN} i_B + u_{CN} i_C , \qquad (3)$$

where u_{AN}, u_{BN} and u_{CN} and i_A, i_B and i_C are the output voltaje and current in each leg in the ac side, respectively.

A novel Balanced Sinusoidal Source Current (BSSC) control strategy is proposed. The objective is that the charger injects (or demand) into the grid a current without harmonics, in phase with the positive sequence fundamental component of the phase-to-neutral grid voltage, attaining a unity displacement power factor.

The inverter reference currents, $i_{inv,ref}$, are obtained multiplying a constant, K, by the unit vector of the positive sequence fundamental component of the phase-to-neutral grid voltage, as shown in

$$i_{inv,ref} = K \vec{u}_{s_1}^{+} = \frac{P_s}{\left| u_{s_1}^{+} \right|} \vec{u}_{s_1}^{+} , \qquad (4)$$

where P_s is the three-phase active power injected or demanded from the grid, $\vec{u}_{s_1}^{+}$ is unit vector of the positive

sequence fundamental component of the phase-neutral grid voltage and $\left| u_{s_1}^{+} \right|$ is its module.

The unit vector is calculated as

$$\vec{u}_{s_1}^{+} = \frac{\begin{bmatrix} u_{AN_1}^{+} \\ u_{BN_1}^{+} \\ u_{CN_1}^{+} \end{bmatrix}}{\sqrt{u_{AN_1}^{+}{}^2 + u_{BN_1}^{+}{}^2 + u_{CN_1}^{+}{}^2}} , \qquad (5)$$

where $u_{AN_1}^{+}$, $u_{BN_1}^{+}$ and $u_{CN_1}^{+}$ are the instantaneous positive sequence fundamental components of the phase-neutral grid voltage.

According to (5), equation (4) can be solved as shown in

$$i_{inv,ref} = \frac{P_s}{u_{AN_1}^{+}{}^2 + u_{BN_1}^{+}{}^2 + u_{CN_1}^{+}{}^2} \begin{bmatrix} u_{AN_1}^{+} \\ u_{BN_1}^{+} \\ u_{CN_1}^{+} \end{bmatrix} . \qquad (6)$$

The inverter output power is equal to the inverter input power, so

$$P_s = \overline{U_{dc} \cdot I_{dc}} , \qquad (7)$$

where U_{dc} and I_{dc} are the output voltage and current of the DC/DC converter (input voltage and current of the inverter), equation (6) is finally solved (and therefore, the inverter reference current obtained) as

$$i_{inv,ref} = \frac{\overline{U_{dc} \cdot I_{dc}}}{u_{AN_1}^{+}{}^2 + u_{BN_1}^{+}{}^2 + u_{CN_1}^{+}{}^2} \begin{bmatrix} u_{AN_1}^{+} \\ u_{BN_1}^{+} \\ u_{CN_1}^{+} \end{bmatrix} . \qquad (8)$$

The block diagram of the control strategy of the inverter is displayed in Fig. 10. An Autoadjustable Synchronous Reference Frame (ASRF) has been used to obtain the positive sequence fundamental component of the grid voltage [23].

V. TRACKING TECHNIQUES

The switching signals of both converters are generated with a synchronous hysteresis band controller. At every sample time, the measured current is compared with the reference current, (2) or (8), depending on the converter.

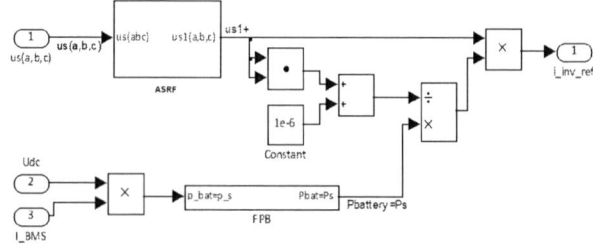

Fig. 10. Block diagram of the control strategy of the inverter.

978-1-4244-8806-3/11 $26.00 © 2011 IEEE   374

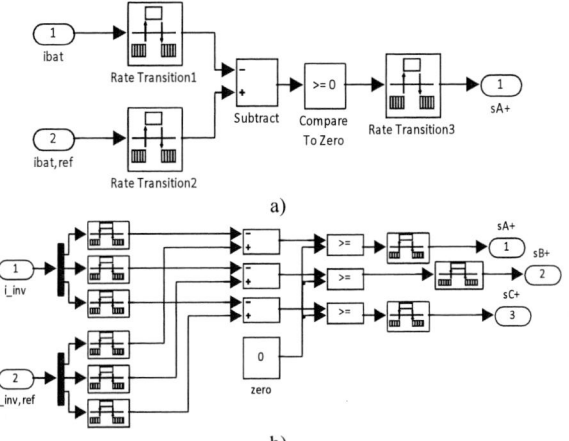

a)

b)

Fig. 11. Block Diagram of the tracking technique for a) DC/DC Converter b) Phase A of the inverter.

The block diagram of the tracking technique for both converters is shown in Fig. 11.

VI. SIMULATION RESULTS

The complete system simulated, using MATLAB-Simulink is shown in Fig. 12. The parameters values are summarized in Table I.

1. Case 1: Sinusoidal source voltage

a) Charge mode

In Fig. 13(a), the phase A of the grid voltage is shown in the upper subplot. In the lower subplot, the reference current and the real current demanded to the grid are displayed.

b) Discharge mode

In Fig. 13(b), the generation of the reference current and the measured current are shown for the discharging mode.

Fig. 13 shows that the reference current is correctly followed in both charge and discharge modes and the obtained THD_{is} is 8,1% and 9,4%, respectively.

2. Case 2: Voltage source with harmonic distortion. $HDv3 = 20\%$ (3rd harmonic); $HDv5 = 20\%$ (5th harmonic)

In this case, harmonic distortion is included in the grid voltage. It allows to check the performance of the charger when the mains present disturbances. As the reference current

TABLE I
PARAMETERS OF THE SIMULATION MODEL

RMS grid voltage	230 V
Grid frequency	50 Hz
U_{dc}	1380 V
$L_{DC} = L_{AC}$	50 mH
$R_{DC} = R_{AC}$	5 Ω
I_BMS	10 A

is obtained from the voltage positive sequence fundamental component, no harmonic distortion will affect the reference current generation. To carry out the simulation, the parameters remain equal as in Case 1. The generation of the reference and measured currents are shown for

a) Charge mode: in Fig. 14(a).

b) Discharge mode: in Fig. 14(b).

The effect of harmonic distortion can be seen in the voltage (upper subplot in Fig: 14(a) and Fig. 14(b)). The lower subplots show that the reference current generated is sinusoidal without harmonic distortion and it is correctly followed. The obtained THD_{is} for both the charge and discharge mode is 8,4% and 8,6%, respectively.

3. Case3: Voltage source with imbalance (zero sequence and negative sequence of the fundamental component: 20%)

The performance of the charger is checked under unbalanced grid voltage conditions. No disturbances will affect the reference current generation. The generation of the reference and measured currents are shown for

a) Charge mode: in Fig. 15(a).

b) Discharge mode: in Fig. 15(b).

The effect of imbalance can be seen in the voltage (upper subplot in Fig: 15(a) and Fig. 15(b)). The lower subplots in Fig. 15 show that the reference current generated is sinusoidal without imbalance and it is correctly followed. The obtained THD_{is} for both the charge and discharge mode is 8,4% and 8,6%, respectively.

VII. CONCLUSIONS

A three-phase bidirectional battery charger for electric vehicles has been presented in this paper. The charger operates with a novel Balanced Sinusoidal Source Current control strategy, demanding or injecting into the grid a balanced and sinusoidal current, making the car to become a

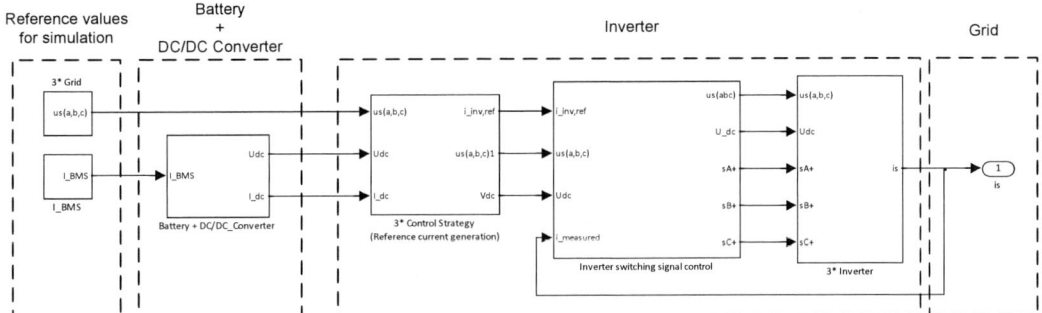

Fig. 12. Block diagram of the control strategy of the inverter.

978-1-4244-8806-3/11 $26.00 © 2011 IEEE

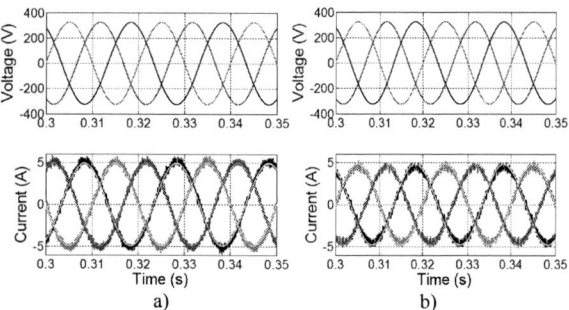

Fig. 13. Evolution of the ideal voltage and the current into the bidirectional inverter's output during (a) battery's charge (b) battery's discharge.

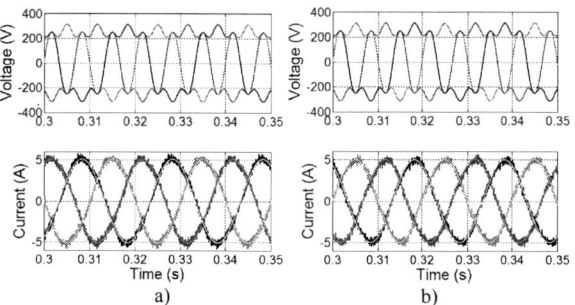

Fig. 14. Evolution of the distorted voltage and the current into the bidirectional inverter's output during (a) battery's charge (b) battery's discharge.

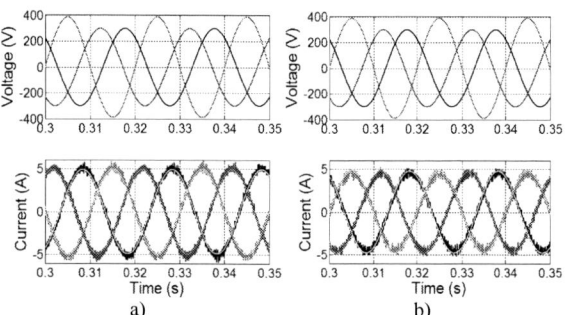

Fig. 15. Evolution of the unbalanced voltage and the current into the bidirectional inverter's output during (a) battery's charge (b) battery's discharge.

smart vehicle contributing to the so-called Smart Grid. Simulation results under ideal and disturbed source voltages are presented to test the operation of the proposed charger.

ACKNOWLEDGMENT

This work was supported by the Spanish Ministry of Science and Innovation under the research project PSE-370000-2009-22, co-financed by FEDER.

REFERENCES

[1] M. El Chehaly, O. Saadeh, C. Martinez, G. Joos, "Advantages and applications of vehicle to grid mode of operation in plug-in hybrid electric vehicles". *Electrical Power & Energy Conference*, pp. 1-6. Oct. 2009.

[2] Sikai Huang, D. Infield, "The potential of domestic electric vehicles to contribute to Power System Operation through vehicle to grid technology". *Universities Power Engineering Conference*, pp. 1-5. Sep. 2009.

[3] Y. Ota, H. Taniguchi, T. Nakajima, K.M. Liyanage, A. Yokoyama, "An

Autonomous Distributed Vehicle-to-Grid Control of Grid-Connected Electric Vehicle". *Industrial and Information Systems*, pp. 414-418. Dec. 2009.

[4] K. H. Chao, P. Y. Chen, C. H. Cheng, "A Three-Level Converter with Output Voltage Control for High-Speed Railway Tractions", *IEEE Conference of Industrial Electronics Society*, pp. 1793–1798, Nov. 2007.

[5] S. Jaganathan, W. Gao, "Battery Charging Power Electronics Converter and Control for Plug-in Hybrid Electric Vehicle", *IEEE Conference Vehicle Power* and *propulsion*, pp. 440–447, Sep. 2009.

[6] I. Cvetkovic, T. Thacker, D. Dong; G. Francis, V. Podosinov, D. Boroyevich, F. Wang, R. Burgos, G. Skutt, J. Lesko, "Future Home Uninterruptible Renewable Energy System with Vehicle-to-Grid Technology", *IEEE Energy Conversion and Exposition*, pp. 2675–2681, Sept. 2009.

[7] Lixin Tang; Gui-Jia Su. "A Low-Cost, Digitally-Controlled Charger for Plug-In Hybrid Electric Vehicles", *IEEE Energy Conversion Congress and Exposition*, pp. 3923–3929, Sept, 2009.

[8] M. Kisacikoblu, B. Ozpineci, L. Tolbert, "Examination of a PHEV Bidirectional Charger System for V2G Reactive Power Compensation", *IEEE Applied Power Electronics Conference and Exposition*, pp. 458–465, Feb. 2010.

[9] X. Zhou, S. Lukic, S. Bhattacharya, A. Huang, "Design and Control of Grid-connected Converter in Bi-directional Battery Charger for Plug-in Hybrid Electric Vehicle Application", *IEEE Vehicle Power and Propulsion Conference*, pp. 1716–1721, Sept. 2009.

[10] X. Zhou, G. Wang, S. Lukic, S. Bhattacharya, A. Huang, "Multi-Function Bi-directional Battery Charger for Plug-in Hybrid Electric Vehicle Application", *IEEE Energy Conversion Congress and Exposition*, pp. 3930–3936, Sept. 2009.

[11] Y. J. Lee, A. Khaligh, A. Emadi, "Advanced Integrated Bidirectional AC/DC and DC/DC Converter for Plug-In Hybrid Electric Vehicles", *IEEE Journals Vehicular Technology*, vol. 58, pp. 3970–3980, Oct. 2009.

[12] R. Rao, S. Vrudhula, D. Rakhmatov, "Battery Modeling for Energy Aware System Design", *Computer*, vol. 36. NO. 12, pp. 77–87. Dec. 2003.

[13] J. Zhang, S. Ci, H. Sharif, M. Alahmad, "An Enhanced Circuit-Based Model for Single-Cell Battery". *Power Electronics Conference and Exposition (APEC) 2010 Twenty-fifth Annual IEEE*, pp. 672–675. 2010.

[14] D. Rakhmatov, S. Vrudhula, D. A. Wallach, "A Model for Battery Lifetime Analysis for Organizing Applications on a Pocket Computer". *IEEE Journal Very Large Scale Integration (VLSI) Systems*, vol. 11, issue 6, pp. 1019–1030, Dec. 2003.

[15] F. Cao, A. Charkey, K. Williams, "Thermal Behavior and End-Of-Life Characteristics of the Nickel-Zinc Battery". *Energy Conversion Engineering Conference and Exhibit, 2000. (IECEC) 35th Intersociety*, vol. 2, pp. 985–994, Jul. 2000.

[16] J. Newman, K. E. Thomas, H. Hafezi, D. R. Wheeler, "Modeling of Lithium-Ion Batteries". *Journal of Power Sources*, pp. 838–843, 2003.

[17] P. Rong, M. Pedram, "An Analytical Model for Predicting the Remaining Battery Capacity of Lithium-Ion Batteries", ". *IEEE Journal Very Large Scale Integration (VLSI) Systems*, vol. 14, pp. 441–451, May 2006.

[18] M. Zheng, B. Qi, X. Du, "Dynamic Model for Characteristics of Li-Ion Battery on electric Vehicle", *4th IEEE Conference Industrial Electronics and Applications*, pp. 2867–2871, May 2009

[19] O. Tremblay, L. A. Dessaint, A. L. Dekkiche, "A Generic Battery Model for the Dynamic Simulation of Hybrid Electric Vehicles", *IEEE Conference Vehicle Power and Propulsion*, pp. 284–289, Jun. 2008.

[20] R. Kroeze, P. Krein, "Electrical Battery Model for Use in Dynamic Electric Vehicle Simulations", *IEEE Conference Power Electronics Specialists*, pp. 1336-1342, Aug. 2008.

[21] M. Chen, G. Rincón-Mora, "Accurate Electrical Battery Model Capable of Predicting Runtime and *I-V* Performance", *IEEE Journals Energy Conversion*, vol. 21, issue 2, pp. 504–511, Jun. 2006.

[22] M. Ortúzar, J. Moreno, J. Dixon, "Implementation and Evaluation of an Ultracapacitor-Based Auxiliary Energy System for Electric Vehicles", *IEEE Transactions on Industrial Electronics*, vol. 54, issue 4, pp. 2147-2156, Jul. 2007.

[23] M. Milanés-Montero, E. Romero-Cadaval, V. Miñambres-Marcos, F. Barrero-González, "Novel Method for Synchronization to Disturbed Three-Phase and Single-Phase Systems", *IEEE International Symposium on Industrial Electronics*. Jun. 2004.

Closed Current Loop Regulator for Electric Kart DC Motor Torque Stabilization

Kristaps Vitols, Nadav Reinberg, Ilya Galkin, Alvis Sokolovs
Riga Technical University
ktx@inbox.lv, snadar4@gmail.com, gia@avene.eef.rtu.lv, alvis@eef.rtu.lv

Abstract-**PID regulator has been adapted to control DC motor current of an electric cart. Both simulation and experiments were carried out to evaluate controller operation during step-change of control signal. The aim is to stabilize motor torque and monitor motor current to protect power converter from dangerous current overshoots.**

I.   INTRODUCTION

Motor control has always been an important task, when precise amount of power or torque has to be produced. With the conventional car which has an internal combustion engine most regulation is done by the driver. The response time of internal combustion engine is rather large, compared to an electric motor. Electrical motor torque response can be in the range of 1 millisecond, while internal combustion engine response time can be about hundred times longer [1]. Nowadays the engine is controlled via an on-board computer, but couple of decades ago, most of the control job was really done by the driver. If the car is equipped with an electric motor, torque response is much faster, thus if the driver presses the accelerator pedal too much then the car "jumps". To avoid such behavior motor torque must be limited within the safe operation zone by the on-board computer.

There are other problems arising as well. Electrical motors have large starting or inrush currents. These currents produce the high starting torque, which is beneficial. On the other hand high current can damage motor windings or semiconductor switches of power converter which delivers power from batteries to the motor. In order to avoid damage, current must be kept within safe operation levels for both motor and converter. Since motor torque is proportional to the motor current, both can be controlled with the same regulator. Because current measurement transducers are simpler and commonly used than torque sensors, the regulation parameter is armature current. Input signal of the controller is the desired current level, or current command which is proportional to the desired torque.

The proposed regulation system is part of the electric kart (Fig. 1), which is built as a student project for educational purposes. Kart chassis has been chosen for particular project due to its relatively inexpensive platform and compact size if compared to a full size car. The small dimensions and lack of outer shell gives easy access to all parts if any improvements are necessary.

Fig. 1. The kart with installed motors and batteries.

II.   THE KART

The discussed current regulator is part of a more complex system. Two permanent magnet DC (PMDC) motors are used to achieve traction and drive the kart. Each motor is connected to one of the rear wheels through pulley sheave-belt coupling. The coupling reduces motor speed by ratio 3:1. Because each motor has its own power converter the speed and torque of individual wheel can be controlled separately, hence electronic differential can be implemented. Electrical energy is stored in two lead-acid battery packs. Each pack is used to supply individual converter. Parameters of motors, batteries and motor power converters are summarized in Table I [2].

Weight of the kart is approximately 230 kilograms. Assuming that average driver weights around 80 kilograms the total weight of the kart is 310 kilograms. From known parameters such as tire diameter, coupling ratio and motor speed, the maximal speed can be is calculated 91 km/h. If both motors are operated at their nominal torque, theoretically it takes approximately seven seconds to achieve top speed from a stand still position. The average time of kart operation in steady state driving mode is estimated 15 minutes. Experimental performance test have not been done, because the kart is not yet fully functional.

TABLE I
DRIVE PARAMETERS

Motor	
Type	PMDC
Power	7 kW
Voltage	72 A
Speed	3300 RPM
Torque	20 Nm
Battery pack	
Type	Lead-Acid
Voltage	72 V
Energy	1.6 kWh
Max discharge current	300 A
Max charge current	6 A
Weight	38.4 kg
Power converter	
Function	Drive and regenerative mode
Switch type	MOSFET
Max current	171 A
Max voltage	150 V
Switching frequency	40 kHz

In order to achieve good driving characteristics, motor torque has to be controlled. Since DC motor torque is proportional to motor current, sufficient control can be achieved using closed current loop regulator. By regulating current, the system can be protected from undesirable over current which can harm power converter. Each converter is operated by individual regulator circuit. Both regulators are controlled by main control board, which calculates necessary torque to achieve desired performance. One of the main board tasks is the implementation of electronic differential function implementation since there is no mechanical one. Precise torque control is essential in order to achieve good performance of electronic differential.

III. THE REGULATOR

As concluded in previous paragraph, the main control board generates current command signal which is fed to the current regulator. Thus the direct task of the current regulator is to achieve the defined motor current.

DC motors typically have very high starting currents and armature current can increase rapidly if the applied voltage increases. Specific motor is equipped with additional series connected inductor with inductance of 200µH that increases the total load inductance, thus decreasing the rate of change of armature current (ΔI) from 0.89A/µs to 0.23A/µs.

Maximal permissible continuous motor current is 100A, but maximal permissible converter current is 171A. Motor can operate with increased current for relatively long time, but semiconductor switches can be damaged if their rated current is exceeded and no sufficient cooling is provided. In particular case there is a "safe zone" (I_{Limit}) of 71A. The minimum time (t) in which the safe zone will be breached can

be calculated according to (1) if the current change time is known. In this particular setup, the minimal time is around 300 microseconds, during which motor current must be read and appropriate PWM signal for switches must be calculated, in order to limit current value to allowable level.

$$t = I_{limit} / \Delta I. \qquad (1)$$

The regulator also has to keep current ripple within certain limits. Generally current ripple depends upon load inductance and switching frequency. Increasing one or both will decrease ripple. Motor inductance cannot be changed since motor is a closed system, but overall load inductance can be increased with an addition inductor as it is done in this particular case. Decoupling capacitor of the converter sets the allowable limit for current ripple, in particular case it is limited to 10.6 amps. In order to increase its lifetime the current ripple should be smaller. Since regulator sets the switching frequency, the current ripple can be controlled easily. In order to maintain low current ripple in particular system, commutation frequency of each converter is chosen to be 40 kHz. Such frequency is rather high for hard switching converter, but development of present converter can give useful hints for future projects when soft switching converters are investigated. Advanced DC/DC converters with MOSFET switches can operate at up to 200 kHz [7].

Relevant parameters for the regulator performance evaluation are the response and settling time of output signal or controlled parameter. In ideal case the controlled parameter – load current, would momentarily reach desired value and settle instantly. In practice the time necessary to reach desired value depends not only on the load characteristics but also on regulator performance. There might be current overshoots due to overregulation, and it usually takes some time for controllable parameter to settle to a desired value. Smaller current overshoot and settling time are indicators of well tuned regulator.

There are several types of regulators that can solve particular tasks. In this paper proportional (P), proportional-integral (PI) and proportional-integral-derivative (PID) controllers are discussed.

The most simple control method is a proportional (P) controller. This control method assumes amplification of input and output signal error by a certain coefficient – proportional gain. If the gain value is too large, the output will be sensitive to the input value, however if the gain value is too low, the output might not be entirely controllable. The gain value must be tuned to an optimum value in order to achieve the best performance. Because each output value is proportional to the error, it might be impossible to achieve ideal operation of the regulator. P controller can be characterized by relatively high overregulation and slow settling time of controlled parameter.

P regulator combined with an integral part forms a *PI* regulator. *I* part has its own gain value and it ensures calculation of output value with respect to the previous error

978-1-4244-8806-3/11 $26.00 © 2011 IEEE

value. As a result regulator response time increases and it can reach the desired value in shorter time.

By addition of *D* part the regulator forms a full PID controller which allows increase of the response time even more. Thus the desired value is achieved faster. As a drawback of PID regulator the high sensitivity to small input signal changes can be mentioned. Even measured noise can destabilize regulation system. Besides PID regulator there are three gain coefficients that must be tuned by implementation of different tuning procedures.

IV. SIMULATION

In order to check regulators compatibility with the load it is common practice to perform computer aided Simulation. In this case simulation of DC motor current regulation is done in MatLab Simulink. Power part of the model is built of a 72V DC voltage source, an ideal MOSFET with intrinsic diode, freewheeling diode and a DC motor model (Fig. 2) [4]. Parameters of the real motor are used in simulation. The motor model consists of two parts: electrical, that consists of armature resistance, inductance and back-EMF; and mechanical, that consists of active resistance equivalent to friction and static torque. Motor and load inertia is determined by inductance of mechanical part. Back-EMF value of electrical part is determined by the current in the mechanical part and it is proportional to motor speed. A voltage source simulates motor torque which is proportional to the armature current. Control unit (Fig. 3) contains P, PI or PID controller (Fig. 4) that depending on the motor armature current feedback provides duty cycle for PWM generator. Controller gain values were found using Ziegler-Nichols Tuning Rules [2] and adjusted to optimum value with trial-and-error method. Controller can be easily modified to be a P, PI or PID regulator by adding necessary gain loops. Step response test was carried out in order to test regulator performance.

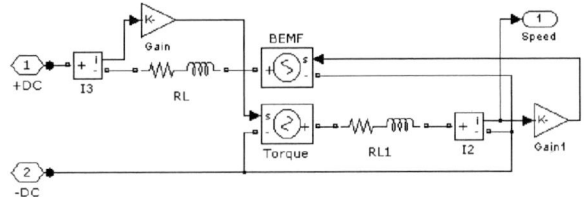

Fig. 2. PMDC motor model in Simulink.

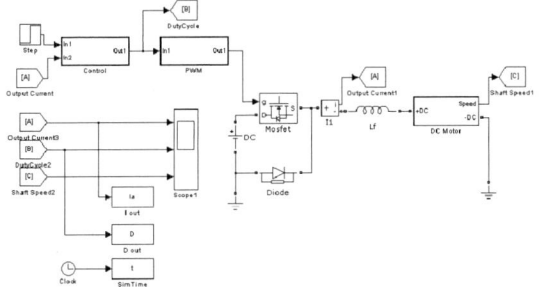

Fig. 3. Simulation setup in Simulink environment.

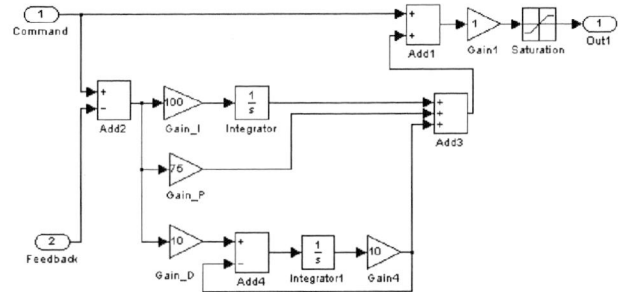

Fig. 4. PID control block model in Simulink.

First simulation was done with proportional control. Due to the low initial back-EMF of the motor armature current increased to a 130A in 0.8ms. However current decaying was slow – it took 0.45 seconds to reach 100 amps value (Fig. 5).

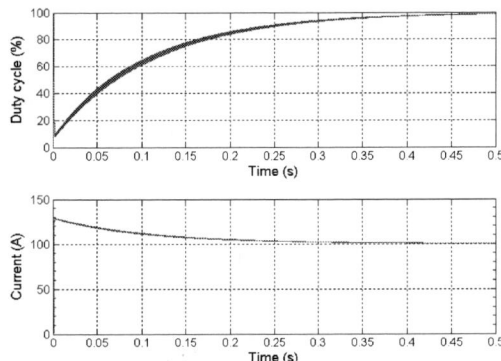

Fig. 5. Simulated P regulator step response.

Second simulation with PI control has shown current overshoot value of 110 amps in 0.4 milliseconds and much faster transient settling time – 10 milliseconds (Fig. 6).

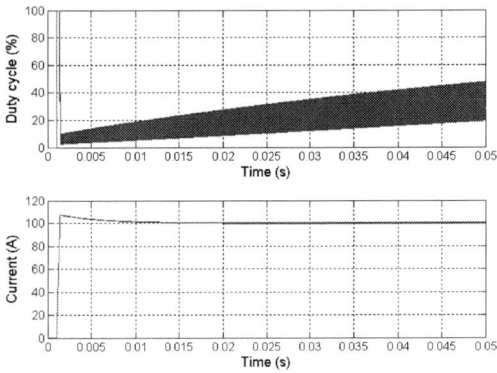

Fig. 6. Simulated PI regulator step response.

Third simulation was done with PID controller. As expected this control method has shown the best results – almost ideal transient without any overshoot, current reaches 100 amps in 0.3 milliseconds (Fig. 7). However this control method requires more calculation operations and more important, it can lead to system instability. Generally

978-1-4244-8806-3/11 $26.00 © 2011 IEEE   379

derivative part is used, when there is a need to control processes that happen very quickly. When controlling a motor, most processes, like increase of load, does not need fast response [6].

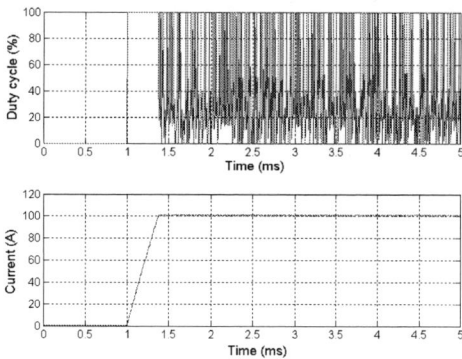

Fig. 7. Simulated PID regulator step response.

In particular application PI control is considered because of satisfactory transient current shape compared to P control and better stability compared to PID control.

V. EXPERIMENTS

Since the kart project is intended for educational purposes, students should be able to examine electric motors, motor drives and regulators. The regulator is commonly implemented using a microcontroller. Microcontroller part is show in Fig. 8 inside dashed square. This project is linked with another project, in which students learn MSP430 microcontroller programming and implementation. In order to improve project cooperation, the current regulator is implemented using the same educational platform: MSP430F1232 16-bit microcontroller with 8kB program memory. Processor is running at 8MHz which is its maximal frequency. For analog signal reading controller is equipped with an eight input 10-bit ADC. The regulator input signal V_{Iref} is provided from a potentiometer that is connected to the accelerator pedal. Motor current I_{MOTOR} is measured with HAIS 100-P current sensor and that value is used for feedback. Microcontroller performs conversion of measurements and writes values in to memory. This process takes approximately 30µs. Calculation of appropriate duty cycle and generation of PWM signal takes approximately 55µs. Program run time is reduced by using built-in peripheral Timer_A which generates PWM signal automatically. Total execution time to set new duty cycle is approximately 85µs. During this time motor current can increase by 20A. If compared to allowed limit of 71A then present performance of microcontroller is satisfactory.

To evaluate regulator performance, tests similar to the simulation were carried out. Motor was loaded with a synchronous three phase AC generator. Setup layout can be seen in Fig. 9.

Due to the power limitations of the synchronous generator, maximal DC motor current could not be achieved. In most cases the desired current level was set to 40 amps.

The first experiment was done with P regulator and multiple gain values to achieve the desired result. With 30 amp input command, current overshoot reached approximately 50 amps. It took 300 milliseconds to settle to the desired value (Fig. 10).

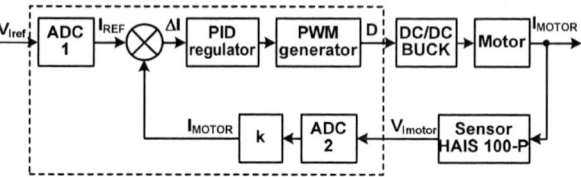

Fig 8. Experimental regulator.

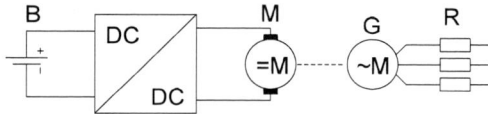

Fig. 9. Experimental setup.

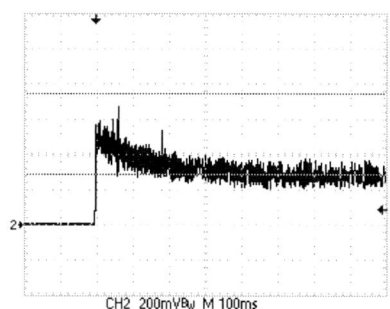

Fig. 10. P regulator step response, 20A/div, 100ms/div.

To improve regulator operation, the integral part was added. Initially both gain values were determined according to the Ziegler-Nichols Tuning Rules. Further experimentation with tuning yielded better results for the particular system. With 50 amp input command, there was almost no current overshoot. It can be seen in Fig. 11 that the current rise time is approximately 8 milliseconds. The difference of simulated and experimentally obtained current rise time is 7.6ms that in case of mechanical system is negligible. This performance makes PI regulator sufficient for the motor torque control [3].

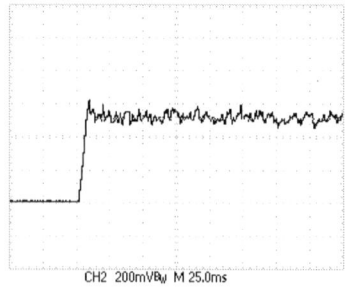

Fig. 11. PI regulator step response, 20A/div, 25ms/div.

978-1-4244-8806-3/11 $26.00 © 2011 IEEE

Full PID regulator was tested to further examine regulator possibilities. All gains were adjusted experimentally. The step response to the input current command of 50A is shown in Fig. 12. There was no motor current overshoot and current reached desired value in approximately 20ms. The slow current rise time as well as the inadequate shape of the steady state current indicates that there is a problem with appropriate PWM generation. Most likely the microcontroller has not enough computational power to perform all calculations. Fast changes of motor current result in controller produced PWM signal for a current that has already changed to another value. To obtain simulated PID regulator response, the controller program code has to be optimized.

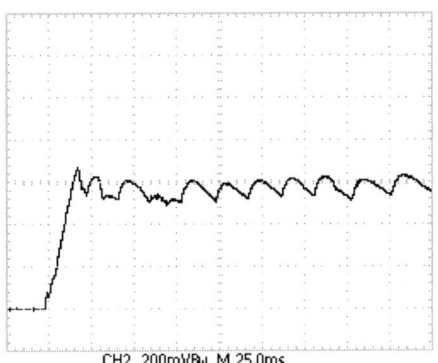

CH2 200mVBw M 25.0ms

Fig. 12. PID regulator step response, 20A/div, 25ms/div.

VI. CONCLUSION

Electric vehicle motor torque has to be controlled using appropriate regulator. Torque regulation improves overall system reliability. Additional task for the regulator is to limit motor current in order to protect power convertor from too high current overshoots.

Evaluation of P, PI and PID regulator performance with the specific DC motor by means of Simulink model was done. It can be concluded that P regulator is not sufficient to control the motor. PI regulator gives decent performance. The best result can be achieved if PID regulator is used.

Experimental results approve that the PI regulator can be used in particular electric kart application. Step response analysis showed no significant current overshoot, and current rise time was high enough to achieve decent dynamic performance of the kart. Although PID has the largest potential, its implementation in microcontroller was not fully successful. The experiment shoved some stability problems, most likely due to software and hardware issues. To use PID regulator, it should be precisely tuned, to avoid any instability. Most processes in motor control take relatively long time, which is why PID controller is not an absolute necessity. In similar manner regenerative braking regulator implementation is aim of future work [7]. Because of low maximal battery charge rate, limited to six amps, Power converter must be improved, to control small current values more precisely.

REFERENCES

[1] M. Mizan, "IPMSM Control for Electric Vehicle with Separate Wheel Drives", Proceedings, vol. 6, pp. 274-277, [EPE-PEMC, Latvia, 2004].

[2] K. Vitols, N. Reinberg, A. Sokolovs, I. Galkin, "Drive Selection for Electric Kart // EPE-PEMC 2010, Macedonia, Ohrid, 6.-8. September, 2010.

[3] K. Ogata, "Modern Control Engineering", Pearson Education, pp. 577-658, 2010.

[4] R. C. Dorf, "Modern Control Systems", 10th edition, Pearson Education International, 2005.

[5] DC Motor Model, http://www.ecircuitcenter.com/Circuits/dc_motor_model/DCmotor_model.htm, 2010.

[6] H. Miyazaki, T. Ohamae, "Driving Stability for Electric Vehicle with Independently Driven Two Wheels in case of Inverter Failure", EPE 2005, Germany.

[7] A. Kawamura, M. Pavlovsky, Y. Tsuruta, "State-of-the-Art High Power Density and High Efficiency DC-DC Choper Circuits for HEV and FCEV Applications", EPE-PEMC 2008, pp. 7-20, Poland, 2008.

Comparative Study of Steady-State Performance of LED Drivers at Different Modulation Techniques

Ilya Galkin, Irena Milashevski, Oleg Teteryonok

Riga Technical University, Tallinn University of Technology, Riga Technical University

gia@eef.rtu.lv, renamilashevski@gmail.com, tetervenoks@eef.rtu.lv

Abstract – **The paper deals with current and light regulators for luminaries with Light Emitting Diodes. Several topologies of the regulators with typical control formula (buck, boost and buck-boost) and several control methods (pulse-width and frequency modulation) are analyzed from the point of view of their control performance. Special attention is paid to the non-linearity of the proposed solutions and their accuracy in practically available control platforms. The control solutions are analyzed analytically and through simulations as well as confirmed experimentally. The conclusions about the most prospective solution are made.**

I. INTRODUCTION

Lighting is one of the most energy consuming industrial and household technologies. According to [1] about 19% of the electrical energy produced over the world is spent to lighting. Therefore it is quite reasonable to improve efficiency of lighting systems that can be done in two basic ways: utilization of the lighting technologies that produce more light per power unit and making lighting systems intelligent – provide lighting exactly when and where it is required.

Utilization of Light Emitting Diodes (LEDs) provides an excellent opportunity to use both of them. On the one hand modern LEDs have efficiency of several tens lumens per watt [2] that is comparable with high pressure sodium lamps (about 140lm/W) [3]. On the other hand it is possible to effectively adjust light produced by a LED lamp with no negative impact on the lamp itself.

Amount of light produced by a LED lamp is proportional to the current of its LED elements. This brings forward two light control methods: 1) fluent regulation mode when LED current remains constant for the same value of the control command; 2) pulse regulation mode when the current is either maximal or zero but its average value is constant for the same value of the control command. The third light regulation method is possible with LEDs because their power is rather small. For this reason LED lamp usually includes a number of LEDs and it is possible to divide them into groups and control each group separately utilizing some kind of Pulse Code Modulation (PCM). This method, however, ensures lesser dimming levels and lower accuracy of lighting and is not preferable for this reason. The second mentioned method is also undesirable because the light produced by the LED follows its current at a very high rate ([4] and [5]) which leads to flickering and stroboscopic effects. Therefore the fluent regulation mode is preferable.

The fluent regulation of LED current is possible with a current regulator. However, design and implementation of such regulators is not an easy task due to the more complicated schematics. For this reason various DC choppers are usually discussed as the power supplies for LED luminaries [6]: buck, boost, buck-boost and Cuk [7]. Each of these converters may be controlled in different ways – Pulse Width Modulation (PWM), Frequency Modulation (FM) etc.

The particular schematic and control method have strong impact on the efficiency and accuracy of the light regulation as well as on the weight and size of the regulator. Previously authors have studied the impact of the control method of a buck converter on its losses [8]. This research tries to find the best solution form the point of view of the control performance.

II. GENERAL ASSUMPTIONS AND CRITERIA FOR COMPARISON

A set of criteria that relates to the regulation curve (Fig. 1) of an LED lamp has been chosen in order to form a clear base for comparison of drivers and modulation methods. The meaning and importance of these criteria are explained below.

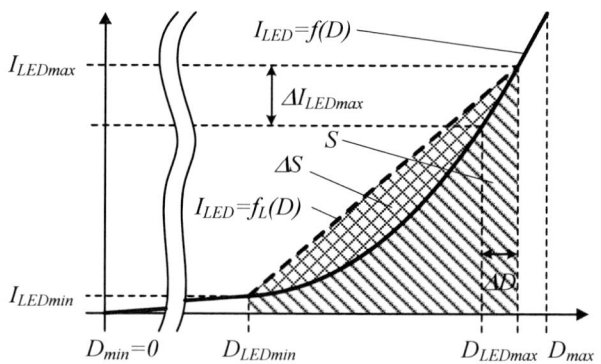

Fig. 1. General definitions for comparison of control performance.

The first one is nonlinearity of the regulation curve. The nonlinearity has a strong effect on the stability and dynamic performance of the control system. It is an integral criterion that is equal to the root-mean-square declination ΔS of the regulation curve from a line connecting its border points to the root-mean-square value S of the curve itself:

$$NL = \frac{\Delta S}{S}. \qquad (1)$$

In (1) ΔS is defined as

$$S = \sqrt{\frac{1}{D_{LED\max} - D_{LED\min}} \int_{D_{LED\min}}^{D_{LED\max}} f^2(D)dD} , \qquad (2)$$

but S – as follows

$$\Delta S = \sqrt{\frac{1}{D_{LED\max} - D_{LED\min}} \int_{D_{LED\min}}^{D_{LED\max}} \left[f(D) - f_L(D)\right]^2 dD} . \qquad (3)$$

In (1)…(3) D is duty cycle of operation of the chosen converter. This control parameter depends on a physical parameter (pulse width or period) that is variable with the particular modulation method. Its values at the ends of the regulation curve are obvious $D_{min}=0$ and $D_{max}=1$ while those, corresponding to the minimal and maximal values of LED current – D_{LEDmin} and D_{LEDmax}.

The second parameter for comparison reflects practical utilization of the duty cycle. It can be defined as the span of active values of this parameter:

$$D_{SPAN} = \frac{D_{LED\max} - D_{LED\min}}{D_{\max} - D_{\min}} = D_{LED\max} - D_{LED\min} . \qquad (4)$$

Another parameter is the practical inaccuracy of the duty cycle. It may be defined as a ratio of technically achievable inaccuracy ΔD of the duty cycle to its practical span:

$$\Delta D_P = \frac{\Delta D}{D_{LED\max} - D_{LED\min}} = K_{II} \cdot \Delta D , \qquad (5)$$

where $K_{II}=1/(D_{LEDmax}-D_{LEDmin})$ is inaccuracy increase ratio. This coefficient expresses the impact of the converter and LEDs V-A curve on the tolerance of regulation. Technical inaccuracy ΔD is defined differently for different modulation techniques and has to be analyzed for each of them.

The last (but not the least one) parameter is the relative inaccuracy of LED current:

$$\Delta I_{LED}\% = \frac{\Delta I_{LED\max}}{I_{LED\max} - I_{LED\min}} \cdot 100\% , \qquad (6)$$

where ΔI_{LEDmax} is changes of LED current corresponding to the changes of control parameter ΔD in the worst point of the regulation curve.

III. NONLINEARITY AND PRACTICAL OF DUTY CYCLE

The nonlinearity of regulation curve of a LED lamp and practical utilization of duty cycle depend only on regulation curve of the chosen converter and Volt-Ampere curve of utilized LEDs. At the same time, modulation technique has

no direct impact on them. In this section these parameters are estimated for different converters and with different values of the input voltage. This estimation has been made analytically and also based on experimental data.

A. Approximation of V-A Curve

Analytical estimation of the regulation curve requires expressions for output voltage of the dimming converter and Volt-Ampere curve of LED series. The last one does not exist in analytical form but has been measured experimentally (for 7 series connected LEDs type W724C0 produced by Seoul Semiconductor). Based on the experimental data a polynomial approximation of the curve has been obtained. In this work approximation with a 3rd order polynomial has been utilized:

$$I_{LED} = f(V_{LED}) = a_0 + a_1 \cdot (V_{LED} - V_0) + ... \qquad (7)$$
$$... + a_2 \cdot (V_{LED} - V_0)^2 + a_3 \cdot (V_{LED} - V_0)^3 .$$

where V_0 is the voltage of the working point, but coefficients $a_0...a_3$ depend on the choice of the working point. Usually only a workspace is approximated, but not the whole characteristic. In this work V_{LED} domain is [17…23,5V] and the corresponding I_{LED} is from the range [0, 3A]. For this range suitable working point is $V_0=20.2V$ that gives coefficients $a_0=557$, $a_1=442$, $a_2=89.6$ and $a_3=1.92$. These values of coefficients provide accuracy of about 2% that is sufficient at the preliminary stage. More accurate and valuable results are obtained experimentally.

B. Impact of regulators

Steady state formula of buck converter can be found from learning and scientific literature. Taking into account that the output voltage is applied to LEDs it can be written as:

$$V_{LED} = f(V_{IN}, D) = D \cdot V_{IN} . \qquad (8)$$

Applying (8) to (7) produces:

$$I_{LED} = f(V_{LED}) = a_0 + a_1 \cdot (D \cdot V_{IN} - V_0) + ... \qquad (9)$$
$$... + a_2 \cdot (D \cdot V_{IN} - V_0)^2 + a_3 \cdot (D \cdot V_{IN} - V_0)^3 .$$

In a similar way applying of steady state formula of the boost converter

$$V_{LED} = f(V_{IN}, D) = \frac{V_{IN}}{1 - D} \qquad (10)$$

produce

$$I_{LED} = f(V_{LED}) = a_0 + a_1 \cdot (\frac{V_{IN}}{1 - D} - V_0) + ... \qquad (11)$$
$$... + a_2 \cdot (\frac{V_{IN}}{1 - D} - V_0)^2 + a_3 \cdot (\frac{V_{IN}}{1 - D} - V_0)^3 ,$$

978-1-4244-8806-3/11 $26.00 © 2011 IEEE

but those of the buck-boost

$$V_{LED} = f(V_{IN}, D) = \frac{D \cdot V_{IN}}{1-D} \qquad (12)$$

produces the following

$$I_{LED} = f(V_{LED}) = a_0 + a_1 \cdot \left(\frac{D \cdot V_{IN}}{1-D} - V_0\right) + \dots \qquad (13)$$

$$\dots + a_2 \cdot \left(\frac{D \cdot V_{IN}}{1-D} - V_0\right)^2 + a_3 \cdot \left(\frac{D \cdot V_{IN}}{1-D} - V_0\right)^3.$$

Equations (9), (11) and (13) have been utilized to calculate nonlinearity of regulation curve (Table I), practical utilization of duty cycle (Table I, see also Fig. 2) and increase of inaccuracy with particular converter.

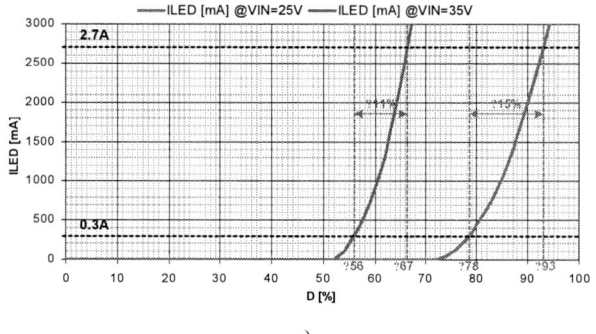

a)

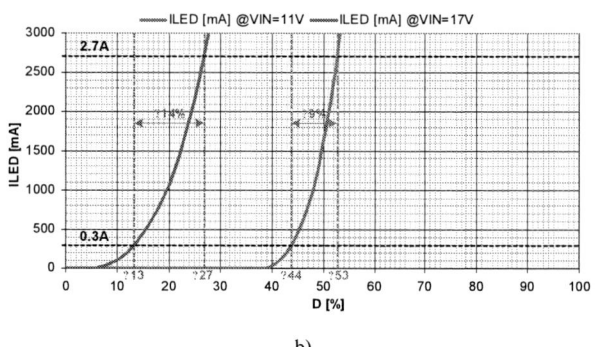

b)

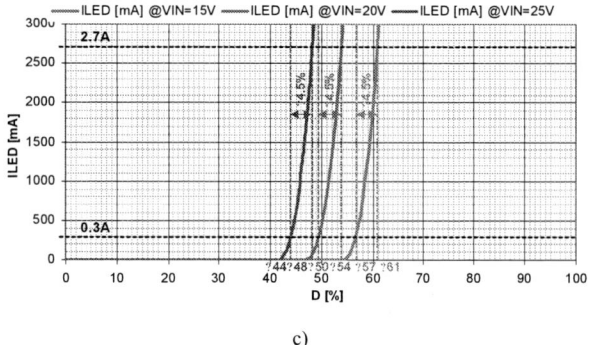

c)

Fig. 2. Regulation curve of a LED lamp containing 7 series connected LEDs W724C0 made by Seoul Semiconductor and fed by a DC/DC converter: a) buck; b) boost; c) buck-boost.

TABLE I
CALCULATED NONLINEARITY AND ACCURACY OF LED LUMINARIES
(FOR THE WORKING RANGE OF CURRENT 0.3…2.7A)

	Buck		Boost		Buck-Boost		
	25V	35V	11V	17V	15V	20V	25V
NL	16%	16%	20%	22%	20%	18%	18%
D_{SPANa}	15%	11%	9%	14%	5%	5%	5%
K_{II}	6.7	9.1	11.1	7.1	20	20	20

IV. ACCURACY OF MODULATION TECHNIQUE

There are two basic modulation modes: pulse width modulation (PWM - Fig. 3-a) and frequency modulation (FM). Besides that two kinds of frequency modulation can be introduced: constant pulse frequency modulation (CPFM - Fig. 3-b) and constant pause frequency modulation (CZFM - Fig. 3-c). Accuracy of the duty cycle and LED's current obtained with these modes are quite different. Their analysis is presented below.

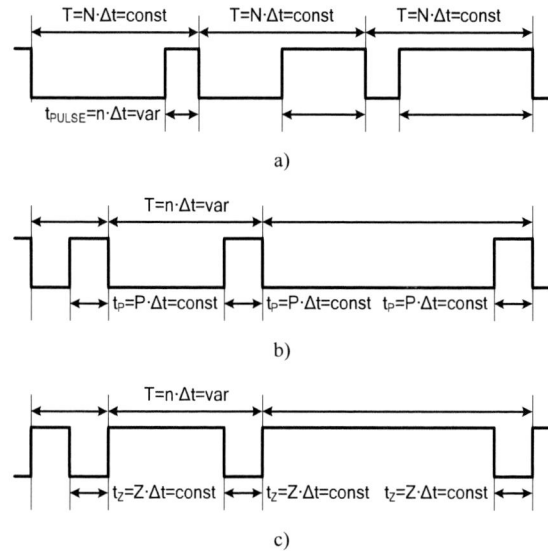

Fig. 3. Modes of pulse modulation: a) pulse width modulation; b) constant pulse frequency modulation; c) constant pause frequency modulation.

A. Pulse Width Modulation

At PWM modulation period T is a constant which is proportional to minimal countable time step (elementary step) Δt and number of such steps N that together produces $T = N \cdot \Delta t$. At the same time pulse width is variable $t_P = P \cdot \Delta t$, where n – is an integer number that is defined by the control system depending on the control conditions. Then the duty cycle is a function of n:

$$D(n) = \frac{t_P}{T} = \frac{n}{N}. \qquad (14)$$

Inaccuracy of duty cycle can be defined as a difference of its two closest values:

$$\Delta D = \frac{n+1}{N} - \frac{n}{N} = \frac{1}{N}. \qquad (15)$$

From (15) is obvious that the inaccuracy at PWM is inversely proportional to the number of elementary time steps in the modulation period. For instance at N=100 the inaccuracy is 1%. In practice N is limited by the performance of control system and cannot be chosen infinitely high.

B. Constant Pulse Frequency Modulation

At CPFM pulse duration is a constant proportional to the elementary step Δt and number of such periods P ($t_P = P \cdot \Delta t$). At FM period is variable. It can be expressed as $T = n \cdot \Delta t$, where n – is an integer number that defined by the control. Then the duty cycle can be expressed a function of n:

$$D(n) = \frac{t_P}{T} = \frac{P}{n}. \qquad (16)$$

At CPFM the inaccuracy of duty cycle is also function of n. If it is defined as a difference of two closest values of the duty cycle then:

$$\Delta D(n) = \frac{P}{n} - \frac{P}{n+1} = \frac{P}{n^2 - n}. \qquad (17)$$

The bigger is n, the smaller is duty cycle. Its minimal value is equal to 0 and it is achieved at infinity n:

$$D_{MIN} = \lim_{n \to \infty} \frac{P}{n} = 0 \qquad (18)$$

The same regards also inaccuracy of the duty cycle at such values:

$$\Delta D_{@DMIN} = \lim_{n \to \infty} \left(\frac{P}{n} - \frac{P}{n+1} \right) = 0. \qquad (19)$$

Some numerical values calculated with (16) and (17) for $P=1$ are given in Table II. The last row in this table shows that FM provides much higher than PWM inaccuracy of the duty cycle at comparable length of period.

TABLE II
DUTY CYCLES OBTAINED WITH CPFM AT LONG PERIODS

n	D(n)	ΔD(n)
10	10%	0.91%
20	5%	0.24%
100	1%	0.01%

In CPFM mode higher values of the duty cycle are obtained at lower values of n. The minimal value of this index is $P+1$. Then the maximal value of the duty cycle may be fond as:

$$D_{MAX} = \frac{P}{n_{MIN}} = \frac{P}{P+1}. \qquad (20)$$

The inaccuracy of the duty cycle then can be found as the difference of its maximal value and the next smaller value:

$$\Delta D_{@DMAX} = \frac{P}{P+1} - \frac{P}{P+2} = \frac{P}{P^2 + 3P + 2}. \qquad (21)$$

The values of the maximal duty cycle calculated with (20) and (21) are given in Table III.

TABLE III
DUTY CYCLES OBTAINED WITH CPFM AT SHORT PERIODS

P	D_MAX(n=P+1)	ΔD@_DMAX(n=P+1)
1	50%	17%
5	83%	12%
10	91%	8%

Table II and Table III show that CPFM provides higher values of the duty cycle at high frequencies when number of counted elementary steps Δt is low. The accuracy of duty cycle at such conditions is extremely low. At the same time this modulation method provides also high accuracy of the duty cycle, but only in the range of its low values if number of counted elementary steps is high. This phenomenon contradicts with nature of DC/DC converters feeding LED elements that mostly require higher accuracy in the range of higher values of duty cycle. Therefore control of this hardware is rater complicated in CPFM mode.

C. Constant Pause Frequency Modulation

This mode assumes that pause between two neighbor impulses remains constant. It is generated as a short series of the elementary time steps with total length $t_{PAUSE} = Z \cdot \Delta t$, where Z is number of the steps. At the same time the number of elementary time steps in period n is a variable whose value is defined by regulation loop and this period still can be expressed as $T = n \cdot \Delta t$. Then the duty cycle can be defined as:

$$D(n) = \frac{t_{PULSE}}{T} = \frac{T - t_{PAUSE}}{T} = \frac{n - Z}{n} = 1 - \frac{Z}{n}. \qquad (22)$$

The inaccuracy of duty cycle then is:

$$\Delta D(n) = \left(1 - \frac{Z}{n+1} \right) - \left(1 - \frac{Z}{n} \right) = \frac{Z}{n^2 + n}. \qquad (23)$$

At such approach the bigger is n, the bigger is duty cycle. It achieves value 1 at infinity n:

$$D_{MIN} = \lim_{n \to \infty} \left(1 - \frac{Z}{n} \right) = 1 \qquad (24)$$

Inaccuracy of the duty cycle at such values is:

$$\Delta D_{@DMIN} = \lim_{n \to \infty}\left(\frac{N}{n} - \frac{N}{n+1}\right) = 0 . \quad (25)$$

Numerical values calculated with (22) and (23) for $Z=1$ are given in Table IV.

TABLE IV
DUTY CYCLES OBTAINED WITH CZFM AT LONG PERIODS

n	D(n)	ΔD(n)
10	90%	0.91%
15	93%	0.42%
100	99%	0.01%

In CPFM mode lower values of the duty cycle are obtained at lower values of n. Then the minimal duty cycle is:

$$D_{MAX} = 1 - \frac{Z}{n_{MIN}} = 1 - \frac{Z}{Z+1} = \frac{1}{Z+1} . \quad (26)$$

The inaccuracy of the duty cycle then can be found as:

$$\Delta D_{@DMAX} = \left(1 - \frac{Z}{Z+2}\right) - \left(1 - \frac{Z}{Z+1}\right) = \frac{Z}{Z^2+3Z+2} . \quad (27)$$

The values of the minimal duty cycle calculated with (26) and (27) are given in Table V.

TABLE V
DUTY CYCLES OBTAINED WITH CZFM AT SHORT PERIODS

Z	$D_{MIN}(n=Z+1)$	$\Delta D_{@DMIN}(n=Z+1)$
1	50%	17%
5	17%	12%
10	9%	8%

Table IV and Table V prove that CZFM provides higher values of the duty cycle at lower frequency with the higher accuracy that corresponds to the requirements of DC/DC converters feeding LED elements. Lower values of the duty cycle are valid at higher frequency with lower accuracy, but this drawback is less important for the discussed application.

V. ANALYSIS OF EXPERIMENTAL VERIFICATION

A series of experiments has been conducted in order to check the above listed analytical assumptions. First of the all static regulation curve of the LED luminaries have been measured (Fig. 4) and the corresponding parameters found (Table VI). Comparison of these results with purely analytically calculated (Fig. 2 and Table I) proves the expected tendencies.

At the same time there is a certain difference that could be explained with impact of the parasitic parameters of elements of real converters. This especially regards boost converter

with high transfer gain (with 11V input). The higher output voltage is necessary the bigger is the difference between calculated and measured duty cycle. That is why the measured span in the case of boost and buck-boost converters is higher than calculated. Increase of the span, in turn, has impact on the corresponding V-A curves making the total relative non-linearity lower.

TABLE VI
MEASURED NONLINEARITY AND ACCURACY OF LED LUMINARIES
(FOR THE WORKING RANGE OF CURRENT 0.3…2.7A)

	Buck		Boost		Buck-Boost		
	25V	35V	11V	17V	15V	20V	25V
NL	12%	12%	9%	14%	7%	8%	11%
D_{SPANa}	17%	11%	16%	19%	8%	7%	7%
K_{II}	5.9	9.1	6.3	5.3	12.5	14.3	14.3

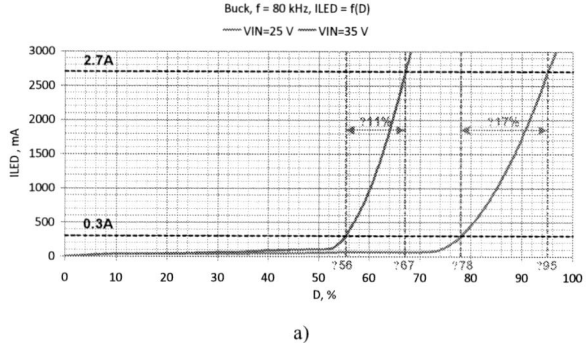

a)

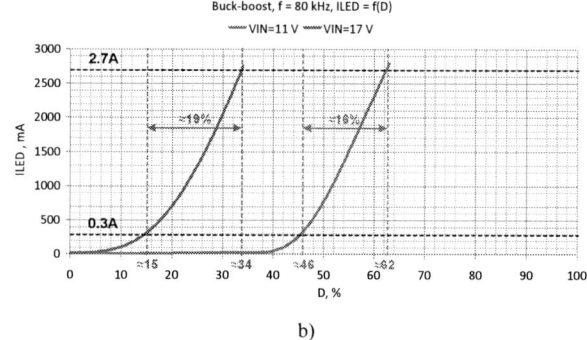

b)

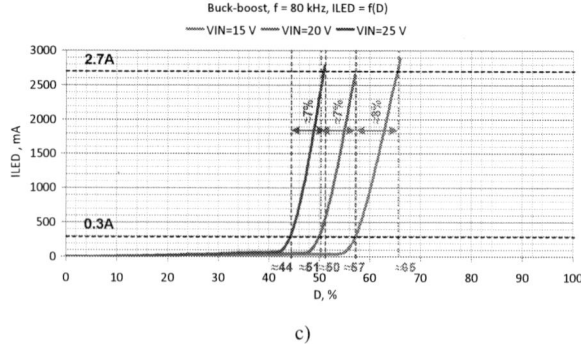

c)

Fig. 4. Measured regulation curve of the above described LED lamp fed by a DC/DC converter: a) buck; b) boost; c) buck-boost.

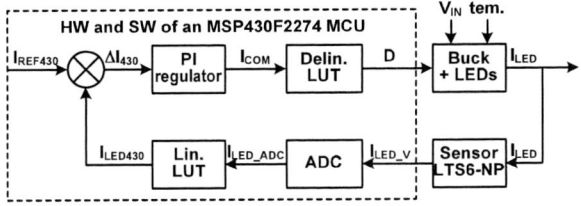

Fig. 5. Functional diagram of the test-bench for control parameters testing composed of a buck converter, sensor and control system with PI regulator.

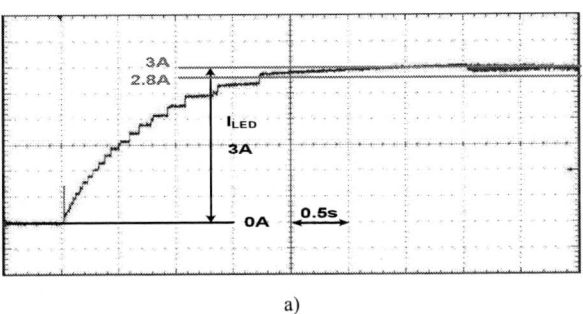

a)

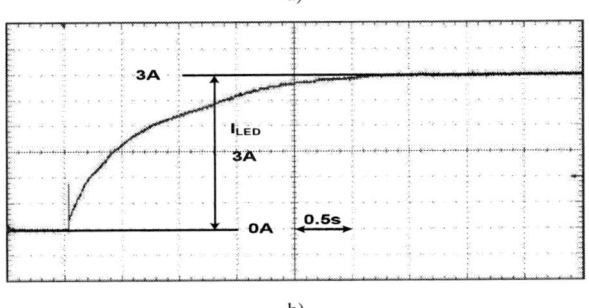

b)

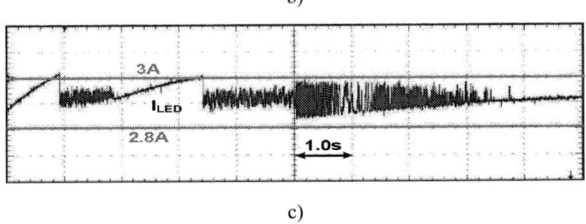

c)

Fig. 6. Influence of modulation technique on the performance of control system: a) startup step response at PWM (Δt=125ns; N=100); b) startup step response at CZFM (Δt=125ns; Z=1); c) steady-state operation at PWM.

At the next stage the impact of modulation technique has been investigated. As it has been previously shown CZFM technique provides more accurate regulation than PWM. This is especially well seen in the startup of LED lamp operating in a closed loop (Fig. 5). CZFM (Fig. 6-a) obviously produce less current steps than PWM (Fig. 6-b). Significant inaccuracy also leads to unavoidable static error. This, in turn, initiates low frequency oscillations between closest available current levels (red lines in Fig. 6) during steady-state operation (Fig. 6-c) if the reference current level is different.

VI. CONCLUSIONS

In the given paper the impact of dimming converter and its pulse modulation technique on its control performance have

been investigated. The results of analytical estimation correspond (taking into account differences between real lamp and its mathematical model) to those experimentally obtained. This correspondence validates the proposed assumptions and the conclusions drawn from them.

The type of dimming converter has a significant effect on the overall control performance. The nonlinearity of regulation curve in the case of boost converter is more significant due to the higher nonlinearity of the converter itself. It must also be noted that the nonlinearity of the V-A curve of LEDs has the most significant contribution in the total nonlinearity of the system.

All the discussed converters show narrow span of active (practically useful) values of the duty cycle. In the case of buck (especially with the input voltage equal to maximal operating voltage of LEDs) and boost (especially with the input voltage equal to minimal operating voltage of LEDs) the practical span is acceptable, but in the case of buck-boost converter – it is too narrow.

The choice of modulation technique is also important. In general FM shows better accuracy than PWM. However, since in the case of FM accuracy is not a constant, specific FM type must correspond to the converter and to the working point. In the case of CZFM the accuracy is higher at higher values of the duty cycle. This makes CZFM suitable for a LED lamp fed from buck converter at high current working point. At the same time a LED lamp with boost converter and low current working point requires CPFM.

It must also be noted that wide frequency band in case of FM may have negative effect on other parameters of the lamp – weight and size of reactive components, loses, EMC etc. These topics, however, require a special research on them.

REFERENCES

[1] P. Waide, S. Tanishima., Light's Labor's Lost. International Energy. 9 rue de la Federation. 75739 Paris Cedex 15: IEA Publications, 2006.

[2] Seoul Semiconductors, W724C0 datasheet, 2008, available electronically at www.seoulsemicon.com, last checked on August 13, 2010.

[3] U.S. DoE., High-Intensity Discharge Lamps Analysis of Potential Energy Savings. available electronically at www1.eere.energy.gov/.../pdfs/ hid_energy_savings_report.pdf. last checked on August 13, 2010.

[4] Schmid, M.; Kuebrich, D.; Weiland, M.; Duerbaum, T.; , "Evaluation on the Efficiency of Power LEDs Driven with Currents Typical to Switch Mode Power Supplies," Industry Applications Conference, 2007. 42nd IAS Annual Meeting. Conference Record of the 2007 IEEE , vol., no., pp.1135-1140, 23-27 Sept. 2007

[5] Suzdalenko, A.; Galkin, I.; , "Choice of power and control hardware for smart LED luminary," Electronics Conference (BEC), 2010 12th Biennial Baltic , vol., no., pp.331-334, 4-6 Oct. 2010

[6] van der Broeck, Heinz; Sauerlander, Georg; Wendt, Matthias; , "Power driver topologies and control schemes for LEDs," Applied Power Electronics Conference, APEC 2007 - Twenty Second Annual IEEE , vol., no., pp.1319-1325, Feb. 25 2007-March 1 2007

[7] de Britto, J.R.; Demian, A.E.; de Freitas, L.C.; Farias, V.J.; Coelho, E.A.A.; Vieira, J.B.; , "A proposal of Led Lamp Driver for universal input using Cuk converter," Power Electronics Specialists Conference, 2008. PESC 2008. IEEE , vol., no., pp.2640-2644, 15-19 June 2008

[8] I. Galkin, L. Bisenieks, A. Suzdalenko, "Impact Of Pulse Modulation Method Of Led Dimmer For Street Lighting On Its Efficiency", Proceedings of the 4th European DSP Education and Research Conference (EDERC2010), pages 160-164, Nice (France), December 1-2, 2010.

Space Vector Modulation with Reduced Switching Losses for Motor Drive Inverters

Mikhail Egorov and Valery Vodovozov
Department of Electrical Drives and Power Electronics
Tallinn University of Technology (TUT)
Tallinn, Estonia
e-mail: edrive@narod.ru

Abstract − **The paper presents simulation and experimental studies of a motor drive inverter with space vector modulation. A novel technique of current-dependent clamping for inverter legs is proposed. This discontinuous modulation method decreases the temperature by an average 12 % thanks to the commutation frequency mitigation and additionally by 10 % due to the current reduction. The benefits of the new control upon the six-step, pulse-width, and continuous space vector switching patterns are described.**

Keywords-power electronic converter; inverter; motor drive; modulation; power loss; IGBT

I. INTRODUCTION

The growth in industrial motor drives over the past 10 years has exceeded 25 % resulting from the increased demand for efficient reliable process control, power flexibility, process improvement, and control complexity. This progress requires significant effort in mitigating numerous technical problems associated with new control modes [1].

Advances in the power technologies estimate that about 50 to 60 % of electric power in industrialized countries is flowing through electronic systems and the percent is growing up. Additional cost of power electronics can be recovered by saving energy. According to the EPRI estimation [2], 15 % of grid energy can be saved easily by widespread but economical applications of power electronics. Saving energy not only provides the direct economic benefit, but helps preserving the dwindling fossil and nuclear fuel reserves. Indirectly, it helps mitigation of environmental pollution, such as global warming problems.

The major driving force in this field is the need for increased power density. This is explained by currently enlarging requirements for higher power output along with simultaneous reducing of system size and maintaining low system cost [3]. The losses of motor drive depend on multiple factors, such as power module voltage and current, supply voltage of the power cascade, average and rms currents of IGBT, forward converter impedance and voltage drop, modulation index, phase shift between the current and voltage waves, fundamental current amplitude, reverse recovery pick current and reverse recovery time of the back diodes, etc.

Among them, the switching frequency plays an extremely important role. It is an important issue to realize the higher power density and drive control characteristics. An increase or reduction of the losses may be effectively estimated based on the number of commutations required for different switching patterns and on energy at the switching instants [4].

Increase in the output power density basically comes from the reduction of volume of reactive components as well as the size of the heat sinks, which are inversely proportional to the switching frequency. Over the last decades, the performance of power electronics has been driven by improvements in fast switching semiconductor components. This leads to the substantial reductions of structural capacitances and inductances associated with device and system-level packaging. To reduce the negative impact of modulation and to enlarge the range of linear operation, the switching frequency of power converters is to be raised also [5]. Among other advantages of high operating frequencies are the lack of audible noise and the dynamic performance improvement [6].

Along with such tendency, it appears at present that in some high frequency power processing technologies, fundamental limits are being reached that will not be overcome without a radical change in the design and implementation of power systems. At the same time, while electronic power processing has steadily moved toward achieving the best output waveforms, high frequency operation doesn't reduce the total harmonic content but only shifts the harmonics having significant amplitude to higher frequency bands. In power converters, the frequency of the first significant harmonic is proportional to the number of pulses per alternation therefore the higher the number of pulses, the easier the filtering.

Besides, increase of the switching frequency means:

- need in more expensive semiconductor devices
- growth of power losses
- heightened electromagnetic interferences (EMI)
- excess ware of moving parts of driven equipment
- failure of the cable and motor winding insulation

One of the ways to improve the situation refers to optimization of modulation processes by generation effective switching patterns. In this paper the models and algorithms are studied from the viewpoint of searching the economical relation between the switching frequencies and duration of transition intervals in alternating current (ac) drive inverters.

II. SPACE VECTOR MODULATION

The three main modulation techniques are often compared: low-frequency simple six-step modulation, high-frequency pulse-width modulation (PWM), and progressive space vector modulation (SVM).

Referring to the common-mode three-phase bridge inverter shown schematically in Fig. 1, in the six-step operation the switching of the three inverter legs (**VT1-VT4, VT2-VT5, VT3-VT6**) supplied by the direct current (dc) voltage U_d is phase-shifted by 120°. Each phase $L1$, $L2$, $L3$ is under the current during half a modulating period and closed during another half-period. A specific phase is switched from the positive pole to the negative one being alternately in series with the remaining two parallel-connected phases. A simplified model given in Fig. 2, *a*, may be applied to represent a switching logic of the three-phase inverter circuit shown in Fig. 1. Each load terminal here assumes a potential defined by the control. For the reversing switches there are 8 different switching combinations that may be designated by the binary variables 100, 110, 010, 011, 001, 101, 111, and 000, which indicate whether the switch is in the top (1) or bottom (0) position, thus defining all possible switching states shown in Fig. 2, *b*. During the modulating period, a phase voltage may be equal to $\pm 2U_d / 3$, $\pm U_d / 3$, or zero depending on which transistors are switched on.

Clearly, to produce the typical six-step output voltage, only one pair of the inverter switches needs to change its state: **VT2** with **VT5**, **VT1** with **VT4**, or **VT3** with **VT6**. The six-step switching pattern is easier to implement compared to all the other control algorithms. It does not require high switching speed and can produce the line-to-line fundamental voltage of 15.5% higher amplitude than the supply dc voltage level. However, this algorithm is the source of significant harmonic distortion as its waveforms are the low frequency rectangles. Taking into account both the non-linear dependence of the

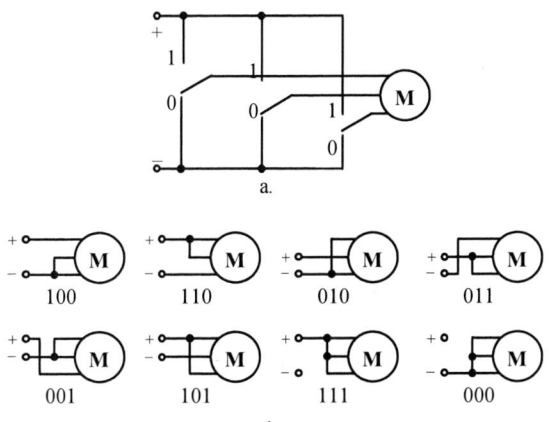

Fig. 2. Replacement circuits of motor drive

fundamental harmonic on the duty cycle and the adverse influence on the induction motors of the harmonics of order three and multiples of three [7], the six-step inverters have found occasional use in motor drives.

Far more popular, especially for the converters with fast switching IGBTs, are the sinusoidal PWM techniques. As compared to the six-step technique, the PWM circuit output presents the chain of high frequency pulses, the duration of which is modulated to obtain the necessary specific waveform on the constant modulating period. PWM is valuable for drive performance in respect to voltage and current harmonics, torque ripple, acoustic noise emitted from an induction motor and also electromagnetic interference.

Today, SVM is considered as the most sophisticated method for generating a fundamental sine wave providing a higher voltage to the motor and lower total harmonic distortion as compared to both the six-step and PWM methods. The possibility to get the line-to-line voltage as high as U_d is an important advantage of the SVM algorithm. Due to the higher line-to-line voltage amplitude, the torque generated by the motor is higher resulting in the better dynamic response of the motor. By increasing of output voltage, the user can design the motor control system with reduced current rating, keeping the same power capacity. It helps to drop inherent conduction loss of the inverter.

To proceed from the six-step modulation to SVM, each of the binary variables 100, 110, 010, 011, 001, 101, 111, and 000, is associated with a particular space vector $U_0...U_7$ the states of which correspond to the definite space position of the resultant motor voltage vector. The desired three-phase voltages at the output of the inverter could be represented by an equivalent space vector u^* rotating in the counterclockwise direction at a constant angular frequency for steady state operating condition. The magnitude of the space vector is related to the magnitude of the output voltage and the modulating time this vector takes to complete one revolution is the same as the fundamental time period of the output voltage. Figure 3, *a*, illustrates the

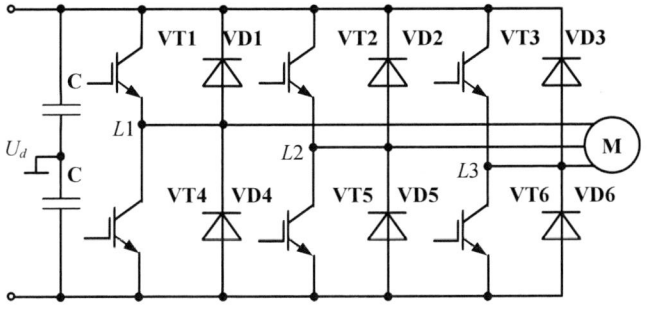

Fig. 1. Circuit diagram of inverter-fed motor drive

978-1-4244-8806-3/11 $26.00 © 2011 IEEE

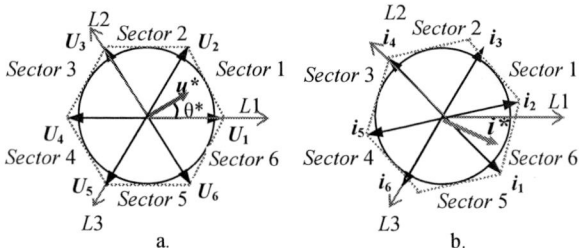

Fig. 3. Voltage and current space vectors

switching states and corresponding voltage vectors of the three-phase inverter shown in Fig.1.The vector set includes six active voltage space vectors U_1 to U_6 corresponding to the switching states 100, 110, 010, 011, 001, 101, and two zero voltage space vectors U_0, U_7 keeping with 111 and 000.

In SVM, vector $u*$ is treated through adequate timing of adjacent non-zero and zero space vectors. It is composed by a switching sequence comprising the neighbor space vectors $U_1...U_6$, while filling up the rest of the time interval with zero vectors U_0 or U_7 during the voltage alternation. As a result, the end of the vector $u*$ travels along the hexagon or stops. The modulating period 2π consists of six sectors, each including some fixed sampling intervals T_c. To move between the neighbor vectors U_i and U_{i+1}, the switching sequence of pulses U_i and U_{i+1} has to be generated, the time durations of which are consequently t_i and t_{i+1}, that is

$$u* = \frac{1}{T_c} (t_i U_i + t_{i+1} U_{i+1} + t_0 U_{0,7}), \qquad (1)$$

where U_i is the magnitude of the first space vector; U_{i+1} is the magnitude of the space vector valid in the next T_c interval; $U_{0,7} = 0$; t_i and t_{i+1} are the time durations for the two adjacent vectors that are to be computed in each T_c in real time or beforehand with keeping in memory; t_0 is the zero vector duration. The solution for t_i, t_{i+1} and t_0 results in

$$t_i = \frac{\sqrt{3}u*}{2U_d} T_c \sin\left(\frac{\pi}{3} - \theta*\right), \; t_{i+1} = \frac{\sqrt{3}u*}{2U_d} T_c \sin\theta*$$

$$t_0 = T_c - t_i - t_{i+1}, \; \theta* = T_c N \qquad (2)$$

where N is the number of the sampling periods counted from the sector starting point. While traveling between the adjacent sectors, the computations repeat. In fact, this technique produces an average of three voltage space vectors U_i, U_{i+1}, and U_0 (U_7) over a sampling interval T_c.

The reference signal $u*$ comes to the input of a conventional SVM controller in the form of modulation index and phase angle synchronized by the sampling clock. The output of this modulator presents the control signals for switching IGBT devices.

Knowing the values of t_i, t_{i+1} and t_0 gives the way to construct the switching patterns for the consecutive periods of T_c.

In continuous space vector modes, both U_0 and U_7 are used in the same sampling cycle. In this strategy, the time t_0 of the total zero voltage vector is divided equally between U_0 (111) and U_7 (000), being symmetrically distributed at the start and end of the sampling period. As opposite to the six-step modulation where the switching sequence of the inverter legs is given by $U_1 \rightarrow U_2 \rightarrow U_3 \rightarrow U_4 \rightarrow U_6 \rightarrow U_6 \rightarrow ...$ within the modulating periods, in continuous SVM the switching sequence of the upper legs of the inverter shown in Fig. 1 is given by $U_0 \rightarrow U_i \rightarrow U_{i+1} \rightarrow U_7 \rightarrow U_7 \rightarrow U_{i+1} \rightarrow U_i \rightarrow U_0$. The scheme has three switch-ons and three switch-offs within a switching cycle.

An idea of the discontinuous modulation is based on the assumption that one phase is clamped by 60° to the lower or upper level of the dc voltage [8]. While the reference vector occupies the first sector, the phase voltage of $L1$ obtains the maximum as compared to other phases. Switch **VT1** of the phase $L1$ keeps in conduction all the time whereas phases $L2$ and $L3$ modulate. As opposed to the continuous modulation, it gives only one zero state of **VT1** per sampling period hence only one vector – U_0 or U_7 – is used in the switching cycle. Thus, one of the main advantages of the method is the reduction of the number of commutations from 6 to 4 in a period T_c, providing a 33% decrease in the switching losses.

III. CURRENT-DEPENDENT CLAMPING OF INVERTER LEGS

Clearly, the character of the load voltage transients has no direct equivalence with the reference signals because of electromagnetic processes in an inverter-motor system,. The degree of such discrepancy depends on the load inductances and resistances as well as on the mode of equipment operation, thus resulting in additional voltage distortion and reduced use of supply power.

As the switching losses are proportional to the magnitude of the current being switched, it would be advantageous to avoid switching of the inverter leg carrying the highest current. This is possible in most cases because all adjacent vectors differ in the state of switches in only one leg. To this end, an effective discontinuous SVM method may be proposed.

Derive the phase current I_L as an integral of the phase voltage U_L that depends on the circuit electromagnetic time constant T_e [9]:

$$I_L = \frac{1}{L} \int U_L dt, \; I_{max} = \frac{2}{3L} \cdot \frac{T_e}{6} U_d \qquad (3)$$

Next, derive the load voltages and currents through the dc supply voltage U_d using Fig. 2:

$$U_{L1} = \frac{U_d}{3}, U_{L2} = \frac{U_d}{3}, U_{L3} = -\frac{2U_d}{3}, I_{L1} = \frac{I_{max}}{2} +$$

$$+ \frac{U_d}{3}t, I_{L2} = -I_{max} + \frac{U_d}{3}t, I_{L3} = \frac{I_{max}}{2} - \frac{2U_d}{3}t \qquad (4)$$

978-1-4244-8806-3/11 $26.00 © 2011 IEEE

Consider the current vector trajectory as a continuous function during each time interval for which the voltage is constant. Since the voltage space vector changes its discrete positions at each 60 degrees, the current space vector trajectory results close to hexagonal as presented in Fig. 3, *b*.

An electromagnetic time constant of the motor drive $T_e = \dfrac{L}{R}$ can be calculated beforehand on the basis of the motor data using the methodology [10] for the induction motor drive having the stabilized rotor flux linkage:

$$R = R_1 + k_2^2 R_2, \quad L = \sigma L_1, \quad \sigma = 1 - k_1 k_2, \quad k_1 = \frac{L_{12}}{L_1}, \quad k_2 = \frac{L_{12}}{L_2},$$

$$L_{12} = \frac{3}{2} \cdot \frac{X_{12}}{314}, \quad L_1 = \frac{X_1}{314} + L_{12}, \quad L_2 = \frac{X_2}{314} + L_{12}, \qquad (5)$$

where index 1 designates the stator parameters, index 2 indicates the rotor parameters, and $X_1, X_2, X_{12}, R_1, R_2$ are the rated values obtained from the motor data sheet or from an online identification procedure.

Now, the current vector position that depends on the time constant T_e may be identified and stored in a lookup table before modulation. In the beginning of each modulation sector, the control system calculates and selects from the lookup table the phase, which is expected to pass the highest current, and clamps an IGBT switch of this phase by 60° alternately to the lower and upper levels of the dc voltage thus preventing the high current commutation. An example of the prohibited switches counted in respect to Fig. 3 is presented in Table I.

TABLE I. IGBT SWITCHES WITH HIGHEST CURRENT

	I sector	II sector	III sector	IV sector	V sector	VI sector
Non-switching	VT1	VT3	VT2	VT1	VT3	VT2

Using an opposite way, this approach can be effectively implemented by sensing the maximum phase currents before each sector processing rather than its calculation and keeping in a lookup table. To this end, the current sensors of the power switches should be used. By comparison the three phase current values, the highest one is selected by the control circuit for exclusion from the commutation process.

IV. EXPERIMENTATION AND SIMULATION

The power converter model designed in the *Matlab/Simulink* environment was accomplished with the induction motor,

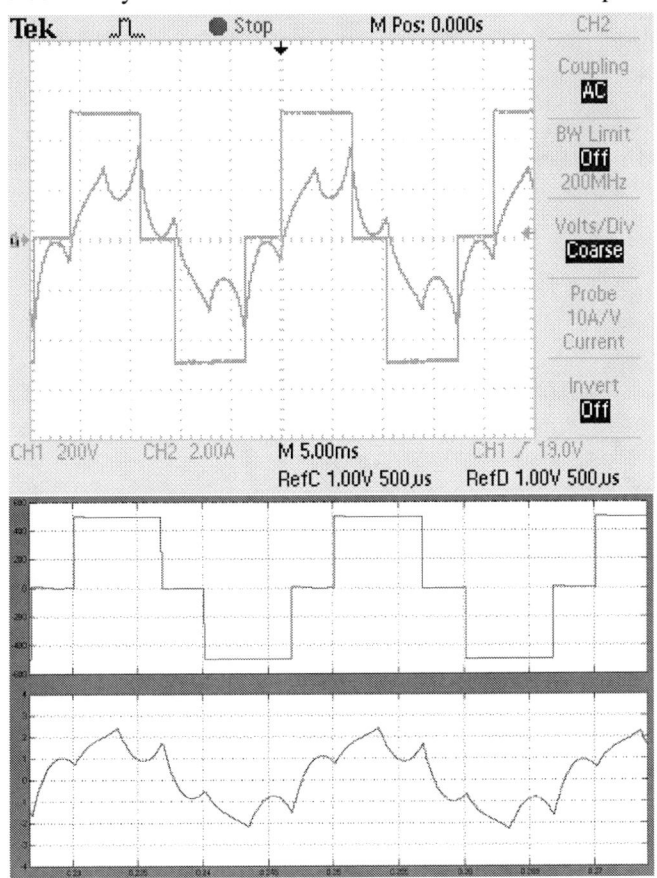

Fig. 4. Experimental and simulation traces of six-step modulation

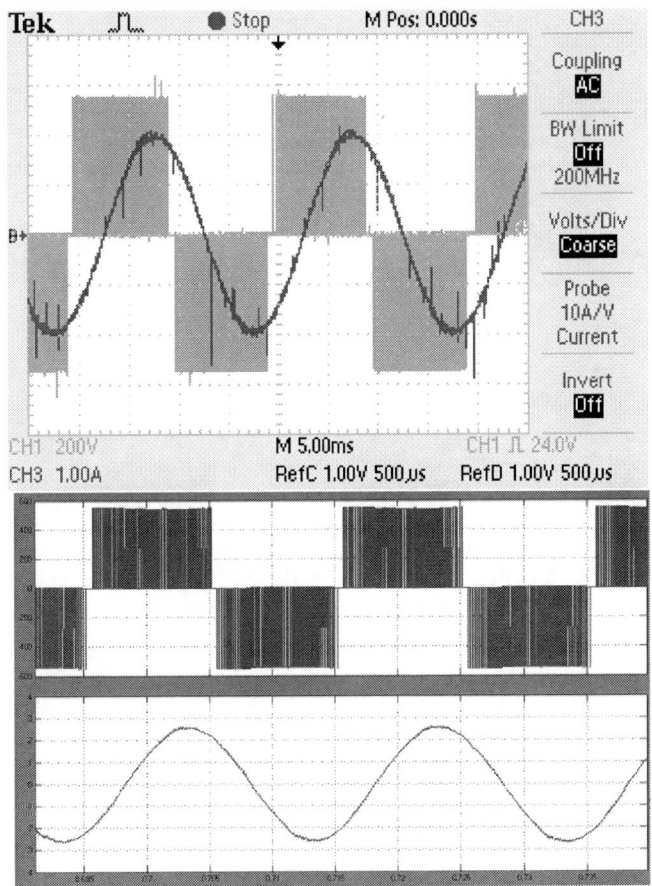

Fig. 5. Experimental and simulation traces of PWM

loading device, and independent bridge-connected IGBT simulators. Simulation of an induction motor drive was conducted for the six-step, sinusoidal PWM, and SVM continuous and discontinuous techniques with the output frequency 50 Hz under the load 2.8 Nm.

To obtain sufficient resolution, the modulating period was divided into six sectors with 15 sampling periods T_c of 12 points each. All switching patterns were calculated at the beginning of each sampling period, based on the value of the referenced voltage and current vectors. Therefore, u^* were updated every sampling interval. The inverter modulation frequency was 6 kHz, sampling frequency 134.4 kHz, dc link voltage 600 V, maximum modulation index 1.

Validation of the simulated characteristics generated by different toolboxes is an important stage of the comparative analysis. To verify the proposed algorithm, to correct a model, to test, and to identify the results of a particular drive application, an experimental setup was developed. An experimental workplace consists of two induction motor drives – the testing drive and the loading one. The three-phase bridge inverter was built on IGBT modules IRG4PH40KD, 1200 V, 15 A with fast recovery diodes. Parameters of the testing motor used for experimentation were as follows: rated power 0.55 kW, voltage 400 V, current 1.6 A, speed 1350 r/min, torque 3.8 Nm, moment of inertia 0.003 kgm².

The traces of waveforms for the six-step modulation are presented in Fig. 4. Here, experimental and simulated phase-to-phase voltage and phase current traces are given on the top and bottom screens respectively.

The timing diagrams of PWM are displayed in Fig. 5.

Figures 6 and 7 represent the waveforms of the continuous and discontinuous SVM accordingly with the latter resulting from the proposed algorithm. The reduced switching numbers are obvious in these traces.

Comparative experimental and simulation results are presented in Tables II and III.

TABLE II.   EXPERIMENTAL AND SIMULATION THD

No	Modulation	Voltage THD, %	Current THD, %
1	Experimental six-step	30.9	35.3
2	Simulation six-step	30.62	27.48
3	Experimental PWM	38	0.9
4	Simulation PWM	30.16	1.32
5	Experimental continuous SVM	44	0.9
6	Simulation continuous SVM	67.4	2.06
5	Experimental discontinuous SVM	53	2.2
6	Simulation discontinuous SVM	52.65	2.44

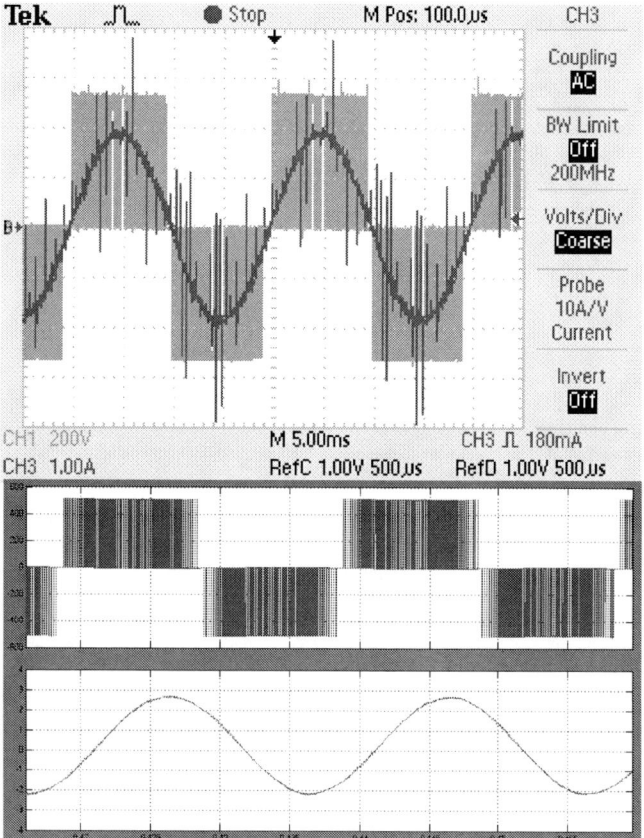

Fig. 6. Experimental and simulation traces of continuous SVM

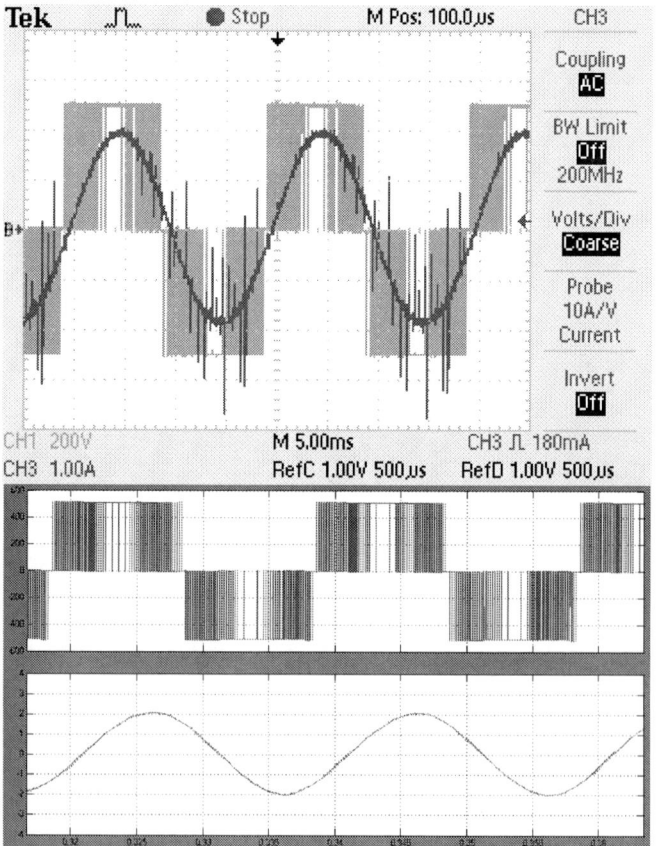

Fig. 7. Experimental and simulation traces of discontinuous SVM

TABLE III.   IGBT COMMUTATIONS IN A SINGLE SECTOR

No	Modulation	Averaged number of commutations	Switching frequency, kHz
6	Six-step algorithm	3	0.05
1	Sinusoidal PWM	60	6
2	Continuous SVM	60	6
4	Discontinuous SVM	40	4

To proceed to the converter loss analyses, take into account that the saturation voltage caused the power losses depends on the current values. Appropriate diagrams are usually presented in the manufacturer data sheets along with the transistor resistances. IGBT conduction intervals, transients, and voltage drops are defined by the circuit topology and temperature. Due to these obstacles, the specialized software was involved for the power loss and heat counting. Today, the most effective toolboxes are proposed by Semicron, Eupec/Infineon, Mitsubishi Electric and International Rectifier. Using the Semikron online package *Semisel*, ver. 3.1.1.3 [11], the group of 18 IGBT modules was explored and compared, such as Skiip, MiniSkiip, Semitrans, Skim, Semitop, and Semix. Figure 8 presents an averaged relationship between IGBT junction temperature, switching frequency, and phase currents in regard to the operational conditions of the discovered products. Similar results were obtained with the toolboxes *Melcosim* of Mitsubishi Electric and *IGBT selector tool* of International Rectifier.

Particularly, it has been found that the proposed discontinuous modulation method decreases the temperature on an average by 12 % thanks to the commutation frequency mitigation and additionally by 10 % due to the current reduction.

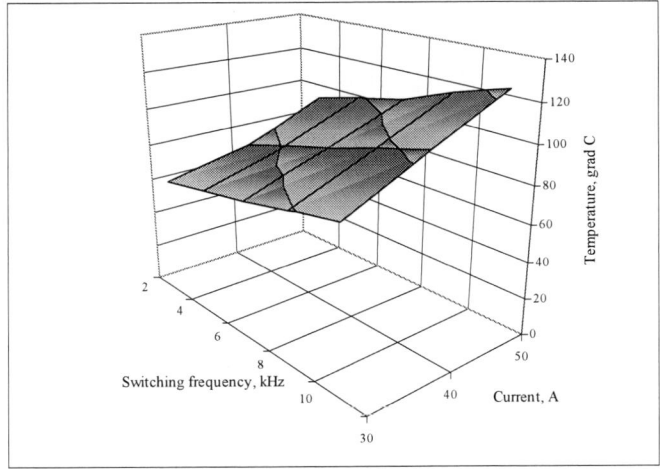

Fig. 8. Temperature-frequency-current correlation of Semikron modules

V.   CONCLUSION

Three main modulation techniques were compared in the paper: low-frequency simple six-step modulation, high-frequency PWM, and progressive SVM. The influence of the modulation method on the voltage and current harmonics, torque ripple, acoustic noise emitted from the motor and electromagnetic interference is described. A space vector based approach that decreases switching losses has been proposed in this paper based on the current-dependent clamping. The converter behavior was improved by decreasing the commutation frequency. Advantages of this method are illustrated in the steady-state mode of the converter performance in an ac drive application. Experimental and simulation results revealed an improvement in the motor drive performance of the discontinuous SVM operation when compared to the six-step modulation, PWM and conventional SVM techniques. The proposed category of the modulation techniques can be effectively applied in motor drives working in the open loop voltage/frequency control modes as well as in the vector controlled drives.

ACKNOWLEDGMENTS

This research work has been supported by Estonian Ministry of Education and Research (Project SF0140016s11), Estonian Science Foundation (Grant ETF8020) and Estonian Archimedes Foundation (project "Doctoral School of Energy and Geotechnology II").

REFERENCES

[1]   R. J. Kerkman, G. L. Skibinski and D. W. Schlegel, "AC drives: Year 2000 (Y2K) and beyond," *Applied Power Electronics Conference and Exposition APEC'99*, Dallas, TX, vol.1, 1999, pp. 28-39

[2]   B. K. Bose, "Power electronics and motor drives recent progress and perspective," *IEEE Transactions on Industrial Electronics*, Vol. 56, No. 2, 2009, pp. 581-588.

[3]   G. F. Labrique, "Developments in power converters," *Integrating Research, Industry and Education in Energy and Communication Engineering MELECON'89*, Lisbon, Portugal, 1989, pp. 59 – 64

[4]   N. Mohan, T. M. Undeland, and W. P. Robbins, *Power Electronics: Converters, Applications, and Design*, Hoboken, NJ, John Wiley & Sons, 2003, 802 p.

[5]   H. Ohashi, "Power electronics innovation with next generation advanced power devices," *The 25th International Telecommunications Energy Conference INTELEC'03*, 2003, pp. 9-12

[6]   E. Gubia-Villabona,   P. Sanchis-Gurpide,   O. Alonso-Sadaba, A. Lumbreras-Azanza, and L. Marroyo-Palomo, "Simplified high-frequency model for AC drives," *28th Annual Conference of the Industrial Electronics Society IECON'02*, v. 2, pp. 1144 - 1149

[7]   D. W. Novotny and T. A. Lipo, *Vector Control and Dynamics of AC Drives*, NY, Oxford University Press, 1996, 440 p.

[8]   K. Xing, F. C. Lee, D. Borojevic, Z. Ye, and S. Mazumder, "Interleaved PWM with discontinuous space-vector modulation," *IEEE Transactions on Power Electronics*, 14 (5), 1999, pp. 906-917

[9]   D. O. Neacsu, "Space vector modulation – An introduction," *The 27th Annual Conference of the IEEE Industrial Electronics Society IECON'01*, Colorado, USA, 2001, pp. 1583-1592.

[10]   V. Vodovozov and D. Vinnikov, *Electronic Systems of Motor Drive*, Tallinn, TTU, 2008, 248 p.

[11]   *Semikron Semisel*, Available at: http://semisel.semikron.com/Circuit.asp

Quasi-Z-Source Inverter Based Bi-Directional DC/DC Converter: Analysis of Experimental Results

J. Zakis[1], D. Vinnikov[1], I. Roasto[1], L. Ribickis[2]

[1] Department of Electrical Drives and Power Electronics, Tallinn University of Technology
Ehitajate tee 5, 19086 Tallinn (Estonia), Phone:+372 6203709,
e-mail: janis.zakis@ieee.org
[2] Institute of Industrial Electronics and Electrical Engineering, Riga Technical University
Kronvalda blvd. 1, Riga, LV-1010 (Latvia)
Phone number:+371 67089415, e-mail: leonids.ribickis@rtu.lv

Abstract-**This paper presents simulation and experimental results of a quasi-Z-source inverter (qZSI) based bi-directional DC/DC converter for supercapacitor (SC) interfacing. The bi-directional capability provides both the charging and discharging of the SC. During the charging mode the converter operates as a conventional VSI performing the voltage buck function, but during the discharging mode the converter operates as a qZSI based step-up DC/DC converter performing the voltage boost function. The proposed converter discussed was tested in both operating modes - SC charging and discharging. Theoretical assumptions are proved by the PSIM simulation results and verified experimentally on a 500 W laboratory prototype. The most important results obtained are outlined.**

I. INTRODUCTION

Supercapacitor (SC), one of the most promising high power storage element, is widely used in different industrial applications. But one of the most challenging applications of a SC is energy buffering in distributed power generation systems powered by fuel cells (FC) and/or photovoltaic (PV).

Since residential power supply systems with FC and/or PV cannot respond fast at short high power demand spikes, a short term controllable support system with a SC should be introduced. The support system should be connected at the supported system (SS) DC bus. In most cases SS side DC bus voltage (U_{SS}) should be 600 V for residential applications, which is enough to modulate 3-phase (400 VAC) or 1-phase (230 VAC) sine wave.

Since SC properties do not allow switching it directly to the SS DC bus, it is necessary to elaborate a suitable power electronic interface which could provide reliable energy transfer between the SC and the SS [1]. Moreover, the interface converter should be able to provide bi-directional power flow in order to charge or discharge the SC.

The aim of this paper is to verify theoretical results obtained by the power electronics group of Tallinn University of Technology in 2009 [2]. For this purpose the novel interface converter model for SC applications was analyzed and tested. The general power circuit of the proposed converter is presented in Fig. 1. It consists of a voltage-source half-bridge inverter (VSI) at the supported energy system side, a medium frequency isolation transformer and a voltage-fed quasi-Z-source inverter (qZSI) on the SC side. In our previous research [2] we used full-bridge VSI on the SS side but now we used the half-bridge VSI in SS side in order to reduce the transformer turns ratio. Also, in the places of the cascaded qZS-network [3]-[6], the single stage qZS-network was used.

In general, the operation modes of the proposed bi-directional converter can be subdivided as the charging mode and the discharging mode of the SC [2].

During the charging of the SC, the converter operates in the voltage buck mode and the charging process of the SC is controlled by the SS side half-bridge VSI. In the energy recovery mode, the supported energy system is being supplied from the SC. To ensure the constant DC-link voltage (U_{DC}) level in the conditions of the discharging of the SC the qZSI with the shoot-through PWM control is used to boost the decreasing voltage of the SC [7].

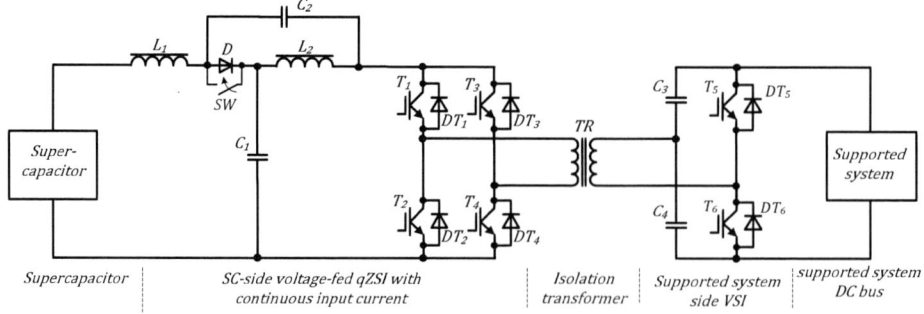

Fig. 1. General power circuit of the proposed bi-directional DC/DC converter for SC interfacing.

978-1-4244-8806-3/11 $26.00 © 2011 IEEE

II. OPERATION MODES OF THE PROPOSED CONVERTER

As mentioned above, the proposed interface converter operates in two modes - SC charging and discharging. The operation modes will be described in more detail below.

A. Charging Mode of the SC

Charging mode of the discussed converter can be realized if the SS DC bus voltage (U_{SS}) is at its rated value. If it is so, then the SC is being charged up to its nominal voltage (U_{SC}). In other words, we can say that the converter operates in the buck mode performing power transfer from the SS DC-bus to the SC. The general power circuit configuration in the charging mode is shown in Fig. 2.

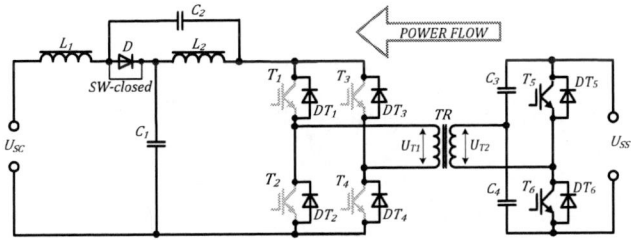

Fig. 2. General power circuit of the interface converter in the charging mode.

In this operation mode the qZSI transistors ($T_1...T_4$) are not conducting, but anti-parallel diodes ($DT_1...DT_4$) operate as full bridge rectifier. The diode D of the qZS-network is shortened with SW (contactor) and passive elements (L_1, L_2, C_1, and C_2) of the qZS-network perform as a low-pass filter.

The SC charging process can be regulated by adjusting the SS side half-bridge VSI transistor (T_5, T_6) duty cycle D. Neglecting losses in the circuit elements, the voltage of the SC (U_{SC}) during the charging mode is

$$U_{SC} = \frac{D \cdot U_{SS}}{n}, \quad (1)$$

where U_{SS} is the supported system DC bus voltage, D is the duty cycle of the VSI switches ($T_5...T_6$) and n is the turns ratio of the isolation transformer:

$$n = \frac{U_{T2}}{U_{T1}}, \quad (2)$$

where U_{T1} and U_{T2} are the amplitude voltages of the primary (SC side) and the secondary (SS side) windings of the isolation transformer, respectively.

B. Discharging Mode of SC

If the SS DC bus voltage drops during the operation, the proposed interface converter should provide the energy recovery function. It is achieved by the discharging of the SC. The power circuit configuration in the discharging mode is shown in Fig. 3. During the discharging mode, the switch SW is opened and power flows from the SC to the SS. The power flow is controlled by the qZSI power transistors ($T_1...T_4$). The SS side half-bridge VSI diodes DT_5 and DT_6 of the power modules (T_5 and T_6) together with capacitors C_3 and C_4 act as a voltage doubler rectifier (VDR).

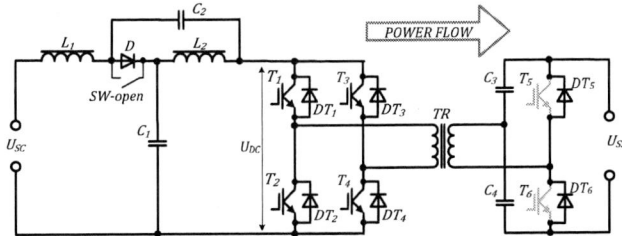

Fig. 3. General power circuit of the interface converter in the discharging mode.

It is well known that during the discharging of the SC its voltage level decreases. To provide the DC-link voltage (U_{DC}) in the rated level of the converter the SC voltage (U_{SC}) should be preregulated. For that reason we should use the major advantage of the qZSI, i.e. the voltage boost capability by implementing special switching state of qZSI transistors - shoot-through state [7]–[9]. Shoot-through state is simultaneous conduction of both transistors of one inverter leg transistors realized in order to boost the converter input voltage (U_{SC}). It should be mentioned that this operating state is forbidden in traditional VSI because of short circuiting of DC capacitors.

While discharging of the SC, the converter must ensure stabilized voltage on the supported system DC-bus (U_{SS}) that could be evaluated as

$$U_{SS} = 2 \cdot n \cdot U_{DC}. \quad (3)$$

The proposed converter is intended for operation conditions when the SC voltage (U_{SC}) decreases up to half of the nominal ($0.5U_{SC}$). Despite SC voltage decrease, the qZSI should provide voltage preregulation in order to ensure the desired DC-link voltage (U_{DC}) level of the interface converter within the whole operating interval. The SC voltage boost can be achieved by changing the shoot-through duty cycle D_S

$$U_{DC} = U_{SC} \frac{1}{1 - 2D_S}, \quad (4)$$

where $D_S = \dfrac{t_S}{T}$,

where t_S is the duration of shoot-through and T is the period. The maximal shoot-through duty cycle ($D_{S,max}$) should be associated with the minimal SC voltage:

$$D_{S,max} = \frac{1}{2} \cdot \left(1 - \frac{U_{SC,min}}{U_{DC}}\right). \quad (5)$$

The minimal shoot-through duty cycle ($D_{S,min}$) is associated with the rated operating voltage of the SC:

$$D_{S,min} = \frac{1}{2} \cdot \left(1 - \frac{U_{SC}}{U_{DC}}\right). \quad (6)$$

III. SIMULATION RESULTS

To verify the analysis [2] a simulation model was developed in simulation software PSIM. Both operation modes - SC charging and discharging (Figs. 2 and 3) were studied. The parameters of the SC stack of two Maxwell Technologies SC modules (PCM14014) connected in series with a total capacitance value of 85 F, rated voltage 28.8 V and equivalent series resistance 20 mΩ were taken as an example for the simulations. General operating and component ratings for the proposed converter are summarized in Table I.

TABLE I
GENERAL OPERATION AND COMPONENT RATINGS FOR THE PROPOSED CONVERTER

Parameter	Value
Rated power of converter P	500 W
Capacitance of capacitors C_1, C_2	730 µF
Capacitance of capacitors C_3, C_4	220 µF
Capacitance of supercapacitor stack C_{SC}	70 F
Transformer turns ratio, n	1:1
Transformer operation frequency, f_{TR}	15 kHz
Charging mode	
Supported system voltage, U_{SS}	60 V
SC-voltage, U_{SC}	28.8 V
Duty cycle of VSI switches, D	0.49
Discharging mode	
Rated SC voltage, U_{SC}	28.8 V
Minimal SC voltage, $U_{SC,min}$	14.4 V
Desired DC-link voltage amplitude, U_1	30 V
Duty cycle of qZSI switches in active states, D_A	0.35
Duty cycle of qZSI switches in shoot-through states, D_S	0.02...0.26

A. Charging Mode

It was assumed that during the charging mode, the SC is charged up to 28.8 V. The supported system side VSI was operated without dead time at the highest possible duty cycle D=0.49. SC is never discharged fully empty. It is considered empty when the terminal voltage of the SC has decreased approximately to half of nominal. That is automatically also initial voltage while charging.

Also, wire active resistances and inductances were taken into account. The operation frequency of the isolation transformer during the charging mode was 15 kHz.

Fig. 4 shows the charging process of the SC in the interval of 40 s.

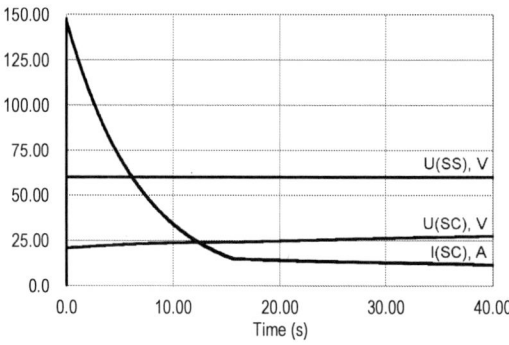

Fig. 4. Operating waveforms of the proposed converter during the SC charging process: SS DC bus voltage (U_{SS}), SC stack current (I_{SC}) and SC stack voltage (U_{SC}).

It can be seen that the SS system DC bus voltage U_{SS} remains the same, but the SC current I_{SC} decreases proportionally to the SC voltage U_{SC} growth.

Fig. 5 shows the voltage waveforms of the SS system DC bus U_{SS}, transformer U_{TR}, DC-link voltages U_{SS} (a) and transformer voltage U_{TR} and current I_{TR} (b) Since the half-bridge VSI was used in the SS side, the transformer U_{TR} and the DC-link voltage U_{DC} amplitude (30 V) is half of the SS DC bus voltage U_{SS} amplitude (60 V).

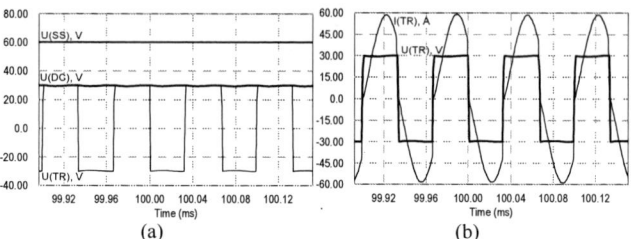

(a) (b)

Fig. 5. Simulation results of the proposed converter during the SC charging mode: SS side voltage (U_{SS}), transformer voltage (U_{TR}), DC-link voltage (U_{DC}) (a) and transformer voltage (U_{TR}), transformer current (I_{TR}) (b).

Finally, Fig. 6 shows that the SC is being charged up to the rated voltage U_{SC} if the interface converter SS side voltage U_{SS} is at the predefined value (60 V).

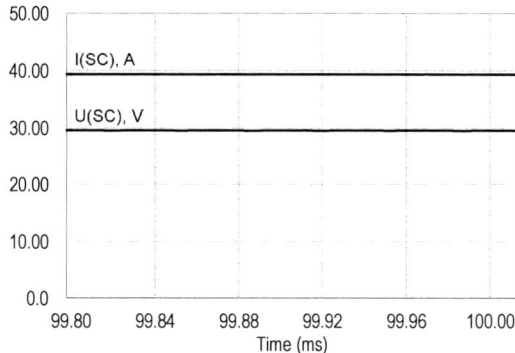

Fig. 6. Simulation results of the proposed converter during the SC charging mode: SC voltage (U_{Sc}) and SC current (I_{SC}).

B. Discharging Mode

The circuit configuration in Fig. 3 was used in simulations. The simulations were made at two extremes when the SC voltage U_{SC} is at its rated value (28.8 V) and at half of the rated value (14.4). It means that in order to provide constant DC-link voltage U_{DC} also, the minimal shoot-through duty cycle $D_{S,min}$ and maximal shoot-through duty cycle $D_{S,max}$ of the qZSI should be introduced. $U_{SC,max}$ should be associated with $D_{S,min}$=0.02 and $U_{SC,min}$ should be associated with $D_{S,max}$=0.26. The operation frequency of the isolation transformer during the discharging mode is 15 kHz.

Operation with 20 % of full load

First simulations were made with U_{SC}=14.4 V and U_{SC}=28.8 V at a load resistance of 36 Ω that provides the converter power of 100 W.

978-1-4244-8806-3/11 $26.00 © 2011 IEEE

Fig. 7 shows two conditions when the SC voltage U_{SC} is nominal (Fig. 7*a*) and when the SC voltage is half of the nominal $0.5U_{SC}$ (Fig. 7*b*). It is obvious that to provide the constant power rating of the converter the SC current I_{SC} have to increases (Fig. 7*b*).

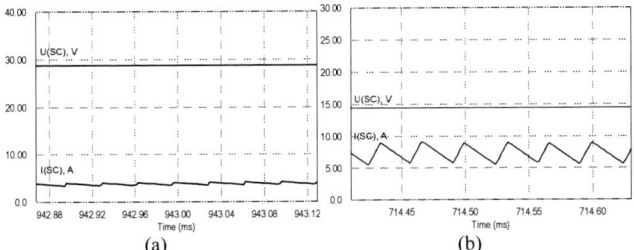

(a)   (b)

Fig. 7. Simulation results of the proposed converter during the SC discharging mode: SC stack voltage (U_{SC}) and current (I_{SC}) at nominal (28.8 V) (a) and at half of nominal (14.4 V) (b) voltage.

The nominal SC voltage U_{SC} and minimal shoot-through duty cycle of the qZSI (0.02) can provide the nominal DC-link U_{DC}, transformer U_{TR} and SS side DC bus U_{SS} voltage amplitudes (Fig. 8*a*). Despite the operating voltage of the SC U_{SS} decrease up to half of the nominal $0.5U_{SC}$ the DC link voltage as well as U_{TR} and U_{SS} can be kept on the predefined level by the implementation of the maximal shoot-through duty cycle (0.26) (Fig. 8*b*).

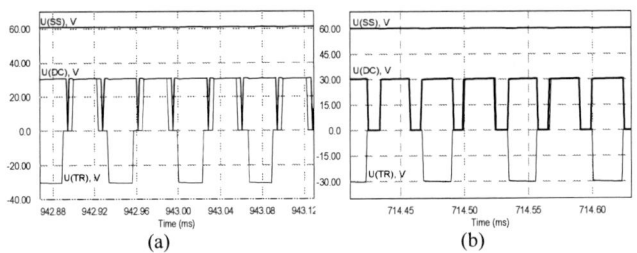

(a)   (b)

Fig. 8. Simulation results of the proposed converter during the SC discharging mode: DC-link voltage (U_{DC}), transformer voltage (U_{TR}), SS DC bus voltage (U_{SS}), at SC stack nominal (28.8 V) (a) and half of nominal (14.4 V) (b) voltage.

Irrespective of the fact that the SC U_{SC} voltage can decrease the SS DC bus voltage U_{SS} and current I_{SC} are the same in the case of nominal and half of the nominal voltage of the SC (Fig. 9*a* and *b*).

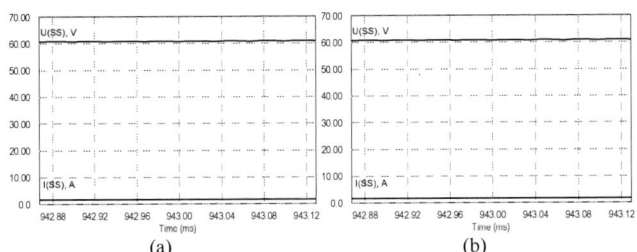

(a)   (b)

Fig. 9. Simulation results of the proposed converter during the SC discharging mode: SS side voltage (U_{SS}) and current (I_{SS}) at SC stack nominal (28.8 V) (a) and half of nominal (14.4 V) (b) voltage.

Operation with a full load

The same converter topology was used for the simulations with a load resistance of 7.2 Ω, which provides the 500 W load of the proposed converter. Also, in this case two experiments at the nominal (U_{SC}=28.8 V) and half of the nominal ($0.5U_{SC}$=14.4 V) value of the SC voltage were made. Fig. 10 shows that the SC current I_{SC} is higher at the power rating of 500 W than in the case of 100 W, particularly in the case of higher voltage boost.

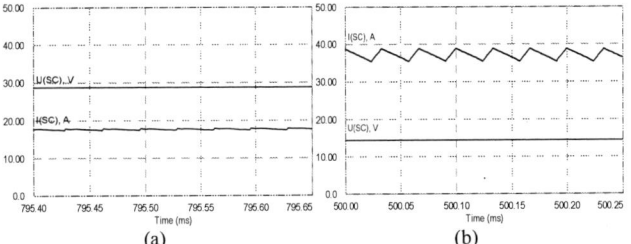

(a)   (b)

Fig. 10. Simulation results of the proposed converter during the SC discharging mode: SC stack voltage (U_{SC}) and current (I_{SC}) at nominal (28.8 V) (a) and half of nominal (14.4 V) (b) voltage.

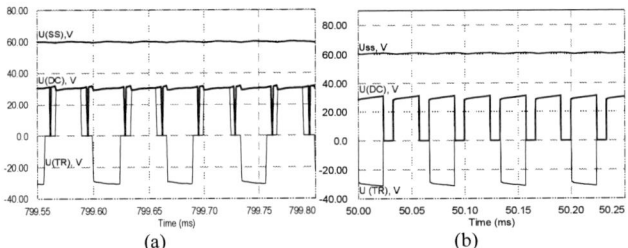

(a)   (b)

Fig. 11. Simulation results of the proposed converter during the SC discharging mode: DC-link voltage (U_{DC}), transformer voltage (U_{TR}) SS DC bus voltage (U_{SS}), at SC stack nominal (28.8 V) (a) and half of nominal (14.4 V) (b) voltage.

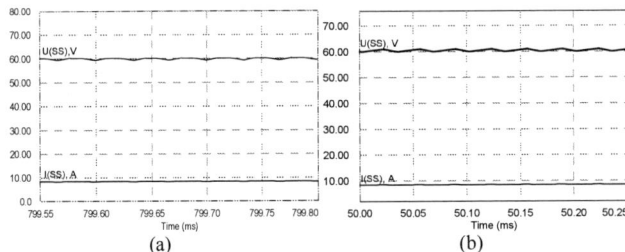

(a)   (b)

Fig. 12. Simulation results of the proposed converter during the SC discharging mode: SS side voltage (U_{SS}) and current (I_{SS}) at SC stack nominal (28.8 V) (a) and half of nominal (14.4 V) (b) voltage.

Despite the higher load power the variation of the SC voltage does not make it difficult to keep U_{DC}, U_{TR} and U_{SS} on the predefined level (Fig. 11*a* and *b*). Also, U_{SS} and I_{SS} in both SC voltage extremes are the same (Fig 12*a* and *b*).

IV. EXPERIMENTAL RESULTS

All the theoretical and simulation results were verified on the laboratory setup of the qZSI based bi-directional DC/DC converter. The operation and element ratings of the experimental setup are listed in Table I. The SC stack consists

978-1-4244-8806-3/11 $26.00 © 2011 IEEE   397

of a series connection of two Maxwell Technologies SC modules (PCM14014) with a total capacitance value of 85 F, rated voltage U_{SC}=28.8 V and equivalent series resistance of 20 mΩ.

A. Charging Mode

At the very beginning of SC charging process the SC current I_{SC} can reach high values. To overcome this issue the SC stack was charged up to 12.5 V.

Fig. 13 shows that during the charging process the SS DC bus voltage U_{SS} is constant and the SC stack voltage U_{SC} increases up to the nominal value (28.8 V). At the beginning of charging process the SC stack current I_{SC} (Fig. 13) is around 135 A.

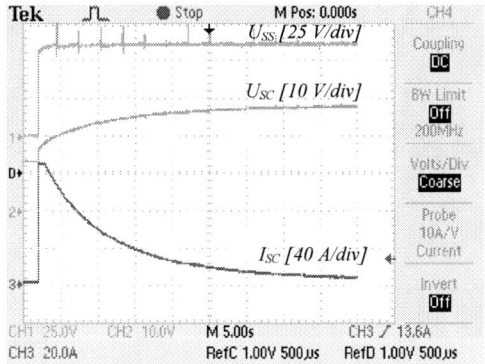

Fig. 13. Experimental waveforms of the charging process of the SC: SS DC bus voltage (U_{SS}), SC stack voltage (U_{SC}) and SC stack current (I_{SC}).

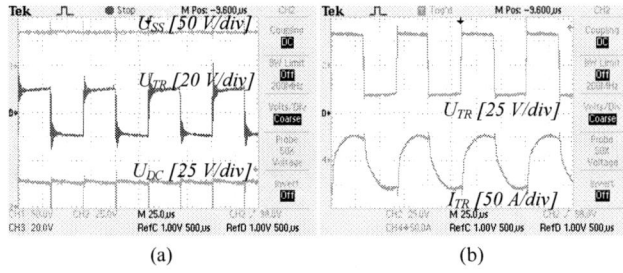

(a)   (b)

Fig. 14. Experimental waveforms during the SC charging mode: SS side voltage (U_{SS}), transformer voltage (U_{TR}), DC-link voltage (U_{DC}) (a) and transformer voltage (U_{TR}), transformer current (I_{TR}).

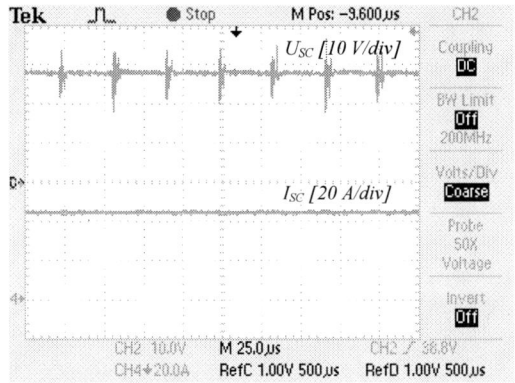

Fig. 15. Experimental waveforms during the SC charging mode: SC voltage (U_{SC}) and SC current (I_{SC}) during steady state.

Figs. 14 and 15 show that experimental waveforms of the proposed converter are in full agreement with the same waveforms that were made in simulation software.

B. Discharging Mode

The experimental verification of the proposed converter was carried out to repeat the theoretical results obtained in the previous section.

Operation with 20 % of full load

The practical experiments at the power rating of 100 W of the proposed converter are presented in Figs. 16 - 18. As pointed out above, if the converter is supplied with the nominal or half of the nominal SC voltage (U_{SC} ...0.5U_{SC}) (Fig. 16a and b) it can ensure the desired U_{DC}, U_{TR} and U_{SS} at rated values (Fig. 17a and b) by the regulation of D_{S} of the qZSI.

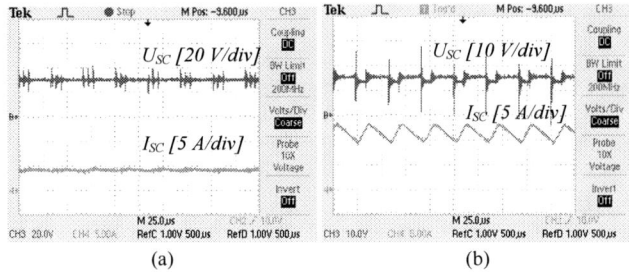

(a)   (b)

Fig. 16. Experimental waveforms of SC stack voltage (U_{SC}) and current (I_{SC}) at nominal (28.8 V) (a) and half of nominal (14.4 V) (b) value.

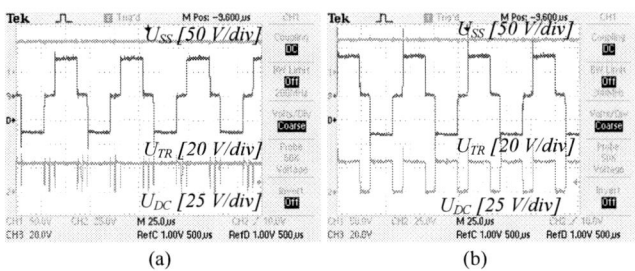

(a)   (b)

Fig. 17. Experimental waveforms of SC stack voltage (U_{SC}), transformer voltage (U_{TR}) and DC-link voltage (U_{DC}) at SC stack nominal (28.8 V) (a) and half of nominal (14.4 V) (b) value.

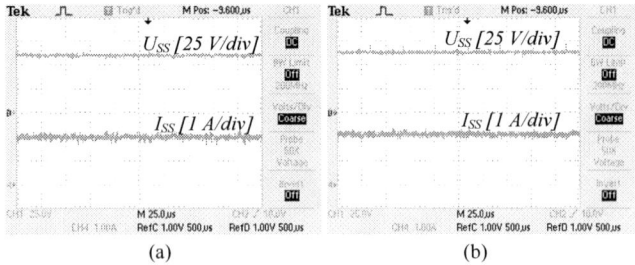

(a)   (b)

Fig. 18. Experimental waveforms of SS side voltage (U_{SS}) and current (I_{SS}) at SC stack nominal (28.8 V) (a) and half of nominal (14.4 V) (b) value.

It can be concluded that despite the SC voltage U_{SC} drop the desired SS side DC bus voltage U_{SS} level (Fig. 18a and b) can provide the same I_{SS} at equal load conditions.

Operation with a full load

The second group of practical experiments was carried out at the power rating of the proposed converter of 500 W. Also, in this case all the results are in agreement with theoretical assumptions and simulation results, i.e. appropriate voltage boost can compensate the reduced SC U_{SC} (Fig. 19a and b) in order to obtain the desired voltage level on the DC-link, transformer and SS bus (Fig. 20a and b).

As a result of comparing the experimental waveforms in the conditions of 100 W and 500 W loading it can be concluded that the operation of the proposed converter is sufficiently adequate and it can successfully provide demanded voltage gain on the SS DC bus when it is necessary.

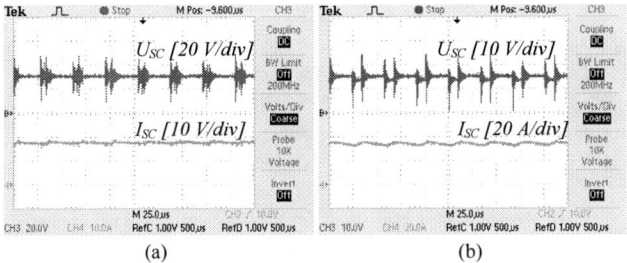

(a)   (b)

Fig. 19. Experimental waveforms of SC stack voltage (U_{SC}) and current (I_{SC}) at nominal (28.8 V) (a) and half of nominal (14.4 V) (b) value.

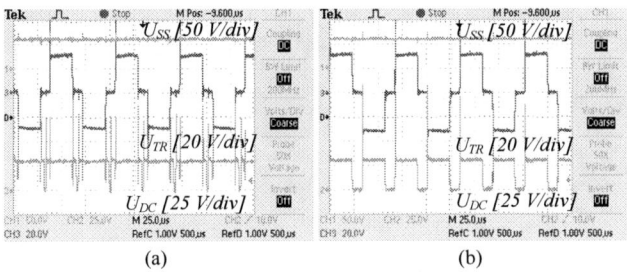

(a)   (b)

Fig. 20. Experimental waveforms of SC stack voltage (U_{SC}), transformer voltage (U_{TR}) and DC-link voltage (U_{DC}) at SC stack nominal (28.8 V) (a) and half of nominal (14.4 V) (b) value.

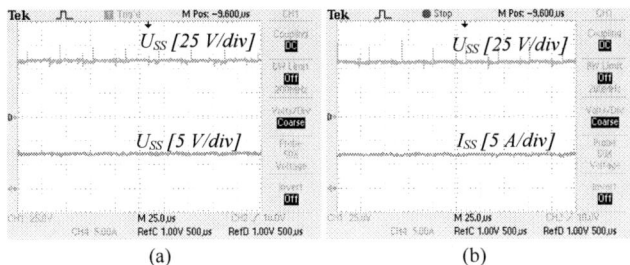

(a)   (b)

Fig. 21. Experimental waveforms of SS side voltage (U_{SS}) and current (I_{SS}) at SC stack nominal (28.8 V) (a) and half of nominal (14.4 V) (b) value.

V.   CONCLUSIONS

This paper describes a qZSI based bi-directional DC/DC converter for supercapacitor interfacing. The proposed converter was simulated in both operation modes. Also, an experimental verification was conducted on a 500 W laboratory prototype and compared with simulation results.

The most important generalizations and recommendations result in the following:

- to increase the efficiency of the proposed converter the cross-section of the conductors in the SC side should be properly selected;
- since some of the alternative energy sources (FC) are sensitive to the output current ripple, special attention should be paid to the selection of qZS-network inductors for the conditions of maximal boost;
- to escape additional input voltage boost on the account of semiconductor loss, power switches with the lowest forward voltage drop capability and qZS-network diodes with fast recovery time should be used;
- to reduce the high inrush current during the charging mode of the SC soft starters should be used.

The worst case efficiency of proposed converter is 87%.

It can be concluded that the proposed SC interfacing converter is a suitable and competitive solution for short term high-power energy storage devices in FC and/or PV powered residential energy supply systems.

ACKNOWLEDGMENT

This research work was supported by Estonian Ministry of Education and Research (Project SF0140016s11), Estonian Science Foundation (Grant ETF8538) and European Social Fund's Researcher Mobility Program "Mobilitas" (Grant no MJD42).

REFERENCES

[1] Nowak, M.; Hildebrandt, J.; Luniewski, P., "Converters with AC transformer intermediate link suitable as interfaces for supercapacitor energy storage", IEEE 35th Annual Power Electronics Specialists Conference PESC'04, vol. 5, pp. 4067-4073, 20-25 June 2004.

[2] D. Vinnikov, I. Roasto, J. Zakis, „New bi-directional DC/DC converter for supercapacitor interfacing in high-power applications", in Proc. EPE-PEMC 2010, Vol. 11, pp. 38-43.

[3] Vinnikov, D.; Roasto, I., "Quasi-Z-Source-Based Isolated DC/DC Converters for Distributed Power Generation", IEEE Transactions on Industrial Electronics, vol. 58, no. 1, pp. 192-201, Jan. 2011.

[4] Strzelecki, R., Adamowicz, M., "Boost-buck inverters with cascaded qZ-type impedance networks", Electrical Review, vol. 86, no. 2, pp. 370-375, 2010.

[5] Vinnikov, D., Roasto, I.; Strzelecki, R., Adamowicz, M., "Performance Improvement Method for the Voltage-Fed qZSI with Continuous Input Current", in Proc. of 15th IEEE Mediterranean Electrotechnical Conference MELECON'2010, pp. 1459-1464, April 25-28, 2010.

[6] Vinnikov, D., Roasto, I.; Strzelecki, R., Adamowicz, M., "Two-Stage Quasi-Z-Source Network Based Step-Up DC/DC Converter", in Proc. of IEEE International Symposium on Industrial Electronics ISIE'2010, pp. 1143-1148, July 4-7, 2010.

[7] Vinnikov, D.; Roasto, I.; Zakis, J.; Strzelecki, R. "New Step-Up DC/DC Converter for Fuel Cell Powered Distributed Generation Systems: Some Design Guidelines", in journal "Electrical Review" ISSN 0033-2097, Vol. 86., Nr 8. pp. 245-252, 2010.

[8] J. Anderson, F.Z. Peng, "Four quasi-Z-Source inverters", in Proc. of IEEE Power Electronics Specialists Conference PESC'2008, pp. 2743-2749, 15-19 June 2008.

[9] Yuan Li; Anderson, J.; Peng, F.Z.; Dichen Liu, "Quasi-Z-Source Inverter for Photovoltaic Power Generation Systems", in Proc. of IEEE Applied Power Electronics Conference and Exposition APEC'2009, pp. 918-924, Feb. 15-19, 2009.

Comparison of two Power Electronic Schemes for 3 kW Li-Ion Battery Charger

Alexander Suzdalenko and Ilya Galkin
Riga Technical University
Aleksandrs.Suzdalenko@RTU.lv, gia@avene.eef.rtu.lv

Abstract- Battery management systems are getting of high importance, mainly due to growing interest in electrical vehicles. It can be concluded that the amount of electrical energy that is going to be transformed to different forms by means of power electronics is growing. In context of energy efficiency and stability two battery charger power circuits are presented, calculated and simulated. The results are discussed and plans for the future work are defined.

I. INTRODUCTION

The major problem of renewable energy sources is their unpredictability. Energy flow needs to be smoothed by electrical energy storage devices such as batteries, flywheels, super capacitors, etc. It should be concluded in this context, that transformation of electrical energy into another form by means of power electronics is one of the areas with the most explicit challenge - to transform electrical energy as efficiently as possible.

Moreover, the usage of distributed power storages, including batteries, sooner or later will also facilitate the introduction of MicroGrid and SmartGrid concept. That allows managing consumer electricity demand by means of information transfer about prices for electrical energy at present time, meaning that power plants will work at more constant load.

II. DESCRIPTION OF PROTOTYPE

Battery management system is definitely going to be significant challenge that will be used not only in households to ensure the stability of local electricity supply, but also for charging electrical vehicles, which will inevitably emerge in the nearest future. As it might be expected, the amount of transformed energy in charging applications is expected to be noticeable, that is why this problem should be studied and an effective battery charging system should be developed.

The count of semiconductor elements in the current path should be minimized in order to eliminate its conduction loses and to achieve high efficiency, as well as to implement advantageous converter topologies realizing zero voltage or/and zero current switching, to eliminate switching loses.

Figure 1 shows preliminary configuration of the planned battery charger, which consists of diode bridge, resonant DC/DC converter with implemented PFC control algorithm. The algorithm is controlled by microcontroller with appropriate configuration of IO ports. Also it is planned to make HMI by connecting keyboard and display to the charger system.

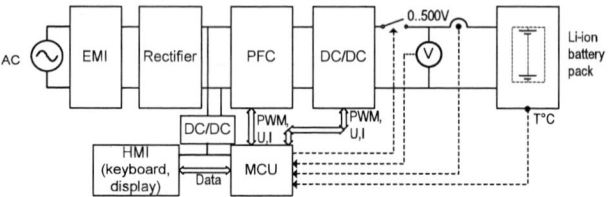

Fig..1. Functional diagram of the preliminary Li-ion charger

This is a very challenging task, because of the very high requirements for the efficiency and at the same time it should not be an expensive solution, because potentially it should be used in mass production. That is why advanced power converter topologies should be implemented in this task. Preliminary analysis revealed the advisability of resonant rectifier topologies because of combined power factor correction with a DC regulation schemes in single stage [2, 6]. This assumes that the switching loses will be minimized as well as component count will be decreased, in comparison with traditional boost active PFC topology and DC-DC regulator implemented in two stages. There are many of resonant converter topologies which should be simulated and analyzed before practical implementation.

MCU is used to control the charger. It implements several functions at a time: monitors parameters of a battery (U, I, temperature), controls DC/DC converter by programmed algorithm and supports human machine interface (HMI). That is why it should be chosen with adequate computing power and required peripheral modules.

HMI module is planned to be realized with external keyboard and display for interaction with user. The display allows to configure the charger by changing parameters of battery (C [kWh], U_{nom} [V], U_{max} [V], I_{cont} [A]). It will be possible to choose one of several pre-programmed configurations for different batteries or different battery technologies to make the charger as versatile as possible.

It is planned to implement the "constant current constant voltage" (CCCV) algorithm as the main charging algorithm.

III. CHOICE OF POWER CONVERTER TOPOLOGY

Battery charger consists of several energy conversion stages: 1) AC/DC rectifying, 2) power factor correction stage and 3) DC/DC conversion for adjusting charger's output voltage to battery voltage. It is possible to combine some of these stages into one by utilizing specialized scheme topologies, for example one of the PFC boost family converters, like full bridge, half bridge, bridgeless for

978-1-4244-8806-3/11 $26.00 © 2011 IEEE   400

implementing AC rectifying and power factor correction in a single stage. These topologies can also adjust DC link voltage in some ranges, allowing to provide desirable DC voltage level for the next conversion stage. Considering, that battery voltage can also be smaller than charger's input voltage, for this reason a buck converter is obligatory to use with AC/DC+PFC scheme, adjusting its charging voltage to control battery charging current.

There are known configurations of converter topologies, which realize resonance in power flow, allowing to make switching in Zero Voltage (ZVS) or in Zero Current (ZCS) moments, called Quasi-Resonant converters (QRC) [1, 4]. The basic idea is to use additional reactive elements in parallel or in series with switching semiconductors, which defines the switching mode.

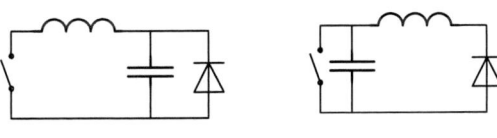

a   b

Fig.2 Quasi-resonant switches. (a) Zero-current. (b) Zero-voltage

The configuration of reactive element provides either ZCS for active switch and ZVS for diode, or vice versa in (2.b), providing desirable switching conditions either for transistor or for diode.

There are also known enhanced quasi-resonant converters that were previously mentioned, which realize ZVS or ZCS for both semiconductors, called multi-resonant converters [1].

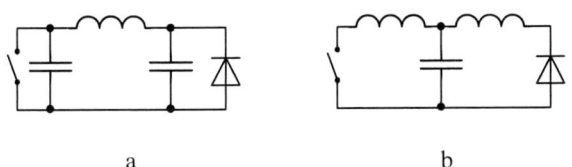

a   b

Fig.3. Multi-resonant switches. (a) Zero-voltage. (b) Zero-current

IV. DESCRIPTION OF SIMULATION MODELS

Two power converter topologies have been chosen for the charger system to compare its profitability and reveal advantages and disadvantages. First solution is simple one - diode bridge rectifier, boost for PFC and buck for DC/DC regulation. The second solution is chosen with respect to defined requirements (minimized number of semiconductor elements in a current path, as well as soft switching). That is why the half bridge rectifier has been chosen as main AC/DC rectifier, which also realizes PFC feature, the DC/DC regulator – resonant buck converter, with minimized switching loses [3].

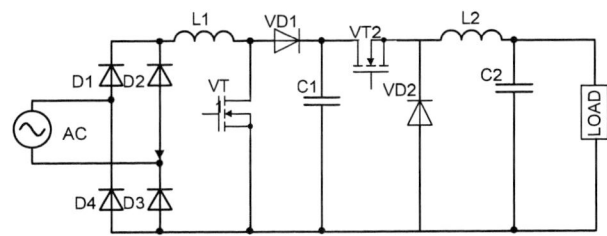

Fig.4. Simple charger – diode rectifier, boost for PFC and Buck for DC/DC regulation

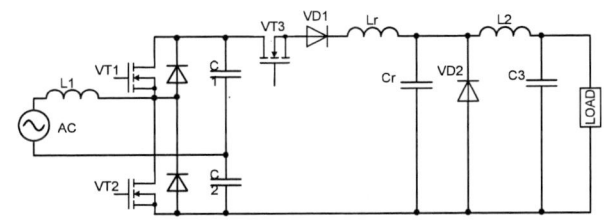

Fig.5. Advanced charger – Half bridge for AC/DC+PFC and zero current switching Buck for DC/DC

A. Simple configuration

The disadvantages of this circuit should be mentioned as relatively big semiconductor count in the current path (4 at a moment), hard switching of semiconductor elements. These are potential power losses, which is unwanted in this application. The simulation model is presented in Figure 6.

B. Advanced configuration

Half bridge converter has known advantages comparing with the previous topology – there is only one semiconductor element in the current path at a time. Secondly, it operates as voltage doubler that is why the DC link voltage is high enough even at 110V grids. It operates this way - lower transistor VT2 (see Fig.5) commutates the input voltage through the C2 capacitor during positive half period of input voltage. Inductor's energy is accumulated during transistor on state, but after transistor is switched off, the energy is commutated through diode, which is integrated into transistor VT1, boosted current then flows through capacitor C1. During negative input's voltage half period the upper transistor SW1 is driven, while lower transistor is represented only as a diode.

The correction of power factor is realized with the line follower algorithm with PI regulator (see. Fig.7), which adopts the amplitude of the consumed current to maintain dc link voltage close to reference value, but due to sinusoidal current consumed from the input, DC link voltage is also sinusoidal with certain offset value, which in ideal is equal to reference value.

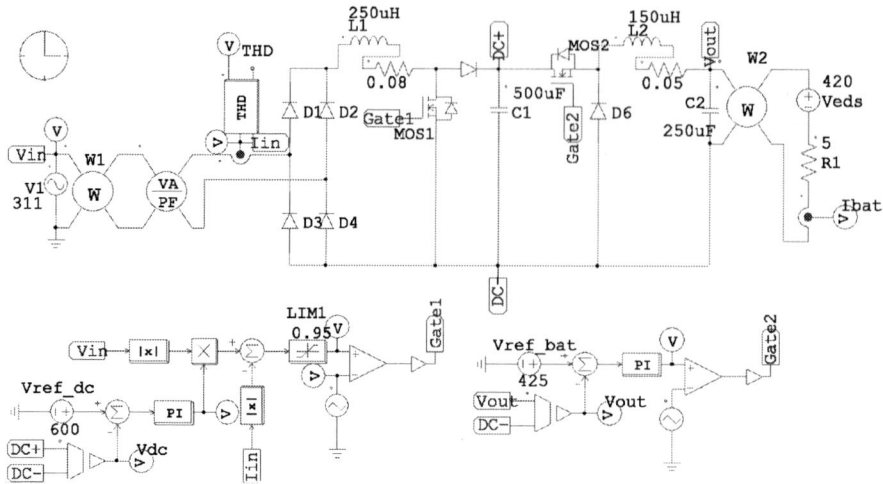

Fig.6. Simple charger model in PSIM simulation program

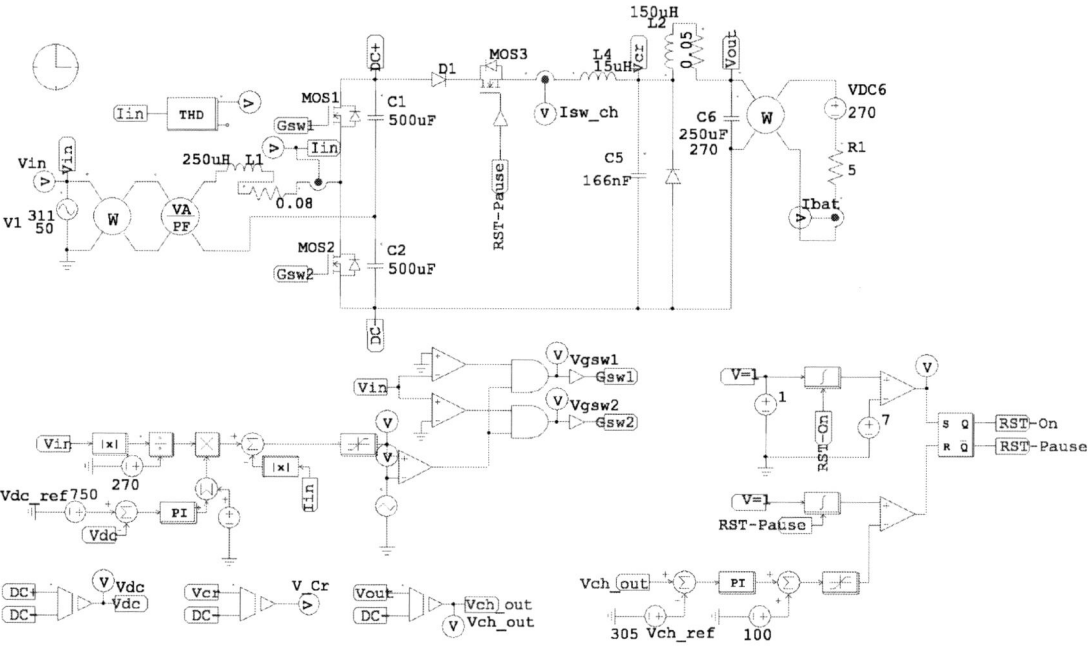

Fig.7. Advanced charger model in PSIM simulation program

The resonant contour can be calculated, by using following formulae [5]:

$$E_{\tan k} = \frac{P_{out}}{f_r} = \frac{1}{2}L_r \cdot i_L^2 = \frac{1}{2}C_r \cdot V_{pri}^2 \qquad (1)$$

Resonance frequency of the LC contour can be calculated from formula below:

$$f_r = \frac{1}{2\pi\sqrt{L_r C_r}} \qquad (2)$$

After choosing the *Pout* and *Vpri*, it is possible to calculate the capacitance of Cr:

$$C_r = \frac{2P_{out}}{V_{pri}^2 \cdot f_r} = \frac{2 \cdot 3000}{600^2 \cdot 10^6} = 166[nF] \qquad (3)$$

Value of resonance inductor can be found using (2) equation below:

$$L_r = \frac{1}{C_r(2\pi f_r)} = \frac{1}{1.66 \cdot 10^{-9}(2\pi 10^6)2} = 15 \cdot [\mu H] \qquad (4)$$

Buck regulator is connected to the DC bus. The main switch operates at Zero Current switching, due to serially connected inductor (see Fig. 8). It is driven by frequency modulation (FM) with constant pulse time, where PI regulator adopts the pause time of the working period to maintain certain voltage at the output

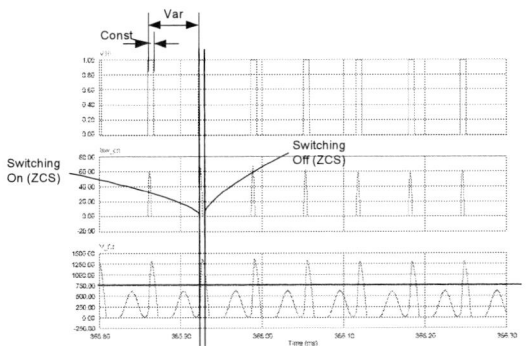

Fir.8 Zero Current switching periods

The output capacitor is chosen in relation with fallowing formula:

$$C_o = I_{out\_max} \frac{T_{off}}{C_{ripple}} = 7 \cdot \frac{10 \cdot 10^{-6}}{1} = 70[\mu F] \Rightarrow 125[\mu F] \quad (5)$$

V. RESULTS

The following table (Table I) presents the simulated results at input voltage 311V (peak).

TABLE I
SIMULATION RESULTS AT 311V INPUT VOLTAGE

Vout/Iout	Efficiency		THD		PF	
	A.	B.	A.	B.	A.	B.
270v@7A	96.5	94.6	12.4	2.2	0.992	0.998
320V@7A	96.6	95.1	11.8	1.5	0.993	0.999
370V@7A	96.7	95.1	11.3	1.3	0.993	0.999
420V@7A	96.8	95.0	9.6	-	0.994	0.999
420V@4A	96.8	94.9	13.2	5.4	0.990	0.999
420V@1A	95.7	92.1	21.2	19	0.955	0.987

A – simple solution; B – advanced solution

The next table (Table II) shows similar measurements at 155V input voltage (peak).

TABLE II
SIMULATION RESULTS AT 155V INPUT VOLTAGE

Vout/Iout	Efficiency		THD		PF	
	A.	B.	A.	B.	A.	B.
270v@7A	92.8	91.7	8.6	4	0.994	0.998
320V@7A	92.3	92.1	8.2	5	0.995	0.998
370V@7A	92.8	91.7	7.9	4.7	0.995	0.998
420V@7A	91.4	91.1	7.9	3.9	0.995	0.998
420V@4A	93.7	92.2	9.4	4.8	0.993	0.998
420V@1A	94.8	88.6	21.9	11.8	0.975	0.992

A – simple solution; B – advanced solution

VI. CONCLUSIONS

In this paper two power electronic topologies have been presented for implementation of 3 kW battery charger system. Charger schematics has been studied, calculated and simulated in PSIM6.0 software with the 0.1 μs time step during 0.7 s. During this period PI regulators reached steady-state values, so the output values were constant or were fluctuated at a constant offset and with constant amplitude.

The simple solution consisted of diode rectifier, boost PFC and buck regulator is relatively simple from point of view of control – it has only two easy controllable (with PWM) switches. On the other hand, it has big number of semiconductor elements in the current path, if to compare it with second examined solution. Besides that all of the switching elements in the first solution are hard switching, in contrast to realized Zero Current switching approach in the second solution. Despite these conceptual ideas the simulation results showed that "advanced" topology is not the best because of the higher losses occurring in the AC/DC converter stage – the switches in half-bridge topology are hard switched, at the same time the voltage applied to the transistors also is higher which requires additional safety initiatives. At the same time, THD values of simple solution are noticeably higher and PF values are also slightly worse. Additionally, it is possible to eliminate switching losses in both examples by using additional circuits to achieve ZCS or ZVS switching mode, which is planned to be done in the nearest future. It is also planned to present additional charger circuit examples with galvanic isolation.

REFERENCES

[1] W.A. Tabisz, F.C.Y. Lee, "Zero-voltage-switching multiresonant technique-a novel approach to improve performance of high-frequency quasi-resonant converters," *IEEE Transactions on Power Electronics.* vol. 4, no. 4, October 1989., pp. 740-741.

[2] H. Benqassmi; J.-C. Crebier, J.-P. Ferrieux, "Comparison between current-driven resonant converters used for single-stage isolated power-factor correction," *IEEE Transactions on Industrial Electronics.* vol. 47, no. 3, June 2000., pp. 518-524.

[3] R.L. Steigerwald, "A Comparison of Half-Bridge Resonant Converter Topologies," *IEEE Transactions on Power Electronics.* vol. 3, no. 2, April 1988., pp. 174-182.

[4] B.Andreycak. "Zero Voltage Switching Resonant Power Conversion". *Application note U-318* [Online [Accessed October 1, 2010].

[5] M.Braun: Power Supplies Cookbook, ELSEVIER, Burlington, UK, 2001, 265pp.

[6] R. Liu, I. Batarseh, and C. Q. Lee, "Resonant power factor correction circuits with resonant capacitor-voltage and inductor-current-programmed controls," in *Proc. IEEE PESC'93*, Seattle, WA, June 20–24, 1993, pp. 675–680

Power Losses Calculation Methodology to Evaluate Inverter Efficiency in Electrical Vehicles

Josep Pou, Daniel Osorno, Jordi Zaragoza, Salvador Ceballos[*], and Carles Jaen

Technical University of Catalonia, Department of Electronic Engineering, Catalonia, Spain, e-mail: *josep.pou@upc.edu*

[*]Tecnalia Research & Innovation, Energy Unit, Basque Country, Spain, e-mail: *salvador.ceballos@tecnalia.com*

Abstract–Nowadays, electrical vehicles (EVs) are of special interest and fuel engines are starting to be substituted by electrical motors. Besides the batteries and the motor, the power electronic devices are important parts in an EV, especially the inverter. This ac/dc power converter drives the electrical motor. Several tools may be needed to achieve an optimal design of the inverter. This paper presents a simplified methodology to estimate power losses in a two-level inverter made up with isolated gate bipolar transistors (IGBTs). This methodology is based on the IGBT manufacturer datasheet; therefore, the knowledge of internal parameters used in other more complex methodologies is not required. The model is implemented in Matlab-Simulink and allows simulation of different power devices, modulation techniques, and operating conditions very easily. Some results are obtained and validated with commercial electronic software and also with the simulation tool Semisel, provided by the manufacturer of the IGBTs.

Index Terms – Electric vehicle (EV), inverter, power losses, conduction losses, switching losses.

I. INTRODUCTION

Nowadays, the most popular vehicle is the fuel car. However, the world is being witness of one of the biggest energy changes in history. Low reserves of oil are forcing this change towards plugged-in hybrid electric vehicles (PHEVs) and fully electric vehicles (EVs)[1],[2]. Fuel engines will be replaced by electric motors fed from battery cells through power electronic converters.

The efficiency of PHEVs is much higher (up to 80%) than the fuel vehicles, which is only about 20% [3]. Moreover, since oil reserves are running out, in the long term EVs will probably be the best successors of the fuel vehicle.

The inverter is the element that drives the electrical motor in an EV. This power converter is usually made up of IGBTs and fed from a dc-bus. The dc-bus can come either directly from battery cells or through a dc-dc boost converter. In the last case, the battery provides lower voltage and it is boosted to a proper value required by the inverter to drive the motor.

The inverter has power losses that can be divided into two different types: conduction losses and switching losses. Conduction losses are produced when the switch is in the on-state, and switching losses appear when the switch is turned on and off. Subsequently, switching losses are proportional to the switching frequency of the power devices. Estimation of the total inverter power losses is needed to select proper power devices. The maximum junction temperature of the IGBTs should not be exceeded at any time.

Our research group is a partner of the Spanish Project called CITYELEC - Electrification Systems for Urban Mobility (PSE-370000-2009-004) [4]. This project has received economic support from the European Union. Our team has worked on the on board electrical power system. Two different EV motion configurations are considered:

- A stand-alone motion that is made up by a sole motor (Fig. 1) providing mechanical energy to the wheels through a transmission, and
- A second architecture that consists of two motors (Fig. 2) located inside the wheels [5],[6]. This option avoids any mechanical transmission because the driving torque is applied directly to the front wheels. Each motor and electrical drive is rated at less than half of that in the stand-alone configuration.

In the case of the stand-alone scheme, the motor rated power is 15 kW and the maximum current is 225 A. On the other hand, in the in-wheel configuration, the power rating is 5.5 kW and the maximum current is 80 A. In both cases, the type of motor selected for the design is a permanent magnet synchronous machine (PMSM).

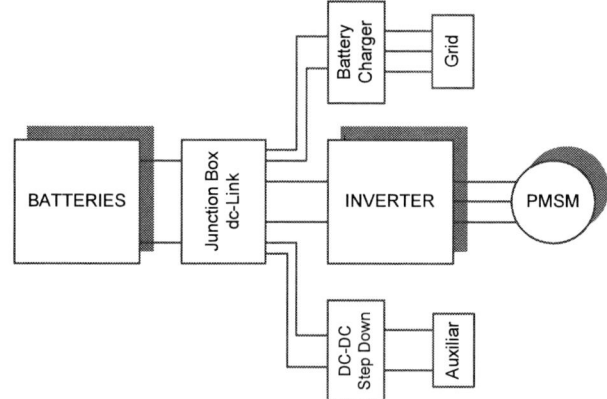

Fig. 1. EV; stand-alone configuration.

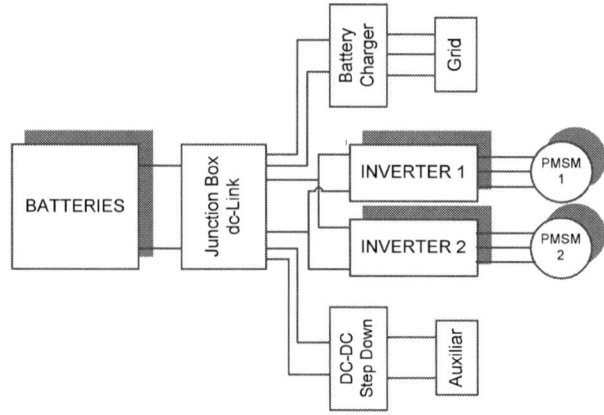

Fig. 2. EV; in-wheel configuration.

II. IGBT MODELS

There are different ways of modeling IGBTs for the calculation of power losses. However, some complex models require many IGBT parameters to perform the calculations.

A common model of an IGBT is shown in Fig. 3. It contains two devices, a metal-oxide-semiconductor field-effect transistor (MOSFET) and a bipolar junction transistor (BJT). Although this model seems to be very simple, it is very difficult to implement because the MOSFET and the BJT must be previously modeled, and many parameters are needed.

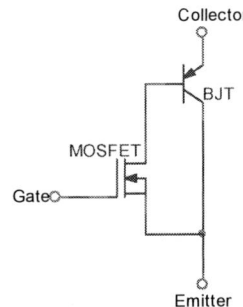

Fig. 3. Precise IGBT model – Not too detailed.

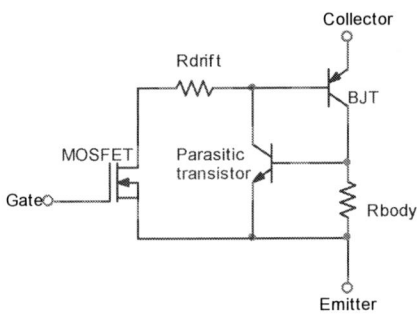

Fig. 4. Precise IGBT model – Medium detailed.

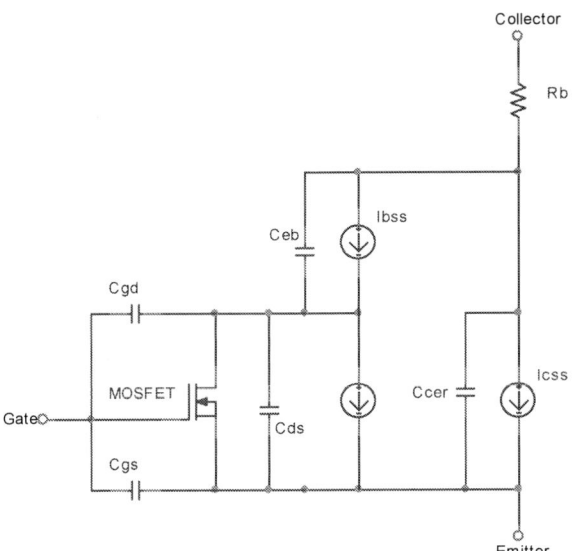

Fig. 5. Precise IGBT model – Very detailed.

Fig. 4 shows a more detailed model than the previous one, including a parasitic transistor and resistors to model the internal behavior of the MOSFET and the BJT. But the designer has the same problem; the parameters of the internal devices are not given in the datasheets.

Fig. 5 shows an IGBT scheme that can be easily implemented by simulation software [7],[8]. However, still a large list of IGBT parameters are needed, for instance, the "gate-drain capacitance" or the "collector-emitter redistribution capacitance".

Consequently, it is not easy for the designer to estimate the inverter power losses. Therefore, sometimes the inverter is over-sized to avoid temperature break down of the switches. This paper shows a way of modeling inverter power losses making use of only information provided in the IGBT datasheets [9],[10]. The proposed algorithm provides not only relatively accurate information, but it shows power losses for the entire operating conditions of the inverter.

Since practical operation conditions of an EV can make the electrical machine perform both, as a motor and as a generator, all possible output current angles have to be considered.

The instantaneous current and the voltage drop on the power devices are needed to calculate the conduction losses. On the other hand, the switching losses are calculated at any IGBT switching on and off events, adding up a calculated amount of energy at each switching step. This model only uses the specific IGBT data provided by the datasheet. No other parameters needed in standard models are required in this approach.

III. CONDUCTION MODEL

For the calculation of conduction losses, the voltage drops on the IGBT and the reverse diode under the on-state condition are needed. For this, the model is represented by a voltage source and a series resistor (Fig. 6). The characteristic values are obtained from the device datasheet using the V_{CE} vs. I_C curve. It is important to consider the junction temperature to choose the correct curve. The gate voltage is selected as 15 V (or the voltage provided by the switch drive), and the worst case temperature of 150 °C is considered.

For this EV application, the SEMIX151GD066HD IGBT is selected from Semikron. Fig. 7 shows the IGBT conduction curves given in the datasheet.

For a proper linearization, the maximum operating output current should be considered. In this example, the values extracted from the IGBT V_{CE} vs. I_C curve are given in Table I.

The same procedure is applied to the diode conduction characteristic. Fig. 8 shows the diode datasheet graph with the linearized curve. Table I also shows the characteristic values obtained from linearization of the diode.

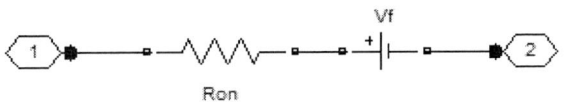

Fig. 6. On-state IGBT model.

978-1-4244-8806-3/11 $26.00 © 2011 IEEE

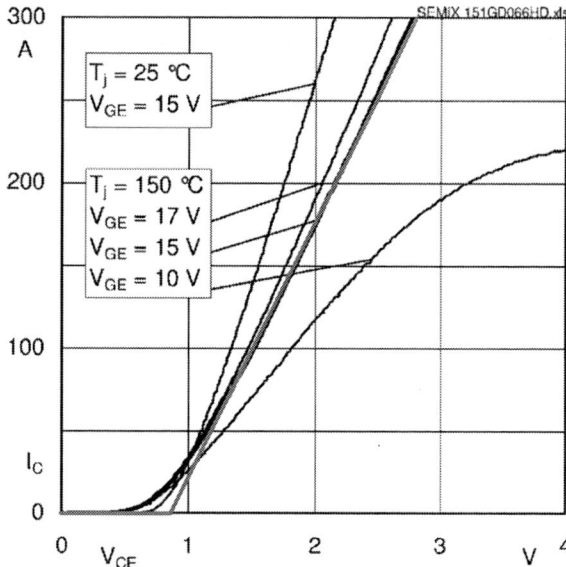

Fig. 7. SEMIX151GD066HD IGBT - V_{CE} vs. I_C curves.

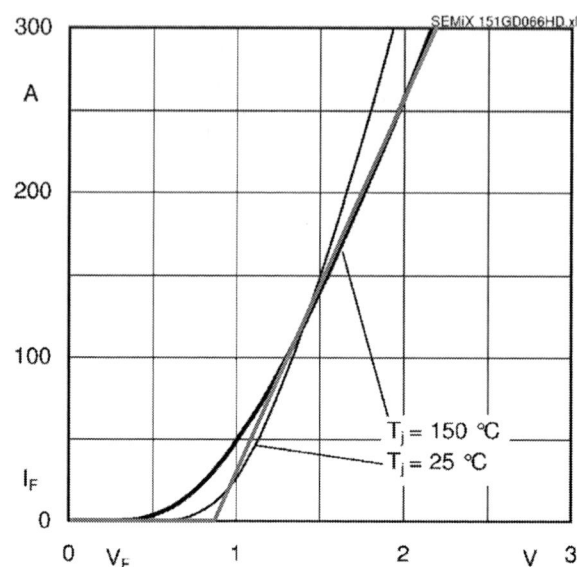

Fig. 8. Diode within the SEMIX151GD066HD - V_F vs I_F curve linearized.

Although the IGBT model used for the conduction losses calculation is very simple, in practice, more sophisticated models are not required because they would provide similar results.

IV. SWITCHING MODEL

Switching losses are calculated using a polynomial approximation of the energy vs. current charts curves. This procedure consists on obtaining values from the curves and then adjusting a second-order polynomial approximation by a proper math tool, such as the Polyfit function in Matlab. As these curves are given for a specific voltage value (the voltage that the IGBT has to stand in the off state), it is necessary to consider the real dc bus voltage of the inverter [3]. As in the conduction model, the junction temperature is another parameter that should be taken into account.

Fig. 9 shows the IGBT switching approximation giving three different points. Datasheets usually show the three switching curves in the same graph, but others show the turn-on (E_{on}) and turn-off (E_{off}) IGBT energy in a graph, and the diode recovery energy (E_{rr}) in another graph.

Table II shows the characteristic points selected from the datasheet curves. Applying the Polyfit function to the captured data, the switching losses approximation is obtained, as Table III shows. In this table, the V_{off} value is the dc voltage that the IGBT has to stand when it is in the off state. The switching losses are assumed proportional to this voltage.

The switching losses fitting curves should be checked with the original curves to ensure they are accurate approximations. This checking is performed by placing the approximated curves obtained by Polyfit on the same graph of the manufacturer datasheet, as Fig. 10 shows.

Summarizing, the power-losses model of the IGBT is obtained from the manufacturer datasheet. Conduction (V_{CE} vs. I_C) and switching curves (E vs. I_C) are needed. The

TABLE I
PARAMETERS FROM THE IGBT AND DIODE CONDUCTION CURVES

Linearized conduction losses	Value
V_T (IGBT threshold voltage)	0.85 V
R_T (IGBT conduction slope)	5.86 mΩ
V_D (Diode threshold voltage)	0.8 V
R_D (Diode conduction slope)	4.76 mΩ

conduction losses are linearized by a dynamic resistor (reverse slope) and a threshold voltage. Switching losses are approximated by a second-order polynomial using the Polyfit function of Matlab. Three points are extracted from the curves. The switching approximation should be validated with the original graph to ensure the second-order approximation is accurate enough.

V. POWER LOSSES CALCULATION

In this work, the power losses calculation algorithm is performed by two Matlab files; the main file (.m structure) and a Simulink program (.mdl file), which is activated from the main file. The Simulink program generates the phase control signal (s_a for phase a) in accordance with the selected modulation strategy. This control signal defines the state of the upper IGBT of the leg as follows;

$$s_a = \begin{cases} 1 \text{ if the upper switch is the on state, and} \\ 0 \text{ otherwise.} \end{cases}$$

Two modulation strategies are programmed; sinusoidal pulsewidth modulation (SPWM) and space-vector modulation (SVM). SPWM is used in this study to compare the proposed algorithm with PSIM and Semisel (from Semikron), because the last one can only work with SPWM.

Some inputs of the Matlab-Simulink program are the relative phase of the output current and the dc-link voltage (V_{dc}). The output current is imposed by a sinusoidal current source in which both, its rms value and relative phase are defined. The V_{dc} voltage is the dc-link voltage of the inverter.

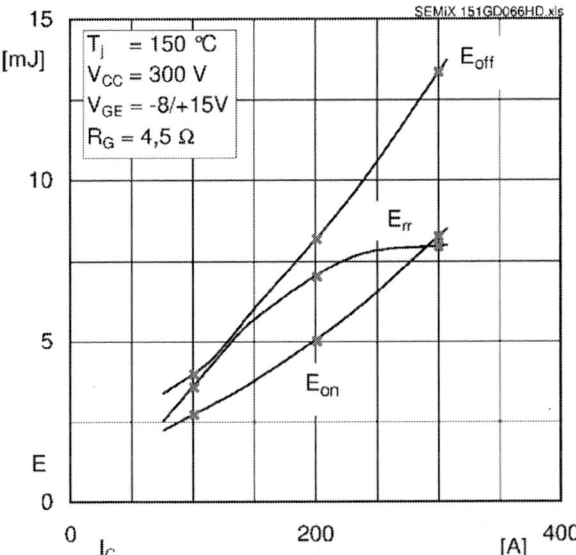

Fig. 9. SEMIX151GD066HD IGBT – Switching losses approximation.

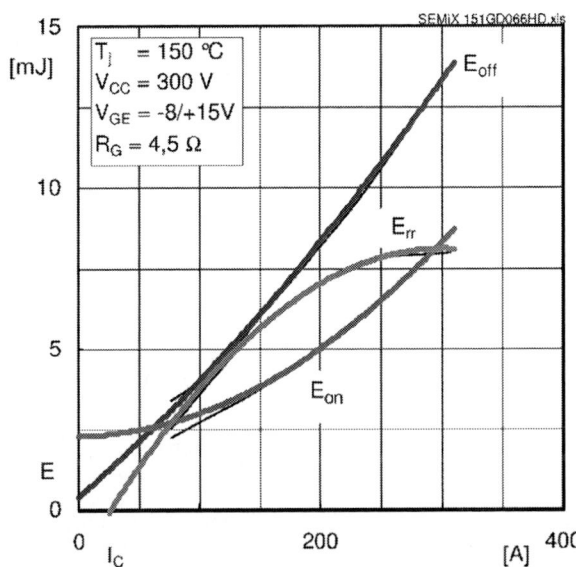

Fig. 10. SEMIX151GD066HD IGBT – Verifying the switching losses approximation.

TABLE II
E_{on}, E_{off} AND E_{rr} POINTS

Curve	Current, I_C (A)	Energy (mJ)
Eon	[100, 200, 300]	[4, 8.3, 13.3]
Eoff	[100, 200, 300]	[3, 5, 8.3]
Err	[100, 200, 300]	[3.75, 7, 8.1]

TABLE III
E_{on}, E_{off} AND E_{rr} APPROXIMATION

Switching	Value
Eon (µJ)	$\frac{V_{off}}{300}\left(35000\,I_C^2 + 32.5\,I_C + 400\right)$
Eoff (µJ)	$\frac{V_{off}}{300}\left(65000\,I_C^2 + 5\cdot10^5\,I_C + 0.0023\right)$
Err (µJ)	$\frac{V_{off}}{300}\left(-107500\,I_C^2 + 64.75\,I_C - 0.00165\right)$

Power losses are calculated in four groups; the transistor conduction losses (1), the diode conduction losses (2), the transistor switching losses (3), and the diode switching losses (4), as follows:

$$P_{Tcond} = \frac{1}{T}\int_0^T |i_a|\left(V_T + r_T\,|i_a|\right)dt_{(s_a=1)\&\&(i_a>0)\|(s_a=0)\&\&(i_a<0)} \quad (1)$$

$$P_{Dcond} = \frac{1}{T}\int_0^T |i_a|\left(V_D + r_D\,|i_a|\right)dt_{(s_a=1)\&\&(i_a<0)\|(s_a=0)\&\&(i_a>0)} \quad (2)$$

$$P_{Tsw} = \frac{1}{n}\sum_{j=1}^n E_{on\,(i_a>0)\&\&(turn\_on)\|(i_a<0)\&\&(turn\_off)} +$$

$$+ \frac{1}{n}\sum_{j=1}^n E_{off\,(i_a<0)\&\&(turn\_on)\|(i_a>0)\&\&(turn\_off)} \quad (3)$$

$$P_{Dsw} = \frac{1}{n}\sum_{j=1}^n E_{rr\,(i_a>0)\&\&(turn\_on)\|(i_a<0)\&\&(turn\_off)} \quad (4)$$

Fig. 11 shows the Matlab-Simulink implementation of the conduction losses calculation based on (1) and (2). Fig. 12 shows the switching losses model for the transistor based on (3) and (4).

The model provides the losses of one leg; therefore, since the same switching patterns, dc-link voltage, and ac currents are applied to the other phase legs, the inverter total losses are obtained by multiplying by three those results. All the losses calculation algorithms are integrated in a block, as Fig. 13 shows.

VI. VALIDATING THE POWER LOSSES MODEL

The power losses calculation methodology is applied to an inverter made up with the power device SEMIX151GD066HD (dual IGBT). The operating parameters are the following:

- Voltage bus (V_{dc}): 450 V
- I_a (rms): 137 A
- Power factor (PF): 0.82
- Carrier frequency (f_c): 8 kHz
- Modulation index (m): 0.85

Table IV shows some results obtained using the proposed Matlab-Simulink algorithm. The power losses are given for the entire inverter and for individual power devices in this table.

The results obtained with the proposed methodology are compared with other results from:

- a PSIM inverter model [5],[11], and
- the Semisel program.

The Semisel program is an online application powered by Semikron which allows simulating different topologies, but using only one modulation strategy, SPWM. Fig. 14 shows the configuration page. Table IV shows the results obtained from the Simsel program.

978-1-4244-8806-3/11 $26.00 © 2011 IEEE   407

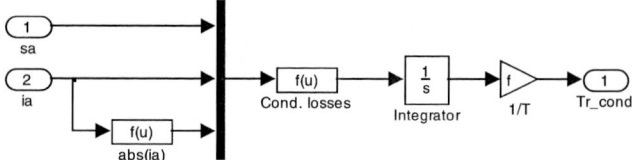

Fig. 11. Matlab - Simulink: Conduction losses model.

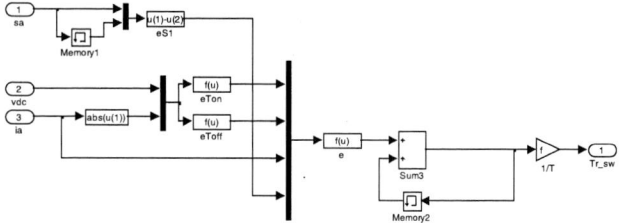

Fig. 12. Matlab-Simulink: Switching losses model.

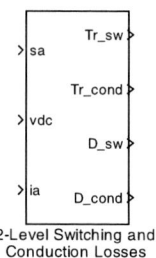

2-Level Switching and Conduction Losses

Fig. 13. Matlab-Simulink losses model.

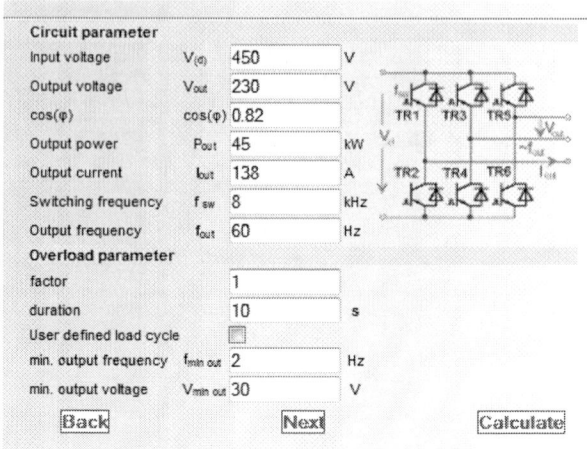

Fig. 14. Semisel – Two-level inverter: Circuit design.

TABLE IV
POWER LOSSES. SIMULATION RESULTS

Power Losses (W)		IGBT Cond. Losses	IGBT Sw. Losses	Diode Cond. Losses	Diode Sw. Losses	Total Power Losses
Proposed Matlab-Simulink	Complete Inverter	502.7	312.2	124.6	157.1	1,096.4
	Single Switch	83.8	52.0	16.7	26.2	182.7
PSIM	Complete Inverter	485.8	304.0	120.6	159.6	1070.1
	Single Switch	81.0	50.7	20.1	26.6	178.4
Semisel	Complete Inverter	486	306	120	96	1008
	Single Switch	81	51	20	16	168

The power losses results show that the proposed model differs on a maximum of 8.9% in comparison with the application provided by the manufacturer (Semisel), while there is only a 2.4% difference when compared with PSIM.

An important feature of the proposed Matlab-Simulink algorithm is its flexibility. It is very simple to modify the power devices, modulation strategy, operating conditions, etc. Furthermore, with the proposed Matlab-Simulink model, it is possible to sweep the modulation index (m) from 0 to 1 and change the power factor (PF) from -180° to 180° during simulation. Fig. 15 shows the power losses obtained for the entire inverter operating conditions keeping constant the rms value of the output current (137 A).

VII. CONCLUSION

The proposed power losses estimator presented in this paper shows good results when compared with other simulation programs; such as PSIM and Semisel (Semikron). An interesting feature of the proposed Matlab-Simulink model is that the data required for the power losses calculation can be found in the datasheet of the semiconductor manufacturer. Furthermore, unlike many simulation programs, the one proposed here allows to obtain a wide range of results in a 3-D representation for different modulation indexes and current phases. The proposed methodology has been implemented in Matlab-Simulink environment, but it could be implemented in other mathematical software.

The next step in this research is to add a thermal module into the proposed algorithm, so that the parameters of the power switches change with the temperature during simulation. Consequently, temperatures values could be given, and then more accurate power losses results could be obtained.

The main objective of this work is to have a tool to select optimal power devices, modulation strategies, thermal sink, etc. specifically in EV applications. Nevertheless, the proposed methodology and the models can be used for the design of other power electronic converters and applications.

Furthermore, the data obtained from the proposed methodology can be saved on a lookup table and have it available to be used in an EV simulation platform. The power losses could be estimated from a continuous model using the stored data and requiring only little processing. That kind of model would be able to operate in large simulations and provide the power losses while the simulation is running.

AKNOWLEDGEMENTS

This work was supported by the Ministerio de Ciencia e Innovación of Spain under Project CITYELEC - Electrification systems for urban mobility (PSE-370000-2009-004), and by the AGAUR of the Generalitat de Catalunya under Project EOLO (2009 CTP 00005).

978-1-4244-8806-3/11 $26.00 © 2011 IEEE

REFERENCES

[1] J. Larminie and J. Lowry, Eds., *Electric Vehicle Technology Explained.* John Wiley & Sons, Ltd, 2003.

[2] F. L. Mapelli, D. Tarsitano, and M. Mauri, "Plug-In Hybrid Electric Vehicle: Modeling, Prototype Realization, and Inverter Losses Reduction Analysis," *IEEE Tran. Ind. Electron.,* vol. 57, no. 2, pp. 598-607, Feb. 2010.

[3] J. Pallisé (Coordinator), F. Astals, I. Cairó, J. Comellas, F. Martínez, J. Pallisé, and J. Serra, "Diagnosis and perpectives of the electrical vehicle in Catalonia," (in Catalan), 2009. Available: http://www15.gencat.cat/cads/AppPHP/images/stories/publicacions/informesespecials/2010/versi_ie_10_web.pdf

[4] Project CITYELEC: Electrification systems for urban mobility. Available: http://www.cityelec.es/ and http://www.cityelec.upc.edu/

[5] H. Gong, F. Chai, Y. Pei, and S. Cheng, "Research on torque performance for permanent-magnet in-wheel motor," in *Proc. IEEE Vehicle Power and Propulsion Conference (VPPC'08),* 3-5 Set. 2008, Harbin, China, pp. 1-4.

[6] S. Dong-joo and K. Byung-il, "Multi-objective optimal design for in-wheel permanent magnet synchronous motor," in *Proc. IEEE International Conference on Electrical Machines and Systems (ICEMS'09),* 15-18 Nov. 2009, Tokyo, Japan, pp. 1-5.

[7] Jr. A. R. Hefner, "A dynamic electro-thermal model for the IGBT," *IEEE Tran. Ind. Applic.,* Vol. 30, No. 2, pp. 394-405.

[8] V. Kumar Khanna, *The Insulated Gate Bipolar Transistor. IGBT. Theory and Design,* IEEE Press. Wiley-Interscience, 2003.

[9] J. Zaragoza, J. Pou, S. Ceballos, E. Robles, P. Ibañez, and J. L. Villate, "A Comprehensive Study of a Hybrid Modulation Technique for the Neutral-Point-Clamped Converter," *IEEE Tran. Ind. Electron.,* vol. 56, no. 2, pp. 294-304, Feb. 2009.

[10] Semikron, "Semikron Application Manual." Available: http://web.fel.zcu.cz/fel/kev/SOV/Text/Semikron%20-%20Application_Manual/Index.html

[11] Powersim Inc., "IGBT Loss Calculation Using the Thermal Module," March 2007. Available: http://www.powersimtech.com/manual/Tutorial%20IGBT%20Loss%20Calculation.pdf

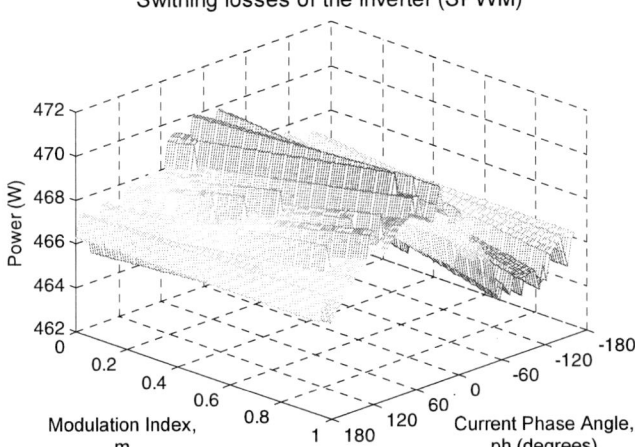

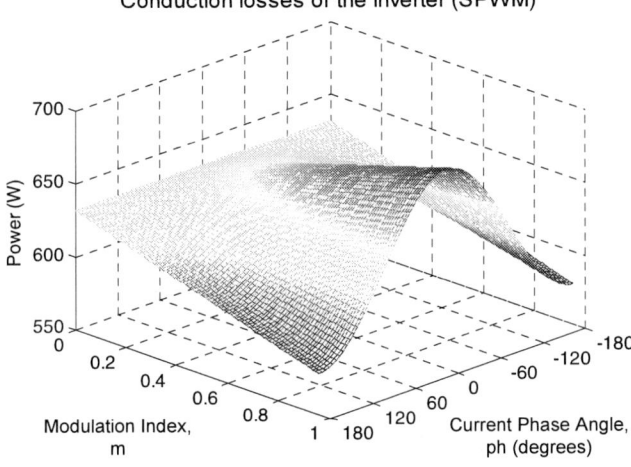

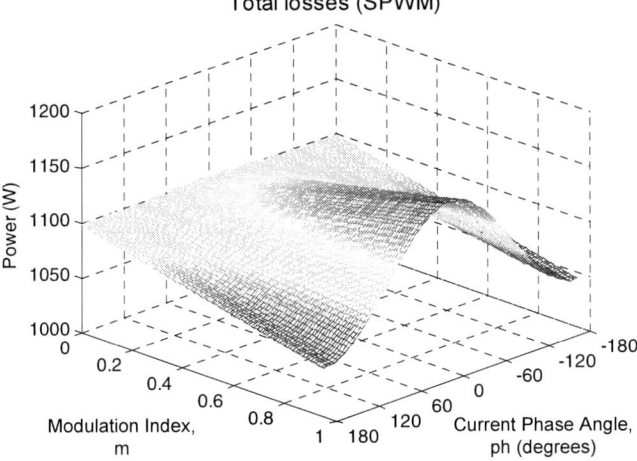

Fig. 15. Matlab-Simulink model . Inverter wide-range power losses analysis.

Perspectives of Improvement of AC Power Transmission Based on Achievements of Modern Power Electronics

Yevgeny I. Sokol, Yury P. Goncharov, Vladimir V. Ivakhno, Vladimir V. Zamaruev, Sergey Y. Krivosheev,
Alexander P. Lastovka, Alexander Y. Ivanov, Marina A. Chernetchenko
National Technical University "KhPI", Ukraine lastovka@kpi.kharkov.ua

Abstract – **possibility of development of AC power transmission is analyzed based on ideology of electronic systems of secondary power supplies with intermediate high-frequency link. Means of compensation of negative effects due to influence of electric power lines as distributed capacitance and inductance objects are considered.**

I. INTRODUCTION

Nowadays power electronics is quite widely used in AC power transmission (FACTS, for example). However, those applications do not so far deal with fundamentals of such transmission, which have been formed at the end of XIX century and used three-phase sinusoidal low-frequency (50-60 Hz) AC as a power carrier. Application of that basic signal was explained by possibility of direct connection of induction motors to the mains that had no alternative at that time. However, there are several drawbacks of the drive system with direct connection of induction motor to the mains:
– there are no effective means of wide-range regulation of revolution frequency, and at the same time, according to known data, about 70 % of all drives should be designed regulated;
– reactive power, comparative with the active one, is consumed from the mains;
– start-up currents reach sevenfold values of the nominal current.

Achievements of power electronics have allowed to create a system of frequency-regulated drive, in which the motor is connected to the mains via intermediate DC link, and in the nearest future such system will replace the traditional system with direct motor connection. However, there will be no solid reasons for monopolistic use of three-phase sinusoidal low-frequency AC in power transmission systems (PTS) and a possibility of application of alternative basic signals appears. As such alternative a single-phase high-frequency square-wave current (meander), that has a leading position in the modern system of secondary power supplies with the intermediate high-frequency link [1, 2].

Application of that alternative basic signal instead of three-phase sinusoidal low-frequency AC has several advantages:
– design of power lines is simplified and it is possible to use simple coaxial cables to obtain, as in DC transmission, a double benefit in energetic parameters due to the fact that for ideal meander the ratio K_{arms} between amplitude and RMS

values of the voltage and current equals a unity, which is $\sqrt{2}$ for a sinusoid;
– circuitry of rectifiers and inverters is simplified because they become single-phase ones with double-decreased rated power of the switches due to the same K_{arms} ratio;
– dimensions of transformers and power filters are reduced due to increased frequency;
– control and regulation are significantly improved due to application of quick-response converters in the main energetic channel.

Fig. 1-2 shows examples of PTS structures with alternative basic signal, which are used in the following analysis. They can be embedded into existing PTS without its major change.

In traditional distribution 220/380 V networks with low

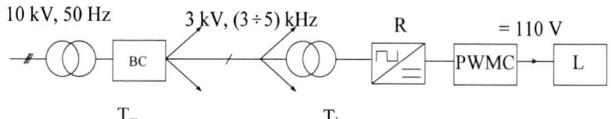

Fig. 1. Alternative solution of low-voltage (220/380 V, 50 Hz) distribution system: T_m – traditional mains step-down transformer; BC – basic converter, consisting of three-phase rectifier and single-phase inverter; T_i – step-down transformer in intermediate high-frequency link; R – rectifier; PWMC – step-down PWM converter with power factor correction functions.

voltage and high current at assigned power there is a high consumption of conductive material in cable power lines and losses reach 10 %. Application of the intermediate high-frequency link (Fig. 1) allows to avoid that by increase of voltage in the cable power lines up to 3 kV, and to install a compact step-down transformer T_i with a rectifier and a PWM converter on power distribution units of a single consumer. The latter allows to stabilize voltage at the customer which improves quality of consumed power. Low distance between the power distribution unit and the consumer allows to reduce voltage up to 110 V or less which increases electric safety in comparison with traditional 220 V. Shift to DC coincides with existing trend of increase of DC share in public consumption and allows to get rid of negative effects associated with AC application at the last conversion stage (light blinking, acoustic noise etc).

Another example is a distributed system of power supply of electrified railroad (Fig. 2). In parallel with a main contact system CS a single-phase cable is mounted in the ground or on the same poles and a voltage with magnitude 50 kV, increased frequency 300 Hz and a waveform close to a square

978-1-4244-8806-3/11 $26.00 © 2011 IEEE

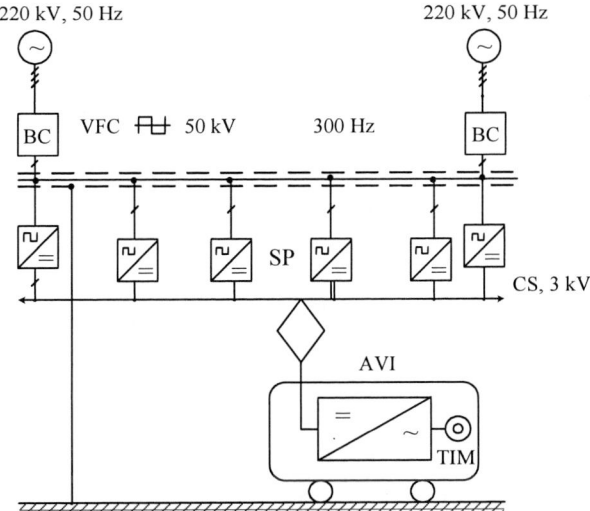

Fig. 2. Distributed system of power supply of electrified railroad: BC – basic converter; VFC – high-voltage high-frequency cable; SP – sub-supply point; CS – DC contact system; AVI – autonomous voltage-source inverter; TIM – traction induction motor.

is applied. Contact system is connected to the cable with a certain pitch via sub-supply points SP. If it is a DC CS with a standard 3 kV voltage the sub-supply points are transformers with rectifiers and filters.

The advantages are as follows:
– reduced material consumption in the cable and energy losses due to better use of insulation;
– the same effect is reached in transformers and filters of sub-supply points due to increased frequency and square waveform of the input voltage;
– multiple reduction of number of connections with the feeding network;
– phase loads of the mains are balanced due to filter in the intermediate DC link of BC and due to the fact that ratio of frequencies of voltages in the cable and in the mains equals three;
– expanded possibilities of control of power supply system which can be used, for instance, for limiting balancing currents flowing through CS and the mains.

However, it is obvious that a lot of work is required to get rid of drawbacks of alternative basic signal. The most significant differences in comparison with traditional secondary power supply systems are brought by power lines as the objects with distributed capacitance and inductance. Possible resonant effects and impulse current loads on converters due to power line capacitance recharge are associated with that. The main goal of the work consisted in finding ways to overcome those obstacles.

Major deviations of the current waveforms from the square may be caused by PTS own inductances. However, for cable power lines their values are 10-times less in comparison with overhead power lines, and a role of magnetic dissipation fluxes in transformers can be reduced by changing

proportions of volumes of coils and cores, that coincides with intentions to obtain low losses from eddy currents in winding conductors at increased frequencies.

II. RESONANT PROPERTIES OF POWER LINES AS A LOAD OF CONVERTER

Let us use equations for a power line as a distributed parameters object [3]. With distributed active resistance equal to zero we obtain:

$$u_1 = u_2 ch(p^*) + \rho i_2 sh(p^*), \quad i_1 = i_2 ch(p^*) + \frac{u_2}{\rho} sh(p^*), \quad (1)$$

where u_1, u_2, i_1, i_2 – input and output voltages and currents of power line; $\rho = \sqrt{L/C}$ – impedance; $p^* = p\sqrt{LC}$ – complex disturbance frequency in relative units; L and C – aggregate values of distributed inductance and capacitance in the power line.

In open circuit in (1) we consider $i_2 = 0$ and then for input operator conductance we obtain:

$$Y_1 = \frac{i_1}{u_1} = \frac{1}{\rho} \cdot \frac{e^{2p^*} - 1}{e^{2p^*} + 1}. \quad (2)$$

Poles of input conductance, turning the denominator (2) into 0:

$$2p^* = 2j\omega^* = j(\pi + 2q\pi) \text{ or } \omega^* = (2q+1)\pi/2, \quad (3)$$

where ω^* – relative resonant frequencies; $q = 0,1,2...$ – integer number.

Thus, in the open circuit a resonance appears on odd-numbered harmonics of the main resonant frequency:

$$\omega_p = \frac{\pi}{2\sqrt{LC}}. \quad (4)$$

Using similar assumptions for another marginal case $u_2 = 0$ (short circuit), we find that resonance appears on even-numbered harmonics of the main resonant frequency. The lowest resonant frequency $2\omega_p$ is twice higher than in the open circuit which is a positive feature of the short circuit.

Equation (4) for the main resonant frequency becomes more obvious if we represent aggregate L and C as functions of specific parameters per a unit of length l of the power line and consider relationship [3] between specific parameters:

$$L = L_s l, \quad C = C_s l, \quad L_s C_s = \sqrt{\mu_0 \varepsilon \varepsilon_0},$$

$$\text{then } \omega_p = \frac{\pi}{2l\sqrt{\mu_0 \varepsilon \varepsilon_0}}, \quad (5)$$

where μ_0 and ε_0 - magnetic and dielectric permeability of vacuum; ε - relative permeability of cable insulator.

III. SHAPE OF BASIC SIGNAL

It is obvious that ideal meander is inacceptable for practical application in the considered systems because it contains wide range of slowly-decreasing according to hyperbolic law $1/\omega$ harmonics. High-frequency part of that range will overlap a zone of power line resonant frequencies, leading to resonant oscillations. The ideal meander is also inacceptable due to skin effect in conductors which interferes a wire communications and other factors.

Real basic signal can be considered as a sequence of lengthy flat tops and relatively short edges, whose shapes can vary. With sinusoidal edges the frequency spectrum of such signal can be at first approximation considered limited by magnitude of frequency of signal on edges:

$$f_e = \frac{1}{2t_e} = \frac{f_1}{t_e^*}, \quad t_e^* = \frac{t_e}{T/2}, \quad (6)$$

where t_e^* - relative length of edges; $f_1 = 1/T$ - frequency of the first harmonic.

If the lowest resonant power line frequency exceeds width of the spectrum f_e of the basic signal not less than in two times, the resonant effects can be explained as a process of impact excitation of resonant oscillations due to derivative discontinuity in junction points of edges with flat tops. From that point the sinusoidal edge is better than the linear one because the signal itself and its first derivative are continuous unlike linear signal that has its first derivative discontinuity.

The mechanism of impact excitation can be effectively suppressed if to use infinitely differentiable time function ("flat" meander) as a basic signal. A quite simple method of its microprocessor realization is done on the base of integration of "bell"-shaped time function.

$$x = X_m e^{-at^2}, \quad a = \left(\frac{\pi}{t_e}\right)^2, \quad X_m = \frac{\pi}{0.89 t_e}, \quad (7)$$

where t_e, according to (6), duration of sinusoidal edge with equal tilt in zero transition point.

Fig. 3 explains an algorithm of ideal meander forming. For that the algebraic sum of two shifted "bells" is formed according to Fig. 3a, and the integration of that signal at assigning of initial value equal to unity is performed (Fig. 3b). For periodic repetition of the signal Fig 3b, the current time is reset after each period T (Fig. 3c).

IV. PARTITION OF LONG POWER LINE ONTO AUTONOMOUS SECTIONS

At increase of length of the power line its lowest resonant frequency decreases and inevitably enters a zone of working harmonics of the basic signal. Then the normal operation of

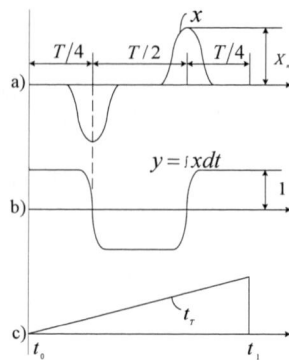

Fig. 3. Algorithm of ideal meander forming

the power line without special control techniques becomes impossible due to appearance of resonant oscillations of high amplitude. For instance, in the structure (Fig. 2) the reasonable length l between neighboring basic converters BC makes about 150 km. With $\varepsilon = 2.3$, that corresponds polyethylene insulation in the cable, from (5) with consideration of $\mu_0 = 1.256 \cdot 10^{-6}$ H/m and $\varepsilon_0 = 8.85 \cdot 10^{-12}$ F/m, we obtain $f_p = \omega_p / (2\pi) = 330$ Hz, which is close to accepted main frequency 300 Hz.

The principle of overcoming those obstacles consists in partition of the long line onto small sections, working according to Chapter 2 terminology in short circuit at both inputs (Fig. 4). To realize that in the connection points of the

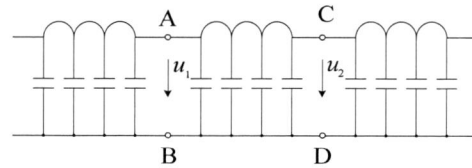

Fig. 4. Partition of long power line onto sections, without non-controlled dynamic interaction

sections the converters are installed that do not have resonant process components in their output voltages u_1, u_2... Then it is the same for them as a short circuit between points of pairs AB, CD... As follows from Chapter 2, the short circuit doubles the lowest resonant frequency and resonant effects take place autonomously that leads to n-times increase of resonant frequencies, where n – number of sections.

In the circuit (Fig. 2) the short circuit is provided by converters of sub-supply points SP. In more general case those can be special converters-correctors, that do not transfer active power and, consequently, acting as active filters. At the long power line it makes sense to use them for compensation of capacitive current consumed at the section eliminating additional losses at its transfer along the line to the feeding converter.

The short circuit practically can't be completely realized due to existence of filters of harmonics with PWM frequency in converters (Fig. 5). With application of traditional $L_f C_f$ – filter the short circuit is quite simple to provide only at u_1'

converter commutator terminals, not at u_1 line terminals. That is why dividing inductance L_f of the power filter should be minimized. Principally, the short circuit could be realized

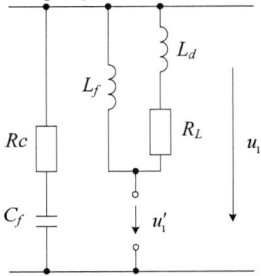

Fig. 5. Explanation of influence of power filters of PWM frequency

by control techniques directly at AB terminals, but that requires high PWM frequency for oscillatory components processing. If not to do that due to low quickness of modern power GTO devices, the mechanism of short circuit realization consists in the following.

The formed voltage meander u_1' is synchronized with voltage in the power line only by frequency. That can be done by alignment of time points t_0, t_1... in Fig. 3 with moments of voltage u_1 zero transitions and their consequent shift on time interval $\approx T/4$. The latter allows to keep time points on the flat top of the meander, where formed signal does not rapidly change. All other parameters of the meander are formed by relatively inertial regulators, not responding to high frequency of resonant oscillations. Particularly, regulation of the phase shift between u_1 and u_1' allows to control active component of the consumed current. Regulation of meander amplitude u_1' changes the reactive component, which it makes sense to keep equal to 0. Finally, by unbalance by positive and negative current amplitudes we can compensate aperiodic component of the current that arises due to inaccuracy of control algorithm.

The disadvantage of application of u_1 as a reference signal for synchronizing by frequency is a low inductive impedance of L_ϕ, separating voltage sources u_1 and u_1', that decreases stability of the mentioned current regulators. To get rid of such disadvantage it is possible to use a voltage from the neighboring sub-supply SP as a reference signal. But it is needed to identify it with the help of indirect sensor, acting by signals from the actual sensors in the given sub-supply SP and by known parameters of the power line between it and neighboring SP.

As an additional mean for suppression of resonant oscillations it is possible to use RC- and RL-dumping circuits, as shown in Fig. 5. Losses in the resistors RC and RL, as given estimation shows, are measured by thousandths of entire converter power. It is possible to use equivalent resistors on a base of low-power converters with high-frequency PWM, returning dumped energy to the feeding source.

V. SIMULATION RESULTS

Both structures (Fig. 1 and Fig. 2) were modeled using «Matlab-Simulink». For the circuit (Fig. 1) a structure with two series modules with three-level design of a rectifier and an inverter in each module [4] was used as a basic converter. It allowed to obtain 9 levels of output voltage and to form it by amplitude-impulse technique without PWM at frequency of switching equal to the frequency of intermediate link of 4 kHz. The cable length was supposed not exceeding 2 km, which limits minimum main resonant frequency at about 25 kHz. That allows not to use additional converters-correctors and to use only traditional step-down PWM. The PWM frequency is 200 kHz, which can be realized on modern low-voltage FETs.

For the structure (Fig. 2) the same conversion blocks were used but with additional PWM in one of four regulation zones. In converters of sub-supply points all kinds of regulators of consumed current parameters were used, described in Chapter 3. A voltage from the neighboring sub-supply SP was used as a reference signal for synchronization by frequency. To estimate the efficiency of the use of "flat" meander, the shift to continuous models of conversion commutators was done, which excludes impact of PWM interference. Fig. 6a shows a cable voltage in one of the

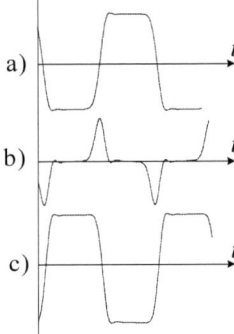

Fig. 6. Processes in a model of the railroad power distribution system: a- cable voltage; b – SP current under no active load; c – SP current under active load

intermediate points which contains voltage overshoot with magnitude only 2 %, while with sinusoidal edges the overshoot value is about 5 times higher. Fig. 6b shows input current of the SP, that has no active load. It contains only impulses of capacitive cable current which take place during meander edges. Fig. 6b represents input current of the SP, under nominal active load with capacitive cable current. Amplitude of aggregate capacitive current in the given structure due to long length of cable power line is 5 times higher than amplitude of nominal active current of the basic converter. However, that fact does not lead to significant increase in amplitude current load of the SP because due to peak character of a load railroads have a big rated power reserve of sub-supply points relative to rated power of the basic converter.

Due to the short circuit, not ideal because of the presence of a filtering inductor with a specific inductive impedance of

978-1-4244-8806-3/11 $26.00 © 2011 IEEE

about 3 % relatively to an impedance of rated load of the SP, the resonant frequencies of series sections increase from 330 Hz to about 4 kHz and that is why cannot be traced.

VI. PERSPECTIVE STRUCTURE OF THE PTS, DESIGNED ON A BASE OF THE CONSIDERED CONTROL TECHNIQUES

It is shown in Fig. 7 and, obviously, represents own authors' opinion. Three-phase low-frequency sinusoid must

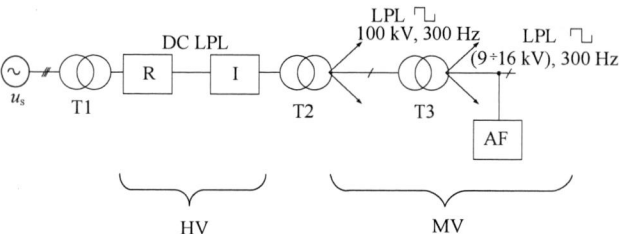

Fig. 7. Perspective structure of PTS: HV – high voltage level; MV – medium voltage level; AF – active filter; LPL – long power line

retain in parts of the system with rotating electric machines – generators and motors. With high voltage levels, starting from 300 kV, it makes sense to use DC power lines.

With long line of the DC power lone it compensates expenses on high-voltage converters: rectifier R and inverter I in Fig. 7. Besides that, it solves all problems associated with distributed capacitance of high-voltage cables. For voltages of 100 kV and less it makes sense to use single-phase cable power lines, transferring high-frequency voltage with quasi-square waveform.

DC application in this part of the system is not efficient because at each change of a voltage level two converters are required – rectifier and inverter, each on full rated power. Single-phase meander has almost the same energetic parameters than a DC, but differs by possibility of direct transformation.

In nodes of the system the converters-correctors should be installed according to ideology described in Chapter 4, which accept part of capacitive currents and provide the short circuit for free components of the process. As a result, that complex system can be split onto a number of simple components with no possibility of non-controlled dynamic interaction. That allows to exclude an overlap of resonant frequencies zone and a zone of working frequencies of the basic signal.

At the outputs of the single-phase 9-16 kV power lines either rectifiers of high-voltage drives are installed or, under low voltage, structures (Fig. 1) with substitution of three-phase voltage u_s source by single-phase 9-16 kV power lines, that is equivalent to traditional voltage levels of 6-10 kV.

CONCLUSIONS

1. Design of DC power line on a base of ideology of electronic systems of secondary power supplies with replacement of three-phase sinusoid of the basic signal (carrier of transferred power) by single-phase high-frequency DC with square waveform is proposed. This ideology can be used in separate parts of PTS as well as in the entire system. The energetic benefits are comparable with those, obtained in secondary power supplies with intermediate high-frequency link.

2. Reasonable form of the basic signal of alternative PTS ("flat" meander) is found, which provides a property of infinite differentiability to the basic signal, like with traditional sinusoid. Thus negative effects are minimized, associated with a possibility of appearance resonant effects and interferences for control and communication devices.

3. A technique for control of parts of the system with alternative basic signal is proposed, consisting in partition of a complex system onto simple components, with no possibility of non-controlled dynamic interaction. Thus, the resonant effects, associated with PTS as the object with distributed capacitance and inductance are removed.

4. Simulation results in «Matlab-Simulink» have proved simulation veracity of the given work.

REFERENCES

[1] R. Severns and G. E. Bloom, Modern Dc-to-Dc Switchmode Power Converter Circuits, New York: Van Nostrand Reinhold, 1985.

[2] Marty Brown Power supply cook book. Elsevier Science Burlington, MA 01803, 2005

[3] Ионкин П.А., Даревский А.И., Кухаркин Е.С. и др. Теоретические основы электротехники, Т-1, под ред. Ионкина П.А. Высшая школа, Москва, 1978, 544 с.

[4] Сокол Е.И., Гончаров Ю.П., Ивахно В.В и др. Преобразователь для связи между питающей энергосистемой переменного тока и сетевым районом с нетрадиционными параметрами электроэнергии. Технічна електродинаміка, тем. випуск «Силова електроніка і енергоефективність» Київ, ч. 2, 2010, с. 56-60

[5] Gunnar Asplund, Lennart Carlsson, HVDC – ABB from pioneer to world leader. //ABB Review – 4/2008, pp. 59-64.

[6] Rahul Chokhawala, Powering Platforms – Connecting oil and gas platforms to mainland power grids. //ABB Review – 1/2008, pp. 52-56.

Parallel resonant inverter with auxiliary AC/DC converter used for induction heating

Jan Mućko *

* University of Technology and Life Sciences, Institute of Electrical Engineering, Bydgoszcz, Poland

mucko@utp.edu.pl

Abstract – this article describes the parallel resonant inverter with the auxiliary AC/DC converter that limits commutation overvoltage and enables the energy of parasitic inductances to be returned to the feeding circuit. The use of the auxiliary AC/DC converter enables the connection of the HF transformer between the inverter output and a parallel resonant circuit as well as the operation of the inverter in the frequency range, ensuring ZVS type switching. The efficiency of the system was also improved.

Keywords - **current source inverter, parallel resonant inverter, induction heating, ZVS, auxiliary converter.**

I. INTRODUCTION

The operational principles of the parallel resonant inverter with voltage bi-directional (VB) switches (with diodes connected in series with transistors), at negligible parasitic inductances between the inverter and the resonant circuit (Fig. 1a), are generally well-known [1, 2]. For systems of this type, it is possible to use soft commutated switches that turn on at zero voltage (ZVS) or turn off at zero current (ZCS). The type of switches that may be used is determined by their switching frequency f_s (Fig. 1b, 1c). If the switching frequency is greater than the damped resonant frequency f_{dr} of the resonant circuit, ZCS switches may be used. If, on the other hand, $f_s < f_{dr}$, ZVS switches may be used. The damped resonant frequency (if the resonant tank quality factor is sufficiently large) is approximately equal to $f_0 = 1/2\pi\sqrt{L_r C_r}$. The role of VB-ZCS switches may be served by SCR thyristors, which have been used for parallel resonant inverters for a long time now. On the other hand, the use of VB-ZVS switches, intended for frequencies higher than by ZCS switches, has posed some considerable difficulties. These difficulties were related to overvoltage created by parasitic inductances between the inverter and the resonant circuit during commutation processes.

In many applications, for electrical safety reasons, the inductor has to be separated from the mains. Transformers are used for that purpose. Places where a transformer may be connected to the circuit of this power converter are shown in Fig. 2a. Most commonly, reference sources [1, 3] describe the arrangements with the transformers connected at places A or B. If the transformer is connected on the mains side (point A) the overall dimensions are large due to low mains frequency (50 Hz).

With the transformer connected at B, the inverter voltage matches the inductor voltage and the transformer is supplied

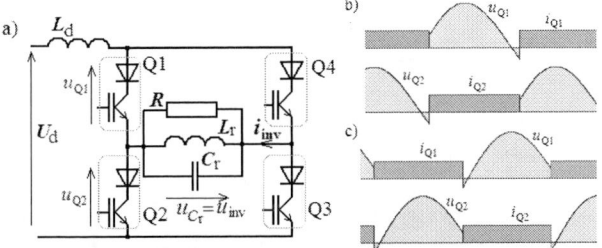

Fig. 1. The schematic diagram of a parallel resonant inverter by neglecting of parasitic inductance between the inverter and the resonant circuit (a) and waveforms of current and voltage of Q1 and Q2 switches for: b) $f_s < f_{dr}$, ZVS switching possible, c) $f_s > f_{dr}$, ZCS switching possible.

with the voltage of an increased frequency (tens – hundreds kHz). Consequently, the mass of both the winding and the transformer core may be reduced. Since the winding of the transformer constitutes an element of the resonant circuit, the circulation of current of high values in this circuit significantly heats up the winding. Water cooling of the transformer is necessary. This requires a more complex design and increased overall dimensions of the transformer. For water cooling of the primary winding, the separation should be provided by means of an additional transformer at place A.

With the transformer connected at C, the inverter voltage matches the inductor voltage and the transformer is supplied with the voltage of an increased frequency. The mass of both the winding and the transformer core is reduced. The transformer winding is not an element of the resonant circuit and the values of current passing through it are considerably

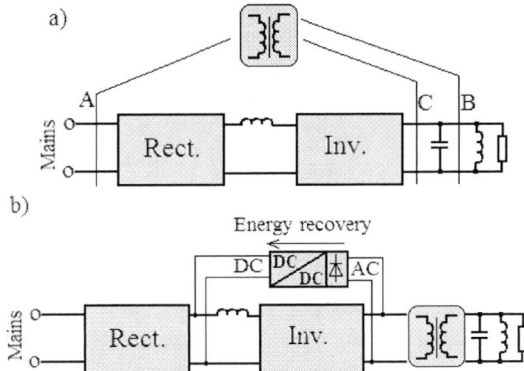

Fig. 2. Schematic diagram of the induction heating system with the current source inverter and the parallel resonant circuit: a) points where the transformer may be connected; b) arrangement with the auxiliary AC/DC converter and the transformer connected between the inverter and the resonant circuit.

978-1-4244-8806-3/11 $26.00 © 2011 IEEE

lower than with the transformer connected at B. In this case, the transformer operation should be considered even without water cooling. Therefore, connection of the transformer at C offers certain advantages in terms of its overall dimensions, the mass and the simplicity of the whole system. However, parasitic leakage inductances of the transformer result in overvoltages, and consequently to losses of energy in semiconductor elements and overvoltage suppressors. The losses become greater with an increase of parasitic inductances and commutated current. In order to limit overvoltages and decrease the consequential losses, an auxiliary AC/DC converter was used. In such an arrangement, unlike in the case of standard overvoltage suppressors, energy is not lost but returned into the feeding circuit of the inverter. The auxiliary AC/DC converter consists of two units: a diode rectifier (AC/DC) and a resonant DC/DC converter [4, 5]. The DC/DC converter transmits the energy to the feeding circuit of the inverter only at the specified value of overvoltage equal to U_{lim}.

The schematic diagram of the induction heating system, including a current source inverter (with VB switches), and a transformer connected between the inverter and a parallel resonant circuit is shown in Fig. 3. In this diagram, the transformer is represented only by leakage inductances (L_{Tr}). The parameters and values of the secondary side of the transformer are referred to the primary side.

II.   OPERATION PRINCIPLE AND ANALYSIS

Figure 4 shows waveforms of current and voltage in the system, including commutation processes, for various switching frequencies of the transistors. Figure 4a illustrates a situation when the commutation occurs before the capacitor C_r voltage wave "goes through zero" ($f_s=1.05 \cdot f_0 > f_{dr}$). Then, conditions exist for the ZCS commutation. Turning off at the ZCS conditions enforces the presence of an inductive snubber at turning on; this role may be served by leakage inductances of the transformer. Figure 4b illustrates a situation when the commutation occurs around the point where the capacitor C_r voltage wave "goes through zero" ($f_s=1.0 \cdot f_0 \approx f_{dr}$). Conditions for ZVS and ZCS commutation are not fulfilled. Figure 4c

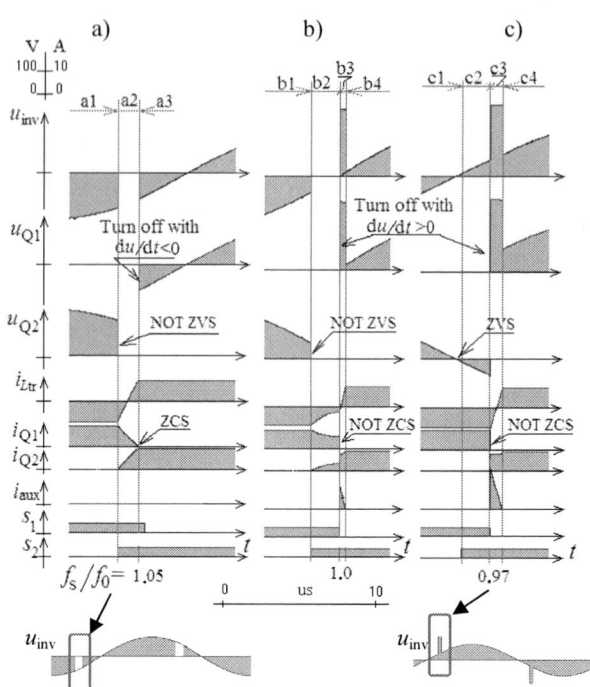

Fig. 4. Waveforms of current and voltage in the inverter for: a) f_s=1.05·f_0>f_{dr}, b) f_s=1.0·f_0≅f_{dr}, c) f_s=0.97·f_0<f_{dr},

illustrates a situation when the commutation occurs after the capacitor C_r voltage wave "goes through zero" ($f_s=0.97 \cdot f_0 < f_{dr}$). Then, conditions exist for ZVS turning on. The presence of capacitative snubber at turning off is recommended. The role of this type of snubber may be served by self-capacity of switches and input capacity of the auxiliary AC/DC converter.

The use of the auxiliary AC/DC converter is beneficial, regardless of the chosen strategy of the switch control. The converter protects the system against overvoltages when these occur for both ZCS and ZVS switching.   If the ZCS control is used, a sudden change of the damped resonant frequency (caused by, for example, introduction of a load or reaching the Curie temperature) may result in a change of the phase shift between the output current and voltage of the inverter and generation of overvoltage.

Table 1 shows equivalent circuit diagrams of the inverter for time intervals indicated in Figs. 4a - 4c. The conditions fulfilled by current and voltage values of the system in these intervals were also specified. Switching at the frequency $f_s \cong f_{dr}$ (Fig. 4b) is undesirable since ZVS or ZCS switching is not possible and it may occur only in transient states. What follows, an analysis of operation of the system for conditions illustrated in Figs. 4a and 4c ($f_s > f_{dr}$ and $f_s < f_{dr}$), is presented.

For the adequate high quality factor of the resonant circuit, the shape of voltage u_{Cr} of the capacitor C_r may be regarded as sinusoidal. Further investigation will relate only to such situations when the switching frequency is slightly different from the damped resonant frequency ($f_s \approx f_{dr} \approx f_0$). The input

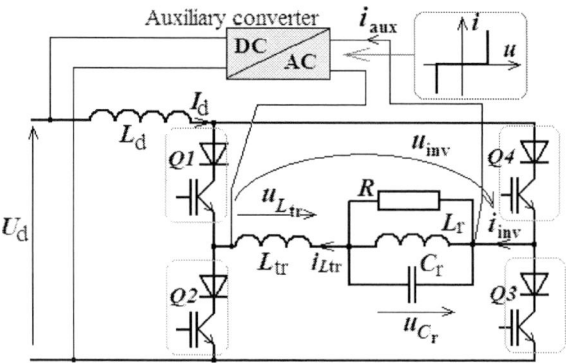

Fig. 3. Schematic diagram of the inverter with the auxiliary AC/DC converter.

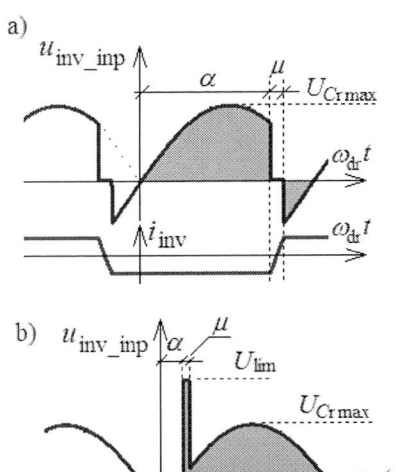

Fig. 5. Waveforms of input voltage and output current of the inverter for various switching frequencies of transistors: a) $f_s > f_{dr}$, b) $f_s < f_{dr}$.

voltage of the inverter, when commutation is not taking place, equals voltage $\pm u_{Cr}$ (Fig. 5). The output voltage during the commutation equals zero at $f_s > f_{dr}$ or voltage U_{lim} (limited by the auxiliary converter) when $f_s < f_{dr}$.

TABLE I.
EQUIVALENT CIRCUIT DIAGRAMS OF THE INVERTER FOR VARIOUS SWITCHING FREQUENCIES IN SPECIFIC SWITCHING OPERATION MODES.

a) $f_s = 1.05 \cdot f_0 > f_{dr}$	
Interval a1 $s_1=1, s_2=0, s_3=0, s_4=0,$ $i_{inv}=i_{Ltr}=-I_d<0, u_{Cr}<0, i_{aux}=0$	U_d (circuit)
Interval a2, begins when: $s_2, s_4\ 0\rightarrow 1$, $s_1=1, s_2=1, s_3=1, s_4=1$ $-I_d \leq i_{inv}=i_{Ltr}\leq I_d, u_{Cr}<0, i_{aux}=0$	U_d (circuit)
Interval a3, begins when: $i_{inv}=i_{Ltr}=I_d$, $s_1=x, s_2=1, s_3=x, s_4=1, (x=0$ or $1)$ $i_{inv}=i_{Ltr}=I_d, u_{Cr}\leq 0, i_{aux}=0$	U_d (circuit)
b) $f_s = 1.0 \cdot f_0 \cong f_{dr}$	
Interval b1 $s_1=1, s_2=0, s_3=1, s_4=0,$ $i_{inv}=i_{Ltr}=-I_d<0, u_{Cr}<0, i_{aux}=0$	U_d (circuit)
Interval b2, begins when: $s_2, s_4\ 0\rightarrow 1$, $s_1=1, s_2=1, s_3=1, s_4=1,$ $-I_d \leq i_{inv}=i_{Ltr}\leq I_{23}< I_d, u_{Cr}<0$ or $u_{Cr}\geq 0, i_{aux}=0,$	U_d (circuit)
Interval b3 begins when: $s_1, s_3\ 1\rightarrow 0$, $s_1=0, s_2=1, s_3=0, s_4=1$ $I_{23}<=i_{Ltr}<I_d, i_{inv}=I_d, u_{Cr}<0$ or $u_{Cr}\geq 0, i_{aux}>0$	U_d (circuit)
Interval b4, begins when: $i_{Ltr}=I_d$, $s_1=0, s_2=1, s_3=0, s_4=1$ $i_{inv}=i_{Ltr}=I_d, u_{Cr}>0, i_{aux}=0$	U_d (circuit)
c) $f_s = 0.97 \cdot f_0 < f_{dr}$	
Interval c1 $s_1=1, s_2=0, s_3=1, s_4=0, i_{inv}=i_{Ltr}=-I_d, u_{Cr}\geq 0,$ $i_{aux}=0$	U_d (circuit)
Interval c2, begins when: $s_1, s_3\ 1\rightarrow 0$, $s_1=0, s_2=1, s_3=0, s_4=1,$ $-I_d< i_{Ltr}<I_d, i_{inv}=I_d, u_{Cr}>0, i_{aux}>0$	U_d (circuit)
Interval c3, begins when: $i_{Ltr}=I_d$, $s_1=0, s_2=1, s_3=0, s_4=1$ $i_{inv}=i_{Ltr}=I_d, u_{Cr}>0, i_{aux}=0$	U_d (circuit)

The method for determining the amplitude of the sinusoidal voltage of the resonant capacitor is shown in Fig 5. The mean input voltage of the inverter equals the feeding voltage:

$$U_d = \frac{1}{\pi}\int_0^\pi u_{inv\_inp}(\omega_{dr}t)\,d\omega_{dr}t, \tag{1}$$

where $\omega_{dr}=2\pi f_{dr}$. Therefore, for $f_s > f_{dr}$

$$U_{Cr\,max} = -\frac{\pi U_d}{[\cos(\alpha+\mu)+\cos\alpha]} \tag{2}$$

and for $f_s < f_{dr}$

$$U_{Cr\,max} = \frac{\pi U_d}{[\cos(\alpha+\mu)+\cos\alpha+U_{lim}\mu]}. \tag{3}$$

The commutation angle μ depends on the value of the commuted current, inductance L_{Tr} and the voltage on this inductance during commutation. It can be determined using the following relation:

$$\Delta i = \frac{1}{L_{Tr}}\int_0^{\mu/\omega_{dr}} u_{LTr}(t)dt = 2I_d. \tag{4}$$

During the commutation, for $f_s > f_{dr}$

$$u_{LTr} = U_{Cr\,max}\sin(\omega_{dr}t). \tag{5}$$

On the basis of equation (5):

$$\mu = \arccos\left(\cos\alpha - \frac{2I_d L_{Tr}\omega_{dr}}{U_{Cr\,max}}\right)-\alpha. \tag{6}$$

For $\alpha = \pi-\mu$ the commutation ends when $u_{Cr}=0$. With linear approximation of the sine function, the angle is:

$$\mu \approx \sqrt{\frac{4I_d}{\omega_{dr}U_{Cr\,max}}}. \tag{7}$$

For frequency $f_s < f_{dr}$ the voltage on the inductance L_{Tr} is

$$u_{LTr}=U_{lim}-U_{Cr\,max}\sin(\omega_{dr}t). \tag{8}$$

Assuming a short duration of commutation in relation to the period of the output voltage of the inverter, the commutation angle may be determined using the approximate relation (9). For $\alpha \approx 0$, the voltage $u_{Cr} \approx 0$ and then the angle is described with the equation (10).

$$\mu \approx \frac{2I_d \omega_{dr} L_{Tr}}{U_{lim} - U_{Cr\,max}\sin\alpha}, \qquad \mu \approx \frac{2I_d \omega_{dr} L_{Tr}}{U_{lim}} \qquad (9), (10)$$

The feeding current I_d may be estimated on the basis of equal values of power in the DC circuit and AC circuit:

$$U_d I_d = \frac{1}{\eta}\frac{U_{Cr\,max}^2}{2R} \qquad (11)$$

where: η - estimated efficiency of the system.

Considering the shape of the current i_{aux} (Figs. 3 and 4c), it is possible to determine energy E_{rec} (for each commutation) and power P_{rec} returned to the feeding circuit by the auxiliary AC/DC converter:

$$E_{rec} = U_{lim} I_d \frac{\mu}{\omega_{dr}}, \qquad P_{rec} = 4I_d^2 L_{Tr} f_s \qquad (12), (13)$$

Formulas (12) and (13) have been devised with an assumption that the commutation begins at the angle $\alpha \approx 0$.

III. INVESTIGATION RESULTS

In order to verify the above assumptions, a laboratory model was built with the transformer placed between the inverter and the parallel resonant circuit (Fig. 6). For the recovery of energy stored in leakage inductances of the transformer, the AC/DC converter was used. It was constructed as a resonant DC/DC converter [4, 5] (with the constant transformation coefficient) with the rectifier at input. Figure 7 shows oscillograms that illustrate waveforms of voltage and current in the system, for: U_d=192 V, I_d=7.58 A, direct current returned by the AC/DC converter I_{drec}=0.84 A,

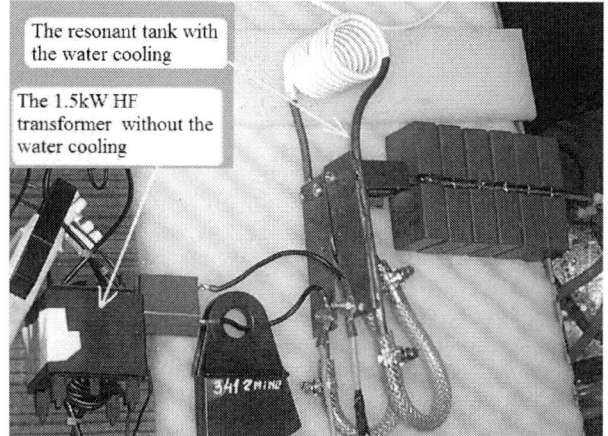

Fig.6. Elements of the output circuit of the prototype current source inverter used for induction heating.

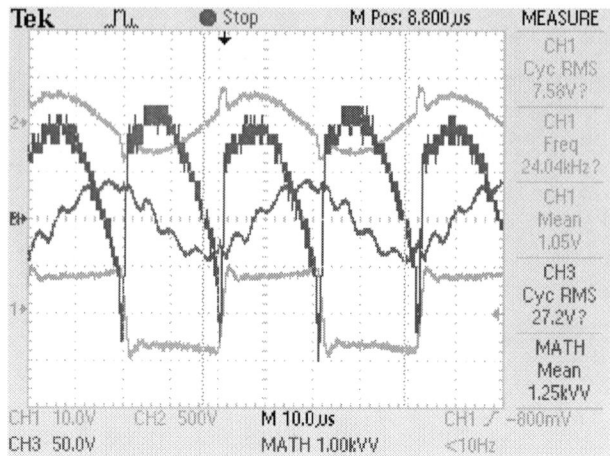

Fig. 7. Waveforms of instantaneous currents, voltage and power in the experimental system: CH1 = i_{LTr} 10 A/div, CH2 = u_{inv} 500 V/div, CH3 = inductor current 650 A/div, MATH = CH1*CH2 = 1kW/div.

U_{lim}=2*U_d, f_s= 24 kHz. Power delivered to the primary winding of the transformer reached approx. 1250W; power returned to the feeding circuit – P_{rec} = 160W and power at the inverter input $U_d \cdot I_d$ = 1455W. The efficiency of the inverter with standard overvoltage suppressor, under the very same conditions as shown in the oscillogram in Fig. 7, was approx. 85%. When the standard suppressor was replaced with the energy recovery converter, efficiency improved to approx. 95%.

IV. CONCLUSION

The arrangement of the current source inverter (with auxiliary AC/DC converter) described herein enabled connection of the transformer between the inverter (with transistors operating as ZVS switches) and the parallel resonant circuit. The enhanced efficiency of the inverter with the auxiliary AC/DC converter was due to the fact that instead of being lost in standard overvoltage suppressors, the energy was returned to the feeding circuit. The transformer (separating the inductor from the feeding circuit) connected between the inverter and the resonant tank, generated less heat than a similar transformer within the resonant circuit. Positive results of the experiments show that the theoretical assumptions were correct. In the future, prototype systems of much greater power will be built and tested. The design of the inverter (with the auxiliary AC/DC converter) described herein may be used as an alternative to systems with parallel resonant circuit supplied by the voltage source inverter through additional inductive elements as described in the reference sources [6, 7].

REFERENCES

[1] E.J. Dede, J. Jordan, et al., "Conception and Design of a Parallel Resonant Converter for Induction heating", *6th Applied Power Electronics Conference and Exposition, (APEC)*, 1991, pp. 38-44.
[2] T. Citko, H. Tunia, B. Winiarski: „*Resonant Circuits in the Power Electronics*" („*Układy rezonansowe w energoelektronice*"), Wydawnictwa Politechniki Białostockiej, 2001r.

[3] J. Jordan, J.M. Magraner, et al., "Short-Circuit Critical Frequency for Induction Heating Parallel Resonant Inverters", *13th European Conference on Power Electronics and Applications (EPE)*, 2009, pp. 1-9.

[4] J. Mućko, H.G. Langer, J.Ch. Bendien, "A Novel Resonant DC to DC Converter with High Power Density and High Efficiency", *3rd European Conference on Power Electronics and Applications (EPE)*, Aachen, 1989, pp. 1467...1471.

[5] J. Mućko: "The control methods of series resonant inverters which make possible the simultaneous work of transistors as the ZVS and the "almost ZCS" switches", *Electrical Review*, ("Metody sterowania szeregowego falownika rezonansowego zapewniające jednoczesną komutacje ZVS i "prawie ZCS", *Przegląd Elektrotechniczny*) No 6/2010, pp. 137-142.

[6] A. Schönknecht, R. De Doncker: "Novel Topology for Parallel Connection of Soft-Switching High-Power High-Frequency Inverters", *IEEE Transactions on Industry Applications*, Vol. 39, 2003, pp. 550 - 555

[7] M. A. Dzieniakowski, J. Fabianowski, R. Ibach, "LCL-Load Modular Converter For Induction Heating", *13th Conference EPE-PEMC*, 2008, pp. 2082 - 2086

Sensorless position identification of permanent magnet motor by zero voltage vector

Janusz Wiśniewski, Włodzimierz Koczara
Warsaw University of Technology (Warsaw, Poland)
wisniewj@isep.pw.edu.pl, koczara@isep.pw.edu.pl

Abstract- **The paper presents a simple sensorless method, which can be used to determine poles position of a permanent magnet axial flux motor. Described method is related to impact of a back electromotive force on current shape produced when a zero voltage vector is applied. A goal of this method is poles position detection for speed range from 2% of nominal speed of the motor. The goal of zero voltage vector method is fulfilling gap between method for initial position detecting e.g. PIPCRM and simple back electromotive force estimation method. Presented method has been tested on experimental set using 40 kVA / 16 poles / 3000 rpm axial flux permanent magnet motor.**

1. INTRODUCTION

The control system of the permanent magnet motor using maximum torque strategy requires information about the poles position. Modern control systems have eliminated a mechanical rotor position sensor, replacing it with special algorithms which calculates the poles position from the measured electrical quantities. Such methods called as sensorless and they are field of research for many researchers [1], [2], [3]. In recent publications [4], the method PIPCRM (Position Identification by Parallel Current Rate Measurement) has been presented. That method can determine the poles position of the permanent magnet machine at standstill or at low speed. The PIPCRM method is based on measurement of the stator current changes during stator core saturation. Saturation of the magnetic circuits of stator coils causes that their self inductance decreases and consequently an increase of stator current. In previous publications, the dependence of phase currents as a function of rotor position has been presented . The PIPCRM method can be applied to approximately 5% of rated speed of permanent magnet motor. However, the saturation of the stator core is a kind of nuisance. Therefore, it is proposed to use a method of poles position determined by a zero voltage vector. This method can shorten the duration of position measurement time and eliminates the necessity for a saturation of the stator core. It can be said that zero voltage method can be expansion of the PIPCRM method.

2. FUNDAMENTALS OF THE METHOD

Describing the method, it is required to introduce a mathematical model of the permanent magnet motor. The basic model of permanent magnet machines is quite sufficient to describe the method idea. It is not necessary to introduce the model which takes into account the magnetic saturation and the saliency of the motor.

Thus, a simple mathematical model is used and it is presented as follows:

$$U_S = R_S \cdot I_S + L_S \cdot \frac{d}{dt} I_S + E \qquad (1)$$

where:

$$U_S = \begin{bmatrix} u_{SA} \\ u_{SB} \\ u_{SC} \end{bmatrix}; I_S = \begin{bmatrix} i_{SA} \\ i_{SB} \\ i_{SC} \end{bmatrix}; L_S = \begin{bmatrix} L_S & M & M \\ M & L_S & M \\ M & M & L_S \end{bmatrix}; E = \begin{bmatrix} e_A \\ e_B \\ e_C \end{bmatrix}$$

u_{SA}, u_{SB}, u_{SC}	— phase voltage
i_{SA}, i_{SB}, i_{SC}	— phase current
R_S	— resistance of the stator windings
L_S	— self inductance of the stator windings
M	— mutual inductance of the stator windings

Assumed that stator resistance and self inductance of each phase are constant, mutual inductance between phases are equal to each other.

Voltage source inverter, supplying a three phase permanent magnet motor, can generate eight output voltage vectors. All possible vectors are presented in Fig. 1. Each voltage vector is obtained by different switches combination. These vectors can be divided in two groups. First of them is group containing an active vectors, marked as V_1 to V_6. Second group contains two zero voltage vectors.

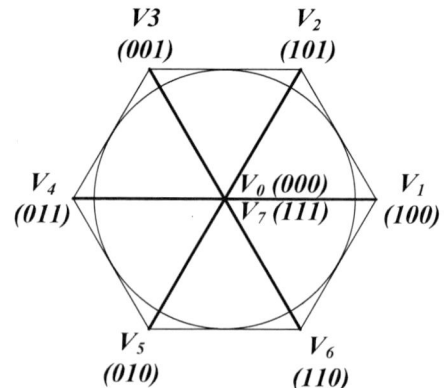

Fig. 1. Diagram of the hexagon of possible switching vectors for a three leg voltage source inverter.

Presented method uses only zero voltage vectors, marked as '000' or '111'. They correspond to states of the converter switches, which are showed in Fig. 2 and 3.

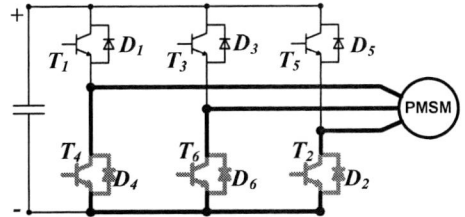

Fig. 2. Diagram showing the states of the switches when the vector '000' is produced by voltage source inverter.

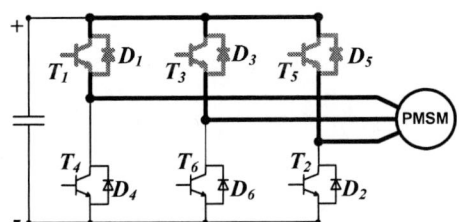

Fig. 3. Diagram showing the states of the switches when the vector '111' is produced by voltage source inverter.

When zero voltage vector is applied to stator windings, the voltage equation of the permanent magnet motor will be written as follows:

$$0 = U_S = R_S \cdot I_S + L_S \cdot \frac{d}{dt} I_S + E \qquad (2)$$

Neglecting stator resistance and writing for each phases, (2) can be expressed by:

$$i_{SA} = \int \left(\frac{-e_A}{L_{SA}}\right) \cdot dt \qquad (3)$$

$$i_{SB} = \int \left(\frac{-e_B}{L_{SB}}\right) \cdot dt \qquad (4)$$

$$i_{SC} = \int \left(\frac{-e_C}{L_{SC}}\right) \cdot dt \qquad (5)$$

The zero voltage vector method is applied when the rotor of the permanent magnet motor is already in motion. Based on laboratory tests, this speed was determined as about 2% of the rated speed. The method takes advantage of small values of electromotive force to determine the rotor position. The concept of the method is using the electromotive force generated in the stator coils when rotor speed is a non-zero. Then, if zero voltage vector is supplied, the stator windings will be short-circuited. The stator current depends on the transient values of electromotive forces. In case of switch the '111' vector on, the current will flow by the top transistors and diodes of the converter. All possible directions of the stator currents are shown in Fig. 4.

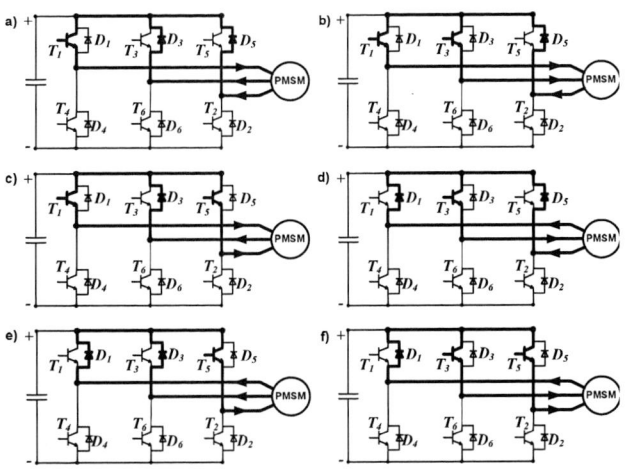

Fig. 4. Possible directions of the stator currents when applying zero voltage vector and non-zero speed.

To determine the position of the rotor, presented method uses the measured stator current at the time when zero voltage vector is applied. In the method, the stator current is proportional to transient value of electromotive force. It can be said that electromotive force is indirectly used to determine the poles position. Since the back electromotive force is function of the poles position φ, it can be calculated using stator currents and (6).

$$\varphi = atan\left(\frac{I_A - \frac{1}{2} \cdot I_B - \frac{1}{2} \cdot I_C}{\frac{\sqrt{3}}{2} \cdot I_B + \frac{\sqrt{3}}{2} \cdot I_C}\right) \mp \frac{\pi}{2} \; [rad] \qquad (6)$$

where:

I_A, I_B, I_C — stator currents measured at the end of zero voltage vector, called as current responses.

Electromotive force vector angle is delayed by π / 2 relative to the position of electrical axis of the magnetic flux excitation. Therefore, depending on the direction of rotation, the calculated value of *atan* must be increased or reduced by π / 2.

In the developed method, the zero voltage vector is held by a specified time t_{TEST}. Usually, measurement time t_{TEST} is longer than one period of the carrier wave of the PWM modulator. During t_{TEST} stator current increases as a result of short-circuit the stator coils by the power electronics converter. The current flow, caused by applying the zero voltage vector to the stator coils, is shown in Fig. 5. The figure indicates the current response also. Figure 6 shows the stator currents, analogous to Fig. 5, but at the different position of the rotor.

Dependence of the stator current from the poles position, for the given direction of rotation is shown in Fig. 7. Waveforms, which are shown in Figure 8, will be shifted by 180 electrical degrees in the case of opposite speed direction.

978-1-4244-8806-3/11 $26.00 © 2011 IEEE

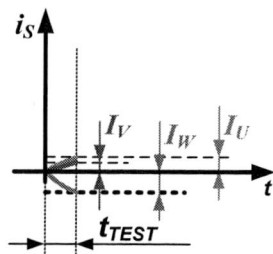

Fig. 5. Theoretical current responses (at zero voltage vector) for given rotor position.

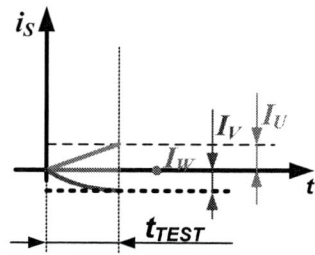

Fig. 6. Theoretical current responses (at zero voltage vector) for rotor position equals to 60 electrical degrees.

At given speed, the zero voltage vector produces the stator current, which causes that permanent magnet motor provides electromagnetic torque. Specifically, it is braking torque. However, the current value used to determine the rotor position is typically about 1% of rated current of the permanent magnet motor.

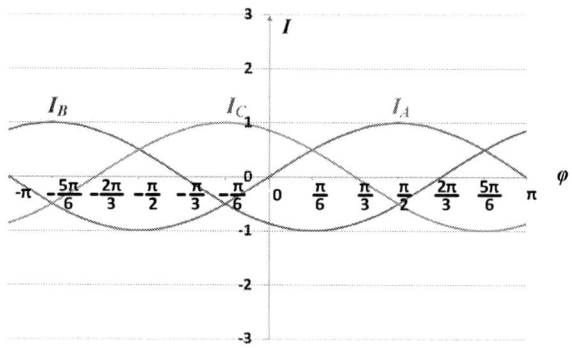

Fig. 7. Theoretical current responses (at zero voltage vector) as a function of rotor position for the positive direction of rotation.

3. LABORATORY RESULTS

Zero voltage vector method has been tested in laboratory set. All results have been carried out using 40kW / 16 poles / 3000 rpm axial flux permanent magnet motor, which is shown in Fig. 8. The sensorless control system has been implemented using DSP SHARC combined with FPGA processor. Such drive system has been designed to operate at standstill and at low speed.

Fig. 8. Laboratory 40kW / 16 poles / 3000 rpm axial flux permanent magnet motor..

An achievement of researches is previously presented PIPCRM method [5]. This method can be used to detect only an initial rotor position. Combining the PIPCRM method with zero voltage vector method, the speed range has been extended [6].

Both methods require control called as Detect and Drive [5]. System control which was realized by DSP processor is presented in Fig. 9. The Detect and Drive is labeled as 'Control - Measurement selection' in Fig. 9.

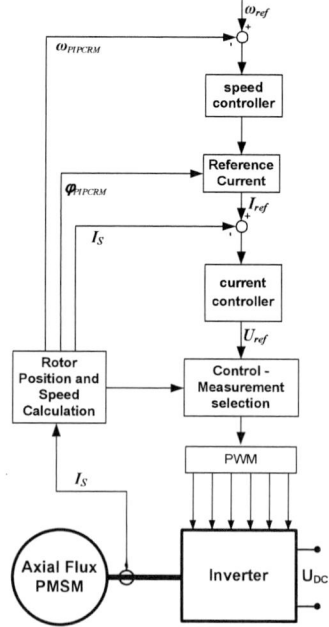

Fig. 9. Block diagram of the sensorless control system

The Detect and Drive (D&D) switches between position measurement and driving torque production. That characteristic control algorithm has three steps. First is zero-current stage. Load currents are switched off. When currents fall down to zero, next stage will occur. In that part of the D&D procedure, test voltage vector is applied to the stator windings. Next, phase currents are measured and rotor position is calculated. Last part of algorithm is load-current stage. Goal of that stage is driving. It means that torque producing currents are set with respect to the calculated electrical angle. Time sequence of Detect and Drive procedure is showed in Fig. 10.

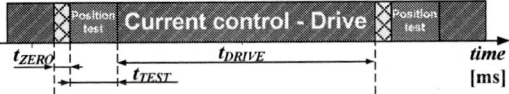

Fig. 10. Time sequence of the Detect and Drive control

Results of the rotor position test by zero voltage vector are shown in Fig. 11 and 12. The presented waveforms were carried out for same rotor position but different speed. Fig. 11 shows stator currents at 12.6 rad/s, Fig. 12 at 21 rad/s. It can be seen that time of the position test is lower than 200 μs. In comparison to the PIPCRM position test [5], it is five time less.

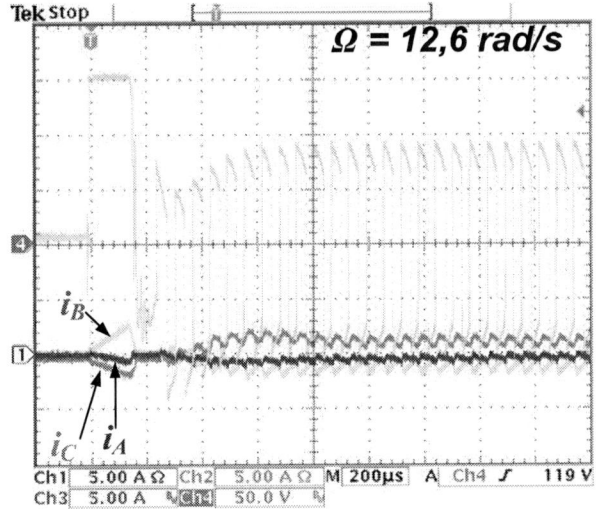

Fig. 11. Experimental results: The voltage and phase current responses i_{SU}, i_{SV}, i_{SW}.

Low speed operation, with different speed, is shown in Fig. 13 (125 rpm) and Fig. 14 (212 rpm). Fig. 14 shows waveforms, which were carried out at 7% of nominal speed of used permanent magnet motor. However, other laboratory tests show that method can be used up to 15 % of nominal speed. Since driving torque is produced between position measurements only, therefore lower time of position measurement allows average electromagnetic torque to be increased.

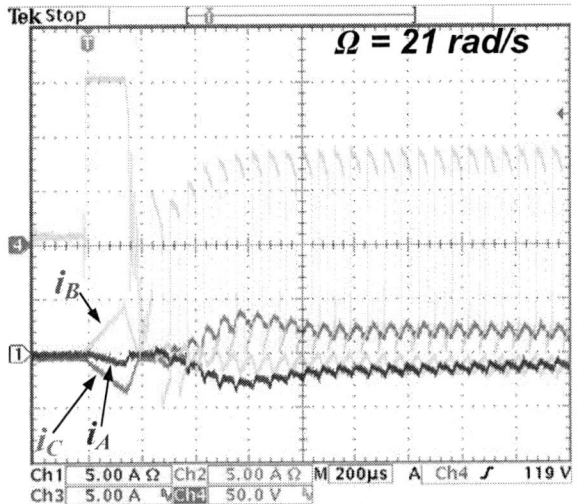

Fig. 12. Experimental results: The voltage and phase current responses i_{SU}, i_{SV}, i_{SW}.

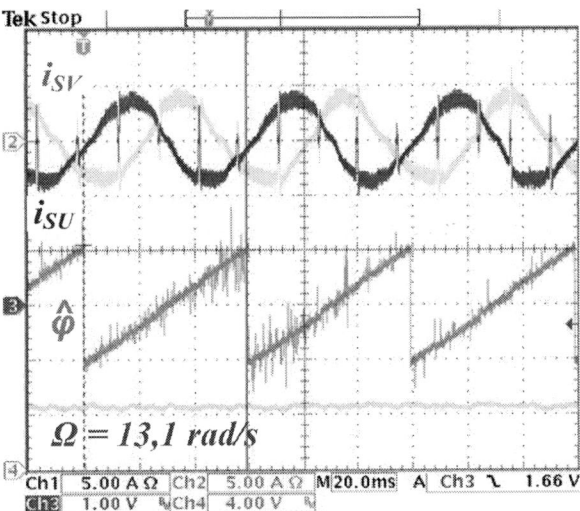

Fig. 13. Experimental results: Low speed operation: From the top: phase current i_{SU} and i_{SV}, calculated rotor position φ and speed Ω = 125 rpm

4. CONCLUSIONS

The paper describes the method of poles position identification at low speed. Presented method can be used at control system working from 1-2% up to about 15 % of nominal speed of used Axial Flux Permanent Magnet Motor. Moreover it is possible to combine zero voltage vector method with the PIPCRM method, what was done and tested in laboratory set. Results show that both methods combined together can be used to provide startup of the axial flux permanent magnet motor from standstill. Some results from sensorless startup are presented in Fig. 15. In that case the load torque was set to 100 Nm

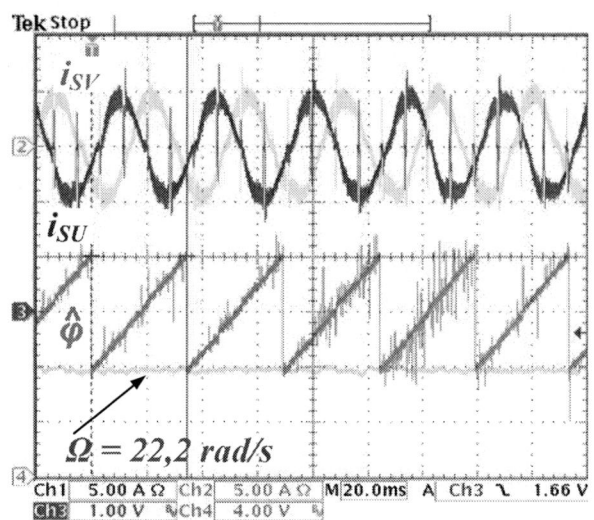

Fig. 14. Experimental results: Low speed operation: From the top: phase current i_{SU} and i_{SV}, calculated rotor position φ and speed Ω = 212 rpm

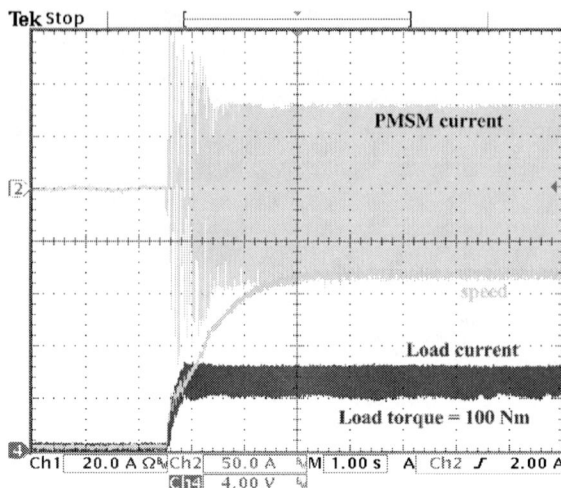

Fig. 15. Experimental results: Sensorless startup of the permanent magnet motor from standstill to 350 rpm with 100 Nm load. [5]

Achieved top speed makes possibility to apply a back electromotive force estimator. It will be in scope of further researches.

REFERENCES

[1] Zhaofeng Li, A. Kawamura, "Investigation on resolution of initial position estimation for IPMSM", *14th International Power Electronics and Motion Control Conference*, EPE-PEMC 2010, Ohrid, Republic of Macedonia, 6-8 September 2010

[2] P. Brandstetter, O. Francik, P. Simonik, "AC drive with permanent magnet synchronous motor and sensorless control", *14th International Power Electronics and Motion Control Conference*, EPE-PEMC 2010, Ohrid, Republic of Macedonia, 6-8 September 2010

[3] E. de M Fernandes, A.C. Oliveira, C.B. Jacobina, A.M.N. Lima, "Comparison of HF signal injection methods for sensorless control of PM synchronous motors", Twenty-Fifth Annual IEEE Applied Power Electronics Conference and Exposition, APEC 2010, Palm Springs, CA, 21-25 February 2010

[4] J. Wiśniewski, W. Koczara: "Control of Axial Flux Permanent Magnet Motor by the PIPCRM Method at Standstill and at Low Speed", *13th International Conference on Power Electronics and Motion Control*, EPE-PEMC 2008, Poznań, Poland, 1-3 September 2008

[5] J. Wiśniewski, W. Koczara: "The Sensorless Rotor Position Identification and Low Speed Operation of the Axial Flux Permanent Magnet Motor Controlled by the Novel PIPCRM Method", *IEEE 39th Power Electronics Specialists Conference*, PESC 2008, Rhodes, Greece, 15-19 June 2008

[6] J. Wiśniewski, W. Koczara: "Sensorless Control of Axial Flux Permanent Magnet Motor at Standstill and at Low Speed", *Electrical Review*, ISSN 0033-2097, R. 85 No 7/2009, pp. 177-181

Optimized Cabin Power Supply with a +/- 270 V DC Grid on a Modern Aircraft

J. Brombach[1], T. Schröter[2], A. Lücken[1] and D. Schulz[1]

[1]Electrical Power Systems, Helmut-Schmidt-University, D - 22043 Hamburg, Germany

[2]Airbus Operations GmbH, D- 21129 Hamburg, Germany

brombach@hsu-hh.de

Abstract- **This paper deals with an onboard power supply of +/- 270 V DC (HVDC, high voltage DC). Conventional aircraft grids have a main voltage level of 115 V AC with a variable frequency of 360...800 Hz. In the future, fuel cells might replace the conventional auxiliary power unit (APU) to optimize the eco efficiency. With that technology HVDC-grids can be integrated in an aircraft. Changing to a HVDC-grid has a lot of advantages. Higher voltage means lower cable weight. Furthermore, the electrical converter architecture can be optimized. Especially the converters inside the loads can be build much lighter when using a +/- 270 V DC supply. This paper gives an overview of the electrical architecture on a modern short- and midrange aircraft and the effect of a +/- 270 V DC power grid thus focusing on the converter architecture in a modern aircraft. Concluding all benefits are summarized.**

I. INTRODUCTION

A major activity of the aviation industry is the reduction of emissions in combination with cost efficiency. On the one hand eco-friendly aircraft save the airlines money because of the forecasted rising fuel prices in the future, on the other hand environmental pollution will become increasingly a problem for human society. The main part of development is focused on reducing weight and because of that directly reduced fuel consumption. The electrical systems use only 0.2 % of the whole engine power [1] at cruise, so savings in the electricity consumption have only a small effect on aircraft fuel consumption. Weight savings on the other hand have a positive "snowball effect". For example, saving one kilogram in the equipment means also a possible weight reduction of the aircraft structure and the engine by an additional 600 g. Saving of 1.6 kg of the aircraft weight translates into lower fuel consumption and an extended performance.

In conclusion saving weight has an impact on aircraft performance or − at the stable performance level − a huge impact on eco efficiency and fuel consumption. Saving one kilogram of weight will reduce costs by approximately 4500 $ [2] at a short- and midrange aircraft over a 20 year period of operation.

As part of the project "cabin technology and multifunctional fuel cell systems" the institute of Electrical Power Systems of the Helmut-Schmidt-University is dealing with a high-efficiency onboard power supply.

A conventional aircraft grid is based on a main bus voltage of 115 V AC (360...800 Hz). Some newer developments [3] [4] use 230 V AC. Recent concepts try to integrate a +/- 270 V DC voltage level into the aircraft [3] [4].

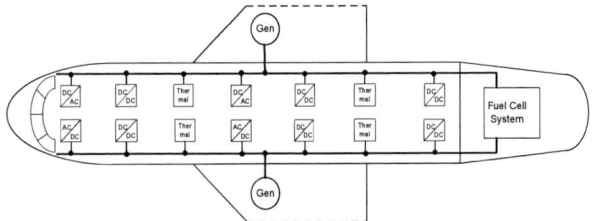

Fig. 1. Aircraft onboard power grid

In addition to the higher main bus voltage the traditional auxiliary power unit (APU) could be replaced by a multifunctional fuel cell system to reduce pollution during ground operation. It has many advantages over the traditional APU [5] [6].

Integrating such a fuel cell system makes an adaption of the primary aircraft grid necessary. For that a +/- 270 V DC voltage level is an possible option in the future as well. Fig. 1 shows a fuel cell fed grid on modern aircraft with the different load types. This paper deals with the use and integration of this new voltage level in the cabin and cargo system.

Most loads in a modern aircraft are supplied with internal converters. This paper focuses on the converter architecture. The effect of a higher voltage level on the weight of the cable systems is presented in other papers [7].

II. CONVERTERS IN THE AIRCRAFT

In conventional aircraft the chain of converters is small and consists normally of one or two converters. Fig. 2 shows loads which are connected to the AC grid. The only converter here is the switching power supply in the loads. Loads which are supplied by 28 V DC could have two converters: The central transformer rectifier unit and one possible DC/DC converter in the load.

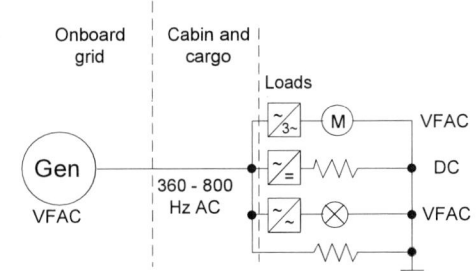

Fig. 2. Converters in conventional aircraft power supply system

III. CHANGE TO +/- 270 V DC GRID

On a modern aircraft the main bus could have a voltage of +/- 270 V DC. Especially, if a fuel cell system would be integrated, the step to a high voltage DC (HVDC) grid is recommended. Fig. 3 shows a grid with HVDC and conventional cabin and cargo systems which have an operating voltage of 115 V 400 Hz AC. A main converter feeds the cabin and cargo grid. The HVDC grid is either being fed by a fuel cell system with a DC/DC converter or being fed by the engine driven generators over a rectifier unit (AC/DC-converter).

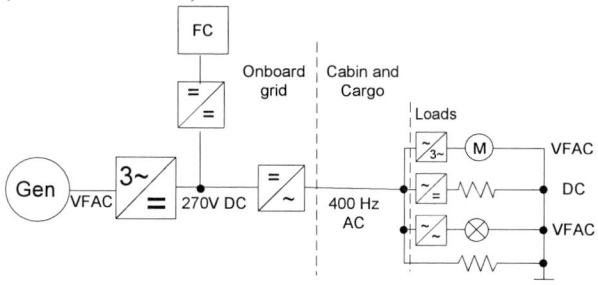

Fig. 3. Converters in a modern aircraft power supply system with conventional cabin and cargo loads

Fig. 4 shows the architecture with a DC-supply for the cabin and cargo systems without a 400 Hz AC converter.

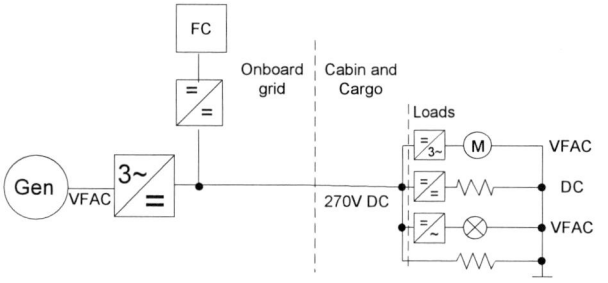

Fig. 4. Converters in a modern aircraft power supply system with adapted cabin and cargo loads

IV. TYPES OF CONVERTERS IN AN AIRCRAFT

The type of the particular converter in an aircraft depends on the supplying voltage. Especially converters which have an AC input or output voltage need filters or PFCs (power factor corrections) to reduce the harmonics which are produced by a rectifier or the inverter. Fig. 5 shows the input current of a conventional full-wave rectifier. The problem is that the stabilizing capacitor inside the rectifier is only charged at the voltage maximum. The harmonic current characteristic of this kind of rectifier is poor. In conventional energy supply grids these rectifiers are only allowed up to a power of 75 W [8].

In aircraft applications all loads have to abide fix harmonic current rules, too. The ratios of allowed harmonic currents are referenced to the fundamental component of the load current. Therefore all loads have to use filters to eliminate the harmonics.

To reduce the input harmonics active or passive PFC can be used. Especially active PFCs can reduce the harmonic

currents significantly. Mainly in "weak" aircraft grids with high grid impedance (compared to conventional power grids) it is particularly important to reduce harmonic currents to ensure the voltage THD (total harmonic distortion) stays in the allowed range

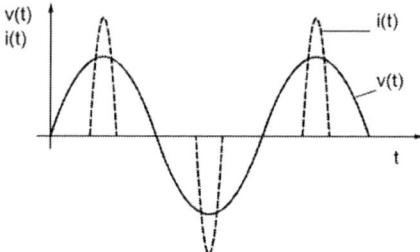

Fig. 5. Input current of a conventional full-wave rectifier with sinusoidal input voltage [9]

The active PFCs work with a boost converter behind the full wave rectifier. The controller of the boost converter controls the input current so that it is proportional to the input voltage. The drawback of this system is the complexity and the system weight. The boost converter has to lift the voltage from nearly zero to the link voltage of the converter.

Fig. 6 shows block diagrams of the four possible converter types onboard an aircraft. In addition the power density of all blocks and the whole converters are given. The weights are based on supplier specification and technical literature [9] and for a converter power smaller than 5 kW.

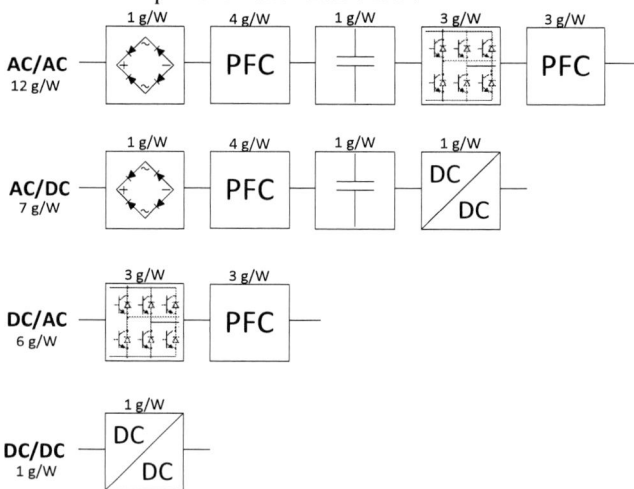

Fig. 6. Power density of different converter types for aircraft application with a power less than 5 kW

A. AC/AC-converter

The AC/AC-converter is the type with the most complex topology. In addition to the rectifier, the inverter and the DC-link, two PFCs are installed to keep the harmonics in the allowed range. The overall power density is approximately 12 g/W. The output PFC is simpler because the output load defines the filter and not the harmonic current requirement of the grid. Furthermore the PFC output does not need a boost converter like the input PFC to guarantee the sinusoidal current at the output.

978-1-4244-8806-3/11 $26.00 © 2011 IEEE   426

B. AC/DC-converter

The AC/DC-converter has on the input side the same components as the AC/AC converter. But on the output side the inverter is replaced by a DC/DC converter to stabilize the voltage on a defined level. The power density of this converter type for aircraft application is 7 g/W. For comparison an AC/DC converter for a mobile computer has a power density of approximately 5 g/W, without overload ability.

C. DC/AC-converter

The DC/AC converter has on the output side the same components as the AC/AC-converter. A DC-link and a rectifier with a PFC are not necessary. That leads to a higher power density of 6 g/W.

D. DC/DC-converter

DC/DC-converters are very small and light because of electronic components with a high power density. A DC/DC-converter only needs a small number of electronic components which leads to an average power density of 1 g/W.

E. Converters for power greater than 5 kW

For architectures with converters greater than 5 kW higher power densities are given. Especially the rectifier (AC/DC-converter) have, compared to the other ones an over proportional high power density. The respective densities are shown in table I.

TABLE I
POWER DENSITY OF ONBOARD AIRCRAFT CONVERTERS GREATER THAN 5 KW

Converter type	Power density[1]
AC/AC	1.43 kg/kW
AC/DC	0.51 kg/kW
DC/AC	0.95 kg/kW
DC/DC	0.17 kg/kW

[1]supplier data

V. ANALYSIS OF THE LOADS

This chapter deals with the ELA (electrical load analysis) of the cabin and cargo systems on a modern short- and midrange aircraft. For these studies all loads in an aircraft power supply system have been manually allocated by their characteristics like internal voltage or power characteristic at voltage drops.

A. Load demand of the cabin and cargo systems

The maximum overall power demand of the cabin- and cargo-system of a modern short- and midrange aircraft is approximately 105 kW. The following calculations are based on that power demand.

Fig. 7 shows the percentage distribution of the cabin and cargo load types weighted to power. It can be seen that constant power and ohmic loads are almost equally distributed. The load type is quite important, especially in fuel cell grids. The load current characteristic is essential for the size of the entire energy supply system.

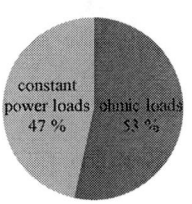

Fig. 7. Load types in the cabin and cargo area weighted to power

The power consumption of constant power loads is independent from the input voltage (1). The load current is inversely proportional to the input voltage.

Ohmic loads have the contrary characteristics. The load current is proportional to the input voltage (2).

$$P = const. \tag{1}$$

$$I \sim \frac{1}{U}$$

$$P = U \cdot I \tag{2}$$

$$P \sim U,$$

$$I \sim U$$

In a nutshell the load current is constant and not independent of the input voltage.

B. Types of loads on the aircraft

Fig. 8 shows the quantitative distribution of the internal voltage of all loads. The number of the different kinds of loads is in the same range.

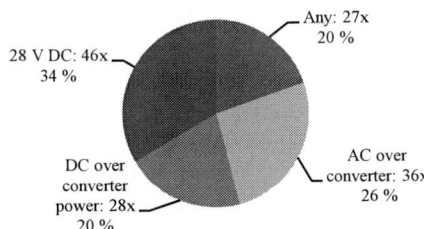

Fig. 8. Internal voltage of the load types in the cabin and cargo area weighted by the quantity

Half of the loads are DC loads such as In-flight entertainment systems or other microelectronics. The other half are thermic loads, AC motoric loads or lights.

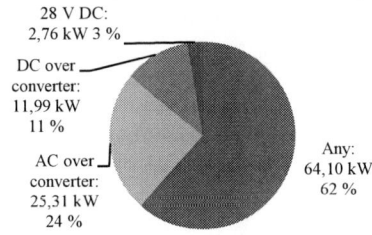

Fig. 9. Internal voltage of the load types in the cabin and cargo area weighted to power

Fig. 9 shows the distribution of the internal voltage of all loads weighted to power. The relative part of loads which can operate with all kind of voltages (e.g. thermal loads) is the greatest one. These loads should be connected to the voltage of the main bus bar. The rest of the loads operate with AC or DC. The key is to reduce the weight of the converters here.

VI. OPTIMIZING THE CABIN AND CARGO SYSTEMS WITH A +/- 270 V POWER SUPPLY

This chapter deals with possible optimization of the three given architectures. The basis upon which all weight calculations are made is the power of the cabin and cargo system. The part of the converter weight for other aircraft systems is not part of this consideration for an easier comparability of the cabin systems. The absolute weight for the considered system is calculated correctly because of the nearly constant power density.

A. State of the art – 115 V AC supply

The converter architecture of a conventional AC grid is simple. The internal voltage and the power densities (Fig. 6) are being used to calculate the weight of the converters. In a modern short- and midrange aircraft all converters would have a weight of approximately 400 kg.

B. Change to +/- 270 V DC grid

If the main bus bar voltage would change to +/- 270 V DC there are two possibilities for the cabin and cargo systems. The first one is to supply the loads with 115 V 400 Hz AC with a main DC/AC-converter. The second possibility is to adapt the loads to the new main bus bar voltage.

Changing the architecture would not affect the weight of the AC/DC-converter at the engine generator and the weight of the DC/DC converter of the fuel cell system. Table II shows the power densities of the main converters. Oversizing of 20 % is taken into consideration.

TABLE II
WEIGHT PROPORTION OF THE CABIN AND CARGO SYSTEMS ON THE MAIN CONVERTERS IN THE AIRCRAFT

Converter type	Location	Weight[1]
AC/DC	Generator	66 kg
DC/DC	Fuel cell	22 kg
DC/AC	Cabin entry	125 kg

[1]supplier data

The internal converters of the cabin and cargo loads would weigh 170 kg. Compared to the 400 kg of a conventional cabin, the DC voltage has a weight benefit.

C. Optimization potential of +/- 270 V supply on the aircraft converter architecture.

The key is the overall converter weight of the conventional architecture. In this example the weight is approximately 400 kg. A change of the main bus bar voltage to +/- 270 V with a conventional 115 V 400 Hz AC cabin would increase the overall weight to 624 kg (+224 kg) because of the aircraft converters.

The overall weight of the cabin and cargo load converters with a +/- 270 V DC supply is 258 kg. Compared to the standard AC system a saving of 142 kg seems possible.

Compared to an aircraft with +/- 270 V DC main buses and a conventional cabin, weight savings up to 366 kg seem possible on a normal short- and midrange aircraft.

VII. CONCLUSION

In conclusion a +/- 270 V DC system has a lot of benefits. Especially the loads have a big potential for weight savings on a modern aircraft. Half of the loads (weighted on power 38 % of the loads) operate with internal switching power supplies. These add 400 kg to the weight because of the required components for the power quality (PFCs). A DC supply could reduce the weight of these devices by 230 kg. In addition a conventional 115 V 400 Hz AC cabin needs a main converter if the main bus bar is DC. Additionally a supply with high voltage DC makes the main converter obsolete and saves up to 125 kg. The reduction of cable weights is given in [7], but not the focus of that paper.

Summarizing a +/- 270 V DC supply for the cabin and cargo systems is highly advisable.

ACKNOWLEDGEMENT

The work "cabin technology and multifunctional fuel cell systems" has been supported by Airbus and the German Federal Ministry of Education and Research (support code: 03CL03A).

REFERENCES

[1] A. Reichert, "Decrease of the Bidirectional Load Flow on Modern Airliners by Improvement of the Primary Electrical Power Grid," *diploma thesis (in German).* Helmut-Schmidt University, Hamburg, Germany, 2010.

[2] F. Heider, "Economical Comparison between Standardized Cable Architecture and Cable Architecture Especially Designed for Low-Cost-Carrier Based on Modern Short- and Midrange Aircraft" *bachelor thesis (in German).* Helmut-Schmidt University, Hamburg, Germany, 2010.

[3] Boeing, *AERO - 787 No-Bleed Systems.* http://www.boeing.com/commercial/aeromagazine/articles/qtr_4_07/article_02_3.html (accessed November 24, 2010).

[4] Heuck, K. and Dettmann, K.-D., *Electrical Power Systems,* 8. Edition (in German) Wiesbaden: Vieweg, 2010. http://www.gbv.de/du/services/toc/bs/479445842.

[5] Rajashekara, K., Grieve, J., and Daggett, D., "Hybrid Fuel Cell Power in Aircrafrt: A feasibility study for onboard power generation using a combination of solid oxide fuel cells and gas turbines," *IEEE Industry Application Magazine,* vol. 14, no. 3, pp. 54–60, 2008.

[6] K. A. Friedrich, "Fuel Cell Systems in aviation (in German)", *BWK das Energie- Fachmagazin,* VDI Verlag, Germany, April 2010.

[7] Brombach, J., Lücken, A., Schulz, D., and Schröter, T., *Structural and Functional improvement of the electrical energy distribution of modern aircraft.* (in German) Hamburg: 59. Deutscher Luft- und Raumfahrtkongress der DGLR, 2010

[8] DIN EN 61000-3-2, *current harmonics <16 A,* Electromagnetic interference (EMI).

[9] Schlienz, U., *Switching Power Supplies and their periphery:* 2. Edition (in German), Wiesbaden: Vieweg, 2007. http://dx.doi.org/10.1007/978-3-8348-9222-5.

Supply System Based on the 18-pulse Diode Rectifier with Coupled Reactors and Integrated with the Series Active Power Filter

P. Mysiak[1], R. Strzelecki[1], H. Tunia[2], K. Zymmer[3]

Gdynia Maritime University[1], Faculty of Electrical Engineering, Department of Ship Automation,81-87 Morska Str., 81-225 Gdynia, Poland, mysiak@am.gdynia.pl, rstrzele@am.gdynia.pl

Kielce University of Technology[2], Faculty of Electrical Engineering, Automatics and Computer Science, Chair of Power Engineering Electronics, 7 Al. Tysiąclecia Państwa Polskiego Str., building D, 25-314 Kielce, Poland, ene@tu.kielce.pl

Electrotechnical Institute[3], Department of Power Converters, 28 Pożaryskiego Str., 04-703 Warsaw, Poland, npm@iel.waw.pl

Abstract - The article presents the principle of operation and selected results of simulation and laboratory tests of the 18-pulse converter system with direct voltage output, cooperating with a small series active power filter (S-APF). The presented system makes it possible to reduce, especially in autonomous supply networks, undesired higher harmonics in the network current. The 18-pulse nature of operation of the rectifier is reached using a set of coupled three-phase network reactors (CDT and CTR). Simultaneous use of the magnetically coupled reactors and the small active power filter provides opportunities for easy and cheap reduction of the supply current THD coefficient, down to the values acceptable by clean power converters (CPC) conditions.

Index Terms - Active filters, multi-pulse converters, coupled reactors, power conditioning.

I. INTRODUCTION

The problem of cooperation of power electronics converters, including rectifying systems, with the supply network, in particular their negative effect onto the network is becoming more and more important due to increasing power and more frequent application of these devices. This negative effect, expressed, for instance, by the value of the supply current deformation coefficient THD_i, can be minimised in many ways, among which the multi-pulse technique, the modulation technique, and the active filter technique can be named.

Although it gives excellent results, the modulation technique is not used in large-power systems, due to limited possibilities to reach high-frequency connections of the semiconductor power connectors and increased connection losses, as well as voltages stresses which intensify ageing of the insulation of electric devices [1], [2], [3], [4], [5].

Active filters provide opportunities for full compensation of electric current deformation, but they have a complicated structure (measuring systems, control systems) and are rather expensive, especially in case of large power applications [1], [4], [6], [7], [8], [9].

In large-power systems, a most reasonable solution seems to be the use of the multi-pulse technique [10]. It secures the generation of multi-stepwise waveform of the current taken from the supply source, with the number of steps equal to the number of pulses q in the rectified voltage time-history counted over the supply voltage period. Increasing q decreases the current deformation coefficient THD_i when the receiver reveals characteristics of a current source. Comparable THD_i values are also obtained when the direct current circuit reveals voltage source characteristics, when assuming that supply source reveals current source characteristics. The level of THD_i is strongly affected by the supply source reactance. For very small reactance, THD_i increases considerably, especially when $q \leq 12$ [11], [12].

The multi-pulse technique can be executed in the following versions:

- parallel or series connection of basics systems (for instance, three-phase bridge systems) with relevant phase shift of supply voltages [3], [5], [11], [13],
- use of multi-pulse autotransformers [14], [15], [16], [17],
- parallel connection of basic systems with the use of coupled reactors [12], [18], [19], [20], [21].

Parallel connection of basic systems is in common use. For instance, the 12-pulse system is realised by parallel connection of two three-phase bridge systems using a reactor for de-coupling those systems, supplied from two different secondary windings of the transformer with relevant phase shift. However, this solution is rather expensive. The rating power of the electromagnetic systems exceeds the direct current power [2], [10].

As a rule, the rating power of the autotransformers does not exceed 70% of the direct current power. Using transformers is a favourable solution when the direct current circuit reveals the current source characteristics [3], [14], [16], [17].

In the situation when the direct current circuit reveals the voltage source characteristics, a very attractive solution is the use of coupled three-phase reactors for parallel connection of three-phase bridge systems. In this case the rating power of the electromagnetic systems is much smaller than in all above discussed systems [3], [11], [12]. Unfortunately, when the systems with coupled reactors are used, the deformation of the symmetry and sinusoidal shape of the supply voltage

978-1-4244-8806-3/11 $26.00 © 2011 IEEE

results in considerable deformations of the network current. In those cases a reasonable solution is the use of an additional small, low-power series active filter [1], [4], [8], [22].

Ref. [11] presents the design and basic relations required for designing the 18-pulse diode rectifier working in the system of coupled three-phase reactors, and characteristic for parallel operation of three three-phase bridge rectifiers. Such a system can be considered a compromise between the quality of the electric current taken from the supply network (THD≈7%) and relatively small complexity of the rectifier structure.

If we take into account the internal impedance of the supply source, then at full load of the rectifier the waveform of the voltage passed to the system has a stepwise shape. The analysed converter with constant-voltage output is generally designed for cooperation with an autonomous low-voltage supply network. The presented system provides opportunities for remarkable reduction of undesired higher harmonics, mainly of an order of 11, 13, 17 and 19, in the network current. The 18-pulse operation of the rectifier is obtained using a coupled three-phase reactor (CDT) for preliminary current division, and a set of coupled network reactors (CTR). Along with the reduction of higher harmonics in the supply current to about 7%, another effect of the rectifier's operation is the power coefficient reaching the level close to one. It is also noteworthy that the apparent power and the dimensions of the magnetic circuit are small. The article presents the results of the laboratory tests of the 20kVA system, complemented by selected results of simulation and laboratory tests concerning possible cooperation of the rectifier with an active power series filter (S-APF) to improve remarkably the shape of the waveform of the supply line current.

II. 18-PULSE DIODE RECTIFIER WITH CDT&CTR

Figure 1 presents a schematic diagram of the 18-pulse diode rectifier characteristic for parallel operation of three three-phase bridge converters, which operates at the leading load [11], [19]. The three-phase supply network is represented by the source voltage e and the line impedance Z_s. Moreover, reactors of inductance L_d are connected in series to the supply source, to reduce, in general, the amplitude of higher harmonics of the currents which are taken from the network by the 18-

pulse rectifier. The basic assumption in the adopted concept of the presented rectifier system is the construction of three three-dimensional vectors of three-phase voltages supplying the bridge rectifiers, which are shifted by 20° with respect to each other. When this condition is met, the three-phase supply voltage v_K, which has the sinusoidal shape at no-load running, takes the 18-step shape at nominal load. The voltages v_K measured between the terminals K and the neutral point 0 can be interpreted as the quantities created as a result of cyclic switching the current voltage V_{DC} on (Fig.1) via the diodes of three bridge rectifiers. The 18-pulse waveform of the voltages v_K (Fig.2) is only obtained when the diode conduction angles in particular rectifiers are equal to π, and when the conduction intervals (18 intervals) are symmetrically distributed along the supply voltage period. In this case each of the three-phase bridge rectifiers generates a three-level voltage at the alternating-current terminals. The phase voltages measured at terminals $D2$ (with respect to the neutral point 0) are by 20° ahead of the relevant phase voltages measured at terminals $D1$. In turn, the phase voltages measured at terminals $D3$ are phase-delayed by 20° with respect to the relevant phase voltages measured at terminals $D1$. As a consequence, CTR executes 40° phase shift between two output voltage systems. The applied electromagnetic systems CDT and CTR make it possible to obtain the required division of the electric current, taken from the supply source, on particular bridge rectifiers. The three three-phase current systems, which supply these rectifiers reveal the same rms value of the phase currents and an approximately sinusoidal shape. Like for voltages, the system currents $P2$ are by 20° ahead of the system currents $P1$, while the system currents $P3$ are phase delayed by 20° with respect to the rectifier $P1$.

The effect of the action of the coupled three-phase reactors is converting the three-phase voltage of the supply source into the nine-phase voltage. The reactors comprise six separate magnetic cores, on which relevant windings are wound. Selection of the number of winding turns and their connections results from the required voltage phase shift (CTR system) and required preliminary electric current division (CDT system).

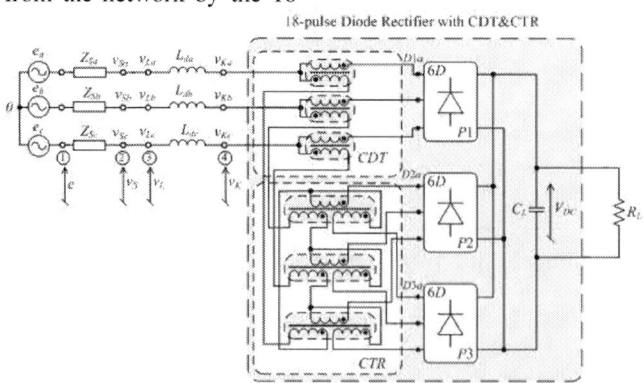

Fig. 1. Supply system based on the 18-pulse diode rectifier with coupled three-phase reactors (CDT&CTR).

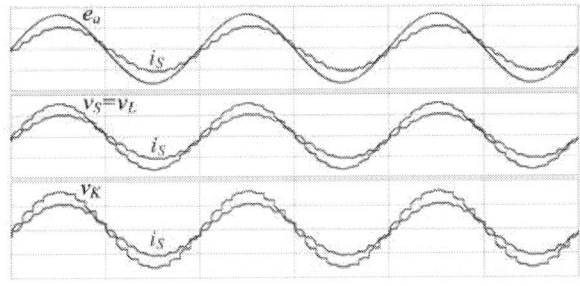

Fig. 2. Typical waveforms of the input voltages and current (200 V/div, 40 A/div)

The simplified model of the examined system, shown in Fig.3, provides opportunities for correct qualitative assessment of the effect of its parameters and supply voltage disturbances on the deformation of the input current.

Fig. 3. Simplified model of the system shown in Fig.1

Although it assumes full symmetry of the system, this model can also be applied for assessing the effect of non-symmetry of supply voltages. This non-symmetry is a source of additional deformations of voltages v_K.

A. Investigation of the Mains Parameters Effect

The input current harmonic content coefficient is defined acc. to [22] by the formula:

$$THD(i_S) = \frac{1}{I_{(1)}} \sqrt{\sum_{h=2}^{\infty} I_{(h)}^2} \qquad (1)$$

where:

$$I_{(h)} = \left| \frac{E(jh) - V_K(jh)}{Z_S(jh) + jX_d} \right|, \quad h = 1,2,3,\dots \quad (2)$$

is the harmonic of an order of h of the base current, and $X_d = \omega L_d$ is the reactance of the network reactor. Hence, assuming the linearity of the network and reactor impedance L_d:

$$Z_S(jh) = R_S + jhX_{S(1)}, \quad X_d = hX_{d(1)}$$

we get:

$$THD(i_S) = \frac{1}{I_{(1)}} \sqrt{\sum_{h=2}^{\infty} \frac{\left| E(jh) - V_K(jh) \right|^2}{R_S^2 + h^2 \left(X_{S(1)} + X_{d(1)} \right)^2}} \quad (3)$$

Since the short-circuit power of the network is:

$$S_{short} = E_{(1)}^2 \Big/ \sqrt{\left(R_S^2 + X_{S(1)}^2 \right)}$$

then if $R_S^2 \ll X_S^2$ then $E_{(1)}^2 \big/ S_{short} \approx X_{S(1)}$. In this case the relation (3) takes the form:

$$THD(i_S) \approx \frac{S_{short}}{I_{(1)}} \sqrt{\sum_{h=2}^{\infty} \frac{\left| E(jh) - V_K(jh) \right|^2}{h^2 \left(E_{(1)}^2 + S_{short} X_{d(1)} \right)^2}} \quad (4)$$

Formula (4) describes the coefficient $THD(i_S)$ as a function of both the inductance L_d of the network reactor and the short-circuit power S_{short} of the supply source. Since:

$$P_{rec} \approx I_{(1)} E_{(1)}$$

then the dependence of $THD(i_S)$ upon the receiver power P_{rec} (when assuming the linearity of CDT&CTR) is given by the formula:

$$THD(i_S) \approx \frac{S_{short} E_{(1)}}{P_{rec}} \sqrt{\sum_{h=2}^{\infty} \frac{\left| E(jh) - V_K(jh) \right|^2}{h^2 \left(E_{(1)}^2 + S_{short} X_{d(1)} \right)^2}} \quad (5)$$

Formulas (4) and (5) are confirmed by functional relations shown in Figs. 4 and 5, which were determined from detailed simulation investigations.

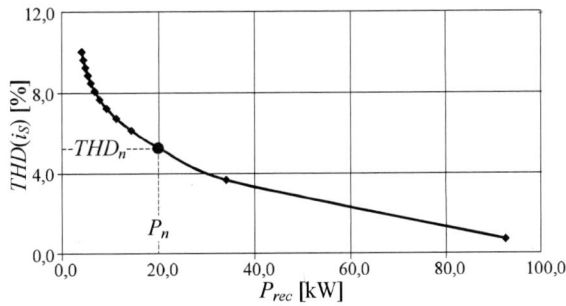

Fig. 4. $THD(i_S)$ versus load power P_{rec} for the linear CDT&CTR.

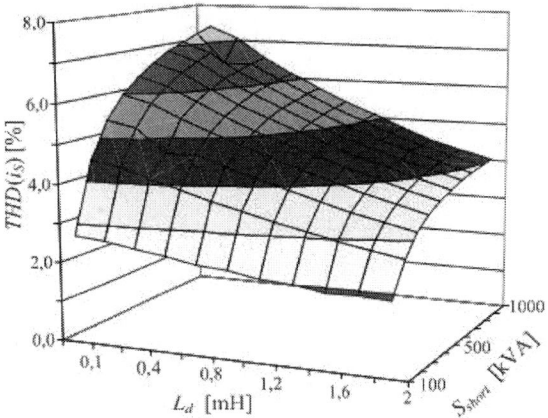

Fig. 5. Variation of $THD(i_S)$ depending on short-mains power S_{short} and line reactor L_d.

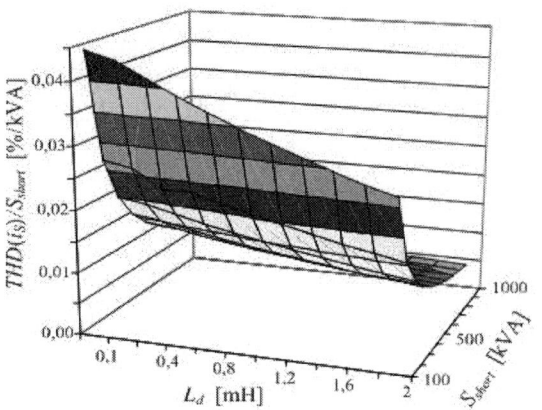

Fig. 6. Dependence of the $THD(i_S)/S_{short}$ ratio versus short-mains power S_{short} and line reactor L_d.

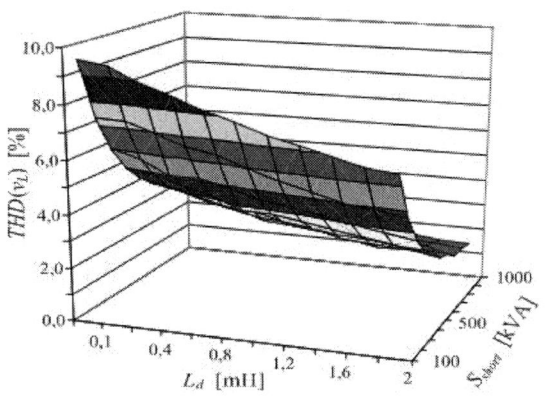

Fig. 7. Variation of $THD(v_L)$ depending on short-mains power S_{short} and line reactor L_d.

It is noteworthy that the coefficient $THD(i_S)$ is not most appropriate for assessing the action of the receiver/ deformed current source onto other receivers connected to the same supply line. A better criterion, taking into account the effect of the short-circuit power, is the ratio:

$$\frac{THD(i_S)}{S_{short}} \approx \frac{1}{I_{(1)}} \sqrt{\sum_{h=2}^{\infty} \frac{|E(jh) - V_K(jh)|^2}{h^2 \left(E_{(1)}^2 + S_{short} X_{d(1)} \right)^2}} \ , (6)$$

Figure 6 shows the ratio (6) variations vs. inductance L_d and power S_{short}, determined from the simulations. The trends of these variation are similar to those observed for the rectifier input voltage harmonic content coefficient (Fig.7). This justifies formulating a conclusion about the adequacy of the assessment of the action of the examined receiver onto other line receivers using the $THD(i_S)/S_{short}$ ratio.

B. Power Voltage Disturbance Effect

The characteristics of the system shown in Fig.1 also strongly depend on the disturbances of the supply voltage: its harmonic deformations and non-symmetry. The appearance of those disturbances, even in a normative form, results in remarkable increase of

deformation of the electric current i_S. This effect can be explained using the model shown in Fig.3. In particular, it is noticeable that the effect of the deformation of the supply voltage on the deformation of the electric current is larger for lower harmonics, which is connected with the filtering action of the reactor L_d and impedance Z_S. This is also confirmed by the changes, recorded in the simulations, of the coefficient $THD''(i_S)$ (in relation to $THD(i_S)$ representing the sinusoidal voltage) as functions of the amplitude of a single harmonic of an order of h, see Fig.8. Initial slower increase of the $THD''(i_S)/THD(i_S)$ ratio results from relatively small changes of voltage harmonics v_K at rectifier's terminals (Fig.4.).

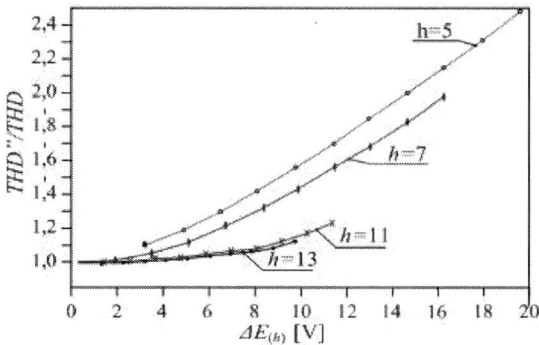

Fig. 8. Relative increase of $THD(i_S)$ versus the amplitude of the single power voltage harmonic

The change of the voltage v_K is also a direct cause of deformation of the current i_S in case of non-symmetric supply voltage. This non-symmetry changes the diode conduction intervals in all elementary rectifiers of the examined system (Fig.1). As a result, after summing up of the component voltages by the reactors CDT&CTR, the shapes of the voltages v_K differ from the sinusoid. The dominating harmonic in their spectrum is the third harmonic of the current i_S.

In order to assess more precisely the effect of supply non-symmetry on the deformation of the current i_S, simulation investigations were carried out in which the non-symmetrical voltage was obtained in the way shown in Fig.9. The simulation results included, among other relations, changes of the $THD''(i_S)$ coefficient (in relation to its value $THD(i_S)$ for the sinusoidal and symmetrical voltage) as a function of the amplitude ΔE and phase γ of the booster voltages, shown in Fig.10. Noticeable is the much larger effect of the amplitude non-symmetry than that of the phase non-symmetry.

Based on the above presented results we can conclude that the multi-pulse rectifiers with coupled reactors, including the examined 18-pulse system, are good solutions of "clean power converter" type only for the symmetrical and sinusoidal supply voltage [11]. When the supply voltage does not meet these two conditions, even if the disturbances are within normative limits, it results in remarkable deformations of the network current. In those cases it is advisable to use an additional small series active filter [4], [8].

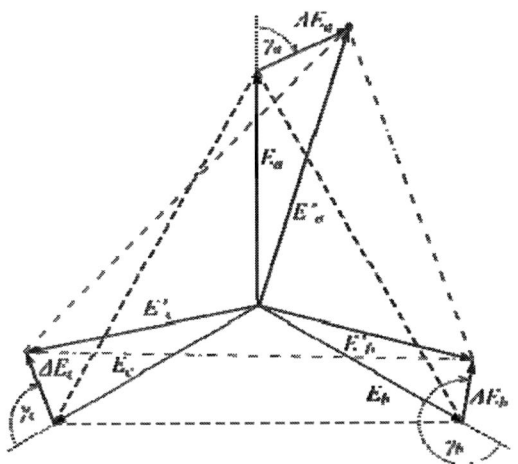

Fig. 9. Phase variation of the power voltage during the tests of the power voltage unbalance effect

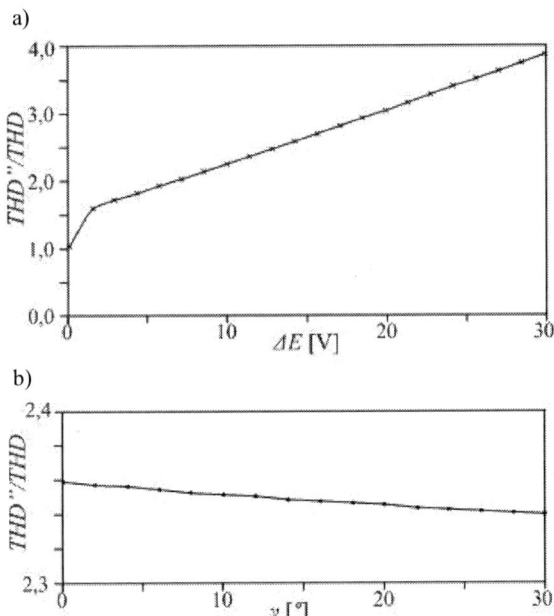

Fig. 10. Relative increase of $THD(i_S)$ in cases of power voltage unbalance: a) changes of the additional voltage amplitude $\Delta E_a = -\Delta E_c$ for $\gamma=0$; b) changes of the additional voltage phase γ for $\Delta E=30$ V.

III. INTEGRATED SYSTEM

Figure 11 shows the AC/DC supply system constructed using an 18-pulse rectifier with CDT&CTR and integrated with a series active power filter (S-APF) of relatively small power [4], [8], [11], [22]. The applied S-APF system comprises three single-phase bridge inverters (with booster transformers Tr and filtering condensers C_F) which were connected to the rectifier output using DC circuits, as was shown in Fig.12. This system provided opportunities for implementation of the 3-level PWM algorithms, and did not require controlling the active power flow between S-APF and the rectifier, due to the fact that this flow is relatively small and naturally tends to self-stabilising reduction.

A. Control Method

The S-APF system in the supply system shown in Fig.11 is controlled by blocking the undesired harmonic components. For those harmonics the system should reveal very large impedance (equal to infinity in theory), while for the required component – the basis harmonic – very small impedance (zero in theory). Therefore the booster voltage v_F in each phase (Fig.11) should be proportional to the undesired components $\tilde{i}_S$ of the electric current i_S in the same phase. This principle of control is given by the formula:

$$\boldsymbol{v}_F = K \cdot \tilde{\boldsymbol{i}}_S \qquad (7)$$

where K is the proportionality coefficient in $[\Omega]$. In the 3-phase system we take into account that:

$$\boldsymbol{v}_F = \begin{bmatrix} v_{Fa(ref)}(j\omega) \\ v_{Fb(ref)}(j\omega) \\ v_{Fc(ref)}(j\omega) \end{bmatrix}, \tilde{\boldsymbol{i}}_S = \boldsymbol{G}(j\omega) \cdot \begin{bmatrix} i_{Sa}(j\omega) \\ i_{Sb}(j\omega) \\ i_{Sc}(j\omega) \end{bmatrix}$$

and

$$\boldsymbol{G}(j\omega) = \begin{bmatrix} 1 & 0 \\ \dfrac{1}{2} & \dfrac{\sqrt{3}}{2} \\ \dfrac{1}{2} & -\dfrac{\sqrt{3}}{2} \end{bmatrix} \cdot \boldsymbol{H}(j\omega) \cdot \begin{bmatrix} 1 & -\dfrac{1}{2} & -\dfrac{1}{2} \\ 0 & \dfrac{\sqrt{3}}{2} & -\dfrac{\sqrt{3}}{2} \end{bmatrix}$$

where $\boldsymbol{H}(j\omega)$ is the transmittance of the high-pass filter (HPF) which separates the blocked network current harmonics in the α-β coordinate system. This transmittance determines basic blocking abilities of the S-APF system. In 3-phase systems, reasonable options are the $\boldsymbol{H}(j\omega)$ realisations making use of the Clark-Park transformation.

In the simplest case shown in Fig.13, HPF is fully executed in the rotating coordinates d-q based of the difference of the constant components extracted by the 1-st order low-pass filters (LPF) and the non-filtered signals. For this solution, $\boldsymbol{H}(j\omega)$ in the α-β coordinate system acc. to [23] has the form:

$$H(j\omega) = \dfrac{\begin{bmatrix} \omega_C + j\omega & -\omega_S \\ \omega_S & \omega_C + j\omega \end{bmatrix} \cdot \begin{bmatrix} j\omega & \omega_S \\ -\omega_S & j\omega \end{bmatrix}}{\omega_S^2 + (\omega_C + j\omega)^2} \qquad (8)$$

where: ω_S – network frequency, ω_C – LPF cut-off frequency. Unfortunately, a disadvantage of this solution is the appearance of remarkable through couplings between the filtered harmonic current components $i_{S\alpha}$, $i_{S\beta}$.

A most favourable transmittance $\boldsymbol{H}(j\omega)$ can be obtained by extracting only the constant components using LPF in the d-q coordinate system. Based on those components and making use of the inverse Park transformation wedetermine the sinusoidal waveforms

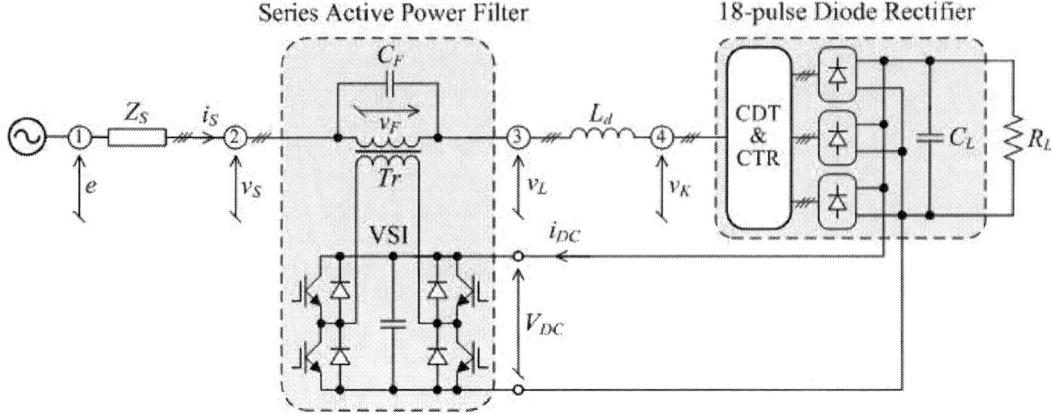

Fig. 11. Supply system integrated with the series active power filter.

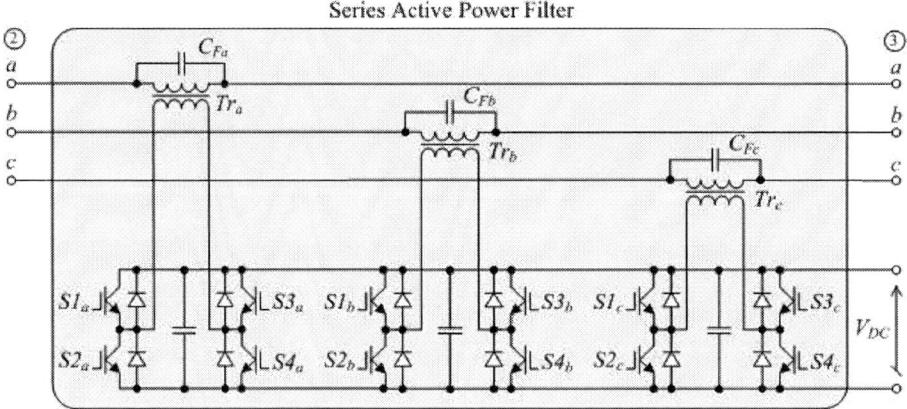

Fig. 12. Application circuits of the series active power filter.

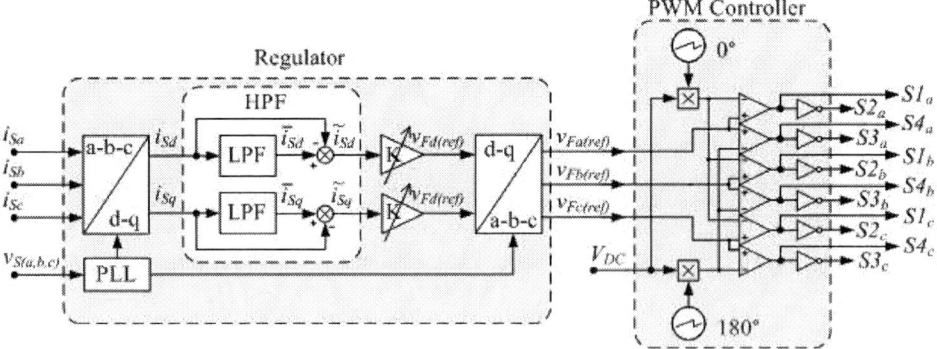

Fig. 13. Basic control system of the series active power filter

in the α-β coordinates. The difference between the obtained sinusoidal waveforms and those unfiltered, in the α-β coordinates, is the result of the filtration: extraction of higher harmonics. In this case the transmittance $H(j\omega)$ is given acc. to [23] by the formula:

$$H(j\omega) = 1 - \frac{\begin{bmatrix} \omega_C + j\omega & -\omega_S \\ \omega_S & \omega_C + j\omega \end{bmatrix} \cdot \omega_C}{\omega_S^2 + (\omega_C + j\omega)^2} \qquad (9)$$

A common disadvantage of the transmittances $H(j\omega)$ given by formulas (8) and (9), and connected with possible non-symmetry of the supply voltage of the system shown in Fig.11, is partial non-filtering of the negative sequence component of the basic harmonic. This results in partial blocking of this component, which excessively increases the overall power of the S-APF system. A reasonable solution here is the realisation of the transmittance $H(j\omega)$ in the way similar to that given by formula (9), but done separately for the positive and

978-1-4244-8806-3/11 $26.00 © 2011 IEEE   434

negative sequence components. We can write this division as:

$$H(j\omega) = 1 - D_p(j\omega) - D_n(j\omega) \qquad (10)$$

where acc. to [23]:

$$D_p(j\omega) = \frac{\begin{bmatrix} \omega_C + j\omega & -\omega_S \\ \omega_S & \omega_C + j\omega \end{bmatrix} \cdot \omega_C}{\omega_S^2 + (\omega_C + j\omega)^2} \qquad (11)$$

$$D_n(j\omega) = \frac{\begin{bmatrix} \omega_C + j\omega & \omega_S \\ -\omega_S & \omega_C + j\omega \end{bmatrix} \cdot \omega_C}{\omega_S^2 + (\omega_C + j\omega)^2} \qquad (12)$$

Hence, after placing (11) and (12) into (10) acc. to [23] we get:

$$H(j\omega) = \frac{\begin{bmatrix} 1 & 0 \\ 0 & 1 \end{bmatrix} \cdot \left[(j\omega)^2 + \omega_S^2 - \omega_C^2 \right]}{\omega_S^2 + (\omega_C + j\omega)^2} \qquad (13)$$

It results from formula (13) that the components $i_{S\alpha}$ and $i_{S\beta}$ are filtered in the same way. Consequently, when we use HPF with transmittance (13), the booster voltages v_F in each phase are given by the relation:

$$v_{F(a,b,c)} = K \cdot \frac{\left[(j\omega)^2 + \omega_S^2 - \omega_C^2 \right]}{\omega_S^2 + (\omega_C + j\omega)^2} \cdot i_{S(a,b,c)}$$

B. Simulation Results

Figures 14 and 15 show sample simulation results for the supply system shown in Fig.11. The simulations were performed using the package PSIM Version 8.0.4. The presented results illustrate the action of the system with the S-APF system switched off and on, for the supply voltage deformation (Fig.14) and supply voltage non-symmetry (Fig.15). In both cases the basic S-APF control system was used (Fig.13). Using the simplest transmittance HPF (8) was dictated by the main goal of the simulation, which was assessing the functionality and features of the proposed solution, without any attempt to optimise dynamic and size parameters – a goal planned for the nearest future.

Table I collects the basic parameters of the examined system (Fig.11), which were recorded in the simulation. These parameters describe the action of the system in the conditions shown in Figs.14 and 15, and refer to the situation before and after switching S-APF on. Along with the average $THD_{AV}(i_S)$ values, the presented data include the $THD(i_S)$ values calculated for each phase, and the overall powers of the receiver S_{REC} and the series active power filter S_{APF}. It is noteworthy that in each case the overall power S_{APF} is approximately equal to only 10% of the overall power S_{REC}.

a)
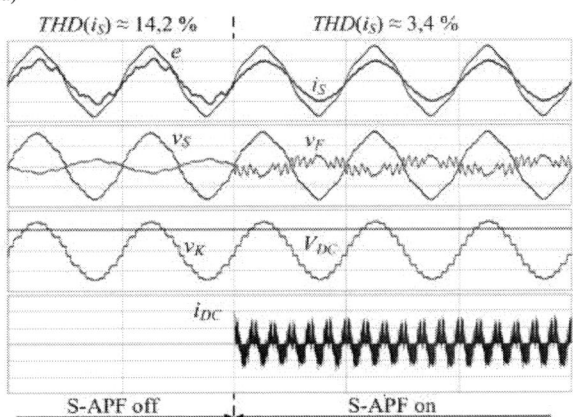
b)
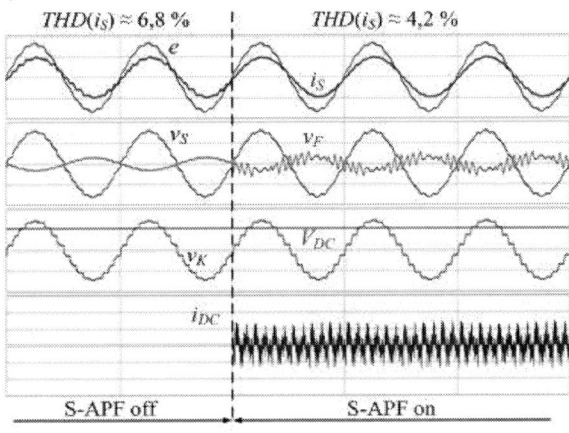

Fig. 14. Voltages and currents in the supply system (Fig.11) with S-APF switched off and on, for supply voltage deformation: a) 5-th harmonic of 6% in magnitude, b) 13-th harmonic of 3% in magnitude ($e, v_S, v_K \rightarrow 200$ V/div; $v_F \rightarrow 100$ V/div; $V_{DC} \rightarrow 400$ V/div; $i_S \rightarrow 50$ A/div; $i_{DC} \rightarrow 5$A/div).

TABLE I
BASIC PARAMETERS OF THE INTEGRATED SUPPLY SYSTEM (FIG.11), BEFORE AND AFTER SWITCHING S-APF ON

Parameter	Conditions of system operation, acc. to Figures			
	Fig. 14a	Fig. 14b	Fig. 15a	Fig. 15b
S_{REC} [kVA]	18,3	18,15	18,1	20,5
S_{APF} [kVA]	0/1,872	0/1,653	0/1,7	0/1,89
S_{APF}/S_{REC}[%]	0/10,2	0/9,1	0/9,3	0/9,2
$THD(i_{Sa})$ [%]	14,1/3,4	6,4/4,6	12,7/3,6	9,1/2,8
$THD(i_{Sb})$ [%]	14,1/3,4	6,4/4,6	16,9/3,8	10,5/2,8
$THD(i_{Sc})$ [%]	14,1/3,4	6,4/4,6	15,6/3,8	8,6/2,7
$THD_{AV}(i_S)$ [%]	14,1/3,4	6,4/4,6	15,1/3,8	9,1/2,8

IV. EXPERIMENTAL RESULTS

The basic goal of the experimental investigations was practical verification of the efficiency of operation of the proposed system (Fig.11) in steady-state conditions. The investigations were performed for the case of S-APF

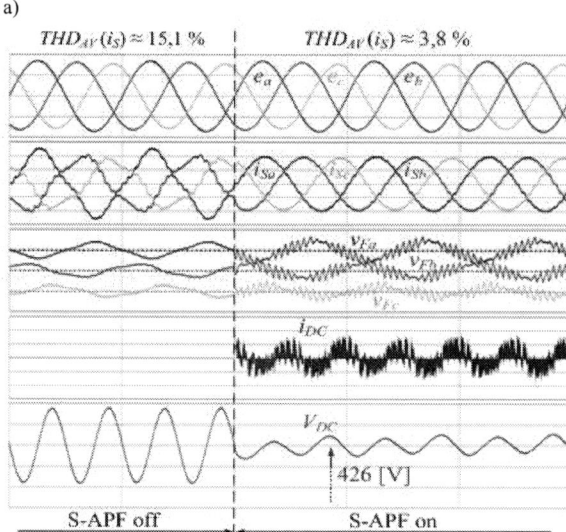

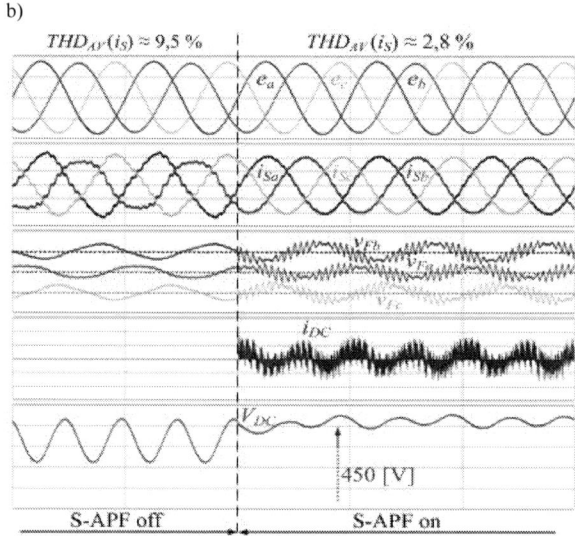

Fig.16.   View of the laboratory model of the supply system shown in Fig.11.

Fig. 15. Voltages and currents in the supply system (Fig.11) with S-APF switched off and on, for supply voltage non-symmetry: a) ΔE_a= ΔE_b=30 [V], ΔE_c=0 [V], γ_a=γ_c=0°, γ_b=180°; b) ΔE_a= ΔE_b=30 [V], ΔE_c=0 [V], γ_a=γ_c=γ_b=0° ($e \rightarrow$ 200 V/div; $i_S \rightarrow$ 20 A/div; $v_F \rightarrow$ 100 V/div; $i_{DC} \rightarrow$ 5 A/div; $V_{DC} \rightarrow$ 2 V/div).

integration with the 18-pulse rectifier of power S_{REC}=20 [kVA], and an additional network reactor L_d = 0,6 [mH].

Figure16 shows a laboratory model of the supply system constructionally integrated with S-APF localised at the bottom of the cubicle. In the system the booster transformers Tr (Fig.12) with the transformation ratio 1:12 were used. Therefore, taking into account the voltage $V_{DC} \approx$ 500 [V], the amplitude of the booster voltages u_F did not exceed 45V. The phase current waveforms i_{Sa}, i_{Sb}, i_{Sc} recorded in nominal conditions of this model operation and their spectra before and after switching S-APF on, are shown in Fig.17. Figures 18 and 19 show the same data, but for 75% and 125% of the nominal load. Their analysis clearly confirms high efficiency of the proposed solution for improving the shape of the network current. It is also noticeable that almost half of the volume of the assembly cubicle

(Fig.16) is occupied by auxiliary connecting elements. This indicates the potential for more compact constructions of similar systems, revealing good mass and size parameters.

As the supplement to the results shown in Figs.17-19, Table II collects the measured phase shifts angles φ_{Sa}, φ_{Sb}, φ_{Sc} of the currents i_{Sa}, i_{Sb}, i_{Sc} with respect to the phase voltages v_{Sa}, v_{Sb}, v_{Sc}, and the rms values I_{Sa}, I_{Sb}, I_{Sc} of these currents, for three different loads: 100%, 125% and 75 % of the nominal load, and for S-APF switched off and on. The presented data also include average values $\cos\varphi_{AV}$ =($\cos\varphi_a$ +$\cos\varphi_b$ +$\cos\varphi_c$)/3. These results indicate an additional effect of the action of the S-APF system, which is levelling of the rms values and phase shift angles of the network currents. Moreover, the recorded $\cos\varphi_{AV}$ values confirm the energy related advantages of the examined solution.

TABLE II

PHASE SHIFT ANGLES AND RMS VALUES OF THE SUPPLY CURRENTS I_S, AND THE $\cos\varphi_{AV}$ IN THE EXPERIMENTAL SETUP, FOR 75%, 100 % AND 125% OF THE NOMINAL LOAD

S-APF	φ_a [°] I_{Sa} [A]	φ_b [°] I_{Sa} [A]	φ_b [°] I_{Sb} [A]	$\cos\varphi_{AV}$
Load 75%				
On	20/23,7	18/23,5	19/23,3	0,946
Off	18/24,6	11/23,9	16/21,9	0,966
Load 100%				
On	16/29,3	16/29,5	17/29,7	0,959
Off	11/28,2	8/31	13/31,1	0,983
Load 125%				
On	15/33,8	14/33,7	15/33,5	0,967
Off	11/37,8	5/37,3	8/34,7	0,991

By confirming the results of the simulation calculations, the presented results of the laboratory measurements make it possible to conclude that also in real conditions the series active power filter can remarkably reduce higher harmonics of the network

978-1-4244-8806-3/11 $26.00 © 2011 IEEE   436

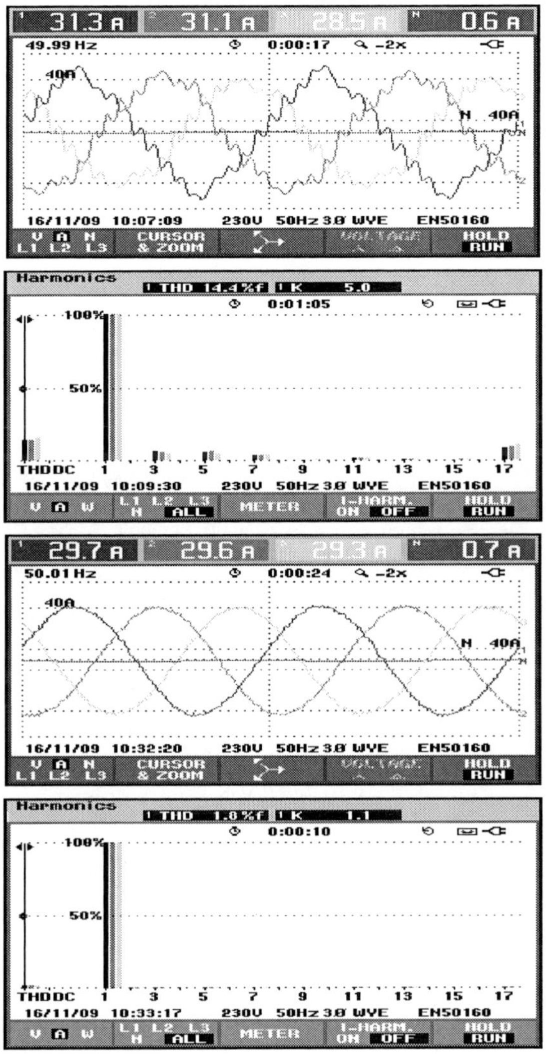

Fig. 17. Oscillograms and spectra of network currents $i_{Sa,b,c}$ in the system of power S_{REC}=20 kVA (nominal load) with reactor L_d =0,6 [mH], before and after switching S-APF on.

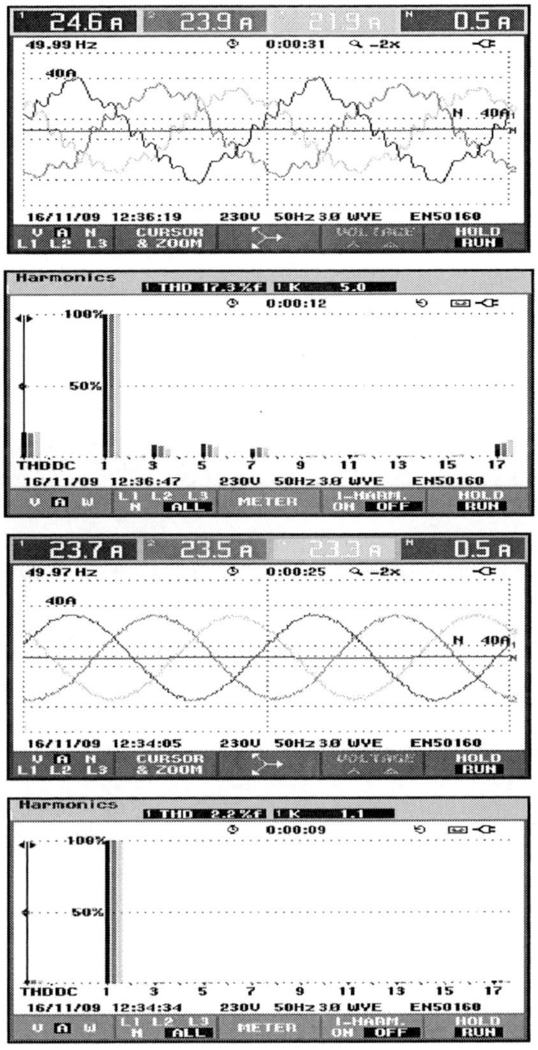

Fig. 18. Oscillograms and spectra of network currents $i_{Sa,b,c}$ in the system of power S_{REC}=15 kVA (75% of nominal load) with reactor L_d =0,6 [mH], before and after switching S-APF on.

current of the 18-pulse converter with CDT&CTR. S-APF is efficient in cases of nominal load and in under- or overload states. Moreover, the integration of the rectifier with S-APF in the examined supply system makes it possible to simplify the structure of the converter, by resignation from the network reactor L_d, and simultaneous remarkable reduction of the negative effect on the supply network.

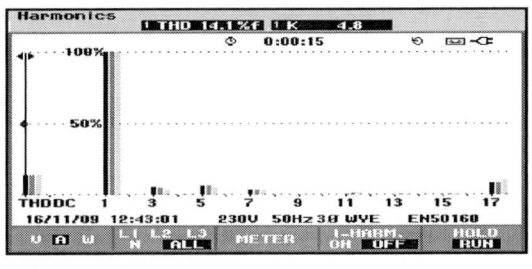

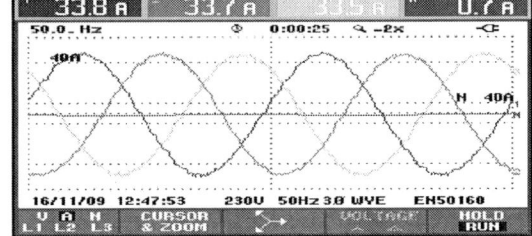

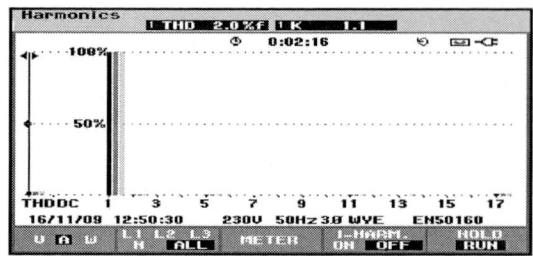

Fig. 19. Oscillograms and spectra of network currents $i_{Sa,b,c}$ in the system of power S_{REC}=25 kVA (125% of nominal load) with reactor L_d =0,6 [mH], before and after switching S-APF on.

V. CONCLUSION

The application of the 18-pulse converter with the system of coupled reactors cooperating with a low-power series active power filter, seems to be an interesting solution for the problem of clean AC-DC energy conversion, due to:

- simplification of the converter system resulting from possible elimination of the network reactor,
- small overall power of the both systems [22], decreasing their costs,
- remarkable reduction of the content of higher harmonics in the supply current waveform,
- small susceptibility of the system to supply asymmetry and load changes,
- potential for construction of low-cost and high-reliability supply systems.

The advantages resulting from the application of the presented filter system refer to efficient minimisation of the negative action of the multi-pulse converter onto the supply network, in various supply conditions. We can assume that the 18-pulse rectifier in the configuration with a series active power filter compose a converter system working in conditions close to CPC.

REFERENCES

[1] Fujita H., Akagi H.: An approach to harmonic current-free ac/dc power conversion for large industrial loads: The integration of a series active filter with a double-series diode rectifier," *IEEE Trans. Ind. Appl.*, vol. 33, no. 5, pp. 1233–1240, 1997.

[2] Paice D. A.: Clean Power Electronic Converters. Paice & Associates Inc., 2004.

[3] Paice D. A.: Power Electronic Converter Harmonics: Multipulse Methods for Clean Power. IEEE, New York, 1996.

[4] Strzelecki R., Supronowicz H.: Filtration of the harmonic in AC supply systems (in Polish), *Adam Marszałek Publishing House, Poland*, 215 p., 1997/1999 (ed.1/ed.2).

[5] Wei L., Guskov N. N., Lukaszewski R. A., Skibinski G. L.: Mitigation of Current Harmonics for Multipulse Diode Front-End Rectifier Systems. IEEE Transactions On Industry Applications, vol. 43, no. 3, May/June, 2007, pp.787-797.

[6] Pan Z., Peng F.Z., Wang S.: Power Factor Correction Using a Series Active Filter. IEEE Transactions On Power Electronics, v. 20, n. 1 (2005), 148 - 153.

[7] Peterson M., Singh B.N., Rastgoufard P.: Active and Passive Filtering for Harmonic Compensation. 40th Southeastern Symposium on System Theory, New Orleans, USA, March 16-18, 2008, pp. 188 - 192.

[8] Strzelecki R., Supronowicz H.: Power factor in AC supply systems and improvements methods (in Polish), *Publishing House of the Warsaw University of Technology*, Poland, 2000.

[9] Strzelecki R., Supronowicz H.: Improving the filtering characteristics of the passive filters by use of active filter. Arch. of El. Eng., Vol. XLIV, No l, 1995, pp. 109-122.

[10] Oguchi K., Hoshi N., Kubota T., Namatame T.: Harmonic Draining Transformer-Coupled Boost-Type Rectifier Systems with Sinusoidal Input Currents. Proc. European Conference Power Electronics EPE2007, Aalborg 2007.

[11] Mysiak P.: Multi-pulse diode rectifier with current harmonic blocking reactors, *Publishing House of the Gdynia Maritime University*, Poland, 2010.

[12] Mysiak P.: Multipulse Diode Converters - Frequency Domain Analysis of Operation of the Applied Coupled Three-Phase Reactor. 6th International Conference-Workshop, Compatibility And Power Electronics (CPE 2009), Badajoz, Spain, May 20-22, 2009.

[13] Chivite-Zabalza F.J., Forsyth A.J.: A Passive 36-Pulse AC-DC Converter With Inherent Load Balancing Using Combined Harmonic Voltage And Current Injection. IEEE Transactions On Power Electronics, vol. 22, no. 3, May 2007, pp. 1027-1035.

[14] Choi S., Enjeti Prasad N., Pitel Ira J.: Autotransformer configurations to enhance utility power quality of high power AC/DC rectifier systems. IEEE PESC conf., 1995.

[15] Singh B., Bhuvaneswari G., Garg V.: An Improved Power-Quality 30-Pulse AC-DC for Varying Loads. IEEE Transactions On Power Delivery, vol. 22, no. 2, April 2007, pp. 1179 - 1187.

[16] Singh B., Bhuvaneswari G., Garg V.: Scott-Connected Autotransformer Based Multipulse AC-DC Converters for Power Quality Improvement In Vector Controlled Induction Motor Drives. Proc. Int. IEEE Conf. PEDS2005, 2005, pp.1491 - 1496.

[17] Singh B., Garg V., Bhuvaneswari G.: A Novel T-Connected Autotransformer-Based 18-Pulse AC-DC Converter for Harmonic Mitigation in Adjustable-Speed Induction-Motor Drives. IEEE Transactions On Industrial Electronics, vol. 54, no. 5, Oct. 2007, pp. 2500-2511.

[18] Mysiak P.: A 24-pulse diode rectifier with coupled three-phase reactor. Journal of the Chinese Institute of Engineers, Vol. 30, No. 7, pp. 1189-1204 (2007), ISSN 0253-3839, National Taiwan University of Science and Technology.

[19] Mysiak P.: Eighteen-pulse diode rectifier with three-phase coupled reactors. Archives of Electrical Engineering, Vol. L, No 1, Warsaw, 2001.

[20] Singh B., Gairola S., Singh B.N., Chandra A., Al-Haddad K.: Multipulse AC-DC Converters for Improving Power Quality: A Review. IEEE Transactions On Power Electronics, vol. 23, no. 1, January 2008, pp. 260- 261.

[21] Oguchi K., Maeda G., Hoshi N., Kubota T.: Coupling Rectifier Systems with Harmonic Cancelling Reactors. IEEE Ind. Appl. Magazine, July/August 2001, pp. 53-63.

[22] Akagi H., Watanabe E.H., Aredes M.: Instantaneous power theory and applications to power conditioning, *John Wiley&Sons*, 2007.

[23] Peng F.Z., Akagi H., Nabae A: Compensation characteristics of the combined system of shunt passive and series active filters" *IEEE Trans. on Ind. Appl.*, vol.29, no.1, pp.144-151, 1993.

978-1-4244-8806-3/11 $26.00 © 2011 IEEE

High-Voltage LED-Based Efficient Lighting using a Hysteretic Controlled Boost Converter

A. Leon-Masich, H. Valderrama-Blavi, J.M. Bosque, A. Cid-Pastor, and L.Martínez-Salamero.

GAEI Laboratory, Rovira i Virgili University. Av. Països Catalans 26, 43007 Tarragona, CAT-Spain

antonio.leon@urv.cat, hugo.valderrama@urv.cat

Abstract—Car industries are especially interested on decreasing the power rating of the electrical sub-system to reduce wire weight and fuel consumption. Using LED devices, a typical 50W halogen-based car spot-light, could be replaced by a LED array of 8-12W. To improve the array life-time and simplify light dimming, all LED devices are series-connected to a high-voltage current-source driver. This driver, based on a single-stage boost converter has a very high voltage gain. Supplied from a 12 V car-battery, and limited only by the commercially-available switching devices, our prototype reaches 900V. Hysteretic control is decisive to achieve the required extremely high duty ratios, circa 98%, as well as to guarantee DCM-CCM border converter operation to reduce switching losses.

I. INTRODUCTION

CURRENTLY, lighting applications based on LEDs are becoming more popular. Their superior longevity, low maintenance requirements, and high-efficiency (lumen/Watt) are the main cause of their commercial success. Among these new applications we find: LCD backlighting, automobiles, traffic lights, general-purpose lighting.

Efficient-lighting is also possible using gas-discharge lamps. These lamps have different light spectrums and color rendering depending mainly on the gas mixture, and internal pressure. Table I compares diverse lamp types. As can be seen, various lam parameters have an important variability or value dispersion. Efficiency data dispersion comes directly from technical reviews and official data sheets from manufacturers. Operation voltage dispersion instead, deals directly with lamp power rating. For a given discharge lamp type, from 75 W to 400 W range, as higher be the power rating, higher the voltage.

TABLE I
EFFICIENCY AND VOLTAGES OF DIVERSE LAMPS

Lamp Type	Efficiency (Lumen/W)	Strike Voltage (V)	Operation Voltage (V)
Incandescent Bulb	10-16		Any
Car Halogen Bulb	20-24		12-14
Mercury Vapor	60-80	180	90-130
Compact Fluorescent	40-90	180	90-110
Metal Halide (Pulse Start)	70-110	3-6 kV	90-130
Metal Halide (Probe Start)	70-110	0.5-0.7 kV	90-130
High Press. Sodium Vapor	130	5 kV	45-180
Car Xenon Lamp	50	7-10 kV	80
White LED (Phosphor)	150		3.2-3.6

Discharge lamps main drawback is that complex electronic ballasts are required to create and maintain the discharge arc. High pressure lamps require high voltage sparks to ignite the lamp. Thus, pulse start lamps require 3-6 kV during 1-2 µs, and probe start lamps, with a third starting-auxiliary electrode, require 0.5-0.75 kV during 250-300 µs. Nevertheless, fluorescent, and other low pressure lamps, as mercury vapor, require sparks below grid voltage, simplifying greatly the ballasts. As no converter to ignite the arc is required, ballast only takes care of controlling lamp current, once started.

Light emitting diodes (LED's) have a higher efficiency and life-time that many discharge lamps, work always at ambient temperature, and have less complicated drivers.

According to that, LED systems should play an important role in automobile and general purpose lighting. For instance, according to Table I, a 50 W car halogen bulb should produce the same light than a LED-based 8 W lamp. The main problem is that high-power devices are still less efficient than 20 mA-5mmø conventional devices. Thus, to make a powerful light-spot, the combination of small devices in a LED array is still required, and the most common technique.

Different matrix architectures can be considered to create a LED array, from the most extreme form-factor cases, where all the LEDS would be connected in series or in parallel, to available commercial lamps and literature works proposing intermediate designs where M strings, each one with N LEDs in series, are connected in parallel.

The main advantage of a pure-series array is that all devices operate naturally at the same current. No special circuit, strategy or additional control is required to balance currents among parallel branches. The main drawback is also evident. A light spot with high number of LEDs should be supplied at high voltage. This option, analyzed in this work, is attractive for its simplicity, avoids balancing problems of parallelized structures, and implies an interesting challenge in high-voltage devices, applications, and high gain step-up conversion, especially for automotive and portable applications.

In this paper we present a single-stage boost converter capable to increase the voltage up to 900 V, from a 12V car-battery, as shown in the results given at Section IV.

This output voltage is sufficient to supply 280 white-LEDs in series at their nominal current, 20 mA. The only limiting factor is the switching devices breakdown. Promising silicon-carbide devices will help to increase the output voltage.

The key point of the whole system, protected under patent, is the hysteretic controller which has two important tasks. By one side, hysteretic control makes feasible converter duty-rations above 98%, unaffordable with PWM technologies. By other side, assures converter operation at DCM-CCM border, eliminating the diode turn-off losses.

The remaining part of this paper is organized as follows. At Section II a quick review about LED fundamentals and driver circuits is presented. Next, at Section III the converter operation and control are discussed, and finally at Section IV some experimental and simulation results are given.

II. WHITE LED'S AND DRIVERS

A. Review on LED Basics.

A Light emitting diode (LED) is a PN junction especially designed to emit light when is polarized.

When the PN junction is polarized directly, the potential barrier is reduced, and the electron excess form the N-side conduction band, begin to cross the junction starting the recombination process with the holes excess at the P-side valence band. Two types of recombination processes take place simultaneously, competing among themselves. Some drops release energy as light (photon transition), other produce heat (phonon transition). By setting appropriately the carrier life-times of both transition types, radiative τ_r and phonon τ_p, the photonic quantum efficiency η_Q of a certain material can be optimized (1).

$$\eta_Q = \frac{1/\tau_r}{(1/\tau_r) + (1/\tau_p)} = \frac{1}{1 + (\tau_r/\tau_p)} \quad (1)$$

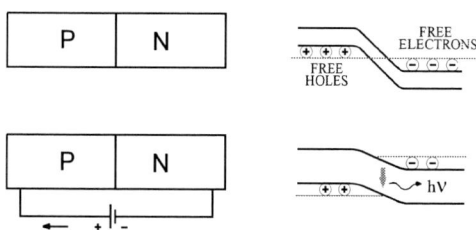

Fig. 1. LED Operation Principle

The energy released during the recombination transient is determined by the energy band-gap, between the initial and final states. Therefore the minimum emitted wavelength (2) or color, is set by the gap of the material (E_g=Ec–Ev), while other radiations with higher wavelength, or less energy, can be achieved by incorporating impurities on the material.

$$E_g = \frac{hc}{\lambda_o} \quad \rightarrow \quad P(\lambda_o) = I_d \eta_Q E_g \quad (2)$$

LEDs of different color have different voltage drops, but surprisingly, the optical power depends on both, light color and device current (2). Consequently, brightness-control or light-dimming can be achieved regulating the device current.

Similarly, as optical power is color-dependent, to compare the features of diverse devices, a good solution is use always the same test current I_d. For instance, with 5 mm diameter LEDs, this current is usually 20 mA.

B. White lighting using LED technology

LED devices light-color is related to the semiconductor band-gap energy. As a result, pale hues, mixed tones, and wide-spectrum lights, as white, cannot be realized by simply tuning the band-gap. Two different methods to implement white LEDs appear in the literature.

First method, color mixing, consists on placing in a single bulb various devices of different colors to achieve a white hue. Diverse color combinations [1] with diverse rendering and efficiency performances have been tested: blue and yellow to create a dichromatic source; red, green, and blue (RGB) to obtain a tri-chromatic light; and blue, green, yellow, and red LEDs to realize a tetra-chromatic white source.

Neglecting the devices different aging-rates, causing a color temperature shift, the real problem of this method is to regulate adequately the current of each device to control the global hue. Besides, at short distances, instead of a white light beam, we see the different colors of each device.

The second method is based on coating a single color LED with a special phosphor layer, implementing a wavelength conversion. For instance, blue LEDs coated with yellowish phosphor deliver a 6,500 K white light. By adding a second red phosphor layer, a warmer 3,200 K white can be achieved. Similarly, a near-UV LED coated with an RGB phosphor gives a 4,000 K white light. Blue LEDs are preferred because frequency shift is lower and, the final device is more efficient.

TABLE II
WHITE LED'S COLOR RENDERING

λ_A (nm)	λ_B (nm)	λ_C (nm)	λ_D (nm)	Render (%)
480	580			16.2
480	540	615		89.4
460	525	590	640	95
480	YAG: Ce^{3+} Phosphor Coating			81
405	RGB White Phosphor Coating			95

Fig. 2 depicts different light spectra. Black trace shows the standard illuminant D55, equivalent to 5,500 K noon-light, usually introduced as reference to compare other light sources.

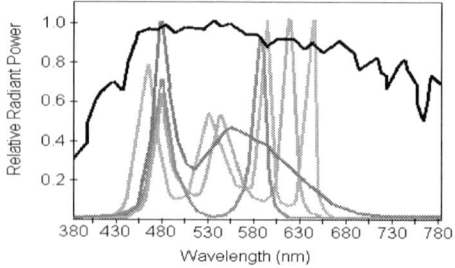

Fig 2. White-LED's and D55 spectra

Pale color traces (violet, blue, and green) show the spectra given by color mixing techniques, and finally red trace shows the spectra given by a blue LED coated by a yellow phosphor.

C. Review on LED Driver Circuits.

As explained before most common technique of achieving a considerable light-output in a LED-based spot is placing a high number of small power LED's in a series-parallel array, where for practical purposes the number of parallelized branches is minimized, favoring the device series connection.

The simplest driver can be found in the LED spots adapted to E14 and E27 standard fittings. These lamps include up to 3 parallelized branches of 20-26 series devices. Although there is nothing to balance the branches currents, its simplicity, low size, and efficiency make this driver remarkable. Realize that the capacitor is the key part, capable of blocking all the grid voltage to supply a single 3V LED, with negligible power dissipation as capacitor voltage lags the current 90°.

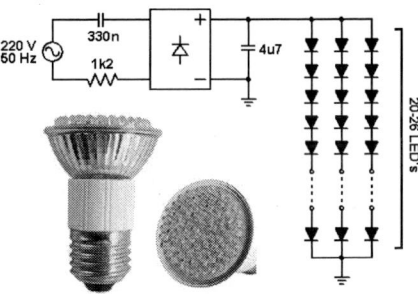

Fig 3. Commercial LED-based spot and driver circuit

To adapt the previous driver to an equivalent DC voltage supply, the 330n capacitor should be replaced by a big-sized power resistance, or a DC-to-DC buck-based converter.

$$P_{LED's} = \underbrace{[3.2x26]}_{V_L} x \underbrace{[0.02x3]}_{I_L} = 5.0\,W$$
$$P_R = \underbrace{[220 - 3.2x2.6]}_{V_R} x \underbrace{[0.02x3]}_{I_R} = 8.2\,W \quad (3)$$

To improve current balancing, many LED drivers include a linear current regulator per branch [1]. This solution has a poor operating efficiency because the voltage drop across the linear regulators cannot be minimized. In fact, the current regulator input voltage V_o must be adjusted for the worst case. Feeding all the branches with the same current, the worst case is given by the highest voltage drop branch, operating at the lowest temperature, when LED's forward voltage is higher.

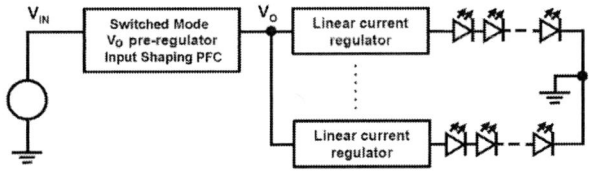

Fig 4. Simplest strategy for current balancing

To increase the driver efficiency, a very common solution in the literature is to make the current regulators input-voltage V_o adaptative [2-4]. Another option [5] to increase efficiency uses switched-mode current sources to balance LED branches, but the component cost can be excessive.

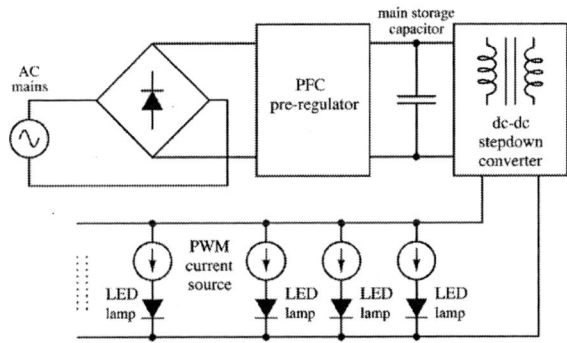

Fig. 5. Driver using PWM current sources

Different converters are proposed as voltage pre-regulators. Among them, boost [3-4-7], buck [6], or galvanic isolation converters like forward or flyback [6]. Concerning converter control, PWM technique is the most used, although this techniques is not practical useful to achieve extreme duty ratios, around 98%, as required in our boost driver.

D. High Voltage- Gain Converters

As can be seen in previous section, balancing parallelized currents without losing efficiency is a complex issue.

Conversely, if we decide to organize the LED array in a pure series configuration, with a single current branch, the lamp driver may require a high output voltage. Besides, if the LED-lamp must be supplied from a low voltage source, like in a car-battery light-spot, then a high voltage-gain converter will be required.

Once again, magnetic coupling [9-10], seems to be an additional advantage if a high voltage-gain ratio is required, as in push-pull, forward, flyback, Cuk, or tapped boost switch converters [11-12]. Not only the voltage gain given by turns-ratio N, thus reducing the switch duty cycle, is important. Besides, the switching devices at the primary side only must block a little fraction (1/N) of the converter output voltage, compared to conventional boost converter. Nevertheless, the transformer must be designed carefully to reduce leakage inductances. Stray inductors cause high-voltage speaks during the MOSFET switch-off transient, reducing dramatically converter efficiency unless an active clamp or non-disipative snubber be used to recover those losses.

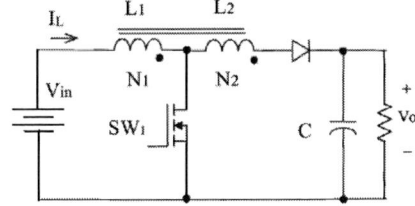

Fig 6. Tapped boost switch converter

978-1-4244-8806-3/11 $26.00 © 2011 IEEE   441

High voltage-gain converters can also be achieved by cascading diverse elementary boost stages, performing quadratic and cubic boost converters, or combining switched capacitor [13-14] stages with boost-derived converters. Anyway, all these last options consider much more devices and complicated circuits than the solution presented here.

III. BOOST CONVERTER OPERATION

In this section we will analyze how to operate the boost converter to obtain the fewest as possible losses. Fig. 7 depicts the converter circuit, as well as its hysteretic controller.

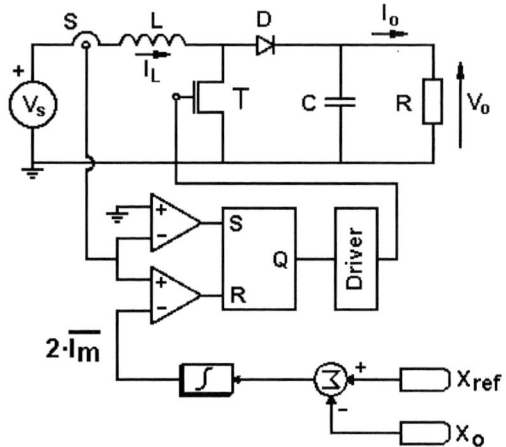

Fig. 7. Boost converter with hysteretic controller

A. DCM-CCM boundary operation

In order to achieve, the highest output voltage as possible using commercial devices, we have made our converter using conventional silicon technology. Although emerging silicon carbide (SiC) technology is very promising, at this point still is difficult to find shottky diodes over 600V, and therefore an hyperfast bipolar diode from IXYS has been used instead. Table III depicts the switching devices used.

TABLE III
SWITCHING DEVICES AT EXPERIMENTAL PROTOTYPE

INFINEON CoolMos **IPW90R120C3**	900 V	36 A	0.12 Ω	270 nC	36 €
	V_{BD}	I_{MAX}	R_{on} - V_f	Q_g - t_{rr}	cost
DSEP 30-12CR IXYS HiperDynFredt	1.2 kV	30 A	1.5 V	20 ns	3.2 €

Shottky diodes are majority-carrier devices, no charge is stored during their ON-state, and very fast turn-OFF without reverse recovery peak is possible. Nevertheless, bipolar diodes are minority-carrier devices, and some charge Q_f according to the ON-state current, and carrier life-time is stored inside. Then a reverse-recovery peak I_{rr} appears during the diode turn-off to empty the stored charge. This process causes switching losses, unless diode turn-OFF be at zero current, imposing inductor DCM operation.

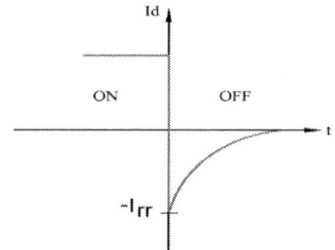

Fig 8. Idealized turn-OFF of a bipolar diode

By other hand, working at deep DCM conditions, to preserve the input average current in order to satisfy the load requirements, the input current peaks would be much higher (fig. 9), increasing all devices RMS currents, and therefore, the conduction losses of mosfet and diode. In fact, as deeper is the DCM mode, higher is the current stress, and therefore the conduction losses. Consequently, the DCM-CCM boundary is the working zone with the lowest conduction losses, but keeping diode turn-OFF at zero current.

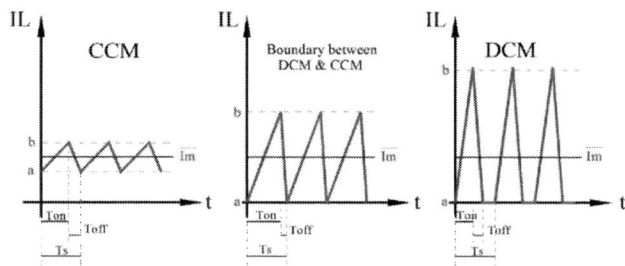

Fig. 9. Inductor current at CCM, border, and DCM

The conduction losses and RMS currents of inductor, mosfet, and diode have been calculated for all conduction modes, and given at Table IV. To find the expressions, the high voltage-gain hypothesis is assumed. At these conditions, the OFF-state, where the diode conducts to charge the output capacitor, has a negligible duration. That is, $T_{OFF} \ll T_{ON}$, and then $T = T_{ON} + T_{OFF}$, or equivalently $D = D_{ON} + D_{OFF} \approx D_{ON}$.

TABLE IV
CONDUCTION LOSSES AND RMS CURRENTS

		Inductor	MOSFET	Diode
RMS Current	DCM	$\dfrac{2}{\sqrt{3D}}\overline{I_m}$	$\dfrac{2\sqrt{D_{ON}}}{\sqrt{3D}}\overline{I_m}$	$\dfrac{2\sqrt{D_{OFF}}}{\sqrt{3D}}\overline{I_m}$
	Boundary	$\dfrac{2}{\sqrt{3}}\overline{I_m}$	$\dfrac{2\sqrt{D_{ON}}}{\sqrt{3}}\overline{I_m}$	$\dfrac{2\sqrt{D_{OFF}}}{\sqrt{3}}\overline{I_m}$
	CCM	$\overline{I_m} < \sqrt{\overline{I_m^2} + \dfrac{(b-a)^2}{12}} < \dfrac{2}{\sqrt{3}}\overline{I_m}$	$I_{L(RMS)} \cdot \sqrt{D_{ON}}$	$I_{L(RMS)} \cdot \sqrt{D_{OFF}}$
Conduction Losses		$\dfrac{4}{3}R_L\overline{I_m^2}$	$\dfrac{4}{3}R_{ON} \cdot D_{ON}\overline{I_m^2}$	$\dfrac{2}{\sqrt{3}}V_F\overline{I_m}\sqrt{D_{OFF}}$

To calculate the averaged input current I_m, the only data required is the switching period T_S, because at DCM-CCM boundary, and also assuming high gain, $D_{ON} \approx 1$.

$$\overline{I_m} \approx \frac{V_S}{2L}D_{ON} \cdot T_S \approx \frac{V_S}{2L} \cdot T_S \qquad (4)$$

978-1-4244-8806-3/11 $26.00 © 2011 IEEE   442

$$\overline{I_m} = \frac{P_o}{\eta V_S} \Rightarrow P_o = \frac{\eta V_S^2 T_S}{2L} \qquad (5)$$

Neglecting the efficiency η effect, the input current I_m, and the output power P_o are both proportional and directly controllable by adjusting the switching period T_S (5). In fact, the converter behaves like a Loss Free Resitor, because the input current I_m is also proportional to the supply voltage V_S.

B. Converter Control.

As can be seen at fig. 5, the flip-flop activates the switch when the inductor current is zero, and turns-OFF the mosfet when the inductor current reaches $2 \cdot I_m$. This is equivalent to force the average input current I_m to the appropriated value, to fulfill the output requirements.

Thus, the hysteretic controller has two tasks. By one side assures that the converter operation at DCM-CCM boundary, by other side it assures that the switch be at ON-state, only the minimum time to deliver the expected output power. As higher is the output power Po, lower is the switching frequency, optimizing the switching losses.

Nevertheless, the most important is that any extreme duty-ratio can be realized, making feasible very high voltage-gains. In classical PWM systems, such extreme duty ratios are not possible due to different causes. Common integrated-circuit PWM controllers limit the duty-ratio range internally. Sometimes, switching noise makes the comparison between the control signal and the triangular carrier very unstable or chaotic, loosing progressively more switching cycles as higher is the duty-ratio attempted. Duty ratios over 95% are practicably unfeasible. And, finally, sometimes the devices rise and fall times make unaffordable such duty ratios if the switching frequency is constant, and cannot be reduced.

Control-loop is closed using an integrator controller. At steady-state conditions, the integrator gives the appropriate reference $2 \cdot I_m$ to assure a zero steady-state error between a reference variable X_{ref}, and its measured value X_o. Reference variable X_{ref} can be an either an output voltage reference, or in the case of a LED-based lamp, a 20 mA current reference.

$$2 \cdot \overline{I_m} = k \int (X_{REF} - X_O) \cdot dt \qquad (6)$$

By applying the sliding motion theory to the system presented here, it can be easily proved that the inductor current will track $2 \cdot I_m$, the steady-state zero error of the output variable, $X_{ref} - X_O = 0$, as well as the system stability. Detailed formulas are not given for paper brevity.

IV. EXPERIMENTAL AND SIMULATION RESULTS

To demonstrate the feasibility of the presented, low losses, single-stage, high output voltage converter, some preliminary simulations were carried out by means of PSIM program.

Fig. 10 depicts these results. Inductor current saw-tooth waveform is clearly appreciated IN Fig. 10c. The steady-state output voltage reaches 650 V (Fig 10b), as expected from an array of 204 standard-size white LEDs. The output current is 20 mA (Fig. 10a), corresponding to the rated current of 5 mmø standard devices. Finally, the nonlinear characteristic of the

diodes can be seen, because the output current only begins growing when LED's voltage have reached around 600 V.

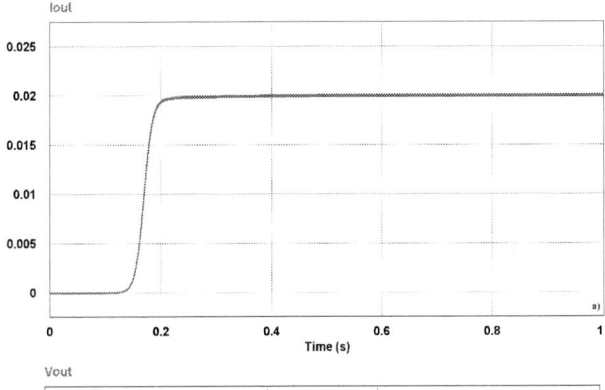

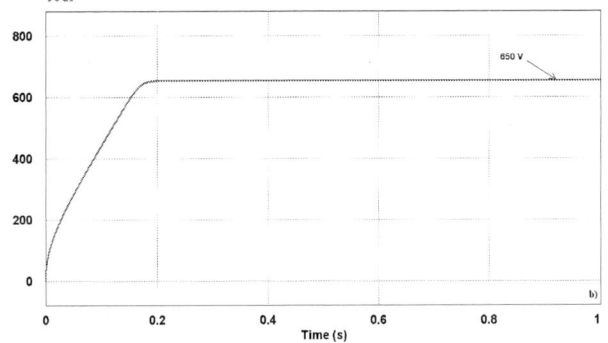

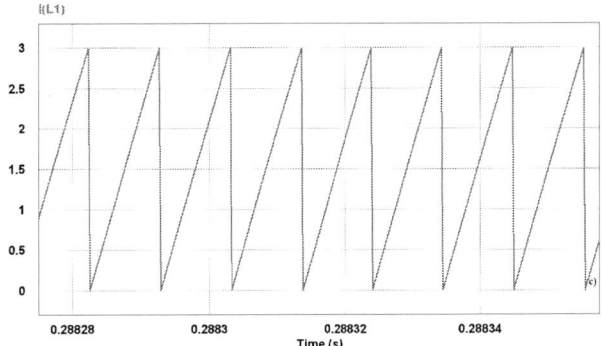

Fig. 10. PSIM Simulated Results

Finally, we have built the boost converter prototype to be supplied from a 12 V car-battery. First, we connected a 204 LED light-spot made at our laboratory. Subsequently, we connected diverse resistive loads with different input voltages to check the robustness of the converter and control. The converter performance at 12 V can be appreciated in Table V.

TABLE V
EXPERIMENTAL MEASUREMENTS

Load	Vin(V)	Iin(A)	Vout(V)	Iout(mA)	η(%)
10 kΩ	12.09	0.475	199	20.019	69.37
20 kΩ	12.05	0.878	398	20.003	75.25
30 kΩ	12.02	1.297	597	20.011	76.76
40 kΩ	12.04	1.733	795	20.031	76.90
43 kΩ	12.03	1.884	849	20.008	75.13
43 kΩ	11.99	2.298	920	21.672	72.36

Fig. 11. LED-based Spot-light and Boost converter

Diverse oscilloscope captions proving the good system performance are given at Fig. 12. In all, the boost converter is always delivering the same output current (Ch4), but diverse output voltages have been tested (Ch3). The caption from Fig 12b, with an output voltage of 643 V, is related to the LED-based spot-light test. The figure 12.c shows the Boost converter working with extreme voltage gain, approximately 100, with an output resistance of 45 kΩ and supplied at 9 V (Ch1). The captions show the inductor current in the boundary between DCM & CCM (Ch2). Realize that each output voltage implies a different switching frequency.

V. CONCLUSIONS

A high voltage-gain single stage-boost converter has been simulated and realized achieving an extreme voltage gain about 100. The system is protected under patent P20131281. As expected, hysteretic control is the key issue to minimize switching losses as well as achieving the extreme duty-ratios required to those voltage gains.

The main problem of increasing the voltage gain and the converter power rating comes mainly from the lack of high-performance available devices. For instance, mosfets are probably appropriate to operate at high-voltages, but have too much resistance to be a good choice to operate at low input voltage, especially if the output power is increased. In this sense, recently emerging silicon carbide devices, especially mosfets, and diodes (shottky and JBS), could help to improve the converter efficiency, power rating and voltage gain.

VI. ACKNOWLEDGEMENTS

This work has been partially sponsored by the Spanish Ministry of Research and Science, under grants: DPI2009-14713-C03-02, and Consolider RUE CSD2009-00046.

VII. REFERENCES

[1] Z. Lei, G. Xia, L. Ting, G. Xiaoling, L. Q. Ming, S. Guangdi," *Color rendering and luminous efficacy of trichromatic and tetrachromatic LED-based white LEDs*", Elsevier Microelectronics Journal, 2006.

[2] Yuequan Hu, Milan Y. Jovanovic (2008). "*A Novel LED Drive with Adaptative Drive Voltage*", Delta Power Electronics laboratory, Research triangle Park, NC 27709, USA.

[3] M.Nishikawa, Y.Ishizuka, H.Matsuo, K.Shigematsu. "*An LED Drive Circuit with constant-Output-Current Control and Constant Luminance Control*", INTELEC Telecomunications Energy Conference, pp 1-6, December 4, 2006..

[4] O.Ronat, P.Green, S.Ragona. "*Accurate Current Control to Drive High Power LED Strings*", IEEE Appec 2006, pp 376-380, March, 19, 2006.

[5] M.Nishikawa, Y.Ishizuka, H.Matsuo, K.Shigematsu. "*An LED Drive Circuit with constant-Output-Current Control and Constant Luminance Control*", INTELEC Telecomunications Energy Conference, pp 1-6, December 4, 2006..

[6] X.Qu, S.-C.Wong, C.-K.Tse. "*Non Cascading Structure for Electronic Ballast design for Multiple LED Lamps with independent Brightness control*", IEEE transactions on power electronics, VOL 25, NO.2, Frebruary 2010.

[7] C.-Y Wu,T.-F.Wu, J.-R.Tsai. "*Multistring LED backlighting driving system for LCD Panels with Color Seqüential Dysplay and area Control*". IEEE Transactions on industrial electronics, VOL.55, NO.10, October 2008.

[8] P.Klimczak. S.Munk-nielsen. "*Comparative Study on Paralleled vs Scaled Dc-dc Converters in High Voltage Applications*", IEEE-EPE-PEMC 2008, International Power Electronics an Motion Control Conference.

[9] R.Jong-Wai, C.Youlin, R.Yon Duan, Y.R-Chang. "*High-Efficiency DC-DC Converter with High Voltage Gain and Reduced Switch Stress*", IEEE Transactions on Power Electronics, VOL.54,NO.1, Frebruary 2007.

[10] I.Lourd, D.Dah-Chuan, V.G Agelidis. "*High Gain Switched-Coupled Inductor Boost Converter*". International Conference on Power Electronics and drive Systems (PEDS), pp 423-428, January 19, 2010.

[11] D. A. Grant, Y. Darroman and J. Suter, "Synthesis of Tapped-Inductor Switched-Mode Converters," IEEE Trans. On Power Electron, vol. 22, no. 5, Sep. 2007.

[12] C.-S.Leu, P.-Y-Huang. "*A novel voltaje doubler Rectifier for High Output Voltage Applications*", IEEE Power Electronics Conference (IPEC), pp 2082-2085, June 31, 2010.

[13] F.L.Luo, H.Ye, M.H.Rashid. "*Multiple Lift Push-Pull Switched-Capacitor Luo-Converters*". IEEE Power Electronics Specialists Conference, VOL.2, NO. 415-420, November 2, 2002.

[14] F.L.Luo, H.Ye. "*Positive Output multiple-lift push-pull switched-capacitor Luo-Converters*".IEEE transactions on industrial electronics, pp 594-602, June 1, 2004.

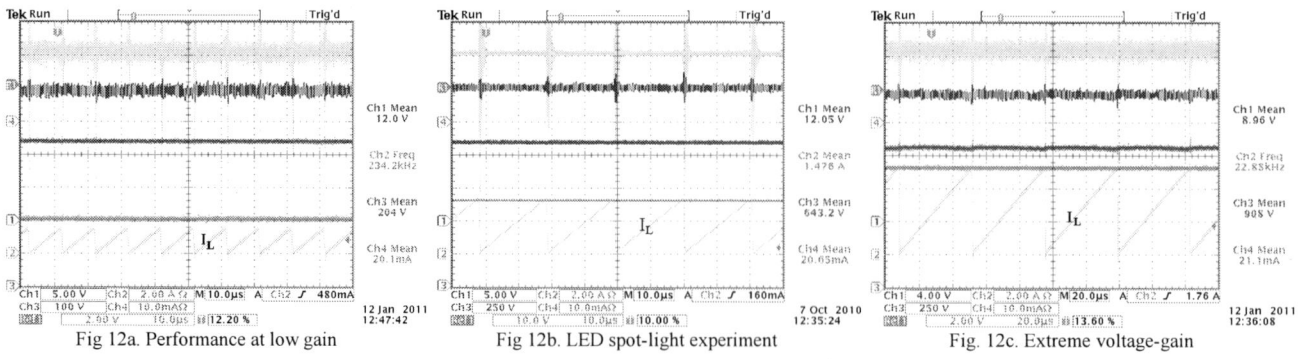

Fig 12a. Performance at low gain   Fig 12b. LED spot-light experiment   Fig. 12c. Extreme voltage-gain

Motor Cable Effect on the Converter Fed AC Motor Common Mode Current

Jaroslaw Luszcz
Gdansk University of Technology (Gdansk, Poland)
jlusz@ely.pg.gda.pl

Abstract- **Analysis of conducted EMI in AC motor drives fed by pulse width modulated voltage converters requires to consider parasitic capacitances in converters, motor windings and feeding cables to be taken into account. Motor voltage transients and related common mode currents are significantly correlated with resonance effects occurring in load circuits. The levels of intensity of these phenomena depend noticeably on frequency dependant impedance – frequency characteristics of converter load, motor windings and feeding cable. An analysis of frequency converter load impedance characteristics allows for identification and determination of frequency ranges in which the foremost contributions to EMI noise generation have voltage ringing phenomena associated with the load parasitic capacitances. This paper presents a method to model an AC motor with a feeding cable in conducted EMI frequency range up to *30 Mhz*. Distributed parasitic capacitances of AC motor windings are modelled as a ladder circuit. The proposed circuit model allows for an analysis of the influence of the motor feeding cable parameters on common mode currents generated in AC motor drive system, particularly in AC motor itself. The simulation results obtained based on the proposed ladder circuit model are verified by the experimental tests which were carried out for an exemplary adjustable speed AC motor drive application.**

I. INTRODUCTION

Voltage source, pulse width modulated (PWM) inverters commonly applied in contemporary adjustable speed drives (ASD) are a cause of accompanying undesirable high frequency (HF) side effects in the powered AC motor and supplying grid [1]–[6]. Increase of pulse width modulation carrier frequency and decrease of converter transistors' switching time intensify existing HF problems in various ways. One of the foremost problems related to these phenomena is generation of high frequency stray current pulses flowing through drive components - due to high levels of output voltage steeps (*dV/dt*) - and unavoidable parasitic capacitances [7]–[9]. HF stray currents, flowing in conducting components of the ASD due to the existence of parasitic inter-capacitances, are not limited by standard regulations related to EMC of power electronic converters. Nevertheless the foremost consequences of the stray HF current flow, especially through the parasitic capacitances, is the generation of HF voltage components appearing in all conductive elements of the ASD system. These HF voltage components, especially common mode (CM) voltages, are highly hazardous and are the fundamental source of conducted EMI emission noise of power electronic converters. Detailed modelling of an ASD system in a wide

frequency range for analyzing all interactions between a PWM voltage converter and an AC motor with cable is challenging and requires the use of complex methods and broadband models [10]–[13].

One of the main problems to model inverter feed AC motor drives in a wide frequency range is accurate identification of parasitic parameters of all drive components, especially in HF range. For the conducted EMI frequency range, usually up to *30 MHz*, the predominant model identification problem is related to parasitic capacitances of the system components. These capacitances induce various HF objectionable effects such as: local resonances in parasitic sub-circuits and increased flow of harmful CM currents affiliated to them [3], [5], [14]–[20].

Increased CM currents generated on the output side of the frequency converter due to the extraordinary resonance effects in the feeding cable and the AC motor windings, propagate to all the surrounding circuits and return back to the supply grid through the frequency converter. This paper presents a method for analyzing the influence of motor feeding cable on the motor CM currents flowing through parasitic capacitances between motor windings, stator and rotor. The proposed identification method of the load broadband circuit model parameters is based on the load impedance frequency characteristics analysis [9]. Determination of the most meaningful resonance frequencies observed on the load impedance characteristic allows for a ladder circuit model configuration arrangement with the appropriate number of rungs. Using this method the reasonable complexity level of the model can be kept, what is essential for its parameter identification difficulty level.

II. GENERATION OF COMMON MODE CURRENTS IN ASD

Analysis of CM currents propagation in the ASD is mostly related to the load side of frequency converter, which is usually voltage source PWM controlled inverter with IGBT switches. In this type of inverter, the output voltage stress (*dV/dt*) due to switching transient is usually high due to the bipolar DC bus voltage commutation during transistor switching time. Nevertheless, contemporary line side PWM controlled converters can also have significant effect on the CM currents generation but in this paper only the ASD with a line side diode rectifier has been analyzed. This simplification allows observing more clearly motor cable effects on the evaluated phenomena by minimizing simultaneous influence of line side converter.

978-1-4244-8806-3/11 $26.00 © 2011 IEEE

In a typical ASD configuration with AC motor fed by shielded cable presented in Fig. 1 four major CM current loops can be emphasized:

- the first one $I_{CM,Cable}$, which is the smallest loop and allow for CM current flow between energized motor cable wires and cable shield connected to output side of converter due to cable internal capacitances,
- the second one I_{Shield}, which carry most of motor CM current flowing through winding capacitances back to grounded converter chassis through motor cable shield grounded on both ends by converter and motor stator,
- the third one $I_{CM,C}$, which carry the part of motor CM current through ground connection between converter and motor other then motor cable shield, usually it is protective earth installation (PE) characterized by equivalent impedance of motor grounding connection $Z_{PE,M}$ and converter grounding connection $Z_{PE,C}$,
- the fourth one $I_{CM,Cin}$ which is the largest loop and conduct the remain part of motor CM current through the electric power grid impedance Z_G.

Presented in Fig. 1 CM currents loops shows that motor feeding cable and its parameters are essential for CM currents generation and sharing between motor windings and feeding cable. According to presented in Fig. 1 circuit diagram the total CM current of converter is a sum of CM currents of motor windings and feeding cable (1). Nevertheless, the distributed character of parasitic capacitances cause many resonance interactions between a motor winding and a feeding cable, what complicate the analysis of CM currents.

$$I_{CM,Cout} = I_{CM,M} + I_{CM,Cable} \qquad (1)$$

The example of detailed CM current path is presented in Fig. 2, where equivalent lumped parasitic capacitances of motor, cable and converter are indicated. Analysis of the CM currents in ASD is problematic because of difficulty to identify impedance-frequency characteristics of the sub-circuits, especially the power grid impedance, converter to ground impedance and the converter load impedance within the analysed frequency range. Difficulties with determination of inverters load CM impedance are associated with parasitic

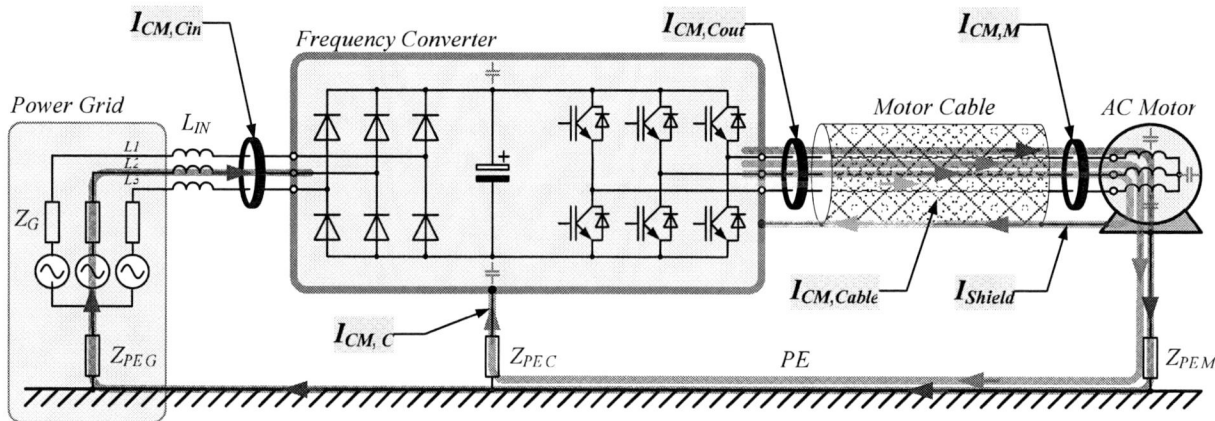

Fig. 1 The most significant common mode currents loops in ASD with shielded motor cable.

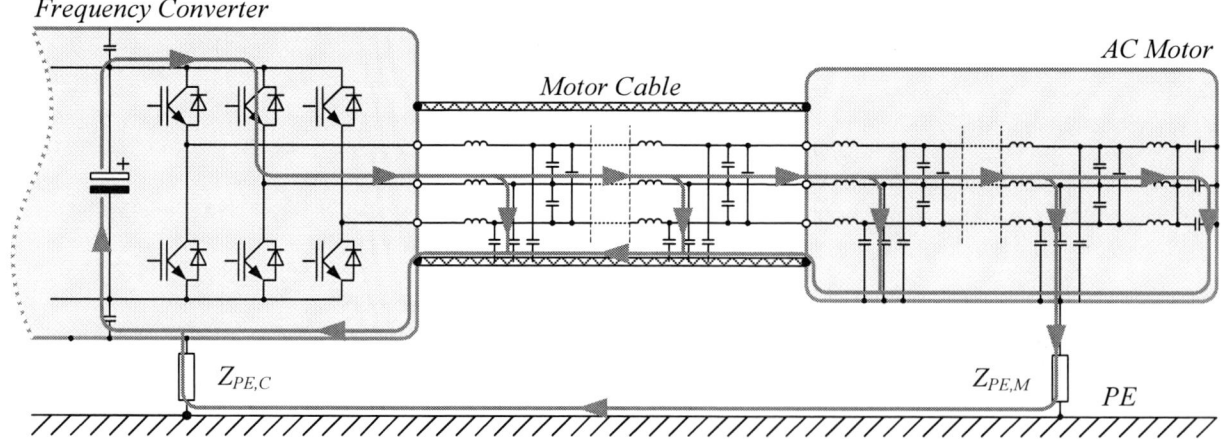

Fig. 2 The detailed example of foremost possible common mode current paths between converter and AC motor windings

capacitances of the motor windings and the feeding cable with reference to the ground. The parasitic capacitances are essential for the converters load CM impedance variations. Distributed character of the parasitic capacitances jointly with rapid voltage changes induces transmission line effects in the motor feeding cable and motor windings. The impedance mismatch between the inverter outputs, the feeding cable and the motor windings usually appears and complicates the analysis.

III. MODELLING OF THE CONVERTER LOAD IN THE CONDUCTED EMI FREQUENCY RANGE

Broadband modelling of AC motor windings with a feeding cable is complicated because of distributed parasitic capacitances which should be determined. Complex and unequal distribution of these parasitic capacitances along the motor windings make identification process particularly difficult. An effortlessness of the identification of the parameters of the AC motor windings model is the main motivation for developing rationally simplified models.

A. Modelling of AC Motor Windings

A simplified model of all distributed capacitances in AC motor windings in a wide frequency range can be formulated based on a ladder circuit model with an appropriate number of rungs - adequate for the given frequency range and expected accuracy. This simplification can be effectively used in a frequency range in which lumped circuit models are not accurate enough and detailed determination of distributed parasitic parameters of the evaluated system is too complicated. General configuration of the proposed circuit model with the N – step ladder circuit is presented in Fig.3.

The CM impedance $Z_{M(CM)}$ of motor windings as a relation between the winding voltage at the motor terminal A and the total CM current I_{CM} flowing through the distributed winding to the ground capacitances C_{gk} can be formulated based on the proposed circuit model as follows (2).

$$Z_{M(CM)}(s) = \frac{u_A(s)}{\sum_{k=0}^{N} i_{C_{gk}}(s)} \tag{2}$$

Assuming that the winding self capacitances C_{s1}, C_{s2}, ... , C_{sN} are usually considerably smaller then the winding to ground capacitances C_{g0}, C_{g1} ,..., C_{gN}, the evaluated

transfer impedance can be determined analytically based on the presented circuit model with using formula (3).

$$Z_{M(CM)}(s) \approx \cfrac{1}{sC_{g0} + \cfrac{1}{sL_1 + \cfrac{1}{sC_{g1} + ... + \cfrac{1}{sL_N + sC_{gN}}}}} \tag{3}$$

The reflection coefficient characteristic $\Gamma(s)$ at motor windings terminals for the given source impedance $Z_0=50\ \Omega$ can be determined using formula (4). The example of the simulated reflection characteristic of the evaluated AC motor is presented in Fig. 4.

$$\Gamma(s) = \frac{Z_{T(CM)}(s) - Z_0}{Z_{T(CM)}(s) + Z_0} \tag{4}$$

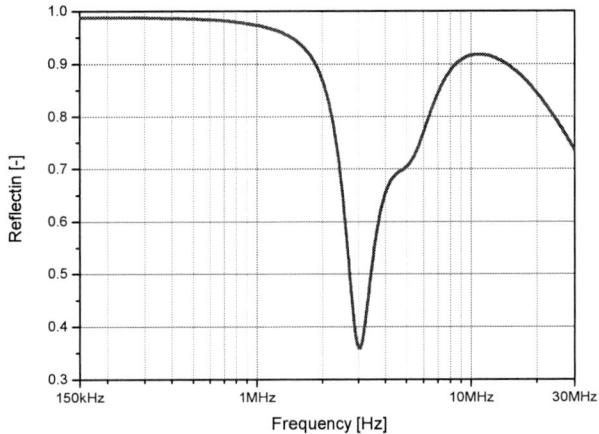

Fig. 4 Reflection characteristic of the evaluated AC motor windings

B. Modelling of a Feeding Cable

A classic AC Motor feeding cable can be represented as three conductor lines referenced to the ground and in a shielded cable which are commonly used for feeding AC motors in ASD. The cable shield represents the reference ground and equalizes conductors to ground parasitic capacitances, so the symmetric circuit model can be implemented as shown in Fig. 5.

Length and impedance related parameters of the AC motor feeding cable are essential for the ASD general performance and its particular influence on the conducted and radiated EMI emissions. Commonly used lengths of AC motor cables are from few meters up to few hundreds of meters long and are divided into two classes: short and long cables.

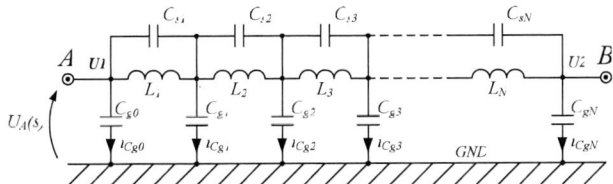

Fig. 3 Distributed parasitic capacitances between motor's windings and ground, simplified lumped representation by N – step LC ladder circuit.

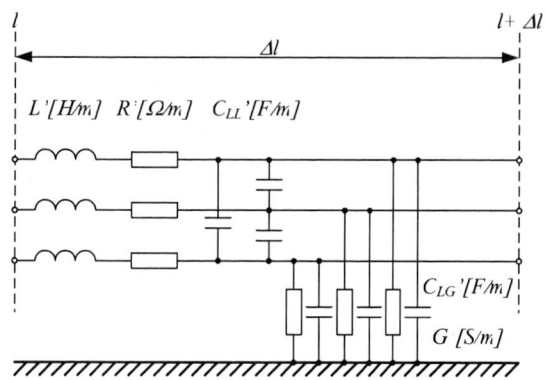

Fig. 5 Classic lossy transmission line circuit model of AC motor feeding cable.

A critical cable length is defined for various typical ASD applications which mean that for the feeding cable longer than the critical length adverse effects are expected to appear intensively and some protective methods should be applied. The critical length depends on spectral characteristics of transmitted signals and pulse propagation velocity. Therefore, the critical cable length can be referenced to the shortest wavelength of transmitted signals.

Theoretically, a critical cable length parameter, which allows to estimate the intensity level of exposure of the ASD to reflection phenomena in the AC motor cable, is commonly calculated based on the converter output voltage rise time t_r. The wavelength related to the voltage rise time can be calculated using formula (5) where v is signal propagation velocity in the evaluated cable.

$$\lambda = t_r \cdot v \qquad (5)$$

The signal propagation velocity in typical power cables is related to the relative dielectric permeability of cable insulation ε_r (6) and also correlated with propagation factor k_p, which can be determined based on the "per meter" parameters of cable: resistance $R'[\Omega/m]$, inductance $L'[H/m]$, capacitances C'_{LL}, C'_{LG} [F/m] and conductance $G'[S/m]$ according to formula (7) where inductance L_0 and capacitance C_0 are equivalent per unit length parameters [21].

$$v = \frac{1}{\sqrt{L_0 C_0}} = \underbrace{\frac{1}{\sqrt{\mu_0 \varepsilon_0}}}_{c_0} \cdot \frac{1}{\sqrt{\varepsilon_r}} \qquad (6)$$

$$k_p(s) = \sqrt{\left(R_0 + sL_0\right)\left(G_0 + sC_0\right)} \qquad (7)$$

In general, the signal propagation velocity in power cables is approximately two times smaller than the light speed c_0, for the reason that dielectric permeability for typical cables insulations varies from 3 to 8, so obtained velocity is usually within the range of 40-60 % of the light speed. The characteristic cable impedance Z_C Eq. 8 allows determining the real reflection coefficient Γ at the motor terminals using Eq. 9 where Z_M is the motor windings input impedance.

$$Z_C(s) = \sqrt{\frac{\left(R_0 + sL_0\right)}{\left(G_0 + sC_0\right)}} \approx \sqrt{\frac{L_0}{C_0}} \qquad (8)$$

$$\Gamma(s) = \frac{Z_M(s) - Z_C(s)}{Z_M(s) + Z_C(s)} \qquad (9)$$

For the given frequency of transmitted signal, the critical cable length can be correlated to the wavelength determined by consideration of the real velocity of the signal propagation. The minimum length of the cable for which the reflection effects can be noticeably increased is $\lambda/4$.

Parameters per length of presented type of transmission line model for particular cable can be determined by measuring the cable CM impedances in open circuit Z_{OC} and short circuit Z_{SC} configurations. [3]

IV. THE FEEDING CABLE IMPACT ON THE MOTOR CM CURRENTS

High frequency interactions between an AC motor and a feeding cable can be analyzed by simulation. The proposed circuit model of an AC motor with feeding cable is presented in Fig. 6, where the AC motor is represented as an N – rung LC ladder circuit whereas the feeding cable is represented as a lossy transmission line. The foremost benefit of using circuit models is its relatively low complexity level in relation to the achieved accuracy. A reasonable balance kept between model simplicity and their adequacies allows for significant reduction of its parameters identification efforts and decrease radically computational overheads of the simulation analysis. An identification of parameters of the circuit model of the AC motor with feeding cable can be completed based on the terminal impedance measurement. Based on this model various simulations have been carried out using the motor and the feeding cable parameters which are identified experimentally. Examples of simulation results obtained for the tested AC motor an AC motor with the feeding cable are presented in Fig. 7.

Presented simulation results allow investigating the influence of the feeding cable on the CM impedance changes. Particularly, the observed decrease of the frequencies for which the CM impedance of the converter loads approaches lowest values can be successfully analyzed for different feeding cable lengths.

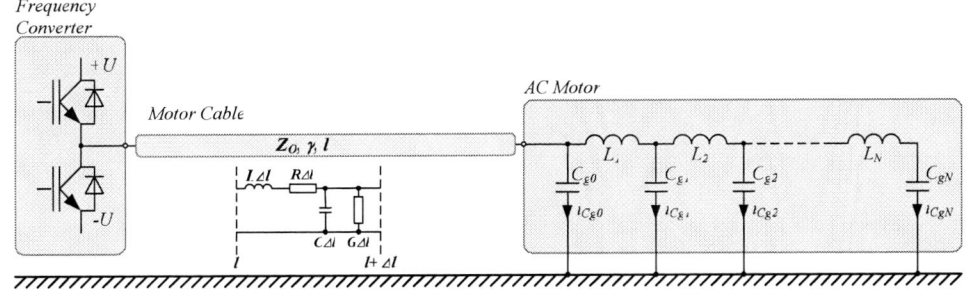

Fig. 6 Circuit model of AC motor winding with feeding cable for broadband CM currents analysis – one phase representation.

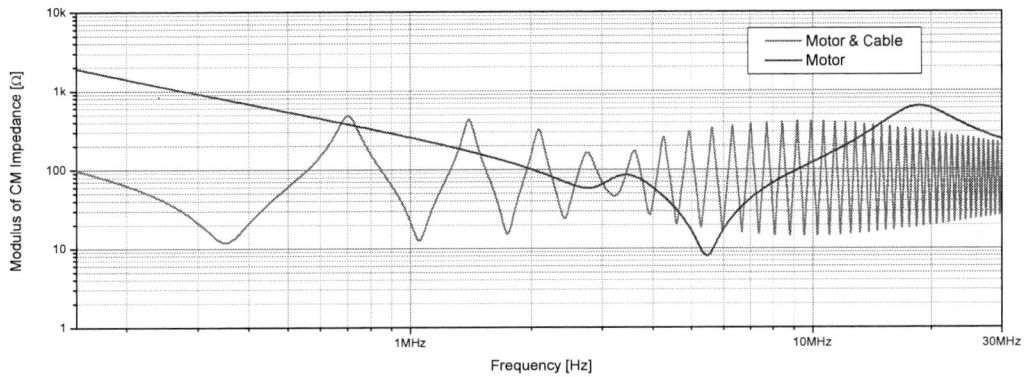

Fig. 7 Example of simulation results of common mode impedance of evaluated AC motor and AC motor with feeding cable.

V. EXPERIMENTAL VERIFICATION

Experimental investigation of CM current generation and propagation has been done for a typical ASD with voltage source inverter (VSI) feeding AC motor, both of rated power *7.5 kW*. A four pole, *50 Hz* AC motor with no mechanical load, was connected to the converter's output terminals by a three wire shielded cable. Two lengths of feeding cables were used: the first one was very short (less than *0.2 m*) and the second one was relatively long - *100 m*. In both cases the cable shield was connected at both ends to the ground: one of them to converter ground terminal and second one to the AC motor grounding terminal.

Configuration of the experimental test bench is presented in Fig. 1 where the AC motor CM currents $I_{CM,M}$ have been measured for both lengths of AC motor feeding cable. The CM current probe has been inserted onto the very short unshielded section of feeding cable, as closely as possible to the motor winding terminals. Additionally the input side of the evaluated converter was feed through a line impedance stabilization network (LISN) in order to minimize power grid impedance influence on the obtained measurement results.

Measurement results calculated as a ratio between AC motor CM currents recorded for the motor with short and long feeding cable is presented in Fig. 8. Based on the obtained measurement results, in the evaluated frequency range CISPR (*150kHz-30MHz*), few characteristic frequency sub ranges related to the analyzed feeding cable influence on the AC motor CM currents can be determined.

In the frequency range below *2 MHz* an increase of motor CM current was remarked, especially for the frequency ranges (around *350 kHz*, *1 MHz* and *1.07 MHz*) correlated closely with the feeding cable impedance characteristic. The maximum increase approximately 15 dB was recorded in frequency about *350 kHz*. A decrease of the motor CM current, caused by feeding cable, was noticed for the frequencies higher than *2 MHz*. The maximum attenuation effect of feeding cable (about *-20 dB*) in the frequency range about *4-6 MHz* is correlated with the AC motor reflection characteristic presented in Fig. 3 and winding CM impedance presented in Fig. 7.

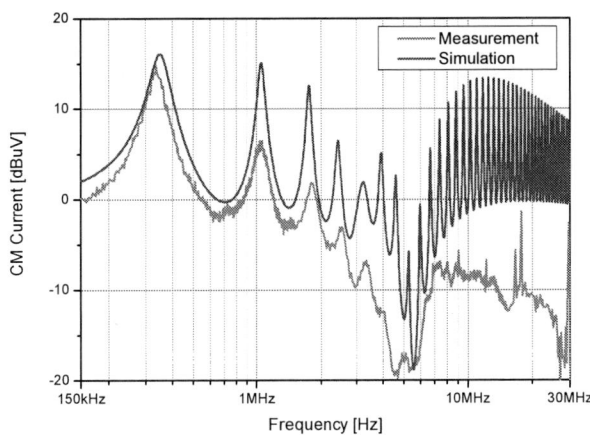

Fig. 8 Measurement and simulation results of the influence of feeding cable on the CM current of AC motor.

For frequency range above 6 MHz the measured CM current attenuation effect of motor feeding cable became smaller, correspondingly with of AC motor windings reflection characteristic (Fig. 4).

Simulation result obtained based on the proposed simplified circuit model are presented in comparison to the experimental results in Fig. 8. Presented comparison shows that the investigated circuit model characterizes accurately frequency ranges in which the AC motor CM currents are noticeably increased due to the feeding cable resonances. Nevertheless, the obtained by simulation levels of increase are noticeably higher in comparison to measurement results. It is a result of numerous simplifications assumed in model development, especially related to dielectric losses in cable and motor winding. The frequency dependence of dielectric losses where not sufficiently represented in the model, therefore the simulation results are overestimated in high frequency range. In the frequency range above the motor winding main resonance, in the investigated case above ~4 MHz, the obtained by simulation magnitudes of CM currents are not adequate enough due to underestimated losses in the model, however the multi resonance response of the converters load in that frequency range is clearly illustrated by the model.

VI. CONCLUSIONS

The main objective of the presented work was to investigate the influence of the AC motor feeding cable on the motor CM currents magnitudes. Detailed modelling of high frequency interactions occurring at output side of frequency inverter, motor cable and AC motor windings require broadband model's, which are enormously complex to parameterize, because of many parameters which have to be determined, usually based on experimental measurement. The proposed simplified circuit model of the converter load is relatively uncomplicated and its parameters are relatively effortless to determine. Experimental verification of the simulation results, which have been done for the investigated ASD based on this model, shows that the capability of such model can be sufficient enough for analysis motor feeding cable influence on the motor CM current. Results of experimental and simulation investigations confirm that motor feeding cable has significant influence on the generation of AC motor CM current, which are which are essential for EMI emission of ASD. Simplified broadband circuit modelling of converters load allows determining characteristic frequency ranges in which motor cable influence is mostly noticeable. In the frequency range above the main resonance frequency of motor winding more precise determination of cable influence on CM current emission require more advanced models.

REFERENCES

[1] S. Ogasawara, H. Akagi H.; Analysis and Reduction of EMI Conducted by a PWM Inverted-Fed AC Motor Drive System Having Long Power Cables. Proc. of IEEE 31st Annual Power Electronics Specialists Conference (PESC'00), Vol. 2, 2000, pp. 928–933.

[2] J. Adabi, F. Zare, G. Ledwich, A. Ghosh, "Leakage Current and Common Mode Voltage Issues in Modern AC Drive Systems", AUPEC, Perth, Australia, Dec 2007.

[3] J. Łuszcz, K. Iwan, "AC motor transients and EMI emission analysis in the ASD by parasitic resonance effects identification", European Conference on Power Electronics and Applications, EPE 2007.

[4] F. Zare, "Modelling of Electric Motors for Electromagnetic Compatibility Analysis", AUPEC 2006, Melbourne, Australia, Nov 2006.

[5] A. Said, K. Al-Haddad, "A new approach to analyze the overvoltages due to the cable lengths and EMI on adjustable speed drive motors", Power Electronics Specialists Conference, 2004. PESC 04. 2004 IEEE 35th Annual, Volume 5, 20-25 June 2004 Page(s):3964 - 3970 Vol. 5.

[6] T. Weidinger, "Elimination of increased excitation of common-mode oscillations in electrical drive systems with active front end and long motor cables", Power Electronics and Motion Control Conference, 2008. EPE-PEMC 2008.

[7] F. Zare, "High frequency model of an electric motor based on measurement results", Australian Journal of Electrical & Electronics Engineering (AJEEE), Vol 4, No 1, 2008, page 17-24.

[8] L. Arnedo, K. Venkatesan, "High frequency modelling of induction motor drives for EMI and overvoltage mitigation studies", Electric Machines and Drives Conference (IEMDC'03), Volume 1, 1-4 June 2003 Page(s): 468 - 474.

[9] J. Łuszcz, I. Moson, "AC Motor Windings Circuit Model for Common Mode EMI Currents Analysis", The 5th International Conference CPE 2007, Compatibility in Power Electronics, May 29 - June 1, 2007, Gdansk, Poland.

[10] A.F. Moreira, T.A. Lipo, G. Venkataramanan, S. Bernet, "High-frequency modelling for cable and induction motor over voltage studies in long cable drives", IEEE Transaction on Industry Application vol.38, no.5, pp.1297–1306, 2002.

[11] N. Hanigovszki, J. Poulsen, G. Spiazzi, F. Blaabjerg, "An EMC evaluation of the use of unshielded motor cables in AC adjustable speed drive applications", Power Electronics Specialists Conference, 2004. PESC 2004. Volume 1, 20-25 June 2004 Page(s):75 - 81 Vol.1.

[12] J. Łuszcz, "Motor Cable as an Origin of Supplementary Conducted EMI Emission of ASD", 13th European Conference on Power Electronics and Applications (EPE 2009), 8 - 10 September 2009, Barcelona, Spain.

[13] J. Łuszcz, "Motor Cable Influence on the Converter Fed AC Motor Drive Conducted EMI Emission" The 5th International Conference CPE 2009, Compatibility and Power Electronics, May 20 - 22, 2009, Badajoz, Spain.

[14] A. Muetze, A. Binder, "Calculation of Circulating Bearing Currents in Machines of Inverter-Based Drive Systems", IEEE Transactions on Industrial Electronics, Volume 54, Issue 2, April 2007 Pages:932 – 938.

[15] H. Akagi, S. Tamura, "A passive EMI filter for elimination both bearing current and ground leakage current from an inverter-driven motor" IEEE Trans. Power Electronics, vol. 21, no. 5, pp. 1459–1469, Sep. 2006.

[16] J. Adabi, F. Zare, A. Ghosh, "End-winding Effect on Shaft Voltage in AC Generators", 13th European Conference on Power Electronics and Applications (EPE 2009), 8 - 10 September 2009, Barcelona, Spain.

[17] A. Kempski, R. Strzelecki, R. Smolenski, Z. Fedyczak, "Bearing current path and pulse rate in PWM-inverter-fed induction motor" PESC. 2001 IEEE, Volume: 4 , 17-21 June 2001 Pages:2025 - 2030 vol. 4.

[18] G. Costabile, B. Vivo, L. Egiziano, V. Tucci, M. Vitelli, L. Beneduce, S. Iovieno, A. Masucci, "An accurate evaluation of electric discharge machining bearings currents in inverter-driven induction motors", European Conference on Power Electronics and Applications (EPE 2007) 2-5 Sept. 2007.

[19] A. Muetze, A. Binder; Calculation of motor capacitances for prediction of discharge bearing currents in machines of inverter-based drive systems. 2005 IEEE International Conference on Electric Machines and Drives, Page(s): 264 – 270.

[20] N. Hanigovszki, J. Landkildehus, G. Spiazzi, F. Blaabjerg; An EMC evaluation of the use of unshielded motor cables in AC adjustable speed drive applications. IEEE Transactions on Power Electronics, Volume: 21 , Issue: 1, 2006 , Page(s): 273 - 281

[21] C.R. Paul, Introduction to Electromagnetic Compatibility. New York: John Wiley & Sons, 1992.

Enabling Technologies for Matrix Converters in Aerospace Applications

L Empringham, *Member, IEEE*, L de Lillo, *Member, IEEE*, S Khwan-On, C Brunson,
P W Wheeler, *Member, IEEE*, J C Clare, *Member, IEEE*

Abstract—The matrix converter is a direct AC-AC power converter topology that does not contain an intermediate DC-link stage. This paper presents different enabling technologies to enhance the suitability of the matrix converter topology for aerospace applications. The use of silicon carbide power devices to improve efficiency and reduce system weight together with an improved, high speed fault detection technique are presented. Results from different prototype matrix converters to validate the concepts are also presented.

I. INTRODUCTION

THE matrix converter is a direct AC-AC power converter topology. It consists of a matrix of bi-directional switches arranged in such a way so that any input phase can be connected to any output phase. As in the majority of cases, a three input, three output converter will consist of a 'matrix' of nine bi-directional switches, hence the term 'matrix converter'. The gating of the devices is modulated to achieve the desired output voltage. A small input filter is necessary to reduce the line side switching harmonics of the system.

The matrix converter first appeared in the literature in a book by Gyugi and Pelly[1] and in a journal publication by Daniels and Slatery[2], in 1976, these both presented the direct AC-AC power circuit concept. In these publications it was also described as a 'force commutated cycloconverter'. Since then, as with many power converter topologies, work has concentrated on the derivation of modulation and control strategies. In comparison to a DC-Link voltage source inverter, the modulation of the switches in the matrix converter is more difficult since three changing voltages can be used for the modulation. The main characteristics of the matrix converter are listed below:

- No Bulky DC-Link capacitor
- No large input boost reactor
- Sinusoidal input and output currents
- 86% voltage transfer ratio
- Controllable input displacement factor
- Bidirectional power flow

Figure 1 shows a schematic diagram of a three input, three output matrix converter. A detailed review of the characteristics of the matrix converter is described in[3]. Since this type of converter is completely reversible and bi-directional, the input and output can be interchanged. In this example the voltage-stiff supply is defined as the input and the current-stiff motor is defined as the output but this can be reversed depending on the application. Whilst at first glance, the matrix converter seems to be a more complicated concept when

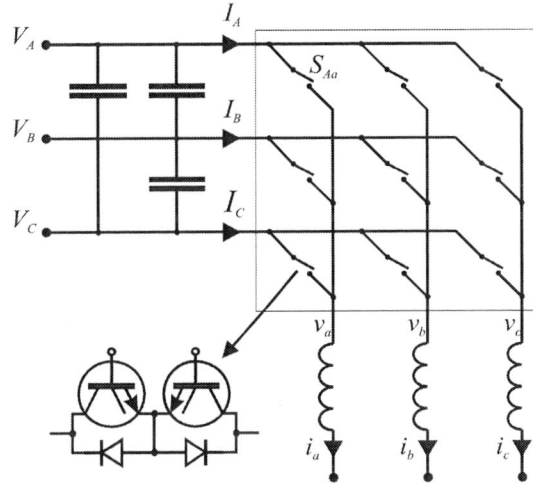

Fig. 1. Schematic representation of a three input phase, three output phase, direct matrix converter

compared to a voltage source inverter (VSI), it has been shown in[4] that the VSI can be thought of is a subset of the matrix converter and techniques to modulate and control one can be used on the other.

Due to the lack of bulky passive components, the matrix converter offers a more compact and power dense solution for motor drive applications when compared to traditional VSI topologies. This is an important aspect for industrial sectors such as the aircraft industry.

This paper will describe the matrix converter and discuss some enabling technologies which will result in further improvements in weight reduction or efficiency and improved robustness, reliability and fault detection capabilities of this topology. Results from different prototype converters are presented to validate the concepts presented.

II. BI-DIRECTIONAL SWITCH ARRANGEMENTS

The main component of the matrix converter is the bi-directional semiconductor switch. A single device that can both conduct current in each direction and block voltage in both directions is currently not commercially available although some work has been reported into the monolithic bidirectional switch(MBS) in[5][6].

Bi-directional switches must therefore be constructed from available unidirectional devices. The most used switching device for matrix converter applications is currently the IGBT although other devices have been used in the past such as

978-1-4244-8806-3/11 $26.00 © 2011 IEEE

the BJT, MTO, MCT and the MOSFET. The reverse blocking IGBT is also gaining favor. Figure 2 shows some typical arrangements.

The diode bridge arrangement uses only one active device but the conduction losses are higher than other arrangements since the current flows through one switch and two diodes. The common emitter and common collector arrangements allow the current direction to be controlled, this allows greater flexibility when performing the commutation of current between input phases. Arrangement d) is for devices that can block voltage in both directions such as the RB-IGBT or MTO.

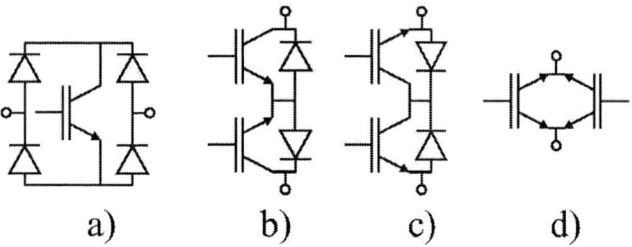

Fig. 2. Possible bi-directional switch arrangements: a) Diode bridge (DB) arrangement b) Common emitter (CE) arrangement c) Common collector (CC) arrangement, d) Reverse Blocking IGBT (RB-IGBT) arrangement

The simple diode bridge arrangement requires only one gate drive per bidirectional switch but suffers the highest conduction losses of all the arrangements. The fact that there is only one controllable device per switch also means that advanced commutation techniques which remove the need for snubbers cannot be used such as those described in [7][8]. It does however offer simpler control and less gate drive requirements than the other options.

The reverse blocking IGBT arrangement together with the Common Collector configuration only require 6 isolated gate drive power supplies in contrast to the nine required by the Common Emitter arrangement. Since the Common emitter arrangement is modular in structure and gate drive power arrangement, i.e. one power supply per bi-directional switch, it may be better for high power converters where the physical size of the converter becomes significant and the stray inductance of busbars may prevent the use of the six isolated power supply methods.

III. SiC DEVICES IN MATRIX CONVERTER APPLICATIONS

Wide bandgap semiconductor materials such as Galium Nitride and Silicon Carbide (SiC) for use in power devices have received renewed interest in the last few years. The use of SiC offers some potential advantages over Silicon (Si). Potential advantages are itemized below:

- Higher temperature operation - Reduced heatsink requirements
- Higher blocking voltages - Higher power density
- Reduced Switching times - higher efficiency
- Higher Current density chips - smaller packages

These advantages however cannot be fully exploited using presently available technology. Faster switching times will degrade EMI performance and present industrial packaging technology limits the operating temperature of the devices. Higher blocking voltages with smaller chip sizes causes increased problems for corona and insulation systems.

The higher blocking voltages of SiC devices mean that devices with low voltage structures, such as schottky barrier diodes, can be constructed to withstand higher voltages than the equivalent silicon part. 1200V Silicon Carbide schottky barrier diodes can be used as the antiparallel diode in place of commercial Si devices with standard Silicon IGBT's. The advantage is that there is little or no reverse recovery loss associated with the commutation process[9].

A direct matrix converter will suffer higher conduction losses when compared to a rectifier - DC-Link - Inverter, or back to back inverter topology since the current path is always through an IGBT and a Diode. The switching losses are however lower since the commutation voltage is always lower than the equivalent voltage source converter[10][11]. This means that beyond a certain frequency, the overall losses of a matrix converter will be less than that of the VSI. In aerospace applications, size and weight is important, so any reduction in the size of passive components by increasing switching frequency may be advantageous and as such, the reduced losses at higher switching frequencies gives the matrix converter a further advantage.

The use of SiC schottky barrier diodes in matrix converter applications has the potential to further reduce the losses associated with the commutation process. This allows the switching frequency to be increased for the same overall efficiency. It is often not clear however what improvement can be made when looking at manufacturers data sheets since the turn on energy of the IGBT is increased by the reverse recovery of the diodes in the circuit. One way of determining any potential benefit is to directly implement two matrix converters with different diodes and to measure any improvement.

IV. 100kW SiC-Si HYBRID MATRIX CONVERTER CONSTRUCTION AND EVALUATION

A 100kW matrix converter prototype has been constructed to determine the potential improvements SiC diodes may have on switching losses[12]. The three phase to three phase converter is constructed using a stack of three power modules, each containing three bi-directional switches. The first version of the prototype used all silicon devices rated at 1700V, 600A. The power module was constructed by Dynex Semiconductor using a standard E-type package. The internal bus-bar structure was modified to create the necessary common emitter arrangement of bi-directional switches. A second set of modules have also been implemented using 1200V, 500A Silicon Carbide antiparallel diodes. Figure 3 shows a photograph of the Dynex DIM600EZM17-E000 described above.

The forward voltage drop against load current of both types of module were measured at different temperatures. It was seen that both the SiC schottky barrier diodes and the silicon diodes exhibited very similar characteristics at 25 degrees C. This was due to the implementation of the SiC devices. Each

978-1-4244-8806-3/11 $26.00 © 2011 IEEE

Fig. 3. The DYNEX, E-type, 600A, 1700V, 3x bi-directional switch module

500A diode was constructed using ten 50A SiC diode chips. Figure 4 shows the characteristics of the diodes at 100 degrees C.

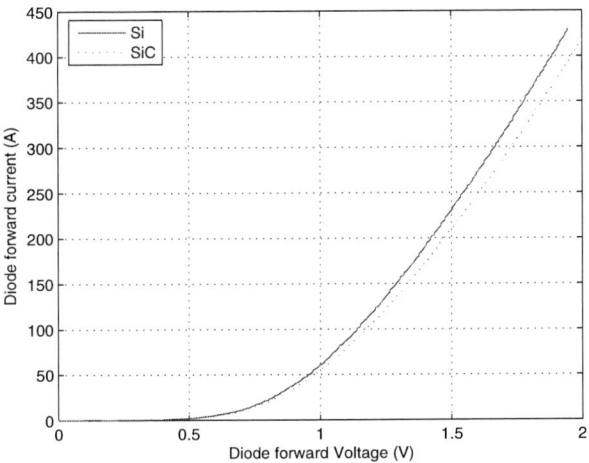

Fig. 4. Forward voltage drop versus load current for Si and SiC diodes at 100 degrees Celsius

A. Loss Measurement

The measurement of electrical losses in power electronic switching circuits is not an easy task. The input power to the converter is relatively simple to determine but even modern high bandwidth power analyzers do not sample fast enough to capture the high speed switching edges with the necessary resolution. It is for this reason that a calorimetric technique of directly determining the losses was preferred.

The cooling system of the 100kW prototype matrix converter used an individual water jacket on each power module. The coolant flow rate, liquid inlet temperature and liquid outlet temperature were all measured. These measurements were then used, together with the specific heat capacity of the water, to directly calculate the losses in the power module according to equation 1.

$$P_{Loss} = 4186 f \Delta T \qquad (1)$$

Where: P_{Loss} is the power loss, 4186 is the specific heat capacity of water, f is the flow rate in L/s and ΔT is the change in temperature between the inlet and the outlet.

Loss to the ambient through the case of the power module was neglected and will be justified later in the paper. Figure 5 shows a photograph of the top of the converter, highlighting the thermal measurement points.

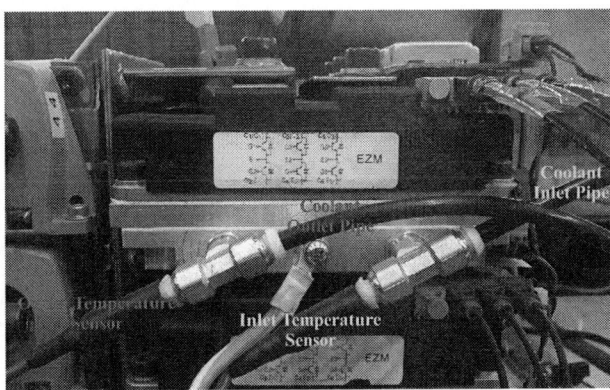

Fig. 5. Photograph of one module of the completed matrix converter showing liquid inlet and outlet temperature measurement

B. Loss Comparison Results

The matrix converter drives a 6000RPM Permanent Magnet Synchronous Machine which is in turn coupled to a loading rig. For the following results, the system was run at 80kW and supplied from a 400Hz, 235V (L-N) aircraft supply. Each converter, All Si and hybrid SiC-Si, were operated at different switching frequencies until thermal stability was reached. The results of the tests can be seen in figure 6. It can be clearly seen that the removal of the reverse recovery in the antiparallel diodes has a significant impact on the overall losses of the system.

Any loss to the ambient air from the converter will not be measured by the loss measurement system, only the heat conducted to the coolant was measured. The coolant temperature used was chosen to keep the overall temperature of the power module as close to the ambient as possible. Nevertheless, with the coolant temperature and flow rates fixed to be the same for the two converters, inevitably, the surface temperature of the all silicon modules was some 20 degrees higher than the hybrid SiC-Si module. This means that some extra loss in the all silicon module was not measured. This implies that the efficiency profile of the all Si converter shown in figure 6 should be slightly lower. This re-emphasizes the advantage gained by using SiC diodes.

V. INCREASING THE MATRIX CONVERTER RELIABILITY

The reliability of matrix converters is often thought of as being less than that of more traditional industrial drive topologies such as the VSI or the back to back PWM rectifier - inverter because of the increase in the number of semiconductors used.

978-1-4244-8806-3/11 $26.00 © 2011 IEEE

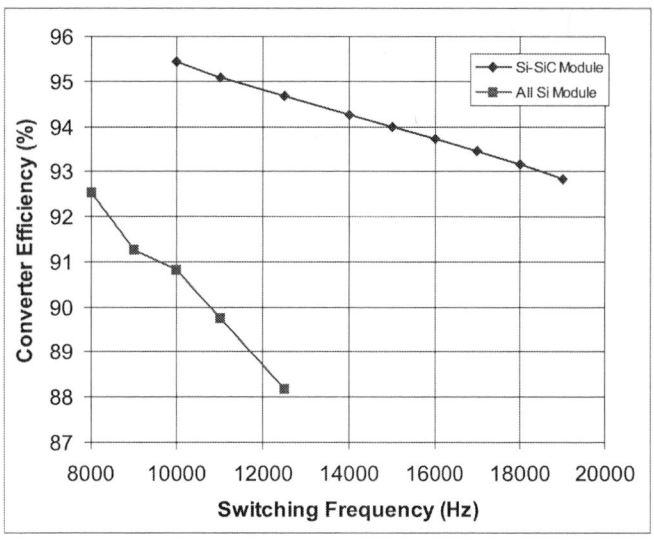

Fig. 6. Graph of switching frequency versus converter efficiency for both Si and SiC-Si matrix converters

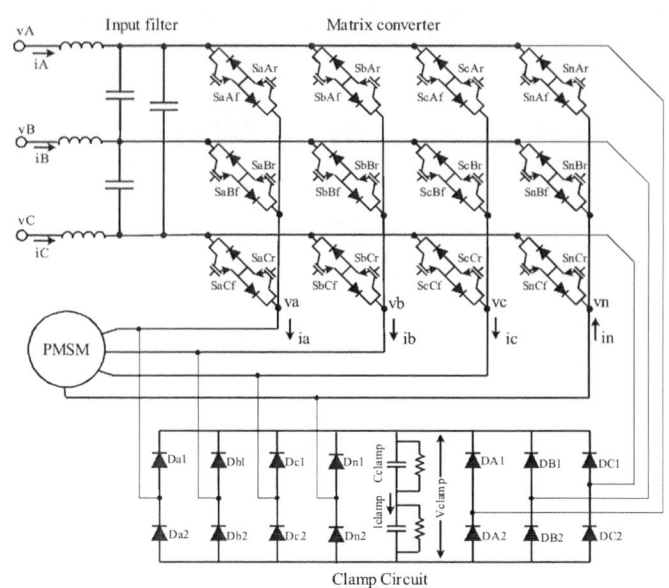

Fig. 7. Four phase fault tolerant motor drive system

Textbook reliability studies have shown however in [13] that this may not be the case. If a matrix converter using a typical bi-directional switch arrangement such as the one shown in figure 2(b) or (c) is assumed, 18 IGBT's and 18 diodes are needed together with associated gate driver circuitry, compared to the 6 IGBT's and 12 diodes of the rectifier-VSI drive and 12 diodes and IGBT's of the Back-to-Back VSI converter. The failure rate for the matrix converter is increased due to the increased number of devices but the voltage stress to which the matrix converter devices are submitted is much reduced compared to the VSI topologies. The IGBT blocking voltage would typically be a around 590V for the rectifier-VSI and 750V for the back to back VSI for a 400V line-line system, whereas, devices in a matrix converter would only be subjected to a half wave sinusoidal voltage with peak input line to line magnitude.

Theoretical studies are interesting but practical solutions are needed to convince the aircraft industry that the matrix converter is a viable option with regards to motor drive applications. Increasing the robustness or the drive systems ability to continue to operate, even under faulted conditions would be a key advantage. Several methods have been suggested in order to improve the availability of the drive circuit after a fault occurs. One potential solution is shown in figure 7[14].

The drive system is operated as a standard three phase to three phase converter under normal operating conditions. If a fault in the converter or motor occurs and drive to one of the motor phases is lost, the fourth phase in the matrix converter is activated and drives the neutral point of the motor. The modulation and control strategy of the converter are then changed to compensate for the fact that a two phase machine is now being driven with the aim of reducing resultant torque ripple and to continue the operation of the motor. Many strategies for remedial action have been discussed in the literature but few deal with actual detection of the faults within the converter.

Failures in the matrix converter circuit typically lead to either a device staying on when it should turn off, creating a short circuit of the input supply, or a device remaining off, creating an open circuit of the inductive load. If a short circuit is created, the drive should normally be disabled to prevent further damage.

The matrix converter can be protected from open circuits of the load using a diode bridge clamp circuit.

A. Clamp circuit

Since there are no free-wheeling paths within the matrix converter circuit for the inductive load current, the protection of the main power devices due to over-voltages needs to be addressed. Over-voltages can occur due to an open-circuit of the load.

The classical method of protecting the matrix converter is to use a diode bridge arrangement on the input and output of the converter[15]. This arrangement provides a path for the inductive load current. The size of the capacitor is typically small and can be calculated based on the load inductance, initial clamp voltage and final clamp voltage:

$$W_{Load} = \frac{1}{2}L(i_a^2 + i_b^2 + i_c^2) = \frac{3}{4}L\hat{I}^2 \qquad (2)$$

The maximum load energy is calculated in (1) where W_{Load} is the energy stored in the inductance of the motor, L is the equivalent line inductance of the motor and i_a, i_b, and i_c are the line currents of the motor. The change in energy of the clamp capacitor from its initial voltage to its final voltage can be used to calculate the size of capacitor and is shown in (2):

$$W_{Load} = \frac{1}{2}C_{Clamp}(V_{MAX}^2 - V_{INI}^2) \qquad (3)$$

where: C_{Clamp} is the capacitance of the clamp capacitor, V_{MAX} is the maximum allowable clamp voltage, V_{INI} is the initial clamp voltage

The size of the clamp capacitor can be minimized with the use of an active dissipation device[16][17]. An active clamp of this type can also improve the reliability and robustness of the matrix converter power circuit. In the event of failures which repeatedly generate an open circuit of the load, the energy transferred to the clamp circuit can be dissipated and the converter may continue to operate, albeit with a potentially poorer power quality.

B. Open-Circuit Failure Detection

An open circuit failure in the matrix converter could be caused by a number of different problems:

- IGBT Failure
- Corresponding diode failure
- IGBT / Diode bonding / packaging failure
- Gate drive failure
- Gate drive isolated power supply failure
- DSP - Gate drive communication failure

Methods which use measurements of the input and output voltages together with knowledge of the switching state of the matrix converter to determine failures down to the switch level have been reported[18][19]. Here, modulated error signals are generated from the information. Due to noise and sampling issues, these signals are integrated over several switching periods in order to determine where the problem has occurred. This technique takes time to determine the location of the failed device and three output voltage sensors, not normally necessary for the control of the converter, need to be added.

C. Use of the Clamp Circuit

The proposed, high speed fault diagnosis method uses the clamp circuit to detect faults. Since any inductive load current must flow through the clamp circuit during an open circuit failure, measurement of the current in the clamp circuit can be used to detect faults. Since positive and negative load current flows the same way through the clamp capacitor, absolute load current measurements are used in the calculations.

Error signals are calculated for each of the devices in the circuit. The switching state together with the clamp current and corresponding load current are used to determine the error signal. The error signal for a device will be large if it is gated and its corresponding load current is the same as the clamp current. Clearly, if the load current for a phase is flowing through the clamp circuit then the matrix converter devices that are supposed to be on, are faulty.

A 7.5kW, four phase matrix converter as illustrated in Figure 7 has been constructed to test the proposed method. Figure 8 shows the absolute load currents and the clamp current as sampled by the controller DSP. The drive system was deliberately run at low load since it is most difficult to determine a fault when all of the load currents are low and of similar magnitude. It can still be clearly seen that the magnitude of the clamp current corresponds to one of the absolute load currents.

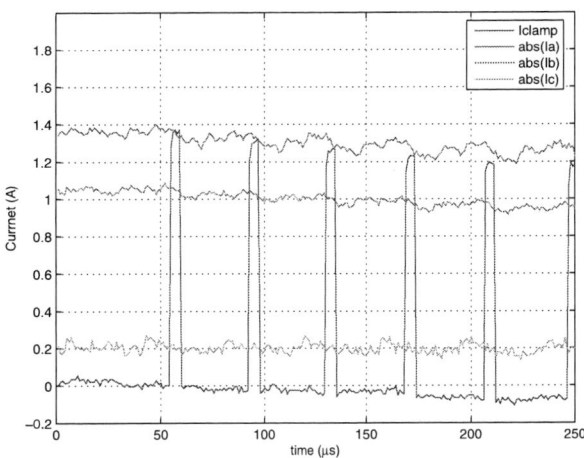

Fig. 8. Absolute measured load currents and clamp current during open circuit failure

The integrated error signals calculated for the matrix converter can be seen in Figure 9. The integrator is reset after every PWM period. It can be seen that the faulty device can be determined within a single controller PWM period. This method provides a fast localization of the faulted device. This will minimize the impact of the system when changing to a remedial control strategy.

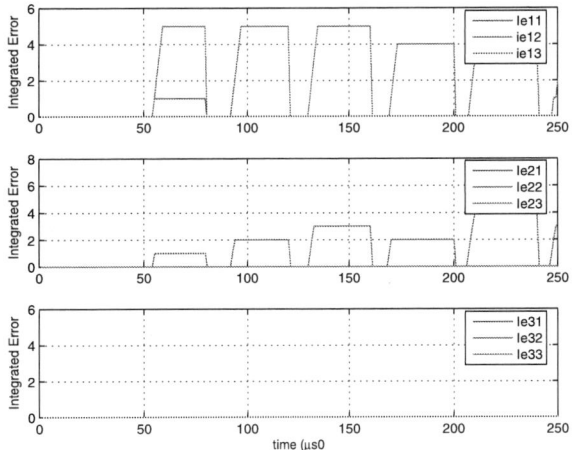

Fig. 9. Integrated error signals for each device in the converter

VI. CONCLUSIONS

The matrix converter offers a viable alternative to standard DC-link based VSI topologies for motor drive applications. It may not be the best option for all applications but the potential advantages of weight savings, sinusoidal input and output waveforms and controllable input power factor irrespective of the load are beneficial attributes when considered for more electric aircraft applications.

This paper has presented a high speed fault detection technique to enable the detection of open-circuit faults within the power circuit. Correct identification of which IGBT has failed to operate can be reliably performed within one PWM cycle of the converter. This leads to the minimum disturbance to the load and input power system when the converter changes to a remedial control strategy to compensate for the failure. Slower detection techniques may cause undue stress on the converter, load and power system while the detection process is performed. The addition of one extra current sensor within the existing clamp circuit is the only extra hardware that is needed to implement this technique.

The reported potential advantages of Silicon Carbide power devices may not all be practically realizable using present technology but the use of SiC schottky barrier diodes in place of standard Si diodes has been shown to reduce the losses associated with the commutation process. This can benefit the converter in two ways, either in an improved efficiency, or in an increase in switching frequency and hence a reduction of the passive component mass.

It can be seen from the results in figure 6 that an all silicon converter using the operating conditions previously discussed, switching at 8kHz has the same efficiency as a hybrid SiC-Si matrix converter operating at 19kHz. This could have a significant impact on the size and weight of the passive filter components.

REFERENCES

[1] L. Gyugi and B. Pelly, *Static power frequency changers: theory, performance and applications.* John Wiley and Sons, 1976.

[2] A. Daniels and D. Slattery, "New power convertor technique employing power transistors," *Electrical Engineers, Proceedings of the Institution of*, vol. 125, no. 2, pp. 146 –150, 1978.

[3] P. Wheeler, J. Rodriguez, J. Clare, L. Empringham, and A. Weinstein, "Matrix converters: a technology review," *Industrial Electronics, IEEE Transactions on*, vol. 49, pp. 276 –288, Apr. 2002.

[4] Holmes D.G and Lipo T.A, "Implementation of a Controlled Rectifier Using AC-AC Matrix Converter Theory," *IEEE Transactions on Power Electronics*, vol. 7, pp. 240 – 250, Jan. 1992.

[5] R. Sittig and F. Heinke, "Monolithic bidirectional switch, part i: Device concept," *Solid State Electronics*, vol. 44, pp. 13 87– I3 92, August 2000.

[6] F. Heinke and R. Sittig, "Monolithic bidirectional switch, part ii: Simulation of device characteristics," *Solid State Electronics*, vol. 44, pp. 1393–1398, August 2000.

[7] N. Burany, "Safe control of four-quadrant switches," in *Industry Applications Society Annual Meeting, 1989., Conference Record of the 1989 IEEE*, pp. 1190 –1194 vol.1, Oct. 1989.

[8] J. Oyama, T. Higuchi, E. Yamada, T. Koga, and T. Lipo, "New control strategy for matrix converter," in *Power Electronics Specialists Conference, 1989. PESC '89 Record., 20th Annual IEEE*, pp. 360 –367 vol.1, June 1989.

[9] C. Johnson, M. Rahimo, N. Wright, D. Hinchley, A. Horsfall, D. Morrison, and A. Knights, "Characterisation of 4h-sic schottky diodes for igbt applications," in *Industry Applications Conference, 2000. Conference Record of the 2000 IEEE*, 2000.

[10] S. Bernet, S. Ponnaluri, and R. Teichmann, "Design and loss comparison of matrix converters, and voltage-source converters for modern ac drives," *Industrial Electronics, IEEE Transactions on*, vol. 49, pp. 304 –314, Apr. 2002.

[11] M. Apap, J. Clare, P. Wheeler, M. Bland, and K. Bradley, "Comparison of losses in matrix converters and voltage source inverters," in *Matrix Converters, IEE Seminar on (Digest No. 2003/10100)*, pp. 4/1 – 4/6, 2003.

[12] L. Empringham, P. Wheeler, and J. Clare, "Power density improvement and robust commutation for a 100 kw si-sic matrix converter," in *Power Electronics and Applications, 2009. EPE '09. 13th European Conference on*, pp. 1 –8, 2009.

[13] P. Wheeler, J. Clare, L. de Lillo, K. Bradley, M. Aten, C. Whitley, and G. Towers, "A comparison of the reliability of a matrix converter and a controlled rectifier-inverter," in *Power Electronics and Applications, 2005 European Conference on*, pp. 7 pp. –P.7, 0-0 2005.

[14] S. Khwan-on, L. de Lillo, P. Wheeler, and L. Empringham, "Fault tolerant four-leg matrix converter drive topologies for aerospace applications," in *Industrial Electronics (ISIE), 2010 IEEE International Symposium on*, pp. 2166 –2171, 2010.

[15] P. Nielsen, F. Blaabjerg, and J. Pedersen, "New protection issues of a matrix converter: design considerations for adjustable-speed drives," *Industry Applications, IEEE Transactions on*, vol. 35, no. 5, pp. 1150 –1161, 1999.

[16] L. Empringham, L. de Lillo, P. Wheeler, and J. Clare, "Matrix converter protection for more electric aircraft applications," in *IEEE Industrial Electronics, IECON 2006 - 32nd Annual Conference on*, pp. 2564 – 2568, 2006.

[17] J. Andreu, J. De Diego, I. de Alegria, I. Kortabarria, J. Martin, and S. Ceballos, "New protection circuit for high-speed switching and start-up of a practical matrix converter," *Industrial Electronics, IEEE Transactions on*, vol. 55, no. 8, pp. 3100 –3114, 2008.

[18] S. M. A. Cruz, M. Ferreira, and A. J. M. Cardoso, "Output error voltages - a first method to detect and locate faults in matrix converters," in *Proc. 34th Annual Conf. of IEEE Industrial Electronics IECON 2008*, pp. 1319–1325, 2008.

[19] S. M. A. Cruz, M. Ferreira, and A. J. M. Cardoso, "Diagnosis of open-circuit faults in matrix converters," in *Proc. 13th European Conf. Power Electronics and Applications EPE '09*, pp. 1–11, 2009.

Power Electronic Grid-Interface for Renewable Ocean Wave Energy

Marian P. Kazmierkowski and Marek Jasiński
Institute of Control and Industrial Electronics
Warsaw University of Technology
Warsaw, POLAND
mpk@isep.pw.edu.pl, mja@isep.pw.edu.pl,

Abstract - **This paper presents information on ocean wave energy converters and power electronics grid-interface. In the introduction a basic terms and methods of ocean wave energy capture are discussed. Further several most important ocean wave energy conversion prototypes are briefly described. The generators and power electronics solutions for Power Take Off (PTO) system are presented on the example of Wave Dragon MW ocean wave energy converter. The Wave Dragon MW captures power from the ocean waves by means of low-head Kaplan turbines and converts it into rotating mechanical power. Problems which can appear in mechanical power to electrical power conversion in Wave Dragon MW can be expected to be similar as in wind turbine. However, subject of the mechanical energy conversion from ocean waves to electrical energy is not well identified and further research should be carried out. Additionally, specific problems of AC-DC-AC grid-interfacing converters under grid voltage distortions - including grid impedance estimation, higher harmonics and voltage dips compensation - are discussed. Some simulated and experimental oscillograms that illustrate properties of the presented systems are shown.**

I. INTRODUCTION

Recently, development of renewable energy sources (RES), including offshore wave energy, arises from the requirement to increase the security of supply, reduce greenhouse effect, emission of carbon dioxide and acid rain gases. The oceans cover 75% of the world surface and as such ocean energy is a global resource. There are different forms of renewable energy available in the oceans:

- waves,
- currents,
- thermal gradients,
- salinity gradients,
- the tides,
- and others.

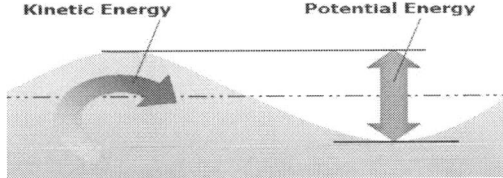

Fig. 1. Energy of sea waves [9]

Ways to exploit these high energy densities resources are being investigated worldwide. The power in a wave is proportional to the square of the amplitude and to the period of the motion (Fig. 1). Long period (~7-10 s), large amplitude (~2 m) waves exceeds energy fluxes about 40-50 kW per meter width of oncoming wave.

High ocean wave power resources are located along the Western Europe coast. Higher levels of wave power can be found only in the southern part of South America and in the Antipodes.

The available wave power resource in the area of north-eastern Atlantic including the North Ocean and in the Mediterranean achieves about 290 GW and 30 GW, respectively (Fig. 2). The potential of world wide ocean wave energy contribution is estimated to be about 10% (2 000 TWh/year) of the total energy consumption.

Among the main difficulties in wave power generations are:

- difficulty to obtain maximum efficiency because of irregularity in wave direction, amplitude and phase,
- problem of coupling irregular slow motion of wave to electrical generators,
- very high loading in the case of extreme weather conditions (hurricanes) over 100 times higher as average conditions.

However, among important advantages are:

- environmental compatibility,
- free of pollution energy conversion,
- low visual and acoustic impact.

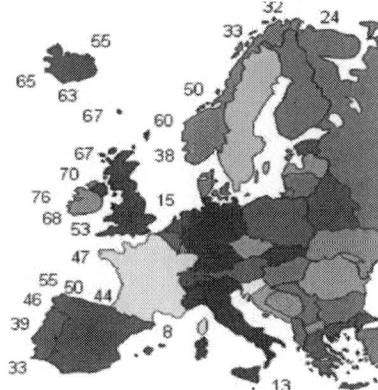

Fig. 2. Available sea wave power resources in Europe [9]

978-1-4244-8806-3/11 $26.00 © 2011 IEEE

II. OVERVIEW OF OCEAN WAVE ENERGY CONVERTERS

Recently, the ocean wave energy conversion has been developed strongly, particularly in EU countries, Canada, Japan, USA, India, China and others. In contrast to other RES the number of concepts for wave energy conversion is very wide. Although over 1000 wave energy conversion techniques are patented worldwide, the apparent large number of concepts for wave energy converters can be classified within a few basic types: Oscillating Water Columns, overtopping devices, heaving devices (floating or submerged), pitching devices and surging devices. Below some selected European pilot wave plant are listed [9].

A. Oscillating Water Columns (OWC)

The power plant is located at the island of Pico, on the Azores (Portugal). It is a shoreline OWC equipped with a 400 kW Wells turbine. This plant was built between 1995 and 1999, under the co-ordination of Instituto Superior Técnico, Lisboa (Portugal) and has been co-funded by the EC (*http://www.ist.utl.pt*). Currently an commercial project is developed.

Fig. 3. Prototype of 400 kW OWC plant at the Pico island (Portugal)

B. Pelamis

The Pelamis, developed by Ocean Power Delivery Ltd (OPD, UK), *http://www.oceanpd.com*, is a semi-submerged, articulated structure composed of cylindrical sections linked by hinged joints.

Fig. 4. Pelamis commercial scale prototype

The wave-induced motion of these joints is resisted by hydraulic rams, which pump high-pressure oil through hydraulic motors via smoothing accumulators. The hydraulic motors drive electrical generators to produce electricity.

C. Archimedes Wave Swing (AWS)

The AWS device has been developed by Teamwork Technology BV (Nederland), *http://www.waveswing.com*, the rights now owned by AWS Ocean Energy LTD (UK), *http://www.awsocean.com*, consists of a hollow, pressurized steel structure, the upper part of which is initiated to heave motions by the periodic changing of hydrostatic pressure beneath a wave. Being submerged, the device is characterized by low visual and acoustic impact. New model AWS II is under development.

Fig. 5. Prototype of 2 MW AWS energy converter

D. Wavebob

The Wavebob energy converter (Fig. 6) has been developed by Wavebob Ltd (UK), *http://www.clearpower.ie*. The device comprises a wave energy absorber and a hydraulic power take-off system driving synchronous alternators.

Fig. 6. Prototype of 500 kW Wavebob energy converter

E. Wave Dragon

The Wave Dragon (Fig. 7 and 8) is an offshore overtopping device developed by a group of companies led by Wave Dragon ApS (DK), *http://www.wavedragon.net*. It utilizes a patented wave reflector design to focus the wave towards a ramp and fill a higher-level reservoir. Electricity is produced by a set of low-head Kaplan turbines.

The development work is to a large extent built on the concept: use proven technologies when going offshore. The plant consists of two wave reflectors focusing the incoming waves towards a ramp, a reservoir for collecting the overtopping water and a number of hydro turbines for converting the pressure head into mechanical power. Generators connected with turbines convert mechanical

Fig. 7. Prototype of Wave Dragon in fjord Nissum Bredning [13]

energy into electrical power which is controlled by power electronics AC/DC/AC converters.

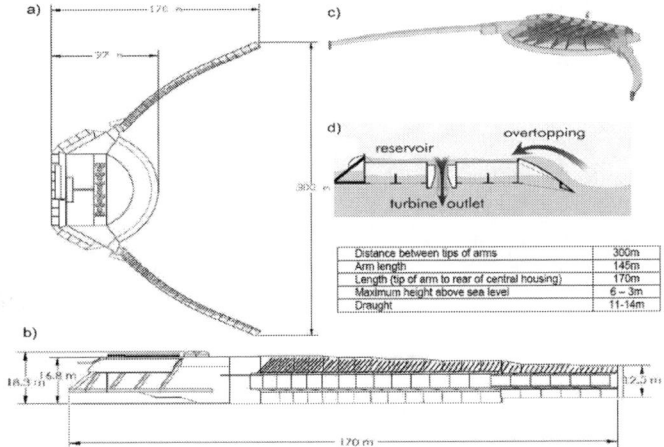

Fig. 8. Wave Dragon MW [1]

The utilization of the overtopping principle as opposed to power absorption via moving bodies means that the efficiency grows with the size of the converter. This means that only practical matters set limits for the size of this WEC. Additionally Wave Dragon due to its large size can act as a floating foundation for MW wind turbines, thus adding a very significant contribution to annual power production at a marginal cost [2]. Wave Dragon (WD) has been developed during the last nine years. Grid connected prototype presented in Fig. 7 (a scale 1:4.5 of a North Ocean production plant) of the WD MW is presently being tested in a Danish fjord *Nissum Bredning*. Every activity is focused on one goal: to produce electricity with the highest efficiency in the lowest possible costs – and in an environmental friendly and reliable way [2].

The only way to achieved this goal (from power electronics point of view) is choosing proper generating system and power electronics converter with robust and accuracy control methodology. Therefore, 16 parallel units of full controlled AC/DC/AC converters with PMSG are considered to be implemented (Fig. 9). However, well designed control schemes (estimators, controllers' parameters in control loops etc.) for

grid side AC-DC converter as well as for generator side DC-AC converter have a major impact on final parameters of the RES.

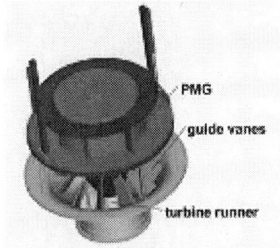

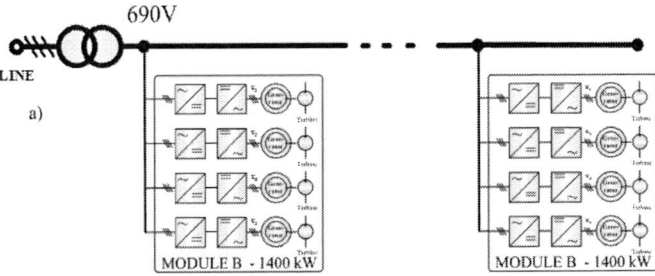

Fig. 9. The system is based on 16 PM generators coupled with low-head Kaplan turbines (upper view) and each generator is controlled by one dedicated converter [7]

III. DIRECT POWER AND TORQUE CONTROL OF AC/DC/AC CONVERTER WITH PERMANENT MAGNET SYNCHRONOUS GENERATOR

Developed by Warsaw University of Technology, Direct Power Control with Space Vector Modulation (DPC-SVM) and Direct Torque Control with Space Vector Modulation (DTC-SVM) [6] are very promising for control of an AC/DC/AC converter. When both algorithms are joined together for control of the AC/DC/AC converter connecting electrical machine and supply grid the *Direct Power and Torque Control with Space Vector Modulation* is obtained – *DPTC-SVM* [4,5].

A. Direct Torque Control with Space Vector Modulation – DTC-SVM

The DTC-SVM control scheme joins the switching table based DTC and FOC features in one control structure as in Fig. 10. The command electromagnetic torque M_{ec} of generator is delivered from outer PI speed controller. Then, M_{ec} and command stator flux Ψ_{Sc} amplitudes of generator are compared with estimated actual values of torque M_e and stator flux Ψ_S in generator. The torque e_M and flux e_ψ errors are fed to two PI controllers. The output signals are the command stator voltage components U_{Syc}, and U_{Sxc} in stator flux coordinate system respectively. Further, these signals are transformed into $\alpha\beta$ stationary coordinates using γ_{Ψ_S} flux angle position. Obtained stator voltage vector $\mathbf{U}_{Sc}$ is delivered to SVM which generates duty cycles vector $D_2(D_{2A}, D_{2B}, D_{2C})$ for the DC-AC converter generator side.

Proper design of the PI torque and flux controller parameters is given in [6].

978-1-4244-8806-3/11 $26.00 © 2011 IEEE

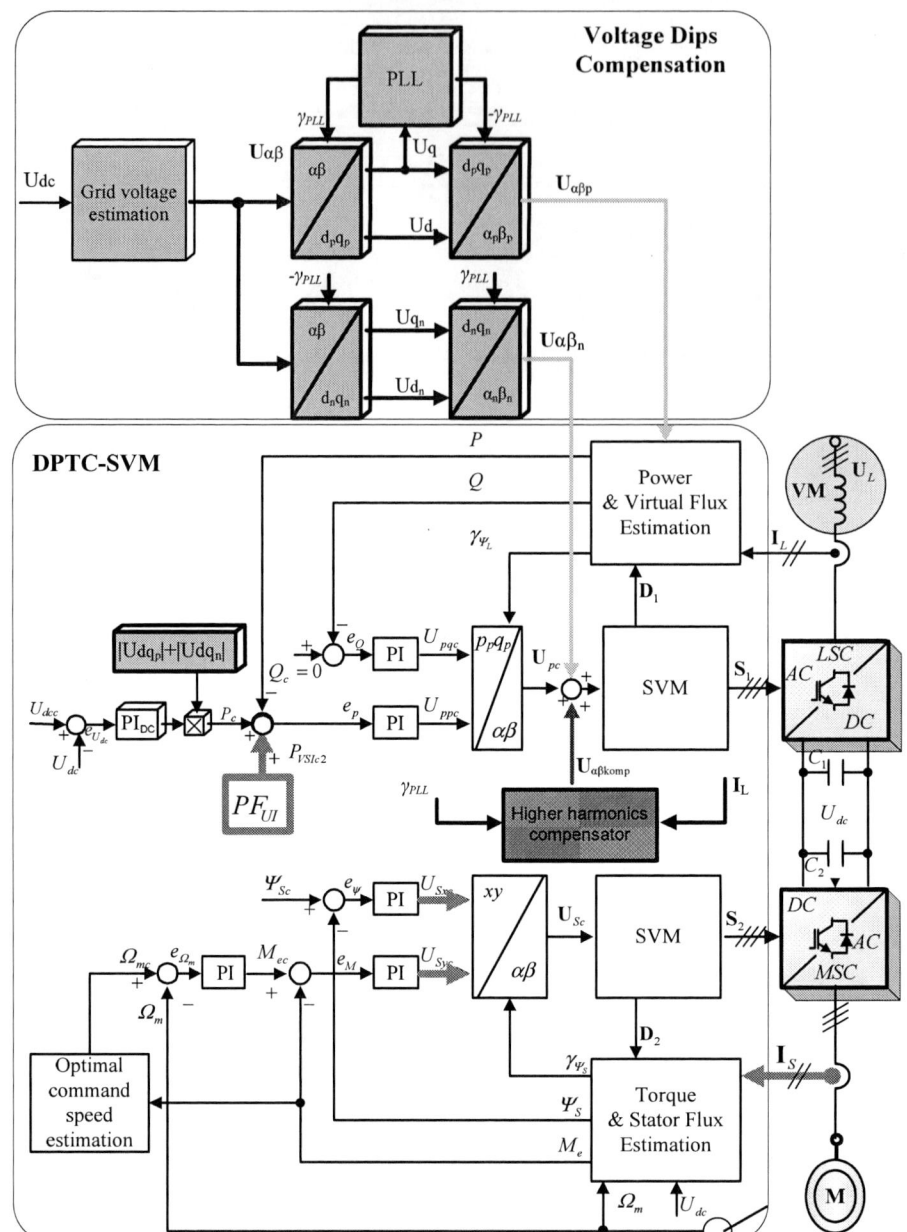

Fig. 10. Direct Power and Torque Control-Space Vector Modulated (DPTC-SVM) with active filter power feedforward (PF$_{UI}$), higher harmonics and voltage dips compensation. Where VM - virtual machine, SVM - space vector modulator.

B.   Direct Power Control with Space Vector Modulation – DPC-SVM

Direct power control with space vector modulation – DPC-SVM guarantees high dynamics and static performance via an internal power control loops. This method joins the concept of hysteresis based DPC and virtual flux (VF) oriented control (V-FOC) [6]. The DPC-SVM with constant switching frequency uses closed active and reactive power control loops (Fig. 10). The command active power P_c are generated by outer DC-link voltage controller, whereas command reactive power Q_c is set to zero for unity power factor operation.

These values are compared with the estimated P and Q values, respectively and calculated errors e_p and e_Q are delivered to PI power controllers. Voltages generated by power controllers are DC quantities, which after transformation to stationary $\alpha\beta$ coordinates, using γ_{Ψ_L} VF position angle, the voltages are delivered to SVM block. The proper design of the power controller parameters is very important especially in respect to line side power quality [7].

C.   Active power feedforward – PF

In spite of very good dynamics behaviors of DPTC-SVM scheme, the stabilization of the DC-link voltage can be improved. Therefore, active power feedforward – PF$_{UI}$ from generator side DC-AC converter to grid side AC-DC converter was introduced. The PF$_{UI}$ deliver information about machine states directly to active power control loop of the AC-DC. Thanks to faster control of power flow between generator and grid, the fluctuation of the DC-link voltages will be significantly reduced. So, the life time of the DC-link capacitors can be enlarged.

Estimation of this power is quite difficult, because the parameters of the machine and states of power switches are needed. Therefore, simplified power estimator based on command stator voltage U_{Sc} and actual stator current can be used [3]:

$$P_{LSC} = \frac{3}{2}\left(I_{Sx}U_{Sxc} + I_{Sy}U_{Syc}\right) \qquad (1)$$

D. Grid parameter estimation

In order to estimate the grid impedance, the method proposed in [1, 10] has been applied. Basically, the voltage drop U_{Grid} over the grid impedance is used to estimate its value. Therefore, a step in the grid power is required to eliminate the unknown actual grid voltage from the equations. Since the real power of the grid side converter controls the DC-Link voltage, the reactive power is increased to provide a current step without influencing the active power control. Note that this is not possible whenever the current rating of the inverter is already reach by active current, e.g. in full power operation.

As can be seen in Fig. 11, there is a voltage drop on the grid impedance denoted U_{Grid}. There is also a voltage drop on the known grid side inductor of the used LCL-filter titled U_{L1}. The measured Voltage U_{Meas} will therefore show a drop whenever the current is increased.

A three phase PLL is used to transfer the three phase voltages in a two phase synchronous reference frame, where U_q is always zero. The voltage and currents in synchronous coordinates dq are low pass filtered by 3rd order Chebychev filter to remove ripple which will distort and influence measurement accuracy. The values are measured twice at different grid current. The difference that results from this is used to calculate the grid impedance according to equations Eq. (2) and Eq. (3). This block scheme is shown in Fig. 12.

For the LCL filter parameters L_2=361 µH, L_1=288 µH, R_1=R_2=2 mOhms, C=84.9 µF. and the grid impedance L_G=192 µH and R_G=2mOhms. The step in reactive power is shown in Fig. 13.

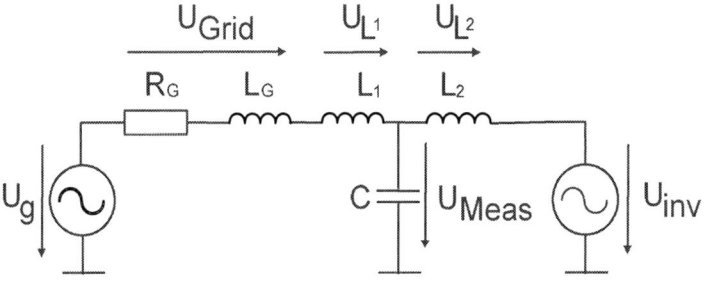

Fig. 11: Voltage drop on grid impedance U_{GRID} used for grid impedance estimation.

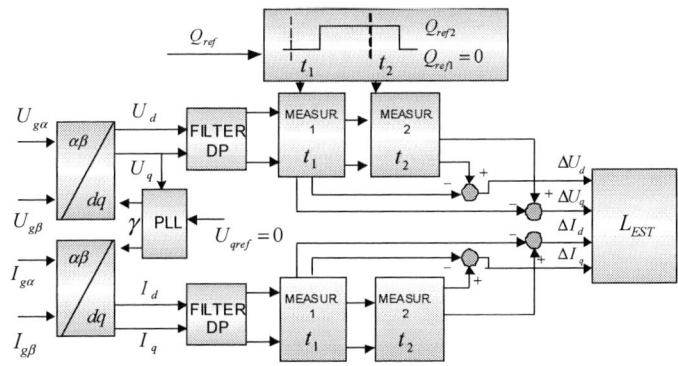

Fig. 12. Algorithm of grid impedance estimation according to E$_{qs}$. (2) and (3).

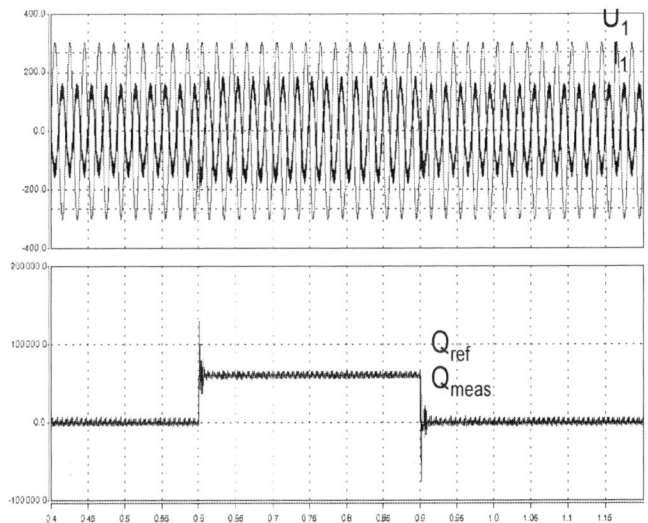

Fig. 13: Step in reactive power used for grid parameter estimation Q = 0 → 60 kVA → 0

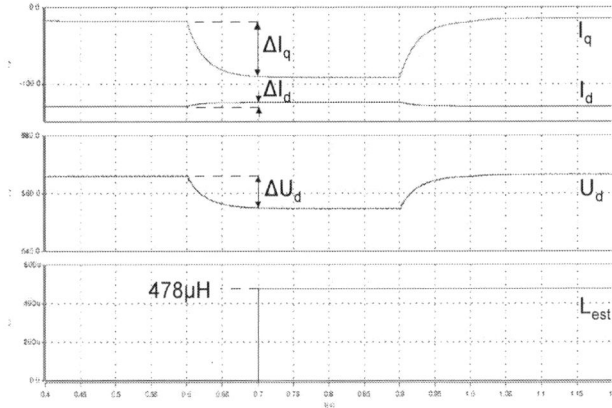

Fig. 14: Measured and filtred grid currents and voltages during step in reactive power used for grid parameter estimation

The measured and filtered currents and voltages are shown in Fig. 14. During test, the quadrature component of the measured voltage stays zero, due to the used PLL. The inductance of the grid is very well estimated to be L_{est}=478 µH, compared to L_G+L_1=480µH.

978-1-4244-8806-3/11 $26.00 © 2011 IEEE   461

V. VOLTAGE DIPS AND HARMONIC COMPENSATION

For voltage dips and harmonic compensation the control scheme has to be expanded as shown by dashed blocks. Typical test considers a 60% non-symmetrical voltage dip in phase "a". In Fig.15 operation in generating mode with 100ms dip is presented. It can be seen that after 70 ms from the sag beginning actual powers return to command values. Small 50W (3%) oscillation of active power are caused by little current unbalance in phase "b".

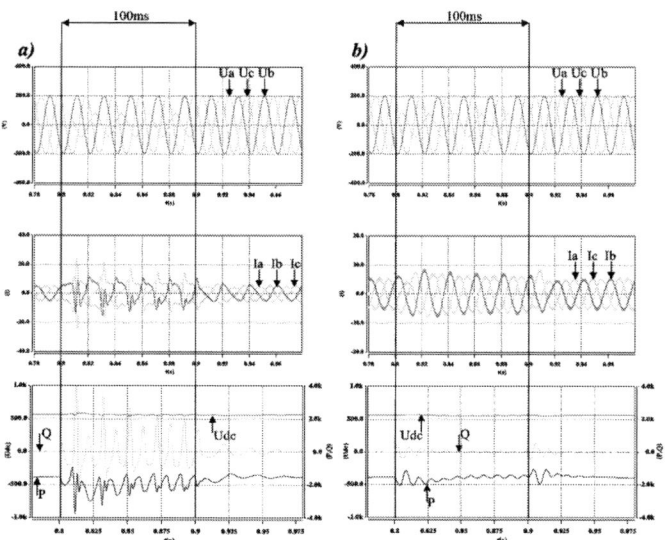

Fig. 15. Simulations results in generating mode: under 60% voltage sag in phase "a": a) for classical control structure, b) with grid voltage compensation

Experimental results for higher harmonic compensation are presented in oscilograms of Fig. 16.

Finally, in Fig. 17 grid phase currents, powers and DC voltage waveforms during voltage dip are presented: a) classical control – compensation is off (during 30% voltage dip), b) grid voltage dip (60%) compensation is on. As it can be observed compensation assures stable operation during voltage dip and power level is the same as before the dip.

IV. CONCLUSIONS

The paper has reviewed development of ocean wave energy converters with special focus on Wave Dragon MW device. Power train consists of 16 PM generators with AC-DC-AC grid interfacing converters. The AC-DC-AC grid interfacing converter system based on the Direct Power and Torque Control with Space Vector Modulation (DPTC-SVM) algorithm has been described. Also, problem of grid impedance estimation higher harmonic and voltage dips compensation has been illustrated with experimental oscilograms.

V. ACKNOWLEDGEMENT

Dr M. Jasinski is grateful for the support of the Centre of Advanced Study, Warsaw University of Technology.

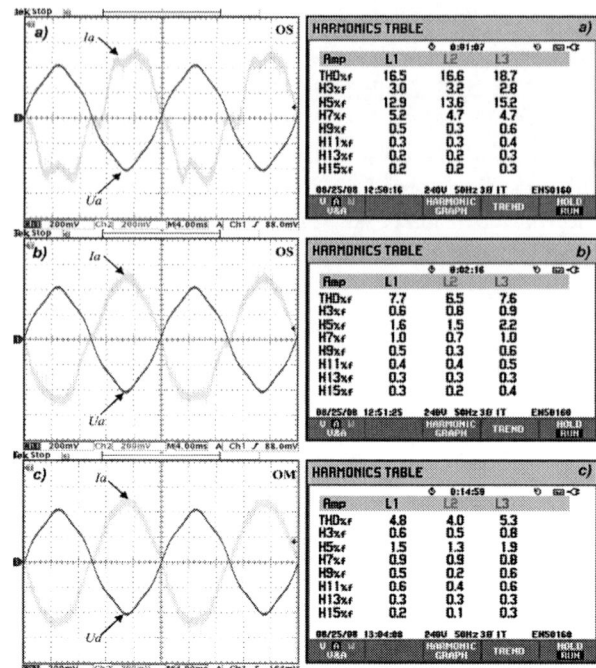

Fig. 16 Harmonics compensation process: a) before compensation, b) with compensation and power controllers tuned based on symmetry optimum, c) with compensation and power controllers tuned based on modulus optimum; I – 2A/div, U-100V/div

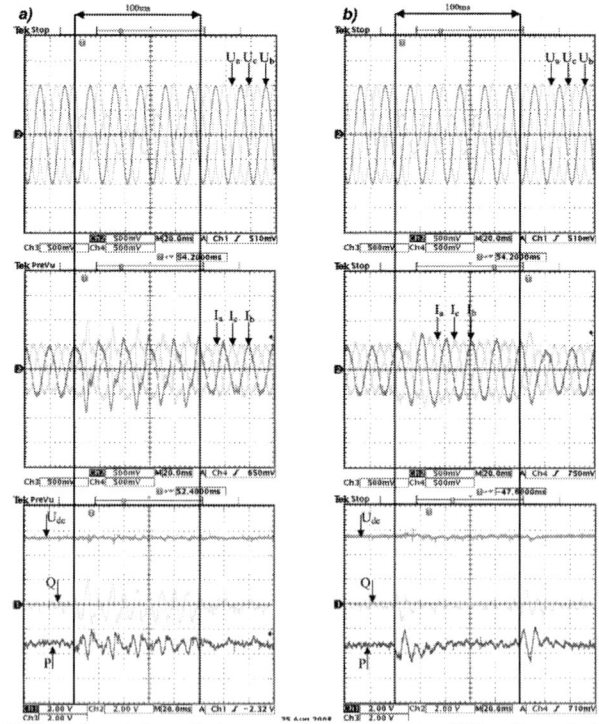

Fig. 17 Experimental results in generating mode under 60% voltage dip in phase "a": a) 30% for classical control, b) 60% for control with sag compensation I – 5A/div, U-100V/div, P,Q – 1kW/div, Udc-200V/div.

REFERENCES

[1] M. Citabaru, R. Teodorescu, P. Rodriguez, A. Timbus, F. Blaabjerg, "On-line Grid Impedance Estimation for single-phase grid connected systems using PQ variations" In Proc. IEEE-PESC '07, Orlando, Florida 2007

[2] L. Christiansen, E. Friis-Madsen, J. P. Kofoed, "Worlds Larges Wave Energy Project 2007 in Wales", *PowerGen 2006*.

[3] H. Hur, J. Jung and K. Nam, "A Fast Dynamics DC-link Power-Balancing Scheme for a PWM Converter-Inverter System," *IEEE Trans. on Industrial Electronics*, vol. 48, No. 4, August 2001, pp. 794–803.

[4] M. Jasinski, *Direct Power and Torque Control of AC/DC/AC Converter-Fed Induction Motor Drives*, Warsaw University of Technology, Ph.D. Thesis, Warsaw, Poland, 2005.

[5] M. Jasinski et. al: "Control of AC-DC-AC Converter for Multi MW Wave Dragon Offshore Energy Conversion System", in Proc. Of the IEEE – Intern. Symp. On Industrial Electronics ISIE'07, 2007, pp. 2685-2690.

[6] M. Kazmierkowski, R. Krishnan, and F. Blaabjerg, *Control in Power Electronics*, Academic Press, 2002, pg. 579.

[7] M. P. Kazmierkowski, M. Jasinski and H. Ch. Soerensen: „Ocean Waves Energy Converter – Wale Dragon MW", Electrotechnical Review, vol. 84, 2007, No. 2, 2008, pp.1-13.

[8] M. P. Kazmierkowski, M. Jasinski, „Power Electronics for Sea Wave Renewable Energy" International Conf. OPTIM 2010,

[9] *Ocean Energy Conversion in Europe*, Centre for Renewable Energy Sources (CRES), 2006

[10] K. Rothenhagen, M. Jasiński, M.P. Kazmierkowski, "Grid Connection of Multi-Megawatt Clean Wave Energy Power Plant under Weak Grid Condition", in Proc. of International Conf. EPE-PEMC, pp. 1927-1933, Poznan, Poland, 2008.

[11] D. Swierczynski, *Direct Torque Control with Space Vector Modulation (DTC-SVM) of Inverter-Fed Permanent Magnet Synchronous Motor Drive*, Warsaw University of Technology, Ph.D. Thesis, Warsaw, Poland, 2005.

[12] P. Vas, *Sensorless Vector and Direct Torque Control*, Oxford University Press.

[13] *www.wavedragon.net*

FPGA-Based Implementation by Direct Torque Control of a PMSM Machine

Badre BOSSOUFI[1)2)], Mohammed KARIM [1)], IONIȚĂ Silviu[2)], Ahmed LAGRIOUI[1)]

[1)] Laboratory of Data processing, Imagery and Analyzes Numerical (LIIAN)
Faculty of Sciences Dhar El Mahraz 1796 Fez-Atlas, Morocco
[2)] Center of Modeling and simulation of the systems, Faculty of Electronics, Communications and Computers,
University of PITEȘTI, Romania
Badre_isai@hotmail.com, karim_lessi@yahoo.fr, silviu.ionita@upit.ro, lagrioui71@gmail.com

Abstract -- In this paper, we present a new contribution of FPGAs (Field-Programmable Gate Array) for control of electrical machines. A detailed description of the structure of direct torque control for PMSM drive, a bench test was realized by a prototyping platform, the experimental results obtained show the effectiveness and the benefit of our contribution and the different stages of implementation for the control FPGA.

Key words: FPGA, Direct Torque Control (DTC), Permanent Magnet Synchronous Machine (PMSM), reusability, DSP, ADC.

I. INTRODUCTION

The speed performance of new components and the flexibility inherent of all programmable solutions give today many opportunities in the field of digital implementation for control systems. This is true for software solutions as microprocessor or *DSP* (Digital Signal Processor). However, specific programmable hardware technology such as Field Programmable Gate Array (*FPGA*) can also be considered as an especially appropriate solution in order to boost performances of controllers [1-3]. Indeed, these generic components combine low cost development, thanks to their re- configurability, use of convenient software tools and more and more significant integration density [4, 5].

The *FPGA* technology is now used by an increasing number of designers in various fields of application such as signal processing [6], telecommunication [7], video [8], embedded control systems [9], and electrical control systems [10]. This last domain, i.e. the studies of control of electrical machines, will be presented in this paper. Indeed, these components have already been used with success in many different applications such as Pulse Width Modulation (*PWM*) [11], control of induction machine drives [12-14] and multimachine system control. This is because the *FPGA*-based implementation of controllers can efficiently answer current and future challenges of this field.

The principal advantages of the digital solutions are as follows:

- High flexibility of changing structures of control;
- Immunity against disturbances;
- No problems of variations of control parameters.

With technological advancement, increased integration of *FPGA* devices is increasing. Nowadays, the density of *FPGA* components can achieve the equivalent of *10* million logic gates with switching frequencies of around *500 MHz*. This allows the implementation of complex algorithms control in their entirety with a small period of time to load.

The inherent parallelism of *FPGA* components offers the possibility to run several algorithms in parallel control and configure them according to the defined criteria. Dynamic configuration between the algorithms control has as objective to select the appropriate algorithms depending on your point of operation. It may be useful also to ensure continuous operation in case of faults (sensors, switches ...).

In this paper, a new contribution for the *FPGA-Based* implementation of controls electrical. This approach is based on concept modularity and reusability.

This paper presents the realization of a platform for *DTC* control of *PMSM* using *FPGA* based controller. This realization is especially aimed for future high performance applications. In this approach, not only the architecture corresponding to the control algorithm is studied, but also architecture and the ADC interface and *RS232 UART* architecture.

Considering the complexity of the diversity of the electric control devices of the machines, it is difficult to define with universal manner a general structure for such systems. However, by having a reflexion compared to the elements most commonly encountered in these systems, it is possible to define a general structure of an electric control device of machines which is show in *Fig.1*:

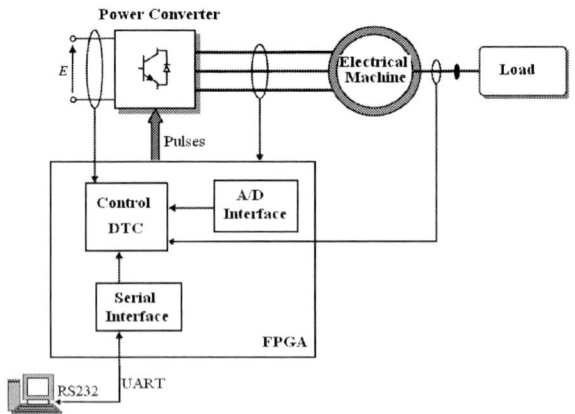

Fig.1. Architecture DTC Control

The concept of implementation is defined as being the introduction of a functionality given on a physical support. Within the framework of control of electric machines, the functionality to be introduced constitutes the algorithm of control, whose objective is to control the development state of mechanical or electric variables of the electric machine (flux, power, torque, speed...). As for the physical support, it constitutes the target of implementation. The latter can be of analogical or numerical nature.

According to the application, consider several algorithms of control can be used such as the current control algorithms, of active or reactive power, speed, position...etc. the structure of the these control algorithms generally comprises an internal loop of regulation of the current. Finally it is often the most difficult to implement because it generally constitutes the most complex part of the control algorithm.

The other control loops are relatively much simpler to implement. This is why, within the framework of this work, one will particularly be interested in the current control technique implementation of the electric machines "Direct Torque Control (*DTC*)"

II. DIRECT TORQUE CONTROL STRUCTURE

In this paper, we apply the command on a machine type *PMSM* (Permanent Magnet Synchronous Motor), which consists of three stator windings and a rotor magnet. This motor is described by the following equation (Voltage, Flux, Torque...) [2],

$$u_{sd} = r_s.i_{sd} + \frac{d\Phi_{sd}}{dt} - \omega.\Phi_{sq} \tag{1}$$

$$u_{sq} = r_s.i_{sq} + \frac{d\Phi_{sq}}{dt} - \omega.\Phi_{sd} \tag{2}$$

$$\Phi_{sd} = L_d.i_{sd} + \Phi_f \tag{3}$$

$$\Phi_{sq} = L_q.i_{sq} \tag{4}$$

$$Ce = \frac{3}{2} p \left[\phi_f.I_q + (L_d - L_q).I_d.I_q \right] \tag{5}$$

$$Ce - C_r = J.\frac{d\Omega}{dt} + f.\Omega \tag{6}$$

Where Ω is the rotation's speed, p the Number of pairs of poles, J the moment of inertia, f the Coefficient of viscous friction, C_r the resistive torque, Φ_f the flux produced by the permanent magnet, L_{sd} and L_{sq} the d-q axis stator inductance, V_{sd} and V_{sq} the d-q axis stator voltage, r_s the stator winding resistance and C_e the electromagnetic torque,

In the middle of 1980's Depenbrock and Isao Takahashi proposed Direct Torque Control for electrical machines [9, 11], more than one decade later. Idea the *DTC* basic for electrical motor is slip control [2]. In the 1990's, *DTC* for PMSM was developed [7, 12, 13]. The DTC control a magnet synchronous motor Standing is based on the direct determination of sequence of commands applied to switches a

voltage inverter. This strategy is based generally on the use of comparators hysteresis whose role is to control the amplitudes stator flux and electromagnetic torque, as in [8, 14]. The *Fig.2* is a typical *DTC* system. It includes estimators flux and torque electromagnetic, flux and torque hysteresis controllers and a switching table.

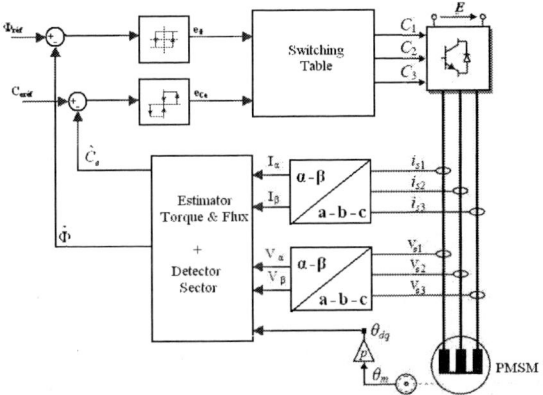

Fig.2. DTC Control Principle

The *Fig.2* presented the chosen control. The characteristics memorized for an optimal torque control strategy generates the d-q component of the vector stator current. Then, an *abc-to-αβ* transformation generates the flux and torque electromagnetic. Finally, these generated parameters are compared to the references and hysteresis controllers to allow the determination of the switching table status C_1, C_2 and C_3 of the voltage inverter.

III. DEVELOPMENT OF THE IMPLEMENTATION
A. FPGA devises

There are several manufacturers of *FPGA* components such: Actel, Xilinx and Altera...etc. These manufacturers use different technologies for the implementation of *FPGAs*. These technologies are attractive because they provide reconfigurable structure that is the most interesting because they allow great flexibility in design.

Nowadays, *FPGAs* offer the possibility to use dedicated blocks such as *RAMs*, multipliers wired interfaces *PCI* and *CPU* cores.

The architecture designing was done using with *CAD* tools. The description is made graphically or via a hardware description language high level, also called *HDL* (Hardware Description Language). Is commonly used language *VHDL* and Verilog. These two languages are standardized and provide the description with different levels, and especially the advantage of being portable and compatible with all *FPGA* technologies previously introduced.

The *Fig.3* summarizes the different steps of programming an FPGA. The synthesizer generated with *CAD* tools first one Netlist which describes the connectivity of the architecture. Then the placement-routing optimally place components and performs all the routing between different logic. These two

978-1-4244-8806-3/11 $26.00 © 2011 IEEE

steps are used to generate a configuration file to be downloaded into the memory of the *FPGA*. This file is called bitstream. It can be directly loaded into *FPGA* from a host computer.

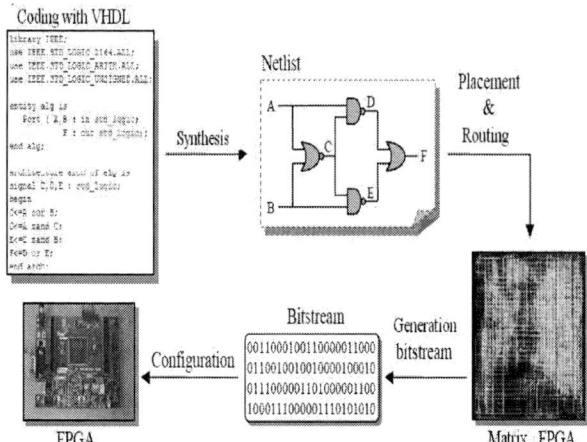

Fig.3. Programming FPGA devisees

In this paper an *FPGA XC3S500E* Spartan3E from Xilinx is used. This *FPGA* contains 400,000 logic gates and includes an internal oscillator which issuer a 50MHz frequency clock. The map is composed from a matrix of 5376 slices linked together by programmable connections (*Fig.4*).

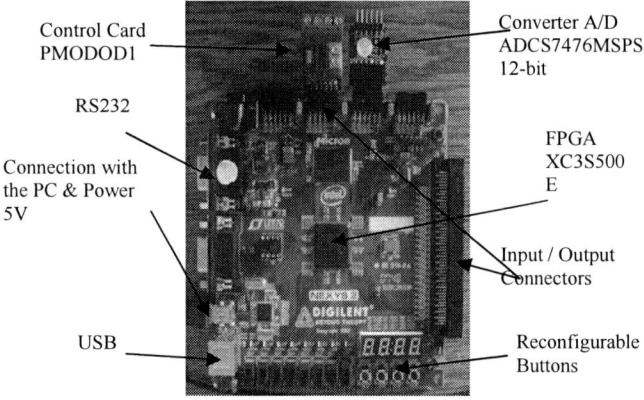

Fig.4. The FPGA XC3S500E Spartan

B. Simulation Procedure

The simulation procedure begins by verifying the functionality of the control algorithm by trailding a functional model using Simulink's System Generator for Xilinx blocks. For this application, the functional model consists in a Simulink timeis discretired model of the *DTC* algorithm associated with a voltage inverter and PMSM model. The *Fig.5* gives a global view of the functional model.

The *Fig.5* shows in detail the programming of the control shown in *Fig.2* (*DTC* Control) in the S*YSTEM GENERATOR* environment from Xilinx; we will implement it later in the memory of the *FPGA* for the simulation of *PMSM*.

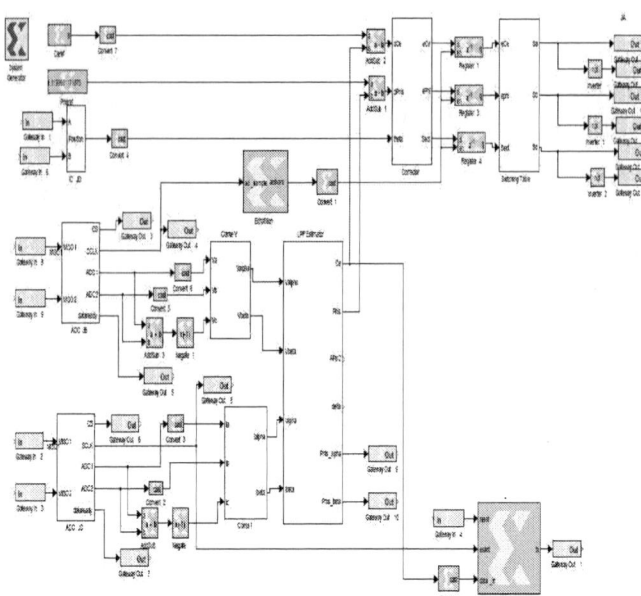

Fig.5. Functional Model DTC from SYSTEM GENERATOR

The description of the different modules is detailed below.

- Two blocks of coordinates transformation: the Clark transformation (*abc-to-αβ*);

- The block switching table is the most important, because it can provide control pulses to the *IGBT* voltage inverter in the power section from well-regulated voltages.

- The block estimator torque and flux, corrector hysteresis and detector sector.

- The block encoder interface *IC* allows the adaptation between the *FPGA* and the acquisition board to iniquity the rotor position of the *PMSM*;

- The *ADC* interface allows the connection between the *FPGA* and the analog-digital converter (*ADCS7476MSPS 12-bit A / D*) that will be bound by the following two Hall Effect transducers for the acquisition of the stator currents machine;

- Block "Timing" which controls the beginning and the end of each block, which allows the refresh in the voltages reference V_{10}, V_{20} and V_{30} at the beginning of each sampling period;

- The *RS232* block allows signal timing and recovery of signals viewed, created by another program on Matlab & Simulink to visualize the desired output signal.

The second step of the simulation is the determination of the suitable sampling period and fixed point format. The *Fig.6* gives the specification model of the *abc-to-αβ* (Clark) transformation.

For example, we present the construction of Block Clark's Transformation in the system generator environment from Xilinx, which is characterized by the following system (7):

978-1-4244-8806-3/11 $26.00 © 2011 IEEE

$$\begin{bmatrix} V_\alpha \\ V_\beta \end{bmatrix} = \frac{2}{3} \begin{bmatrix} 1 & -\dfrac{1}{2} & -\dfrac{1}{2} \\ 0 & \dfrac{\sqrt{3}}{2} & -\dfrac{\sqrt{3}}{2} \end{bmatrix} \begin{bmatrix} V_a \\ V_b \\ V_c \end{bmatrix} \qquad (7)$$

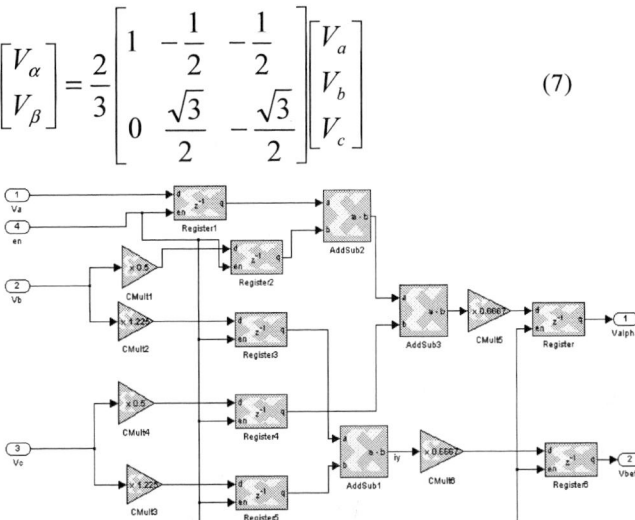

Fig.6. Clark Transformation Model

The specification model is then used for the definition of the corresponding Data Flow Graph *DFG*. *Fig.7* showed the *DFG* corresponding to the Clark transformation module.

C. Optimization of the consumed resources

The Algorithm Adequation Architecture (*A3*) ensures this application, in order to generate a new graph called Factorized *FDFG* (Data Flow Graph). This method is used to generate optimized hardware architecture for each module. If an operator is used several recovery, its factorization via the *A3* methodology consists in using it only one time. *Fig.8* showed the *FDFG* of the Clark transformation module.

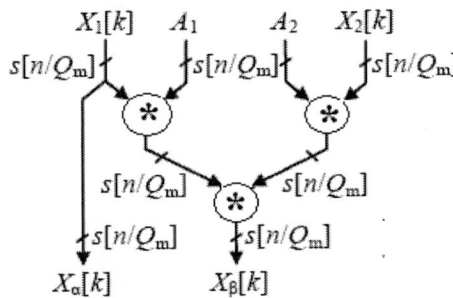

Fig.7. DFG of the Clark Transformation

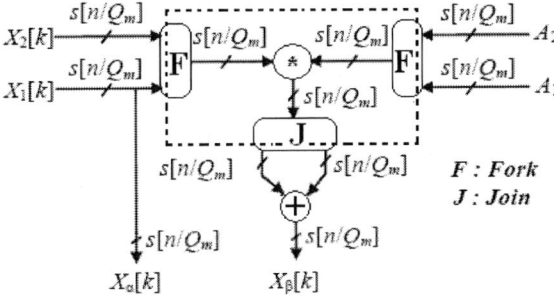

Fig.8. FDFG of the Clark Transformation

D. Modular design of the control architecture

To design the control architecture to implement a datapath and control unit is defined for each module as shown in *Fig.9*. The data path is mainly made up of operators and registers. The registers can provide the technical Pipeline which consists on decomposing the performance of an algorithm in different sequences according to the data dependency. The communication between the registers and operators can be achieved through data buses, multiplexers and demultiplexers. Concerning the control unit, its role is to ensure the flow of data according to very specific sequences within the data path. The control unit is other than a finite state machine (*FSM*: Finate State Machine). *Fig.9* and *Fig.10* present two different architectures *DFG* and *FDFG* processing module Clark.

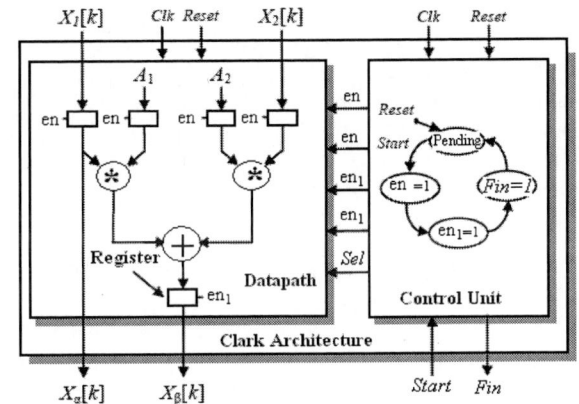

Fig.9. DFG Clark transformation module architecture

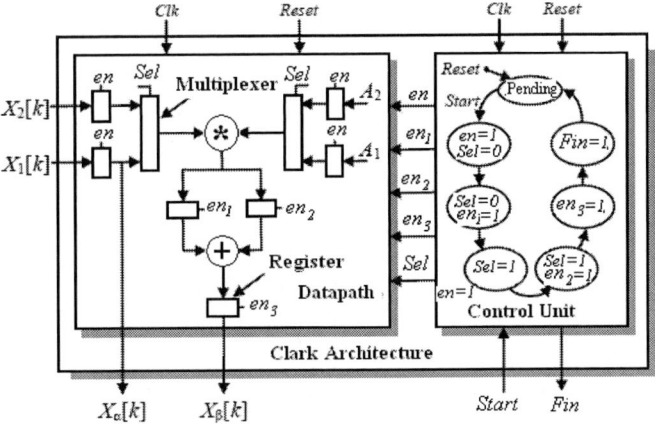

Fig.10. FDFG Clark transformation module architecture

IV. EXPERIMENTAL SET-UP

For this work, the used *FPGA* target is a *XC3S500E Spartan3E* the firm Xilinx, the *FPGA* based hardware control system includes the *DTC*, an *ADC* interface and *IC* interface in one *FPGA* chip. *Fig.11* presents the corresponding implemented architecture.

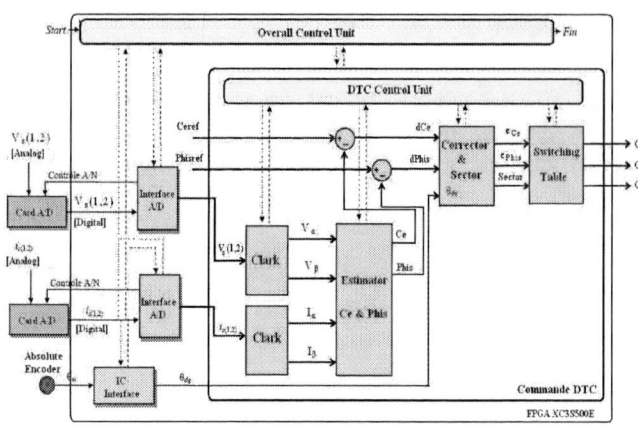

Fig.11. FPGA based hardware DTC

The control unit for architecture ensures implanted control module of the *ADC* interface, the encoder interface and the command encoder *DTC*. The *Fig.12* shows the timing diagram for the mode of operation of these four modules. At the beginning of each sampling period, the *A/D* interface module and encoder interface are activated simultaneously. Then, after a delay conversion from analog to digital $t_{A/D}$, the control module *DTC* is activated. This is driven by its own control unit and allows you to refresh the reference torque and flux after a computation time equal to $t_c r$. Once activated, the control unit control module *DTC* activates the first two modules of the transformation of Clark (*123-αβ*) that will calculate the components $i_{s\alpha}$ and $i_{s\beta}$ stator current vector and the components $V_{s\alpha}$ and $V_{s\beta}$ stator voltage vector. The computation time of this transformation is equal to t_C. When the modules of this transformation indicate the end of the calculation, the estimator module flow and electromagnetic torque is activated. It has the same computation time equal to test and calculates the flux and torque estimated.

Subsequently, when the estimator module indicates the end of the calculation, the correction module of hysteresis and the detector area is activated to calculate the error of the torque and stator flux and the area *N*. This module is characterized by a computation time equal to t_{cr}. Finally, when the errors of flux and torque and sector *N* are calculated, the module of the switching table is activated. The latter has a running time equal to t_{table} and generates the control signals C_1, C_2 and C_3 for controlling the switches of the voltage inverter.

The following table shows the performance of computing time and resource consumption, obtained during the implementation of the control *DTC* architecture. The resources consumed are obtained for a fixed point format *20/Q18*. The total computing time of t_{DTC}, in the *DTC* squeal module is equal to 0.99μs. By adding the analog to digital conversion time $t_{A/D}$, total time T_{ex} architecture equals 3.43μs.

TABLE I
FPGA Performances

Module	Latency	Time Calculation
Interface A/D	120	$t_{A/D}=2.4$ μs
Interface A/D	120	$t_{A/D}=2.4$ μs
IC Interface	2	$t_{Cod}=0.04$ μs
123-αβ	18	$t_C = 0.34$ μs
Estimator Flux & Torque	12	$t_{Est} = 0.25$
Corrector & Sector	14	$t_{Cr} = 0.22$
Switching Table	8	$t_{table} = 0.18$μs
$t_{DTC} = t_C + t_{Est} + t_{Cr} + t_{table}$		$t_{DTC} = 0.99$ μs
Run time $T_{ex}=T_{A/D} + t_{DTC}$		$t_{ex} =3.43$ μs
Resources Consumed	Number of Slices	1850 de 5376 (34%)
	Wired Multipliers	12 de 16 (65%)
	Memory RAM	6%

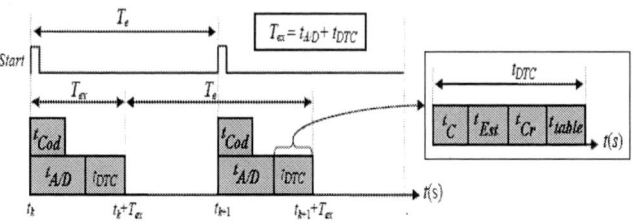

Fig.12. Time diagram of the control structure DTC

To test the *FPGA* based controller, a prototyping platform for the control of a Permanent magnet Synchronous Machine was assembled.

Fig.13. Prototyping platform control

The *Fig.14* shows the experimental results obtained during the implementation of the *DTC* control, it presents the state of control signals for the switches of the inverter voltage in the area in which there is the reference voltage vector. These results are similar to those presented in the theories. Furthermore, the control signals generated from the *FPGA* board will be filtered before being injected into the voltage inverter.

The above figures show that the phases are balanced and demonstrate the proper functioning of the Switching table.

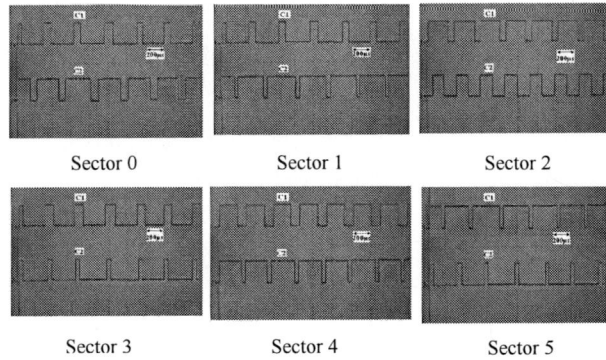

| Sector 0 | Sector 1 | Sector 2 |

| Sector 3 | Sector 4 | Sector 5 |

Fig.14. Switching states of control signals C_1 and C_2 for the DTC Control

The *Fig.14* and *Fig.12* show that control system satisfy the basic requirements of the control strategy and validate therefore the good functionality of the system. In fact, It can be noted that:

- The switching frequency is limited to the sampling frequency of the control algorithm to guarantee safe operation of the semiconductor power devices.

- The switching frequency increases weakly when the stator current vector magnitude decreases.

Implementing the *DTC* control in *FPGA* has the drawbacks:

- The switching frequency is variable. It is limited to half the sampling frequency of the control algorithm and maximum at very low speed.

- The zero voltage vectors are not applied.

At the hardware level, the execution time of the control architecture is of the order of several microseconds, which allows a better control of current, including a *THD* lower.

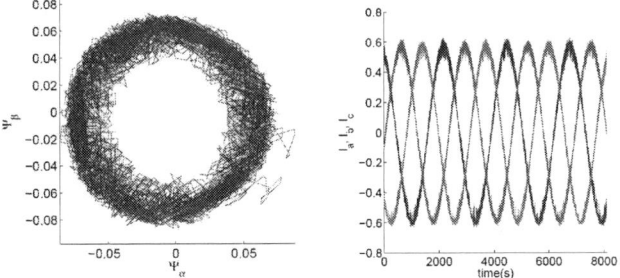

Fig.15. d-axis and q-axis current in the PMSM and stator flux locus for DTC PMSM

In Fig.15 the experimental results *DTC* of *PMSM* with the *FPGA* platform are shown. Clearly the constant set values for stator flux linkage magnitude and torque result, because of the hysteresis controllers in the *DTC* scheme, in d-axis and q-axis current components that are bounded within hysteresis limits. The hysteresis control is visible in the stator flux locus plot as well. Update frequency for this implementation is 20 kHz. All results were extracted from the *FPGA* by the ChipScope tool of Xilinx.

CONCLUSION

In this work, we presented the *FPGA* implementation of Direct Torque Control architecture for permanent magnet synchronous machine (*PMSM*). The results obtained show the benefits of an appropriate methodology that allows the creation of a library of reusable modules optimized. This work provides a new benefit to our laboratory work and a new route for the development and implementation of other control architectures.

ACKNOWLEDGMENT

We thank all those who contributed to make this work, including my teachers in the Center of Modeling and simulation of the systems, University of Pitesti from Romania and in my home laboratory *LIIAN* from Morocco, and all my friends for their support

REFERENCES

[1] M.W. NAOUAR, « Commande numérique à base de composants FPGA d'une machine synchrone », algorithmes de contrôle du courant, *Thèse de Doctorat, Ecole Nationale d'Ingénieurs de Tunis et l'Université de Cergy Pontoise*, Tunis, 2007.

[2] B.BOSSOUFI, M.KARIM, S.IONITA, A.LAGRIOUI, "Performance Analysis of Direct Torque Control (DTC) for Synchronous Machine Permanent Magnet (PMSM) " *in Proc. IEEE-SIITME'2010*, pp. 275-280, 23-26 Sep 2010, Pitesti, Romania.

[3] NAOUAR M.-W. ; MONMASSON E.; SLANW-BELKHODIA I. "FPGA-based torque controller of a synchronous machine" *in Proc. IEEE-ICIT'04*, pp. 8-10, Hammamet, Tunisia, Dec. 2004.

[4] E. MONMASSON, and M. Cirstea "FPGA Design Methodology for Industrial Control Systems – A Review," *IEEE Trans Ind. Electron..*, vol.54, no. 4, pp.1824-1842, August. 2007.

[5] Y.A. Chapuis, J. P. Blonde, and F. Braun, "FPGA Implementation by Modular Design Reuse mode to Optimize Hardware Architecture and performance of AC Motor Controller Algorithm," *in Proc. EPE-PEMC Conf.*, Sept. 2004.

[6] L. Charaabi, "Conception des architectures matérielles dédiées à la commande de systèmes électriques", *Thèse de Doctorat, Ecole Nationale d'Ingénieurs de Tunis*, 2006.

[7] Y. Kebbati, Y.A. Chapuis and F. Braun, "Reuse methodology in FPGA/ASIC Digital Integration Solution for Vector Control of Motor Drives", *in Proceeding of IEEE International Symposium on Signal, Circuit and Systems (SCS'2001)*, PP.333-33-, Romania, 2001.

[8] X. Lin-Shi, F. Morel, A. M. Llor, B. Allard and J.M. Retif "Implementation of Hybrid Control for Motor Drives," *IEEE Trand. Ind Electron.*, vol.54, no.4, pp.19446-1952, August. 2007.

[9] H. T. moon, H. S. Kim and M. J. Youn, "A Discret Time Predictive Current Control for PMSM" *IEEE Trans. Power Electronic.*, vol.18, no.1, pp. 464-472, Janvier. 2003.

[10] H.J. Lee; S.K. Kim; Y.A. Kwon; S.J. Kim, "ASIC design for DTC based speed control of Induction motor," *in Proc. IEEE ISIE'01 Conf.*, 2001, pp. 956 –961

[11] F. Ricci, H. Le-Huy, "An FPGA-based rapid prototyping platform for variable-speed drives," *in Proc. IEEE IECON'02 Conf.*, 2002, pp. 1156- 1161.

[12] K. Tazi, E. Monmasson and J-P. Louis, "The use of FPGAs to build an AC drive specific hardware- software library. The case of Park's transformation," *in Proc. EPE-PEMC'2000 Conf.*, 2000, vol. 7 pp.181-185.

[13] E. Monmasson; B. Robyns; E. Mendes, B. De Fornel, "Dynamic reconfiguration of control and estimation algorithms for induction motor drives," *in Proc. IEEE ISIE Conf.*, 2002, pp. 828 –833.

[14] Takahashi and T. Noguchi, "A new quick-response and high-efficiency control strategy of an induction machine," *in IEEE Trans. Ind. Applicat.*, vol. 22, pp. 820-827, Sept./Oct. 1986.

Magnetic Sensor Coil Shape Geometry and Bandwidth Assessment

Lauri Kütt[1], Muhammad Shafiq[2]
[1]Tallinn University of Technology, Estonia
[2]Aalto University School of Electrical Engineering, Finland
lauri.kutt@ttu.ee; muhammad.shafiq@aalto.fi

Abstract- **There are quite many scientific examples out there regarding on-line measurements of fast pulses and fast partial discharge phenomena [1] [2] [3] using magnetic sensor coils. In these examples, most attention has been paid on the results obtained with specific design of magnetic sensor. There is though a lack of actual design consideration of the sensor regarding its target bandwidth and sensitivity. In current paper the focus is put on the design aspects of such a magnetic loop sensor. Geometrical shape of the magnetic sensor loop is considered and coils with different shapes are put into tests to determine their real parameters.**

I. INTRODUCTION

Signal processing systems capabilities are increasing and this opens up new possibilities for performing systems diagnostics are emerging which previously have been not available for on-line analysis. These are provided by observing different phenomena occurring at higher speeds and shorter durations. Such new methods include for example partial discharge measurements and isolation diagnostics on electric motors, electric power transmission, switchgear etc. Capabilities of such diagnostics system could be traced to two aspects – speed of sensor and speed of processing unit. Today, processing units speed is basically as fast as required, as new processors are getting ever faster and cheaper. This means that sensor speed becomes essential limiting factor.

Basically a magnetic field sensor for detecting very fast events has two requirements – it has to have as wide bandwidth as possible and it has to have as high sensitivity as possible. Also for a practical system for on-line measurements means ability to operate in harsh environments. In addition, the price of such a sensor should be as small as possible.

Comparing different technologies available, there is rather simple solution. The magnetic loop current sensor is very cheap and robust device, having also quite high bandwidth. In this paper focus is put on developing a faster sensor, which would allow detecting very fast pulses and short high-speed events on energized wire. This sensor application is only for the diagnostics, while RMS mains frequency values are not observed at all. Different shapes of sensors are discussed in relation to sensitivity and its possible bandwidth.

II. PRINCIPLES

Operation of magnetic field sensor relies on the Biot-Savart law, which sets a relation between magnetic field around an object and distance of observation from the object [4].

$$\vec{H} = \frac{1}{4\pi} \int_a^b \frac{I \cdot dl' \times \vec{R}}{4\pi \left| R \right|^2}, \tag{1}$$

where H – magnetic field strength,
I – current through conductor length dl',
R – vector from point of dl' to observation point.

Magnetic field strength around the wire can be expressed as

$$H = \frac{I}{2\pi r}, \tag{2}$$

where r – distance of observation point from wire.

Flux density around the wire can be thus expressed as

$$B = \mu \cdot H = \frac{\mu \cdot I}{2\pi r}, \tag{3}$$

where μ - relative magnetic permeability.

When such flux density is passing through some area, the area holds magnetic flux of

$$\Phi = \int_A \vec{B} \cdot \vec{da}, \tag{4}$$

where da – normal vector of the smallest piece of area A.

Considering loop with n turns of wire area, sum of flux through all the loops is considered flux linkage and it is the sum of flux through all turns

$$\psi = \sum_{i=1}^{n} \Phi_i, \tag{5}$$

where n – number of loops in coil.

When all turns of wire have same orientation towards the magnetic field and same area, this can be simplified to

$$\psi = n \cdot \Phi. \tag{6}$$

Effective output of such magnetic loop is actually brought by Faraday's law

$$e = -\frac{d\psi}{dt}. \tag{7}$$

978-1-4244-8806-3/11 $26.00 © 2011 IEEE

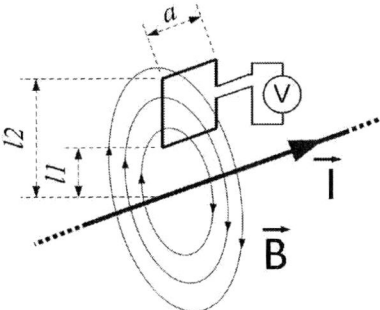

Fig. 1. Basic current measurement set-up with loop near the current-carrying wire.

Application of using a small rectangular loop to sense magnetic field has been presented in figure below. Loop is rectangular-shaped, its lower edge being at a height of l_1 and its higher edge being at l_2 distance from the wire.

Considering simple, single-turn loop, it is visible, that the flux density at distance l_1 is significantly higher than flux density at distance l_2. To sum the flux density along the loop, integration has to be carried out. This case is simple because of rectangular loop.

The integration will start at distance l_1 and end at distance l_2. As the distance l is changing, so is also the magnetic flux density, with the relation of $1/l$.

$$\Phi = \int_A B \cdot da = \frac{\mu_0 I}{2\pi} \int_{l_1}^{l_2} \frac{1}{l} a \, dl = \frac{\mu_0 a I}{2\pi} \ln \frac{l_2}{l_1}. \quad (8)$$

Making appropriate substitutions, the output of magnetic loop sensor would be

$$e = -\frac{n \mu_0 a}{2\pi} \ln \frac{l_2}{l_1} \frac{di}{dt}. \quad (2)(9)$$

This formula in essence holds all the factors that make up the magnetic sensor coil sensitivity.

1) The smaller the distance from wire to coil l_1, the higher the output.
2) The larger the distance from wire to coil far end l_2, the higher the output.
3) More turns in coil means more output.
4) Higher relative magnetic permeability leads to greater output.
5) Most important in our application would be the rate of current change of the observed application.

Note that formula (9) is intended only for the rectangular shaped sensor coils. For different shaped coils, different formulae for calculating the magnetic flux have to be used. Still, referencing to the magnetic field strength formula (2), it is clear that the further away the point is from the wire, the smaller is the magnetic field. This means that to achieve most sensitivity, most of the sensor coil effective loop area should be situated as close to the wire as possible.

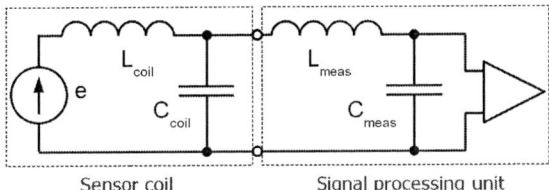

Fig. 2. Resonance circuit of the sensor coil and measurement system. L_{coil} and C_{coil} are coil self-inductance and self-capacitance respectively; L_{meas} and C_{meas} are measurement signal processing unit input inductance and capacitance repectively.

III. SENSOR COIL BANDWIDTH

Every piece of wire holds some inductance. Also, every piece of wire holds some capacitance. This applies also to every magnetic loop sensor, which has some value of self-inductance and self-capacitance. These two are one of the main factors that determine the essential operating bandwidth of the magnetic loop sensor. Other important factors are input inductance and capacitance of the signal processing unit. A schematic of components determining resonance are shown in Fig. 2.

If the connecting wires from coil to signal processing unit are short, L_{meas} can be neglected and resonance frequency of such a system can be calculated as

$$f = \frac{1}{2\pi \sqrt{L_{coil}(C_{coil} + C_{meas})}}. \quad (10)$$

There are a variety of different approaches to coil capacitance and inductance assessment. Most widely cited and used have been works of Wheeler for inductance calculations [5]. Practical testing and assessment of capacitance has been conducted by Medhurst for the circular coils [6]. Some newer work is focused on theoretical sides of capacitance also [7]. However, as none of formulas provided above provide reasonably accurate inductance and capacitance values authors of this paper will not refer to these. Also, none of the formulas provide reference to different shape of the loop vs. inductance and capacitance. Instead, general theoretical considerations are used from the papers mentioned to draw some conclusions on the general capacitance and inductance expected dynamics considering coil shapes.

Capacitance of a coil, made up of loops will provide the result that capacitance of a loop is extremely small; major part of the capacitance is made up of turn-to-turn capacitance [7]. This is valid especially, when turns pitch is smaller than the dimension of the loop. For calculation of turn-to-turn capacitance, a straightforward formula has been provided

$$C_{tt} \cong \pi D, \quad (11)$$

where D is diameter of circular coil.

A common formula for a circular loop length is

$$P_{loop} = \pi \cdot D. \quad (12)$$

978-1-4244-8806-3/11 $26.00 © 2011 IEEE   471

This means that turn-to-turn capacitance is expected to be in reference to single turn loop wire length. Conclusion is rather straightforward – in order to provide smallest capacitance, the wire length in loop should be minimized.

As for sensor coil inductance, attention should be paid to the formula on inductance [5]

$$L = \mu_0 n^2 a \left[0,48 \ln\left(1 + \pi\frac{a}{b}\right) + 0,52 arcsh\left(\pi\frac{a}{b}\right) \right], \quad (13)$$

where a – coil radius,
n – number of turns in coil,
b – coil axial length.

One of the most essential effects is provided by number of turns in coil. This should be kept as low as possible. Second, a linear factor in inductance would be provided by the radius. Third part is a factor, made up of coil radius divided by coil length. The latter one has rather interesting effect on the sensor coil design, as it states that the sensor coil should be as long as possible to keep the inductance low. There is only one way of having the coil in a uniform field around a wire with maximum length and this is having the coil wound in torus. Basically it refers that Rogowski type sensor coils should have smallest inductance for the application.

IV. PRACTICAL TESTING

For testing the coils in practice, several different coils were made. All the coils prepared for testing were made to have as close sensitivity as possible, to provide a comparison between shape and bandwidth. Calculations were carried out with a presumption of isolation of 30 mm from the wire being measured (dia. 3 mm). All coils were prepared with the same type of wire dia. 0.18 mm and same winding pitch of 8 mm. Every coil has 4 turns.

Coil geometrical shapes used for testing have been presented in Fig. 3 and were following:

1) *Circular coil*, $r = 50$ mm. Circular coil was the largest, as it does have rather small area near the wire itself, where the magnetic field is strongest.
2) *Oval coil* 1, with edge radius $r = 12.5$ mm and centre section length $l = 75$ mm.
3) *Oval coil* 2, with edge radius $r = 7.5$ mm and centre section length $l = 125$ mm.
4) *D-shape coil*, with edge radius $r = 12.5$ mm, centre section length $l = 70$ mm and D-height $hD = 40$ mm.
5) *Rectangular coil* 1, with length $l = 55$ mm and side height $h = 50$ mm.
6) *Rectangular coil* 2, with length $l = 105$ mm and side height $h = 20$ mm.

Tests were carried out using pulse signal from desktop signal generator HP33220A, which was feeding a line with 50 ohm load. From the line, coils under test were placed perpendicular to the wire, with 30 mm spacing towards the wire. Coils were connected using low-capacitance (2.2 pF) active differential probes to LeCroy Wavesurfer 24Xs oscilloscope. To determine coil parameters, resonant oscillations

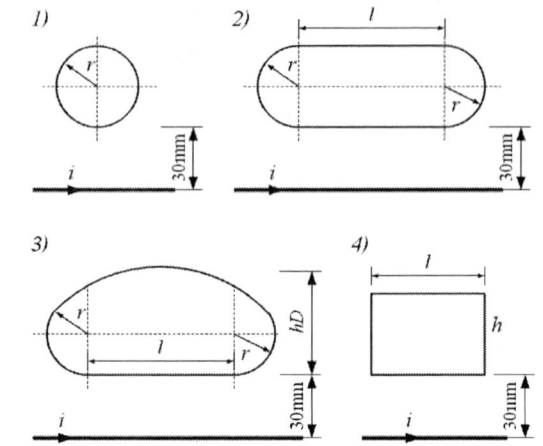

Fig. 3. Geometry of different coils used for testing. 1 – circular coil; 2 – oval coils; 3 – D-shape coils; 4 – rectangular coils. Coil parameters: r – radius; l – coil length; hD – D-height; h – rectangular coil height.

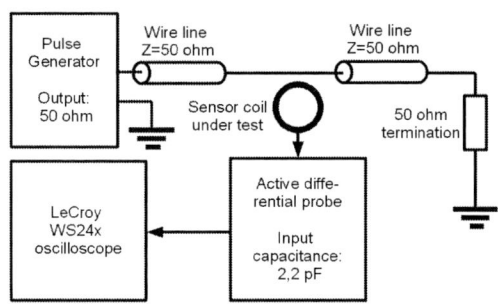

Fig. 4. Coil parameters testing set-up.

of the step response were observed on the oscilloscope. Schematic of the testing set-up used is presented in Fig. 4.

Coil characteristics, determined from the measurement sequence, have been stated in Table I.

TABLE I
COIL MEASUREMENT RESULTS AND PARAMETERS

Coil	Inductance (µH)	Capacitance (pF)	Resonant frequency (MHz)
Circular coil	2.71	1.34	83.6
Oval coil 1	1.26	1.14	132.7
Oval coil 2	1.46	1.28	116.4
D-shape coil	1.47	1.1	125.3
Rectangular coil 1	1.37	1.02	134.5
Rectangular coil 2	1.33	1.08	132.7

It rather well turns out that the circular coil, which also was the largest in its geometry, has the lowest resonance frequency. This is because this coil has the largest inductance and largest capacitance of the tested shapes. This implies, that using circular cross-section is not advised, if highest bandwidth is targeted.

To make conclusions on coils capacitance values, length of a wire on single loop of coil were calculated. This was then used to find out the capacitance per length value for each of the coils tested. Results of these calculations have been given in Table II.

TABLE II
COIL CAPACITANCE ANALYSIS

Coil	Wire length (m)	Capacitance (pF)	Capacitance per single loop length (pF/m)
Circular coil	0.314	1.34	4.26
Oval coil 1	0.229	1.14	4.99
Oval coil 2	0.297	1.28	4.31
D-shape coil	0.197	1.1	4.93
Rectangular coil 1	0.210	1.02	4.86
Rectangular coil 2	0.250	1.08	4.32

An interesting trend emerges. The coils, which have larger dimension (1, 3, 6) tend to have quite uniform capacitance per length value approximately 4.30 pF/m; other coils which are somewhat more compact (2, 4, 5) tend to have a value approximately 4.90 pF/m. The difference itself in total is rather small (12%) which would state rather well, that overall capacitance is in quite good relation to the loop wire length.

For the inductance, such clear relation to shape cannot be established so well. This would be disclosed in future papers.

V.   DISCUSSION AND CONCLUSIONS

With the results above, it seems rather proved that when sensor coil shapes are discussed, their capacitance is quite clearly related to their wire length. This provides a good starting point for design of magnetic loop sensors that would be as fast as possible with highest sensitivity.

Results of these tests show that there is major effect on the shape of the sensor coil to its performance. Circular-shaped coils might be simplest and cheapest to manufacture, however the size needed to have the same sensitivity as other shapes brings larger inductance and capacitance values. Using more flat designs of the sensor coils provides 1.6 times higher bandwidth than the circular design.

During this investigation tracing of the actual measured capacitance and inductance values provides somewhat surprising results, as none of the formulas available from other research and studies could explain the inductance and capacitance values of these coils. For example, formula [6] provided almost 5 times higher capacitance than in reality

while approach in [7] provided capacitance value that was around 2 times less than the measured capacitance. This implies that work on the real capacitance formula for such coils will continue.

Inductance values here have been calculated from the testing results. Work will also continue to provide the inductance formulas for different shapes of coils.

For the researchers, these results proved to be more optimistic than expected. The resonance frequency values acquired proved to be even higher than expected. On one side, this is due to smaller capacitance values of the real coils, than were indicated by the previously used formulas. As the bandwidth is improved, this provides also more possibilities for the range of measurements.

ACKNOWLEDGMENT

This research was supported by European Social Fund's Doctoral Studies and Internationalisation Programme DoRa.

Authors thank Estonian Archimedes Foundation (Interdisciplinary project "Optimal energy conversion and control in Smart and Microgrids" within the framework of Doctoral School of Energy and Geotechnology-II) for support of this study.

REFERENCES

[1]   G. M. Hashmi, M. Lehtonen, A. Elhaffar, "Modeling of Rogowski Coil for On-line PD Monitoring in Covered-Conductor Overhead Distribution Networks", 19th International Conference on Electricity Distribution (CIRED07), Vienna, Austria, Paper No. 0207, 2007.

[2]   G. Robles, M. Argueso, J. Sanz, R. Giannetti, B. Tellini, Identification of parameters in a Rogowski coil used for the measurement of partial discharges, Instrumentation and Measurement Technology Conference, 2007 (IMTC 2007). Warsaw, Poland; pp 1-4.

[3]   V. Dubickas, H. Edin, High-Frequency Model of the Rogowski Coil With a Small Number of Turns, IEEE Transactions on Instrumentation and Measurement, Volume: 56, Issue:6, Dec. 2007, pp. 2284-2288.

[4]   N. Ida, Engineering Electromagnetics (Second Edition), Springer New York, 2004.

[5]   H. A. Wheeler, Inductance Formulas for Circular and Square Coils, Proceedings of the IEEE, Vol. 70, No. 12, December 1982, pp 1449-1450.

[6]   R. G. Medhurst, H.F. Resistance and Self-capacitance of single-layer solenoids, Wireless Engineer, March 1947, pp 80-91.

[7]   G. Grandi, M. K. Kazimierczuk, A. Massarini, U. Reggiani, Stray Capacitances of Single-Layer Solenoid Air-Core Inductors, IEEE Transactions on Industry Applications, Vol. 35, No. 5, September/October 1999, pp 1162-1168.